PLANT BREEDING
Mendelian to Molecular Approaches

PLANT BREEDING

Mendelian to Molecular Approaches

Editors

H K Jain

M C Kharkwal

Kluwer Acdemic Publishers

BOSTON DORDRECHT LONDON

Narosa Publishing House

New Delhi Chennai Mumbai Kolkata

A C.I.P. catalogue record for the book is available from the Library of Congress

ISBN 1-4020-1981-5

Co-published by Kluwer Academic Publishers
P.O. Box 17, 3300AA Dordrecht, The Netherlands
with Narosa Publishing House, New Delhi 110002, India

Sold and distributed:
In North, Central and South America by
Kluwer Academic Publishers, 101 Philip Drive
Norwell, MA 02061, U.S.A., Europe and Rest of the World by
Kluwer Academic Publishers, P.O. Box 322
3300 AH Dordrecht, The Netherlands

In India, Bangladesh, Nepal, Pakistan and Sri Lanka by Narosa Publishing House
22 Daryaganj, Delhi Medical Association Road, New Delhi 110 002, India

FOREWORD

The Indian Society of Genetics and Plant Breeding was established in 1941 in recognition of the growing contribution of improved crop varieties to the country's agriculture. Scientific plant breeding had started in India soon after the rediscovery of Mendel's laws of heredity. The Indian Agricultural Research Institute set up in 1905 and a number of Agricultural Colleges in different parts of the country carried out some of the earliest work mostly in the form of pure-line selections. In subsequent years, hybridization programmes in crops like wheat, rice, oilseeds, grain legumes, sugarcane and cotton yielded a large number of improved cultivars with significantly higher yields.

A turning point came in the 1960s with the development of hybrids in several crops including inter-specific hybrids in cotton. And when new germplasm with dwarfing genes became available in wheat and rice from CIMMYT and IRRI, respectively, Indian plant breeders quickly incorporated these genes into the genetic background of the country's widely grown varieties with excellent grain quality and other desirable traits. This was to mark the beginning of modern agriculture in India as more and more varieties were developed, characterized by a high harvest index and response to modern farm inputs like the inorganic fertilizers. India's green revolution which has led to major surpluses of food grains and other commodities like sugar and cotton has been made possible by the work of one of the largest groups of plant breeders working in a coordinated network.

The Indian Society of Genetics and Plant Breeding with a membership of over 2000 is a representative body of this large community of agricultural scientists in the country. The Society has been fostering close interactions between its members located in different parts of India as well as all over the world. As a part of this effort, the Society organizes periodic symposia on topics of current interest. The present book "Plant Breeding: Mendelian to Molecular Approaches" is the outcome of an international symposium entitled "Hundred Years of Post-Mendelian Genetics and Plant Breeding - Retrospect and Prospects". The symposium marked the Diamond Jubilee Celebrations of the Society and coincided with the Centenary of rediscovery of Mendel's laws of biological inheritance. The focus of the symposium was on hundred years of plant breeding as a tribute to Gregor Johann Mendel.

As the world is already celebrating the Golden Jubilee of the discovery of structure of DNA in the year 2003, I hope that documentation of plant breeding researches in the past century in this book will demonstrate the immense economic impact of the science of genetics and its application in agriculture.

I would like to place on record our sincere thanks to the previous Executive Council, which was instrumental in organizing the symposium. On behalf of the present Executive Council, I like to express my deep appreciation especially to Dr. Mangala Rai, past President and Dr. M.C. Kharkwal, Organizing Secretary of the International Diamond Jubilee Symposium, who worked hard to make the event successful. We owe our thanks to the Editors for their untiring efforts in the preparation of this book. I am sure the book will be of value to students, teachers and practitioners of plant breeding.

M. Mahadevappa
President, ISGPB

PREFACE

The idea of preparing this book grew out of a symposium which the Indian Society of Genetics and Plant Breeding had organized to mark the rediscovery of Mendel's laws of biological inheritance. Mendel's work has had its greatest economic impact in the rapid development of the science of plant breeding. Improved varieties of crop plants currently help to feed a human population of more than six million people. One hundred years later following major advances in molecular genetics, plant breeding, with the availability of new tools of modern biotechnology, is again expected to make major contributions to world agriculture.

The participants in the symposium were asked to write critical reviews on a number of selected topics in genetics, biotechnology and plant breeding. The concepts and methodology of plant breeding supported by advances in classical and molecular genetics received special attention in the preparation of the book. The purpose has been to document the development of the science of plant breeding beginning with some of the landmark discoveries.

Plant breeding in the 21st century will be characterized by an increasing integration of the tried and tested classical methods with the newer techniques of modern biotechnology. The book points to the kind of integration which will be taking place. First and foremost, plant breeders of the future will be much better placed in the field of analysis of genetic diversity which they will seek for their hybrdization programmes. Microsatellites and many other loci known for their polymorphic DNA sequences have led to the development of elegant techniques for the assessment of genetic diversity available in different plant populations. Marker assisted selection should offer new opportunities to pick up traits of interest in segregating populations. Already, plant breeders are interested in the marking and manipulation of quantitative trait loci (QTLs) of value for higher yields and durable resistance to biotic and abiotic stresses. Transgenic crop varieties are being released with increasing frequency despite some of the controversy surrounding them. The safety concerns must obviously receive serious attention considering that some of the genes of interest and the proteins they code for diverged from the ancestoral populations of humans millions of years ago. However, the newly found ability to transfer genes across taxonomic barriers of all kinds provides such a powerful recombinant DNA technology that it is bound to increase the range of genetic variability available to plant breeders.

Our high-yield agriculture has come a long way since the domestication of the wild progenitors of the present day crop plants. Mendel's work paved the way for their scientific improvement. The identification and incorporation of plant-type genes in the 1960s led to the genetic reconstruction of traditional cultivars to fit them into a more modern agriculture. The high yielding varieties of various crops now available would be further improved with the emergence of more efficient techniques of plant breeding.

We are grateful to the contributors of the various chapters for their positive response and to the members of the Programme Committee for their assistance in determining the contents of this book. The support received from many colleagues in the preperation of the final manuscript is thankfully acknowledged.

H.K. JAIN
M.C. KHARKWAL

CONTENTS

ABBREVIATIONS

ADP - Adenosine diphosphate
AFLP - Amplified fragment length polymorphism
AFP - Antifreeze protein
AGI - The *Arabidopsis* genome initiative
AIDS - Acquired immune deficiency syndrome
AMMI - Additive main effects and multiplicative interaction
AP-PCR - Arbitrary primer PCR
APR - Adult plant resistance
ARS - Autonomously replicating sequences
AS-PCR - Allele specific PCR
ATP - Adenosine triphosphate
BAC - Bacterial artificial chromosome
BCTV - Beet curly top virus
BGMV - Bean golden mosaic virus
BIPs - Biparental progenies
BLB - Bacterial leaf blight
BLUP - Best linear unbiased prediction
BMV - Brome mosaic virus
BSA - Bulked segregant analysis
Bt - Bacillus thuringiensis
BYDV - Barley yellow dwarf virus
CAP - Catabolite activator protein
CAPS - Cleaved amplified polymorphic sequences
CBD - Convention on biological diversity
CBF - C-repeat binding factors
CCA - Clone-by-clone approach
CCD - Charged coupled device
CCMV - Cowpea chlorotic mottle virus
cDNA - Complementary DNA
CEPH - Centre for studies on human polymorphism
CGRFA - Commission on genetic resources for food and agriculture
CHA - Chemical hybridizing agents
CHIAS - Chromosome image analyzing system
CIM - Composite interval mapping
CMS - Cytoplasmic male sterility
CRP - Cyclic AMP receptor protein
DAF - DNA amplification fingerprinting
DGGE - Denaturing gradient gel electrophoresis
DH - Doubled haploid
DNA - Deoxyribonucleic acid
dNTP - Deoxy-nucleotide tri-phosphate
DRE -Drought responsive element
ds RNA - Double stranded RNA
DUS - Distinctness, uniformity and stability
ECFs - Extended chromatin fibers
ECR - Extended chromosomal region
EDFs - Extended DNA fibers
EDV - Essentially derived varieties
ELC - Expression-linked copy
ELISA - Enzyme-linked immunosorbent assay
EMS - Ethyl methane sulphonate
ESAG - Expression site-associated gene
EST - Expressed sequence tags
FACS - Fluorescence activated cell sorter
FDR - First division restitution
FISH - Fluorescence *in-situ* hybridization
GATT - General Agreement on Tarrif and Trade
GEI - Genotype by environment interaction
GISH - Genomic *in-situ* hybridization
GLC - Gas-liquid chromatography
GMO - Genetically modified organisms
GPC - Grain protein content
GRU - Genetic resources unit
GUS - β-glucuronidase gene
HI - Harvest index
HPLC - High pressure liquid chromatography
HSPs - Heat shock proteins
HVR - Hyper variable repeats
HUGE - Human genome project
IDA - International depository authority
IEDC - Induced embryonic determined cells
IPR - Intellectual property rights
IRAP - Inter retrotransposon amplified polymorphism
IRGSP - International rice genome sequencing project
ISSR - Inter simple sequence repeats
ITMI - International triticeae mapping initiative
kb - kilobase or kilobase pairs
LAI - Leaf area index
LINEs - Long interspersed nuclear elements
LRR - Leucine rich repeats
LTR - Long terminal repeats
MAALs - Monosomic alien addition lines
MAB - Molecular marker assisted breeding
MARs - Matrix attachment regions
MAS - Molecular marker assisted selection
MBC - Map based cloning

McFISH - Multicolour FISH
MITEs - Miniature inverted repeat transposable elements
MME - Mixed model equation
MRD - Modified Roger's distance
mRNA - Messenger RNA
MS - Murashige and Skoog medium
MTAs- Material transfer agreements
NHP - Non-histone protein
NILs - Near isogenic lines
NMR - Nuclear magnetic resonance
ORF - Open reading frame
PAC - Plant arfiticial chromosome
PAGE - Polyacrylamide gel electrophoresis
PAP - Protein amount polymorphism
PBRs - Plant breeders rights
PCR - Polymerase chain reaction
PEDC - Pre-embryonic determined cells
PEG - Polyethylene glycol
PFGE - Pulse-field gel electrophoresis
PGFRA - Plant genetic resources for food and agriculture
PGR - Plant genetic resources
PIC - Prior informed consent
PTGS - Post-transcriptional gene silencing
PVPA - Plant variety protection act
QPM - Quality protein maize
RAP - RNA amount polymorphism
RAGE - Reversion analysis of gene expression
RAMP - Random amplified microsatellite polymorphism
RAPD - Random amplified polymorphic DNA
RBIP - Retrotransposon based insertional polymorphism
RCS - Rice centromeric sequence
rDNA - Ribosomal DNA
REMAP - Retrotransposon microsatellite amplified polymorphism
REML - Restricted maximum likelihood
RFLP - Restriction fragment length polymorphism
RGA - Resistance gene analogue
RGP - Rice genome project
RILs - Recombinant inbred lines
RNA - Ribonucleic acid
RNAi - Interfering RNA
RNase - Ribonuclease
RUE - Radiation use efficiency
RWC - Relative water content
SAMPL - Selective amplification of microsatellite polymorphic loci
SAGE - Serial analysis of gene expression
SAR - Scaffold attachment region
SCARs - Sequence characterized amplified regions
SDR - Second division restitution
SDRF - Single dose restriction fragment
SIM - Simple interval mapping
SINEs - Short interspersed nuclear elements
SITL - Simply inherited trait loci
SITM - Simply inherited trait markers
SMA - Single marker approach
SNPs - Single nucleotide polymorphisms
SRAP - Sequence-related amplified polymorphism
SSAP - Sequence specific amplified polymorphism
SSCP - Single strand conformational polymorphism
SSLP - Simple sequence length polymorphism
SSRs - Simple sequence repeats
STMs - Sequence tagged microsatellites
STS - Sequence tagged sites
T-DNA - DNA of Ti plasmid transferred to plant
TGMS - Temperature sensitive genetic male sterility
TGMV - Tomato golden mosaic virus
TGS - Transcriptional gene silencing
Ti - Tumor inducing plasmid of *Agrobacterium tumefaciens*
TIR - Terminal inverted repeats
TLC - Thin layer chromatography
TLP - Thaumatin-like protein
TMV - Tobacco mosaic virus
TNV - Tobacco necrosis virus
tRNA - Transfer RNA
TRIPs - Trade related (aspects of) intellectual property
TTC - Tripple test cross
UPOV - Union le protection de obtention vegetale (International Union for Protection of New Varieties of Plants)
UTR - Untranslated region
VAM - Vesicular arbuscular mycorriza
VNTR - Variable number of tandom repeats
WGSA - Whole genome shotgun approach
WIPO - World intellectual property organization
WTO - World trade organization
WUE - Water use efficiency
YAC - Yeast artificial chromosome

Plant Breeding - Mendelian to Molecular Approaches
H. K. Jain and M. C. Kharkwal (eds.)

1

Hundred Years of Genetics: Revisiting some of the Landmarks

H.K. Jain[1]

Abstract

The human genome sequence report published in the year 2000 followed the rediscovery of Mendel's laws of biological inheritance a hundred years earlier. The intervening period has been marked by major discoveries which have become landmarks in the history of genetics. The present paper recounts some of these landmark discoveries. The treatment is not exhaustive; rather an attempt has been made to show how one important finding followed another in rapid succession, as important questions were asked at each stage. This only indicates the intense interest of a large number of scientists from different disciplines as the young science started to unfold its potential as an integrating force in biology.

Introduction

A remarkable feature of genetics is that within a relatively short span of a hundred years, one can see its beginning, and, one might say, its end, with the completion of the Human Genome Project. The latter part of this statement, however, is not true because many new discoveries remain to be made particularly in the field of gene expression and development. There are whole new fields which remain to be fully explored such as neurogenetics. At the same time, it would probably be true to say that some of the most important questions which arose from Mendel's (1866) discovery have already been answered and the focus is now shifting to new applications in the field of medicine and agriculture.

Mendel's Lasting Legacy

Mendel's own work in some ways accounts for this rapid advancement. Mendel's discovery of genes which he called 'factors' showed them to be robust and discreet entities that could be traced through generations of sexual reproduction without contamination and loss of structure or function. These findings gave rise to the concept of particulate nature of genetic material very different from the idea of blending inheritance that had prevailed in earlier years. Today, it is possible to isolate individual genes and to clone, sequence and patent them, and put them in the market place for sale. All this bears testimony to the essence of Mendel's findings.

Mendel's discovery raised many questions and answers and these have become milestones in the development of genetics in the last hundred years. I propose to recount

[1]Centre for Science Writing, 40 Surya Niketan, New Delhi 110 092, India

some of these milestones and I would like to begin with a number of discoveries made during 1910 and 1920, which established the chromosome theory of heredity. The chromosome theory lies at the heart of classical genetics.

Chromosome Theory of Heredity

The very first question following the rediscovery by the three European biologists, Correns, Tschermak and De Vries, in 1900, related to the physical presence of the Mendelian factors or genes as Johannsen (1909) called them a few years later. The first answer to this question was derived from purely empirical studies. Boveri, in Germany and Sutton (1902), a graduate student in the USA, observed the striking parallelism in the inheritance of genes and the transmission of chromosomes during sexual reproduction. Both recognized independently that Mendelian inheritance could be explained in terms of behaviour of chromosomes in the course of meiotic cell division. They proposed on the basis of their observations that genes should be present on the chromosomes. The proof for the chromosome theory of heredity came from the experiments of Morgan (1910), Bridges (1916), Sturtevant (1913) and others who organized a particularly productive school of research on *Drosophila* in the early years of the 20th century. In the course of one of his experiments, Morgan found that the gene for eye colour in *Drosophila* followed the course of transmission of the X-chromosome [The sex determining chromosomes in insects had been reported earlier by McClung (1901), Wilson, 1905 and others]. Morgan had crossed a white-eyed male fly with a red-eyed female and he observed that in the first generation both male and female flies were red-eyed. However, in the reciprocal cross when a white-eyed female fly was crossed with a red-eyed male, the results were quite different. All the female flies were red-eyed and all the males were white-eyed. These results could be best understood if it was supposed that the gene for eye colour was located on the X-chromosome. The white-eyed female parent must transmit one of its X-chromosomes to her sons, which is able to express this recessive character in a hemizygous condition. However, when the female parent is red eyed, it transmits one of its X-chromosomes both to its sons and to its daughters and, therefore, both would show the red eye character. Morgan demonstrated that sex-linked traits like the white eye colour in *Drosophila* follow a criss-cross inheritance. The male transmits his sex linked recessive traits to his grandsons only through his daughters, never to or through his sons.

Further evidence which provided strong support for the physical presence of genes on chromosomes came with the observation of some exceptional flies in the F_1 generation. In a cross when white-eyed female flies were mated with red-eyed males, about one fly in 3000 showed an unexpected eye colour, white in females and red in males. These observations seemed to contradict Morgan's hypothesis and if an acceptable explanation was not found, the chromosome theory would have received a serious setback. However, the *Drosophila* workers were so convinced about the validity of their theory that they made a bold prediction. Bridges (1916) suggested that the exceptional white-eyed female flies must have received both of their X-chromosomes from their mother, instead of one from each parent and the exceptional red eyed males must have received their single X-chromosome not from their mother, but from their father. Bridges believed that in the course of meiotic cell division, the paired X-chromosomes had shown non-disjunctional separation. A cytological examination showed that the exceptional white-eyed females did indeed carry in their body cells three sex chromosomes, XX and Y.

Discovery of Linkage

A second potential roadblock in the march of genetics in those early years came when linked inheritance was observed. The exception to the principle of independent segregation of genes was described by the British geneticist Bateson (1909), while he was studying segregation for petal colour in sweetpeas. Bateson proposed that gametes with the parental gene combinations must have been formed in a disproportionately large number because of abnormal meiotic cell divisions.

However, there was little evidence of such abnormality. It was Morgan (1916) who came out with the correct explanation and proposed the theory of linkage on the basis of his own experiments on *Drosophila*. Morgan argued that genes located on the same chromosome would be expected to show linked inheritance. While this made sense, the problem was that the observed linkage was almost always incomplete; recombinant individuals did appear in the segregating population although with a reduced frequency. It is at this point that Morgan proposed that the maternal and paternal chromosomes exchange segments during meiosis. In other words, they crossover. The first suggestion of this kind had been made earlier by De Vries purely on the basis of his cytological observations. Morgan found in this observation an explanation for partial linkage. Morgan went on to propose that the amount of recombination between linked genes is a function of the distance separating them in the chromosome. This suggestion was to lead to the concept of linear arrangement of genes on the chromosomes and to the construction of first linkage maps in *Drosophila*. The problem with linkage maps has always been the relative absence of marker genes. This problem was to be solved some 70 years later when repetitive nucleotide sequences of varying length were found to be dispersed all along the length of chromosomes. But more of it later.

Cytology Provides Proof of Crossing Over

Cytologists, meanwhile, from their observations on meiotic cell division could study crossing over in the form of structures which they called chiasma. The chiasmata marked the point where the maternal and paternal chromatids had crossed over. Darlington in the 1930s proposed the partial chiasmatype hypothesis suggesting that the exchange involved not the whole chromosomes but chromatids from the two homologous chromosomes. The genetic confirmation of this theory was to come later from the observations of Lindergren (1993) in *Neurospora*. The advantage in *Neurospora* is that the products of a single meiosis in the form of four chromatids can be recovered in a single ascus. The linearly arranged ascospores in this body can be scored for segregation of spore colour and for their arrangement. Lindergren observed that the spores were arranged in groups of twos rather than groups of fours, following crossing over. This could be best understood if the exchange involved chromatids rather than the whole chromosomes.

Another remarkable experiment by Creighton and McClintock (1931) confirmed that physical exchange of segments between maternal and paternal chromosomes is associated with the recombination of linked genes. To demonstrate this Creighton and McClintock identified marker chromosomes in a variety of maize. In these chromosomes, one of the two homologous partners could be identified by size difference in both of its arms. They discovered that in a doubly heterozygous plant, crossing over of genes was associated with exchange of segments between the two chromosomes. As the genes were recombined, the marker segments at the end of the chromosomes crossed over.

Cytology and genetics came together giving rise to the discipline of cytogenetics when Bovari and Sutton (1902) proposed the chromosome theory of heredity. The new discipline gained strength from studies like those of Creighton and McClintock (1931) and earlier from those of Bridges (1916), who proposed non-disjunctional separation of X-chromosomes in female meiosis to explain the occurrence of exceptional white-eyed female *Drosophila* flies in a cross involving white-eyed females and red-eyed males.

Quantitative Inheritance: Gene Interactions

A number of other important discoveries were made during this early period of classical genetics. These included the development of the multiple factor hypothesis by East (1916) to explain the inheritance of quantitative characters, which showed continuous type of variation in a segregating population. This hypothesis was later developed into the polygene theory by Mather (1943), who also talked of effective factors made up of tightly linked blocks of genes, which natural selection will favour for their adaptive value. One wonders whether our present concept of quantitative trait loci (QTL) comes close to the definition of effective factors.

Also, multiple alleles were discovered at many loci; the locus for self-incompatibility in plants, for example, may have hundreds of alleles. Similarly, the genes for blood group antigens were found to occur in large number of allelic forms. The simple concept of one-gene one-character was discarded when it was found that interactions of genes in determining many traits is quite common. Epistatic gene interactions were widely recognized in quantitative genetic analysis.

First Genetic Transformation Experiments

The most important question in the evolving concept of gene in the early 1940s related to their chemical and physical structure. With the chromosome theory of heredity fully established, every one wanted to know what the genes were made up of. One suggestion was that the genes in the chromosomes are made up of proteins. The reason for this was simple. With 20 different amino acids, it was possible to account for thousands of different genes, with variation in the sequence of aminoacids in the polypeptide chains. Some remarkable experiments carried out during this period however, showed conclusively, that genes are made up of DNA and not proteins. The first set of experiments related to genetic transformation demonstrated in the *Diplococcus pneumoniae*, a bacterium, which causes pneumonia. Two different strains of this bacterium could be identified by their cell wall. In one of the strains the cell wall formed a capsule and these were of the virulent kind. In the other strain, the cell wall did not form a capsule and these bacteria did not have the property of virulence. As early as 1928, Griffith had injected mice with a mixture of dead cells of the virulent and living cells of the avirulent strain. Against all expectations, the injected mice developed pneumonia and died. From the heart of the dead mice, Griffith was able to recover living virulent becteria with capsulated cell walls similar to those of the heat-killed strain used in injection. It was clear that the dead virulent cells had in some way transferred the property of virulence and capsule formation to the avirulent strain.

With genetic transformation demonstrated in this way, the question arose as to what was the nature of the specific substance which the dead cells of the virulent strain had transferred to the living cells of the avirulent strain, which brought about the transformation? The question was answered by Avery, MacLeod and McCarty (1944) 16 years later. The significance of Griffith's findings had been lost during those years, possibly because this was not a high priority question at that stage. In their experiment, they substituted dead cells of the capsule forming virulent strain with DNA extract of these cells. The results were similar to those obtained by Griffith (1928). It was clear that it was the DNA in the becterial cells which brought about genetic transformation.

Phage DNA Carries Information for Infection and Multiplication

More direct evidence showing that DNA is the genetic material came from the classic experiment of Hershey and Chase (1952) who carried out their experiment on Phage T2, a virus which infects *E. coli* as a host for its multiplication. They labelled the protein coat of the infecting virus particles with radioactive sulphur and labelled their DNA with radioactive phosphorus. They observed that while the protein coat of the virus, failed to enter the bacterial cells, the labelled DNA could be recovered in the infected cells as the virus multiplied. It was concluded that it is the DNA which carried information for infection and for the production of fresh virus particles.

In retrospect, it seems natural that DNA constitutes the genetic material and not proteins. Alexander Rich (1995) proposed that in the evolution of life, the parental nucleotide molecule was initially an RNA polymer. This ribopolynucleotide chain, in turn, evolved into a double stranded DNA molecule with specialized functions like the template activity. Proteins with their 20 amino acids are better suited to be the workhorse of life processes with their structural ability to form different architectures.

Gene's Multiple Mutational and Recombinational Sites

The 1940s also saw some remarkable advances in the understanding of the physical structure of the gene. The *Drosophila* geneticists were beginning to find evidence of genetic recombination between different alleles of a gene. The concept so far had been that the gene was not only a unit of biological inheritance but also of genetic recombination. So deep-rooted was this concept that when inter-allelic recombination was first observed, the *Drosophila* geneticists called them as pseudoalleles. While they could be resolved by crossing over, in complementation tests, they appeared to be alleles of the same gene. However, the concept of pseudoalleles had a short history, for soon extensive studies on microorganisms showed that different alleles of a gene recombined quite commonly and all that was needed to detect recombination within a locus was the ability to score a very large segregating population.

It was the work of Benzer (1955) on the fine structure of rII locus in phage T4 which showed clearly that a gene may have many different mutational sites and that these could be resolved readily through the classic hybridization experiments not very different from those of Mendel. Benzer made excellent use of the fact that a segregating population running into millions could be scored because efficient screening techniques were available to pick up the rare wild type recombinants. Benzer could identify as many as 300 different mutational sites in the rII locus and could map them in a linear order much in the same way as the *Drosophila* geneticists had mapped a large number of genes on a chromosome, using the 3-point cross. Benzer could measure a map distance as small as 0.01 per cent between two tightly linked mutational sites in the rII locus. Realising that his discovery was in total contradiction to the older concept of the gene, Benzer coined the terms recon, muton and cistron to describe the different kinds of units within a gene. The new terminology, however, did not last long as continued advances led to a much better understanding of gene structure, both in chemical and physical terms. It was not considered wise to replace the widely popular term gene with Benzer's terminology of cistron.

DNA Structure and Semi-conservative Replication

As genetics completed 50 years of its foundation, the focus shifted to an understanding of the structure of the DNA molecules. The 1953 paper of Watson and Crick and a parallel paper published by Wilkins *et al.*, provided the answer. Three properties of DNA duplex molecule were to lay the groundwork for a revolution in biological sciences, which was not fully anticipated at that time. These three properties are the complementary pairing of purine and pyrimidine bases in the two chains of the duplex, the random sequence of bases and their linear arrangement along the length of the molecule. The Chargaff (1950) rule was formulated based on a recognition of the quantitative equivalence of purines and pyrimidines in the DNA molecules from a wide variety of organisms. Watson and Crick, in their paper, pointed out that their model of DNA should help to explain how genes may be replicated and how mutations may arise.

Watson and Crick pointed out that their model of DNA had the characteristics of a template. The two strands of the DNA duplex on uncoiling could serve as templates for laying down of new strands with complementary base pairing. This model of replication was described as semi-conservative, because one of the DNA strands is conserved in the newly formed duplex molecule.

Evidence in support of the semi-conservative replication of DNA came from an experiment of Meselson and Stahl (1958). They studied the replication of DNA in *E. coli* bacteria. At the start of their experiment they made the parental DNA in the bacterial cells heavy by growing the cells repeatedly on a medium containing N-15 atoms of nitrogen. The bacterial cells with their heavy DNA were made to produce the first, second and subsequent generation cycles on a normal medium containing N-14 atoms. These progenies were examined for their DNA in a cesium chloride density gradient in an ultracentrifuge. They found that the first generation bacteria yielded DNA which occupied a position midway between the heavy and the normal DNA. The second generation cells yielded two kinds of DNA molecules - half labeled and completely unlabeled. The semi-conservative replication of DNA was later demonstrated in replicating chromosomes by Taylor and his colleagues (1957).

How far have we come following the landmark discovery of the structure of the DNA molecule? The answer to this question can be found in the proceedings of a symposium organized by the New York Academy of Sciences in 1993 to commemorate 40 years of DNA research. The proceedings show that once the structure of DNA had been discovered, everything seemed to fall in place and progress in genetics, in particular, and molecular biology in general, was very rapid.

Mutations as Misprints in Replicating DNA

In addition to the mechanism of replication of genes, the newly discovered structure of DNA helped to explain how mutations arise as misprints during the replication of DNA. Transitions and transversions were recognized by Ernest Freeze (1959) in the late 1950s as the simplest kind of base pair substitution changes, which can result in phenotypically visible mutations. It was clear that a single base pair change in DNA could alter the functioning of a gene, which is now well recognized as in the case of several haemoglobin mutant proteins. The other molecular mechanism of gene mutation involved the addition or deletion of a nucleotide base pair, resulting in what came to be known as frameshift mutations. Witkin (1969) showed how DNA repair mechanisms keep mutation rates under check.

While on the subject of mutations, it is important to recognize the pioneering contributions of Muller (1927) whose earlier work on *Drosophila* helped to lay down some of the basic concepts. Muller developed elegant techniques for scoring mutations in *Drosophila* and in determining their rates. It was he who showed that genes generally show a mutation rate of 10^{-5}-10^{-6} and introduced the concept of generation time while considering mutation rates in different organisms. Above all, he showed for the first time in 1927 that mutations could be induced in *Drosophila* through treatment with X-rays.

Mutation: Not a Directed Process

The DNA structure also suggested that mutations would occur in a random manner, because basically they are unpredictable mistakes during the process of replication. This suggestion helped to answer a very fundamental question in biology - whether mutations arose as an adaptive response of the organism to the environment. An elegant experiment by Luria and Delbruck in 1943 provided convincing evidence in support of random, non-adaptive nature of mutations. In their experiment Luria and Delbruck counted the number of mutant individuals showing drug resistance in different cultures of wild type bacteria. The different cultures were expected to show a great deal of variation for resistant cells, depending on the random occurrence of mutations at different stages of culture growth. However, if mutations arose as an Luria adaptive response when the cultures were exposed to an antibiotic, the variation in the number of resistant cells will be small. Luria and Delbruck (1943) found that the different bacterial cultures showed large variation, much larger than that expected, as a result of adaptive response to the antibiotic. They concluded that mutations for drug resistance had occurred irrespective of exposure to the drug.

Gene mutations are the ultimate source of all variation, without which evolution is not possible. With mutations arising as random events following mispairing of DNA bases, should one say that all living organisms including humans are a product of mistakes in the replicating DNA? Strictly speaking this is true, but one has to remember that natural selection picks up only that variant which adapts to its environment. So there is a great deal of rejection and only the fit survive.

Gene Function

The third important discovery in the 1940s helped to answer the questions as to how genes determine the phenotype. The answer was that genes help to bring about transformations in biochemical pathways by catalyzing the different steps in the synthesis and breakdown of metabolites. Beadle and Tatum (1941) proposed the one-gene one-enzyme hypothesis, based on their studies on *Neurospora*. It should be emphasized that Beadle, a maize geneticist, carried out what was basically a classical genetic analysis, inducing mutations in the fungus, making crosses and studying segregation. The only thing which was different was that the phenotype in this case related to nutritional requirement on a minimal medium. That is where Tatum, a microbiologist became an important collaborator. The one-gene one-enzyme hypothesis was later revised as one-gene one-polypeptide or one-gene one-function. It is interesting to recall here that as early as 1902, soon after the rediscovery of Mendel's work, Garrod, a British physician, was talking of genes, metabolism and enzymes based on his study of the pedigree of persons showing alkaptonuria. Garrod published his book "Inborn Errors of Metabolism" in 1909. Beadle has commented that scientists at that time were not mentally prepared for a discovery of this magnitude so far ahead of its time.

Insights into DNA's Variation

The 1960s were marked by several important sets of discoveries, some of them unexpected and rather surprising. The unexpected findings related to a number of properties of DNA, which had significant implications for the evolution and organization of genes in the genomes. Cytologists had observed a great deal of variation in the chromosome number of different species and now it was found that the DNA content in the genomes of different organisms showed a surprisingly large variation, not all of which could be understood in terms of the complexity of the different groups. This variation became known as the C-value paradox. For example, it was observed that the humans and the common toad have very similar content of DNA in their nuclei. This finding was followed by a discovery by Britten and Kohne in 1968 which showed that a large proportion of DNA was made up of repetitive sequences with no apparent function. They reached this conclusion on the basis of their studies on the denaturation and renaturation kinetics of eukaryotic DNA. Another discovery, made many years later, having a bearing on DNA content was the split nature of genes, which were found to be made up of exons and introns. While the genes in bacteria had only the coding sequences called exons, genes in eukaryotes had coding sequences separated by relatively large non-coding sequences described as introns.

The linkage maps constructed by the *Drosophila* workers in earlier years had indicated an organization of genes in the chromosomes marked by order and neatness. But now it was becoming clear that the formation of genes in the DNA had followed a different course. Geneticists still continue to argue about the relative merits of the two theories proposing early and late occurrence of introns in the genes. Early here means 4 billion years ago, while late means one billion years.

The function of the repetitive sequences continues to be controversial but they have proved to be of great value as molecular markers. The microsatellites, in particular, showing a great deal of allelic variation and distributed all along the length of the chromosomes, have helped to discover a large number of human genes associated with various diseases. It was the finding of repetitive sequences of varying length and the discovery of restriction enzymes by Hamilton Smith in 1970 that opened up a whole new world of analysis of genetic diversity and the development of DNA fingerprinting techniques. Hundreds of restriction enzymes evolved by bacteria as a defence mechanism have become molecular scissors for a wide range of studies. including those relating to genetic engineering.

The Genetic Code

The real stories of the 1960s, however, are the deciphering of the genetic code and the understanding of gene regulation. It is interesting to recall here that the general properties of the genetic code were described in a major paper by Crick and his colleagues (1961) purely on the basis of a genetic analysis. Crick and his collaborators made excellent use of the rII locus, whose fine structure had been described earlier by Benzer. The trick they employed was to induce frameshift mutations in this locus, using acridine dyes, which were known to act through the addition or deletion of a base pair. They found that while a plus or a minus mutation will render the gêne inactive, if a double mutant combining a plus mutation with a minus mutation was produced, the wild type phenotype was restored. So they had a group of base addition mutants and another group of base deletion mutants which when combined together would suppress the mutant phenotype. From these observations Crick and his colleagues proposed a triplet code and concluded that it was 'comma free'.

The direct understanding of the genetic code had to wait for the contributions of organic chemists, among them Khorana and his colleagues (1967) and Nirenberg and his collegues (1964) in particular. I would like to refer to the work of Khorana who pioneered the advent of synthetic messengers with a pre-determined repeating nucleotide sequence, in cell free systems. It was the extension of this technique which helped to determine the codons for all the different amino acids. Their work also showed that the code is highly degenerate.

The understanding of the genetic code shifted the focus on the transfer and translation of genetic information leading to the development of the polysome model of protein synthesis. I would like to refer to only one or two important experiments of this period. As a working hypothesis, a one-gene one-ribosome relationship was proposed. However, Brenner and his colleagues (1965) in the course of a multiple labelling experiment showed that the ribosomes do not carry the message of genes and that they had a different function. In the course of their experiment they made the ribosomes of *E. coli* heavy and then infected these cells with T2 virus. The newly synthesized messenger RNA was found to go to the pre-existing heavy ribosomes. No new ribosomes were produced following infection.

The other important discovery relates to the identification of transfer RNA molecules. Crick had proposed that there would be need for an adaptor molecule which will take the amino acids to the ribosomes. Following this prediction, transfer RNA molecules were soon discovered for each of the 20 amino acids.

Colinearity

An interesting demonstration of the colinearity of nucleotide bases in the DNA and amino acids in the proteins was established during this period, first by Sarabhai and colleagues (1964), who studied mutations in the gene directing the synthesis of head protein of phage T4. The site of mutation showed complete correspondence with the site of termination of the protein chain. In another experiment, Yanofsky *et al.*, in 1966, mapped a number of mutational sites in the A gene, which directed the synthesis of the enzyme tryptophan synthetase in *E. coli*. The different mutations were found to result in substitution of one amino acid by another. This was perhaps the first experimental demonstration of amino acid substitutions in a protein as a result of gene mutations

Regulation of Gene Activity

Even as advances were being made in the understanding of the genetic code, Jacob and Monod came out with a paper in 1961, which described some of the major concepts underlying our understanding of gene regulation and eventually that of development, differentiation and morphogenesis. For the first time, two kinds of genes were recognized - the structural genes and the regulatory genes. A significant finding was the discovery of constitutive mutations, both at the regulatory locus and in the operator region. Soon the repressor protein itself was discovered by Gilbert and Muller-Hill (1965). Of particular interest was the organization of functionally related genes such as the nine genes involved in histidine biosynthesis into a single operon in *E. coli*. They were coregulated.

The discovery of and the homeobox provided important insights into our understanding of development and differentiation. The *Drosophila* geneticists had earlier reported some of the homeotic mutations but they were not sure of their significance. The fact that homeobox is ancient and has been conserved in large groups of organisms for millions of years is obviously of great evolutionary interest. The humans are not so unique after all. In fact, as the human genome sequence was to show later, most of its DNA has been conserved from the DNA of apes like Chimpanzee and Orangutan.

Genes in Evolution and Crop Improvement

Even as rapid advances were being made in the development of the new science, genetics was having its impact on things practical and profound. Perhaps the greatest economic impact of the rediscovery of mendelian principles was in providing a scientific basis for the theory and practice of plant breeding. The laws of segregation and independent assortment showed that the parental genes could be recovered in populations of hybrid progenies intact and uncontaminated in many diverse combinations. Plant breeders could select genotypes from these populations that showed the best expression for characters of economic interest to them. With continued selection and evaluation they could evolve an improved variety marked by a changed frequency of alleles at different loci.

A more profound contribution was in providing for the first time a scientific basis for Darwin's theory of evolution (1859). The theory demanded continued availability of heritable variation for populations to evolve leading to the formation of new races and species. Gene mutations and genetic recombination were now recognized as the basic source of this variation. It is this genetic variation in Mendelian populations which provides the raw material for evolutionary change. Dobzhansky (1970) has described a Mendelian population as a community of sexually reproducing cross fertilized individuals who share in a common gene pool. Fisher (1930), Haldane (1932) and Wright (1951) in a series of papers described how a number of factors operating individually and collectively would change gene frequencies in a Mendelian population. These factors included gene and chromosomal mutations, natural selection, migrations resulting in gene flow and random fluctuations arising from sampling errors due to small size of the population. A population which is under pressure from these factors continues to attain new frequency of allels at different loci. Over a period of millions of year the genetic structure of the population would be greatly modified giving rise to a new race and subsequently to a new species.

The role of these factors in directing evolution of new species has been described in Fisher's *The Genetical Theory of Natural Selection* (1930), *Haldane's Causes of Evolution* (1932) and *Wright's Evolution and the Genetics of Populations* (1978).

It should be clear that both the development of new crop varieties through breeding and the evolution of new species in the course of Darwinian evolution are similar processes involving gene frequency changes in populations. The essential difference is that in crop improvement, the plant breeder replaces the role of natural selection. Also ofcourse, Darwinian evolution has a much longer time span extending over several million years with significant changes in the environment. Plant breeders create an artificial environment with the application of fertilizers, pesticides, irrigation and other inputs of modern agriculture.

Other Important Discoveries

I have focused mainly on the evolution of the gene concept - its structure, function and regulation. While doing so I have left out many important discoveries, for example, the mechanism of sex determination, including the discovery of the SRY locus in the Y-chromosome for maleness, and the X-chromosome inactivation in female diploid cells. Above all, I have left out important discoveries in the field of microbial genetics including the discovery of sexuality in bacteria by Lederberg and Tatum (1946). Making crosses between different multiple mutants of *E. coli*, they were the first to recover wild type recombinants which could not possibly arise through reverse mutation, considering the number of loci involved. We all know how important a role *E. coli* has played in genetic analysis. I have also not considered chromosome rearrangements and polyploidy, both of which have been important in evolution. The humans, for example, in the course of their evolution have lost two of their chromosomes through telocentric fusion. The great apes like chimpanzee and gorilla, their close cousins, have a diploid number of 48 chromosomes. As for chromosome doubling, allopolyploidy has played a very important role in the evolution of higher plants including some plants of great economic value to man, like wheat and upland cotton. Major advances have been made in the field of cytogenetics, specially with the development of new techniques like Fluorescent *in-situ* hybridization (FISH).

Powerful DNA Technology

I would like to conclude this paper with the more recent advances resulting in the development of a powerful DNA technology. The 1980s saw the emergence of genetic engineering which is now beginning to make a significant impact, especially in the field of medicine. In agriculture, transgenic varieties are now being produced in increasing number and in countries like the USA they cover millions of hectares of land. They relate mostly to herbicide resistance which is not of great interest in many developing countries. However, transgenic plants will be increasingly developed for the production of chemical compounds of medicinal and nutritional value. This is already happening in the case of transgenic animals.

Human Genome Sequence, Transposing Elements and Repeats

Finally, the closing years of the last century saw the completion of the Human Genome Project, one of the greatest adventures in the history of science. The draft sequence of nucleotides in the human genome was reported in *Nature* (Human Genome Sequencing Consortium, 2001) and in *Science* (Venter *et al.*, 2001). It is difficult to recount many of the important findings reported in the two papers. A number of them continue to be widely discussed and debated. I would like to refer to four of them. First, the draft shows that the coding sequences in the human genome constitute 1.1 % of the total DNA, whereas the repeat sequences account for nearly 50 %. Introns, larger than the exons, take up about 24 % of genome's DNA.

The repeats have always been something of a mystery. It would now appear that many of the human repeat sequences, particularly the long interspersed and the short interspersed repeat elements are derived through the activity of the transposing elements. Even the unique sequences of the genes with their coding function may, in the ultimate analysis, owe their origin to transpon-derived repeats. Over millions of years, they have diverged so much that they no longer appear as repeats.

We have always asked the question how large genomes evolved from the self-replicating molecules when life first originated. We now have the answer. The picture which emerges is that eukaryotic genomes are far more dynamic than we had thought earlier. The transposing elements provide a powerful instrument for genetic alterations of various kinds including gene mutations, chromosome rearrangements and the proliferation of repetitive sequences. When McClintock (1951) first reported transposing elements in maize, they appeared to be an exception to the rule. Now that exception appears to be the rule. In 1980, Jain published a short paper proposing that the repetitive sequences should be described as incidental DNA, a byproduct of the process of mutability, which natural selection will always favour. Orgel and Crick (1980), Doolittle and Sapienza (1980) and other authors had described the repetitive sequences as selfish DNA. What we now find is that many repeats are closely associated in their generation with transposable elements, which remain a powerful instrument of mutations in the genome.

The second interesting finding is that some of the families of repeat sequences were born tens of millions of years ago and they survive now in different species and in different populations of the same species. In this way they provide us with a valuable resource for the study of phylogenetic relationships.

Third, the genome sequence shows the presence of more than 1.4 million single nucleotide polymorphisms (SNPs). These SNPs have already become of great interest as markers for identification of genes for diseases in humans.

Fourth, there has been much debate that the number of genes in the human genome are no more than 39,000. This debate however, is no longer considered so important. Many scientists have argued that a great deal of diversity of genetic significance is created at the post-transcription and post-translation stages. It is the ability of proteins to form different architectures that is more important. The one-gene one-function hypothesis has been discarded. Gene remains as a unit of genetic information, but that information can be presented in many different forms through alternate splicing, exon shuffling and other means.

The human genome sequence with the coding exons scattered in a vast desert of introns, repeat sequences of diverse kind and pseudogenes may appear a bit messy. This, however, is understandable. Mutability, after all, is the most important attribute for genomes to keep evolving and to remain adaptive in changing environments. Natural selection, therefore, has provided a in-built mechanism, like the activity of transposing elements for mutability, even if it comes with a price.

References

Avery O.T., MacLeod C.M. and McCarty M. 1944. Studies on the chemical nature of the substance inducing transformation of pneumococcal types: Induction of transformation by a deoxyribonucleic acid fraction isolated from *pneumococcus* type III. *J. Exp. Med.*, **79**: 137.

Bateson W. 1909. *Mendel's Principles of Heredity*. Cambridge University Press, Cambridge, U.K.

Beadle G.W. and Tatum E.L. 1941. Genetic control of biochemical reaction in *Neurospora, Proc. Natl. Acad. Sci.*, **27**: 499.

Benzer S. 1955. Fine structure of a genetic region in bacteriophage. *Proc. Natl. Acad. Sci.* **41**: 344.

Botstein D., White R.L., Sholnick M. and Davies R.W. 1980. Construction of a genetic linkage map in man using restriction fragment polymorphism. Amer. J. Human Genetics, **32**: 314.

Boveri T. 1902. Uber mehrpolige Mitosen als Mittel zur Analyse der Zelkerns. *Verh. phys. Med. Gesellsch.* Wurzburg. **35**: 67.

Brenner S., Stretton A.O.W. and Kaplan S. 1965. Genetic code: The "non-sense" triplets for chain termination and their suppression. *Nature*, **206**: 994.

Bridges C.B. 1916. Nondisjuction as a proof of the chromosome theory of heredity. *Genetics*, **1**: 107.

Britten R.J. and Kohne D.E. 1968. Repeated sequences in DNA. *Science*, **161**: 359.

Chargaff E. 1950. Chemical specificity of the nucleic acids and mechanism of their enzymatic degradation. *Experientia*, **6**: 201.

Correns C. 1900. G. Mendel Regel uber das Verhalten der Nachkommenschft der Rassenbastarde. *Ber. deutsch. botan. Gesellsch*, **18**: 158. English translation available in The Birth of Genetics, published as a supplement to 1950 volume of *Genetics*, 33-41.

Creighton M.R. and McClintock B. 1931. A correlation of cytological and genetical crossing over in *Zea mays*. *Proc. Natl. Acad. Sci.*, **17**: 485.

Crick F.H.C., Barnett L, Brenner S. and Watts-Tobin R.J. 1961. General nature of the genetic code for proteins. *Nature*, **192**: 1227.

Darlington C.D. 1965. Recent Advances in Cytology. Churchill, London.

Darwin C. 1859. *The Origin of Species*, John Muray, London, U.K.

De Vries H. 1900. *Sur la loi de disjoinetion des hybrids*, C. R. Acad. Sci. (Paris), **130**: 845. (See English translation in The Birth of Genetics, Suppl. *Genetics*, 1950, volume 30-32.

Dobzhansky Th. 1970. *Genetics of the Evolutionary Process*. Columbia University Press, New York.

Doolittle W.F. and Sapienza. 1980. Selfish genes, the phenotype paradigm, and genome evolution. *Nature*, **284**: 601-603.

East E.M. 1916. Studies on size inheritance in *Nicotiana*. *Genetica*, **1**: 164.

Fisher R.A. 1930. *The Genetical Theory of Natural Selection*. Clarendon Press, Oxford, U.K.
Freeze E. 1959. The difference between spontaneous and base-analogue induced mutations of phage T4. *Proc. Natl. Acad. Sci.*, **45**: 622.
Freese E. 1963. Molecular Mechanisms of Mutations. *In*: *Molecular Genetics*, Part 1, (ed.) J. H. Taylor, Academic Press, New York, pp. 207.
Gilbert W. and Muller-Hill B. 1965. Isolation of the lac repressor. *Proc. Natl. Acad. Sci.*, **58**: 2415.
Garrod A.E. 1909. *Inborn Errors of Metabolism*. Oxford Univ. Press, Oxford, U.K.
Griffith F. 1928. The significance of pneumococcal types. *J. Hygiene*, **27**: 113.
Haldane J.B.S. 1932. *The Causes of Evolution*, Longman, London.
Hershey A.D. and Chase M. 1952. Independent functions of viral protein and nucleic acid in growth of bacteriophage. *J. Gen. Physiol.*, **36**: 39.
International Human Genome Sequencing Consortium. 2001. Initial sequencing and analysis of the human genome. *Nature*, **409**: 860.
Jacob F. and Monod J. 1961. Genetic regulatory mechanisms in the synthesis of proteins. *J. Mol. Biol.*, **3**: 318.
Jain H.K. 1980. Incidental DNA. *Nature*, **288**: 647.
Johannsen W. 1909. *Elemente der Exakten Erblichkeitslehre*, Fisher, Jena 516 pp. Classic Papers in Genetics. (ed.) Peters J.A., Prentice Hall, 1959.
Khorana H.G., Bochi H., Ghosh H., Gupta H., Jacob F., Kossel T.M., Morgan R., Narang R., Ohtsuka S.A. and Wells R.D. 1967. Polynucleotide synthesis and the genetic code. *Cold. Symp. Harb. Symp. Quant. Biol.*, **31**: 39.
Lederberg J. and Tatum E.L. 1946. Gene recombination in *E. coli*, *Nature*, **158**: 558.
Lindergren C.C. 1993. The genetics of *Neurospora* III. Pure bred stocks and crossing over in *N. crassa*, *Bull. Torrey Bot. Club*, **60**: 133.
Luria S. E. and Delbruck M. 1943. Mutation of bacteria from virus sensitivity to virus resistance. *Genetics*, **28**. 491.
Mather K. 1943. Polygenic inheritance and natural selection. *Biol. Rev.*, **18**: 32.
McClintock B. 1951. Chromosome organization and gene expression. *Cold Spring Harb. Symp.*, **16**: 13-47.
McClung C.E. 1901. Notes on the accessory chromosome. *Anat. Anz.*, **20**: 220.
Mendel G.J. 1866. Versuche uber Pflanzenhybriden. Verh. *Naturforsch.* Ver. Brunn., **4**: 3-47. Following English translation are available: *J. Roy. Hort. Soc.*, **26**; Mendel's Principles of Heredity, Gabriel and Fogel, 1955; Classic Papers in Genetics, (ed.) Peters J. A., Prentice Hall, 1959.
Meselson M. and Stahl F.W. 1958. The replication of DNA in *E. coli*. *Proc. Natl. Acad. Sci.*, **44**: 671.
Morgan T.H. 1910. Sex limited inheritance in *Drosophila*. *Science*, **32**: 120.
Morgan T.H. and Bridges C.B. 1916. Sex-linked inheritance in *Drosophila*. *Carnegie Inst. Wash.* Publ. No. **237**: 1.
Muller H.J. 1927. Artificial transmutation of the gene. *Science*, **66**: 84.
Muller H.J. 1962. Studies in Genetics. *The Selected Papers of H. J. Muller*. Indiana University Press, with a foreword by Joshua Lederberg and a tribute to Muller as a Teacher by G. Pontecurvo.

Nirenberg M.W. and Matthei J.H. 1961. The dependence of cell-free protein synthesis in *E. coli* upon naturally occurring or synthetic polyribonucleotides. *Proc. Natl. Acad. Sci.*, **47**: 1588.

Nirenberg M.W. and Leder P. 1964. RNA codewords and protein synthesis: The effect of trinucleotides upon the binding of sRNA to ribosomes. *Science*, **145**: 1399.

Orgel L.E. and Crick F.H.C. 1980. Selfish DNA: The ultimate parasite. *Nature*, **284**: 604-607.

Rich Alexander. 1995. The nucleic acids. A backward glance. *In*: *DNA: The Double Helix.* (ed.) Donald A. Chambers, The New York Academy of Sciences, pp. 97-142.

Sarabhai A., Stretton A.D.W., Brenner S. and Bolle A. 1964. Colinearity of the gene with polypeptide chain. *Nature*, **201**: 13.

Smith H.O. 1979. Nucleotide sequence specificity of restriction endonucleases. *Science*, **205**: 455-462.

Sturtevant A.H. 1913. The linear arrangement of sex-linked factors in *Drosophila* as shown by their mode of association. *Exp. Zool.*, **14**: 43.

Sutton W.S. 1902. The Chromosomes in heredity. *Biol. Bull.*, **4**: 231.

Taylor J.H., Woods P.S. and Hughes W.I. 1957. The organisation and duplication of chromosomes as revealed by autoradiographic studies using tritium labelled thymidine. *Proc. Natl. Acad. Sci.*, **45**: 122.

Tschermak E. 1900. Ueber kunstliche Kreuzung bei *Pisum sativum.* Zeits. *Landw. Verschsw. Oesterr.*, **3**. Heft 5. English translation in *The Birth of Genetics*, supplement to 1950. (volume **35**) of Genetics.

Venter J.C. *et al.* 2001. The sequence of the human genome. *Science*, **291**: 1304.

Watson J.D. and Crick F.H.C. 1953a. A structure for deoxyribose nucleic acid. *Nature*, **171**: 737.

Watson J.D. and Crick F.H.C. 1953b. Genetical implications of the structure of deoxyribonucleic acid. *Nature*, **171**: 964.

Wilkins M.H.F., Stokes A.R. and Wildon H.R. 1953. Molecular structure of deoxypentose nucleic acids. *Nature*, **171**: 738.

Wilson E.B. 1905. The chromosomes in relation to the determination of sex in insects. *Science*, **22**: 500-502.

Witkin E.M. 1969. The role of DNA repair and recombination in mutagenesis. Proc. XII Intern. Congr. *Genetics*, **3**: 225.

Wright S. 1951. Genetic structure of populations. *Ann. Eugenics*, **15**: 323-354.

Wright S. 1978. *Evolution and the Genetics of Populations*, Vol. 4., Variability within and among natural population. The University of Chicago Press, Chicago.

Yanofsky C. 1960. The tryptophan synthetase systems. *Bact. Rev.*, **24**: 221.

Yanofsky C., Horn V. and Thorpe D. 1964. Protein structure relationships revealed by mutational analysis. *Science*, **146**: 1593.

Yanofsky C., Cox E.C. and Horn V. 1966. The unusual specificity of an *E. coli* mutator gene. *Proc. Nat. Acad. Sci.*, **55**: 274-281.

Plant Breeding - Mendelian to Molecular Approaches
H. K. Jain and M. C. Kharkwal (eds.)

2

A Century of Advances in Plant Breeding Methodologies

M.C. Kharkwal[1] and Darbeshwar Roy[2]

Abstract

Plant breeding in the first half of the twentieth century which started with the rediscovery of Mendel's work, mainly involved development of pure-line(s), clones, hybrids, synthetics and composites with higher yield, stability, better quality and resistance to biotic and abiotic stresses. Following the early success, significant contributions were made in the development of semi-dwarf high yielding varieties of wheat by Borlaug and his colleagues at CIMMYT, Mexico, and of rice by Beachell and his colleagues at IRRI, Philippines. Also, high yielding hybrids were developed in maize, sorghum, pearl millet, sunflower, cotton, rice, pigeonpea and in other crops. Efforts are currently also underway to develop hybrids based on genetically engineered male-sterility systems and apomixis. The twentieth century also saw significant developments in mutation breeding, quantitative genetics and other areas relevant to plant breeding, which helped in a better understanding of the genetic architecture of a trait for formulating suitable breeding strategies. Advances in molecular genetics in the last twenty years have opened up altogether new possibilities in analyzing genetic diversity, fingerprinting genotypes, and using molecular marker assisted selection. The last decade of the twentieth century also saw the development of transgenics, the genetically engineered varieties in a number of crop plants for commercial cultivation. This paper takes a journey in the world of plant breeding through the 20th century.

Introduction

The twentieth century was an eventful century for the development and growth of the science of genetics and plant breeding. It started with the great landmark in the history of biological sciences – the elucidation and rediscovery of Mendel's laws of inheritance, simultaneously, but independently by Correns, De Vries and Tschermak in 1900. Studies of Spillman (1901) on wheat, provided further support to the basic "laws" of Mendel. With Mendel's laws taken as a fresh and exciting breakthrough in biology, the next few decades witnessed series of developments in the subjects of genetics, cytogenetics and

[1]Indian Agricultural Research Institute, New Delhi 110 012, India
[2]G.B. Pant University of Agriculture and Technology, Pantnagar 263 145, India

plant breeding. In 1900, Bateson introduced the terms "allelomorph, homozygote, heterozygote and F_1 and F_2", and in 1906 he coined the term "genetics". The phenomenon of linkage in sweetpeas was described by Bateson and Punnett in 1902, and in the same year De Vries proposed his mutation theory of evolution. In 1903, Johannsen proposed the pure-line theory. The importance of selection in the improvement of wheat, oats and barley was emphasised by Nilsson-Ehle in Sweden in 1901 and in the same year Hayes introduced the centegener progeny test. Biffen's report in 1902, that resistance to stripe rust in wheat was due to a single recessive gene was, the first explanation of the nature of inheritance of disease reaction. The next few years saw spectacular developments in various disciplines related to practical/applied plant breeding, and some new and sophisticated theoretical breeding procedures and methodologies were proposed. These advances helped the breeders in better understanding of the genetic architecture of different traits, which in turn helped in formulating suitable breeding methods in crops differing in breeding systems. Thus, the gap between the expected and the observed response narrowed considerably.

Plant breeders' efforts have been focused around improving yield, resistance to biotic and abiotic stresses, quality and stability of genotypes/varieties/hybrids of economic plants. Although a very large number of plant breeders engaged in various crops the world over have made significant contributions towards improvement of economic plants by way of developing high yielding varieties and hybrids with resistance to diseases, insect-pests and having wider adaptability, based on application of standard breeding methodologies, the work of only a few are considered in this paper, because of their wider significance and applications.

Origin of Agricultural Crops

The early works of De Candolle, who published the book "Origin of Cultivated Plants" in 1882, and Russian geneticist and plant geographer, N. I. Vavilov (1926), led to the concept of centres of origin of crop plants, as also the centres of diversity. Vavilov (1926, 1951) collected plants from all over the world and identified regions where crop species and their wild relatives coexisted with greater genetic diversity. He recognised eight such centres of diversity and two or three subsidiary centres of diversity for different crops, which since then have been slightly modified with respect to distinct areas and crops in some centres. The centres of diversity were also considered as centres of origin of crop plants, and thus each crop was assigned to a specific centre. The centres of diversity were confined to tropics and sub-tropics between 20° and 45° latitude north and south of equator. There were some notable gaps; Africa (except Ethiopia), North America and Australia were not represented in the centres designated by Vavilov. Now it is well known that Australia has unique diversity for soybean and cotton; North America for sunflower (*Helianthus annus* and *H. tuberosus*), and West Africa for *Oryza glaberrima* and *Dioscorea* spp.

Harlan (1971) hypothesised 'centres' and 'non-centres' for the origin of crop plants to explain the lack of Vavilovian patterns of variation for many crops. He proposed that 'centres' are narrow areas where crops originated in a specific time and from where these spread to other parts of the world. The 'non-centers' of agricultural origin, on the other hand, represent broad geographical areas where crops were domesticated over a long period of time. He delineated three major centres and three non-centres for the origin of crop species. These centres have served as the sources of genes for resistance to diseases and insects/pests and thus contributed to practical plant breeding. Increased plant breeding activity from the beginning of 20th century led to rapid development and spread of high yielding varieties, replacing numerous genetically variable and important land races and other valuable variants.

Concerned with the preservation of land races which were heterogeneous, served as stores of genetic variability, and remained nearly stable for a long time on account of their inherent capabilities, Sir Otto Frankel coined the term 'genetic resources' in 1968 and called upon the plant breeders to be aware of the gradual loss of germplasm and to take urgent measures to conserve the genetic resources of crop plants prior to a point of no return. Realization of the importance of germplasm, particularly the land races, in the development of high yielding varieties and as sources of resistance to biotic and abiotic stresses led to the establishment of gene banks for long-term and short-term storage of germplasm. The rapid developments in the field of molecular biology also led to the development of *in vitro* techniques for storage of tissues, cells and nucleic acid, and the genome or the useful genes through cryopreservation.

Vavilov's Law of Homologous Series: Vavilov studied exhaustive world collections of crop plants and established important relationships between the related species of genera, based on which he propounded the "Law of Homologous Series" and concluded that the "species and genera that are genetically closely related are characterized by similar series of variations with such regularity that by knowing the series of forms within the limits of one species, we can predict the occurrence of parallel forms in other species and genera".

Harlan and DeWet (1971) gave the concept of 'gene pools' for effective and judicious use of germplasm. They classified whole genetic variation at different levels as primary, secondary and tertiary gene pools, which has now been narrowed down to a single gene pool, as virtually no barrier exists for gene flow even across the plant kingdom through the use of genetic engineering.

Breeding Methods

Plant breeders working with self-pollinated crops have employed conventional methods of breeding such as 'pure-line' (Johannsen, 1903), 'bulk' (Nilsson-Ehle, 1909), 'backcross' method (Harlan and Pope, 1922) and 'pedigree' (Love, 1927). Harrington (1937) proposed 'mass pedigree' method to overcome some deficiencies in the pedigree method. Later, Goulden (1939) and Brim (1966) proposed 'single seed descent' method of handling segregating populations, which provides the breeder an opportunity to practice selection among homozygous lines. In the field of practical plant breeding, pedigree selection has played an important role in the development of high yielding varieties in self-pollinated crops. As the need of intermating between individuals in segregating generations was felt to extract superior recombinant pure breeding lines, Jensen (1970) proposed 'diallel selective mating', which provides opportunity for additional recombination. It was developed for crop species where it is difficult to make crosses and only few seeds per cross are available. Composite cross bulk population (Suneson, 1956) approach was proposed to develop locally-adapted, high-yielding varieties. In order to breed a variety with higher stability or wider adaptibility, 'disruptive selection' (Mather, 1956, and Thoday, 1972), which has evolutionary implications, is widely employed. For allowing recombination to take place freely among individuals in self-pollinated crops and considering the problem associated with hand emasculation and pollination, Brim and Stuber (1973) proposed the use of male sterility in such crops for practising 'recurrent selection'.

Attempts have also been made to increase yield by changing the relative contribution of different yield components; such attempts have usually failed, because increase in one component tends to be accompanied by reduction in another (Adams, 1967). Donald (1968) proposed the concept of crop ideotype, i.e. breeding a variety with those characteristics which contribute the most towards high yield under a wide range of climatic conditions and management practices. Plants with these features generally exhibit improved level of photosynthesis, growth and grain production. Although an increase in harvest index (HI) is believed to result in increased yield in most cereals, increasing harvest index alone was not enough (Jain, 1986). When HI is improved either by keeping the biomass constant or by improving it simultaneously, only then increase in yield would be realised. For instance, reducing tillering would lead to reduced biomass/unit area, so that even a very high HI may not lead to any increase in grain yield. That is why uniculm types are no longer used and even in the new plant types (NPT) of rice and wheat, the focus is on moderate tillering. Breeders are now engaged in designing crop varieties by way of tailoring the different traits to suit our needs.

Theory of Heterosis

The term 'heterosis' for hybrid vigour was introduced by Shull (1914) as shorthand for awkward expression as "Stimulation of Heterozygosis". Jones, in 1917, gave his well-known explanation of heterosis. Various theories such as dominance, which explains hybrid vigour due to the operation of favourable dominance in two or more of its components (Bruce, 1910; Keeble and Pellew, 1910 and Jones, 1917), overdominance (East, 1936 and Hull, 1952) and epistasis (Power, 1934 and Jinks, 1955) were proposed to explain hybrid vigour. True overdominance has been shown not to have demonstrable role in hybrid vigour and spurious overdominance can arise from a linkage disequilibrium without or with epistasis (Gardner, 1963, and Jinks, 1981). Epistasis also includes additive × dominance and dominance × dominance interactions, which can not be fixed in homozygous state. Fixation of heterosis in homozygotes is possible only with greater role of dominance in heterosis manifestation and not of epistasis although fixation of heterosis in heterozygous state through apomixis will involve epistatic interactions. The fact that the better parent heterosis is associated with duplicate type of epistasis and in the absence of epistasis, heterosis appears mainly due to dispersed directionally dominant genes, it is often possible to produce inbred lines superior in performance to heterotic F_1 parent. Heterosis and inbreeding depression were described as related phenomenon. The concept of coefficient of inbreeding was given by Wright (1921) and Malecot (1948). Haldane (1942), Bartlett (1946) and Fisher (1950) also proposed methods for calculation of inbreeding coefficient.

Hybrid Breeding

Shamel (1905), was perhaps the first to study yields of two inbred lines of corn and their hybrids for three generations. However, even earlier it was recognized that hybrids yield higher in comparison to open-pollinated variety (OPV). Following the realization of superiority of hybrids, Shull (1908, 1909), and East (1908), who proposed the dominance theory of heterosis also proposed the production of single cross hybrids. Because of the problems, such as poor vigour and seed production of inbred lines associated with single cross hybrid seed production, later Jones (1918) suggested the use of double cross hybrids. The problem then breeders faced was of selecting a few lines, out of a large number of available inbred lines, which could give superior hybrids. Davis (1927) proposed a screening technique called top cross test, in which inbred lines are crossed with a common broad base tester, preferably, with an open pollinated variety (OPV) and lines producing hybrids with unsatisfactory performance were discarded, thus eliminating undesirable inbred lines right in the early phase of hybrid development programme. As the number of double crosses increased with the increase in the number of inbred parents, Jenkins (1934) proposed and compared four methods for predicting the performance of double cross hybrids from the use of single cross hybrid performance, so that only a selected number of double cross hybrids could be produced, which resulted in saving of time and resources.

Jenkins (1935) and Sprague (1946) showed that the yielding ability of inbred lines could be determined before they approach homozygosity, i.e. the potential of lines to be productive in crosses could be determined in early generations early testing and consequently the lines lacking such ability could be discarded in early generation of inbreeding. This would substantially reduce the number of unproductive lines.

The top cross test for preliminary screening of inbred lines and prediction methods for reducing the number of double cross hybrids for testing were widely used and contributed a lot to the efficiency of varietal development programmes. Over the years inbred lines have been developed through different methods, such as pedigree, backcross, convergent technique (Richey, 1927), gamete selection (Stadler, 1944) and the problems associated with single cross hybrids have been overcome. Single cross hybrids in maize are being commercially grown in many countries. The best linear unbiased prediction procedure developed by Henderson (1975) has been found useful for routine identification of superior single crosses prior to field testing (Bernardo, 1996).

Improved Populations

As the hybrid seed needs to be replaced by farmers every season, research efforts were directed towards development of improved open pollinated populations, i.e., composites and synthetics (Hayes and Garber, 1919; Jenkins, 1940). While a composite is a mixture of genotypes from different sources which is maintained through open-pollination, a synthetic variety is a population developed by *inter se* crossing of a set of good combiner inbred lines with subsequent maintenance through open-pollination. The components of synthetics are inbreds or clones and thus synthetics can be periodically reconstituted or synthetisized. Improved populations are obtained through various intra- and inter-population improvement methods. The recurrent selection schemes developed included mass selection, family selection, ear-to-row selection (Hopkins, 1899), S_1 or S_2 selection, recurrent selection for *gca* (Jenkins, 1940), for *sca* (Hull, 1945), reciprocal recurrent selection for full-sib (Comstock *et al.*, 1949), modified ear to row selection (Lonnquist, 1964), half-sib (Hallauer and Eberhart, 1970) and some modifications of reciprocal recurrent selection (Paterniani, 1973). These methods of population improvement are widely used by maize breeders and a number of improved open pollinated varieties have been developed in India, the USA and other countries. Also, synthetics have been developed and released as commercial varieties in maize and some forage crops.

Male Sterility

The commercial exploitation of hybrid vigour or heterosis depends, besides other factors, on how easily one can produce hybrid seeds and this is where male sterility, self-incompatibility, apomixis and chemical hybridizing agents (CHAs) (McRaes, 1985) played an important role. Therefore, the landmarks in hybrid breeding included not only the proposal of development of single, double and three-way cross hybrids, but also the discoveries of genetic male sterility, cytoplasmic male sterility and fertility restorer genes. East and Mangelsdorf (1925) reported gametophytic self-incompatibility and, later, Gerstel (1950), Hughes and Babcock (1950) and Mather (1950) reported sporophytic self-incompatibility. Jones and Davis (1944) discovered Cytoplasmic-genetic (C) type of male sterility in onion. The discovery of cytoplasmic male sterility in maize by Rhoades (1933) provided the basis needed for the subsequent production of hybrids in sorghum, wheat, and other plants. In maize, different cytoplasmic male sterility sources *viz*., T, C and S cms (Rhoades, 1933 and Duvick, 1959) were used to produce hybrids. In sunflower, the genetic male sterility, cytoplasmic male sterility and fertility restorer genes were discovered in 1965, 1969 and 1970, respectively. Leclercq (1966) reported the discovery of cytoplasmic male sterility in the progeny of interspecific crosses of sunflower. Meyer and Meyer (1965) created cytoplasmic male sterility by placing *hirsutum* cotton genome in the cytoplasm of wild species, *G. anomalum*. In pearl millet [*Pennisetum glaucum* (L.) R.Br.], Tift 23-A, which is of USA origin, is the source of male sterility. In India, a number of high yielding hybrids were developed in maize, sorghum, pearl millet, sunflower and cotton through manual emasculation and pollination using mechanical detasselling. In pigeonpea [*Cajanus cajan* (L.) Millsp.], an often cross-pollinated crop, a hybrid ICPH-8 was developed by ICRISAT scientists, using genetic male sterility system. Similarly, self-incompatibility mechanism has been used to produce hybrids in members of cruciferae family. In rice (*Oryza sativa* L.), hybrids based on three line system, using cytoplasmic-genetic male sterility have been developed in China, India, Vietnam, Egypt and some other countries. Two-line rice hybrids based on environmental (temperature and photo-period) sensitive genetic male sterility have also been developed in China. The first superfine grained aromatic rice hybrid Pusa RH-10, developed at IARI, has been released in the year 2001 for commercial cultivation in India.

Development of Apomictic Varieties

The problem of replacing hybrid seed every season can be overcome by fixing the hybrid vigour through the use of apomixis, a genetically controlled reproductive mechanism without the union of gametes that can be used to propagate a genotype through seed. Once apomixis is transferred to a hybrid variety, it would breed true to type. Obligate apomixis can also provide a means of seed production of aneuploids, polyploids, structural hybrids and other chromosomal aberration that might otherwise be sterile. Apomixis has been reported in grain sorghum, pearl millet and in a number of perennial forage grasses. The 'law of homologous series' (Vavilov, 1951) sugests that apomixis should exist or could be developed in other cereal crops also. Taliaferro and Bashaw (1966) outlined the procedure for transferring genes for apomixis. Genes controlling apomixis have been found in wild relatives of some cultivated crops (Hanna, 1995). Transfer of apomixis to important crops would make it possible to develop true breeding hybrids, eliminating the labour intensive and costly process of seed production, and thus reducing the seed cost and making hybrid seed available at ease. Apomictic cultivars have been developed in Kentucky blue grass (*Poa pratensis*) and Buffelgrass (*Cenchrus ciliaris* L.). Research efforts are underway to develop apomictic hybrids in maize, pearl millet and other crops. Genes for apomixis have been transferred from *Tripsacum* into maize at CIMMYT and from *P. squamulatinum* to pearl millet, and some laboratories are already working on the identification, cloning and transfer of gene(s) responsible for apomixis. However, this effort has not been successful so far in grain crops (Rutger, 1992; Savindan, 2000).

Breeding for Resistance to Biotic and Abiotic Stresses

One way of ensuring stability in yield of a variety is to make it resistant against diseases and pests. The resistance/susceptibility trait is the outcome of interaction between host/plant and pathogen/insect. Pathogens/insects have co-evolved with crop species. In general, resistance is dominant to susceptibility. In 1902, Biffen's report that resistance to stripe rust in wheat was due to a single recessive gene, was the first explanation of the nature of inheritance of disease reaction. Early efforts to improve the stem rust resistance in wheat began with Freeman's work shortly after, in 1904. Jones and Gilman released a yellows-resistant cabbage in 1915. Hayes *et al.* (1920) successfully transferred stem rust resistance from durum to bread wheat in the famous Marquis × Iumillo cross. Painter (1951) first described the different resistance mechanisms such as non-preference, antibiosis and tolerance in plants against insects. Flor (1956), working with flax rust fungus, *Melampsora lini*, proposed classical gene-for-gene hypothesis, which states that for each gene conditioning resistance in host, there is a corresponding gene in the parasite conditioning pathogenicity. Gene-for-gene hypothesis was a major break-through in the genetics of disease resistance. This gene-for-gene relationship was shown to be functioning in case of wheat–Hessian fly (*Mayetiola destructor*) system by Hatchett and Gallun (1970). Van der Plank (1963, 1968, 1982), Robinson (1969) and Nelson (1975) introduced the concept, where horizontal (general) or race-nonspecific or non-hypersensitive or uniform or polygenic resistance is distinguished from vertical (specific) or race-specific or hypersensitive or monogenic/oligogenic resistance. Van der Plank (1963, 1968, 1982) gave the concepts of vertifolia effect, weak/strong resistance gene, stabilizing selection and modelling of epidemics, which are kept in mind while carrying out resistance breeding programme. Leonard (1969) gave the concept of race competition, fitness and adaptibility of a race to a particular host variety. Parlevliet (1978, 1979) studied the components of polygenic resistance and Wolfe *et al.* (1976) applied the gene-for-gene hypothesis in the field conditions, to find out the good combination of genes to breed for disease resistance. Person *et al.* (1976) and Leonard (1977) developed models to predict the fitness of pathogen population in susceptible, resistant and a mixture of resistant and susceptible varieties. Ellingboe (1981), in his models for studying the specificity of resistance/susceptibility, proposed that incompatibility is the active response.

The plant breeder and pathologists would like to prevent the occurrence of epidemics and evolution of a super or complex race of an insect or pathogen. Various strategies such as resistance breeding, gene deployment, pyramiding of genes, development of multilines (Jensen, 1952; Borlaug, 1953 and Browning and Frey, 1969) or mixtures (Jensen, 1952 and Barett, 1987), use of durable resistance (Johnson, 1984), and tolerance instead of resistance. A number of multilines were developed in the USA and India. The first commercial multiline wheat cultivars were Miramer-63,and Miramer-65. Oat multilines E68 and M68 were released for commercial cultivation in the USA by Iowa State University. In India, Kalyansona and Sonalika multilines (KSML3 and KML 7406) were developed. In North America, gene deployment strategy for disease resistance was designed in case of brown rust in Oat and black rust in wheat, by constructing *Puccinia* path showing spore movement. In India also, in case of black (stem) rust in wheat, *Puccinia* path was explored.

The abiotic stresses in the form of drought, salinity, extreme temperature, quality and quantity of water and nutrient inbalances are equally devastating. Millions of hectares of crop lands have been abandoned because of salinization, scarcity of water and other factors. Plant breeders, agronomists and plant physiologists have initiated joint efforts to breed for abiotic stresses in a big way, with the establishment of phytotrons and rain-out shelters towards the end of twentieth century.

Mutation Breeding

De Vries (1901) is credited with the discovery of mutations, which he described as sudden heritable changes in the organisms that caused relatively large effects on the phenotype. He coined the term 'mutation' and also referred to the new types of radiations (gamma rays and X-rays), discovered earlier by scientists like Roentgen (1895), Becquerel (1896) and Pierre and Madam Curie (1897) that might be applied to create genetic variation in plant populations. However, the use of radiations for generating novel genetic variability could take off only after the discovery of mutagenic action of X-rays demonstrated on *Drosophila* by Muller in 1927, and in maize, barley and wheat by Stadler in 1928 and 1930. The practical potential of radiation as a plant breeding tool was later shown by Delaunay (1931) and Sapehin (1936) in wheat. The X-ray treatment to two-rowed barley by Nilsson-Ehle (1948) in Sweden produced mutants having characters of dense ears and very stiff straw. Such mutants later on named as 'erectoides' by Gustafsson (1954) could withstand high doses of fertilizers and gave high yield. The Swedish programme covered several crop species and generated a lot of information on mutation breeding. Oehlker (1943), Auerbach and Robson (1946), in Britain, and Rapoport (1946, and 1948), in Russia demonstrated mutagenic effects of chemicals like mustard gas and several other alkylating agents to be highly mutagenic. Since then a number of chemical mutagenic agents have been discovered that can increase the frequency of artificially induced mutations. In USA, mutation breeding was employed with some success by Konzak (1954) and Frey (1955) on oats, and by Gregory (1955) on peanuts. Extensive work and significant contribution in understanding of basic and applied nature of mutation phenomenon on cereal and legume crops have been made in Germany (Gaul *et al.*, 1969; Gottschalk and Wolff, 1983). During 1950-1970s, several countries including China, India, Japan, the Netherlands, Russia and the USA took up the task of crop improvement through systematic mutation breeding approaches and reported spectacular accomplishments of mutation breeding in evolving several superior crop varieties (Kharkwal, 1996; Kharkwal *et al.* 2001, 2003).

By the beginning of 21st century, over 2252 varieties of crop plants including cereals, oilseeds, pulses, vegetables, fruits, fibres and ornamental plants developed through mutation breeding have been released for commercial cultivation in different countries (Maluszynski *et al.*, 2000). In India, sustained efforts for crop improvement through induced mutations at some of the major research centers like Indian Agricultural Research Institute (IARI), New Delhi; Bhabha Atomic Research Centre (BARC), Mumbai; Tamil Nadu Agricultural University (TNAU), Coimbatore; National Botanical Research Institute (NBRI), Lucknow and several others, have resulted in the development and release of about 308 mutant varieties of more than 50 crop species for commercial cultivation (Kharkwal *et al.*, 2001; 2003).

Breeding for Quality

Besides carbohydrates, proteins and fats are also important constituents of our diet and so the geneticists/breeders also paid attention for their improvement. The genetic manipulation for erucic acid (which is under control of two loci with multiple alleles) and glucosinolates free rape oil (Stefansson *et al.*, 1961; Downey, 1964) had helped in markedly increasing the production of rape in the western world. Mertz *et al.* (1964) reported that *opaque-2* (*o2*) gene doubled the content of lysine and tryptophan in the endorsperm mutant of maize. Nelson *et al.* (1965) and Nelson (1967) reported two mutants, *opaque-2* (*o2*) and *floury-2* (*fl2*), with high lysine and tryptophan contents. *Floury-2* improves protein quality but it also depresses yield. The incorporation of *opaque-2* has provided unique populations, inbreds and hybrids in maize with superior quality with double the lysine content than normal maize (Nelson, 1969). Initially the strain had low yield, soft grain with chalky appearance that was unacceptable to consumers and was more susceptible to ear rot, a fungal disease. Vasal and his colleagues (1979) at CIMMYT had made selection of favourable modifiers and developed hard endosperm *opaque-2* (*o2*) population, equal in yield with their normal counterparts. Intensive efforts by CIMMYT led to the development of several Quality Protein Maize (QPM) populations and heterotic pools by large-scale backcrossing of improved *o2* donor stocks followed by recurrent selection to accumulate modifiers to improve kernel vitreousness, hardness and other desirable traits, while retaining nutritional superiority (Vasal, 1994). In India, two quality protein maize (QPM) cultivars, *Shaktiman-1* and *Shaktiman-2*, have been released for commercial cultivation in 2002 (Anonymous, 2002).

High lysine mutants in barley were reported by Munck *et al.* (1971) and Ingversion *et al.* (1973). High lysine mutants in sorghum were reported by Singh and Axtell (1973) and Mohan and Axtell (1975). The recurrent selection experiment at the University of Illinois resulted in increase of oil content in maize from 4.7 to 17 per cent, although grain yield was reduced drastically. In rice, germplasm lines having high iron, zinc and provitamin A content have been identified at IRRI and are being used in breeding programmes. Attempts have also been made to increase protein content in cereals like wheat, but these are rare examples of success. Jain *et al.* (1976) have shown that selection for higher percentage of protein is often associated with shrivelled grains and lower seed yield. They proposed that selection should be made for the absolute amount of protein in the seed combined with high seed weight.

Cytogenetics in Plant Breeding

Plant breeders manipulate gene(s) whereas cytogeneticists work at genome or whole chromosome or chromosomal segments level. Sears (1954) developed a complete series of nullisomics, monosomics, trisomics and tetrasomics. These aneuploids provided the basic material for studies related to breeding, genetics, cytogenetics and evolution of wheat (Gupta, 1995). Sears (1956) proposed a scheme for transfer of gene for brown rust (*Lr9*) resistance from *Aegilops umbellulata* to hexaploid wheat using pollen irradiation. He could transfer a chromosome segment containing gene for resistance through segmental substitution. Sears and Okamoto (1958), and Riley and Chapman (1958) are credited for the fundamental discovery of *Ph* (Pairing homoeologous) gene that is located on the long arm of 5B chromosome. The *Ph* gene suppresses the homoeologous pairing. This discovery had obvious application in transfer of gene(s) from wild species to cultivated wheat.

Balanced tertiary trisomics (BTT) technique for development of male sterile line for use in hybrid seed production was proposed by Ramage (1965) in barley, and by Khush and Rick (1967) in tomato. However, this technique had certain shortcomings such as weak plants, reduced female fertility and low pollen production, which aggravated the problem of poor pollen dispersal. O'Mara (1940) developed a method for producing alien addition lines for transferring characters from two species or genera by addition of one or more chromosomes from one species to the full diploid complement of the other species. Law (1972) proposed a method for producing inter-varietal (entire) chromosome substitution. The purpose of the substitution was to determine the effect of individual chromosomes and after demonstrating that disease resistance or some other desirable trait is conditioned by a certain chromosome, to substitute the desired chromosome into an otherwise acceptable variety.

The conventional methods of plant breeding take longer time to develop homozygous pure breeding lines. Haploid produced from F_1 or F_2 individuals followed by diploidization using colchicine can quickly produce homozygous lines. Guha and Maheshwari (1964) developed haploids from pollen grains in *Datura inoxia*, whereas Kasha and Kao (1970) produced haploids in *Hordeum vulgare* ($2n = 2x = 14$) through chromosome elimination, following inter-specific cross with *H. bulbosum*. Chase (1952) outlined the production of haploids from diploids in maize.

Induced polyploids have been a success in fruits, root and forage crops. A triploid, produced as a result of crossing between diploid and tetraploid genotypes has resulted in the development of seedless watermelon that was grown commercially (Kihara, 1951). Also, triploid sugarbeets with higher sugar production capacity have been recommended for commercial cultivation (Matsumura, 1953). Triploid banana (seedless) and autotetraploid rye grass, *berseme* and alfalfa have been produced for commercial cultivation. Translocations are common varietal distinguishing characteristics in a number of plant species (Burnham, 1956) and are important features of species differentiation. Translocations have also been used for chromosome mapping (Rhoades, 1955). Darlington and Mather (1950) suggested that balanced lethals have evolved as a means of maintaining structural hybridity even in self-fertilizing species like *Oenothera lamarckiana*. Barbara McClintock (1951) discovered transposable elements in maize. The *Ac-Ds* system transposon based techniques are being used effectively for transposon mutagenesis, molecular tagging of genes and genetic transformation in a variety of plant materials.

Synthesis of New Crops

Blakeslee and Avery (1937) showed that chromosome doubling can be induced in plants by treatment with colchicine, which enabled the plant breeders to combine genomes from different species or genera which has resulted in synthesis of new crops. *Raphanobrassica* ($2n = 4x = 36$), an allotetraploid was developed as a result of doubling of the chromosomes of F_1 between *Raphanus sativus* ($2n = 2x = 18$), the radish and *Brassica oleracea* ($2n = 2x = 18$), the cabbage by the Russian botanist, Karpechenko (1927). The amphidiploid, unfortunately, combined the non-commercial features of both, the above ground parts of the radish and the roots of the cabbage.

Triticale is a well known example of a man-made plant. Hexaploid and octaploid triticales were produced as a result of doubling of the chromosome number in the F_1 of tetraploid *T. durum* ($2n = 4x = 28$) and hexaploid. *T. aestivum* ($2n = 6x = 42$) wheat with rye, *Secale cereale* ($2n = 2x = 14$). Rimpau (1891) obtained the first fertile triticale amphidiploid in 1888. Although the first triticales were inferior and unproductive, subsequent breeding efforts resulted in forms which were promising from an agronomic view point (Müntzing, 1935, 1939). The objective behind this intergeneric cross was to enrich the high protein content of wheat with the lysine of rye and to combine the high yield of wheat with the resilience of the rye, which is resistant to drought, tolerates acidity, low temperature and is resistant to frost. The first dwarf, fertile, and partial photoperiod insensitive triticale strain, *Armadillo*, was developed by CIMMYT. In 1968 and 1969 other varieties were released for commercial cultivation in Hungary, Spain and Canada (for a review, see Gupta and Priyadarshan, 1982). The first amber colour grain triticale variety suitable for human consumption - DT 46, developed by plant breeders at IARI, New Delhi, was released for commercial cultivation in 1994.

Developments in Related Fields of Plant Breeding

Plant breeding became more precise due to the availability of knowledge from other fields of science such as genetics, including biometrical/quantitative genetics and molecular genetics, statistics, plant physiology, plant pathology, entomology, etc. Information generated in these fields have helped in understanding the genetic architecture of a trait and the physiological or biochemical processes leading to the development of a trait. This helped formulating a more precise breeding strategy and also in precise evaluation of the genotype(s). Overall such information have increased the efficiency of selection, which a breeder practices in the field.

Quantitative Genetics

Darwin (1859) first emphasized the importance of small cumulative changes in evolution and Galton (1889) distinguished between continuous variability and stepwise variability and developed the concept of quantitative analysis of continuous variation that is characteristic of quantitative traits. Soon after the rediscovery of Mendel's laws in 1900 by Correns, De Vries, and Tschermak, controversy arose over the nature of inheritance of quantitative traits, which ended in 1906 when Yule proposed that the inheritance of continuous variation that is characteristic of quantitative traits can be explained in Mendelian fashion if number of genes controlling the trait(s) are assumed to be large. Further, Johannsen (1903, 1906) working with *Phaseolus* demonstrated that although the phenotypic variation is continuous, the underlying genetic variation is discrete and continuous phenotypic variation is actually due to the influence of environment. Based on his observations, he proposed the "pure line theory". Johannsen's pioneering work paved the way to a greater understanding of those processes, by which the genotype and the environment jointly regulate the development of a character. The multifactorial hypothesis put forward independently by Nilsson-Ehle (1909) and East (1916) and the Mather's polygene concept (1949) provided the genetic basis for explaining continuous variation exhibited by quantitative traits. Sax (1923) and Rasmusson (1935) showed the linkage of polygenes with Mendelian factors. Thoday (1961) suggested a method for detection and location of polygenes.

Plant breeders manipulate variation. The use of statistics in genetics and plant breeding started with the use of chi-square test for goodness of fit of data on qualitative traits to segregation ratio suggested by Harris in 1912. Soon it also became possible to partition total variation into genetical and environmental variation (R.A. Fisher, 1918). Later Fisher, Immer and Tedin (1932) gave the genetical interpretation of third degree statistics in the study of quantitative variation. The genetical variation was further partitioned into additive, dominance and interaction (epistasis) variation (Wright, 1935, Mather, 1949; Falconer, 1960). Sprague and Tatum (1942) coined the terms general (*gca*) and specific combining ability (*sca*) and Griffing (1956) proposed methods for estimating them. This analysis is widely used by plant breeders for selecting inbred parents for development of hybrids. Thereafter, basic generation (Mather, 1949; Mather and Jinks, 1982) and multiple mating designs such as BIPs, NC I, NC II, NC III, Diallel, TTC and partial diallel were developed to work out the genetical architecture of quantitative traits (Comstock and Robinson, 1952; Hayman, 1954; Kempthorne, 1957; Kempthorne and Curnow, 1961 and Kearsey and Jinks, 1968). The effects of linkage and epistasis on means, variances and covariances were also studied (Mather, 1938; Cavalli, 1952; Cockerham, 1954; Hayman and Mather, 1955; Opsahl, 1956; Van der Veen, 1959; and Schnell, 1961). All these developments greatly enhanced the ability of plant breeders in manipulating genetic variation.

The population can be characterized by gene frequencies and genotypic frequencies which determine the genetic structure of a population. Hardy and Weinberg (1908) independently, proposed Hardy-Weinberg equilibrium. The different population improvement methods also became available, which aimed at changing gene and genotypic frequencies by recurrent selection. Wright (1931) showed the importance of random genetic drift in determining the genetic structure of population and Mayr (1954) proposed 'founder principle' for the evolution of a new population. The stochastic forces guide the breeder to select a sample of particular size from a population for either conservation or using it as base population in recurrent selection program. Kimura (1970) proposed the neutral theory of molecular evolution and Crow and Kimura (1970) proposed a method for estimating the number of alleles per locus and sampling theory of alleles, which helped in formulating sample size to be used for collection and conservation of plants.

Simultaneously researches were on to formulate different methods of selection - including individual, family and within family selections and also those for prediction of response to selection (Falconer, 1960). Selection criteria were also outlined, which included single trait selection, independent culling, and selection index (Fisher, 1936; Smith, 1936; Hazel, 1943). Wright (1921) developed path analysis, which has helped breeders in choosing trait(s) from a set of correlated traits for yield improvement.

Genotypes respond differently in different environments. Fisher (1926) suggested the use of factorial design in the field experimentation for investigating G × E interactions. Use of factorial design led to the separation of G × E effects. The power of the analysis of variance for investigating G × E interactions was demonstrated by Immer, Hayes and Powers (1934). The objective before plant breeders always is to develop a high yielding variety with stability or general adaptability. Sprague (1963) recognized the importance of G × E interactions and cited it as one of the factors limiting further improvement in maize. He called for greater attention of geneticists and breeders to the problem of G × E interaction. Finlay and Wilkinson (1963) proposed the regression coefficient b as a measure of stability, whereas Breese (1969) advocated the use of $S^2{}_d$, (deviation from regression), for characterizing the stability of a genotype. Eberhart and Russell (1966) and Perkins and Jinks (1968) proposed models and methods for estimating these two parameters of stability. A concept of coadapted gene complex or super gene and its role in the adaptation of populations to their environments was also suggested by Mather (1956).

Like in other sciences, prediction has value in practical plant breeding. Jinks and Pooni (1976) gave a method for predicting the proportion of superior recombinant pure breeding lines derived from the F_1 of the two pure breeding parents. Panse (1940), Knott (1979), and Jinks and Pooni (1982) gave a theory for use of early generation selection, i.e. selection based on performance of F_3 families in pedigree selection scheme and found selection to be effective. Finny (1958) and Simmonds (1985) presented principles underlying multistage selection in clonally propagated crops.

Estimation of Diversity

Plant breeders choose diverse parents/populations and make crosses for extracting superior recombinant inbred lines [pure line(s)] or more heterotic population(s). Several procedures of multivariate analyses such as D^2 statistics, canonical analysis and factor analysis etc., for the grouping of varieties and measurement of diversity have been extensively used. Mahalanobis (1936) proposed a method called D^2 analysis for estimating the genetic distance between lines or populations based on measurement of quantitative traits, which is widely used by plant breeders. Anderson (1957) suggested the use of Metrogylph analysis, a semigraphical approach to measure genetic diversity among a large number of accessions for their initial categorization. Nei (1972, 1973) proposed a method for estimating the degree of similarity between two populations, which is also being used for estimation of genetic similarity and dissimilarity using molecular (banding patterns) data.

Field Designs

The foundations for the design and analysis of field experiments using randomization and the power of the analysis of variance was laid by Fisher, (1925, 1935, 1936). The idea of randomization is also credited to Fisher. Various field designs such as CRD, Augmented RBD, Latin Square, Split plot, Lattice, Alpha Lattice and Individual Plant Randomization, etc. were developed with the following objectives: (a) to control and estimate the environmental sources of variation, (b) to precisely estimate the means of different entries in the evaluation trial and (c) to work out the package of practices for a variety, for realizing the maximum yield potential (Yates, 1937; Fisher, 1950; Cochran and Cox, 1957; Federer, 1967; Panse and Sukhatme, 1978). Other field designs such as Rhombic, Honeycomb and Moving blocks were also proposed for assessing genotypes in segregating population while practising selection (Fasoulas, 1973, 1987).

Biotechnology in Crop Improvement

Three major biotechnological approaches for crop improvement include (a) the use of molecular marker assisted selection (MAS); (b) the production of genetically modified (GM) or the so called transgenic crops, and (c) the use of tissue culture for micro propogation and other purposes.

Marker Assisted Selection: The close association between molecular markers and target traits has been demonstrated in many cases. This facilitates 'marker assisted selection'. With the aid of molecular marker technology, one can now identify, locate and estimate the genetic effects of individual Quantitative Trait Loci (QTLs) (Dudley, 1993). Although a number of workers have proposed models for estimating gene effects and other parameters, the interval mapping approach developed by Lander and Botstein (1989) is considered as the pioneering work in the field of QTL mapping. Isozymes, the biochemical markers, have shown linkages with some of the important agronomical traits in various crops such as tomato, e.g. the association between acid phosphatase locus, *Aps-1* and *Mi*, the gene conferring nematode resistance (Rick and Fobes, 1974). With the advent of molecular markers such as hybridization based Restriction Fragment Length Polymorphism (RFLPs) and Polymerase Chain Reaction (PCR) based RAPDs, AFLPs, SNPs, mini-and micro-satellites (Williams *et al.*, 1990; Rafalski and Tingey, 1993 and Vos *et al.*, 1995) plant breeders can now not only fingerprint genotypes, but can also practise marker-aided selection.

Transgenic Crops: Conventional gene transfer resulting from sexual crosses has mostly been restricted to organisms belonging to the same species. Advent of genetic engineering technologies enlarged the scope of crop improvement as gene transfer circumvented sexual barrier. Gene(s) from even bacteria can be transferred to plants through genetic engineering. The *Agrobacterium* mediated plant transformation method developed in 1980s still remains the most convenient, and can be used both for dicot and monocot species, although initially it could be used only for dicot species. The other important method of gene transfer through micro projectile gun was developed by Stanford in 1990. Developments in the field of genetic engineering have led to the development of transgenics in several crop species. Genetically engineered cotton with *Bt*-toxin gene (*Bt*-cotton) that shows resistance against lepidoptera pests is a popular example. Similarly, genetically engineered potato with increased nutritive value has been developed (Chakraborty *et al.*, 2000). Genetically engineered plants with herbicide tolerance or resistance to diverse biotic stresses, including viruses, are already in commercial cultivation in some developed as well as in some developing countries. Transgenic varieties with enhanced nutritional quality are also receiving increasing attention (Potrykus, 1993; Ye *et al.*, 2000). The next few decades would also witness a higher level of commercial cultivation of transgenics in developing countries like India.

Genetic engineering approaches such as programmed cell death (Ellis and Horvitz, 1986) for controlling diseases and *barnase-barstar* system, for developing the genetically engineered male sterility and fertility restorer system for hybrid seed production in *Brassicas* have been proposed and are being used (Mariani *et al.*, 1992).

Tissue Culture: The demonstrations of hormonal control of differentiation (Skoog and Miller, 1957) and totipotency of single plant cell (Steward *et al.*, 1958) laid the foundation for clonal propagation and micro-propagation of plants through tissue culture techniques. Tissue culture techniques have provided breeders with the additional opportunity of carrying out selection at the individual cell level against biotic and abiotic stresses. These have also provided opportunities to produce somatic hybrids, including cybrids, and in overcoming problems associated with inter-specific or inter-generic crosses. *In vitro* propagation also facilitated large scale multiplication of derived genotypes in long duration plants such as date palm. Tissue culture has also revolutionized the horticultural industry. Rapid multiplication through the use of synthetic seeds using somatic embryos in crops such as potato also offered another highly useful application. The *in vitro* method of germplasm storage has also become an important component in long-term preservation of plant genetic resources.

Genomics

The term "genome" for a basic (haploid) chromosome set was proposed by Winkler in 1916. More recently, the term 'genome' has been used to refer to haploid DNA content (1C) of an organism or organelle. The study of genomes (1C DNA) of biological entities, popularly described as genomics, has emerged as a sub-specialization in biology. Genomics encompasses the structural and functional aspects of the hereditary material composing a complete genome. Beginning with the discovery of the chromatin, the study of genome moved from the genetic mapping of few genes/mutations by Morgan (1911) and Kihara (1917, 1924) based on chromosome pairing behaviour to complete sequencing and functional analysis of the whole genome, thereby bringing genomics to the forefront of biology. The first foray into genomics was a proposal to use DNA technology to extend Sturtevant's (1913) original concept of genetic mapping. Instead of tracing visible mutations as had been done in fruit flies, David Botstein and colleagues, working at the Whitehead Institute of Biomedical Research, USA, proposed in 1980 that one could construct a genome map, i.e. a complete genetic map of all the human chromosomes by following the inheritance of DNA sequence variations, termed DNA polymorphism

(Botstein *et al.*, 1980). Each polymorphism could be used to locate a DNA sequence marker at a specific site thus giving the genetic map of chromosome. One could then localize genes causing specific human diseases by matching their inheritance patterns with those of the landmarks (DNA markers) on these genetic maps.

The first success using this strategy came in 1983, when the gene causing Huntington's disease was shown to map at the tip of the short arm of human chromosome 4 (Gusella *et al.*, 1983). A more elaborate vision aiming at study of genome at a very fine scale giving a nucleotide-by-nucleotide account was initiated in the year 1985. The entire human genome was proposed to be sequenced providing a complete catalogue of every human gene. Although, on its face, the proposal seemed a logistic impossibility, decades of hard work by large groups of researchers at different laboratories of the world transformed this vision into a reality (Venter *et al.*, 2001 and Collins *et al.* 2001). Mapping and sequencing of the whole genome of an organism today constitute Structural Genomics, while assigning biological function(s) to the DNA sequences using a variety of tools and techniques is Functional Genomics.

The genome of more than 60 living organisms have now been completely sequenced. The first bacterial genome to be sequenced was that of *Haemophilus influenzae* in the year 1995 (Fleishmann *et al.*, 1995). In the year 1998, the genome of the first multicellular organism – the 97 million base (Mb) DNA sequence of the round worm, *Caenorhabditis elegans* was published (CSC, 1998). The first plant genome to be completely sequenced is that of the weed *Arabidopsis thaliana*, a wild relative of mustard (TAGI, 2000). Now the genome sequencing of the first crop species, rice (*Oryza sativa*) is also completed with high degree of accuracy (http://rgp.dna. affrc.go.jp).

The sequence of the whole genome provides a wealth of information on the genome structure, its evolution pattern and relationship with the other genomes. It also permits analysis of biological processes/pathways at the whole genome level. Concerted efforts are now being made to derive meaning from the sequence data, so that new and novel genes are identified, isolated and utilized in crop improvement. A variety of gene prediction packages based on the structure and functions of known genes are available, that are aiding the process of gene identification. Mutants created by transposon, retrotransposon and T-DNA insertions, are facilitating the process (Springer, 2000). Another landmark development is the availability of micro-arrays, also known as DNA chips. In this approach, random cDNAs from libraries or predicted genes from the genome sequences are placed (arrayed) on a glass slide as microscopic dots by using a robot called arrayer. Thousands of these spots so created on a single slide (that are now called micro-arrays) are then hybridized with fluorescent-labeled cDNAs synthesized from mRNA of different tissues or treatments. Different fluorescent lables are used for different treatments. When hybridized, differentially expressed genes are easily identified based on the colour pattern of the spots. Using this approach, several novel genes are simultaneously being identified (Kawasaki *et al.*, 2001) and subsequently validated in genetic transformation experiments to see if the mutant phenotypes are complemented.

The science of genomics is now a well-established field in biology, that is advancing the frontiers of our knowledge of gene structure and function at the whole genome level. With this development, the study of biology will never be the same. New genes isolated through genomics approaches are expected to solve many intractable problems that looked insurmountable in the past. The application of biotechnology is likely to provide the much needed food and nutritional security to the masses without harming the environment.

Some Significant Achievements in Plant Breeding

The twentieth century, the most happening century for the science, has witnessed spectacular developments in the field of genetics and plant breeding. Significant improvements were achieved in a very large number of crops world over. However, achievements made in top three crops of global importance - wheat, rice and sugarcane only are mentioned here in brief as landmarks of the last century.

Wheat: Although breeders have developed a number of high yielding varieties in different self-pollinated crops following conventional methods of plant breeding, the most significant contribution was that of Borlaug and his colleagues at CIMMYT, Mexico. They utilized the dwarfing genes *Rht-B1* and *Rht-D1* available in the variety Norin-10 and transferred this much desired feature to spring wheat background. This dwarfing source can be traced back to a Japanese dwarf variety 'Daruma' which was crossed with a US winter wheat cultivar 'Fultz'. A 'Daruma-Fultz' derivative was crossed with another variety 'Turky Red' to develop Norin-10. Borlaug's group developed dwarf spring wheat types *viz*., Lerma Rojo 64 and Sonora 64, which along with Mayo 64 and Sonora 63 were introduced in India in 1963. Lerma Rojo and Sonora 64 were released for commercial cultivation in 1965. These varieties formed the basis of wheat revolution in India. Simultaneously, advanced generation segregating populations were obtained from CIMMYT in 1963 and, thereafter, a number of high yielding varieties such as Kalyansona (Sonalika-227), Sonalika (RR-21), Chhoti Lerma and Safed Lerma were developed. Kalyansona and Sonalika were released in 1966 and 1967, respectively, for commercial cultivation and are still popular. Chhoti Lerma and Safed Lerma were popular in peninsular India. Borlaug bred varieties that could respond to inputs such as fertilizer, insecticide, weedicide, irrigation, farm machanisation and were well adapted to the tropical and sub-tropical wheat growing areas of the world. By 1980, 80 % of the wheat crop in south and southeast Asia was planted with high yielding wheat varieties.

The term "Green Revolution", was coined by William S. Gaud, former Administrator of the United States Agency for International Development, in a speech to the Society of International Development in 1968, to signify the phenomenal gains in food grain production achieved in 1966 to 1968 in India and Pakistan through the introduction of semi-dwarf/dwarf high yielding varieties of wheat. In India, the productivity of wheat and rice rose from 0.8 and 0.9 t/ha in 1963 to 2.0 and 1.7 t/ha, in 1983 respectively, whereas in China it rose from 1.0 t/ha in wheat and 2.0 t/ha in rice to 2.5 and 4.7 t/ha, respectively during the same period.

Rice: Green revolution in rice occurred during 1960s and 1970s as a result of development of IR-varieties by Beachell and his colleagues at IRRI, Philippines (Khush *et al.*, 2001). The rice variety, Taichung Native 1 (TN1), that was introduced in India in 1962 was developed as a result of selection from the cross as of Chinese variety *Dee-geo-wu-gen* (*Dee-geo* means short-leg in Chinese), a spontaneous dwarf mutant with a variety carrying a recessive semi-dwarfing gene named sd_1 and Tsai-Yuan-Chung, a traditional tall variety from Taiwan. Variety IR-8 was developed from a cross of Peta with *Dee-geo-wu-gen* (DGWG) and released for cultivation in Bangladesh, India, Thailand and Columbia in 1966. Further crossing between Peta and DGWG resulted in the development of IR-20 which was released in India in 1969. Subsequently, a variety Jaya was developed in India from the cross of TN1 and T141. IR-24, a variety free of chalky grains, was developed from the cross of IR-8 and Sigadis to correct the defect of IR-8, which had chalky grains, and thus IR 24 was widely accepted. Later, IRRI developed IR-36, a high yielding dwarf variety having wider adaptability and multiple resistance to biotic/abiotic stresses. IR-36 was released for commercial cultivation in several countries resulting into improved productivity. By 1970s, half of the Asian rice growing area was planted with semidwarf varieties. As of today, about 90 % of the dwarf varieties under cultivation in tropical Asia are based on DGWG dwarfing gene.

Sugarcane: After wheat and rice, sugarcane is the most important crop in the world. A vegetatively propagated crop, which is cultivated in more than 100 countries, meets over 70% of the total sweetener requirement. The important cane producing countries are India, Brazil, Cuba, China, Thailand, Mexico, Australia and USA. In recent years, India has emerged as the largest producer of sugar, although Brazil has larger area under sugarcane and also higher total cane production. The present day commercial sugarcane cultivars are complex inter-specific hybrids incorporating genes from cultivated and wild species of *Saccharum*. For centuries, clones belonging to cultivated species of *S. barberi* and *S. sinense* were grown in India and China and clones belonging to *S. officinarum* were grown in the Pacific and other parts of the world. The two wild species of sugarcane are *S. Spontaneum* and *S. robustum*. The two species *S. barberi* and *S. sinense*, were under cultivation in India until the advent of hybrid sugarcane cultivars, which were not only hardy and tolerant to environmental stresses, but also had fairly good sucrose content. They were, however, very low yielding due to thin stalks, poor flowering and sterility. On the other hand, *S. officinarum*, that was cultivated in pacific region, had all the desirable attributes of a commercial sugarcane such as thick stemmed colourful canes with high sugar yield, good sucrose and purity, low fibre and starch content, high individual cane weight, easy trashing and moderately high yields under optimal cultural conditions. It has, however, low tolerance to biotic and abiotic stresses. Owing to their majestic appearance, need for good management, and a lack of their capacity to survive on their own in nature these were called as 'noble canes' in Java. Sugarcane breeding was initiated soon after the following two landmark discoveries in the genetic development of sugarcane, (a) in 1889 the independent discovery of the occurrence of true seed and its germination by Soltwedel, in Java (Indonesia) and by Harrison and Bovell in Barbados, and (b) the discovery of the beneficial effects of interspecific hybridization. Three species of *Saccharum* i.e., *S. officinarum, S. barberi* and *S. spontaneum* were used in the production of improved varieties. Hybrids among these three species were systematically crossed with a series of 'noble' canes for improvement in quality. The method of backcrossing of F_1 derived from cross *S. officinarum* × *S. spontaneum* with noble canes to dilute/eliminate the undesirable characters/influence of *spontaneum* (wild parent) was termed as 'Nobilization' by breeders in Java.-Later, attention was directed towards inter-varietal crosses involving products of hybridization of successive generation. The early breeding work in Java culminated in the development of 'wonder cane' POJ 2878. In India, the nobilization of *S. spontaneum* was initiated by Barber (1916) at Coimbatore soon after the success of the first inter-specific hybrid, Co 205 (*S. officinarum* × *S. spontaneum*) which replaced *S. barberi*. Later, hybrid clones like POJ 213 obtained from Java were introduced into early hybridization programme. These early efforts culminated in the development of commercially important cane varieties such as Co 290, Co 312 and Co 313 in sub-tropical India and Co 419, the 'wonder cane' of tropical India (Venkataraman, 1938). In fact, the early cane varieties produced in Java and India through hybridization with wild species have gone into the constitution of many commercial cultivars later produced in all the major sugarcane growing countries. The dramatic success of the earlier hardy 'Co' varieties in sub-tropical India as a result of 'nobilization' process is an outstanding achievement that has resulted in an increase of more than 300% in cane production, 500% in sugar production and 100% in yield during the last seven decades.

Conclusion

Increased productivity through genetic improvement of crop plants has had a significant impact on world agriculture. The efforts of plant breeders and other scientists engaged in crop improvement activity, directly or indirectly, during the past century has helped in generating food for about 6 billion people on this planet, thereby disproving the doomsday prediction of thinkers like Malthus (1798) about inevitability of world famine.

Plant breeding has followed the general pattern of introduction, selection and hybridization. Introduction has been crucial for world agriculture because of which many of the world's crops are now cultivated outside the region of their domestication. Selection has exploited the natural genetic variation. Planned hybridization and induced mutations have helped in generating the additonal desired variability. Moreover, plant breeders have improved selection process by using quantitative genetics to better understand the phenotype, genotype and environmental interactions. Currently, molecular techniques are being used to further enhance the selection of quantitative traits. Indeed, plant breeding has to be an endless pursuit as the perfect variety has never been bred. Present varieties are merely more desirable than their predecessors.

Plant breeders are working to extend the phenomenal success of Green Revolution by intensifying selection, developing hybrid varieties in more crops, and increasing the range of plant functions through mutation and transgenic breeding. There is a long wish-list of plant breeders to incorporate desirable traits in crop varieties. To mention a few, these include increased biomass production, increased harvestable yield, desired nutritional and processing quality, resistance to various biotic and abiotic stresses, response to inputs, production economics, etc. Plant breeding will continue to play a crucial role in crop improvement in 21st century because the needs are many, the targets are very high, but the natural resources like aerable land and water are fast dwindling due to the exponential rise in global population, urbanization and environmental degradation of agricultural land. Methodologies developed and used for plant breeding during the last century have proven their worth and contributed significantly to crop improvement and advancement of agriculture in the world. During this process, they have been refined a lot and are expanding tremendously and rapidly with great deal of precision. The hands of modern plant breeders have been further strengthened particularly after the spectacular advances made in molecular biology and addition of powerful tools of molecular techniques to their kits. The success of the past century augurs well regarding the potential of the future of plant breeding in meeting the daunting challenge of providing food for everyone and thus creating a hunger-free world through sustainable agriculture.

References

Adams M.W. 1967. Basis of yield component compensation in crop plants with special reference to the fieldbean (*Phaseolus vulgaris*). *Crop Sci.*, **7**: 505-510.

Allard R.W. 1999. *Principles of Plant Breeding*. 2nd Ed. John Wiley and Sons, New York.

Allard R.W. and Bradshaw A.D. 1964. Implications of genotype-environment interactions in applied plant breeding. *Crop Sci.*, **4**: 503-508.

Anderson E. 1954. A semi-graphical method for the analysis of complex problem. *Proc. Natl. Acad. Sci.* Wash., **43**: 923-927.

Anonymous. 2001. *Production Technology of Quality Protein Maize*. Directorate of Maize Research (ICAR), Pusa Campus, New Delhi 110 012. pp 14.

Auerbach C. and Robson J.M. 1946. Chemical production of mutations. *Nature*, **157**: 302.

Avery O.T., MacLeod C.M. and McCarty M. 1944. Studies on the chemical nature of the substance inducing transformation of pneumococcal types: Induction of transformation by a deoxyribonucleic acid fraction isolated from *Pneumococcus* type III. *J. Exp. Med.*, **79**: 137-158.

Barber C.A. 1916. *Studies in Indian sugarcane*, No. 2. Mem. Dept. Agric. India, Bot. Ser., **8**: 103-199.

Barett J.A. 1978. A model for epidemic development in varietal matures. *In*: *Plant Disease Epidemiology*. Scott & A. Bainsbridge, Blackwells.

Bartlett M.S. 1946. The use of transformations. *Biometrics*, **1**: 39-53.

Bateson W. 1902. *Mendel's Principles of Heredity: a defence*. Cambridge, U.K.

Bateson W. 1909. *Mendel's Principles of Heredity*, Camb. Univ. Press, Cambridge, U.K.

Bateson W. and Punnett R.C. 1905. Experimental studies in the physiology of heredity. *Reports to the Evolution Committee of the Royal Society*. II. Harrison and Sons, London.

Bernardo R. 1996. Best linear unbiased prediction of maize single cross performance. *Crop Sci.*, **36**: 50-36.

Biffin R.H. 1905. Mendel's laws of inheritance and wheat breeding. *J. Agric. Sci.*, **1**: 4-48.

Blakeslee A.F. and Avery A.G. 1937. Methods of inducing chromosome doubling in plants by treatment with colchicine. *Science*, **86**: 408.

Borlaug N.E. 1953. New approach to the breeding of wheat varieties resistant to *Puccinia graminis*. *Phytopathology*, **43**: 463.

Botstein D., White R.W., Sholnick M. and Davies R.W. 1980. Construction of a genetic linkage map in man using restriction fragment polymorphism. *Amer. J. Human Genetics*, **32**: 314-331.

Breese E.L. 1969. The measurement of significance of genotype environment interactions in grasses. *Heredity*, **24**: 27-44.

Brim C.A. 1966. A modified pedigree method of selection in soybean. *Crop Sci.*, **6**: 220.

Brim C.A. and Stuber C.W. 1973. Application of genetic male sterility to recurrent selection schemes in soybean. *Crop Sci.*, **13**: 528-530.

Browning J.A. and Frey K.J. 1969. Multilines cultivars as a means of disease control. Annu. Rev. *Phytopathology*, **7**: 355-82.

Bruce A.B. 1910. The Mendelian theory of heredity and the augementation of vigor. *Science*, **32**: 627-628.

Burnham C.R. 1956. Chromosomal interchanges in plants. *Bot. Rev.*, **22**: 419-552.

Cavalli L.L. 1952. An analysis of linkage in quantitative inheritance. *In*: *Quantitiative Inheritance*, (eds.) E.C.R. Reeve and C.H. Waddington, HMSO, London. pp. 134-144.

Chakraborty S., Chakraborty N. and Datta A. 2000. Increased nutritive value of transgenic potato by expressing a non-allergenic seed albumin gene from *Amaranthus hypochondriaus*. *Proc. Natl. Acad. Sci.*, USA, **97**: 3724-3729.

Chase S.S. 1952. Production of homozygous diploids of maize from monoploids. *Agron. J.*, **44**: 263-267.

Chrispeels M.J. and Sadava D.E. 1994. *Plant, Genes and Agriculture*. Jones and Bartlett Publishers, Boston.

Cochran W.G. and Cox G.M. 1950. *Experimental designs*, John Wiley and Sons, New York.

Cockerham C.C. 1954. An extension of the concept of partitioning of hereditary variance for analysis of conveniance among relatives when epistasis is present. *Genetics*, **19**: 859-882.

Collins F. *et al.* 2001. Initial sequencing and analysis of the human genome. *Nature*, **409**: 860-921.

Comstock R.E. and Robinson. H.F. 1952. Genetic parameters, their estimation and significance. *Proc. VI. Intl. Grassland Congr.*, **1**: 284-291.

Comstock R.E., Robinson H.F. and Harvey P.H. 1949. A breeding procedure designed to make use of both general and specific combining ability. *Agron. J.*, **41**: 360-367.

Correns C. 1900. Mendels Regal über dasVerhalten der Nach-Kommenschaft Rassen Bastarde. *Ber. Deutsch. bot. Ges.*, **18**: 158-168.

Crow J.F. and Kimura M. 1970. *An Introduction to population Genetics Theory*. Harper and Row, New York.

Dalrymple D. 1986. *Development and spread of high yielding rice varieties in developing countries*. Washington D.C. USAID.

Darlington C.D. and Mather K. 1950. *The Elements of Genetics*. The McMillan Company, New York.

Darwin C. 1859. *The origin of species*. John Murray, London.

Davis R.L. 1927. Report of the plant breeder. *Rep, Puerto Rico Agric Expt. Stn*. pp. 14-15.

De Candolle A. 1882. *Origin of cultivated plants*. Hafner, New York.

Delaunay L.N. 1931. Resultate eines dreijahrigen Rontgen Versuchs mit Weizen. *Der Züchter*, **3**: 129-137.

De Vries H. 1901. *Die Mutations – Theorie*. I, Von Veit, Leipzig, 648 p.

Donald C.M. 1968. The breeding of crop ideotype. *Euphytica*, **17**: 385-403.

Downey R.K. 1964. A selection of *Brassica campestris* L. containing no erucic acid in its seed oil. *Can. J. Plant Sci*., **44**: 295.

Dudley J.W. 1993. Molecular markers in plant improvement: Manipulation of genes affecting quantitative traits. *Crop Sci*., **33**: 660-668.

Duvick D.N. 1959. The use of cytoplasmic male-sterility in hybrid seed production. *Economic Bot*., **8**: 167-195.

Duvick D.N. 1965. Cytoplasmic male sterility in corn. *Adv. Genet*., **13**: 1-56.

East E.M. 1916. Studies on size inheritance in *Nicotiana. Genetics*, **1**: 164-176.

East E.M. 1908. *Inbreeding in corn*. Connecticut Agric. Expt. Stn. 1907. pp. 419-428.

East E.M. 1936. Heterosis. *Genetics*, **21**: 375-397.

East E.M. and Mangelsdorf A.J. 1925. A new interpretation of the hereditary behaviour of self-sterile plants. *Proc. Natl. Acad. Sci*., USA, **11**: 116-183.

Eberhart A.A. and Russell W.A. 1966. Stability parameters for comparing varieties. *Crop Sci*., **6**: 36-40.

Ellingboe A.H. 1981. Changing concepts in host-parasite genetics. *Ann. Rev. Phytopathol*., **19**: 125-143.

Elliott Fred C. 1958. *Plant breeding and cytogenetics*, McGraw Hill, New York.

Ellis H. and Horvitz H.R. 1986. Genetic control of programmed cell death in the nematode *C. elegance. Cell*, **44**: 817-829.

Falconer D.S. 1960. *Introduction to quantitative genetics*. Longman, London.

Fasoulas A.C. 1973. *A New Approach to Breeding Superior Yielding Varieties*. Aristotelian Univ. of Thessaloniki, Publication **3**: 114 pp. Thessaloniki, Greece.

Fasoulas A.C. 1987. A moving block evaluation technique for improving the efficiency of pedigree selection. *Euphytica*, **36**: 477-478.

Federer W.T. 1961. Augmented designs with one way elimination of heterogeniety, *Biometrics*, **17**: 444.

Finlay K.W. and Wilkinson G.N. 1963. The analysis of adaptation in a plant breeding programme. *Aust. J. Agr. Res*., **14**: 742-754.

Finny D.J. 1958. 1958. Plant selection for yield improvement. *Euphytica*, **7**: 83-106.

Fisher R.A. 1918. The correlation between relatives on the supposition of Mendelian inheritance. *Trans. R. Soc. Edinb*., **52**: 399-433.

Fisher R.A. 1926. The arrangement of field experiments. *Journal of the Ministry of Agriculture*, **33**: 503-513.

Fisher R.A. 1935. *The Design of Experiments* (5th edition). Oliver and Boyd. Edinburgh.

Fisher R.A. 1936. The use of multiple measurements in taxonomic problems. *Annals of Eugenics*, **7**: 179-188.

Fisher R.A. 1950. *Statistical methods for research workers* (11th edition). Oliver and Boyd, Edinburgh.

Fisher R.A., Immer F.R. and Tedin O. 1932. The genetical interpretation of statistics of the third degree in the study of quantitative inheritance. *Genetics*, **17**: 107-124.

Fleishmann R. *et al.* 1995. Whole genome random sequencing and assembly of *Haemophillus influenza* RD. *Nature*, **269**: 495.

Flor H.H. 1956. The complemenary genic systems in flax and flax rust. *Adv. Genet.*, **8**: 29-64.

Frankel O.H. 1970. Genetic conservation in perspective. p. 469-49. *In*: *Genetic Resources in Plants - their exploration and conservation*, (eds) Frankel O.H. and E. Bennet. IBF Handbook No. 11, Blackwell, Oxford and Edinburg.

Freeman E.M. and Johnson E.C. 1909. The loose smuts of wheat and barley. *USDA Bureau Plant Ind. Bull.* 152. 48 pp.

Frey K.J. 1955. Agronomic mutations in oats induced by X-ray treatment, *Agron. J.*, **47**: 207-210.

Galton F. 1889. *Natural Inheritance*. Macmilan and Co., London.

Gardner C.O. 1963. Estimates of genetic parameters in cross fertilizing plant and their implications in plant breeding. pp. 225-252. *In*: *Statistical genetics and plant breeding*. Special publ. 982, NAS-NRC, Washington, D.C.

Gaul H., Ulonska E., Winkel C.Z. and Braker G. 1969. "Micro-mutations influencing yield in barley - studeis over nine generations". *In*: *Induced mutations in plants*, Proc. Symp. Pullman, 1969, IAEA, Vienna, pp. 375-398.

Gerstel D.V. 1950. Self incompatibility studies in Gauyule. II. Inheritance. *Genetics*, **35**: 282-286.

Gottschalk W. and Wolff G. 1983. Induced Mutations in Plant Breeding, Monograph on *Theoretical and Applied Genetics* 7. Springer-Verlag, Berlin, Heidelberg, pp, 323-327.

Goulden C.H. 1939. Problems in plant selection. *In*: *Proc. 7th Int. Genet. Congr.*, (ed) R.C. Burnett, Edinburgh, Cambridge Univ. Press, pp. 132-133.

Gregory W.C. 1955. X-ray breeding of peanuts (*Arachis hypogaea*). *Agron. J.* **47**: 396-399.

Griffing B. 1956. Concept of general and specific combining ability in relation to diallel crossing systems. *Austr. Journ. Bio. Sci.*, **9**: 463-493.

Guha S. and Maheshwari S.C. 1964. *In vitro* production of embryos from anthers of *Datura*. *Nature*, **204**: 497.

Gupta P.K. 1995. *Cytogenetics*. Rastogi & Co., Meerut, India

Gupta P.K. and Priyadarshan P.M. 1982. *Triticale - present status and future prospects*. Advances in Genetics. **21**: 253-345.

Gusella J.F., Wexler N.S., Conneally P.M., Naylor S.L. *et al.* 1983. A polymorphic DNA marker genetically linked to Huntington's disease. *Nature*, **306**: 234-238.

Gustafsson A. 1954. Swedish mutation work in plants. Background and present organization. *Acta Agr. Scand.*, **4**: 361-364.

Haldane J.B.S. 1942. *New Paths in Genetics*, Harper and Bros., New York.

Hallauer A.R. and Eberhart S.A. 1970. Reciprocal full-sib selection. *Crop Sci.*, **10**: 315-316.

Hallauer A.R. and Miranda J.B. 1981. *Quantitative genetics in maize breeding*. Iowa State Univ. Press/Ames.

Hanna W.W. 1995. Use of apomixis in cultivar development. *Adv. Agron.*, **54**: 334-350.

Hardy G.H. 1908. Mendelian proportions in a mixed population. *Science*, **28**: 49.

Harlan H.V. and Pope M.N. 1922. The use and value of backcrosses in small grain breeding. *J. Hered.*, **13**: 319-322.

Harlan J.R. 1971. Agricultural origins: Centres and non-centres. *Science*, **174**: 468-474.

Harlan J.R. and DeWet J.M.J. 1971. Towards a rational classification of cultivated plants. *Taxon*, **20**: 509-517.

Harrington J.B. 1937. The Mass Pedigree Method in the Hybridization Improvement of Cereals. *J. Amer. Soc. Agron.* **29**: 379-384

Harrington J.B. 1952. *Cereal breeding procedures.* Food and Agric. Org., U.N. Paper 28, Rome, Italy.

Harris J.A. 1912. A simple test of the goodness of fit of Mendelian ratios. *Am. Naturalist*, **46**: 741-745.

Hatchett J.H. and Gallun R.L. 1970. Genetics of the ability of the Hessian fly *Mayetiola destructor*, to survive on wheats, having different genes for resistance. *Ann. Entomol Soc. Amer.*, **63**: 1400-1407.

Hayes H.K. and Garber R.J. 1919. Synthetic production of high protein corn in relation to breeding. *J. Amer. Soc. Agron.*, **11**: 309-319.

Hayes H.K., Parker J.H. and Kurzweil C. 1920. Genetics of rust resistance in crosses of varieties of *Triticum vulgare* with varieties of *T. durum* and *T. dicoccum. J. Agri. Res.*, **19**: 523-542.

Hayman B.I. 1954. The theory and analysis of diallel crosses. *Genetics*, **39**: 789-809.

Hayman B.I. and Mather K. 1955. The description of genic interactions in continuous variation. Biometrics, **11**: 69-82.

Hazel L.N. 1943. The genetic basis of constructing selection index. *Genetics*, **28**: 476-490.

Henderson C.R. 1975. Best linear unbiased estimation and prediction under a selection model. *Biometrics*, **31**: 423-427.

Henshaw G.G. 1979. Tissue culture and germplasm storage. *Int. Assoc. Plant Tissue Culture*, Newsletter, **28**: 2-7.

Hopkins G.G. 1899. Improvement in the chemical composition of the corn kernel. *Illionois Agric. Exp. Stn. Bull.*, **55**: 205-240.

Hughes M.B. and Babcock E.B. 1950. Self-incompatibility in *Crepis foetida* L. subsp. *rhoedaifolia. Genetics*, **35**: 570-580.

Hull F.H. 1945. Recurrent selection for specific combining ability. *Agron. J.*, **37**: 134-145.

Ingversion J., Koie B. and Doll H. 1973. Induced seed protein mutant of barley. *Experientia*, **29**: 1151-52.

Jain H.K. 1986. Eighty Years of Post Mendelian Breeding for Crop Yield: Nature of selection pressure and future potential, *Indian J. Genet.*, **46**(Suppl.): 30-53.

Jain H.K., Singhal N.C. and Austin A. 1976. Breeding for higher protein yields in bread wheat. Experimental approach and phenotypic markers. *Z. Pflanzenzuchtg*, **77**: 110-111.

Jenkins M.T. 1934. Methods of estimating the performance of double crosses in corn. *J. Amer. Soc. Agron.*, **26**: 199-204.

Jenkins M.T. 1935. The effect of inbreeding and of selection within inbred lines of maize upon the hybrids made after successive generation of selfing. *Iowa State J. Sci.*, **9**: 429-50.

Jenkins M.T. 1940. The segregation of genes affecting yield of grain in maize. *J. Soc. Agron.*, **32**: 55-63.

Jensen N.F. 1952. Intra-varietal diversification in oat breeding. *Agron. J.*, **44**: 30-34.

Jensen N.F. 1970. A diallel selective mating system for cereal breeding. *Crop Sci.*, **10**: 629-635.

Jinks J.L. 1955. A survey of genetical basis of heterosis in a variety of diallel crosses. *Heredity*, **9**: 223-238.

Jinks J.L. and Mather K. 1955. Stability in the development of heterozygotes and homozygotes. *Proc. Roy. Soc. B.*, **143**: 561-578.

Jinks J.L. and Pooni H.S. 1976. Predicting the properties of recombinant inbred lines derived by single seed descent. *Heredity*, **36**: 253-66.

Jinks J.L. and Pooni H.S. 1982. Predicting the properties of pure breeding lines extractable from a cross in the presence of linkage. *Heredity*, **49**: 265-270.

Johannsen W.L. 1903. Über Erblickeit in Populationen und in reinen Leinen. Gustav Fischer, Jena. (Heredity in populations and pure lines. *In*: *Classic Papers in Genetics.* (ed.) J.A. Peters. Prentice-Hall, Englewood Cliffs, New Jersy, 1959, pp. 20-26)

Johannsen W.L. 1926. *Elemente der exacten Erblichkeitslehre.* Gustav Fischer, Jena.

Johnson R. 1984. A critical analysis of durable resistance. *Annu. Rev. Phytopathol.*, **32**: 309-330.

Jones D.F. 1917. Dominance of linked factors as a means of accounting for heterosis. *Proc. Natl. Acad. Sci.*, **3**: 310-312.

Jones D.F. 1918. The effects of inbreeding and crossbreeding upon development. *Connecticut Agric. Exp. Stn. Bull.* **207**: 5-100.

Jones H.A. and Davis G.N. 1944. Inbreeding and heterosis and their relation to the development of new varieties of onions. *U.S.D.A. Tech. Bull.*, **874**: 1-28.

Jones L.R. and Gilman J.C. 1915. The control of cabbage yellows through disease resistance. *Wis. Agr. Exp. Sta. Res. Bull.*, **38**: 1915.

Karpechenko G.D. 1927. Polyploid hybrids of *Raphanus sativus* L. × *Brassica oleracea* L. *Bull. Appl. Bot., Genet. Pl. Breed.*, **17**: 305-410.

Kasha K.J. and Kao K.N. 1970. High frequency haploid production in barley (*Hordeum vulgare* L.) *Nature*, **225**: 874-876.

Kawasaki S., Borchert C., Deyholes M., Wang H., Brazille S., Kawai K., Galbraith D. and Bohnert H.J. 2001. Gene expression profiles during the initial phase of salt stress in rice. *Plant Cell*, **13**: 889-905.

Kearsey M.J. and Jinks J.L. 1968. A general method of detecting additive, dominance and epistatic variation for metrical traits. II. Application to inbred lines. *Heredity*, **24**: 45-47.

Keeble F. and Pellew C. 1910. The mode of inheritance of stature and of time of flowering in peas (*Pisum sativum*). *J. Genetics*, **1**: 47-56.

Kempthorne O. and Curnow R.N. 1961. The partial diallel cross. *Biometrics*, **17**: 229-250.

Kharkwal M. C. 1996. Accomplishments of mutation breeding in crop improvement in India. *In*: *Isotopes & Radiations in Agriculture and Environment Research.* (eds.) Sachdev M.S., Sachdev P. and Deb D.L. Indian Society for Nuclear Techniques in Agriculture and Biology. New Delhi. pp. 196-218.

Kharkwal M.C., Pawar S.E. and Pandey R.N. 2001. Seventy five Years of Research on Induced Mutations with special reference to Crop Improvement in India. *In*: *Applications of Radioisotopes and Radiation Technology in the 21st Century.* (eds.) N. Ramamurty, M. Ananthakrishnan and A.N. Nandakumar, Proc. of NAARRI International Conference December 12-14, 2001, Mumbai. pp. 230-235.

Kharkwal M.C., Pandey R.N. and Pawar S.E. 2003. Mutation Breeding for Crop Improvement. pp 601-645. *In*: *Plant Breeding - Mendelian to Molecular Approaches.* (eds.) Jain H.K. and Kharkwal M.C., Narosa Publishing House, New Delhi, India.

Khush G.S. and Rick C.M. 1967. Studies on the linkage map of chromosme 4 of the tomato and on transmission of induced deficiencies. *Genetica*, **38**: 74-94.

Khus G.S., Coffman W.R. and Beachell H.M. 2001. The history of rice breeding: IRRI's contribution, pp. 117-135. *In*: *Rice Research and Production in the 21st Century*. (ed.) Rockwood W.G.: Symposium honouring Robert F. Chandler Jr., Los Banos (Philippines); IRRI. 224 p.

Kihara H. 1917. Über cytologiesche studien bei einigen Getreidearten. Mit. 1. Spezies-Bastard der Weizens und Weigen roggenn-bastard. *Bot. Mag. Tokyo*, **32**: 17-19.

Kihara H. 1924. Cytologishe und genetische studien bei wichtigen Getreidarten mit besonderer Rücksicht auf das verhalten der chromosomen und die sterilität in den Bastarden. *Mem. Coll. Sci.* Kyoto Imp. Univ., **1**: 1-200.

Kihara H. 1951. Triploid watermelons. *Proc. Am. Soc. Hort. Sci.*, **58**: 217-230.

Kimura M. 1970. The length of time required for selectively neutral mutant to reach fixation through random frequency drift in a finite population. *Genet. Res.* Cambridge, **15**: 131-133.

Knott D.R. 1979. Selection for yield in wheat breeding. *Crop Sci.*, **15**: 295-299.

Konzak C.F. 1954. Stem rust resistance in oats induced by nuclear radiation. *Agron. J.*, **46**: 538-540.

Lander E.I. and Botstein D. 1989. Mapping Mendelian factors underlying genetic traits using RFLP linkage maps. *Genetics*, **121**: 185-199.

Law C.N. 1972. *Intervarietal substitution*. Plant Breeding Institute Report. 1972.

Leclercq P. 1966. Une sterilite Male Utilisable pour la production d' hybrides simples de tournesol. *Annales de l'Amelioration des plantes*, **16**: 135-144.

Leonard K.J. 1969. Selection in heterogeneous populations of *Puccinia graminis* f. sp. *avenae*. Phytopathology, **59**: 1851-1857.

Leonard K.J. 1977. Selection pressures and plant pathogens. Annals of the *New York Acad. of Sci.*, **287**: 207-222.

Lonnquist J. H. 1964. Modification of the ear-to-row procedure for the improvement of maize populations. *Crop Sci.*, **4**: 227-228.

Love H.H. 1927. A programme for selecting and testing small grains in successive generations following hybridization. *J. Amer. Soc. Agron.*, **19**: 705-712.

Mahalanobis P.C. 1936. On the generalized distance in statistics. *Proc. Natl. Inst. Sci. India*, **2**: 49-55.

Malecot G. 1948. *Les mathematiques de 1 Heredete*, Masson et cie. Paris.

Malthus T.R. 1798. *An essay on the Principle of Population*. Oxford Univ. Press. Reprinted by McMillan & Co., London, 1926.

Maluszynski M., Nichterlein K., Van Zantan L. and Ahloowalia B.S. 2000. Officially released mutant varieties—the FAO/IAEA database. *Mutation Breeding Review*. **12**: 1-84.

Mariani C., Gossele V., De Beuckeleer M., De Block M., Goldberg R.B., De Greef W. and Leemans J. 1992. A chimeric ribonuclease-inhibitor gene restorers fertility to male sterile plants. *Nature*, **357**: 384-387.

Mather K. 1938. *The Measurement of Linkage in Heredity*. Methuen, London.

Mather K. 1950. The genetic architecture of heterostyly in *Primula sinensis*. *Evolution*, **4**: 340-352.

Mather K. 1956. Response to selection. *Cold Spring Harbour Symp. on Quant. Biology*, **20**: 158-165.

Mather K. and Jinks J.L. 1982. *Biometrical Genetics*, 3rd edn. Chapman and Hall, London.

Matsumura S. 1953. (ed), "*Improvement of sugarbeet by means of triploidy*." No. 14 Science-Sha Tokyo, Japan.

Mayr E. 1954. Change of genetic environment and evolution. *In*: *Evolution as a process* (eds.) J.S. Huxley, A.C. Hardy and E.B. Ford. pp. 156-180. Allen and Unwin, London.

McClintock B. 1951. Chromosome organization and gene expression. *Cold Spring Harbour Symp. Quant. Biol.*, **16**: 13.

McRaes D.H. 1985. Advances in chemical hybridization. *In*: *Plant Breeding Reviews*. (ed.) J. Janick. Vol. 3. Avi Publishing Company, West Port, Connecticut, pp. 169-191.

Mendel G. 1865. Versuche über Pflanzen-Hybriden. *Verh. Naturforschung Ver.* In Brünn Verh., **4**: 3-47.

Mertz E.T., Bates L.S. and Nelson O.E. 1964. Mutant gene that changes protein composition and increases lysine content of maize endosperm. *Science*, **145**: 279-80.

Meyer V.G. and Meyer J.R. 1965. Cytoplasmically controlled male sterility in cotton. *Crop Sci.*, **5**: 444-448.

Mohan D.P. and Axtell J.D. 1975. DES induced high lysine mutants in sorghum. Pap. presented at *Ninth Biennial Grain Sorghum Res. and Util. Cong*. Lubbock, Tex., 4-6 Mar. 1975.

Morgan T.H. 1911. An attempt to analyse the constitution of the chromosomes on the basis of sex limited inheritance in *Drosophila. J. Exp. 2001*, **11**: 365.

Muller H.J. 1927. Artificial transmutation of gene. *Science*, **66**: 84-87.

Mullis K.B. 1990. The unusual origin of the polymerase chain reaction. *Scientific American*, **262**: 56-65.

Mullis K.B., Faloona F., Scharf S., Saiki R., Horn G. and Erlich H. 1986. Specific enzymatic amplification of DNA *in vitro*: The polymerase chain reaction. *Cold Spring Harbor Symp. Quanti. Bio.*, **51**: 263-273.

Munck L., Karlsson K.E. and Hagberg A. 1971. Selection and characterization of a high protein lysine variety from the world barley collection. pp. 544-58. *In*: *Barley Genetics II*. (ed.) Nilan R., Pullman, Wash.

Müntzing A. 1935. Triple hybrids between rye and two wheat species. *Hereditas*, Lund, **20**: 137-160.

Müntzing A. 1939. Studies on the properties and the ways of production of rye-wheat amphidiploids. *Hereditas*, Lund., **25**: 387-430.

Nei M. 1972. Genetic distance between populations. *Amer. Naturalist*, **106**: 283-291.

Nei M. 1973. Analysis of gene diversity in subdivided populations. *Proc. Natl. Acad. Sci.* (USA), **70**: 3321-3323.

Nelson O.E. 1967. A mutant gene affecting protein synthesis in the maize endosperm. *Genet.*, **21**: 209-230.

Nelson O.E. 1969. Genetic modification of protein quality in plant. *Advance. Agron.*, **21**: 171-194.

Nelson O.E., Mertz E.T. and Bates L.S. 1965. Second mutant gene affecting the amino-acid pattern of maize endosperm protein. *Science*, **150**: 1469-1470.

Nelson R.R. 1975. Horizontal resistance in plants: Concepts, controversies and applications, pp. 1-20. *In*: *Proc. of the Seminar on horizontal resistance to blast disease of rice*, CIAT Publ. Ser. C.E. 9, Cali, Colombia.

Nilsson-Ehle H. 1908. Surles travaux de selection × due troment et del avoine excute's a Svalof 1900-1912. *Bull. Mens. Renseign. Ag.*, Malad. Plantes IV, **6**: 3-12.

Nilsson-Ehle H. 1909. *Kreuzungsuntersuchungen an Hafer und weisen*. Lunds Univ. Aarskr. N.F. Afd. Ser. 2. Vol. **5**(2): 1-22.

Nilsson-Ehle H. 1948. The future possibilities of Swedish barley breeding. *Svalöof 1886-1946*, pp. 113-126.

Oehlker F. 1943. Chromosome mutation in meiosis by chemicals. *In*: *Mutation Research - Problems, Results and Prospects*, (ed.) C. Auerbach (1976), Champman and Hall, U.K.

O'Mara J.G. 1940. Cytogenetic studies on Triticinae. I. A method for determining the effect of individual *Secale* chromosomes on *Triticum*. *Genetics*, **25**: 401-408.

Opsahl B. 1956. The discrimination of interaction and linkage in continuous variation. *Biometrics*. **12**: 415-432.

Painter R.H. 1951. *Insect resistance in crop plants*. McMillan Co., New York.

Panse V.G. 1940. The application of genetics to plant breeding 2. The inheritance of quantitative characters and plant breeding. *J. Genet*., **40**: 283-302.

Panse V.G. and Sukhatme P.V. 1978. *Statistical Methods for Agricultural Workers*, ICAR, New Delhi.

Parlevliet J.E. 1978. Race-specific aspects of polygenic resistance of barley to leaf rust, *Puccinia hordei* relation during epidemic development. *Phytopathology*, **67**: 776-778.

Parlevliet J.E. 1979. The multiline approach in cereals to rusts: Aspects, problems and possibilities. *Indian J. Genet*., **39**: 22-29.

Paterniani E. 1967. Inter population improvement. Reciprocal recurrent selection variations. *Maize*, **8**: CIMMYT, Mexico.

Paterniani E.A. 1973. Recent studies on heterosis. *In*: *Agricultural Genetics*. (ed.) R. Moav. pp. 1-22. Natl. Counc. Res. Div. Jerusalem.

Perkins J.M. and Jinks J.L. 1968a. Environmental and genotype-environmental components of variability. III. Multiple lines and crosses. *Heredity*, **23**: 339-356.

Perkins J.M. and Jinks J.L. 1968b. Environmental and genotype-environmental components of variability. IV. Non-linear interactions for multiple inbred lines. *Heredity*, **23**: 525-535.

Person C. Groth J.V. and Mylyk O.M. 1976. Genetic changes in host-parasite populations. *Ann. Rev. Phytopathology*. **14**: 177-188.

Plaisted R.L. and Paterson L.C. 1959. A technique for evaluating the ability of selections to yield consistently in different locations or seasons. *Amer. Potato. J*., **36**: 381-385.

Potrykus I. 1993. Gene transfer to plants: Approaches and available techniques. pp. 126-137. *In*: *Plant Breeding - Principles and Prospects*, (eds.) M.D. Hayward, N.O. Bosemark and I, Ramagosa. Champman and Hall, London.

Powers Le Roy. 1934. The nature of interaction of genes differentiating habit of growth in a cross between varieties of *Triticum vulgare*. *J. Agr. Res*., **49**: 163-172.

Powers Le Roy. 1951. Gene analysis by partitioning method when interactions of genes are involved. *Bot. Gaz*. **113**: 1-23.

Rafalski J.A. and Tingey S.V. 1993. Genetic diagnostics in plant breeding RAPDs, microsatellites and machines. *Trends Genet*., **9**: 275-279.

Ramage R.T. 1965. Balanced tertiary trisomics for use in hybrid seed production. *Crop Sci*., **5**: 177-178.

Rapport I.A. 1946. Carbonyl compunds and the Chemical mechanism of mutation. C.R. *Doklady Acad. Sci*., USSR, **54**: 65.

Rapoport I.A. 1948. Alkylation of the gene molecule. *Doklady Acad. Sci*., USSR, **59**: 1183-1186.

Rasmusson J.M. 1935. Studies on the inhertiance of quantitative characters in *Pisum*. I. Preliminary note on the genetics of flowering. *Hereditas*, **20**: 161-180.

Rhoades M.M. 1933. The Cytoplasmic inheritance of male-sterility in *Zea mays*. *J. Genetics*, **27**: 71-93.

Rhoades M.M. 1955. The Cytogenetics of Maize. *In*: *Corn and Corn Improvement*. Academic Press, New York. pp. 123-219.

Richey F.D. 1927. The convergent improvement of selfed lines of corn. *Amer. Naturalist.*, **61**: 430-49.

Rick C.M. and Fobes J.F. 1974. Association of an allozyme with nematode resistance. *Rep. Tomato Genet. Coop.* No. **24**: 25.

Riley R. and Chapman V. 1958. Genetical control of cytologically diploid behaviour of hexaploid wheat. *Nature*, **182**: 713-715.

Rimpau W. 1891. Kreuzungsprodukte Landwirtschaftlicher Kulturpflanzen. *Landwirtschaftl. Jahrb.*, **20**: 335-371.

Robinson R.A. 1969. Disease resistance terminology. *Rev. Appl. Mycol.*, **48**: 593-600.

Roy Darbeshwar. 2000. Plant Breeding-Analysis and Exploitation of Variation. *Narosa, New Delhi.*

Rutger J.N. 1992. Searching for apomixis in rice. *In*: Proc. Apomixis Workshop. Atlanta, G.A., pp. 36-39.

Sapehin A.A. 1930. Rontgen-mutationen beim Weizen (*Triticum vulgare*). *Der Zuchter*, **2**: 257-259.

Sapehin A.A. 1936. X-ray mutants in soft-wheat, *Bull. Appl. Bot. Genet. Pl. Breed.* Ser. II., **9**: 3-37.

Savindan Y. 2000. Apomixis: Genetics and Breeding. *Plant Breeding Reviews*, **18**: 13-83.

Sax K. 1923. The association of size differences with seed-coat pattern and pigmentation in *Phasiolus vulgaris*. *Genetics*, **8**: 552-560.

Schnell F.W. 1961. Some general formulations of linkage effects in inbreeding, *Genetics*, **46**: 947-957.

Sears E.R. 1954. The aneuploids of common wheat. *Mo. Agr. Exp. Stn. Res. Bull.*, **572**: 1-58.

Sears E.R. 1956. The transfer of leaf rust resistance from *Aegilops umbellulata* to wheat. *Brook haven Symposia in Biology*, **9**: 1-22.

Sears E.R. and Okamoto M. 1958. Intergenomic relationship in hexaploid wheat. *Proc. X. Int. Cong. Genet.*, Montreal. Can. Univ. of Toronto. Press. **2**: 258-259.

Shamel A.D. 1905. The effect of inbreeding in plants. *USDA Yearbook*. Pp. 377-392.

Shull G.H. 1908. The composition of a field of maize. *Am. Breeders' Assoc. Rep.*, **4**: 296-301.

Shull G.H. 1909. A pure line method of corn breeding. *Am. Breeders' Assoc. Rep.*, **5**: 51-59.

Shull G.H. 1914. Duplicated genes for capsule form in *Bursa bursa pastoris*. *Z. Ind. Abstr. Ver.*, **12**: 97-149.

Simmonds N.W. 1979. *Principles of Crop Improvement*. Longman, London, U.K.

Simmonds N.W. 1985. Two stage selection strategy in plant breeding. *Heredity*, **55**: 393-399.

Singh R. and Axtell J.D. 1973. High lysine mutant gene (*hl*) that improves protein quality and biological value of grain sorghum. *Crop Sci.*, **13**: 535-539.

Skoog F. and Miller C.O. 1957. Chemical regulation of growth and organ formation in plant tissues cultivated *in vitro*. *In*: *Biological Action of Growth Substances*. Symp. Soc. Exp. Biol., **11**: 118-131.

Smith. 1936. A discriminant function for plant selection. *Ann. Eng.*, **7**: 240-250.

Spillman W.J. 1909. The Hybrid wheats. *Wash. Agr. Exp. Sta. Bull.*, 89.

Sprague G.F. 1946. Early testing of inbred lines of corn. *J. Am. Soc. Agron.*, **38**: 108-17.

Sprague G.F. 1963. Orietation and objectives. *Statistical Genetics and Plant Breeding*, (eds.) W.D. Hanson and H.F. Robinson. Natl. Acad. Sciences. NRC Publication 982, IX-XV.

Sprague G.F. and Tatum L.A. 1942. General *vs.* specific combining ability in single crosses of corn. *J. Amer. Soc. Agron.*, **34**: 923-932.

Springer P.S. 2000. Gene Traps: Tools for Plant Development and Genomics. *Plant Cell.*, **12**: 1007-1020.

Stadler L.J. 1928. Genetic effects of X-rays in maize. *Proc. Natl. Acad. Sci.*, USA, **14**: 69-75.

Stadler L.J. 1930. Some genetic effects of X-rays in plants, *J. Hered.*, **21**: 3-19.

Stadler L.J. 1944. Game selection in corn breeding. *J. Am. Soc. Agron.*, **36**: 988-89.

Stanford J.C. 1990. Biolistic plant transformation. *Physiologica Planterum.*, **79**: 206-209.

Stefansson B.R., Houghen F.W. and Bowney R.K. 1961. Note on the isolation of rape plants with seed oil free from erucic acid. *Can. J. Plant Sci.*, **41**: 213-215.

Steward F.C., Mapes M.O. and Mears K. 1958. Growth and organized development of cultured cells, II: Organization in cultures grown from freely suspended cells. *Amer. Jour. Bot.*, **45**: 705-708.

Sturtevant A.H. 1913. The linear arrangement of six sex linked factors in *Drosophila*, as shown by their mode of association. Jour. Exp. Zool., **14**: 43-59. The first paper on gene mapping. (Reprinted in the *Classic Paper in Genetics.* (ed.) Peters J.A. (1959), Prentice Hall, Englewood Cliffs, N.J.

Suneson C.A. 1956. An evolutionary plant breeding method. *Agron. J.*, **48**: 188-191.

The *Arabidopsis* Genome Initiative (TAGI). 2000. Analysis of the genome sequence of the flowering plant *Arabidopsis thaliana*. *Nature*, 408: 796-815.

Taliafero C.M. and Bashaw E.C. 1966. Inheritance and control of obligate apomixis in breeding Buffelgrass (*Cenchrus ciliaris* L.). Crop Sci., 6: 473-476.

Tanksley S.D. and Rick C.M. 1980. Isozyme linkage map of tomato : applications in genetics and plant breeding. *Theor. Appl. Genet.*, **57**: 161-170.

Tschermack E. Von. 1900. Über Kunstliche Kreuzung bei *Pisum sativum*, *Ber. der Bot. Ges.*, **18**: 232-239.

Thoday J.M. 1961. Location of polygenes. *Nature*, **191**: 368-370.

Thoday J.M. 1972. Disruptive selection. *Proc. Royal. Soc. London. B.*, **182**: 109-143.

Van der Plank J.E. 1963. *Plant Diseases. Epidemics and Control.* Academic Press, New York.

Van der Plank J.E. 1968. *Disease resistance in plants.* Academic Press, New York and London.

Van der Plank J.E. 1982. *Host-pathogen interactions in plant diseases.* Academic Press, New York.

Van der Veen. 1959. Test of non-allelic interaction and linkage for quantitative characters in generations derived from two diploid pure lines. *Genetics*, **12**: 415-432.

Vasal S.K. 1994. High quality protein corn. p. 80-121. *In*: *Specialty corns*, (ed.) A.R. Hallauer.

Vasal S.K., Villegas E. and Bauer R. 1979. Present status of breeding quality protein maize. pp. 127-48. *In*: *Proc. Symp. On seed protein Improvement in cereals and grain legumes.* Neuherberg, Fed. Repub. Ger. 4-8 Sept. 1978.

Vasil I.K. 1978. Plant tissue culture and crop improvement. *Int. Assoc. Plant Tissue Culture Newsletter*, **26**: 2-10.

Vavilov N.I. 1926. Studies on the origin of cultivated plants. *Bulletin of Applied Botany and Plant Breeding*. **16**: 1-245.

Vavilov N.I. 1951. *The origin, variation, immunity and breeding of cultivated plants*. Translated from Russian by K.S. Chester. Chron. Bot., **13**: 1-366.

Venkataraman T.S. 1938. Hybridization in and with the genus *Saccharum*. *Proc. Ind. Sci. Congr.*, **25**: (Part 2, Sec. 9): 267-284.

Venter J.C. *et al.* 2001. The sequence of the human genome. *Science*, **291**: 1304-1351.

Vos P., Hogers R., Bleeker M., Reijans M., Van der Lee T., Hornes M., Frijters A., Pot J., Peleman J., Kuiper M. and Zabeau M. 1995. AFLP: A new technique for DNA fingerprinting. *Nucleic Acids Res.*, **23**: 4407-4414.

Weinberg W. 1908. Über den nachweis der Vererbung beim Menschen. (Translated into English and printed in collection of Boyer S.H. (1963) and of Corwin H.O. and Jenkins J.B. 1976).

Williams J.G.K., Kubelik A.R., Livak K.J., Rafalski J.A. and Tingey S.V. 1990. DNA polymorphisms amplified by arbitrary primers are useful as genetic markers. *Nucleic Acids Res.*, **18**: 6531-6535.

Winkler H. 1916. Über die experimentelle Erzeugung von Pflanzen mit abweichenden chromosomenzahlen. *Zeitschr. Bot.*, **8**: 417-531.

Wolfe J.A., Barrett J.A., Shattock R.C., Shaw D.S. and Whitebread R. 1976. Phenotype-phenotype analysis: Field application of the gene-for-gene hypothesis in host-pathogen relations. *Ann. Appl. Bio.*, **82**: 369-374.

Wright S. 1921. System of mating. *Genetics*, **6**: 111-178.

Wright S. 1931. Evolution in Mendelian population. *Genetics*, **16**: 97-159.

Wright S. 1935. The analysis of variance and the correlation between relatives with respect to diviations from an optimum. *J. Genetics*, **30**: 243-256.

Yates F. 1937. *The design and analysis of factorial experiments*. Imperial Bureau of Soil Science, Harpenden.

Ye X., Al-Babili S., Kloti A., Zhang J., Lucca P., Bayer P. and Potrykus I. 2000. Engineering the provitamin A (β-carotene) biosynthetic pathway into (carotenoid free) rice endosperm. *Science*, **287**: 303-305.

Yule G.V. 1906. On the theory of inheritance of quantitative compound characters on the basis of Mendel's laws. *Third Int. Conf. Genet.*, 140-142.

Plant Breeding - Mendelian to Molecular Approaches
H. K. Jain and M. C. Kharkwal (eds.)

3

Plant Breeding Science and Practice in the Twentieth Century: Some Landmarks

N.G.P. Rao[1]

Abstract

This paper attempts a brief analysis of the growth of the science of plant breeding and some of its landmarks. Advances in our understanding of evolutionary and population genetics have significantly influenced the theory and practice of plant breeding. Pure line selections among the domesticated land races of crop plants in the various eco-geographic regions and selections from crosses between elite lines yielded cultivars, which helped to increase agricultural production during the 20th century. Induced mutations, chromosome manipulations and distant hybridization involving wild species and genera, generated lot of interest but their direct contribution to improved cultivars has been limited. Transgenics to date have succeeded largely in incorporating relatively simply inherited traits. The breeding procedures have become more and more reductionist as compared to an integrated individual as the unit of selection under evolution. Among the landmarks in plant breeding are: pure line selections in land races and selections from elite crosses in almost all crops; inter-specific cultivars of sugarcane grown across continents which became feasible because of their vegetative propagation. Recent improvements for yield and fibre characters in *G. arboreum* cotton resistant to sucking pests and tolerant to boll worms has been a significant development; other significant developments include exploitation of wider germplasm sources including dwarfing genes as in rice, wheat and sorghum to develop semi-dwarf, high yielding varieties with marked changes in adaptability, input responses, leading to more productive cropping systems like that of rice-wheat; exploitation of heterosis employing male sterility and other systems in crops like maize, sorghum, pearl millet, cotton, sunflower and castor. The performance of hybrid sorghum, pearl millet, cotton and sunflower in rainfed agriculture has been a major contribution.

Introduction

The first principles of plant breeding were enshrined in the process of evolution of species and origin of crop plants in nature. Plant breeding is man made evolution. Except for the rates and direction of genetic change the principles of evolution in nature are valid for plant breeding. Nature is the first plant breeder.

[1]HIG (OG) Block 10, Flat 4, Baghlingampally, Hyderabad 500 044, India

The origin of settled agriculture is lost in antiquity. Subsequently, with the passage of time, increasing need of food and fibre prompted man to rendering cultivated forms more distinct from their wild ancestors. Land races developed during migration and domestication under diverse eco-geographic situations as an adaptive response through both natural and human selection. Farmers are the second generation plant breeders. Farmers are known to select novel forms even today.

Sexuality of plants was known and influenced crop hybridization in the pre-Mendelian era. Tobacco hybrids of Koelreuter, progeny tests of Vilmorin and improvement of sugar content in sugar beet from 6 to 16 % by mass and pedigree selection, and novel fruit and ornamental plants derived from hybridization are early examples of human endeavour from Europe. Plant breeding has come to be known as the Art of Plant Breeding. The artistic aspect of plant breeding continues inspite of application of science.

The enumeration of Mendel's laws (1865) and their rediscovery independently by Correns, De Vries and Tschermak in 1900 provided the genetic basis for plant breeding. Subsequent developments in various branches of genetics, cytogenetics, population genetics, quantitative genetics and molecular genetics enhanced our knowledge and opportunities to genetically modify crop plants. Allied sciences - plant physiology, biochemistry, entomology, plant pathology, statistics, etc., further influenced breeding efforts and procedures for multi-trait improvement and breeding efficiency. The era of science of plant breeding is the journey from phenomics to genomics.

Plant breeding is now both a public welfare and market oriented economic activity. Public and private sectors are involved. Globalization, market economy, plant variety rights, IPR, genetically modified organisms (GMOs) bring in ethical, social, commercial and legal implications.

Evolution: Significance to Plant Breeding

Evolutionary changes have been taking place in life forms since the origin of life on earth, resulting in diversity of species. Evidences from different disciplines indicate that early agricultural economies must have developed during late Mesolithic and Neolithic ages (15,000-10,000 BC), independently in different parts of the world (Harlan, 1992).

The principles of evolution emanating from Darwinism and neo-Darwinism are important to present day plant breeding. Darwin's (1859) theory of origin of species by means of natural selection and the preservation of the favoured ones focused on the dynamic processes and changes in plant populations leading to the formation of species, intra-specific variants like races, ecotypes and even higher taxa.

Huxley's (1942) evolutionary synthesis and Mayr's (1991) discussion attempt to reconcile Darwin's theory of natural selection with Mendelian hereditary principles. The consequence is the emergence of "neo-Darwinism." The Darwinian theory postulated that selection targeted individuals. In the reductionist genetic hypothesis independent genes play a larger independent role in fitness and evolution and involve largely allelic frequencies in populations.

The early works of De Candolle (1886) and Vavilov (1926) led to the concept of centres of diversity of crop plants as also the centres of origin. Vavilov recognised 12 such centres. Tropical West Africa which did not find a place in Vavilov's map has now emerged as an important area of diversification for several crops. Also the lack of coincidence between the areas of wild ancestors, areas of domestication and evolutionary diversification in several instances warranted changes in the geography of crop evolution. Harlan (1992) preferred to designate the centres of origin as 'Ecological Regions of Diversity'.

Genetics, Population and Adaptation Processes : Basis for Plant Breeding

The story of twentieth century plant breeding science is the travels and travails from Mendelian factors to heredity molecules; from genotype to phenotype through the environment and refinements in tools and breeding plans to meet agricultural needs of ever increasing mankind in the twenty-first century.

Genetic Processes

The genetic studies based on natural populations provided the scientific basis for plant breeding. Mendel (1865) hypothesized that traits were transmitted by factors and follow a predictable pattern of inheritance. About the same period East (1908) and Shull (1909), observed inbreeding depression on selfing and restoration of vigour on crossing in maize. This led to the development of concept of heterozygote advantage and the role of dominance, 'overdominance' and gene interactions in heterosis and led to its utilisation in cross pollinated crops. The concept was later extended to self-fertilised crops. First observations on the role of cytoplasm in inheritance were also observed during this period.

The occurrence of spontaneous mutations in nature during evolution was an important source of variation. Gene mutations or point mutations are sudden changes at a major locus and are generally recessive in nature. Mutations are alterations in base sequence in the DNA in situations that lead to altered polypeptides and hence phenotypic effects. The word mutation was conceived by De Vries (1900) from his observations on *Oenothera*, which later proved to be a structural heterozygote.

Both ionising and non-ionising radiations induce mutations. Several chemicals which are alkylating agents like sulphur mustard, nitrogen mustard, epoxides, ethyleneimines, sulphates and sulphonates, diazoalkenes and nitrous compounds have been used to induce mutations. Several induced mutants have been reported, but only very few went into large scale cultivation. Aruna castor from Indian Agricultural Research Instute (IARI) centre at Hyderabad and TG Groundnut mutants from Bhabha Atomic Research Centre, (BARC) at Trombay entered into large scale cultivation in India.

Apomixis is emerging as a potential tool for improving grain crops in future. To-date apomixis does not appear to have been transmitted from wild to cultivated crops. Two approaches being pursued are direct transfer of apomixis and gene isolation, or both. Two molecular markers, RFLP and RAPD, were reported to be linked to apospory in *Pennisetum*. A review of the current state of apomixis in relation to genetics and plant breeding will be found in Savindan (2000).

Cytological and genetical studies on *Drosophila* and maize by Morgan (1911), Muller (1927), Sturtevant (1913) and McClintock (1951) led to the location of genes on specific chromosomes, operation of linkages between genes on a chromosome which led to the chromosome theory of heredity and development of genetic and cytological maps. The 1930-1950 period was dominated by cytogenetic researches in all crops. The studies involved chromosome pairing and structural changes in chromosomes, particularly in wide crosses and were related to wild gene transfer. Studies on changes in chromosome number led to the development of autopolyploids and allopolyploids and the use of aneuploids like trisomics, monosomics and nullisomics in gene location and developing substitution lines. Sears developed genetic stocks to locate genes for rust resistance in wheat and were used extensively in breeding for rust resistance. Chromosome 5B in wheat also attracted attention because it has genes controlling homeologous pairing. Cytogenetic studies laid particular emphasis in gene transfer from wild species and genera. The most successful example is that of sugarcane where crosses of *S. officinarum* × *S. spontaneum* resulted in many commercial varieties widely cultivated. Efforts at inducing translocations have resulted in some genes transfer across species as in the case of leaf rust from *Aegilops* to bread wheat. Frey (1984) reported that introgression of wild genes into cultivars enhances variability for yield considerably and hence offers scope for yield improvement. But there seem to be no examples of developing high yield cultivars through wild gene incorporation. There was considerable interest in synthesizing amphi-diploids and allopolyploids as in cotton, *Brassica* etc. They did provide proof of their origin, but no where near natural amphi- diploids. Synthetic *Triticale* and *Raphano-brassica* generated considerable interest but could not become commercial.

Several colchicine induced autopolyploids with doubled chromosome numbers have been synthesised in many crops and again there are no examples of commercialisation. Autopolyploid grape only raised some expectations. The synthesised autopolyploids were no more successful than the diploids. Haploids occur in nature. Haploids have also been induced through wide hybridization and through anther culture. The techniques of producing haploids and doubled haploids have been improved (Raina, 1997). However in commercial breeding the conventional inbreds are ruling.

Plant breeding has now entered the era of molecular biology. Discoveries of the structure of DNA and genetic code gave molecular basis to the Mendelian 'factors' or 'genes'. The DNA from one organism can function in another and leads to transformation systems and transgenics. The availability of alternative transformation systems such as *Agrobacterium*, Biolistics, Silicon fibre, Electroporation, etc., enabled development of transgenics. Traits like herbicide resistance, insect resistance and virus resistance from bacteria and viral pathogens have been incorporated. First generation transgenic crops of maize, cotton, potatoes, soyabeans, etc., made commercial impact in USA, Canada, Argentina and some other countries. Quality trait improvement has also potential as in the case of golden rice, quality protein gene of *Amaranthus* isolated in India, etc. The release of transgenics (genetically modified organisms) raised issues of environment, bio-safety and led to regulatory and legislative measures. Concerns have also been expressed on genetically modified food crops. India is yet to develop transgenics with the exception of *Bt* cotton on which trials are being conducted. Transgenics may have a great potential in altering amino acid profile of oils and orientation to end products for industrial processing and nutritional attributes. The transgenics derived till today largely involve simply inherited traits like herbicide/insect/disease resistance and they are also subject to breakdowns due to changes in pathogen or environment, as has been the case with traditional methods of major gene transfer.

The DNA markers have emerged as an important tool for selection. Among the marker assisted selections (MAS), the first marker system developed was the RFLP (Restriction Fragment Length Polymorphism). This was followed by the RAPD (Randomly Amplified Polymorphic DNA), SSRs (Simple Sequence Repeats) and Microsatellite based markers.

The genetic mapping of QTLs with DNA markers provides a quantitative perspective. These tools are expected to be more precise and effective compared to traditional selection criteria. The DNA markers are significant in basic studies, but they are yet to reach a stage when they could replace present approaches.

The era of functional genomics, a new phase of genome analysis has arrived. Massive relational data bases and computational approaches are taking their places alongside banks of automated sequences, and arrays of biochips as the tools of this new science. Comparative genomics is offering unparalleled opportunities to link biological kingdoms, enabling crossing species and genera. Genome based data bases are available and the success of the genome projects depends on the availability and use of the data bases.

Population Processes

Fisher's *Genetical theory of natural populations*, Wright's *Evolution in Mendelian populations* and Haldane's *The causes of evolution* laid the foundations for population genetic-theory. The Hardy-Weinberg law concerning equilibrium population under random mating enumerates the role of mutation, migration and selection in diploid random mating populations in changing gene frequency.

Unlike simply inherited traits most economic traits are governed by a large number of genes called polygenes with minor effects relative to environmental variation, result in measurable continuous variation and hence such traits are known as metric traits. An analytical approach to plant breeding based on quantitative genetic theory revolves around the concepts of phenotype, genotype, environment and genotype × environment interactions. The basic model is:

$$P = G + E + (GE) \quad \text{and} \quad \sigma^2 P = \sigma^2_G + \sigma^2_E + \sigma^2_{GE}$$

Plant breeding is based on this model amplifying genetic, environmental and G × E interaction components and maximising gain from selection through appropriate breeding plans. This basic model has been adopted to replicated, randomised, field experiments conducted over single/multiple locations/years/environments, etc.

Sprague and Tatum (1942) coined the words 'general combining ability' (gca) and 'specific combining ability' (sca). Parent offspring regressions, graphic methods like Vr-Wr graphs, methods of partitioning the mean from segregating generations, mating designs like diallel and its modifications have been more conveniently used for estimating the genetic parameters, particularly in self-pollinated crops. Comstock and Robinson, 1952 developed I, II and III designs and partitioned the genetic variation into various components in the open pollinated populations.

The estimation and understanding the role of genetic variances and the findings that intermating results in dissipating repulsion phase linkages besides changes in gene frequencies led to development of efficient breeding plans, such as pedigree and mass pedigree methods. The intra- and inter-population improvement procedures, composites, synthetics, gene pools and high yield inbreds from diverse populations to maximise heterosis were developed.

Adaptation Processes

Another aspect of genetic parameters pertains to the estimation of g × e interaction variances and regression methods of Finlay and Wilkinson (1963); and Eberhart and Russell (1966) enhanced our understanding of the concepts of adaptation and adaptability of cultivars, and phenotypic stability. The g × e studies led to demarcating adaptation zones, variety testing procedures and the relative significance of the various g × e interactions in variety development plans. A review of genotype-environment interactions and their significance for cultivar development will be found in *Kang* (1998). The changes in adaptation concept is a landmark of the century. Irrespective of the tools used in crop breeding, adaptation processes will have a permanent place in cultivar development.

Plant physiologists (Wallace and Yan, 1998) refer to the whole system crop physiology and the constant capacity plant system hypothesis. The focus has been on system output and its near fully integrated components of yield-biomass and biomass per day, yield and yield per day, days to flower and maturity and harvest index.

Adaptation and productivity are complexly inherited traits and are much affected by the environment. Allard (1996) observed that population behaviour can be explained on the basis of selecting chromosomal segments, in which the marker loci reside and that marker assisted dissection of the genetic basis of adaptation is feasible. Thus, discretely inherited markers often provide breeders with effective ways of identifying, tracking and incorporating regions of chromosome with favourable effects on adaptation into elite materials. Changes in crop duration, dry matter production and distribution and crop phenology have reduced seasonal yield failures and fluctuations and conferred greater stability of performance across years and geographies (Rao, 1982; Rao *et al.*, 1989).

The genetics of resistance to plant diseases and insect pests received attention from early 1900s. Plant defence mechanisms for disease resistance-avoidance, resistance, tolerance and durable resistance have been exemplified. Breakdown of simply inherited resistances due to new pathogen races has been a frequent occurrence. To get over this, the concept of components of partial resistance equivalent of quantitative resistance has been used to move towards durable resistance. Such examples will be found in case of barley leaf rust, late blight of potato, bacterial leaf blight of rice, etc. (Parlevliet, 1992). Similarly, in case of insect pests, the resistance mechanisms - non-preference, antibiosis and tolerance have been used in breeding programmes.

Plant modifications to tolerate environmental stresses such as drought and water-use efficiency, soil toxicities and beneficial host - soil-microbe interactions are also features of adaptation. Water-use efficiency reflects in greater dry matter production or economic yield per unit of water in a given evaporative environment. This has been estimated from gas exchange by measuring the ratio of CO_2 fixed to water transpired. C_3 plants tend to have lower water-use efficiency than C_4 photosynthesis, as in maize or sorghum. In case of drought resistance, dehydration, avoidance and tolerance have been the major mechanisms. Improved root depth, cuticle thickness and increased production indicate tolerance to dehydration (Boyer, 1996).

Toxic soil conditions such as salinity, aluminium toxicity, etc., have been successfully handled by genetic manipulation of correlated physiological traits. Plant soil-microbe interactions in the rhizosphere are significant. They may enhance growth or provide protection from pathogenic micro-organisms. Symbiotic *Rhizobia* and non-symbiotic or associative BNF by *Azotobacter*, *Azospirillum*, etc., contribute to soil fertility. The role of nodulins, flavonoids and plant lectins in BNF is being investigated. With molecular genetic approaches, the efforts could lead to more success.

With limitations on cultivable land and water, there is considerable emphasis on crop diversification, crop intensification through multiple cropping and on inter-crops, with the twin objectives of promoting productivity per day and sustainability of production. In this context, competition between crops in space and time assumes importance. Differences in intra- and inter-plant competition within a species, competition between species for water, nutrients and sunlight is critical. Plant densities and duration, and architecture becomes significant in adaptation to cropping systems. The stability of cropping systems across years and seasonal fluctuations also need to be considered. Rao *et al.* (1981) and Rao (1991) reported these aspects in relation to sorghum based cropping systems. A review of competitive ability and plant breeding has been done by Fasoula (1997).

Some Significant Landmarks in Plant Breeding

The plant breeding history in India could be considered under two broad periods:

1. Pre-green Revolution (before 1965)
2. Green Revolution (1965-2000)

Pre-Green Revolution Period

By and large, the native tropical cultivars are characterized by tallness, photosensitivity, relative lateness compared to duration of rainy season, low harvest indices, low economic yields, local adaptation and low response to increasing fertilizer and population levels. Yields in India were not only low, but vulnerable to climatic fluctuations resulting in food shortages and even famines in drought prone areas.

(*a*) The Significant Role of Land Races

Cultivars selected from locally domesticated land races, spontaneous mutants and some induced mutants, and products of hybridization, largely between closely related cultivars and a few involving distant hybridization, dominated the agricultural scene of the pre-green revolution era. Consequent to these plant breeding efforts and increasing fertilizer use and cultivation practices during the first 20 years after independence, crop production doubled from 51 to 108 m tonnes between 1950 and 1970 (Table 1). A few landmarks in plant breeding during this period are focussed as follows.

Table 1. Progress of production of some agricultural commodities in India (M.tons)

Crop/Commodity	1950-1951	1960-1961	1970-1971	1980-1981	1990-1991	1999-2000
Foodgrains	50.82	82.02	108.42	129.59	176.39	201.56
Rice	20.58	34.58	42.22	53.63	74.29	88.55
Wheat	6.46	11.00	23.83	36.31	55.14	70.10
Coarse cereals	15.38	23.74	30.55	29.02	32.70	29.36
Grains legumes	8.41	12.70	11.82	10.93	14.26	13.55
Oilseeds	5.16	6.98	9.63	9.37	18.61	21.54
Cotton (m. bales)	3.04	5.60	4.76	2.01	9.84	10.45
Jute (m. bales)	3.31	5.26	6.19	8.16	9.23	9.53
Sugarcane	57.05	110.00	126.37	154.25	241.05	292.64
Potato	1.66	2.72	4.81	9.67	15.21	24.64
Tobacco	0.26	0.31	0.36	0.48	0.56	0.70
Coconut (m. nuts)	35.82	46.39	60.75	57.20	97.30	149.25

Wheat: Some of the prominent land races from which pure line selections were made included: *Sharbati, Dara, Karachi choice, Lal Kanthi, Lal pissi*, etc., in bread wheat; *Bansi, Malwi, Khandwa, Hara, Jalelia*, etc., in *durum* and *Khapli* etc., in *dicoccum*.

Systematic pure line selection in land races of wheat was initiated in 1904 at Pusa, Bihar by Howards (1910) and later pursued by B.P. Pal and his colleagues. Pusa 4 (NP 4), a selection from *Mundia* (awnless) was known for its quality and was awarded first prize in several international exhibitions (1916-1920). It was also reported to have been cultivated in South Africa, Rhodesia, Australia, Hungary, etc. Several promising selections were made in the local land races and released for cultivation at Layalpur, Kanpur, Nagpur, Niphad and Parbhani, etc., to suit local needs.

Hybridization between land races for incorporation of specific traits was the next step and several hybrid derivatives were cultivated in different parts of the country. NP 52 and a series of NP wheats from Pusa; PbC 518, PbC 591, etc., from Punjab; Niphad 4, etc., from Maharashtra were well known in different parts of the country.

The use of exotic varieties in hybridizations for combining yield and rust resistance formed the next important stage in wheat improvement. Varieties like Thatcher, Federation, Gabo, Hofed, Timstein, Chinese and Kenyan in the bread wheat group; Gaza and Arabian durums and some exotic dicoccums were used for improving the respective species. The NP 700 and 800 series wheats, kenphad wheats, etc., were the result of hybridization and selection.

This period was also marked by the use of cytogenetic approaches, distant hybridization, translocations, monosomics, nullisomics, radiations, etc., in wheat improvement. The transfer of leaf rust resistance from *Aegilops* sp. to wheat by Sears (1956) and his colleagues using ionising radiations to produce translocations between wheat and *Aegilops* carrying a gene for leaf rust resistance is cited as a classical example. The *Lr.9* was found resistant to several wheat rust pathotypes across the world. Knowledge of the genetic control of chromosome pairing in wheat enabled transfer of genes from alien species to wheat. In wide crosses, progenies lacking chromosome 5B have high level of pairing and recombination.

Another very significant contribution in wheat improvement pertains to the development of knowledge about the wheat rusts and their management. The identification of 45 *Sr* genes for stem rust, 47 *lr* genes for leaf rust and *29 lr* genes for stripe rusts and their manipulation provided an ideal example of pathologist-breeder cooperation in containing rust epidemics. Critical epidemiological studies of wheat rusts, the genetics of rust resistance aided breeders to successfully develop resistant cultivars, gene deployment to create genetic diversity in large geographic areas, furnishing a strategy for rust control and management. The works of Howard (1928), Pal (1966) and Sawhney (1998) summarize wheat-breeding efforts in India.

Rice: Pure line selection in land races of agro-ecological regions and agro-ecosystems began in early 1900s and continued till the sixties as the main method for the genetic improvement of rice. Some inter-varietal hybridization was also practised for incorporation of specific traits. Before 1960s, of the 450 improved varieties selected in India, only 25 were of hybrid origin. Thus, selection in land races was the main avenue for cultivar improvement under various situations, which ruled rice fields in various states. Particular mention may be made of GEB 24 selected at Coimbatore, which combined moderate yields, good grain quality and wider adaptation. Also, these land race selections provided the mainstay for securing resistance for various biotic and abiotic stresses and were used as donors in later years. Mention may be made of varieties like TKM 6 for stem borer resistance; PTB 21 for tungro virus resistance; SR 26B for salt tolerance; *Jala-magna* and FR 43B for deep water and flood prone conditions. The list of such donors from land race selections will be large and could be retrieved from rice germplasm catalogues and breeders.

The *indica-japonica* hybridization program to improve yield levels of *indica* rice was initiated in several countries. Success during the pre-green revolution period was limited. ADT 27 developed at Coimbatore was a successful example. The Korean breeders seem to have had more success with *indica-japonica* crosses. *Mahsuri* a derivative from *indica-japonica* crosses of Malaysia and introduced into India combined high yields, good grain quality and wider adaptation and was extensively cultivated. Several variants of *Mahsuri* were selected in various parts of India. Besides, *Mahsuri* provided the parental material in later breeding programs. The adaptation of *Mahsuri* in India is a landmark. Gangadharan (1985) and Siddiq (2001) provide a comprehensive account of rice breeding researches.

(*b*) The Success of Interspecific Crosses in Sugarcane

The establishment of the Sugarcane Breeding Station at Coimbatore, India during 1912 was a significant step, which laid the basis for sugarcane improvement and a stable sugar industry not only in India but across many countries. Mention should be made of three sugarcane species which played a vital role in developing commercial hybrid canes, *Saccharum officinarum* (2n = 80), *S. barberi* (2n = 81, 82, 83, 91-92, 105) and *S. spontaneum* (2n = 40 to 128). Barber (1916) at Coimbatore crossed a *S. officinarum* clone *vellai* with locally available *S. spontaneum*, which gave rise to commercially accepted Co. 205 and replaced *S. barberi*. In India intercrossing of elite derivatives of the basic crosses between the three species led to development of varieties like Co 290, Co 312, Co 313 for sub-tropical India and Co 419 for tropical India, which became to be known as the 'Wonder Cane of India'. The breeding philosophies of almost all sugarcane growing countries were on similar lines. Sugarcane is one example where hybridization with wild species led to the development of widely cultivated varieties across the world. This example of sugarcane is certainly a landmark in the early Plant Breeding History of India. Barber (1916) and Sir T.S. Venkatraman (1938) pioneered these breeding efforts.

Green Revolution Period

(*a*) Exploitation of Dwarfing Genes in Wheat and Rice

Exploitation of dwarfing genes in wheat and rice initiated the green revolution. The dwarf forms furnished lodging resistance, superior harvest indices and better utilization of environmental resources. Plant modifications due to major changes at a few loci associated with pleiotropic effects led to changes in form, function and adaptation. The discovery of the Norin-10 dwarfing genes in wheat and the *Dee-geo-wu-gen* (DGWG) dwarfing genes in rice led to marked plant modification in wheat and rice. *Rht-B1* and *Rht-D1* of Norin source which are partially dominant, largely contributed to advances in wheat. In rice, the *Sd1* gene from the *Dee-geo-wu-gen* (DGWG) was the main contributor. A comprehensive review of the dwarfing genes will be found in Milach and Federizzi (2001).

The Semi-dwarf Wheats: *Gains* was probably the first semi-dwarf variety released in USA during 1962 using Norin-10 dwarfing genes. This line subsequently entered the CIMMYT wheat breeding program. CIMMYT introduced several breeding populations carrying the dwarfing genes into the programs of North America, Europe and Asia, which initiated transformation of the wheat scenario.

In India, the bulk introduction of the Mexican semi dwarf spring wheats - Sonora 64, Lerma Rojo took place in 1965-66. The initial introductions because of red grain colour were not acceptable to the Indian consumer, for whom the preference is amber coloured grain. Selection for amber grain in later introductions led to the development of popular varieties like Kalyan sona and Sonalika released in 1967, which spread rapidly and paved the way for further improvement in yield, quality and disease resistance. Thus, Kalyan sona and Sonalika may be considered as landmark in wheat improvement. Subsequent breeding efforts at IARI and various agricultural universities of India led to a number of varieties including HD series, reflecting improvements in yield, quality and disease resistance. Semi-dwarf *durum* and *dicoccum* wheats have also been developed. Similarly, all the wheat growing states developed semi-dwarf high yield wheat varieties.

Semi-dwarf Rice: The era of short-statured *indica* varieties in rice began with the advent of Taichung Native-1, Taichung 65, Tainan 3, etc. But it is the development of IR 8 at the International Rice Research Institute and its introduction into several national research programs of Asia that heralded the transformation of the rice scene. IR 8 has been mentioned as the miracle rice and is certainly a landmark in rice breeding. A number of IR varieties have been developed at IRRI and introduced into several National Research Programs, which became useful directly or indirectly by furnishing parental material and by release of several new varieties in the respective states/countries. Similarly, the rice variety Jaya, developed in India, set a benchmark for its yield level in the mid-late maturing group. A number of medium duration and mid-early varieties attained the yield levels of Jaya. What is more significant is the improvement of grain quality to suit local conditions at high yield levels. The aromatic rice, *Basmati*, characterized by super fine grain, pleasant aroma, extra elongation of kernel and soft texture of cooked-rice rendering it a delicacy, is in demand in India and abroad. Pusa Basmati, Pusa Sugandh, semi-dwarf Basmati rice varieties have high yields compared to traditional Basmatis. A review of rice breeding will be found in *Siddiq* (2001).

(*b*) Exploitation of Heterosis

Maize: Historically, commercial exploitation of heterosis began with maize in USA during 1930 through double cross hybrids. Extensive quantitative genetic studies involving open pollinated varieties of maize and population improvement methods by exploiting additive genetic variance led to development of high yield inbreds, which in turn promoted commercial single-cross hybrids. Single-cross hybrids were cultivated since 1960 and today almost the entire hybrid corn in USA is single-cross. The quantitative genetic studies during 1950s and 60s at the Universities of North Carolina, Nebraska and Iowa constituted a landmark in global maize improvement and provided the quantitative genetic base for crop improvement. Hybrid maize in India became a commercial success from the beginning of the green revolution era and today, double-cross, three-way-cross, and single-cross hybrids are cultivated extensively. Quality protein maize has also been developed and released.

Sorghum: The discovery of cytoplasmic genetic male sterility in sorghum enabled development of commercial sorghum hybrids in USA and later in other countries, notably India. Commercial sorghum hybrid CSH 1, based on male sterile combine kafir 60, was released in 1964 and stood the test of time. Yellow endosperm short statured derivatives from African × American dwarfs provided the male parents. The development of restorers like CS 3541, a dwarf zerazera, and 296A, a *milo* based sterile involving *karad* a local Indian variety marked significant steps in breeding CSH 5 and CSH 9, which were widely adapted and stable across rainfed environments. Converted zerazeras like CS 3541 and 296 A are extensively employed in several breeding programs of private and public sectors and are landmarks in parental development.

Pearl Millet: About the same time in 1960, the availability of male steriles in pearl millet led Indian breeders to develop commercial hybrids. The modified HB 3 (BJ 104) and HB 4 (BK 560) became widely adapted with some tolerance to downy mildew. Extensive work on downy mildews had led to the release of downy mildew resistant hybrids and populations.

Cotton: India is the first country to grow cotton hybrids on a commercial scale. Cotton Hybrid-4 developed in Gujarat during 1970 and the demonstration that hybrid seed could be produced on a commercial scale through hand emasculation and pollination by C.T. Patel initiated the hybrid cotton era. A number of intra-*hirsutum* hybrids and a few *hirsutum* × *barbadense* hybrids rule over 30 % of the cotton area in India.

Castor: Gujarat also took the lead in the development of castor hybrids, which turned Gujarat into the number one castor growing state in India.

Rice: China took the lead in developing rice hybrids. The discovery of 'Wild Abortive' or WA and the WA-based CMS lines led hybrid rice to be commercially feasible. The first hybrid was released in China during 1976. During the last decade India made extensive efforts to develop commercial rice hybrids. The first Indian basmati scented quality rice hybrid, Pusa RH-10, developed by IARI, scientists was released in the year 2001 for commercial cultivation in India.

The advent of commercial hybrids in various crops led to the growth of private sector investments for research and development of hybrids in various crops. The contribution of hybrids of sorghum, pearl millet, cotton and sunflower in improving yields in rainfed agriculture is a significant landmark (Rao, 1991). During 1970s the growth rates of rainfed sorghum and pearl millet were even higher than wheat and rice. A comprehensive review of 'Heterosis and plant breeding' will be seen in Stuber (2000).

(*c*) Species Change in Cotton

Before 1950s the entire cotton in India was *desi* diploid (*G. arboreum* and *G. herbaceum*). Currently the American tetraploid cottons and their hybrids constitute over 70-80 %, a near reversal of the situation after the visit of S.C. Harland, a renowned cotton scientist/consultant from UK during 1950.

It was believed that diploid *desi* cottons are short-stapled, low yielding and non-responsive to irrigation and that their replacement with American tetraploids was the only solution to enhance yield and quality. That long staple *arboreums* could be bred was demonstrated by Rao (1960) and Qureshi and Rao (1973). Recent work in Andhra

Pradesh, Maharashtra and Karnataka and multilocation trials have consistently proved that *arboreum* cottons out yield American cottons and their hybrids under rainfed as well as irrigated situations. The fibre properties including staple length, strength, fibre maturity are equal or superior to the currently cultivated American cottons and their hybrids. Besides, they are resistant to sucking pests and more tolerant to boll worms. *Arboreum* improvement is likely to have significant consequences on the future scenario of Indian cottons and initiate a trend towards more areas under *arboreums* and also provide an alternate scenario for the pest management processes of cotton.

(*d*) Introduction of New Crops in India

Soybean: The introduction of soybean into India during 1890 and 1932 was not of much consequence. But the introduction of soybean varieties Brag and Clark 63 by the Jawaharlal Nehru Krishi Viswavidyalaya, Jabalpur initiated efforts to adopt and improve soybean in Indian agriculture. Madhya Pradesh leads in soybean cultivation and it is rapidly spreading into Maharashtra, Andhra Pradesh, Karnataka and other states, rendering soybean a major constituent of several rainfed agricultural systems. Export of soya meal has become an important source of Indian agricultural exports. The area increased from 0.5 m. ha in 1979-80 to > 6.0 m. ha in 1999-2000 and the production rose from 0.25 m.t to 6.6 m.t during same period.

Sunflower: The initial introduction of sunflower did not strike root. But later as the edible oil situation became critical, with the introduction of EC 68414 and EC 68415, the early maturing variety Morden and sunflower hybrids, changed all this and sunflower got established as a commercial crop in Karnataka and other states. Refining sunflower oil is also a major industry. The area increased from 0.12 m. ha in 1979-80 to > 2.00 m. ha in 1999-2000 and the production increased from 0.07 m.t to 1.17 m.t during the same period.

Outlook

The decade of 1990s witnessed declining yield growth rates of almost all crops and experimental yields appeared plateauing. The twenty first century is an era of free trade and open markets. Comparative and competitive advantages become primary concerns in the global village. Quality criteria and standards and consumer needs of both domestic and export sectors will be of prime concern. This may warrant drastic changes in the crops scenario. Added to this, cropland and water will be in short supply. There are forewarnings of climate change leading to global-warming, changes in rainfall patterns and so on. This demands for new strategies.

In tropics, time will become an important resource. To maximize per day productivity, crop intensification has to be optimized commensurate with resources. Short season varieties including grain legumes, which provide opportunities for crop intensification and crop diversification may even meet challenges of climate change. Problems associated with short season crops may have to be solved through grain drying and other means.

The concept of food security is changing. In addition to productivity concerns micronutrients have to be incorporated in staple food grains. The food grain demand may double. Agriculture has to become more inclusive and address the entire food chain from primary production to processing, marketing and consumer needs. End product orientation and designer crops will become breeding objectives. Cost cutting and cost efficiency will warrant greater precision. Specialized crop specific/system specific zones or focal areas with processing and market infrastructure may have to emerge.

The host-pathogens relationships are continuously evolving. The pest-disease scenario and host-pest interactions will change over time. Holistic approaches towards broad based resistances rather than single traits will be needed.

The breeders genetic tool kit is rapidly changing. The advent of molecular tools seem to have reflected in the neglect of field work and traditional tools. We should realize that there is no substitute for field selection, population and adaptation processes and integrated plant systems. The experimental approaches may be reductionist, but the holistic approach to integrated plant systems should not be overlooked. The plant breeder has more challenging tasks and he needs to remain close to earth.

References

Allard R.W. 1996. Evolution of adaptedness in plants. *Euphytica*, **92**: 1-11.

Allard R.W. 1999. *Principles of Plant Breeding*. John Wiley and Sons Inc., New York.

Barber C.A. 1916. *Studies in Indian sugarcane* Vol. 2. Kem. Dept. Agric. Ind. Bot. Sec. 7. pp. 1-106.

Ben Miflin. 2000. Crop improvement in the 21st century. *J. Expt. Bot.*, **51**: 1-8.

Boyer J.S. 1996. Advances in drought tolerance in plants. *Advances in Agronomy*, **56**: 187-217.

Comstock R.E. and Robinson H. F. 1952. Estimation of average degree of dominance. *In*: *Heterosis*, (ed.). Gowen J. W. Iowa State Univ. Press, Ames, Iowa pp. 494-516.

Correns C. 1900. Mendelel's Regel Uber das Verhalten der Vachkommens auf der Rassenbastarde. *Ber. Deutsh. Bot. Ges.*, **17**: 158-168.

Darwin C. 1859. The origin of species by means of natural selection or the preservation of favoured races in the struggle for life. *Amer. Ed.*, New York 1868.

De Candolle A. 1886. *Origin of cultivated plants*. Hafner, New York (reprint of 2nd edn. 1959).

De Ponti O.M.B. and Mollema C. 1992. Emerging breeding strategies for insect resistance. *In*: '*Plant Breeding in the 1990s*' (eds.) H.T. Stalker and J.P. Murphy. pp. 323-346.

De Vries H. 1901. *Die mutations - Theorie*. I. Von Veit, Leipzig. 648p.

Dudley J.W. 1997. Quantitative genetics and plant breeding. *Adv. Agronomy*, **59**: 1-21.

East E.M. 1908. Inbreeding in corn. *Connecticut Agric. Expt. St.*, 419-428.

Eberhart S.A. and Russel W.A. 1966. Stability parameters for comparing varieties. *Crop Sci.*, **6**: 36-40.

Fasoula D.A. 1997. Competitive ability and plant breeding. *Plant Breeding Reviews*, **14**: 89-138.

Fasoula D.A. and Fasoula V.A. 1997. Gene action and plant breeding. *Plant Breeding Reviews*, **15**: 315-374.

Finlay K.W. and Wilkinson G.N. 1963. The analysis of adaptation in plant breeding programme. *Australian J. Agric. Res.*, **14**: 742-54.

Fisher R.A. 1926. The arrangement of field experiments. *Journal of the Ministry of Agriculture*, **33**: 503-513.

Frey K.J. 1984. Breeding Approaches for increasing cereal crop yields. *In*: *Cereal Production*, (ed.) E.J. Gallagher, Butterworths, London. pp. 47-68.

Gangadharan C. 1985. Breeding. pp. 73-109. *In*: *Rice Research in India*. ICAR, New Delhi.

Haldane J.B.S. 1942. *New Paths in Genetics*, Harper and Bros., New York.

Harlan J.R. 1992. *Crops and Man*. Madison, Wisconsin USA.

Howard A. and Howard G.L.C. 1910. *Wheat in India: its production, varieties and improvement*, Thacker Spink & Co., Calcutta, 128 pp.

Howard A. and Howard G.L.C. 1928. The improvement of Indian wheat. A brief summary of investigation carried out at Pusa from 1905 to 1924 including an account of the Pusa hybrids. *Bull. Pusa Agric. Res. Inst.*, 171-26.

Huxley J.R. 1942. *Evolution: The modern synthesis*. Allen and Unwin, London.

ICRISAT 1981. *International workshop on intercropping*, (ed.) R.W. Willey, pp. 1-401.

Kang M.S. 1998. Using genotype-by-environment interaction for crop cultivar development. *Advances in Agronomy*, **62**: 199-252.

Mayr E. 1991. *One long argument*. Harvard University press, Cambridge.

Mendel G.J. 1865. Vensuche Uber Aflangenhybriden. English translation by W. Batesen in *J. Roy Hort. Soc.*, **26**: 1-32, 1901.

Milach S.C.K. and Federizzi L.C. 2001. Dwarfing genes in plant improvement. *Advances in Agronomy*, **73**: 35-61.

McClintok B. 1951. Chromosome organization and genetic expression. *Cold Spring Harbor Symp. Quant. Biol*.; **16**: 13-47.

Morgan T.H. 1911. An attempt to analyse the constitution of the chromosomes on the basis of sex limited inheritance in *Drosophila*, *J. Exp*., 2001, **11**: 365.

Muller H.J. 1927. Artificual transmutation of gene. *Science*, **66**: 84-87.

Pal B.P. 1966. *Wheat*. Indian Council of Agricultural Research, New Delhi, India. Pp. 370.

Parlevliet J. 1992. *Selecting components of partial resistance in plant breeding in the 1990s*, (eds.) H.T. Stalker and J.P. Murphy. pp. 281-302.

Patel C.T. 1971. Hybrid 4 - a new hope towards self-sufficiency in cotton in India. *Cot. Dev.*, **1**: 1-6.

Qureshi M.R.H. and Rao N.G.P. 1973. Breeding gaorani cottons *G. arboreum* race *indicum*) of a new plant architecture, larger yield and superior fibre attributes. *Indian J. Genet.*, **34A**: 849-853.

Rao N.G.P. 1960. Studies in breeding long staple gaorani cottons. *Ind. Cot. Gr. Rev.*, **14**: 19-22.

Rao N.G.P. 1982. Transforming traditional sorghum in India. *In*: *Sorghum in the eighties*. ICRISAT pp. 39-60.

Rao N.G.P. 1991. Advancing Dryland Agriculture: Plant Breeding accomplishments and perspective. Second B.P. Pal Memorial Lecture. *Indian J. Genet.*, **51**: 145-173.

Rao N.G.P. and Narayana L.L. 1968. Apomixis in grain sorghum. *Indian J. Genet.*, **30**: 230-236.

Rao N.G.P., Rana B.S. and Tarhelkar P.P. 1981. Stability, Productivity and Profitability of some intercropping systems in dryland agriculture. *In*: *Proc. Int. Workshop on Intercropping* (ed.) R.W. Willey. 292-298, ICRISAT, Patancheru.

Rao N.G.P., Rao G.R.K. and Acharya H.S. 1989. Yield stability of sorghum and millet across climates. *In*: *Climate and Food security*, IRRI, Philippines pp. 165-186.

Rao V.S. 2001. Wheat. *In*: *Breeding Field crops*, (ed.) V.L. Chopra. Oxford and IBH Publishing Co. Pvt. Ltd., New Delhi.

Raina S.K. 1997. Double haploid breeding in cereals. *Plant Breeding Reviews*, **15**: 141-86.

Savidan Y. 2000. Apomixis: Genetics and Breeding. *Plant Breeding Reviews*, **18**: 13-83.

Sawhney R.N. 1998. Genetic analysis of wheat rust resistance and strategies to breed for durable resistance. *Indian J. Genet.*, **58**: 391-416.

Sears E.R. 1956. The transfer of leaf rust resistance from *Aegilops umbellulata* to wheat. *Brookhaven Symp. Biol.*, **9**: 1-21.

Shull G.H. 1909. *Rept. American Breeders Association*, **5**: 296-301.

Siddiq E.A. 2001. Rice. *In*: *Breeding Field Crops*, (ed.) V.L. Chopra. Oxford and IBH Publishing Co. Pvt. Ltd., New Delhi. pp 1-85.

Sprague G.F. and Tautum. 1942. General and specific combining ability in single crosses of corn. *J. Am. Soc. Agron.*, **34**: 923-932.

Srinivasan T.V. 1989. Sugarcane. *In*: *Plant Breeding-Theory and Practice*, (ed.) V.L. Chopra. Oxford and IBH Publishing Co. Pvt. Ltd., New Delhi. pp. 253-273.

Stuber C.W. 2000. *Heterosis in plant breeding*, pp. 228-251.

Sturtevant A.H. 1913. The linear arrangement of sex-linked factors in *Drosophila* as shown by their mode of association. *Exp. Zool.*, 14: 43.

Tarhalkar P.P. and Rao N.G.P. 1981. Genotype - Plant density considerations in the development of efficient intercropping system for sorghum. *In*: *Proc. Int. Workshop on Intercropping*, pp 35-40 (ed.) R.W. Willey. ICRISAT, Patancheru 502 324, A.P., India.

Vavilov N.I. 1926. Studies on the origin of cultivated plants. *Bulletin of Applied Botany and Plant Breeding*, **16**: 1-245.

Venkatraman T.S. 1938. Hybridization in and within the genus *Saccharum*. *Proc. Ind. Science Congress* 25 (Part-II Section 9) pp. 267-284.

Wallace D.H. and Yan. 1998. *Plant Breeding and Whole System Crop Physiology*. CAB, UK.

Wright S. 1921. Systems of mating. *Genetics*, **6**: 111-178.

Wright S. 1968. *Evolution and genetics of populations*. University of Chicago Press, Vol. 1.

Plant Breeding - Mendelian to Molecular Approaches
H. K. Jain and M. C. Kharkwal (eds.)

4

Chromosome Manipulations for Crop Improvement

D.S. Brar[1] and H.S. Dhaliwal[2]

Abstract

The parallelism established between Mendel's laws of inheritance and chromosomal theory of inheritance during 1900-1904 laid the foundation of cytogenetics. Later, a series of cytogenetic stocks such as monosomics, trisomics, haploids, autotetraploids, synthetic and extracted alloploids, alien chromosome addition and substitution lines were developed in several crops. Advances were also made in molecular cytogenetic techniques such as chromosome banding, chromosome image analyzing systems, genomic *in situ* hybridization (GISH), fluorescence *in situ* hybridization (FISH) including fiber-FISH, micro-dissection of chromosomes, flow sorting of chromosomes, and, more recently, in developing artificial plant chromosomes. Cytogenetic stocks have been characterized genetically and cytologically and are being 'used in construction of molecular, genetic, physical and comparative maps. Several breeding lines have been developed and varieties released through haploidy breeding. Autotetraploids have been produced in numerous crops. A large number of interspecific hybrids and amphiploids have been produced. Alloploid breeding involving crosses between natural diploid/polyploids × synthetic alloploids and manipulation of alien addition and substitution lines have met with great success. Several useful genes have been transferred utilizing cytogenetical tools for resistance to diseases, insects, abiotic stresses, CMS sources and quality traits in wheat, oats, cotton, rice, *Brassica* and other horticultural species. Genes for cyst nematode and BYDV resistance have been introgressed through tissue culture induced chromosomal exchanges. FISH has been employed in numerous studies to characterize alien segments. Integration of chromosome manipulation techniques and molecular cytogenetic techniques with structural and functional genomics is essential to solve some of the plant breeding problems.

Introduction

Gregor Mendel (1865) established that heredity is particulate; alleles exist in two or more alternate forms, out of which only one allele of a gene is transmitted in each gamete. Different alleles of a gene do not blend when they come in contact in zygotes, but retain

[1]International Rice Research Institute, Metro Manila, Philippines
[2]Punjab Agricultural University, Ludhiana 141 004, India

their identity and integrity from generation to generation. Mendel's discoveries were ignored from 1865 to 1900. Cytologists made a succession of discoveries until 1900. Based on these discoveries, Sutton and Boveri (1903) independently proposed the chromosome theory of heredity.

A parallelism between Mendel's laws of inheritance and the behaviour of chromosomes during meiosis laid the foundation of 'cytogenetics'. In 1882 Flemming described the process of mitosis. Waldeyer coined the term chromosome in 1882. Several reviews are available on chromosome manipulations involving different kinds of numerical and structural changes and their application in genetic and breeding research (Burnham, 1962; Khush, 1973; Mujeeb-Kazi and Jewell, 1985; Brar and Khush, 1986; Singh, 1993; Gupta, 1999). A large number of cytogenetic stocks have been produced, characterized and used in construction of genetic maps, gene location, manipulation of pairing controlling mechanism, alien introgression and in developing improved germplasm with new characteristics. Notable examples include development of improved cultivars through haploid breeding in barley, rice, *Brassica*, potato; tetraploid varieties in forages, ornamentals, rye, triploid seedless watermelon, triploid sugarbeet, synthetic and extracted allopolyploids such as Triticale, *Brassica* and transfer of several useful alien genes in wheat, oats, cotton, tobacco and other plant species. During the last three decades, major advances have been made in chromosome research, particularly in chromosome image analyzing system, chromosome banding, flourescence *in situ* hybridization (FISH), flow sorting of chromosomes, uptake of isolated chromosomes, micro-dissection of chromosomes, computer assisted chromosome analysis and possibilities to construct plant artificial chromosomes (PAC). Similarly, advances in molecular and cellular biology have accelerated chromosome research to develop comprehensive cytological maps, genetic maps, physical maps, comparative genetic maps, genomic differentiation and repattering of karyotypes, physical location of centromeres, characterization of alien chromosome segments, and physical location of genes on chromosomes using BAC-FISH including fibre FISH techniques. These advances have provided new opportunities for precise manipulation of chromosomes, aimed at the genetic enhancement of crops. Some of the chromosome manipulations have been summarized in Tables 1-5.

Haploid Breeding

Haploids are important to develop homozygous lines from a segregating population much faster than by any other technique to fix recombinants and they serve as mapping populations to locate genes governing economic traits. Among the many methods used to produce haploids, the most common include: (*a*) isolation of haploids from natural populations, (*b*) anther culture, (*c*) interspecific crosses, (*d*) genetic selection technique and (*e*) mutant gene (Table 1). In the first method, a number of genetic stocks have been found to give increased frequency of haploids. Such genetic stocks have been extensively used to isolate haploids in corn, flax, *Brassica*, etc. The second method (anther culture)

Table 1. Some examples on improved germplasm developed through different chromosome manipulation techniques in crop plants

Technique	Method used	Crop(s)	Improved germplasm and/or new cytogenetic stocks developed
Haploid breeding	Anther culture	Rice, wheat, tobacco, *Brassica*	Several improved varieties and elite breeding lines have been developed
		Rice etc.	Doubled haploids (mapping populations) developed and several genes mapped on rice chromosomes
	Genetic selection	Maize	doubled haploids produced
	Chromosome elimination through interspecific crosses:		
	Aegilops caudata (alien cytoplasm) × *T. aestivum* (Salmon strain)	Wheat	30 % progenies are haploid
	H. vulgare × *H. bulbosum*	Barley	Mingo, Gwylyn, Rodeo cultivars released
	Oat × maize	Oat	Oat-maize addition lines developed
	Wheat × pearl millet	Wheat	Wheat haploids produced
	Wheat × maize	Wheat	Wheat haploids produced
	Durum wheat × maize	Durum wheat	Durum wheat haploids produced
	Interploid matings	Potato	Several genes for pest resistance, frost tolerance and tuber formation under high temperature transferred into potato cultivars
Autoploid breeding	Chromosome doubling	Forage	Alsike clover
		Forage	Svalov's ulva
		Forage	Pusa Giant Berseem
		Grapes	Pierce
		Strawberry	Tetraploid strawberry
		Turnip	Sirius
		Radish	Siberia 1, Siberia 2
		Brassica	Tetraploid toria
		Ornamentals	*Zizania, Antirrhinum, Chrysanthemum*
Triploid breeding	Intercrosses (4n × 2n)	Watermelon	Seedless watermelon
	4n × 2n	Sugarbeet	High yielding "Polybeets"
Alloploid breeding	Synthetic and extracted alloploids,	Triticale wheat, *Brassica*, cotton, tobacco	Number of cultivars released; useful genes introgressed into elite breeding lines
	Irradiation of alien addition lines	Wheat, oats	Many useful alien genes transferred into commercial cultivars
	Induced homoeologous pairing	Wheat, oats	Many useful alien genes transferred
Multiple chromosomal translocations	2n (normal karyotype) × 2n (homozygous chromosome translocations)	Watermelon	Seedless watermelon
Induced chromosomal exchanges among distant genomes via tissue culture	Tissue culture of rye grass: *Lolium multiflorum* (2n) × *L. perenne* (4n)	Rye grass	Chromosomal exchanges obtained among *Lolium* multiflorum × L. *perenne*
	Tissue culture of wheat-rye amphiploids	Wheat	Cereal cyst-nematode resistance introgressed from rye
	Tissue culture of disomic alien addition line (2n = 44)	Wheat	BYDV resistance transferred from *Thinopyrum intermedium*
	Tissue culture of wheat × *Dasypyrum villosum* hybrid	Wheat	Chromosomal translocations induced among wheat-*Dasypyrum* genomes
	Tissue culture of wheat-barley hybrid	Wheat	Increased chromosome associations between wheat and barley

has been extensively used worldwide to produce haploids in several plant species. A number of promising cultivars in tobacco, rice, wheat, and *Brassica* have been released using anther culture derived dihaploids. The third method involving interspecific crosses has proved to be an important procedure to produce haploids in barley, wheat, and potato. Hagberg and Hagberg (1980) reported a mutant gene '*hap*' that could produce haploids in barley. Plants homozygous for the *hap* gene produce progeny that includes 10-14 % haploids. The practical usefulness of this method is lacking. Alien cytoplasm (*Aegilops caudata* × *T. aestivum*) can result into high frequency (30 %) of haploids (Kihara and Tsunewaki, 1962).

Interploid mating in potato: Haploids in cultivated potato (*S. tuberosum* 2n = 48) offer the advantage of disomic rather than tetrasomic inheritance. This has made gene transfer possible from numerous wild and cultivated tuber bearing 24 chromosome *Solanum* species. Haploids are obtained in potato through interploid matings of *S. tuberosum* (2n = 48) × Phureja'(2n = 24). Haploid frequency can be increased 10-15 folds by selection of seed parent such as Merrimack and WIS. Ag. 231. Similarly selection of superior "pollinators" 1.1, 1.3 and 1.22 selections of Phureja (PI 225681) can have profound effect on haploid frequency. These haploids have been successfully crossed with 23 of the 24-chromosome tuber bearing species. Inter-haploids show regular chromosome pairing. These dihaploids are produced mainly as a result of union of the two polar nuclei of the *S. tuberosum* with the restitution sperm nucleus of the male parent *S. phureja* and or through aberrant cytokinesis. Two important features include the production of haploids and formation of 2n gametes. Both first division restitution (FDR) and second division restitution (SDR) occur in potato in such interploid matings. The 2x male parent produces 2n pollen, resulting in almost exclusively 4x progeny, since a severe triploid (3x) block occurs as a result of abnormal development of the pentaploid endosperm associated with 3x embryos. Selected 4x clones from 4x × 2x crosses have been released as cultivars. Tuber yield of many Phureja-haploid F_1 hybrids was substantially higher than their parents. These dihaploids of potato which are cross-compatible with the diploid species have been utilized for transferring disease resistance, frost resistance and tuber formation under high temperature conditions (Swaminathan, 1951; Peloquin *et al.*, 1964). Watanabe *et al.* (1996) reported several pest resistant, diploid potato germplasm through interploid matings in *Solanum*.

Haploid via chromosome elimination: In certain interspecific crosses, an entirely different mechanism operates, where chromosomes of one parent get selectively eliminated after fertilization, resulting into the production of haploid plants (Table 1). Some of the classical examples include elimination of *bulbosum* chromosomes in the cross of *H. vulgare* × *H. bulbosum*; elimination of maize chromosomes in wheat × maize and oat × maize crosses. Interestingly, haploids from such crosses can be produced through sexual hybridization followed by embryo rescue. Ho and Kasha (1975) made crosses of primary trisomics of barley with tetraploid *bulbosum* and located genes governing chromosome elimination on chromosomes 2 and 3. Some high yielding varieties (Mingo, Rodeo, Gwylan) have been released through *bulbosum* method. Barley doubled haploids were found to be no different from homozygous lines developed by pedigree or SSD methods for agronomic traits (Park *et al.*, 1976). Thus, this method seems to have good potential in barley breeding.

Laurie *et al.* (1990) reviewed production of haploids from wheat-maize crosses. Evidence for hybrid origin of embryo was confirmed from 4C nuclear DNA content of parents-wheat, maize and the hybrid embryos. Zygotes at metaphase confirmed the hybrid origin of embryos. At anaphase, the maize chromosome did not appear to be attached to the spindle and tended to lag behind the wheat chromosomes. The 2-celled embryos had 4-10 maize chromosomes at metaphase, while anaphase and telophase preparations showed lagging chromosomes. Metaphase from 4-celled embryos had none to 5 maize chromosomes. Further analysis showed maize chromosomes were effectively eliminated within three cell division cycles. The results from wide crosses (wheat × maize) indicate that fertilization occurs with high efficiency. This may either mean that cell recognition signals between gametes are not important or that such signals are conserved.

Haploids have also been produced from crosses of *T. durum* × maize (Almouslem *et al.*, 1998). Jauhar *et al.* (2000) observed seed set on synthetic haploids of durum wheat produced from the crosses of durum wheat × maize. The durum cultivars had *Ph1* gene, their haploids mostly formed univalents and had irregular meiosis. However, some haploids (2.75 seeds) set viable seeds. The seeds had viable embryo probably resulting from fusion of unreduced male and female gametes. It is hypothesized that lack of pairing may be pre-requisite for the occurrence of meiotic restitution and hence chromosome doubling. Dogramaci-Altuntepe and Jauhar (2001) reported durum wheat substitution haploids from crosses of durum × maize. Fluorescence *in situ* hybridization (FISH) analysis showed that the substitution haploids consisted of 7A-genome, 6B genome and 1D-genome chromosomes. Inagaki and Hash (1998) obtained haploids from crosses of bread wheat, durum wheat and triticale with pearl millet. Chromosome elimination can also occur under the influence of alien cytoplasm. Kihara and Tsunewaki (1962) reported that in the cross of *Ae. caudata* and *T. aestivum* (strain Salmon), 30 % progenies were haploids. These findings on production of haploids from diverse cross-combinations indicate that mechanism of chromosome elimination should be operative in other crops as well. There is a need to identify such chromosome elimination systems for producing haploids, particularly in crops where anther culture response is poor or restricted to selected genotypes.

Autoploid Breeding

The technique of artificial doubling of chromosome number through colchicine treatment has been extensively used in large number of crop plants. In comparison with diploids, induced autotetraploids are often 'gigas' for vegetative characteristics. However, due to genic imbalance and multivalent formation at meiosis, such autotetraploids show reduced seed fertility. The reduction in seed fertility is common in the initial stages, however, subsequent selection can improve seed fertility and reduce multivalent formation. Inspite of many efforts to improve fertility, autotetraploids are not popular in crops where seed is the commercial product. In seed crops, a few tetraploid rye varieties have been released such as Tetra Petkus, Fourex, Dubbel Stahl and Steel have been released in Germany, Sweden, and USA. The tetraploid rye has larger kernel size, superior ability to emerge under adverse conditions and possesses higher protein content. A large number of autotetraploids have been produced in many forage crops. Some of these have been released as varieties in forage crops, ornamentals and fruit trees; notable examples include Alsike and red clovers, including Svalof's Ulva and Pusa Giant Berseem. A few tetraploid varieties have been released in turnips, spinach, radish, Pierce grapes, snap-dragon, *Zizania, Antirrhinum,* and *Chrysanthemum* (Table 1).

The seed sterility in autotriploids has been used as an advantage in crops where seedlessness is a desirable character. The example of triploid "seedless" watermelons is well-known. The cross 4n (female) × 2n (male) produces viable triploid seeds. But in the triploid progenies less than 1 per cent of the gametes are viable, hence triploids show high seed sterility and produce no true seed. Triploid sugarbeet is another example of success of triploid breeding. It yields larger tubers and more sugar per unit area. Induced autotetraploidy has not been very successful to achieve goal as expected from the gigas nature. However, chromosome doubling has been quite successful in the improvement of allopolyploids.

Synthetic Amphiploids

A number of crop plants such as wheat, cotton, *Brassica,* oats are natural allopolyploids. Numerous amphiploids (Triticale, *Raphanobrassica, Festuca-lolium, Pennisetum glaucum* × *P. purpureum, Brassica* spp., *Gossypium* spp.) have been produced by intercrossing two or more distantly related taxa followed by chromosome doubling to restore the chromosomal balance. Several amphiploids from hybrids involving cultivated and wild relatives have been produced during the last 50 years, especially after the discovery of the action of colchicine. In many cases, such amphiploids show varying levels of fertility. Such amphiploids lack genomic harmony, show meiotic instability, and several other undesirable features resulting from nuclear-cytoplasmic interactions.

Numerous amphiploids have been produced, but examples of success of the man-made amphiploid are too few. Triticale (AABBDDRR) is an amphiploid of *Triticum* spp. (AABBDD), and *Secale cereale* (RR). It combines good agronomic traits of *S. cereale* which are lacking in *T. aestivum* and enable its cultivation in non-traditional wheat areas. More than 100 varieties have been released; however, triticale is still not very popular, as it has poor grain characteristics and also suffers from infertility despite the fact that it has a perfect chromosome balance. None of the synthetic amphiploids produced in the past such as *Raphanobrassica, B. campestris* × *B. nigra, B. campestris* × *B. oleracia, Oryza sativa* × other *Oryza* spp. have shown promise. However, such synthetic amphiploids have been used successfully in crosses with the natural diploids or polyploids to extract desirable alloploids.

Extracted Alloploids

As discussed in the previous section, incorporation of a complete alien genome may bring in several undesirable characteristics. Thus the alternative approach is to extract stable alloploids from crosses of synthetic amphiploids with natural allopolyploids or related wild relatives. Triticale is the classical example for demonstrating the usefulness of extracted alloploids. hexaploid triticales are comparatively better in terms of meiotic stability and seed yield. The extracted triticales derived from the progenies of (1) octaploid triticale × hexaploid triticale and (2) hexaploid wheat × hexaploid triticale are much better. They are superior to the primary triticales in meiotic stability, grain yield, and grain quality. That superiority is probably due to the recombination of A and B genomes of wheat and substitution of some of the rye genome chromosomes with the D genome chromosomes of bread wheat. According to Gupta (1984), 29 out of 41 released varieties of triticale have R/D chromosome substitution. The most popular hexaploid triticale are "Armadillo" strains carrying 2R-2D chromosome substitution. Crosses of 8x triticale with Armadillo resulted in the production of Maya 2 Armadillo from which Mapache was selected and released as Cananea-79 in Mexico.

The extracted alloploids have become popular to introgress genes from synthetic alloploids to natural allopolyploid. *Brassica* is another example where improved lines have been derived from the crosses of natural allopolyploids with the synthetic allopolyploid. Cotton is another good example of the transfer of fiber strength of *G. thurberi* introgressed into *G. hirsutum* through *arboreum-thurberi* synthetic allotetraploid. Similarly, promising introgression has been achieved from *sylvestris* chromosomes into *tabacum* by way of synthetic allopentaploid. The Vavilov Institute has developed a series of 42-chromosome amphiploids by crossing 4x wheat with eincorn wheats and doubling the chromosome number of F_1s. Over 30 allopolyploids were intercrossed with common wheat cultivars and promising lines were extracted.

A number of genes for disease and insect resistance and tolerance to abiotic stresses and improved quality characteristics have been introgressed into crops through crosses with wild relatives. The sterile F_1's are backcrossed with the recurrent parent to introduce desirable alien genes and develop chromosome stable lines. Numerous examples on introgression of alien genes are available in wheat, cotton, rice, barley, sunflower, tomato and others (Brar and Khush, 1986; Khush and Brar, 1989; Brar and Khush, 1997 Friebe *et al.*, 1996, 1999).

Aneuploid Breeding

A large number of aneuploid stocks such as monosomics, nullisomics, primary trisomics, secondary trisomics, tertiary trisomics and compensating trisomics have been produced (Table 2). These stocks have been characterized and employed in genetic analysis for mapping major genes as well as genes governing quantitatively inherited traits (see for review Khush, 1973 Singh 1993, Gupta 1999). Gene mapping using aneuploid stocks is based on modification of genetic ratios involving crosses of aneuploids with normal stocks having normal chromosome complement.

Table 2. Development of cytogenetics stocks and their application in genetic and breeding research

Cytogenetic stock	Crop(s)	Application in genetic and breeding research
Monosomics	Wheat, *Nicotiana* cotton, oats	Several genes mapped on specific chromosomes and genetic maps constructed Alien chromosome substitution lines developed Development of chromosome substitution lines
Primary trisomics	*Datura*, tomato, barley, maize, pearl millet	Genes assigned on specific chromosomes and linkage groups established
Secondary trisomics	Tomato, *Datura*, rice	Genes assigned to specific arms of chromosomes
	Rice	Centromeres mapped on molecular map of rice
Tertiary trisomics	Tomato, barley	Genes assigned to specific segments of chromosomes Production of hybrid barley, 'Hembar' variety released
Compensating trisomics	Tomato	Genes assigned to specific arms of chromosomes
Translocations	Maize, barley, pearl millet, rice	Genes assigned to specific chromosome arms and chromosome segments Breakpoints on chromosomes identified
	Barley	Intercrossing of translocations produced duplication of defined chromosome segments in barley
A-B translocations	Maize	Genes assigned to chromosome arms
Induced chromosome deficiency	Tomato	Many genes mapped on different chromosome regions of tomato
Deletion stocks	Wheat	Construction of physical maps and mapping of genes in wheat

Monosomics (2n-1): Monosomics have been reported in several polyploid plant species such as wheat, oats, *Nicotiana*, cotton, etc. These have been extensively studied in wheat. Sears (1954) was the first to discover that 21 different chromosomes of wheat fall into seven homoeologous groups of three distinct genomes. This was based primarily on the ability of each tetrasome to compensate for the nullisome of each of the other two chromosomes of the same group. The tetrasomes compensate for the deleterious effect of the nullisomes (Sears *et al.*, 1996). Using monosomics, numerous genes have been mapped in wheat and chromosome substitution lines developed. Due to lack of survival in diploid species, use of monosomics is mainly restricted to polyploid species.

Primary Trisomics (2n+1): Primary trisomics have one extra intact chromosome in addition to the normal diploid chromosome complement. Primary trisomics are one of the most extensively studied aneuploids in assigning genes to specific chromosomes in diploid species such as *Datura*, maize, barley, tomato, and rice. These have also been used to construct molecular map in rice (McCouch *et al.*, 1988).

Secondary and Telo-Trisomics: In these trisomics, the extra chromosome is isochromosome for one of the chromosome arms of the complement. Secondary trisomics or telo-trisomics have been reported in maize, barley, tomato, and rice. These trisomics have been used to map only a few genes on specific arms of barley, tomato and rice chromosomes (Singh *et al.*, 1996b).

Telo-trisomics: Rhoades (1936) discovered the first telotrisomic in *Zea mays*. Since then, telotrisomics have been developed in a number of species including *Datura, Nicotiana, Triticum, Secale, Lycopersicon* and rice. Singh *et al.* (1996a) used secondary and telotrisomics for mapping of centromeres on the genetic map of rice. Cheng *et al.* (2001) reported a complete set of telotrisomics in rice covering all the 24 chromosome arms. A rice centromere BAC clone was used as a marker probe in FISH analysis to verify the telocentric nature of extra chromosome in telotrisomic stocks of rice.

Tertiary Trisomics: In tertiary trisomics, extra chromosome consists of two arms of two non-homologous chromosomes. After a gene has been assigned to a specific chromosome by use of primary trisomics, it is possible to delimit it further to its respective chromosome arm by testing the segregation in the progeny of primary trisomics. Tertiary trisomics have been used to map a few genes on chromosome arms of barley and tomato.

Balanced Tertiary Trisomics and Hybrid Barley Production: Ramage (1965) produced balanced tertiary trisomics for the production of F_1 hybrid barley. A pair of recessive male sterility alleles are present on normal chromosome, while the extra chromosome (interchanged chromosome) carried a dominant allele linked to the translocation point. A variety of barley - Hembar, was released using such trisomics derived from the segmental interchange T_2-7d induced by X-rays in Bonus barley. The system avoids the need for fertility restorer genes in the pollen parent as in wheat. However, the balanced tertiary trisomics are generally weak and do not produce sufficient pollen to pollinate the male-sterile diploids for increasing the seed of male-sterile parent. Hence, the system could not be used on commercial scale for practical barley breeding.

Compensating Trisomics: Compensating trisomics are those in which one chromosome is missing but is compensated for by two other modified chromosomes. On account of their unique segregation pattern, they can be used advantageously over other trisomics in indexing a gene on an arm of a specific chromosome. Khush and Rick (1967) located several genes in tomato using these trisomics.

Alien Chromosome Addition Lines

One of the major barriers in introducing alien genes into cultivars, is the lack of recombination between chromosomes of the parental species. Also, incorporation of the complete alien genome could cause genetic and physiological disturbances. To circumvent these limitations, addition or substitution of single chromosome and incorporation of small alien chromosome segments or genes through manipulation of pairing controlling mechanism seem to be advantageous.

As early as 1940, O'Mara described a method for the addition of single chromosome pairs of the donor species into the recipient species. Alien addition lines have been produced in wheats, oats, *Nicotiana*, cotton, maize, rice, potato, sugarbeet, etc. (Table 3).

Table 3. Some examples on alien chromosome addition lines in crop plants

Recipient species	Alien chromosome(s) added (donor species)
Cereal crops	
Triticum aestivum	*Secale cereale*
T. aestivum (*S. cereale* cytoplasm)	*Secale cereale*
Triticum aestivum	*Haynaldia villosa*
Triticum aestivum	*Aegilops umbellulata*
Triticum aestivum	*Aegilops comosa*
Triticum aestivum	*Aegilops variabilis*
Triticum aestivum	*Aegilops sharonensis*
(*Ae. sharonensis* cytoplasm) *T. aestivum*	*Aegilops sharonensis*
(*Ae. longissima* cytoplasm) *T. aestivum*	*Aegilops longissima*
Triticum aestivum	*Agropyron elongatum*
Triticum aestivum	*Agropyron intermedium*
Triticum durum	*Secale cereale*
Triticum durum	*Agropyron elongatum*
Triticum durum	D-genome *(Triticum aestivum)*
Triticum durum	*Aegilops umbellulata*
Triticum durum	*Aegilops squarrosa*
Avena sativa	*Avena strigosa*
Avena sativa	*Avena hirtula*
Avena sativa	*Avena barbata*
Avena sativa	*Zea mays*
Oryza sativa	*Oryza officinalis*
Oryza sativa	*Oryza minuta*
Oryza sativa	*Oryza latifolia*
Oryza sativa	*Oryza australiensis*
Oryza sativa	*Oryza brachyantha*
Oryza sativa	*Oryza granulata*
Zea mays	*Tripsacum dactyloides*
Other crops	
Nicotiana tabacum	*Nicotiana glutinosa*
Gossypium hirsutum	*Gossypium stocksii*
Gossypium hirsutum	*Gossypium anomalum*
Gossypium hirsutum	*Gossypium sturtianum*
Solanum tuberosum	*Lycoperson esculetum*
Brassica campestris	*Brassica oleracea*
Raphanus sativus	*Brassica oleracea*
Beta vulgaris	*Beta procumbens*
Beta vulgaris	*Beta corollifera*

Gerstel (1945) developed mosaic resistance *N. tabacum* and added 11 chromosome pair from *Nicotiana glutinosa*. Corn-*Tripsacum* monosomic addition lines were produced from crosses involving multiple recessive tester stock of corn (Galinat, 1977). Monosomic alien addition lines (MAALs) have been produced in rice carrying extra chromosome from six wild species, (Brar and Khush, 1997). Dhaliwal *et al.* (1987) used addition lines to locate genes for Karnal bunt (*Neovossia indica*) resistance on 6R and 7R of rye and 1H and 6H of barley. Garriga-Caldere *et al.* (1998) identified 7 of the 12 chromosome addition lines of tomato in the background of potato.

Recently, from the crosses of oat × maize, a complete set of MAALs representing each of 10 maize chromosome in oat has been developed (Kynast *et al.*, 2001). Disomic addition ($2n = 6x + 2 = 44$) lines have been established for maize chromosomes 1, 2, 3, 4, 6, 7, and 9. RFLP markers have been used to characterize MAALs of maize (Riera-Lizarasu *et al.*, 1996). The oat-maize addition lines have become important cytogenetic stocks for (1) isolation of chromosome specific probes, (*b*) flow sorting, (*c*) microdissection of chromosomes, (*d*) physical mapping and selective isolation and mapping of cDNA on specific chromosome, and (*e*) analysis of expression of maize genes.

Ananiev *et al.* (1997) constructed a cosmid library from oat-maize addition lines. A multiprobe (mixture of labeled fragments) of highly repetitive maize specific sequences was used to selectively isolate cosmid clones containing maize genomic DNA. Ananiev *et al.* (1998) used McSIa, maize centromere-associated sequence as a probe to isolate cosmid clones from genomic library of chromosome 9 addition line. Analysis of six cosmid clones containing centomeric DNA revealed a complex organization. The McSIa sequence was found to consist of a long terminal repeats of a retrotransposon-like repeated elements. Riera-Lizarazu *et al.* (2000) produced radiation hybrids through irradiation of oat-maize chromosome 9 addition line. These radiation hybrid derivatives are good sources of region-specific DNA for cloning of genes. Okagaki *et al.* (2001) mapped 400 sequences to maize chromosomes using PCR and DNA from oat-maize addition lines. These cytogenetic stocks would be important in combination with data on positional cloning to identify candidate genes for various agronomic traits. These alien chromosome addition lines are important sources of genetic variation carrying useful genes. However, none has become popular as a commercial variety because of the instability of the alien chromosome and incorporation of undesirable characters associated with the alien chromosome.

Proposed Method for Producing Hybrid Wheat via Alien Addition Lines: A novel scheme of producing F_1 hybrid seed of wheat by exploiting addition lines was suggested by Driscoll (1972). The scheme involves three lines designated as X ($2n = 44$), Y ($2n = 43$), and Z ($2n = 42$). The recessive male sterility allele (*ms*) is present on wheat chromosome, while dominant allele (*Ms*) is present on the homoeologous alien chromosome. The Z line is male sterile while X and Y lines are male fertile. The main drawback with the addition lines is that they are generally less stable and as such rapidly revert to euploid state. Pollen with addition chromosome is less competitive than the euploid pollen, in fertilizing the egg cell. The X, Y, Z system could not become practical because of limitations of chromosome instability of the addition lines and seed multiplication of the desired lines due to insufficient pollination.

Alien Chromosome Substitution Lines

Alien chromosome substitution refers to the replacement of one or more pairs of chromosomes of the cultivated parent with an equal number of pairs of an alien species. Methods for producing substitution lines using monosomic series have been discussed by Sears (1953), Unrau *et al.* (1956), and Khush (1973). Substitution lines have been studied extensively in wheat. Alien chromosome substitution lines have been developed in *Triticum, Nicotiana*, and *Gossypium* species. Alien chromosomes have been substituted in wheat with *Aegilops, Secale, Agropyron*, and barley; in oats, with *Avena barbata* and in *N. tabacum* with *N. glutinosa* and *N. plumbaginifolia*. In tobacco, substitution lines were developed to transfer resistance to tobacco mosaic and black-shank diseases (Gerstel, 1945). Weique - a commercial wheat cultivar with a pair of substituted chromosomes of *Agropyron intermedium* incorporating resistance to stem rust has been developed (Wienhues, 1965). In oats, substitution lines have been produced to transfer mildew resistance of *Avena barbata* to *A. sativa* (Thomas *et al.*, 1975). Using molecular markers, a series of overlapping chromosome substitution lines have been developed in rice, *O sativa*, carrying chromosome segments of *O. glaberrima* (Doi *et al.*, 1997).

Although a number of substitution lines have been produced in wheat, only a few have become commercial varieties. In many cases, the alien chromosome does not compensate fully for the substituted chromosome and sometimes carries undesirable genes along with the desirable ones. Some of the wheat cultivars developed in Germany and Eastern Europe are spontaneous substitutions, in which 1R of rye has been substituted for 1B of wheat (Zeller, 1973). Rajaram *et al.* (1983) reported that several cultivars developed at CIMMYT in which a segment of 1B of wheat substituted by a segment of 1R of rye, have wider adaptability and resistance to *Septoria tritici*. Several wheat cultivars around the world have wheat-rye translocation (1BL.IRS).

Intervarietal Chromosome Substitution: Sears (1953) developed series of aneuploid stocks in wheat which have been extensively used to produce intervarietal as well as inter specific and intergeneric chromosome substitution lines. Intervarietal chromosome substitution lines have been developed by Law *et al.* (1981) between Cappelle-Desperez and Chinese Spring; Cappelle-Desperez and Koga-II and Vilmorin-27 and Cappelle-Desperez. Intervarietal chromosome substitutions have been used in genetic analysis of agronomic traits. Pairs of Hope chromosomes were substituted with their homologues in Chinese Spring. The results indicated that for yield, ear number, and height, the estimates between chromosome interactions were nearly all positive. By contrast, within chromosome, interactions were almost all negative. Intervarietal chromosomal substitutions have provided a wealth of information on genetics of quantitative traits in wheat, *Nicotiana*, and cotton.

Induced Homoeologous Pairing and Alien Gene Transfer

In wheat (AABBDD 2n = 6x = 42), there is no intergenomic pairing between chromosomes belonging to the A, B, and D genomes. Okamoto (1957) discovered that chromosome 5B of wheat carries a gene (or genes) that suppresses the pairing of homoeologous chromosomes. In the absence of this gene (*Ph*, pairing homoeologous), the homoeologous chromosomes of the A, B and D genomes pair with each other and also with their homoeologues from related species and genera. Jauhar (1975) observed genetic control of diploid-like meiosis in hexaploid tall fescue similar to wheat. In one of the monosomic line (90-64-7), multivalents were observed indicating genetic control of homoeologous pairing. Manipulating *Ph* locus can induce homoeologous pairing and obtain recombination between the chromosomes of wheat and other alien species which otherwise do not pair. Besides, other pairing regulators located on chromosomes 3A, 3D and 4D are also known in wheat. Various wild relatives such as *Aegilops speltoides* and *Avena longiglumis* have genes that promote homoeologous pairing in F_1 hybrids. B-chromosomes in some strains also suppress pairing, very much like *ph* (Dover and Riley, 1972). Three methods have been used for inducing homoeologous and to transfer agronomically important alien genes into commercial cultivars (Table 4).

Wild Species for Inducing Homoeologous Pairing: Wild species such as *Aegilops speltoides* and *Avena longiglumis* accession Cw57 are known to induce homoeologous pairing in wheat and oats, respectively. Riley *et al.* (1968) used an addition line of *Aegilops comosa* having an extra chromosome pair (2M) for transferring rust resistance from *Ae. comosa* to wheat. This disomic addition line (2n = 44) was crossed with *Aegilops speltoides* (2n = 14). In this 29-chromosome hybrid, the 2M chromosome of *Ae. comosa* paired with the 2D chromosome of wheat because of suppression of *Ph* activity. The homoeologous pairing occurred because the dominant allele of *Ae. speltoides* neutralizes the effect of *Ph*. A wheat line having the rust resistance gene from *Ae. comosa* was selected and named "Compair". Sears (1972) transferred resistance to brown rust from *Ag. elongatum* to wheat using this technique (Table 4).

The Cw57 strain of *A. longiglumis* has a gene or genes that neutralize the effect of the gene of *A. sativa* responsible for the suppression of homoeologous pairing. *A. longiglumis* has been successfully used to transfer mildew resistance from *A. barbata* and *A. prostrata* to the cultivated oat *A. sativa* (Thomas *et al.*, 1980; Griffiths and Thomas, 1983).

Aghaee-Sarbarzeh *et al.* (2000) made crosses of Chinese Spring (wheat) and Chinese Spring carrying *Ph* gene (CS *Ph*) with three accessions of *Ae. kotschyi* (UUSS), one accession of *Secale cereale* (RR) and one amphiploid of *Triticum durum* × *Ae. umbellulata* (AABBUU). Meiotic analysis showed significant increase in chiasma frequency in all the crosses with CS *Ph* over those with normal wheat. The results indicated the effectiveness of *Ph* gene transferred from *Ae. speltoides* in 6x wheat in inducing homoeologous recombination and thus transfer of alien genetic variation with least linkage drag.

Table 4. Some examples on chromosome manipulations used for developing improved germplasm in cereals

Crop (recipient)	Donor species	Method	Character transferred
Bread wheat (*Triticum aestivum*)			
	Triticum durum	1	Stem rust resistance
	T. dicoccum	1	Stem rust resistance
	T. timopheevi	1	Cytoplasmic male sterility
	T. monococcum	1	Stem rust (*Sr21*) resistance
	T. monococcum	1	Stem rust (*Sr35*) resistance
	Aegilops ventricosa	1	Alpha amylase and eye spot resistance
	Ae. ventricosa	2	Eye spot resistance
	Ae. umbellulata	3	Leaf rust (*Lr9*) resistance
	Agropyron elongatum	3	Stem rust (*Sr24, Sr26*) resistance
	Ag. elongatum	3	Leaf rust (*Lr19*) resistance
	Ag. intermedium	3	Leaf rust (*Lr38*) resistance
	Ag. elongatum	3	Wheat streak mosaic virus resistance
	Th. intermedium	3	Barley yellow dwarf virus resistance
	Secale cereale	3	Leaf rust (*Lr25*) resistance
	Secale cereale	3	Powdery mildew (*Pm7, Pm17*) resistance
	Secale cereale	3	Hessian fly (*H25*) resistance
	Secale cereale	3	Green bug resistance
	Ae. comosa	4	Leaf rust resistance
	Ag. elongatum	4	Leaf rust resistance
	Ae. ovata	4	Leaf rust resistance
	Ae. triuncialis	4	Leaf rust resistance
	Ag. intermedium	4	Streak mosaic virus resistance
	Thinopyrum intermedium	4	Barley yellow dwarf virus resistance
	T. dicoccoides	4	High protein content
	Ae. umbellulata	4	High molecular weight glutenin
Cultivated oat (*Avena sativa*)			
	A. barbata	3	Mildew resistance
	A. strigosa	3	Crown rust resistance
	A. barbata	4	Mildew resistance
	A. prostrata	4	Mildew resistance
Maize (*Zea mays*)	*Zea diploperennis*	1	BLB, BSDM and Rhizactonia resistance

1=Interploid matings; 2=Bridging species; 3=Radiation induced alien translocations; 4=Induced homoeologous pairing

Dhaliwal *et al.* (2002) used two substitution lines of hexaploid wheat cv. WL711, with 5 U of *Ae. triuncialis* and 5 M of *Ae. ovata* carrying rust resistance gene. These lines were crossed with Chinese Spring wheat carrying the *Ph* gene and backcrossed to wheat variety WL711. The rust resistance gene(s) from both the *Aegilops* substituted chromosomes were transferred to wheat with reduced alien chromatin. Microsatellite marker analysis showed chromosomal exchanges between the substituted alien chromosome and wheat chromosomes through induced homoeologous pairing.

Use of Nullisomic 5B System: In the absence of chromosome 5B, pairing takes place between chromosomes of wheat and those of alien species. Sears (1973) produced several wheat lines carrying resistance to leaf rust from *Agropyron elongatum* using a nulli-5B system. As a result of induced homoeologous pairing, several wheat-*Agropyron* transfers were obtained (Sears 1972, 1973).

Use of Induced Ph Mutant: The *Ph* mutant for inducing homoeologous pairing has not been used extensively for the transfer of alien genes. Wall *et al.* (1971) isolated a *Ph* mutant after ethyl methane sulphonate (EMS) treatment. However, the level of homoeologous pairing in this mutant is not as high as in nulli-5B. Another *Ph* mutant was isolated by Sears (1977) by X-raying normal pollen. The irradiated pollen was used to pollinate monosomic 5B plants carrying the marker gene (Hairy-Neck). Of the 438 M_1 plants tested for pairing mutations, mostly by crossing to *T. kotschyi*, one mutation was obtained that appeared to be a deficiency of *Ph*.

Wang *et al.* (1977) used a *Ph* mutant and induced recombination between chromosome 4B of wheat and *Agropyron intermedium* chromosome carrying genes for resistance to streak mosaic virus. Khushnir and Halloran (1984) used *Ph1b* and Ph_2 mutants to obtain homoeologous recombination and transfer of genes for high protein from *T. turgidum/T. dicoccoides* to *T. aestivum*. Giorgi and Cuozzo (1980) isolated a mutant in tetraploid wheat variety, Cappelli, showing increased homoeologous pairing with *Ae. longissima* and *Secale cereale*. Giorgi and Barbera (1981) crossed a *Ph* mutant of durum wheat with four tetraploid *Aegilops species* (*Ae. cylindrica, Ae. kotschyi, Ae. columnaris*, and *Ae. triuncialis*). The chromosome pairing was high in all these crosses as compared to that in the control hybrids. Islam and Shepherd (1992) produced wheat-barley hybrids using the *Ph1b* mutants and observed recombination between wheat and barley utilizing isozyme markers. Gill and Gill (1996) developed PCR based marker specific for the *Ph1b* deletion.

Sherman *et al.* (2001) produced wheat-barley recombinants involving chromosome 4 and 7. Wheat-barley disomic addition lines were crossed with Chinese Spring wheat carrying the *Ph1b* mutant to promote homoeologous pairing. PCR markers were used to characterize barley introgression. Progeny testing showed 9 recombinants involving chromosome 4 and 11 recombinants involving barley chromosome 7. Xin *et al.* (2001), using *Ph* mutant, induced homoeologous recombination between wheat and *Thinopyrum*. The wheat -*Th. intermedium* translocation lines, YW642, and YW643 showed good resistance to BYDV. GISH showed that some of these lines are homozygous wheat - *Th.*

intermedium translocation lines. The chromosome segments of *Th. intermedium* were transferred to the distal end of wheat chromosomes.

Dvorak and Dubcovsky (1995) made genetic analysis of chromosomes produced by targeted recombination of two pairs of related homoeologous chromosomes in the absence of *Ph1*. In the presence of *Ph1*, crossing over occurred frequently in the homologous region, but was very low in the homoeologous region. In the absence of *Ph1*, the frequencies of crossing over in the homoeologous region were similar to the frequencies of crossing over between wheat 1A homologous. Frequencies of crossing over between 4B and 4D were reduced in all intervals in the absence of *Ph1*.

Irradiation of Cytogenetic Stocks for Introgression of Alien Genes

The incorporation of the complete alien genome as in amphiploids, or addition or substitution of a single alien chromosome has not met with much success. The method consists of irradiating the pollen or seed of the monosomic alien addition lines followed by recovery of the translocation of alien chromosome segments in the successive progenies, either genetically or cytologically or by both. Sears (1956) used this elegant technique for transferring a small chromosome segment carrying leaf rust resistance from *Aegilops umbellulata* to 6B chromosome of common wheat. Since then, many genes have been transferred through this method (Table 4). A number of improved strains carrying useful alien genes (Transfer, T4, Agatha, and Transec) for resistance to various diseases of wheat have been developed using this technique with 6A chromosome of wheat. The most successful transfer involves translocation of *Agropyron elongatum* chromosome, carrying resistance to wheat stem rust, to chromosome 6A of wheat (Knott, 1961). The stem rust resistance from *Ae. elongatum* has been incorporated into Australian wheat cultivars such as Eagle, Kite, Jabiru, and Avocet (Fisher and Martin, 1974). Riley 67, a commercial variety of soft red winter wheat produced at Purdue University, possesses leaf rust resistance from *A. umbellulata*.

Aung and Thomas (1976) successfully transferred the gene for mildew resistance from tetraploid *A. barbata* into the cultivated oat by means of an induced translocation (Table 4). The translocation involved the long arm of the shortest chromosome of *A. sativa* and the short arm of an *A. barbata* chromosome, which carries the gene for mildew resistance. A segment carrying *Pel5* gene for crown rust resistance was transferred from *A. strigosa* to *A. sativa* using thermal neutrons.

Crasta *et al.* (2000) irradiated seeds of monosomic alien addition line carrying *Thinopyrum intermedium* (group 7) chromosome carrying resistance to BYDV and analyzed M_2 population. *Thinopyrum* chromosome specific repetitive elements and RFLPs were used to identify wheat-*Thinopyrum intermedium* translocation lines. The BYDV resistance was located on the distal end of the introgressed *Th. intermedium* chromosome.

Enhancing Chromosomal Exchanges and Gene Transfer Among Distant Genomes

The distant hybrids frequently lack chromosomal exchange due to strong divergence among genomes. The karyotypic changes arising through tissue culture resulting from chromosome breakage and translocations have been reported in several plant species (Table 1). Amongst these karyotypic changes, translocations are most important to introgress useful alien genes into crop plants. Orton (1980) reported that somaclonal variants obtained from tissue culture of a sterile hybrid (*H. vulgare* × *H. jubatum*) had enhanced bivalent formation, as compared to the original hybrid which was asynaptic. Ahloowalia (1983) produced 4000 plants from tissue culture of triploid rye grass *Lolium multiflorum* (2x) × *L. perenne* (4x). The regenerants showed chromosomal translocations and other rearrangements. Lapitan *et al.* (1984) reported a high degree of chromosome structural changes in amphiploids of wheat × rye hybrids regenerated from tissue culture. The karyotype analysis by C-banding of the 10 amphiploids showed three wheat × rye, one wheat × wheat translocations, seven deletions, and five amplifications of heterochromatin bands of rye chromosomes. Larkin *et al.* (1989) and Banks *et al.* (1995) obtained a number of translocation stocks with BYDV resistance from a disomic addition line, which carries a pair of 7 Ai-1 chromosomes from *Thinopyrum intermedium*. Similarly, resistance to cereal cyst nematode was introgressed from rye (Table 1).

Dahleen (1999) observed enhanced pairing among tissue culture regenerates of *H. vulgare* × *Elymus canadensis*. Li *et al.* (2000) reported intergeneric chromosomal translocations between *Triticum* and *Dasypyrum*. These translocations were characterized through GISH, both at callus stage as well in regenerated plants. Molnar-Lang *et al.* (2000) multiplied hybrids between winter wheat × winter barley through tissue culture and produced backcross after pollination with wheat. Meiotic analysis of the hybrids showed 1.59 chromosome arm associations per cell. However, the number of chromosome associations per cell increased to 4.72 after *in vitro* culture in hybrids.

The method seems to be promising for introgression of alien genes through tissue culture of species, which show limited homoeologous pairing. Tissue culture cycle using explants of wide hybrids (F_1s, chromosome addition, and/or substitution lines), including somatic hybrids could enhance the frequency of genetic exchange between alien and cultivated genomes. Novel traits can be transferred through exchange of chromosome segments via tissue culture. Irradiation of the protoplast fusion products would enhance the exchange of genetic material between the genomes of alien and cultivated species. Numerous interspecific and intergeneric somatic hybrids already produced through tissue culture in crop plants could be used to derive desired translocations among distant genomes.

Structural Changes in Chromosomes

A large number of cytogenetic stocks involving chromosomal interchanges (translocations), inversions, deletions and duplications have been produced in maize, barley, rice, etc. As an example, in barley alone, cytologically defined stocks include over 1000 reciprocal translocations, inversions, deletions and duplications. These stocks have been used extensively in establishing linkage group and mapping of genes in barley, maize and other crops. McClintock (1950)'s pioneer work on breakage fusion bridge cycle and variegation in maize is a landmark in the history of cytogenetics. The AC-Ds system and the mutable loci analysis in maize have resulted into new understanding of transposable elements and jumping genes.

Translocations: Among the different types of structural aberrations, translocations have been extensively employed in genetic analysis of traits and in mapping of genes in maize and barley (Burnham, 1962; Singh, 1993). Multiple chromosomal interchange stocks can be used to produce seedless fruit and vegetable crops. Translocations have been used for duplication of defined segments of chromosomes 6 and 7 in barley, which seem to be promising for enhanced yield (Hagberg and Hagberg, 1987). Translocations are being used for producing duplications in the short arm of chromosome 5 and 6, which carry genes for mildew resistance and alpha-amylase activity, respectively. The duplications of V-*v* factor can be used commercially for establishing permanent heterosis in barley.

A-B Translocations: B chromosomes, commonly referred to as supernumerary or accessory chromosomes, have been reported in many plant species. These have been extensively studied in rye and maize. B chromosomes have a minute short arm and are highly heterochromatic - the centromere, followed by euchromatin and then heterochromatic regions that compose most of the long arm. B chromosome has the unusual property of nondisjunction at the second pollen mitosis. In all of the other stages of mitosis, its behaviour is identical to the normal A chromosomes. The deficiency-duplication created by the B-A translocations has been used to map genes. Roman and Ullstrup (1952) located *Hm* gene on long arm of chromosome 1 in maize using B-A translocations.

B chromosomes contain a heterogeneous cluster of related specific sequences. The misdivision derivatives will localize this cluster to the functional centromere. The B-specific sequence has the $(CCCTAAA)_n$ motifs, which appear to be characteristic of centromere regions in a number of organisms. There is homology to the maize knob, which can act as a type of facultative centromere when abnormal chromosome 10 is present. Reduction of the B-specific sequences indicate that they are indeed derived from the centromere.

Synthetic Complex Translocation Stocks for Production of Homozygous Lines and Seedlessness

Multiple interchanges involving all the chromosomes of the haploid complement have been produced in many diploid crops for gametic selection and establishing homozygous lines. Burnham (1946) suggested use of multiple interchanges for developing homozygous lines and called it as "*Oenothera*" method of gamete selection. Such complex translocations occur naturally in *Oenothera*. Using recurrent irradiation and intercrossing procedures, complex interchange stocks have been synthesized in *Triticum monococcum*, *Hordeum vulgare*, *Pennisetum glaucum* and *Zea mays* (Brar *et al.*, 1973; Gupta, 1999).

However, the method could not become popular to develop homozygous lines, because of high sterility of complex interchange stocks. The technique has been applied with advantage for commercial production of F_1 seedless watermelons (Sakaguchi and Nishimura, 1969).

Induced Chromosome Segmental Deficiency: Marker gene can be located by use of induced deficiency method. A line homozygous for recessive alleles is pollinated with X-rayed pollen of a line homozygous for the corresponding normal allele. Plants of mutant phenotypes are selected in the progeny for cytological study. Khush (1973) mapped many genes in tomato using this method.

Comparative Maps: Structural Differentiation of Genomes and Identification of Orthologous Loci

It has become possible to study genomic relationships across crossability barriers involving divergent species and genera using molecular markers. Such analysis is difficult through classical cytogenetic techniques. Potato and tomato genomes show extensive colinearity, and karyotypes differ only by a few structural chromosomal rearrangements (Bonierbale *et al.*, 1988). Of the 12 chromosomes, the order of loci for 9 chromosomes was identical, while in three chromosomes, intrachromosomal rearrangements were found. Three appeared to be paracentric inversions with one break point at or near the centromere. Ahn *et al.* (1993) found extensive homoeologues in several regions of the genomes of wheat, rice and maize. Comparative genome mapping in rice, maize, sorghum, wheat, barley, foxtail millet and sugarcane into a single synthesis shows that gene order is highly conserved between species within the grass family, although amount and organization of repetitive DNA sequences have diverged considerably (Devos and Gale, 1997).

Because of strong crossability barriers, hybrids between *Oryza meyeriana* and *O. ridleyi* complexes are difficult to produce, hence assigning genomes based on meiotic pairing could not be carried out. Thus, alternative approach based on total genomic DNA hybridization and molecular divergence was used and two new genomes, GG and HHJJ have been assigned to *O. meyeriana* and *O. ridleyi* complexes respectively (Aggarwal *et al.*, 1997). Brar *et al.* (1996) detected introgression of small alien chromosome segments from distant genomes (FF and GG) of *Oryza* from species which show limited homoeologous pairing.

Jena *et al.* (1994) constructed comparative RFLP map of *O. sativa* and wild species, *O. officinalis* (CC). The linkage order on different chromosomes of *O. officinalis* was mostly conserved relative to *O. sativa*, but some chromosomal rearrangements were detected. Nine of the 12 chromosomes of rice were homosequential to those of wild species. Kennard *et al.* (1999) developed comparative maps of wild rice (*Zizania palustris* 2n = 2x = 30) and *O. sativa* (2n = 2x = 24). Although the genomes of two species differ in DNA content, co-linear markers were found for 11 of the 12 linkage groups of rice. Comparative maps offer new possibilities to identify and isolate orthologous loci from diverse plant systems.

Chromosomal Image Analyzing System: Imaging methods have been developed to characterize plant chromosomes and develop quantitative cytological maps (Fukui and Ijima, 1991). This system has become an important supplement to conventional cytogenetic techniques in chromosomal analysis.

Chromosome Banding for Characterization of Karyotypes and Alien Introgression

The development of chromosome banding techniques has provided additional tool to identify individual chromosomes. Giesma C-banding has been used in chromosome identification and in analyzing the evolutionary relationships (Gill and Kimber, 1974) and to detect introgressed alien chromosome segments (Table 5). The banding techniques are used for identification of chromosome segments that consist of either GC or AT rich regions, or for constitutive heterochromatin. In barley, all chromosomes can be identified by using C- or N-banding techniques, which reveal blocks of heterochromatin as dark staining regions. Friebe *et al.* (1992) identified *Agropyron intermedium* chromosomes based on banding in wheat × *Ag. intermedium* partial amphiploid and derived alien addition lines. The chromosomal constitution of eight radiation-induced, rust-resistant wheat-*Agropyron intermedium* derivatives were analyzed by C-banding and GISH. Jiang *et al.* (1993) analyzed wheat germplasm carrying resistance to wheat streak mosaic virus from *Agropyron elongatum*. Jiang *et al.* (1994) detected introgression from *Elymus trachycaulus* chromatin into wheat through chromosome banding. Endo and Gill (1996) identified 436 deletion by C-banding in wheat. Deletion chromosomes were stably transmitted to the progeny.

***In Situ* Hybridization: Karyotypic Changes and Alien Introgression**

One of the major advances in chromosome research is the development of *in situ* hybridization techniques. Classical cytogenetic techniques in combination with molecular markers and FISH have enhanced the precision on characterization of changes in karyotype, determining structural differences in chromosomes, differentiation among distant genomes and detection of introgressed alien genes and chromosome segments. Both mitotic and meiotic chromosomes have been used in *in situ* hybridization analysis. *In situ* hybridization involves hybridization of DNA or RNA probes to the cytological preparation and allows the visualization of specific nucleotide sequences directly on chromosomes. The method was developed by Gall and Pardue (1969). Since then, isotopic probes were used extensively to map both repetitive and low copy DNA sequences. Rayburn and Gill (1985) extended non-isotopic protocols for *in situ* hybridization. Now, fluorescence *in situ* hybridization (FISH) has become the protocol of choice for *in situ* hybridization. Various procedures such as genomic *in situ* hybridization (GISH) including repetitive DNA, BAC clones, etc. can be used. Numerous reports are available on location of repetitive DNA sequences on chromosomes of wheat, barley, tomato, etc. BAC and YAC libraries have increased the efficiency of physical mapping of single copy and small sequences on chromosomes. Jiang *et al.* (1995) physically mapped 40 kb BAC clones containing *Xa21* gene for bacterial blight resistance on chromosome 11 of rice (Table 5). Recently, major advances have been made to precisely characterize the genes through extended DNA fiber-FISH.

Table 5. Some examples on the application of chromosome banding and *in situ* hybridization techniques in crop plants

Technique	Application	Reference
C-banding	Identified individual rye chromosomes	Gill and Kimber (1974)
C-banding`	Identified D-genome chromosomes in wheat	Dhaliwal *et al.* (1990)
C-banding	Identified 436 deletions in wheat	Endo and Gill (1996)
C-banding	Characterized translocations and pericentric inversions in wheat	Kawahara and Taketa (2001)
C-banding	Introgressed *teosinte* into maize	Sachan and Tanaka (1977)
C-banding + *in situ* hybridization	Identified partial amphiploids and chromosome addition lines in wheat	Friebe *et al.* (1992)
C-banding + *in situ* hybridization	Non-homoeologous wheat-*Agropyron* translocation resistant to leaf rust identified	Friebe *et al.* (1993)
C-banding + *in situ* hybridization	Identified wheat-*Agropyron* translocations carrying resistance to wheat streak mosaic virus	Jiang *et al.* (1993)
GISH	Characterized autosyndetic and allosyndetic pairing among A and E genome of *Oryza*	Abbasi *et al.* (2000)
GISH	Characterized chromosome associations between wheat and barley	Molnar-Lang *et al.* (2000)
GISH	Characterized monosomic alien addition lines in *Beta vulgaris*	Gao *et al.* (2001)
GISH	Characterized potato + tomato somatic hybrids	Gavrilenko *et al.* (2001)
GISH	Characterized monosomic allien addition lines in tomato	Haider Ali *et al.* (2001)
GISH	Characterized wheat-*Thinopyrum* amphiploids, and translocated chromosomes	Cai *et al.* (2001)
GISH	Identified introgression of *festuca* chromosome segments in *Lolium multiflorum*	Morgan *et al.* (2001)
FISH	Identified homologue in *B. napus*	Snowdon *et al.* (2002)
FISH	Physical location of centromeres in cereals	Jiang *et al.* (1996)
BAC + FISH	Mapping of *Xa21* gene on chromosome 11 of rice	Jiang *et al.* (1995)
BAC + FISH	Identified individual chromosomes in potato	Dong *et al.* (2000)
YAC + FISH	Mapped hs^1 *Pro*–1 gene for nematode resistance	Desel *et al.* (2001)
FISH	Identified substitution haploids with 7A, 6B, 1D chromosomes	Dogramaci-Aluntepe and Jauhar (2001)
FISH	Mapping of rDNA loci and *Arabidopsis* BAC in *Brassica*	Ziolkouski and Sadowski (2002)

GISH can effectively detect alien introgression (Table 5). Schwarzacher *et al.* (1992) identified *Leymus multicaulis*, *Thinopyrum bessarabicum* and *Hordeum vulgare* and *Secale cereale* in alien addition, substitution and translocation lines of bread wheat. Mukai *et al.* (1993) analyzed breakpoints in the wheat-rye translocation lines resistant to Hessian fly, through *in situ* hybridization. Zhong *et al.* (1999) analyzed introgression line of tomato carrying chromosome 6 region of *L. peruvianum* containing the nematode resistance gene *Mi*. Using *Mi*-BAC clones and FISH on pachytene chromosomes, the *Mi* was located at the border of the euchromatin and heterochromatin regions in the short arm (6S) and Aps-1 in the pericentromeric heterochromatin of the long arm close to the euchromatin. Wang *et al.* (2000) analyzed introgression lines derived from cross of *T. aestivum* × goat grass and found introgressed alien chromosome segments.

Pickering *et al.* (2000) used sequential GISH and FISH to analyze derived recombinant lines in barley. FISH was employed with a short oligonucleotide sequence as probe located introgression on the long arm of barley chromosome 2H. Jauhar and Peterson (2000) produced intergeneric F_1 hybrids between durum wheat (*Triticum turgidum* AABB) and *Thinopyrum junceiforme* ($J_1J_1J_2J_2$). Segmental exchanges between alien chromosomes were confirmed through FISH. Some of the derived lines showed introgression for scab resistance from *Thinopyrum* to wheat.

Garriga-Caldere *et al.* (1999) used GISH to analyze somatic hybrid (potato+tomato) and MAALs of potato carrying chromosomes 1 and 8 of tomato. A low level of allosyndetic pairing, 1.1% and 1.3 % respectively, was observed for monosomic additions of 1 and 8.

In many species, such as rice and potato which have small sized somatic chromosomes, it is not possible to distinguish chromosomes based on morphology and banding pattern. It has been found that large-insert DNA clones can be conveniently used as chromosome specific DNA markers. Dong *et al.* (2000) demonstrated that FISH signals derived from BACs can be used as chromosome-specific cytogenetic DNA markers for chromosome identification in potato.

A major improvement in FISH detection and sensitivity has been achieved with the extended DNA fiber. The accessibility of probe to the target DNA is no longer hindered by condensed, intact chromosome structures, thus a strong detection for FISH signals is possible.

Zhong *et al.* (1996) mapped telomeric repeat of *Arabidopsis thaliana* and tomato specific repeat, TGRI, containing 162 bp motif on pachytene chromosomes and extended DNA fibers from interphase nuclei of tomato. Kjima *et al.* (1999) used extended DNA fiber-FISH isolated from both *indica* and *japonica* rices. The results showed that copy number of 5S rDNA was 1.3 times more in *indica* than *japonica*, and telomere sequences in *indica* rice were three times longer than that of *japonica*. Centromere specific probes have become available. One of the probes 745 bp repetitive DNA clone (pSau 3A9) isolated from sorghum was located through FISH in centromeric regions of all cereals.

However, it shows absence of such DNA sequences homologous to pSau3A9 in dicot species, suggesting rapid divergence of centromere related sequences compared to telomere related sequences in plants (Jiang *et al.*, 1996).

Characterization of Homoeologous Pairing Through FISH: It is difficult, if not impossible, to distinguish autosyndetic and allosyndetic pairing in hybrids of species using classical cytogenetic techniques. However, FISH offers the advantage to precisely characterize pairing among and within genomes. This is particularly useful when chromosomes of parents are of similar size and lack diagnostic cytological markers. Begona *et al.* (1995) compared homoeologous pairing in wheat-rye derivatives using C-banding and GISH. The homoeologous pairing detected by C-banding and GISH was 2.5 and 9.2 %, respectively. This shows underestimation of wheat/rye associations by C-banding. Abbasi *et al.* (1999) observed autosyndetic and allosyndetic pairing through GISH among A and E genomes of *Oryza*.

Flow Sorting and Chromosome Mediated Gene Transfer

Flow cytometry has become a useful technique to characterize materials with different ploidy levels. The technique has been used to characterize n + n, 2n + n backcross progenies derived from the crosses of maize × apomictic *Tripsacum*. Flow cytometry has been used to sort chromosomes in *Haplopappus*, *Petunia*, tomato, wheat, etc. This technique enables rapid quantitative karyotyping and the sorting of individual chromosomes. In wheat, from the suspension of intact wheat chromosomes, only chromosome 3B could be identified on flow karyotypes; remaining chromosomes formed three composite peaks and could be sorted only as groups (Vrana *et al.*, 2000).

Sorted chromosomes can be used to develop chromosome addition lines and for chromosome mediated gene transfer. Such sorted chromosomes can also be used for cloning. Refined procedures are available to isolate chromosomes from cultured cells. Szabados *et al.* (1981) devised a technique for the uptake of isolated chromosomes by protoplasts of other species. Mitotic protoplasts from synchronized cell cultures were used for isolation of chromosomes. The uptake of isolated chromosomes into the protoplasts of recipient species (wheat, parsley, and maize) was induced by polyethylene glycol (PEG) treatment. The method looks promising but has not been used to introduce isolated single chromosomes.

Micro-Dissection of Chromosomes

The microdissection procedure was first used on *Drosophila* polytene chromosomes (Scalenghe *et al.*, 1981). Since then attempts have been made on microdissection of chromosomes of barley, wheat, rye, oat, etc. The method involves isolation of small copies of a specific chromosome or chromosome region or arm using a micromanipulator. The method requires very little starting material; the chromosomes could be isolated from a single slide prepared from a single root tip meristem. Cheng *et al.* (2001) used chromosome microdissection in rice. The extra chromosome arms of 2n+.5S+.5S, and 2n+.5L were successfully microdissected from the prometaphase cells with a microneedle controlled by a Leitz micromanipulator. Primer pairs from STS and microsatellites on different chromosome arms were selected to amplify the PCR products from both microdissected chromosomes and control samples. The results revealed that the amplified DNAs from the microdissected chromosome arms came from extra chromosome arms 5S and 5L.

Artificial Plant Chromosomes

During the last decade, major progress has been made to construct yeast artificial chromosomes (YAC) and bacterial artificial chromosomes (BAC). Both BAC and YAC have been extensively used in physical mapping of plant genomes. Recently, emphasis has been placed to construct artificial human and plant chromosomes. Harrington. (1997) and Rosenfeld (1997) constructed artificial human chromosomes. Long synthetic arrays of alpha satellite DNA were combined with telomeric DNA and genomic DNA to generate artificial chromosomes in human HT1080 cells. The resulting linear micro-chromosomes contain exogenous alpha satellite DNA, are mitotically and cytogenetically stable in the absence of selection for up to 6 months in culture, bind centromere proteins specific for active centromeres, and are one-fifth to one-tenth the size of endogenous human chromosomes. Future research would lead to the construction of plant artificial chromosomes (PAC). Three basic prerequisites are necessary: (*a*) origin of replication for chromosome duplication, (*b*) a centromere to allow the division and migration along the microtubules of duplicated chromosomes into daughter cells, and (*c*) telomere at the ends to confer stability. PAC will be useful to understand the regulation of gene expression including potential regulatory elements and to conserve information carried by genes, while regulating their expression, since these structures are established in a similar manner as in chromosomes. It is expected that in future, artificial plant chromosomes would become important to move genes across distant plant species.

References

Abbasi F.M., Brar D.S., Carpena A.L., Fukui K. and Khush G.S. 1999. Detection of autosyndetic and allosyndetic pairing among A and E genomes of *Oryza* through genomic *in situ* hybridization. *Rice Genet. Newsl.*, **16**: 24-25.

Aggarwal R.K., Brar D.S. and Khush G.S. 1997. Two new genomes in the *Oryza* complex identified on the basis of molecular divergence analysis using total genomic DNA hybridization. *Mol. Gen. Genet.*, **254**: 1-12.

Aghaee-Sarbarzeh M., Singh H. and Dhaliwal H.S. 2000. *phl* gene derived from *Aegilops speltoides* induces homologous chromosome pairing in wide crosses of *Triticum aestivum*. *J. Hered.*, 536-5163.

Ahloowalia B.S. 1983. Spectrum of variation in somaclones of triploid rye grass. *Crop Sci.*, **23**: 1141-1147.

Ahn S., Anderson J.A., Sorrells M.E. and Tanksley S.D. 1993. Homoeologous relationship of rice, wheat and maize chromosomes. *Mol. Gen. Genet.*, **241**: 483-490.

Almouslem A.B., Jauhar P.P., Peterson T.S., Bommineni V.R. and Rao M.B. 1998. Haploid durum wheat production via hybridization with maize. *Crop Sci.*, **38**: 1080-1087.

Ananiev E.V., Lizarazu O.R., Rines H.W. and Phillips R.L. 1997. Oat-maize chromosome addition lines: A new system for mapping the maize genome. *Proc. Natl. Acad. Sci.*, USA, **94**: 3524-3529.

Ananiev E.V., Phillips R.L. and Rines W.H. 1998. Chromosome-specific molecular organization of maize (*Zea mays* L.). centromeric regions. *Proc. Natl. Acad. Sci.*, USA, **95**: 13073-13078.

Aung T. and Thomas H. 1976. Transfer of mildew resistance from the wild oat, *Avena barbata* into cultivated oat. *Nature*, **260**: 603-604.

Banks P.M., Larkin P.J., Bariana H.S., Lagudah E.S., Appels R., Waterhouse P.M., Brettell R.I.S., Chen X., Xu H.J., Xin Z.Y., Qian Y.T., Zhou X.M., Cheng Z.M. and Zhou G.H. 1995. The use of cell culture for subchromosomal introgressions of barley yellow dwarf virus resistance from *Thinopyrum intermedium* to wheat. *Genome*, **38**: 395-405.

Begona F.C., Benavente E. and Orellana J. 1995. Meiotic pairing in wheat-rye derivatives detected by genomic *in situ* hybridization and C-banding: A comparative analysis. *Chromosoma*, **103**: 554-558.

Bonierbale M.W., Plaisted R.L. and Tanksley S.D. 1988. RFLP maps based on a common set of clones reveal modes of chromosomal evolution in potato and tomato. *Genetics*, **120**: 1095-1103.

Boveri Th. 1902. Über mehrpolige Mitosen als Mittel Zur Analyse der Zellkerns. *Verh. Phys. - med. Ges.*, **35**: 67-90.

Brar D.S., Dalmacio R., Elloran R., Aggarwal R.K., Angeles R. and Khush G.S. 1996. Gene transfer and molecular characterization of introgression from wild *Oryza* species into rice. *In*: *Rice Genetics III*. International Rice Research Institute, Manila, Philippines, pp. 477-486.

Brar D.S., Minocha J.L. and Gill B.S. 1973. Translocations in *Pennisetum typhoides* involving all the chromosomes in one complex. *Curr. Sci.*, **42**: 653-654.

Brar D.S. and Khush G.S. 1986. Wide hybridization and chromosome manipulation in cereals. pp. 221-263. *In*: *Handbook of Plant Cell Culture*, Vol. 4. Techniques and Applications. (eds). D.A. Evans, W.R. Sharp, P.V. Ammirato. MacMillan Publ. Co., New York.

Brar D.S. and Khush G.S. 1997. Alien introgression in rice. *Plant Mol. Biol.*, **35**: 35-47.

Burnham C.R. 1946. An "*Oenothera*" or multiple translocation method of establishing homozygous lines. *J. Amer. Soc. Agron.*, **38**: 702-707.

Burnham C.R. 1962. *Discussions in Cytogenetics*. Burgess Publishing Company, pp. 375.

Cai X., Jones S.S. and Murray T.D. 2001. Molecular cytogenetic characterization of *Thinopyrum* genomes conferring perennial growth habit in wheat-*Thinopyrum* amphiploids. *Plant Breed.*, **120**: 21-26.

Cheng Z.K., Yan H.H., Yu H.X., Tang S., Jiang J., Gu M. and Zhu L.H. 2001. Development and applications of a complete set of rice telotrisomics. *Genetics*, **157**: 361-368.

Crasta O.R., Francki M.G., Bucholtz D.B., Sharma H.C., Zhang J., Wang R.C., Ohm H.W. and Anderson J.M. 2000. Identification and characterization of whet-wheatgrass translocation lines and localization of barley yellow dwarf virus resistance. *Genome*, **43**: 698-706.

Dahleen L.S. 1999. Tissue culture increases meiotic pairing of regenerants from barley × canada wild rye hybrids. *J. Heredity*, **90**: 265-269.

Desel C., Jung C., Cai D., Kleine M. and Schmidt T. 2001. High-resolution mapping of YACs and the single-copy gene *HsI*, *Pro-I* on *Beta vulgaris* chromosomes by multi-color fluorescence *in situ* hybridization. *Plant Mol. Biol.*, **45**: 113-122.

Devos K.M. and Gale M.D. 1997. Comparative genetics in the grasses. *Plant Mol. Biol.*, **35**: 3-15.

Dhaliwal H.S., Friebe B., Gill K.S. and Gill B.S. 1990. Cytogenetic identification of *Aegilops squarrosa* chromosome additions in durum wheat. *Theor. Appl. Genet.*, **79**: 769-774.

Dhaliwal H.S., Gill B.S. and Waines J.G. 1977. Analysis of induced homoeologous pairing in a *Ph* mutant wheat × rye hybrid. *J. Hered.*, **68**: 207-209.

Dhaliwal H.S., Multani D.S. and Singh B.B. 1987. Identification of rye and barley chromosomes carrying resistance to Karnal bunt of wheat (*Neovossia indica*). *Cereal Res. Comm.*, **15**: 191-194.

Dhaliwal H.S., Singh H. and William M. 2002. Transfer of rust resistance from *Aegilops ovuta* into bread wheat (*Triticum aestivum* L.) and molecular characterization of resistant derivatives. *Euphytica*, **126**: 153-159.

Dogramaci-Altuntepe M. and Jauhar P.P. 2001. Production of durum wheat substitution haploids from durum × maize crosses and their cytological characterization. *Genome*, **44:** 137-142.

Doi K., Iwata N. and Nishimura Y. 1997. The construction of chromosome substitution lines of African rice (*Oryza glaberrima* Steud.) in the background of *japonica* rice (*O. sativa* L.). *Rice Genet. Newsl.*, **14**: 39-40.

Dong F., Song J., Naess S.K., Helgeson J.P., Gabhardt C. and Jiang J. 2000. Development and applications of a set of chromosome-specific cytogenetic DNA markers in potato. *Theor. Appl. Genet.*, **101**: 1001-1007.

Dover G.A. and Riley R. 1972. The prevention of pairing of homoeologous meiotic chromosomes of wheat by a genetic activity of supernumerary chromosomes of *Aegilops*. *Nature*, **240**: 159-161.

Driscoll C.J. 1972. X, Y, Z system of producing hybrid wheat. *Crop Sci.*, **12**: 516-517.

Dvorak J. and Dubcovsky J. 1995. Recombination between homoeologous chromosomes in wheat in the absence of the Ph1 locus. pp 64-75. *In*: *Classical and Molecular Cytogenetic Analysis*. (eds.) Raupp W.J. and Gill B.S., Kansas State University, Manhattan.

Endo T.R. and Gill B.S. 1996. The deletion stocks of common wheat. *J. Hered.*, **87**: 295-307.

Fisher J.A. and Martin R. 1974. Kite, Condor and Egret - three new wheats. *Agric. Gaz. NSW*, **85**: 10-13.

Flemming W. 1982. Introduction Cited in *Cytology and Cytogenetics* by C. P. Swanson (1957). Prentice Hall Inc., New Jersey, pp. 1-13.

Friebe B., Hatchett J.H., Gill B.S., Mukai Y. and Sebesta E.E. 1991. Transfer of Hessian fly resistance from rye to wheat via radiation-induced terminal and intercalary chromosomal translocations. *Theor. Appl. Genet.*, **83**: 33-40.

Friebe B., Mukai Y., Gill B.S. and Cauderon Y. 1992. C-banding and *in situ* hybridization analyses of *Agropyron intermedium*, a partial wheat × *Ag. intermedium* amphiploid, and six derived chromosome addition lines. *Theor. Appl. Genet.*, **84**: 899-905.

Friebe B., Jiang J., Gill B.S. and Dyck P.L. 1993. Radiation-induced nonhomoeologous wheat-*Agropyron intermedium* chromosomal translocations conferring resistance to leaf rust. *Theor. Appl. Genet.*, **86**: 141-149.

Friebe B., Jiang J., Raupp W.J., McIntosh R.A. and Gill B.S. 1996. Characterization of wheat-alien translocations conferring resistance to diseases and pests: Current status. *Euphytica*, **91**: 59-87.

Friebe B., Kynast R.G., Hatchett J.H., Sears R.G., Wilson D.L. and Gill B.S. 1999. Transfer of rye derived Hessian fly resistance genes *H21* and *H25* from bread wheat into durum wheat. *Crop Sci.*, **39**: 1692-1696.

Fukui K. and Ijima K. 1991. Somatic chromosome map of rice by imaging methods. *Theor. Appl. Genet.*, **81**: 589-596.

Galinat W.C. 1977. The origin of corn. *In*: *Corn and Corn Improvement*, (ed.) G.F. Sprague.

Gall J.G. and Pardue M.L. 1969. Formation and detection of RNA-DNA hybrid molecules in cytological preparation. *Proc. Natl. Acad. Sci.*, USA, **63**: 378-383.

Gao D., Guo D. and Jung C. 2001. Monosomic addition lines of *Beta corolliflora* Zoss in sugarbeet: Cytological and molecular marker analysis. *Theor. Appl. Genet.*, **103**: 240-247.

Garriga-Caldere F., Huigen D.J., Angrisano A., Jacobsen E. and Ramanna M.S. 1998. Transmission of alien tomato chromosomes from BC_1 to BC_2 progenies derived from backcrossing potato (+) tomato fusion hybrids to potato: The selection of single additions for seven different tomato chromosomes. *Theor. Appl. Genet.*, **96**: 155-163.

Garriga-Caldere F., Huigen D.J., Jacobsen E. and Ramanna M.S. 1999. Prospects for introgressing tomato chromosomes into the potato genome: An assessment through GISH analysis. *Genome*, **42**: 282-288.

Gavrilenko T., Thieme R. and Rokka V.M. 2001. Cytogenetic analysis of *Lycopersicon esculentum* (+) *Solanum tuberosum* somatic hybrids and their androgenetic regenerants. *Theor. Appl. Genet.*, **103**: 231-239.

Gerstel D.U. 1945. Inheritance in *Nicotiana tabacum* II. The addition of *Nicotiana glutinosa* chromosomes to tobacco. *J. Hered.*, **36**: 197-206.

Gill B.S. and Kimber G. 1974. Giemsa C-banding karyotype of rice. *Proc. Natl. Acad. Sci.*, USA, **71**: 1247-1249.

Gill K.S. and Gill B.S. 1991. A DNA fragment mapped within the submicroscopic deletion of *Ph1*, a chromosome pairing regulator gene in polyploid wheat. *Genetics*, **129**: 257-259.

Giorgi B. and Cuozzo L. 1980. Homoeologous pairing in a *ph* mutant of tetraploid wheat crossed with rye. *Cereal Res. Comm.*, **8**: 485-490.

Giorgi B. and Barbera F. 1981. Use of mutants that affect homoeologous pairing for introducing alien variation in both durum and common wheat. *In*: *Induced mutations - a tool in plant research*. IAEA, Vienna, pp. 37-47.

Griffiths N. and Thomas H. 1983. Gene transfer. *In*: *Annual Report* for 1982 Welsh Plant Breeding Station, Aberstwyth, UK, pp. 145-146.

Gupta P.K. 1984. Role of loss of the rye heterochromatin in improvement of hexaploid triticale. *The Nucleus*, **27**: 100-107.

Gupta P.K. 1999. *Cytogenetics*. Rastogi Publications, Meerut, India, pp. 418.

Hagberg A. and Hagberg G. 1980. High frequency of spontaneous haploids in the progeny of an induced mutation in barley. *Hereditas*, **93**: 341-343.

Hagberg A. and Hagberg G. 1987. Some vigorous and productive duplications in barley, pp. 423-426. *In*: *Barley Genetics*, (eds). V. S. Yasuda and T. Konishi. Okayama, Japan.

Haider Ali S. N., Ramanna M. S., Jacobson E. and Visser R. G. F. 2001. Establishment of a complete series of a monosomic tomato chromosome addition lines in the cultivated potato using RFLP and GISH analyses. *Theor. Appl. Genet.*, **103**: 687-695.

Harrington J.J. 1997. Formation of *de novo* centromeres and construction of first-generation human artificial chromosomes. *Nature Genetics*, **15**: 345-355.

Ho K.M. and Kasha K.J. 1975. Genetic control of chromosome eliminating during haploid formation in barley. *Genetics*, **81**: 263-275.

Inagaki M.N. and Hash C.T. 1998. Production of haploids in bread wheat, durum wheat and hexaploid triticale crossed with pearl millet. *Plant Breeding*, **117**: 485-487.

Islam A.K.M.R. and Shepherd K.W. 1992. Production of wheat-barley recombinant chromosomes through induced homocologous pairing. I. Isolationof recombinants involving barley arms 3HL and 6HL. *Theor. Appl. Genet.*, **83**: 489-494.

Jauhar P.P. 1975. Genetic control of diploid like meiosis in hexaploid tall fescue. *Nature*, **254**: 595-597.

Jauhar P.P. and Peterson T.S. 2000. Hybrids between durum wheat and *Thinopyrum junceiforme*: Prospects for breeding for scab resistance. *Euphytica*, **00**: 1-10.

Jauhar P.P., Dogramaci-Altuntepe M., Peterson T.S. and Almouslem A.B. 2000. Seedset on synthetic haploids of durum wheat: Cytological and molecular investigations. *Crop Sci.*, **40**: 1742-1749.

Jena K.K., Khush G.S. and Kochert G. 1994. Comparative RFLP mapping of a wild rice, *Oryza officinalis*, and cultivated rice, *O. sativa*. *Genome*, **37**: 382-389.

Jiang J., Friebe B., Dhaliwal H.S., Martin T.J. and Gill B.S. 1993. Molecular cytogenetic analysis of *Agropyron elongatum* chromatin in wheat germplasm specifying resistance to wheat streak mosaic virus. *Theor. Appl. Genet.*, **86**: 41-48.

Jiang J., Friebe B. and Gill B.S. 1994. Recent advances in alien gene transfer in wheat. *Euphytica*, **73**: 199-212.

Jiang J., Morris K.L.D. and Gill B.S. 1994. Introgression of *Elymus trachycaulus* chromatin into common wheat. *Chromosome Res.*, **2**: 3-13.

Jiang J., Gill B.S., Wang G.L., Ronald P.C. and Ward D.C. 1995. Metaphase and interphase fluorescence *in situ* hybridization mapping of the rice genome with bacterial artificial chromosomes. *Proc. Natl. Acad. Sci.*, USA, **92**: 4487-4491.

Jiang J., Nasuda S., Dong F., Scherrer C.W., Woo S.S., Wing R.A., Gill B.S. and Ward D.C. 1996. A conserved repetitive DNA element located in the centromeres of cereal chromosomes. *Proc. Natl. Acad. Sci.*, USA, **93**: 14210-14213.

Kawahara T. and Taketa S. 2001. Fixation of translocation 24.4B infers the monophyletic origin of Ethiopian tetraploid wheat. *Theor. Appl. Genet.*, **101**: 705-710.

Kennard W., Phillips R., Porter R. and Grambacher A. 1999. A comparative map of wild rice (*Zizania palustris* 2n = 2x = 30). *Theor. Appl. Genet.*, **99**: 793-799.

Khush G.S. and Rick C.M. 1967. Studies on the linkage map of chromosome 4 of the tomato and on transmission of induced deficiencies. *Genetica*, **38**: 74-94.

Khush G.S. 1973. *Cytogenetics of aneuploids*. Academic Press, New York, pp. 301.

Khush G.S. and Brar D.S. 1989. Wide hybridization in plant breeding. *In*: *Plant Breeding and Genetic Engineering* (A.H. Zakri ed.) SABRAO, Malaysia, pp. 141-188.

Khushnir W. and Halloran G.M. 1984. Transfer of high kernel weight and high protein content from wild tetraploid wheat to bread wheat using homologous and homoeologous recombination. *Euphytica*, **33**: 249-256.

Kihara H. and Tsunewaki K. 1962. Use of alien cytoplasm as a new method of producing haploids. *J. Genet.*, **37**: 310-313.

Kjima K., Ohmido T., Hirose T., Mitsui T. and Fukui K. 1999. Visualization of the DNA stretches in a rice genotype by extended DNA fiber-FISH. *In*: *Plant and Animal Genome VII*, San Diego, USA, pp. 83.

Knott D.R. 1961. The inheritance of rust resistance. VI. The transfer of stem rust resistance from *Agropyron elongatum* to common wheat. *Can. J. Plant Sci.*, **41**: 109-123.

Kynast R.G., Riera-Lizarazu O., Vales M.I., Okagaki R.J., Maquieira S.B., Chen G., Ananiev E.V., Odland W.E., Russell C.D., Stec A.O., Livingston S.M., Zaia H.A., Rines H.W. and Phillips R.L. 2001. A complete set of maize individual chromosome additions to the oat genome. *Plant Physiol.*, **125**: 1216-1227.

Lapitan N.L.V., Sears R.G. and Gill B.S. 1984. Translocations and other karyotypic structural changes in wheat × rye hybrids regenerated from tissue culture. *Theor. Appl. Genet.*, **68**: 547-554.

Larkin P.J., Banks P.M., Bhati R., Brettell R.I.S., Davies P.A., Ryan S.A., Scowcroft W.R., Spindler L.H. and Tanner G.J. 1989. From somatic variation to variant plants: Mechanisms and applications. *Genome*, **31**: 705-711.

Laurie D.A., O'Donoughue L.S. and Bennett M.D. 1990. Wheat × maize and other wide sexual hybrids: Their potential for genetic manipulation and crop improvement. pp. 95-126. *In*: *Gene manipulation in plant improvement II*: (ed.) J.P. Gustafson. 19th Stadler Genetics Symposium. Plenum Press, New York.

Law C.N., Snape J.W. and Worland A.J. 1981. Intraspecific chromosome manipulation. *Phil. Trans. R. Soc.*, Lond., **292**: 509-518.

Li H.J., Guo B.H., Li Y.W., Du L.Q., Jia X. and Chu C.C. 2000. Molecular cytogenetic analysis of intergeneric chromosomal translocations between wheat (*Triticum aestivum* L.) and *Dasypyrum villosum* arising from tissue culture. *Genome*, **43**: 756-762.

McClintock B. 1950. The origin and behaviour of mutable loci in maize. *Proc. Natl. Acad. Sci.*, USA, **36**: 344-355.

McCouch S.R., Kochert G., Yu Z.H., Wang Z.Y., Khush G.S., Coffman W.R. and Tanksley S.D. 1988. Molecular mapping of rice chromosomes. *Theor. Appl. Genet.*, **76**: 815-829.

Mendel G. 1865. *Versuche über Pflanzen-Hybriden*. Verh. Naturf Orshung Ver. In Brüm Verh., **4**: 3-47.

Molnar-Lang M., Linc G., Logojan A. and Sutka J. 2000. Production and meiotic pairing behaviour of new hybrids of winter wheat (*Triticum aestivum*) × winter barley (*Hordeum vulgare*). *Genome*, **43**: 1045-1054.

Morgan W.G., King T. P., Koch S., Harper J.A. and Thomas H. 2001. Introgression of chromosomes of *Festuca arundinacea* var. *glaucescens* into *Lolium multiflorum* revealed by genomic *in situ* hybridization. *Theor Appl. Genet.*, **103**: 696-701.

Mujeeb-Kazi A. and Jewell D.C. 1985. CIMMYT's wide cross program for wheat and maize improvement. *In*: *Biotechnology in International Agricultural Research*. International Rice Research Institute, Manila, Philippines, pp. 219-226.

Mukai Y., Fribe B., Hatchett M., Yamamoto M. and Gill B.S. 1993. Molecular cytogenetic analysis of radiation-induced wheat-rye terminal and intercalary chromosomal translocations and the detection of rye chromatin specifying resistance to Hessian fly. *Chromosama*, **102**: 88-95.

Okagaki R.J., Kynast R.G., Livingston S. M., Russell C.D., Rines H.W. and Phillips R.L. 2001. Mapping maize sequences to chromosomes using oat-maize chromosome addition materials. *Plant Physiol.*, **125**: 1228-1235.

Okamoto M. 1957. *Asynaptic effect of chromosome V*. Wheat Information Service 5:6.

O'mara J.G. 1940. Cytogenetic studies on Triticinae. I. A method for determining the effect of individual *Secale* chromosomes on *Triticum. Genetics*, **25**: 401-408.

Orton T.J. 1980. Haploid barley regenerated from callus cultures of *Hordeum vulgare* × *H. jubatum. J. Hered.*, **71**: 780-782.

Park S.J., Walsh E.J., Reinbergs E., Song L.S.P. and Kasha K.J. 1976. Field performance of doubled haploid barley lines in comparison with lines.

Peloquin S.J., Hougas R.W. and Gabert A.C. 1964. Haploidy as a new approach to the cytogenetics and breeding of *Solanum tuberosum*. p. 21-28. *In*: *Chromosome manipulations and plant genetics*. (eds.) R. Riley and K.R. Lewis, Oliver and Boyd Ltd., London.

Pickering R.A., Malyshev S., Kunzel G., Johnston P.A., Korzun V., Menke M. and Schubert I. 2000. Locating introgressions of *Hordeum bulbosum* chromatin within the *H. vulgare* genome. *Theor. Appl. Genet.*, **100**: 27-31.

Rajaram S., Maan E.E., Ortiz-Ferrara G. and Mujeeb-Kazi A. 1983. Adaptation, stability and high yield potential of certain 1B/1R CIMMYT wheats. *In*: *Sixth Intern. Wheat Genet. Symp.*, Kyoto, Japan, pp. 613-621.

Ramage R.T. 1965. Balanced tertiary trisomics for use in hybrid seed production. *Crop Sci.*, **5**: 177-178.

Rayburn A.L. and Gill B.S. 1985. Use of biotin labeled probes to map specific DNA sequences on wheat chromosomes. *J. Hered.*, **76**: 78-81.

Rhoades M.M. 1936. A cytogenetic study of a chromosome fragment in maize. *Genetics*, **21**: 491-502.

Riera-Lizarazu O., Rines H.W. and Phillips R.L. 1996. Cytological and molecular characterization of oat × maize partial hybrids. *Theor. Appl. Genet.*, **93**: 123-125.

Riera-Lizarazu O., Vales M.I., Ananiev E.V., Rines H.W. and Phillips R.L. 2000. Production and characterization of maize chromosome 9 radiation hybrids derived from an oat-maize addition line. *Genetics*, **156**: 327-339.

Riley R., Chapman V. and Johnson R. 1968. Introduction of yellow rust resistance of *Aegilops comosa* into wheat by genetically induced homoeologous recombination. *Nature*, **217**: 383-384.

Roman H. and Ullstrup A.J. 1952. The use of A-B translocations to locate genes in maize. *Agron. Jour.*, **43**: 450-454.

Rosenfeld M. 1997. Human artificial chromosomes get real. *Nature Genetics*, **15**: 333-335.

Sachan J.K.S. and Tanaka R. 1977. Variation and pattern of C-banding in *Zea* chromosomes. *Nucleus*, **20**: 61-64.

Sakaguchi S. and Nishimura Y. 1969. Breeding seedless watermelon by using induced chromosome translocations. *Jarq.*, **3**: 18-22.

Scalenghe F., Tureo E., Edstrom J.E. and Pirotta V. 1981. Microdissection and cloning of DNA from a specific region of *Drosophila melanogaster* polytene chromosomes. *Chromosama*, **82**: 205-216.

Schwarzacher T., Anamthawat-Jonsson K., Harrison G.E., Islam A.K.M.R., Jia J.Z., King I.P., Leitch A.R., Miller T.E., Reader S.M., Rogers W.J., Shi M. and Heslop-Harrison J.S. 1992. Genomic *in situ* hybridization to identify alien chromosomes and chromosome segments in wheat. *Theor. Appl. Genet.*, **84**: 778-786.

Sears E.R. 1953. Nullisomic analysis in common wheat. *Amer. Nat.*, **87**: 245-252.

Sears E.R. 1954. The aneuploids of common wheat. *Res. Bull., Mo. Agr. Exp. Stn.*, 572, 59pp.

Sears E.R. 1956. The transfer of leaf rust resistance from *Aegilops umbellulata* to wheat. Brookhaven. *Symp. Biol.*, **9**: 1-22.

Sears E.R. 1966. Nullisomic-tetrasomic combinations in hexaploid wheat. p. 29-45. *In*: *Chromosome manipulations and plant genetics.* (eds.) R. Riley and K.W. Lewis, Oliver and Boyd Ltd., London, London.

Sears E.R. 1972. Chromosome engineering in wheat. *Stadler Genet. Symp.*, **4**: 23-38.

Sears E.R. 1973. *Agropyron*-wheat transfers induced by homoeologous pairing. *In*: 4th Intern. *Wheat Genet. Symp.*, Columbia. pp. 191-199.

Sears E.R. 1977. An induced mutant with homoeologous pairing in common wheat. *Can. J. Genet. Cytol.*, **19**: 585-593.

Sherman J.D., Smith L.Y., Blake T.K. and Talbert L.E. 2001. Identification of barley genome segments introgressed into wheat using PCR markers. *Genome*, **44**: 38-44.

Singh K., Ishii T., Parco A., Huang N., Brar D.S. and Khush G.S. 1996a. Centromere mapping and orientation of the molecular linkage map of rice (*Oryza sativa* L.). *Proc. Natl. Acad. Sci.*, **93**: 6163-6168.

Singh K., Multani D.S. and Khush G.S. 1996b. Origin, characterization, and use in determining the orientation of chromosome map. *Genetics*, **143**: 517-529.

Singh R.J. 1993. *Plant Cytogenetics.* CRC Press, Boca Raton, pp. 391.

Snowdon R.J., Friedrich T., Friedt W. and Kohler W. 2002. Identifying the chromosomes of the A-C-genome diploid *Brassica* species. *B. rapa* (Syn. *campestris*) and *B. oleracea* in their amphidiploid *B. napus. Theor. Appl. Genet.*, **104**: 533-538.

Sutton W.S. 1903. The chromosomes in heredity. *Biol. Bull.*, **4**: 231-251.

Sutton W.S. and Boveri T. 1903. *An appraisal. Cited in Cytogenetics* by C.P. Swanson, T. Merz, W.J. Yong (1967). Prentice Hall Inc., New Jersey, pp. 1-7.

Swaminathan M.S. 1951. Notes on induced polyploids in the tuber-bearing *Solanum* species and their crossability with *S. tuberosum. Amer. Potato Jour.*, **28**: 472-489.

Szabados L., Hadlaczky G. and Dudits D. 1981. Uptake of isolated plant chromosomes by plant protoplasts. *Planta*, **151**: 141-145.

Thomas H., Leggett M. and Jones I.T. 1975. The addition of a pair of chromosomes of the wild oat, *Avena barabta* (2n = 28) to the cultivated oat, *A. sativa* L. (2n = 42). *Euphytica*, 24: 717-724.

Thomas H., Powell W. and Aung T. 1980. Interfering with regular mieotic behaviour in *Avena sativa* as a method of incorporating the gene for mildew resistance from *A. barbata. Euphytica*, **29**: 635-640.

Unrau J., Person C. and Kuspira J. 1956. Chromosome substitution in hexaploid wheat. *Can. J. Bot.*, **34**: 629-640.

Vrana J., Kubalakova M., Simkova H., Chihalikova J., Lysak M.A. and Dolezel J. 2000. Flow sorting of mitotic chromosomes in common wheat (*Triticum aestivum* L.). *Genetics*, **156**: 2033-2041.

Wall A.M., Riley R. and Chapman V. 1971. Wheat mutants permitting homoeologous meiotic chromosome pairing. *Genet. Res.*, **18**: 311-328.

Wang R.C., Liang G.H. and Heyne E.G. 1977. Effectiveness of *ph* gene in inducing homoeologous pairing in *Agrotrichum. Theor. Appl. Genet.*, **51**: 139-141.

Wang W.N., Hang A., Hansen J., Burton C., Mallory-Smith C.A. and Zemetra R.S. 2000. Visualization of A- and B-genome chromosomes in wheat (*Triticum aestivum* L.) x jointed goat grass (*Aegilops cyclindrica* Host) backcross progenies. *Genome*, **43**: 1038-1044.

Watanabe K.N., Orrillo M., Vega S., Golmirzaie A.M., Perez S., Crusado J. and Watanabe J.A. 1996. Generation of pest resistant, diploid potato germplasm with short-day adaptation from diverse genetic stocks. *Breed. Sci.*, **46**: 329-336.

Wienhues A. 1965. Cytogenetische Untersuchungen über die chromosomale Grundlage der Rostre-sistenz der Weizensortie Weique. *Zuchter*, **35**: 352-354.

Xin Z.Y., Zhang Z.Y., Chen X., Lin Z.L., Ma Y.Z., Xu H.J., Banks P.M. and Larkin P.J. 2001. Development and characterization of common wheat-*Thinopyrum intermedium* translocation lines with resistance to barley yellow dwarf virus. *Euphytica*, **119**: 161-165.

Zeller F.J. 1973. 1B/1R wheat-rye chromosome substitutions and translocations. *In*: *5th Intern. Wheat Genet. Symp.*, Columbia, Missouri, pp. 209-222.

Zhong X.B., Bodeau J., Fransz P.F., Williamson V.M., Van Kammen A., De Jong J.H. and Zabel P. 1999. FISH to meiotic pachytene chromosomes of tomato locates the root-know nematode resistance gene *Mi-1* and the acid phosphatase gene *Aps-1* near the junction of euchromatin and pericentromeric heterochromatin of chromosome arms 6S and 6L, respectively. *Theor. Appl. Genet.*, **98**: 365-370.

Zhong X.B., Fransz P.F., Van-Wennekes E.J., Zabel P., Van Kammen A. and De Jong J.H. 1996. High resolution mapping by fluorescence *in situ* hybridization to pachytene chromosomes and extended DNA fibres. *Plant Mol. Biol. Rep.*, **14**: 232-242.

Ziolkowski P.A. and Sadowski J. 2002. FISH-mapping of rDNAs and *Arabidopsis* BAC on pachytene complements of selected *Brassicas*. *Genome*, **45**: 189-197.

Plant Breeding - Mendelian to Molecular Approaches
H. K. Jain and M. C. Kharkwal (eds.)

5

Advances in Molecular Cytogenetics: Potential for Crop Improvement

P.K. Gupta[1] and M.K. Dhar[2]

Abstract

Advances made over the last two decades in techniques and approaches used in molecular cytogenetics research and the results obtained so far have been briefly reviewed. These include the use of chromosome banding, fluorescence *in situ* hybridisation (FISH) and multicolour FISH (McFISH), genomic *in situ* hybridisation (GISH), flow cytometry, pulse field gel electrophoresis (PFGE), microdissection and microcloning, among others. These tools have been used both for identification of individual chromosomes and for physical localization of DNA sequences on individual chromosomes. Construction of genetic and physical maps of crop plants involving molecular markers as well as genes for agronomic traits has been discussed briefly. The use of these maps in comparative genomics, both at the macro- and micro-colinearity levels and for map based cloning has been briefly reviewed. The development and use of bacterial artificial chromosomes (BACs), yeast artificial chromosomes (YACs) and the future prospects of the development of plant artificial chromosomes (PACs) have also been discussed. The progress made in discovery of new genes through reverse genetics and functional genomics approach has also been presented.

Introduction

Cytogenetics is an area of research that relates structure, behaviour and function of chromosomes with the genes that these chromosomes carry (Gupta, 1995). Since many of these genes are relevant to crop improvement programmes, advances in cytogenetics research have always been considered important for crop improvement. In recent years, research in this discipline has been greatly influenced by molecular tools that have been extensively utilized for crop biotechnology, leading to the emergence of a new area of cytogenetics research called "molecular cytogenetics". The latest development in the field of crop biotechnology, which would influence cytogenetics in a big way, is the emergence of genomics as the new area of research, which should prove extremely useful for future crop improvement programmes. In this connection, a landmark is the recent publication of complete sequence of *Arabidopsis* genome on December 14, 2000 (TAGI, 2000 and the

[1]Ch. Charan Singh University, Meerut 250 004, India
[2]University of Jammu, Jammu 180 006, India

three accompanying papers). The follow-up studies in this weed and several other small genomes (rice, sorghum, *Medicago truncatula*, etc.) will certainly lay the foundation for future crop improvement programmes.

Molecular cytogenetics involves an understanding of the structural and functional organization of genome within the chromosomes, including physical localization of genes on chromosomes (physical mapping) using *in situ* hybridization or deletion mapping. This would also facilitate isolation and/or manipulation of genes, which are important for crop improvement. For instance, with the availability of molecular maps (both genetic and physical) of all major crops and the discovery of macro- and microsynteny conservation in many cases, it is now possible to clone mapped genes through the techniques involving map based cloning and/or microdissection/microcloning.

Research in the area of molecular cytogenetics also provides useful information for understanding relatedness of a crop with its wild relatives, that make an important genetic resource for the improvement of this crop. For instance, in any programme on wide hybridization undertaken with the objective of alien gene transfer, it is important to undertake molecular cytogenetic analysis not only of the crop in question, but also of the wild species to be used for alien gene transfer. This will not only help to determine the feasibility of such alien gene transfer, but will also enhance the efficiency of such a transfer. The alien gene transfer may or may not involve manipulation of a diploidizing system like *Ph* system of wheat, so that a molecular analysis of *Ph* system in wheat has been an active area of research. In still other cases, it may also be necessary to analyze the degree of recombination in different regions of the chromosome that carry the gene of interest, so that we know whether or not the gene for the targeted trait is located in a region of high recombination. Advances in molecular cytogenetics have influenced in the past and will continue to influence in future, our understanding of the structural organization, behaviour and function of chromosomes and also that of the genes they carry. This knowledge base is important for devising strategies for manipulating genes that are important for crop improvement. The present paper aims to review the recent developments in this paper.

Chromosome Identification/Sorting: Differential Banding, CHIAS and Flow Cytometry

Identification of individual chromosomes is the first step of cytogenetics research in any crop. Until late 1960s and early 1970s, precise chromosome identification could be undertaken only in those species where differences in size and morphology of chromosomes could be resolved either at mitotic metaphase or at pachytene of meiosis. However, in crops like barley and bread wheat, individual chromosomes do not differ significantly in size and morphology, and pachytene analysis is not feasible. In crops such as these and also in other crops, chromosome banding techniques developed in 1970s (Vosa, 1970), and CHIAS (chromosome image analysing system), developed in 1980s (Fukui, 1986) provided reliable methods for chromosome identification and karyotype preparation. The banding technique that was originally developed for human chromosomes (Caspersson *et al.*, 1969), was later applied to several plant and animal systems. Among crop plants, wheat and rye were the first to be used for preparation of C-banded karyotypes (Gill and Kimber, 1974 a, b). Thereafter, the technique was used extensively to identify chromosomes in a number of other plant species (Friebe *et al.*, 1995). The banding technique also resolved intraspecific variations in DNA organization within chromosomes, in the form of variations in banding patterns of individual chromosomes. For instance, C-banding polymorphism in various cultivars of wheat were described by Friebe and Gill (1994) and that in 185 accessions of *Triticum araraticum* were described by Badaeva *et al.*, (1994). Among chromosomes of a number of crops, the banding technique also allowed detection of structural rearrangements that are not uncommon among different cultivars of a crop. This variation resolved by banding technique thus challenged the concept of constancy of chromosome structure and karyotype within individual species, and suggested variation in the organization of DNA within individual chromosomes of a species. Therefore, utilizing popular cultivars for individual crops, standard karyotypes and nomenclature for C-bands were proposed in several crops including wheat, barley and rye, to bring about uniformity in karyotype description (Gill *et al.*, 1991.

During mid-1980s, with the availability of computer technology, it also became possible to analyse chromosomes using the chromosome image analysing system (CHIAS), which would digitize chromosome images and would conduct analysis of a set of chromosomes within 25 minutes (Fukui, 1986). Several improvements in this technique were made, which led to the development of second and third generation CHIAS systems (CHIAS II and CHIAS III), which became available during 1990s and proved very useful for chromosome research. CHIAS II and CHIAS III also allowed measurements of several new parameters including condensation pattern (CP) and density volume (DV), which can prove useful in the identification of individual chromosomes (Fukui, 1987; Fukui and Nakayama, 1998).

During 1980s and 1990s, another technique that became popular for chromosome research was flow cytometry, which facilitated estimations of genome size, flow karyotyping and sorting of individual chromosomes. For instance, sorting of all the 42 arms of bread wheat chromosomes could be undertaken using ditelocentric stocks and sorting of all the 10 maize chromosomes could be achieved using oat-maize addition lines (Arumuganthan *et al.*, 2000). Flow sorting of individual chromosomes has also been achieved in several other plant species (De Laat and Blaas, 1984; Conia *et al.*, 1987; Arumuganathan and Earle, 1991; Wang *et al.*, 1992; Lucretti *et al.*, 1993; Lee *et al.*, 1996). For instance, flow cytometry has been used to isolate specific chromosomes from protoplasts in *Petunia* hybrida (Conia *et al.*, 1987) and from root tips in maize (Lee *et al.*, 1996) and wheat (Lee *et al.*, 1997). The sorted chromosomes, or the DNA derived from individual chromosomes sorted out as above, can be used for a variety of purposes, including physical mapping of specific loci and preparation of chromosome specific libraries. Other uses of flow cytometry include identification of chromosome addition lines, polyploids and aneuploids.

Physical Localization of DNA Sequences on Chromosome Using *in situ* Hybridization

(*a*) Fluorescence *in situ* Hybridization (FISH)

In situ hybridization (ISH) technique was initially developed, using radioactive probes, by Gall and Pardue (1969). The technique is useful in locating genes or DNA sequences on chromosomes. Although the original technique was highly sensitive, it was time consuming and cumbersome. These limitations were overcome by the development of non-radioactive *in situ* hybridization technique (Langer-Safer *et al.*, 1982) that was initially used for detection of repetitive DNA sequences. However, several improvements in the technique have been made. In plants, Rayburn and Gill (1985) were the first to apply non- radioactive *in situ* hybridization to plants. They used biotin-labelled DNA probes and the hybridization sites were detected by enzymatic reactions using horse radish peroxidase or alkaline phosphatase conjugated avidin/streptavidin. Subsequently, fluorescence *in situ* hybridization (FISH) using fluorochromes for signal detection became more common. This system offered many advantages. The foremost among these advantages included ability to label several DNA probes with different haptens, which can be detected simultaneously by different fluorochromes.

Although, many probe labelling and fluorescent reagents are currently available, essentially haptens in the form of combinations of biotin, digoxigenin and fluorescein labelled nucleotides are used. This technique, described as multicolor FISH (McFISH) was initially used in both humans (Ried *et al.*, 1992) and plants (Mukai, 1996) to detect simultaneously seven different probes. Later, in humans, as many as 24 different colours, one for each human chromosome, could be developed (Schrock *et al.*, 1996). Multicolour FISH not only allows identification of individual chromosomes using one or more repetitive sequences simultaneously, but also enables physical mapping of sequences of interest. Several repetitive DNA sequences have actually been cloned from a variety of plant species and used for FISH. Thus depending upon the distribution of repetitive sequences in a particular genome, FISH helps both in conventional genome analysis in polyploid species (McIntyre *et al.*, 1990; McNeil *et al.*, 1994) and identification of individual chromosomes both in diploid and polyploid species. For instance, repetitive DNA sequence pAs1 (derived from *Aegilops tauschii*) permits identification of D genome chromosomes (Rayburn and Gill, 1985; 1986), while *pSc119.2* (derived from rye) mainly hybridizes to B genome chromosomes of bread wheat (Mukai *et al.*, 1993). The D genome clone pAs1 and the barley clone pHvG38 were also used for two-colour FISH to identify the entire chromosome complement of bread wheat, as an alternative to C-banding (Pederson and Langridge, 1997). FISH was also used for the identification of individual maize chromosomes in oat-maize addition lines (Kynast *et al.*, 2000). Besides this, 5S and 18S-26S rDNA sequences were also used for identification of chromosomes carrying these sequences.

The resolution of FISH has improved considerably through capturing images by highly sensitive charge coupled device (CCD) camera and processing them through digital image analysis systems, discussed above.

In situ hybridization can be conducted either on condensed metaphase chromosomes or on extended pachytene or interphase chromosomes. Since the resolution on highly condensed metaphase chromosomes is usually restricted to 2 Mbp, *in situ* hybridization to the less condensed pachytene chromosomes is often used without losing the ability to identify individual chromosomes (Shen *et al.*, 1987; Zhong *et al.*, 1996). Further improvement in resolution up to 100 kb has been achieved in humans by applying FISH to interphase nuclei (Trask *et al.*, 1989) which, however, was at the expense of individual chromosome recognition.

(*b*) Genomic *in situ* Hybridization (GISH)

GISH is a modification of *in situ* hybridization technique (Le *et al.*, 1989; Schwarzacher *et al.*, 1989) and proved useful in genome analysis of several polyploid species. In a hybrid, it also proved useful in discriminating chromosomes belonging to two related parental species (e.g. hybrids between *Hordeum vulgare* and *H. bulbosum*). GISH involves labelling of total genomic DNA of one species and using it as a probe, in presence of excess DNA of the other species. The unlabelled excess DNA of one species first blocks the availability of DNA sequences of this species for hybridization, so that the labelled DNA gets hybridized to the chromosomes of the other species only. The technique was successfully utilized in several species including wheat (Mukai *et al.*, 1993), *Aegilops*, oats (Chen and Armstrong, 1994; Jellen *et al.*, 1994), tobacco (Kenton *et al.*, 1993), rice, cotton, etc.

(*c*) Fiber FISH

Recently, dramatic progress has been made in physical mapping through the development of fluorescence hybridization to extended chromatin fibers (ECFs) or extended DNA fibers (EDFs) that was first reported in early 1990s (Heng *et al.*, 1992). Shortly after, a variety of related methods was reported (Fidlerova *et al.*, 1994; Gerdes *et al.*, 1994; Haaf and Ward, 1994; Houseal *et al.*, 1994; Parra and Windle, 1993; Wiegant *et al.*, 1992). The fibre technique proved highly versatile in permitting physical fine mapping of YACs, cosmids, lambda and plasmid clones, thus permitting direct mapping of DNA fragments ranging from a few kilobases (Florijn *et al.*, 1995) to several hundred kilobases (Haaf and Ward, 1994). High resolution physical mapping by extended DNA fiber FISH has also been successfully achieved in *Arabidopsis thaliana* and tomato (Fransz *et al.*, 1996).

(*d*) Identification of Alien Chromosome Segments Using FISH and GISH

A number of studies have been conducted in wheat, where using FISH and GISH, presence of DNA sequences from related wild and cultivated alien species were shown to be present in the wheat genome (for reviews see, Mukai, 1996; Mukai and Yamamoto, 1998). Among recent studies, for instance, Wang *et al.* (1999) used C-banding, GISH and RFLP analyses for cytogenetic characterization of wheat - *Elymus tsukushiense* introgression lines. *E. tsukushiense* is a distant relative of wheat and an excellent source for resistance to wheat scab. An addition line NAU702 was identified and found to have a high level of resistance to scab. This could prove highly useful in developing improved germplasm for scab resistance breeding. Several other studies involved the use of FISH or GISH for detection of a whole alien chromosome or segments in wheat germplasm.

Molecular Maps, Comparative Genome Analysis and Colinearity

Molecular genetic/physical maps are now available for all major crops and have been utilized for comparative genomic studies of chromosomes in several groups of plants including Brassicaceae, Poaceae, Fabaceae and Solanaceae. Recently, individual maize chromosomes have been added to oats and used for preparation of molecular maps of individual maize chromosomes (Kynast *et al.*, 2001; Okagaki *et al.*, 2001). Availability of deletion stocks in several crops including wheat and maize also facilitated the preparation of physical maps of chromosomes. A novel radiation hybrid technology, earlier used in humans, was also used for the preparation of physical map of maize chromosome 9 (Okagaki *et al.*, 2000; Riera-Lizarazu *et al.*, 2000). Recent developments of techniques for manipulation and cloning of large DNA fragments such as pulse field gel electrophoresis (PFGE), yeast artificial chromosome (YAC) and bacterial artificial chromosome (BAC) have also provided powerful tools to generate physical maps of crop plant genomes. For instance, PFGE has been used for isolation of large DNA molecules representing small chromosome fragments in crops like tomato (Ganal *et al.*, 1989) and sugarbeet (Jung *et al.*, 1990). Similarly, BAC clones have been successfully utilized for mapping in several crops including rice, cotton and sorghum (Jiang *et al.*, 1995; Hanson *et al.*, 1995; Zwick *et al.*, 1998). Despite the availability of these tools, *in situ* hybridization techniques in plants have mainly been used for mapping repetitive DNA sequences and multicopy gene families. Mapping of low or single copy sequences has proved difficult, although on occasions these unique sequences have also been mapped, particularly with the help of fibre FISH, discussed above.

The studies involving genetic and physical mapping in several groups of plants as above suggested that in majority of cases, along the length of individual chromosomes, not only the gene content and order (synteny), but also the intragenic and intergenic DNA sequence polymorphisms have largely been conserved in related species (Draye *et al.*, 2001; Dubcovsky *et al.*, 2001). However, segmental duplications, deletions and rearrangements have also been shown to play a role in the evolution of these related taxa (Devos and Gale, 2000; Heslop-Harrison, 2002; Bennetzen, 2000; Paterson *et al.*, 2000). For instance, selective elimination of genomic sequences of wheat and its related species during allopolyploidization has recently been demonstrated in artificially synthesized F_1 hybrids and raw allopolyploids (Shaked *et al.*, 20001; Ozkan *et al.*, 2001).

Organization of repetitive DNA sequences including those in ribosomal DNA (rDNA) has been studied within individual chromosomes and genomes. Special attention has been paid to characterization and isolation of telomeric and centromeric sequences, to assist in efforts being made to assemble artificial chromosomes in some plant species like rice (Nonomura and Kurata, 2000; see later). Distribution and organization of transposons (including MITES, LTR and non-LTR-retrotransposons, LINES and SINES) and microsatellites (SSRs), and methylation of DNA sequences are the other areas of research, which received attention of cytogeneticists and molecular biologists (Gupta, 2000). In addition to the above, comparative analysis of molecular genetic maps also helped in predicting the location of genes controlling key agronomic traits, the only exception being the disease resistance genes, which undergo changes much faster (Leister *et al.*, 1998). Among a number of genes that have already been studied through comparative genome analysis, two important examples include (*i*) gibberillin-insensitive dwarfing genes, which have been found to be orthologous between wheat and maize (Peng *et al.*, 1999), and (*ii*) the QTLs controlling shattering, which have been found to be orthologous among several members of the grass family including foxtail millet, maize, sorghum, rice and pearl millet (Paterson *et al.*, 1995; Wang *et al.*, 2001).

Several other important conclusions that have been made on the basis of comparative genomics include the following: (*i*) Maize is a tetraploid with $x = 5$ and retrotransposons make upto 50-80 % of maize genome (in other plant genomes also, retrotransposons make a sizable fraction of the genome). (*ii*) About two-thirds of *Arabidopsis* genome is duplicated suggesting that it is a segmental polyploid, if not a true tetraploid like maize. (*iii*) In cereals, extensive rearrangements took place between homoeologous chromosomes. (*iv*) Rice genome, being an anchor grass genome, has been divided into 20-35 blocks, which can be used for reconstructing any of the grass genome. (*v*) There are specific gene rich regions in all chromosomes studied, particularly in cereals. For instance, in wheat chromosomes, it was shown that very few genes are located in the proximal one-third of each chromosome arm, and that there is a high density of genes in the distal regions. Within the gene rich regions themselves, the distribution of genes could differ, even being more in rice than in wheat.

Microcolinearity in DNA Sequences of Small Genomic Regions

The studies on molecular maps across related plant species (as above), though revealed significant conservation of gene content, gene order and gene homology, had their own limitations. For instance, only up to about one marker per 10 centimorgan genetic distance is available for comparative analysis, thus making it difficult to analyse small deletions, duplications and inversions involving only a few centimorgans. Moreover, many genes have multiple homologs within the same genome, thus making it difficult to establish orthology between genes from different but related species (Bennetzen, 2000). In view of these limitations, in recent years, instead of mapped markers, DNA sequences representing small regions of genomes have been used for comparative analysis, sometimes resolving what is described as microcolinearity (Bennetzen, 2000; Feuillet and Keller, 1999; Doyle and Gaut, 2000; Table 1; see later for more details). Although very few plant genomic sequences have been used for such a study, some patterns have been delineated on the basis of available sequences.

Comparative analysis of gene sequences and repetitive DNA sequences has shown that they evolve at different rates. However, it has been shown that the gene sequences are more conserved relative to repetitive DNA, although disease resistance genes have been shown to change rapidly. It has also been shown in several studies that the gene density differs and is related with genome size. For instance, at the *Sh2/A1* region, the intergenic regions were seven times longer in maize than in the orthologous regions in rice and sorghum. Similarly, in a 280 kb region of the *Adh1/u22* locus, in maize the genes were separated by more than 120 kb containing 10 retroelement families, but in sorghum the orthologous genes were only 50 kb apart.

Table 1. Large genomic fragments sequenced in some cereal species

	Crop and species	Fragment length sequenced	Surrounding region of gene/locus	Reference
1.	Bread wheat			
	(*Triticum aestivum*)	16 kb	*Lrk10*	Feuillet and Keller (1999)
		23 kb	*Lrk10*	Stein *et al.* (2000)
2.	Diploid wheat			
	(*Triticum tauschii*)	16 kb	*SBE-1*	Rahman *et al.* (!997)
	Triticum monococcum	211 kb		Wicker *et al.* (2001)
	Triticum monococcum	180 kb	*Vrn1*	Dubcovsky *et al.* (2001)
	Triticum monococcum	112kb, 125 kb, 151 kb	*X1*	Li and Gill (2002)
	Triticum monococcum	45 kb, 125 kb	*Sh2*	Li and Gill (2002)
3.	Barley			
	(*Hordeum vulgare*)	160 kb	*HvLrk*	Feuillet and Keller (1999)
		60 kb	*mlo*	Panstruga *et al.* (1988)
		204 kb	*Mla*	Wei *et al.* (1999)
		66 kb	*rar1*	Shirasu *et al.* (2000)
		90 kb	*Vrn1*	Dubcovsky *et al.* (2001)
4.	Rice			
	(*Oryza sativa*)	340 kb	*Adh1-Adh2*	Tarchini *et al.* (2000)
		50 kb	*BAC 3615*	Dubcovsky *et al.* (2001)
		150 kb	*chr 1/6*	Rice Genome Project (http://www.staff.or.jp/GenomeSeq.html)
5.	Sorghum			
	(*Sorghum vulgare*)	78 kb	*Adh1*	Tikhonov *et al.* (1999)
6.	Maize			
	(*Zea mays*)	218 kb	*Adh1*	Tikhonov *et al.* (1999)
		140 kb	*A1/Sh2*	Civardi *et al.* (1994)
		78 kb	*Zein*	Llaca and Messing (1998)

It has also been shown that microcolinearity may occur not only between related species but also between more distantly related species. For instance, in the *adh* region, colinearity is greater between closely related maize and sorghum than between either of these two species and the distantly related rice (Tikhonov *et al.*, 1999). In contrast to this, wheat shows greater colinearity with distantly related barley than with the closely related rye (Devos *et al.*, 1993). Sequenced segments of tomato and *Arabidopsis* have also been compared, and it was shown that large-scale duplications followed by selective gene loss created a network of synteny between the two species (Ku *et al.*, 2000).

DNA sequences have also been utilized for evolutionary studies, where phylogenetic trees could be prepared on the basis of similarities between sequences of individual genes like 5S rRNA, 18S rRNA, *rbcL* and several others. Already, by 1980s, large sequence data and sophisticated computer based analytical tools had become available for this purpose for redrawing the phylogenetic tree of life, which has recently been interpreted once again (Woese, 2000). However, in most of these earlier studies, sequences of one or two genes were utilized at a time, but the sequences of individual genes like 18S rRNA and *rbcL* did not find the most parsimonious trees. Therefore, more recently DNA sequences of plastid genes *rbcL* and *atpB* and the nuclear 18S rDNA of 560 species of angiosperms and seven non-flowering plants were utilized together for preparing a well-resolved phylogenetic tree (Solttis *et al.*, 1999). Similarly, a study of DNA sequences of five mitochondrial, plastid and nuclear genes (*atp1*, *matR*, *atpB*, *rbcL*, 18S rDNA) of 105 species belonging to 103 genera and 63 families suggested that *Amborella*, Nymphaeales and Illicales-Trimeniaceae-*Austrobaileya* represent the first stage of Angiosperm evolution, with *Amborella* being sister to all other angiosperms (Qui *et al.*, 1999). Almost the same conclusions were also drawn through a study of duplicate phytochrome genes, *PHYA* and *PHYC* (Mathews and Donoghue, 1999).

Map Based Cloning

Both genetic and physical maps have proved useful for map based cloning of specific genes. An estimation of the physical distance (in terms of kilobases) between two tightly linked molecular markers flanking the gene of interest, may prove useful for this purpose. Several studies have clearly demonstrated the existence of "gene rich" regions and uneven recombination frequencies along the length of chromosomes, so that it is difficult to determine such distances by conventional genetic analysis. In such cases, BAC clones and contigs consisting of a number of cosmids hybridizing with one or more flanking markers may prove useful. Flanking molecular markers can also be used for FISH at interphase nuclei, to estimate the physical distance between the markers (in terms of proportion of chromosome length) and the position of these markers (in terms of proximity to the centromere). This was successfully done initially in humans (Trask *et al.*, 1989), and later in several crop species. Thus besides other approaches, FISH mapping can also prove to be a valuable tool for map-based cloning in plants. This approach has been used by Jiang *et al.* (1995) to determine the relationship among three BAC clones in rice. Several genes, particularly those imparting disease resistance in a variety of crop plants have been isolated using such an approach of map based cloning.

Chromosome Microdissection and Microcloning

The techniques of chromosome microdissection and microcloning are new to plant systems. These were initially developed in 1981 on *Drosophila* polytene chromosomes and later used extensively to analyze human chromosomes (see Kao, 1996, for a review). During the last decade the techniques have been used effectively in several plant genera including *Hordeum* (Fukui *et al.*, 1991; Schondelmaier *et al.*, 1993), *Triticum* (Albani *et al.*, 1993; Vega *et al.*, 1994; Liu *et al.*, 1997), *Avena* (Chen and Armstrong, 1994), *Zea* (Stein *et al.*, 1998), *Secale* (Sandery *et al.*, 1991; Zhou *et al.*, 1999) and *Beta* (Jung *et al.*, 1992).

Individual chromosomes as well as chromosome segments representing satellites, chromosome arms and centromeres could also be dissected out and utilized for a variety of purposes including development of probes for RFLPs and tagging of important genes. Microdissected barley chromosomes carrying translocations were also used for PCR mediated physical mapping of translocation breakpoints (TBs), which can now be used as physical landmarks (Sorokin *et al.*, 1994; Kuenzel and Korzun, 1996; Kuenzel *et al.*, 2000). Sex chromosomes in *Melandrium* could also be dissected for developing chromosome specific probes. Special cytogenetic stocks such as monosomics, ditelocentrics and telotrisomics (Schondelmaier *et al.*, 1993) and addition lines (Jung *et al.*, 1992) can also be used for microdissection and microcloning in plant systems.

Genomics and Whole Genome Sequencing

During mid- and late 1990s, a new area of biology emerged in the form of genomics and 'reverse genetics', which involved the use of high throughput approaches for the study of the genome as a whole (both structure and function), rather than studying individual genes. One of the major activities in this area involved sequencing of whole genomes in some model organisms including *Arabidopsis* and rice as model plant systems. Two approaches have been used for whole genome sequencing, the BAC to BAC approach (utilizing the physical maps) and the whole genome shotgun (WGS) approach, recently used for human genome. In the past, assembly of genomic contigs using BAC clones has been the primary strategy for sequencing the whole genome of *Arabidopsis* and also for the ongoing efforts of sequencing the rice genome.

As mentioned above, *Arabidopsis* genome has now been completely sequenced (TAGI, 2000), utilizing the 'BAC to BAC' or 'map first' approach. More than 90 % of the genome that has been sequenced (~118 Mb) has brought some interesting features to light. For instance, it has been shown that much (~two-thirds) of the *Arabidopsis* genome is duplicated, but only 19 % of ESTs fall in these duplicated regions. Substantial proportion of centromeric DNA of each of the five chromosomes has also been sequenced, resolving some interesting features. In chromosome 2, sequence corresponding to 75 % of the mitochondrial genome was also noticed near the centromere, suggesting lateral transfer of this DNA from mitochondrion genome to the nuclear genome.

Rice genome is also being sequenced with the efforts of public as well as private sectors. An International Rice Genome Sequencing Project (IRGSP) was created for this purpose in 1998, with a target to complete rice genome sequencing by the year 2008 using BAC to BAC approach. Meanwhile, Monsanto and Syngenta (two private companies) independently used whole genome shotgun (WGS) approach for rice genome sequencing and produced rough drafts of rice genome shotgun sequences in the years 2000 and 2001, respectively. Two more drafts of rice genome were published later.

In view of this competition between public and private sectors, IRGSP has advanced the date of targeted completion of rice genome sequencing from 2008 to 2004. India has also joined this international effort, so that as a part of international collaboration, a part of chromosome 11 of rice genome is also being sequenced at two different laboratories in New Delhi (Delhi University, South Campus and IARI). The sequences of rice chromosomes 1 and 4 have already been published (Sasaki *et al.*, 2002; Feng *et al.*, 2002).

Artificial Chromosomes for Crop Plants

Construction of artificial chromsomes in plants and ability to introduce them in plant cells will perhaps revolutionize chromosome research in plant systems. It will tell us about the minimum essential elements that are needed for the maintenance, transmission and function of chromosomes within plant cells. For the purpose of crop improvement, it will also allow plant breeders to introduce within plant cells, chromosome segments that are as long as several kb or even Mb long. Construction of rice artificial chromosome is being attempted, utilizing the rice centromeric sequences (RCS), rice telomeric sequences and the different marker sequences (*HPT* gene, GFP-tag and *GUS*) along with the elements of yeast artificial chromosome (YAC) (Nonomura and Kurata, 2000)

Conclusions

The paper is a brief overview of the rebirth of plant cytogenetics and progress made in plant molecular cytogenetics during the last two decades of the 20th century, where different molecular tools have been used for cytogenetics research. These molecular tools include recombinant DNA technology, the polymerase chain reaction and several non-gel based high throughput approaches that are being developed. Tools of genomics research developed recently are also a component of molecular cytogenetics research, which will certainly dominate the early years of 21st century.

Utilizing the approaches of genomics, reverse genetics and comparative genomics in model plants and major crop plants, molecular cytogenetics in future will also facilitate the discovery and isolation of many genes of agronomic importance, which are not amenable to the conventional Mendelian approach of genetic analysis. The results of majority of these molecular cytogenetic studies have been shown to be relevant to crop improvement programmes, so that in future these will be extensively utilized in what is popularly described as precision breeding.

References

Albani D., Cote M.J., Armstrong K.C., Chen Q., Segal A. and Robert L.S. 1993. PCR amplification of microdissected wheat chromosome arms in a simple tube reaction. *Plant J.*, **4**: 899-903.

Arumuganathan K. and Earle E.D. 1991. Nuclear DNA content of some important plant species. *Plant Mol. Biol. Rep.*, **9**: 208-218.

Arumuganathan K., Champoux J., Li L. and Gill K.S. 2000. Flow-cytometric sorting of individual chromosomes and chromosome arms of corn and wheat. *In*: Proc. 6th Intern. Congr. *Plant Mol. Biol.*, Abstract No. S29-2, Quebec (June 18-24, 2000).

Badaeva E.D., Badaev N.S., Gill B.S. and Filatenko A.A. 1994. Intraspecific karyotype divergence in *Triticum araraticum (Poaceae)*. *Pl. Syst. Evol.*, **192**: 117-145.

Bennetzen J.L. 2000. Comparative sequence analysis of plant nuclear genomes: Microcolinearity and its exceptions. *The Plant Cell*, **12**: 1021-1029.

Burke D.T., Carle G.F. and Olson M.V. 1987. Cloning of large segments of exogenous DNA into yeast by means of artificial chromosome vectors. *Science*, **236**: 806-812.

Caspersson T., Zech L., Modest E.J., Foley G.E., Wagh U. and Simonsson E. 1969. Chemical differentiation with fluorescent alkylating agents in *Vicia faba* metaphase chromosomes. *Exp. Cell Res.*, **58**: 128-140.

Chen Q. and Armstrong K. 1994. Genomic *in situ* hybridization in *Avena sativa*. *Genome*, **37**: 607-612.

Civardi L., Xia Y., Edwards K.J., Schnable P.S. and Nikolau B.J. 1994. The relationship between genetic and physical distances in the clones a1-sh2 interval of the *Zea mays.* L. genome. *Proc. Natl. Acad. Sci.*, USA, **91**: 8268-8272.

Conia J., Bergounioux C., Perennes C., Muller P., Brown S. and Gadal P. 1987. Flow cytometric analysis and sorting of plant chromosomes from *Petunia hybrida* protoplasts. *Cytometry*, **8**: 500-508.

De Laat A.M.M. and Blass J. 1984. Flow-cytometric characterization and sorting of plant chromosomes. *Theor. Appl. Genet.*, **67**: 463-467.

Devos K.M., Atkinson M.D., Chinoy C.N., Harcourt R.L., Koebner R.M.D., Liu C.J., Masojc P., Xie D.X. and Gale M.D. 1993. Chromosomal rearrangements in the rye genome relative to that of wheat. *Theor. Appl. Genet.*, **85**: 673-680.

Devos K.M. and Gale M.D. 2000. Genome relationships: The grass model in current research. *The Plant Cell*, **12**: 637-646.

Doyle J.F. and Gaut B.F. 2000. *Plant Molecular Evolution*. Kluwer Academic Publishers. Dordrecht, The Netherlands.

Draye X., Lin Y.R., Qian X.Y., Bowers J.E., Burow G.B., Morrell P.L., Peterson D.G., Presting G.G., Ren S.X., Wing R.A. and Paterson A.H. 2001. Towards integration of comparative genetic, physical, diversity and cytomolecular maps for grasses and grains, using the sorghum genome as a foundation. *Plant Physiol.*, **125**: 1325-1341.

Dubcovsky J., Ramakrishna W., SanMiguel P.J., Busso C.S., Yan L., Shiloff B.A. and Bennetzen J.L. 2001. Comparative sequence analysis fo collinear barley and rice bacterial artificial chromosomes. *Plant Physiol.*, **125**: 1342-1353.

Feng Q. *et al.* 2002. Sequence and analysis of rice chromosome 4. *Nature*, **420**: 312-316.

Feuillet C. and Keller B. 1999. High gene density is conserved at syntenic loci of small and large grass genomes. *Proc. Natl. Acad. Sci.*, USA, **96**: 8265-8270.

Florijn R.J., Bonden L.A.J., Vrolijk H., Wiegant J., Vaandrager J.W., Bass F., Den Dunnen J.T., Tanke H.J., Van Ommen G.J.B. and Raap A.K. 1995. High resolution DNA mapping and colour bar-coding of large genes. *Hum. Mol. Genet.*, **4**: 831-836.

Fidlerova H., Senger G., Kost M., Sanseau P. and Sheer D. 1994. Two simple procedures for releasing chromatin from routinely fixed cells for fluorescence *in situ* hybridization. Cytogenet. *Cell Genet.*, **65**: 203-205.

Fransz P.F., Stam M., Montijn B.M., Hoopen R.T., Wiegant J., Kooter J.M., Oud O. and Nanninga N. 1996. Detection of single-copy genes and chromosome rearrangements in *Petunia hybrida* by fluorescence *in situ* hybridisation. *Plant J.*, **9**: 767-774.

Friebe B. and Gill B.S. 1994. C-band polymorphism and structural rearrangements detected in common wheat (*Triticum aestivum*). *Euphytica*, **78**: 1-5.

Friebe B., Endo T.R. and Gill B.S. 1995. Chromosome banding methods. *In*: *Plant Chromosomes: Laboratory methods*, (ed.) K. Fukui. CRC Press, Boca Raton, Florida, USA.

Fukui K. 1986. Standardization of karyotyping plant chromosomes by a newly developed chromosome image analyzing system (CHIAS). *Theor. Appl. Genet.*, **72**: 27-32.

Fukui K., Minezawa M., Kamisugi Y., Yanagisawa T., Fujishita M. and Sakai F. 1991. Microdissection of barley chromosome by the cell workstation. *Barley Genet.*, **6**: 272-276.

Fukui K. and Nakayama S. 2000. Imaging: An indispensable tool for modern chromosome research. *In*: *Genetics and Biotechnology in Crop Improvement*, (eds.) P. K. Gupta *et al.* pp. 38-51. Rastogi Publications, Meerut, India.

Gall J.G. and Pardue M.L. 1969. Formation and detection of RNA - DNA hybrid molecules in cytological preparations. *Proc. Natl. Acad. Sci.*, USA, **63**: 378-383.

Ganal M.W., Young N.D. and Tanksley S.D. 1989. Pulsed field gel electrophoresis and physical mapping of large DNA fragments in the Tm-2a region of chromosome 9 in tomato. *Mol. Gen. Genet.*, **215**: 395-400.

Gerdes M.G., Carter M.C., Moen J.P.T. and Lawrence J.B. 1994. Dynamic changes in the higher - level chromatin organization of specific sequences revealed by *in situ* hybridization to nuclear halos. *J. Cell Biol.*, **126**: 289-303.

Gill B.S. and Kimber G. 1974a. The Giemsa C-banded karyotype of rye. *Proc. Natl. Acad. Sci.*, USA, **71**: 1247-1249.

Gill B.S. and Kimber G. 1974b. Giemsa C-banding and the evolution of wheat. *Proc. Natl. Acad. Sci.*, USA, **71**: 4086-4090.

Gill B.S., Friebe B. and Endo T.R. 1991. Standard karyotype and nomenclature system for description of chromosome bands and structural aberrations in wheat (*Triticum aestivum*). *Genome*, **34**: 830-839.

Gupta P.K. 1995. Recent trends in cytogenetics. *In*: *Perspectives in Cytology and Genetics*, (eds.) G. K. Manna and S. C. Roy, **8**: 633-642.

Gupta P.K. 1998. Plant cytogenetics and plant breeding in the twenty first century. *In*: *Perspectives in Cytology and Genetics*, (eds.) G. K. Manna and S. C. Roy. **9**: 65-78.

Gupta P.K. 2000. Chromosome and genome research in plants: Some recent developments. *The Nucleus*, **43**: 94-113.

Gupta P.K., Roy J.K. and Prasad M. 2001. Single Nucleotide Polymorphisms (SNPs) in Plants: A new paradigm in molecular marker technology and DNA polymorphism detection. *Curr. Sci.*, **80**: 524-535.

Haaf T. and Ward D.C. 1994. High resolution ordering of YAC contigs using extended chromatin and chromosomes. *Hum. Mol. Genet.*, **3**: 629-633.

Hanson R.E., Zwick M.S., Choi S., Islam-Faridi M.N., McKnight T.D., Wing R.A., Price H.J. and Stelly D.M. 1995. Fluorescent *in situ* hybridization of a bacterial artificial chromosome. *Genome*, **38**: 646-651.

Heng H.H.Q., Squire J. and Sui L.C. 1992. High resolution mapping of mammalian genes by *in situ* hybridization to free chromatin. *Proc. Natl. Acad. Sci.*, USA, **89**: 9509-9513.

Heslop-Harrison J.S. 2000. Comparative genome organization in plants: From sequence and markers to chromatin and chromosomes. *The Plant Cell*, **12**: 617-635.

Houseal T.W., Dackowski W.R., Landes G.M. and Klinger K.W. 1994. High resoltuion mapping of overlapping cosmids by fluorescence *in situ* hybridization. *Cytometry*, **15**: 193-198.

Jellen E.N., Gill B.S. and Cox T.S. 1994. Genomic *in situ* hybridization differentiates between A/D and C genome chromatin and detects intergenomic translocations in polyploid oat species (genus *Avena*). *Genome*, **37**: 613-618.

Jiang J., Gill B.S., Wang G.L., Ronald P.C. and Ward D.C. 1995. Metaphase and interphase fluorescence *in situ* hybridization mapping of the rice genome with bacterial artificial chromosomes. *Proc. Natl. Acad. Sci.*, USA, **92**: 4487- 4491.

Jung C., Kleine M., Fischer F. and Herrmann R.G. 1990. Analysis of DNA from a *Beta procumbens* chromosome fragment in sugar beet carrying a gene for nematode resistance. *Theor. Appl. Genet.*, **79**: 663-672.

Jung C., Claussen U., Horsthemke B., Fischer F. and Herrmann R.G. 1992. A DNA library from an individual *Beta patellaris* chromosome conferring nematode resistance obtained by microdissection of meiotic metaphase chromosome. *Plant Mol. Biol.*, **20**: 503-511.

Kao F.T. 1996. Chromosome microdissection and microcloning: Application to genome analysis. *In*: *Methods of genome analysis in plants*, (ed.) P.P. Jauhar. CRC Press, Boca Raton, Florida, USA.

Kenton A., Parokonny A.S., Gleba Y.Y. and Bennett M.D. 1993. Characterization of the *Nicotiana tabacum* L. genome by molecular cytogenetics. *Mol. Gen. Genet.*, **240**: 159-169.

Kuenzel G. and Korzun L. 1996. Physical mapping of cereal chromosome with special emphasis on barley. *In*: *5th Intern. Oat Conf. and 7th Intern. Barley Genetics Symp.*, (eds). Scoles G., Rossnagel B., Pp. 197-206. University Extension Press, University of Saskatchewan, Canada.

Kuenzel G., Korzun L. and Meister A. 2000. Cytological integrated physical restriction fragment length polymorphism maps for the barley genome based on translocation breakpoints. *Genetics*, **154**: 397-412.

Ku H.M., Vision T., Liu J. and Tanksley S.D. 2000. Comparing sequenced segments of the tomato and *Arabidopsis* genomes: Large-scale duplication followed by selective gene loss creates a network of synteny. *PNAS*, USA, **97**: 9121-9126.

Kynast R.G., Odland W.E., Okagaki R.J., Riera-Lizarazu O., Maquieira S.M., Russel C.D., Rines H.W. and Phillips R.L. 2000. Complete set of maize individual-chromosome additions in oat. *CSSA Abstracts*, Division C-7, page 188.

Kynast R.G., Riera-Lizarau O., Vales M.I., Okagaki R.J., Maquieira S.M., Chen G., Ananiev E.V., Odland W.E., Russell C.D. and Stec A.O. 2001. A complete set of maize individual chromosome additions to the oat genome. *Plant Physiol.*, **125**: 1216-1227.

Langer-Safer P.R., Levine M. and Ward D.C. 1982. Immunological method for mapping genes on *Drosophila* polytene chromosomes. *Proc. Natl. Acad. Sci.*, USA, **79**: 4381-4385.

Le H.T., Armstrong K.C. and Miki B. 1989. Detection of rye DNA in wheat-rye hybrids and wheat translocation stocks using total genomic DNA as a probe. *Plant Mol. Biol. Rep.*, **7**: 150-158.

Lee J.H., Arumuganathan K., Kaeppler S.M., Kaeppler H.F. and Papa C.M. 1996. Cell synchronization and isolation of metaphase chromosomes from maize (*Zea mays* L.) root tips for flow cytometric analysis and sorting. *Genome*, **39**: 697-703.

Lee J.H., Yen Y., Arugumunathan K. and Baenziger P.S. 1997. DNA content of wheat chromosomes at interphase estimated by flow cytometry. *Theor. Appl. Genet.*

Leister D.M., Kurth J., Laurie D.A., Yano M., Sasaki T., Devos K.M., Graner A. and Schulze-Lefert P. 1998. Rapid reorganization of resistance gene homologues in cereal genomes. *Proc. Natl. Acad. Sci.*, USA, **95**: 370-375.

Li W. and Gill B.S. 2002. The colinearity of *Sh2/A1* orthologous region in rice, sorghum and maize is interrupted and accompanied by genome expansion in the triticeae. *Genetics*, **160**: 1153-1162.

Liu B., Segal G., Vega J.M., Feldman M. and Abbo S. 1997. Isolation and characterization of chromosome-specific DNA sequences from a chromosome arm genomic library of common wheat. *Plant J.*, **11**: 959-965.

Llaca V. and Messing J. 1998. Amplicons of maize zein genes are conserved within genic but expanded and constricted in intergenic regions. *Plant J.*, **15**: 211-220.

Lucretti S., Dolezel J., Shubert I. and Fuchs J. 1993. Flow karyotyping and sorting of *Vicia faba* chromosomes. *Theor. Appl. Genet.*, **85**: 665-672.

Mathews S. and Donoghue M.J. 1999. The root of angiosperm phylogeny inferred from duplicate phytochrome genes. *Science*, **286**: 947-950.

McIntyre C.L., Pereira S., Moran L.B. and Appels R. 1990. New *Secale cereale* (rye) DNA derivatives for the detection of rye segments in wheat. *Genome*, **33**: 635-640.

McNeil D., Lagudah E.S., Hohmann U. and Appels R. 1994. Amplification of DNA sequences in wheat and its relatives: the Dgas44 and R350 families of repetitive sequences. *Genome*, **37**: 320-327.

Mukai Y. 1996. Multicolor fluorescence *in situ* hybridization: A new tool for genome analysis. *In*: *Methods of genome analysis in plants*, (ed.) P. P. Jauhar. CRC Press, Boca Raton, Florida, USA, pp 181-192.

Mukai Y. and Yamamoto M. 1998. Application of multicolor fluorescence in situ hybridization to plant genome analysis. *In*: *Genetics and Biotechnology in Crop Improvement*, (eds.) P. K. Gupta, S. P. Singh, H. S. Balyan, P. C. Sharma and B. Ramesh. Rastogi Publications, Meerut, India.

Mukai Y., Friebe B., Hatchett M., Yamamoto M. and Gill B.S. 1993. Molecular cytogenetic analysis of radiation-induced wheat-rye terminal and intercalary chromosomal translocations and the detection of rye chromatin specifying resistance to Hessian fly. *Chromosoma*, **102**: 88-95.

Nonomura K.I. and Kurata N. 2000. Construction of rice artificial chromosome. *In*: *Proc. 6th Intern. Congr. Plant Mol. Biol.*, Abstract No. S 03-60, Quebec (June 18-24, 2000).

Okagaki R.J., Kynast R.G., Odland W.E., Russel C.D., Livingston S.M., Rines H.W. and Phillips R.L. 2000. Mapping maize chromosomes using oat-maize radiation hybrid lines. *CSSA Abstracts*, Division C-7, page 188.

Okagaki R.J., Kynast R.G., Livingston S.M., Russell C.D., Rines H.W. and Phillips R.L. 2001. Mapping maize sequences to chromosomes using oat-maize radiation hybrid lines. *CSSA Abstracts*, Division C-7, 188.

Ozkan H., Levy A.A. and Feldman M. 2001. Allopolyploidy-induced rapid genome evolution in the wheat (*Aegilops-Triticum*) group. *Plant Cell*, **13**: 1735-1747.

Panstruga R., Busches R., Piffanelli P. and Schulze-Lefert P. 1998. A contiguous 60 kb genomic stretch from barley reveal momecular evidence for gene islands in monocot genome. *Nucleic Acids Res.*, **26**: 1056-1062.

Parra I. and Windle B. 1993. High resolution visual mapping of stretched DNA by fluorescent hybridization. *Nature Gen.*, **5**: 17-21.

Paterson A.H., Lin Y.R., Li Z., Schertz K.F., Doebley J.F., Pinson S.R. M., Liu S.C., Stansel J.W. and Irvine J.E. 1995. Convergent domestication of cereal crops by independent mutations at corresponding genetic loci. *Science*, **269**: 1714-1718.

Paterson A.H., Bowers J.E., Burow M.D., Draye X., Elsik C.G., Jiang C.X., Katsar C.S., Lan T.H., Lin Y.R., Ming R. and Wright R.J. 2000. Comparative genomics of plant chromosomes. *Plant Cell*, **12**: 1523-1539.

Pedersen C. and Langridge P. 1997. Identification of the entire chromosome complement of bread wheat by two-color FISH. *Genome*, **40**: 589-593.

Peng J.R., Richards D.E., Hartley N.M., Murphy G.P., Devos K.M., Flintham J.E., Beales J., Fish L.J., Worland A.L., Pelica F., Sudhakar D., Christou P., Snape J.W., Gale M.D. and Harberd N.P. 1999. 'Green revolution' genes encode mutant gibberellin response modulators. *Nature*, **400**: 256-261.

Qui Y.L., Lee J., Bernasconl-Quadronl F., Solttis D.E., Solttis P.S., Zanis M., Zimmer E.A., Chen Z., Savolainen V. and Chase M.W. 1999. The earliest angiospersm: Evidence from mitochondrial, plastid and nuclear genomes. *Nature*, **25**: 404-406.

Rahman S., Abrahams S., Abbott D., Mukai Y., Samuel M., Morell M. and Appels R. 1997. A complex arrangement of genes at a starch branching enzyme I locus in the D-genome donor of wheat. *Genome*, **40**: 465-474.

Rayburn A.L. and Gill B.S. 1985. Use of biotin-labelled probes to map specific DNA sequences on wheat chromosomes. *J. Hered.*, **76**: 78-81.

Rayburn A.L. and Gill B.S. 1986. Molecular identification of the D-genome chromosomes of wheat. *J. Hered.*, **77**: 253-255.

Ried T., Baldini A., Rand T.C. and Ward D.C. 1992. Simultaneous visualization of seven different DNA probes by *in situ* hybridization using combinatorial fluorescence and digital imaging microscopy. *Proc. Natl. Acad. Sci.*, USA, **89**: 1388-1392.

Riera-Lizarazu O., Vales M.I., Ananiev E.V., Rines H.W. and Phillips R.L. 2000. Production and characterization of maize chromosome 9 radiation hybrids derived from an oat-maize addition line. *Genetics*, **156**: 327-339.

Sandery M.J., Forster J., Macadam S.R., Blunden R., Jones R.N. and Brown S.D.M. 1991. Isolation of a specific common to A- and B-chromosomes of rye (*Secale cereale*) by microcloning. *Plant Mol. Biol. Rep.*, **9**: 21-30.

Sasaki T. ***et al.*** 2002. The genome sequence and structure of rice chromosome 1. *Nature*, **420**: 312-316.

Schondelmaier J., Martin R., Jahoor A., Houben A., Graner A., Koop H.U., Herrmann R.G. and Jung C. 1993. Microdissection and microcloning of the barley (*Hordeum vulgare* L.) Chromosome 1HS. *Theor. Appl. Genet.*, **86**: 629-636.

Schrock E., Du Manoir S., Veldman T., Schoell B., Wienberg J., Ferguson-Smith M.M.A., Ning Y., Ledbetter D.H., Bar-Am I., Soenksen D., Garini Y. and Ried T. 1996. Multicolor spectral karyotyping of human chromosomes. *Science*, **273**: 494-497.

Schwartz D.C. and Cantor C.R. 1984. Separation of yeast chromosome - sized DNA by pulse field gradient gel electrophoresis. *Cell*, **37**: 67-75.

Schwarzacher T., Leitch A.R., Bennett M.D. and Heslop-Harrison J.S. 1989. *In situ* localization of parental genomes in a wide hybrid. *Ann. Bot.*, **64**: 315-324.

Shaked H., Kaushkush K., Ozkan H., Feldman M. and Levy A.A. 2001. Sequence elimination and cytosine methylation are rapid and reproducible responses of the genome to wide hybridisation and allopolyploidy in wheat. *Plant Cell*, **13**: 1749-1759.

Shen D.L., Wang Z.F. and Wu M. 1987. Gene mapping on maize pachytene chromosomes by *in situ* hybridization. *Chromosoma*, **95**: 311-314.

Shirasu K., Schulman A.H., Lahaye T. and Schulze-Lefert P. 2000. A contiguous 66-kb barley DNA sequence provides evidence for reversible genome expansion. *Genome Res.*, **10**: 908-915.

Solttis P.S., Solttis D.E. and Chase M.W. 1999. Angiosperms phylogeny inferred from multiple genes as a tool for comparative biology. *Nature*, **25**: 402-404.

Sorokin A., Marthe F., Houben A., Pich U. and Graner A. 1994. Polymerase chain reaction mediated localization of RFLP clones to microisolated translocation chromosomes of barley. *Genome*, **37**: 550-555.

Stein N., Ponelies N., Musket T., McMullen M. and Weber G. 1998. Chromosome microdissection and region specific libraries from pachytene chromosomes of maize (*Zea mays* L.). *Plant J.*, **13**: 281-289.

Stein N., Feuillet C., Wicker T., Schlagenhauf E. and Keller B. 2000. Sub genome chromosome walking in wheat: a 450-kb physical contig in *Triticum monococcum* L. spans the *Lr10* resistance locus in hexaploid wheat (*Triticum aestivum* L.). *Proc. Natl. Acad. Sci.*, USA, **97**: 13436-14441.

TAGI. 2000. Analysis of the genome sequence of the flowering plant *Arabidopsis thaliana*. *Nature*, **408**: 796-815.

Tarchini R., Biddle P., Wineland R., Tingey S. and Rafalski A. 2000. The complete sequence of 340 kb of DNA around the rice *Adh1-Adh2* region reveals interrupted colinearity with maize chromosome 4. *Plant Cell*, **12**: 381-391.

Tikhonov A.P., Sanmiguel P.J., Nakajima Y., Gorestein N.M., Bennetzen J.L. and Avramova Z. 1999. Colinearity and its exceptions in orthologous *adh* regions of maize and sorghum. *Proc. Natl. Acad. Sci.*, USA, **96**: 7409-7414.

Trask B., Pinkel D. and Van den Engh G. 1989. The proximity of DNA sequences in interphase cell nuclei correlated to genomic distance and permits ordering of cosmids spanning 250 kilobase pairs. *Genomics*, **5**: 710-717.

Vega J.M., Abbo S., Feldman M. and Levy A.A. 1994. Chromosome painting in plants: *In situ* hybridization with a DNA probe from a specific microdissected chromosome arm of common wheat. *Proc. Natl. Acad. Sci.*, USA, **91**: 12041-12045.

Vosa C.G. 1970. Heterochromatin recognition with fluorochromes. *Chromosoma*, **30**: 366-371.

Wang M.L., Atkinson M.D., Chinoy C.N., Devos K.M. and Gale M.D. 1992. Comparative RFLP-based genetic maps of barley chromosome 5 (1H) and rye chromosome 1R. *Theor. Appl. Genet.*, **84**: 339-344.

Wang R.L., Stec A.O., Hey J., Lukens L. and Doebley J. 1999. The limits of selection during maize domestication. *Nature*, **398**: 236-239.

Wang Z.M., Le Thierry D'Ennequin M., Panaud O., Gale M.D., Sarr A. and Devos K.M. 2001. Trait mapping in foxtail millet. *Science*, **290**: 2114-2117.

Wei F., Gobelman-Werner K., Morroll S.M., Kurth J., Mao L., Wing R.A., Leister D.M., Schulze-Lefert P. and Wise R.P. 1999. The *Mla* (powdery mildew) resistance cluster is associated with three NBS-LRR gene families and suppressed recombination within a 240-kb DNA interval on chromosome 5S (1HS) of barley. *Genetics*, **153**: 1929-1948.

Wicker T., Stein N., Albar L., Feuillet C., Schlagenhauf E. and Keller B. 2001. Analysis of contiguous 211 kb sequence in diploid wheat (*Triticum monococcum* L.) reveals multiple mechanisms of genome evolution. *Plant J.*, **26**: 307-316.

Wiegant J., Kalle W., Mullenders L., Brookes S., Hoovers J.M.N., Dauwerse J.G., Van Ommen G.J.B. and Raap A.K. 1992. High resolution *in situ* hybridization using DNA halo preparations. *Hum. Mol. Genet.*, **1**: 587-591.

Woese C.R. 2000. Interpreting the universal phylogenetic tree. *PNAS*, USA, **97**: 8392-8396.

Zhong X., De Jong J.H. and Zabel P. 1996. Preparation of tomato meiotic pachytene and mitotic metaphase chromosomes suitable for fluorescence *in situ* hybridization (FISH). *Chromosome Res.*, **4**: 24-28.

Zhou Y., Hu Z., Dang B., Wang H., Deng X., Wang L. and Chen Z. 1999. Microdissection and microcloning of rye (*Secale cereale* L) chromosome 1R. *Chromosoma*, **108**: 250-255.

Zwick M.S., Islam-Faridi M.N., Czeschin Jr. D.J., Wing R.A., Hart G.E., Stelly D.M. and Price H.J. 1998. Physical mapping of the liguleless group in *Sorghum bicolor* using rice RFLP-selected sorghum BACs. *Genetics*, **148**: 1983-1992.

Plant Breeding - Mendelian to Molecular Approaches
H. K. Jain and M. C. Kharkwal (eds.)

6

Transposable Elements and Epigenetic Mechanisms: Significance and Implications

K.C. Upadhyaya[1] and B.M. Prasanna[2]

Abstract

Our understanding of the dynamic potential of the genetic material in living organisms is largely a consequence of the discovery and genetic characterization of plant transposable elements. The ubiquitous presence of transposons, and their significance in relation to structure, organization, and evolution of plant genomes are now well-recognized. Molecular characterization of major plant transposons, particularly in maize, led to isolation of an array of genes influencing plant development, biotic stress resistance and other economically important traits in various crop species, through homologous/heterologous transposon tagging strategies. Retrotransposons are also being utilized as molecular tools in DNA fingerprinting, genetic linkage mapping, phylogenetic studies, and molecular breeding in crop plants. Concurrent with these significant developments, 'epigenetics', the study of heritable changes in gene expression that occur without a change in DNA sequence, has developed as an important frontier in genetics research. Intensive research in recent years provided a better understanding of how epigenetic mechanisms regulate plant transposons. Besides transposon regulation, epigenetic control of gene expression is recognized as a fundamental feature of plant development, particularly flowering and seed development. A detailed understanding of salient epigenetic mechanisms, such as DNA methylation, paramutation, genomic imprinting and gene silencing, should lead to practical solutions and novel strategies for future plant improvement programmes. The purpose of this review is to highlight the significance and implications of transposon function and major epigenetic phenomena in relation to genome evolution, gene regulation and crop improvement.

Introduction

Molecular stability is considered as the hallmark of a genetic material. Although the genome is innately stable, it does provide opportunities for variations to creep in, particularly through mutations and recombination. Pioneering discoveries in the last five

---

[1]Jawaharlal Nehru University, New Delhi 110 067, India
[2]Indian Agricultural Research Institute, New Delhi 110 012, India

decades have led to a clear appreciation of the fact that genomes are dynamic with several interactive networks operating simultaneously for regulation of gene function. Elegant analysis by Barbara McClintock during the 1950s revealed the occurrence of transposable genetic elements in maize, providing the first experimental basis for the dynamic nature of the genome. Transposable elements are genetic entities capable of moving from one location to another within a genome. McClintock referred to these elements as 'controlling elements', as she demonstrated that these elements not only affect the expression of genes in which they were inserted, but also influence other genes located in the vicinity (McClintock, 1951). Studies on transposable genetic elements were largely confined to a very few laboratories till they were also discovered in bacteria in the late 1960s (Shapiro, 1969). Since then transposons have been shown to exist in almost all the organisms, including prokaryotes and eukaryotes, in which they have been looked for. It is now well recognized that transposons are ubiquitous components of all the genomes, from the bacteria to the humans, and contribute significantly to the dynamic nature of the genome itself.

Transposable elements in maize have been initially discovered and extensively analyzed through studies on a range of variegated phenotypes, resulting from frequent transpositional activity in somatic tissues. Anthocyanin biosynthesis has served as an excellent platform for analyzing transposition. For instance, insertion of an element in one of the genes involved in production of anthocyanin pigments may disrupt the functionality of the gene. Transposition of the element out of this target gene in a particular cell results in functional restoration of the gene, leading to the production of anthocyanin pigment in all the descendants of that cell. Thus, the maize kernels display patches of dark pigmentation or colouration (red, purple, bronze, etc.) against a colourless (yellow or white) background. Transposition in germinal tissues could lead to fully pigmented kernels in the progeny (Fig. 1). In case of *Antirrhinum majus* (snapdragon) plants, somatic transposition often leads to variegated flowers containing dark red spots or patches against a white petal background. Although best-studied using model systems such as anthocyanin biosynthesis, transposon activity can virtually influence any part of the genome, and, therefore, can have considerable implications to gene structure, function and regulation. Rapid strides have been made in the last few decades in 'transposon biology', leading to a better understanding of the significance of transposition in gene regulation and genome evolution. Simultaneously, advances have also been made in analysis of epigenetic mechanisms other than transposition, in both plants and animal systems.

'Epigenetics' was originally defined by Waddington (1956) as all those 'interactions of genes with their environment that bring the phenotype into being'. The term 'epigenetics' is now most often used to denote heritable, often reversible, changes in gene expression that do not involve changes in gene sequence, but 'changes in access to the underlying genetic information' (Holliday, 1987). Phenomena that point to an epigenetic mode of inheritance have been studied in diverse organisms, from bacteria and yeasts, to plants,

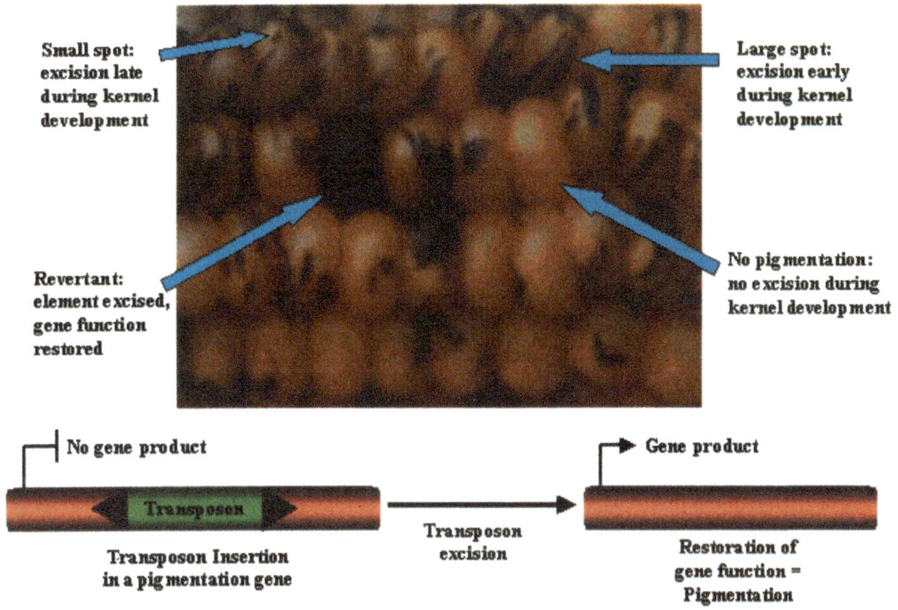

Fig. 1. **Transposon insertion in *R*, a gene regulating anthocyanin pigmentation in maize, leading to a mutable allele, *R-marbled* (*R-mb*) with variegated kernel phenotype. Revertant (pigmented) sector in the aleurone result from excision of the transposable element from a single cell during the clonal development of maize endosperm.**

invertebrates, and mammals. Epigenetic regulatory strategies assume particular significance to the plants in the context of developmental plasticity and environmental interactions. Certain epigenetic strategies might have also been elaborated and favoured by plants, because of the opportunity to select, and to transmit to the next generation, metastable states that are established in the somatic tissues throughout the development (Walbot, 1996). It is now well recognized that genetic and epigenetic variation arising during ontogeny can be a potentially important source of transmissible variation.

Among various epigenetic phenomena in plants, transposition, paramutation, genomic imprinting and gene silencing have attracted considerable attention, particularly in the recent years. These phenomena not only provide vital clues for understanding how genes are regulated, but also have significant influence over the growth and development of plants. Epigenetic control of gene expression can be considered from the standpoint of normal development that requires differential, but stable, expression of genes in various cell types. Many epigenetic effects, however, are observed in unusual circumstances, providing new insights into mechanisms underlying differential gene expression and inter-allelic communication. The present article reviews the current status of plant transposable elements and epigenetic phenomena. Our major aim is to discuss the significance and implications of salient epigenetic mechanisms, particularly in relation to crop improvement.

Transposable Elements and their Molecular Properties

Transposable elements are classified into two broad categories depending on their mode of transposition. The class I elements or retrotransposons transpose via an RNA intermediate, in a 'copy-and-paste' reaction catalyzed by element-encoded reverse transciptase. The class II elements or bacterial type transposons move through a DNA intermediate, through a 'cut-and-paste' reaction catalyzed by element-encoded transposase. Class I transposons constitute retrotransposons with long terminal repeats (LTRs), retrotransposons without LTRs, LINEs (long interspersed nuclear elements) and SINEs (short interspersed nuclear elements). LTR retrotransposons contain long terminal repeats, a *gag* open reading frame (ORF), and a polymerase (*pol*) ORF that encodes protease, endonuclease, reverse transcriptase and RNase H. By sequence homology and the order of the internal domains, the LTR retrotransposons can be further categorized into *Ty1/copia* and *Ty3/gypsy* families of retrotransposons (see Grandbastien, 1992; Kumar and Bennetzen, 1999, for reviews).

The class II transposable elements, simply called as 'transposons', possess terminal inverted repeats (TIR) consisting of a small number of nucleotides; occasionally the TIRs may be a few hundred bases (such as *Mutator* of maize). They may occur as a two-element system, comprising autonomous and non- autonomous elements. For instance, *Ac-Ds* (*Activator- Dissociation*), the first transposon system characterized by Barbara McClintock, refers to *Ac*, the autonomous element that can mediate its own transposition, and *Ds*, a non-autonomous element that cannot transpose by itself, but requires the presence of an *Ac* element in the genetic background for its transposition. The non-autonomous elements almost invariably are deletion derivatives of their cognate functional autonomous elements. *Spm* (*Suppressor-mutator*) is another transposon system analyzed by McClintock. Peter Peterson simultaneously studied the same element, but named it differently as *En-I* (*Enhancer-Inhibitor*) (see Peterson, 1987, for review). The integration site of the mobile elements, whether retroelement or transposon, is flanked by a direct repeat (target site duplication) (Fig. 2).

DNA transposable elements were well-analyzed in plants, primarily because they gave rise to altered phenotypes at a high frequency in both somatic as well as germinal tissues Fig. 2. McClintock called the affected genes as 'mutable genes', since their mutation frequencies (both forward and reverse) are much higher than the normal mutation events associated with a gene. Excellent reviews are available on mobile elements with an account of the genetic and molecular findings of the last five decades (Fedoroff, 1983, 2000; Nevers *et al.*, 1985; Peterson, 1987; Wessler, 1988; Carpenter *et al.*, 1988; Prasanna, 1991; Upadhyaya, 1995; Walbot, 2000). We shall only highlight here some important features of transposable elements in plants and their utility in gene tagging.

DNA Transposable Elements or Transposons: *Ac* was the first element isolated from the *waxy* locus of maize and characterized at the molecular level (Fedoroff *et al.*, 1983). Subsequently, several transposable elements from many other maize loci have been

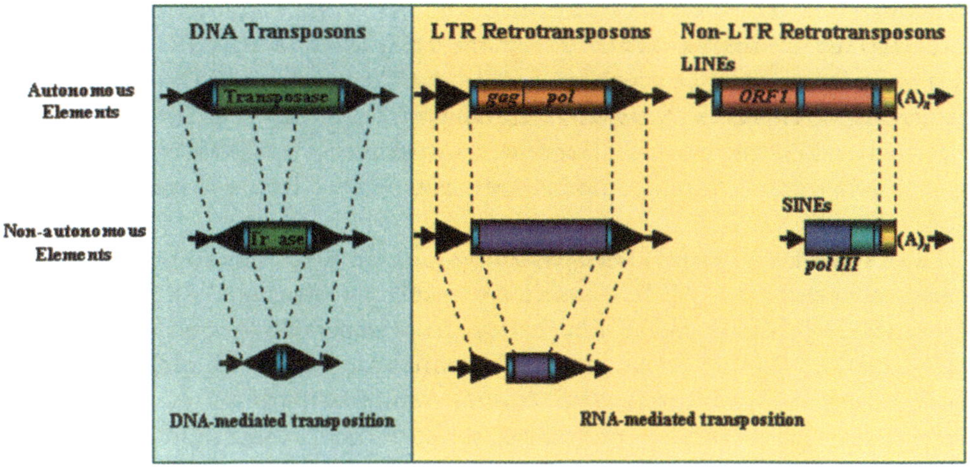

Fig. 2. Basic structural features of DNA transposons and retrotransposons in eukaryotes. Transposon insertion results in duplication of a short genomic sequence (indicated by arrows)at the insertion site, which can be variable in size and/or sequence among transposon families. LTR retrotranposons have long terminal repeats (black triangles) in direct orientation. Autonomous LTR retroelements contain at least two genes, *gag* and *pol*. Non-LTR retrotransposons (LINEs and SINEs) terminate by a simple sequence repeat, usually poly (A).

isolated and characterized. The *Ac* element is 4563-bp long, has 11-bp imperfect TIRs, and causes an 8-bp target site duplication upon integration. The ends of the transposon contain sequences that are essential both for its transcription and transposition. DNA sequence of the *Ac* element consists of five exons and four introns, encodes a transcript of 35,000 nucleotides having an open reading frame (ORF) of 2421 bp that corresponds to 807 amino acids. The ORF protein (the putative transposase) has been shown to bind to the sequences located at the termini. The *Ds* elements are a heterogeneous lot, with varying degree of deletions of the internal sequences; but all the *Ds* elements contain 11-bp intact TIRs (Becker and Kunze, 1997).

Spm is an 8.3-kb element with terminal inverted repeats of 13-bp, and causes a 3-bp target site duplication. The primary *Spm* transcript is alternatively spliced, yielding mRNA sequences with several open reading frames, some with extensive overlaps. Four large transcripts have been identified and designated as *tnpA, tnpB, tnpC* and *tnpD*. The *tnpA* transcript is the shortest and most abundant. The TnpA and TnpB proteins are the only ones for which functions are known and both are required for transposition (Masson and Fedoroff, 1989; Frey *et al.*, 1990). Besides its role in transposition, TnpA also works both as a positive and a negative regulator of the *Spm* promoter (Gierl *et al.*, 1988; Schlappi *et al.*, 1994).

An array of transposable elements from maize, *Antirrhinum*, *Arabidopsis*, tobacco and petunia have been isolated and characterized in recent years. Some of them fall in the category of autonomous elements, while a large majority constitutes the non-autonomous category. A 3.3-kb autonomous element, *Tag1* from *Arabidopsis* has recently been characterized. Its most abundant 2.3-kb transcript encodes 86-kDa transposase that belongs to the hAt or *Ac* super family (Liu and Crawford, 1998). Analogous to the regulation of *P*-element in *Drosophila*, the transposition of *Tag1* is differentially regulated in germinal and somatic tissues. Germinal transposition is dependent on the 5′ promoter region, while somatic transposition is dependent on the 5′-intron sequences (Liu *et al.*, 2001).

There is another group of transposable elements called "miniature inverted repeat transposable elements" (MITEs), which are found to be associated with genes in several plant species, including maize, rice, barley, green pepper and *Arabidopsis* (Casa *et al.*, 2000; Le *et al.*, 2000). Although MITE families are numerous and diverse, all are distinguished by several structural features reminiscent of non-autonomous DNA elements, such as absence of open reading frame or coding capacity and the presence of short (usually 11-14 bp) TIRs. MITEs are small (only 100-500 bp), have high copy number (1000-15,000 per haploid genome), and show preference for insertion into 2 to 3-bp targets that are rich in A and T sequences.

Recently, a distinct and previously unknown category of DNA transposable element, named *Helitrons*, has been identified in the genomes of *Arabidopsis* and *Caenorhabditis elegans* (Kapitonov and Jurka, 2001; Feschotte and Wessler, 2001). Interestingly, this element can transpose via rolling circle replication. Autonomous *Helitrons* encode 5′ → 3′ DNA helicase and nuclease/ligase, similar to those encoded by known rolling circle replicons. These elements do not contain terminal repeats, but transpose precisely between 5′-A and T-3′ with no modification of the AT target sites.

By comparing the basic structural features, transposons could be grouped into distinct families. For example, both *Ac-Ds* elements of maize and *Tam3* of *Antirrhinum* have 11-bp TIRs and 8-bp target site duplication. Similarly, *Spm* of maize, *Tgm1* of soybean, *Tam1* and *Tam2* of *Antirrhinum* have 13-bp TIR and 3-bp target site duplication. Thus, on the basis of structural criteria, these two groups of elements (*Ac/Tam3* and *Spm/Tgm1/Tam1/Tam2*) would constitute two distinct families, though the members of the same family are present in different plant genera. This indicates the probable horizontal movement of transposons across plant genera, and suggests that members of a single transposon family might have evolved from a common progenitor.

Retrotransposons: The Class I elements or retrotransposons are mobile genetic elements that are ubiquitous throughout the plant kingdom. These elements constitute a major portion of the nuclear genomes (in some cases as high as 50-70 % of the total DNA) of plants (Kumar and Bennetzen, 1999). They are distributed as interspersed repeats almost throughout the length of all the chromosomes. Retrotransposons are indeed the most abundant and widespread class of transposable elements in plants. These elements may or may not possess long terminal repeats (LTRs).

The LTR retrotransposons encode a number of proteins, specified by three major genes, *gag, pol* and *int*. These genes are all transcribed as a part of a single transcript, and are translated into a polyprotein that is cleaved into functional polypeptides by a *pol*-encoded protease. The *pol* gene encodes reverse transcriptase and RNaseH activities that are required for replication/transposition of the retrotransposon, whereas *int* encodes an integrase that allows the DNA copy of the element to insert at a new chromosomal location. In some cases, *gag, pol* and *int* proteins are encoded within a single translational reading frame, but in other cases, two or more reading frames may be present. This leads to a requirement for frame shift or translational reinitiation to make appropriate peptides downstream of the change in frame, and thereby, may lead to a lower level of these downstream proteins relative to those present upstream.

The LTR retrotransposons are further sub-divided into *Ty1-copia* and *Ty3-gypsy* families that differ from each other in both the order of encoded gene products and in their degree of sequence similarities. Both *copia* and *gypsy* groups of retroelements are widely distributed in the plant kingdom and are commonly found in high copy numbers (up to a million copies per haploid genome), particularly in plants with large genomes. In *copia*-like retrotransposons, the endonuclease domain is positioned 5′ to the reverse transcriptase domain, while in the *gypsy*-like retrotransposons; it is at the 3′ end of the reverse transcriptase domain. The *Ty3- gypsy* retrotransposons encode the same functions as do *Ty1-copia* family of retrotransposons. Examples of the *Ty1-copia* group of retrotransposons are BARE1 (barley), Bs1, *Opie*, PREM-1 (maize), SIRE-1 (soybean), *Tnt1, Tto1* (tobacco), *Tos17* (rice), *panzee* (chickpea); and those of the *Ty3-gypsy* family are *Athila* (*Arabidopsis*), *cereba* (barely), *cinful, Grande-1, Zeon* (maize), RIRE3 (rice), *cyclops-2* (pea) etc. (see Kumar and Bennetzen, 1999, for a review).

Non-LTR retrotransposons, LINEs and SINEs are also found in high copy numbers (up to 250,000) in various plant species studied so far. The recently sequenced *Arabidopsis* genome reveals that despite the small size of the genome, *Arabidopsis* has as many as 2109 Class I elements comprising all the categories of retroelements, that is, *copia* and *gypsy* types of LTR retrotransposons, non-LTR retrotransposons, LINEs and SINEs (*Arabidopsis* Genome Initiative, 2000).

The LINE elements are relatively simpler than LTR retrotransposons. LINEs have *gag* and *pol* genes, but lack an identified integrase. The *gag* genes of LINEs usually specify a protein with endonuclease activity that may be involved in integration of LINE DNA into different chromosomal locations. The small retrotransposons, called SINEs, are different from other classes of retrotransposons, in that they do not encode any *trans*-acting transposition functions. All known SINEs are derived from RNA polymerase III products that appear to have evolved the ability to be efficiently replicated and integrated by utilizing proteins encoded by LINEs and/or LTR retrotransposons.

Retroelements and C-value Paradox: Plants have a 1000-fold variation in their genome size, ranging from 125 Mb of *Arabidopsis* to 120,000 Mb (120 billion base pairs) of the ornamental lily, *Fritillaria* (40 times the size of human or maize genome). Even species from the same plant family can exhibit striking differences in genome size, although the total number of genes might not be substantially different between them; for example, the rice genome (430 Mb) is approximately 12 times smaller than the genome of diploid wheat (5700 Mb). Retrotransposons play an important role in determining the size of a plant genome. The big genome size of maize (3200 Mb versus 125 Mb of *Arabidopsis*) contains approximately 300,000 copies of retrotransposons that comprise more than 50 % of the nuclear genome. The LTR-retrotransposons are mainly responsible for the vast differences in genome sizes between plant species (Bennetzen, 2000). In spite of a small genome size, the *Arabidopsis* genome contains 13-14 % as transposon and retrotransposon sequences (*Arabidopsis* Genome Initiative, 2000). Analysis of 340-kb of contiguous DNA sequence around the *Adh1-Adh2* locus in rice reveals that almost 15 % of the DNA is represented by LTR retrotransposons (Tarchini *et al.*, 2000). Similarly, study of a 211-kb contiguous genomic region of *Triticum monococcum* orthologous to the *Lr10* leaf rust resistance locus in hexaploid wheat reveals an overall gene diversity of one gene per 42 kb (as compared to one gene per 4-5 kb for the *Arabidopsis*), and 70 % of these sequences comprise several classes of transposons and retroelements (Wicker *et al.*, 2001). Thus, the presence of a vast number of retroelements in the genome may explain the so-called 'C-value paradox', that is, the size of an organism's genome may not be correlated with its obvious complexity.

Evolutionary Implications of Transposable Elements

Analysis of empty donor sites of transposons, both in homologous and heterologous plant systems, revealed that excision of the element leaves behind an entire or a part of the target site duplication, leading to 'transposon footprints'. Imprecise excision of transposons might also cause deletions, inversions as well as translocations of small segments of the genome. The functionality of the revertant gene retaining such altered structural features would largely depend on whether or not such sequence changes could be tolerated in the protein encoded by that gene. It is now well demonstrated that excision events in the germinal tissue can result in a full spectrum of allelic variability, from null to fully functional. Thus, excision and integration of a transposable element may generate fine genetic variability, possibly resulting in the evolution of a repertoire of different kinds of proteins (see Peterson, 1987, for a review).

Transposable elements are known to cause gene inactivation by being inserted within a gene or in its vicinity or by causing an alteration in transcript processing and/or stability. In some cases, insertion of transposon or retrotransposon alters the spatial and temporal pattern of gene expression, or the structure of resultant protein (Wessler, 1988; Flavell *et al.*, 1994; Kunze *et al.*, 1997). Computer-assisted searches have shown that many wild type genes contain retrotransposon and transposon sequences in their regulatory regions, suggesting that they have contributed regulatory units to the normal plant genes. The elements can rearrange genomes either by acquiring a portion of another gene and then amplifying it throughout the genome or by creating intron-less pseudogenes (Kumar and Bennetzen, 1999). These rearrangements can serve as raw material for the possible evolution of new genes. The existence of solo LTRs in many plants (Vicient *et al.*, 1999) shows that unequal recombination between the LTRs of a single retrotransposon has also occurred. It appears that retrotransposons and transposons provide useful structural components to the plant genome for evolution, for new functions and restructuring of the genome, by providing sites for recombinational enzymes.

An innate stability of the genome has been developed over evolutionary time-scale. Sequences of many key genes are conserved across diverse plant genomes. The preservation of 'synteny', that is, many genes remaining grouped together at the same relative positions in the genome irrespective of the genome size, suggests the segmental stability of genomes. In spite of such islands of genomic stability, the remarkable fluidity of the genome is made amply clear by comparison of genomes of sorghum and maize. A comparison of orthologous genomic regions of maize and sorghum reveals that the intergenic space in maize is seven-fold larger than in sorghum. The sequence expansion is accounted for by the presence of retroelements (Chen *et al.*, 1997). Based on such observations, it was suggested that genetic obesity of maize is a direct consequence of amplification of retrotransposon sequences from 120 Mb to 240 Mb, that occurred 1-3 million years ago, a short period in evolutionary time (SanMiguel *et al.*, 1998). There is also evidence that retrotransposons can be eliminated from chromosomes by intrachromosomal recombination. Here, LTR sequences act as sites for recombinational enzymes, and the DNA sequences between the LTRs are eliminated (Kalendar *et al.*, 2000). Thus, retroelements can serve as means of both chromosomal (genome) shrinkage as well as expansion.

Regulation of Transposon and Retrotransposon Activity

McClintock observed that the *Spm* element could change to a number of different states in which it was either silent or alternated between active and inactive phases during development. She described different developmental patterns of transposon activity and reported that a given developmental pattern could be heritable (see Fedoroff, 2000, for a review). It was observed that elements could remain silent for a number of generations and then might return to an active form at a low frequency. An inactive element could be activated by the presence of an active element, suggesting the existence of transposon-encoded epigenetic activators. McClintock (1951) concluded that transposable elements are often maintained in genomes in an inactive yet genetically competent state. Fedoroff (2000) termed this state as 'cryptic state'.

Since retrotransposons cannot transpose without an RNA template for reverse transcription, the simplest way to control their activity would be via regulation of transcription initiation. Many retrotransposons show unique patterns of developmental and/or environmental regulation. Many of the plant retrotransposons studied so far are transcriptionally activated by various biotic and abiotic stresses (Wessler, 1996; Mhiri *et al.*, 1997; Beguiristain *et al.*, 2001). Expression of the tobacco *Tnt1* and *Tto1* retrotransposons is greatly enhanced by several abiotic stresses including protoplast isolation, cell culture, wounding, methyl jasmonate, CuCl3 and salicylic acid (Takeda *et al.*, 1998, 1999). Similarly, biotic stresses, such as infection by viral, bacterial and fungal pathogens, can activate the transcription of these elements. In contrast to *Tnt1* and *Tto1*, transcription of *Tos17* in rice is induced only by tissue culture. Only a fraction of retrotransposon population in plants appear to be transcriptionally active, as revealed by a recent survey of EST databases that indicates that approximately 1.2 % of the total sequences represent retrotransposon complementary sequences (Vicient *et al.*, 2001).

There is a strong evidence that retroelements may confer certain selective advantage on the system, possibly to withstand adverse environmental stresses. Analysis of BARE1 element of barley in "Evolutionary Canyon" Mount Carmel in Israel indicates that plants grown at the top of the canyon have three times more retroelements than the plants grown near the bottom of the canyon. Plants grown at higher elevation apparently gained more copies of retroelements. Kalendar *et al.* (2000) have speculated that a larger genome achieved through the ample presence of transposons might help plants to cope up with more stressful high and dry areas of canyon. This is consistent with the suggestion that sunlight, likely to be more plentiful at higher elevation of evolutionary canyon, may be an important environmental agent involved in genomic restructuring (Moffat, 2000), as it has been shown earlier that shorter wavelength UV light can activate transposons, such as *mutator* in the maize pollen (Walbot, 2000).

Application of Transposons and Retrotransposons

Transposon Mutagenesis and Gene Tagging: Transposons are versatile molecular tools for isolation and identification of target genes that display mutant phenotype when inactivated. Mutant alleles when tagged with a transposon can be isolated by using an element-complementary sequence as a probe. Transposons have extensively been used for tagging genes in plants. This has been facilitated by isolation and characterization of functional transposable elements, mainly from maize (*Ac, Spm, Mu*), *Antirrhinum* (*Tam3*), and more recently from *Arabidopsis* (*Tag1*). However, presence of active mobile elements and their molecular characterization has not been a limiting factor in utilizing transposon-induced mutations as tools to tag genes in diverse plant species, as well-characterized functional elements from maize were shown to efficiently transpose in heterologous plant systems (reviewed by Haring *et al.*, 1991; Sundaresan, 1996; Walbot, 1992, 2000).

The maize transposons, *Ac/Ds* and *Spm* have been mainly utilized for transposon mutagenesis. This is largely because of the fact that genetic and molecular properties of these transposons are relatively well understood, so that the timing, range and frequency of transposition can be experimentally controlled. Since transposon-induced mutations are inherently unstable, two-element systems have been developed, with one element providing transposase and the other being a transposase-defective receptor element

(Sundaresan, 1996). Transposase encoded by *Ac* and *Spm* is required to mobilize non-autonomous elements, *Ds* and *dSpm*, respectively. Both these elements are designed to carry distinct phenotypic markers for easy identification of plants carrying one or both the elements. A variety of phenotypic excision assays have been developed using marker/reporter genes (such as kanamycin, GUS, streptomycin, luciferase and green fluorescence protein) containing transposon inserts, where excision could be monitored by restoration of marker/reporter gene activity (reviewed by Pereira, 1998). For transposon tagging, transposase genes and non-autonomous elements are separately introduced in the plant by *Agrobacterium*-mediated transformation or by particle bombardment. Transformants with single integration events are selected. Mutagenesis is initiated by crossing homozygous plants that have non-autonomous elements with homozygous plants possessing transposase genes. In the F_1 progeny, the non-autonomous element will transpose to new chromosomal locations. To recover stable germinal insertions, the F_1 progeny plants are self-pollinated, and the F_2 plants that inherit a transposed non-autonomous element are selected, eliminating those plants with the autonomous element inherited through segregation. The absence of transposase gene from the mutant lines is essential to maintain stability. Since both *Ac* and *Spm* have a preference to move to linked sites, it is desirable to start with a large F_1 population to maximize the chances of obtaining a mutational spectrum covering the entire genome. A number of genes from several plant species have been isolated using the above strategy (Sundaresan, 1996).

Modified *Ac/Ds* based systems, called 'enhancer traps' and 'gene traps' have been developed for facilitating transposon-induced mutagenesis and identification of genes involved in a variety of plant functions (Sundaresan *et al.*, 1995). Gene traps and enhancer traps are reporter genes that are not normally expressed unless they are integrated near or within a gene. Enhancer trap reporter gene has a minimal promoter that is only expressed when inserted near *cis*-acting chromosomal enhancers. Gene trap reporter gene has no promoter, so that reporter gene expression can only occur when it gets inserted within a transcribed chromosomal gene, creating a transcriptional fusion.

Following transposon mutagenesis, mutation affecting the gene of interest has to be identified and sequences flanking the transposon are to be isolated. The flanking sequences will be part of the gene in which the transposon has been integrated, affecting the functionality of the gene. These sequences are then used as a probe to screen the genomic or cDNA library to isolate the gene. First the mutant plants are to be identified by checking the co-segregation of the tag (transposon) with the mutant phenotype in progeny plants. Two types of strategies have often been used to isolate the flanking sequences. The first strategy utilizes Southern hybridization with a transposon-derived probe. The other strategy is based on PCR to amplify DNA sequences flanking the transposon. The flanking sequences can easily be amplified from a single or low copy insertion of transposon sequence using inverse PCR, with the outward primers designed from the sequences of the element. However, where multiple insertions are present or an endogenous transposon is being utilized for tagging, it becomes quite challenging to isolate the gene-specific flanking sequences. For this purpose, several techniques collectively called "transposon display" based on AFLP have been designed recently, that essentially involve restriction digestion and adapter ligation followed by PCR amplification (Van der Broeck *et al.*, 1998; Frey *et al.*, 1998; Yephremov and Saedler, 2000). After digestion of DNA isolated from the identified plant, adapters are ligated, and then the flanking region is amplified using the primers specific to the transposon and the adapter. A comparison of the amplicon (fragment) display pattern with that obtained from the non-targeted plant helps in identifying the amplified fragment containing the desired sequences.

It was initially considered that since retrotransposons move through RNA intermediates, they are not suitable for transposon mutagenesis. Recent research by Hirochika and coworkers (2001) disproved this assumption as they successfully utilized the rice retrotransposon, *Tos17* for mutagenesis and gene tagging of several agronomically important genes. The features of retrotransposons that make them suitable genetic tools for gene tagging in plants are as follows:

(*a*) Retrotransposon-mediated insertion mutations are stable as these elements transpose in a replicative mode;
(*b*) Transposition target sites are usually unlinked to the site of the original copy, making it relatively easy to generate a large collection of random insertions for saturation mutagenesis;
(*c*) Transposition can be activated by abiotic and biotic stress conditions;
(*d*) High mutagenicity due to preferential transposition into genes;
(*e*) Low copy numbers, thereby facilitating the identification of the retrotransposon insertion responsible for a specific mutation; and
(*f*) Endogenous retrotransposons that are ubiquitous in plants and may be active in many species (Hirochika, 2001).

Tos17 is highly active under tissue culture conditions and preferentially inserts into or near genes. It has been effectively utilized to generate a large collection of random insertions for saturation mutagenesis for cloning a number of genes in rice (Sato *et al.*, 1999; Hirochika, 2001). Additionally, *Tos17* has been used in a reverse genetic approach to identify null mutant lines for the *OSH15* gene (*Oryza sativa* homeobox; a *knotted*-type homeobox gene), demonstrating that *OSH15* has a role in rice internode development (Sato *et al.*, 1999).

Significance of Transposon Activity in Breeding Programmes: Although the properties of transposons were well-characterized using mutable genes related to characters such as anthocyanin pigmentation, it is instructive to keep in view that transposons are capable of influencing expression of a number of genes involved in different pathways, including those that are of interest to the breeders. Active maize mobile elements were detected in several maize populations that are often used by breeders in USA, such as BSSS (Iowa Stiff Stalk Synthetic), Hays Golden, Hallauer's Long Ear (BSLE), Lancaster, Kolkmeier, and Illinois Reverse High Protein Selection (RHPC21) (Peterson, 1988). Prasanna and Sarkar (1997) reinforced the pervasiveness of genetically active mobile elements, such as *Ubiquitous* (*Uq*), in North-Eastern Himalayan (NEH) maize lines, specific breeding populations used in Indian maize breeding programmes, as well as in some popular inbred lines such as CM111. While the presence of active mobile elements in the maize populations can be of some value, since plant transposons by virtue of imprecise excision from the target genes can generate genetic variability that can be further selected upon by the breeders, the same can be a major bottleneck in maintaining the purity of the inbred lines, particularly those used in hybrid seed production.

Retrotransposons as Tools in Molecular Breeding: Retrotransposons are now being utilized as molecular tools in DNA fingerprinting, genetic linkage mapping, phylogenetic studies, and molecular breeding. The following properties of retrotransposons make them particularly suitable candidates for generating molecular markers in a variety of crop plants:

(*i*) Presence in high copy number in heterogeneous populations;

(*ii*) Dispersed throughout the genome;

(*iii*) Insertion of retrotransposon into new genomic sites occurs without losing the parental copies;

(*iv*) The consequences of retrotransposition could range from the alteration of a few hundred to a few thousand bases at the site of insertion; and

(*v*) Most retrotransposon insertions are irreversible; therefore, changes remain relatively fixed, which is a good attribute for analyzing phylogenetic relationships (Kumar and Hirochika, 2001).

Moreover, retroelements are frequently observed in regions adjacent to known plant genes. Since these elements consist of LTRs with a highly conserved terminus, this property can be exploited for primer design in the development of retrotransposon-based markers (Kumar and Bennetzen, 1999). The above features have also aided in use of retroelements as DNA markers to study biodiversity in maize, pea and barley (Purugganan and Wessler, 1995; Ellis *et al.*, 1998; Kalendar *et al.*, 1999), and to generate genetic linkage maps in barley, oat and pea (Kumar *et al.*, 1997; Waugh *et al.*, 1997; Ellis *et al.*, 1998; Yu and Wise, 2000).

Several molecular marker techniques based on retroelements have emerged during the last few years. These are briefly outlined below:

(*i*) ***S-SAP (Sequence-Specific Amplified Polymorphism)***: SSAP is a multiplex amplified fragment length polymorphism (AFLP)-like technique that displays individual retrotransposon insertions as bands on a sequencing gel. Fragments are amplified by PCR, using one primer designed from the conserved terminus of the LTR and another based on the presence of a nearby restriction endonuclease site.

(*ii*) ***IRAP (Inter-Retrotransposon Amplified Polymorphism)***: It is a dominant, multiplex marker system that examines variation in retrotransposon insertion sites. IRAP fragments between two retrotransposons are generated by PCR, using outward-facing primers annealing to LTR target sequences. Fragments are separated by high-resolution agarose gel-electrophoresis (Kalendar *et al.*, 1999).

(*iii*) ***REMAP (Retrotransposon-Microsatellite Amplified Polymorphism)***: It is a dominant, multiplex marker system that examines variation in retrotransposon insertion sites. REMAP fragments between retrotransposons and microsatellites are generated by PCR, using one primer based on a LTR target sequence and another based on a microsatellite or simple sequence repeat (SSR) motif. Fragments are separated by high-resolution agarose gel-electrophoresis (Kalendar *et al.*, 1999).

(*iv*) ***RBIP (Retrotransposon Based Insertional Polymorphism)***: It is a codominant marker system that uses PCR primers designed from the retrotransposon and its flanking DNA to examine insertional polymorphisms for individual retrotransposons. Presence or absence of insertion is investigated by two PCRs, the first using one primer from the retrotransposon and the other from the flanking DNA, the second using primers designed from both flanking regions. Polymorphisms are detected by simple agarose gel-electrophoresis or by dot hybridization assays. A drawback of the method is that sequence data of the flanking regions are required for primer design. But, a major advantage is that RBIP does not necessarily require a gel-based detection system and can be easily adapted to automated, gel-free procedures in order to increase the sample throughput (Flavell *et al.*, 1998).

Molecular breeding is dependent on reliable genetic linkage mapping data and markers linked to agronomic traits. SSAP markers have also been used to construct linkage maps in barley (Manninen *et al.*, 2000), oat (Yu and Wise, 2000) and pea (Ellis *et al.*, 1998), using LTR-specific sequences (Pearce *et al.*, 1999). Several quantitative trait loci (QTLs) have been mapped into seven linkage groups in barley using BARE-1-SSAP analysis. In barley, retrotransposon-based monitor systems, such as IRAP and REMAP, in conjunction with SSAP, have been used to map a locus conferring resistance to the 'net blotch' disease caused by the fungus *Pyrenophora teres*, on to chromosome 6H. The *syn19* and *piz* loci conferring resistance to *Magnarporthe grisea* in pea, the *Hero* locus conferring resistance to potato cyst nematode in tomato are the other salient examples of effective utilization of this strategy (see Kumar and Hirochika, 2001, for review).

Utility of Retrotransposons in Analyses of Biodiversity and Phylogeny: Biodiversity and phylogenetic studies are essential for preserving both landraces and related wild species. Retrotransposon-based markers are especially suitable for studying phylogenetic relationships and genetic diversity within and between species, as active retrotransposon family, for instance, produces new insertions in the genome leading to polymorphism. These new insertions can then be detected and used to establish the temporal sequences of insertion events, thereby helping to determine phylogenies. Such genetic properties of retrotransposons have been exploited to study biodiversity and phylogeny in the genera *Brassica, Hordeum, Oryza* and *Pisum* (see Kumar and Hirochika, 2001, for review). A multi-retrotransposon approach has been used recently to estimate phylogenetic relationships between different species in legumes (Pearce *et al.*, 2000) and cereals (Gribbon *et al.*, 1999; Kalender *et al.*, 2000).

Paramutation and 'Allelic Communication'

Paramutation refers to an interaction between two alleles of a gene, resulting in a heritable and directed change in one of the alleles. This epigenetic modification, which can be maintained through one or more meiotic generations, was first discovered by R. A. Brink (1956). Brink (1958) coined the term 'paramutation' and defined it as a "directed, heritable change in the expression of one allele brought about by an earlier association with another allele". He recognized the potentially reversible nature of this event as compared to classical mutations caused by nucleotide changes. Significantly, paramutation is a rare exception to the principle of segregation or 'purity of gametes' proposed by Mendel, which asserts that two alleles in a heterozygote will be transmitted to the next generation without 'contamination' during gamete formation. Alleles sensitive to altered expression were termed 'paramutable', and alleles inducing the change as 'paramutagenic'. Alleles sensitive to paramutation are referred to as 'paramutant', and are designated with an apostrophe (for instance, *B'*, *Pl'*, *R'*, etc. in maize). Many alleles at loci displaying paramutation may not be sensitive to this phenomenon; such alleles have been referred to as either 'neutral' or 'non-paramutagenic' (Brink, 1973).

Paramutation in Plants - Salient Examples: Brink's discovery of paramutation was first made in specific alleles at the maize *R* (*red*) locus, particularly with the paramutable *R-r:standard* (*R-r:std*) haplotype, that conditions anthocyanin pigmentation of various plant parts and the aleurone layer of the endosperm. In *R-r:std*, aleurone pigmentation is considerably sensitive to paramutation than plant pigmentation (Brink and Mikula, 1958). In the most extensively studied *r1* haplotypes, neither the plant nor aleurone expression is markedly reduced in F_1 plants. Consequently, silencing of paramutable *r1* haplotypes typically is monitored by crossing the F_1 plants with null, recessive testers and analyzing kernel pigment levels in the progeny. The phenotype is strongest if the F_1 is used as male. The fact that many, but not all, paramutable *r1* haplotypes also undergo genomic imprinting (discussed later), increases the sensitivity of this assay. Paramutagenic *r1* haplotypes, *R-stippled* (*R- st*) and *R-marbled* (*R-mb*), are distinguished by their characteristic patterns of anthocyanin pigmentation in the aleurone (Prasanna and Sarkar, 1996).

Soon after Brink's discovery, Coe (1959) discovered another example of paramutation at the *b1* (*booster*) locus in maize. A functional *b1* allele is an activator of the anthocyanin pathway. *B′* confers light purple pigmentation, while *B-I* confers dark purple pigmentation. The heterozygotes (*B′*/B-I) are light purple, which implies that *B′* is dominant and *B-I* is recessive. However, when we cross the heterozygotes (*B'*/B-I × *B'*/B-I), all the plants are still light purple. While expectation based on Mendel's principles should indicate that one quarter of the plants would be dark purple (*B-I/B-I*), no dark purple plants were seen in the progeny. In contrast to the paramutation at *r1* locus, at the *b1* and *pl1* loci, newly altered paramutagenic alleles (*B'* and *Pl'*) are strongly paramutagenic, and induce heritable silencing (paramutagenicity) of specific paramutable alleles (Coe, 1966; Hollick *et al.*, 1995).

In all these cases, two assays are routinely used to monitor paramutation: (a) the ability of a paramutagenic allele to cause a heritable reduction in the expression of a paramutable allele; and (b) heritable alteration of the paramutant allele into a paramutagenic allele. Particularly with respect to the second property, some intriguing differences were observed among various loci in maize displaying paramutation (reviewed by Chandler *et al.*, 2000). Paramutable alleles of *b1* and *pl1* can undergo spontaneous paramutation (in the absence of a paramutagenic allele), whereas paramutagenic alleles at *r1* cannot. Also, at *r1*, the extent of paramutagenicity obtained by a paramutant haplotype depends on the nature of the crosses. Besides the above differences, paramutant alleles at *r1* and *pl1* can revert back to their original states after several sexual generations, but paramutant alleles at *b1* cannot.

All the four maize loci (*r1, b1, pl1* and *p1*) that were intensively analyzed with respect to paramutation encode transcription factors that activate the biosynthesis of coloured flavonoid pigments (reviewed by Hollick *et al.*, 1997). Paramutation at *r1* and *p1* correlates with increased DNA methylation in the transcribed or promoter proximal regions, whereas the DNA methylation levels do not change within the comparable regions of *b1* and *pl1* after paramutation.

Paramutation was also observed at the *nivea* locus in *Antirrhinum majus* and was well-analyzed in crosses of a highly variegated line 53 (carrying *Tam1* transposable element) with the stable white line 44 (containing the *Tam2* element). Line 44 is paramutagenic, while line 53 is paramutable (Upadhyaya *et al.*, 1985). It was also demonstrated that the *Tam2* elements at the *nivea* locus is not required for paramutagenicity of line 44; rather factor(s) in the genetic background of line 44 must be responsible for this phenomenon (Sommer *et al.*, 1988).

Mutations Influencing Paramutation: Isolation and characterization of mutants altered in establishment and/or maintenance of paramutation will be crucial for understanding the mechanisms underlying paramutation. One recessive mutation, *mop1-1* (*mediator of paramutation*), has been recently identified that can specifically influence alleles of loci in maize, such as *r1, b1* and *pl1*, that participate in paramutation, suggesting that paramutation at these loci is mechanistically related, in spite of the differences in phenomenology. Seedling-based screens have also identified a novel class of trans-acting *rmr* (*required to maintain repression*) loci that allow heritable restoration of high levels of *pl* gene action. Several *rmr* genes are proposed to play an important role in growth and development. Efforts are currently underway to clone these *rmr* genes by transposon-tagging and candidate-gene approaches. It is likely that additional genes participating in paramutation shall be identified in the near future.

Mechanism(s) Underlying Paramutation: Two interesting aspects of paramutation are: (i) the ability of alleles to sense each other (or to be sensed), which establishes particular expression states; and (ii) the ability to maintain these expression states through development and transmit them to the next generation. Paramutation, thus, provides an excellent system for studying mechanisms involved in establishing and maintaining expression states that are both meiotically and mitotically heritable, besides exploring allelic communication. The stable, heritable inactivation or activation of particular genes is crucial during development of multi-cellular organisms to maintain determined gene expression states.

While allelic interactions and communication among chromosomes have been observed in a variety of organisms, little is currently known about how alleles communicate. Genetic and molecular observations suggest that paramutation reflects alterations in chromatin structure that are sensitive to chromosome pairing interactions (reviewed by Henikoff and Comai, 1998). An alternative model involving DNA/RNA/protein interactions was discussed by Chandler and coworkers (2000). Mutagenesis experiments are also currently being carried out in some laboratories for isolation of mutants that fail to paramutate and mutants that have lost the ability to induce paramutation. Analysis of these mutants may allow us to locate specific site(s) necessary for paramutation and lead to ideas about the mechanisms involved. An eventual understanding of paramutation would explain how alleles interact in the nucleus to influence the regulation of each other, how heritable expression states are established, and how such expression states are maintained through numerous cell divisions and transmitted to the next generation.

Potential Role(s) of Paramutation: The number of genes for which paramutation has been well-documented still remains small, and mostly explored in maize and *Antirrhinum* among the plants. Does this mean paramutation is confined to very few loci or are such studies too limited to identify this phenomenon in a greater number of loci across various species? Hollick *et al.* (1997) pointed out that paramutation was easy to observe in genes regulating pigmentation; other genes may be similarly affected, but not as readily seen. Chandler *et al.* (2000) favour this hypothesis, and suggest that sensitivity of anthocyanin pigmentation to the expression level of the regulatory genes has enabled subtle, but reproducibly heritable changes in gene expression to be detected.

If the above hypotheses were true, what could be the potential role for paramutation? It has been previously postulated that paramutation and other examples of homology-dependent gene silencing might reflect an important cellular mechanism for protection against invasive DNA. It is now well established that the genomes of many species, particularly eukaryotes, have a large component of dispersed repetitive DNA in the form of transposable elements. A consequence of targeting and inactivating invasive sequences such as transposable elements and viruses would be to reduce mutation rates and ectopic recombination between repeated sequences (Matzke *et al.*, 1996; Yoder *et al.*, 1997; Matzke *et al.*, 2000). Chandler *et al.* (2000) recently discussed three other potential roles for paramutation, in terms of: (*i*) possible association with recombination mechanisms; (*ii*) establishment and maintenance of chromatin domain boundaries; and (*iii*) adaptive mechanism for transmission of gene expression states.

Imprinting and 'Functional Hemizygosity'

There are two important assumptions implicit in Mendelian inheritance: (*i*) allelic identity remains the same irrespective of its transmission through the male or female gametes; (*ii*) a 'structurally intact' normal gene has the full potential for normal function. Genomic imprinting challenges both these assumptions. Firstly, it turns out that a gene that comes from a specific sex (male or female) may not be identical in function to the same gene that is inherited from the opposite sex. 'Genomic imprinting' is, thus, defined as a reversible modification of DNA that causes differential expression of maternally or paternally inherited genes, although the gene sequence remains the same. In other words, the phenotype elicited from a locus is differentially modified by the sex of the parent contributing that particular allele. This process ultimately results in a functional difference between the genetic information contributed by each parent, and the imprinted loci can be thus considered 'functionally hemizygous'.

Imprinting usually has the following four important characteristics:

(*i*) It is imposed during gametogenesis;
(*ii*) It suppresses gene transcription;
(*iii*) It is reimposed following DNA replication and remains restricted to one chromosome; and
(*iv*) It is potentially reversible (in animals, if the offspring is of the same sex, the imprinted gene remains in the same state in that individual's gametes, but if the offspring is of a different sex, imprinting is reversed).

Imprinting is Ubiquitous: Although imprinting has been more intensively analyzed in animals, probably the first convincing evidence for imprinting came from the studies at the *R* locus in maize (Kermicle, 1970). The standard *R-r* allele shows anthocyanin pigmentation in both plant parts and aleurone layer of the endosperm, while the *r* allele causes colourless kernels when homozygous. If *rr* is crossed to *RR*, the resulting kernels are solidly coloured when *RR* is the female parent, but mottled when *RR* is used as the male parent. Kermicle (1970) demonstrated that this difference between the products of reciprocal crosses was due to preferential expression of the *R* allele when maternally derived. Endosperm was solidly coloured whenever it possessed a maternal copy of *R*, but mottled even when it possessed two paternal copies of *R* (with one extra copy of *R* that can be experimentally provided using a B chromosome in maize) without a maternal copy. Since extra paternal copies of the allele could not substitute for a maternal copy, the study provided strong evidence that dosage of *R* is not the cause for mottling, but the *R* allele is 'imprinted' when it is transmitted through the pollen parent.

Imprinting-like phenomena have been observed in a wide range of phyla in both the plant and animal kingdoms (Kermicle and Alleman, 1990; Messing and Grossniklaus, 1999; Grossniklaus *et al.*, 2001). Manifestations of this gamete-of-origin dependent modification were reported in different plants, scale insects (coccids), *Drosophila*, fishes, marsupials, rodents, yeast, humans, etc., suggesting that either there is strong evolutionary conservation of this phenomenon or that it evolved independently in different genera and species. It is not yet clear whether a similar molecular mechanism is involved in all these cases.

Effects of Imprinting on Seed Development: In contrast to the influence of genomic imprinting in the embryo development in mammals, the endosperm is the organ that appears to be strongly influenced by the imprinting process in flowering plants. Classic experiments by Lin (1982, 1984), and his hypothesis of parental imprinting of maize genome, are important milestones in our understanding of genetic mechanisms controlling normal morphogenesis of endosperm. The hypothesis clearly pointed out the possible unequal contribution of maternal and paternal genomes to seed development. In maize, this aspect has been well analyzed through studies on the 'small-kernel' effect. Reciprocal translocations between autosomes and B-chromosomes can be used to generate endosperms that are deficient for a paternal copy of the segment translocated on to the B-chromosome. For specific chromosomal regions (such as 1L, 1S and 10L), paternal deficiency caused the development of small kernels (Beckett, 1978; Lin, 1982; Birchler and Hart, 1987). These studies confirmed the association between paternal gene expression and larger kernel size, and maternal gene expression and smaller kernel size. Because segments of several chromosomes cause small-kernel effects in maize, imprinting is believed to have an influence over the functioning of a substantial number of loci (Matzke and Matzke, 1993).

Very few genes that are involved in seed development and display imprinting effects have so far been isolated. The first imprinted gene that is essential for proper seed development to be isolated and characterized (using transposon mutagenesis) was the *MEDIA* (*MEA*) gene in *Arabidopsis*. In higher plants, seed development requires maternal gene activity in the haploid (gametophytic) as well as diploid (sporophytic) tissues of the developing ovule. Genomic imprinting regulates the *mea* locus, with only maternal *MEA* alleles being active during early seed development (Vielle-Calzada *et al.*, 1999). The *MEA* gene encodes a SET-domain protein of the *Polycomb* group (of *Drosophila*) that regulates cell proliferation by exerting a gametophytic maternal control during seed development. Seeds carrying a mutant *mea* allele abort and exhibit cell proliferation defects in both embryo and endosperm. Paternally inherited *MEA* alleles are transcriptionally silent in both the young embryo and endosperm. Mutations at another locus, *decrease in DNA methylation1* (*ddm1*) are able to rescue *mea* seeds by functionally reactivating paternally inherited *MEA* alleles during seed development. Therefore, it appears that the maintenance of the genomic imprint at the *mea* locus requires zygotic *DDM1* activity.

Because *DDM1* encodes a putative chromatin-remodelling factor, chromatin structure is likely to be interrelated with genomic imprinting in *Arabidopsis* (Vielle-Calzada *et al.*, 1999).

Recent studies in plants clearly highlight that the mother's influence on seed development precedes that of the father's, contradicting the long-held assumption that genes inherited from both the male and the female parents influence seed development equally from the onset. The studies indicate that the early stages of embryonic development in plants are under exclusive maternal jurisdiction, as in flies, worms and amphibians, although the mechanisms employed might be different. Paternal genetic material appears to remain 'silent' for several days after fertilization, unlike the maternal component that activates immediately (Vielle-Calzada *et al.*, 2000). These findings could have significant implications for ágriculture in general, and plant breeding and genetic engineering in particular, as the seed is economically the most important plant part and plays a vital role in agriculture.

Possible Mechanism of Imprinting: In mammals, where imprinting has been studied in greater detail than in plants, the phenomenon involves addition of a methyl group to position 5′ of cytosine and occurs most frequently at CpG dinucleotides. Methylation is one of the ways in which an unwanted gene is turned off; however, it is not yet clear how cells decide which genes to methylate. Significantly, 'housekeeping genes' are seldom methylated, even though they have CpG dimers in their promoter regions. Furthermore, methylation of DNA residues can be stably preserved through replication process by the action of "maintenance methylases", which use the replicated hemimethylated DNA as a template (Razin and Riggs, 1980). It has, therefore, been suggested that DNA methylation may generate an epigenetic marking responsible for the imprinting phenomenon.

Why does Imprinting Occur?: Despite significant advances in research on imprinting, particularly in mammals, nobody knows for sure the exact reason(s) for the occurrence of imprinting. There have been several models/theories regarding the evolutionary origin or advantages of genomic imprinting, but so far none have been subjected to or withstood rigorous testing. Holliday (1987) suggested that the function of genomic imprinting is to produce functional haploid gene sets in order for the cell's regulatory machinery to be able to 'fine tune' itself on just a single gene copy. It was further implied that imprinting would also serve to prevent possible detrimental cross-talk between maternal and paternal transcripts generated by developmentally important regulatory loci. Barlow (1993) suggested that genome imprinting originated from the use of DNA methylation as a host defense mechanism.

Many theories on imprinting were based primarily on genomic imprinting in mammals, and very few were proposed to explain the origin or evolutionary advantages of genomic imprinting in plants. One school of thought is that genomic imprinting plays an important role during endosperm development in plants, which is a possible checkpoint for proper sexual reproduction. Imprinting is also thought to be the main reason why mammals are the only group of vertebrates in which successful parthenogenesis is known (Surani, 1987; Solter, 1988).

Gene Silencing

It is now becoming evident that plants, fungi and animals have evolved surveillance mechanisms that recognize foreign nucleic acid sequences including those of infecting viruses, transposons and retrotransposons, and other foreign sequences and silence them transcriptionally or post-transcriptionally. These epigenetic silencing mechanisms that rely on recognition of nucleic acid sequences have been identified in plants, fungi, *Drosophila* and *C. elegans* (Fagard *et al.*, 2000; Matzke *et al.*, 2001; Waterhouse *et al.*, 2001). In plants, the process was observed, when an additional copy of the endogenous gene was introduced in order to upregulate the expression of that gene. The results were dramatically opposite; instead of upregulation, both the endogenous and transgenes were silenced. This phenomenon of gene silencing has been referred to as 'cosuppression' (Jorgensen, 1995).

Gene silencing as a consequence of insertion of a transgene, either in the sense or in the antisense orientation, is a sequence-dependent phenomenon. Analysis of the cosuppressed and virus-resistant plants (due to presence of transgene in the form of viral gene sequences) revealed that in both the cases, the transgenes were highly transcribed in the nucleus, but the steady-state levels of their RNAs in the cytoplasm were very low. This led to the proposal that the transgene mRNA was somehow perceived by the cell as unwanted and sequence-specific degradation was induced. Thus, in the cosuppressed and virus-resistant lines, not only the transgene mRNAs but also the mRNA from the homologous endogenous gene and invading viruses RNA (homologous to the transgene) were degraded. This was interpreted to be due to post-transcriptional gene silencing (PTGS), resulting from the degradation of RNA transcribed from the homologous host gene and transgene in the nucleus. When PTGS affects transgenes sharing homology with an RNA virus, plant becomes resistant to the infection by this virus. This phenomenon was referred to as 'RNA-mediated virus resistance'.

Another form of gene silencing is transcriptional gene silencing (TGS) that involves inhibition of transcription. Both TGS and PTGS lead to reduction in accumulation of gene transcripts, the former by inhibiting transcription by hypermethylation, and the latter by degrading the synthesized transcript. This epigenetic silencing may persist over many cell divisions or plant generations. Recent studies indicate that both TGS and PTGS are mechanistically and possibly functionally related. TGS is associated with the hypermethylation of promoter sequences and PTGS is associated with hypermethylation of the transcribed or coding sequences. Hypermethylation can spread within promoter regions or within transcribed regions, but spreads to a lesser extent from promoter to adjacent transcribed regions and from transcribed to adjacent promoter regions.

In the past decade, there has been considerable research into transgene-mediated virus resistance, cosuppression, antisense suppression and TGS in plants. Intensive research has also been carried out on RNA interference (RNAi) in *Drosophila*, *C. elegans*, mammals and quelling in fungi. These seemingly disparate phenomena have produced pieces of a jigsaw puzzle which, when put together, begin to reveal the existence and characteristics of a natural defense system in plants against viruses and transposable elements. Many of the details and ratifications are yet to come in, but the current picture is that of an elegant system that can recognize invading viruses and transposable elements and marshal the plant's defenses against them.

There is strong evidence that double-stranded RNA (dsRNA) plays a dual role in plant gene silencing by initiating both the RNA degradation step of PTGS or RNA-mediated DNA methylation. Although the ability of dsRNA molecules triggering the mechanism of degradation of homologous RNAs (called RNAi, RNA interference) was recognized first in the worm, *C. elegans*, it became apparent on the basis of common molecular features and gene products that PTGS in plants also involves dsRNA. Studies in plants and *Drosophila* have elucidated that silencing is accompanied by the accumulation of small RNA (21-25 nucleotides) in sense and antisense orientation that are homologous to the silenced locus. Silencing can be triggered locally rather than spread through the organism, in plants and *C. elegans* and between nuclei in heterokaryons in *Neurospora* via a mobile silencing signal. These observations suggest conservation of a mechanism across kingdoms and possibly derived from an ancestral mechanism directed against invading nucleic acids (Fagard *et al.*, 2000).

Phenomena such as PTGS in plants, quelling in fungi, and RNAi in animals, require a set of related proteins (SGS2, QDE-1 and EGO-1, respectively). Several other classes of gene products are also required for PTGS, quelling and RNAi (see Matzke *et al.*, 2001, for review). The identities of these proteins suggest possible features common to the underlying molecular mechanisms involved in silencing, the synthesis and amplification of dsRNA, and targeting of mRNA. The cellular RNA-directed RNA polymerase plays a central role in PTGS, and is required even in the RNAi system, where the process is induced by exogenously supplied dsRNA. Mutation in these genes affects silencing mechanism as has been shown by isolation and characterization of a mutation in *AGO1* gene of *Arabidopsis* (similar to *QDE-2* of *Neurospora*) that is impaired in PTGS (Fagard *et al.*, 2000).

Interrelationships among Transposons, DNA Methylation, Paramutation and Gene Silencing

DNA methylation may have evolved as an epigenetic means of containing the spread of transposons in the host genomes. Transposons and retrotransposons, in spite of constituting a large fraction of plant genomes, do not appear to transpose freely. Transposition of the elements is apparently under strict regulatory control(s); consequently, majority of the elements are in cryptic state (Fedoroff, 2000). DNA methylation appears to play a vital role in regulating transposon activity. The transposable element sequences are usually heavily methylated, and this epigenetic state of silencing could be maintained for a number of generations. The cryptic elements could be activated by a variety of genomic stress factors including UV, tissue culture etc., besides other conditions which can induce demethylation. Researchers demonstrated that at least two distinct kinds of transposons in *Arabidopsis* are activated and start transposing, specifically in the context of reduced DNA methylation (*ddm1* mutants) (Miura *et al.*, 2001; Singer *et al.*, 2001). Similarly, the activity of retrotransposons increases in *ddm1* plants (Hirochika *et al.*, 2000).

Induction of methylation is mechanistically considered similar to TGS, as it can occur following production of dsRNA (Waterhouse *et al.*, 2001). Evidence that dsRNA is involved in transposon silencing comes from the reactivation of transposons in the PTGS and RNAi defective mutants and presence of ~21-nucleotide dsRNAs complementary to transposons in the PTGS extracts (Waterhouse *et al.*, 2001). Thus, transposition may be regulated by methylation as well as degradation of transposase transcript.

Interactions between homologous transposable elements also share some features with paramutation (McClintock, 1965; Krebbers *et al.*, 1987; Martienssen, 1998). Expression of a gene containing a transposon, or expression of the transposable element itself, can be heritably changed in response to *trans*-interactions with other such elements elsewhere in the genome. Recently, allelic interactions between transposons that activate excision and alter the gap repair mechanism have been described (Van Houwelingen *et al.*, 1999).

Interaction between transgenes at ectopic, non-allelic position, resulting in heritably altered expression was later reported by several researchers (Matzke *et al.*, 1994; Vaucheret *et al.*, 1998), following the initial discovery in 1990s, of an allelic interaction leading to heritable changes in the expression of a transgene in *Petunia* (Meyer *et al.*, 1992). These non-allelic, but clearly homology-dependent interactions are often referred to as 'paramutation-like' (reviewed by Hollick *et al.*, 1997). Where expression levels have been tested (*b1* and *pl1*), paramutation is more similar to TGS than to PTGS. Generally, PTGS is not meiotically heritable, whereas paramutation at *b1, pl1* and *r1* loci in maize as well as several examples of TGS are meiotically heritable.

Paramutation at *b1* is most similar to TGS of the *a1* transgene in *Petunia*, as both are associated with changes in chromatin structure (Van Blokland *et al.*, 1997). However, the coding and promoter-proximal regions are extensively methylated in the *a1* transgene, but not in *b1*. These results combined with some recent models suggesting that PTGS and TGS might have features in common (reviewed by Wolffe and Matzke, 1999), suggest that an understanding of paramutation in maize could reveal important mechanistic information on transgene silencing in plants, which has enormous implications to successful commercialization of transgenic crops.

Conclusions

A paradigm that transposons are genome builders and restructuring agents, which was initially brought about by pioneering discoveries in maize by Barbara McClintock, is now becoming firmly established. Transposons are inherently mutagenic; but, some of their properties, like imperfect excision (in case of many plant transposable elements) leading to allele and protein variability, exon/promoter shuffling, and enhanced ectopic recombination, might have had profound effects on genome evolution and evolutionary history. Transposable elements, particularly retrotransposons, appear to play an important role in genome expansion as well as shrinkage. Retrotransposons are also regulated in response to biotic as well as abiotic stresses, suggesting their possible involvement in responses of the plants to various stresses. Transposable elements are also now serving as powerful molecular tools for isolation of plant genes, including several genes influencing diverse agronomically important characters in various plant species. Also, retroelements are now beginning to be utilized as molecular tools for DNA fingerprinting and phylogenetic analysis.

With the completion or near-completion of genome sequencing projects in a wide variety of organisms, including bacteria, viruses, plants and animals, the next major challenge shall be to understand how genes function and how they are regulated. Achieving this goal requires a closer examination of how epigenetic controls are imposed on genes, and a better understanding of how such controls are maintained and reset during development and sexual reproduction of a plant. Knowledge of epigenetic controls such as imprinting, DNA methylation and gene silencing is also important from the plant breeding and plant genetic engineering viewpoints, and consequently in analyzing stability of performance of both conventionally bred and transgenic cultivars.

References

Arabidopsis Genome Initiative. 2000. Analysis of the genome sequence of the flowering plant *Arabidopsis thaliana*. *Nature*, **408**: 796-815.

Barlow D.P. 1993. Methylation and imprinting: from host defense to gene regulation? *Science*, **260**: 309-310.

Becker H.A. and Kunze R. 1997. Maize Activator transposase has a bipartite DNA binding domain that recognizes subterminal sequences and the terminal inverted repeats. *Mol. Gen. Genet.*, **254**: 219-230.

Beckett J.B. 1978. B-A translocations in maize. I. Use in locating genes by chromosome arms. *J. Hered.*, **69**: 27-36.

Beguiristain T., Grandbastien M., Puigdomenech P. and Casacuberta J.M. 2001. Three *Tnt1* subfamilies show different stress-associated patterns of expression in tobacco. Consequence of retrotransposon control and evolution in plants. *Plant Physiol.*, **127**: 212-221.

Bennetzen J.L. 2000. Transposable element contributions to plant gene and genome evolution. *Plant Mol. Biol.*, **42**: 251-269.

Birchler J.A. and Hart J.R. 1987. Interaction of endosperm size factors in maize. *Genetics*, **177**: 309-317.

Brink R.A. 1956. A genetic change associated with the *R* locus in maize which is directed and potentially reversible. *Genetics*, **41**: 872-889.

Brink R.A. 1958. Paramutation at the *R* locus in maize. Cold Spring Harbor Symp. *Quant. Biol.*, **23**: 379-391.

Brink R.A. 1973. Paramutation. *Annu. Rev. Genet.*, **7**: 129-152.

Brink R.A. and Mikula B. 1958. Plant color effects of certain anomalous forms of the *R-r* allele in maize. *Genetics*, **54**: 899-910.

Carpenter R., Hudson A., Robbins T., Almeida J., Martin C. and Coen E. 1988. Genetic and molecular analyses of transposable elements in *Antirrhinum majus*. *In*: *Plant Transposable Elements*. (ed.) O.E. Nelson, Plenum Press, New York, pp. 69-81.

Casa A.M., Brouwer C., Nagel A., Wang L., Zhang Q., Kresovich S. and Wessler S.R. 2000. The MITE family Heartbreaker (Hbr): Molecular markers in maize. *Proc. Natl. Acad. Sci.*, USA, **97**: 10083-10089.

Chandler V.L., Eggleston W.B. and Dorweiler J.E. 2000. Paramutation in maize. *Plant Mol. Biol.*, **43**: 121-145.

Chen M.P., San Miguel P., De Oliviera A.C., Woo S.S., Zhang H., Wing R.A. and Bennetzen J.L. 1997. Microcolinearity in the *sh2*-homologous regions of the maize, rice and sorghum genomes. *Proc. Natl. Acad. Sci.*, USA, **94**: 3431-3435.

Coe E.H. Jr. 1959. A regular and continuing conversion-type phenomenon at the *B* locus in maize. *Proc. Natl. Acad. Sci.*, USA, **45**: 828-832.

Coe E.H. Jr. 1966. The properties, origin, and mechanism of conversion-type inheritance at the *B* locus in maize. *Genetics*, **53**: 1035-1063.

Ellis T.H, Poyser S.J., Knox M.R., Vershinin A.V. and Ambrose M.J. 1998. Polymorphism of insertion sites of *Ty1*-copia class retrotransposons and its use for linkage and diversity analysis in pea. *Mol. Gen. Genet.*, **260**: 9-19.

Fagard M., Boutet S., Morel J.B., Belleni C. and Vaucheret H. 2000. AGO1, QDE-2 and RDE-1 are related proteins required for post-transcriptional gene silencing in plants, quelling in fungi, and RNA interference in animals. *Proc. Natl. Acad. Sci.*, USA, **97**: 11650-11654.

Fedoroff N.V. 1983. Controlling elements in maize. *In*: *Mobile Genetic Elements*. (ed.). J. Shapiro Academic Press, New York, pp. 1-63.

Fedoroff N.V. 2000. Transposons and genome evolution in plants. *Proc. Natl. Acad. Sci.*, USA, **97**: 7002-7007.

Fedoroff N.V., Wessler S. and Shure M. 1983. Isolation of the transposable maize controlling elements *Ac* and *Ds*. *Cell*, **35**: 235-242.

Feschotte C. and Wessler S.R. 2001.Treasures in the attic: rolling circle transposons discovered in eukaryotic genomes. *Proc Natl. Acad. Sci.*, USA, **98**: 8923-8924.

Flavell A.J., Pearce S.R. and Kumar A. 1994. Plant transposable elements and the genome. Curr. Opin. Genet. Dev., *4: 838-844.*

Flavell A.J, Knox M.R., Pearce S.R. and Ellis T.H. 1998. Retrotransposon-based insertion polymorphisms (RBIP) for high throughput marker analysis. *Plant J.*, **16**: 643-650.

Frey M., Reinecke J., Grant S., Saedler H. and Gierl A. 1990. Excision of the *En/Spm* transposable elements of *Zea mays* requires two-element encoded proteins. *EMBO J.*, **9**: 4037-4044.

Frey M., Stettner C. and Gierl A. 1998. A general method for gene isolation in tagging approaches: amplification of insertion mutagenised sites (AIMS). *Plant J.*, **13**: 717-721.

Gierl A., Lutticke S. and Saedler H. 1988. TnpA product encoded by the transposable element *En-1* of *Zea mays* is a DNA binding protein. *EMBO J.*, **7**: 4045-4053.

Grandbastien M.A. 1992. Retroelements in higher plants. *Trends Genet.*, **8**: 103-108.

Gribbon B.M., Pearce S.R., Kalendar R., Schulman A.H., Paulin L., Jack P., Kumar A. and Flavell A.J. 1999. Phylogeny and transpositional activity of *Ty1*-copia group retrotransposons in cereal genomes. *Mol. Gen. Genet.*, **261**: 883-891.

Grossniklaus U., Spillane C., Page D.R. and Kvhler C. 2001. Genomic imprinting and seed development: endosperm formation with and without sex. *Curr. Opin. Plant Sci.*, **4**: 21-27.

Haring M.A., Rommens C.M.T., Nijkamp H.J.J. and Hille J. 1991. The use of transgenic plants to understand transposition mechanisms and to develop transposon strategies. *Plant Mol. Biol.*, **16**: 449-461.

Henikoff S. and Comai L. 1998. *Trans*-sensing effects: the ups and downs of being together. *Cell*, **93**: 329-332.

Hirochika H., Okamoto H. and Kakutani T. 2000. Silencing of retrotransposons in *Arabidopsis* and reactivation by the *ddm1* mutation. *Plant Cell*, **12**: 357-369.

Hirochika H. 2001. Contribution of the *Tos17* retrotransposon to rice functional genomics. Curr. Opin. *Plant Biol.*, **4**: 118-122.

Hollick J.B., Dorweiler J.E. and Chandler V.L. 1997. Paramutation and related allelic interactions. *Trends Genet.*, **13**: 302-308.

Hollick J.B., Patterson G.I., Coe E.H. Jr., Cone K.C. and Chandler V.L. 1995. Allelic interactions heritably alter the activity of a metastable maize *pl1* allele. *Genetics*, **141**: 709-719.

Holliday R. 1987. Inheritance of epigenetic defects. *Science*, **238**: 163-170.

John R.M. and Surani M.A. 1996. Imprinting genes and regulation of gene expression by epigenetic inheritance. *Curr. Opin. Cell Biol.*, **8**: 348-353.

Jorgensen R.A. 1995. Cosuppression, flower color patterns, and metastable gene expression states. *Science*, **268**: 686-691.

Kalendar R., Grob T., Regina M., Suoniemi A. and Schulman A. 1999. IRAP and REMAP: two new retrotransposon-based DNA fingerprinting techniques. *Theor. Appl. Genet.*, **98**: 704-711

Kalendar R., Tanskanen J., Immonen S., Nevo E. and Schulman A.H. 2000. Genome evolution of wild barley (*Hordeum spontaneum*) by BARE-1 retrotransposon dynamics in response to sharp microclimatic divergence. *Proc. Natl. Acad. Sci.*, USA, **97**: 6603-6607.

Kapitonov V.V. and Jurka J. 2001. Rolling-circle transposons in eukaryotes. *Proc. Natl. Acad. Sci.*, USA, **98**: 8714-8719.

Kermicle J.L. 1970. Dependence of the *R-mottled* aleurone phenotype in maize on mode of sexual transmission. *Genetics*, **66**: 69-85.

Kermicle J.L. and Alleman M. 1990. Gametic imprinting in maize in relation to the angiosperm life cycle. *Development* (suppl.), 9-14.

Krebbers E., Hehl R., Piotrowiak R., Lonnig W., Sommer H. and Saedler H. 1987. Molecular analysis of paramutant plants of *Antirrhinum majus* and the involvement of transposable elements. *Mol. Gen. Genet.*, **214**: 25-327.

Kumar A. and Bennetzen J.L. 1999. Plant retrotransposons. *Annu. Rev. Genet.*, **33**: 479-532.

Kumar A. and Hirochika H. 2001. Applications of retrotransposons as genetic tools in plant biology. *Trends Plant Sci.*, **6**: 127-134.

Kumar A., Pearce S.R., McLean K., Harrison G., Heslop-Harrison J.S., Waugh R. and Flavell A.J. 1997. The *Ty1-copia* group of retrotransposons in plants: genomic organisation, evolution, and use as molecular markers. *Genetics*, **100**: 205-217.

Kunze R., Saedler H. and Lonning W.E. 1997. Plant transposable elements. *Adv. Bot. Res.*, **27**: 331-470.

Le Q.H., Wright S., Yu Z. and Bureau T. 2000. Transposon diversity in *Arabidopsis thaliana*. *Proc. Natl. Acad. Sci.*, USA, **97**: 7376-7381.

Lin B.Y. 1982. Association of endosperm reduction with parental imprinting in maize. *Genetics*, **100**: 475-486.

Lin B.Y. 1984. Ploidy barrier in endosperm development in maize. *Genetics*, **107**: 103-115.

Liu D. and Crawford N.M. 1998. Characterization of putative transposase mRNA of *Tag1*, which is ubiquitously expressed in *Arabidopsis* and can be induced by *Agrobacterium*-mediated transformation with *dTag1* DNA. *Genetics*, **149**: 393-401.

Liu D., Wang R., Galli M. and Crawford N. M. 2001. Somatic and germinal excision activities of the *Arabidopsis* transposon *Tag1* are controlled by distinct regulatory sequences within *Tag1*. *Plant Cell*, **13**: 1851-1863.

Manninen O., Kalendar R., Robinson J., Schulman A.H. 2000. Application of *BARE*-1 retrotransposon markers to map a major resistance gene for net blotch in barley. *Mol. Gen. Genet.*, **264**: 325-334

Martienssen R. 1998. Chromosomal imprinting in plants. *Curr. Opin. Genet. Dev.*, **8**: 240-244.

Masson P. and Fedoroff N. 1989. Mobility of the maize *Suppressor-mutator* element in transgenic tobacco cells. *Proc. Natl. Acad. Sci.*, USA, **86**: 2219-2223.

Matzke M.A. and Matzke A.J.M. 1993. Genomic imprinting in plants: parental effects and trans-inactivation phenomena. *Annu. Rev. Plant Physiol. Plant Mol. Biol.*, **44**: 53-76.

Matzke M.A., Matzke A.J.M. and Eggleston W.B. 1996. Paramutation and transgene silencing: a common response to invasive DNA? *Trends Plant Sci.*, **1**: 382-388.

Matzke M.A., Mette M.F. and Matzke A.J.M. 2000. Transgene silencing by the host genome defense: implications for the evolution of epigenetic control mechanisms in plants and vertebrates. *Plant Mol. Biol.*, **43**: 401-415.

Matzke A.J.M., Neuhuber F., Park Y.D., Ambros D.F. and Matzke M.A. 1994. Homology-dependent gene silencing in transgenic plants, epistatic silencing loci contain multiple copies of methylated transgenes. *Mol. Gen. Genet.*, **244**: 219-229.

Matzke M.A., Matzke, A.J.M., Pruss G.J. and Vance V.B. 2001. RNA-based silencing strategies in plants. *Curr. Opin. Genet. Dev.*, **11**: 221-227.

McClintock B. 1951. Chromosome organization and genic expression. *Cold Spring Harbor Symp. Quant. Biol.*, **16**: 13-47.

McClintock B. 1965. The control of gene action in maize. *Brookhaven Symp. Biol.*, **18**: 162-184.

McClintock B. 1984. The significance of responses of the genome to challenge. *Science*, **226**: 792-801.

Messing J. and Grossniklaus U. 1999. Genomic imprinting in plants. Results Probl. *Cell Differ.*, **25**: 23-40.

Meyer P., Linn F., Heidmann I., Meyer Z.A.H., Niedenhoff I. and Saedler H. 1992. Endogenous and environmental factors influence 35S promoter methylation of a maize *A1* gene construct in transgenic petunia and its colour phenotype. *Mol. Gen. Genet.*, **231**: 345-352.

Mhiri C., Morel J.B., Vernhettes S., Casacuberta J.M., Lucas H., Grandbastien M.A. 1997. The promoter of the tobacco *Tnt1* retrotransposon is induced by wounding and by abiotic stress. *Plant Mol. Biol.*, **33**: 257-266.

Miura A., Yonebayashi S., Watanbe K., Toyama T., Shimada H. and Kakutani T. 2001. Mobilization of transposons by a mutation abolishing full DNA methylation in *Arabidopsis*. *Nature*, **411**: 212-214.

Moffat A.S. 2000. Transposons help sculpt a dynamic genome. *Science*, **289**: 1455-1457.

Nevers P., Shepherd N.S. and Saedler H. 1985. Plant transposable elements. *Adv. Bot. Res.*, **12**: 103-203.

Pearce S.R, Stuart-Rogers C., Knox M.R., Kumar A., Ellis T.H. and Flavell A.J. 1999. Rapid isolation of plant *Ty1*-copia group retrotransposon LTR sequences for molecular marker studies. *Plant J.*, **19**: 711-717.

Pearce S.R., Knox M., Ellis T.H., Flavell A.J. and Kumar A. 2000. Pea *Ty1*-copia group retrotransposons: transpositional activity and use as markers to study genetic diversity in *Pisum*. *Mol. Gen. Genet.*, **263**: 898-907.

Pereira A. 1998. Heterologoüs transposon tagging systems. *In*: *Transgenic Plant Research*, (ed.). K. Lindsey, Harwood Academic Publishers, UK, pp. 91-108.

Peterson P.A. 1987. Mobile elements in plants. CRC Crit. Rev. *Plant Sci.*, **6**: 105-208.

Peterson P.A. 1988. Transposons in maize and their role in corn breeding progress. *Proc. 43rd Annu. Corn Sorghum Res. Conf.*, pp. 51-71.

Prasanna B.M. 1991. Transposable genetic elements in maize. *In*: *Maize Genetics Perspectives*, (eds.) K.R. Sarkar, J.K.S. Sachan and N.N. Singh. Indian Society of Genetics and Plant Breeding, New Delhi, pp. 113-120.

Prasanna B.M. and Sarkar K.R. 1996. The *R* locus in maize - A versatile marker system for probing differential gene expression. *Indian J. Genet.*, **56**: 229-247.

Prasanna B.M. and Sarkar K.R. 1997. Active transposable elements in Indian maize germplasm and assorted genetic testers. *Curr. Sci.*, **73**: 464-469.

Purugganan M.D. and Wessler S.R. 1995. Transposon signature-species-specific molecular markers that utilize a class of multiple-copy nuclear-DNA. *Mol. Ecol.*, **4**: 265-269.

Razin A. and Riggs A.D. 1980. DNA methylation and gene function. *Science*, **210**: 604-610.

San Miguel P., Gaut B.S., Tikhonov A., Nakajima Y. and Bennetzen J.L. 1998. The paleontology of intergene retrotransposons of maize. *Nat. Genet.*, **20**: 43-45.

Sato Y., Sentoku N., Miura Y., Hirochika H., Kitano H. and Matsuoka M. 1999. Loss-of-function mutations in the rice homeobox gene *OSH15* affect the architecture of internodes resulting in dwarf plants. *EMBO J.*, **18**: 992-1002.

Schlappi M., Raina R. and Fedoroff N. 1994. Epigenetic regulation of the maize transposable element system: Novel activation of amethylated promoter by TnpA. *Cell*, **77**: 427-437.

Shapiro J.A. 1969. Mutations caused by the insertion of genetic material into the galactose operon of *Escherichia coli*. *J. Mol. Biol.*, **40**: 93.

Singer T., Yordan C. and Martienssen R. 2001. Robertson's *Mutator* transposons in *A. thaliana* are regulated by the chromatin-remodeling gene, decrease in DNA methylation. *Genes Dev.*, **15**: 591-602.

Solter D. 1988. Differential imprinting and expression of maternal and paternal genomes. *Annu. Rev. Genet.*, **22**: 127-146.

Sommer H., Hehl R., Krebbers E., Piotrowiak R., Lonning W.E. and Saedler H. 1988. Transposable elements in *Antirrhinum majus*. *In*: *Plant Transposable Elements*, (ed.). O.E. Nelson, Plenum Press, New York, pp. 227-236.

Sundaresan V. 1996. Horizontal spread of transposon mutagenesis: new uses for old elements. *Trends Plant Sci.*, **1**: 184-190.

Sundaresan V., Springer P., Volpe T., Howard S., Jones J.D., Dean C., Ma H. and Martienssen R. 1995. Patterns of gene action in plant development revealed by enhancer trap and gene trap transposable elements. *Genes Dev.*, **9**: 1797-1810.

Surani M.A.H. 1987. Evidences and consequences of differences between maternal and paternal genomes during embryogenesis in the mouse. *In*: *Experimental Approaches to Mammalian Embryonic Development*, (eds.) J. Rossant, and R.A. Pedersen, Cambridge University Press, London, pp. 401-435.

Takeda S., Sugimoto K., Otsuki H. and Hirochika H. 1998. Transcriptional activation of the tobacco retrotransposon *Tto1* by wounding and methyl jasmonate. *Plant Mol. Biol.*, **36**: 365-376.

Takeda S., Sugimoto K., Otsuki H. and Hirochika H. 1999. A 13-bp cis-regulatory element in the LTR promoter of the tobacco retrotransposon *Tto1* is involved in responsiveness to tissue culture, wounding, methyl jasmonate and fungal elicitors. *Plant J.*, **18**: 383-393.

Tarchini R., Biddle P., Wineland R., Tinguey S. and Rafalski A. 2000. The complete sequence of 340 kb of DNA around the rice *Adh1-Adh2* region reveals interrupted collinearity with maize chromosome 4. *Plant Cell*, **12**: 381-391.

Upadhyaya K.C. 1995. Transposable elements and transposon tagging in plants. *In*: *Genetic Research and Education: Current Trends and the Next Fifty Years*, (eds.) B. Sharma *et al.* Indian Society of Genetics Plant Breeding, New Delhi, pp. 1168-1187.

Upadhyaya K.C., Sommer H., Krebbers E. and Saedler H. 1985. The paramutagenic line niv44 has a 5-kb insert, *Tam2*, in the chalcone synthase gene of *Antirrhinum majus*. *Mol. Gen. Genet.*, **199**: 201-207.

Van Blokland R., Lohuis M. and Meyer P. 1997. Condensation of chromatin in transcriptional regions of an inactivated plant transgene: evidence for an active role of transcription in gene silencing. *Mol. Gen. Genet.*, **257**: 1-13.

Van der Broeck D., Maes T., Saur M., Zethof J., De K.P., D'Hauw M., Van M.M. and Gerats T. 1998. Transposon display identifies individual transposable elements in high copy number lines. *Plant J.*, **13**: 121-129.

Van Houwelingen A., Souer E., Mol J. and Koes R. 1999. Epigenetic interactions among three dTph1 transposons in two homologous chromosomes activate a new excision-repair mechanism in Petunia. *Plant Cell*, **11**: 1319-1336.

Varmuza S. and Mann M. 1994. Genomic imprinting - defusing the ovarian time bomb. *Trends Genet.*, **10**: 118-123.

Vaucheret H., Beclin C., Elmayan T., Fuerbacj F., Godon C., Morel J.B., Mourrain P., Palauqui J.C. and Vernhettes S. 1998. Transgene-induced gene silencing in plants. *Plant J.*, **16**: 651-659.

Vicient C.M., Suoniemi A., Anamthawat-Jonsson K., Tanskanen J. 1999. Retrotransposon *BARE-1* and its role in genome evolution in the genus *Hordeum*. *Plant Cell*, **11**: 1769-1784.

Vicient C.M., Jddskeldinen M.J., Kalendar R. and Schulman A.H. 2001. Active retrotransposons are a common feature of grass genomes. *Plant Physiol.*, **125**: 1283-1292.

Vielle-Calzada J.P., Baskar A. and Grossniklaus U. 2000 Delayed activation of the paternal genome during seed development. *Nature*, **404**: 91-94.

Vielle-Calzada J.P., Thomas J., Spillane C., Coluccio A., Hoeppner M.A. and Grossniklaus U. 1999. Maintenance of genomic imprinting at the *Arabidopsis medea* locus requires zygotic DDM1 activity. *Genes Dev.*, **13**: 2971-2982.
Vongs A., Kakutani T., Martienssen R. and Richards E.J. 1993. *Arabidopsis thaliana* DNA methylation mutants. *Science*, **260**: 926-1928.
Waddington C.H. 1956. *Principles of Embryology*. Allen and Unwin, London.
Walbot V. 1992. Reactivation of *Mutator* transposable elements of maize by ultraviolet light. *Mol. Gen. Genet.*, 1992. **234**: 353-360.
Walbot V. 1996. Sources and consequences of genotypic and phenotypic plasticity in flowering plants. *Trends Plant Sci.*, **1**: 27-32.
Walbot V. 2000. Saturation mutagenesis using maize transposons. *Curr. Opin. Plant Biol.*, **3**: 103-107.
Waterhouse P.M., Wang M.B. and Lough T. 2001. Gene silencing as an adaptive defence against viruses. *Nature*, **411**: 834-842.
Waugh R., McClean K., Flavell A.J., Pearce S.R., Kumar A., Thomas B.T. and Powell W. 1997. Genetic distribution of BARE1 retrotransposable elements in the barley genome revealed by sequence-specific amplification polymorphism (S-SAP). *Mol. Gen. Genet.*, **253**: 687-694.
Wessler S.R. 1988. Phenotypic diversity mediated by the maize transposable elements *Ac* and *Spm*. *Science*, **242**: 399-405.
Wessler S.R. 1996. Turned on by stress. Plant retrotransposons. *Curr. Biol.*, **6**: 959-961.
Wicker T., Stein N., Albar L., Feuillet C., Schlagenhauf E. and Keller B. 2001. Analysis of a contiguous 211 kb sequence in diploid wheat (*Triticum monococcum* L.) reveals multiple mechanisms of genome evolution. *Plant J.*, **26**: 307-316.
Wolffe A.P. and Matzke M.A. 1999. Epigenetics: regulation through repression. *Science*, **286**: 481-486.
Yephremov A. and Saedler H. 2000. Display and isolation transposon flanking sequences starting from genomic DNA or RNA. *Plant J.*, **21**: 495-505.
Yoder J.A., Walsh C.P. and Bestor T.H. 1997. Cytosine methylation and the ecology of intragenomic parasites. *Trends Genet.*, **13**: 335-340.
Yu G. X. and Wise R.P. 2000. An anchored AFLP- and retrotransposons-based map of diploid *Avena*. *Genome*, **43**: 736-749.

Plant Breeding - Mendelian to Molecular Approaches
H. K. Jain and M. C. Kharkwal (eds.)

7

Genetics of Host-Pathogen Interaction and Breeding for Durable Resistance

Ravi P. Singh[1], S. Rajaram[1], R.G. Saini[2], J. Huerta-Espino[3] and M. William[1]

Abstract

Disease control to achieve stable food production has been one of the challenges to wheat pathologists, geneticists and breeders during the 20th century. An understanding of disease epidemiology, genetic basis of host-pathogen interaction, search for resistance genes, and development of cultivars with built-in disease resistance to a number of important diseases has reduced the occurrence of large-scale epidemics that were common in the first half of the 20th century. Improper use of the major, race-specific type of resistance to control rapidly evolving pathogens (e.g., the rusts of wheat) has led to boom-and-bust cycles that make it necessary to replace cultivars a short time after their release. Durable resistance is based on interaction of minor, additive genes. In case of leaf and yellow rusts of wheat, accumulation of 4-5 slow rusting genes results in a high level of resistance that approaches immunity. This is the type of resistance needed in the 21st century to ensure food security.

Introduction

Diseases and pests have posed major threats to stable production of many food and fiber crops for centuries. The science of crop phytopathology, through the understanding of pathogen epidemiology and utilisation of genetic resistance, has helped prevent or reduce the disease epidemics that were common until 30 or 40 years ago. Wheat has historically been attacked by the rust pathogens. However, modern, semi-dwarf cultivars protected by built-in genetic resistance are now grown successfully under higher input conditions in disease-prone environments. This success can be attributed to numerous independent research findings that have led to improved genetic control of diseases and pests. For each 25-year period of the 20th century, we would like to illustrate a few such discoveries that have greatly impacted resistance breeding in most field crops, especially wheat.

[1]International Maize and Wheat Improvement Center (CIMMYT), Mexico
[2]Punjab Agricultural University, Ludhiana 141 004, India
[3]International Maize and Wheat Improvement Center (CIMMYT), Chapingo, Edo. de Mexico

1900-1925: Initiation of Resistance, Inheritance. Studies and Determination of Pathogenic Specialisation

Determination of the genetic basis of plant disease resistance began soon after the rediscovery of Mendel's laws of segregation and independent assortment in 1901. The first results were reported by Rowland Harry Biffen in 1905 in Cambridge, UK. He studied the F_1 and F_2 populations from a cross of yellow (stripe) rust (caused by *Puccinia striiformis*) resistant cultivar 'Rivet' with susceptible 'Red King'. The F_1 plants were susceptible, and he observed that 195 plants of the F_2 generation were infected with the disease, while 64 were rust-free. This segregation conformed to a 3:1 ratio, as could be predicted from Mendelian genetics, and indicated that resistance was recessive and susceptibility dominant. Biffen classified plants in three categories early in the disease development phase, indicating incomplete dominance of susceptibility. He also found that resistance and susceptibility were independent of other plant characteristics (Biffen 1907). Subsequently, many examples of Mendelian inheritance of disease resistance were documented, though there was also some confusion due to physiologic specialisation in the pathogen population, the influence of environment on the expression of resistance, and the presence of minor-gene-based resistance.

Following two devastating stem rust epidemics in North America in 1904 and 1916, E. C. Stakman and his co-workers at the University of Minnesota, USA, were able to show a biologic form (race) of *Puccinia graminis* (Stakman and Piemeisel, 1917). Soon afterwards, wheat cultivars differentiating stem rust and leaf rust populations were identified (Table 1) and a system of nomenclature based on the resistance/susceptibility

Table 1. Differential cultivars used internationally in the identification of physiologic races in *Puccinia graminis* and *Puccinia triticina*

Puccinia graminis[1]		*Puccinia triticina*[2]	
Cultivars	Known gene(s)[3]	Cultivars	Known gene(s)[4]
Little Club	*LC*	Malakof	*Lr1*
Marquis	*7b, 18, 19, 20, X*	Hussar	*Lr11*
Reliance	*5, 16, 18, 20*	Mediterranean	*Lr3*
Kota	*7b, 18, 19, 28, Kt2*	Democrat	*Lr3*
Arnautka	*9d, a*	Carina	*Lr2b, B*
Mindum	*9d, a, b*	Brevit	*Lr2c, B*
Spelmar	*9d, a, b*	Webster	*Lr2a*
Kubanka	*9g, c*	Loros	*Lr2c*
Acme	*9g, d*		
Einkorn	*21*		
Vernal	*9e*		
Khapli	*7a, 13, 14*		

[1]Stakman *et al.*, (1962); [2]Mains and Jackson (1926) and revised by Johnston and Browder (1966); [3]Roelfs and Martens (1988); [4]Samborski and Dyck (1968)

response pattern of these cultivars was developed in the USA and adopted in other countries (Stakman and Levine, 1922; Mains and Jackson, 1926). Stakman also developed the infection type scale for characterizing resistance as early as 1919. Although somewhat modified, this scale is still widely used. For 36 years Stakman was extensively engaged in not only determining various aspects of wheat rust but also training numerous scientists who later made major impacts.

1925-1950: More Profound Understanding of Epidemiology and Generation of Resistant Cultivars Through Hybridisation

This quarter century was highlighted by the definition of physiologic races, pathogen life cycles, and epidemiology led by great scientists such as E. C. Stakman (stem rust), K. S. Chester (leaf rust), and their co-workers in North America, K. C. Mehta in India, W. L. Waterhouse in Australia, and J. H. Craigie in Canada, among others.

Several methods for conducting disease surveys, collecting field samples, trapping spores and tracing spore movement were developed under the guidance of Stakman. He and his co-workers were able to show the northward movement of stem rust in North America, where rust survived south of Texas in northern Mexico and then spread to the Mississippi Basin in the USA and the Prairie Provinces of Canada in a time span of about 2 months (Stakman and Harrer, 1957). Influenced by the knowledge of the role of alternate hosts in initiating early infections, Stakman personally led a campaign to eradicate susceptible *Berberis* and *Mahonia* species in North America (Stakman *et al.*, 1934). He conducted extensive field surveys annually to monitor disease severity and prevalence, and obtained uredinospore collections for determining physiologic races (Stakman *et al.*, 1929). This information was then used in identifying resistant parents for breeding purposes (Harrer *et al.*, 1944). Today, surveys to monitor numerous diseases and pests of several crops are conducted all over the world.

K. C. Mehta studied rust epidemiology in India and found that the alternate hosts present in Indian hills did not play a significant role in the life cycles of stem and leaf rusts. He also determined that rust inoculum survived in summer in the Himalayas in the north and the Nilgiri and Pulney hills, and then was blown to different wheat-growing areas in the plains (Mehta, 1929; 1933).

W. L. Waterhouse initiated similar work on cereal rusts in Australia in 1919. He demonstrated variation within physiologic races, identified on international differential series, by using additional cultivars (Waterhouse, 1929). This demonstrated that races were not necessarily homogeneous units but carried combinations of similar pathogenic (avirulence/virulence) characteristics. He also began breeding for resistance and successfully developed resistant cultivars that carried single resistance genes (Waterhouse, 1930). He was later joined by I. A. Watson at the University of Sydney, which became a pioneer rust research laboratory that has contributed crucial findings in the area of host-pathogen genetics and bred numerous successful rust resistant cultivars in Australia.

The benefits of interspecific hybridisation for rust resistance breeding were observed in the USA during this period. Although H. K. Hayes had begun interspecific hybridisation in 1914 by crossing hard red spring wheat 'Marquis' with durum wheat 'Imuillo' and selected a resistant spring wheat 'Marquillo' in 1918 (Hayes *et al.*, 1920), the successful rust resistant cultivar 'Thatcher' could be released in 1934 only after further hybridisation (Hayes *et al.*, 1936). Thatcher carries a combination of stem rust resistance genes that, to this day, provides useful resistance in North America.

E. S. McFadden of South Dakota, USA, crossed 'Yaroslav' emmer with Marquis and developed stem rust resistant cultivars 'Hope' and 'H-44' (McFadden, 1930). Stem rust resistance from Hope was back-crossed to Thatcher in Minnesota, leading to the development of 'Newthatch' (Ausemus *et al.*, 1944). These wheats led to the foundation of durable stem rust resistance globally, commonly known now as the '*Sr2*-complex' (Rajaram *et al.*, 1988).

1951-1975: Understanding the Scientific Basis of Resistance and its Utilisation in Breeding

Tremendous research advances were made during this period by a number of scientists, not only in the understanding of host-pathogen genetics but also in breeding for disease resistance. Only a few important contributions are highlighted.

H. H. Flor was the first person to initiate simultaneous inheritance studies involving the pathogenicity in flax rust (*Melampsora lini*) and resistance in its host flax (*Linum usitatissimum*) (Flor, 1942; 1947). These studies led to the formulation of the 'Gene-for-Gene Relationship' concept (Flor, 1956) which states that "for each gene conditioning rust reaction in the host, there is a specific corresponding gene conditioning pathogenicity in the parasite." As explained in Table 2, resistance is expressed only when the host cultivar carries resistance alleles and the pathogen race possesses avirulence alleles. In another three situations, the interaction phenotype is susceptibility. Flor's work received more attention when Person (1959) provided a theoretical analysis of the gene-for-gene relationship and concluded that such relationships should occur as a general rule in host-parasite systems as a result of selection pressure during evolution. Soon after this publication, the gene-for-gene concept was demonstrated or hypothesised in a number of host-pathogen systems, including the rusts (Day, 1974; Luig and Watson, 1961; Loegering and Powers, 1962).

Table 2. Explanation of the "gene-for-gene" hypothesis presented by Flor (1956)

	Pathogen	
Host	Avirulence allele P_1P_1	Virulence allele p_1p_1
	Interaction phenotype	
Resistance allele (R_1R_1)	Resistant	Susceptible
Susceptiblity allele (r_1r_1)	Susceptible	Susceptible

Assumptions: Resistance in the host and avirulence in the pathogen are dominant.

The International Spring Wheat Rust Nursery Program was initiated in 1950 by B. B. Bayles and R. A. Rodenhiser of United States Department of Agriculture, Agricultural

Research Services (USDA-ARS), Beltsville, and operated continuously until the mid-1980s. The purposes of this program were: (1) to find new genes or combinations of genes in wheat, which condition field resistance to rust throughout the world, and (2) to test new varieties and promising selections of wheat developed by plant breeders and pathologists for resistance to rusts. The germplasm and information generated were thus made available to the global wheat community. This nursery was the foundation of numerous other international nurseries and global cooperation to achieve resistance to rust diseases. Germplasm distributed through this nursery was widely used in breeding programs worldwide, including the one led by N. E. Borlaug in Mexico.

I. A. Watson and co-workers in Australia conducted a very dynamic program that studied the evolution of new pathogen races in Australia and used this information in breeding stem rust resistant cultivars carrying single resistance genes. This contributed to the phenomenon of "boom-and-bust" disease cycles and made it clear that utilisation of this philosophy in breeding would need further refining. The concept of using combinations of resistance genes arose from this experience. Watson was firmly convinced that the development of stable, durable resistance in wheat cultivars requires the use of many distinct resistance genes (McIntosh and Smith-White, 1989). However, resistance breakdowns in a stepwise manner lead to a similar problem, as demonstrated by the use of single genes for resistance. The philosophy adopted in Australia, therefore, was to stay ahead of the pathogen through thorough knowledge of the variation in the pathogen including the prediction of the future changes in pathogenicity. These studies were supplemented by the simultaneous search for new resistance genes within wheat and related species. To meet this demand, numerous resistance genes were searched globally, catalogued and transferred from wild relatives to wheat. The first catalogue of gene symbols in wheat was prepared and presented in 1973 by McIntosh, who has updated it several times (McIntosh *et al.*, 1998).

Borlaug used stem rust resistance derived from McFadden's Hope in Mexico and in 1950 released a resistant wheat cultivar, Yaqui 50. Since then the combination of stem rust resistance genes present in this cultivar has become the backbone of durable resistance known to be present in the semi-dwarf wheats from CIMMYT. The adaptation of semi-dwarf wheats to numerous environments during 1960s was partly associated with the durable stem rust resistance they carried (Rajaram *et al.*, 1988). Stem rust epidemics also have become a nonexistent phenomenon wherever cultivars carrying Hope-derived resistance are grown.

The multilineal approach to disease control was proposed (Jensen, 1952; Borlaug, 1953) and researched during this period, and cultivars developed. However, breeders have since realised that although multiline breeding can control disease, it is a very conservative and slow approach in regard to yield, because new varieties may rapidly supersede the recurrent parent. Further problems also arose in the multiplication of each component and the replacement of components that became susceptible. For these reasons, the approach did not succeed.

Working with late blight of potato in Mexico, J. S. Niederhauser and colleagues demonstrated the presence of "partial resistance," a term Niederhauser first used in some cultivated varieties and wild species of potato (Niederhauser *et al.*, 1954). In field trials he found that these varieties show a degree of resistance that is exhibited equally toward all races of the pathogen. These varieties remained green longer than varieties without partial resistance.

The conceptual delineation of resistance into two types, vertical and horizontal, was put forward by J. E. Vanderplank (1963). He defined vertical resistance as being effective against some races but ineffective against others; horizontal resistance was defined as being evenly spread against all races of the pathogen. In relation to horizontal and vertical resistance, pathogens can have variability for corresponding aggressiveness and virulence, respectively. Vanderplank also pointed out that both types of resistance can, and often do, coexist. This delineation provoked numerous debates from the followers of the gene-for-gene relationship, but is now widely accepted. Several other synonymous terms have been used since then, *viz.* race-specific and race-nonspecific resistance.

R. M. Caldwell of Purdue University, USA, elucidated the importance of breeding for general resistance to plant diseases (Caldwell, 1968). Experience and adequate testing in nature have shown that general resistance confers enduring, stable protection against a pathogen or disease. General resistance is such that no natural variants of a pathogen are able to compensate for the restrictions to their penetration, development, or dispersion that such resistance imposes. This can usually be determined by prolonged testing. Working with leaf rust of wheat, Caldwell (1968) emphasised that slow rusting is a form of general or horizontal resistance. He stated that such resistance to rusts involves exclusion of the fungus, limitation of the pustule size without hypersensitivity, or possibly slow growth and development of the fungus. We now know that these components of slow rusting (longer latent period, reduced infection frequency and smaller uredinia, reduced duration and quantity of spore production) are important features of slow rusting resistance. The joint action of these host characters may drastically slow down a disease epidemic to the point of insignificance. J. E. Parlevliet (1975) used the term 'partial resistance' to characterise slow rusting to leaf rust in barley. According to him, partial resistance is a form of incomplete resistance characterised by a reduced rate of epidemic development despite a high or susceptible infection type. Parlevliet (1976; 1986) also showed that partial resistance to leaf rust involves interaction of minor genes, and that the components of resistance may be under pleiotropic genetic control.

R. Johnson of Cambridge, England, described the presence of durable resistance to yellow rust in winter wheat cultivar Cappelle Desprez. This moderate level of adult plant resistance had remained effective for over 20 years in the UK when such resistance was recognised as durable. Durable resistance as defined by Johnson (1978) is that which has remained effective in a cultivar during its widespread cultivation for a long sequence of generations or period of time in an environment favourable to a disease or pest. This term has received wide acceptance and is popularly used.

1975-2000: Better Understanding of the Types of Resistance and Their Utilisation in Crop Improvement

This period is highlighted by the identification, chromosomal localization, transfer from alien sources, and attempt to pyramid these race-specific genes in crop improvement. Designated genes for resistance to the three rust diseases are listed in Table 3. Several other race-specific genes have not yet been catalogued but are present in wheat germplasm or related species and genera. To give an example, 49 resistance genes have been catalogued for resistance to leaf rust (McIntosh *et al.*, 1998), indicating high genetic diversity for resistance. Genes *Lr38* and *Lr39*, which confer the hypersensitive type of adult-plant resistance, were recently designated by R. G. Saini. These are the first genes identified by a scientist in India. Except for two genes, all others confer the hypersensitive type of resistance, and virulence for a majority of these hypersensitive resistance genes has been reported. This means useful genetic diversity is much less than what would appear from the number of catalogued genes. The *Puccinia triticina* (causal pathogen of leaf rust of wheat) population in Mexico has acquired new virulence within 3 years on an average from the time a new semi-dwarf cultivar was released (Singh and Dubin, 1997). A similar phenomenon has been observed in several diseases of wheat and other crops. This experience has prompted CIMMYT's wheat breeding program to adopt strategies that should provide longer lasting resistance.

Table 3. Origin of designated genes for rust resistance in wheat[1]

Origin	Resistance genes		
	Leaf rust	Stem rust	Yellow rust
Triticum aestivum	1, 2a, 2b, 2c, 3, 3ka, 3bg, 10, 11, 12, 13, 14b, 15, 16, 17a, 17b, 18, 20, 22b, 27, 30, 31, 33, 34, 46, 48, 49	*5, 6, 7, 8, 9a, 9b, 9f, 15, 16, 18, 19, 20, 23, 28, 29, 30, 41, 42*	*1, 2, 3, 4, 6, 10, 11, 12, 13, 14, 15, 16, 18, 19, 20, 21, 22, 23, 25, 27, 29*
T. turgidum	14a, 23	*2, 9d, 9e, 9g, 11, 12, 13, 14, 17*	*7, 30*
T. monococcum		21, 22, 35	
T. timopheevi		36, 37, 40	
T. speltoides	*28, 35, 36*	*32, 39*	
T. spelta	*44, 47*		*5*
T. tauschii (syn. *Aegilops tauschii*)	*21, 22a, 32, 39, 40, 41, 42, 43*	*33, 45*	*24, 28*
T. comosum		*34*	*8*
T. ventricosum	*37*	*38*	*17*
Agropyron elongatum	*19, 24, 29*	*24, 25, 26, 43*	
T. umbellulata	*9*		
Secale cereale	*25, 26, 45*	*27, 31*	*9*
Agropyron intermedium	*38*	*44*	
T. araraticum		40	
Haynaldia villosa			26

[1]For further details see McIntosh *et al.* (1998).

The globalisation of CIMMYT's International Screening Nurseries occurred during this period, following the example of USDA's International Spring Wheat Rust Nursery. Nurseries specific for important diseases were created, and improved breeding lines from CIMMYT and other breeding programs that carried resistance to a specific disease, or group of diseases, were distributed for international evaluation. The concept of ACI (average coefficient of infection across locations) arose due to the lack of information on the genetic basis of resistance in these lines (Rajaram *et al.*, 1988). A high degree of phenotypic diversity for disease resistance was observed in these nurseries. Lines showing low average disease values across locations were selected for further crossing (Table 4). Although a single resistance gene effective globally may be selected, this methodology has helped develop globally resistant germplasm in a short time. Many lines have been found to carry either combinations of multiple race-specific genes or combinations of minor, additive genes (Singh and Rajaram, 1991; 1992).

Table 4. Genetic diversity in 280 advanced lines of the 24th International Bread Wheat Screening Nursery classified by average coefficients of infection (ACI) for three rusts in International Multilocation Testing

Disease	No. of locations	No. of entries in ACI classes					
		0-5	5.1-10	10.1-20	20.1-30	30.1-40	>40
Leaf rust	39	168	78	32	2	0	0
Stripe rust	16	26	62	104	71	15	2
Stem rust	15	162	52	60	4	2	0

Breeding to achieve durable slow rusting resistance to leaf rust in wheat was undertaken in CIMMYT during the early 1970s by S. Rajaram. He selected plants and lines in segregating populations that would show up to 20 % disease severity under high disease pressure, when the susceptible check cultivars would have shown complete susceptibility. This major thrust toward accumulating slow rusting genes in CIMMYT germplasm resulted in the release in Mexico of cultivars such as Torim 73, Cocoraque 75, Pavon 76 and Nacozari 76 in the mid-1970s. The moderate resistance level they carried has remained durable. Since then the understanding of the genetic basis of durable resistance to rust diseases in wheat germplasm of CIMMYT origin has been enhanced and a simple breeding methodology formulated. Similar progress has been made for resistance to a number of other diseases of wheat.

The resistance genes involved in durable resistance of recent wheats bred at CIMMYT are expected to give resistance to many cultivars that will be grown by farmers in many countries in the 21st century. Studies have shown that about 82 % of wheat cultivars grown today in developing countries are derived from CIMMYT germplasm, and almost 58 % of these are direct releases from germplasm distributed through different nurseries (Heisey *et al.*, 1999). Hence it is important to discuss advances in the understanding of the genetic basis of durable resistance to rust diseases in wheat and CIMMYT's approach to utilising such resistance in wheat improvement.

Sr2 and Other Minor and Major Genes for Durable Resistance to Stem Rust

Stem rust resistance gene *Sr2*, in addition to other unknown minor genes derived from cultivar Hope, provided the durable resistance foundation in Rockefeller-Mexican spring wheat germplasm developed by Borlaug. Cultivar Yaqui 50, released in Mexico during the 1950s, and other *Sr2*-carrying wheats released since then have stabilised the stem rust situation in Mexico. In recent studies we have not observed any change in stem rust pathotypes in Mexico. Released in 1960 in the Indian subcontinent and subsequently grown on millions of hectares, the cultivar Sonalika has also remained resistant. When present alone, the *Sr2* gene confers slow rusting that may not be adequate under heavy disease pressure, but does provide adequate resistance in combination with other major or minor genes. The *Sr2* gene can be identified by its genetic link to pseudo-black chaff or brown necrosis phenotype seen on the glumes and below the nodes of wheat plants. Unfortunately, not much is known about the other genes and their interactions in the *Sr2* complex.

Knott (1988) has shown that adequate levels of multigenic resistance to stem rust can be selected by accumulating approximately five minor genes. In his studies the genes were different from *Sr2*. It is likely that similar genes are present in CIMMYT germplasm, but more research is needed to document their existence.

Lr34 and Other Minor Genes for Durable Resistance to Leaf Rust

The South American cultivar Frontana is considered one of the best sources of durable resistance to leaf rust (Roelfs, 1988). The Mexican-Rockefeller Program first used the variety in the 1950s. Later derivatives such as Penjamo 62, Torim 73 and Kalyan/Bluebird showed slow rusting characteristics possibly derived from Frontana. Genetic analysis of Frontana and several CIMMYT wheats possessing excellent slow rusting resistance to leaf rust worldwide has indicated that such adult plant resistance is based on the additive interaction of *Lr34* and two or three additional slow rusting genes (Singh and Rajaram, 1992). In Mexico, leaf rust severity on most cultivars can be related to the number of slow rusting genes they carry (Table 5). When susceptible cultivars display 100 % leaf rust severity, cultivars with only *Lr34* display approximately 40 % severity; cultivars with *Lr34* and one or two additional minor genes display 10-15 % severity; and cultivars with *Lr34* and two or three additional genes display 1-5 % severity. Leaf rust may further increase to unacceptable levels on cultivars carrying only *Lr34*, or *Lr34* and one or two additional genes. However, cultivars with *Lr34* and two or three additional genes show a stable response in all environments tested so far, with final leaf rust ratings lower than 10 %. Some cultivars carrying *Lr34* and two or three additional genes are listed in Table 5. Several Indian cultivars also carry different levels of slow rusting resistance that is either due to the presence of *Lr34* or other genes. Some of the cultivars are HD2009, HD2135, HD2189, HI977, HS86, HUW37, IWP72, PBW65, Raj2184, etc. The presence of *Lr34* is indicated by the presence of leaf tip necrosis in adult plants, which is closely linked with it (Singh, 1992a).

Table 5. Some seedling susceptible bread wheats that carry good adult plant resistance to leaf rust in Mexico and other countries

Genotype(s)	Usual leaf rust response[1]	Additive genes[2] for resistance
Jupateco 73S	100S(N)	Highly susceptible
Jupateco 73R	50MSS	*Lr34*
Nacozari 76	30MSS	*Lr34* + 1 gene
Sonoita 81, Bacanora 88, Rayon 89	20MSS	*Lr34* + 1 or 2 genes
Frontana, Parula, Trap, Tonichi 81	10MSS	*Lr34* + 2 or 3 genes
Chapio, Tukuru, Kukuna, Vivitsi	1MSS	*Lr34* + 3 or 4 genes
Pavon 76	40MSS	*Lr46* + 1 gene
Genaro 81, Attila	40MSS	2 genes
Amadina	5MSS	4 genes

[1]Leaf rust response evaluated in Mexico has two components: % severity based on the modified Cobb scale (Peterson *et al.*, 1948) and reaction based on Roelfs *et al.* (1992). The reactions are: MSS = moderately susceptible to susceptible, i.e., medium to large sized uredia without chlorosis or necrosis; S = susceptible, i.e. large uredia without chlorosis or necrosis; N = necrotic leaves following high leaf

Slow rusting can be characterised in greenhouse experiments by evaluating latent period, uredial number (or infection frequency, or receptivity), uredinial size and inoculum production under quantitative inoculation. Characterization of 27 bread wheats of CIMMYT origin by Singh *et al.* (1991) indicated that all components measured were phenotypically diverse (Table 6). The area under disease progress curve of these wheat lines in the field ranged from 1 to 50 % of the very susceptible check cultivar Morocco (Table 6). Singh *et al.* (1991) also reported the likelihood of pleiotropic genetic control of slow rusting components because of highly significant positive or negative phenotypic correlation among latent period, uredinial number and uredinial size. Assuming that the same gene controls several slow rusting components, it can be hypothesised that perhaps only a few genes with additive effects could retard disease progress to a rate that final disease level remains at an acceptable low level.

Table 6. Range of variability for components of slow rusting resistance to leaf rust observed in 28 bread wheats (given as % of susceptible check variety Morocco)

Component	Variability range
Latent period	+14 to +49
Uredial number	–42 to –98
Uredial size	–34 to –78
Area under the disease progress curve	–50 to –99

The heterogeneous Mexican wheat cultivar Jupateco 73 was reselected for the presence and absence of *Lr34* by Singh (1992a). These isogenic Jupateco 73R (*Lr34* present) and Jupateco 73S (*Lr34* absent) genotypes and those of Thatcher developed by Dyck (1987) have yielded useful information on the nature of slow rusting resistance. Our studies using the Jupateco pair have shown that *Lr34* affects all three components of slow rusting, i.e. it increases latent period and decreases uredinial number and uredinial size (Table 7). The effect was more pronounced in post seedling growth stages, although measurable differences also occurred in the seedling stage. Temperature can also influence the expression of resistance conferred by the gene *Lr34* (Singh and Gupta, 1992).

Table 7. Comparison of the three components of slow rusting resistance to leaf rust in seedling and flag leaves of near-isogenic *Lr34* Jupateco 73 reselections tested at 15 °C

Genotype	Latent period				Uredial size	
	(Days)		Uredia/cm^2		(mm^2)	
	Seedling	Flag	Seedling	Flag	Seedling	Flag
Jupateco +*Lr34*	13.8	18.0	21	6	0.21	0.07
Jupateco -*Lr34*	12.8	12.7	47	30	0.38	0.27

Rubiales and Niks (1995) studied the infection process and indicated that slow rusting resistance due to *Lr34* was based on reduced rate of haustorium formation in the early stages of infection, in association with no, or relatively little, plant cell necrosis. Electron microscopic studies by Alvarez-Zamorano (1995) on Jupateco 73 isolines have shown an accumulation of unknown electro-dense substances in the cells of *Lr34* lines near the site, where haustorial mother cells try to dissolve the cell wall of mesophyll cells for the formation of haustoria. It would appear that the accumulation (cell wall apposition) causes a thickening of cell wall, which reduces the establishment of the haustorial tube. If haustoria are formed, slow mycelial growth may be due to restricted movement of fungus from one cell to another by a similar phenomenon. Alvarez-Zamorano (1995) also observed structural change in the *Lr34* line leading to invagination, or contraction of the cell wall, which may delay the completion of the infection process. These observations indicate a mechanism for *Lr34*-based slow rusting different from hypersensitivity, which is associated with race-specific genes. Because pathogen isolates can vary for aggressiveness (Lehman and Shaner, 1996), it may be difficult to differentiate pathogenic variation for increased capability to overcome slow rusting resistance of this type from aggressiveness.

Slow rusting resistance to leaf rust is common in spring wheat germplasm. Our studies have shown that at least 10-12 slow rusting genes are involved in the adult plant resistance of CIMMYT wheats. We have also identified lines, such as Amadina (Table 8), where *Lr34* is absent, but whose level of resistance is high. We therefore believe that durable resistance is feasible even in the absence of *Lr34*. This is the case of Pavon 76 (Table 8), where we have identified a new gene *Lr46* for slow rusting in chromosome 1BL. Genes other than *Lr34* and *Lr46* have not yet been named, though work is currently underway at CIMMYT to identify their chromosomal locations, so that they may be catalogued.

Table 8. Some seedling susceptible bread wheats that carry good adult plant resistance to stripe rust in field trials in Mexico and other countries

Genotype(s)	Usual yellow rust response[1]	Additive genes[2] for resistance
Jupateco 73S	100MS	Moderately susceptible
Jupatco 73R	50M	*Yr18*
Parula, Cook, Trap	15M	*Yr18* + 2 genes
Tonichi 81, Sonoita 81, Yaco	10M	*Yr18* + 2 or 3 genes
Chapio, Tukuru, Kukuna, Vivitsi	1M	*Yr18* + 3 or 4 genes
Amadina	30M	3 genes
Pavon 76, Attila	20M	3 genes

[1]Yellow rust response data from Mexico has two components, % severity based on modified Cobb scale (Peterson *et al.*, 1948) and reaction based on Roelfs *et al.*, (1992). The reactions are M = moderately resistant to moderately susceptible, sporulating stripes with necrosis and chlorosis; and S = sporulating stripes without chlorosis or necrosis; [2]Minimum number estimated from genetic analysis.

Yr18 and Other Minor Genes for Durable Resistance to Stripe Rust

In recent studies, Singh (1992b) and McIntosh (1992) have indicated that the moderate level of durable adult plant resistance to stripe rust of the CIMMYT-derived US wheat cultivar Anza and winter wheats such as Bezostaja is controlled in part by the *Yr18* gene. This gene is completely linked to the *Lr34* gene. The level of resistance it confers is usually not adequate when present alone. However, combinations of *Yr18* and 2-4 additional slow rusting genes result in adequate resistance levels in most environments (Singh and Rajaram, 1994). Cultivars carrying such *Yr18* complexes are listed in Table 8. Genes *Lr34* and *Yr18* occur frequently in germplasm developed at CIMMYT and in various countries. Using Jupateco 73 near-isogenic reselections, studies at CIMMYT have shown that the gene *Yr18* also increases latent period while decreasing infection frequency and length of infection lesions (stripes) to stripe rust in greenhouse experiments (Table 9). The conclusion again was that these components were under pleiotropic genetic control.

Table 9. Comparison of the three components of slow rusting resistance to stripe rust in seedling and flag leaves of near-isogenic *Yr18* Jupateco 73 reselections tested at 15 °C

Genotype	Latent period (Days)	Infection frequency (stripes/cm^2)	Length of stripe (mm)
Jupateco +*Yr18*	20.1	0.7	12.5
Jupateco -*Yr18*	15.9	7.1	47.7

Because stripe rust can develop systemically, it is different from the other two rusts, where every new pustule develops from a new infection. The epidemiology of stripe rust is also different from that of the other two rusts. Johnson (1988) presented examples of adult plant resistance genes that are race-specific in nature. It is difficult to distinguish such resistance from the resistance conferred by genes of race-nonspecific nature based on the adult plant infection type. Low disease severity to stripe rust is most often associated with at least some reduction in infection type. However, we have observed that in the case of potentially durable slow rusting resistance, the first uredinia to appear are moderately susceptible to susceptible. Subsequent growth of fungal mycelium causes some chlorosis and necrosis; therefore, the final infection type is usually rated as moderately resistant-moderately susceptible. Durability of such resistance can be expected if the cultivar's low disease severity is due to the additive interaction of several (4-5) partially effective genes.

Genetic Linkage/Pleiotropism of Resistance Genes Involved in Slow Rusting to Different Rust Pathogens

The genetic linkage between slow rusting genes *Lr34* and *Yr18* was described earlier. More recently it has been shown that durable stem rust resistance gene *Sr2* is closely linked to minor gene *Yr30* conferring yellow rust resistance (Singh *et al.*, 2000). Quantitative trait locus (QTL) analysis of slow rusting resistance to leaf rust and yellow rust in two recombinant inbred populations at CIMMYT has shown that several QTLs confer resistance to both leaf and yellow rust (Table 10). As shown in Table 10, disease specific QTLs were also present for both leaf and yellow rusts, indicating that close genetic linkage or pleiotropism is not a rule. Slow rusting leaf rust resistance gene *Lr46* was linked to a gene for slow rusting yellow rust resistance, recently designated as *Yr29*. Functional aspects of slow rusting genes may be better understood once they are cloned. Because the same, or closely linked, minor slow rusting genes confer resistance to more than one rust disease, generating multiple rust resistance germplasm should be simpler than previously thought.

Table 10. QTLs for slow rusting, additive genes involved in resistance to leaf and yellow rust diseases of wheat mapped by evaluating RILs from crosses of susceptible wheat 'Avocet S' and resistant 'Pavon 76' and 'Parula' for three years at field sites in Mexico

			Disease severity reduction (%)		
Cultivar	Location	Marker	Leaf rust	Yellow rust	Named genes
Pavon 76	1BL	*Wms259*	35	27	*Lr46,Yr29*
	4B	*Wms495*	18	15	
	6A	*Wms356*	14	18	
	6B	*PaggMcaa*	-	18	
	3BS	*PacgMcgt*	-	11	*Yr18,Sr2*
Parula	7DS	*Ltn*[1]	56	46	*Lr34,Yr18*
	7B or 7D	*Pcr156*	29	-	
	1BL	*Wms259*	15	16	*Lr46,Yr29*
	Unknown	*PaagMcta*	22	14	

[1]Leaf tip necrosis, a morphological marker linked to gene *Lr34*.

Crossing and Selection Procedures Employed to Achieve Resistance Based on Additive Interactions of Slow Rusting Genes

It is often thought that selection for resistance based on additive minor genes is difficult. However, at CIMMYT certain measures aimed at enhancing the accumulation of such genes are being taken. These measures are as follows:

(*i*) Selection of parents that lack effective major genes and have moderate to good levels of slow rusting resistance to local rust pathotypes. Such parents are easily identified by testing them at the seedling stage in the greenhouse and as adult plants in the field using the same pathotype. Parents of interest should show susceptibility at the seedling stage and slow rusting in the field. Cultivars known to have durable resistance are also included.

(*ii*) Maintenance of genetic diversity. Parents having different sets of additive genes based on available information are used in crossing. If such information is not available, parents of diverse origins or diverse pedigrees are selected for crosses.

(*iii*) Establishment of high disease pressure in the breeding nursery with chosen rust pathotypes. Spreader rows are planted at optimum distance and artificially inoculated to ensure homogeneous disease spread of desired rust pathotypes in the plot. Susceptible and slow rusting checks are included to assess disease pressure.

(*iv*) Selection of plants with low to moderate terminal disease severity in F_2 and F_3; and from F_4 onwards, selection of plants or lines with low terminal severity. Because adequate resistance levels require the presence of 3-5 additive genes, the level of homozygosity from the F_4 generation onwards is usually sufficient to identify plants or lines that combine adequate resistance with good agronomic features. Moreover, selecting plants with low terminal disease severity under high disease pressure means that more additive genes may be present in those plants.

(*v*) Maintenance of leaf tip necrosis or mild pseudo-black chaff phenotypes. Because leaf tip necrosis is linked to durable resistance genes *Lr34* and *Yr18*, and pseudo black chaff is linked to *Sr2*, these traits are useful morphological markers.

(*vi*) Conduct multilocational testing. As discussed earlier, multilocational testing of useful advanced lines can indicate the effectiveness and stability of resistance across environments. Based on the results, new lines are identified for future crossing.

(*vii*) Genetic analyses of selected lines. To confirm the presence of resistance based on additive genes, important lines are genetically analysed.

Following the methodology described above, we have successfully combined high levels of resistance (comparable to near-immunity) to leaf and yellow rusts with high grain yield potential in wheat lines such as Chapio, Tukuru, Kukuna and Vivitsi (Tables 5 and 8) (Singh *et al.*, 2000). Genetic analysis of such resistance has shown that at least 4 or 5 minor, additive genes conferred resistance to both leaf and yellow rusts. These wheat lines could be released directly for cultivation or be used in future breeding programs.

Resistance to Karnal Bunt

The search for resistance to Karnal bunt (caused by *Tilletia indica* Mitra) at CIMMYT began in the early 1980s. Some bread wheat cultivars were reported to be resistant in India. An extensive search for resistance at CIMMYT has identified four principal sources of resistance: Indian, Chinese, and Brazilian wheats, and synthetic wheats produced at CIMMYT. Resistance appears to be based on a few partially dominant or partially recessive genes and is additive (Fuentes-Davila *et al.*, 1995). Synthetic wheats have derived their resistance from both the *T. durum* and *T. tauschii* parents, and some have shown immunity to the pathogen in repeated tests. It is interesting to note that Karnal bunt does not occur in China or Brazil, but germplasm developed for scab resistance was found to carry Karnal bunt resistance as well. Several high yielding advanced lines are now available, which carry good levels of Karnal bunt resistance. Since Karnal bunt resistance is known to be long-lasting, any discussion on the type of resistance is less pertinent.

Resistance to *Septoria tritici* Blotch

Breeding for resistance to *Septoria tritici* blotch (caused by *Septoria tritici*) at CIMMYT started in the early 1970s. The susceptibility of early Green Revolution semid-warf cultivars became evident when they were grown in North Africa. Steady progress has been made since then. Currently, several high yielding semi-dwarf wheats with good blotch resistance are available. Resistance in these wheats is derived from Argentinean, Brazilian, Russian, West European and Chinese sources. The main problems encountered in early breeding work were to break the association of resistance with lateness and tallness present in the above sources. Two high rainfall sites, Toluca (Mexico State) and Patzcuaro (Michoacan State), are used in Mexico for *Septoria tritici* resistance breeding.

Some high yielding, semi-dwarf resistant lines are Milan (resistance derived from a French source), Corydon (Brazilian source), Catbird (Chinese source), and Bobwhite (Russian source). Efforts are being made to combine these resistances. Genetic studies conducted on CIMMYT wheats indicate that between five to eight genes are operating in resistance to septoria tritici blotch depending on the source population (Briceno, 1992; Jlibene *et al.*, 1992; Matus-Tejos, 1993). From two to three genes are generally needed to confer acceptable levels of resistance, and genes have predominantly additive effects. The selection methodology, therefore, is similar to that described for combining minor, additive genes for resistance to leaf and yellow rusts.

Some synthetic wheats (*T. turgidum*/*T. tauschii*) developed at CIMMYT have shown excellent resistance that appears to be leading towards immunity to the disease. These sources offer new genetic diversity for resistance coming from durum wheat and/or *T. tauschii*. Attempts are being made to transfer this resistance to semi-dwarf wheats and to combine it with other sources currently present in semi-dwarf wheats.

Resistance to *Fusarium* Head Scab

Scab, caused by *Fusarium* spp., is a major production constraint in the humid and semi-humid wheat areas of subtropical countries. *Fusarium graminearum* Schwabe [perfect stage *Gibberella zeae* (Schw.) Petch] predominates in wheat growing areas of China and North and South America. CIMMYT has been involved in breeding for resistance to this disease since 1985. Sources of scab resistance have been divided into three groups according to their geographic origin: China and Japan, Argentina and Brazil, and Eastern Europe.

Progress in breeding for scab resistance has been largely due to the recent expansion of collaboration between Chinese and CIMMYT scientists. The dual objective was to introduce high-yield genes from new CIMMYT materials into resistant Chinese materials for those areas in China where direct CIMMYT materials had not adapted well. Most foreign germplasm introductions into China occurred in the 1950s and 1960s, and more recent materials had not been used in the Chinese breeding programs. In addition the aim was to utilise Chinese scab resistance in CIMMYT's global breeding program. In the past 10 years, Chinese researchers have shared with us more than 500 cultivars and advanced lines, many of them carrying high levels of scab resistance. More than 10,000 materials have been distributed from CIMMYT, either in the form of regular International Nurseries or materials selected by Chinese scientists in CIMMYT's fields in Mexico. The Chinese cultivars that best combined with CIMMYT materials to transmit scab resistance are Sumai#3, Ning7840, Shanghai#5, Yangmai#6, Suzhoe#6, Wuhan#3 and Chuangmai 18. The following new germplasm for Chinese agriculture has evolved from the joint shuttle breeding effort:

- Sichuan Province: Chuanmai 25 (SW2089; Genaro 81 cross); SW5193, SW89-1862 (Veery cross); SW89-5422 (Alondra cross); SW90-1648 (Seri 82 cross); Catbird (Bagula cross made at CIMMYT).
- Jiangsu Province: Ningmai 7 (Shanghai 4-23B-0Y); Ningmai 8, Ning 9338 (Yangmai 158/Kauz); Ning 9341 (Yang87-158/Fasan); Ning8675/Catbird; Ning 9350 (Catbird); Ning 9415 (Shanghai 7//Parula/Veery#6).
- Heilongjiang Province: Longmai 19, Ke92-779 (Roller cross).

Presently, throughout China 5-7 million hectares are cultivated to new varieties carrying CIMMYT germplasm in their pedigrees, representing about 25 % of the total Chinese wheat area.

Chinese sources are probably the best currently available and should be combined with other sources of resistance. Our genetic analyses indicate that a few additive genes confer resistance in Chinese and Brazilian wheats, and that genes present in Chinese sources are different from those in Brazilian materials (Singh *et al.*, 1995; Van Ginkel *et al.*, 1996). Some synthetic wheats have recently been identified whose moderate resistance must be derived from *T. tauschii*, as the *T. turgidum* parents used in generating the synthetics are highly susceptible (Gilchrist *et al.*, 1997). These sources should add new genetic diversity, which is crucial to enhance resistance levels currently present in hexaploid wheat. Because genes for scab resistance are additive, a careful crossing and selection scheme should allow combinations of several genes leading to high levels of resistance and reduced accumulation of fusarium toxins in the grain.

Resistance to Spot Blotch and Tan Spot

The first crosses to incorporate spot blotch (caused by *Bipolaris sorokiniana*) resistance into CIMMYT wheats were made about 20 years ago. These crosses involved moderately resistant cultivars, such as BH1146 from Brazil. However, the level of resistance in progenies was inadequate when tests were carried out in Poza Rica, Mexico, CIMMYT's ME5 testing site. In the mid-1980s, wheat genotypes carrying scab resistance and obtained from the Yangtze River Valley of China showed varying levels of spot blotch resistance when tested in Poza Rica. These Chinese lines included Suzhoe 1 to 10, Wuhan 1 to 3, Shanghai 1 to 8, and certain Ningmai and Yangmai lines. About the same time, the wide crossing program at CIMMYT produced resistant lines that contain *Thinopyrum curvifolium* in their pedigree (Villareal *et al.*, 1995). Some of these lines and their derivatives are showing good resistance and appear promising in Bangladesh, lowland Bolivia, and Nepal. Resistance in wheats such as Sabuf, Chyria 1 and Cugap appears to be controlled by two to three genes (Velazquez-Cruz, 1994), whereas Longmai 10 and Yangmai 6 may carry polygenic resistance with high narrow-sense heritabilities (Sharma *et al.*, unpublished data).

A few synthetic wheats developed at CIMMYT also carry resistance derived from *T. tauschii* accessions. A key problem in selecting for spot blotch resistance is the negative correlation of disease severity with heading date and plant height (Duveiller and Gilchrist, 1994). Therefore, care must be taken if short types with early maturity are required. The current strategy followed at CIMMYT is to combine resistances from these diverse sources. Identification of highly resistant lines from such crosses indicate that resistance is additive.

Tan spot (caused by *Drechslera tritici-repentis*) resistance is not widely dispersed in CIMMYT germplasm, but moderate resistance is known to occur (Rees and Platz, 1992). Newer CIMMYT lines such as Milan, Attila, Corydon, Tinamou, and some Chinese wheats and their derivatives, such as Luan, are also reported to carry high to moderate resistance (Diaz de Ackermann and Kohli, 1998). Tan spot is increasing in areas where reduced tillage practices are being combined with stubble retention. CIMMYT has an ongoing project to search for new and better sources of resistance to tan spot for these areas.

Tolerance and Resistance to BYDV

Tolerance to barley yellow dwarf virus in cultivar Anza and several other CIMMYT wheats is due to the gene *Bdv1* (Singh *et al.*, 1993), which is linked to durable leaf and stripe rust resistance genes *Lr34* and *Yr18* (Singh, 1993). Presence of this gene does not reduce virus titres but does cause slow yellowing of plants. Gene *Bdv1* is widespread in CIMMYT wheats because *Lr34* and *Yr18* occur in a large number of CIMMYT wheats. CIMMYT's highland field location at Toluca has endemic presence of BYDV. CIMMYT lines such as Milan show much higher levels of tolerance/resistance than Anza and are likely to carry genes in addition to *Bdv1*. Recent studies at CIMMYT show that several QTLs are involved in the tolerance mechanism (L. Ayala, unpublished data). Wheat lines developed through Australian-Chinese collaboration and carrying a chromosome 7DL translocation from *Thinopyrum intermedium* have shown true resistance to BYDV (lower virus titres). However, these sources show high symptoms in Mexico despite their low titres. CIMMYT's program is attempting to combine this resistance with other tolerance genes and with high yield potential.

Priorities for the 21st Century: The Next 25 Years

We believe that much has been achieved in the last 100 years in terms of an improved understanding of epidemiology, disease resistance concepts, and host-pathogen interactions. To a certain extent, this knowledge has been applied towards achieving better crop protection and more efficient productivity. However, the degree to which it has been applied towards achieving science-based durable disease resistance has been less than satisfactory and mostly sporadic. The challenge in the next 25 years should be to make our food crops durably resistant to diseases and pests. Though this would still be based on knowledge acquired through conventional plant breeding, quantitative genetics and epidemiology, recent advances in molecular biology may contribute to achieving better resistance to diseases and pests.

We visualise that several cultivars grown in the coming years will be protected against diseases through the utilisation of race-specific, major genes. Strong regional collaboration in monitoring pathogenic virulence should make it possible to give early warning of forthcoming disease epidemics and, consequently, the government will have enough time to reduce or replace vulnerable cultivars.

Issues related to sharing plant genetic resources and proprietary rights must be fully defined to allow all breeding programs access to global germplasm. It would not be fair to seek access to outside germplasm and yet restrict the use of one's own germplasm to outsiders. We must remember that the Green Revolution would not have occurred in India, or would have been delayed, if the current restrictive regulatory guidelines for germplasm exchange had been in place in the 1960s and 1970s, when massive amounts of wheat seed were imported from Mexico. We would like to urge caution when defining germplasm for breeding programs, which should be freely available to all, and a crop variety suitable for farmers and market needs, to which proprietary rights could be sought. We should stress that in the latter case, the genes in the variety should still be available to all for breeding purposes.

Infusion of advanced technologies through a well defined human resource development program is important to continually update scientists at different levels of their careers. Opportunities for all scientists to be able to travel abroad and work together with scientists of advanced research institutions in their respective disciplines are essential.

References

Alvarez-Zamorano R. 1995. Patogenesis de *Puccinia recondita* Rob. ex Desm. f. sp. *tritici* y la resistencia en trigo. Ph.D. thesis. Colegio Postgrad., Montecillos, Mexico. 76 pp.

Ausemus E.R., Stakman E.C., Hanson E.W., Geddes W.F. and Merrit P.P. 1944. Newthatch wheat. *Minn. Agric. Exp. Stn., Tech. Bull.*, 166.

Biffen R.H. 1905. Mendel's law of inheritance and wheat breeding. *J. Agric. Sci.*, **1**: 4-48.

Biffen R.H. 1907. Studies in the inheritance of disease resistance. *J. Agric. Sci.*, **2**: 109-128.

Borlaug N.E. 1953. New approach to the breeding of wheat varieties resistant to *Puccinia graminis tritici*. *Phytopathology*, **43**: 467 (Abstr.).

Briceno F.G. 1992. Inheritance of resistance to *Septoria* leaf blotch in selected spring bread wheat genotypes. MS thesis. Oregon State Univ., Corvallis, USA. 86 pp.

Caldwell R.M. 1968. Breeding for general and/or specific plant disease resistance. *In*: *Proceedings of the 3rd Int. Wheat Genetics Symp.*, (eds.) K.W. Finlay and K.W. Shephard, p. 263-272. Australian Academy of Sciences, Canberra, Australia.

Day P.R. 1974. *Genetics of Host-Parasite Interactions.* Freeman, San Fransisco.

Diaz de Ackermann M. and Kohli M.M. 1998. Research on *Pyrenophora tritici-repentis* tan spot of wheat in Uruguay. *In*: *Helminthosporium Blights of Wheat: Spot Blotch and Tan Spot.* (eds.) E. Duveiller, H.J. Dubin, J. Reeves and A. McNab, p. 134-141. CIMMYT, Mexico.

Duveiller E. and Gilchrist L. 1994. Production constraints due to *Bipolaris sorokiniana* in wheat: current situation and future prospects. *In*: *Wheat in Heat Stressed Environments: Irrigated, Dry Areas and Rice-Wheat Farming Systems.* (eds.) D.A. Saunders and G.P. Hettel, p. 343-352. CIMMYT, Mexico.

Dyck P.L. 1987. The association of a gene for leaf rust resistance with the chromosome 7D suppressor of stem rust resistance in common wheat. *Genome*, **29**: 467-469.

Flor H.H. 1942. Inheritance of pathogenicity in *Melampsora lini*. *Phytopathology*, **32**: 653-669.

Flor H.H. 1947. Inheritance of reaction to rust in flax. *J. Agric. Sci.*, **74**: 241-262.

Flor H.H. 1956. The complementary genic systems in flax and flax rust. *Adv. Genet.*, **8**: 29-54.

Fuentes-Davila G., Rajaram S. and Singh G. 1995. Inheritance of resistance to Karnal bunt (*Tilletia indica* Mitra) in bread wheat (*Triticum aestivum* L.). *Plant Breeding*, **114**: 250-252.

Gilchrist L., Rajaram S., Mujeeb-Kazi A., Van Ginkel M., Vivar H. and Pfeiffer W. 1997. *Fusarium* scab screening program at CIMMYT. *In*: *Fusarium Head Scab: Global Status and Future Prospects.* (eds.) H.J. Dubin, L. Gilchrist, J. Reeves, A. McNab, CIMMYT, Mexico.

Harrer J.G., Loegering W.Q. and Stakman E.C. 1944. Relationship of physiologic races of *Puccinia graminis tritici* to wheat improvement in southern Mexico. *Phytopathology*, **34**: 1002 (Abstr.).

Hayes H.K., Parker J.H. and Kurtzweil C. 1920. Genetics of rust resistance in crosses of varieties of *Triticum vulgare* with varieties of *T. durum* and *T. dicoccum*. *J. Agric. Res.*, **19**: 523-542.

Hayes H.K., Ausemus E.R., Stakman E.C., Bailey C.H., Wilson H.K., Bamberg R.H., Markley M.C., Crim R.F. and Levine M.N. 1936. Thatcher wheat. *S. Bull. Minn., Agric. Exp. Stn.*, 325.

Heisey P.W., Lantican M.A. and Dubin H.J. 1999. Assessing the benefits of international wheat breeding research: An overview of the global wheat impacts study. *In*: *World Wheat Facts and Trends. Global Wheat Research in a Changing World: Challenges and Achievements.* (ed.) P. L. Pingali, Mexico, D.F.: CIMMYT, 1998-99. Pp 19-26.

Jensen N.F. 1952. Intervarietal diversification in oat breeding. *Agron. J.*, **44**: 30-34.

Jlibene M., Gustafson J.P. and Rajaram S. 1992. A field disease evaluation method for selecting wheat resistant to *Mycosphaerella graminicola. Plant Breed.*, **108**: 26-32.

Johnson R. 1978. Practial breeding for durable resistance to rust diseases in self pollinating cereals. *Euphytica*, **27**: 529-540.

Johnson R. 1988. Durable resistance to yellow (stripe) rust in wheat and its implications in plant breeding. *In*: *Breeding Strategies for Resistance to the Rusts of Wheat*, (eds.) N.W. Simmonds and S. Rajaram, p. 63-75. CIMMYT, Mexico.

Johnston C.O. and Browder L.E. 1966. Seventh revision of the international register of physiologic races of *Puccinia recondita* f. sp. *tritici. Plant Dis. Reporter*, **50**: 756-760.

Knott D.R. 1988. Using polygenic resistance to breed for stem rust resistance in wheat. *In*: *Breeding Strategies for Resistance to the Rusts of Wheat*, (eds.) N.W. Simmonds and S. Rajaram, p 39-47, CIMMYT, Mexico.

Lehman J.S. and Shaner G. 1996. Genetic variation in latent period among isolates of *Puccinia recondita* f. sp. *tritici* on partially resistant wheat cultivars. *Phytopathology*, **86**: 633-641.

Loegering W.Q. and Powers H.R. 1962. Inheritance of pathogenicity in a cross of physiological races 111 and 36 of *Puccinia graminis* f. sp. *tritici. Phytopathology*, **52**: 547-554.

Luig N.H. and Watson I.A. 1961. A study of inheritance of pathogenicity in *Puccinia graminis* var. *tritici. Proc. Linn. Soc. N.S.W.*, **86**: 217-229.

Mains E.B. and Jackson H.S. 1926. Physiologic specialization in the leaf rust of wheat, *Puccinia triticina* Erikss. *Phytopathology*, **16**: 89-120.

Matus-Tejos I.A. 1993. Genetica de la resistencia a *Septoria tritici* en trigos harineros. M.S. thesis. Colegio Postgrad., Montecillos, Mexico. 82 pp.

McIntosh R.A. 1992. Close genetic linkage of genes conferring adult-plant resistance to leaf rust and stripe rust in wheat. *Plant Pathol.*, **41**: 523-527.

McIntosh R.A., Hart G.E., Devos K.M., Gale M.D. and Rogers W.J. 1998. Catalogue of gene symbols for wheat. (ed.) A. E. Slinkard, *Proc. 9th Int. Wheat Genetics Symp.*, 2-7 Aug. 1998, Saskatoon, Canada. Vol., **5**: 1-235.

McIntosh R.A. and Smith-White S. Irvine Armstrong Watson 1914-1986. 1986. *Historical Records of Australian Science*, **7**: 405-415.

McFadden E.S. 1930. A successful transfer of emmer characteristics to *vulgare* wheat. *J. Amer. Soc. Agron.*, **22**: 1020-1034.

Mehta K.C. 1929. Annual recurrence of rusts on wheat in India. (Presidential Address). *Proc. Indian Sci. Congr.*, 16, pp. 199-223.

Mehta K.C. 1933. Rusts of wheat and barley in India. A study of their annual recurrence, life-histories and physiologic forms. *Indian J. Agric. Sci.*, **3**: 939-962.

Niederhauser J.S., Cervantes J. and Servin L. 1954. Late blight in Mexico and its implications. *Phytopathology*, **44**: 406-408.

Parlevliet J.E. 1975. Partial resistance of barley to leaf rust, *Puccinia hordei*. I. Effect of cultivar and development stage on latent period. *Euphytica*, **24**: 21-27.

Parlevliet J.E. 1976. Partial resistance of barley to leaf rust, *Puccinia hordei*. III. The inheritance of the host plant effect on latent period in four cultivars. *Euphytica*, **25**: 241-248.

Parlevliet J.E. 1986. Pleiotropic association of infection frequency and latent period of two barley cultivars partially resistant to barley leaf rust. *Euphytica*, **35**: 267-272.

Person C. 1959. Gene-for-gene relationship in host:parasite systems. *Can. J. Bot.*, **37**: 1101-1130.

Peterson R.F., Campbell A.B. and Hannah A.E. 1948. A diagrammatic scale for estimating rust intensity of leaves and stem of cereals. *Can. J. Res. Sect. C.*, **26**: 496-500.

Rajaram S., Singh R.P. and Torres E. 1988. Current CIMMYT approaches in breeding wheat for rust resistance. *In*: *Breeding Strategies for Resistance to the Rust of Wheat.* (eds.) N.W Simmonds and S. Rajaram, CIMMYT, D.F.: Mexico. Pp 101-118.

Rees R.G. and Platz G.J. 1992. Tan spot and its control - some Australian experiments. *In*: *Advances in Tan Spot Research.* (eds.) L.J. Francl, J.M. Krupinsky and M.P. Mc Mullen, Proc. of the 2nd Int. Tan Spot Workshop, Fargo, ND, USA: North Dakota Agri. Expt. St. p 1.

Roelfs A.P. 1988. Resistance to leaf rust and stem rust in wheat. *In*: *Breeding strategies for Resistance to the Rusts of Wheat*, (eds.) Simmonds N.W. and S. Rajaram, p. 10-22. CIMMYT, Mexico.

Roelfs A.P. and Martens J.W. 1988. An interrnational system of nomenclature for *Puccinia graminis* f. sp. *tritici*. *Phytopathology*, **78**: 526-533.

Roelfs A.P., Singh R.P. and Saari E.E. 1992. *Rust diseases of wheat: Concepts and methods of disease management.* CIMMYT, Mexico, D.F. 81pp.

Rubiales D. and Niks R.E. 1995. Characterisation of *Lr34*, a major gene conferring nonhypersensitive resistance to wheat leaf rust. *Plant Dis.*, **79**: 1208-1212.

Samborski D.J. and Dyck P.L. 1968. Inheritance of virulence in wheat leaf rust on the standard differential wheat varieties. *Can. J. Genet. Cytol.*, **10**: 24-32.

Singh R.P. 1992a. Association between gene *Lr34* for leaf rust resistance and leaf tip necrosis in wheat. *Crop Sci.*, **32**: 874-878.

Singh R.P. 1992b. Genetic association of leaf rust resistance gene *Lr34* with adult plant resistance to stripe rust in bread wheat. *Phytopathology*, **82**: 835-838.

Singh R.P. 1993. Genetic association of gene *Bdv1* for tolerance to barley yellow dwarf virus with genes *Lr34* and *Yr18* for adult plant resistance to rusts in bread wheat. *Plant Dis.*, **77**: 1103-1106.

Singh R.P., Burnett P.A., Albarran M. and Rajaram S. 1993. *Bdv1*: a gene for tolerance to barley dwarf virus in bread wheat. *Crop Sci.*, **33**: 231-234.

Singh R.P. and Dubin H. J. 1997. Sustainable control of wheat diseases in Mexico. *In*: *Memorias de 1er Simposio Internacional de Trigo* (Proc. 1st International Wheat Symposium, 7-9 April, 1997, Cd. Obregon, Sonora, Mexico.

Singh R.P. and Gupta A.K. 1992. Expression of wheat leaf rust resistance gene *Lr34* in seedlings and adult plants. *Plant Dis.*, **76**: 489-491.

Singh R.P., Ma H. and Rajaram S. 1995. Genetic analysis of resistance to scab in spring wheat cultivar Frontana. *Plant Dis.*, **79**: 238-240.

Singh R.P., Nelson J.C. and Sorrells M.E. 2000. Mapping *Yr28* and other genes for resistance to stripe rust in wheat. *Crop Sci.*, **40**: 1148-1155.

Singh R.P., Payne T.S. and Rajaram S. 1991. Characterisation of variability and relationship among components of partial resistance to leaf rust in CIMMYT bread wheats. *Theor. Appl. Genet.*, **82**: 674-680.

Singh R.P. and Rajaram S. 1991. Resistance to *Puccinia recondita* f. sp. *tritici* in 50 Mexican bread wheat cultivars. *Crop Sci.*, **31**: 1472-1479.

Singh R.P. and Rajaram S. 1992. Genetics of adult-plant resistance to leaf rust in 'Frontana' and three CIMMYT wheats. *Genome*, **35**: 24-31.

Singh R .P. and Rajaram S. 1994. Genetics of adult plant resistance to stripe rust in ten spring bread wheats. *Euphytica*, **72**: 1-7.

Singh R.P., Huerta-Espino J. and Rajaram S. 2000. Achieving near-immunity to leaf and stripe rusts in wheat by combining slow rusting resistance genes. *Acta Phytopathlogica Hungarica*, **35**: 133-139.

Stakman E.C. and Harrer J.G. 1957. *Principles of Plant Pathology*. Ronald Press, New York.

Stakman E.C. and Levine M.N. 1922. The determination of biologic forms of *Puccinia graminis* on *Triticum* spp. *Minn. Agr. Exp. Sta. Tech. Bul.*, 8.

Stakman E.C., Stewart D.M. and Loegering W.Q. 1962. Identification of physiologic races of *Puccinia graminis* var. *tritici*. U.S. *Agric. Res. Serv.*, E-617 (rev.).

Stakman E.C., Levine M.N., Cotter R.U. and Hines L. 1934. Relation of barberry to the origin and persistence of physiologic forms of *Puccinia graminis*. *J. Agric. Res.*, **48**: 953-969.

Stakman E.C., Levine M.N. and Wallace J.M. 1929. The value of physiologic form surveys in the study of the epidemiology of stem rust. *Phytopathology*, **19**: 951-959.

Stakman E.C. and Piemeisel F.J. 1917. A new strain of *Puccinia graminis*. *Phytopathology*, **7**: 73 (Abstr.)

Vanderplank J.E. 1963. *Plant Diseases: Epidemics and Control*. Academic Press New York and London.

Van Ginkel M., Van der Schaar W. and Zhuping Y. 1996. Inheritance of resistance of scab in two wheat cultivars from Brazil and China. *Plant Dis.*, **80**: 863-867.

Velazquez-Cruz C. 1994. Genetica de la resistencia a *Bipolaris sorokiniana* en trigos harineros. M.S. thesis. Colegio Postgrad., Montecillos, Mexico. 84 pp.

Villareal R.L., Mujeeb-Kazi A., Gilchrist L. and Del Toro E. 1995. Yield loss to spot blotch in spring bread wheat in warm non-traditional wheat production areas. *Plant Dis.*, **79**: 893-897.

Waterhouse W.L. 1929. Australian rust studies. I. *Proc. Linn. Soc. N.S.W.*, **54**: 615-680.

Waterhouse W.L. 1930. Australian rust studies. III. Initial results of breeding for rust resistance. *Proc. Linn. Soc. N.S.W.*, **55**: 596-636.

Plant Breeding - Mendelian to Molecular Approaches
H. K. Jain and M. C. Kharkwal (eds.)

8

Genetic Improvement for Abiotic Stress Responses

Anil Grover[1], Avnish Kapoor[1], D. Kumar[2], H.E. Shashidhar[3] and S. Hittalmani[3]

Abstract

Very often discussion on crop improvement against abiotic stresses employing conventional or traditional breeding and unconventional biotechnological approaches (i.e. molecular breeding and genetic engineering) is carried out at separate platforms. The use of molecular markers and transgenic techniques are just the means to an end of using breeding methods for production of stress tolerant crops. We include both the conventional and unconventional tools of crop improvement under the same umbrella in this paper. The conventional breeding approach takes a longer time to yield results. The selection of stress tolerance through use of DNA markers can enormously help in marker-aided selection. Genetic engineering can make a small but definite contribution to the breeding program. The recent upsurge in the science of structural genomics is leading to an accumulation of a huge wealth of literature on nucleotide sequences. Functional genomics is the science of understanding how the genome works through a control on the expression of genes. The analysis of proteins through proteomics approach is the most direct approach to define the gene function. Stress biotechnology research looks at microarray chip technology and proteomics research with great deal of optimism. The stress biotechnology research is limited by the non-availability of useful genes that would lead to desired genetic change. Therefore, any method that may enable identification, isolation and cloning of novel abiotic stress responsive genes will be a positive step.

Abiotic Stress and Crop Yield

Abiotic stresses are caused by several factors including supra-optimal (high temperature stress) or sub-optimal temperatures (low temperature stress, cold stress or freezing stress), excess of water (submergence stress, water-logging stress or flooding stress) or water deficit (drought stress or water stress or desiccation stress), increased salt levels (salt stress), increased chemicals (chemical stress such as metal stress, pH stress), increased light incidence (light stress) and increased levels of pollutants (pollution stress). The

[1]University of Delhi South Campus, New Delhi 110021, India
[2]Central Arid Zone Research Institute, Jodhpur 342 003, India
[3]University of Agricultural Sciences, Bangalore 560 065, India

oxidative stress results when conditions of water stress, salt stress or temperature stress are accompanied by high light stress. Often the stress conditions are compounded in the farmer's field due to the occurrence of more than one stress at the same time.

The relative damage due to abiotic stress is variable, depending upon the crop and the environment. Sub-marginal and marginal agricultural lands, characteristically depending on rainwaters for cultivation, may be frequently exposed to stringent drought situations and drought associated strains. For instance, below average rains resulting in above average heat may considerably affect the crop productivity in rainy as well as in winter season in the entire north-west India, as being practically experienced year after year. The arid and semi-arid regions are drastically influenced by the soil moisture deficit situations. Out of the total of 1474 m ha arable lands world over, unirrigated rainfed area of 1247 m ha constitutes 84 %.

Drought in any part of the world may considerably fluctuate price structure. For instance, drought in North America reduced world cereal supplies in 1988 and led to higher world price: soft wheat price rose from around \$140 t^{-1} in 1987 to \$200 t^{-1} in mid-June of 1988 and maize price rose from around \$140 t^{-1} in the beginning of 1988 to almost \$160 t^{-1} at the end of the year.

A temperature higher than the optimum decreases the growth rate, maturity period and the metabolic processes, resulting in reduced plant productivity. High temperature stress is a major cause for unstable and low yields of nearly 1 t ha^{-1}, which is far less than the potential yield of 7-8 t ha^{-1} in peanuts and soybeans, and 4 t ha^{-1} in chickpea and pigeonpea. Similarly, wheat yield in northern India is reduced by almost 4 % for every 1 ^{o}C rise in mean temperature over the range 12.2 -27.5 ^{o}C. *Brassica* species, if planted at more than 30 ^{o}C mean temperature, particularly in arid zone, may show a high degree of seedling mortality, leading to sub-optimal and patchy plant stand.

Low temperature affects agricultural crops to a considerable extent in many parts of the world. Though generally mild in northern India, cold shocks cause extensive damage to fruits, vegetables, food crops, ornamental and industrial crops. Magnitude of damage, however, depends on the crop tolerance potential, severity of the frost, wind velocity and certain management practices and natural factors like air humidity, snow, soil conductivity, water reservoirs, etc. In general, the frosts in the northern plains are fewer in December, common in January (especially in second half) and are rare after mid-February. Among the leguminous crops, field pea, chickpea and soybean are mainly damaged by cool climate in India, Mediterranean regions, North Africa and South Europe. The oilseed *Brassica* and potato are susceptible to frost and a single night of frost can ruin them entirely in Northern plains and hilly regions.

Significant progress has been made in development of crops tolerant to biotic stress factors. This has been achieved both by conventional as well as unconventional (through transgenic technology and using molecular markers) ways. However, breeding against biotic factors is a continuous battle, as the pathogens/pests keep evolving. As the factors contributing to abiotic stresses are not subjected to any change due to the evolutionary [H. El] pressure, resistance bred against abiotic stress factor is going to be of stable nature. However, this positive statement about abiotic stress factors is associated with several other negative factors. Different abiotic stress factors mostly operate together under natural conditions. The response of plants to abiotic stresses is highly variable due to this compound nature of abiotic stress factors. The cell, tissue and plant responses to abiotic stress factors involve several mechanisms simultaneously. Thus, it is essentially a multi-genic response involving several biochemical and physiological attributes at the same time. In fact, the knowledge on biochemistry and physiology of plant abiotic stresses is still not completely understood. This comes in the way of using targeted approach for breeding for abiotic stress tolerance. Further, the intensity of abiotic stresses is by and large on the rise due to intensive farming being practiced world-over, or due to other man-made reasons.

On a note of optimism, there are several reports in which partial success has been achieved against abiotic stresses. We take up the issues related to development of abiotic stress tolerant crops in this paper. We will essentially summarize the different genetical approaches (i.e. conventional breeding, molecular breeding, genetic engineering and genomics) that are being followed towards improvement of major crops against abiotic stresses.

Physiological, Biochemical and Genetical Aspects of Major Abiotic Stresses

Resistance to stresses can be manifested, identified, measured and studied at cellular, tissue, organ, whole plant, or at the level of the crop community. Any response that constitutes resistance to be practically usable in breeding, needs to be manifested as an advantage in survival during stress or recovery after alleviation of stress, thus resulting in better crop stand in the field. This enhanced crop stand is likely to translate to higher grain productivity by improved harvest index. Next, we take up the different manifestations of the stress effects.

Drought

The effects of drought on crop plants are complex and variable and are greatly accentuated by a number of interacting factors. The onset of drought in general has been observed to reduce germination, emergence, hypocotyl length, water uptake and the mobilization of dry matter reserves even at thc early growth stages. At the early seedling growth stage, high root: hypocotyl length ratio for instance observed in *Brassica napus* under drought leads to effective extraction of soil moisture (Richards and Thurling, 1978). In addition to differential accumulation of carbohydrates, drought stress causes a marked reduction in leaf area. Soil moisture availability at the seed germination, anthesis and at post- anthesis growth stages may determine to what extent production potential of crops would be expected in drought-prone situations.

The yield losses in field crops may be attributed to inadequate water availability at the critical sensitive growth stages (Boyer, 1996). Crops therefore differ in their susceptibility of growth stages to water stress. Stress exposed plants immediately lower down relative water content (RWC) of their leaves; leaf water potential (ψl), osmotic potential, and turgor pressure (TP) may also be decreased unless plants have finally adapted to drought by opening of stomata again and improving the water potential. The withholding of water at different growth stages in *B. juncea* and rape revealed that stress associated reduced leaf expansion rate due to water stress is related to RWC, ψl and TP. It is indicated that low leaf expansion rate at the early stage and accelerated leaf senescence accompanied by mortality of older leaves at later stage decreased total leaf green area. Thus, improved grain yield of plants under water stress may be due to adequate soil moisture availability at the critical growth stage, uniform leaf expansion rate, and differential accumulation of carbohydrates and physiological adaptation by maintaining turgor pressure.

Drought Tolerance

Plant tolerance to soil moisture deficit is an important adaptive feature and this is achieved in some of the following ways: proportional and sequential growth retardation, diversion of carbohydrates to required portions, physiological potentiality and synthesis and accumulation of certain biochemical compounds. However, drought tolerance mechanisms are not simply confined to the above means only. Their further contribution comes from cell membrane and individual cell to whole plant, biochemically involving synthesis and accumulation of specific proteins, osmolytes and other complex processes. Morphologically, plants adapt to drought by early flowering, decreased plant height, reduction in leaf area and dry weight, and increased leaf conductance.

In the process of physiological adaptation, maintenance of TP in response to decreased cell water potential is an important mechanism. Osmotic adjustment involves accumulation of certain low molecular weight organic solutes intracellularly, which are believed to restore TP. The organic solutes (osmolytes) that generally accumulate are polyols, amino-acids (such as proline) and amino-acid derivatives (such as glycinebetaines).

At the cellular level, integrity and stabilization of cell membrane structure as well as repair mechanisms have been observed to operate in plants. It is shown that drought stress induces changes in protein pattern in root system of rape. For instance, 13 new polypeptides of low molecular weight were accumulated in the tap-root upon exposure to drought stress, 12 out of which were present in short tuber-sized roots. The reversal of these induced proteins during rehydration suggested their involvement in drought tolerance. Similarly, specific drought-induced protein BnD22 was shown to accumulate in *B. napus* leaves adapted to progressive osmotic stress induced by salt and drought stresses. It is suggested that this protein is involved in decreased protease activity in the drought-adapted leaves, thus contributing to delayed leaf senescence (Ilani *et al.*, 1997). More details on the biochemical and molecular aspects of water stress can be found in Grover (2000).

Selection Indices for Drought Tolerance

For a successful selection and evaluation of large number of germplasm lines and population or segregating materials, consideration of phenological and morphological traits showing direct correlation with high drought tolerance concomitant with high productivity are desired. Such parameters are few and those available are inconvenient, time consuming, unreliable and not universal. Of many selection indices reported by a number of workers in pulses and oilseed *Brassica*, the most common and convenient ones include fresh and dry weight of whole plant, seed germination, days-to-flowering, leaf expansion, 1000-seed weight, seed/pod, seed yield, chlorophyll stability index, stomatal conductance, osmoregulation, turgor pressure, leaf water potential, transpirational cooling, anthesis/ accumulation of solutes and specific proteins (Kumar *et al.*, 1987; Yadav *et al.*, 1990). It is suggested that leaf area extension and dry matter partitioning at the anthesis stage and early flowering by and large may prove to be useful morphological parameters, whereas maintenance of high turgor pressure and stomatal conductance and transpirational cooling are the useful physiological drought tolerance parameters. The biochemical parameters need to be further worked upon for making them time-effective and convenient indices for fast screening of large number of genetic materials.

High Temperature Stress

High temperature injury varies considerably with the growth stages, highest injury occurring at the flowering stage whereas susceptibility to injury decreases towards maturity. High temperature stress causes accelerated plant development and consequently reduces both growth and grain yield (Blum, 1988). At the cellular level, high temperature results in metabolic disturbances, depletion of respiratory substrates and reduction in photosynthetic activities. It also leads to denaturation of proteins, damage of cellular structures and inactivation of enzymes. At maturity, high temperature inhibits accumulation of starch into grain, thereby reducing crop productivity.

High temperature (35/40 °C) affects flower morphology by increasing hypanthium length and causing stigma exertion in groundnut, hence the chances of successful fertilization are reduced. The most sensitive phenological bud stage to heat stress in groundnut was 5 days before opening of flowers (Talwar *et al.*, 1999). The poor seed set in rapeseed mustard under high temperature is due to slow growth of pollen tube and its failure to fertilize, causing sterility. At the biochemical level, high temperature resulted in accumulation of amino-acids in the leaves; the most active amino-acids accumulated in groundnut were proline, glutamine, serine and arginine. Proline and glycinebetaine have protective roles on pollen germination and pollen tube growth under heat stress at 45 °C in groundnut. Heat-stressed rape has lower endogenous GA_3, IAA and zeatin contents than in control, while ABA and ethylene contents were increased significantly. Heat shock also increases lipid peroxidation and membrane damage in rape (Zhou *et al.*, 1999). Impact of heat stress may be compounded for yield loss due to drought stress, nutrient stress, diseases and insect pest infestations. High temperature induced yield losses could also be due to accelerated senescence, reduction in photosynthetic rate and increased respiration.

High Temperature Tolerance

High temperature tolerance is a complex phenomenon, because its effects are compounded in association with drought effects. However, plants can resist high temperature by way of avoidance and tolerance. Avoidance mechanism enables the plants to keep their temperature lower than the ambient temperature through transpirational cooling, reflection of solar radiation through increased leaf hairiness and wax deposition. High temperature tolerance mechanisms generally operate in situations when tissue temperature is still higher and yet the plants function. This is achieved by three ways: biomembrane saturation, synthesis of specific isozymes of enzymes such as catalase, peroxidase, superoxide dismutase and finally by protection of biomembranes, molecules, organelles and maintaining their function, where heat shock proteins (HSPs) may play a crucial role.

In specific studies, the accumulation of proline content in tolerant cultivar increased with increasing temperature in anthers and there was surge in translocation of proline from anther to pollen just before anthesis, thus pollen tube grew better in tolerant than susceptible cultivars (Ye *et al.*, 1996). Similarly, accumulation of proline in the leaves in response to high temperature appeared as a possible mechanism to suppress heat stress induced damage in Brassica and Chinese cabbage. It is indicated that heat-stable proteins may contribute to desiccation tolerance in groundnut seeds and the heat-stable polypeptides (9-15.5 kDa) at 45-50 days after planting may confer heat tolerance of groundnut storage proteins which are normally heat-labile (Yang *et al.*, 1998). Generally, high enzymatic activities in tolerant cultivars at supra-optimal temperatures may lead to heat tolerance. For instance, the activities of superoxide dismutase (SOD), catalase and ascorbic acid peroxidase (ASP) increased in tolerant Chinese cabbage but the activities of SOD and ASP decreased in susceptible cultivars under high heat stress temperatures. Morphologically, Chinese cabbage appears to tolerate high temperature by having longer and thicker leaves with thicker palisade cells, more leaves/plant, higher chlorophyll a/b ratio and early bolting.

Studies dealing with relative heat tolerance potential of legumes by testing membrane stability and photosystem (PS II) function in leaves at critical temperatures, have indicated that there are clear differences in relative injury (RI) at and above 40 °C. Exposing tissues to critical temperatures (38-48 °C for chickpea, 45-56 °C for pigeonpea and soybean and 48-58 °C for groundnut) showed distinct differences in the temperature that caused 50 % RI. Heat killing temperatures in chickpea, pigeonpea, soybean and groundnuts were 44.3°C, 49.9, 50.8 and 54 °C, respectively. Heat killing time was much longer in groundnut (139 min), than in soybean (51 min), pigeonpea (47 min) or chickpea (41 min). These food legumes therefore could be ranked from heat tolerant to sensitive in following order: groundnut, soybean, pigeonpea and chickpea. Damage to cell membrane through electrolyte leakage was less and recovery was quicker in groundnut than other legumes.

Selection Indices for High Temperature Tolerance

Plant responses to high temperature are diverse and include protein denaturation, changes in lipid composition, reduction in membrane stability, efficiency of photosynthesis, cessation of cytoplasmic streaming and reduced yield due to high degree of sterility. Thus, these responses are much diversified and involve morphological, biochemical and physiological traits depending upon day/night temperatures and their lengths and the crop species.

In view of these factors, following screening techniques are suggested against high temperature stress across the regions and crop species:

(*a*) Electrolyte leakage that reflects membrane dysfunction leading to increased permeability and leakage of electrolytes, which in turn reduces photosynthetic and mitochondrial activities and ability of plasmalemma to retain solutes and water. Specific studies have shown that heat tolerance of a cultivar can be assessed by measuring electrical conductance of the leachate.

(*b*) Electron transfer from photosystem II (PSII) is a rapid, reliable, non-destructive and a cheaper technique that can detect injury even before visible symptoms appear. Measurement of chlorophyll fluorescence can give quantitative assessment of inhibition or damage to electron transfer. This method for instance has been employed to screen potato for heat tolerance.

(*c*) Assessment of the temperature induction and high stringency lethal stress levels which involves the optimization of stress induction temperature at which the stress responsive genes are expressed.

The genetic variability in gene expression following induction temperature is shown to be responsible for acquired thermotolerance on exposure to severe lethal temperature in a population (Kumar *et al.*, 1999). Heat stress induces expression of specific gene families called heat shock genes (*hsps*) which lead to the synthesis of a new set of proteins called heat shock proteins (HSPs). These proteins are highly conserved and found nearly in all the organisms. These proteins induced within 5 min of exposure, are synthesized from newly-transcribed mRNAs. HSPs are believed to prevent accumulation of aberrant proteins induced following high temperature or other stresses. In soybean, both high and low molecular weight HSPs protected soluble proteins from heat denaturation. Repair of heat damaged/ denatured protein is essential for both survival and recovery from heat stress. The HSP 104 of yeast which is essential for survival under high temperature, functions by reactivation of heat damaged proteins. Similarly, HSP 60 from etiolated *Avena sativa* seedlings stimulated the refolding of chemically-denatured phytochrome to a photoactive form. Proteolysis of denatured proteins is a strategy used by cells to prevent accumulation of denatured proteins. Ubiquitins which target the damaged proteins for proteolysis are coded by a multigene family and show heat induction in higher plants. Overall synthesis and localization of specific HSPs trigger several important physiological and biochemical functions including the maintenance of membrane integrity and chaperoning of proteins (more details on HSPs and related aspects can be seen in Singla *et al.*, 1997; Agarwal *et al.*, 2001; Katiyar-Agarwal *et al.*, 2001).

(*d*) Field screening of germplasm can be done by growing lines in above average high temperatures and assessment of growth related parameters.

Breeding for high temperature tolerance suffers from lack of convenient and rapid selection parameters. Heat tolerant wheat Hindi 62 in Central India, where soil temperature reaches 35-40 °C at the time of sowing, has capability to germinate and exhibits high amylase activity compared to susceptible genotypes. Also at seedling stage, it has tendency for faster accumulation of HSP16.9, HSP17.3 and HSP26.4. Thus these HSPs may be important components of heat stress tolerance during germination in wheat (Viswanathan and Khanna-Chopra, 1996). Pollen germination indicates a good indicator of heat tolerance in groundnut. A surge of proline translocation towards the pollen grains is considered to be an index of heat tolerance in some studies (Talwar *et al.*, 1999). Superoxide dismutase activity was found to be more stable in heat tolerant than susceptible genotypes under high temperature in some species.

The heat killing temperature, electrolyte leakage rate and reduction in chlorophyll content have been used as physiological indices to identify heat tolerant genotypes in Chinese cabbage. The heat tolerant genotypes of Chinese cabbage contained lower amount of Malonidixldehyde (MDA) than heat sensitive genotype. Therefore MDA content in the leaves after heat stress exposure can also be taken as an indicator for heat tolerance identification. SDS gel analysis of Youchu-4, a high yielding genotype, indicated that heat storage proteins SH-1 and SH-2 were enhanced by heat or desiccation tolerance and there was association between HSPs and tolerance to heat, drought and virus resistance in Youchu-4 (Que *et al.*, 1997). Heat tolerant genotype of *B. chinensis* showed longer and thicker leaves, thicker palisade tissues, more leaves/plant, taller and narrow plants and higher chlorophyll a/b ratio. Early bolting cultivars were more heat tolerant than late bolting types. Leaf cells of tolerant chinese cabbage had more water content, high stomatal density, faster transpiration and lower degree of protein degradation than susceptible genotype and cell membrane permeability increased slightly under high temperature. The above morphological, physiological and biochemical parameters are available in different crops, which can help screen large number of germplasm; however, their feasibility and use must be standardized in respect of crop concerned and the situations.

Genetic analysis in Chinese cabbage indicated that the ability to form heads under high temperature is simply inherited. Thus, population breeding has been used to improve the yields of some heat tolerant open pollinated varieties of this crop (Opena and Kuo, 1994). However, expression of heat tolerance traits in the interspecific somatic and sexual hybrids between cabbage × Chinese cabbage and Chinese kale × Chinese cabbage was intermediate between that of parents. The relation between heat tolerance and phosphoglucomutase (PGM) dominant loci was studied with tolerant and sensitive cultivars of Chinese cabbage. Based on this study, it was concluded that PGM-2 locus could be used as a genetic marker of heat tolerance in this crop (Zheng *et al.*, 1998). A heat shock responsive gene designated BocHT-1 was isolated following differential screening of the cDNA library prepared from the cabbage plants grown at high temperature (35 °C) for 6 weeks. Sequence analysis of this gene revealed that it encoded a putative protein that was related to 22 kDa protein induced from drought and salt stress, a member of the family of proteinase inhibitors (Padmanaban *et al.*, 1999).

Low Temperature Stress

Cold climate may induce a variety of effects of adverse nature on plants. The seedling emergence rate is reduced at low temperatures. The exposure of cotyledons to low temperature were observed to delay the flowering by about 5-7 days in linseed. Low temperature may prove harmful for root extension, mean radius, sphere, area and length of root hairs. Leaf and epicotyl dry matter were less affected by low temperature, but leaf area expansion showed greatest reduction. The cold treatment also induces changes in soluble proteins and enzyme activities. The peroxidase activity was inhibited following low temperature and inhibition of this enzyme activity affected seed metabolism. During chilling, decreased values of leaf expansion rates, net CO_2 uptake rate, photosynthetic rate and stomatal conductance have been observed in chickpea. Low night temperature (10-15 °C) reduces translocation of newly-synthesized assimilates. Leguminous plants at low temperature (<10 °C) show reduced nodule formation and nitrogen fixation. The impact of chilling has been observed to be more detrimental during night. In sunflower, cold shock (5-8 °C) for 2 h dropped absorption flux rapidly while transpiration remained constant, thereby inducing water deficit.

Very young and nearly mature *Brassica* seeds are, in general, less affected by freezing than the seeds at intermediate stage of development. Results on low temperature tolerance of *Brassica* revealed that vegetative and bud initiation stages were fairly tolerant to low temperature injury (Chhabra and Ohlsson, 1990), but *Brassica* was most sensitive to freezing injury at early or middle pod development stage and hardened as the maturity progressed. Low temperature tolerance was high in sunflower varieties with short growth (Ungaro *et al.*, 1985). Since germination is delayed at low temperature, screening of sunflower genotypes in petri dishes at 3-5 °C revealed better low temperature tolerance in the varieties, showing less than 11 days to germinate. The intracellular ice formation is the major cause of freezing stress-induced death. The freezing conditions also lead to increased leaf water potential and decreased turgor pressure due to soil moisture deficit situations.

Low Temperature Tolerance

The freezing tolerance is induced by short exposure to pre- hardening sub-lethal temperature in *Brassica* at –6 to –10 °C for 18 h. The pre-hardening treatment of plants to sub- freezing temperature leading to freezing avoidance is a common phenomenon in field crops. Sub-freezing temperature related decrease in leaf water potential caused by change in turgor or osmotic potential may be related to such acquired freezing tolerance in plant tissues. The sunflower cells treated with mefluidide or ABA attained the LT_{50} of –17.5 °C or –18.5 °C, respectively, against –10 °C for controls. Such increase in freezing tolerance of mefluidide-treated cells was paralleled by a 4 to 25 fold increase in ABA (Johnson-Flanagan *et al.*, 1991). Similarly, salt induced cold hardiness involves the synthesis of cold and ABA-responsive proteins and the alteration in protein synthesis is mediated by ABA-induced salt stress in potato (Ryu *et al.*, 1995). Salt stress has been found to increase low temperature tolerance in some herbaceous species also.

Selection Indices for Low Temperature Tolerance

Screening for low temperature tolerance in laboratory has been done generally by luminescence method, modified luminescence method, biochemical luminescence intensity and tetrazolium reduction method. These methods are by and large reliable and effective. Cold tolerance study carried out on 31 short duration pigeonpea cultivars in arid Hissar at 14 °C revealed that root length can be used as a criterion for cold tolerance (Kumar *et al.*, 1995). In field trials, *desi* type pigmented strains showed better germination and cold tolerance than non-pigmented *kabuli* type chickpea (Auld *et al.*, 1988). In case of soybean, a linear correlation between seminal root growth and seedling emergence rate was noted at 6.5-7.0 °C (Balashov and Sherepitko, 1984). Thus, fast rate of emergence and fast growth of seminal root were observed effective growth parameters for selecting cold tolerant soybean lines. Analysis of 209 barley lines for low temperature tolerance revealed high proline content in low temperature tolerant barley than in barley which was susceptible to low temperature (Dobslaw and Bielka, 1988). In *Picea abies* seedling, height and bud flushing were strongly genetically correlated with low temperature desiccation damage score. Early flushing families containing relatively short seedlings were less damaged than late flushing tall seedlings (Danusevicius *et al.*, 1999). Low temperature induced injury was in general less in *Brassica* cultivars having higher cell sap dry matter. Following a night low temperature at Ludhiana, most resistant lines showed reduced electrolyte leakage compared to susceptible ones (Bagga *et al.*, 1987).

Reports available from USSR indicated limited values on extent of genetic coefficient of variation (18 %) and estimates of narrow sense heritability (12-36 %) for seed germination of chickpea at 10 °C. Heritability estimates for germination of common bean at 8 °C (35 %) were limited in USA situations (Dickson, 1971). Interestingly, information on cold tolerance of beans at different stages of germination, seedling vigour and days to flowering at 5-10 °C revealed that tolerance at these stages was inherited independently (Dickson and Petzoldt, 1987). Thus, improvement in cold tolerance through breeding could be achieved in respect of critical sensitive stage, having high heritability estimates. For instance, late vegetative stage in chickpea and flower initiation stage in field pea are important growth stages, which deserve cold tolerance. Genetic analysis in barley has shown that the low temperature tolerance depends on several loci. The broad sense heritability for rapid germination and growth of swede rape at low temperature were estimated to be about 60 % for *napus* and 90 % for *campestris*, indicating complex inheritance (possibly polygenic) for these traits at 10 °C (Acharya *et al.*, 1983).

Production of Abiotic Stress Tolerant Plants Through Conventional Breeding Approach

Breeding for Drought Tolerance

Seed based technology involving improved water use efficiency and enhanced drought tolerance of plants appears reliable and convenient. Thus, improved drought tolerance capabilities of a plant, as a result of better water use efficiency and physiological potential, deserves consideration for enhanced and sustained productions, particularly under resource-constraint situations. The objective of breeding is the development of germplasm that combines high yield potential and drought resistance under water-stress conditions. Isolation and search of heritable genetic variances related to the drought tolerant character calls for effective and reliable screening procedures. It appears difficult to spell out specific breeding strategy for tolerance to drought in crop plants due to both controllable and uncontrollable factors influencing the final expression.

Broadly, resistance to drought is achieved in three ways: escape, avoidance and dehydration tolerance (Ludlow and Muchow, 1990). Under severe drought situations, breeding for drought may not be very successful. However, such efforts may be useful in stabilizing yield production under limited rainfall situations. The experience till date indicates that breeding for drought escape followed by drought avoidance has by far been the most useful and effective in field crops, because it favours higher yields *vis-a-vis* tolerance mechanism under soil moisture stress situations. For selection, phenological traits normally take precedence over the physiological and biochemical traits, because the former are easily identifiable, measurable and transferable. As a whole, no deliberate breeding attempts for developing drought tolerant varieties under truly dry conditions appear to have been initiated. Some of the drought tolerant varieties have been developed by usual breeding under optimum conditions, or screening of germplasm in drought-prone situations.

Breeding for drought tolerance also depends on the particular drought environment for which the cultivar has to be bred. Depending on the environment, plant type and growth habit varies. Crop plants may generally encounter the following three types of environments: (1) stored moisture environment, in which traits such as rapid uptake of soil moisture, water conservation in plant tissues, increased metabolic efficiency, rapid growth and deep root systems are required; (2) variable moisture environment, in which plants with high photosynthetic efficiency, sensitive stomata, dense root system, indeterminate growth habit and rapid osmotic adjustment may perform better; and (3) optimal moisture environment, in which plants with extensive root development, sensitive stomata to moisture deficit, rapid osmotic adjustment and indeterminate growth habit may perform well.

From a large number of reports, it is evident that genetic variations, estimates of heritability and genetic advance for grain yield are by and large higher under optimum conditions and low under-stress situations. Thus, following two schools for breeding crops in unfavourable situations are being floated.

(1) Adaptation to specific environment, which implies that plants adapted to optimum conditions, may not be equally adapted to moisture stress conditions. Plants having improved adaptive traits (like longer and well-developed root system) in stress environment may not have so in non-stress environment. However, genotypes adapted to specific situations tend to have extreme expression of traits. Heritability for yield would be extremely low; therefore, greatly reducing selection efficiency. Selection made for medium or even for low yield may result in high drought tolerance.

(2) Adaptation to variable environment, which implies that stability and yield in dry conditions are independent. Selection for high yield may be done in optimum environment and these genotypes may give higher yields in stress environment also. Genotypes with wide adaptation tend to have traits with intermediate expression. Selection for yield and developmental characters is initially done in a wide range of stress and non- stress environments. Average yield over all the environments is used as a criterion for selection to identify superior types.

The physiological approach for breeding drought tolerance is to combine two (stress and non-stress) environments for drought resistance breeding. The underlying philosophy is that yield potential and drought resistance are independent traits, governed by different gene blocks and hence can be improved independently. An example of using cultivar wax as a morphological index is as follows:

Large number of crosses are made between the varieties having high cuticular wax and high productivity. The plants are grown next year in stress environment and selection is made for high cuticular wax and no attention is given for agronomic improvements. Next year the plants with high cuticular wax layer are grown in optimum environment to allow maximum expression of genetic variation for grain yield. Several cycles of selection are made for high yield and quality and lines are identified for high yield. Finally high yielding lines are tested in stress environment in which the variety is to be grown under natural conditions.

Breeding for High Temperature Tolerance

It is often necessary to breed crop varieties having high temperature tolerance potential towards critical growth stages, for sustained production, particularly under resource-constraint farming systems. Moreover, tolerance to high temperature at the seedling stage in *Brassicas* and chickpea will provide longer planting time, increasing the cropped area and utilization of conserved soil moisture with late rains. Similarly, high temperature tolerant pigeonpea will help its early sowing in the summer, enabling timely sowing of wheat and making success of pigeonpea-wheat rotation. Inspite of the great need, breeding for heat tolerance has not taken off so far in any of the crop plants. A few heat tolerant genotypes available in literature for practical purpose owe their origin from selection of germplasm or recommended varieties under natural or laboratory conditions only. Availability of screening techniques like more number of pods per peduncle, higher fertility and accumulation of proline in pollen grains before anthesis are some of the simple traits which can help push the efforts in this direction. A more favourable factor is that, by and large, heat tolerance potential appears to be governed by few major genes. The limited progress made so far for heat tolerance in most crops is mainly by empirical approach.

In spite of the great surge in physiological research on the resistance and contributing traits, its applications in genetics and breeding are only modest. Furthermore, a negative feature as regards the selection for end product of grain yield is its low heritability in hot and dry environment (Blum, 1988). Therefore, development of screening method based on traits involved in heat response is strongly desired. By and large, number of pods/peduncle, number of floral buds/peduncle and decreased pollen sterility may be useful traits for selecting high temperature tolerant variants. Involvement of major gene in high temperature tolerance with a partial dominance have been observed. Dickson and Petzoldt (1987) also observed that heat tolerance was a dominant character but had low heritability (0-14 %) in beans. Thus, due to low heritability, its incorporation appears difficult. Marrow and Hall (1992) on the basis of pods/peduncle and proportion of tolerant plants in F_1, F_2 and backcross progenies indicated that heat tolerance in cowpea was conferred by a single dominant gene but narrow sense heritability was only 26 %. In view of these genetic information, a pedigree method of breeding with selection in F_2 is suggested. Overall, selection or improvement through conventional breeding for heat tolerance in crop plants may be attempted on heat tolerance based simple traits rather than directly on yields.

Breeding for Low Temperature Tolerance

Main obstacle in breeding for cold tolerance is the limited genetic variations for this trait and lack of suitable cold tolerant criteria. Salt tolerance has been found to increase low temperature tolerance, which indicates the possibility of mutual inclusive breeding. In a specific study, nine salt-induced proteins were identified after 24 h of salt treatment in potato, at which time cold hardiness increased by 3 °C. Comparison of these proteins revealed that five of the salt-induced proteins were also induced by cold treatment and seven were also induced by ABA treatment. These results suggested that a subset of proteins induced by cold- and ABA-treatments were related to salt stress. Thus, possibility for simultaneous breeding for cold and salt stress existed in potato. The synthesis and role of cold regulated protein 14 (COR 14) during the development of cold hardening has been investigated in oats. The cold tolerant cultivar showed higher level of COR14 during cold acclimatisation than the susceptible cultivar (Cattivelli *et al.*, 1995). Relation in the fluidity of specific cell membrane with low temperature tolerance provides new possibility towards understanding mechanism of low temperature tolerance and breeding strategies. The cold tolerance evaluation of chickpea germplasm at ICARDA indicated no significant correlation between growth maturity and grain weight traits. This work shows that the possibility of developing low temperature tolerant line varying in these traits exist in chickpea.

Production of Abiotic Stress Tolerant Plants Through Molecular Breeding Approach
Genetic and phenotypic variability for several components of drought resistance has been well documented over the past few decades (O'Toole and Bland, 1987). Natural selection under low-moisture stress with further selection by generations of farmers of the target habitat has resulted in evolution of traditional land races, which have several desirable traits. Farmers tend to grow these land races inspite of their low grain yield potential. These in fact, serve as invaluable starting points for understanding the plant type requirement of the habitat, local adaptability, taste and cooking preferences. These can be used to develop mapping populations. Markers identified in such mapping populations will be of direct application to breeding.

Drought resistance, manifested by certain varieties and land races, is a culmination of interactions involving several morphological, physiological, phenological and biochemical traits. Each one of these traits is inherited in a complex manner and thus could be further divided into more components for convenience of genetic studies and breeding purposes. In order to enhance drought resistance, breeders rely on any trait, which would be (directly, indirectly, or even remotely) associated with enhanced survival during stress and/or recovery consequent to alleviation of stress. Stability of grain yield, and increased grain yield are desirable attributes too. Utility of a trait enhancing drought resistance has to manifest at whole plant and crop community levels. In view of the vast quantum of variations expressed for each attribute, the magnitude of expression of each component trait and its ability to blend with other casually or causally related traits would decide its ultimate utility. Further, each trait and, in fact, each locus (governing a complex trait) inherits in a Mendelian fashion and has its characteristic genetic parameters. These help characterize the locus with reference to its sensitivity to environment, heritability, intragenic and intergenic interactions, and possible multiple alleles at the locus. These are just a few of the considerations for utility of the trait in molecular breeding for enhanced drought resistance.

Genetic studies and plant breeding for relatively complex traits has experienced a renaissance with the advent of molecular marker technology. With reference to traits with complex inheritance pattern, marker technology has contributed towards increased confidence and better understanding of the underlying inheritance mechanisms. For example, inter-trait correlation and regression values can now be explained by marker (and possibly gene) co-location with positive and negative correlations also being indicated (Leberton *et al.*, 1995; Hemamalini *et al.*, 2000). Digenic interactions can be discerned and quantified (Yadav *et al.*, 1997). Two genes governing a trait may not show-up in an analysis, but may appear when studied for interactions. Epistatic interactions can now be quantified enabling pyramiding of genes.

Mapping and tagging has enabled dissection of the complex trait into its logical components, which are easier to study, comprehend and possibly select for. Many examples of identification of the location of genes associated with drought resistance governed by quantitative trait loci (QTL) are available in sorghum, maize, pearl millet, rice and wheat. With the availability of advanced QTL analysis software, complex analysis can be performed. Thus, while tagging QTL is common, an increasing number of studies are now aimed at addressing fairly tricky genetic and biological phenomena like distinctions between tight linkage and pleiotropy (Leberton *et al.*, 1995), distinctions between coupling phase and repulsion phase when two genes are linked, intergenic and intragenic interactions, environmental sensitivity of genes, etc. Of the various types of markers available for mapping, plant breeders seek molecular markers based on polymerase chain reaction (PCR) that are closely associated with genes controlling enhanced drought resistance and grain yield for practical molecular breeding endeavour. The molecular allele should co-segregate with the trait and also have favourable influence (negative or positive, depending on the phase of linkage) on traits that happen to be congruent (at the locus) and still have no negative influence on other loci of the same (or other) traits. Following the pioneering work of Champoux *et al.* (1995), several reports involving tagging root morphology have appeared (Ray *et al.*, 1996; Hemamalini, 1997; Price *et al.*, 1997; Yadav *et al.*, 1997). With this a consensus on traits associated with drought resistance in rice is emerging. A consolidated view of all the QTL associated with drought resistance is available in the rice genes database (*www.ricegenes.org*).

Genetic compulsions of selecting for complex traits and marker- assisted selection strategies with special reference to enhanced drought resistance have been discussed in several research publications. Drought resistance and grain yield are negatively associated (Venuprasad, 1999). Blum (1980) suggested that it would be easy to select for drought resistance simply by selecting for *low* yields. The genetic basis of grain yield and drought resistance needs to be understood, if we plan to combine the two important agronomic traits together. Molecular marker assisted tagging and selection is expected to help in this goal. As it is not desirable to have a high degree of drought resistance with a low level of grain yield, all breeding programs need to be able to enhance both these traits within the constraints imposed by the environment. It is well known that moisture availability is directly correlated with accumulation of dry matter (Venuprasad, 1999). Tagging root morphology and grain yield (spatially separated) in the same mapping population under contrasting moisture regimes has enabled identification of loci for the two complex traits simultaneously (Venuprasad, 1999). The allelic contribution of the deep rooted and lower yielding parent (Azucena) in enhancing performance of the progeny is documented (Venuprasad, 1999). Transgressants from this study were used for a bulk segregant analysis for maximum root length. Of the several markers tried, two markers (one SSR and one RAPD marker) were found to cosegregate with the trait.

Of the several characters associated with drought resistance, there is an emerging consensus that root morphological traits are the most important ones that are likely to give the maximum advance. The molecular markers are being used for selection of desirable root characters along with conventional selection for phenology and whole plant characters.

Production of Abiotic Stress Tolerant Plants Through Plant Biotechnology Approach

Transgenic Approach for Crop Improvement

Plant molecular biology has made a spectacular growth in the past two decades. It offers several new and novel opportunities for strategic research on production of improved plant types, exploiting the transgenic and non-transgenic approaches. Embryo-culture enables breeders to attempt wide crosses between varieties that are not possible to be hybridized by sexual crosses. This tissue culture based method has proved to be of great help for a number of different applications, including developing of resistance against biotic stress factors. Embryo rescue technique has enabled transfer of genes for resistance to brown planthopper, whitebacked planthopper, bacterial blight and blast from wild species into elite breeding lines of rice. Anther culture allows faster stabilization of breeding lines. Homozygosity can be bred in a rapid way by anther culture. Early generation selection for recessive traits is a major advantage of plants derived from anther culture over plants obtained by conventional inbreeding. Low-temperature tolerance is one of the traits introduced in cultivated *indica* rice varieties by anther culture.

Methods of genetic engineering involve direct introduction of a small number of either foreign or same species plant genes. This method has the following advantages over the conventional breeding-based approach: (1) It is useful for transferring alien genes (such as genes derived from bacteria, insects, fish, etc.) into crops which cannot be sexually accomplished. (2) It is useful in re-introduction of the genes in the same species after *in vitro* modifications. These modifications can be with respect to coding region of the gene or with respect to regulatory regions. In the latter case, it is possible to change the level of expression of the desired gene by cloning suitable promoters. (3) It allows introduction of one or two well-characterized genes to the background of other desired traits. With this approach, there is no need for the extensive back- crossing that is often done in conventional hybridization to achieve the desired genetic make-up.

Plant genetic engineering owes its growth to (*a*) plant tissue culture and (*b*) plant molecular biology disciplines. The methods involved in tissue culture and regeneration response have been standardized for a large number of crops. The protocols to introduce the desired gene in crops have been perfected to a great deal (Birch, 1997; Hansen and Wright, 1999). A range of different vector systems has been constructed for introduction of the transgenes in both dicots and monocots. The methods for stable genetic transformation (using a variety of approaches such as *Agrobacterium tumefaciens*-based method, direct DNA uptake by electroporation and particle gun-based approach) as well as for regulation of the introduced transgene (using a variety of different promoters such as constitutive or induced promoters) have been optimized to a great deal (Grover *et al.*, 1999). Transgenic plants over expressing a number of different genes have been produced, leading to insect, viral, fungal pathogen resistance and herbicide tolerance (Grover and Minhas, 2000).

Abiotic Stress Tolerant Transgenic Plants

Murata *et al.* (1992) and Tarczynski *et al.* (1993) made major contributions in stress molecular biology research by producing transgenic plants for enhanced low temperature stress and salt stress tolerance, respectively. Following these two initial reports, transgenics showing tolerance to salt stress, water stress, oxidative stress and low and high temperature stress have been produced employing various means in the past 6-7 years (see Table 1; also see Grover *et al.*, 1993; Grover *et al.*, 1995; Singla *et al.*, 1997; Pareek *et al.*, 1997; Grover *et al.*, 1998a, Grover *et al.*, 1998b, Grover, 1999, Grover *et al.*, 1999, Minhas and Grover, 1999, Katiyar-Agarwal *et al.*, 1999, Grover and Minhas, 2000, Grover *et al.*, 2000, Grover, 2000, Grover *et al.*, 2001a, Grover *et al.*, 2001b, Katiyar-Agarwal *et al.*, 2001 for detailed information on varied aspects of raising of abiotic stress tolerant transgenics). To obtain these transformants, genes encoding different structural proteins have been employed (Table 1).

In recent years, appreciation is growing for use of regulatory genes as a more effective approach than the use of structural genes for producing stress tolerant plants. This is based on the observations that single regulatory gene may lead to altered expression of a number of different downstream structural genes, leading to a wide arrayed altered responses (Grover *et al.*, 1999). Regulatory machinery involved in synthesis and control of transacting factors has emerged as the focal point for controlling expression of stress-responsive genes. For the regulation of Cold-Regulated (COR) genes, the transcription factor CBF1 (CRT/DRE binding factor 1) is implicated to be the gene regulator. Jaglo-Ottosen *et al.* (1998) have produced transgenic *A. thaliana* plants that overexpress CBF1. The transformed line exhibiting higher levels of CBF 1 transcript also showed greater amounts of different *cor* (cold-regulated genes) transcripts and showed increased tolerance of plants to freezing stress tolerance. This observation has been further confirmed with other specific examples on use of transacting factors (Table 1).

Most of the early work on expression of the transgenes was carried out using constitutive promoters. In some of the experiments, the increased tolerance to abiotic stresses obtained was found to be associated with certain negative effects on phenotype of the transformed plants. Kasuga *et al.* (1999) recently overexpressed Dehydration Responsive Element Binding Protein IA (DREBIA) cDNA under the control of *rd29A* promoter, which is a stress-induced promoter. The over-expression of the DREBIA cDNA in transgenic plants in this experiment gave rise to minimal negative effects on plant growth, while providing an even greater tolerance to stress conditions than did expression of the gene from the CaMV promoter. The stress-induced promoters corresponding to low temperature, water stress, salt stress, anaerobic stress and high temperature stress are at various stages of experimentation and optimization for use in biotechnology research (Grover *et al.*, 2001).

Table 1. Selected examples on production of abiotic stress tolerant transgenic plants using structural and regulatory genes

Gene	Protein	Source	Cellular role(s)	Trans host	Stress type	References
Regulatory Genes						
ABI3	Absicisic acid induced protein	*A. thaliana*	Transcription factor	*A. thaliana*	Cold	Parcy *et al.*, 1994, Tamminen *et al.*, 2001
Alfin1	Member of Zn finger family of proteins	*M. sativa*	Transcription factor	*M. sativa*	Salinity	Winicov & Bastola, 1999
Ala1	Aminophospholipid ATPase1	*A. thaliana*	P-type ATPase	*A. thaliana*	Cold	Gomes *et al*, 2000
At-dbf 2	Cell cycle regulated phosphoprotein	*A. thaliana*	Kinase	*A. thaliana*	High temperature, Salinity, Cold and Osmotic stress	Lee *et al.*, 1999
At-hsf 1	Heat shock transcription factor	*A. thaliana*	Transcription factor	*A. thaliana*	High temperature	Lee *et al.*,1995
Cbf1 and cbf3	CRT/DRE binding factor	*A. thaliana*	Transcription factor	*A. thaliana*	Cold	Jaglo-Ottosen *et al.*, 1998, Gilmour *et al.*, 2000
Cnb1	Calcineurin	*S. cerevisiae*	Ca^{2+}-binding protein	*N. tabacum*	Salinity	Pardo *et al.*, 1998
Dreb	DRE binding protein	*A. thaliana*	Transcription factor	*A. thaliana*	Cold and Dehy-dration	Kasuga *et al.*, 1999, Liu *et al.*, 1998
Oscdpk7	Calcium dependent protein kinase	*O. sativa*	Protein kinase	*O. sativa*	Drought and Salinity	Saijo *et al.*, 2000
Scof1	Soybean cold inducible factor1	*G. max*	Transcription factor	*A. thaliana* and *N. tabacum*	Cold	Kim *et al.*, 2001
Structural Genes						
Afp	Antifreeze protein (AFP)	*Synthetic*	Inhibits ice growth and recrystalli-zation	*S. tuberosum*	Cold	Wallis *et al.*, 1997
Atnhx1	Na^+/H^+ transporter	*A. thaliana*	Vacuolar Na^+/H^+ antiporter	*A. thaliana*	Salinity	Apse *et al.*, 1999
Bet A	Choline dehydrogenase	*E. coli*	Glycinebet-aine biosynthesis	*N. tabacum*	Salinity	Lilius *et al.*, 1996

Bet B	Betaine aldehyde dehydrogenase	*E. coli*	Glycinebetaine biosynthesis	*N. tabacum*	Salinity	Holmstrom *et al.*, 1994
Cod A	Choline oxidase	Arthrobacter globiformis	Glycinebetaine biosynthesis	*A. thaliana*	Cold and Salinity	Hayashi *et al.*, 1997; Alia *et al.*, 1998; Sakomoto *et al.*, 2000
Cor15a	Cold regulated gene	*A. thaliana*	Promotes freezing tolerance	*A. thaliana*	Cold	Artus *et al.*, 1996
Gly 1	Glyoxylase	*B. juncea*	Converts 2-oxoaldehydes into 2-hydroxy acids	*N. tabacum*	Salinity	Veena *et al.*,1999
Gpat	Glycerol 3-phosphate acyltransferase	*Cucurbita* sp.	Fatty acid unsaturation	N. tabacum	Cold	Murata *et al.*, 1992
Gpat		*A. thaliana*	O. sativa		Cold	Yokoi et al., 1998
Hal1	Regulation of Ion transport	*S. cerevisiae*	Regulation of K-transport	*L. esculentum*	Salinity	Gisbert *et al.*, 2000
Hsp 17.7	Heat shock protein	*D. carota*	Heat shock protein	*D. carota*	High temperature	Malik *et al.*, 1999
Hsp101	Heat shock protein	*A. thaliana*	Heat shock protein	*A. thaliana*	High Temperature	Queitsch *et al.*, 2000
Hva1	Lea protein	*H. vulgare*	Unknown	*O. sativa*	Drought and Salinity	Xu *et al.*, 1996
Imt1	Myo-inositol-o-methyl transferase	*M. crystallinum*	D-Ononitol biosynthesis	*N. tabacum*	Drought and Salinity	Sheveleva et al., 1997
Msalr	NADPH-dependent Aldose/aldehyde reductase	*M. sativa*	Detoxification	*N. tabacum*	Drought	Oberschall *et al.*, 2000
MtlD	Mannitol-1 phosphate dehydrogenase	*E. coli*	Mannitol metabolism	*N. tabacum*	Salinity	Tarczynski *et al.*, 1993
MtlD		*E. coli*		*A. thaliana*	Salinity	Thomas *et al.*, 1995
MtlD		*E. coli*		*N. tabacum*	Oxidative stress	Shen *et al.*, 1997
P5cs	O^1-pyrroline 5-carboxylate synthase	*V. aconitifolia*	Proline biosynthesis	*N. tabacum*	Drought	Kishor *et al.*, 1995, Hong *et al.*, 2000
Sac B	Levan sucrase	*B. subtilis*	Fructan biosynthesis	*N. tabacum*	Cold	Pilon-Smits *et al.*, 1995
Sod	Superoxide dismutase	*N. plumbaginifolia*	Superoxide Dismutase	*M. sativa*	Cold	McKersie *et al.*, 1993
Tps1	Trehalose 6-phosphate synthase	*A. thaliana*	Trehalose biosynthesis	*N. tabacum*	Drought and salinity	Holmstrom *et al.*, 1996

Physiological, genetic and biochemical approaches have yielded a great deal of information about several signal transduction pathways in plants. Recently the potentiality of using signal transduction component genes for the improvement of stress tolerance has been demonstrated through over expression of Ca^{++}/calmodulin-dependent protein phosphatase, calcineurin (Table 1). The manipulation of specific activities of signal reception and transduction might prove useful in controlling multiple genes through single gene transfers. To achieve this, more basic research is needed to unravel mechanisms of perceptions and transduction of stress signals from the outside of the cell to the cell nuclei, where the transcriptional induction/activation of the specific gene takes place (Grover *et al.*, 1999).

Production of Abiotic Stress Tolerant Plants Through Integrated Participatory Molecular Breeding Approach

Abiotic stress habitats are innately dynamic and fragile. The variability in the edaphic, biotic, climatic factors of the environment coupled with the social and economic status of the farmers make them highly variable. Unlike the irrigated ecosystems, they cannot be altered by the farmers to suit the crop to be planted, as it is low-external input agricultural system. It is not uncommon to describe the stressful habitats as a population of habitats.

In view of the diversity and dynamism of the habitats for which breeding has to be done, there is a need to change the selection strategy from that which worked wonders in the irrigated ecosystems. The new concept of participatory varietal evaluation and participatory varietal selection envisages trying the breeding material in the farmers' fields with their active involvement. The farmers input with a few nudges from the breeders (mostly regarding the stability and breeding behaviour of lines selected by farmers) is expected to be invaluable, as the farming is of subsistence nature and hardly any produce reaches the market. The traditional accessions are known to possess very good taste, keeping quality and nutrition. The material selected by the breeders or molecular biologists needs to be subjected to the test by farmers for possible adoption. Participatory plant breeding approach provides the opportunity to accomplish this.

An integrated molecular marker assisted breeding and participatory varietal selection is being done in Patna region of Bihar and in Nepal. A DFID funded project involves KRIBHCO, a local NGO of farmers for the testing and selection of QTL introgressed lines. The farmers are encouraged to select interesting material based on their requirements. The lines are subjected to molecular marker analysis and phenotyping for root morphology. Interesting material is in the pipeline.

Epilogue

Water stress, salt stress and low and high temperature stresses adversely affect growth as well as production and partitioning of the biomass, eventually reducing yield of the crops. The yield of all major crops including wheat, rice, oilseed brassicas, chickpea, pigeonpea and cotton can appreciably increase if genes controlling high-level abiotic stress tolerance can be incorporated and expressed. Admittedly, production of abiotic stress tolerant crops has not been a highly successful venture as yet, as far as field-level advantage is concerned. However, the damage to crop production caused by abiotic stresses is too severe to be ignored. This statement particularly holds true for the developing countries where agriculture is not as organized a sector as is in the developed countries. Therefore, even though it is difficult to breed abiotic stress tolerant crops and there is not much success to show, there is every reason to continue and strengthen effort towards understanding how crops respond to abiotic stresses and mount resistance against these stress factors. There is a need to employ all possible tools of crop improvement for production of crops tolerant against abiotic stresses.

Acknowledgements

AG is thankful to the Department of Biotechnology, Government of India (BT/AB/05/02/91-Vol IV) and National Agricultural Technology Project (28(1)/99-NATP/CGP-I/33) for the financial support to his laboratory. Shashidhar and SH thank the Rockefeller Foundation, New York, USA for the financial support (RF98001 #671 and 698).

References

Acharya S.N., Dueck J. and Downey R.K. 1983. Selection and heritability studies on canoda/rapeseed for low temperature germination. *Candian J. Pl. Sci.*, **63**: 377-384.

Agarwal M., Katiyar-Agarwal S., Sahi C., Gallie D.R. and Grover A. 2001. *Arabidopsis thaliana* Hsp100 protein: Kith and Kin. *Cell Stress and Chaperone*, **6**: 219-224.

Alia H., Hayashi A., Sakomoto and Murata N. 1998. Enhancement of the tolerance of *Arabidopsis* to high temperature by genetic engineering of the synthesis of glycine betaine. *Plant J.*, **16**: 155-16.

Apse M.P., Aharon G.S., Snedden G.S.W.A. and Blumwald E. 1999. Salt tolerance conferred by overexpression of a vacuolar Na+/H+ antiport in *Arabidopsis. Science*, **285**: 1256-1258.

Artus N.N., Uemura M., Stepnokus P.L., Gilmour S.J., Lin C. and Thomashow M.F. 1996. Constitutive expression of the cold regulated *Arabidopsis thaliana* COR15a gene affects both chloroplast and protoplast freezing tolerance. *Proc. Natl. Acad. Sci.*, USA. **93**: 13404-13409.

Auld D.L., Bettis B.L., Crock J.E. and Kephart K.D. 1988. Planting dates and temperatures effects on germination emergence and seed yield of chickpea. *Agron. J.*, **80**: 909-14.

Bagga S.K., Bagga S.S., Labana K.S. and Dhillon S.S. 1987. Screening for frost tolerance in Indian mustard (*B. juncea* L.) *In*: *7th Intern Rapeseed Congress*, Poznan, Poland. 11-14 Mar. 1987.

Balashov T.N. and Sherepitko V.V. 1984. Identification of genetic resources of soybean in respect of cold resistance at the seed germination stage. *In*: *Crop response to abiotic stresses* (ed.) D. Kumar, Vol 1. (p. 509) (Scientific Publishers, Jodhpur.)

Birch R.G. 1997. Plant transformation: problems and strategies for practical application. Ann. Rev. Pl. Physiol. *Pl. Mol. Biol.*, **48**: 297-326.

Blum A. 1988. Plant breeding for stress environment. CRC Press Inc. Boca Raton, Florida.

Blum A. 1980. Genetic improvement of drought adaptation. p. 450-452. *In*: *Adaptaiton of plants to water and high temperature stress*, (eds.) N.C. Turner and P.J. Kramer, Wiley- Interscience, New York.

Boyer J.S. 1996. Advances in drought tolerance in plants. *Adv. Agron.*, **56**: 187-218.

Cattivelli L., Crosatti C. and Rizza F. 1995. Increasing in membrane stability and COR-14 (Cold regulated protein) accumulation associated with cold-hardening in oats. *J. Genet. & Pl. Breed.*, (Italy) **49**: 333-338.

Champoux M.C., Wang G., Sarkarung S.S., Mackill D.J., O'Toole J.C., Huang N. and McCouch S.R. 1995. Locating genes associated with root morphology and drought avoidance in rice via linkage to molecular markers. *Theor. Appl. Genet.*, **90**: 969-981.

Chhabra M.L. and Ohlsson I. 1990. A comparison of ontogenesis variation in cold hardiness of *B. juncea. Acta Agriculturae Scandinarica*, **40**: 231-235.

Danusevicius D., Jonsson A. and Eriksson G. 1999. Variation among open pollinated families of *Picea abies* (L.) KARST in response to simulated frost desiccation. *Silvae-Genetica*, **4**: 3-4.

Dickson M.H. 1971. Breeding beans for improved germination under unfavourable low temperature conditions. *Crop Sci.*, **11**: 848-850.

Dickson M.H. and Petzoldt R. 1987. Inheritance of low temperature tolerance in beans at several growth stages. *Hort. Sci.*, **22**: 481-483.

Dobslaw S. and Bielka S. 1988. Studies to establish frost tolerance raking in winter barley by means of protein accumulation . II. Use of protein accumulation as a criterion for establishing the frost tolerance of breeding material. *Archivfur-Zuchtungsforschung*, **18**: 255-260.

Gilmour S.J., Selbot A.M., Salazar M.P., Everard J.D. and Thomashow M.F. 2000. Overexpression of the *Arabidopsis* CBF3 transcriptional activator mimics multiple biochemical changes associated with cold acclimation. *Plant Physiol.*, **124**: 1854-1865.

Gisbert C., Rus A.M., Bolarin M.C., Lopez-Coronado M., Arrillaga I., Montesinos C., Caro M., Serrano R. and Moreno V. 2000. The yeast HAL1 gene improves salt tolerance of transgenic tomato. *Plant Physiol.*, **123**: 393-402.

Gomes E., Jakobsen M.K., Axelsen K.B., Geisler M. and Palmgram M.G. 2000. Chilling tolernace in *Arabidopsis* involves ALA1, a member of new family of putative aminophospholipid translocases. *Plant Cell*, **12**: 2441-2453.

Grover A., Sanan N. and Sahi C. 1998a. Genetic engineering for high-level tolerance to abiotic stresses through over-expression of transcription factor genes: The next frontier. *Curr. Sci.*, **75**: 178-179.

Grover A. and Minhas D. 2000. Towards production of abiotic stress tolerant transgenic rice plants: issues, progress and future research needs. *Proc Indian Natn Acad Sci.*, **B66**: 13-32.

Grover A., Kapoor A., Sahi C., Agarwal S., Katiyar-Agarwal S., Agarwal M. and Dubey H. 2000. Abiotic stress responsive genes and promoters in plants. p. 1-39. *In*: *Biotechnological interventions for abiotic stress management in rainfed crops*, Proceedings Andhra Pradesh Netherlands project workshop, Central Res. Inst. for Dryland Agri., Hyderabad.

Grover A., Hossain M.A., Huq M.E., McGee J.D., Peacock W.J., Dennis E.S. and Hodges T.K. 1995. Studies on the alterations of Pdc gene expression in transgenic rice. p. 911-921. *In*: *Fragile lives in fragile ecosystems.* Proceedings of the International Rice Research Conference, Manila, Philippines.

Grover A., Kapoor A., Satya Lakshmi O., Agarwal S., Sahi C., Katiyar-Agarwal S., Agarwal M. and Dubey H. 2001a. Understanding molecular alphabets of the plant abiotic stress responses. *Curr. Sci.*, **80**: 206-216.

Grover A., Kapoor A., Katiyar-Agarwal S., Agarwal M., Sahi C., Jain P., Satyalakshmi O., Agarwal S. and Dubey H. 2001. Experimentation in biology of plant abiotic stress responses. *Proc. Indian Natl. Acad. Sci.*, **B67**: 189-214.

Grover A., Pareek A. and Maheshwari S.C. 1993. Molecular approaches for genetically engineering plants tolerant to salt stress. *Proc. Indian Natn. Acad. Sci.*, **B59**: 113-127.

Grover A., Pareek A., Singla S.L., Minhas D., Katiyar S., Ghawana S., Dubey H., Agarwal M., Rao G.U., Rathee J. and Grover A. 1998b. Engineering crops for tolerance against abiotic stresses through gene manipulation. *Curr. Sci.*, **75**: 689-696.

Grover A., Sahi C., Sanan N. and Grover A. 1999. Taming abiotic stresses in plants through genetic engineering: current strategies and perspective. *Plant Sci.*, **143**: 101-111.

Grover A. 1999. A novel approach for raising salt tolerant transgenic plants based on altering stress signalling through Ca2+/calmodulin-dependent protein phosphatase calcineurin. *Curr. Sci.*, **76**: 136-137.

Grover A. 2000. Water stress responsive proteins/genes in crop plants p. 397-408. *In*: *Probing Photosynthesis: Mechanism, Regulation and Adaptation*, (eds.) M. Yunus, U. Pathre, and P. Mohanty, Taylor, and Francis.

Hansen G. and Wright M.S. 1999. Recent advances in the transformation of plants. *Trends Pl. Sci.*, **4**: 226-231.

Hayashi H., Mustardy A., Deshnium P., Ida M. and Murata N. 1997. Transformation of *Arabidopsis thaliana* with *cod A* gene for choline oxidase; accumulation of glycinebetaine and enhanced tolerance to salt and cold stress. *Plant J.*, **12**: 33- 142.

Hemamalini G.S. 1997. Molecular mapping of low moisture stress induced response in rice (*Oryza sativa* L.) root at peak vegetative stage. M.Sc. (Agri) Thesis submitted to the Univ. Agril Sciences, GKVK, Bangalore 560 065, India.

Hemamalini G.S, Shashidhar H.E. and Hittalmani S. 2000. Molecular marker assisted tagging of root morphological traits under two contrasting moisture regimes at peak vegetative stage in rice (*Oryza sativa* L.). *Euphytica*, **112**: 69-78.

Holmstrom K.O., Welin B., Mandal A., Kristiansdottir I., Teeri T.H., Lamark T., Strom A.R. and Palva E.T. 1994. Production of the *Escherichia coli* betaine-aldehyde dehydrogenase, an enzyme required for the synthesis of the osmoprotectant glycine betaine, in transgenic plants. *Plant J.*, **6**: 749-758.

Holmstrom K.O., Mantyl E., Welin B. and Palva E.T. 1996. Drought tolerance in tobacco. *Nature*, **379**: 683-684.

Hong Z., Lakkineni K., Zhang Z. and Verma D.P.S. 2000. Removal of feedback inhibition of Δl-Pyrroline-5-Carboxylate synthetase results in increased proline accumulatin and protection of plants from osmotic stress. *Plant Physiol.*, **122**: 1129-1136.

Ilani G., Nespoulous C., Huet J.C., Vartanian N. and Pernollet J.C. 1997. Characterization of BND22, a drought induced proteins expressed in *B. napus* leaves. *Phytochem*, **45**: 1-8.

Jaglo-Ottosen K.R., Gilmour S.J., Zarka D.G., Schabenberger O. and Thomashow M.F. 1998. *Arabidopsis* CBF1 overexpressing induces COR genes and enhances freezing tolerance. *Science*, **280**: 104-106.

Johnson-Flanagan A.M., Huiwen Z., Thiagarahjah M.R. and Saini H.S. 1991. Role of abscisic acid in the induction of freezing tolerance in *Brassica napus* suspension-cultured cells. *Pl. Physiol*, **95**: 1044-1048.

Kasuga M., Liu Q., Miura S., Yamaguchi-Shinozaki K. and Shinozaki K. 1999. Improving plant drought, salt and freezing tolerance by genetic transfer of a single stress-inducible transcription factor. *Nature Biotech.*, **17**: 287-291.

Katiyar-Agarwal S., Agarwal M. and Grover A. 1999. Emerging trends in agricultural biotechnology research: use of abiotic stress-induced promoter to drive expression of a stress resistance gene in the transgenic system leads to high level stress tolerance associated with minimal negative effects on growth. *Curr. Sci.*, **77**: 1577-1579.

Katiyar-Agarwal S., Agarwal M., Gallie D.R. and Grover A. 2001. Search for the cellular functions of plant Hsp100/ Clp family proteins. *Crit. Rev. in Plant Sci.*, **20**: 277-295.

Kim J.C., Lee S.H., Cheong Y.H., Yoo C.M., Lee S.I., Chun H.J., Yun D.J., Hong D.J., Lee S.Y., Lim C.O. and Cho M.J. 2001. A novel cold inducible zinc finger protein from soybean, SCOF-1 enhances cold tolerance in transgenic plants. *Plant J.*, **25**: 247-259.

Kishor K.P.R., Hong Z., Miao G.H., Hu C.A.A. and Verma D.P.S. 1995. Overxpression of A'-pyrroline-5-carboxylate synthetase increases proline production and confers osmotolerance in transgenic plants. *Plant Physiol.*, **108**: 1387-1394.

Kumar A., Singh D.P. and Singh P. 1987. Genotypic variation in response of *Brassica* species to water defecit. *J. Agric. Sci. Camb.*, **109**: 615-18.

Kumar G., Krishinaprasad B.T., Saritha M., Gopalakrishna R., Mukhopadhyay K., Ramamohan G. and Udayakumar M. 1999. Enhanced expression of heat-shock proteins in thermo-tolerant lines of sunflower and their progenies selected on the basis of temperature induction response. *Theor. Appl. Genet.*, **99**: 359- 367.

Kumar U., Shukla A. and Pant K.C. 1995. Cold tolerance in pigeonpea seedling characteristics. *Crop Res.*, (Hisar). **9**: 105-113.

Leberton C., Lazic-Jancic V., Steed A., Pekic S. and Quarrie S.A 1995. Identification of QTL for drought responses in maize and their use in testing causal relationship between traits. *J. Exp. Bot.*, **46**: 853-865.

Lee J.H., Hubel A. and Schoffl F. 1995. Derepression of the activity of genetically engineered heat chock factor causes constitutive synthesis of heat shock proteins and increased thermotolerance in transgenic *Arabidopsis. Plant J.*, **8**: 603-612.

Lee J.H., Montagu M.V. and Verbruggen N. 1999. A highly conserved kinase is an essential component for stress tolerance in yeast and in plant cells. *Proc. Natl. Acad. Sci.*, USA. **96**: 5873-5877.

Lilius G., Holmberg N. and Bulow L. 1996. Enhanced NaCl stress tolerance in transgenic tobacco expressing bacterial choline dehydrogenase. *Biotechnology*, **14**: 177-180.

Liu Q., Kasuga M., Sakuma Y., Abe S., Miura S., Yamaguchi-Shinozaki K. and Shinozaki K. 1998. Two transcription factors, DREB1 and DREB2, with an EREBP/AP2 DNA binding domain separate two cellular signal transduction pathways in drought- and low-temperature responsive gene expression respectively, in *Arabidopsis. Plant Cell*, **10**: 1391-1406.

Ludlow M.M. and Muchow R.C. 1990. Critical evaluation of the possibilities for modifying crops for high production per unit of precipitation. *Adv. Agron.*, **43**: 107-153.

Malik M.K., Solvin J.P., Hwang C.H. and Zimmerman J.L. 1999. Modifies expression of a carrot small heat shock protein gene, *Hsp 17.7* results in increased or decreased thermotolerance. *Plant J.*, **20**: 89-99.

Marrow K.O. and Hall H.E. 1992. Inheritance of heat tolerance during pod set in cowpea. *Crop Sci.*, **32**: 912-18.

McKersie B.D., Chen Y., Beus M., Bowley S.R., Bowler C., Inze D., D'Halluin K. and Botterman J. 1993. Superoxide dismutase enhances tolerance of freezing stress in transgenic Alfalfa (*Medicago sativa* L). *Plant Physiol.*, **103**: 1155-1163.

Minhas D. and Grover A. 1999. Transcript levels of genes encoding various glycolytic and fermentation enzymes change in response to abiotic stresses. *Plant Sci.*, **146**: 41-51.

Murata N., Ishizaki-Nishizawa O., Higashi S., Hayashi S., Tasaka Y. and Nishida I. 1992. Genetically engineered alteration in the chilling sensitivity of plants. *Nature*, **356**: 710-713.

Oberschall A., Deak M., Torok K., Sass L., Vass I., Kovacs I., Feher A., Dudits D. and Hovarth G.V. 2000. A novel aldose/aldehyde reductase protects transgenic plants against lipid peroxidation under chemical and drought stress. *Plant J.*, **24**: 437-446.

Opena R.T. and Kuo C.G. 1994. Breeding for tropical adaptation in heading Chinese cabbage - achievements and continuing challenges. *Pl. Breed. Abst.*, **64**: 467-71.

O'Toole J.C. and Bland W.L. 1987. Genotypic variation in crop plant root system. *Adv. Agron.*, **41**: 91-145.

Padmanaban A., Yangihara S., Annamalai P. and Yajima M. 1999. Molecular study on the function of heat shock responsive gene in *Brassica oleracea* var. *capitata* L. Proceed. JIRCAS-Working Report 1999 No. 14: 31-45. Workshop on *Heat Tolerance of Crops*, Okinawa, 7-9 Oct. 1997.

Parcy F., Valon C., Raynal M., Gaubier-Comella P., Delseny M. and Giraudat J. 1994. Regulation of gene expression programs during seed development: roles of the ABI3 locus and of endogenous abscisic acid. *Plant Cell*, **6**: 1567-1582.

Pardo J.M., Reddy M.P. and Yang S. 1998. Stress signaling through Ca2+/ calmodulin-dependent protein phosphatase calcineurin mediates salt adaptation in the plants. *Proc. Natl. Acad. Sci.*, USA. **95**: 9681-9683.

Pareek A., Singla S.L. and Grover A. 1997. p. 365-391. *In*: *Strategies for Improving Salt Tolerance in Higher Plants*, (eds.) P.K. Jaiswal, R.P. Singh, and A. Gulati, Oxford & IBH Publishing, New Delhi 1997.

Pilon-Smits E.A.H., Ebskamp M.J.M., Paul M.J., Jeuken M.J.W., Weisbeek P.J. and Smeekens S.C.M. 1995. Improved performance of transgenic fructan-accumulating tobacco under drought stress. *Plant Physiol.*, **107**: 125-130.

Price A.H., Tomas A.D. and Virk D.S. 1997. Genetic dissection of root growth in rice (*Oryza sativa* L.) A hydroponic screen. *Theor. Appl. Genet.*, **95**: 132-142.

Queitsch C., Hong S.W., Vierling E. and Lindquist S. 2000. Heat shock protein 101 plays a crucial role in thermotolerance in *Arabidopsis*. *Plant Cell*, **12**: 479-492.

Que Q., Zhu Y.G., Lin J.X. and Yang W.Q. 1997. Some characteristics of high yielding soybean line Youchu 4. *Acta Agronomica Sinica*, **23**: 296-300.

Ray J.D., Yu L., McCouch S.R., Champoux M.C., Wang G. and Nguyen H.T. 1996. Mapping quantitative trait loci associated with root penetration ability in rice. *Theor Appl Genet.*, **92**: 627-933.

Richards R.A. and Thurling N. 1978. Variation between and within species of rapeseeds (*B. campestris* and *B. napus*) in response to drought stress. II. growth and development under natural drought stresses. *Australian J. Agric. Res.*, **29**: 479-490.

Ryu S.B., Costa A., Xin Z. and Li P.H. 1995. Induction of cold hardiness by salt stress involves synthesis of cold and abiscisic acid-responsive proteins in potato (*Solanum commersonni* Dum). *Pl. Cell Physiol.*, (Japan) **36**: 1245-1251.

Saijo Y., Hata S., Kyozuka J., Shimamoto K. and Izui K. 2000. Over-expression of a single Ca^{2+}-dependent protein kinase confers both cold and salt/drought tolernace on rice plants. *Plant J.*, **23**: 319-327.

Sakomoto A., Valverde R., Alia H.H., Chen and Murata N. 2000. Transformation of *Arabidopsis* with the codA gene for choline oxidase freezing tolerance in plants. *Plant J.*, **22**: 449-453.

Shen B., Jensen R.G. and Bohnert H.J. 1997. Increased resistance to oxidative stress in transgenic plants by targeting mannitol biosynthesis to chloroplasts. *Plant Physiol.*, **113**: 1177-1183.

Sheveleva E., Chmara W., Bohnert H.J. and Jensen R.J. 1997. Increased salt and drought tolerance by D-Ononitol production in transgenic *Nicotiana tabaccum* L. *Plant Physiol.*, **115**: 1211-1219.

Singla S.L., Pareek A. and Grover A. 1997. High temperature. p.101-127. *In*: *Plant Ecophysiology*, (ed.) M.N.V. Prasad, John Wiley & Sons.

Talwar H.S., Takeda J., Yashima S. and Senboku T. 1999. Growth and photosynthetic responses of groundnut genotypes to high temperature. *Crop Sci.*, **39**: 460-466.

Tamminen I., Makela P., Heino P. and Tapio Palva E. 2001. Ecotopic expression of ABI3 gene enhances freezing tolerance in response to abscisic acid and low temperature in *Arabidopsis thaliana*. *Plant J.*, **25**: 1-8.

Tarczynski M.C., Jensen R.G. and Bohnert H.J. 1993. Stress protection of transgenic tobacco by production of osmolyte mannitol. *Science*, **259**: 508-510.

Thomas J.C., Sepahi M., Arendall B. and Bohnert H.J. 1995. Enhancement of seed germination in high salinity by engineering mannitol expression in *Arabidopsis thaliana*. *Plant Cell Environ.*, **18**: 801-806.

Ungaro M.R.G., Toledo N.M.P. and Gobbo C.R.L.S. 1985. Trials of cultivars of sunflower. *Revista de Agricultura*. Brazil. **60**: 193-215.

Veena V.S. Reddy and Sopory S.K. 1999. Glyoxylase I from *Brassica juncea*: molecular cloning and it's overexpression confer tolerance in transgenic tobacco under stress. *Plant J.*, **17**: 385-395.

Venuprasad R. 1999. QTL mapping of grain yield and its association with root morphology in rainfed lowland rice (*Oryza sativa* L.). M.Sc (Agri) Thesis submitted to the Univ. Agril Sciences, GKVK, Bangalore 560 065, India.

Viswanathan C. and Khanna-Chopra R. 1996. Heat shock proteins - role in thermotolerance of crop plants. *Curr. Sci.*, **71**: 225-284.

Wallis J.G., Wang H. and Guerra D.J. 1997. Expression of a synthetic antifreeze protein in potato reduces electrolyte release at freezing temperatures. *Plant Mol. Biol.*, **35**: 323-330.

Winicov I. and Bastola D.R. 1999. Transgenic overexpression of the transcription factor *Alfin1* enhances expression of the endogenous MsPRP2 gene in alfalfa and improves salinity tolerance of the plants. *Plant Physiol.*, **120**: 473-480.

Xu D., Duan Z., Wang B., Hong B., David Ho T.H. and Wu R. 1996. Expression of a late embryogenesis abundant protein gene, *HVA1*, from barley confers tolerance to water deficit and salt stress in transgenic rice. *Plant Physiol.*, **110**: 249-257.

Yadav R.S., Courtois B. and Huang N. 1997. Mapping genes associated with root morphology in a doubled haploid mapping populaiton of *indica-japonica* cross in rice. *Theor. Appl. Genet.*, **94**: 619-632.

Yadav S.K., Kumar A. and Singh D.P. 1990. Microenvironment, water relations and yield of mustard cultivars in uniirrigated and irrigated soils. *Beitrage Zur Tropischem Landwirtschaft und Vetrinarmedizin*, **28**: 399-406.

Yang X.Q., Jiang X.C., Fu J.R., Yang X.Q., Jiang X.C. and Fu J.R. 1998. Heat stable proteins and acquisition of desiccation tolerance I peanut seeds. *Acta Botanica Sinica*, **40**: 337-42.

Ye C.L., Key Y.Q., Chen W., Ye C.L., Ke Y.Q. and Chen W. 1996. A study on the physiology of heat tolerance in Chinese cabbage-II. Metabolism of water proteins in leaves and heat tolerance. *Jour. Fujian Agril. Univ.*, **25**: 490-493.

Yokoi S., Hogashi S., Kishtiani S., Murata N. and Toriyama K. 1998. Identification of the cDNA for *Arabidopsis* glycerol-3-phosphate acetyl transferase (GPAT) confers unsaturation of fatty acids and chilling tolerance of photosynthesis on rice. *Mol. Breeding*, **4**: 269-275.

Zheng X.Y., Wu G., Wang Y.J., Zheng X.Y., Wu G.S. and Wang Y.J. 1998. Study on relationship between heat tolerance and phosphoglucomutase loci of Chinese cabbage. *Acta Horticulturae Sinica*, **25**: 52-257.

Zhou W., Leul M. and Zhou W.J. 1999. Uniconazole-induced tolerance of rape plants to heat stress in relation to change in hormonal levels, enzyme activities and lipid peroxidation. *Pl. Growth Regulation*, **27**: 99-104.

Plant Breeding - Mendelian to Molecular Approaches
H. K. Jain and M. C. Kharkwal (eds.)

In-vitro Approaches to Crop Improvement

P.S. Ahuja[1] and P.B. Kirti[2]

Abstract

Propounding of the cell as a basic structural unit of life which is potentially totipotent and the definitive experimental demonstration of totipotency culminated in focusing wide applications of plant tissue, cell and protoplast cultures in the broad areas of plant propagation, virus elimination leading to disease free plants, plant improvement, and the *in vitro* production of secondary metabolites of plants. In addition, the techniques help to resolve fundamental questions in developmental biology. This paper focuses on *in vitro* approaches to crop improvement. In addition to a broad historic perspective, implications of plant cell cultures in inducing genetic variations, underlying mechanisms of induction and its utility to the plant breeders is discussed. The pre-twentieth century era defined the expanse, rhythm, morphology and anatomy of life forms, whereas Mendel laid the foundation for the twentieth century research into inheritance patterns to be looked into qualitatively and quantitatively within life forms as factors later known as genes. History of *in vitro* technology only addresses to issues of how actually DNA and as a consequence the genes can be physically manipulated. *In vitro* approaches have facilitated the mobility of genomes and genes across genera and kingdoms but the laws of their inheritance over sexual generations remain the same.

Introduction

Curiosity and interest in plant regeneration developed 225 years ago when Henri-Louis Duhamel du Monceau (1700-1782), a French General Inspector of Navy and an Agronomist, discovered wound healing and bud formation following ring shaped debarking of an elm tree. An explanation to this phenomenon was found in the 'Cell Theory' independently expressed by Schleiden and Schwann during 1830's (see cross ref Brachet, 1961). This theory propounded that the cell is a basic structural unit of life which is potentially totipotent. The cell theory emerged as a unifying principle in biology and a reductionist experimental approach evolved to prove the concept of totipotency. In the early part of this century, a German botanist named Gottlieb Haberlandt and his students delved for 20 years but could only establish excised root tips *in vitro*. Their work provided the basic theoretical concept of cell culture. Philip R. White (1934) in United States used

[1]Institute of Himalayan Bioresource Technology, Palampur 176 061, India
[2]University of Hyderabad, Hyderabad 500 046, India

the above procedure to arrive at an optimal nutritional medium that could sustain indefinite growth of excised tomato roots *in vitro*. Consequently, the placement of cambial tissues of carrot and tobacco on a medium containing the auxin and indole acetic acid enabled indefinite growth of tissues in culture. This aspect was simultaneously and independently reported by Nobecourt and Gautheret in France and White in the U.S. in the 1939.

Experimental Demonstration of Totipotency in Plants

The discovery of kinetin by Folke Skoog and the finding that adenine stimulates cell division led to the classical experiments of Skoog (1944) and Miller (1957) showing that root and bud initiation in callus cultures were conditioned by a balance between auxins and kinetin. This laid the basis to the definitive experimental demonstration of totipotency by Steward and his group from callus and free cells. For further parity towards zygotic behaviour of somatic cells, Reinert (1959) reported the formation of bipolar embryos from callus cultures. In 1964, Guha and Maheshwari for the first time showed that gametic cells in anthers could be induced to develop into complete plants. The large scale isolation of plant protoplasts using hydrolytic enzymes by Edward Cocking of Britain in 1960 and the consequent result of regeneration of protoplasts to plants by the Japanese worker Takebe in 1971 affirmed the complete dimension of totipotency.

These findings culminated in focusing wide applications of plant tissue, cell and protoplast cultures in the broad areas of plant propagation, virus elimination leading to disease free plants, plant improvement, and the *in vitro* production of secondary metabolites of plants. In addition, the techniques help to resolve fundamental questions in developmental biology. This paper focuses on *in vitro* approaches to crop improvement.

Importance of Genetic Variations and the Role of *in vitro* Technology

The year 1859 and later the re-discovery of Mendel's laws in 1900 by De Vries, Correns and Tschermak can be considered as indelible landmarks for biologists. During 1859, Charles Darwin released first edition of his book "*On the Origin of Species*". The impact of Darwinian theory of evolution on scientific thought in the "premolecular" era has been of enduring importance to man particularly in understanding about his own present form and the different forms of life prevailing around him. The book changed the concept of a static species to that of a dynamically changing population of varying individuals differing in their reproductive fitness in specific environmental niches. These differences lead to the survival of the fittest, thereby lending credence to the expanse of variability within and between species and across genera. Importance of variability stimulated explorers such as Vavilov to trot the globe and author the major treatise of diversity of plants and delineate the centers of origin of major crop plants. Mendel's Laws of segregation and independent assortment of alleles gave a fundamental basis to the management of variability. Thus, in programmes of genetic improvement and breeding of either animals or plants, variation forms the basis of selection and Mendelian inheritance patterns aid in the efficient utilization of variation. Both Darwinism and Mendelism have endured the test of time and are internalised in the study of biology and particularly the science of genetics for over a century now.

A plant breeder utilizes variability to improve the existing potential of crops by (*a*) Direct selection, (*b*) Hybridization, (*c*) Hybridization followed by selections in specific patterns, (*d*) Polyploidization, and (*e*) Mutations. *In vitro* technology helps the plant breeders in each of these facets by the techniques of (*a*) Somaclonal variations, (*b*) *In vitro* selection for biotic and abiotic stresses, (*c*) Embryo/ovule/anther cultures, (*d*) Haploidy/Polyploidy, (*e*) Somatic cell hybridization/cybridization/chromosome transfer, (*f*) Transgenics, and (*g*) Conservation of germplasm.

Tremendous advancements have been made in utilizing these *in vitro* approaches for crop improvement. The paper is not intended to be exhaustive. It describes the salient achievements under each of the techniques listed above.

Somaclonal Variations

Somaclonal variation was first detected among progeny of tissue culture raised plants that were expected to be genetically identical (see reviews of Larkin and Scowcroft, 1981, 1983; Orton, 1984; Ahloowahlia, 1986; Larkin, 1987; Sun and Zheng, 1990; Peschke and Phillips, 1992; Kaeppler and Phillips, 1993 Kaeppler *et al.*, 2000). Somaclonal variation is defined as genetic and phenotypic variation among tissue culture raised plants originating from somatic tissues or callus derived from an explant. Somaclonal variations can be stably established either somatically or meiotically. Somatically stable variation is not heritable and manifests in primary regenerants. Heritable somaclonal variations segregate as qualitative and are quantitative traits in several seed-derived generations and are gainfully used in plant improvement programmes.

The potential usefulness of such somaclonal variation as described by Larkin and Scowcroft (1981) for plant improvement first became apparent in sugarcane. In this crop, differences in plants arising from callus cultures were observed for morphological, cytological and biochemical traits. Plants with increase in cane sugar yield and resistance to eye-spot disease could be evolved. In potato, improved tuber shape, colour and uniformity of tubers and resistance to the killer disease of late blight could be screened from tissue cultured plants. In tomatoes, variations with increased solids and resistance to *Fusarium* race 2 could be detected. In India, it has been possible to develop shattering resistance in the oilseed crop *Brassica campestris* by the group at IARI, Delhi. A variety Pusa Jaikisan in *Brassica juncea*, mustard which was shattering resistance was developed through this technique (Katiyar and Chopra, 1995). In Citronella Java. (*Cymbopogon winterianus*) a variety developed from somaconal variation, named CIMAP/Bio-13, has a better herb yield and higher oil content in the range of 1, 2-1, 4 % (Mathur *et al.*, 1988). Importantly, single gene mutations following Mendelian inheritance patterns arise out of *in vitro* cultures at reasonably high frequencies.

An estimate of 1-3 % somaclonal variation per generation cycle is conservative and some reports have estimated somaclonal variations to be greater than 10 % per cycle

(Larkin *et al.*, 1987). Regardless of the rate, somaclonal variation will remain a valuable tool for the plant geneticist to introduce variation into a breeding programme.

Mechanism of Somaclonal Variations: Reasons for somaclonal variations are largely attributed to pre-existing variation in parental tissues, chromosomal abnormalities, sequence variation, transposon activation and gene silencing. Ploidy changes and chromosome rearrangements have been commonly found among tissue culture raised plants, particularly after extended sub-culture periods, largely as a consequence of endoreduplication. However, chromosome breakage is the most prevalent cytological change (see review Benzion *et al.*, 1986; Kaeppler *et al.*, 1998). Translocations followed by inversions and insertions/deletions are also evident. In maize, these are largely observed between heterochromatic knobs and the centromere, or within the centromeric heterochromatin in oats. This has led to the hypothesis that replication of heterochromatin is delayed in tissue culture, leading to chromosomal abnormalities (Johnson *et al.*, 1987).

Sequence variation arising in somaclones is reviewed by Kaeppler *et al.*, 1998; Olhoft and Phillips, 1999). Sequence variation has been detected by RAPDs/RFLPs/AFLPs and analysis of single gene changes both in terms of Mendelian ratios as well as by molecular characterizations at the gene sequence level. In this regard, tissue culture-derived mutant *Adhl* alleles were found to be due to two independent point mutations, both A-to-T transversions (Brettel and Dennis, 1991; Dennis *et al.*, 1987)

DNA methylation has been another significant cause for somaclonal variations, particularly for quantitative traits. Phillips *et al.* (1994) proposed that variation caused by DNA methylation conditions chromatin changes which affect quantitative traits by modulating effects of multiple loci, by altering timing of replication, and base changes through a process similar to RIP in fungi (Selker and Stevens, 1985). DNA methylation patterns have been shown to vary among somaclones and their progeny in a number of studies. Hypomethylation is more frequently observed than hypermethylation and this is extensively elaborated by Olhoft (1996). In an interesting study by LoSchiavo *et al.* (1989) it was observed that increasing levels of kinetin decreased levels of methylation but with increasing concentrations of auxin 2,4-D, there was an increase in the level of methylation. DNA methylation is also regulated by the developmental timing.

Transposon activation was demonstrated by Peschke and Phillips (1991) through activity of *Spm/En* and by Peschke *et al.* (1987) through activity of *Ac* among progenies derived from regenerants of maize. This was perhaps due to genomic shock, which is induced during the tissue culture process. Genomic shock is known to activate transposable elements in other systems (McClintock, 1984). None of the maize regenerants in the above mentioned study with activity of the transposable elements had any cytogenetic abnormality. Hence, the genomic shock produced in culture is not by chromosome breakages but due to genome stresses that lead to instability of repeated sequences, late replication of heterochromatin and also the activation of transposable elements.

Gene silencing is shown to be another reason for somaclonal variations. Oono (1985) reported a dwarf mutant of rice from regenerants via callus cultures and this mutation was maintained in selfed progenies. The treatment of the dwarf mutant with a DNA methylation inhibitor, 5-deoxyazacytidine restored the normal character. This event indicated that the dwarf mutant was a result of a reversible epigenetic silencing event. Recently, a series of white cob mutants (somaclones) were analyzed (Kaeppler *et al.*, 2000) and a *p* locus observed to be the transcription factor activated the pigment pathway in which the mutations occurred (Chopra *et al.*, 1996). None of the mutants had large deletions as resolved with restriction enzyme analysis. These were as a result of transposition or intragenic recombination. Southern analysis with methylation-sensitive restriction enzymes revealed hypermethylation of the *p* mutants. It was suggested that mutants at *p* locus were unexpectedly frequent due to epigenetic silencing. This result has implication in triggering mutation events at specific loci through pre-determined external stimuli in cell cultures leading to differential epigenetic patterns as opposed to normal developmental processes of fertilization and plant development.

Directed Uses of Somaclonal Variations in Crop Improvement: In most studies of somaclonal variation, the variation is occurring randomly. Recent findings however, suggest that triggering of mutation events at specific loci is due to epigenetic silencing under specific growth conditions on standardized tissues. Even within its largely random induction of variation, the importance of somaclonal variations for crop improvement is clearly established and it is important to use this more efficiently. The amount of variation varies within a species from clone to clone, the culture duration, use of mutagenic agents in cultures and selection pressures for toxins or herbicides. Such an approach for generating variations is specifically suitable for vegetatively propagated older clones that have accumulated a large number of mutant cells.

In sexually propagated plants, useful traits could be induced in adapted cultivars and hence reduce timelag in crop improvement programmes. Furthermore, the phenomena of chromosomal rearrangements could be gainfully exploited by breeders in introgression of specific genes from wild relatives in modern cultivars. Development of isogenic lines without the conventional problems associated to linkage is an important consequence in screening somaclonal variations.

Creating important genetic stocks such as monosomics, ditelosomics with common backgrounds can be an important attainable objective. These stocks would be of wide application in development of substitution lines or for fundamental studies in assigning genes to specific chromosomes, particularly in polyploid species. Facilitation of selection of desirable variations for disease resistance or stress tolerance during the culture phase is another important potential of *in vitro* cultures.

In vitro techniques offer opportunities to select for resistance to biotic and abiotic stresses by facilitating imposition of uniform stress at the cell level. The use of saline media for screening salinity resistant/tolerant cell lines or regenerants has been successful for selecting salt tolerant rice (Reddy and Vaidyanath, 1986). Pathogen toxins are known to be useful as reported for *Helminthosporium maydis* in maize and resistance against *Phoma lingam* in mustard. This technique has also been useful in selection of herbicide tolerant varieties (Swanson *et al.*, 1988).

Embryo Culture

Ever since Hanning (1904) first cultured embryos, the technique has played a vital role in wide hybridization and has served as an asset in accessing wide gene pools by plant breeders in their crop improvement programmes. Perhaps embryo culture has been the first of *in vitro* techniques that was used in plant breeding programmes and was accepted by plant breeders with little or no resistance and to advantage. Laibach (1925) first developed an interspecific hybrid following embryo rescue after crossing *Linum perenne* and *L. austriacum*. This outcome led to a concerted effort in elucidating the technique and culture medium requirements for an extremely wide range of interspecific and intergeneric crosses as is evident from Table 1, which highlights some of the recent developments in crop plants.

In addition to wide hybridization, the emryo culture technique helps in breakage of seed dormancy and reduction of the breeding cycle. Dormancy of seeds is caused by several factors including endogenous inhibitors, specific light requirements, low temperatures, dry storage needs and embryo immaturity (Raghavan, 1977). These limitations are largely overcome by embryo excision and culture as was reported in Iris (Randolph, 1945) where the breeding cycle was shortened from years to months.

Embryo culture between species and genera hybrids has been successfully employed for the production of monoploids and haploids. Kasha and Kao (1970) first developed a technique to produce barley haploids by crossing *Hordeum vulgare* with *H. bulbosum*, with the latter necessarily being the pollen parent. On culturing the hybrid embryos of this cross, the complement of chromosomes from *H. bulbosum* gets selectively eliminated and the resultant is a haploid of cultivated barley. The growth conditions of *H. vulgare*, the pollination and subsequent time of embryo rescue and culture procedure are critical in determining the throughput of successful events. Some of the interesting crosses that have been attempted via embryo cultures are presented in Table 1.

Anther and Pollen Culture for Production of Haploids, Inbreds and Mutations

Since the discovery of naturally occurring haploid plants of *Datura stramonium* (Blakeslee *et al.*, 1922), haploids have attracted the attention of geneticists, plant breeders and embryologists. Anthers and pollen cultures have been obviously attractive to plant breeders for developing haploids and dihaploids or homozygous plants. The function of pollen of higher plants is the fertilization of an egg cell. The pollen grains are highly specialized structures consisting of a vegetative cell and the generative cell, which gives rise to two male gametes, one of which fuses with the egg to form the zygote and the other with a secondary nucleus to form a primary endosperm nucleus. In nature, abnormal development of embryo sac like pollen grains are known to occur in some plant species such as *Hyacinthus* (Stow, 1934) and *Leptomeria* (Mansi Ram, 1959), which indicated their capacity for cell division. La Rue (1954) conceived the idea of culturing pollen of higher plants in order to examine their growth potential. In gymnosperms, *Ginkgo biloba, Taxus* sp., *Ephedra foliata* and *Pseudo-larix amabilis* yielded multi-cellular structures under culture conditions (cf Narayanaswamy, 1994). Guha and Maheshwari (1964, 1966) for the first time reported the development of haploid embryoids and plantlets in anther culture of *Datura innoxia*. This breakthrough generated interest in the production of haploids *in vitro*. To date such embryos have been recorded from anther and microspore culture of over 200 species in a large number of families (Maheshwari *et al.*, 1982; Heberle-Bors 1985; Hu and Huang 1987; Keller *et al.* 1987).

Some of the factors to be taken into consideration are as follows :

(*a*) Genotype as shown for *Lycopersicon esculentum* and *Arabidopsis thaliana* (Gresshoff and Doy, 1972), rice (Guha-Mukherjee 1973), etc. In rice, *japonica* sub-species are more productive than the *indica* sub-species. The anthers of African rice (*Oryza glaberrima*) produced pollen callus with a frequency of 43 %, but it was only 3-5 % for reciprocal hybrids from *O. glaberrima* and *O. sativa*. Low yield of anther cultures of some varieties, especially Hsien rice, and high rate of albino pollen plantlets are the two main problems which hamper the application of technique of anther culture for broad range of rice genotypes. However, some fine varieties or strains have been selected from pollen plants and evaluated through field trials.

(*b*) Stage of microspore development is important and uninucleate microspores are most suitable for induction of androgenesis.

(*c*) Donor plant physiology and the growth of plant material in controlled environment is critical for better response. In *Brassica* pollen, embryogenesis is stimulated in anthers subjected to 30 °C for 24 h or 40 °C for just 1 hour. The pretreatment of floral buds of *Brassica hirta* at 4 °C in sucrose solution enhanced the intensity of callusing as well as the frequency of embryogenic anthers (Bajaj and Mohapatra, 1984).

(*d*) Culture medium and conditions are responsible and most commonly used basal media for anther culture (White, 1943; Murashige and Skoog, 1962; Nitsch and Nitsch, 1969) with addition of varying concentrations of growth regulators and sucrose. Higher concentrations of sucrose (6-12 %) have yielded better results in barley (Clapham, 1973), and rice (Chen, 1978). Activated charcoal and iron in the medium play a very important role. The subject has been reviewed at length by a number of workers (Maheshwari *et al.*, 1982; Dunwell, 1986; Hu and Yang, 1986, Bajaj, 1990; Ferrie *et al.*, 1995).

Anther culture is easy to perform as compared to isolated microspore culture. But, microspore culture offers a number of potential advantages over anther culture, particularly in relation to genetic transformation studies. The simplest method is shed

Table 1. Some important examples of embryo rescue

Species	Purpose	Media	Referance
A. Cereals			
Wheat (*Triticum aestivum*) × Maize	Haploids (chromosome elimination)	MS basal medium with 0.05mg BA/1	Kammholz *et al.*, 1996
- *Triticum aestivum* (6x) × Triticale (6x) - *T. monococcum* (2x) *H. jabatum* (4x) - *T. aestivum* (6x) × Hybrids of [*H. jabatum* (4x) × *Secale cereale* (4x)]	Reciprocal intergeneric hybridization to obtain hybrid progenies and cytological importance	--	Wojciechowska and Pudelska, 1994
Wheat × Rye	Primary Triticale production	Taira and Larter (TL) medium	Sikka, 1993; Aganfonov *et al.*, 1990; Lazar *et al.*,1987
Wheat × Teosinte	Haploid production	--	Ushiyama *et al.*, 1991
Triticum × *Aegilops*	Hybridization and transfer of disease resistance for *Puccinia graminis,P. recondita* and *Erysiphe graminis*	Modified MS medium	Valkoun *et al.*, 1984
Hordeum vulgare × *H. bulbosum*	Haploid production	--	Subramanyam and Kishore, 1999; Pickering, 1988
Hordeum vulgare × *Secale cereale*	Hybrid production and cytological studies	--	Wojciechowska and Pudelska, 1992
Hexaploid *Tritordeum* [Amphidiploid *Hordeum chilense* × *T. turgidum*] × Durum wheat	Hybridization and fertility restoration studies	MS medium with various concentrations of 2, 4-D	Barcelo *et al.*, 1989
Barley × Wheat	Hybridization studies	--	Pershina *et al.*, 1988; Xu *et al.*, 1986
Hordeum vulgare × *Triticum aestivum*	Hybridization studies	--	Chen *et al.*, 1984
Elymus cilaris × *Triticum aestivum*	Regeneration and standardization of media	MS medium and various growth regulators	Wu *et al.*, 1995
Elymus canadensis × *H. vulgare*	Winter hardiness, drought tolerance	--	Dahleen and Joppa, 1992
Oryza sativa cv. *Japonica* × O.s. cv. *Indica*	Hybridization for cold tolerance	MS medium supplemented with 0.25 mg 2, 4-D	Sharma *et al.*, 1999
Oryza sativa × a. *O. officinalis* b. *O. australiensis* c. *O. eichingeri*	Breeding and pathology programs for rice improvement	--	IRRI Reporter, 1985

(*Table 1 continued*)

B. Pulses			
Vigna unguiculata × *V. vexillata*	To incorporate viral pathogen resistance	MS medium and growth regulators	Gomathinayagam *et al.*, 1998
Phaseolus coccineus × *P. vulgaris*	Media standardization for embryo rescue	Gamborg's medium and growth regulators	Mergeai *et al.*, 1997
Cajanus cajan × *C. platycarpus*	Improvement of pigeon pea	MS liquid medium + 0.5 mg BA	Mallikarjuna and Moss, 1995
Interspecific hybridization of *Lathyrus sativus*	Free from neurotoxins	--	Misra *et al.*, 1994
Leucaena leuoceplala × *L. diversifolia*	Improvement	MS + 2mg 2, 4-D + 0.1mg BA	Toruan Mathius, 1992
Vigna radiata × *V. glaberescens*	Pest and viral disease resistance	--	Chen *et al.*, 1989
Phaseolus vulgaris × *P. acutifolius*	Hybridization and cytological studies	Basal medium with BA	Obukosia *et al.*, 1988
Vigna pubescens × *V. unguiculata*	Insect resistance	Modified MS medium	Fatokun and Singh, 1987
Medicago sativa × *M. rupestris*	Hybridization	L2 without hormones	McCoy, 1985
Vigna mungo x *V. radiata*	Hybridization	MS + 1mg IAA + 0.2mg KN	Gosal and Bajaj, 1983
- *Vigna radiata* × *V. angularis* - *V. umbellate* × *V. angularis* - *V. angularis* × *V. umbellate*	Hybridization	LS	Chen *et al.*, 1983
Arachis hypogaea × *A. villosa*	Hybridization	MS + 40mg IAA + 200mg KN	Bajaj *et al.*, 1982
- *Glycine tomentella* × *G. max* - *G. canescens* × *G. max*	Hybridization	EC1 + nurse endosperm	Brou'e *et al.*, 1982
Trifolium sarosiense × *T. pratense*	Hybridization	LIH (basal salts and vitamins of L2)	Phillips *et al.*, 1982
Phaseolus coccineus × *P. acutifolius*	Hybridization		Alvarez *et al.*, 1981
Trifolium ambiguum × *T. repens*	Hybridization	TF_2, EC_1, B×N, + nurse endosperm	Williams and De Lautour, 1980

(*Contd.*)

(*Table 1 continued*)

Species	Purpose	Media	Referance
C. Oil seeds			
Brassica napus × *B. oleracea*	Evolve oilseed rape with *B. oleracea* cytoplasm	--	Starzycki *et al.*, 1999
Brassica juncea × *Crambe abysinica*	*B. juncea* rich in erucic acid	--	Wang Youping and Luo Peng, 1998
- *Brassica campestris* var. *trilouilaris* × *Sinapis eurgida* and a series of intergeneric crosses	Synthesis of artificial allopolyploids	--	Aliya *et al.*, 1998
Arachis hypogea and incompatible hexaploid × diploid sp	Wide hybridization technique and media standardization	B5 + 0.5-1.0mg Picloram	Ozias Atkins, 1989; Stalker and Eaveda, 1988
Brassica juncea × *B. hirta*	Hybridization and media standardization	Basal medium + IAA + KN + CH	Mohapatra and Bajaj, 1987
D. Vegetables			
Cucumis sativus × *C. hystrix*	To develop bridge species between *C. melo* and *C. sativus*	MS medium	Chen *et al.*, 1997
Cucurbita pepo × *C. martinezii*	Powdery mildew and cucumber mosaic virus resistance in *C. pepo*	MS + 0.01mg IAA + 0. 1mg KN	Metwally *et al.*, 1996
Cucumis sativus × 9 African sp. (2n=24) - *C. anguria* - *C. dispsaceus* - *C. prophetarun* - *C. zeyheri*	For analysis of relatedness and improvement of crossability in *Cucumis* L. and to induce fruits containing no viable seeds in *C. melo* × *C. sativus*	--	Raamsdonk *et al.*, 1988
Hakuran [*Brassica campertris* × *B. oleracea*] × -*B. campestris-B. juncea-B. oleracea*	The introduction of useful genes from A-genome species to C-genome species using Hakuran as a bridge species	MS medium	Yamagishi *et al.*, 1984

pollen technique, whereby anthers are allowed to float on the surface of liquid medium and eventual dehiscence of anther releases the microspores into the medium. This method has been employed in a number of species such as *Nicotiana tabacum* (Sunderland and Roberts, 1977), *Datura innoxia* (Tyagi *et al.*, 1979), *Oryza sativa* (Chen *et al.*, 1980), *Hordeum vulgare* (Sunderland and Xu, 1982; Kohler and Wenzel, 1985; Powell *et al.*, 1988) and wheat (Wei, 1982).

The embryo yield in anther culture is lower as compared to microspore culture. In *Brassica napus*, microspore culture showed 10 times more efficient embryogenesis (Kieffer *et al.*, 1993). Some of the advantages of microspore culture are as follows:

(*i*) Microspores can be easily isolated by maceration of whole buds followed by filtration and centrifugation. Even shed pollen technique can be employed.
(*ii*) The culture density for optimum response can be easily adjusted.
(*iii*) Potentially embryogenic microspores can be separated by cell sorting or gradient centrifugation.
(*iv*) Biochemical and physiological studies of embryogenesis can be easily carried out.
(*v*) Sometimes anther wall callus overgrows the microspore derived embryos. This problem can be overcome by isolated microspore culture.
(*vi*) Isolated microspore cultures are convenient for gene transfer and mutagenesis.

Applications of pollen plants: Microspore population is a rich source of genetic variation as a result of gametic recombination that serves as a basis for cultivar improvement, and haploids from these can be readily diplodized and screened. Till date haploids have been produced in a large number of agricultural crops such as wheat, rice, maize, barley, potato, brassica, asparagus, tobacco, etc. Haploids due to their homozygous nature are of great importance, especially for the induction of mutations as recessives are easily detectable (Keller *et al.*, 1987; Morrison and Evans, 1988). Many of our crop plants are highly heterozygous and do not breed true to type. The conventional method of selfing and back crossing is a time consuming process. Through anther/pollen culture haploids could be induced, from which homozygous dihaploids could be developed by the doubling of chromosome number.

It is most important in haploid breeding to make suitable combinations of sexual cross. "If the requisite genes are not available in the breeder stock, neither inbreeding nor haploidy will put them there" (Chase, 1974). The occurrence of gametic variation consequent to hybridization among microspore derived embryos may allow the selection of useful traits without the need for mutagenesis (Morrison and Evans, 1988). This procedure has been used for selection of variants for disease resistance, increased alkaloid content and increased protein levels in a large number of plants. (Voorrips and Visser, 1990; Witherspoon *et al.*, 1991). Yellow mosaic virus-resistant *Hordeum vulgare* lines were produced by anther culture (Foroughi-Whr and Friedt, 1984).

Haploid embryos are potentially useful as recipients for foreign genes, especially where there is a high frequency of embryogenesis. Microspore-derived embryos generally exhibit a high regeneration potential. Also, chromosome doubling results in duplication of introduced trait and the homozygote can be evaluated. In microspore embryogenic system, individual cells can be transformed and transformants can be recovered without the complication of chimeras. So far, most of the transformation experiments have been with microspore derived embryo segments with transformant recovery through secondary embryogenesis. Gene transfer to individual microspores prior to embryo development is potentially more advantageous. For transformation of pollen embryos, *Agrobacterium* mediated gene transfer technique has been used most frequently and fertile homozygous transgenic plants have been recovered (Swanson and Erickson, 1989; Oelck *et al.*, 1991; Huang, 1992; Sangwan *et al.*, 1993). The use of biolistic technique in combination with desiccation and inhibition of DNA resulted in stable transformants in *Brassica napus* microspore derived embryos (Chen and Beversdorf, 1994).

In *Brassicas*, storage lipid and protein biosynthesis and accumulation have been extensively studied (Taylor *et al.*, 1990, 1993; Wiberg *et al.*, 1991). These embryos are rich in lipid biosynthetic enzymes and are useful for *in vitro* screening for oil quality (Taylor *et al.*, 1993). This system has been utilized for studies of chilling tolerance, metabolism of chlorophyll during seed degreening, and glucosinolate metabolism (cf. Palmer and Keller, 1997).

Future Potential: It seems that embryogenic microspores occur in association with gametophytic ones in some species, (Heberle-Bors, 1985). However, microspore stage of development is crucial to determine which path it will follow (Sunderland and Huang, 1987). The use of *in vitro* culture manipulations to control gametophytic or sporophytic development from binucleate pollen of *Nicotiana tabacum* suggests a lack of predetermination. (Benite-Moreno *et al.*, 1988; Zarsky *et al.*, 1992). Attempts are now being made to understand the biochemical and physiological aspects of pollen embryogenesis *in vitro* with studies of gene expression and biosynthesis of specific proteins (Pechan *et al.*, 1991; Reynolds and Kitto, 1992; Binarova *et al.*, 1993; Garrido *et al.*, 1993).

The potential of haploid cells for somatic hybridization has not been fully exploited. A protoplast system would be a valuable tool for recovery of a large number of plants from mutagenized embryos or in gametic-somatic fusions.

Pollen embryos have been produced only in a small number of species. Now it is possible to identify androgenic response of microspores by marker proteins (Vergue *et al.*, 1993). Moreover, very young microspores or megasporocytes can be cultured *in vitro*. Therefore, attempts should be made to analyse the factors regulating the switch from gametophytic to sporophytic development.

The haploid method will be most successful in practice if those who are masters of the haploid technique are also masters of plant breeding (Chase, 1974).

Somatic Embryogenesis

Embryogenesis is a unique process and represents a critical stage in the development of a plant at which its basic organization and body plan are first initiated (Kaplan and Cooke, 1997). Plant embryos arise not only from fertilized or un-fertilized ovule cells but also from somatic or vegetative cells. Ever since the sensational discovery of somatic embryogenesis from secondary phloem cells of carrot by Steward *et al.* in 1958 and also by Reinert, 1958, somatic embryogenesis has become an important biotechnological tool.

Cells Amenable to Somatic Embryogenesis: The actual basis underlying the process of somatic embryogenesis is that every cell of a plant is 'totipotent' and has the potential to behave as a zygotic embryo armed with the ability to germinate into a whole new plant. However, only a few cells respond to the manipulations made by tissue culturists for the induction of somatic embryogenesis. Embryogenesis can be easily induced in cells of zygotic embryos and these are called as the competent or 'pre-embryonic determined cells' or PEDC. However, there are a variety of cells which are well differentiated and can be made to behave as PEDC when they are subjected to major manipulations and are called as the 'induced embryonic determined cells' (IEDC) or the 'potentially competent cells' (PCC). There are highly differentiated cells that do not respond to culture manipulations. These are the non-competent cells.

Factors Affecting Somatic Embryogenesis: A large number of factors have been exploited by tissue culturists in inducing embryogenesis in the IEDC cells of a wide variety of both Gymnospermous and Angiospermous plants. A large number of reviews (Zimmerman, 1993; Dodeman *et al.*, 1997) and books (Ammirato, 1983; Jain *et al.*, 1999) indicate that these range from the age or physiological status of the plant, season of explant collection, part of the plant to be collected as explant up to the orientation of the chosen explant on the culture medium. The culture medium is also governed by different factors like the different strengths and composition of major and minor nutrients, vitamins, reduced nitrogen, other nitrogen sources, carbon sources, growth hormones etc. However, the recent studies by different workers (Chasan, 1993; Liu *et al.*, 1993; Cooke and Cohen, 1993; Zimmerman, 1993) reported that the 'auxins' are the most important 'plant growth regulators' required for the induction of somatic embryogenesis in most of the plants. The continued presence of high concentrations of auxins in the 'somatic embryogenesis induction medium' is believed to help the pro-embryonic masses in the synthesis of all the gene products necessary for the completion of the globular stage of embryogenesis (Zimmerman, 1993). Subsequent development takes place either on a totally auxin free medium, or a medium that contains very low concentrations of auxins. Zimmerman, 1993, explained that while some specific genes and their products were required for the completion of the globular stage, inactivation of other genes and their products was required for further development of the embryos into the bipolar, heart and torpedo stages. Thus, for the embryogenesis programme to proceed further, removal or reduction of auxins was an absolute requirement for the inactivation of the inhibitory genes coupled with the expression of a new set of genes and synthesis of gene products that stimulate further development (Borkird *et al.*, 1986; Zimmerman, 1993).

Stages of Somatic Embryogenesis and Hormonal Basis: The process of embryogenesis is characterized by well marked stages during ontogeny, *viz.* the pro-embryonic masses, the globular stage, the bipolar stage, the heart stage, the torpedo stage and the germination stage. Of these, the formation of the globular stage followed by acquisition of bipolar symmetry and later the proper maturity of the embryos for normal germination are the most important. Two important PGRs are known to play a major role. While auxins are important in the earlier stages, ABA is important for the later stages.

Once, the globular stage is formed, the embryos begin to synthesize their own auxins via an alternate pathway (Michalczuk *et al.*, 1992a,b) and the need to affect its polar transport becomes important for normal morphogenesis (Schiavone and Cooke, 1987; Liu *et al.*, 1993, Cooke and Cohen, 1993). Liu *et al.* (1993) used inhibitors of polar transport of auxins and *PIN 1* mutants of *Arabidopsis* to prove that the lack of polar transport of auxins resulted in the formation of collar-like cotyledons and cylindrical embryos.

Another important step in the process of embryogenesis is the maturation of the embryos, during which ABA induces a signal transduction mechanism for a cascade of events (Galau *et al.*, 1986; Hakman *et al.*, 1990). This leads to the expression of certain genes in developmental pathway responsible for maturation and germination during which synthesis of storage reserves occur. Also the developing embryos at this phase acquire desiccation tolerance. Inadequate food reserves and non-desiccation generally leads to abnormal or precocious germination of the somatic embryos (Merkle, 1995).

Biochemical and Molecular Basis: Recently, there has been a greater understanding of the biochemical and molecular basis of somatic embryogenesis (Kaplan and Cooke, 1997). Three glycosylated extracellular proteins, *viz.* EP-1, EP-2 and EP-3 are involved in the globular stages of somatic embryogenesis. While EP-1, a S-glycoprotein with S like receptor protein kinase related to serine/threonine kinase is known to be involved in the initial embryo related signal transduction, EP-2, a lipid transfer protein is involved in the dermatogen differentiation and promotes the formation of pro-embryogenic masses. Generally this protein is used as a marker for initiating the developments in embryogenesis and is mostly located in the epidermal layers. Lastly, EP-3 protein or endochitinase induces the EP-2 genes and is important in rescuing the embryo arrest at the globular stage. Lately, Zhongsen and Thomas (1998), isolated the 'Embryo Specific Zinc Finger Protein' or '*PEI1*' required for the heart stage embryo formation in *Arabidopsis*, by using the latest gene isolation strategy virtual subtraction. The genes encoding the 'late embryogenesis protein' or the '*Lea*' genes are important for the normal maturation and germination of the embryos. A whole gamut of other genes belonging to DC series like the DC3, DC5, DC1.2, 2.26, 2.15, 3.1, DC4.2, DC7.1, DC9.1, DC10.1 DC8, DC49, DC59 have also been isolated. Although these genes are known to be present in the different developmental stages of somatic embryogenesis, their exact function is yet to be determined (Zimmerman, 1993).

Uses of Somatic Embryos: Somatic embryos have been extensively exploited by plant biotechnologists for varied purposes, as they are discrete propagules with distinct poles, i.e. root and the shoot apices, and have the ability to germinate into complete plants. Somatic embryogenesis has also proved to be an effective method for mass propagation in comparison to other methods. Not only can the somatic embryos be directly delivered to the green houses or fields, encapsulated as artificial seeds, but they also have a lower unit cost with respect to labour, space and acclimatization. Somatic embryos of plants like *Borago officinalis*, cocoa and Jojoba have also proved to be important sources of secondary metabolites, like precursors of prostaglandins, high quality waxes, lipids, linolenic acids and oleo-palmito stearin. Somatic embryos of other crops have also proved to be important sources of important alkaloids.

Besides having the potential for unravelling complex processes of embryogenesis, somatic embryos help in overcoming the problems of seeds like indeterminate growth, restriction to particular seasons and seed shattering in the fields.

Somatic embryos are extremely important for crop improvement through genetic manipulation methods, involving gene delivery into embryos by *Agrobacterium*, biolistic or other direct transfer methods. Plant transformation via direct somatic embryogenesis is limited by the problems of chimera formation because direct embryos are of multicellular origin. The lack of callus phase precludes the opportunity to preferentially propagate transformed cells before the formation of somatic embryos. However, the process of 'recurrent or repetitive or secondary embryogenesis' circumvents the problem of chimerism by allowing the recovery of completely transformed cells to differentiate as somatic embryos from transformed sectors within a primary somatic embryo. Recurrent cycles of direct embryogenesis, thus, effectively substitute for the callus phase found in indirect embryogenic systems, thereby making repetitive embryogenesis a powerful method for transgenic production. Moreover, the embryos formed during repetitive embryogenesis generally originate from single cells, and continued cycling of chimeric secondary embryos on selective agents eventually result in embryos consisting of entirely transformed cells (Polito *et al.*, 1989). A series of important transgenics have been developed by utilizing the somatic embryo pathway and some important examples are carrot (Scott and Draper, 1987), English walnut (McGranahan *et al.* 1988; 1990) and oilseed rape (Swanson and Erickson, 1989).

Limitations: Somatic embryos despite having tremendous potential for mass propagation, germplasm exchange, and gene transfer are limited by problems like recalcitrant nature of some species, loss of embryogenic potential even under permissive conditions, reduced fertility, chances of somaclonal variation and problems of low frequency induction and normal maturation and germination.

Protoplasts and Somatic Hybrids

The phenomena of pre- and post-zygotic incompatibility, which limit the scope of sexual hybridization are overcome to a large extent in somatic hybridization, i.e., hybridization as a result of fusion of protoplasts of two distantly related or unrelated plants. This fusion procedure, in contrast to the natural zygote formation, provides opportunity for genetic recombination of nuclei, and organelle DNAs, and segregation occurs at nuclear, mitochondrial and chloroplast levels. Both recombination and segregation give rise to a host of innumerable combinations between the nucleus, mitochondria and the chloroplasts. Such changes give rise to variations of the kind which are not feasible in nature, or by conventional breeding methodology. The use of protoplast fusions is visualized in the production of cybrids which contain the nuclear genome of one parent and the cytoplasmic genome of the second parent. These can be of particular advantage in the transfer of cytoplasmic male sterility, which is important for developing hybrid seeds in plant breeding. Somatic hybridization studies in *Brassicas* have led to most productive results by way of producing male sterile lines and interspecific hybrids of commercial importance (Kirti *et al.*, 1995; Pental *et al.*, 1991). The technique can also be employed for generating triploids (Pental *et al.*, 1988).

For a successful programme in somatic hybridization, it is important to establish the following:

1. Standardized protocol for the isolation, culture and regeneration of plants from protoplast. The important concerns of protoplast isolation culture and regeneration are:
   - (*i*) *Source of protoplasts*: Tissues should be available from controlled growth conditions in terms of temperature, light and nutrition.
   - (*ii*) *Isolation Procedure*: The enzyme combination, osmotic pressure of the solution, incubation time and temperature and separation protocol need standardization.
   - (*iii*) *Culture procedure*: Medium of culture, density of culture, type of culture i.e. liquid layer, liquid over agar or agarose or in agarose or agar.
   - (*iv*) *Dilution procedure*: Down regulation of density and osmotic pressure without affecting growth.
   - (*v*) *Regeneration of plantlets*: Plant growth regulation using optimal concentration of auxins and cytokinins.
2. Suitable fusion techniques that do not affect the viability and regeneration potential of the protoplast system are employed. These are:
   - (*i*) Adequate yields of protoplasts to be fused.
   - (*ii*) Fusion procedure: (*a*) PEG, (*b*) PEG+High pHCa++; (*c*) Electrofusion.
3. Selection mechanisms for distinguishing hybrids from parents are the most important technical requirement of a somatic hybridization programme (Pental *et al.*, 1984).

Following fusion, a high proportion of parental protoplasts are mixed with a low population of heterokaryons. One of the ways to find hybrid colonies amongst the parental types is "good fortune" (Carlson, 1973). However, more precise and powerful selection procedures are used to recover rare hybrids by utilizing physiological and biochemical capabilities, and also by mechanical isolation using a micro-manipulator. Some selection strategies are outlined below in the Table 2.

Table 2. Outline of selection schemes employed to identify hybrid colonies in fusion experiments

		Selection Mechanism against parent A/B				
		Parent A		Hybrid/Cybrid		Parent B
Scheme I	(a)	Media not suitable	+	media suitable green callus	–	media not suitable
	(b)	media not suitable	+	green callus and plant regeneration	+	green callus - lack of plant regeneration
Scheme II	(a)	media not suitable	+	green callus	+	white callus (albino mutation)
	(c)	+ yello-green callus	+	dark green callus	+	white callus (albino)
Scheme III		+ white calus (albino)	+	green callus	+	white callus (albino)
Scheme IV	(a)	– drug sensitivity	+	green callus/drug resistant	+	white callus (albino and drug resistant)
	(b)	– drug sensitivity	+	green callus/drug resistant	+	anti metabolite (drug resistant)
	(c)	– drug sensitivity	+	green callus/drug resistant	+	irradiation (drug resistant)
	(d)	– drug sensitivity	+	green callus/drug resistant	+	cytoplast (drug resistant)
	(e)	– drug sensitivity	+	green callus/drug resistant	+	auxotrophic mutant and drug resistance
	(f)	– drug sensitivity	+	green callus/drug resistant	+	media not suitable and drug resistance
Scheme V		– antimetabolite	+	green callus	–	irradiation
Scheme VI		– drug sensitivity	+	green callus	–	drug resistant pollen tetrads
Scheme VII	Isolation of Single Heterokaryons Using Micromanipulator					
Scheme VIII	Fluorescence Activated Cell Sorting					
Scheme IX	Callus/Plant Regeneration Stage - PCR Screening					

Successful genetic exchange between species, genera and tribes have been possible through synthesis of: (*a*) amphiploids by symmetric or complete genomic complementation, (*b*) triploids by gameto-somatic fusions, (*c*) aneuploids by asymmetric or partial complementation, (*d*) cybrids by nucleo-cytoplasmic complementation.

Important areas where the technique holds promise are: (*a*) production of male sterile lines, (*b*) improved rootstocks in horticultural plants and (*c*) improved medicinal plants with better and high productivity of secondary metabolites.

Cybrids/Asymmetric Hybrids: The transfer of cytoplasmic DNA and the unidirectional elimination of chromosomes have resulted in a number of cybrids of agricultural importance, especially the male sterile plants. Such plants are very useful for the production of commercial F_1 hybrids (hybrid seed), as the need for the time-consuming and labour-intensive manual emasculation is avoided, e.g. potato (Binding *et al.*, 1982; Pert *et al.*, 1990a,b); *Brassica* (Pelletier *et al.*, 1983; Glimelius *et al.*, 1991); rice (Kyozuka *et al.*, 1989); tomato (Melzer and O' Connell, 1992); carrot (Ichikawa *et al.* 1987); tobacco (Medgyesy, 1990; Galun, 1993); citrus (Vardi *et al.*, 1989; Grosser *et al.*, 1992); chicory (*Cichorium intybus* cv. Magdebourg) - (Dubreucq *et al.*, 1999).

The most important results of cybridization, as summarized by Galun (1993), are given below:

1. When the donor-recipient protoplast-fusion is performed between species of the same genus then chloroplasts can be transferred, in some genera (e.g. *Nicotiana*) without causing detectable morphological pigmentation abnormality. But in other genera, especially when the phylogenetic distance between the fusion partners is great, such abnormalities may occur. In intergeneric fusions such chloroplast transfers may either not be possible at all, or such abnormalities may become evident.
2. By the use of appropriate genetic markers and due selection, the transfer of plastome components (i.e. after plastome recombination) can be achieved. This enables the transfer of plastome traits even among very distant species of different genera.
3. Pure chondriomes of one fusion partner are rarely transferred to a cybrid. Many studies indicate that once two types of mitochondria are housed in the fused protoplast, they may exchange DNA fragments. This should not worry us with respect to the practical use of cybridization (e.g. to induce male sterility), because those chondriome traits in which we are interested can be transferred by cybridization. Moreover, this transfer of only parts of a donor's chondriome has an advantage, because less nuclear genome/chondriome incompatibility is expected. Thus, for chondriomes, the phylogenetic distance of the fusion partners is probably less crucial than in the transfer of plastomes. Obviously a selectable marker in the chondriome (e.g. oligomycin resistance) can facilitate intergeneric transfer of chondriome components.
4. Sorting-out of plastomes and, even more so, the stabilization of the chondriome, in the fusion derivatives may require many cell-divisions and thus not be achieved in the respective cybrid plants. These sorting-outs and fixations may even not reach completion in the sexual progeny of the cybrid. Consequently sexual progenies of cybrids with mixed chloroplast populations and "segregating" mitochondria were reported.

Various methods evolved for the production of cybrids are:

1. Inactivation or removal of the nucleus of one parent by X-irradiation (Zelcer *et al.*, 1978) or by centrifugation through a Percoll gradient (Maliga *et al.*, 1982).
2. Fusion of normal sub-protoplasts or microprotoplasts (Bilkey *et al.*, 1982).

3. Elimination of chromosomes of one parent during proliferation of hybrid cells (Gleba and Hoffmann, 1980)
4. Use of mutants and anti-metabolites (Aviv and Galun, 1988; Bourgin *et al.*, 1986).

Problems of Somatic Hybridization

1. Lack of an efficient regeneration system or a selection method can be a hindrance in its progress.
2. Somatic hybrids between diverse species and genera can be sterile and unbalanced due to choromosomal load. Adequate thought needs to be given to nuclear cytoplasmic ratios for choice of parents in a somatic hybridization programme.
3. Chimeric calli develop when nuclei do not fuse after cell fusion.
4. The result of a somatic fusion programme can not be predicted and hence it is a speculative approach in many of the cases. However, in cybrids where cytoplasmic transfer is involved the procedure is more definitive.
5. It is often assumed that somatic hybridization is possible *in vitro* for two species which are incapable *in vivo*. It should be appreciated that plants are strongly inclined to retain their integrity by chromosome elimination, both *in vivo* and *in vitro*.

***In vitro* Systems and their Application in Chloroplast Transformation**

Efficient regeneration systems from mesophyll protoplasts and further knowledge of chloroplast genome and its functioning have paved the way for chloroplast transformation and production of 'transplastomes'. Chloroplasts are intracellular organelles of plant cells that have a genome and a protein synthetic machinery that is distinct from the nuclear genome. Characteristics like multiple copies of chloroplast genome and their ability to contribute towards maternal inheritance in a large way, have made 'chloroplast transformation' an attractive method for plant genetic modification.

Based on the fact that functioning of chloroplasts depends upon an integrated nucleo-cytoplasmic system, two approaches were developed for chloroplast transformation, *viz.* (*i*) targeting of foreign proteins into chloroplasts using native transit peptide 5'sequences at the amino terminal part of cytosolic precursors and (*ii*) the transformation of the chloroplast genome itself. Chloroplast transformation requires a detailed *in vivo* study of chloroplast gene sequence and their expression.

Advantages of chloroplast transformation or transplastome formation include: (*i*) enhanced expression of the transgene by several folds due to their presence in several thousand copies, (*ii*) maternal inheritance to seed progeny without the problems of gene escape through pollen transfer, (*iii*) site directed or targeted integration of transgenes as compared to random nuclear transformation, (*iv*) no dilution of the transgenes due to the absence of gene segregation in progenies and (*v*) an economical process, wherein maternally inherited genes like resistance to the herbicide (atrazine) or production of vaccines can be of value.

These advantages have not only facilitated the development of several chloroplast vectors like PHD203-GUS, pHD407, pZS148, pZS134, pZS197 and the pPRV series but also enabled some successful transformation experiments (Table 3). Generally the chloroplast vectors have been constructed by either fusing the transgene with the transit peptide of the small subunit of Rubisco and *NPTII* gene of Tn5 or by fusing it to the transit peptide of chlorophyll a/b binding protein and to *GUS* gene of *E. coli*.

Table 3. Examples of successful chloroplast transformations

S. No.	Organism/Plant	Gene used	Method of transformation	Reference
	Direct transformation			
1.	*Chlamydomonas reinhardtii*		Microprojectile	Boynton *et al.*, 1988
2.	*Nicotiana tabacum*	trnI	Microprojectile	Svab *et al.*, 1990
3.	*Nicotiana tabacum*	cat	Microprojectile	Daniell *et al.*, 1990
4.	*Nicotiana tabacum*	gus	Microprojectile	Ye *et al.*, 1990
5.	*Nicotiana tabacum*	gus	PEG mediated	Golds *et al.*, 1993

Selection of transplastomes are generally done on antibiotics, namely spectinomycin and streptomycin that allow the transplastomes to remain green as compared to the non-transformants which turn white. The genes resistant to the antibiotics were obtained from *Nicotiana tabacum* line *SPC2* due to mutations in the 16S rRNA encoding ribosomal DNA of the chloroplast genome. A silent mutation was further induced in this region for new restriction sites. This finding is vital to chloroplast transformation, because the plastid genome exists in large number of identical copies and is further amplified by several thousand copies of the genome, thereby making 'non-lethal' selection an imperative parameter. The other popular selection strategy employs RNA editing, where C is changed to U as in psbF mRNA, where the codon for serine is changed to that for phenylalanine.

Although chloroplast transformation has been successful, it has been restricted only to some lower organisms like *Chlamydomonas*, yeasts and to the model plant 'tobacco'. This is because the mechanisms governing the regulation of chloroplast genes *in vivo* in higher plants is still largely unknown and the complete chloroplast DNA sequence data of only three plants namely liverworts, tobacco and rice have been worked out at present.

Application of *in vitro* Techniques for Conservation of Germplasm

There is an imperative need to conserve the potential of diversity of life, so as to attain productivity and adaptation against known and unknown threats. *In vitro* techniques help to efficiently conserve germplasm resources that are not easily preserved by seeds or other naturally prevailing propagules, both as live cultures maintained at room or low temperature or as cryopreserved tissues at supra-low temperatures of –196 °C in liquid nitrogen. Techniques for their preservation and, importantly, their rejuvenation from these cold preserved tissues have been outlined for a very large number of plant species.

In *ex situ* conservation of germplasm, seed is the most convenient organ to be stored, but the method is not feasible for crops producing recalcitrant seeds (*Shorea robusta, Azadirecta indica, Mangifera indica, Coffea arabica, Camellia sinensis, Quercus* sp., etc.), or for those crops which are propagated vegetatively e.g., *Citrus, Pyrus, Prunus*, etc.

The only storage places for such crops could be field or green houses. At the same time such collections are more susceptible to disease and environmental changes and in comparison to seed banks, cannot be used for long-term storage. In order to overcome these problems, *in vitro* methods for long term storage have been developed for small explants such as apical meristems, embryogenic callus, mutant cells, embryos, cotyledons, hypocotyls, etc.

Various methods have been devised to reduce the growth rate and subsequent reduction of sub-culturing frequency, which saves money and reduces the contamination rates also.

Growth rate can be restricted by employing the following methods:

(*i*) Reducing the mineral salt concentration of medium
(*ii*) Use of growth retardant such as abscisic acid
(*iii*) Use of sucrose free medium, which helps in delaying the sub-culture time
(*iv*) Use of osmotic agents such as mannitol, which raises the osmolarity of the medium
(*v*) An overlay of mineral oil and silicon oil, which restricts the growth of cultures
(*vi*) By inducing dormancy or quiescence, a natural phenomenon in which the plant part or a propagule suspends its growth for a short period, to avoid the adverse environmental conditions by modulating certain factors such as photoperiod, temperature, phytohormones, water potential, etc.
(*vii*) Storing at lower temperatures (4-10 °C) with exposure to light

Advantage of restricted growth technique is that the same basic facilities that are used for micropropagation can be utilized, and the cultures can be readily switched to rapid multiplication when required. Disadvantage is the high cost of labour and space and risk of somaclonal variations.

The most extensively used procedure for long term storage of cultures is cryo-preservation, i.e. storage at ultra-low temperature. Success in cryo-preservation was achieved in carnation, potato and subsequently within a range of other herbaceous and woody species. In this technique, the following factors influence the efficiency of the procedure:

Most of the inspiration of *in vitro* technology in plant cells is motivated by developments made with mammalian cells. The same was the case on cryopreservation. Plant cells and tissues were first exposed to ultra-low temperatures by Sakai in 1960, when he demonstrated that very hard mulberry twigs could withstand freezing in liquid nitrogen after dehydration mediated by extra-organ freezing. A decade later it was shown that tissue cultured cells of flat could sustain freezing to –50 °C after treatment with dimethylsulphoxide (DMSO) (Quatrano, 1968). Later cell cultures of carrot (Latta, 1971) were successfully retrieved after cryopreservation. The developments in plant cell tissue and organ cultures in the last three decades have culminated in evolving procedures that essentially involves, slow dehydrative cooling followed by rapid immersion in liquid nitrogen, rapid thawing, washing and recovery (Kartha, 1985; Withers and Engelman, 1997). Today, there are about 37,600 accessions conserved *in vitro* worldwide.

Tissue culture gene banks are a direct application of these results in conserving genetic variation for crop improvement. National Bureau for Plant Genetic Resources (NBPGR), Pusa, New Delhi is a practical example of this in India. However, it is necessary to mention that *in vitro* conservation should not be seen as a replacement of conventional *in situ* and *ex situ* approach to germplasm conservation.

References

Agafonov I.S., Vilibekova E.N., Vidibekov M.D., Bondarenko D.N., Urmomov V.N., Gorbunov D.L. and Gordov V.T. 1990. Promising methods of producing useful genotypes of agricultural crops resistant to diseases and extreme environmental factors. *Saratov*, 12-14-Sentyabrya, 1990, 4-5.

Ahloowalia B.S. 1986. Limitations to the use of somaclonal variation in crop improvement. P. 14-27. *In*: *Somaclonal variation and crop improvement*, (ed.) J. Semal. Martinus Nijhoff, Boston.

Aliya M., Kato M. and Kukihara F. 1998. Production of intergeneric hybrids between *Brassica* and *Sinapis* species by means of embryo rescue techniques. *Euphytica*, **103**: 123-130.

Alvarez M.N., Ascher P.D. and Davis D.W. 1981. Interspecific hybridization in *Euphaseolus* through embryo rescue. *Hort. Science*, **16**: 541-543.

Ammirato P.V. 1983. Techniques for propagation and breeding. *In*: *Handbook of plant cell culture.* P. 82-123. (ed.) D.A. Evans, W.R. Sharp, P.V. Ammirato and Y. Yamada. Vol. 1, Macmillan Publishing Comapany.

Aviv D. and Galun E. 1988. Transfer of cytoplasmic organelles from an oligomycin-resistant Nicotiana cell suspension into tobacco protoplasts yielding oligomycin-resistant cybrid plants. *Mol. Gen. Genet.*, **215**: 128-133.

Bajaj Y.P.S. 1990. *In vitro* production of haploids and their use in cell genetics and plant breeding. P 3-44. *In*: *Haploids in crop improvement.* I. *Biotechnology in Agriculture and Forestry*, (ed.) Y.P.S. Bajaj. Springer-Verlag, Berlin.

Bajaj Y.P.S. 1994. Somatic hybridization-A rich source of genetic variability. P. 3-32. *In*: *Somatic Hybridization in crop improvement I.* Vol. 27, *Biotechnology in Agriculture and Forestry*, (ed.) Y.P.S. Bajaj. Springer-Verlag, Berlin.

Bajaj, Y.P.S. and Mohapatra D. 1984. *In vitro* induction of genetic variability in rapeseed mustard. *Annual oilseed workshop*, Haryana Agric. Univ., Hissar.

Bajaj Y.P.S., Kumar P., Singh M.M. and Labana K.S. 1982. Interspecific hybridization in the genus *Arachis* through embryo culture. *Euphytica*, **31**: 365-372.

Barcelo P., Vazquez A. and Martin A. 1989. Somatic embryogenesis and plant regeneration from *tritordeum*. *Pl. Breed.*, **103**: 235-240.

Benite-Moreno R.M., Macke F., Hauser V, Alwen A. and Hebrle-Bors E. 1988. Sporophytes and male gemetophytes from cultured immature tobacco pollen. P 137-142. *In*: *Sexual Reproduction in Higher Plants*, (eds.) M. Cresti, P. Gori and E. Pacini. Springer-Verlag, Berlin.

Benzion G., Phillips R.L. and Rines H.W. 1986. Case histories of genetic variability *in vitro*: oats and maize. P. 435-448. *In*: *Cell Culture and Somatic Cell Genetics of Plants*, Vol. 3, (ed.) I. K. Vasil. Academic Press, New York.

Bilkey P.C., Davey M.R. and Cocking E.C. 1982. Isolation, origin and properties of enucleate plant microspores. *Protoplasma*, **110**: 147.

Binarova P., Straatman K., Hause B., Hause G. and Van Lammeren A.A.M. 1993. Nuclear DNA synthesis during the induction of embryogenesis in cultured microspores and pollen of *Brassica napus* L. *Theor. Appl. Genet.*, **87**: 9-16.

Binding H., Jain S.M., Finger J., Mordhorst G., Hehls R. and Gressel J. 1982. Somatic hybridization of an atrazine resistant biotype of *Solanum nigrum* and *Solanum tuberosum*. Part 1. Clonal variation in morphology and in atrazine sensitivity. *Theor. Appl. Genet.*, **63**: 273-277.

Blakeslee A.F., Belling J., Franham M.E. and Bergner A.D. 1922. A haploid mutant in the Jimson weed, *Datura stramonium*. *Science* (Washington), **55**: 646-647.

Borkird C., Choi J.H. and Sung Z.R. 1986. Effect of 2,4-dichlorophenoxyacetic acid on the expression of embryogenic programme in carrot. *Plant Physiol.*, **81**: 1143-1146.

Bourgin J.P., Missionier C. and Goujard J. 1986. Direct selection of cybrids by streptomycin and valine resistance in tobacco. *Theor. Appl. Genet.*, **72**: 11-14.

Boynton J.E., Gilham E.H., Hosler J.P., Jones A.R., Randolph-Anderson B.L., Robertson D., Klein T.M., Shark K.B. and Sanford J.C. 1988. Chloroplast transformation in *Chlamydomonas* with high velocity microprojectiles. *Science*, **240**: 1534-1538.

Brachet J. 1961. The living cell. P. 5-17. *In*: *The living cell. Readings from Scientific American.* (ed.) Kennedy D. Stanford University. Freeman W.H. & Company, San Francisco & London.

Brettell R.I.S. and Dennis E.S. 1991. Reactivation of a silent Ac following tissue culture is associated with heritable alterations in its methylation pattern. *Mol. Gen. Genet.*, **202**: 365-372.

Brou'e P., Douglass J., Grace J.P. and Marshall D.R. 1982. Interspecific hybridization of soybeans and perennial *Glycine* species indigenous to Australia via embryo culture. *Euphytica*, **31**: 715-724.

Carlson P.S. 1973. The use of protoplasts for genetic research. *Proc. Natl. Acad. Sci.*, **70**: 589-602.

Chaplin J.F., Berk L.G., Gooding A.V. and Powell N.T. 1980. Registration of NC744 tobacco germplasm (Reg No GP18). *Crop Sci.*, **20**: 677.

Chasan R. 1993. Embryogenesis: New molecular insights. *Plant Cell*, **5**: 597-599.

Chase S.S. 1974. Utilization of haploids in plant breeding: Breeding diploid species. *In*: *Haploids in Higher plants, advances and potential.* (ed.) Kasha K.J., Univ Guelph, Guelph, pp 211-230

Chen C.C. 1978. Effect of sucrose concentration on plant production in anther culture of rice. *Crop Sci.*, **18**: 905-906

Chen C.C., Tsay H. and Huang C. 1986. Rice (*Oryza sativa* L.) : Factors affecting androgenesis pp. 123-138. *In*: *Biotechnology in Agriculture and Forestry*, Vol 2, (ed.) Y.P.S. Bajaj, Springer-Verlag New York.

Chen H.K., Mok M.C., Shanmugasundaram S. and Mok D.W.S. 1989. Interspecific hybridization between *Vigna radiata* (L.) Wilczeh and *V. glabrescens*. *Theor. Appl. Genet.*, **78**: 641-647.

Chen J.F., Staub J.E., Tashiro Y., Isshiki S. and Miyazaki S. 1997. Successful interspecific hybridization between *Cucumis sativus* L. and *C. hystrix* Chakr. *Euphytica*, **96**: 413-419.

Chen J.L. and Beversdorf W.D. 1994. A combined use of microprojectile bombardment and DNA inhibition enhances transformation frequency of canola (*Brassica napus* L.). *Theor. Appl. Genet.*, **88**: 187-192.

Chen N.C., Baker L.R. and Honma S. 1983. Interspecific crossability among four species of *Vigna* food legumes. *Euphytica*, **32**: 925-937.

Chen X., Du Z.H., Zhang W.X., Yin F.Y., Xu H.J. and Zhu Z.Q. 1984. Barley × wheat hybrids and regenerates. *Acta Agronomica Sinica*, **10**: 65-71.

Chen Yin, Jui-Feng Wang., Tian W., Zuo Q., Zheng S., Lu D. and Zhang G. 1980. Studies on pollen culture *in vitro* and induction of plantlets in *Oryza sativa* subsp. Keng. *Acta-Gent Sin.*, **7**: 46-54.

Chopra P.S., Athma P. and Peterson T. 1996. Alleles of the maize P gene with distinct specificities encode *Myc*-homologous proteins with C-terminal replacements. *Plant Cell*, **8**: 1149-1158.

Clapham D. 1973. Haploid *Hordeum* plants from anthers *in vitro*. *Z. Pflanzenzuecht*, **69**: 142-155.

Cocking E.C. 1960. A method for the isolation of plant protoplasts and vacuoles. *Nature* (London), **187**: 927-929.

Cooke T.J. and Cohen J.D. 1993. The role of auxin in plant embryogenesis. *The Plant Cell*, **5**: 1494-1495.

Dahleen L.S. and Joppa L.R. 1992. Hybridization and tissue culture of *Hordeum vulgare* × *Elymus canadensis*. *Genome*, **35(6)**: 1045-1049.

Daniell H., Vivekananda J., Nielson B.L., Ye G.N., Tewari K.K. and Sanford J.C. 1990. Transient foreign gene expression in chloroplasts of cultured tobacco cells after biolistic delivery of chloroplast vectors. *Proc. Natl. Acad. Sci.*, USA. **8**: 88-92.

Darwin C. 1859. *In*: *The origin of species by means of Natural Selection*, 6th edn (Murray, London, 1872).

De Buyser J., Henry Y., Lonnet P., Hertzog R. and Hespel A. 1987. Florin: a doubled haploid wheat variety developed by the anther culture method. *Plant Breeding*, **98**: 53-57.

Dennis E.S., Brettell R.I.S. and Peacock W.J. 1987. A tissue culture induced *Adh1 null* mutant of maize results from a single base change. *Mol. Gen. Genet.*, **210**: 181-183.

Djehadi C., Royo C., Nachit M.M., N-di-Fonzo and Arans J.L. 2000. Obtaining embryos by interspecific crossing between durum wheat and *Aegilops*. *In*: *Durum wheat improvement in the Mediterranean region: new challenges*. Seminar proceedings, Zaragoza, Spain. 12-14 April, No.40, 271-273.

Dodeman V.L., Ducreux G. and Kreis M. 1997. Zygotic embryogenesis versus somatic embryogenesis. *J. Exp. Bot.*, **48**: 1493-1509.

Duhamel du Monceau H.L. 1756. La Physique des Arbres, ou II Est Trait e de I'Anatomie des plantes et de I'economie Vegetale pour Servir d'Introduction au Trait e complet des Bois et des Forests. P.H.L. Guerin Pub. (cited by Gautherete 1985, p. 52).

Dunwell J.M. 1986. Pollen, ovule and embryo cultureas tools in plant breeding. *In*: *Plant Tissue Culture and its Agricultural Applications*. (eds.) L. A. Withers and P. G. Aldersons.

Fatokun C.A. and Singh B.B. 1987. Interspecific hybridization between *Vigna pubescens* and *V. unguiculata* (L.) Walp through embryo rescue. *Plant cell, Tissue and Organ Culture*, **9**: 229-233.

Ferrie A.M.R., Palmer C.E. and Keller W.A. 1995. *In vitro* embryogenesis in plants. pp. 309-344. *In*: *Plant Embryogenesis*, (ed.) T. A. Thorpe. Dordrecht, Kluwer Acad. Publ.

Foroughi-Whr B. and Friedt W. 1984. Rapid production of recombinant barley yellow mosaic virus resistant *Hordeum vulgare* lines by anther culture. *Theor. Appl. Genet.*, **67**: 377-382.

Galau G.A., Hughes D.W. and Dure L. 1986. Abscisic acid induction of cloned cotton late embryogenesis-abundant (Lea) mRNAs. *Plant Mol. Biol.*, **7**: 155-170.

Galun E. 1993. Cybrids - an introspective review. *IAPTC Newslett.*, **70**: 2-10.

Garrido D., Eller N., Hebrle-Bors E. and Vicente O. 1993. *De novo* transcription of specific mRNAs during the induction of tobacco pollen embryogenesis. Sex. *Plant Reprod.*, **6**: 40-45

Gautheret R.J. 1939. Sur la possibilite de realiser la culture indefinie des tissus de tubercules de carotte. C.R. Hebd. *Seances Acad. Sci.*, **208**: 118-120.

Gleba Y.Y. and Hoffmann F. 1980. 'Arabidobrassica'. a novel plant obtained by protoplast fusion. *Planta*, **149**: 112-117.

Glimelius K., Fahlesson J., Landgren M., Sjodin C. and Sunberg E. 1991. Gene transfer via somatic hybridization in Plants. *TIBTECH*, **9**: 24-31.

Golds T., Maliga P. and Koop H. U. 1993. Stable plastid transformation in PEG treated protoplasts of *Nicotiana tabacum. Bio/Technology*, **11**: 95-97.

Gomathinayagam P., Ganesh Ram S., Rathinaswamy R. and Ramasamy N.M. 1998. Interspecific hybridization between *Vigna unguiculata* (L.) Walp and *V. vexillata* (L.) A. Rich. through *in vitro* embryo culture. *Euphytica*, **102**: 203-209.

Gosal S.S. and Bajaj Y.P.S. 1983. Interspecific hybridization between *Vigna mungo* and *Vigna radiata* through embryo culture. *Euphytica*, **32**: 129-137.

Gresshoff P.M. and Doy C.H. 1972. Development and differentiation of haploid *Lycopersicon esculentum* (tomato). *Planta*, **107**: 161-170.

Grosser J.W., Gnitter Jr F.G., Sesto F., Deng X.X. and Chandler J.L. 1992. Six new somatic *Citrus* hybrids and their potential for cultivar improvement. *J. Am. Soc. Hortic. Sci.*, **27**:169-173.

Guha S. and Maheshwari S.C. 1964. *In vitro* production of embryos from anthers of *Datura. Nature*, **204**: 497.

Guha S. and Maheshwari S.C. 1966. Cell division and differentiation of embryos in the pollen grain of *Datura in vitro. Nature*, **212**: 97-98.

Guha-Mukherjee S. 1973. Genotype difference in the *in vitro* formation of embryoids from rice pollen. *J. Exp. Bot.*, **24**: 139-144.

Gupta H.S., Bhattacharyajee V. and Pattnayak A. 1996. Transfer of cytoplasmic male sterlity in *indica* rice through protoplast fusion. *International - Rice- Research Notes*, **21**: 33-34.

Haberlandt G. 1902. Kulturversuche mit isollierten pflanzenzellen. Sitzungsber, Akad. Wiss. Wien., *Math-Naturwiss*. Kl., Abt. 1., **111**: 69-92.

Hakman I., Stable P., Engstrom P. and Ericksson T. 1990. Storage protein accumulation during zygotic and somatic embryo development in *Picea abies* (Norway spruce). *Physiol. Plant.*, **80**: 441-445.

Hanning E. 1904. Physiology of plant embryos. I. the culture of cruciferous embryos outside the embryo sac. *Bot. Ztg.*, **62**: 46-81.

Heberle-Bors E. 1985. Isolated pollen culture in tobacco: plant reproductive development in a nut shell. Sex. *Plant Reprod.*, **2**: 1-10.

Hu H. 1986. Wheat: Improvement through anther culture. p. 55-72. *In*: *Biotechnology in Agriculture and Forestry*, Vol 2, (ed.) Y.P.S. Bajaj. Springer-Verlag New York,.

Hu H. and Huang B. 1987. Application of pollen derived plants to crop improvement. *Intl. Rev. Cytol.*, **107**: 397-420.

Hu H. and Yang H. 1986. *Haploids of higher plants in vitro*. Springer-Verlag, Berlin, Heidelberg.

Huang B. 1992. Genetic manipulation of microspores and microspore-derived embryos. *In vitro Cell Devel. Biol.*, **281**: 53-58.

Huang B., Swanson E.B., Baszcyznski C.L., MacRae W.D., Bardour E., Armavil V., Wobe L., Arnoldo M., Rozakis S., Westecott M., Keats R.F. and Kemble R. 1991. Application of microspore culture to canola improvement.P. 298-303. *In*: *Proc. of the 8th International Rapeseed Congress*, (ed). D.I. Mcgregor. Canola Council of Canada.

Ichikawa H., Tanno-Sueraga L. and Imamura J.1987. Selection of *Daucus* cybrids based on metabolic complementation between X-irradiated *D. capillifolius* and iodoacetamide- treated *D. carrota* by somatic cell fusion. *Theor. Appl. Genet.*, **74**: 746-752.

IRRI Report. 1985. Breeding and pathology programs apply biotechnology to rice improvement. *IRRI Report*, **3**: 1-3.

Jain S. M., Gupta P. K. and Newton R. J. 1999. *Somatic embryogenesis in woody plants*. Vol. 4 & 5, Kluwer Academic Publishers, London.

Johnson S.S., Phillips R.L. and Rines H.W. 1987. Possible role of heterochromatin in chromosome breakage induced by tissue culture in oats (*Avena sativa* L.). *Genome*, **29**: 439-446.

Kaeppler S.M. and Phillips R.L. 1993. DNA methylation and tissue culture-induced variation in plants. *In vitro Cell Dev. Biol.*, **29**: 125-130.

Kaeppler S.M., Phillips R.L. and Olhoft P. 1998. Molecular basis of heritable tissue culture-induced variation in plants. P. 465-484. *In*: *Somaclonal Variation and Induced Mutations in Crop Improvement*. Current Plant Science and Biotechnology in Agriculture, Vol. 32, (ed.) Jain *et al.*, Kluwer Academic Publishers, Dordrecht, Netherlands.

Kaeppler S.M., Kaeppler H.F. and Rhee Y. 2000. Epigenetic aspects of somaclonal variation in plants. *Plant Mol. Biol.*, **43**: 179-188.

Kammholz S.J., Sutherland M.W. and Banks P.M. 1996. Improving the efficciency of haploid wheat production mediated by wide crossing. *SABRAO Journal*, **28**: 37-46.

Kaplan D.R. and Cooke T.J. 1997. Fundamental concepts in the embryogenesis of dicotyledons: A morphological interpretation of embryo mutants. *The Plant Cell*, **9**: 1903-1919.

Kartha K.K. 1985. *Cryo-preservation of plant cells and organs*. CRC Press, Boca Raton, FL.

Kasha K.J. and Kao K.N. 1970. High frequency haploid production in barley (*Hordeum vulgare* L.). *Nature* (London), **225**: 874-876.

Katiyar R.K., Chopra V.L. 1995. A somaclone of *Brassica juncea* is proccessed into a variety and released for commercial cultivation in India. *Cruciferae Newsletter*, **17**: 92-93.

Keller W.A., Arnison P.G. and Cardy B.K. 1987. Haploids from gametophytic cells:recent developments and future prospects. P 233-241. *In*: *Plant Tissue and Cell Culture*, (eds.) C.E. Green, D.A. Somers, W.P. Hackers and D.D. Biesboer, New York.

Kieffer M., Fuller M.P., Chauvin T.E. and Schlesser A. 1993. Anther culture of kale (*Brassica oleracea* L. con var. Acephala. D. C. Alef). *Plant Cell Tissue and Organ Culture*, **33**: 303-313.

Kirti P.B., Banga S.S., Prakash S. and Chopra V.L. 1995. Transfer of Ogu cytoplasmic male sterility to *Brassica juncea* and improvement of a male sterile line through somatic cell fusion. *Theor. Appl. Genet.*, **91**: 517-521.

Kyozuka J., Kaneda T. and Shimamoto K.1989. Production of cytoplasmic male sterile rice (*Oryza sativa* L.) by cell fusion. *Bio/Technology.*, **7**: 1171-1174.

Laibach F. 1925. Das Taubwerden von Bastardsmen und die kunsliche Aufzucht fruh. absterbender Bastardembryonen. *Z. Bot.*, **17**: 417-459.

Larkin P.J. 1987. Somaclonal variation: history, method, and meaning. *Iowa State J.*, **61**: 393-434.

Larkin P.J. and Scowcroft W.R. 1981. Somaclonal variation: a novel source of variability from cell cultures for plant improvement. *Theor. Appl. Genet.*, **60**: 197-214.

Larkin P.J. and Scowcroft W.R. 1983. Somaclonal variation and crop improvement. P. 289-314. *In*: *Genetic Engineering of Plants: An Agricultural Perspective*, (eds.) T. Kosuge *et al.*, Plenum, New York.

Latta R. 1971. Preservation of suspension cultures of plant cell by freezing. *Can. J. Bot.*, **49**: 1253-1254.

Lazar M.D., Chen T.H.H., Scoks C.J. and Kartha K.K. 1987. Immature embryo and anther culture of chromosome addition lines of rye in Chinese spring wheat. *Plant Science Irish Republic*, **51**: 77-81.

Liu C.M., Xu Z.H. and Chua N.H. 1993. Auxin polar transport is essential for the establishment of bilateral symmetry during early embryognenesis. *The Plant Cell*, **5**: 621-630.

LoSchiavo F., Pitto L., Giuliano G., Torti G., Nuti-Ronchi V., Marazatti D., Vergara R., Orselli S. and Terzi M. 1989. DNA methylation of embryogenic carrot cell cultures and its variations as caused by mutation, differentiation, hormones and hypomethylating drugs. *Theor. Appl. Genet.*, **77**: 325-331.

Maheshwari S.C., Rashid A. and Tyagi A.K. 1982. Haploids from pollen grains: retrospect and prospect. *Amer. J. Bot.*, **69**: 865-879.

Maliga P., Lorz H., Lazar G. and Nagy F. 1982. Cytoplast-protoplast fusion for interspecific chloroplast transfer in Nicotiana. *Mol. Gen. Genet.*, **185**: 211-215.

Mallikarjuna N. and Moss J.P. 1995. Production of hybrids between *Cajanus platycarpus* and *Cajanus cajan*. *Euphytica*, **83**: 43-46.

Mathur A., Ahuja P.S., Pandey B., Kukreja A.K. and Mandal S. 1988. Screening and evaluation of somaclonal variations for quantitative and qualitative traits in an aromatic grass, *Cymbogon winterianus* Jowitt. *Plant Breeding* 1988. 321-334.

Mc Coy T.J. 1985. Interspecific hybridization of *Medicago sativa* L. and *M. rupestris* M.B. using ovule-embryo culture. *Can. J. Genet. Cytol.*, **27**: 238-245.

McClintock B. 1984. The significance of responses of the genome to challenge. *Science*, **226**: 792-801.

McGranahan G.H., Leslie C.A., Uratsu S., Martin L.A. and Dandekar M.A. 1988. *Agrobacterium* mediated transformation of walnut somatic embryos and regeneration of transgenic plants. *Bio/Technol.*, **6**: 800-804.

McGranahan G.H., Leslie C.A., Uratsu S., Martin L.A. and Dandekar M.A. 1990. Improved efficiency of the walnut somatic embryo gene transfer system. *Plant Cell Rep.*, **8**: 512-516.

Medgyesy P. 1990. Selection and analysis of cytoplasmic hybrids. P. 287-316. *In*: *Plant Cell Line Selection.* (ed.) Dix P.J. VCH, Weinheim.

Melzer J.M. and O' Connell M.A. 1992. Effect of radiation dose on the production and the extent of asymmetry in tomato asymmetric somatic hybrids. *Theor. Appl. Genet.*, **83**: 337-339.

Mendel G. 1866. "Versuche uber pflanzenhybriden" (Available in the original German in J. Hered. 42: 1-47. English translation under the title "*Experiments in plant hybridization*" Harvard University Press, Cambridge, MA).

Mergeai G., Schmit V., Lecomte L. and Baudoin T.P. 1997. Development of an *in vitro* culture technique for immature *Phaseolus* embryos. *Biotechnology Agronomy, Society and Environment*, **1**: 49-58.

Merkle S.A. 1995. Strategies for dealing with limitations of somatic embryogenesis in hardwood trees. *Plant Tissue Culture and Biotechnolgy*, **1**: 112-121.

Metwally E.I., Haroun S.A. and El-Fadly G.A. 1996. Interspecific cross between *Cucurbita pepo* L and *Cucurbita martinezii* through *in vitro* embryo culture. *Euphytica*, **90**: 1-7.

Michalczuk L., Ribnicky D.M., Cooke T.J. and Cohen J.D. 1992b. Regulation of indole-3-acetic acid biosynthetic pathways in carrot cell culture. *Plant Physiol.*, **100**: 1346-1353.

Michalczuk L., Cooke T.J. and Cohen J.D. 1992a. Auxin levels at different stages of somatic embryogenesis. *Phytochemistry*, **31**: 1097-1103.

Misra M., Addis G. and Narayan R.K. 1994. Methods for callus induction and differentiation of *Lathyrus sativus* and embryo rescue in interspecific crosses by tissue culture. *J. Agrl. Soci.*, Univ. Wales, **74**: 129-144.

Mohapatra D. and Bajaj Y.P.S. 1987. Interspecific hybridization in *Brassica juncea* × *Brassica hirta* using embryo rescue. *Euphytica*, **36**: 321-326.

Morrison R.A. and Evans D.A. 1988. Haploid plants from tissue culture: new varieties in a shortened time frame. *Biotechnology*, **6**: 684-690.

Murashige T. and Skoog F. 1962. A revised medium for rapid growth and bioassays with tobacco tissue cultures. *Physiol. Plant.*, **15**: 473-497.

Narayanaswami S. 1994. *Plant Cell and Tissue Culture*, Tata McGraw-Hill Publishing Company Limited, New Delhi

Nitsch J.P. and Nitsch C. 1969. *Haploid plants from pollen grains.* Science (Washington), **163**: 85-87.

Nobecourt P. 1939. Sur la perennite et l'augmentation de volume des cultures de tissus vegetaux. *C.R. Seances Soc. Biol. Ses Fil.*, **130**: 1270-1271.

Obukosia S.D., Waithaka K., Gupta V.K. and Kimani P.M. 1988. *In vitro* embryo culture of interspecific hybrids between the Kenyan lines of *Phaseolus vulgaris* L. and *Phaseolus acutifolius* A. Gray. *Acta Horticulturae*, **218**: L221-230.

Oelck M.M., Phan C.V., Eckes P., Donn G., Rakow G. and Keller W.A. 1991. Field resistance of canola transformnts (*Brassica napus* L.) to ignite (Phosphinothricin). P. 293-297. *In*: *Proc. of the 8th International Rapeseed Congress*, (ed). D.I. Mcgregor, Canola Council of Canada.

Olhoft P.M. 1996. DNA methylation pattern changes induced by maize tissue culture. M.S. thesis, University of Minnesota, St. Paul, MN.

Olhoft P.M. and Phillips R.L. 1999. Genetic and epigenetic instability in tissue culture and regenerated progenies. P. 111-148. *In*: *Plant Responses to Environmental Stresses: From Phytohormones to Genome Reorganization*, (ed.) H.R. Lerner, Marcel Dekker, New York.

Oono K. 1985. Putative homozygous mutants in regenerated plants of rice. *Mol. Gen. Genet.*, **198**: 377-384.

Orton T.J. 1984. Genetic variation in somatic tissues: method or madness? *Adv. Plant Path.*, **2**: 153-189.

Ozias Atkins P. 1989. Plant regeneration from immature embryo of peanut. *Plant Cell Reports*, **8**: 217-218.

Palmer C.E. and Keller W.A. 1997. Pollen embryos. P. 392-422. *In*: *Pollen Biotechnology for Crop Production and Improvement.* (eds.) K.R. Shivanna and Sawhney V.K. Cambridge University Press, USA

Pechan P.M., Bartels D., Brown D.C.W. and Schell J. 1991. Messenger RNA and protein changes associated with induction of *Brassica* microspore embryogenesis. *Planta*, **184**: 161-165.

Pelletier G., Primard C., Vedel F., Chetrie P., Rousselle R. and Renard M. 1983. Intergeneric cytoplasmic hybridization in cruciferae by protoplast fusion. *Mol. Gen. Genet.*, **191**: 244-250.

Pental D., Mukhopadhay A. and Pradhan A.K. 1991. A model for the transfer of nuclear and organelle genes from wild species to allopolyploid crop *Brassicas*. *Physiol. Plant*, **82**: 425.

Pental D., Mukhopadhay A., Grover A. and Pradhan A.K. 1988. A selection method for the synthesis of triploid hybrids by fusion of microspores protoplasts (n) with somatic cell protoplasts (2n). *Theor. Appl. Genet.*, **76**: 237-243.

Pental D., Hamill J.D. and Cocking E.C. 1984. Somatic hybridization using a double mutant of *Nicotiana tabaccum*. *Heredity*, **53**: 79-83.

Pershina L.A., Numerova O.M., Belova L.I., Devyatkina E.P. and Shumnyi V.K. 1988. Feasibility in barley × wheat hybrids, *H. geniculatum*. All. × *T. aestivum* L., their regenerants and hybrid progeny of back-crosses to *T. aestivum* L. *Cereal Research Communications*, **16**: 157-163.

Pershina L.A., Numerova O.M., Belova L.I., Devyalkina E.P. and Shumnyi V.K. 1990. Production and characterization of *Hordeum vulgare* × *Triticum timopheevi* Zhuk. hybrids and their progeny. *Tsitologiya-I-Genetika*, **24**: 33-37.

Pert A., Aviv D. and Galun E. 1990a. Protoplast fusion derived *Solanum* cybrids: application and phylogenetic limitations. *Theor. Appl. Genet.*, **79**: 632-640.

Pert A., Aviv D. and Galun E. 1990b. Protoplast fusion derived CMS potato cybrids: potential seed parents for hybrid true potato seeds. *J. Hered.*, **81**: 438-442.

Peschke V.M. and Phillips R.L. 1991. Activation of the maize transposable element *Suppressor-mutator* (Spm) in tissue culture. *Theor. Appl. Genet.*, **81**: 90-97.

Peschke V.M. and Phillips R.L. 1992. Genetic implications of somaclonal variation in plants. *Adv. Genet.*, **30**: 41-75.

Peschke V.M., Phillips P.L. and Gengenbach B.G. 1987. Discovery of transposable element activity among progeny of tissue culture-derived maize plants. *Science*, **238**: 804-807.

Phillips G.C., Collins G.B. and Taylor N.L. 1982. Interspecific hybridization of red clover (*Trifolium pratense* L.) with *T. Sarosiense* Hazsl. Using *in vitro* embryo rescue. *Theor. Appl. Genet.*, **62**: 17-24.

Phillips R.L., Kaeppler S.M. and Olhoft P.M. 1994. Genetic instability of plant tissue cultures: breakdown of normal controls. *Proc. Natl. Acad. Sci.*, USA, **91**: 5222-5226.

Pickering R.A. 1988. The attempted transfer of disease resistance from *Hordeum bulbosum* L., to *H. vulgare* L. *Barley-Genetics-Newletter*, **18**: 5-8.

Polito V.S., McGranahan G., Pinney K. and Leslie C. 1989. Origin of somatic embryos from repetitively embryogenic cultures of walnut (*Juglans regia*): Implications for *Agrobacterium* -mediated transformation. *Plant Cell Reports*, **8**: 219-221.

Quatrano R.S. 1968. Freeze-preservation of cultured flax cells utilizing DMSO. *Plant Physiol.*, **43**: 2057-2061.

Raamsdonk L., Van W.D. and Den-Nijs A.P.M. 1988. Analysis of relatedness and improvement of crossability in the genus *Cucumis* L. *Prophyta*, **42**: 155-156.

Raghavan V. 1977. 3. Applied aspects of embryo culture. P. 375, *In*: "*Applied and Fundamental Aspects of Plant Cell, Tissue and Organ Culture*" (eds.) J. Reinert and Y.P.S. Bajaj, Springer-Verlag, Berlin and New York.

Randolph L.F. 1945. Embryo culture of Iris seed. *Bull. Am. Iris Soc.*, **97**: 33-45.

Reddy P.J. and Vaidyanath K. 1996. *In vitro* characterization of salt stress effects and the selaection of salt tolerant plants in rice (*Oryza sativa* L.). *Theor. Appl. Genet.*, Berlin, W. Ger. Springer International. Feb., **71**: 757-760.

Reinert J. 1959. Uber die Kontrolle der Morphogenese und die Induktion von Advientiveem bryonen an Gewebekuluren aus Karotten. *Planta*, **58**: 318-333.

Reynolds T.L. and Kitto S.L. 1992. Identification of embryoid-abundant genes that are temporarily expressed during pollen embryogenensis in wheat anther cultures. *Plant Physiol.*, **100**: 1744-1750.

Sakai K. 1960. Survival of the twigs of woody plants at –196 °C. *Nature*, **185**: 395-394.

Sangwan R.S., Ducrocq C. and Sangwan-Norreel B.S. 1993. *Agrobacterium* mediated transformation of pollen embryos in *Datura innoxia* and *Nicotiana tabacum*: production of transgenic haploid and fertile homozygous dihaploid plants. *Plant Sci.*, **95**: 99-115.

Schiavone F.M. and Cooke T.J. 1987. Unusual patterns of somatic embryogenesis in the domesticated carrot.: Developmental effects of exogenous auxins and auxin transport inhibitors. *Cell Differ.*, **21**: 53-62.

Scott R.J. and Draper J. 1987. Transformation of carrot tissues derived pro-embryogenic suspension cells: A useful model system for gene expression studies in plants. *Plant Mol. Biol.*, **8**: 265-274.

Selker E.U. and Stevens J.N. 1985. DNA methylation at asymetric sites is associated with numerous transition mutations. *Proc. Natl. Acad. Sci.*, USA, **82**: 8114-8118.

Sharma D., Konwar B.K. and Deka P.C. 1999. Application of embryo rescue in *japonica* × *indica* rice hybridization. *Oryza*, **36**: 35-37.

Sikka T. and Immoren A. 1993. Comparison of callus culture with embryo culture of different lines of embryo rescue for primary triticale production. *Euphytica*, **70**: 185-190.

Skoog F. 1944. Growth and organ formation in tobacco tissue culture. *Am. J. Bot.*, **31**: 19-24.

Skoog F. and Miller C.O. 1957. Chemical regulation of growth and organ formation in plant tissues cultured *in vitro*. P. 118-131. *In*: *The Biological Action of Growth Substances*, (ed.) Porter H.K., Symposia of the society for Experimental Biology No. 11, Cambridge University Press, Cambridge.

Stalker H.T. and Eaveda M. A. 1988. Ovule and embryo culture of *Arachis hypogaea* and interspecific hybrids. *Peanut Science*, **15**: 98-104.

Starzycki M., Starzycka E., Krzymanski J. and Mutuszcza M. 1999. Alloplasmic winter oilseed rape obtained by interspecific crossing within the family Brassicaceae. *K-Rosliny-Oleiste*, **20**: 43-49.

Steward F.C., Mapes M.O. and Smith J. 1958. Growth and organized development of cultured cells. I. Growth and sivision of freely suspended cells. *Am. J. Bot.*, **45**: 693-703.

Steward F.C. 1958. Interpretations of the growth from free cells to carrot plants. *Am. J. Bot.*, **45**: 709-713.

Subramanyam N.C. and Kishor P.B.K. 1997. Haploids from wide hybridization in grass genera: Progress and perspectives. Plant tissue culture: Emerging trends. *In*: *Proceedings of a symposium* held at Hyderabad, India, 29-31 January, pp 11-21.

Sun Z.X. and Zheng K.L. 1990. Somaclonal variation in rice. P. 288-325. *In*: *Biotechnology in Agriculture and Forestry*, Vol. 3, (ed.) Y.P.S. Bajaj, Springer-Verlag, Berlin.

Sunderland N. and Xu Z.H. 1982. Shed pollen culture in *Hordeum Vulgare*. *J. Exp. Bot.*, **33**: 1086-1095.

Sunderland N. and Huang B. 1987. Ultrastructural aspects of pollen dimorphism. *Intl. Rev. Cytol.*, **107**: 175-220.

Sunderland N. and Roberts M. 1977. New approach to pollen culture. *Nature* (London), **270**: 236-238.

Svab Z., Hajdukiewicz P. and Maliga P. 1990. Stable transformation of plastids in higher plants. *Proc. Natl. Acad. Sci.*, USA., **87**: 8526-8530.

Swanson E.B. and Erickson L.R. 1989. Haploid transformation in *Brassica napus* using an octopine producing strain of *Agrobacterium tumefaciens*. *Theor. Appl. Genet.*, **78**: 831-835.

Swanson E.B., Herrgesell M.J., Arnoldo M., Sipell D.W. and Wang R.S.C. 1989. Microspore mutagenesis and selection: canola plants with field tolerance to the imidazolinones. *Theor. Appl. Genet.*, **78**: 525-531.

Swanson E.B. and Erickson L.R. 1989. Haploid transformation in *Brassica napus* using an octopine-producing strain of *Agrobacterium tumefaciens*. *Theor. Appl. Genet.*, **78**: 831-835.

Swanson E.B., Coumans M.P., Brown G.L., Patel J.D. and Beversdorf W.D. 1988. The characterization of herbicide tolerant plants in *Brassica napus* L. after *in vitro* selection of microspores and protoplasts. *Plant Cell Reprod.*, **7**: 83-87.

Takebe I., Labib G. and Melchers G. 1971. Regeneration of whole plants from isolated mesophyll protoplasts of tobacco. *Naturwissenschaften*, **58**: 318-320.

Taylor D.C., Ferrie A.M.R., Keller W.A., Giblin E.M., Pass E.U. and Mackenzie S.L. 1993. Bioassembly of acyllipids in microspore-derived embryos of *Brassica campestris*. *Plant Cell Reprod.*, **12**: 375-384.

Taylor D.C., Weber N., Underhill E.W., Pomeroy M.K., Keller W.A., Scowcroft W.R., Wilen K.W., Moloney M.M. and Holbrook L.A. 1990. Storage protein regulation and lipid accumulation in microspore embryos of *Brassica napus* L. *Planta*, **181**: 18-26.

Toruan Mathius N. 1992. Propagation of *Leucaena* sp. by *Leucaena* sp. by the tissue culture technique. *Menara-perkebunan*, **60**: 20-26.

Turner J. and Facciotti D. 1990. High oleic acid *Brassica napus* from mutagenized microspores. P. 24. *In*: *Proc. 6th Crucifer Genetics Workshop*, (eds.) J.R. McFerson, S. Kresovich, and S.G. Dwyer, USDA-ARS, Geneva NY.

Tyagi A.K., Rashid A. and Maheshwari S.C. 1979. High frequency production of embryos in *Datura innoxia* from isolated pollen grains by combined cold treatment and serial culture of anthers in liqud medium. *Protoplasma*, **99**: 11-17.

Ushiyama T., Shimizu T. and Kuwabara T. 1991. High frequency haploid production of wheat through intergeneric cross with teosinte. *Jap. J. Breed.*, **1**: 353-357.

Valkoun J., Dostal I., Kucerova D., Novak F.J., Havel L. and Dolezel J. 1984. Employment of embryo culture in distant hybridization of wheat. *Plant tissue and cell culture application to crop improvement*, 461-462.

Vardi A., Arzee-Gonen P., Frydman-Shani A., Bleichman S. and Galun E. 1989. Protoplast fusion and vertification by mitochondrial-DNA restriction profiles. *Theor. Appl. Genet.*, **75**: 51-58.

Vergue P., Riccardi F., Beckerf M. and Dermas C. 1993. Identification of a 32-KDa anther marker protein for androgenic response in maize, *Zea mays* L. *Theor. Appl. Genet.*, **86**: 843-850.

Voorrips R.E. and Visser D.L. 1990. Doubled haploid lines with clubroot resistance in *Brassica oleracea*. P. 40. *In*: *Proc. 6th Crucifer Genetics Workshop*, (eds.) J.R. McFerson, S. Kresovich, and S.G. Dwyer. USDA-ARS, Geneva, NY.

Waara S. and Glimelius K. 1995. The potential of somatic hybridization in crop breeding. *Euphytica*, **85**: 217-233.

Wang Youping and Luo Peng. 1998. Intergeneric hybridization between *Brassica* species and *Crambe abyssinica*. *Euphytica*, **101**: 1-7.

Wei Z.M. 1982. Pollen callus culture in *Triticum aestivum*. *Theor. Appl. Genet.*, **63**: 71-73.

White P.R. 1934. Potentially unlimited growth of excised tomato root tips in a liquid medium. *Plant Physiol.*, **9**: 585-600.

White P.R. 1939. Potentially unlimited growth of excised plant callus in artifical nutrient. *Am. J. Bot.*, **26**: 59-64.

Wiberg E., Rahlen L., Hellman M., Tillberg E., Glimelius K. and Stymne S. 1991. The microspore-derived embryo of *Brassica napus* L. as a tool for studying embryo-specific lipid biogenesis and regulation of oil quality. *Theor. Appl. Genet.*, **82**: 515-520.

Williams E.G. and De Lautour G. 1980. The use of embryo culture with transplanted nruse endosperm for the production of interspecific hybrids in pasture legumes. *Bot. Gaz.*, **141**: 252-257.

Withers L.A. and Engelman F. 1997. *In vitro* conservation of plant genetic resources. *In*: *Biotechnology in Agriculture*. (ed.) Altman A., Marcel Dekker, New York.

Witherspoon W.D. Jr., Wernsman E.A., Gooding G.V. Jr. and Rufty R.C. 1991. Characterization of a gameto-clonal variant controlling virus resistance in tobacco. *Theor. Appl. Genet.*, **81**: 1-5.

Wojciechowska B. and Pudelska H. 1992. Intergeneric hybrids of *Hordeum vulgare* L. × *Secale cereale* L. regenerated from embryo callus culture. *Genetica-Polonica*, **33**: 87-96.

Wojciechowska B. and Pudelska H. 1994. Wide hybridization in the Triticeae involving *Hordeum, Triticum, Triticale* and *Hordeum* × *Secale*. *Genetica-Polonica*, **35**: 221-230.

Wu B.H. and Wang Z. 1995. Callus induction, growth and plant regeneration from globular embryos of the F_1 hybrid *Roegneria ciliaris* × *Triticum aestivum* cv. J11. *Acta-Agronomica-Sinica*, **21**: 373-376.

Xu S.L., Zhung Y.Q. and Xu J. 1986. Genetic variation in clonal lines of barley-wheat hybrids and their back cross progeny. *J. Agr. Sci.*, **2**: 20-25.

Ye G.N., Daniell H. and Sanford J.C. 1990. Optimization of delivery of foreign DNA into higher plant chloroplasts. *Plant Mol. Biol.*, **15**: 809-819.

Yin K.C., Hsu C., Chu C.Y., Pi F.Y., Wang S.T., Liu T.Y., Chu C.C., Wang C.C. and Sun C.S. 1976. A study of the new cultivar of rice raised by haploid breeding method. *Sci. Sinica.*, **19**: 227-242.

Zarsky V., Garrido D., Rihova L., Tupy J., Vicente O. and Heberle-Bors E. 1992. Derepression of the cell cycle by starvation is involved in the induction of tobacco pollen embryogenesis. Sex. *Plant Reprod.*, **5**: 189-194.

Zelcer A., Aviv D. and Galun E. 1978. Interspecific transfer to cytoplasmic male sterility by fusion between protoplast of normal *Nicotiana sylvestris* and X-ray irradiated protoplast of male sterile *N. tabacum*. *Z. pflanzenphysiol*, 90: 397-407.

Zimmerman J.L. 1993. Somatic embryogenesis: A model for early development in higher plants. *Plant Cell*, **5**: 1411-1423.

Zhongsen L. and Thomas T.L. 1998. *PEI1*, an Embryo-specific Zinc finger protein gene required for heart-stage embryo formation in *Arabidopsis*. *The Plant Cell*, **10**: 383-398.

Plant Breeding - Mendelian to Molecular Approaches
H. K. Jain and M. C. Kharkwal (eds.)

10

Molecular Markers in Improvement of Wheat and *Brassica*

M. Lakshmikumaran[1], T. Mohapatra[2], V.S. Gupta[3] and P.K. Ranjekar[3]

Abstract

Recent developments in DNA-based marker technology have opened up newer avenues for the studies on plant molecular genetics. The new marker systems have enabled researchers to construct genetic maps and to examine directly the genotypes of plant species rather than phenotypes. These markers have also provided means to address problems related to genome structure and evolution, understand the genetic basis of morphological variation, and study the genomic distribution of genes and the pattern of inter- generic/inter-specific gene flow. Genetic maps based on molecular markers have been constructed for a number of plant species such as *Arabidopsis thaliana, Brassica species, Oryza sativa, Triticum aestivum, Zea mays, Hordeum vulgare, Pisum sativum,* to name a few. The most important and practical application of these maps has been in generating markers linked to both qualitative and quantitative traits of agronomic importance and employing them in marker assisted breeding. The focus of this review is to summarise recent developments in the area of DNA marker technology with special reference to wheat (*Triticum aestivum*) and *Brassica* coenospecies and highlight its applications in specific areas of plant breeding.

Introduction

Recent developments in DNA marker technology have opened up new avenues for studies in genetics and for the breeding of new cultivars. Technologies such as Restriction Fragment Length Polymorphism (RFLP), Random Amplified Polymorphic DNA (RAPD), Sequence Tagged Site (STS), Amplified Fragment Length Polymorphism (AFLP) and more recently Single/Simple Nucleotide Polymorphism (SNP) allow geneticists and breeders to look directly at the genotype of a plant. Molecular or DNA based markers offer many advantages over conventional morphological markers. They are developmentally stable, detectable in all tissues, unaffected by environmental conditions, insensitive to epistatic or pleiotropic effects, and provide a choice of co-dominant or dominant markers.

[1]Tata Energy Research Institute, New Delhi 110 003, India
[2]Indian Agricultural Research Institute, New Delhi 110 012, India
[3]National Chemical Laboratory, Pune 411 008, India

DNA-markers can be classified broadly into two categories, namely hybridization-based (RFLPs) and Polymerase Chain Reaction (PCR)-based (RAPD, STS and AFLP).

Hybridisation Based Markers (RFLP)

The great bulk of genetic variation at the nucleotide level may not be visible at the phenotypic level. It is this variation that is exploited to generate RFLP. RFLPs were first developed for human genome analysis (Botstein *et al.*, 1983) and then adapted for plant genome analysis. Nucleotide sequence variation in DNA resulting from point mutations occasionally leads to loss or gain of restriction endonuclease site. DNA alterations involving larger regions are the results of insertion, deletion, inversion or translocation events and can result in changes in the distribution of several restriction endonuclease sites simultaneously. The size distribution of DNA fragments affected by such changes will be different from that of an unaffected individual, thus resulting in a RFLP pattern. RFLP between two or more individuals may be detected in the DNA by digestion with an appropriate restriction enzyme, followed by electrophoresis and Southern blotting. The Southern blot is then probed with a labelled nucleic acid fragment that is homologous to the affected region. Single or low copy sequences in the genome are ideal to be used as RFLP probes and the source may vary from cDNA sequence to anonymous sequences such as non-coding parts of the gene. Nuclear RFLP probes have been shown to be stably inherited, segregated in a Mendelian fashion and exhibiting co-dominance. Co-segregation of a trait and the RFLP banding pattern data indicate a linkage between the two. These probes may thus be linked to a region by following its inheritance and segregation in a mapping population.

PCR Based Markers

Polymerase Chain Reaction (PCR) is an extremely powerful and sensitive technique, which is used for *in-vitro* amplification of specific target site in a DNA sample. In a typical target PCR, prior sequence information of the site to synthesise site specific oligonucleotide primers is required, in order to amplify the site. At the start of the reaction, the double stranded DNA is denatured by heating and the resulting single stranded template DNA is then annealed to the oligonucleotide primers that are complementary to the sites on either side of the target locus. Using DNA polymerase to synthesise the complementary strand, the entire cycle is repeated 25-40 times, which leads to more than a million-fold amplification in the copy number of the initial target sequence. Modification of this basic PCR technique has led to development of many classes of molecular markers as briefly described here.

RAPD

This represents one of the first methods based on PCR, which was developed simultaneously by Welsh and McClelland (1990) and Williams *et al.* (1990). Basically genomic DNA with single short (usually ten nucleotide in length) primer of arbitrary nucleotide sequence is used in PCR, and the amplification products are electrophoresed and visualized by ethidium bromide staining. Each amplification product is derived from a region of the genome that contains DNA segments (500-2500 bp) with homology to the primers and should be present on the opposite strand of DNA for amplification. Use of decamer primers and low annealing temperatures (37-40 °C) ensures several sites distributed randomly in the genome to bind to a single random primer and gives rise to several amplification products. RAPD polymorphism is the result of either a nucleotide change that alters the primer-binding site, or of an insertion/deletion within the amplifiable region and scored as presence or absence of a particular band. RAPD markers are usually dominant in nature and alleles for a particular locus cannot be differentiated. For mapping purpose, such dominant markers are converted into specific PCR based markers known as Sequence Characterized Amplified Region (SCAR). (Paran and Michelmore, 1993).

In view of its simplicity, relative ease to perform, and no need of prior sequence information, RAPD has been a method of choice for assessing genetic variability, estimating genetic relatedness, fingerprinting of plant genomes, genetic mapping and tagging of genes.

EST/STS/CAPS

With rapid advances in sequencing of genomes, a vast amount of sequence data from complete and partial sequencing projects have been generated. Sequencing of cDNA clones from both ends generates Expressed Sequence Tags (ESTs), whereas, any part of the genome flanked by known sequence would be considered as Sequence Tagged Sites (STS). However, most of these loci, both STS and sites with ESTs have not been localised on linkage maps. The sequences from the ESTs and STSs can be used to design primers, and with the help of PCR technique, amplify the specific genomic region. Amplified products generating polymorphism between individuals can be scored. If no polymorphism is obtained on amplification, then the amplified product is digested with restriction enzymes and scored for polymorphism which is termed as Cleaved Amplified Polymorphic Sequences (CAPS). This is performed on individuals of a segregating population so that the resultant polymorphism scored can then be mapped on a linkage group. EST sequences have been used to map co-dominant ecotype specific PCR markers on *Arabidopsis* genome by Konieczny and Ausubel (1993).

AFLP

Amplified Fragment Length Polymorphism (AFLP) technique has been originally designed to fingerprint genomic DNA samples (Vos *et al.*, 1995), which could be used as a source of genetic markers to generate linkage maps and as markers that are linked to a trait or locus of interest. It is based on selective amplification of a subset of restricted DNA fragments using the PCR procedure. Basically DNA is digested with a combination of two restriction endonucleases, followed by ligation of double stranded restriction-site-specific adapters to the ends of digested DNA to generate template for PCR amplification. The sequence of adapters in combination with restriction site sequence is used to design primers, to add selective nucleotides at the 3' end. PCR leads to amplification of only a selective set of DNA fragments, where the nucleotides flanking the restriction sites match the selective nucleotide of the PCR primer. The primers can be radioactively labelled, or one of the dNTPs during PCR is radiolabeled to facilitate autoradiography of fragments after electrophoresis, on denaturing polyacrylamide gel. Each experiment, on an average, yields about 50-100 bands in the size range of 50-350 bp, depending on the enzyme/primer combination. AFLP has widespread applications including assessment of genetic diversity studies (Hongtrakul *et al.*, 1997), construction and saturation of linkage maps (Becker *et al.*, 1995), and tagging of genes (Voorrips *et al.*, 1997).

Applications of DNA Markers in *Triticum aestivum*

Wheat is one of the major food crops of the world, supplying about 55 % of the carbohydrate consumed world-wide. It is an allo-hexaploid containing three related genomes, A, B and D. The haploid DNA content of hexaploid wheat is approximately 1.7×10^{10} bp (Arumuganathan and Earle, 1991). The average size of a wheat chromosome is about 810 Mb, which is almost twice the size of the haploid rice genome. Such a large wheat genome has evolved through natural cross hybridisation and polyploidy, with about 80 % of the genome consisting of repetitive DNA. This has been the reason for the delay in use of molecular markers in wheat. RFLP was the first marker system to be used in wheat genome mapping. Initially, using low copy sequences as probes in RFLP analysis, chromosome specific partial molecular maps were constructed in wheat (Chao *et al.*, 1989; Devos *et al.*, 1993; Nelson *et al.*, 1995). Subsequently, whole genome maps were developed using RFLP markers, in spite of low level of polymorphism (Table 1). Liu and Sunewaki (1991) constructed a 1800 cM RFLP map covering 197 loci using 66 F_2 plants derived from *T. aestivum* × *T. spelta*. Devos and Gale (1993) also developed a map of hexaploid wheat having several interesting features. Some of the probes detected multiple loci, which mapped to the three homeologous chromosomes in the same linear order.

Table 1. Molecular maps in hexaploid wheat

Mapping population	Marker type	Number of loci	Reference
F_2 population (*T. aestivum* cv. 'Chinese spring' × *T. spelta* var. *deutamelianum*)	RFLP	197	Liu and Sunewaki (1991)
Doubled haploids ('Chinese Spring' × 'Courtot')	RFLP SSR	264 2	Cadalen *et al.* (1997)
F_2 population ('W7984' × 'Opata85')	SSR	279	Roder *et al.* (1998)
Doubled haploids ('Garnets' × 'Saunders'	AFLP SSR	620 42	Penner *et al.* (1998)
F_2 population ('Chinese Spring' × 'Synthetic')	SSR	53	Stephenson *et al.* (1998)

To circumvent the problems of low level of polymorphism detected by RFLP and develop more robust and fast PCR based markers, Devos *et al.* (1995) developed two SSR markers based on sequences already available in the database. These markers were genome specific and detected high levels of variation. Roder *et al.* (1998) developed SSR based map of wheat that consisted of 279 loci. Most of these markers show locus specific amplification and reveal high level of allelic variation, and thus are highly useful in mapping and tagging of agronomically important genes.

Mapping of Genes for Rust Resistance

Rust is one of the diseases of great economic importance, because the productivity of wheat all over the world is affected by this disease, which is caused by fungus *Puccinia* species. Development of genetic resistance to rust is the most efficient, cost effective and eco-friendly approach to overcome the losses caused by rusts. So far, over 100 rust resistance genes have been identified in wheat and related species and some of them have been introgressed into wheat (McIntosh *et al.*, 1995). However, due to selection pressure and evolution, new and virulent races of the fungus appear, which stresses a need to develop durable resistance, which can be achieved by pyramiding seedling and adult plant resistance genes. However, the selection of genotypes carrying two or more genes using traditional host parasite interaction is very time consuming and often not possible, due to lack of isolates with specific virulence and difficulty in identifying one resistance gene in the presence of another gene.

With the utilization of newer and improved DNA markers, more than 50 different loci have been tagged, and majority of these loci represent the sites of disease resistance genes (Table 2) (Landridge and Chalmers, 1998; Naik *et al.*, 2001). Leaf rust resistance gene *Lr*1 (Feuillet *et al.*, 1995) is an example where RFLP marker was converted into allele specific STS marker. Here, probe PSR 567 was found to be completely linked with the gene in one population with 156 F2 plants and the marker was assigned to the single copy locus on chromosome 5DL. Twenty-seven breeding lines containing *Lr*1 resistance gene in different background showed the same band as *Lr*1/6* Thatcher when hybridized with pTAG621. There was no recombination between the gene and the marker in the second F_2 population with 165 plants. As stated above, though the RFLP marker was converted into STS marker, the marker allele was also found to be present in 50 % of the lines not containing *Lr*1. The reason postulated by the authors is the possible origin of the resistance genes as a result of a point mutation from the locus carrying the marker allele (Feuillet *et al.*, 1995).

Sayfarth *et al.* (1999) reported RFLP marker linked with the adult plant resistance gene, *Lr*35. The polymorphic band BCD 260/0.9 was sequenced to construct the STS primer and was used to amplify DNA from NIL, susceptible parent and the population. The product of STS amplification was again digested with tetra-cutter restriction enzyme *Dde I* and the fragments were separated on the gel to obtain a 450 bp dominant marker, called CAPS (Cleaved Amplified Polymorphic Sequence), specific to the resistance. Helguera *et al.* (2000) identified a RFLP marker Xabc465 linked to the leaf rust resistance gene *Lr*47 and converted it to STS marker, which was completely linked to the gene.

Table 2. List of resistance genes tagged in hexaploid wheat

Resistance Genes	Marker system	Reference
Lr1	PCR, RFLP	Feuillet *et al.* (1995)
Lr3	RFLP	Parker *et al.* (1998)
Lr9	RAPD, PCR	Schachermayr *et al.* (1994)
Lr10	RFLP/STS	Schachermayr *et al.* (1997)
Lr19	RFLP	Autrique *et al.* (1995)
Lr20	RFLP	Parker *et al.* (1998)
Lr23	RFLP	Nelson *et al.* (1997)
Lr24/Sr24	RAPD, PCR	Schachermayer *et al.* (1995)
Lr25	RAPD/DGGE	Procunier *et al.* (1995)
Lr27	RFLP	Nelson *et al.* (1997)
Lr28	RAPD/PCR	Naik *et al.* (1998)
Lr29	RAPD/DGGE	Procunier *et al.* (1995)
Lr32	RFLP	Autrique *et al.* (1995)
Lr34	RFLP	Nelson *et al.* (1997)
Lr35	ISSR/PCR	Gold *et al.* (1999), Sayfarth *et al.* (1999)
Lr37	RGA clone	Seah *et al.* 2001
Lr47	RFLP/STS	Helguera *et al.* (2000)
Yr15	RAPD/RFLP	Sun *et al.* (1997)
Yr17	RGA clone	Seah *et al.* (2001)
YrH52	Microsatellite	Peng *et al.* (1999)
Sr5	RFLP	Parker *et al.* (1998)
Sr9e	RFLP	Parker *et al.* (1998)
Sr22	RFLP	Paull *et al.* (1994)
Sr36	RFLP	Parker *et al.* (1998)
Sr38	RGA clone	Seah *et al.* (2001)
Sr39	ISSR	Gold *et al.* (1999)
Pm1	RAPD/RFLP	Ma *et al.* (1994), Hartl *et al.* (1995), Hu *et al.* (1997)
Pm2	RFLP	Ma *et al.* (1994), Hartl *et al.* (1995), Mohler & Jahoor (1996)
Pm3	RFLP	Hartl *et al.* (1993), Ma *et al.* (1994)
Pm4	RAPD	Li *et al.* (1995)
Pm18	RFLP	Hartl *et al.* (1995)
Pm21	RAPD	Qi *et al.* (1996)
T10	RFLP/SCAR	Procunier *et al.* (1997)
Bt-10	RAPD	Demeke *et al.* (1996)
Pch-2	RFLP	Delapena *et al.* (1996)
H3, H5, H6, H9, H, 10, H11, H12, H13, H14, H16, H17	RAPD/DGGE	Dweikat *et al.* (1994, 1997)
H21	RAPD	Seo *et al.* (1997)
Cre1	RFLP/STS	Williams *et al.* (1994)
Cre3/Ccn-D1	RAPD/PCR	Eastwood *et al.* (1994)
Wms1	RAPD/STS	Talbert *et al.* (1996)

William *et al.* (1997) identified two RAPD markers linked with QTLs associated with the *Lr*34 gene and converted them into RFLP markers, to overcome the repeatability problems. DNA bulk segregation analysis was conducted on the population of 77 recombinant inbred lines (RILs) using 400 RAPDs. The polymorphic bands were mapped, and loci separated by 2 cM were found with 1:1 observed segregation in each of the marker and the leaf tip necrosis loci (Ltn), which was considered either as very tightly linked trait, or pleiotropic with *Lr*34 (Singh, 1992).

A similar approach was used for tagging the *Lr*24 gene, where the RAPD marker $CPJ09_{550}$ was converted into RFLP that mapped into translocated fragment from *Agropyron elongatum* on the 3DL in wheat (Schachermayr *et al.*, 1995). Two tightly linked RAPD markers were identified for two alien genes introgressed into wheat viz., *Lr*25 (from rye) and *Lr*29 (from *A. elongatum*) (Procunnier *et al.*, 1995). Similarly, another RAPD marker was identified for *Yr*15, a stripe rust resistance gene from *T. dicoccoides* (Sun *et al.*, 1997).

Mapping of Genes for Quality Traits in Wheat

Improvement of quality of wheat for the end product use is one of the important mandates of wheat breeding all over the world. Wheat is used in preparation of different food products like various types of breads, biscuits, cookies and chapatis. The quality of these end products depends on specific characters like grain protein quality and content, grain hardness etc. These quality traits are quantitative in nature and are governed by a number of genes dispersed on various chromosomes in wheat genome. Identification of markers linked to these traits helps in pyramiding different genes controlling these characters through MAS, and will ultimately guide the breeder to construct beneficial allelic combinations for development of superior cultivars.

Grain Protein Content

Grain Protein Content (GPC) is one of the important quality traits and plays a major role in end product quality of wheat. It is a complex quantitative trait and is influenced by environment. Halloran (1975) and Konzak (1977) have shown that a few genes with major effects, with many other genes with minor effect govern the trait. The reports by Kuspira and Unrau (1975) and Diehl *et al.* (1978) have, however, suggested quantitative nature of GPC trait with several genes distributed throughout the genome. Molecular studies carried out on tetraploid wheat (Blanco *et al.*, 1996; Joppa *et al.*, 1997) and hexaploid wheat (Prasad *et al.*, 1999) have also suggested polygenic nature of this trait with presence of several QTLs on different chromosomes. Recently, Dholakia *et al.* (2001) have demonstrated environmentally stable as well as specific loci associated with GPC and have shown the utility of DNA markers in studying G × E interactions affecting GPC.

Seed Size and Shape

The end use quality of wheat is also greatly influenced by seed size (Campbell *et al.*, 1999), which can be measured indirectly by thousand kernel weight (TKW). Generally, bold grain with attractive and uniform shape, size and amber colour fetches a higher price in the market. Thus, development of cultivars fulfilling these requirements of global market has become a major target of wheat breeding programs. Genetic analysis of seed size in wheat is complicated because of its quantitative nature (Giura and Saulescu, 1996).

During the last two decades, several attempts have been made to understand the genetic basis of seed size in wheat. Blanco *et al.* (1996) have studied the heritability of this trait, while Giura and Saulescu (1996) have reported that chromosomes 6D and 4A are associated with high TKW, whereas 5B and 5D are associated with low TKW. Petrovic and Worland (1988) have shown that TKW is associated with chromosome 5D, while Halloran (1976) have reported that chromosome 4B is associated with seed size. Recently, Campbell *et al.* (1999) have reported that chromosomes 1A, 1B, 3B, and 7A, have the loci, which control TKW. Ammiraju *et al.* (2001) have reported presence of three seed size QTLs on chromosomes 2DL, 6BL and 1DS using ISSR markers. Gupta *et al.* (1999) have reported chromosomes IAS to be associated with TKW using STMS markers. Ammiraju *et al.* (2003) have further dissected the seed size and shape into its components, such as seed length, width and density, and have shown ten markers to be associated with seed length and seed width and three markers for factor form density. The same study has also demonstrated the chromosome regions 1D, 5B, 6A, and 6B for seed length, 2D and 6B for seed width and 1A and 2D for form density. Thus dissection of important traits by QTL analysis will prove to be useful for the wheat breeders, to develop better cultivars that have improved economic value.

Kernel Hardness

Kernel hardness (KH) is yet another important quality parameter of wheat grain, which has a profound effect on milling, baking as well as end use qualities. Hard wheat flour is generally used for baking bread, while soft wheat flour is used for making pastries, biscuits and cakes (Tippless *et al.*, 1994). The hardness of wheat is influenced by many factors such as protein content (PC) with direct effect (Bushuk, 1998), and hectolitre weight (HW) and thousand kernel weight (TKW) with indirect contribution (Pomeranz and Williams, 1990). Many researchers have studied the genetic control of KH. Symes (1969) first demonstrated that hardness is governed by one major gene (*Ha*), located on short arm of 5D (Mattern *et al.*, 1973; Law *et al.*, 1978). Later, Baker (1977) and Pomeranz and Williams (1990) reported presence of two major and three minor genes controlling the expression of kernel hardness. Majority of the molecular studies have been restricted to the *Ha* locus. Sourdille *et al.* (1996) reported presence of a major locus *Xmta* 9 linked closely to the gene *ha* on 5DS and 7 other minor loci distributed on different chromosomes. Genes for puroindoline a (*pin*-a), puroindoline b (*pin*-b) and grain softness protein (*Gsp*-1), closely linked to *Ha* (soft) locus, have been shown to be associated with the expression of grain softness (Dubcovsky *et al.*, 1999, Turner *et al.*, 1999). Recently, Galande *et al.* (2001) attempted to decipher this trait into constituents like NIR, PC, HW and TKW and reported chromosome 2B and 6B to be associated with KH and PC, thus demonstrating the complexity of this trait.

Molecular Markers in *Brassica coenospecies*

The *Brassicas* are an important group of crops yielding edible oils and vegetables. This group comprises of six cultivated species. *B. campestris/rapa* (2n = 20, AA), *B. nigra* (2n

= 16, BB) and *B. oleracea* (2n = 18, CC) are diploids. *B. juncea* (2n = 36, AABB), *B. napus* (2n = 38, AACC) and *B. carinata* (2n = 34, BBCC) are digenomic tetraploids, which evolved in nature following hybridization between the constituent diploid species. Discovery of high levels of inter- as well as intra-specific DNA polymorphism in RFLP profiles obtained with random genomic DNA clones as probes by Figdore *et al.* (1988) encouraged molecular mapping of *Brassica* genomes and tagging of genes for several useful traits.

Development of Linkage Maps

Molecular maps have been developed for several species such as soybean, eucalyptus, coffee, sugarcane, cotton, *Brassica, Arabidopsis*, oat, sorghum, sunflower, poplars, rice, maize, and wheat using RFLPs, minisatellites, microsatellites, RAPD, AFLP, EST/STS and CAPS. The map construction in *Brassica* crops started with the use of isozyme markers (Arus and Orton, 1983; Chen *et al.*, 1989). However, due to their limited availability, and environmental influence, the progress was not significant. Later, the development of DNA-based marker systems such as RFLP and RAPD provided for a large number of markers, which enabled construction of genome maps in the cultivated *Brassica* species (Table 3).

Table 3. Molecular linkage maps constructed in *Brassica* species

Species	Marker type	Reference
B. oleracea	RFLP	Slocum *et al.* (1990), Kianian and Quiros (1991), Kianian and Quiros (1992), Landry *et al.* (1992), Bohuon *et al.* (1996)
	RFLP/AFLP	Voorips *et al.* (1997)
	EST	Lan *et al.* (2000)
	Microsatellite	Saal *et al.* (2001)
	SRAP*	Li and Quiros (2001)
B. campestris	RFLP	Song *et al.* (1991), Chyi *et al.* (1992), Teutonico and Osborn (1994)
B. nigra	RFLP/RAPD	Truco and Quiros (1994)
	RFLP	Lagercrantz and Lydiate (1995)
B. napus	RFLP	Landry *et al.* (1991), Ferreira *et al.* (1994), Uzuonova *et al.* (1995)
B. juncea	RFLP RAPD	Sharma *et al.* (1994), Cheung *et al.* (1997), Axelson *et al.* (2000), Mohapatra *et al.* (2002), Sharma *et al.* (2002)

* Sequence-related amplified polymorphism (SRAP)

Selfed population at F_2 or later stages (F_3, F_4 or Recombinant Inbred Lines) have been used to construct genetic maps in *B. oleracea* (Landry *et al.*, 1992, Kianian and Quiros, 1991), *B. nigra* (Truco and Quiros, 1994), *B. campestris/B. rapa* (Chyi *et al.*, 1992; Kole *et al.*, 1997), *B. napus* (Hoenecke and Chyi 1991; Landry *et al.*, 1991) and *B. juncea* (Sharma *et al.*, 1994; Mohapatra *et al.*, 2002; Sharma *et al.*, 2002). Backcross mapping population has been used to generate maps for *B. nigra* (Lagercrantz and Lydiate, 1995) and Doubled haploid (DH) mapping population has been used to construct maps of *B. oleracea* (Voorrips *et al.*, 1997; Li and Quiros, 2001; Saal *et al.*, 2001), *B. napus* (Ferreira *et al.*, 1994; Uzunova *et al.*, 1995; Foisset *et al.*, 1996) and *B. juncea* (Cheung *et al.*, 1997; Axelson *et al.*, 2000). Apart from these, a combination of mapping populations has also been used. Ramsay *et al.* (1996) and Kearsay *et al.* (1996) have used backcross individuals of DH lines to generate maps of *B. oleracea.*

The molecular linkage maps in *Brassica coenospecies* have greatly facilitated study of genome organisation and evolution pattern. Prior to the development and use of molecular markers, genome relationships were being established based on chromosome pairing behaviour in the inter-specific hybrids. A common assumption was that the n = 8 (*B. nigra*), 9 (*B. oleracea*) and 10 (*B. campestris*) cultivated species of *Brassica* evolved in an ascending dysploid series, from a common primitive genome by natural hybridization and polyploidy resulting in aneuploidy and chromosomal rearrangements. These diploid *Brassica* species thus were considered secondary polyploids (Prakash and Hinata, 1980). Use of RFLP markers to construct linkage maps revealed at a finer scale the structure and evolution of the *Brassica* genomes, strongly supporting the above assumption. A common set of DNA probes used to construct linkage maps in different species enabled comparison of genomes. Besides, detection of duplicate/multiple segregating loci with a single probe greatly facilitated understanding the extent of relationship among the chromosomes of a diploid species. Extensive sequence duplication to the extent of the 50 % of the loci was evident (Truco and Quiros, 1994) in all the three diploid genomes (A, B and C). In general, these duplications were found distributed on more than one chromosome supporting the role of translocation in genome evolution. Reversion of the order of the markers in different linkage groups was also clearly evident (Song *et al.*, 1991; Kianian and Quiros, 1992) thereby suggesting contribution of inversion events to the structuring of *Brassica* genomes.

Molecular mapping has also made it possible to compare the genomes of *Brassica* with that of *Arabidopsis thaliana.* Kowalski *et al.* (1994) found extensive rearrangements between RFLP maps of *A. thaliana* and *B. oleracea,* although islands of conserved gene order were identified. In general, the single copy genes in *Arabidopsis* were found in multiple copies in *Brassica,* and the physical distance between genes was also larger in *Brassica* for conserved chromosome segments (Sadowski *et al.*, 1996). *A. thaliana* is a wild relative belonging to the same family and has a smaller ($n = 5$) simple genome, which has been completely sequenced. The genomic information and tools available in the *Arabidopsis* database should aid characterisation of *Brassica* genomes, eventually leading to genetic enhancement of this group of crops yielding edible oil and vegetables.

Mapping Genes of Economic Importance

During the last decade, a number of agriculturally and economically important genes from crop and tree species have been mapped and tagged using molecular markers, which is a prerequisite for any marker-assisted-breeding and selection programme. The maps of *Brassica* species have been used to tag various traits (Table 4) and some of these are discussed below.

Table 4. List of some of the traits mapped in *Brassica* species

Species	Characters	Reference
B. napus	Resistance against *Leptosphaeria maculans*	Ferreira *et al.* (1995a) Dion *et al.* (1995)
	Resistance against *Albugo candida*	Ferreira et al. (1995c)
	Vernalization requirement and flowering time	Ferreira *et al.* (1995b)
	Nuclear fertility restorer gene	Delourme *et al.* (1994), Jean *et al.* (1997)
	Linolenic acid concentration	Hu *et al.* (1995); Tanhuanpää *et al.* (1995); Thormann *et al.* (1996); Hu *et al.* (1999)
	Seed coat colour	Van Deynze *et al.* (1995)
	Erucic acid synthesis	Ecke *et al.* (1995); Jourden *et al.* (1996c); Thormann *et al.* (1996), Barret *et al.* (1998b); Fourmann *et al.* (1998)
	Seed aliphatic glucosinolate content	Toroser *et al.* (1995)
	Linoleic/Oleic acid	
	Dwarf BREIZH (*Bzh*) gene	Hu *et al.* (1999); Barret *et al.* (1998a);
B. campestris	Oleic acid concentration	Tanhuanpää *et al.* (1996)
	Resistance to *Albugo candida* race2	Kole *et al.* (1996)
B. oleracea	Resistance to *Plasmodiophora brassicae*	Figdore *et al.* (1993); Voorips *et al.* (1997)
	Resistance to *Xanthomonas campestris*	Camarago *et al.* (1995)
B. juncea	Oil content	Sharma *et al.* (1999)
	White rust resistance	Cheung *et al.* (1998); Prabhu *et al.* (1998); Mukherjee *et al.* (2001)
	Oleic acid level	Sharma *et al.* (2002)

Disease Resistance

Molecular markers have been generated for genes conferring resistance to *Leptosphaeria maculans* in *B. napus* by various workers. The resistance locus *LmFr*1 was linked to markers cDNA 011 and cDNA 110 (Dion *et al.*, 1995), and localized onto the Linkage Group 6 (LG6) (Ferreira *et al.*, 1995a). Loci *pb-3* and *pb-4* conferring resistance to *Plasmodiophora brassicae* in *B. oleracea* were identified and linked to RFLP and AFLP markers (Voorrips *et al.*, 1997). Similarly, Figdore *et al.* (1993) also identified markers 14a on LG1, marker 48 on LG4 and 177b on LG9 linked to clubroot resistance (resistance to *Plasmodiophora brassicae* wor. Race 7) in *B. oleracea*. *ACA1* locus, conferring resistance to *Albugo candida* race 2 in *B. rapa* was flanked by markers ec2b3a and wg6c1a and mapped to LG4 (Kole *et al.*, 1996). The white rust resistance locus in *B.*

juncea has been mapped in three independent studies. Cheung *et al.* (1998) identified one co-segregating dominant RFLP marker 140a, which mapped to LG7 in *B. juncea* and designated the locus as *Acr*. In the same year, white rust resistance locus $Ac2_1$ present in an Eastern European source that was effective against a Canadian isolate of the pathogen was mapped using RAPD markers (Prabhu *et al.*, 1998). Recently, Mukherjee *et al.* (2001) mapped a locus designated as *Ac2*(t) effective against Indian isolate of the white rust pathogen. QTLs controlling resistance to *Xanthomonas campestris* were mapped on LG1 and LG9 in *B. oleracea* (Camarago *et al.*, 1995).

Vernalization Requirement

The loci controlling the vernalization requirement and flowering time in *B. napus* were mapped and a number of markers were identified to be linked to both these traits on LG9 (Ferreira *et al.*, 1995b). Two genomic regions controlling the plant habit (annual/biennial) in *B. rapa* were identified, which corresponded to the genomic segments carrying genes related to the flowering time in *B. napus* (Teutonico and Osborn, 1995; Ferreira *et al.*, 1995b). In *B. oleracea*, a total of five QTLs were detected for flowering time and reproductive traits like S-locus *slg6*, responsible for incompatibility response and petiole length (Camarago and Osborn, 1996). Homologues of the CO gene determining flowering time in *A. thaliana* were identified in *B. nigra* in a fine scale comparative mapping analysis (Lagercrantz *et al.*, 1996).

Oil Content and Fatty Acid Composition

The oil content in the seed and its quality as defined by the constitution of fatty acid profile are the traits of commercial importance in oilseed *Brassica* species. Both oil quantity and quality have been studied using molecular markers. Ecke *et al.* (1995) mapped three loci on different linkage groups in *B. napus* using RFLP markers. In *B. juncea*, Sharma *et al.* (1999) identified three loci based on segregation of RAPD markers in a recombinant inbred population. A number of studies have been undertaken to generate markers for fatty acids such as linolenic acid, linoleic acid, oleic acid, palmitic acid and erucic acid. Two RAPD markers, $K\text{-}01_{1100}$ and 25a were generated and linked to the linolenic acid concentration (Hu *et al.*, 1995; Tanhuanpää *et al.*, 1995). In a recent study, RAPD markers linked to oleic, linolenic and linoleic acid were identified in *B. napus* (Hu *et al.*, 1999). The RAPD marker linked to the linolenic acid content was converted to a co-dominant SCAR marker (Hu *et al.*, 1999). The two alleles detected by the SCAR marker corresponded to low linolenic acid concentration (allele A) and high linolenic acid concentration (allele B). Markers linked to genomic regions controlling linolenic acid concentration in *B. napus* and corresponding to the *fad3* (omega-3-desaturase) gene in *A. thaliana* (Arondel *et al.*, 1992) were also identified (Jourden *et al.*, 1996 a,b; Thormann *et al.*, 1996). In another study, a single QTL containing six markers associated with oleic, palmitic and linoleic acid content was detected in *B. rapa* (Tanhuanpää *et al.*, 1996). Sharma *et al.* (2002) recently mapped two major QTLs influencing oleic acid level in *B. juncea* using both single factor analysis of variance and interval mapping. These two loci were located in 10.6 and 14 cM marker intervals respectively, and together explained 32.2 % of the trait variance.

Erucic acid loci have been linked to molecular markers by Ecke *et al.* (1995), Thormann *et al.* (1996), Jourden *et al.* (1996c) and Barret *et al.* (1998b) using BSA or RFLP analysis in *B. napus*. In each of the studies, two QTL were detected. These QTL have been positioned on LG 6 and LG12 (Ecke *et al.*, 1995) or on LG7 and LG15 (Thormann *et al.*, 1996). In an independent study, two QTL associated with the erucic acid level in *B. napus* were detected (Jourden *et al.*, 1996c) and mapped onto two different loci termed as E_1 and E_2. QTL E_1 and E_2 correspond to the two alleles of the β-Ketoacyl-synthase (KCS) derived from *B. campestris* and *B. oleracea*, the two parental species of *B. napus* and encode the fatty acid elongation 1 (Fae1) protein (Fourmann *et al.*, 1998). In *B. rapa* (Syn campestris) erucic acid loci were linked to RFLP markers (Teutonico and Osborn, 1994).

Seed Glucosinolate Content

Variable number of QTLs has been detected for seed glucosinolate content in *B. napus*. For example, Uzunova *et al.* (1995) detected four QTL, while Toroser *et al.* (1995) detected five QTL. These QTL, GSL-1, GSL-2, GSL-3, GSL-4 and GSL-5 localized on LG20, LG1, LG18, LG4 and LG13, respectively. Of these, GSL-1 and GSL-2 are responsible for low and high glucosinolate content (De Quiroz and Mithen, 1996).

Fertility Restorer Gene

Molecular markers for the nuclear restorer gene of the cytoplasmic male sterile (CMS) systems "Ogura" and "Polima" have been generated. Four RAPD markers OPC-02_{1150}, OPD-02_{1000}, OPF-06_{1200} and OPG-02_{700} were linked to the fertility restorer gene for "Ogura" radish cms system (Delourme *et al.*, 1994; 1998). The *Rfp1* locus responsible for fertility restoration in "Polima" cms was linked to 10 RFLP and one RAPD marker. One probe, cRF1b showed perfect linkage to *Rfp1* and *Rfp2* loci indicating that these could be allelic (Jean *et al.*, 1997). Delourme *et al.* (1994) used BSA strategy to identify RAPD fragments linked to restorer gene *Rfo*, in Ogura radish cms lines of *B. napus*. The RAPD fragments were later converted into SCAR markers for mapping purposes.

Seed Coat Colour

Since yellow/light seeded character is recessive, maternally inherited and influenced by environmental factors, it is desirable that the locus responsible for seed coat colour is linked to molecular markers. The seed coat colour gene has been tagged to various RFLP and RAPD markers. RFLP markers linked to seed coat colour in *B. napus* were identified using the Bulked Segregant Analysis (BSA) approach (Van Deynze *et al.*, 1995). RFLP marker 4NF6a and pLF15 co-segregated with seed coat colour in *B. napus*. RFLP markers associated with this trait were then mapped onto *B. napus* linkage maps (Van Deynze *et al.*, 1995). Seed coat colour trait in *B. campestris* was tagged with RAPD markers using *B. campestris-oleracea* addition lines (Chen *et al.*, 1997). A 3:1 ratio of segregation of brown: yellow seed in *B. rapa* indicated a monogenic control of this trait and was mapped to LG5 (Teutonico and Osborn, 1994). Upadhyay *et al.* (1996) studied segregation of the trait in an F_2 population of *B. juncea* and reported duplicate dominant gene action giving a phenotypic ratio of 15:1. Two RFLP markers flanking one of the interacting loci were identified. In a recent report, the seed coat colour trait was tagged using a combined approach of Bulked Segregant Analysis (BSA) and AFLP in *B. juncea* (Negi *et al.*, 2000). The segregation data of 15:1 for brown:yellow seed coat indicated a two gene control of the trait. Bulks of individuals from a selfed mapping population at the F4 stage were created and analysed using AFLP. Dominant AFLP markers specific to the seed colour trait were identified and converted into a co-dominant SCAR markers after PCR-walking (Negi *et al.*, 2000).

Marker Assisted Breeding (MAB)

Plant breeding creates novel combinations of genes and tailors new crop varieties of great economic value. As a continuing process, it has tremendously contributed in the past and is still contributing in a very significant way to the food and nutritional security in the world. Conventional plant breeding, starting from selection of parents to selection of desirable segregants is, however, largely based on phenotypes. The phenotype being the interaction product of genotype and environment, does not always allow an unbiased and objective evaluation of the genetic worth of a genotype. Molecular markers, which assay genetic variation directly at the level of DNA sequence, precisely identify a genotype and thus can complement many aspects of conventional plant breeding, including selection of the parents for hybridisation, gene introgression and elimination of linkage drag, gene pyramiding and development of improved purelines and hybrids. For this to happen in any crop species, the two important prerequisites are, availability of a framework molecular genetic map and markers tightly linked to the genes/loci affecting the traits of agricultural importance. In the crops described in the preceding sections, these two basic requirements have been largely fulfilled. However, massive efforts are now needed to validate the identified markers and develop easy-to-use protocols for their routine use in wheat and *Brassica* breeding programmes. Although no successful case of MAB has been reported in these two crops, application of molecular markers in the following areas of plant breeding looks highly promising, as has been demonstrated in other crops.

Gene Introgression

All the important cultivated crop species have many wild relatives, serving as rich reservoir of agronomically important genes. In wheat for instance, most of the genes for rust resistance have been derived from wild relatives. It has been reported that introgression of a specific locus from a donor genome, using a backcross breeding programme relying on traditional method of selection, requires a minimum of six generations of backcrosses. At this junction, the resultant progeny constitutes nearly 99 % of recipient genome and the rest from the donor genome. In comparison, marker-assisted selection (MAS) in a backcross-breeding programme would aid in attaining the same desired objective in only three backcrosses. Similarly, the genomic region flanking the target gene from the donor is replaced by the recurrent or recipient genome in only two backcrosses in a MAS programme, whereas the same would require nearly 100 backcrosses in a traditional method (Ribaut and Hoisington, 1998). The use of MAS thereby significantly reduces the time needed to develop crop varieties with desired traits. Besides, it is possible to eliminate linked undesirable regions introgressed from the wild along with the gene of interest, by exercising allele specific selection against the wild species. Simultaneous discovery and introgression of QTLs present in the wild relatives by employing the advance-backcross method for yield and its components in rice (Xiao *et al.*, 1998) and tomato (Eshed *et al.*, 1996) exemplifies the application of molecular markers in breaking yield barriers in crop plants.

Gene Pyramiding

Marker-assisted breeding and selection has a tremendous potential in strategies such as gene pyramiding. The principle of gene pyramiding is based on combining two or more major genes conferring resistance to a particular disease in a single plant genotype. The use of a single major gene may limit the resistance against one or few pathotypes of any causal organism and limit the resistance in the cultivars to a few years. In comparison, deployment of pyramided genes would confer resistance against all prevalent pathotypes and make it more durable. A combined effect of the resistance genes is thought to provide a broad spectrum of resistance by both individual gene action as well as quantitative, collective or additive gene action (Babujee and Gnanamanickam, 2000). However, it may be very difficult and a lengthy procedure to ascertain the number of genes that have been pyramided, using a traditional breeding and selection programme. Therefore, molecular markers that are linked or co-inherited with the individual resistance genes would be used for marker-assisted selection. The presence of multiple markers in a single genotype will indicate that pyramiding of gene has taken place (Williams *et al.*, 1996). A typical example of gene pyramiding is in rice. Lines carrying different combinations of four genes for bacterial leaf blight (BLB) resistance, *Xa4, xa5, xa13* and *Xa21*, could be identified (Huang *et al.*, 1997) using markers. These lines were then used by different workers to develop BLB resistant cultivars. For instance, Singh *et al.* (2001) carried out marker aided pyramiding of these genes and developed BLB resistant rice cultivar PR106. This demonstrated that markers could be efficiently utilised to combine recessive and dominant genes together to develop improved cultivars.

Development of Improved Varieties/Hybrids

Development of improved varieties by recombination breeding and of hybrids by use of improved parental inbreds remains major breeding objective in crop species. Classical breeding programmes mainly rely on phenotypic expression in respect of the complex target traits during pedigree or recurrent selection schemes to attain this objective. Molecular markers linked to the QTLs affecting the target traits should help in following their transmission over generations and thus assist in more precise identification of useful recombinants. Marker aided back-cross transfer of QTLs to create near-isogenic lines has been reported in tomato (Eshed and Zamir, 1995). In maize, through marker aided transfer of yield QTLs from inbred line Tx 303 to B73 and from Oh43 to Mo17, improved inbreds were generated, which were called as 'enhanced' lines. Fifteen enhanced B73 lines were crossed with 18 enhanced Mo17 lines to produce 93 hybrids that were evaluated in replicated field trials at North Carolina, USA. Six of the hybrids exceeded the national check hybrids by two standard deviations or more and two highest yielding enhanced hybrids gave 15 % more yield than the checks (Stuber, 1997). These results thus demonstrated that the marker-facilitated backcrossing could be successfully employed to manipulate the complex traits such as heterosis for grain yield.

Practical Considerations in MAB

Molecular markers are getting increasingly integrated into the conventional plant breeding programmes in almost all crops at various levels. At this stage it is worthwhile to look into various practical aspects and thus the feasibility of MAB for improving crop species.

Choice of Markers

RFLP, being the first marker system to be developed, has been extensively utilised in mapping and tagging of plant genes. It is a robust co-dominant marker system and is ideal for the study of comparative genomics. RFLP technique is, however, laborious and is not amenable to automation and thus cannot be used routinely for screening of a large number of segregants in plant breeding programmes. The RFLP markers identified as gene-tags till date have to be converted to PCR based STS markers, which will allow automation and therefore fast handling of a large number of samples. Among the PCR-based markers, AFLP and RAPD, because of high multiplex ratio, will be of great use in finding markers for genes and development of high-resolution maps in the gene region. For MAB, the AFLP and RAPD bands linked to the genes of interest are also to be converted to STS markers, such as SCARs/CAPS. The STS markers particularly those which assay polymorphism in the SSR (which are known as STMS markers) will be useful in detecting differences between closely related genotypes and thus are the markers of choice for MAB. The single/simple nucleotide polymorphism assays being developed in some plant species, although hold promise in case of crops with very low levels of DNA polymorphism, will have to undergo repeated laboratory tests prior to their routine use.

Cost of Assay

The development and use of molecular markers for genotyping involve high expenses. The cost of development of SNP markers is the highest, followed by the cost for STMS, which involve large-scale DNA sequencing. Once developed as in case of crops like rice and wheat, the genotyping cost for the STMS will be cheaper than that of SNP and AFLP, and will almost equal the cost of RAPD genotyping. Development of well furnished laboratory facility and provision of trained technical support, in addition to regular supplies of consummables will add to the cost of marker assisted plant breeding. Therefore, it would be prudent to choose in the beginning an appropriate trait, which is difficult to handle by conventional means.

Development of Protocols

Genotyping using molecular markers involves three major steps: (*i*) DNA isolation, (*ii*) PCR and (*iii*) DNA fragment separation and visualisation. In order to handle large number of samples as required by high resolution mapping of genes and marker aided selection, all these three steps are to be simplified. Isolation of DNA from thousands of samples would involve cost, labour and time. Although mechanical as well as chemical methods that can save time and labour are becoming available, these involve high cost. Newer protocols therefore need to be developed, which may not even involve DNA isolation. Fragment separation is currently being done by agarose gel electrophoresis (RAPD, SCAR, CAPS and STMS), PAGE (STMS), denaturing sequencing gels (AFLP) and automated fragment analysers using capillaries (AFLP and STS/STMS). Use of agarose gels is less cumbersome and costlier as compared to other systems although the resolution is poor. A combination of STMS markers and agarose electrophoresis will thus be preferred over the other methods. However, a non-gel/capillary based protocol would be ideal, which can be easily handled by specialist plant breeders with little exposure to molecular techniques.

The utility of molecular markers in genome mapping, gene tegging and marker-assisted breeding has been immensely demonstrated. Now it needs to be integrated with the conventional breeding programmes. This would require developing facilities, trained personnel and also a change in mindset of all concerned. Newer technologies are being developed very fast. This is expected to reduce the cost of marker development and genotyping, thereby making marker assisted breeding a routine activity in near future.

References

Ammiraju J.S.S., Dholakia B.B., Jawadekar G., Santra D.K., Gupta V.S., Roder M.S., Singh H., Lagu M.D., Dhaliwal H.S., Rao V.S.P. and Ranjekar P.K. 2003. DNA marker facilitated analysis of grain shape and size in wheat (*Triticum aestivum* L). (*Communicated*).

Ammiraju J.S.S., Dholakia B.B., Santra D.K., Singh H., Lagu M.D., Tamhankar S.A., Dhaliwal H.S., Rao V.S.P., Gupta V.S. and Ranjekar P.K. 2001. Identification of inter-simple sequence repeat markers associated with seed size in wheat. *Theor. Appl. Genet.*, **102**: 726-732.

Arondel V., Lemieux B., Hwang T., Gibson S., Goodman H.M. and Somerville C.R. 1992. Map-based cloning of a gene controlling omega-3 fatty acid desaturation in *Arabidopsis*. *Science*, **258**: 1352-1355.

Arumuganathan E. and Earle E.D. 1991. Nuclear DNA content of some important plant species. *Plant. Mol. Biol. Rep.*, **9**: 208-218.

Arus P. and Orton T.J. 1983. Inheritance and linkage relationships of isozyme loci in *Brassica oleracea*. *J. Hered.*, **74**: 405-412.

Autrique E., Singh R.P., Tanksley S.D. and Sorrells M. 1995. Molecular markers for four leaf rust resistance genes introgressed into wheat from wild relatives. *Genome*, **38**: 75-83.

Axelson T., Bowman C.M., Sharpe A.G., Lydiate D.J. and Lagercrantz U. 2000. Amphidiploid *Brassica juncea* contains conserved progenitor genomes. *Genome*, **43**: 679-688.

Babujee L. and Gnanamanickam S.S. 2000. Molecular tools for characterization of rice blast pathogen (*Magnaporthe grisea*) population and molecular marker-assisted breeding for disease resistance. *Curr Sci.*, **78**: 248-257.

Baker R.J. 1977. Inheritance of kernel hardness in spring wheat. *Crop Sci.*, **17**: 900-920.

Barret P., Deloume R., Foisset N. and Renard M. 1998a. Development of SCAR (sequence characterized amplified region) marker for molecular tagging of dwarf BREIZH (*Bzh*) gene in *Brassica napus* L. *Theor. Appl. Genet.*, **97**: 828-833.

Barret P., Delourme R., Renard M., Domergue F., Lessire L., Delseny M. and Roscoe T.J. 1998b. A rapeseed *FAE1* gene is linked to the E1 locus associated with variation in the content of erucic acid. *Theor. Appl. Genet.*, **96**: 177-186

Becker J., Vos P, Kuiper M., Salamini F. and Heum M. 1995. Combined mapping of AFLP and RFLP markers in Barley. *Mol. Gen. Genet.*, **249**: 65-73.

Blanco A., Giovanni de C., Laddomada B., Sciancalepore A., Simeone R., Devos K.M. and Gale M.D. 1996. Quantitative trait loci influencing grain protein content in tetraploid wheat. *Plant Breed.*, **115**: 310-316.

Bohuon E.J.R., Keith D.J., Parkin I.A.P., Sharpe A.G. and Lydiate D.J. 1996. Alignment of the conserved C genomes of *Brassica oleracea* and *Brassica napus*. *Theor. Appl. Genet.*, **93**: 833-839.

Botstein D., White R.L., Skolnick M. and Davis R.W. 1983. Construction of a genetic linkage map in man using restriction fragment length polymorphisms. *Am. J. Hum. Genet.*, **32**: 314-331.

Bushuk W. 1998. Wheat breeding for end-product use. *Euphytica*, **100**: 137-145.

Cadalen T., Boeuf C., Bernard S. and Bernard M. 1997. An inter-varietal molecular marker map in *Triticum aestivum* and comparison with a map from a wide cross. *Theor. Appl. Genet.*, **94**: 367-377.

Camarago L.E.A. and Osborn T.C. 1996. Mapping loci controlling flowering time in Brassica oleracea. *Theor. Appl. Genet.*, **92**: 610-616.

Camarago L.E.A., Williams P.H. and Osborn T.C. 1995. Mapping of quantitative trait loci controlling resistance of *Brassica oleracea* to *Xanthomonas campestris* pv. *campestris* in the field and green house. *Phytopathology*, **85**: 1296-1300.

Campbell K.G., Bergman C.J., Gaulberto D.G., Anderson J.A., Gorpix M.J., Hareland G., Gulcher R.G., Sorrells M.E. and Finney P.L. 1999. Quantitative trait loci associated with kernel traits in a soft X hard wheat cross. *Crop Sci.*, **39**: 1184-1195.

Chao S.P., Sharp P.J., Worland A.J., Warham E.J., Koebner R.M.D. and Gale M.D. 1989. RFLP-based genetic maps of wheat homoeologous group 7 chromosomes. *Theor. Appl. Genet.*, **78**: 493-504.

Chen B.Y., Heneen W.K. and Simonsen V. 1989. Comparative and genetic studies of isozymes in re-synthesized and cultivated *Brassica napus* L., *B. campestris* and *B. alboglabra* Bailey. *Theor. Appl. Genet.*, **77**: 673-679.

Chen B.Y., Jorgensen R.B., Cheng B.F. and Heneen W.K. 1997. Identification and chromosomal assignment of RAPD marker linked with a gene for seed coat colour in a *Brassica campestris- alboglabra* addition line. *Hereditas*, **126**: 133-138.

Cheung W.Y., Friesen L., Rakow G.F.W., Seguin-Swartz G. and Landry B.S. 1997. An RFLP based linkage map of mustard [*Brassica juncea* (L) Czern and Coss]. *Theor. Appl. Genet.*, **94**: 841-851.

Cheung W.Y., Gugel R.K. and Landry B.S. 1998. Identification of RFLP markers linked to the white rust resistance gene (*Acr*) in mustard (*Brassica juncea* (L.) Czern. and Coss.). *Genome*, **41**: 626-628.

Chyi Y.S., Hoeneke M.E. and Sernyk J.L. 1992. A genetic linkage map of restriction fragment length polymorphism loci for *Brassica rapa* (syn Campestris). *Genome*, **35**: 746-757.

De Quiroz H.C. and Mithen R. 1996. Molecular markers for low-glucosinolate alleles in oilseed rape (*Brassica napus* L.). *Mol. Breed*, **2**: 277-281.

Delapena R.C., Murray T.D. and Jones S.S. 1996. Linkage relations among eyespot resistance gene Pch2 on chromosome 7AL in wheat. *Genome*, **40**: 249-252.

Delourme R., Bouchereau A., Hubert N., Renard M. and Landry B.S. 1994. Identification of RAPD markers linked to a fertility restorer gene for the Ogura radish cytoplasmic male sterility of rapeseed (*Brassica napus* L). *Theor. Appl. Genet.*, **88**: 741-748.

Delourme R., Foisset N., Horvais R., Barret P., Champagne G., Cheung W.Y., Landry B.S. and Renard M. 1998. Characterisation of the radish introgression carrying the *Rfo* restorer gene for the *Ogu*-INRA cytoplasmic male sterility in rapeseed (*Brassica napus* L). *Theor. Appl. Genet.*, **97**: 129-134.

Demeke T., Laroche A. and Gaudet D.A. 1996. A DNA marker for BT-10 common bunt resistance gene in wheat. *Genome*, **39**: 51-55.

Devos K.M. and Gale M.D. 1993. The genetic maps of wheat and their potential application in plant breeding. *Outlook Agric.*, **22**: 93-99.

Devos K.M., Milan T. and Gale M.D. 1993. Comparative RFLP maps of the homoeologous group 2 chromosomes of wheat, rye and barley. *Theor. Appl. Genet.*, **85**: 784-792.

Devos K.M., Bryan G.J., Collins A.J., Stephenson P. and Gale M.D. 1995. Application two microsatellite sequences in wheat storage proteins as molecular markers. *Theor. Appl. Genet.*, **90**: 247-252.

Dholakia B.B., Ammiraju J.S.S., Santra D.K., Singh H., Katti M.V., Lagu M.D., Tamhankar S.A., Rao V.S., Gupta V.S., Dhaliwal H.S. and Ranjekar P.K. 2001. Molecular marker analysis of grain protein content in wheat (*T. aestivum*). *Biochem. Genet.*, **39**: 325-338.

Diehl A.L., Johnson V.A. and Mattern P.J. 1978. Inheritance of protein and lysine in three wheat crosses. *Crop Sci.*, **17**: 391-395.

Dion Y., Gugel R.K., Rakow G.F.W., Seguin-Swartz G. and Landry B.S. 1995. RFLP mapping of resistance to the blackleg disease [casual agent, *Leptosphaeria maculans* (Desm) Ces et de Not.] in canola (*Brassica napus* L). *Theor. Appl. Genet.*, **91**: 1190-1194.

Dubcovsky J., Tranquilli G., Lijavetzky D. and Muzzi G. 1999. Genetic and physical characterization of grain texture-related loci in diploid wheat. *Mol. Gen. Genet.*, **262**: 846-850.

Dweikat I., Ohm H., Mackenzie S., Patterson F., Cambron S. and Rateliffe R. 1994. Association of a DNA marker with Hessian fly resistance gene *H9* in wheat. *Theor. Appl. Genet.*, **89**: 964-968.

Dweikat I., Ohm H., Patterson F. and Cambron S. 1997. Identification of RAPD markers for 11 hessian fly resistance genes in wheat. *Theor. Appl. Genet.*, **94**: 419-423.

Eastwood R.F., Lagudah E.S. and Apples R. 1994. A direct search for DNA sequences tightly linked to cereal cyst nematode resistance genes in *Triticum tauschii*. *Genome*, **37**: 311-319.

Ecke W., Uzunova M. and Weissleder K. 1995. Mapping the genome of rapeseed (*Brassica napus* L) II. Localization of genes controlling erucic acid synthesis and oil content. *Theor. Appl. Genet.*, **91**: 972-977.

Eshed Y. and Zamir D. 1995. Introgression line population of *Lycopersicon pennellii* in the cultivated tomato enables the identification and fine mapping of yield associated QTL. *Genetics*, **141**: 1147-1157.

Eshed Y., Gera G. and Zamir D. 1996. A genome-wide search for wild species allele that increase horticultural yield of processing tomatoes. *Theor. Appl. Genet.*, **93**: 877-885.

Ferreira M.E., Rimmer S.R., Williams P.H. and Osborn T.C. 1995a. Mapping loci controlling *Brassica napus* resistance to *Leptosphaeria maculans* under different screening conditions. *Genetics*, **85**: 13-217.

Ferreira M.E., Satagopan J., Yandell B.S., Williams P.H. and Osborn T.C. 1995b. Mapping loci controlling vernalization requirement and flowering time in *Brassica napus*. *Theor. Appl. Genet.*, **90**: 727-732.

Ferreira M.E., Williams P.H. and Osborn T.C. 1994. RFLP mapping of *Brassica napus* using doubled haploid lines. *Theor. Appl. Genet.*, **89**: 615-621.

Ferreira M.E., Williams P.H. and Osborn T.C. 1995c. Mapping loci controlling resistance to *Albugo candida* in *Brassica napus* using molecular markers. *Phytopathology*, **85**: 218-220.

Feuillet C., Messmer M.M., Schachermayr G. and Keller B. 1995. Genetic and physical characterization of the *Lr1* leaf rust locus in wheat (*Triticum aestivum* L.) *Mol. Gen. Genet.*, **248**: 553-562.

Figdore S.S., Kennard W.C., Song K.M., Slocum M.K. and Osborn T.C. 1988. Assessment of the degree of restriction fragment length polymorphism in *Brassica*. *Theor. Appl. Genet.*, **75**: 833-840

Figdore S.S., Ferreira M.E., Slocum M.K. and Williams P.H. 1993. Association of RFLP markers with trait loci affecting clubroot resistance and morphological characters in *Brassica oleracea* L. *Euphytica*, **69**: 33-44.

Foisset N., Delourme R., Barret P., Hubert N., Landry B.S. and Renard M. 1996. Molecular mapping analysis of *Brassica napus* using isozyme, RAPD and RFLP markers on doubled haploid progeny. *Theor. Appl. Genet.*, **93**: 1017-1025.

Fourmann M., Barret P., Renard M., Pelletier G., Delourme R. and Brunel D. 1998. The two genes homologous to *Arabidopsis FAE1* co-segregate with the two loci governing erucic acid content in *Brassica napus*. *Theor. Appl. Genet.*, **96**: 852-858.

Galande A.A., Kale A., Tiwari R., Ammiraju J.S.S., Santra D.K., Lagu M.D., Rao V.S., Gupta V.S., Misra B.K., Nagarajan S. and Ranjekar P.K. 2001. Genetic analysis of kernel hardness in bread wheat using PCR based markers. *Theor. Appl. Genet.*, 103: 601-606.

Giura A. and Saulescu N.N. 1996. Chromosomal location of genes controlling grain size in a large grained selection of wheat (*Triticum aestivum* L.). *Euphytica*, **89**: 77-80.

Gold J., Harder D., Townley-Smith F., Aung T. and Procunier J. 1999. Development of a molecular marker for rust resistance genes *Sr39* and *Lr35* in wheat breeding lines. EJB Electronic J. Biotech., **2**: 1-6, http://www.ejb.org/content/vol2/issue1/full/1.

Gupta P.K., Balyan H.S., Prasad M., Varshney R.K., Roy J.K., Harjit Singh and Dhaliwal H.S. 1999. Molecular markers for some quality traits in wheat. *In*: *National Symposium on Frontiers of Research in Plant Sciences*. p 14, Dec 2-4, 1999, Calcutta, India.

Halloran G.M. 1975. Genetic analysis of grain protein percentage in wheat. *Theor. Appl. Genet.*, **46**: 79-86.

Halloran G.M. 1976. Genetic analysis of hexaploid wheat, *Triticum aestivum* using intervarietal chromosome substitution lines. *Euphytica*, **25**: 65-71.

Hartl L., Weiss H., Stephen U., Zeller F.J. and Zahoor A. 1995. Molecular identification of powdery mildew resistance genes in common wheat *Triticum aestivum* L. *Theor. Appl. Genet.*, **90**: 601-606.

Hartl L., Weiss H., Zeller F.J. and Zahoor A. 1993. Use of RFLP markers for the identification of alleles of the *Pm3* locus conferring powdery mildew resistance in wheat (*Triticum aestivum* L.). *Theor. Appl. Genet.*, **86**: 959-963.

Helguera M., Khan I.A. and Dubcovsky J. 2000. Development of PCR markers for the wheat leaf rust resistance gene *Lr47*. *Theor. Appl. Genet.*, **100**: 1137-1143.

Hoenecke M. and Chyi Y.S. 1991. Comparison of *Brassica napus* and *B. rapa* genomes based on restriction fragment length polymorphism mapping. *In*: *Rapeseed in a changing world, Proc 8th International Rapeseed Cong.* Saskatoon, Saskatchewan, Canada, **4**: 221-226.

Hongtrakul V., Heustis G.M. and Knapp S.J. 1997. Amplified fragment length polymorphisms as a tool for DNA fingerprinting sunflower germplasm: genetic diversity among oilseed inbred lines. *Theor. Appl. Genet.*, **95**: 400-407.

Hu J., Li G., Struss D. and Quiros C.F. 1999. SCAR and RAPD markers associated with 18-carbon fatty acids in rapeseed, *Brassica napus*. *Plant Breeding*, **118**: 145-150.

Hu J., Quiros C., Arus P., Struss D. and Robbelen G. 1995. Mapping of a gene determining linolenic acid concentration in rapeseed with DNA-based markers. *Theor. Appl. Genet.*, **90**: 258-262

Hu X.Y., Ohm H.W. and Dweikat I. 1997. Identification of RAPD markers linked to the gene Pm1 for resistance to powdery mildew in wheat. *Theor. Appl. Genet.*, **94**: 832-840.

Huang N., Angeles E.R., Domingo J., Magpantay G., Singh S., Zhang G., Kumarvadivel L., Bennet J. and Khush G.S. 1997. Pyramiding of bacterial blight resistance genes in rice: marker assisted selection using RFLP and PCR. *Theor. Apple. Genet.*, **95**: 313-320.

Jean M., Brown G.G. and Landry B.S. 1997. Genetic mapping of nuclear fertility restorer genes for the Polima cytoplasmic male sterility in canola (*Brassica napus* L) using DNA markers. *Theor. Appl. Genet.*, **95**: 321-328.

Joppa L.R., Changheng D., Hart G.E. and Hareland G.A. 1997. Mapping gene(s) for grain protein in tetraploid wheat (*Triticum turgidum* L.) using a population of recombinant inbred chromosome lines. *Crop Sci.*, **37**: 1586-1589.

Jourden C., Barret P., Brunel D., Delourme R. and Renard M. 1996a. Specific molecular marker of genes controlling linolenic acid content in rapeseed. *Theor. Appl. Genet.*, **93**: 512-518.

Jourden C., Barret P., Horvais R., Delourme R. and Renard M. 1996b. Identification of RAPD marker linked to linolenic acid genes in rapeseed. *Euphytica*, **90**: 351-357.

Jourden C., Barret P., Horvais R., Foisset N., Delourme R. and Renard M. 1996c. Identification of RAPD markers linked to loci controlling erucic acid level in rapeseed. *Mol. Breed*, **2**: 61-71.

Kearsay M.J., Ramsay L.D., Jennings D.E., Lydiate D.J., Bohoun E.J.R. and Marshall D.F. 1996. Higher recombination frequency in female compared to male meiosis in *Brassica oleracea*. *Theor. Appl. Genet.*, **92**: 363-367.

Kianian S.F. and Quiros C.F. 1991. Genetic analysis of major multigene families in *Brassica oleracea* and related species. *Genome*, **35**: 516-527.

Kianian S.F. and Quiros C.F. 1992. Generation of *Brassica oleracea* composite RFLP map: linkage arrangements across varios populations and evolutionary implications. *Theor. Appl. Genet.*, **84**: 544-554.

Kole C., Teutonico R., Williams P.H. and Osborn T.C. 1996. Molecular mapping of a locus controlling resistance to *Albugo candida* in *Brassica rapa*. *Phytopath.*, **86**: 367-369.

Kole C., Kole P., Vogelzang R. and Osborn T.C. 1997. Genetic linkage map of a Brassica rapa recombinant inbred population. *J. Heredity*, **88**: 553-557.

Konieczny A. and Ausubel F.M. 1993. A procedure for mapping *Arabidopsis* mutations using co-dominant ecotype-specific PCR based markers. *The Plant J.*, **4**: 403-410.

Konzak C.F. 1977. Genetic control of the content aminoacid composition and processing properties of proteins in wheat. *Adv. Genetics*, **19**: 407-582.

Kowalski S.P., Lan T.H., Feldman K.A. and Paterson A.H. 1994. Comparative mapping of *Arabidopsis thaliana* and *Brassica oleracea* chromosomes reveals islands of conserved organization. *Genetics*, **138**: 499-510.

Kuspira J. and Unrau J. 1957. Genetic analysis of certain characters in common wheat using whole chromosome substitution lines. *Can. J. Plant Sci.*, **37**: 300-326.

Lagercrantz U. and Lydiate D.J. 1995. RFLP mapping in *Brassica nigra* indicates differing recombination rates in male and female meioses. *Genome*, **38**: 255-264.

Lagercrantz U., Putterill J., Coupland G. and Lydiate D.J. 1996. Comparative mapping of *Arabidpsis* and *Brassica*, fine scale genome colinearity and congruence of genes controlling flowering time. *The Plant J.*, **9**: 13-20.

Landgridge P. and Chalmers K.J. 1998. Techniques for marker development. *In*: *Proc. 9th International wheat genetics symposium*, 2-7 August, 1998, Saskatoon, Saskatchewan, Canada.

Lan T.H., DelMonte T.A., Reischmann K.P., Hyman J., Kowalski S.P., McFerson J., Kresovich S.N. and Paterson A.H. 2000. An EST-enriched comparative map of *Brassica oleracea* and *Arabidopsis thaliana*. *Genome Res.*, **10**: 776-788.

Landry B.S., Hubert N., Crete R., Chiang M.S., Lincoln S.E. and Etoh T. 1992. A genetic map for *Brassica oleracea* based on RFLP markers detected with expressed DNA sequences and mapping of resistance genes to race 2 of *Plasmodiophora brassicae* (Woronin). *Genome*, **35**: 409-420.

Landry B.S., Hubert N., Etoh T., Harada J.J. and Lincoln S.E. 1991. A genetic map for *Brassica napus* based on restriction fragment length polymorphisms detected with expressed DNA sequences. *Genome*, **34**: 543-552.

Law C.N., Young C.F., Brown J.W.S., Snape J.W. and Worland A.J. 1978. The study of grain-protein control in wheat using whole chromosome substitution lines. *In*: Seed protein improvement by nuclear techniques. *IAEA*, Vienna 483-502.

Li G. and Quiros C.F. 2001. Sequence-related amplified polymorphism (SRAP), a new marker system based on a simple PCR reaction: its application to mapping and gene tagging in *Brassica*. *Theor. Appl. Genet.*, **103**: 455-461.

Li S., Zhang Z., Wang B., Zhong Z., and Yao J. 1995. Tagging the *Pm4a* gene in NILs by RAPD analysis. *Acta Genet. Sin.*, **22**: 103-108.

Liu Y.G. and Sunewaki K. 1991. RFLP analysis in wheat II. Linkage maps of RFLP sites in common wheat. *Jpn. J. Genet.*, **66**: 617-633.

Ma Z.Q., Sorrells M.E. and Tanksley S.D. 1994. RFLP marker linked to powdery mildew resistance genes *Pm1, Pm2, Pm3* and *Pm4* in wheat. *Genome*, **37**: 871-875.

Martin G.B., Brommonschenkel S.H., Chunwongse J., Frary A., Ganal M.W., Spivey R., Wu T., Earle E.D. and Tanksley S.D. 1993. Map-based cloning of a protein kinase gene conferring disease resistance in tomato. *Science*, **262**: 1432-1436.

Martin G.B., Williams J.G.K. and Tanksley S.D. 1991. Rapid identification of markers linked to a *Pseudomonas* resistance gene in tomato by using random primers and near- isogenic lines. *Proc. Natl. Acad. Sci.*, USA, **88**: 2336-2340.

Marathee J.P. and Gomez-MacPherson H. 2001. Future World Supply and Demand. *In*: *The World Wheat Book, A History of Wheat Breeding*, (eds.) A.P. Bonjean and W.J. Angus, pp 1107-1116. Lavoisier Publishers, Paris, France.

Mattern P.J., Morris R., Schmidt J.W. and Johnson V.A. 1973. Locations of genes for kernel properties in the wheat variety "Cheyenne" using chromosome substitution lines. *In*: *Proc. 4th Int. wheat Genet. Symp.*, Columbus, Missouri, 703-707.

McIntosh R.A., Wellings C.R. and Park R.F. 1995. *Wheat Rusts: An Atlas of Resistance Genes*. Kluwer Academic Publishers, Dordrecht. pp 65-66.

Michelmore R.W., Paran I. and Kesseli R.V. 1991. Identification of markers linked to disease-resistance. *Proc. Natl. Acad. Sci.*, USA, **88**: 9828-9832.

Mohapatra T., Upadhyay A. Sharma A. and Sharma R.P. 2002. Detection and mapping of duplicate loci in *Brassica juncea*. *J. Plant Biochem. Biotech.*, **11**: 37-42.

Mohler V. and Jahoor A. 1996. Allele specific amplification of polymorphic sites for the detection of powdery mildew resistance loci in cereals. *Theor. Appl. Genet.*, **93**: 1078-1082.

Mukherjee A.K., Mohapatra T., Varshney A., Sharma R. and Sharma R.P. 2001. Molecular mapping of a locus controlling resistance to *Albugo candida* in *Brassica juncea*. *Plant Breeding*, **120**: 483-487.

Naik S., Tiwari R., Gill B.S., Gupta V.S., Rao V.S.P., Ranjekar P.K., Nagarajan S. and Keller B. 2001. Utility of molecular markers in breeding rust resistance wheat material. *In*: *"Role of Resistance in Intensive Agriculture"*, (eds) Nagarajan S. and D.P. Singh, pp 205-215. Kalyani Publishers, India, New Delhi.

Naik S., Gill K.S., Rao V.S.P., Gupta V.S., Tamhankar S.A., Pujar S, Gill B.S. and Ranjekar P.K. 1998. Identification of a STS marker linked to the Aegilops speltoides-derived leaf rust resistance gene *Lr28* in wheat. *Theor. Appl. Genet.*, **97**: 535-540.

Negi M.S., Devic M., Delseny M. and Lakshmikumaran M. 2000. Identification of AFLP fragments linked to seed coat colour in *Brassica juncea* and conversion to SCAR marker for rapid selection. *Theor. Appl. Genet.*, **101**: 146-152.

Nelson J.C., Van Deynze A.E., Autrique E., Sorrels M.E., Lu T.H., Merrilino M., Atkinson M.D. and Leroy P. 1995. Molecular mapping of wheat. Homoeologous group 3. *Genome*, **38**: 516-524.

Nelson J.C., Singh R.P., Autrique J.E. and Sorrels M.E. 1997. Mapping genes conferring and suppressing leaf rust resistance in wheat. *Crop Sci.*, **37**: 1928-1935.

Olson M., Hood L., Cantor C.H. and Botstein D. 1989. A common language for physical mapping of the human genome. *Science*, **24**: 1434-1435.

Paran I. and Michelmore R.W. 1993. Development of reliable PCR-based markers linked to downy mildew genes in lettuce. *Theor. Appl. Genet.*, **85**: 985-993.

Parker G.D., Chalmers K.J., Rathjen A.J. and Landgridge P. 1998. Mapping loci associated with flour colour in wheat (*Triticum aestivum* L.). *Theor. Appl. Genet.*, **96**: 1021-1032.

Paull J.G., Pallotta M.A., Landgridge P. and The T.T. 1994. RFLP markers associated with *Sr22* and recombination between chromosome 7A of bread wheat and the diploid species *Triticum boeoticum. Theor. Appl. Genet.*, **89**: 1039-1045.

Peng J.H., Fahima T., Roder M.S., Li Y.C., Daham A., Grama A., Ronin Y.I., Korol A.B. and Nevo E. 1999. Microsatellite tagging of the stripe- rust resistance gene *YrH52* derived from wild emmer wheat, *Triticum dicoccoides*, and suggestive negetive crossover interference on chromosome 1B. *Theor. Appl. Genet.*, **98**: 862-872.

Penner G.A., Zirino M., Kruger S. and Townley-Smith F. 1998. Accelerated recurrent parent selection in wheat with microsatellite markers. *In*: *Proc. 9th Wheat Genet. Symp.*, (ed.) Slikand A.E., Vol. 1, Univ. Extension Press, Univ. of Saskatchewan, Saskatoon, Canada, pp. 131-134

Petrovic S. and Worland A.J. 1988. The use of reciprocal monosomic analysis to detect variation between certain chromosomes of the wheat varieties 'Bersee' and 'Sava'. *In*: *Proc. Intl. Wheat Genet. Symp.*, (eds.) T.E. Miller and R.M.D. Koebner, 7th Cambridge, England, 13-19 July 1988. Inst. Plant. Sci. Res. Cambridge Lab., Cambridge, pp 629-633.

Pomeranz Y. and Williams P.C. 1990. Wheat hardness: its genetic, structural and biochemical background, measurements and significance. *In*: *Advances in Cereal Science and Technology*, AACC, St. Paul, Vol. X: 471-544

Prabhu K.V., Somers D.J., Rakow G.F.W. and Gugel R.K. 1998. Molecular markers linked to white rust resistance in mustard *Brassica juncea. Theor. Appl. Genet.*, **97**: 865-870.

Prakash S. and Hinata K. 1980. Taxonomy, cytogenetics and origin of crop *brassicas*, a review. *Opera Bot.*, **55**: 1-57.

Prasad M., Varshney R.K., Kumar A., Balyan H.S., Sharma P.C., Edvard K.J., Singh H., Dhaliwal H.S., Roy J.K. and Gupta P.K. 1999. A microsetellite marker associated with a QTL for GPC on chromosome arm 2DL of bread wheat. *Theor. Appl. Genet.*, **99**: 341-345.

Procunier J.D., Knox R.E., Bernier A.M., Gray M.A. and Howes N.K. 1997. DNA markers linked to a *T10* loose smut resistance gene in wheat (*Triticum aestivum* L.). *Genome*, **40**: 176-179.

Procunier J.D., Townley-Smith T.F., Fox S., Prashar S., Gay M., Kim W.K., Czarnecki E. and Dyck P.L. 1995. PCR based RAPD/DGGE markers linked to leaf rust resistance genes *Lr29* and *Lr25* in wheat (*Triticum aestivum* L.). *J. Genet. Breed.*, **49**: 87-92.

Qi L.L., Cao M.S., Chen P.D., Li W.L. and Liu D.J. 1996. Identification, Mapping and application of polymorphic DNA associated with resistance gene *Pm21* of wheat. *Genome*, **39**: 191-197.

Ramsay L.D., Jennings D.E., Bohuon E.J.R., Arthur A.E., Lydiate D.J., Kearsey M.J. and Marshall D.F. 1996. The construction of a substitution library of recombinant backcross lines in *Brassica oleracea* for the precision mapping of quantitative loci. *Genome*, **39**: 558-567.

Rao S., Singh G. and Chatterjee M. 2001. Indian Wheat Pool. *In*: *The World Wheat Book, A History of Wheat Breeding*, (eds.) A.P. Bonjean and W.J. Angus, pp 773-816. Lavoisier Publishers, Paris, France.

Ribaut J.M. and Hoisington D. 1998. Marker-assisted selection: New tools and strategies. *Trends Plant Sci.*, **3**: 236-239.

Roder M.S., Korzun V., Wendehake K., Plaschke J., Tixer M.H., Leroy P. and Ganal M.W. 1998. A microsatellite map of wheat. *Genetics*, **149**: 2007-2023.

Saal B., Plieske J., Hu J., Quiros C.F. and Struss D. 2001. Microsatellite markers for genome analysis in Brassica. II. Assignment of rapeseed microsatellites to the A and C genomes and genetic mapping in *Brassica oleracea* L. *Theor. Appl. Genet.*, **102**: 695-699.

Sadowski J., Gaubier P., Delseny M. and Quiros C.F. 1996. Genetic and physical mapping in diploid *Brassica* species of a gene cluster defined in *Arabidopsis thaliana. Mol. Gen. Genet.*, **251**: 298-306.

Sayfarth R., Feuillet C., Schachermayr G., Winzeler M. and Keller B. 1999. Development of molecular marker for the adult plant leaf rust resistance gene *Lr35* in wheat. *Theor. Appl. Genet.*, **99**: 554-560.

Schachermayr G.M., Feuillet C. and Keller B. 1997. Molecular markers for the detection of the wheat leaf rust resistance gene *Lr10* in diverse genetic backgrounds. *Mol. Breed.*, **3**: 65-74.

Schachermayr G.M., Messmer M.M., Feuillet C., Winzeler H., Winzeler M. and Keller B. 1995. Identification of molecular markers linked to the *Agropyrum elongatum*-derived leaf rust resistance gene *Lr24* in wheat. *Theor. Appl. Genet.*, **30**: 982-990.

Schachermayr G.M., Siedler H., Gale M.D., Winzeler H., Winzeler M. and Keller B. 1994. Identification and localization of molecular markers linked to the *Lr9* leaf rust gene of wheat. *Theor. Appl. Genet.*, **88**: 110-115.

Seah S., Bariana H., Jahier J., Sivasithamparam K. and Lagudah E.S. 2001. The introgressed segment carrying rust resistance genes *Yr17, Lr37*, and *Sr38* in wheat can be assayed by a cloned disease resistance gene-like sequence. *Theor. Appl. Genet.*, **102**: 600-605.

Seo Y.W., Johnson J.W. and Jarret R.L. 1997. A molecular marker associated with the H21 hessian fly resistance gene in wheat. *Mol. Breed.*, **3**: 177-181.

Sharma A., Mohopatra T. and Sharma R.P. 1994. Molecular mapping and character tagging in *Brassica juncea.* I Degree, nature and linkage relationship of RFLPs and their association with quantitative traits. *J. Plant Biochem. Biotech.*, **3**: 85-89.

Sharma R., Mohapatra T., Mukherjee A.K., Krishanpal and Sharma R.P. 1999. Molecular markers for seed oil content in Indian mustard. *J. Plant Biochem. Biotech.*, **8**: 61-64.

Sharma R., Aggarwal R.A.K., Kumar R., Mohapatra T. and Sharma R.P. 2002. Construction of RAPD linkage map and localization of QTLs for oleic acid level using recombinant inbreds in mustard. *Genome*, **45**: 467-472.

Singh R.P. 1992. Association between gene *Lr34* for leaf rusts resistance and leaf tip necrosis in wheat. *Crop Sci.*, **32**: 874-878.

Singh S., Sidhu J.S., Huang N., Vikal Y., Li Z., Brar D.S., Dhaliwal H.S. and Khush G.S. 2001. Pyramiding three bacterial blight resistance genes (*xa5, xa13* and *Xa21*) using marker assisted selection into indica rice cultivar PR106. *Theor. Appl. Genet.*, **102**: 1011-1015.

Slocum M.K., Figdore S.S., Kenard W.C., Suzuki J.Y. and Osborn T.C. 1990. Linkage arrangement of restriction fragment length polymorphic loci in *Brassica oleracea. Theor. Appl. Genet.*, **80**: 75-64.

Song K.M., Suzuki J.Y., Slocum M.K., Williams P.H. and Osborn T.C. 1991. A linkage map of *Brassica rapa* (syn campestris) based on restriction fragment length polymorphism loci. *Theor. Appl. Genet.*, **82**: 296-304.

Sourdille M., Perretant R., Charmet G., Leroy P., Gautier, M.F., Jourdier P., Nelson J.C., Sorrells M.E. and Bernard M. 1996. Linkage between RFLP markers and genes affecting kernel hardness in wheat. *Theor. Appl. Genet.*, **93**: 580-586.

Stephenson P., Bryan G., Kirby J., Collins A.J., Devos K.M., Busso C. and Gale M.D. 1998. Fifty new microsatellite loci for the wheat genetic map. *Theor. Appl. Genet.*, **97**: 946-949.

Stuber C.W. 1997. Case history in crop improvement: yield heterosis in maize. *In*: *Molecular dissection of complex traits*, (ed.) Paterson A.H., CRC Press, Boca Raton, New York, pp.197-206.

Sun G.L., Fahima T., Korol A.B., Turpeinen T., Grama A., Ronin Y.I. and Nevo E. 1997. Identification of molecular markers linked to the *Yr15* stripe rust resistance gene of wheat originated in wild emmer wheat, *Triticum dicoccoides*. *Theor. Appl. Genet.*, **95**: 622-622.

Symes K.J. 1969. Influence of a gene causing hardness on the milling and baking quality of two wheats. *Aus.t J. Agric. Res.*, **20**: 971-979.

Talbert L.E., Bruckner P.L., Smith L.Y., Sears R. and Martin T.J. 1996. Development of PCR markers linked to resistance to wheat streak mosaic virus in wheat. *Theor. Appl. Genet.*, **93**: 463-467.

Tanhuanpää P.K., Vilkki J.P. and Vilkki H.J. 1995. Association of a RAPD marker with linolenic acid concentration in the seed oil of rapeseed (*Brassica napus* L). *Genome*, **38**: 414-416.

Tanhuanpää P.K., Vilkki J.P. and Vilkki H.J. 1996. Mapping of a QTL for oleic acid concentration in spring turnip rape (*Brassica rapa* ssp oleifera). *Theor. Appl. Genet.*, **92**: 952-956.

Teutonico R..A and Osborn T.C. 1994. Mapping of RFLP and quantitative trait loci in *Brassica rapa* and comparison to the linkage maps of *B. napus, B. oleracea* and *Arabidopsis thaliana*. *Theor. Appl. Genet.*, **89**: 885-894.

Teutonico R.A. and Osborn T.C. 1995. Mapping loci controlling vernalization requirement in *Brassica rapa*. *Theor. Appl. Genet.*, **91**: 1279-1283.

Thormann C.E., Romero J., Mantet J. and Osborn T.C. 1996. Mapping loci controlling the concentration of erucic and linolenic acids in seed oil of *Brassica napus* L. *Theor. Appl. Genet.*, **93**: 282-286.

Tippless K.H., Kilborn R.H. and Preston K.R. 1994. Bread-wheat quality defined, *In*: *Wheat: Production, properties and quality*, (eds.) Bushuk W., Rasper V.F., Blackie Academic and Professional, Chapman and Hall, Glasgow, 25-35.

Toroser D., Thormann C.E., Osborn T.C. and Mithen R. 1995. RFLP mapping of quantitative trait loci controlling seed aliphatic glucosinolate content in oilseed rape (*Brassica napus* L). *Theor. Appl. Genet.*, **91**: 802-808.

Truco M.J. and Quiros C.F. 1994. Structure and organization of the B genome based on a linkage map in *Brassica nigra*. *Theor. Appl. Genet.*, **89**: 590-598.

Turner M., Mukai Y., Leroy P., Charef B., Appels R. and Rahman S. 1999. The *Ha* locus of wheat: Identification of a polymorphic region for tracing grain hardness in crosses. *Genome*, **42**: 1242-1250.

Upadhyay A., Mohapatra T., Pai R.A. and Sharma R.P. 1996. Molecular mapping and character tagging in Indian mustard (*Brassica juncea*) II. RFLP marker association with seed coat colour and quantitative traits. *J. Plant Biochem. Biotech.*, **5**: 17-22.

Uzunova M., Ecke W., Weissleder K. and Robbelen G. 1995. Mapping the genome of rapeseed (*Brassica napus* L) I. Construction of an RFLP linkage map and localization of QTLs for seed glucosinolate content. *Theor. Appl. Genet.*, **90**: 194-204.

Van Deynze A.E., Landry B.S. and Pauls K.P. 1995. The identification of restriction fragment length polymorphisms linked to seed colour genes in *Brassica napus*. *Genome*, **38**: 534-542.

Voorrips R.E., Jongerius M.C. and Kanne H.J. 1997. Mapping of two genes for resistance to clubroot (*Plasmodiophora brassicae*) in a population of doubled haploid lines of *Brassica oleracea* by means of RFLP and AFLP markers. *Theor. Appl. Genet.*, **94**: 75-82.

Vos P., Hogers R., Bleeker M., Reijans M., Van de Lee T., Hornes M., Frijters A., Pot J., Peleman J., Kuipers M. and Zabeau M. 1995. AFLP: a new technique for DNA fingerprinting. *Nucl. Acids. Res.*, **23**: 4407-4414.

Welsh J. and McClelland M. 1990. Fingerprinting genomes using PCR with arbitrary primer. *Nucl. Acid Res.*, **18**: 7213-7218.

Williams C.E., Wang B., Holsten T.E., Scambray J., De Assis Goes da Silva F. and Ronald P.C. 1996. Markers for selection of the rice *Xa21* disease resistance gene. *Theor. Appl. Genet.*, **93**: 1119-1122.

Williams G.K., Kubelik A.R., Livak K.J., Rafalski J.A. and Tingey S.V. 1990. DNA polymorphisms amplified by arbitrary primers are useful as genetic markers. *Nucl. Acids Res.*, **18**: 6531-6535.

William H.M., Hoisington D., Singh R.P. and Gonzalez-de-Leon D. 1997. Detection of quantitative trait loci associated with leaf rust resistance in bread wheat. *Genome*, **40**: 253-260.

Williams K.J., Fisher J.M. and Landgridge P. 1994. Identification of RFLP markers linked to cereal cyst nematode resistance gene (*Cre*) in wheat. *Theor. Appl. Genet.*, **93**: 463-467.

Xiao J., Li J., Grandilo S., Ahn S.N., Yuan L., Tanksley S.D. and McCouch S.R. 1998. Identification of trait-improving quantitative trait loci alleles from a wild rice relative, *Oryza rufipogon*. *Genetics*, **150**: 899-909.

Plant Breeding - Mendelian to Molecular Approaches
H. K. Jain and M. C. Kharkwal (eds.)

11

Genome Mapping and Map Based Cloning

C. Kole[1] and P.K. Gupta[2]

Abstract

Genome mapping has emerged as a potential tool that provides the complete depiction of the genomes of plants and animals and thereby the means for their further manipulation. It involves elucidation of the nuclear genome of higher plants and animals as well as the smaller cytoplasmic genomes as chloroplasts and mitochondria. It involves basically linkage mapping employing molecular markers and using mainly segregating populations. First generation or primary genetic linkage maps have been constructed in several plant and animal systems, particularly those of fundamental or of economic interests. An array of markers have been used for such purposes, those included mainly isoenzyme, RFLP, RAPD, AFLP and SSR. The mostly used mapping populations included F_2, recombinant inbred lines, backcross, doubled haploid lines and CEPH. The second generation genetic maps, such as high density, high resolution and saturated linkage maps have also been developed in several plants and animals mainly by enriching the primary maps locally or globally. Molecular mapping has paved the way for positioning of simply inherited trait loci (SITL) controlling oligogenic characters and quantitative trait loci (QTL) controlling polygenic characters. These have in turn facilitated marker assisted breeding (MAB) or molecular breeding (MB) for improvement in these traits. It has also recently been possible to mendelize the QTLs for precise monitoring of the gene clusters. Construction of molecular maps of two or more species/genera using a common set of markers and characters have resulted in comparative mapping that provides valuable information on genome homology and thus could illuminate on phylogenetic relationship and evolution in several taxa. Genome mapping has also provided the platform for chromosome walking or chromosome landing for isolating the chromosomal fragment containing a target gene and its (map based) cloning (MBC) for use in genetic transformation.

Introduction

Genome mapping has emerged as a potential tool in life sciences. It stands on the platform built with the ideas inherited from the classical concepts of genetics as well as the recent

[1]Allahabad Agricultural Institute, Allahabad 211 007, India
[2]Ch. Charan Singh University, Meerut 250 004, India

findings of molecular biology and genetic engineering. The strategies used in genome mapping involve traditional as well as the most modern tools of biotechnology. Genome mapping has facilitated enrichment of almost all the branches of life sciences including evolution and phylogenetic relationship; genome organization; and localization, isolation, sequencing and manipulation of genes. In a variety of organisms, it has also provided the benchmarks for marker aided genetic tailoring for desirable traits including the docile simply inherited qualitative traits as well as the complex quantitative traits, which could not be mapped and manipulated by using conventional tools.

The term 'genome' represents the 'repertoire of DNA' present in an organelle: nucleus, mitochondrion or chloroplast. A genome map is the depiction of the hereditary information contained in an organelle that controls all the attributes of an organism and is rightly called as the 'blue print'. Among the three genomes, nuclear genome is the largest and consists of the 'true chromosomes' in higher organisms. The linear alignment of the genes in the chromosomes in this genome facilitates landmarking with the so-called 'genetic markers'. The present article deals with genome mapping in the nuclear genome only.

Genesis of Genome Mapping Concept: A Historical Resume

The concept of genome mapping stems from the 'genetic linkage' phenomena proposed by Morgan (1910) and is a refined product of the 'linkage mapping' strategy of the '*Drosophila* School'. Development of genetic maps in *Drosophila* involved the use of the mutants for perceivable morphological characters, known as 'visible markers' or 'simply inherited trait markers' (SITM). Linkage mapping using these markers was tried in a number of other organisms including bean, maize, *Arabidopsis*, rice etc., to develop partial genetic maps.

Genetic mapping using the SITMs is constrained with several limitations, which have been elaborately delineated in literature. Any two organisms of a species have only a few characters with contrasting expressions (phenotypes). Genetic mapping, therefore, requires a large number of mating experiments over a long period of time. Genes for the morphological traits comprise only a very small proportion of the total number of genes in an organism and thus the genetically active regions cover relatively a small total length of the genome, particularly in higher organisms. Hence, the number of SITMs is too meager to depict an entire genome. Again, mutations for many characters often cause lethality, and loci for such characters can not be used as markers. Another limitation of these markers is the influence of genetic environment, inter-allelic interactions or modifier genes for example, and non-genetic environment, the external growing conditions.

Attempts were also made tq use some unique chromosomal features as landmarks for cytological mapping. This strategy involved induction of deletions by mutagens to correlate these landmarks with the affected characters. This approach was initially successful only in some cases, since the unique features as 'puffs', 'loops' and 'knobs' were available only in some organisms, and also because a deletion often causes lethality.

In the meantime, protein markers particularly isoenzyme markers emerged as another promising tool for genetic mapping (Markert and Moller, 1959). These were used for the development of partial genetic maps in quite a few organisms (Tanksley and Orton, 1983). Their number, however, is too few for developing complete genetic maps and, therefore, protein markers were mainly used for tagging genes of economic value.

The elegant discoveries in the field of molecular genetics in the fifties, sixties and seventies of the twentieth century culminating in the development of the recombinant DNA technology paved the way for using DNA as the source of the most potential molecular markers for complete genome mapping.

Since the development of restriction fragment length polymorphisms (RFLPs) as the first generation DNA markers in 1980 (Botstein *et al.*, 1980) and their use in human genome mapping, RFLPs have been widely used across organisms of both animal and plant systems. Later on, several other DNA markers were described. These are based on the principles of either RFLP and/or of polymerase chain reaction (PCR) (Mullis *et al.*, 1986; Mullis and Faloona, 1987). An array of DNA markers is now available to choose from for a specific organism, population and purpose.

During the last quarter of the twentieth century techniques were also developed for cloning large segments of DNA in vectors (Schmidt *et al.*, 1995) for developing BAC (Shizuya *et al.*, 1992) and YAC (Burke *et al.*, 1987) libraries, and for genetic transformation together with already much used DNA sequencing methods (Maxam and Gilbert 1977; Sanger *et al.*, 1977). These techniques facilitated the construction of physical maps comprised of contiguous overlapping mega base DNA clones spanning large regions of chromosomes. These techniques also represented a major step toward complete sequencing of genomes. While genetic mapping using DNA markers facilitated saturation of any genome with closely spaced landmarks, physical mapping (including whole genome sequencing) provided complete knowledge about the nucleotide sequence of the genome, leaving almost nothing more to know about the physical and chemical organization of a genome.

Genetic mapping although initially started in human genome (Botstein *et al.*, 1980) but has been most effectively and extensively practiced in the plant systems. The number of genetic maps constructed so far in the higher plant species exceeds those in higher animal species. This is mainly because of the ease of raising large segregating populations for mapping in most of the higher plants. In the remaining part of this article, we discuss the achievements and prospects of genome mapping in the higher plants.

Genetic Mapping: General Strategies

Construction of genetic linkage maps employs basically the classical strategy of linkage mapping. It comprises primarily the following component steps.

1. Selection of genetically diverse parents aiming at optimum DNA polymorphism
2. Identification of probes/primers detecting DNA polymorphism in parents
3. Raising of segregating population(s) for assessment on recombination events
4. Linkage analysis and construction of linkage maps

Selection of Parents

Theoretically any two individuals (genotypes) can be used for development of a genetic map. But their genetic differences should spread over the entire genome to facilitate the depiction of landmarks on all the chromosomes, at least at intervals resolvable from cross-over products. From applied point of view, parents differing in some economically important traits should be selected. Generally, two genotypes of a species are selected based on their apparent difference in morphological characters or from the results of various kinds of analysis on genetic distance, using morphological trait values and/or molecular markers. Most of the genetic maps developed so far in plants included two varieties of a species. In some cases adequate polymorphism was not available within a species, so that two species or even genera had to be used. These may be exemplified with the genetic maps in tomato using *Lycopersicon esculentum* and *L. pennelli* (Bernatzky and Tanksley, 1986) and *L. esculentum* and *L. cheemanii* (Paran *et al.*, 1995), rice using *Oryza sativa* and *O. longistaminata* (Causse *et al.*, 1994) and blueberry using *Vaccinium darrowi* and *V. elliottii* (Rowland and Levi, 1994) to cite a few. A compromise has, however, always been made between the extent of polymorphism, the feasibility of raising the segregating population and obviously the application value of the map in tagging and tracking of some economic traits.

For mapping in plants, generally homozygous (pure or inbred) lines are used as parents. Some out-crossing plants such as sugarcane or potato, however, can not sustain selfing (inbreeding). In these cases, heterozygous parents are used and marker analysis is done following 'single dose restriction fragment' (SDRF) procedure (Wu *et al.*, 1992).

Screening of Probes/Primers Detecting Parental Polymorphism

Several types of DNA markers are now available for use in genetic mapping. Out of these, RFLPs have been used almost universally. They still rank first in plant genetic mapping. Some PCR based markers such as random amplified polymorphic DNA (RAPD), arbitrary primer-PCR (AP-PCR), DNA amplification fingerprinting (DAF) and amplified fragment length polymorphism (AFLP) have also been used, mainly as supplement to the RFLPs. Attempts to develop maps based only on RAPDs have also been made in a few cases e.g., blue berry by Rowland and Levi (1994) and *Brassica campestris* by Nozaki *et al.* (1997). Both in animal and plant systems, microsatellite markers (SSRs) later emerged as the most effective marker system since these detect polymorphism even between closely related individuals (Gupta and Varshney, 2000). Isozyme markers have also been used as supplement to RFLPs in quite a few cases. Examples of selected genetic maps in some major economic plants are furnished in Table 1.

Detection of DNA and isozyme markers, their merits, demerits and utility have been described at length by several workers including Lakshmikumaran *et al.* (2003) elsewhere in this book. Some issues relevant to genetic mapping will only be discussed here.

Table 1. Genetic linkage maps of molecular markers in some major economic plants

Plant	Markers	Loci #	Interval (cM)	Coverage (cM)	Authors
Rice	RFLP	135	9.0	1389	McCouch *et al.*, 1988
	IZ,SITL	150	NA	NA	Kinoshita, 1993
	RFLP	726	4.0	1491	Causse *et al.*, 1994
	MS	121	16-20	900	Chen *et al.*, 1997
	AFLP,RFLP	343	8.9	3058	Maheswaran *et al.*, 1997
	RFLP	149	10.9	1630	Jena *et al.*, 1994
Wheat	RFLP	197	9.13	1800	Liu & Tsunewaki, 1991
	RFLP,SSR	266	NA	NA	Cadalen *et al.*, 1997
	SSR,AFLP	662	NA	NA	Penner *et al.*, 1998
Maize	RFLP	113	NA	NA	Helentjaris *et al.*, 1986
	RFLP,IZ,SITL	338	NA	NA	Helentjaris, 1987
	RFLP	62	2.1	NA	Burr *et al.*, 1988
Oat	RFLP	532	2.8	1482	O'Donoughue *et al.*, 1992
Barley	RFLP,IZ,RAPD,SITL	295	4.2	1250	Kleinhofs *et al.*, 1993
Rye	RFLP,IZ,RAPD,SITL	60	6.0	350	Philipp *et al.*, 1994
Sorghum	RFLP	98	10.0	949	Whitkus *et al.*, 1992
	RFLP	190	9.4	1789	Xu *et al.*, 1994
Tomato	RFLP	104	NA	NA	Helentjaris *et al.*,1986
	RFLP,IZ	112	NA	NA	Bernatzky & Tanksley, 1986
	RFLP,IZ	1030	1.2	1276	Tanksley *et al.*, 1992
	RFLP	132	10.0	1209	Paran *et al.*, 1995
Potato	RFLP,IZ	134	8.8	1189	Bonierbale *et al.*, 1988
	RFLP,IZ	977	0.7	684	Tanksley *et al.*, 1992
Lettuce	RFLP,IZ,SITL	53	7.6	404	Landry *et al.*, 1987
P. vulgaris	RFLP,IZ	244	5.0	1200	Vallejos *et al.*, 1992
	RFLP,RAPD,IZ	152	6.5	827	Nodari *et al.*, 1993a
Soybean	RFLP	252	8.5	2147	Diers *et al.*, 1992
Mungbean	RFLP	171	9.0	1570	Menancio-Hautea *et al.*, 1993
Cowpea	RFLP,RAPD,SITL	97	7.0	684	Menancio-Hautea *et al.*, 1993
B. rapa	RFLP	280	6.6	1850	Song *et al.*, 1991
	RFLP	360	5.2	1876	Chyi *et al.*, 1992
	RFLP	139	13.5	1785	Teutonico & Osborn, 1994
	RFLP	144	6.0	890	Kole *et al.*, 1997
B. oleracea	RFLP	258	3.5	820	Slocum *et al.*, 1990
	RFLP	201	5.5	1112	Landry *et al.*, 1992
	RFLP	108	6.9	747	Kianian & Quiros, 1992
B. napus	RFLP	120	11.8	1413	Landry *et al.*, 1991
	RFLP	132	7.7	1016	Ferreira *et al.*, 1994
	RFLP,RAPD,SITL	207	6.9	1441	Uzunova *et al.*, 1995
	RFLP,AFLP,IZ,SITL	480	4.1	2007	Kole, 1997
B. juncea	RFLP	342	6.6	2073	Cheung *et al.*, 1997
	RAPD	114	6.9	790.4	Sharma *et al.*, 2002
	RFLP,AFLP	1029	3.5	1629	Pradhan *et al.*, 2002
Blueberry	RAPD	70	3.3	950	Rowland & Levi, 1994
Sugarbeet	RFLP	413	1.5	621	Hallden *et al.*, 1996
Banana	RFLP,IZ,RAPD	90	10.0	606	Faure *et al.*, 1993
Cotton	RFLP	705	7.1	4675	Reinisch *et al.*, 1994
Arabidopsis	RFLP	320	2.0	630	Reiter *et al.*, 1992

Detection of RFLPs is based upon the differential length of DNA segment between two restriction sites. The differential lengths occur due to insertion or deletion of DNA sequences in between two restriction sites or mutation in the restriction sites. DNA samples of the candidate individuals are digested with a suitable restriction enzyme, the fragments are separated by electrophoresis, denatured to produce single stranded DNA fragments and transferred to a nitrocellulose paper or hybond membrane, called as Southern blots (Southern, 1975). The probe DNA is denatured, labelled (generally with a radiolabelled nucleotide or conjugated enzymes that liberate a colour substrate) and used for hybridization with the Southern blot. The hybridized blot is exposed to an X-ray film varying from 2 hr to 10 days (at –85 °C), depending on the isotopic counts. The developed film (autoradiograph) with the differential positions of the hybridization events is scored for different alleles of a locus. The commonly used DNA probes detect single locus (particularly the cDNAs and cloned genes), albeit multilocus probes are not infrequent.

RFLPs have several merits and find wide usage in genetic mapping. These can be produced in infinite numbers and are so far the best for construction of primary linkage maps with optimum genome coverage. These produce (mostly) codominant fragments and facilitate distinguishing heterozygotes from homozygotes. These are almost free from any sort of interallelic interactions, particularly epistasis, and are hardly influenced by environmental fluctuations. The only limitation with RFLPs involves the need for DNA libraries/probes, the higher amount of DNA for each individual and high initial investments. DNA probes, which include genomic or c-DNA clones (homologous, heterologous or orthologous) and cloned genes are now available commercially, or as generous gifts for most of the common (plant/animal) species.

RAPDs (Williams *et al.*, 1990) are second only to RFLPs in their utility in genetic mapping, particularly in plants. It was the first PCR based marker along with the very similar marker AP-PCR (Welsh and McClelland 1990). RAPDs are detected by using single short arbitrary DNA sequences of 10bp (decamer) as primers to amplify DNA samples (template) of two or more individuals using programmable (PCR) machines (Saiki *et al.*, 1988) to monitor denaturation of template DNA into single strands (95 °C), annealing of primer to complementary 5′-sites on the single DNA strands (37-55 °C) and formation of a new strand (72 °C) mediated by a thermophilic DNA polymerase enzyme (Taq polymerase) using free deoxy-nucleotide tri-phosphates (dNTPs). Priming sites at two complementary strands generally measuring up to a distance of about 2000bp lead to the amplification of intervening segment. Insertion or deletion in the priming site(s) in an individual leads to the failure of amplification. Presence or absence of amplified fragment(s) can be scored tacitly by electrophoresis of PCR products followed by staining of the gel.

The main advantage of RAPDs is the simplicity of the technique associated with speed and cost effectiveness. RAPDs can theoretically be detected for any part of the genome. A single primer often detects more than one locus. The major limitation of these markers is their dominant intra-allelic interaction. For all practical purposes, distinguishing heterozygotes from homozygotes from intensity polymorphism is not that simple. Therefore, use of RAPDs is limited only to segregating populations consisting of homozygous individuals. Another constraint of analysis with these markers is low reproducibility. It is often necessary to repeat an experiment in several ways! Two other PCR-based markers include AP-PCR (Welsh and McClelland, 1990), which involves the use of decamer primers, and DAF (Baura *et al.*, 1992) which makes use of primers, 6-8 bases in length.

For overcoming the demerits of RFLP or RAPD analysis, new DNA markers have been proposed, which include AFLP (Vos *et al.*, 1995) and cleaved amplified polymorphic sequences (CAPs) (Konieczny and Ausubel 1993; Jarvis *et al.*, 1994). The CAPs, however, are not that effective for animal system where individuals used are generally genetically close. AFLPs on the other hand are constrained with the tendency of clustering at certain chromosomal stretches.

In human and other animal systems, the hypervariable repeats (HVR) or 'minisatellites' (Jeffreys *et al.*, 1985) and the simple sequence length polymorphism (SSLP) or 'microsatellites' also called simple sequence repeats or SSRs (Weber and May, 1989; Tautz, 1989) are effective to detect polymorphism even between closely related individuals. SSRs have also been used for enrichment of primary linkage maps in plants.

Mapping Population

Different segregating populations have been used for genetic mapping in plants. These include F_2 (selfed or intercross), backcross (BC), doubled haploid (DH), recombinant inbred (RI) and inter-mated (IM) populations. In animal systems, as in case of mice, RI lines are produced by sib-mating and used for mapping in humans at 'CEPH' (Center for Studies on Human Polymorphism).

Choice of mapping population depends on several considerations including the mode of reproduction and breeding behaviour of the organism, type of markers available, traits to be mapped and map density desired. A back cross (BC) population is raised by crossing an F_1 hybrid with one of its parents. It segregates for the allele inherited from the non-recurrent parent. Any BC individual, therefore, contains one recombinant chromosome in a homologous pair. BC population experiences one cycle of meiosis that leads to very strong linkage among marker alleles inherited from a common parent. In certain interspecific crosses selfing leads to sterility but backcrossing facilitates recovery of fertile progeny. In such cases genetic mapping is done in BC population. This population requires only two generations to be raised. The limitations of using BC populations in genetic mapping include the lowest recombination event per gamete (x), and its further propagation that is achieved only through clonal propagation. Besides, possibility of only two genotypes per locus in this population limits the use of the map to study all kinds of gene action.

F_2 populations are raised through selfing of an F_1 or through intermating between F_1s, thus needing only two generations. The possibility of three genotypes, in all possible combinations of parental alleles makes the framework map highly useful to study all kinds of gene actions. Moreover, both the chromosomes of homologous pair experience recombination events and thereby, provide twice more information than a BC population. F_2 populations have been used in most of the mapping programs till date. F_2 and BC populations should be used for construction of primary genetic linkage maps in an organism studied for the first time. Recently, 'permanent populations' are being used to develop genetic linkage maps that can be enriched infinitely by addition of marker and trait loci. These include DH and RI population. These populations can be grown in multiple locations and seasons, and thereby, facilitate precise estimation of the genetic component of variation from the total phenotypic variation. This advantage makes them highly useful for accurate detection of chromosomal regions controlling the complex quantitative traits.

DH individuals are raised by culture of single pollen grains from F_1 plants, followed by induction of chromosome doubling. A DH population, therefore, can be developed only in those species, which respond to anther culture. Several genetic linkage maps have been constructed using DH lines and used for mapping of several economic traits. DH individuals have two identical homologous chromosomes, each being the same recombinant. Thus, the recombinational information remains the same as in a BC population.

RI population is raised by 'single seed descent' (Brim 1966) method or recurrent intermating among siblings for five to six generations starting from an F_2 population. RI lines are nearly homozygous, never completely homozygous as DHs. Because of several cycles of meiosis experienced, the RI lines have an overall gain in recombinational information over an F_2 population. Development of RI lines in plants requires shorter time as compared to mice. Another advantage of RI lines is the possibility of growing them over a number of locations or seasons that provides the scope for fast enrichment of a map and detection of precise locations of gene clusters controlling complex quantitative traits.

Intermating among different F_2-derived lineage for several generations has recently been a popular practice among plant geneticists, because this strategy facilitates the accumulation of 50 % more recombinational information each generation. Intermated populations are being used in several crops since their first use for high resolution mapping in *Arabidopsis* (Liu *et al.*, 1996).

Linkage Analysis

The probe-enzyme combinations or primers already found to detect polymorphism between the parents are used for marker analysis in the mapping populations. The autoradiographs for a particular marker are scored for the segregating alleles and the individuals are genotyped according to the parental genotypes. Genotyping for codominant and dominant markers should be done differently as per the instruction of the particular program used for linkage analysis. Confusing observations must always be scored as missing and the software could take care of such data. For certain probes multiple fragments produced often create confusion for selecting out the segregating pairs and need critical observation. Presence of more than one fragment in a parent and also in segregating individuals is also not uncommon. In RI population, whatever the number of selfing generations, codominant markers may detect alleles from both the parents in the chromosomal stretches remaining still heterozygous and an observer should restrain to be biased for scoring alleles for a particular parent! The simple PCR based markers are scored from the gel photographs. The dominant bands in DH, BC and to some extent RI populations do not make any problem in scoring and further analysis. In F_2s, however, use of a subset of mapping population (putatively homozygous for morphological markers or known codominant markers) has been found to be more reliable than going for scoring the (PCR based) dominant bands considering intensity polymorphism.

Once all individuals of a mapping population are genotyped using each polymorphic marker, linkage analysis can be done. Several comparable computer programmes can be used for framing the linkage groups. These include Linkage I (Suiter *et al.*, 1983), MAPMAKER v2.0 (Lander *et al.*, 1987), MAPMAKER/EXP 3.0 (Lincoln *et al.*, 1992a), Cri-Map (Weaver *et al.*, 1992), G-Mendel (Liu and Knapp, 1992), JoinMap (Stam, 1993) and MapManager (Manly, 1995). Choice of a program, however, may depend on the kind of computer available (Macintosh/IBM), criteria of assembling marker loci and maps, targeted use of the genetic map besides several other factors. After construction of the genetic map it is worthwhile to align the marker data as per the order in the linkage groups to verify any scoring error.

Genetic Distance of Marker Loci

Linkage mapping is based on the assumption that chance of crossing over between two loci is proportional to their physical distance. As it involves frequency of recombination, the number of individuals in the mapping population becomes a critical factor. The second critical factor is the occurrence of double crossing over. Likelihood of double recombinants is proportional to the square of the recombination distance between two loci that is taken care of by adjustment of recombination fraction to centi Morgan (cM) unit. The two often used algorithms for such adjustment follow mapping functions described by Haldane (1919) and Kosambi (1944). Most of the computer programs for genetic mapping are equipped with this conversion mode. The optimum number of individuals needed for genetic mapping depends on several factors, primarily level of resolution desired, availability of manpower and time. About 1-5 cM marker interval has been obtained in several primary genetic maps using about 50-100 individuals.

Recombination frequency may vary due to genetic, epigenetic and environmental factors. Crossing over is known to be reduced in regions close to centromere and inverted segments. It has been found recently that the repetitive DNA elements are cold to recombination events, and they constitute the major part of a genome! One may therefore question the utility of genetic mapping, as far as the distances of marker intervals are concerned. This differential recombination frequency, however, does not affect the sequence of marker loci in a species that is fairly conserved. Reorganization in a genome often involves structural rearrangement that leads to speciation. It may be exemplified from the chromosomal inversions, leading to new species in *Drosophila* of classical genetics and from rearrangements in taxonomically close cereal species resolved recently by molecular marker analysis (Bennetzen and Freeling, 1993).

Optimum Coverage of a Genetic Map

It is sometimes seen that two or more genetic maps of the same species or very closely related species cover strikingly different distances. This poses the question, 'when a map should be considered as complete'. Depiction of linkage groups of the same number as the haploid number of chromosomes and absence of too many unlinked markers is the primary criteria to assume that a genetic map is complete. An organism with a small genome and less number of chromosomes would obviously span a shorter distance and require lesser number of markers to cover its chromosomes, as compared to an organism with a large genome with many gametic chromosomes. Average recombination fraction between marker loci is another criteria to judge coverage. In general, recombination fraction of 5 % with lower range of variation is thought to be optimum. Addition of new markers particularly in case of RI and DH based maps, to any of the linkage groups in the primary genetic map also provides a positive indication of optimum coverage. Tendency of a marker to show linkage to the markers of another linkage group is obviously a negative indication. More confirmation of completeness can be obtained by using repetitive DNA elements from telomeres as probes and determining the strength of linkage to the terminal markers of the linkage group in the primary genetic linkage map (Ganal *et al.*, 1992; Burr *et al.*, 1992).

Interpretation of a Genetic Linkage Map

A genetic linkage map provides useful first hand information about the nature of genome organization and reorganization. Clustering of marker loci might indicate that they are located near the centromere region. Mapping of a group of markers with preferential skewness in a genomic location is an evidence for its involvement with biological fitness (Kole *et al.*, 1997), and this particular location might contain loci of some deleterious alleles from either of the parents. Detection of duplicated loci in clusters might indicate genome reorganization due to chromosomal aberrations. When such duplications are too frequent, particularly for loci detected with cDNA probes or cloned genes, it may suggest paleoploidy (Kole *et al.*, 1997).

Correlation of Genetic Linkage Groups and Chromosomes

Genetic maps describe just the linkage groups of a genome with molecular markers as the landmarks. It could be useful to assign each individual molecular genetic linkage group to a specific chromosome. The chromosomes of many organisms have been identified using unique features as puffs, loops, knobs and chemically induced bands besides their relative size and position of centromeres. A linkage group can be correlated to any of these chromosomes by two ways. The first one involves the use of aneuploid stocks, substitution lines or additional lines for genetic mapping. This strategy has been used effectively in rice, tomato, cotton and wheat. The second option involves *in situ* hybridization of DNA probes with chromosomes. Repetitive DNA elements as well as single copy DNA probes could be used for this purpose. Establishment of such correlation is useful to integrate the information of location of genes determined by cytological mapping with those localized by molecular mapping. Hybridization with repetitive DNA elements also provides information about individual chromosomes in terms of genetic activity (or inertness!).

Application of Genetic Maps

A genetic map provides not only information of fundamental interest but is also the platform for several applied tools. A framework map can be used for locating genes controlling simply inherited traits and complex quantitative traits, particularly those of interest in agriculture, medicine, industry or in development. Besides, it can be used for further saturation of the targeted regions, for comparative mapping among two or more closely related taxa and also for map based cloning of genes.

Detection of Loci for Simply Inherited Traits

This is generally called as gene mapping or qualitative trait loci mapping. The strategy applied is quite simple. The phenotypic expression of the concerned character is recorded on the individuals of a mapping population; they are converted to genotypes depending upon which parent they resemble, followed by linkage analysis with the marker data already available. In some cases the phenotypic expression may be recorded using metric data or scores. For mapping of the trait locus, these quantified phenotypic data are converted to discrete forms resembling the parents, followed by routine linkage analysis. Literature reveals mapping of an array of such loci in several organisms. A list of mapped genes in plant systems is furnished in Table 2.

An interesting but simple technique is now followed for gene mapping. It involves tagging of the gene first with tightly linked markers following the concept of bulked segregant analysis (BSA) described by Michelmore *et al.* (1991). The DNAs of segregating individuals with two extreme phenotypes are pooled into two groups and used for detection of RAPDs. The bulking theoretically makes mimicry of near isogenic lines that helps in the detection of genomic areas close to the trait loci. It is perhaps the most frequently used single technique in molecular breeding. Once the tightly linked or flanking markers are detected, it is fairly easy to position it on the available map, either by analyzing the DNAs of a mapping population followed by linkage analysis or using randomly selected probes or primers detecting marker loci on the linkage groups to analyze the DNA bulks.

Detection of Quantitative Trait Loci (QTL)

One of the unique contributions of genetic mapping is the possibility of detection of genomic regions controlling quantitative traits. Most of the economic characters are quantitative in nature (Table 3). These characters are highly complex with low to

Table 2. A list of genes mapped in some major crops

Crop	Trait	Gene	Marker	Marker & cM Distance	Chrm	Reference
Rice	Leaf blast resistance	*Pi-1*	RFLP	*NPB181 (3.5)*	11	Yu 1991
		Pi-2(t)	RAPD	*RG64 (2.1)*	6	Hittalmani *et al.*, 1995
		Pi-4(t)	RFLP	*RG869 (15.3)*	12	Yu *et al.*, 1991
		Pi-ta	RFLP	*RZ397 (3.3)*	12	Yu *et al.*, 1991
		Pi-5(t)	RFLP	*RG498, RG788 (5-10)*	4	Wang *et al.*, 1994
		Pi-6(t)	RFLP	*RG869 (20)*	12	Yu, 1991
		Pi-7(t)	RFLP	*RG103 (5-10)*	11	Wang *et al.*, 1994
		Pi-10(t)	RAPD	*RRF6, RRH18*	5	Naqvi *et al.*, 1995
		Pib	RFLP	*RZ123*	2	Miyamoto *et al.*, 1996
	Bacterial leaf blight resistance	*Xa1*	RFLP	*Npb235 (3.3)*	4	Yoshimura *et al.*, 1992
		Xa2	RFLP	*Npb235, Npb197 (3.4, 9.4)*	4	Yoshimura *et al.*, 1992
		Xa3	RFLP	*Npb181, Npb78 (2.3, 3.5)*	11	Yoshimura *et al.*, 1992
		Xa4	RFLP, RAPD	*Npb181, Npb78, (1.7, 1.7)*	11	Yoshimura *et al.*, 1992;1995a,b
		Xa5	RFLP	*RG556 (0-1)*	5	McCouch *et al.*, 1991
		Xa10	RAPD	*OP07 (5.3)*	11	Yoshimura *et al.*, 1995
		Xa13	RFLP	*RZ390, RG136 (0, 3.8)*	8	Yoshimura *et al.*, 1995
		Xa21	RAPD	*Pta818, Pta248 (0-1)*	11	Ronald *et al.*, 1992
	Gall midge resistance	*Gm2*	RFLP	*RG329 (1.3), RG476 (3.4)*	4	Mohan *et al.*, 1994
		Gm4-t	RAPD	$E20_{570}$ *(0.4* from *R1813 & 1.4* from *S1633b*	8	Nair *et al.*, 1997
	Brown plant hopper resistance	*Bph1*	RFLP	*XNpb248*	12	Hirabayashi & Ogawa, 1995
		Bph10 (t)	RFLP	*RG457 (3.7)*	12	Ishii *et al.*, 1994
	Stripe resistance	*Stv-b*	RFLP	*St10 (0)*	11	Hayano-Saito *et al.*, 1998
	Green leaf hopper resistance	glh_{lb1}	RAPD	$OPA19_{320}$ *(27.7)*	N.A.	Padmavathi (2002)
	Fragrance	*Fgr*	RFLP	*RG28 (4.5)*	8	Ahn *et al.*, 1992
	Fertility restorer	*Rf3*	RFLP	*RG532 (0.2)*	1	Zhang *et al.*, 1997
	Photoperiod insensitivity	*Se1*	RFLP	*RG640 (0)*	6	Mackill *et al.*, 1993
		Se3	RFLP	*A19 (5-10)*	6	Maheswaran, 1995
	Semi dwarf	*Sdg(t)*	RFLP	*RZ182 (4.3)*	5	Liang *et al.*, 1994
		Sd1	RFLP	*RG109 (0.8)*	1	Cho *et al.*, 1994
	PGMS	*PMS1*	RFLP	*RG477 (4.3)*	7	Zhang *et al.*, 1994
Wheat	Vernalization response	*Vrn1*	RFLP	*Xrz395 (80* from centromere*)*	5AL	Nelson *et al.*, 1995

	Vrn1	RFLP	*Xwg44 (7.5)*	5AL	Korzun *et al.*, 1997
		STMS	*XgwM186 (23.6)*	5AL	Korzun *et al.*, 1997
	Vrn-A^{m}2	RFLP	*Xgw199 (125* from centromere)	5AL	Dubcovsky *et al.*, 1998
	Vrn-D1	STMS	*Xgwm212 (3.3), Xgwm292 (4.1)*	5D	Snape *et al.*, 1998
Aluminium tolerance	*Alt-BH*	RFLP	*Xbcd1230 (1.1), Xcdo1395 (11.3)*	4DL	Riede & Anderson, 1996
Kernel hardness	*ha*	RFLP	*XksuF11 (5.7)*	2AL	Sourdille *et al.*, 1996
	ha	RFLP	*Xbcd120 (4)*	2DL	Sourdille *et al.*, 1996
	ha	RFLP	*XksuG48 (4.8)*	6DS	Sourdille *et al.*, 1996
Dwarfness	*Rht12*	RFLP	*Xpsr1201 (15.1)*	5AL	Korzun *et al.*, 1997
	Rht12	STMS	*Xgwm291, Xgwm410 (5.4,11)*	5AL	Korzun *et al.*, 1997
	Rht8	STMS	*WMS261 (0.6)*	2DS	Korzun *et al.*, 1998
	Rht-B1	RFLP	*Xpsr144-4b (13.2)*	4BS	Sourdille *et al.*, 1998
			Xfba1-4b (5)		Cadalen *et al.*, 1998
	Rht-D1	RFLP	*Xglk578-4D (2.8),*	4DS,	Sourdille *et al.*, 1998
			Xgfba211-4D (20%)	4BS	Cadalen *et al.*, 1998
Fertility restoration	*Rf4* *Rf3,5DRf*	RFLP	*Xksu48(13.1%), Xcdo442(19.3%), Xcdo786(8.8%)*	6BS, 1BS, 5D	Ma & Sorrells, 1995
Amylose content	*Wx-B1*	RFLP	*Xbcd1738(6.2)*	4A	Araki *et al.*, 1999
Eye spot	*Pch2*	RFLP	*Xpsr121(3.8)*	7D	De la Pena *et al.*, 1996
	Pch2	RFLP	*Xcdo347(11.0), Xwg380(18.8)*	7AL	De la Pena *et al.*, 1996
Powdery mildew resistance	*Pm*	RFLP	*Xwg516*	2B	Rong *et al.*, 1998
	Pm1	RFLP	*WHS178 (3)*	NA	Jahoor, 1998
	Pm2	RFLP	*Xbcd1871 (3.5)*	5D	Ma *et al.*, 1994
	Pm2	RFLP	*Xfba393*	5DS	Nelson *et al.*, 1995
	Pm3b	RFLP	*Xbcd1434 (1.3)*	1AS	Ma *et al.*, 1994
	Pm4a	RFLP	*Xbcd1231-2A (1.5)*	2AL	Ma *et al.*, 1994
	Pm4a	RFLP	*Xcdo678-2A (1.6)*	2AL	Ma *et al.*, 1994
	Pm12	RFLP	*Xpsr10, Xpsr106, Nor2, Xpsr141, Xpsr113* (all tight)	6DS	Jia *et al.*, 1996
	Pm12	RFLP	*Xpsr142, Xpsr149, Xpsr2, Xpsr605, Xpsr154, Xpsr546* (all tight)	6DL	Jia *et al.*, 1996

(*Contd.*)

(*Table 2 Continued*)

Crop	Trait	Gene	Marker	Marker & cM Distance	Chrm	Reference
Wheat	Cereal cyst nematode resistance	*Cre1*	RFLP	*Xglk605 (7.3), Xglk588 (8.4)*	2BL	Williams *et al.*, 1994
		Ccn-D1	RAPD RFLP	*OPE20* (tight) *CSE20-2 (tight)*	2DS	Eastwood *et al.*, 1994
	Wheat Streak mosaic virus resistance	*Wsm1*	STS	*Wg232* (tight)	4L	Talbert *et al.*, 1996
	Hessian fly resistance	*H23*	RFLP	*XksuH4 (6.9), XksuG48a (15.6)*	6D	Ma *et al.*, 1993
		H24	RFLP	*Xcnl BCD451 (5.9), Xcnl CDO482 (5.9), Xksu G48b (12.9)*	3DL	Ma *et al.*, 1993
	Durable stem rust resistance	*Sr2*	RFLP	*Xbcd1871 (3.5)*	5B	Nelson *et al.*, 1995
	Leaf rust resistance	*Lr1*	RFLP	*PSR580 (13.8), PSR567* (tight)	5DL	Feuillet *et al.*, 1995
		Lr3	RFLP	*Xmwg798 (0)*	6BL	Sacco *et al.*, 1998
		Lr13	RFLP	*Xbcd1709 (<7.9), Xpsrs912 (9.1)*	2B	Seyfarth *et al.*, 1998
		Lr32	RFLP	*Xbcd1278 (3.6 ± (2.6), Xcdo395 (6.9 ±-3.6)*	3DS	Autrique *et al.*, 1995
	Stripe rust resistance	*Yr15*	RFLP	*Nor1B (11)*	1B	Sun *et al.*, 1997
			STMS	*WMS533 (4.5)*		Fahima *et al.*, 1997
		YrH52	RFLP STMS	*Nor1 (1.4), NA (0.02-0.35)*	1B	Peng *et al.*, 1999
	Loose smut resistance	*T10*	RAPD RFLP	*UBC353 (14) Xcrc4.2 (14), Xcrc153.2 (10)*	2BL	Procunier *et al.*, 1997
	Septoria nodorum blotch resistance		RAPD	*UBC521 (15), RC37 (13)*	3A	Cao *et al.*, 1998
	Scab resistance	*FHB*	AFLP RFLP	*XeageMetal (16.2), ksuH16 (12.7), Xbcd1331 (7.3), Xcdo1387 (7.4), Xcdo524 (6.9)*	3BS, 2AL, 6BS, 4AL, 6BS	Anderson *et al.*, 1998
Barley	Stem rust resistance	*rpg4*	RAPD		7M	Borovkova *et al.*, 1995

	Powdery mildew resistance	*MiLa*	RFLP	*WG645 (0)*	2	Saghai-Maroof *et al.*, 1994
Mungbean	Bruchid resistance	*Bruchid*	RFLP	*PA882 (3.6)*	8	Young *et al.*, 1992
Soybean	SMV resistance	*Rsv*	RFLP/SSR	*PA186 (1.5), pK644a (2.1) SM176 (0.5)*	NA	Yu *et al.*, 1994
Tomato	*F. oxysporum* resistance	*I2*	RFLP	*TG105 (0.0±4.8)*	11	Sarfatti *et al.*, 1989
	TMV resistance	*Tm2*	RFLP	*TG3 (0.3), CD3 (0.9)*	9	Young and Tanksley, 1989
	Powdery mildew resistance	*Lv*	RAPD	*OP57, OP58, OP218 (1 from CD134)*	12	Chunwongse *et al.*, 1994
		Lv	RFLP	*CT211, CT219 (5.5)*	12	Chunwongse *et al.*, 1994
	Salt tolerance		IZ	*Est3, Prx7, Pgdh 2, pgt1*	1,3,12	Foolad and Jones, 1993
Arabidopsis	Freezing tolerance	*SFR2*		*GAPC (2.3), nga126 (3.6)*	3	Thorlby *et al.*, 1999
		SFR3		*AIG1 (1.2), jcc (0.4), T27K12 (0.4)*	1	Thorlby *et al.*, 1999
		SFR4		*F1119T7 (1.2), mi106 (0.0), nga208 (5.9)*	1	Thorlby *et al.*, 1999
		SFR5		*m219 (2.8), m235 (14.1)*	1	Thorlby *et al.*, 1999
		SFR7		*g3883 (9.5), nga1107 (12.1)*	4	Thorlby *et al.*, 1999
		SFR8		*CDC2A (7.1), BGL1 (8.3)*	3	Thorlby *et al.*, 1999
		SFR9		*LFY3 (6.2), g2368 (3.6)*	5	Thorlby *et al.*, 1999
Mustard	White rust resistance	*Ac21*	RAPD	*WR2 (7), WR1a/3 (1.4)*	NA	Prabhu *et al.*, 1998
		Ac2(t)	RAPD	*$OPBO6_{1000}$ (5.5)*	NA	Mukherjee *et al.* (2001)
Rapeseed	Turnip mosaic virus resistane	*TuRBO1*	RFLP	*pO120b (3.6)*	6	Walsh *et al.*, 1999
	Blackleg	*LEM1*	RFLP	*wg2a3b (1), wg5a1a (1.9)*	6	Figdore *et al.*, 1993
	White rust resistance	*ACA1*	RFLP	*Tg6a12 (12.9)*	9	Ferreira *et al.*, 1995
Indian rape	Erucic acid	*Eru1*	RFLP	*tg1f8 (7.2)*	1	Teutonico & Osborn, 1994
	Seed color	*Yls*	RFLP	*M456b (6.5) ec3c8b (7.8)*	5	Teutonico & Osborn, 1994
	Leaf pubescence	*Pub1*	RF:P	*Ec2b3 (8.4)*	4	Teutonico & Osborn, 1994
	White rust resistance	*ACA1*	RFLP	*Wg6c1a (5.0), ec2b3a (5.5)*	4	Kole *et al.*, 1996
		AC2	RFLP	*Wg6c1a (9.8)*	4	Kole *et al.*, 1997
		Aca1 (AC2+ AC7)	RFLP	*Wg6c1a (5.0), ec2b3a (5.5)*	4	Kole *et al.*, 2002a
	Leaf pubescence	*PUB1*	RFLP	*Ec2b3a (7.9)*	4	Kole *et al.*, 1996
		PUB1	RFLP	*Wg6c1a (14.3)*	4	Kole *et al.*, 1997
	Seed colour	*YLS1*	RFLP	*Wg2c4c (9.7), wg3h8a (11.3)*	5	Kole *et al.*, 1997
	Self-incompatibility	*SLG6a*	RFLP	*$Wg2d_{11}$ (9.9)*	2	Panigrahi (2002)
		SLG6b	RFLP	*$Wg5b_9$ (4.2)*	9	

Table 3. A list of QTLs detected for important characters in some major crops

Crop	Character	QTL #	LG	Reference
Rice	Osmotic adjustment, dehydration tolerance	15	81,3,5,7,8	Lilley *et al.*, 1996
	Root traits related drought resistance	28	2,3,4,6,7,11	Ali, 1999
	Drought avoidance	36	1,2,3,4,5,6,7,9,10,11,12	Courtois *et al.*, 2000
	Yield related characters	26	1,2,4,5,6,8,10,11	Lin *et al.*, 1996
	Plant morphology and RYMV resistance	12	1,2,3,4,7,8,9,12	Albar *et al.*, 1998
	Primary seed dormancy	5	3,5,6,7,12	Lin *et al.*, 1995
	Sheath blight resistance	6	2,3,8,9,12	Li *et al.*, 1995
	Total tiller no., effective tiller no., 100-grain weight and resistance to green leafhopper	1 each	NA	Padmavathi (2002)
	Heading date & Plant height	7	3,4,8,12	Li *et al.*, 1995
	Panicle and grain characteristics	18	1,2,3,4,5,6,7,8,10	Redona & Mackill, 1998
	Heading date	5	6,7,8	Yano *et al.*, 1997
	Yield related attributes	8	1,2,4,6,12	Wu *et al.*, 1996
	Tiller number	15	1,2,3,4,6,7,8,12	Yan *et al.*, 1998
	Fertility restoration	2	10	Tan *et al.*, 1998
	Grain characters	12	1,3,10	Huang *et al.*, 1997
	Root morphology and drought avoidance	15	-	Champoux *et al.*, 1995
	Heterosis	32	-	Xiao *et al.*, 1995
	Agronomic importance	22	-	Lu *et al.*, 1997
	N_2 fixation	4	-	Wu *et al.*, 1995
Wheat	Grain protein content	1	2DL	Prasad *et al.*, 1999, 2003
	Grain weight	1	1AS	Varshney *et al.*, 2000
	Growth characters	14	1BL, 1DL, 2AS, 2DS, 2DL, 4AL, 4DS, 5DL, 7BL	Kulwal *et al.*, 2003
	Haploid formation	2	4B, 1 Unassigned	Torp *et al.*, 1998
	Green plant formation	5	2A,2B,3A,5B,1 Unassigned	Torp *et al.*, 1998
Maize	Grain yield	6	-	Stuber *et al.*, 1992
	Plant height	3	-	Schon *et al.*, 1993
	Ear height, plant height, GDD to anthesis, GDD to silk delay & GDD to silk emergence	23	-	Veldboom *et al.*, 1994
	Agronomic importance (9 Characters)	46	-	Doebly and Stec, 1991
	European corn borer resistance	7	-	Schon *et al.*, 1993
	Plant height	11	1,2,3,4,5,6,7,8,10	Khairallah *et al.*, 1998
	Days to female flowering	13	1,2,3,4,5,6,7,8,10	Khairallah *et al.*, 1998
	SE corn borer resistance	7	3,5,6,8,9	Khairallah *et al.*, 1998
	Anthesis, silking	9	1,3,5,6,7,8,9,10	Khairallah *et al.*, 1998
Barley	Flowering time	8	2 (2H), 3(3H), 4(4H), 7 (5H), 6 (6H), 1(7H)	Laurie *et al.*, 1995
	Plot yield	7	2(2H)S, 2(2H)L, 3(3H)L, 7(5H)S, 6(6H)L, 1(7H)L	Bezant *et al.*, 1997

	Plant grain weight	7	2(2H)S, 7(5H)L, 1(7H)S, 5(1H)S, 3(3H)L, 4(4H)S, 6(6H)S	Bezant *et al.*, 1997
	Thousand grain weight	10	2(2H)L, 3(3H)L, 4(4H)L, 7(5H)L, 2(2H)S, 6(6H)L, 6(6H)S, 1(7H)S, 1(7H)L, 3(3H)S	Bezant *et al.*, 1997
	Ear grain weight	6	5(1H)S, 2(2H)L, 3(3H)L, 7(5H)L, 6(6H)L, 1(7H)L	Bezant *et al.*,1997
	Ear grain number	7	5(1H)S, 7(5H)L, 6(6H)L, 3(3H), 4(4H)L, 2(2H), 6(6H)S	Bezant *et al.*, 1997
P. vulgaris	Nodule number	4	-	Nodari *et al.*, 1993b
	Blight resistance	4	-	Nodari *et al.*, 1993b
Cow pea	Seed weight	4	-	Fatokun *et al.*, 1992
Mungbean	Powdery mildew resistance	3	3,7,8	Young *et al.*, 1993
Potato	Trichome related attributes	12	-	Bonierbale *et al.*, 1994
	Tuber shape	1	-	Van Eck *et al.*, 1994
Tomato	Yield related attributes	58	-	De Vicente & Tanksley, 1993
	Soluble solids	7	-	Paterson *et al.*, 1991
	Fruit mass	11	-	Paterson *et al.*, 1991
	Fruit pH	9	-	Paterson *et al.*, 1991
Sugarbeet	Cercospora leaf spot	5	1,2,3,9	Nilsson *et al.*, 1999
Arabidopsis	Growth, flowering time	2	2	Olds, 1996
B. campestris	Freezing tolerance	6	2,4,5,7,9,10	Teutonico *et al.*, 1995
	Flowering time	7	2,3,5,8	Osborn *et al.*,1997
	Palmitic acid	1	-	Tanhuanpaa *et al.*, 1995
	Self incompatibility	1	2	Nozaki *et al.*, 1997
	NS glycoprotein	1	1	Nozaki *et al.*, 1997
	Bolting time	2	8,9	Nozaki *et al.*, 1997
	Flowering time	4	2,3,10	Kole *et al.*, 2001
	Freezing tolerance	3	1,9	Kole *et al.*, 2002b
	Winter survival	5	1,2,3,7	Kole *et al.*, 2002b
	White rust resistance	2	4,2	Kole *et al.*, 2002a
B. napus	Flowering time	4	9,12,16	Osborn *et al.*, 1997
	Flowering time	6	2,3,8,10,13	Kole *et al.*, 2001
	Freezing tolerance	1	8	Kole *et al.*, 2002b
	Winter survival	16	1,2,3,4,5,6,8,9,17,19	Kole *et al.*, 2002b
	Flowering time	7	2,3,7,8,9,12	Butruille *et al.*, 1999
	Plant height	4	2,3,12,13	Butruille *et al.*, 1999
	Seed weight & oil content	10	1,2,3,5,10,12,17	Butruille *et al.*, 1999
B. oleracea	Flowering time	3	2,6,8	Camargo & Osborn, 1996

moderate heritability. These traits are controlled by polygenes or minor genes, each with small effect. Quantitative characters exhibit continuous variation following normal distribution. The nature of gene action of these traits is conventionally studied by using biometrical approaches applying several models that involve many assumptions. Most of these assumptions however are hardly tenable.

It is now possible to study the correlation of phenotypic trait data of segregating individuals of a population with their marker data to identify the putative genomic locations of the gene clusters controlling the concerned character (Paterson *et al.*, 1988). The concept of interval mapping (Lander and Botstein, 1989) to detect such QTLs in marker intervals using complete framework maps has became a popular strategy of mapping polygenes. Presently a number of excellent computer programmes are available for QTL mapping. These include programmes proposed by Haseman and Elston (1972), Amos and Elston (1989), Lander and Botstein (1989), Goldgar (1990), Haley and Knott (1992), Kruglyk and Lander (1995), Olson (1995), Almassy and Blangero (1998), Ghosh and Majumder (2000a,b), Zeng, 1994. Of these the first such program used most widely is the computer program MAPMAKER/QTL 1.1 (Lincoln *et al.*, 1992b). In this programme, the likelihood positions are determined by scanning the whole genome as described by the complete map, interval wise, by fixing generally a likelihood of odds of presence of a QTL at the precision level of LOD 2 (1 % level).

The programme provides several useful genetic information besides the putative chromosomal locations of the gene clusters. It provides the degree of likelihood of presence of a QTL, contribution of parental alleles to the QTL, the extent of variation explained by a QTL and the nature of gene action depending on the mapping population (additive, dominant or recessive). The programme allows for detection of QTLs with lesser effects by rescanning of the genome by fixing the major ones, verification of linked QTLs whether spurious or true, identification of the most reliable zone or confidential interval of a QTL and formulation of multi-locus model (Zeng, 1994) with several QTLs that provide information on the ultimate picture of the genetics of the character. One can use the linked markers to evaluate the nature of gene action between QTLs such as epistasis. It is also possible to learn about the nature of variation of a character and to transform the data to normal distribution, besides obtaining the values of mean, skewness, kurtosis, variance, etc. of the character. A list of important QTLs detected in some major crops is furnished in Table 3.

High Resolution Mapping

In certain cases the linked markers are too distant from the trait loci to monitor the desirable gene(s) by marker assisted selection (MAS). It becomes necessary to detect tightly linked markers for effective MAS. The strategy followed in most cases involves marker assisted backcross breeding using the markers flanking the gene or QTL and developing a population segregating only in the area around the gene/gene cluster of interest, parallel to the near isogenic lines (NILs) of conventional breeding, and enriching the area with more markers. The RAPD and AFLP markers have been used widely for this purpose besides RFLPs. The population derived in most cases exhibit discontinuous variation for the same quantitative character and a single locus could be detected for it. The complex trait thus is resolved into a single discrete locus and the strategy is called popularly as Mendelization. Because of heterozygosity in a small region only in the genome, high frequency of crossing over in a small window paves the way for saturation mapping and detection of tightly linked, even co-segregating markers. Successful Mendelization has been done for QTLs controlling quality related traits in tomato (Paterson *et al.*, 1988) and a QTL controlling vernalization responsive flowering time in Indian rape (Kole *et al.*, 2001). The proven possibility of such Mendelization of QTLs will hopefully change the face of breeding for quantitative traits in near future in all organisms. It is not overambitious to claim that the often complicated and not-so-precise biometrical approaches will no longer come in the way of breeding programmes.

Comparative Mapping

Studies on evolution and phylogenetic relationship based on similarities and differences between closely related taxa (sub-species, species or genera) are well practiced tools in life sciences. It is of fundamental importance to have information on the genetic origin and evolutionary pathways of domesticated plants and animals. From practical point of view also, information on genome homology of allied taxa is equally important. In conventional breeding, several characters have been introgressed into domesticated plants from allied species and genera, even without any information on the genes involved. For this purpose, information on the cross-breeding behaviour related to cytological proximity and fertility were needed. Traditionally, similarity and differences between taxa were studied by using mainly morphological, cytological and biometrical attributes. Advent of molecular markers now could provide the most accurate information on similarity of chromosomal regions of two or more closely related taxa. Much information has already accumulated on this aspect in the last two decades of the previous century.

Comparative mapping facilitated this kind of study, and it is now possible to study genome homology even between two taxa, which are not cross-compatible. Several such attempts have shown that closely related taxa contain several chromosomal stretches with similar gene order completely conserved, with only a small number of structural rearrangements, most of them being inversions and translocations. This is immensely useful, as information from one taxon can be used for the other for manipulation of the homologous genomic region (Paterson *et al.*, 2000).

Construction of linkage maps for comparative mapping is done by using orthologous set of genetic markers. It is ensured that the probes are preferably from a common ancestor. The second aspect considered is to select DNA probes that will detect single locus in the genome. That is why the cDNA probes are best suited for comparative mapping. It is well known that the expressed DNA sequences slowly evolve as compared to the intergenic repetitive DNA.

The popular method to construct maps is to use cDNAs to detect RFLPs in the target chromosomes. Another potential alternative is to use CAPs developed from cDNAs. These will satisfy the orthology as well as conservativeness. SSRs have also been used in some cases, but less frequently as compared to RFLPs and CAPs. Theoretically, the PCR-based markers such as RAPD, AP-PCR, DAF or AFLP have been developed with one of the objectives of detecting multiple loci, and it becomes their demerit to be used in comparative mapping. The SCARs are similar to microsatellites, for they also are not conserved.

Comparative mapping has been employed in all kinds of crops of all ploidy levels (Table 4). It has revealed conserved gene order, albeit punctuated by minor chromosomal rearrangements, gene or genome duplication as in maize, sugarcane, cotton and Brassicas; subgenomic origin and composition as in cotton and similarity in gene order of closely related taxa as in mungbean and cowpea, and tomato and potato. Extensive conservation of gene repertoire and order has been delineated in the family poaceae (Hulbert *et al.*, 1990, Ahn and Tanksley, 1993; Ahn *et al.*, 1993; Kurata *et al.*, 1994b). In fact the grass family constitutes a single gene system (Bennetzen and Freeling, 1993; Helentjaris, 1993; Shields, 1993).

Detection of QTLs controlling the polygenic traits on framework maps of orthologous markers has also paved the way for studying their homology across taxa. QTL locations for seed size, shattering and flowering behaviour have been compared in sorghum, sugarcane, maize, wheat, barley and rice (Paterson *et al.*, 1995). It also includes QTLs for height and flowering in maize, sugarcane and other grasses (Pereira *et al.*, 1994; Lin *et al.*, 1995) and vernalization requirement, flowering time, winter survival, freezing tolerance and resistance to white rust within different *Brassica* species *inter se* or with *Arabidopsis* (Teutonico and Osborn, 1995; Teutonico *et al.*, 1995; Osborn *et al.*, 1997; Kole *et al.*, 2001/2002a,b).

Comparative mapping of regions with QTLs has also elucidated some basic aspects of evolution. A qualitative trait locus controlling late flowering in *Arabidopsis* has been found to be homologous to a QTL controlling vernalization responsive flowering time in Indian rape (Kole *et al.*, 2001). Presence of duplicated QTLs in several crops including vernalization responsive and independent flowering time (Ferreira *et al.*, 1995; Teutonico and Osborn, 1995; Teutonico *et al.*, 1995), winter survival, freezing time (Kole, 1997; Kole *et al* 2002b.) white rust resistance (Kole *et al.*, 2002a) in oilseed *Brassicas* and shattering in maize clearly indicates the trend of evolution of gene clusters and evidences for chromosomal duplication thousands of years prior to domestication of crop plants.

The homology of genes or gene clusters, as above, has useful practical implications. Closely related crop plants differ due to only a few genes with major effects. It is possible to isolate these unique genes and use them for genetic tailoring of related crops. The information from rodents and other domesticated animals can also be used for searching and cloning QTLs in humans. Comparative mapping therefore will have a high impact in agriculture and medicine in the present century.

Table 4. A list of some comparative mapping studies

Crops	Markers	Phenotypic Traits	References
O. sativa, O. officinalis	RFLP	-	Jena *et al.*, 1994
Sorghum, Rice, Maize	RFLP	Large seed, reduced desarticulation of mature inflorescence, photoperiod insensitivity	Paterson *et al.*, 1995
Maize, Rice	RFLP	-	Ahn & Tanksley, 1993
Wheat, Rice	RFLP		Kurata *et al.*, 1994
Maize, Wheat, Rice	RFLP	-	Ahn *et al.*, 1993, Moore *et al.*, 1995a
Maize, Sorghum	RFLP	-	Hulbert *et al.*, 1990; Whitkus *et al.*, 1992; Berhan *et al.*, 1993, Binelli *et al.*, 1993, Pereira *et al.*, 1994
Rice, Barley, Wheat	RAPD, Isozyme	Smut resistance (PSH1-31)	Chen *et al.*, 1998
Foxtail millet, Rice	RFLP	-	Devos *et al.*, 1998
Rice, Wheat, Maize, Sugarcane, Foxtail millet, Sorghum	RFLP	-	Moore *et al.*, 1995b
Sugarcane, Maize, Sorghum	RFLP	-	Dufour *et al.*, 1996
Rice, Maize, Wheat, Oat	RFLP	-	Van Deynze *et al.*, 1995a
Rice, Barley	RFLP	-	Saghai-Maroof *et al.*, 1996
Maize, Wheat	RFLP	-	Devos *et al.*, 1994
Wheat, Rye	RFLP	-	Devos *et al.*, 1992
Rice, Barley, Wheat	RFLP	Leaf rust resistance *Lr1*, *Lr10*	Gallego *et al.*, 1998
Rice, Barley	RFLP	Barley malting quality, QTLs and *rpG* 4 chromosomes	Han *et al.*, 1998
Potato, Tomato	RFLP, Isozyme	-	Bonierbale *et al.*, 1988, Tanksley *et al.*, 1992
Tomato, Pepper	RFLP	-	Prince *et al.*, 1993
Arabidopsis, B. oleracea	RFLP		Kowalski *et al.* (1994)
B. nigra, B. oleracea, B. rapa	RFLP	-	Lagercrantz & Lydiate, 1996
B. rapa, B. napus, Arabidopsis	RFLP	Resistance to *A. candida*	Kole, *et al.*, 2002a
B. rapa, B. napus	RFLP	QTLs for vernalization requirement, flowering time, winter survival, freezing tolerance	Kole, 1997, Kole *et al.* 2002b
B. rapa, B. napus, Arabidopsis	RFLP	QTLs and genes for flowering time	Osborn *et al.*, 1997, Kole, 1997, Kole *et al.*, 2001 Kole *et al.*, 2001a
B. rapa, B. napus	RFLP	-	Hoenecke & Chyi, 1991
Arabidopsis, B. nigra	RFLP	Flowering time	Lagercrantz *et al.*, 1996
Mungbean, Cowpea	RFLP		Menacio-Hautea *et al.*, 1993
Pea, Lentil	RFLP		Weeden *et al.*, 1992

Marker Assisted Breeding

The most important practical utility of a genetic map is the information on the marker loci linked to or flanking SITLs and QTLs, controlling characters of economic interests. For effective marker assisted selection or breeding, however, marker loci should be tightly linked with RF of <1 %. Details on MAS/MAB are beyond the purview of this article.

Physical Mapping and Map Based Cloning

It is well known that the distance between two genetic markers is deduced from the recombination frequencies. It is worthwhile to know its correspondence to the actual physical distance, the amount of DNA separating the two marker loci. It is generally obtained from the total DNA in the genome and the total coverage of a complete genetic map for each centi-Morgan (cM). However, the DNA quantity per cM varies widely in higher plants from 280 kb in *Arabidopsis* to 7000 kb in barley and even more in some other plants for which information on genome length is yet to come (Paterson, 1996). Interestingly, the DNA amount intervening the orthologous loci in different plants is more or less comparable. It points towards the repetitive DNA elements for the vagaries of correspondence between genetic distance and physical distance between taxa. The relationship between genetic and physical distances varies even within a single genome at different locations. In tomato, the DNA amount varies from 50kb to over 4000kb per cM. This anomaly has been best documented by clustering of many markers near the centromere and it is well known that this region often acts as a crossover suppressor (Khush and Rick, 1967; Tanksley *et al.*, 1992). The relation between genetic distance and physical distance is influenced by introgressed chromatin (Ganal *et al.*, 1989; Paterson *et al.*, 1996) and recombinational hot-spots, besides repetitive DNA. Phenotype based genetic mapping has a resolution limit of 0.1-1.0 cM for a locus. If it corresponds to 28-280 kb in *Arabidopsis*, the variation in higher plants could be easily imagined! Manipulation of DNA fragments of such large sizes is the primary limitation of cloning genes based on mapping positions that invokes for other alternative approaches.

Three strategies are now employed to clone plant genes. These involve candidate gene and mutagenesis approaches besides map-based cloning (MBC). MBC is an effective means to isolate naturally occurring mutations, even if masked by the presence of other mutations or environment. MBC is based on the segregation of alleles in a population and requires a minimum *a priori* information. The effectiveness of MBC is, however, limited in organisms with large genome and depends on other factors as repetitive DNA fractions and polyploidy. MBC needs primarily a high-resolution map of the region that contains the gene linked to, or flanked tightly (preferably <1 % RF) by marker loci, spanning over a one megabase DNA clone or by a contig of several megabase DNA clones. It requires standardized protocols for identification of the transcripts in the megabase DNA clone and identification of the target gene by mutant complementation through transformation studies.

Concept of physical mapping dates back to 1980s (Steinmetz *et al.*, 1982) that involved 'chromosome walking' to obtain the genic DNA corresponding to the target locus by walking from one marker to another. It is based on the principle of walking along the chromosome to identify a series of contiguous DNA clones or contigs. Chromosome walking may involve a particular region to isolate the gene localized by genetic mapping or the entire chromosome, so as to serve as a resource for future. The major criteria in contig assembly are the size of the steps that requires a proper balance between inputs and resolution. Larger the size, faster could the assembly be completed but with lower resolution, for the target genes are in larger segments!

Chromosome walking was practised previously with the 'λ bacteriophage' as the cloning vector that could carry 10-29 kb of exogenous DNA (Steinmetz *et al.*, 1982). It was followed by the 'cosmid' vectors, which could carry up to 40 kb of foreign DNA. These vectors, therefore became popular for assembling physical maps (Coulson *et al.*, 1986). Since larger clones are required for characterization of large genomes as in human and higher crop plants, these vectors were no longer suitable and the age of 'artificial chromosome' vectors emerged. These vectors became immediately popular for physical mapping and contig assembly in various genome projects for plants and animals, including the 'Human Genome Project' (HUGE) (Burke *et al.*, 1987). The simple strategy of inserting a cloned DNA segment into an artificial chromosome and then cloning it in yeast cells made it possible to propagate DNA segments as long as 100 kb ! Yeast artificial chromosomes (YAC) stand as an important milestone for physical mapping in eukaryotes. Cloning of DNA segments of 400-700 kb length in YACs from the earlier 100-150 kb, is a reality! Bacterial artificial chromosome (BAC) surfaced as an alternative to YAC when the ease and time taken for cloning are desired, albeit its capability of replicating slightly smaller segments ranging from 150-350 kb (Shizuya *et al.*, 1992). The size of a 'megabase library' is again a critical factor of cloning a target gene. Generally the number of DNA inserts of 3x equivalence of haploid genomic DNA amount leads to 95 % confidence of picking up a DNA element represented once in the library and 5x leads to 99 % confidence level.

Contig assembly is proceeding in several organisms now. In humans, a DNA library of 33000 clones with an average size of 900 kb has been fingerprinted by several techniques to permit assembly of a set of overlapping cloned genomic DNA fragments that span each of the human chromosomes (Cohen *et al.*, 1993). In *Arabidopsis*, 374 YAC clones have been organized into a physical map of four contigs, which span the vast majority of chromosome 4 (Schmidt *et al.*, 1995). Such YAC clones, comprising a physical map of the genome, can readily be integrated with existing genetic maps, simply by applying mapped DNA markers as labeled probes to the array of YACs. In this manner, new DNA markers might be quickly mapped to a level of resolution which otherwise would require study of several hundred recombinant plants (Liu *et al.*, 1996).

Construction of complete contig maps is an enormously costly affair. It is affordable for small genomes such as the model plants, *Arabidopsis* and rice, but not in higher plants. Moreover, the contig maps in these species can be applied to other crops that share genome homology as the cruciferous vegetables and oil seeds (Kowalski *et al.*, 1994; Osborn *et al.*, 1997; Kole, 1997, 2001) from *Arabidopsis* and crops of the grass family from rice. For other higher plants however, high-density maps with DNA markers could be the viable alternative for gene cloning (Tanksley *et al.*, 1995). Enrichment of a particular chromosomal region with additional markers using DNAs of NILs or pooled DNAs mimicking NILs has proved successful (Giovannoni *et al.*, 1991; Kole, 1997, 2001). A second pooling technique was suggested by Churchill *et al.*, (1993) for high-resolution mapping of specific chromosomal regions. This strategy of local enrichment by high-resolution mapping (HRM) can identify markers close enough to a target gene, to enable direct landing on a megabase DNA containing the gene (Tanksley *et al.*, 1995).

Map based cloning using contig assembly, or chromosome walking/landing has emerged as a potential tool for cloning genes in higher plants, but in several crops, it is constrained with the problems of physically large genomes, large proportion of repetitive DNA fractions and polyploidy which call for an alternative approach as mutagenesis for cloning of important genes.

MBC requires basically (a) localization of the target gene in a small chromosomal region preferably flanked by two DNA markers, and spanned by a one magabase DNA clone or by a contig of several megabase DNA clones; (b) standardized method for identification of the transcripts in the megabase DNA clone; and (c) workable transformation system for introducing foreign DNA into the plant species of interest, allowing identification of the target gene by mutant complementation. The first criterion can be addressed by using a large number of markers and a large mapping population to facilitate HRM to search for megabase DNA clones containing the target gene. For this purpose, available contig maps can be used or local contig maps can be developed by chromosome walking. Isolation of transcripts from megabase DNA clones frequently poses many technical difficulties and requires combination of several approaches (Gardiner and Mural, 1995). In tomato, conventional approach of direct screening of a cDNA library with the megabase DNA clone has been successful (Martin *et al.*, 1994). This approach is not suitable for transcripts with hyper-expression or expression in specific tissue or growth stages. Elution and amplification by PCR (Parimoo *et al.*, 1991; Lovett *et al.*, 1991; Morgan *et al.*, 1992; Fan *et al.*, 1993; Lovett 1994) of the megabase DNA clones selectively bound to cDNAs have provided a solution for such cases. Several cDNA clones from human and other mammals have been isolated by this approach (Gardiner and Mural, 1995). A technique that employs PCR to amplify hypomethylated DNA upstream from an interspersed repetitive element found very frequently in human DNA (Valdes *et al.*, 1994) is another way of effective gene isolation. Future genetic maps are expected to include a large number of transcribed sequences, considering the ongoing

DNA sequencing projects. These will be helpful in searches for candidate genes. At least in rice (Sasaki *et al.*, 1994) and *Arabidopsis* (Hofte *et al.*, 1993, Newman *et al.*, 1994) sequence data base and high-throughput mapping efforts (Kurata *et al.*, 1994a) have culminated in near-comprehensive genetic maps of expressed sequences. This will eliminate the need for region-specific searches in each of these two crops and other species with close genome homology. Transformation of a plant with recessive phenotype having a target gene with dominant phenotype, provides its identity and this mutant complementation approach is the popular means of identification of target genes. The major limitation of this strategy is the difficulty of transformation in several plant species, particularly the monocots. Another limitation of this strategy is the phenomenon of 'position effect' that involves the effects of DNA adjacent to the transformation site and that requires multiple transformation events for confirmation of the function of a candidate gene. MBC, therefore, should be supplemented or substituted with other approaches of gene cloning, depending on the situation. Literature is now quite enriched with the methodologies of physical mapping required for MBC. A list of the genes cloned by using MBC is furnished in Table 5.

Table 5. A list of some genes isolated through map based cloning

Crop	Trait/Product	Gene	References
Arabidopsis	Restriction of Long distance movement of tobacco etch virus	*RTM 1*	Chisholm *et al.*, 2000
	Dual specificity to pathogen	*RPM 1*	Grant *et al.*, 1995
	Fertilization independent seed development	*FIS 1* & *FIS 2*	Luo *et al.*, 1999
	Farnesylation in meristem development	*WIGGUM*	Ziegelhoffer *et al.*, 2000
Rice	Baterial blight Resistance	*Xa 1*	Yoshimura *et al.*, 1996
	Baterial blight Resistance	*Xa 21*	Song *et al.*, 1995
	Blast resistance gene	*Pi-ta2*	Nakamura *et al.*, 1997
	Methyl transferase homolog to *X. oryzaep* v. *oryzae*	*Xor 11*	Choi & Leach, 1994
Tomato	Iron uptake & intercoastal chlorosis	*Chloronerva*	Ling *et al.*, 1999
	Protein kinase gene conferring resistance to *Pseudomonas Syringae*	*Pto*	Martin *et al.*, 1993
	Resistance to pathogen	*Cf. 2* & *Cf. 9*	Dixon *et al.*, 1996
	Sensitivity to fenthion	*Fen*	Martin *et al.*, 1994

Future Road Map of Genome Mapping and Map Based Cloning

Development of genetic linkage maps using almost all kinds of markers in all sorts of mapping populations has been undertaken in all economically important plants. Several other plant species will receive attention in the course of time. These include some major field crops grown in the developing countries, horticultural crops, fodder crops, medicinal, aromatic and plantation crops, and forest trees. Among animal systems, efforts were made in the past two decades mainly on human and mouse. All kinds of cattle and poultry species, birds and forest animals will come to the fore in the near future.

In most of the molecular mapping endeavors in plants, mapping populations have been developed from intervarietal crosses, which releases lower degree of DNA polymorphism, thus giving little scope of mapping genes conferring adaptability and stress resistance. Involvement of distant local land races, sub-species, species or even genera in crossing program could facilitate identification of desirable genes in allied or wild taxa besides paving the way for construction of high-density maps. This will lead to detection of desirable genes in agronomically not so useful local land races and wild species and will also allow us to build new gene combinations with heterotic effects.

In years to come, more attention will be paid to development of infinite mapping populations such as DH and RI for using them in linkage mapping of marker and trait loci on global basis. Exchange of mapping populations and probes will be encouraged more.

Comparative mapping is still limited to a few families including poaceae, cruciferae, solanaceae, fabaceae. Studies of genome homology across species or genera will mitigate the chance of redundant cloning of identical genes. Formation of networks of scientists handling closely related species or genera could facilitate comprehensive comparative mapping.

Most of the agronomically important traits are polygenic in nature and are controlled by QTLs. Information on location, number and gene action of QTLs for many characters in the major crops is now known. However, they need to be Mendelized and assembled into a common genetic background employing high resolution mapping and marker assisted backcross breeding. Similar attempts will be made for development of genotypes with durable resistance for abiotic and biotic stresses.

Physical mapping is a costly and time consuming affair and has been completed in humans and *Arabidopsis* and is nearly complete in rice through global efforts. However, similar attempts will be taken up on international collaborative basis for other plant and animal species particularly those with smaller genome size. Local mapping of short chromosomal regions containing economic genes or gene clusters in different laboratories and their integration in course of time will hopefully become a popular strategy.

The fruits of genome mapping and map based cloning are yet to be realized at the agricultural and medical fields. Marker assisted selection (MAS) and genetic transformation with cloned genes will emerge as the two routine tools of enormous applied significance in developing desirable plant varieties and animal breeds in the next few years.

References

Ahn S., Bollich C.N. and Tanksley S.D. 1992. RFLP tagging of a gene for aroma in rice. *Theor. Appl. Genet.*, **84**: 825-828

Ahn S., Anderson J.A., Sorrells M.E. and Tanksley S.D. 1993. Homeologous relationships of rice, wheat, and maize chromosomes. *Mol. Gen. Genet.*, **241**: 483-490

Ahn S. and Tanksley S.D. 1993. Comparative linkage maps of the rice and maize genomes. *Proc. Natl. Acad. Sci.*, USA, **90**: 7980-7984

Albar L. and Lorieux M. *et al.* (8 authors). 1998. Genetic basis and mapping of the resistance to rice yellow mottle virus. I. QTLs identification and relationship between resistance and plant morphology. *Theor. Appl. Genet.*, **97**: 1145-1154

Ali. 1999. Mapping quantitative trait loci for root traits related to drought resistance in rice (*Oryza sativa* L.) using AFLP markers. Ph.D. Thesis, Texas Tech University, Lubbock, Texas, USA.

Almassy L. and Blangero J. 1998. Multipoint quantitative trait linkage analysis in general pedigrees. *Am. J. Hum. Genet.*, **62**: 1198-1221

Amos C.I. and Elston R.C. 1989. Robust methods for the detection of genetic linkage for quantitative data from pedigrees. *Genet Epidemiol*, **6**: 349-360

Anderson J.A., Waldron B.L., Moreno-Sevilla B., Stack R.W. and Frohberg R.C. 1998. Detection of *Fusarium* head blight resistance QTL in wheat using AFLPs and RFLPs. *In*: *Proc. 9th Int. Wheat Genet Symp.*, (ed.) Slinkard A.E., Univ Saskatchewan, Saskatoon, **1**: 135-137.

Araki E., Miura H. and Sawada S. 1999. Identification of genetic loci affecting amylose content and agronomic traits on chromosome 4A of wheat. *Theor. Appl. Genet.*, **98**: 977-984.

Autrique E., Singh R.P., Tanksley S.D. and Sorells M.E. 1995. Molecular markers for four leaf rust resistance genes introgressed into wheat from wild relatives. *Genome*, **38**: 75-83.

Baura T.J., Gresshoff P.M., Lewis S.A. and Dean R.A. 1992. DNA amplification fingerprinting (DAF) of isolates of four common *Meloidogyne* species, and their host races. *Phytopathology*, **82**: 1095

Bennetzen J.L. and Freeling M. 1993. Grasses as a single genetic system: genome composition, colinearity and compatibility. *Trends Genet.*, **9**: 259-261

Berhan A.M., Hulbert S.H., Butler L.G. and Bennetzen J.L. 1993. Structure and evolution of the genome of *Sorghum bicolor* and *Zea mays*. *Theor. Appl. Genet.*, **86**: 598-604.

Bernatzky R. and Tanksley S.D. 1986. Toward a saturated linkage map in tomato based on isozymes and random cDNA sequences. *Genetics*, **112**: 887-898.

Bezant J., Laurie D.A., Pratchett N., Chojecki J. and Kearsey M. 1997. Mapping QTL controlling yield and yield components in a spring barley (*Hordeum vulgare* L.) cross using marker regression. *Mol. Breed.*, **3**: 29-38.

Binelli G., Gianfrancheschi L., Pe M.E., Taramino G., Busso C., Stenhouse J. and Ottaviano E. 1992. Similarity of maize and sorghum genomes as revealed by maize RFLP probes. *Theor. Appl. Genet.*, **84**: 10-16.

Bonierbale M.W., Plaisted R.L. and Tanksley S.D. 1988. RFLP maps based on a common set of clones reveal modes of chromosomal evolution in potato and tomato. *Genetics*, **120**: 1095-1103.

Bonierbale M.W., Plaisted R.L., Pineda O. and Tanksley S.D. 1994. QTL analysis of trichome-mediated insect resistance in potato. *Theor. Appl. Genet.*, **87**: 973-987.

Borovkova I.G., Steffenson B.J. and Jin Y. *et al.* (8 authors). 1995. Identification of molecular markers linked to the stem rust resistance gene *rpg4* in barley. *Phytopathology*, **85**: 181-185.

Botstein D., White R.L., Skolnick M. and Davis R.W. 1980. Construction of a genetic linkage map in man using restriction fragment length polymorphisms. *Am. J. Hum. Genet.*, **32**: 314-331.

Brim C.A. 1966. A modified pedigree method of selection in soybeans. *Crop Sci.*, **6**: 220.

Burke D.T., Carle G.F. and Olson M.V. 1987. Cloning of large segments of exogenous DNA into yeast by means of artificial chromosome vectors. *Science*, **236**: 806-812.

Burr B., Burr F.A., Matz E.C. and Romero-Severson J. 1992. Pinning down loose ends: Mapping telomeres, and factors affecting their length. *Plant Cell*, **4**: 953-960.

Burr B., Burr F.A., Thompson K.H., Albertson M.C. and Stuber C.W. 1988. Gene mapping with recombinant inbreds in maize. *Genetics*, **118**: 519-526.

Butruille D.V., Guries R.P. and Osborn T.C. 1999. Linkage analysis of molecular markers and quantitative trait loci in populations of inbred backcross lines of *Brassica napus* L. *Genetics*, **153**: 949-964.

Cadalen T., Boeuf C., Bernard S. and Bernard M. 1997. An intervarietal molecular marker map in *Triticum aestivum* L. Em. Thell. And comparison with a map from a wide cross. *Theor. Appl. Genet.*, **94**: 367-377.

Cadalen T. and Sourdille G. *et al.* (9 authors). 1998. Molecular markers linked to genes affecting plant height in wheat using a double-haploid population. *Theor. Appl. Genet.*, **96**: 933-940.

Camargo L.E.A. and Osborn T.C. 1996. Mapping loci controlling flowering time in *Brassica oleracea*. *Theor. Appl. Genet.*, **92**: 610-616.

Camargo L.E.A., Savides L., Jung G., Nienhuis J. and Osborn T.C. 1997. Location of the self-incompatibility locus in an RFLP and RAPD map of *Brassica oleracea*. *J. Hered.*, **88**: 57-60.

Cao W., Huges G.R., Ma H. and Dong Z. 1998. Development of DNA based markers for resistance to *Septoria nodorum* blotch in durum wheat. *In*: *Proc. 9th Int. Wheat Genet. Symp.*, (ed.) Slinkard A.E., Univ Saskatchewan, Saskatoon, **3**: 95-97.

Causse M.A., Fulton T.M. and Cho Y.G. *et al.* (13 authors). 1994. Saturated molecular map of the rice genome based on an interspecific backcross population. *Genetics*, **138**: 1251-1274.

Champoux M.C., Wang G.L., Sarkarung S., Mackill D.J., O'Toole J.C., Huang N. and McCouch S.R. 1995. Locating genes associated with root morphology and drought avoidance in rice via linkage to molecular markers. *Theor. Appl. Genet.*, **90**: 969-981.

Chen X., Temnykh S., Xu Y., Cho Y.G. and McCouch S.R. 1997. Development of a microsatellite framework map providing genome wide coverage in rice (*Oryza sativa* L.). *Theor. Appl. Genet.*, **95**: 553-567.

Chen X.M., Line R.F. and Leung H. 1998. Genome scanning for resistance-gene analogs in rice, barley, and wheat by high-resolution electrophoresis. *Theor. Appl. Genet.*, **97**: 345-355.

Cheung W.Y., Friesen L., Rakow G.F.W., Seguin-Swartz G. and Landry B.S. 1997. A RFLP-based linkage map of mustard [*Brassica juncea* (L.) Czern and Coss]. *Theor. Appl. Genet.*, **94**: 841-851.

Chisholm S.T., Mahajan S.K., Whittam S.A., Yamamoto M.L. and Carrington J.C. 2000. Cloning of the *Arabidopsis RTM1* gene, which controls restriction of long-distance movement of tobacco etch virus. *Proc. Natl. Acad. Sci.*, USA, **97**: 489-494.

Cho Y.G., Eun M.Y., McCouch S.R. and Chae Y.A. 1994. The semidwarf gene, *sd-1* of rice (*Oryza sativa* L.) II. Molecular mapping and marker-assisted selection. *Theor. Appl. Genet.*, **89**: 54-59.

Choi S.H. and Leach J.E. 1994. Identification of the *Xor*II methyltransferase gene and a *vsr* homolog from *Xanthomonas oryzae* pv. *Oryzae*. *Mol. Gen. Genet.*, **244**: 383-390.

Chunwongse J., Bunn T.B., Crossman C., Jiang J. and Tanksley S.D. 1994. Chromosomal localization and molecular-marker tagging of the powdery mildew resistance gene (*Lv*) in tomato. *Theor. Appl. Genet.*, **89**: 76-79.

Churchill G., Giovannoni J.J. and Tanksley S.D. 1993. Pooled-sampling makes high-resolution mapping practical with DNA markers. *Proc. Natl. Acad. Sci.*, USA, **90**: 16-20.

Chyi Y.S., Hoenecke M.E. and Sernyk J.L. 1992. A genetic linkage map of restriction fragment length polymorphism loci for *Brassica rapa* (syn. *campestris*). *Genome*, **35**: 746-757.

Cohen D., Chumakov I. and Wessenbach J. 1993. A first-generation physical map of the human genome. *Nature*, **366**: 698-701.

Coulson A., Sulston J., Brenner S. and Karn J. 1986. Toward a physical map of the nematode *Caenorhabditis elegans. Proc. Natl. Acad. Sci.*, USA, **83**: 7821-7825.

Courtois B., Mclaren G., Sinha P.K., Prasad K., Yadav R. and Shen L. 2000. Mapping QTLs associated with drought avoidance in upland rice. *Mol. Breed.*, **6**: 55-66.

De la Pena R.C., Murray T.D. and Jones S.S. 1996. Linkage relations among eyespot resistance gene *Pch2*, endopeptidase *EP-A1B* and RFLP marker *Xpsr121* on chromosome 7A of wheat. Plant. Breed., **115**: 273-275.

DeVicente M.C. and Tanksley S.D. 1993. QTL analysis of transgressive segregation in an interspecific tomato cross. *Genetics*, 1**34**: 585-596.

Devos K.M., Atkinson M.D., Chinoy C.N., Liu C.J. and Gale M.D. 1992. RFLP-based genetic map of the homoeologous group 3 chromosomes of wheat and rye. *Theor. Appl. Genet.*, **83**: 931-939.

Devos K.M., Chao S.P., Li Q.Y., Simonetti M.C. and Gale M.D. 1994. Relationship between chromosome 9 of maize and wheat homoeologous group 7 chromosomes. *Genetics*, **138**: 1287-1292.

Devos K.M., Wang Z.M., Beales J., Sasaki T. and Gale M.D. 1998. Comparative genetic maps of foxtail millet (*Setaria italica*) and rice (*Oryza sativa*). *Theor. Appl. Genet.*, **96**: 63-68.

Diers B.W., Klein P., Fehr W.R. and Shoemaker R.C. 1992. RFLP analysis of soybean seed protein and oil content. *Theor. Appl. Genet.*, **83**: 608-612.

Dixon M.S., Jones D.A., Keddie J.S., Thomas C.M., Harrison K. and Jones J.D.G. 1996. The tomato *Cf-2* disease resistance locus comprises two functional genes encoding leucine-rich repeat proteins. *Cell*, **84**: 451-459.

Doebly J. and Stec A. 1991. Genetic analysis of the morphological differences between maize and teosinte. *Genetics*, **129**: 285-295.

Dubcovsky J., Lijavetzky A.L. and Tranquilli G. 1998. Comparative RFLP mapping of *Triticum monococcum* genes controlling vernalization requirement. *Theor. Appl. Genet.*, **97**: 968-975.

Dufour P., Grivet L.D., Hont A., Deu M., Trouche G., Glaszman J.C. and Hamon P. 1996. Comparative genetic mapping between duplicated segments on maize chromosome 3 and 8 and homoeologous regions in sorghum and sugarcane. *Theor. Appl. Genet.*, **92**: 1024-1030.

Eastwood R.F., Lagudah E.S. and Appels R. 1994. A direct search for DNA sequences tightly linked to cereal cyst nematode resistance genes in *Triticum tauschii. Genome*, **37**: 311-319.

Fahima T., Chague V., Sun G., Korol A., Ronin Y., Roder M., Grama A. and Nevo E. 1997. Identification and potential use of PCR markers flanking the *Triticum dicoccoides*-derived stripe rust resistance gene *Yr15* in wheat. *In*: *Wheat 5th Int. Cong. Plant Mol. Biol.*, Singapore, Abstr, 249.

Fan W., Wei X., Shukla H., Parimoo S., Xu H., Sankhavaram P., Li Z.B. and Weissman S.M. 1993. Application of cDNA selection techniques to regions of the human MHC. *Genomics*, **17**: 575-581.

Fatokun C.A., Menancio-Hautea D.I., Danesh D. and Young N.D. 1992. Evidence for orthologous seed weight genes in cowpea and mungbean based on RFLP mapping. *Genetics*, **132**: 841-846.

Faure S., Noyer J.L., Horry J.P., Bakry F., Lanaud C. and De Leon D.G. 1993. A molecular marker-based linkage map of diploid bananas (*Musa acuminata*). *Theor. Appl. Genet.*, **87**: 517-526.

Ferreira M.E., Williams P.H. and Osborn T.C. 1994. RFLP mapping of *Brassica* using doubled haploid lines. *Theor. Appl. Genet.*, **89**: 615-621.

Ferreira M.E., Satagopan J.M., Yandell B.S., Williams P.H. and Osborn T.C. 1995a. Mapping loci Controlling Vernalization requirement and flowering time in *Brasica napus*. *Theor. Appl. Genet.*, **90**: 727-732.

Ferreira M.E., Williams P.H. and Osborn T.C. 1995b. Mapping of a locus controlling resistance to *Albugo candida* in *Brassica napus* using molecular markers. *Phytopathology*, **85**: 218-220.

Feuillet C., Messmer M., Schachermayr G. and Keller B. 1995. Genetic and physical characterization of the *Lr1* leaf rust locus in wheat (*Triticum aestivum* L.). *Mol. Gen. Genet.*, **248**: 553-562.

Figdore S.S., Ferreira M.E., Slocum M.K. and Williams P.H. 1993. Association of RFLP markers with trait loci affecting clubroot resistance and morphological characters in *Brassica oleracea* L. *Euphytica*, **63**: 33-44.

Foolad M.R. and Jones R.A. 1993. Mapping salt-tolerance genes in tomato (*Lycopersicon esculentum*) using trait-based marker analysis. *Theor. Appl. Genet.*, **87**: 184-192.

Gallego F., Feuillet C., Messmer M. and Penger A. *et al.* (8 authors). 1998. Comparative mapping of the two wheat leaf rust resistance loci *Lr1* and *Lr10* in rice and barley. *Genome*, **41**: 328-336.

Ganal M.W., Broun P. and Tanksley S.D. 1992. Genetic mapping of tandemly repeated telomeric DNA sequences in tomato *Lycopercicon esculentum*. *Genomics*, **14**: 444-448.

Ganal M.W., Young N.D. and Tanksley S.D. 1989. Pulsed field gel electrophoresis and physical mapping of large DNA fragments in the *Tm-2a* region of chromosome 9 in tomato. *Mol. Gen. Genet.*, **215**: 395-400.

Gardiner K. and Mural R. 1995. Getting the message: identifying transcribed sequences. *Trends Genet.*, **11**: 77-79.

Ghosh S. and Majumder P.P. 2000a. Mapping quantitative trait loci via the EM algorithm and Bayesian classification. *Genet. Epidemiol.*, **19**: 97-126.

Ghosh S. and Majumder P.P. 2000b. An improved procedure of mapping a quantitative trait locus via the EM algorithm using posterior probabilities. *J. Genet.*, **79**: 47-53.

Giovannoni J.J., Wing R.A., Ganal M.W. and Tanksley S.D. 1991. Isolation of molecular markers from specific chromosomal intervals using DNA pools from existing mapping populations. *Nucleic Acids Res.*, **19**: 6553-6558.

Goldgar D.E. 1990. Multipoint analysis of human quantitative genetic variation. *Am. J. Hum. Genet.*, **47**: 957-967.

Grant M.R., Godiard L. and Straube E. *et al.* (8 authors). 1995. Structure of the Arabidopsis *RPM1* gene enabling dual specificity disease resistance. *Science*, **269**: 843-846.

Gupta P.K. and Varshney R.K. 2000. The development and use of microsatellite markers for genetic analysis and plant breeding with emphasis on bread wheat. *Euphytica*, **113**: 163-185.

Haldane J.B.S. 1919. The combination of linkage values and calculation of distance between the loci of linked factors. *J. Genet.*, **8**: 299-309.

Haley C.S. and Knott S.A. 1992. A simple regression method for mapping quantitative trait loci in line crosses using flanking markers. *Heredity*, **69**: 315-324.

Hallden C., Hjerdin A. and Rading I.M. *et al.* (9 authors). 1996. A high density RFLP linkage map of sugar beet. *Genome*, **39**: 634-645.

Han F., Kleinhofs A., Ullrich S.E., Kilian A., Yano M. and Sasaki T. 1998. Synteny with rice: analysis of barley malting quality QTLs and *rpg4* chromosome regions. *Genome*, **41**: 373-380.

Haseman J.K. and Elston R.C. 1972. The investigation of linkage between a quantitative trait and a marker locus. *Behav. Genet.*, **2**: 3-19.

Hayano-Saito Y., Tsuji T., Fujii K., Saito K., Iwasaki M. and Saito A. 1998. Localization of the rice stripe disease resistance gene, *stv-b1*, by graphical genotyping and linkage analyses with molecular markers. *Theor. Appl. Genet.*, **96**: 1044-1049.

Helentjaris T. 1987. A genetic linkage map for maize based on RFLPs. *Trends Genet.*, **3**: 217-221.

Helentjaris T. 1993. Implications for conserved genomic structure among plant species. *Proc. Natl. Acad. Sci.*, USA, **90**: 8308-8309.

Helentjaris T., Slocum M.K., Wright S., Schafer A. and Nienhuis J. 1986. Construction of genetic linkage maps in maize and tomato using restriction fragment length polymorphisms. *Theor. Appl. Genet.*, **72**: 761-769.

Hirabayashi H. and Ogawa T. 1995. RFLP mapping of *Bph-1* (Brown plant hopper resistance genes) in rice. *Breed Sci.*, **45**: 369-371.

Hittalmani S., Foolad M.R., Mew T.V., Rodriguez R.L. and Huang N. 1995. Development of a PCR-based marker to identify rice blast resistance gene, *Pi-2(t)*, in a segregating population. *Theor. Appl. Genet.*, **91**: 9-14.

Hoenecke M. and Chyi Y.S. 1991. Comparison of *Brassica napus* and *B. rapa* genomes based on restriction fragment length polymorphism mapping. *In*: *Proc. 8th Int. Rapeseed Cong.*, (ed.) McGregor D.I., GCIRC, Saskatoon: 1102-1107.

Hofte H., Desprez T. and Amselem J. *et al.* (28 authors). 1993. An inventory of 1152 expressed sequence tags obtained by partial sequencing of cDNAs from *Arabidopsis thaliana*. *Plant J.*, **4**: 1051-1061.

Huang N. and Parco A. *et al.* (9 authors). 1997. RFLP mapping of isozymes, RAPD, and QTLs for grain shape, brown plant hopper resistance in doubled-haploid rice population. *Mol. Breed.*, **3**: 105-113.

Hulbert S.H., Richter T.E., Axtell J.D. and Bennetzen J.L. 1990. Genetic mapping and characterization of sorghum and related crops by means of maize DNA probes. *Proc. Natl. Acad. Sci.*, USA, **87**: 4251-4255.

Ishii T., Brar D.S., Multani D.S. and Khush G.S. 1994. Molecular tagging of genes for brown plant hopper resistance and earliness introgressed from *Oryza australiensis* into cultivated rice, *O. sativa*. *Genome*, **37**: 217-221.

Jahoor A. 1998. Marker assisted breeding in cereals, specially with respect to synteny among loci for mildew resistance. *In*: *Genetics and Biotechnology for crop Improvement*, (ed.) Gupta P.K., Rastogi Pub., Meerut, pp. 237-254.

Jarvis P., Lister C., Szabo V. and Dean C. 1994. Integration of CAPS markers into the RFLP map generated using recombinant inbred lines of *Arabidopsis thaliana*. *Plant Mol. Biol.*, **24**: 685-687.

Jeffreys A.J., Wilson V. and Thein S.L. 1985. Hypervariable minisatellite regions in human DNA. *Nature*, **314**: 4251-4255.

Jena K.K., Khush G.S. and Kochert G. 1994. Comparative RFLP mapping of a wild rice, *Oryza officinalis* and cultivated rice, *Oryza sativa*. *Genome*, **37**: 382-389.

Jia J., Devos K.M., Chao S., Miller T.E., Reader S.M. and Gale M.D. 1996. RFLP-based maps of the homoeologous group-6 chromosomes of wheat and their application in the tagging of *Pm12*, a powdery mildew resistance gene transferred from *Aegilops speltoides* to wheat. *Theor. Appl. Genet.*, **92**: 559-565.

Khairallah M.M., Bohn M. and Jiang C. *et al.* (8 authors). 1998. Molecular mapping of QTL for southwestern corn borer resistance, plant height and flowering in tropical maize. *Plant Breed*, **117**: 309-318.

Khush G.S. and Rick C.M. 1967. Studies on the linkage map of chromosome 4 in tomato and on the transmission of induced deficiencies. *Genetica*, **38**: 74-94.

Kianian S.F. and Quiros C.F. 1992. Generation of a *Brassica oleracea* composite RFLP map: linkage arrangements among various populations and evolutionary implications. *Theor. Appl. Genet.*, **84**: 544-554.

Kinoshita T. 1993. Report of the committee on gene symbolization, nomenclature and linkage group. *Rice Genet. Newsl.*, **10**: 7-39.

Kleinhofs A., Kilian A. and Saghai-Maroof *et al.* (22 authors). 1993. A molecular, isozyme and morphological map of the barley (*Hordeum vulgare*) genome. *Theor. Appl. Genet.*, **86**: 705-712.

Kole C. 1997. Molecular mapping in *Brassica campestris* (syn. *rapa*). Final Project Report submitted to DBT, GOI, Lab. Mol. Biol. Biotech., Orissa Univ Agril Tech, Bhubaneswar, India.

Kole C., Kole P., Vogelzang R. and Osborn T.C. 1997. Genetic Linkage map of a *Brassica rapa* recombinant Inbred population. *J. Hered.*, **88**: 553-557.

Kole C., Quijada P., Michaels A.D., Amasino R.M. and Osborn T.C. 2001. Evidence for homology of flowering-time genes *VFR2* from *Brassica rapa* and *FLC* from *Arabidopsis thaliana*. *Theor. Appl. Genet.*, **102**: 425-430.

Kole C., Teutonico R., Mengistu A., Williams P.H. and Osborn T.C. 1996. Molecular mapping of a locus controlling resistance to *Albugo candida* in *Brassica rapa*. *Phytopathology*, **86**: 367-369.

Kole C., Williams P.H., Rimmer S.R. and Osborn T.C. 2002a. Linkage mapping of genes controlling resistance to white rust (*Albugo candida*) in *Brassica rapa* (syn. *campestris*) and comparative mapping to *B. napus* and *Arabidopsis thaliana*. *Genome*, **45**: 22-27.

Kole C., Thormann C.E., Karlsson B.H., Palta J.P., Gaffney P., Yandell B. and Osborn T.C. 2002b. Comparative mapping of loci controlling winter survival and related traits in oilseed *Brassica rapa* and *B. napus*. *Mol. Breed.*, **9**: 201-210.

Konieczny A. and Ausubel F.M. 1993. A procedure for mapping Arabidopsis mutations using co-dominant ecotype-specific PCR-based markers. *Plant J.*, **4**: 403-410.

Korzun V., Roder M.S., Ganal M.W., Worland M.J. and Law C.N. 1998. Genetic analysis of the dwarfing gene *Rht8* in wheat. Part I. Molecular mapping of *Rht8* on the short arm of chromosome 2D of bread wheat (*Triticum aestivum* L.). *Theor. Appl. Genet.*, **96**: 1104-1109.

Korzun V., Roder M.S,. Worland M.J. and Borner A. 1997. Intrachromosomal mapping for dwarfing (*Rht12*) and vernalization response (*Vrn1*) in wheat by using RFLP and microsatellite markers. *Plant Breed.*, **116**: 227-232.

Kosambi D.D. 1944. The estimation of map distance from recombination values. *Ann. Eugen.*, **12**: 172-175.

Kowalski S.D., Lan T.H., Feldmann K.A. and Paterson A.H. 1994. Comparative mapping of *Arabidopsis thaliana* and *Brassica oleracea* chromosomes reveals islands of conserved gene order. *Genetics*, **138**: 499-510.

Kruglyk L. and Lander E.S. 1995. A nonparametric approach for mapping quantitative trait loci. *Genetics*, **139**: 1421-1428.

Kulwal P.L., Roy J.K., Balyan H.S. and Gupta P.K. 2002. QTL mapping for some growth and leaf chracters in bread wheat. *Plant Science*, **164**: 267-277.

Kurata N. *et al.* (28 authors). 1994a. A 300 kilobase interval genetic map of rice including 883 expressed sequence. *Nature Genet.*, **8**: 365-372.

Kurata N., Moore G., Nagamura Y., Foote T., Yano M., Minobe Y. and Gale M.D. 1994b. Conservation of genome structure between rice and wheat. *Biotechnology*, **12**: 276-278.

Lagercrantz U. and Lydiate D.J. 1996. Comparative genome mapping in *Brassica*. *Genetics*, **144**: 1903-1910.

Lagercrantz U., Putterill J,. Coupland G. and Lydiate D. 1996. Comparative mapping in *Arabidopsis* and *Brassica*, fine scale genome colinearity and congruence of genes controlling flowering time. *Plant J.*, **9**: 13-20.

Lakshmikumaran M., Mohapatra T., Gupta V.S. and Ranjekar P.K. 2003. Molecular Markers in Improvement of Wheat and *Brassica*. *In*: *Plant Breeding: Mendelian to Molecular Approaches*, (eds.) Jain H.K. and Kharkwal M.C., Narosa, New Delhi, 230-256.

Lander E.S. and Botstein D. 1989. Mapping Mendelian factors underlying quantitative traits using RFLP linkage maps. Genetics, **121**: 185-199; and Corrigendum. *Genetics*, **136**: 705.

Lander E.S., Green P., Abrahamson J., Barlow A., Dally M.J., Lincoln S.E. and Newburg L. 1987. MAPMAKER: An interactive computer package for constructing primary genetic linkage maps of experimental and natural populations. *Genomics*, **1**: 174-181.

Landry B.S., Hubert N., Crete R., Chang M.S., Lincoln S.E. and Etoh T. 1992. A genetic map of *Brassica oleracea* based on RFLP markers detected with expressed DNA sequences and mapping of resistance genes to race 2 of *Plasmodiophora brassicae* (Woronin). *Genome*, **35**: 409-420.

Landry B.S., Hubert N., Etoh T., Harada J.J. and Lincoln S.E. 1991. A genetic map of *Brassica napus* based on restriction fragment length polymorphisms detected with expressed DNA sequences. *Genome*, **34**: 543-552.

Landry B.S., Kesseli R.V., Farrara B. and Michelmore R.W. 1987. A genetic map of lettuce (*Lactuca sativa* L.) with restriction fragment length polymorphism, isozyme, disease resistance and morphological markers. *Genetics*, **116**: 331-337.

Laurie D.A., Pratchett N., Bezant J.H. and Snape J.W. 1995. RFLP mapping of five major genes and eight quantitative trait loci controlling flowering time in a winter × spring barley (*Hordeum vulgare* L.) cross. *Genome*, **38**: 575-585.

Li Z.B., Pinson S.R.M., Marchetti M.A., Stansel J.W. and Park W.D. 1995. Characterization of quantitative trait loci (QTLs) in cultivated rice to field resistance to sheath blight (*Rhizoctonia solani*). *Theor. Appl. Genet.*, **91**: 382-388.

Liang C.Z., Gu M.H., Pan X.B., Liang G.H. and Zhu L.H. 1994. RFLP tagging a new semidwarfing gene in rice. *Theor. Appl. Genet.*, **88**: 898-900.

Lilley J.M., Ludlow M.M., McCouch S.R. and O'Toole J.C. 1996. Locating QTL for osmotic adjustment and dehydration tolerance in rice. *J. Exptl. Bot.*, **47**: 1427-1436.

Lin H.X., Qian H.R. and Zhuang J.Y. *et al.* (8 authors). 1996. RFLP mapping of QTLs for yield and related characters in rice (*Oryza sativa* L.). *Theor. Appl. Genet.*, **92**: 920-927.

Lin Y.R., Schertz K.F. and Paterson A.H. 1995. Comparative analysis of QTLs affecting plant height and maturity across the Poaceae, in reference to an interspecific sorghum population. *Genetics*, **141**: 391-411.

Lincoln S.E., Dally M.J. and Lander E.S. 1992a. Constructing genetic linkage maps with MAPMAKER/ EXP 3.0. *Whitehead Institute Technical Reports*, Ed. 3.

Lincoln S., Dally M. and Lander E.S. 1992b. Mapping genes controlling quantitative traits with MAPMAKER/QTL 1.1. *Whitehead Institute Technical Reports*, Ed. 2.

Ling H.Q., Koch G., Baumlein H. and Ganal M.W. 1999. Map-based cloning of *chloronerva*, a gene involved in iron uptake of higher plants encoding nicotianamine synthase. *Proc. Natl. Acad. Sci.*, USA, **96**: 7098-7103.

Liu B.H. and Knapp S.J. 1992. G-Mendel: A program for Mendelian segregation and linkage analysis of individual or multiple progeny populations using log-likelihood ratios. *J. Hered.*, **81**: 407.

Liu S.C., Kowalski S.D., Lan T.H., Feldmann K.A. and Paterson A.H. 1996. Genome-wide high-resolution mapping by recurrent intermating, using *Arabidopsis thaliana* as a model. *Genetics*, **142**: 247-258.

Liu Y.G. and Tsunewaki K. 1991. Restriction fragment length polymorphism analysis of wheat. II. Linkage maps of the RFLP sites in common wheat. *Japan J. Genet.*, **66**: 617-633.

Lovett M. 1994. Fishing for complements: finding genes by direct selection. *Trends Genet.*, **10**: 352-357.

Lovett M., Kere J. and Hinton L.M. 1991. Direct selection: A method for the isolation of cDNAs encoded by large genomic regions. *Proc. Natl. Acad. Sci.*, USA, **88**: 9628-9632.

Lu C., Shen L. and Tan Z. *et al.* (7 authors). 1997. Comparative mapping of QTLs for agronomic traits of rice across environments using a doubled-haploid population. *Theor. Appl. Genet.*, **94**: 145-150.

Luo M., Bilodeau P., Koltunow A., Dennis E.S., Peacock W.J. and Chaudhury A.M. 1999. Genes controlling fertilization-independent seed development in *Arabidopsis thaliana*. *Proc. Natl. Acad. Sci.*, USA, **96**: 296-301.

Ma Z.Q., Gill B.S., Sorrells M.E. and Tanksley S.D. 1993. RFLP markers linked to two Hessian fly resistance genes in wheat (*Triticum aestivum* L.) from *Triticum tauschii* (Coss.) Schal. *Theor. Appl. Genet.*, **85**: 750-754.

Ma Z.Q. and Sorells M.E. 1995. Genetic analysis of fertility restoration in wheat using restriction fragment length polymorphisms. *Crop Sci.*, **35**: 1137-1143.

Ma Z.Q., Sorrells M.E. and Tanksley S.D. 1994. RFLP markers linked to powdery mildew resistance genes *Pm1, Pm2, Pm3* and *Pm4* in wheat. *Genome*, **37**: 871-875.

Mackill D.J., Salam M.A., Wang Z.Y. and Tanksley S.D. 1993. A major photoperiod-sensitivity gene tagged with RFLP and isozyme markers in rice. *Theor. Appl. Genet.*, **85**: 536-560.

Maheswaran M. 1995. Identification of quantitative traits for days to flowering and photoperiod sensitivity in rice (*Oryza sativa* L.). Tamil Nadu Agricultural University, Coimbatore, India (Ph.D. Thesis).

Maheswaran M., Subudhi P.K., Nandi S., Xu J.C., Parco D., Yang C. and Huang N. 1997. Polymorphism, distribution and segregation of AFLP markers in a doubled haploid rice population. *Theor. Appl. Genet.*, **94**: 39-45.

Manly K.F. 1995. New functions in Map Manager, a macro computer program for genomic mapping. *Proceedings of Plant Genome III*, San Diego, CA, 15-19 Jan 1995, p. 61.

Markert C.L. and Moller F. 1959. Multiple forms of enzymes: tissue, ontogenic, and species-specific patterns. *Proc. Natl. Acad. Sci.*, USA, **45**: 753-763.

Martin G.B., Brommonschenkel S.H., Chunwongse J., Frary A., Ganal M.W., Spivey R., Wu T., Earle E.D. and Tanksley S.D. 1994. Map based cloning of a protein kinase gene conferring disease resistance in tomato. *Science*, **262**: 1432-1436.

Martin G.B., Frary A., Wu T., Brommonschenkel S., Chungwongse J., Earle E.D. and Tanksley S.D. 1994. A member of the tomato *Pto* gene family confers sensitivity to fenthion resulting in rapid cell death. *The Plant Cell*, **6**: 1543-1552.

Maxam A.M. and Gilbert W. 1977. A new method for sequencing DNA. *Proc. Natl. Acad. Sci.*, USA, **74**: 560-564.

McCouch S.R., Abenes M.L., Angeles R., Khush G.S. and Tanksley S.D. 1991. Molecular tagging of a recessive gene *Xa-5* for resistance to bacterial blight of rice. *Rice Genet. Newsl.*, **8**: 143-145.

McCouch S.R., Kochert G., Yu Z.H., Wang Z.Y., Khush G.S., Coffman D.R. and Tanksley S.D. 1988. Molecular mapping of rice chromosomes. *Theor. Appl. Genet.*, **76**: 815-829.

Menancio-Hautea D., Fatokun C.A., Kumar L., Danesh D. and Young N.D. 1993. Comparative genome analysis of mungbean (*Vigna radiata* L. Wilczek) and cowpea (*V. unguiculata* L. Walpers) using RFLP mapping data. *Theor. Appl. Genet.*, **88**: 797-810.

Michelmore R.W., Paran I. and Kesseli R.V. 1991. Identification of markers linked to disease resistance genes by bulked segregant analysis: A rapid method to detect markers in specific genomic regions by using segregating populations. *Proc. Natl. Acad. Sci.*, USA, **88**: 9828-9832.

Miyamoto M., Ando I., Rybka K., Kodama O. and Kawasaki S. 1996. High resolution mapping of the *indica*-derived rice blast resistance genes. 1. *Pi-b. Mol. Plant Microbe Interaction*, **9**: 6-13.

Mohan M., Nair S., Bentur J.S,. Prasad Rao U. and Bennet J. 1994. RFLP and RAPD mapping of the rice *Gm-2* gene that confers resistance to biotype 1 of gall midge (*Orseolia oryzae*). *Theor. Appl. Genet.*, **87**: 782-788.

Moore G., Foote T., Helentzaris T., Devos K.M., Kurata N. and Gale M.D. 1995a. Was there a single ancestral cereal chromosome? *Trend Genet.*, **11**: 81-82.

Moore G., Devos K.M., Wang Z. and Gale M.D. 1995b. Cereal genome evolution. *Curr. Biol.*, **5**: 737-739.

Morgan J.G., Dolganov G.M., Robbins S.E., Hinton L.M. and Lovett M. 1992. The selective isolation of novel cDNAs encoded by the regions surrounding the human interleukin 4 and 5 genes. *Nucl. Acids Res.*, **20**: 5173-5179.

Morgan T.H. 1910. Sex-linked inheritance in *Drosophila*. *Science*, **32**: 120-122.

Mukherjee A.K., Mohapatra T., Varshney A., Sharma R. and Sharma R.P. 2001. Molecular mapping of a locus controlling resistance to *Albugo candida* in Indian mustard. *Plant Breed.*, **120**: 483-487.

Mullis K. and Faloona F. 1987. Specific synthesis of DNA *in vitro via* a polymerase catalyzed chain reaction. *Methods Enzymol*, **155**: 335-351.

Mullis K., Faloona F., Scharf S., Saiki R.K., Horn G. and Erlich H. 1986. Specific enzymatic amplification of DNA in vitro: The polymerase chain reaction. *Cold Spring Harbor Symp Quant. Biol.*, **51**: 263.

Nair S., Kumar A., Srivastava M.N. and Mohan M. 1996. PCR-based DNA markers linked to a gall midge resistance gene, *Gm4l* has potential for marker aided selection in rice. *Theor. Appl. Genet.*, **92**: 660-665.

Nakamura S., Asakawa S., Ohmido N., Fukui K., Shimizu N. and Kawasaki S. 1997. Mol Construction of an 800kb contig in the near-centromeric region of the rice blast resistance gene *Pi- ta* using a highly representative rice BAC library. *Gen. Genet.*, **254**: 611-620.

Naqvi N.I., Bonman J.M., Mackill D.J., Nelson R.J. and Chattoo B.J. 1995. Identification of RAPD markers linked to a major blast resistance gene in rice. *Mol. Breed.*, **1**: 341-348.

Nelson J.C., Sorrells M.E., Van Deynze A.E., Lu L.H., Atkinson M.D., Bernard M., Leroy P., Faris J. and Anderson J.A. 1995. Molecular mapping of wheat. Major genes and rearrangements in homoeologous groups 4, 5 and 7. *Genetics*, **141**: 721-731.

Newman T., De Bruijin F.J. and Green P. *et al.* (12 authors). 1994. Genes Galore: A summary of methods for accessing results from large-scale partial sequencing of anonymous *Arabidopsis* cDNA clones. *Plant Physiol.*, **106**: 1241-1255.

Nilsson N.O. and Hansen M. *et al.* (9 authors). 1999. QTL analysis of *Cercospora* leaf spot in sugar beet. *Plant Breed.*, **118**: 327-334.

Nodari R.O., Tscai S.M., Gilbertson R.L. and Gepts P. 1993a. Towards an integrated linkage map of common bean. II. Development of an RFLP-based linkage map. *Theor. Appl. Genet.*, **85**: 513-520.

Nodari R.O., Tscai S.M., Guzman P., Gilbertson R.L. and Gepts P. 1993b. Towards an integrated map of common bean. 3. Mapping genetic factors controlling host-bacterial interactions. *Genetics*, **134**: 341-350.

Nozaki T., Kumarzaki A., Koba T., Ishikawa K. and Ikehashi H. 1997. Linkage analysis among loci for RAPDs, isozymes and some agronomic traits in *Brassica campestris* L. *Euphytica*, **95**: 115-123.

O' Donoughue L.S., Wang Z.M., Roder M.S., Kneen B., Leggett M., Sorells M.E. and Tanksley S.D. 1992. An RFLP based linkage map of oats based on a cross between two diploid taxa (*Avena atlantica* × *A. hirtul*). *Genome*, **35**: 765-771.

Olds T.M. 1996. Genetic constraints on life-history evolution: Quantitative-trait loci influencing growth and flowering in *Arabidopsis thaliana*. *Evolution*, **50**: 140-145.

Olson J.M. 1995. Robust multipoint linkage analysis: an extension of the Haseman-Elston method. *Genet. Epidemiol.*, **12**: 177-193.

Osborn T.C., Kole C., Parkin I.A.P., Sharpe A.G., Kuiper M., Lydiate D.J. and Trick M. 1997. Comparison of vernalization responsive flowering time genes in *Brassica rapa*, *B. napus* and *Arabidopsis thaliana*. *Genetics*, **146**: 1123-1129.

Padmavathi G. 2002. Molecular mapping of genes controlling some agronomic characters in rice. Unpub Ph.D. Thesis, Utkal Univ, Bhubaneswar, India.

Panigrahi J. 2002. Molecular mapping in *Brassica campestris*. Unpub Ph.D. Thesis, Utkal Univ, Bhubaneswar, India.

Paran I., Goldman I., Tanksley S.D. and Zamir D. 1995. Recombinant inbred lines for genetic mapping in tomato. *Theor. Appl. Genet.*, **90**: 542-548.

Parimoo S., Patanjali S.R., Shukla H., Chaplin D.D. and Weissman S.M. 1991. cDNA selection: Efficient PCR approach for the selection of cDNAs encoded in large chromosomal DNA fragments. Proc. Natl. Acad. Sci., *USA, 88: 9623-9627.*

Paterson A.H. 1996. Physical mapping and map-based cloning: Bridging the gap between DNA markers and genes. *In*: *Genome Mapping in Plants*, (ed.) Paterson A.H., Academic Press, RG Landes Co, Austin, pp. 55-62.

Paterson A.H. and Bowers J.E. *et al.* 2000. Comparative genomics of Plant chromosomes. *Plant Cell*, **12**: 1523-1539.

Paterson A.H., Damon S., Hewitt J.D., Zamir D., Rabinowitch H.D., Lincoln S.E., Lander E.S. and Tanksley S.D. 1991. Mendelian factors underlying quantitative traits in tomato: comparison across species, generations, and environments. *Genetics*, **127**: 181-197.

Paterson A.H., Deverna J.W., Lanini B. and Tanksley S.D. 1990. Fine mapping of quantitative trait loci using selected overlapping recombinant chromosomes in an interspecies cross of tomato. *Genetics*, **124**: 735-742.

Paterson A.H., Lander E.S., Hewitt J.D., Peterson S., Lincoln S.E. and Tanksley S.D. 1988. Resolution of quantitative traits into Mendelian factors by using a complete RFLP linkage map. *Nature*, **335**: 721-726.

Paterson A.H., Lin Y.R., Li Z.B., Schertz K.F., Doebley J.F., Pinson S.R.M., Liu S.C., Stansel J.W. and Irvine J.E. 1995. Convergent domestication of cereal crops by independent mutations at corresponding genetic loci. *Science*, **269**: 1714-1718.

Peng J.H., Fahima T., Roder M.S., Li Y.C., Dahn A., Grama A., Ronin Y.I., Korol A.B. and Nevo E. 1999. Microsatellite tagging of the stripe rust resistance gene *YrH52* derived from wild emmer wheat *Triticum dicocoides* and suggestive negative crossover interference on chromosome 1B. *Theor. Appl. Genet.*, **98**: 862-872.

Penner G.A., Zirino M., Kruger S. and Townley-Smith F. 1998. Accelerated recurrent parent selection in wheat with microsatellite markers. *In*: *Proc. 9th Int. Wheat Genet Symp.*, (ed.) Slinkard A.E., Univ Extn Press, Univ Saskatchewan, Saskatoon. **1**: 131-134.

Pereira M.G., Lee M. and Rayapati P.J. 1994. Comparative RFLP and QTL mapping in sorghum and maize. Poster 169. *In*: *2nd Int. Conf. Plant Genome*, New York: Scherago International, Inc.

Philipp U., Wehling P. and Wricke S.G. 1994. A linkage map of rye. *Theor. Appl. Genet.*, **88**: 243-248.

Prabhu K.V., Somens D.J., Rakow G. and Gugel R.K. 1998. Molecular markers linked to white rust resistance in mustard *Brassica juncea*. *Theor. Appl. Genet.*, **97**: 865-870.

Pradhan A.K., Gupta V.S., Mukhopadhyay A., Arumugam N., Sodhi Y.S. and Pental D. 2003. A high-density linkage map in *Brassica juncea* (Indian mustard) using AFLP and RFLP markers. *Theor. Appl. Genet.*, **106**: 607-614.

Prasad M., Varshney R.K., Kumar A., Balyan H.S., Sharma P.C. and Gupta P.K. 1999. A microsatellites marker associated with a QTL for grain protein content on chromosomal arm 2DL of bread wheat. *Theor. Appl. Genet.*, **99**: 341-345.

Prasad M., Kumar N., Kulwal P.L., Roder M., Balyan H.S., Dhaliwal H.S. and Gupta P.K. 2003. QTL analysis for grain protein content using SSR markers and validation studies using NILs in bread wheat. *Theor. Appl. Genet.*, **106**. 659-667.

Prince J.P., Pochard E. and Tanksley S.D. 1993. Construction of a molecular linkage map of pepper, and a comparison of synteny with tomato. *Genome*, **36**: 404-417.

Procunier J.D., Knox R.E., Bernier A.M., Gray M.A. and Howes N.K. 1997. DNA markers linked to a T10 loose smut resistance gene in wheat (*Triticul aestivum* L.). *Genome*, **40**: 176-179.

Redona E.D. and Mackill D.J. 1998. Quantitative trait locus analysis for rice panicle and grain characteristics. *Theor. Appl. Genet.*, **96**: 957-963.

Reinisch A.R., Dong J., Brubaker C.L., Stelly D., Wendel J. and Paterson A.H. 1994. An RFLP map of cotton (*Gossypium hirsutum* × *G. barbadense*): Chromosome organization and evolution in a disomic polyploid genome. *Genetics*, **138**: 829-847.

Reiter R.S., Williams J.G.K., Feldman K.A., Rafalski J.A., Tingey S.V. and Schlonik P.A. 1992. Global and local genome mapping in *Arabidopsis thaliana* by using recombinant inbred lines and random amplified polymorphic DNAs. *Proc. Natl. Acad. Sci.*, USA, **89**: 1477-1481.

Riede C.R. and Anderson J.A. 1996. Linkage of RFLP markers to an aluminium tolerance gene in wheat. *Crop Sci.*, **36**: 905-909.

Ronald P.C., Albano B., Tabien R., Abenes L., Wu K.K., McCouch S.R. and Tanksley S.D. 1992. Genetic and physical analysis of the rice bacterial blight disease resistance locus *Xa21*. *Mol. Gen. Genet.*, **236**: 113-120.

Rong J.K., Millet E., Manisterski J. and Feldman M. 1998. A powdery mildew resistance gene from wild emmer transferred into common wheat and tagged by molecular markers. *In*: *Proc. 9th Int. Wheat Genet. Symp.*, (ed.) Slinkard A.E., Univ. Saskatchewan, Saskatoon, **3**: 148-150.

Rowland L.J. and Levi A. 1994. RAPD-based genetic linkage map of blueberry derived from a cross between diploid species (*Vaccinium darrowi* and *V. elliottii*). *Theor. Appl. Genet.*, **89**: 863-868.

Sacco F., Suarez E.Y. and Naranjo T. 1998. Mapping of the leaf rust resistance gene *Lr3* on chromosome 6B of Sinvalocho MA wheat. *Genome*, **41**: 686-690.

Saghai Maroof M.A., Yang G.P., Biyashev R.M., Maughan P.J. and Zhang Q. 1996. Analysis of the barley and rice genomes by comparative RFLP linkage mapping. *Theor. Appl. Genet.*, **92**: 541- 551.

Saghai Maroof M.A., Zhang Q. and Biyashev R.M. 1994. Molecular marker analyses of powdery mildew resistance in barley. *Theor. Appl. Genet.*, **88**: 733-740.

Saiki R.K., Gelfand D.H., Stoffel S., Scharf S.J., Higuchi R., Horn G.T., Mullis K.V. and Erlich H.A. 1988. Primer-directed enzymatic amplification of DNA with a thermostable DNA polymerase. *Science*, **239**: 487.

Sanger F., Nicklen S. and Coulson A.R. 1977. DNA sequencing with chain-terminating inhibitors. *Proc. Natl. Acad. Sci.*, USA, **74**: 5463-5467.

Sarfatti M., Katan J., Fluhr R. and Zamir D. 1989. An RFLP marker in tomato linked to the *Fusarium oxysporum* resistance gene *I2*. *Theor. Appl. Genet.*, **78**: 755-759.

Sasaki T., Song J., Koga-ban Y. and Matsui E. *et al.* (21 authors). 1994. Toward cataloguing all rice genes: large scale sequencing of randomly chosen rice cDNAs from a callus cDNA library. *Plant J.*, **6**: 615-624.

Schimdt R., West J., Love K., Lenehan Z., Lister C., Thompson H., Bouchez D. and Dean C. 1995. Physical map and organization of *Arabidopsis* chromosome 4. *Science*, **270**: 480-483.

Schon C.C., Lee M., Melchinger A.E., Guthrie W.D. and Woodman W.L. 1993. Mapping and characterization of QTL affecting resistance against second generation European corn borer in maize with the aid of RFLPs. *Heredity*, **70**: 648-659.

Seyfarth R.C., Feuillet C. and Keller B. 1998. Development and characterization of molecular markers for the adult leaf rust resistance genes *Lr13* and *Lr35* in wheat. *In*: *Proc. 9th Int. Wheat Genet. Symp.*, (ed.) Slinkard A.E., Univ Saskatchewan, Saskatoon, **3**: 154-155.

Sharma R., Agrawal R.A.K., Kumar R., Mohapatra T. and Sharma R.P. 2002. Construction of an RAPD linkage map and localization of QTLs for oleic acid level using recombinant inbreds in mustard (*Brassica juncea*). *Genome*, **45**: 467-472.

Shields R. 1993. Pastoral synteny. *Nature*, **365**: 297-298.

Shizuya H., Birren B., Kim U.J., Mancino V., Slepak T., Tachiiri Y. and Simon M. 1992. Cloning and stable maintenance of 300-kilobase-pair fragments of human DNA in *Escherichia coli* using an F-factor-based vector. *Proc. Natl. Acad. Sci.*, USA, **89**: 8794-8797.

Slocum M.K., Figdore S.S., Kennard W.C., Suzuki J.Y. and Osborn T.C. 1990. Linkage arrangement of restriction fragment length polymorphism loci in *Brassica oleracea*. *Theor. Appl. Genet.*, **80**: 57-64.

Snape J.W., Semikhodski A., Sarma R. and Korzun V. *et al.* 1998. Mapping vernalization loci in wheat and comparative mapping with other cereals. *In*: *Proc. 9th Int. Wheat Genet Symp.*, (ed.) Slinkard A.E., Univ Saskatchewan, Saskatoon, **3**: 156-158.

Song K.M., Suzuki J.Y., Slocum M.K., Williams P.H. and Osborn T.C. 1991. A linkage map of *Brassica rapa* (syn. *campestris*) based on restriction fragment length polymorphism loci. *Theor. Appl. Genet.*, **82**: 296-304.

Song W.Y., Wang G.L., Chen H.S., Kim H.S., Pi L.Y., Holsten T., Gardner J., Wang B., Zhai W.X., Zhu L.H., Fauquet C. and Ronald P.A. 1995. A receptor kinase like protein encoded by the rice disease resistance gene *Xa 21*. *Science*, **270**: 1804-1807.

Sourdille P., Charmet G., Trotett M., Tixier M.N., Boeuf C., Negre S., Barloy D. and Bernard M. 1998. Linkage between RFLP molecular markers and the dwarfing genes *Rht-B1* and *Rht-D1* in wheat. *Hereditas*, **128**: 41-46.

Sourdille P., Perretant M.R., Charmet G., Leroy P., Gautier M.F., Jourdier P., Nelson J.C., Sorells M.E. and Bernard M. 1996. Linkage between RFLP markers and genes affecting kernel hardness in wheat. *Theor. Appl. Genet.*, **93**: 580-586.

Southern E.M. 1975. Detection of specific sequences among DNA fragments separated by gel electrophoresis. *J. Mol. Biol.*, **98**: 503-517.

Stam P. 1993. Construction of integrated genetic linkage maps by means of a new computer package: Join Map. *Plant J.*, **3**: 739-744.

Steinmetz M., Minard K. and Horvath S. *et al.* 1982. A molecular map of the immune response region from the major histocompatibility complex of the mouse. *Nature*, **300**: 35-42.

Stuber C.W., Lincoln S.E., Wolff D.W., Helentjaris T. and Lander E.S. 1992. Identification of genetic factors contributing to heterosis in a hybrid from two elite maize inbred lines using molecular markers. *Genetics*, **132**: 823-839.

Suiter K.A., Wendel J.F. and Case J.S. 1983. Linkage-1: a Pascal computer program for the detection and analysis of genetic linkage. *J. Hered.*, **74**: 203-204.

Sun G.L., Fahima T., Korol A.B., Turpeinien T., Grama A., Ronin Y.I. and Nevo E. 1997. Identification of molecular markers linked to *Yr15* stripe resistance gene of wheat originated in wild emmer wheat, *Triticum dicoccoides*. *Theor. Appl. Genet.*, **95**: 622-628.

Talbert L.E., Bruckner, Smith L.Y., Sears R. and Martin T.J. 1996. Development of PCR markers linked to resistance to wheat streak mosaic virus in wheat. *Theor. Appl. Genet.*, **93**: 463-467.

Tan X.L., Vanavichit A., Amornsilpa S. and Trangoonrung S. 1998. Genetic analysis of rice CMS-WA fertility restoration based on QTL mapping. *Theor. Appl. Genet.*, **96**: 994-999.

Tanhuanpaa P.K., Vilkki J.P. and Vilkki H.J. 1995a. Identification of a RAPD marker for palmitic acid concentration in the seed oil of spring turnip rape (*Brassica rapa* ssp. *oleifera*). *Theor. Appl. Genet.*, **91**: 477-480.

Tanksley S.D., Ganal M.W. and Martin G.B. 1995. Chromosome landing: A paradigm for map-based gene cloning in plant species with large genomes. *Trends Genet.*, **11**: 63-68.

Tanksley S.D., Ganal M.W. and Prince J.P. *et al.* 1992. High density molecular linkage maps of the tomato and potato genomes. *Genetics*, **132**: 1141-1160.

Tanksley S.D. and Orton T.J. 1983. *Isozymes in Plant Genetics and Breeding*. Parts 1A and 1B. Elsevier, Amsterdam, Netherlands.

Tautz D. 1989. Hypervariability of simple sequences as a general source for polymorphic DNA markers. *Nucl. Acids. Res.*, **17**: 6463-6471.

Teutonico R.A. and Osborn T.C. 1994. Mapping of RFLP and qualitative trait loci in *Brassica rapa* and comparison to the linkage maps of *B. napus*, *B. oleracea*, and *Arabidopsis thaliana*. *Theor. Appl. Genet.*, **89**: 885-892.

Teutonico R.A. and Osborn T.C. 1995. Mapping loci controlling vernalization, requirement in *Brassica rapa*. *Theor. Appl. Genet.*, **91**: 1279-1283.

Teutonico R.A., Yandell B.S., Satagopan J.M., Ferreira M.E., Palta J.P. and Osborn T.C. 1995. Genetic analysis and mapping of genes controlling freezing tolerance in oil seed *Brassica*. *Mol. Breed.*, **1**: 329-339.

Thorlby G., Veale E., Butcher K. and Warren G. 1999. Map positions of SFR genes in relation to other freezing-related genes of *Arabidopsis thaliana*. *Plant J.*, **17**: 445-452.

Torp A.M., Hansen A.L., Holme I.B. and Anderson S.B. 1998. Genetic markers for haploid formation in wheat anther culture. *In*: *Proc. 9th Int. Wheat Genet. Symp.*, (ed.) Slinkard A.E., Univ Saskatchewan, Saskatoon, **3**: 159-161.

Uzunova M., EckeW., Weissleder K. and Robbelen G. 1995. Mapping the genome of rapeseed (*Brassica napus* L) I. Construction of an RFLP linkage map and localization of QTLs for seed glucosinolate content. *Theor. Appl. Genet.*, **90**: 194-204.

Valdes J.M., Tagle D.A. and Collins F.S. 1994. Islad rescue PCR: A rapid and efficient method for isolating transcribed sequences from yeast artificial chromosomes and cosmids. *Proc. Natl. Acad. Sci.*, USA, **91**: 5377-5381.

Vallejos C.E., Sakiyama N.S. and Chase C.D. 1992. A molecular marker-based linkage map of *Phaseolus vulgaris*. *Genetics*, **131**: 733-740.

Van Deynze A.E., Nelson J.C. and O' Donoughue L.S. *et al.* (9 authors). 1995. Comparative mapping in grasses. Oat relationships. *Mol. Gen. Genet.*, **249**: 349-356.

Van Eck H.J., Jacobs J.M.E., Stam P., Ton J., Stickema W.J. and Jacobsen E. 1994. Multiple alleles for tuber shape in diploid potato detected by qualitative and quantitative genetic analysis using RFLPs. *Genetics*, **137**: 303-309.

Varshney R.K., Prasad M., Roy J.K., Kumar N., Singh H., Dhaliwal H.S., Balyan H.S. and Gupta P.K. 2000. Identification of eight chromosomes and a microsatellite marker on 1AS associated with QTL for grain weight in bread wheat. *Theor. Appl. Genet.*, **100**: 1290-1294.

Veldboom L.R., Lee M. and Woodman W.L. 1994. Molecular marker- facilitated studies in an elite maize population: I. Linkage analysis and determination of QTL for morphological traits. *Theor. Appl. Genet.*, **88**: 7-16.

Vos P., Hogers R., Bleeker M., Reijans M., Van Lee T., Hornes M., Frijters A., Pot J., Peleman J., Kuiper M. and Zabeau M. 1995. AFLP: a new technique for DNA fingerprinting. *Nucleic Acids Res.*, **23**: 4407-4414.

Walsh J.A., Sharpe A.G., Jenner C.E. and Lydiate D.J. 1999. Characterization of resistance to turnip mosaic virus in oilseed rape (*Brassica napus*) and genetic mapping of TURBO 1. *Theor. Appl. Genet.*, **99**: 1149-1154.

Wang G.L., Mackill D.J., Bonman J.M., McCouch S.R., Champoux M.C. and Nelson R.J. 1994. RFLP mapping of genes conferring complete and partial resistance to blast in a durably resistant rice cultivar. *Genetics*, **136**: 1421-1434.

Weaver R., Helms C., Mishra S.K. and Donis-Keller H. 1992. Software for analysis and manipulation of genetic linkage data. *Am. J. Hum. Genet.*, **50**: 1267-1274.

Weber J.L. and May P.E. 1989. Abundant class of human DNA polymorphisms which can be typed using the polymerase chain reaction. *Am. J. Hum. Genet.*, **44**: 388-396.

Weeden N.L., Muehalbauer F.J. and Ladizinsky G. 1992. Extensive conservation of linkage relationships between and total genetic maps. *J. Hered.*, **83**: 123-129.

Welsh J. and McClelland M. 1990. Fingerprinting genomes using PCR with arbitrary primers. *Nucl. Acids. Res.*, **18**: 7213-7218.

Whitkus R., Doebley J.F. and Lee M. 1992. Comparative genome mapping of sorghum and maize. *Genetics*, **132**: 1119-1130.

Williams J.G.K., Kubelik A.R. and Livak K.J. *et al.* 1990. Oligonucleotide primers of arbitrary sequence amplify DNA polymorphisms which are useful as genetic markers. *Nucl. Acids Res.*, **18**: 6531-6535.

Williams K.J., Fisher J.M. and Langridge P. 1994. Identification of RFLP markers linked to the cereal cyst nematode resistant gene (*Cre*) in wheat. *Theor. Appl. Genet.*, **89**: 927-930.

Wu K.K., Burnquist W., Sorrells M.E., Tew T.L., Moore P.H. and Tanksley S.D. 1992. The detection and estimation of linkage in polyploids using single-dose restriction fragments. *Theor. Appl. Genet.*, **83**: 294-300.

Wu P., Zhang G. and Huang N. 1996. Identification of QTL controlling quantitative characters in rice using RFLP markers. *Euphytica*, **89**: 349-354.

Wu P., Zhang G., Ladha J.K., McCouch S.R. and Huang N. 1995. Molecular-marker-facilitated investigation on the ability to stimulate N_2 fixation in the rhizosphere by irrigated rice plants. *Theor. Appl. Genet.*, **91**: 1177-1183.

Xiao J., Li J., Yuan L. and Tanksley S.D. 1995. Dominance is the major genetic basis of heterosis in rice as revealed by QTL analysis using molecular markers. *Genetics*, **140**: 745-754.

Xu G.W., Magill C.W., Schertz K.F. and Hart G.E. 1994. A RFLP linkage map of *Sorghum bicolor* (L.) Moench. *Theor. Appl. Genet.*, **89**: 139-145.

Yan J.Q., Zhu J., He C.X., Benmoussa M. and Wu P. 1998. Quantitative trait loci analysis for the developmental behaviour of tiller number in rice (*Oryza sativa* L.). *Theor. Appl. Genet.*, **97**: 267-274.

Yano M., Harushima Y., Nagamura Y., Kurata N., Minobe Y. and Sasaki T. 1997. Identification of quantitative trait loci controlling heading date in rice using a high-density linkage map. *Theor. Appl. Genet.*, **95**: 1025-1032.

Yoshimura S., Yoshimura A., Iwata N., McCouch S.R., Abenes M.L., Baraoidan M.R., Mew T.W. and Nelson R.J. 1995a. Tagging and combining bacterial blight resistant genes in rice using RAPD and RFLP markers. *Mol. Breed.*, **1**: 375-387.

Yoshimura S., Yoshimura A., Nelson R.J., Mew T.W. and Iwata N. 1995b. Tagging *Xa-1*, the bacterial blight resistance gene in rice, by using RAPD markers. *Breeding Sci.*, **45**: 81-85.

Yoshimura S., Yoshimura A., Saito A., Kishimoto N., Kawase M., Yano M., Nakagahra N., Ogawa T. and Iwata N. 1992. RFLP analysis of introgressed segments in three near-isogenic lines of rice for bacterial blight resistance genes, *Xa-1, Xa-3* and *Xa-4*. *Japan J. Genet.*, **67**: 29-37.

Yoshimura S., Umehara Y., Kurata N., Nagamura Y., Sasaki T., Minobe Y. and Iwata N. 1996. Identification of a YAC clone carrying the *Xa-1* allele, a bacterial blight resistance gene in rice. *Theor. Appl. Genet.*, **93**: 117-122.

Young N.D., Danesh D., Menancio-Hautea D. and Kumar L. 1993. Mapping oligogenic resistance to powdery mildew in mungbean with RFLPs. *Theor. Appl. Genet.*, **87**: 243-249.

Young N.D., Kumar L., Menancio-Hautea D., Danesh D., Talekar N.S., Shanmugasundaraum S. and Kim D.H. 1992. RFLP mapping of a major bruchid resistance gene in mungbean (*Vigna radiata* L. Wilczek). *Theor. Appl. Genet.*, **84**: 839-844.

Young N.D. and Tanksley S.D. 1989. RFLP analysis of the size of chromosomal segments retained around the *Tm-2a* locus of tomato during back-cross breeding. *Theor. Appl. Genet.*, **77**: 353-359.

Yu Z.H. 1991. Molecular mapping of rice (*Oryza sativa* L.) genes via linkage to restriction fragment polymorphism (RFLP) markers. Cornell University, Ithaca, New York (Ph. D. Thesis).

Yu Z.H., Mackill D.J., Bonman J.M. and Tanksley S.D. 1991. Tagging genes for blast resistance in rice via linkage to RFLP markers. *Theor. Appl. Genet.*, **81**: 471-476.

Yu Y.G., Saghai-Maroof M.A., Buss G.R., Maughan P.J. and Tolin S.A. 1994. RFLP and microsatellite mapping of a gene for soybean mosaic virus resistance. *Phytopathology*, **84**: 60-64.

Zeng Z.B. 1994. Precision mapping of quantitative trait loci. *Genetics*, **136**: 1457-1468.

Zhang Q., Liu K.D., Yang G.P., Saghai-Maroof M.A., Xu C.G. and Zhou Z.Q. 1997. Molecular marker diversity and hybrid sterility in indica-japonica rice crosses. *Theor. Appl. Genet.*, **95**: 112-118.

Zhang Q.F., Shen B.S., Dai X.K., Mei M.H., Saghai-Maroof M.A. and Li Z.B. 1994. Using bulked extremes and recessive class to map genes for photoperiod-sensitive genic male sterility in rice. *Proc. Natl. Acad. Sci.*, USA, **91**: 8675-8679.

Ziegelhoffer E.C., Medrano L.J. and Meyerowitz E.M. 2000. Cloning of the *Arabidopsis WIGGUM* gene identifies a role for farnesylation in meristem development. *Proc. Natl. Acad. Sci.*, USA, **97**: 7633-7638.

Plant Breeding - Mendelian to Molecular Approaches
H. K. Jain and M. C. Kharkwal (eds.)

12

Genome-wide Molecular Approaches in Plants: From Structure to Function

Akhilesh K. Tyagi, J.P. Khurana, P. Khurana, A. Mohanty and A.K. Bharti

Abstract

Advent of technologies capable of creating 'genome map' even without knowledge of the phenotype or a product of a gene has tremendously improved our capability to comprehend genomes. Large fragments of DNA (> 100 kb) can be cloned in YAC (yeast artificial chromosome), BAC (bacterial artificial chromosome) and PAC (P1-derived artificial chromosome) vectors, which can be aligned to chromosomes with the help of DNA markers. Clones of contigs can be used to prepare shotgun sub-clones to generate sequence of the entire genome of organisms. Such an approach has been successfully used for *Arabidopsis* genome sequencing, and the International Rice Genome Sequencing Programme has also adopted the same strategy. Another approach involves sequencing the shotgun clones of the entire genome, which can then be assembled by computation. Simultaneously, progress is being made in the functional genomics of plants with more than a million ESTs (expressed sequence tags) available in databases and expression chips representing thousands of genes being put to test. The challenge for the future would be to define the function of genes and give them a place in the regulatory network of the cell. Gene knockouts and gene tags would prove to be of significant value as also computational biology. These developments should eventually pave the way not only for discovery of novel genes but also help in precision breeding.

Introduction

In the beginning of the last century, "factors" of Mendel controlling traits according to the laws of heredity were re-discovered. This was followed by the realization that heredity factors reside in chromosomes in a linear order, which paved the way for discovery of genetic linkage by Morgan and the development of linkage map by Sturtevant. Subsequently, we learned that "factors" embodied in DNA have a precise genetic code and can be read in their entirety. Biochemical and molecular research defined that genes, the contemporary term for "factors", specify organismic shape and function. Many hypotheses evolved by *in vitro* experimentation were confirmed *in vivo* by transgenic approach. Today, we stand to see a hypothesis arising *in silico*, as the result of a plethora

University of Delhi South Campus, New Delhi 110 021, India

of genome-wide information and its analysis by power of computation, and genomes becoming a starting point for the research leading to an experimental canvas akin to "virtual cell". The emerging global scenario in genomics is very exciting, as it is the beginning of an era where diversity at molecular level could be appreciated in terms of genes, proteins and their functions, and one could address the questions related to evolution and dynamic life in real earnest.

Plant genome research has also made great strides in recent years. While *Arabidopsis* genome has already been sequenced, the rice genome is on its way to be sequenced. Marker-based breeding is complementing the conventional approach leading to precision breeding and the function of newly discovered genes is being defined. An attempt has been made in this paper to provide a brief review of the progress in plant genomics with emphasis on *Arabidopsis* and rice. For further details, one may consult other recent reviews and references cited therein (Brent, 2000; Gupta, 2000; Heslop-Harrison, 2000; Lander and Weinberg, 2000; Maheshwari *et al.*, 2001).

Genome Maps

A map of the genome of an organism defines relative position of different features of the genome, which could be recognized by visual scrutiny or analyzed by experimentation. Cytogenetic maps are based on microscope examination of chromosomes. Starting from chromosome number counting, the approaches include chromosome banding, fluorescence *in situ* hybridization (FISH) with gene-specific sequences, and genomic *in situ* hybridization (GISH) with whole genome. While FISH has been used to carry out cytogenetic mapping in *Arabidopsis* as well as rice (Fransz *et al.*, 1998; Fukui and Ohmido, 2000), GISH has been used to determine the D genome in rice (Yan *et al.*, 1999; Fukui and Ohmido, 2000).

In linkage map or genetic map, relative distances of traits or genes and their order on the chromosome are defined on the basis of the frequency of genetic recombination between genes. The genetic distance of each centiMorgan (cM) represents 1 % recombination frequency, but a recombination frequency of more than 50 % reflects independent chromosomal position. Recent developments in molecular marker technology allow construction of a genetic map even without establishing the relationship of a gene to the phenotype or the product. The concept developed in 1980 when Botstein and coworkers realized that "...a set of probes for DNA polymorphism ......should provide a new horizon in .... genetics". Thus, a piece of DNA became a milepost on the road of organism's DNA and road map started to provide information about the location of gene(s) influencing a phenotype. Such DNA markers are normally the result of mutations in single base pair resulting in loss or gain of a restriction site, segmental deletions or additions. Neither these markers are dependent on the developmental state of the plant, nor are they influenced by environment. They can be detected in the form of restriction fragment length polymorphism (RFLP) by using a DNA or EST probe. Other approach utilizes PCR technology to amplify and detect polymorphism at DNA level. Several kinds of markers including random amplified polymorphic DNA (RAPD), microsatellite or simple sequence length polymorphism (SSLP), sequence-tagged sites (STS) and amplified fragment length polymorphism (AFLP) are available (Mohan *et al.*, 1997). A new and rich source of polymorphism is emerging in the form of single nucleotide polymorphisms (SNPs), as a result of nucleotide sequence information (Brookes, 1999).

By making use of variable quality and number of markers, genetic maps have been prepared for *Arabidopsis* (Blanc *et al.*, 2000) as well as rice (McCouch *et al.*, 1988; Nagamura *et al.*, 1997; Harushima *et al.*, 1998; Cho *et al.*, 1998). The latest genetic map of rice includes 3267 markers (Rice Genome Research Program http://rgp.dna.affrc.go.jp/).

The physical map reflects distances between genes in the form of DNA length and serves as a precursor to positional cloning of genes and genomic DNA sequencing. As a matter of fact, the nucleotide sequence of the entire genome represents the ultimate physical map of an organism. A physical map can be anchored to the genetic map by use of DNA markers, which can also be mapped to chromosomes by FISH. To prepare a physical map, large DNA inserts are cloned in vectors like YACs, BACs and PACs (Burke *et al.*, 1987; Shizuya *et al.*, 1992; Ioannou *et al.*, 1994). Subsequently, seed clones are selected with the help of DNA markers, which specify a particular locus on the chromosome. Cross hybridization of DNA sequences from ends of the clones (sequence tag connectors, STC) to new clones, or overlap of fingerprints of clones help extend the physical map in the form of contigs till they merge into each other. Physical maps based on the use of variable cloning vectors have been generated both for *Arabidopsis* (Marra *et al.*, 1999; Mozo *et al.*, 1999) and rice (Kurata *et al.*, 1997; Saji *et al.*, 2001). A relationship between cytogenetic, genetic and physical maps is shown in Fig. 1. Work on development of genome maps of several major plants is already in progress.

An interesting outcome of genome mapping is the analysis of synteny between plant genomes (Bennetzen, 2000). Extensive colinearity has been observed in the rice genome and genomes of other cereals and millets (Gale and Devos, 1998). Such alignments on different maps at orthologous positions are possible even for quantitative trait loci (QTLs). However, colinearity of *Arabidopsis* and rice does not seem to be sufficient to allow map-based prediction and isolation of genes among eudicot-monocot divide (Devos *et al.*, 1999).

Genome-wide Sequencing

The discovery of entire sequences of genomes can serve the purpose of global heritage and resource to understand genetic events, SNPs, and comparative genomics. To obtain information about nucleotide sequence at genome-wide scale, variable numbers of strategies have been adopted. To determine information about expressed component of the genome, mRNAs are converted into cDNA, which can be sequenced from 5′ and 3′ ends. Single-pass sequence information from either end is called as expressed sequence tag (EST). More than 1.2 million ESTs of plants are available as per dbEST release 081701 of

August 2001 from National Center for Biotechnological Information (NCBI). The number of ESTs for major crops is given in Table 1.

Table 1. Number of ESTs for selected plants available in public database

S.No.	Plant	ESTs
1.	*Glycine max* (soybean)	180,830
2.	*Medicago truncatula* (barrel medic)	137,588
3.	*Lycopersicon esculentum* (tomato)	126,562
4.	*Arabidopsis thaliana* (thale cress)	113,330
5.	*Zea mays* (maize)	102,551
6.	*Oryza sativa* (rice)	80,099
7.	*Sorghum bicolor* (sorghum)	76,440
8.	*Hordeum vulgare* (barley)	68,480
9.	*Triticum aestivum* (wheat)	66,191
10.	*Solanum tuberosum* (potato)	48,769
11.	*Gossypium* sp. (cotton)	30,416
12.	*Brassica* sp. (mustard, rape)	2,760
13.	*Mentha piperita* (peppermint)	1,316
14.	*Ricinus communis* (castor bean)	750
15.	*Cicer arietinum* (chickpea)	20

It is estimated that a much larger number of ESTs is also available in private sector. It should, however, be kept in view that *Arabidopsis* gene content has been estimated to be only 25,498 and for rice only about 20,000 non-redundant ESTs have been worked out (The *Arabidopsis* Genome Initiative, 2000; Sasaki and Burr, 2000; White *et al.*, 2000). Such ESTs serve an important resource of markers, reflect expression potential of their source and also help in annotation of the genomic sequence.

Most of the DNA sequencing is based on Sanger's dideoxy chain termination method (Sanger *et al.*, 1977). With the emergence of capillary and cycle sequencing, fluorescent dideoxynucleotides, PCR technology and automation combined with computation, it became possible to sequence the entire genome of organisms. First to be sequenced were microbe genomes, starting with *Haemophilus influenzae*, numbering more than 40 today (Hunkapillar *et al.*, 1991; Lee *et al.*, 1997; Friedman and Meldrum, 1998). Subsequently, genomes of eukaryotes like yeast, *Caenorhabditis* and *Drosophila* were sequenced before human genome sequence was released in February 2001 (International Human Genome Consortium, 2001; Venter *et al.*, 2001). The most commonly used approach for genome level sequencing is the shotgun approach. But, one can take two alternative routes to achieve the objective. In the first alternative, one could prepare a physical map with the help of large DNA fragments cloned in BACs or PACs and shotgun the clones to assemble their DNA sequence. This clone-by-clone sequence can be assembled into larger contigs by over-lapping DNA sequence, ultimately leading to the complete genome sequence (Roach *et al.*, 1999). In the second alternative, the whole genome shotgun route involves shearing the entire genome into smaller fragments to prepare shotgun library before sequencing and its assembly by powerful computers (Venter *et al.*, 1996; 1998). While clone-by-clone approach is relatively slow, it is precise as it gives an assembled sequence aligned with the help of well-defined DNA markers; the whole genome random sequence approach on the other hand, is faster in the beginning but may pose problems in gap filling and assignment at later stages (Fig. 1). Both routes have turned out to be

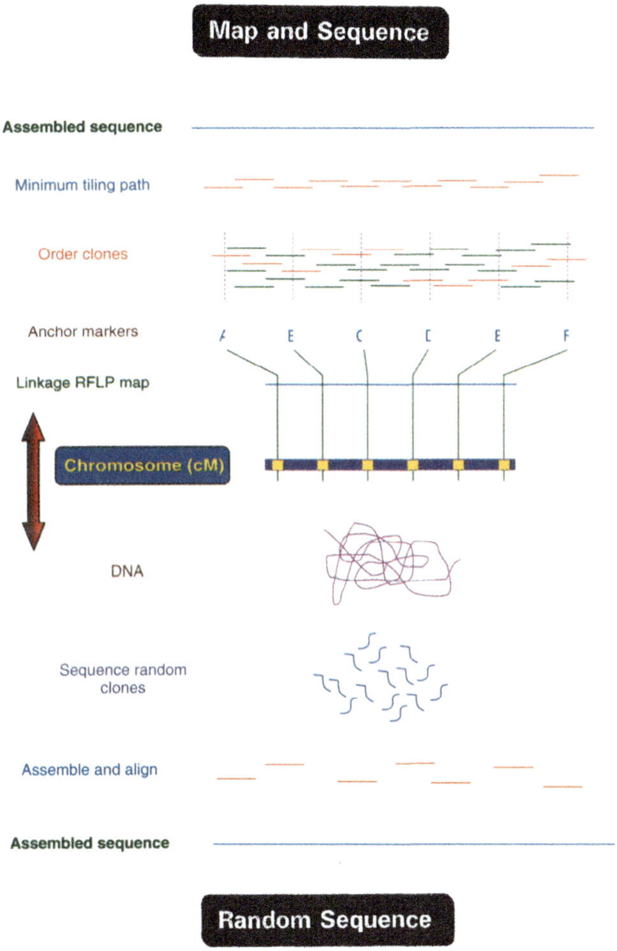

Fig. 1. The genome-wide sequencing approaches. Map and sequence approach involves correlative information about cytogenetic and molecular marker-based linkage map. Anchor markers are used to order clones and select minimum tiling path of over-lapping clones. Shotgun sequencing and assembly of DNA from clones provides the assembled sequence of chromosomes. In random sequence approach, the total DNA is fragmented and cloned before sequencing of clones randomly. The sequence of the entire genome is assembled and aligned chromosome-wise with the help of very powerful computation system.

rewarding, in particular the utility of whole genome shotgun approach was established with *Drosophila* and human genome sequencing. The shotgun approach and high level of accuracy (> 99.99 %) entails 10-12 folds redundancy in sequencing effort.

The *Arabidopsis* Genome Initiative

The plant genome sequencing programs in public sector have followed the clone-by-clone approach. *Arabidopsis thaliana* was selected for its small genome size (~130 Mb). The organized sequencing of *Arabidopsis thaliana* genome was initiated as "The *Arabidopsis* Genome Initiative (AGI)" in 1996 with a large amount of information on genetic and physical mapping as well as genome libraries already available (Pennisi, 2000). The AGI had several institutions from Europe, Japan and USA participating together. While first results of complete sequence of chromosomes 2 and 4 were out in 1999, the sequence of remaining three chromosomes and 115.4 Mb of the whole genome sequence (125 Mb) were out in December 2000 (The *Arabidopsis* Genome Initiative, 2000). The genome was found to contain 25,498 genes (about 15,000 being unique) with a potential to code for 11,000 proteins. About 60 % of the genome has been found to be duplicated. While only 9 % of the genes have been assigned function experimentally, a large number shows similarity with other organisms. Assigning a proper function to these genes would be a big task for future. One full issue of *Plant Physiology* (124, 2000) was devoted to have a look at the *Arabidopsis* genome information and to highlight its future application in plant genomics.

The Rice Genome Initiative

For sequencing the genome of a crop plant, rice has been considered ideal due to the large amount of information about its genetics, mapping and ESTs already available, particularly due to a seven-year effort in the form of Rice Genome Research Programme (RGP) in Japan. In 1997, The International Rice Genome Sequencing Project (IRGSP) was organized with Japan, Korea, USA, China and UK as its members. Subsequently, Taiwan, Canada, Thailand, India, France and Brazil also joined the efforts and were assigned specific chromosomes for sequencing (Sasaki and Burr, 2000; Delseny *et al.*, 2001). Indian effort on chromosome 11 in the form of Indian Initiative for Rice Genome Sequencing (IIRGS) is being made jointly by the Department of Plant Molecular Biology, University of Delhi South Campus, New Delhi and the National Research Centre for Plant Biotechnology (NRCPB), Indian Agricultural Research Institute, New Delhi. All participating countries can utilize common resources like YAC, BAC and PAC libraries for *Oryza sativa* ssp. *japonica*, cultivar Nipponbare, to generate and consolidate physical map of rice genome.

Originally, the rice genome sequencing was expected to be completed in 10 years. However, recent developments in technology have helped advance the possible date of completion significantly. Already, ~50 Mb of finished sequence and another ~100 Mb of phase I/II sequence from 1,055 BAC/PACs is available in public database and 106 Mb has been aligned by markers. Complete phase II sequence of the whole 430 Mb genome is likely to be released by the end of 2002. In 2000, Monsanto made its sequence available to IRGSP for completion and complementing public sector efforts (Barry, 2001). In January 2001, Syngenta also claimed success of its efforts on rice genome sequencing, but full information is yet to become available. These efforts will be followed by gap closure and annotation requiring much more time due to the nature of work involved. These projects would yield a plethora of information on gene content leading to gene discovery, which can pave the way for functional genomics. Further, comparative genomics of rice with other cereals and millets having larger genomes (from ~1,000 to 16,000 Mb), has already shown the potential of global utility of working on rice (Heslop-Harrison, 2000).

Functional Genomics

The progress in structural genomics of plants has led to great excitement in experimental biologists since the vast information is seen as a challenge to understand function of genes at genome-wide scale (Hieter and Boguski, 1997). Several international efforts are also being organized e.g. "The 2010 Project" on *Arabidopsis* (Chory *et al.*, 2000) or IRRI's effort to coordinate rice functional genomics (Fisher *et al.*, 2000). Significant success achieved in recent years promises to understand plant gene function to a level of "Virtual Plant".

The discovery and cloning of genes of importance has been greatly aided by the availability of mutant phenotypes and genome maps. The position of mutation can be mapped by analyzing association or distances between a marker and the mutation. Linked markers on the two sides of mutation help define the genetic interval, which could be refined by making use of new markers and fine-mapping (Lukowitz *et al.*, 2000). Genomic clones within flanking markers are isolated and target gene(s) selected. The gene of interest can be identified subsequently by complementation of mutation or gain-of-function after genetic transformation. Positional cloning has led to identification of several genes like *Pto* in tomato (Martin *et al.*, 1993), *ABI1* and others in *Arabidopsis* (Leung *et al.*, 1994; Lukowitz *et al.*, 2000) as well as resistance genes *Xa*21, *Pi-b* and *Pi-ta* in rice (Delseny *et al.*, 2001). Even Quantitative Trait Loci (QTL) like *Hd1* for heading date have been identified by positional cloning (Yano *et al.*, 2000).

Another approach of isolating a gene of interest or defining function of genes involves their interruption by known DNA sequences, which can serve as probes to isolate the gene. Earlier, native transposons were moved in maize genome to identify genes and their function (Vodkin, 1989). Subsequently, it was found that maize transposons *Ac/Ds* or *En/Spm* could be made to move and tag genes in heterologous systems (Osborne and Baker, 1995). Such systems have been developed to a significant level in *Arabidopsis* as well as in rice (Sussman *et al.*, 2000; Tyagi and Mohanty, 2000). This has resulted in cloning of several genes from *Arabidopsis* (Parinov and Sundaresan, 2000) and from rice (Sato *et al.*, 1999; Tyagi and Mohanty, 2000). In rice, a native retrotransposon *Tos17* has been found to be of special utility for gene tagging, as it is activated only during tissue culture and moves to a new site by making its copy (Hirochika, 1999). Already 30,000 lines carrying 250,000 insertions have been raised. The mutated gene as a result of interruption can be identified by cloning sequences flanking the retrotransposon. Since the system does not require transgenic approach and containment, it may find favour for large-scale open-field screening.

Genes can also be disrupted by T-DNA tagging after transformation with *Agrobacterium* (Krysan *et al.*, 1999). In rice, 22,000 transgenics representing 25,700 tagging events have been reported (Jeon *et al.*, 2000). The tagging technology, however, suffers from the fact that several gene disruptions do not give a visible phenotype due to functional compensation by other genes or members of a multigene family. Further, the number of insertions required to reach saturation levels is huge (> 10^5 to a million) and feasibility of the same remains to be demonstrated.

To know about genes involved in a selected process, one can take advantage of gene expression patterns. Trap vectors have been constructed to gain knowledge about expression patterns of genes or their regulatory elements (Fig. 2). These may be based on simple T-DNA insertion or can involve transposable elements to jump in the genome and provide knowledge about expression by incorporating variable designs (Pereira, 2000; Springer, 2000). The components providing information about expression pattern may include a reporter gene (*gus* or *GFP*) with minimal promoter, which can be activated by an enhancer in the vicinity (enhancer trap). Insertion of a reporter gene alone can be used to know about the action of the promoter (promoter trap) for the interrupted gene. On the other hand, the reporter gene can be made part of the transcribed mRNA from the gene (gene trap) by incorporating splice acceptor site with the reporter gene. While enhancer trap is independent of orientation, promoter trap and gene trap require directional insertion to be effective. These systems have been shown to work in *Arabidopsis* (Sundaresan *et al.*, 1995) as well as in rice (Chin *et al.*, 1999).

Another form of tagging involves activation of a gene by inserting a strong enhancer in the vicinity of the gene (Walden *et al.*, 1994). This would result in ectopic expression of the gene (or gain-of-function) revealing its function, which cannot be determined by gene interruption. Weigel *et al.* (2000) have reported 25,000 activation tags in *Arabidopsis*.

While trap vectors provide information about the hetero-catalytic activity of the gene, one can gain useful insights into the function by analyzing the levels of expression product, i.e. mRNA. Analysis of the transcript levels of all possible genes of an organism would provide global information about its transcriptosome (Duggan *et al.*, 1999). Recent development in the area of microarrays allows rapid analysis of several thousand genes at one time (Lemieux *et al.*, 1998; Brown and Botstein, 1999). Microarrays or DNA chips can be prepared by using several thousand cDNAs or their PCR products for spotting on a solid matrix at high density. Alternatively, unique information from EST database can be used to synthesize small oligonucleotides, representing each EST by photolithographic process (Schaffer *et al.*, 2000). These arrays representing transcriptosome of an organism are hybridized to cDNAs, derived from mRNA of two states, which have been labelled

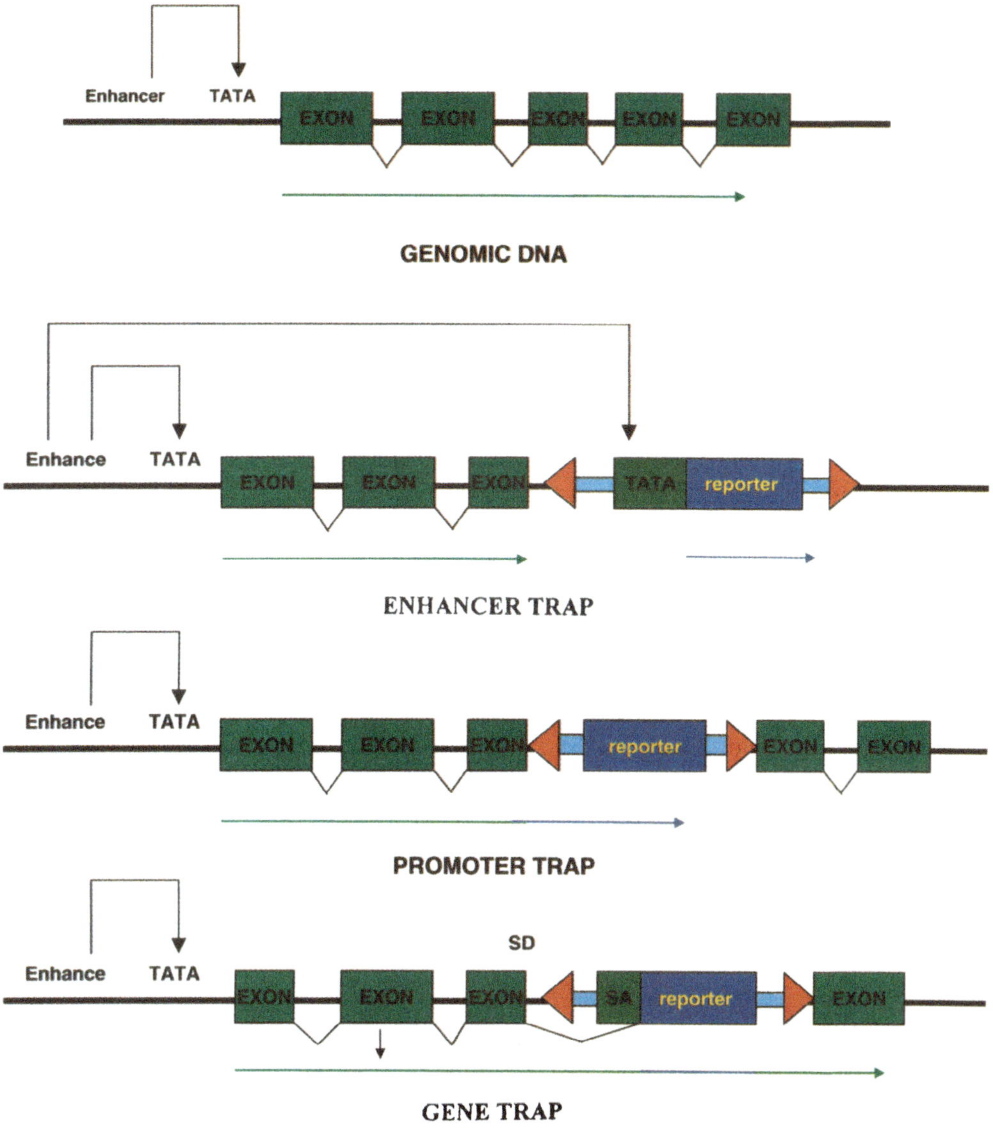

Fig. 2. Enhancer trap, promoter trap and gene trap vectors (Springer, 2000)

differently with fluorescent dyes. This creates a situation like "reverse northern-dot blots". The measurement of fluorescence of labeled cDNAs hybridized to their counter-parts in the microarray is monitored by a high-resolution scanner followed by image analysis, data extraction, and data management. Ultimately one can get information about relative abundance of mRNAs for a full set of genes on the microarray. Affimatrix has made a chip of 8,000 genes from *Arabidopsis* available for expression analysis (Wisman and

Ohlrogge, 2000) and efforts are on to develop a microarray with > 20,000 genes. This is a tremendous progress, keeping in view the modest start from 48 ESTs of *Arabidopsis* (Schena *et al.*, 1995). Microarray analysis of 1300 genes in *Arabidopsis* under drought and cold stresses has resulted in identification of 44 and 19 cDNAs, respectively, specific to these stresses (Seki *et al.*, 2001). In rice too, microarrays with 9,000 ESTs have already been prepared and tested (Yazaki *et al.*, 2000) and attempts are on to improve the scale.

In addition to the above, development of serial analysis of gene expression (SAGE) not only provides qualitative but also quantitative information about expression of genes (Velculescu *et al.*, 1995). The technique makes use of small tags generated from a unique position in mRNAs of a transcriptosome and concatenated to allow cloning and sequencing for analysis. The approach has been used for transcript profiling in seedlings of rice (Matsumura *et al.*, 1999). They could analyze 10,122 tags representing 5,921 expressed genes out of which only 23.1 % matched EST, thereby reflecting on the gap that may exist in EST database and the true nature of transcriptosome. Similar information emerged in our investigations to isolate target-specific cDNAs by differential screening (Tyagi *et al.*, 2000). Several alternatives to improve the potential of the technique have come in the form of RAGE and TALEST and it is hoped that the technique will find wide acceptability in genome-wide expression analysis.

Bioinformatics and Genome Annotation

The emerging genome sequencing information is inundating all other disciplines of biological sciences. But, making sense of all these sequences entails identification and subsequent prediction of functions for various genes in each genome. This is not only a Herculean task but also an interdisciplinary field involving extensive computational biology. As these raw sequences offer little, they place major demands on bio-computational professionals and biologists to interpret genomes (Lewis *et al.*, 2000). This process of interpreting raw sequence data into useful biological information is the process of genome annotation. Although it is important to remember that these are mere predictions, the outcome can be immensely important for functional genomics (Gerstein and Jansen, 2000; Nierman *et al.*, 2000).

Annotation of genomic sequences involves the use of existing information on the presence of genes, ESTs and other peripheral gene expression information in different species. There are two major classes of gene prediction methods - *ab initio* methods and homology-based methods. The *ab initio* methods take into account the complex signals laid out in the sequences, e.g. exons, introns, non-coding 5′ and 3′ exons and alternatively spliced products, especially in the eukaryotes, which is in contrast to the relatively simpler sequences with a single open reading frame and short intergenic regions in the prokaryotes. Thus, *ab initio* methods must combine these attributes in predicting genes. In general, these gene-finding methods are either 'signal sensor' or 'content sensor' based. In broad terms, signal sensor methods exploit information about pertinent sites like splice junctions, start and stop codon, promoter and terminators, etc. The content sensor methods employ extended length information, such as base composition of exons and introns. Most successful methods combine the two (Lewis *et al.*, 2000).

Characterization of homology-based methods has traditionally employed methods such as BLAST or FASTA for determining similarity between sequences being analyzed and the ones in the database at either the nucleic acid level or the predicted protein level. Although these techniques remain invaluable, new variations would extend the capabilities. Features being incorporated are the protein domain based searches and the EST databases. In fact, the large-scale EST sequencing programs initiated with gene discovery in mind, are now becoming an invaluable source of data, both for gene prediction and for confirming models of gene structure.

There are many more challenging problems to be addressed in the near future. The genome is much larger than the sum of its genes. Thus, problems arise in identification of the intergenic regions or in describing the genome as a whole. The alignment of non-coding regions and genome-wide comparisons and automatic genome annotation programs are the challenges of the future. Some of the latter is being attempted at the Oakridge Genome Annotation Channel (http://compbio.ornal.gov/chanel) and ensEMBL (http://ensembl.ebi.ac.uk) is in the pipeline. The Rice Genome Research Program (RGP) launched large-scale genome sequencing in 1998, and developed a new genome database called INE (Integrated rice genome Explorer), to integrate all information on its genome sequencing. Appropriately called the INE (Ine in Japanese means rice!), it was developed with an intergrated map view (Sakata et al., 2000). Its advanced and promising version of *in silico*, automated annotation program was recently developed for annotating the rice genome data (http://RiceGAAS.dna.affrc.go.jp/). This program (Rice GAAS) automatically screens the query sequence with first the GENSCAN, RiceHMM database and then the database for splice site (SplicePredictor), homology search analysis programs like BLAST, profileScan, MOTIF, tRNA gene prediction (tRNAscan-SE), repetitive DNA analysis programs (RepeatMasker, Printrepeats), and automatically moves on to signal scan program (Signal Scan), protein localization (PSORT) and finally for classification and secondary structure prediction of membrane proteins (SOSUI). The near future may thus witness development of various online annotation programs for various genomes to help reduce the time gap between sequencing and annotation.

Conclusions

It is obvious that emerging global scenario about genomics is very exciting. But collation of all data and uncovering relevant information requires much greater deployment of informatics tools. And, this is being called just the beginning of an era where diversity at molecular level could be appreciated in terms of genes, proteins and their functions and we could address the questions related to evolution and dynamic life in real earnest. The logic states that all predictions need a scientific test. The effort to discover sequence of entire genome of crop plants like rice would lead to a global heritage of great value that will eventually help in understanding genetic events in crops of tremendous potential for food security and their manipulation. It would serve as a resource to make inventions, generate new intellectual property and complement efforts of breeders to provide more food in years to come.

Acknowledgements

Our work in genome biology is supported by the Department of Biotechnology, Government of India, New Delhi.

References

Barry, G.F. 2001. The use of the Monsanto draft rice genome sequence in research. *Plant Physiol.*, **125**: 1164-1165.

Bennetzen J.L. 2000. Comparative sequence analysis of plant nuclear genomes: microcolinearity and its many exceptions. *Plant Cell*, **12**: 1021-1029.

Blanc, G., Barakat A., Guyot R., Cooke R. and Delseny M. 2000. Extensive duplication and reshuffling in the *Arabidopsis* genome. *Plant Cell*, **12**: 1093-1101.

Botstein D., White R.L., Skolnick M. and Davis R.W. 1980. Construction of a genetic linkage map in man using restriction fragment length polymorphisms. *Am. J. Hum. Genet.*, **32**: 314-331.

Brent R. 2000. Genomic biology. *Cell*, **100**: 169-183.

Brookes A.J. 1999. The essence of SNPs. *Gene*, **234**: 177-186.

Brown P.O. and Botstein D. 1999. Exploring the new world of the genome with DNA microarrys. *Nat. Genet. Suppl.*, **21**: 33-37.

Burke D.T., Carle G.F. and Olson M.V. 1987. Cloning of large segments of exogenous DNA into yeast by means of artificial chromosome vectors. *Science*, **236**: 806-812.

Chin H.G., Choe M.S., Lee S. H., Park S.H., Koo J.C., Kim N.Y., Lee J.J., Oh B.G., Yi G.H., Kim S.C., Choi H.C., Cho M.J. and Han C. 1999. Molecular analysis of rice plants harboring an *Ac/Ds* transposable element-mediated gene trapping system. *Plant J.*, **19**: 615-623.

Cho Y. G., McCouch S. R., Kniprer M., Kang M.R., Pot J., Groenen J.T.M. and Lun E. 1998. Integrated map of AFLP, SSLP and RFLP markers using a recombinant inbred population of rice (*Oryza sativa* L.). *Theor. Appl. Genet.*, **97**: 370-380.

Chory J., Ecker J. R., Briggs S., Caboche M., Corruzi G. M., Cook D., Dangl J., Grant S., Guerinot M.L., Henikoff S., Martienssen R., Okada K., Raikhel N.V., Somerville C. R. and Weigel D. 2000. Functional genomics and the virtual plant. A blueprint for understanding how plants are built and how to improve them. *Plant Physiol.*, **123**: 423-425.

Delseny M., Salses J., Cooke R., Sallaud C., Regad F., Lagoda P., Guiderdoni E., Ventelon M., Brugidou C. and Ghesquihre A. 2001. Rice genomics: present and future. *Plant Physiol. Biochem.*, **39**: 323-334.

Devos K.M., Beales J., Nagamura Y. and Sasaki T. 1999. *Arabidopsis*-rice: will colinearity allow gene prediction across the eudicot-monocot divide? *Genome Research*, **9**: 825-829.

Duggan D.J., Bittner M., Chen Y., Meltzer P. and Trent J.M. 1999. Expression profiling using cDNA microarrays. *Nat. Genet. Suppl.*, **21**: 10-14.

Fischer K. S., Barton J., Khush G. S., Leung H. and Cantrell R. 2000. Collaborations in rice. *Science*, **290**: 279-280.

Fransz P., Armstrong S., Alonso-Blanco C., Fischer T.C., Torres-Ruiz R.A. and Jones G. 1998. Cytogenetics for the model system *Arabidopsis thaliana*. *Plant J.*, **13**: 867-876.

Friedman N. and Meldrum D. 1998. Capillary tube resistive thermal cycling. *Anal. Chem.*, **70**: 2997-3002.

Fukui K. and Ohmido N. 2000. Rice genome research: An alternative approach based on molecular cytology. *In*: *Genomes*, (ed.) J.P. Gustafson, Kluwer Acedemic/Plenum Publishers, New York, pp. 109-121.

Gale M.D. and Devos K.M. 1998. Comparitive genetics in the grasses. *Proc. Natl. Acad. Sci.*, USA, **95**: 1971-1974.

Gerstein M. and Jansen R. 2000. The current excitement in bioinformatics-analysis of whole-genome expression data: how does it relate to protein structure and function? *Curr. Opin. Struct. Biol.*, **10**: 574-584.

Gupta P.K. 2000. Chromosome and genome research in plants: some recent developments. *The Nucleus*, **43**: 94-113.

Harushima Y., Yano M., Shomura A., Sato M., Shimano T., Kuboki Y., Yamamoto T., Lin S.Y., Antonio B.A., Parco A., Kajiya H., Huang N., Yamamoto K., Nagamura Y., Kurata N., Khush G.S. and Sasaki T. 1998. A high-density rice genetic linkage map with 2275 markers using a single F2 population. *Genetics*, **148**: 479-494.

Heslop-Harrison J.S. 2000. Comparative genome organization in plants: from sequence and markers to chromatin and chromosomes. *Plant Cell*, **12**: 617-635.

Hieter P. and Boguski M. 1997. Functional genomics: It's all how you read it. *Science*, **278**: 601-602.

Hirochika H. 1999. Retrotransposon of rice as a tool for forward and reverse genetics. *In*: *Molecular Biology of Rice*, (ed.) K. Shimamoto. Springer, Tokyo, pp. 43-58.

Hunkapiller T., Kaiser R.J., Koop B.F. and Hood L. 1991. Large-scale and automated DNA sequence determination. *Science*, **254**: 59-67.

International Human Genome Sequencing Consortium. 2000. Initial sequencing and analysis of the human genome. *Nature*, **409**: 860-921.

Ioannou P.A., Amemiya C.T., Garnes J., Kroisel P.M., Shizuya H., Chen C., Batzer M.A. and De Jong P.J. 1994. A new bacteriophage P1-derived vector for the propagation of large human DNA fragments. *Nat. Genet.*, **6**: 84-89.

Jeon J.S., Lee S., Jung K.H., Jun S.H., Jeong D.H., Lee J., Kim C., Jang S., Yang K., Nam J., An K., Han M.J., Sung R.J., Choi H.S., Yu J.W., Choi J.H., Cho S.Y., Cha S.S., Kim S.I. and An G. 2000. T-DNA insertional mutagenesis for functional genomics in rice. *Plant J.*, **22**: 561-570.

Krysan P.J., Young J.C. and Sussman M.R. 1999. T-DNA as an insertional mutagen in *Arabidopsis*. *Plant Cell*, **11**: 2283-2290.

Kurata N., Umehara Y., Tanoue H. and Sasaki T. 1997. Physical mapping of the rice genome with YAC clones. *Plant Mol. Biol.*, **35**: 101-113.

Lander E.S. and Weinberg A. 2000. Genomics: journey to the center of biology. *Science*, **287**: 1777-1782.

Lee L.G., Spurgeon S.L., Heiner C.R., Benson S.C., Rosenblum B.B., Menchen S.M., Graham R.J., Constantinescu A., Upadhya K.G. and Cassel J.M. 1997. New energy transfer dyes for DNA sequencing. *Nucleic Acids Res.*, **25**: 2816-2822.

Lemieux B., Aharoni A. and Schena M. 1998. Overview of DNA chip technology. *Mol. Breed.*, **4**: 277-289.

Leung J., Bouvier-Durand M., Morris P.C., Guerrier D., Chefdor F. and Giraudat J. 1994. *Arabidopsis* ABA response gene *ABI1*: features of a calcium-modualated protein phosphatase. *Science*, **264**: 1448-1452.

Lewis S., Ashburner M. and Reese M.G. 2000. Annotating eukaryote genomes. *Curr. Opin. Struct. Biol.*, **10**: 349-354.

Lukowitz W., Gillmor C.S. and Scheible W.R. 2000. Positional cloning in *Arabidopsis*. Why it feels good to have a genome initiative working for you. *Plant Physiol.*, **123**: 795-805.

Maheshwari S.C., Maheshwari N. and Sopory S.K. 2001. Genomics, DNA chips and a revolution in plant biology. *Curr. Sci.*, **80**: 252-261.

Marra M., Kucaba T., Sekhon M. et al. 1999. A map for sequence analysis of the *Arabidopsis thailana* genome. *Nat. Genet.*, **22**: 265-270.

Martin G.B., Brommonschenkel S.H., Chunwongse J., Frary A., Ganal M.W., Spivey R., Wu T., Earle E.D. and Tanksley S.D. 1993. Map-based cloning of a protein kinase gene conferring disease resistance in tomato. *Science*, **262**: 1432-1436.

Matsumara H., Nirasawa S. and Terauchi R. 1999. Transcript profiling in rice (*Oryza sativa* L.) seedlings using serial analysis of gene expression (SAGE). *Plant J.*, **20**: 719-726.

McCouch S.R., Kochert G., Yu Z.H., Wang Z.Y., Khush G.S., Coffman W.R. and Tanksley S.D. 1988. Molecular mapping of rice chromosomes. *Theor. Appl. Genet.*, **76**: 815-829.

Mohan M., Nair S., Bhagwat A., Krishna T.G., Yano M., Bhatia C.R. and Sasaki T. 1997. Genome mapping, molecular markers and marker-assisted selection in crop plants. *Mol. Breed.*, **3**: 87-103.

Mozo T., Dewar K., Dunn P., Ecker J. R., Fischer S., Kloska S., Lehrach H., Marra M., Martienssen R., Meir-Ewert S. and Altmann T. 1999. A complete BAC-based physical map of *Arabidopsis thaliana* genome. *Nat. Genet.*, **22**: 271-275.

Nagamura Y., Antonio B.A. and Sasaki T. 1997. Rice molecular genetic map using RFLPs and its applications. *Plant Mol. Biol.*, **35**: 79-87.

Nierman W.C., Eisen J.A., Fleischmann R.D. and Fraser C.M. 2000. Genome data: what do we learn? *Curr Opin. Struct. Biol.*, **10**: 343-348.

Osborne B.I. and Baker B. 1995. Movers and shakers: maize transposons as tools for analyzing other plant genomes. *Curr. Opin. Cell Biol.*, **7**: 406-413.

Parinov S. and Sundaresan V. 2000. Functional genomics in *Arabidopsis*: large-scale insertional mutagenesis complements the genome sequencing project. *Curr. Opin. Biotech.*, **11**: 157-161.

Pennisi E. 2000. *Arabidopsis* comes of age. *Science*, **290**: 32-35.

Pereira A. 2000. A transgenic perspective on plant functional genomics. *Transgenic Res.*, **9**: 245-260.

Roach J. C., Siegel A F., Van den Engh G., Trask B. and Hood L. 1999. Gaps in Human Genome Project. *Nature*, **401**: 843-845.

Saji S., Umehara Y., Antonio B.A., Yamane H., Tanoue H., Baba T., Aoki H., Ishige N., Wu J., Koike K., Matsumoto T. and Sasaki T. 2001. A physical map with yeast artificial chromosome (YAC) clones covering 63% of the 12 rice chromosomes. *Genome*, **44**: 32-37.

Sakata K., Antonio B.A., Mukai Y., Nagasaki H., Sakai Y., Makino K. and Sasaki T. 2000. INE: a rice genome database with an integrated map view. *Nucleic Acids Res.*, **28**: 97-101.

Sanger F., Nicklen S. and Coulson A. R. 1977. DNA sequencing with chain-terminating inhibitors. *Proc. Natl. Acad. Sci.*, USA, **74**: 5463-5467.

Sasaki T. and B. Burr. 2000. International rice genome sequencing project: the effort to completely sequence the rice genome. *Curr. Opin. Plant Biol.*, **3**: 138-141.

Sato Y., Sntoku N., Miura Y., Hirochika H., Kitano H., Matsuoka M. 1999. Loss of function mutations in the rice homeobox gene *OSH15* affect the architecture of internodes resulting in dwarf plants. *EMBO J.*, **18**: 992-1002.

Schaffer R., Landgraf J., Perez-Amador M. and Wisman E. 2000. Monitoring genome-wide expression in plants. *Curr. Opin. Biotech.*, **11**: 162-167.

Schena M., Shalon D., Davis R.W. and Brown P.O. 1995. Quantitative monitoring of gene expression patterns with a complementary DNA microarray. *Science*, **270**: 467-470.

Seki M., Narusaka M., Abe H., Kasuga M., Yamaguchi-Shinozaki K., Carninci P., Hayashizaki Y. and Shinozaki K. 2001. Monitoring the expression pattern of 1300 *Arabidopsis* genes under drought and cold stresses by using a full-length cDNA microarray. *Plant Cell*, **13**: 61-72.

Shizuya H., Birren B., Kim U.-J., Mancino V., Slepak T., Tachiiri Y. and Simon M. 1992. Cloning and stable maintenance of 300-kilobase-pair fragments of human DNA in *Escherichia coli* using an F-factor-based vector. *Proc. Natl. Acad. Sci.*, USA, **89**: 8794-8797.

Springer P.S. 2000. Gene traps: tools for plant development and genomics. *Plant Cell*, **12**: 1007-1020.

Sundaresan V., Springer P., Volpe T., Haward S., Jones J.D.G., Dean C., Ma H. and Martienssen R. 1995. Patterns of gene action in plant development revealed by enhancer trap and gene trap transposable elements. *Genes Dev.*, **9**: 1797-1810.

Sussman M.R., Amasino R.A., Young J.C., Krysan P.J. and Austin-Phillips J. 2000. The *Arabidopsis* knockout facility at the University of Wisconsin-Madison. *Plant Physiol.*, **124**: 1465-1467.

The *Arabidopsis* Genome Initiative. 2000. Analysis of the genome sequence of the flowering plant *Arabidopsis thaliana. Nature*, **408**: 796-815.

Tyagi A.K., Khurana J.P., Sharma A.K., Mohanty A., Dhingra A., Raghuvanshi S., Mukhopadhaya A., Gupta V., Anand S., Kathuria H., Bhusan S., Thakur J. and Kumar D. 2002. Organ-specific gene expression and genetic transformation for improvement of rice. *In*: *Proceedings of IVth International Rice Genetics Symposium*, Manila, The Philippines (In Press).

Tyagi A.K. and Mohanty A. 2000. Rice transformation for crop improvement and functional genomics. *Plant Sci.*, **158**: 1-18.

Velculescu V.E., Zhang L., Vogelstein B. and Kinzler K.W. 1995. Serial analysis of gene expression. *Science*, **270**: 484-487.

Venter C.J., Smith H.O. and Hood L. 1996. A new strategy for genome sequencing. *Nature*, **381**: 364-366.

Venter C.J., Adams M.D., Sutton G.G., Kerlavage A.R., Smith H.O. and Hunkapiller M. 1998. Shotgun sequencing of the human genome. *Science*, **280**: 1540-1542.

Venter J. C. *et al.* 2001. The sequence of the human genome. *Science*, **291**: 1304-1351.

Vodkin L. 1989. Transposable element influence on plant gene expression and variation. *In*: *The Biochemistry of Plants*, vol. 15, Academic Press Inc., San Diego, pp. 83-132.

Walden R., Fritze K., Hayashi H., Miklaschevichs E., Harling H. and Schell J. 1994. Activation tagging: a means of isolating genes implicated as playing a role in plant growth and development. *Plant Mol. Biol.*, **26**: 1521-1528.

Weigel D., Ahn J.H., Blazquez M.A., Borevitz J.O., Christensen S.K., Fankhauser C., Ferrandiz C., Kardailsky I., Malancharuvil E.J., Neff M.M., Nguyen J.T., Sato S., Wang Z.Y., Xia Y., Dixon R.A., Harrison M.J., Lamb C.J., Yanofsky M.F. and Chory J. 2000. Activation tagging in *Arabidopsis*. *Plant Physiol.*, **122**: 1003-1013.

White J.A., Todd J., Newman T., Focks N., Girke T., Martinez de Ilarduya O., Jaworski J.G., Ohlrogge J.B. and Benning C. 2000. A new set of *Arabidopsis* expressed sequence tags from developing seeds. The metabolic pathway from carbohydrates to seed oil. *Plant Physiol.*, **124**: 1582-1594.

Wisman E. and Ohlrogge J.B. 2000. *Arabidopsis* microarray service facilities. *Plant Physiol.*, **124**: 1468-1471.

Yan H., Min S. and Zhu L. 1999. Visualization of *Oryza eichingeri* chromosomes in intergenomic hybrid plants from *O. sativa* × *O. eichingeri* via fluorescent *in situ* hybridization. *Genome*, **42**: 48-51.

Yano M, Katayose Y., Ashikari M., Yamanouchi U., Monna L., Fuse T., Baba T., Yamamoto K., Umehara Y., Nagamura Y. and Sasaki T. 2000. *Hd1*, a major photoperiod sensitivity quantitative trait locus in rice, is closely related to the *Arabidopsis* flowering time gene CONSTANS. *Plant Cell*, **12**: 2473-2483.

Yazaki J., Kishimoto N., Nakamura K., Fujii F., Shimbo K., Otsuka Y., Wu J., Yamamoto K., Sakata K., Sasaki T. and Kikuchi S. 2000. Embarking on rice functional genomics via cDNA microarray: use of 3′ UTR probes for specific gene expression analysis. *DNA Res.*, **7**: 367-370.

Plant Breeding - Mendelian to Molecular Approaches
H. K. Jain and M. C. Kharkwal (eds.)

13

Application of Biotechnology to Maize and Wheat Improvement

David Hoisington[1]

Abstract

Today, scientists can take advantage of genes that are derived from various sources, including related and unrelated species, those identified via genetic mapping experiments and most recently from the efforts of functional genomics. Through the application of molecular genetics and genetic engineering, coupled with conventional crossing approaches, these genes can be efficiently incorporated into modern plant varieties. One of the most studied traits at CIMMYT is abiotic stress tolerance, especially tolerance to water-limited conditions. Quantitative Trait Loci (QTL) mapping, has identified several regions of the maize genome involved in the response to water stress. Efforts are underway to identify the underlying genes in these regions and to determine their potential to further improve the water-stress responses in maize and wheat. Candidate gene approaches are also being used employing resistance-like sequences isolated from rice and maize, to find possible homologies with genes conditioning disease resistance in wheat. The possibility of utilizing markers identified for *Lr4/Yr29* and *Lr34/Yr18* in applications in the wheat breeding activities are being explored. While the PCR-based marker systems have allowed more effective and efficient genotyping, DNA-array technology offers a substantially increase the number of genes that can be analysed. Efforts are also underway to develop complete EST databases for many cereals, including maize and wheat. Marker-assisted selection for polygenic trait improvement is in an important transition phase, and this field is on the verge of producing convincing results. Considering the potential for the development of new strategies, the future for polygenic trait improvement through DNA markers and the contribution of this to plant breeding efforts worldwide, appear bright.

Introduction

This presentation will provide an overview of the application of molecular genetics, genetic engineering and functional genomics to two of the major cereals, maize and wheat. It is not intended to provide a comprehensive coverage for either crop, but to give a few concrete examples of how molecular technology has been useful in the genetic dissection and manipulation of important traits in these species.

[1]The International Maize and Wheat Improvement Center (CIMMYT), Mexico

Today, scientists can take advantage of genes that are derived from various sources, including related and unrelated species, those identified via genetic mapping experiments and most recently, from the efforts of functional genomics (the area aimed at understanding the function of all genes in an organism) (Fig. 1). Through the application of molecular genetics and genetic engineering, coupled with conventional crossing approaches, these genes can be efficiency incorporated into modern plant varieties.

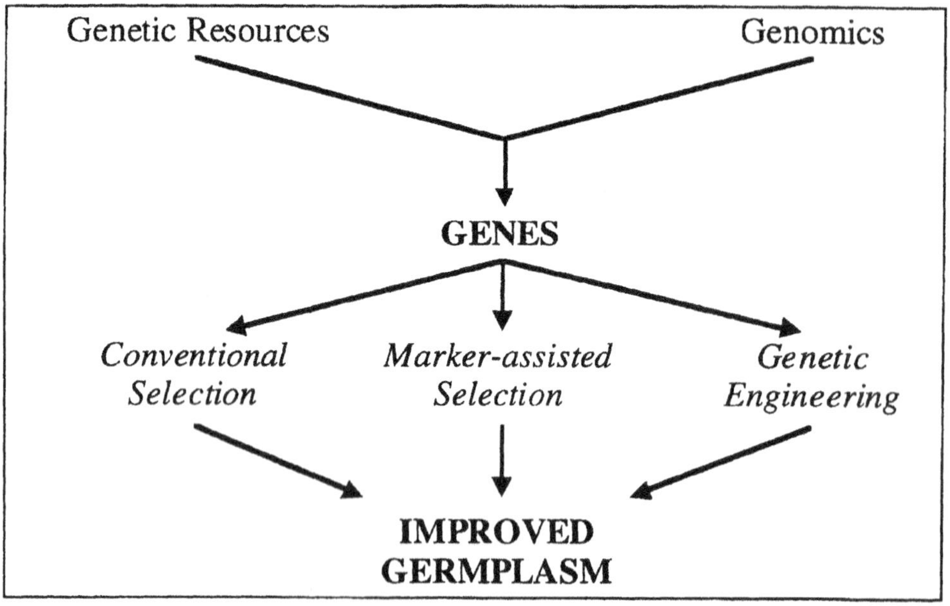

Fig. 1. Plant improvement options in the 21st century

Molecular Marker Technology

Molecular genetics, or the use of molecular techniques for detecting differences in the DNA of individual plants, has many applications of value to crop improvement. Such molecular markers, when very tightly linked to genes of interest, can be indirectly used to select for the desirable allele, and represents the simplest form of marker-assisted selection (MAS), whether used to accelerate the backcrossing of such an allele or in pyramiding several desirable alleles. Markers can also be used to dissect polygenic traits into their Mendelian components or quantitative trait loci (QTL), thus, increasing the understanding of the inheritance and gene action for such traits, and allowing for the use of MAS, as a complement to conventional selection procedures. Molecular markers are also used to probe the level of genetic diversity among different cultivars, within populations, among related species etc. The applications of such evaluations are many, including varietal fingerprinting for identification and protection, understanding relationships among the units under study, efficiently managing genetic resources, facilitating introgression of chromosomal segments from alien species, and even tagging of specific genes. In addition, markers and comparative mapping of various species have been very valuable for improving our understanding of a genome structure and function, and have allowed the isolation of a few genes of interest via map-based cloning.

Molecular marker technology has evolved from hybridization-based detection to new sequence-based systems. Each has their advantages and disadvantages. Restriction fragment length polymorphisms (RFLPs), were the first to be developed (some 15 years ago) and have been widely and successfully used to construct linkage maps of various species, which include maize and wheat. With the development of the polymerase chain reaction (PCR) technology, several marker types emerged. The first of those were RAPD markers (random amplified polymorphic DNA) which quickly gained popularity over RFLPs due to the simplicity and decreased costs of the assay. However, most researchers now realise the weaknesses of RAPDs and use them with much less frequency. Microsatellite markers or simple sequence repeats (SSRs), combine the power of RFLPs (co-dominant markers, reliable, specific genome location) with the ease of RAPDs and have the advantage of detecting higher levels of polymorphism. The AFLP (Amplified Fragment Length Polymorphism) approach takes advantage of the PCR technique to selectively amplify DNA fragments previously digested, with one or two restriction enzymes. Playing with the number of selective bases of the primers and considering the number of amplification products per primer pair, this approach is certainly powerful in terms of polymorphisms identified per reaction. Most recently, systems that detect single base pair changes, termed Single Nucleotide Polymorphisms (SNP), are becoming available. While fairly expensive to develop and requiring sequencing of several alleles, they do detect high levels of polymorphism and can be detected with simple and automated technology.

Molecular Genetics of Drought Tolerance in Maize

Maize was one of the first major crop species for which, a complete molecular marker map was developed (Helentjaris *et al.*, 1986). Since the first publications, many other maps have been produced which are now consolidated into a consensus map by using a 'bin' allocation to chromosome segment (Gardiner *et al.*, 1993). Given the high level of polymorphism found even between highly related lines, this consensus map allows one to rapidly identify possible markers for use in further saturating a region of interest, or for developing alternative (e.g., PCR-based) marker systems. Efforts are underway to develop saturated microsatellite marker maps, and maps composed of AFLP loci are available for a number of maize populations. Most recently, efforts are focused on sequencing alleles at numerous loci to develop the information necessary for SNP analyses.

One of the most studied traits at CIMMYT is abiotic stress tolerance, especially tolerance to water-limited conditions. Plants vary tremendously in their ability to withstand abiotic stresses, both, between species and within populations of a single species. Abiotic stresses limit crop productivity in every season and in every crop worldwide, yet the nature of tolerance is not well characterized. Understanding the mechanisms of tolerance will have a significant impact on crop productivity (Boyer 1982). In general, tolerance to abiotic stresses is associated with a host of morphological and physiological traits. These include, root morphology and depth, plant architecture, variation in leaf cuticle thickness, stomatal regulation, osmotic adjustment, antioxidant capacity, hormonal regulation, desiccation tolerance (membrane and protein stability), maintenance of photosynthesis, and the timing of events during reproduction (Bohnert *et al.*, 1995; Shinozaki and Yamaguchi-Shinozaki, 1996; Bray 1997; Nguyen *et al.*, 1997; Ribaut *et al.*, 2001). The complexity of these responses is not surprising, because plants must be able to tolerate significant variations in temperature and water potential during development. One of the most challenging traits for which to breed among the abiotic stresses, is drought tolerance, due to a large part in its unpredictable nature. Selected materials need to perform well under both, water-limited and well-watered conditions. In addition, establishing optimal environments to select for improved performance under drought is complicated by environmental variation. Furthermore, selection is slowed because there is generally only one dry cycle per year per location.

Loss to drought in the tropics alone is thought to exceed 20 million tons of grain per year, or, around 17 % of well-watered production (Edmeades *et al.*, 1998), reaching up to 60 % in severely affected regions, such as, southern Africa in 1991-92. The magnitude of these losses has made breeding for drought tolerance a major focus of CIMMYT (for review, see Heisey and Edmeades, 1999). Although, impressive progress has been achieved through conventional breeding, it should be kept in mind, that conventional breeding for yield improvement remains time consuming and laborious, since carefully managed field conditions are required. In addition, there is a decrease in the genetic variance and heritability of yield components that parallels an increase in environmental stress (Ribaut *et al.*, 1997b). Considering these limitations to efficient selection, and that only one relatively rain-free crop season per year is available for selection in most tropical countries, the use of molecular genomics could provide a useful tool to complement phenotypic selection.

The construction of QTL linkage maps using segregating populations is routine at CIMMYT. During the past seven years, major efforts have been dedicated to the genetic dissection of drought tolerance components in maize under water-stress, before, during, and after flowering. These efforts have resulted in the identification of QTLs involved in the expression of yield components and secondary morphological traits of interest, such as ASI. As presented earlier, of primary interest are secondary traits that are correlated with yield, and demonstrate segregation with high heritability under water-limited conditions. To date, genetic dissection has been conducted in four different crosses, under different water regimes (WW: well-watered conditions, IS: Intermediate Stress, and SS: Severe Stress conditions) and in several environments (Kenya, Mexico, and/or Zimbabwe).

Much of the past QTL identification has focused on yield components and secondary traits of interest, each important measures for drought tolerance in maize but complex polygenic traits. To more deeply explore, at the genetic level, the maize plant's response under water-limited conditions, it is necessary to identify the QTL involved in the differential expression of the key physiological pathways that induce the drought tolerance phenotype. To achieve this objective, a recombinant inbred line (RIL) population was developed by single seed descent from F_3 families obtained by crossing Ac7643 with Ac7729/TZSRW. The same morphological traits measured with the F_3 families have also been evaluated on this RIL population. RIL families are more suitable than F_3 families for physiological measurements because they are genetically fixed; on the other hand, they are poor material for evaluation of yield components since they usually demonstrate high inbreeding depression. In addition to the physiological parameters measured in-house, such as relative water content, osmotic adjustment, and chlorophyll content, collaborations with other research groups have allowed the evaluation of root growth under hydroponics (Roberto Tuberosa, Bologna University), quantify the ABA content in the ear at the flowering stage (Tim Setter, Cornell University), and evaluate the photosynthetic apparatus and dehydration phenomena under low temperature conditions (Yvan Fracheboud, Institute of Plant Sciences ETH, Zurich). Identification at the same genomic location of QTL related to physiological and morphological traits should be expected, given that changes in physiological pathways have an impact on the plant phenotype. As an example from the first field evaluation, a QTL for chlorophyll content was identified on chromosome 2, near a QTL for ASI (under IS and SS) and grain yield (under IS only). This QTL for chlorophyll content was consistent when measurements were conducted on the ear leaf, and on a young leaf close to the tassel. On chromosome 6, a QTL for relative water content corresponds exactly to a QTL for ASI (under IS and SS) and grain yield (under IS and SS). At the same chromosomic region, the identification of a dehydrin gene (*dhn*1) has also been reported (Campbell and Close, 1997). Since, several physiological pathways involved in the drought tolerance mechanism are well known (e.g., ABA biosynthesis), the characterization of the gene(s) corresponding to identified QTL can be achieved, making the candidate gene approach an attractive option.

Based on the QTL and mapping information, a backcross marker-assisted selection (BC-MAS) project was initiated in 1994. The line P_1 (Ac7643) was used as the drought-tolerant donor line and CML247 was used as the recurrent parent. CML247 is an elite tropical inbred line developed by CIMMYT, with outstanding combining ability and good yield per se under well-watered conditions. It is susceptible to drought, in part, because its ASI is large under drought. Genetic data from a segregating F_2 population derived from the $P_1 \times$ CML247 cross, were combined with F_3 evaluations in the field under different water regimes, to identify QTLs for traits of interest. The QTLs for ASI identified in this cross were quite consistent with those in the original $P_1 \times P_2$ cross. Of the five QTLs originally identified from P1 that conferred a short ASI, only the QTL on chromosome 6 was not detected in the second cross. The QTL on the short arm of chromosome 1 was shifted by 40 cM in the new cross, and the three other QTLs on chromosomes 2, 8, and 10 were in similar positions in both. A new QTL for ASI was detected on the short arm of chromosome 3. These results demonstrate the need to make a new genetic map when the recurrent line is changed in BC-MAS schemes. A single good-quality trial under drought conditions, however, might be sufficient for identifying QTLs of interest, provided, QTL identification has been previously carried out in another cross involving the donor line.

Five genomic regions involved in the expression of a short ASI were selected to be transferred from P_1 into the CML247 genome. The screening of large populations (about 2,000 plants) at each selection cycle during backcrossing, has been possible because of the development of reliable PCR-based markers, used here as preselection tools (Ribaut *et al.*, 1997a). After two BCs and two self-pollinations, the best genotype was fixed from the donor line for the five target regions (12 % of the genome), as well as, for 7 % of the genome outside the QTL regions (Fig. 2). The 70 best BC_2F_3 (i.e., S_2 lines) were identified and crossed with two CIMMYT tester inbreds, CML254 and CML274. These hybrids, as well as, the BC_2F_4 families (S_3 lines) derived from the selected BC_2F_3 plants, were evaluated in 1998, 1999 and 2000, under several water regimes. Results show that

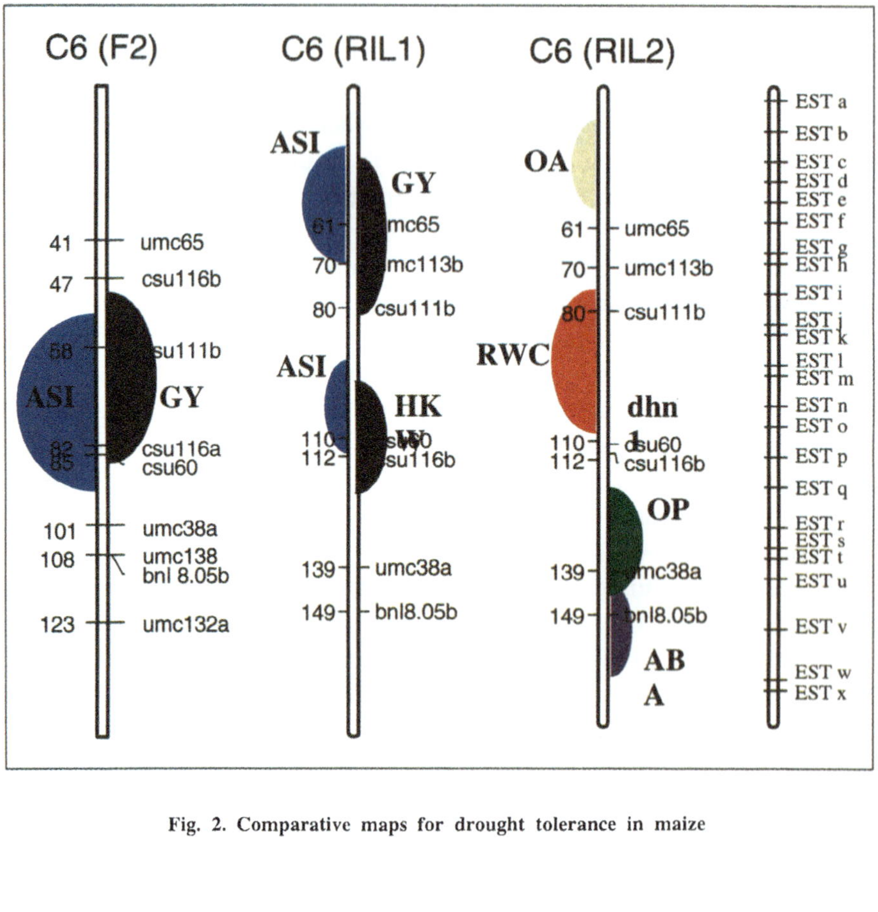

Fig. 2. Comparative maps for drought tolerance in maize

under stress conditions that induce a yield reduction of at least 80 %, the mean of the 70 selected genotypes performed better than the control crossed with CML254 and CML274. In addition, the best genotype among the 70 selected, ($BC_2F_3 \times$ testers) performed two to four times better than the control ones. This difference became less marked when the intensity of stress decreased; for a stress inducing less than 40 % yield reduction, hybrids resulting from the MAS or developed with the "original" version of CML247 performed the same. Although which genotypes performed best depended on the stress intensity, few of the genotypes always performed significantly better than the controls across the six water-limited trials. No yield reduction was observed under well-watered conditions.

Through the recent development of genomic technologies that provide structural and functional information (Habben *et al.*, 1999), gene characterization (i.e., the localization, sequence, and expression framework of a gene) has received a significant boost during the last few years. To date, if one tries to establish a list of candidate genes for drought tolerance based on gene function, hundreds of genes can easily be listed (Skriver and Mundy, 1990; Bray, 1993; Ingram and Bartels, 1996; Cushman and Bohnert, 2000). The questions that now arise are, how to prioritize research aimed at characterizing the genes involved in the drought-tolerance process, and once those genes are characterized, how to identify and efficiently manipulate the elite alleles at those target loci, to improve a given variety. The first question must be addressed principally by the research groups conducting basic genomic research. Of course, establishing such priorities is more or less a function of the available resources and research objectives of a group. The recent discovery of promoter regulatory elements, like DRE (dehydration-responsive element) or ABRE (ABA-responsive element) involved in both dehydration- and low-temperature-induced gene expression in *Arabidopsis* (Shinozaki and Yamaguchi-Shinozaki, 1996), as well as, the identification of transcriptional factors interacting with those promoters (Liu *et al.*, 1998), are exciting developments. The characterization of the genes involved in the initiation phase of the stress response (e.g., genes encoding for stress-induced transcription factors) should be a logical priority, since they represent the "upstream keys" to global genomic responses that might involve hundreds of genes. Moreover, once they have been identified, expression of these key genes should serve as a "timing reference", to identify expression products from downstream genes involved in stress responses. This can be achieved using microarray technology, as described by Chu (1998).

Recent work has shown how different genes can provide new clues to understanding stress tolerance in plants (Shinozaki and Yamaguchi-Shimozaki, 1996). They actively play a role in the biosynthesis of various osmoprotectants (Tarczynski and Bohnert 1993), modification of cell membrane structure (Kodama *et al.*, 1994), and sometimes, code for enzymes that detoxify plant cells. Analyses of the expression of dehydration-induced genes have shown, that at least four independent signal pathways function in response to dehydration. Two are abscisic acid (ABA) dependent (Abe *et al.*, 1997, Chandler and Robertson 1994), and two are ABA-independent. Several stress- induced genes, such as *rd29A*, are induced through the ABA-independent pathway. The *rd29A* gene in particular is responsible for dehydration and cold-induced expression (Yamaguchi-Shinozaki and Shinozaki, 1994). Expression of the *DREB1A* gene, under constitutive promoters such as CaMV 35S, leads to strong increases in tolerance to abiotic stresses; however, it also induces growth retardation under normal growth conditions (Liu *et al.*, 1998), raising concerns about its possible use in plant breeding. However, plants produced using the stress regulated *rd29A* promoter, demonstrated increased tolerance to freezing, salt and water limited stresses, without producing changes in the normal phenotype of the transformed plants. One way to validate these candidate genes, is to introduce them into a genotype via genetic engineering and test if the plant's tolerance to water-stress has been improved. This approach has been followed at CIMMYT, to evaluate and characterize the effect of the *DREB1A* gene under water-limited conditions, when introduced into wheat germplasm.

At the 4-leaf stage, the T_1 plants were subjected to water-limited stress by withholding water. The wheat lines started to show differences in the wilting of the leaves after 10 days without water. Controls started to show water-limited symptoms (loss of turgor and bleaching of the leaves) after 10 days of stress, and severe symptoms (e.g., senescence of all leaf tissue) were evident in the control samples after 15 days without water, while the transgenic plants either showed none, or reduced symptoms. These preliminary results suggest, that the *rd26::DREB1A* gene has a general effect on tolerance to water-limited conditions. Whether this tolerance is attributable to reduced evapotranspiration of the transgenic plants or differential regulation of the water status in the cell (e.g., osmotic adjustment), it clearly needs to be determined through future physiological measurements on both, the control and the transgenic plants.

Wheat: Durable Fungal Resistance

The developments in molecular genetics in wheat have been slower, especially when compared to other crops such as maize, rice or tomato. This is due to the wheat's ploidy level, the size and complexity of its genome, the very high percentage of repetitive sequences and the low level of polymorphism. Far less QTL studies have been reported when compared to other grass species. However, due to the large number of disease and pest resistances controlled by major genes, the mapping of such genes has dominated the research activities in wheat molecular genetics. On the other hand, the hexaploid nature of wheat and its amenity to cytogenetic manipulation have offered unique tools for molecular geneticists of wheat. Such maps have corroborated cytological evidence of major chromosome rearrangements (Devos *et al.*, 1995; Nelson *et al.*, 1995a) and have allowed the comparative mapping among related species (e.g., Ahn *et al.*, 1993; Börner *et al.*, 1998; Devos *et al.*, 1994).

The first RFLP maps were reported by Chao *et al.* (1989) for the group 7 homoeologous chromosomes. Since then, maps for each homoelogous chromosome has been published (Devos *et al.*, 1992; Devos *et al.*, 1993; Xie *et al.*, 1993; Devos *et al.*,

1995; Jia *et al.*, 1994). The Norwich wheat RFLP linkage map has also been published (Gale *et al.*, 1995) and altogether contains over 500 loci. Another important mapping population was developed at CIMMYT by crossing a synthetic (amphihexaploid) wheat (*T. tauschii* × Altar84 durum) to a spring bread wheat cultivar 'Opata85', and was genotyped at Cornell University, USA. The use of such a non-intervarietal cross, resulted in a very dense map (about 1000 RFLP loci) due to the higher polymorphism level (Van Deynze *et al.* 1995; Nelson *et al.*, 1995a, 1995b; Nelson *et al.*, 1995c; Nelson *et al.*, 1995a; Marino *et al.*, 1996). Recently, Röder *et al.* (1998) placed 279 SSR loci on the map also refered to as the ITMI (International Triticeae Mapping Initiative) map.

The existence of numerous sets of wheat *near-isogenic lines* (NILs), differing in the presence or absence of a resistance allele has facilitated the mapping of genes for which such lines exist (e.g., Hartl *et al.*, 1993; 1995 for *Pm1, Pm2*, and *Pm3*; Schachermayr *et al.*, 1994, 1995 for *Lr9* and *Lr24*; Demeke *et al.* 1996 for *Bt-10*; Sun *et al.* 1997 for *Yr15*). Dweikat *et al.* (1997) screened a series of NILs in 'Newton' for Hessian fly resistance alleles using 1600 random 10-mer primers. One to three RAPD markers were identified for each of the 11 genes being tagged, and linkage determined by screening F_2 populations segregating for each individual gene. On the other hand, Feuillet *et al.* (1995) screened 'Thatcher' NILs for *Lr1* (on 5DL) with 37 RFLP probes mapping to group 5 chromosomes and found three to be linked to the gene after testing on F_2 populations between Thatcher and *Lr1*/6*Thatcher. The same approach was used by Williams *et al.* (1996) who found two RFLP markers flanking the *Cre* (*Cre1*) gene on the long arm of 2B.

Bulk segregant analysis (BSA) developed by Michelmore *et al.* (1991) to tag disease resistance genes in lettuce, has been successfully applied in wheat. This approach has been mostly used with RAPDs (e.g., Hartl *et al.*, 1995 for *Pm1* and *Pm2*; Hu *et al.*, 1997 for *Pm1*) although, it is now being used with AFLPs (Goodwin *et al.* 1998; Hartl *et al.*, 1998). Either marker technique is used to screen two bulks of DNA samples from individuals identified in the two opposite tails of a segregating population for a target trait. For a major gene, all loci in the genome should appear to be in linkage equilibrium, except in the region of the genome linked to the target gene. To overcome the problems of limited repeatability of RAPDs, and the fact that repetitive sequences are often amplified (Devos and Gale 1992), Eastwood *et al.* (1994) and William *et al.* (1997) used BSA and RAPDs on DNA enriched for low-copy sequences. In both cases, there was a noted increase in repeatability and levels of polymorphism detected compared with non-enriched DNA. The AFLP technology offers the advantage of the high number of DNA fragments amplified with one primer combination, and the problem of highly repetitive DNA is overcome by using methylation sensitive endonucleases, such as, *Pst*I and *Sse*I.

At CIMMYT, our efforts in breeding for disease resistance in general, and leaf rust resistance in particular, have focused on the use of durable resistance (Van Ginkel and Rajaram, 1993). Such resistance is controlled by a number of minor genes also referred to as adult plant resistance (APR) genes. In order to determine the number and location of these genes, and find tightly linked markers that will enhance the breeding efforts for such resistance, we have been involved in mapping APR loci in the leaf rust resistant cultivars 'Parula' and 'Frontana'. William *et al.* (1997), used BSA on RILs from a cross between Parula and 'Siete Cerros', to identify three RAPD markers associated with two leaf rust resistance loci. Nullisomic-tetrasomic analysis showed that these are located on 7BL and 1BS or 1DS. We have also constructed a genetic linkage map using RFLP, SSR and AFLP markers in a segregating population of Frontana × 'INIA66' in order to map primarily, the durable leaf rust resistance and other important traits that are segregating in the same population. Although the map now includes about 450 marker loci, some gaps still exist and efforts are focused on filling those with the SSR markers that are becoming available (e.g., Röder *et al.*, 1998). With the current map and using composite interval mapping, we have identified five and seven QTL for leaf rust resistance and BYDV tolerance, respectively (Khairallah *et al.*, 1998).

Rust diseases, specifically leaf rust and yellow rust are globally important foliar fungal diseases in wheat. Enhanced resistance to these fungal pathogens plays a major role in the adaptation of improved wheat cultivars, with high yield potential into different agro-climatic regions. *Fusarium* head blight is another fungal disease of global significance. In contrast to the rich genetic resources available for breeding better varieties, molecular genetic tools and a knowledge base for understanding factors conditioning disease resistance, and tools for manipulation of such resistances using biotechnology approaches are lacking in wheat compared to other cereals of global importance. Characterization of genes conditioning durable resistance to rust diseases and *fusarium* head blight remained the focus of activities during the current year.

Although 10-12 slow rusting genes are known to be present in the CIMMYT spring wheat germplasm, only two such genes, namely *Lr34* and *Lr46* have been designated. We have been able to further characterize *Lr46* and have been able to pin point genomic location using, bulked segregant analysis and partial linkage mapping, with the aid of existing linkage maps. It has also been demonstrated that as in the case of *Lr34/Yr18/Bdv1* complex, this is also a complex of genes that have effects on at least leaf rust and yellow rust commonly. Further, markers have also been identified for *Lr34/Yr18* complex (Figs. 3 and 4).

We have been able to characterize several loci conditioning durable resistance in improved CIMMYT wheats such as Pavon 76, Parula and Tonichi. New gene designations have been obtained for two such genes and designate as *Yr29* in chromosome 1BL and *Yr30* in chromsome 3BS. Several quantitative loci conditioning resistance to *fusarium* head blight have also been identified.

We are using candidate gene approaches, using resistance like sequences isolated from rice and maize to find possible homologies with genes conditioning resistance in wheat. We believe that better characterization of gene complexes in wheat that would confer durable resistance to rust diseases would help in better understanding of the functional aspects of disease resistance, not only to the rusts but likely for other pathogens as well, since these genes are known to confer resistance to multiple diseases. We are exploring the possibility of utilizing markers identified for *Lr46/Yr29* and *Lr34/Yr18* in applications in the wheat breeding activities (Figs. 3 and 4).

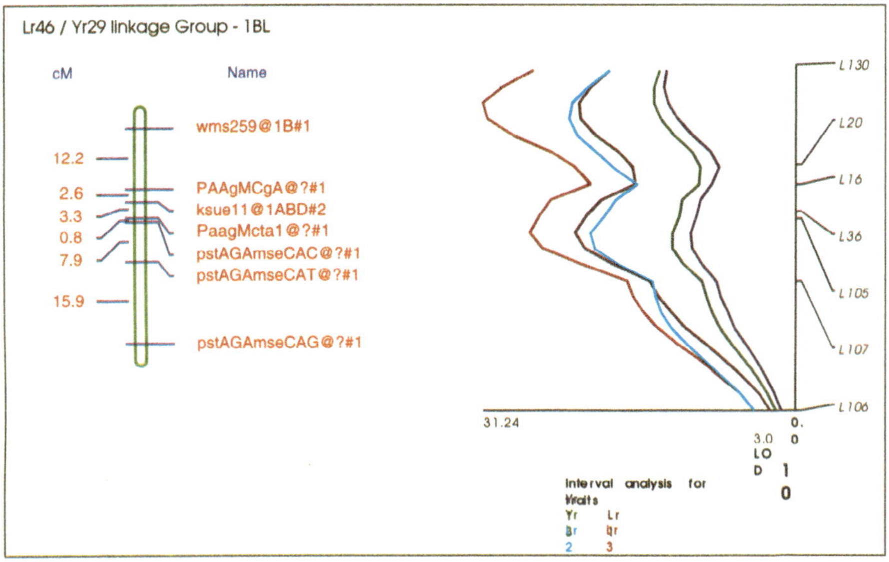

Fig. 3. Linkage analysis of *Lr46/Yr29* in wheat

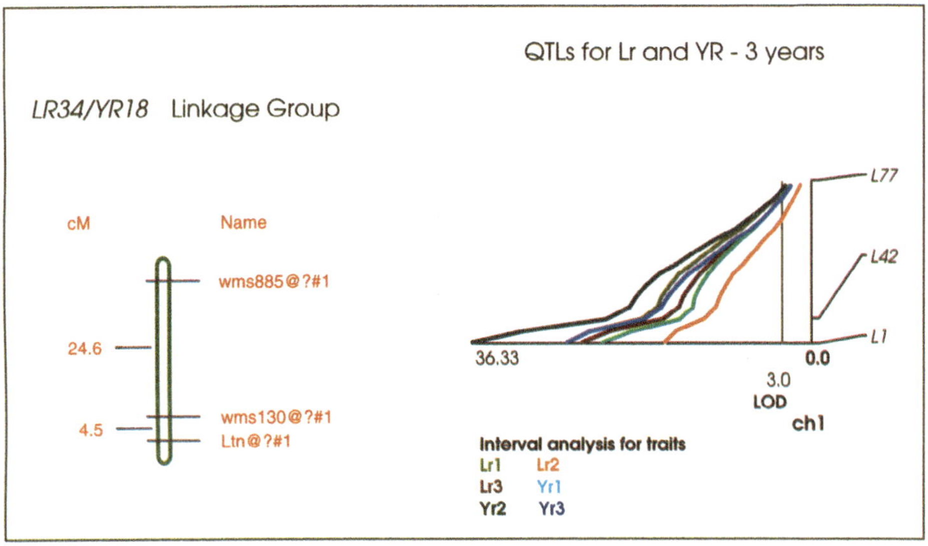

Fig. 4. Linkage analysis of *Lr34/Yr18* in wheat

The Future Application of Biotechnology in Plant Improvement

What is in hold for the future? While always difficult to predict, there are some significant developments in marker technologies and functional genomics worth mentioning. While the PCR-based marker systems have allowed more effective and efficient genotyping, DNA-array technology offers to substantially increase the number of genes that can be analysed (Shalon, 1995; Schena *et al.*, 1995; Shalon *et al.*, 1996). Currently, the cost of the arrayer (to develop the chips containing the desired genes), the array reader (to detect the presence of each gene) and a set of gene sequences (to develop primers to be arrayed), have limited the application of this new technology. Both the arrayer and reader are decreasing in price and this will make this technology available to many laboratories in the near future. In addition, efforts are underway to develop complete EST databases for many cereals including maize and wheat. If this data can remain in the public sector, chips containing a significant number of cereal genes will be produced and used in the not too' distant future.

Marker-assisted selection for polygenic trait improvement is in an important transition phase, and the field is on the verge of producing convincing results. Considering the potential for the development of new strategies (Ribaut and Hoisington, 1998), the future for polygenic trait improvement through DNA markers and the contribution of this to plant breeding efforts worldwide appear bright.

References

Abe H., Yamaguchi-Shinozaki K., Urao T., Iwasaki T., Hosokawa D. and Shinozaki K. 1997. Role of Arabidopsis MYC and MYB homologs in drought- and abscisic acid-regulated gene expression. *Plant Cell*, **9**: 1859-1868.

Ahn S., Anderson J.A., Sorrells M.E. and Tanksley S.D. 1993. Homoeologous relationships of rice, wheat and maize chromosomes. *Mol. Gen. Genet.*, **241**: 483-490.

Bohnert H.J., Nelson D.E. and Jensen R.G. 1995. Adaptations to environmental stresses. *Plant Cell*, **7**: 1099-1111.

Börner A., Korzun V. and Worland A.J. 1998. Comparative genetic mapping of loci affecting plant height and development in cereals. *Euphytica*, **100**: 245-248.

Boyer J.S. 1982. Plant productivity and environment. *Science*, **218**: 443-448.

Bray E.A. 1993. Molecular Responses to Water Deficit. *Plant Physiologist*, **103**: 1035-1040.

Bray E.A. 1997. Plant responses to water deficit. *Trends in Plant Science*, **2**: 48-54.

Campbell S.A. and Close T.J. 1997. Dehydrins: genes, proteins and associations with phenotypic traits. *New Phytol.*, **137**: 61-74.

Chandler P.M. and Robertson M. 1994. A dehydrin cognate protein from pea (*Pisum sativum* L.) with an atypical pattern of expression. *Plant Mol. Biol.*, **26**: 805-816.

Chao S.P., Sharp P.J., Worland A.J., Warham E.J., Koebner R.M.D. and Gale M.D. 1989. RFLP-based genetic maps of wheat homoeologous group 7 chromosomes. *Theor. Appl. Genet.*, **78**: 495-504.

Chu S., DeRisi J., Eisen M., Jutholland J., Botstein D., Brown P.O. and Herskowitz I. 1998. The transcriptional program of sporulation in budding yeast. *Science*, **282**: 699-705.

Cushman J.C. and Bohnert H.J. 2000. Genomic approaches to plant stress tolerance. *Current Opinions in Plant Biol.*, **3**: 117-124.

Demeke T., Laroche A. and Gaudet D.A. 1996. A DNA marker for the BT-10 common bunt resistance gene in wheat. *Genome*, **39**: 51-55.

Devos K.M. and Gale M.D. 1992. The use of randomly amplified DNA markers in wheat. *Theor. Appl. Genet.*, **84**: 567-572.

Devos K.M., Atkinson M.D., Chinoy C.N., Liu C. and Gale M.D. 1992. RFLP based genetic map of the homeologous group 3 chromosomes of wheat and rye. *Theor. Appl. Genet.*, **83**: 931-939.

Devos K.M., Chao S.P., Li Q.Y., Simonetti M.C. and Gale M.D. 1994. Relationships between chromosome 9 of maize and wheat homeologous group 7 chromosomes. *Genetics*, **138**: 1287-1292.

Devos K.M., Dubcovsky J., Dvorák J., Chinoy C.N. and Gale M.D. 1995. Structural evolution of wheat chromosomes 4A, 5A, and 7B and its impact on recombination. *Theor. Appl. Genet.*, **91**: 282-288.

Devos K.M., Miller T.E. and Gale M.D. 1993. Comparative RFLP maps of the homoeologous group 2 chromosomes of wheat, rye and barley. *Theor. Appl. Genet.*, **85**: 784-792.

Dweikat I., Ohm H., Patterson F. and Cambron S. 1997. Identification of RAPD markers for 11 hessian fly resistance genes in wheat. *Theor. Appl. Genet.*, **94**: 419-423.

Eastwood R.F., Lagudah E.S. and Appels R. 1994. A direct search for DNA sequences linked to cereal cyst nematode resistance genes in *Triticum tauschii. Genome*, **37**: 311-319.

Edmeades G.O., Bolanos J., Banziger M., Ribaut J.M., White J.W., Reynolds M.P. and Lafitte H.R. 1998. Improving crop yields under water deficits in the tropics. *In*: *Crop Productivity and Sustainability - Shaping the Future.* (eds.) Chopra V.L., Singh R.B. and Varma A., Proceedings of Second International Crop Science Congress. Oxford and IBH, New Delhi, pp. 437-451.

Feuillet C., Messmer M.M., Schachermayr G. and Keller B. 1995. Genetic and physical characterization of the Lr1 leaf rust resistance locus in wheat (*Triticum aestivum* L.). *Mol. and Gen. Genet.*, **248**: 553-562.

Gale, M.D., Atkinson, M.D., Chinoy, C.N. Harcourt, R.L., Jia, J., Li, Q.Y. and Devos, K.M. 1995. Genetic maps of hexaploid wheat. *In*: *Proceeding of the 8th International Wheat Genetics Symposium.* (eds.) Li Z.S. and Xin Z.I., China Agricultural Scientech Press, Beijing, pp. 29-40.

Gardiner J., Melia-Hancock S., Hoisington D.A., Chao S. and Coe E.H. 1993. Development of a core RFLP map in maize using an immortalized-F_2 population. *Genetics*, **134**: 917-930.

Goodwin S.B., Hu X.Y. and Shaner G. 1998. An AFLP marker linked to a gene for resistance to Septoria tritici blotch in wheat. *In*: *Proc. 9th Intl. Wheat Genet. Symp.*, **3**: 108-110.

Habban J., Helentjaris T., Sun Y. and Zinselmeier C. 1999. Utilizing new technologies to investigate drought tolerance in maize: a perspective from industry. *In*: *Molecular Approaches for the Genetic Improvement of Cereals for Stable Production in Water-Limited Environments.* (eds.) Ribaut J.M. and Poland D., A Strategic Planning Workshop held at CIMMYT, El Batan, Mexico, 21-25 June (1999) Mexico D.F.: CIMMYT, pp. 154-155.

Hartl L., Mori S. and Schweizer G. 1998. Identification of a diagnostic molecular marker for the powdery mildew resistance gene Pm4b based on fluorescently labelled AFLPs. *Proc. 9th Intl. Wheat Genet. Symp.*, 111-113.

Hartl L., Weiss H., Stephan U., Zeller F.J. and Jahoor A. 1995. Molecular identification of powdery mildew resistance genes in common wheat (*Triticum aestivum* L.). *Theor. Appl. Genet.*, **90**: 601-606.

Hartl L., Weiss H., Zeller F.J. and Jahoor A. 1993. Use of RFLP markers for the identification of alleles of the Pm3 locus conferring powdery mildew resistance in wheat (*Triticum aestivum* L). *Theor. Appl. Genet.*, **86**: 959-963.

Heisey P.W. and Edmeades G.O. 1999. Maize production in drought-stressed environment: technical options and research resource allocation. *In*: *World Maize Facts and Trends.* CIMMYT 1997/98

Helentjaris T., Weber T. and Wright S. 1986. Use of monosomics to map cloned DNA fragments in maize. *Proc. Natl. Acad. Sci.*, **83**: 6035-6039.

Hu X.Y., Ohm H.W. and Dweikat I. 1997. Identification of RAPD markers linked to the gene Pm1 for resistance to powdery mildew in wheat. *Theor. Appl. Genet.*, **94**: 832-840.

Ingram J. and Bartels D. 1996. The molecular basis of dehydration tolerance in plants. *Annual Review Plant Physiology and Plant Molecular Biology*, **47**: 377-403.

Jia J., Devos K.M., Chao S.P., Miller T.E., Reader S.M. and Gale M.D. 1994. RFLP-based maps of the homoeologous group-6 chromosomes of wheat and their application in the tagging of *Pm12*, a powdery mildew resistance gene transferred from Aegilops speltoides to wheat. *Theor. Appl. Genet.*, **92**: 559-565.

Khairallah M., Guillén-Andrade H., Alarcon J., Rodriguez C., Ayala L., Henry M., Singh R.P., Jiang C., Sharp P. and Hoisington D. 1998. Mapping of durable resistance to leaf rust and tolerance to BYDV in wheat. *In*: *Proc. 9th Intl. Wheat Genet. Symp.*, **3**: 282-284.

Kodama H., Hamada T., Horiguchi G., Nishimura M. and Iba K. 1994. Genetic enhancement of cold tolerance by expression of a gene for chloroplast w-3 fatty acid desaturase in transgenic tobacco. *Plant Physiology*, **105**: 601-605.

Liu Q., Kasuga M., Sakuma Y., Abe H., Miura S., Yamaguchi-Shinozaki K. and Shinozaki K. 1998. Two transcription factors, DREB1 and DREB2, with an EREBP/AP2 DNA binding domain separate two cellular signal transduction pathways in drought- and low-temperature-responsive gene expression, respectively, in Arabidopsis. *Plant Cell*, **10**: 1391-1406.

Marino C.L., Nelson J.C., Lu Y.H., Sorrels M.E., Leroy P., Lopes C.R. and Hart G.E. 1996. RFLP-based linkage maps of the homeologous group 6 chromosomes of hexaploid wheat (*Triticum aestivum* L. em. Thell.). *Genome*, **39**: 359-366.

Michelmore R.W., Paran I. and Kesseli R.V. 1991. Identification of markers linked to disease-resistance genes by bulked segregant analysis: A rapid method to detect markers in specific genomic regions by using segregating populations. *Proc. Natl. Acad. Sci.*, **88**: 9828-9832.

Nelson J.C., Sorrells M.E., Van Deynze A.E., Lu Y.H., Atkinson M., Bernard M., Leroy P., Faris J.D. and Anderson J.A. 1995a. Molecular mapping of wheat: major genes and rearrangements in homoeologous groups 4, 5, and 7. *Genetics*, **141**: 721-731.

Nelson J.C., Van Deynze A.E., Autrique E., Sorrells M.E., Lu Y.H., Merlino M., Atkinson M.D. and Leroy P. 1995b. Molecular mapping of wheat-homoeologous group 2. *Genome*, **38**: 516-524.

Nelson J.C., Van Deynze A.E., Autrique E., Sorrells M.E., Lu Y.H., Negre S., Bernard M. and Leroy P. 1995c. Molecular mapping of wheat-homoeologous group 3. *Genome*, **38**: 525-533.

Nguyen H.T., Babu R.C. and Blum A. 1997. Breeding for drought resistance in rice: physiology and molecular genetics consideration. *Crop Sci.*, **37**: 1426-1434.

Ribaut J.M. and Hoisington D. 1998. Marker-assisted selection: new tools and strategies. *Trends in Pl. Sci.*, **3**: 236-239.

Ribaut J.M., Edmeades G.O., Perotti E. and Hoisington D. 2001. QTL analyses, MAS results, and perspectives for drought-tolerance improvement in tropical maize. *In*: *Molecular Approaches for the Genetic Improvement of Cereals for Stable Production in Water-Limited Environments.* (eds.) Ribaut J.M. and Poland D., A Strategic Planning Workshop held at CIMMYT, El Batan, Mexico, 21-25 June (1999) Mexico D.F.: CIMMYT, pp. 131-136.

Ribaut J.M., Hu X.Y., Hoisington D. and Gonzalez-de-Leon D. 1997a. Use of STSs and SSRs as rapid and reliable preselection tools in a marker-assisted selection-backcross scheme. *Pl. Mol. Biology Reporter*, **15**: 154-162.

Ribaut J.M., Jiang C., González-de-León D., Edmeades G.O. and Hoisington D.A. 1997b. Identification of quantitative trait loci under drought conditions in tropical maize .I. Yield components and marker-assisted selection strategies. *Theor. Appl. Genet.*, **94**: 887-896.

Röder M.S., Korzun V., Wendehake K., Plaschke J., Tixier M.H., Leroy P. and Ganal M.W. 1998. A microsatellite map of wheat. *Genetics*, **149**: 1-17.

Schachermayr G.M., Messmer M.M., Feuillet C., Winzeler H., Winzeler M. and Keller B. 1995. Identification of molecular markers linked to the Agropyron elongatum-derived leaf rust resistance gene *Lr24* in wheat. *Theor. Appl. Genet.*, **90**: 982-990.

Schachermayr G.M., Siedler H., Gale M.D., Winzeler H., Winzeler M. and Keller B. 1994. Identification and localization of molecular markers linked to the *Lr9* leaf rust resistance gene of wheat. *Theor. Appl. Genet.*, **88**: 110-115.

Schena M., Shalon D., Davis R.W. and Brown P.O. 1995. Quantitative monitoring of gene expression patterns with a complementary DNA microarray. *Science*, **270**: 467-470.

Shalon D. 1995. "DNA microarrays: A new tool for genetic analysis." Ph.D thesis, Stanford University, Stanford, CA.

Shalon D., Smith S.J. and Brown P.O. 1996. A DNA microarray system for analyzing complex DNA samples using two-color fluorescent probe hybridization. *Genome Methods*, **6**: 639-645.

Shinozaki K. and Yamaguchi-Shinozaki K. 1996. Molecular responses to drought and cold stress. *Current Opinions in Biotech.*, **7**: 161-167.

Skriver K. and Mundy J. 1990. Gene expression in response to abscisic acid and osmotic stress. *Plant Cell*, **2**: 503-512.

Sun G.L., Fahima T., Korol A.B., Turpeinen T., Grama A., Ronin Y.I. and Nevo E. 1997. Identification of molecular markers linked to the *yr15* stripe rust resistance gene of wheat originated in wild emmer wheat, Triticum dicoccoides. *Theor. Appl. Genet.*, **95**: 622-628.

Tarczynski M. and Bohnert H. 1993. Stress protection of transgenic tobacco by production of the osmolyte mannitol. *Science*, **259**: 508-510.

Van Deynze A.E., Dubcovsky J., Gill K.S., Nelson J.C., Sorrells M.E., Dvorák J., Gill B.S., Lagudah E.S., McCouch S.R. and Appels R. 1995. Molecular-genetic maps for group 1 chromosomes of Triticeae species and their relation to chromosomes in rice and oat. *Genome*, **38**: 45-59.

Van Ginkel M. and Rajaram S. 1993. Breeding for durable disease resistance in wheat: an international perspective. *In*: *Durability of Disease Resistance*. (eds.) Jacobs T.H. and Parlevliet J.E., Kluwer Academic Press. Dordrecht, the Netherlands, pp. 259-272.

William H.M., Hoisington D., Singh R.P. and González-de-León D. 1997. Detection of quantitative trait loci associated with leaf rust resistance in bread wheat. *Genome*, **40**: 253-260.

Williams K.J., Fisher J.M. and Langridge P. 1996. Development of a PCR-based allele-specific assay from an RFLP probe linked to resistance to cereal cyst nematode in wheat. *Genome*, **39**: 798-801.

Xie D.X., Devos K.M., Moore G. and Gale M.D. 1993. RFLP-based genetic maps of the homoeologous group 5 chromosomes of bread wheat (*Triticum aestivum* L.). *Theor. Appl. Genet.*, **87**: 70-74.

Yamaguchi-Shinozaki K. and Shinozaki K. 1994. A novel cis-acting element in an Arabidopsis gene is involved in responsiveness to drought, low-temperature or high-salt stress. *Plant Cell*, **6**: 251-264.

Plant Breeding - Mendelian to Molecular Approaches
H. K. Jain and M. C. Kharkwal (eds.)

14

Transgenics in Crop Improvement

Swapan K. Datta[1], Niranjan Baisakh[1], Vai Ramanathan[2] and K.K. Narayanan[2]

Abstract

Agricultural biotechnology is no longer a dream but a reality that will dominate the 21st century for agriculture and human well-being. Plant genetic engineering has revolutionized and supplemented conventional plant breeding in crop improvement. Apart from broadening the genetic base, transformation technology has helped improve crop productivity, plant protection, and nutrition. This review summarizes some significant developments and achievements in the field of gene technology that accelerate designing the rice crop to meet the needs of consumers, farmers, industrialists, and other end users.

Introduction

Ever since the advent of agriculture, there has been a constant need to improve crop plants for higher productivity and quality and to satisfy changing human preferences. This need is felt more acutely today, particularly in the developing world, where the population continues to increase. Farmers in South and South-East Asia must produce an extra 6.7 million tons of unmilled rice every year, without fail, just to maintain the current level of nutrition (Hossain *et al.*, 2000); and this task does not become any easier with diminishing land and water resources.

Genetic modification of plants probably began through the selection of better types about 10,000 years ago when human agricultural activities began and useful results were often a product of random or chance events. Following the laws of genetics, plant breeding became a deliberate and predictable activity. Traditional plant breeding methods have been very successful and have helped provide the volume of food required to allow the world population to grow to its current six billion. Breeding efforts have provided us with remarkable diversity among crop species and even some new ones such as triticale. However, recent trends in crop productivity indicate that traditional methods alone will not allow us to keep pace with the growing demand for food, fiber, and fuel. Yield increases have reached a plateau or have fallen below the rate of population increase for many food crops. In India, the yield of the major staple, rice, increased 3-5 per cent annually during the Green Revolution years, but in the latter half of the 1990s annual

[1]International Rice Research Institute, Metro Manila, Philippines
[2]Metahelix Life Sciences, Bangalore 562 153, India

growth in rice production was a mere 1.5 per cent (monthly review of the Indian economy reported in *The Hindu*, 26 January 2000). The country's population during the same period grew at around 1.9 per cent annually. Biotechnology can bolster plant breeding efforts to meet these new challenges in a sustainable way.

Reproductive barriers often limit conventional plant breeding. Developments in plant biology in the last three decades (Goldberg, 2001), such as plant genetic transformation, have opened up new vistas in crop improvement, thereby allowing the transfer of desirable gene(s) across species and genera (overcoming crossability barriers that limit the scope of conventional breeding), to develop transgenic species with novel traits such as in-built tolerance to several biotic and abiotic stresses, improved nutritional quality, enhanced grain filling, etc. Moreover, advances in genetic transformation techniques provide plant breeders with access to new and broader gene pools. Transgenic plants can be considered as the most recent development in our efforts to genetically improve crops.

Genetic Transformation of Plants

Genetic engineering of plants has lagged almost a full decade behind the genetic engineering of microorganisms because of the difficulty in transferring foreign DNA to plant cells and the longer duration of crop plants. Plant transformation was first demonstrated independently in 1983 by research groups in Gent, Belgium, in St Louis, Missouri (USA), and by collaborating groups from Washington University, St Louis, and Cambridge University, UK (Bevan *et al.*, 1983; Herrera-Estrella *et al.*, 1983; Fraley *et al.*, 1983). All three groups transferred and expressed bacterial antibiotic resistance genes using the *Agrobacterium*-mediated method. However, plant transformation became routine in the 1990s, a full decade after genetically engineered human insulin went on sale. After the first report of gene transfer with seed protein phaseolin from beans to sunflower and tobacco (Murai *et al.*, 1983) and a bacterial gene for neomycin phosphotransferase II (*NPT II*) to tobacco a year later (Horsch *et al.*, 1984), plants have been transformed with genes across species and genera, including bacteria, viruses, and animals. With dramatic progress in the improvement of transformation techniques, more than 50 different species of transgenic plants have been produced from both monocots and dicots and some (including rice in China) are under field experiments worldwide (James, 2000).

Many transformation approaches have been tested in the past for their comparative efficiency, including the *Agrobacterium tumefaciens*-mediated transformation system and direct gene transfer system, that is protoplast- and biolistic-mediated transformation methods (Potrykus, 1990a & b; Datta *et al.*, 1990; Fig. 1: Datta *et al.*, 1997; Datta, 1999; Datta and Datta, 2001).

Among the methods available, the *Agrobacterium*-mediated method and the biolistic method are the most widely used.

Agrobacterium tumefaciens is a soil-borne Gram-negative bacterium that is capable of genetically colonizing susceptible host plants. It is capable of transferring any piece of

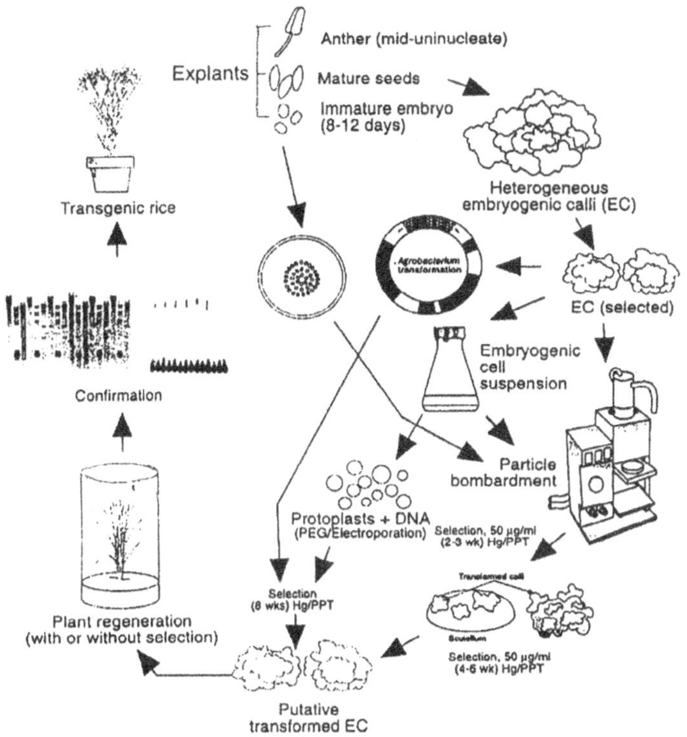

Fig. 1. A schematic presentation of the development of transgenic indica rice using biolistic, *Agrobacterium*, and protoplast methods (Datta *et al.*, 1997)

DNA placed in its T-DNA between a pair of direct repeats called border sequences with the help of a site-specific, strand-specific endonuclease. This feature has been extensively exploited in the genetic transformation of plants. Different strains of *Agrobacterium* have different host ranges and some crop plants, particularly the monocots, are considered recalcitrant to *Agrobacterium* infection. A lot of strategies have been successfully implemented to overcome this recalcitrance. Very often, T-DNA integration occurs in transcriptionally active regions of the plant genome and hence the expression of the transgene becomes a routine phenomenon. Detailed insight into the *Agrobacterium*-mediated DNA transfer process in plant cells is given in the report of Zupan and Zambryski (1995) and Zupan *et al.*, (2000). Several variants of the *Agrobacterium*-mediated transformation protocol have been used to transform *Arabidopsis*, the model plant. Many laboratories routinely transform *Arabidopsis* using the whole-plant or the floral dip method, which are both efficient and easy to practice. A recent modification of the floral dip method, called the floral spray method, might help in expanding this approach to a variety of other plants (Chung *et al.*, 2000).

Biolistic transformation (also referred to as particle gun bombardment or microprojectile bombardment) is carried out by shooting DNA-coated tungsten or gold particles into target tissue (Sanford *et al.*, 1987). The microprojectiles (tungsten/gold particles) can be accelerated with gunpowder, helium, or electric discharge. The advantage of this method is that any tissue can be transformed provided that it can be regenerated through tissue culture into plants. Usually, transformation using this method results in complex patterns of DNA integration compared to T-DNA transfer, which, usually results in precise, low-copy integrations and simple integration patterns (Tinland, 1996; Baisakh *et al.*, 1999, Datta *et al.*, 2000). Co-suppression of the transgene/endogenous gene can occur because of the integration of multiple copies of the transgene (Flavell, 1994). The transfer of long DNA molecules can be a challenge since the molecules can be sheared by the force involved in accelerating the microprojectile, unlike with the *Agrobacterium* method (Hamilton *et al.*, 1996). Generally, the whole plasmid representing the clone of the transgene is bombarded, resulting in the integration of the vector backbone into the plant genome, which is also possible in the *Agrobacterium*-mediated method (Ramanathan and Veluthambi, 1995; Wenck *et al.*, 1997; Baisakh, 2000). Variations of the protocol in which only the transgene is introduced as a linear fragment with bombardment of the minimal expression cassette also exist and they work efficiently (Fu *et al.*, 2000, Datta *et al.*, unpublished). This method comes in handy in transforming plants that are known to be recalcitrant to *Agrobacterium*-mediated transformation. Its utility in transient expression studies is also immense. The method has been quite useful in basic studies on transformation and in modifying gene expression (Napoli *et al.*, 1990). Many commercially available transgenic plants have been developed by biolistic transformation.

Promoters Used in Transgenic Crops

The fate of the introduced gene(s) in the transgenic plant largely depends on the promoter that drives its expression as well as its position in the genome. Promoter sequences upstream of the gene(s) of interest are very important in plant transformation for determining the levels and patterns of the transgene expression. Two major categories of promoters, constitutive and tissue-specific, are used widely.

Constitutive promoters direct the expression of a foreign gene in all plant tissues at all stages of plant development with some variations in the levels across tissues and stages. This group of promoters includes cauliflower mosaic virus 35S (CaMV 35S) promoter (Guilley *et al.*, 1982), rice actin I (*Act*1) promoter (McElroy *et al.*, 1990), maize ubiquitin (*Ubi*1) promoter (Christensen *et al.*, 1992) and maize alcohol dehydrogenase I (*Adh*1, also called *Emu*) promoter (Dennis *et al.*, 1984). Several studies show the non-constitutiveness of the 35S promoter in rice (Benfey and Chua, 1990) and maize *Adh*1 promoter (Zhang and Wu, 1988). A hierarchy of several constitutive promoters was shown on the basis of levels of transient expression of *gus* transgene in rice suspension cell cultures: *Ubi*1 > *Act*1 > *Adh*1 > CaMV 35S (Li *et al.*, 1997). However, such a comparison was artificial without considering several factors and, moreover, CaMV 35S promoter and *Act*1 promoter have been shown to strongly drive the constitutive expression of transgenes in rice (Datta *et al.*, 1990; Lin *et al.*, 1995; Tu *et al.*, 1998a; Datta *et al.*, 1999a & b).

The other group, tissue-specific promoters, drives the spatial and temporal expression of the transgene(s). Such promoters studied and used so far in rice and other monocots include maize phosphoenolpyruvate carboxylase (PEPC) promoter driving green tissue-specific expression (Tada *et al.*, 1991; Datta *et al.*, 1998; Ku *et al.*, 1999), pith-specific promoter (Datta *et al.*, 1998), wound-inducible promoter (Pin 2, Win: Xu *et al.*, 1993), root-specific promoter (*Rcg2*: Kharb *et al.*, 1999), endosperm-specific promoter (*Hordein*: Shewry, 1993; Cho *et al.*, 1999, rice glutelin *Gt1*: Ye *et al.*, 2000; *tubulin*: Lee *et al.*, 1999), pollen specific promoter (Dodds *et al.*, 1999), rice rubisco (*rbcS*) promoter (Kong *et al.*, 1999), stress-inducible promoter (ABA: Jain and Wu, 1999), etc. These promoters are useful for directing the expression of the transgenes only in the tissues where it is required and when it is needed.

The expression of the introduced gene would also vary based on where it gets integrated in the genome. To normalize gene expression and reduce position effect, matrix attachment regions (MARs) have been used in the gene construct for both biolistic and *Agrobacterium*-mediated transformations (Lucca *et al.*, 2001).

Choice of the transformation method depends on various factors such as the purpose, the target plant, equipment availability, and so on. Agrolistics is a transformation procedure recently developed by combining biolistics and the *Agrobacterium* T-DNA processing system (Hansen and Chilton, 1996). The method involving microprojectile bombardment of plant tissues prior to Agro-infection was also found to be efficient in increasing the transformation frequency (Bidney *et al.*, 1992; Datta *et al.*, unpublished data). This combination provides some advantages over both the original methods of transformation in that, by using a non-biological agent, precise, low-copy integration events can be achieved.

Selectable Markers Used in Developing Transgenic Crops

Selection of putative transgenic tissues following transformation irrespective of the method, is the key step for the final recovery of transgenic plants. Dominant selectable markers are an integral part of plant transformation strategies. For this purpose, a selectable marker gene is used, cointegrated in the plasmid fused with the gene of interest, or is harboured on a separate plasmid for cotransformation. Several selective agents and suitable resistance genes have been investigated concurrently with the studies on gene transfer and cell culture.

The most widely used inhibitors are kanamycin, G418, and hygromycin. All of these are aminoglycoside antibiotics that interfere with the translation machinery of prokaryotic and eukaryotic cells. However, all of them can be inactivated by phosphorylation reactions mediated by the products of either the Tn5 neomycin phosphotransferase II (*npt*-II gene, also known as *aph*-II or *neo*) (Herrera-Estrella *et al.*, 1983) or the hygromycin phosphotransferase gene (*hph*, also called *hpt* or *aph*-IV) originally found in *Escherichia coli* (Blochinger and Diddelmann, 1984). The *hph* gene is subsequently modified for plant expression (Waldron *et al.*, 1985) and for further elimination of the internal restriction site to-form cassettes for vector construction (Zheng *et al.*, 1991). Although kanamycin has been used successfully as a selective agent in plant transformation, it has some limitations such as its low efficiency in screening transformed calli and inability of transformed calli to regenerate green plants (Battraw and Hall, 1990). These problems were circumvented by using G418 (Peterhans *et al.*, 1990). Now, the hygromycin B resistance gene is widely used as an efficient selective agent for almost all the transformation methods in several crops, including rice, without any fertility or albino problems (Datta *et al.*, 1990; Lin *et al.*, 1995; Datta *et al.*, 1998; Tu *et al.*, 1998a & b; Baisakh *et al.*, 1999; Datta *et al.*, 1999a & b; Datta *et al.*, 2000).

The bialaphos (also called Basta) resistance gene (*bar*) encodes phosphinothricin acetyltransferase (PAT), and acetylated phosphinothricin is no longer inhibitory to glutamine synthase. Phosphinothricin (PPT) or Basta has been used successfully as a selective agent for several crop plants including rice (Christou *et al.*, 1991; Cao *et al.*, 1992; Datta *et al.*, 1992; Rathore *et al.*, 1993). Hence, the usefulness of a particular resistance marker depends on the characteristics of the selection agent, the resistance gene, and the plant material (Angenon *et al.*, 1994).

The use of all these genes as selectable markers may be perceived to pose a risk for the environmental release of the transgenic products. A recent development is based on the use of selective genes, which give the transformed cells a metabolic advantage over the untransformed cells, which are starved with a concomitant slow reduction in viability. Such a new strategy involves the use of mannose as the selective agent, which, after uptake, is phosphorylated by a hexokinase to an unmetabolized mannose-6-phosphate that accumulates in the cell, resulting in severe growth inhibition (Malca *et al.*, 1967). But, the phosphomannose isomerase gene (*pmi*) allows conversion of mannose-6-phosphate to fructose-6-phosphate, which is readily metabolized. The *pmi* gene as a selectable marker gene has been and is being used for plant transformation including rice (Joersbo *et al.*, 1998; Datta *et al.*, 2000; Lucca *et al.*, 2001).

Similarly, in plant transformation studies, reporter (assessable marker) genes are necessary for rapid detection of DNA introduction. They are usually fused to the plant regulatory sequences *in vitro* and are used to determine when, where, and at what level a regulatory sequence directs gene expression *in vivo*. They can also be used for protein targeting studies if fused to a signal peptide coding sequence. Commonly used reporter genes are the chloramphenicol acetyltransferase gene (CAT) (Knight *et al.*, 1987), the luciferase gene (*LUC*) (Ow *et al.*, 1986), and the β-glucuronidase gene (*GUS*) (Jefferson *et al.*, 1986). Recently, the gene for green fluorescent protein (*gfp*) cloned from the jellyfish *Aequorea victoria* (Chalfie *et al.*, 1994) has also been used in quite a few transformation studies.

Transgenics in Stabilizing Production

A considerable proportion of the crop produce is lost because of biotic and abiotic stresses. Conventional breeding, which has often exploited the natural variability in a species, has produced crop varieties with in-built resistance to several of these stress agents. However, in instances where the natural variability is limited or non-existent altogether, transgenic breeding could be a viable and alternative solution. Transgenic plants that are tolerant to biotic agents such as insect pests and disease agents such as viruses, fungi, and bacteria have been produced, although only insect and virus-resistant transgenic crops have been widely commercialized. Weeds can also significantly reduce crop yields. Transgenic crops with in-built resistance to broad-action weedicides have also been commercialized in many countries. These transgenic crops allow the spraying of the weedicide in a standing crop. The weeds are killed but the crop remains unaffected. This makes weed control more effective and cheaper. It also allows "no-till" cultivation, aiding in soil and water conservation. Abiotic stresses have been more difficult to tackle by transgenic approaches, but some recent developments hold great promise in this area as well.

Insect Resistance

Insect-resistant transgenic crops can be cited as one of the exemplary success stories of agricultural biotechnology. The transgenic *Bacillus thuringiensis* (*Bt*) varieties are in many ways better than *Bt* as a spray formulation. In the *Bt* transgenics, the protein is expressed in all tissues at all times, whereas the effectiveness of the sprays would be affected by a lack of uniform coverage and instability of the *Bt* protein, especially on exposure to sunlight. Considerable progress has been made in developing transgenic crops with resistance to target insect pests over the past decade (Hilder and Boulter, 1999). Though there have been many approaches for incorporating insect resistance in transgenic plants, such plants carrying the insecticidal protein gene from *B. thuringiensis* have by far been the most successful. *Bt* is a soil bacterium that makes crystalline inclusions (cry proteins) during sporulation. These crystals dissolve in the alkaline environment of insect guts and release protoxin molecules that are processed by the gut proteases to give rise to active insecticidal proteins. These proteins interfere with the ion channel pumps and ultimately lead to the death of the insect larva that ingest the crystal. These proteins are quite specific in their host range (determined largely by ligand- receptor interaction) and this fact has been exploited in the development of transgenics tolerant of specific groups of insect pests. More than 50 different cry proteins have been characterized that have different target insect specificity.

The first transgenic tobacco plants with *Bt* were produced in 1987 (Barton *et al.*, 1987; Fischoff *et al.*, 1987; Vaeck *et al.*, 1987). Gene truncation, the use of different promoters, enhancer sequences, and fusion proteins resulted in a significant improvement in the amount of insecticidal protein in the transgenic plants (Perlak *et al.*, 1991). NewLeaf is the brand name of the first *Bt* product to be commercialized in 1995. This is transgenic potato expressing cry3A protein to control colorado potato beetle. The introduction of this product reportedly decreased chemical insecticide use by 40 %. This was followed by the release of pest-resistant transgenic cotton and maize. Subsequently, several *Bt* crops have been released for cultivation and, in 2000, such insect-resistant crops occupied 8.2 million ha globally (James, 2000). It has been estimated that *Bt* cotton alone reduced the use of chemical insecticides in the United States by more than 2 million pounds or nearly one million gallons from 1996 to 1998. The study also found that *Bt* cotton increased yields by 85 million pounds and farmer profits by US$92 million in 1998 (Gianessi and Carpenter, 1999). This is an example of the potential of biotechnology to come up with a solution for combating a problem in a manner that is more environment-friendly. Among the cereals, maize was the first one to be transformed and field-tested with the *Bt* gene and the crop exhibited a high level of resistance to the European corn borer (Koziel *et al.*, 1993). After the first transgenic rice produced in the same year (Fujimoto *et al.*, 1993) with the codon-optimized and truncated *cry* gene, several reports have accumulated recently worldwide about developing transgenic rice carrying a single or fused *cry* genes under different constitutive or tissue-specific promoters that showed 100 percent resistance to stem borers and leaffolder under greenhouse and natural field conditions (Wu *et al.*, 1997; Alam *et al.*, 1998, 1999; Datta *et al.*, 1998; Tu *et al.*, 1998a; Tu *et al.*, 2000a; Ye *et al.*, 2001). Two reports from the International Rice Research Institute, Philippines, are the first-ever reports of transgenic hybrid rice (Shan you 63) as well as an elite *indica* IR72 with fused *Bt* genes that were field-tested (Fig. 2: Tu *et al.*, 2000a; Ye *et al.*, 2001) in China.

There were many challenges to producing the first *Bt* transgenic crops. Because the insecticidal protein gene is bacterial in origin, it is expressed very poorly in plants. Extensive codon optimization was done with many of these native bacterial genes to obtain useful levels of expression in plants. Low expression levels were also addressed by directly transforming the plastids of plants with the gene (De Cosa *et al.*, 2001). Since the plastids have a gene expression machinery similar to that of prokaryotes, the genes could be introduced without extensive modification, and the sheer number of this organelle in a given plant cell would result in very high levels of expression.

However, there are a few concerns regarding the use of *Bt* transgenic crops, the two major ones being: the effect on non-target organisms, and the possibility of the target insects developing resistance to the *Bt* protein. A report in *Nature* (Losey *et al.*, 1999) indicated that monarch butterfly larvae were affected when fed with pollen from *Bt* corn and this was widely and incorrectly interpreted to mean that *Bt* crops threatened non-pest insects. Several follow-up studies showed that the effect of pollen from *Bt* crops was negligible on non-target insects including butterflies under field conditions (Hodgson, 1999). Though *Bt* crops have been under wide cultivation since 1995, there has not been any instance of a pest developing resistance. However, given the experience of the diamondback moth having developed resistance to *Bt* sprays, the development of resistance in the insect cannot be discounted. As a proactive measure, several strategies for insect resistance management have been developed as a package for the cultivation of *Bt* crops. These strategies include refugia (growing a small proportion of the area under a

Fig. 2. Field evaluation of *Bt* hybrid rice in Wuhan province of China against yellow stem borer (Tu *et al.*, 2000a); the line separating the transgenics (with no white head) from the control (with whiteheads, arrow marked)

non-*Bt* crop along with the *Bt* transgenic crop), gene pyramiding, and a high dosage of the protein in the plant to prevent any insects from escaping from the *Bt* field. A recent study (Tabashnik *et al.*, 2000) shows that the natural frequency of the recessive resistance alleles did not increase in spite of the extensive cultivation of *Bt* cotton from 1997 to 1999 and the *Bt* cotton crop remained extremely effective against the pink bollworm.

Besides the *Bt cry* gene, many other candidate genes have been used to develop insect-resistant transgenic plants, such as protease inhibitors (Xu *et al.*, 1996), α-amylase inhibitors (Ishimoto *et al.*, 1996), chitinases (Rigev *et al.*, 1996), lectins (Hilder *et al.*, 1995; Sudhakar *et al.*, 1998), vegetative insecticidal proteins from *Bt* (Estruch *et al.*, 1996), cholesterol oxidases (Shukle and Murdock, 1993; Corbin *et al.*, 1994), and toxins from predators such as mites and scorpions (Barton and Miller, 1991). Recent studies (reviewed in Muthukumar and Narayanan, 1998) show that some of these candidates could be promising and might prove to be an effective alternative to the *Bt* approach.

Virus Resistance

Transgenic crop varieties have been successfully deployed to control viral diseases. One of the classical examples is the success with the genetically engineered papaya, which virtually rescued the papaya industry in Hawaii from the threat of the dreaded ring spot disease (Yeh *et al.*, 1998). The transgenic approach would be more appropriate in situations where sufficient levels of resistance to the virus are not available in the related germplasm or the resistance is difficult to transfer by normal crossing techniques because of either reproductive isolation or linkage with other undesirable traits. The production of virus-tolerant transgenic plants has been based on several approaches. In most instances, a gene coding for the complete viral protein or part of a viral protein has been introduced into the crop by transformation.

Inappropriate expression of a pathogen gene could result in impairment of infection or attenuated symptoms of the disease (Sanford and Johnston, 1985). The first application of this concept was in the development of virus-resistant plants. Many different approaches have been made in an effort to build up resistance to viral pathogens in plants. The underlying mechanisms are different and complex for these approaches and are not clearly understood. Genes encoding structural (coat-protein) and non-structural (replicase, movement proteins) proteins have been effectively used to confer resistance to plants (Beachy *et al.*, 1990; Cooper *et al.*, 1995; Hammond-Kosack and Jones, 1996). Antisense RNA against the coat protein gene and defective interfering transcripts were also shown to be effective in controlling viral infection (Lindbo and Dougherty, 1992; Hammond and Kamo, 1995). Expression of a defective movement protein (TMV) resulted in wide-spectrum resistance to many unrelated viruses (Cooper *et al.*, 1995). Likewise, Tacke *et al.*, (1996) used a defective movement protein from potato leafroll virus (PLRV pr17 which is an open reading frame overlapping the coat protein gene but in a different translation frame) to confer resistance against potato virus X, potato virus Y, and potato leafroll virus in potato plants.

β-1,3-glucanases, known to be involved in plant defense against fungal infection, are induced during the local lesion response to viral infection (Moore and Stone, 1972; Kaufmann *et al.*, 1987; Ward *et al.*, 1991). Their role in viral pathogenesis is not clearly established. Callose is one of the substrates on which the glucanases act. Callose deposition could impede the spread of viral particles. It is tempting to speculate that viruses may be exploiting the host defense against fungi to their advantage. By inducing β-1,3-glucanases, they mediate callose degradation to facilitate their movement. Beffa and Meins (1996) developed transgenic tobacco plants with reduced levels of β-1,3-glucanases by expressing antisense RNA directed against the transcript. These plants exhibited decreased susceptibility to tobacco mosaic virus and tobacco necrosis virus.

Antisense RNA molecules have been successfully deployed in controlling viroid replication in plants. Molecules targeting the minus strand of a viroid seem to be more effective than those against the plus strand (Atkins *et al.*, 1995). Transgenic tomato plants that made antisense RNA against citrus exocortis viroid exhibited a slight reduction in viroid accumulation, whereas antisense RNA against the plus strand actually resulted in increased viroid accumulation.

Ribozymes are short RNA molecules that can catalytically cleave target RNA molecules with high specificity. They have been used extensively in animal cell cultures to suppress gene expression, reduce viral RNA replication, and so on. Their potential has been realized very well *in vitro*. *In vivo*, their ability to perform various functions is being explored. For control of viral replication at the whole-plant level, they hold enormous potential. Edington *et al.* (1993) found that ribozymes targeting the TMV RNA could delay the development of disease symptoms. De Feyter *et al.* (1996), using the same strategy, obtained transgenic plants resistant to TMV infection. These are examples of ribozymes that inhibit viruses with an exclusively cytoplasmic replication cycle. Yang *et al.* (1997) showed that ribozymes could be used effectively to control replication of viroid RNA in the nucleus. Potato spindle tuber viroid has a small RNA genome with a nuclear replication phase. Yang *et al.*, (1997) showed that transgenic potato plants expressing a ribozyme against the PSTVd minus-strand RNA were resistant to the viroid. This is in clear contrast to the antisense RNA approach for the same viroid-host interaction. Atkins *et al.* (1995) could see only a slight reduction in viroid accumulation with antisense RNA against the minus strand, whereas ribozymes directed against the minus strand could effectively reduce viroids to an undetectable level. These ribozymes have only 9-11 nucleotide regions hybridizing to the viroid RNA. Atkins *et al.* (1995) also showed that ribozymes targeting the plus strand were not as effective as the ones targeting the minus strand of the viroid.

The difference in the efficacy of the positive versus negative strand for targeted antisense RNA/ribozymes could be due to the differential accumulation or stability between the plus- and minus-strand RNAs of viroids. Various studies indicate a differential abundance between the plus and minus RNAs (Matousek *et al.*, 1994; Faustmann *et al.*, 1986). Also, the difference observed between the antisense RNA approach of Atkins *et al.* (1995) and the ribozyme approach of Yang *et al.* (1997) may be due to differences in the efficiency of viroid replication in different host plants. When Yang *et al.* (1997) studied the efficiency of ribozymes in controlling RSTVd in tomato plants, they found that the viroid accumulation was significant in tomato plants expressing ribozymes against the minus strand, unlike their observations with transgenic potato. These experiences suggest that the strategy for controlling the pathogen will have to be carefully devised, taking into consideration the nature of biological interaction between the host and pathogen on a case-by-case basis.

Mammalian systems use an interferon-induced antiviral pathway called the 2′, 5′ oligoadenylate system. The system has two components: 2′, 5′ oligoadenylate synthetase and 2′, 5′ oligoadenylate-dependent RNase (RNaseL). Ogawa *et al.* (1996) demonstrated that this system can work in plants as well. Their work showed that dsRNA and replicating intermediates of RNA viruses induce cell death, which does not happen in animal systems.

Resistance to Fungal Diseases

Plant species deploy an assortment of defense responses soon after infection or exposure to fungal infection. These responses involve the biosynthesis and accumulation of pathogenesis-related proteins (PR proteins, Datta and Muthukrishnan, 1999). Besides, plants have resistance genes upstream from the defense responses that recognize the avirulence genes in the pathogens, which otherwise is known as gene-for-gene hypothesis. A good deal of progress has been made in deploying the transgenic strategy to develop transgenic crops resistant to or tolerant to fungal diseases using these resistance (R) genes as well as overexpressing the PR genes. Among the PR genes, the most attractive initial candidates for manipulation of the single-gene defense mechanism approach are the genes encoding chitinases and β-1,3-glucanases, because of their hydrolytic action on chitin and glucan, respectively, the two major structural components of the fungal cell wall. The first report of success was the resistance of transgenic tobacco and *Brassica napus* expressing a bean vacuolar chitinase under the cauliflower mosaic virus 35S promoter against necrotrophic fungus, *Rhizoctonia solani* (Broglie *et al.*, 1991). Similarly, overexpression of the β-1,3-glucanase from soybean in transgenic tobacco led to enhanced resistance against *Alternaria alternata* and *Phytophthora parasitica* var. *nicotianae* (Yoshikawa *et al.*, 1993). Details of the successful reports on several crops overexpressing chitinase (PR-3 group) and glucanase (PR-2 group), cloned from different sources, singly or in combination, as well as several other PR genes of different groups, are documented (Datta *et al.*, 1999b). In rice, the first report on overexpression of a rice chitinase gene showing enhanced resistance against sheath blight fungus *R. solani* came in 1995 (Lin *et al.*, 1995). Thereafter, the regulation, expression, and function of different chitinase genes (Anuratha *et al.*, 1996; Xu *et al.*, 1996; Baisakh *et al.*, 1999, 2001; Datta *et al.*, 2001) have been studied in rice. Recently, a PR-5 gene encoding a thaumatin-like protein (TLP) from rice, was introduced into rice and the high-level constitutive expression of the TLP protein showed enhanced resistance against *R. solani* (Datta *et al.*, 1999a).

Combined expression of multiple PR genes along with the R genes could provide effective and durable resistance in crop breeding (Wenzel, 1985).

Resistance to Bacterial Infection

Considerable research effort has been directed to using resistance genes in crop breeding programs including rice (Khush *et al.*, 1989). Five different classes of R genes cloned from different plant species have been characterized and tested for their efficiency in conferring resistance against bacterial pathogens. Among these, the fifth class representing a map-based cloned *Xa21* gene from rice conferring resistance to *Xanthomonas oryzae* pv. *oryzae* (*Xoo*) (Song *et al.*, 1995; Wang *et al.*, 1996) is the best one studied. *Xa21* encodes a receptor-like kinase consisting of leucine repeat regions (LRRs) in the putative extracellular domain and a serine-threonine kinase in the putative intracellular domain, thus indicating an evolutionary linkage with the other classes of R genes at the molecular level (Song *et al.*, 1995). By virtue of its wide-spectrum resistance (Song *et al.*, 1995; Wang *et al.*, 1996) against bacterial blight (BB), a serious disease of rice, *Xa21* is the first R gene to be transferred into rice. Therefore, International Rice Research Institute is focusing on developing and deploying transgenics with *Xa21* against bacterial blight (Tu *et al.*, 1998b; Baisakh *et al.*, 2000; Datta and Datta, 2002. Song *et al.* (1995) transferred *Xa21* into rice which showed resistance to a single isolate race 6 of BB pathogen. Subsequently, Wang *et al.* (1996) reported on the broad-spectrum resistance of transgenic rice with the *Xa21* gene against 29 diverse isolates from eight different countries, indicating that a single cloned gene is sufficient to confer multi-isolate resistance. However, in both these studies, a *japonica* rice, Taipei 309, was being used. In 1998, two reports came simultaneously from the International Rice Research Institute (Tu *et al.*, 1998b) and the Scripps Research Institute in La Jolla, California (Zhang *et al.*, 1998), that showed the successful production of transgenic *indica* rice with the *Xa21* gene. In both studies, the transgene followed a Mendelian segregation (3:1) pattern in the subsequent selfing generation, indicating a single-locus insertion. Under containment facilities, the transgenics were shown to be resistant to BB pathogen race 4 (Zhang *et al.*, 1998) and races 4 and 6 (Tu *et al.*, 1998b).

The first-ever report of transgenic rice with *Xa21* developed at the International Rice Research Institute and field-tested at Huazhong University, Wuhan, China (Fig. 3: Tu *et al.*, 2000b), showed clearly the resistance of the transgenic homozygous lines to multiple strains and races of BB pathogen under natural field conditions and artificial inoculation. This shows the potential of transgenic breeding for broad-spectrum resistance to bacterial disease(s) with a single cloned R gene. Research is in progress in many laboratories for

Fig. 3. Field evaluation of transgenic IR72/*Xa21* against bacterial blight in China showing the resistant reaction of transgenics with less lesion length than the control plants (Tu *et al.*, 2000b)

the deployment of novel R genes or novel members from the same family (such as *Xa5*, *Xa13*, etc. in the *Xa* family) for durable, multiple, and horizontal resistance.

Herbicide Resistance

Weeds are considered to be one of the major pests of several important crops, including rice, and they cause considerable yield loss. The chemical control of weeds through the application of herbicides poses a threat to human health and the environment (Pandey and Velasco, 1999), apart from the heavy costs involved. Moreover, the herbicides are usually non-selective, that is, they do not distinguish crop plants from weeds. Engineering for herbicide tolerance is a new way of conferring selectivity and enhancing crop safety and production (Oxtoby and Hughes, 1990). Hence, herbicide technology combined with the transgenic approach might give farmers a way to modernize agricultural technology with reduced labour (such as a choice of crops for rotation or double cropping) in an environmentally friendly system.

Mechanisms of herbicide resistance include (1) a modified target site, (2) enhanced detoxification or delayed activation, and (3) alterations in the uptake, translocation, or compartmentalization of the herbicide. The first two mechanisms have been identified mainly in plants. Herbicide resistance genes were isolated for several herbicides with different modes of action. Genes that coded for a herbicide target or detoxification enzymes were transferred into crop plants. The transgenic plants expressing these genes were tolerant to the active ingredients of the herbicides.

Herbicide-tolerant transgenics are now available in crops such as soybean, cotton, rice (Fig. 4), rapeseed, maize, sugarbeet and alfalfa (Rai and Prasanna, 2000).

Many reports from the United States have documented field trials of herbicide-resistant transgenics for diverse crop plants such as rice, wheat, maize, barley, tomato, tobacco, potato, cotton, soybean, rapeseed, peanut, sugarcane, alfalfa, pea, clover, beet, cantaloupe, grape, lettuce, carnation, and kiwi (Snow and Palma, 1997). The common herbicides for which herbicide tolerance has been engineered are glyphosate (Roundup™), phosphinothricin (Basta™, Herbiace™), Bromoxynil (Buctril™), sulfonylurea (Glean™, Oust™), and 2,4-D. Herbicide-resistant seed and related chemical packages are transforming the input supply market and the competitive arena dramatically, especially in the soybean and maize market, where 70 per cent of all area is expected to be growing Roundup Ready soybeans treated with Roundup herbicide or a soon-to-be-marketed generic glyphosate equivalent (Hayenga, 1999).

Tolerance of Drought, Salinity and Cold Stress

Crop productivity is affected or limited because of a variety of environmental factors such as high-/low- temperature stress, salinity, water supply, and so on. Modifying crops to tolerate unfavourable environmental conditions could improve productivity under adverse conditions and could also expand the land area under cultivation. A variety of genes isolated from bacteria, animals, and plants have been tested for their ability to confer stress tolerance to plants. These analyses to a large extent have been carried out in dicot

Fig. 4. Tolerance of transgenic rice for herbicide Basta compared with the susceptibility of the non-transformed control

model systems such as *Arabidopsis* and tobacco. Quite a few studies have tested various genes in plants such as rice, maize, and others. Holmberg and Bulow (1998) have reviewed efforts on improving stress tolerance of plants by the transgenic approach.

Most abiotic stresses eventually cause water deficit. Although a minor water deficit results in a reduction in photosynthesis, major deficits lead to complete inhibition of photosynthesis and the production of reactive oxygen intermediates such as superoxides and peroxides. These intermediates affect membrane integrity and cause severe impairment of several physiological processes and biochemical reactions. Hence, many transgenic approaches involve the use of genes whose products can either scavenge or protect from reactive oxygen species.

The production of transgenic plants that accumulate osmoprotectants (e.g., glycine betaine) and osmolytes has been a very popular approach for studying tolerance to salinity, drought and cold stress. Osmolytes are molecules with low molecular mass and they could be quaternary amines, amino acids, or sugar alcohols. Some plant species that have inherent tolerance of abiotic stresses have been shown to accumulate these osmolytes under stress. These molecules are known to raise the osmotic potential of the cell, which could help it in combating water stress. They could stabilize membranes and other macromolecular structures.

The genes that can potentially confer abiotic stress tolerance fall into three categories and they encode (*a*) enzymes that make protective metabolic products, (*b*) regulatory proteins and (*c*) protective proteins. Studies with some such genes, the expression of which could either lead to accumulation of an osmolyte or make a protein that can scavenge radicals, are discussed below.

Mannitol-1-Phosphate Dehydrogenase (mtlD): Tobacco plants transformed with *E. coli mtlD* showed mannitol accumulation leading to increased tolerance of high salinity of transgenic plants relative to control plants (Tarczynski *et al.*, 1993). Thomas *et al.* (1997) reported enhancement in seed germination under high salinity in transgenic *Arabidopsis* plants. However, Karakas *et al.* (1997) reported that transgenic plants showed a marginal increase in dry weight upon salt stress but no difference in growth could be observed between transgenic and control plants upon drought stress. Targeting mannitol biosynthesis to chloroplasts in transgenic tobacco plants also resulted in increased tolerance of oxidative stress (Shen *et al.*, 1997). Transgenic plants with *mtlD* showed increased mannitol accumulation in seedlings with increased shoot height and higher shoot and root fresh weight under salt stress compared with non-transgenics.

Δ^1-Pyrroline-5-Carboxylate Synthetase: Under water stress, many plants accumulate compatible osmolytes, which could protect components of the cell from damage. Proline, a potent osmolyte, accumulates in water-stressed plants and, when the stress is removed, proline levels are restored and these changes in proline level are tightly regulated (Yoshiba *et al.*, 1997). The enzyme Δ^1-pyrroline-5-carboxylate synthetase (P5CS) catalyzes the conversion of glutamate to Δ^1-pyrroline-5-carboxylate, which is then reduced to proline. Overexpression of a gene encoding for mothbean P5CS in transgenic tobacco plants resulted in the accumulation of proline (Kishor *et al.*, 1995). The transgenic plants had increased biomass and enhanced flower development under salt stress. However, the leaf sap of transgenic tobacco showed decreased osmotic potential during water stress and this puts the relationship between proline accumulation and water-stress tolerance in question. P5CS has successfully been transferred to *Arabidopsis* (Nanjo *et al.*, 1999) and rice (Zhu *et al.*, 1998), where it confers tolerance of water-limiting stress conditions to transgenic plants.

Trehalose-6-Phosphate Synthetase Gene (TPS1): The yeast trehalose-6-phosphate synthetase gene (*TPS1*) was introduced in tobacco to overproduce trehalose, which is a non-reducing disaccharide that can serve as an osmolyte. Transgenic plants accumulating trehalose exhibited multiple phenotypic alterations and improved drought tolerance (Romero *et al.*, 1997). Pilon-Smits *et al.* (1998) introduced bacterial trehalose-6-phosphate synthase (*OtsA*) and trehalose-6- phosphate phosphatase (*OtsB*) in tobacco. The transgenic lines showed better growth under drought stress in terms of dry weight and produced larger leaves. The dry weight of leaves under stress was 85 % higher in transgenics with the maximum trehalose accumulation compared with the control plants. Detached leaves from young well-watered transgenic plants showed better capacity to retain water when air-dried than the wild-type plants. Transgenics performed more efficient photosynthesis under stress. They also accumulated non-structural carbohydrates along with glucose, fructose, and sucrose. They were able to maintain a lower osmotic potential than the wild-type plants, thus helping them fare better under stress. A more recent study with transgenic potato constitutively overexpressing the yeast *TPS1* gene corroborates earlier results correlating trehalose accumulation with drought tolerance. Although the regenerating plantlets suffered abnormalities in culture, they recovered in the greenhouse and gave rise to normal plants that were drought-tolerant (Yeo *et al.*, 2000).

Levansucrase: Fructans are polyfructose molecules that are produced by many plants and bacteria. Owing to their solubility, they may help plants survive under periods of osmotic stress. Pilon-Smits *et al.* (1995) introduced a gene encoding for bacterial fructan (SacB) isolated from *Bacillus subtilis* into tobacco. The transgenic plants performed significantly better under PEG-mediated drought stress than the wild-type tobacco. The drought resistance correlated well with the amount of fructan accumulated. Transgenic plants have been developed from several other crops (potato, tobacco, chicory, barley, etc.) with levansucrase, and those plants grow better under drought stress than their wild-type counterparts (Meer *et al.*, 1994; Ebskamp *et al.*, 1994; Sprenger *et al.*, 1997; Schellenbaum *et al.*, 1999).

Myo-inositol O-methyltransferase (IMT1): The *IMT1*-encoded enzyme leads to the production of the methylated inositol D-ononitol. Expression of a cDNA encoding myo-inositol O-methyltransferase in tobacco during salt and drought stress resulted in the accumulation of methylated inositol (D-ononitol), which in turn conferred tolerance for both stresses. The transgenic plants also recorded higher levels of photosynthesis and CO_2 fixation than the untransformed control plants (Sheveleva *et al.*, 1997; 2000).

Choline Oxidase: Water stress conditions in halophytes and some bacteria result in the accumulation of glycine betaine as an adaptive response. The choline oxidase gene (*codA*) isolated from *Arthrobacter globiformis* that converts choline to glycine betaine via betaine aldehyde was fused with a transit peptide for plastid localization and introduced into *Arabidopsis*. The transgenic plants accumulated glycine betaine (up to 50 μM in the chloroplasts) and showed enhanced tolerance of salt and cold stress (Hayashi *et al.*, 1997; Alia *et al.*, 1998). The comprehensive study carried out by Huang *et al.*, (2000) clearly indicates that mere installation of a betaine biosynthetic pathway is insufficient for conferring cold tolerance to *Arabidopsis*, tobacco, and *Brassica*. Transgenic *Arabidopsis* carrying the *codA* gene showed resistance to light stress and increased accumulation of H_2O_2 and several other stress-related chemicals (Alia *et al.*, 1999). Transgenic rice expressing the *codA* gene in the chloroplast and cytosol returned to normal growth faster than the wild type after an initial growth inhibition under salt and low temperature stress (Sakamoto *et al.*, 1998). These results indicated that the subcellular compartmentalization of the biosynthesis of glycine betaine was a critical element in the efficient enhancement of tolerance of stress in the engineered plants.

Polyamines: Polyamines have been implicated in a variety of stress responses in plants. Andersen *et al.* (1998) showed that transgenic carrot cell lines overexpressing mouse ornithine decarboxylase, which converts ornithine to the diamine putrescine, withstood salt and osmotic stress. The role of polyamines in the stress tolerance of several crops has been reviewed by Galston *et al.* (1997).

Oxidative Stress-related Genes

Stress induces several processes in plants, leading to the generation of reactive oxygen species that cause oxidative stress. Plants have evolved systems to combat oxidative stress with a battery of gene products that aid in quenching the active oxygen species that primarily damage membranes. Examples of some of the enzymes involved in such protective processes are superoxide dismutases, glutathione reductase, glutathione-S-transferase/peroxidase, catalases, glyoxalases, etc. In the following section, some of the studies involving ectopic expression of some of these genes are discussed.

Moran *et al.* (1994) showed major reductions in photosynthesis and transpiration under drought-induced oxidative stress in pea plants. Analysis of four drought-tolerant varieties of tobacco by Van Rensburg and Kruger (1994) revealed that oxidative stress-related genes were induced that were evident from the elevated levels of glutathione reductase, superoxide dismutase, and ascorbate peroxidase, and, to a lesser extent, from catalase activities. They also associated these activities with the integrity of photosynthetic pigments. These observations suggest that abiotic stresses primarily affect plants through oxidative damage. Many groups have developed transgenic plants expressing oxidative stress-induced genes and their results are not easy to interpret. Tepperman *et al.* (1988) developed transgenic tomato and tobacco plants overexpressing Cu/Zn superoxide dismutase (SOD) and failed to see any protection against superoxide toxicity. However, targeting the constitutively overexpressed Mn-SOD into chloroplasts and mitochondria reduced cellular damage (Bowler *et al.*, 1991). Gupta *et al.* (1993) overexpressed Cu/Zn-SOD in tobacco and the transgenic plants retained 90 % photosynthesis under chilling and high light stress. Van Camp *et al.* (1996) reported overexpression of Fe-SOD in tobacco plants. Though they found elevated levels of oxidative stress-related enzymes in the transgenic plants, these were not any more tolerant of salt stress than the control plants. Different groups have used different classes of SOD derived from different plant species. This makes it difficult to draw a general inference. More recent work on SOD suggests that there could be value in pursuing its utility in protecting plants from oxidative stress resulting from abiotic stresses. Van Breusegem *et al.* (1999) developed transgenic maize plants with *Arabidopsis* Fe-SOD expressed under the CaMV 35S promoter and containing the chloroplast targeting the transit peptide signal from pea RBISCO. They observed enhanced methyl viologen (a chemical that induces oxidative stress) resistance and higher growth rates under low temperatures in transgenic maize plants. McKersie *et al.* (2000) developed transgenic alfalfa plants using the same transgene as Van Breusegem *et al.* (1999) and showed that increases in Fe-SOD levels could be correlated with better survival during winter under field conditions. However, the transgenic alfalfa plants did not show any change in photosynthetic oxidative stress tolerance. These workers suggested that elevated levels of Fe-SOD resulted in amelioration of secondary injury symptoms and enhanced the recovery of transgenic plants from low-temperature shocks.

Roxas *et al.* (1997) constitutively overexpressed Nt107 cDNA (which encodes a glutathione-S-transferase with additional peroxidase activity) in tobacco. They found that transgenic seedlings showed better growth than the control plants under cold and salt stress. More work needs to be done to understand the various oxidative stress-related genes and the interaction of their products during stress.

Late Embryogenesis Abundant (LEA) Protein Gene *hva1*

A barley group 3 LEA protein HVA1 was previously characterized from the barley aleurone. This gene, *hva1*, is specifically expressed in the aleurone layers and embryos during late seed development correlating with the seed desiccation stage. The expression of this gene is rapidly induced in young seedlings by abscissic acid and by several stress conditions, including dehydration, salt, and extreme temperatures. Xu *et al.* (1996) produced transgenic rice plants expressing the barley *hva1* gene. It was found that accumulation of barley HVA1 protein in vegetative tissues of transgenic rice plants conferred increased tolerance of water deficit and salt stress and the extent of stress tolerance correlated with the level of HVA1 protein accumulation.

CBF1/CBF3

Many cold-regulated genes have a common upstream regulatory motif called C-repeat. These are coordinately regulated by C-repeat binding factors (CBF). The same motif is also referred to as DRE (drought-responsive element). Transcription factors bind these motifs under stress and activate the expression of genes resulting in tolerance to the stress. Jaglo-Ottosen *et al.*, (1998) found that overexpression of the Arabidopsis transcription factor CBF1 using the CaMV 35S promoter resulted in enhanced freezing tolerance. More recently, Kasuga *et al.*, (1999) demonstrated the utility of expressing a transcription factor in an inducible fashion to obtain stress tolerance in plants. When CBF_3 was constitutively overexpressed, the recovery of transgenic plants was affected. The regenerated plants also displayed stunted growth and other morphological abnormalities. However, when the expression was driven by a cold-inducible rd29A promoter of *Arabidopsis*, Kasuga *et al.*, (1999) obtained healthy transgenic plants that were more tolerant to drought, salt, and low temperature than plants constitutively expressing CBF3. Expression of a master switch gene in an inducible manner to control stress-responsive gene expression could be a valuable approach for engineering stress tolerance in crop plants.

Enhancing Shelf Life

Tomato Fruit Ripening: Tomatoes are usually harvested in the green stage to enable mechanical handling and longer shelf life. This prevents the development of complete flavour that is obtained with vine-ripened fruits. Expression of an antisense polygalacturonase construct results in delayed softening without interfering with the ripening process (Smith *et al.*, 1988; Sheehy *et al.*, 1988). With these transgenics, it is possible to mechanize handling after the fruits are allowed to ripen fully on the vines. Calgene Inc. marketed such transgenic tomatoes under the brand name "Flavr-Savr".

Another approach involved the inhibition of ethylene biosynthesis to delay the ripening process. This was achieved by overexpressing a 1-aminocyclopropane-1-carboxylate (ACC) deaminase isolated from a bacterial source. The enzyme degrades ACC, an intermediate in ethylene synthesis (Klee *et al.*, 1991).

Improving Productivity

The primary goal of crop improvement has been to increase productivity. The first wave of transgenics has successfully addressed traits such as insect and herbicide resistance to sustain the maximum yield potential of the crops. The second wave of transgenic varieties is expected to enhance yield. This could be achieved by improving source strength by essentially increasing net photosynthesis at the leaf level. Photosynthetic losses can occur at many steps: light harvesting, electron transport, carbon assimilation, partitioning of the photosynthate, and respiration. Various approaches that could minimize these losses would be to alter canopy structure, improve light acclimation, improve photoprotection, incorporate features of the C_4 photosynthetic pathway into C_3 crops, alter stomatal responses, alter carbohydrate metabolism, delay leaf senescence, alter metabolite signalling, reduce respiration capacity, and improve nitrogen economy. It is unlikely that achieving one of the above with a transgene would directly improve yield. In the continuum from source strength to sink capacity, all of the above play major roles. Understanding the effect of expression of some useful genes in photosynthesis and partitioning will give clues to the useful combination of alterations that could result in discernible increases in yield. Various efforts along these lines have been described in a recent review (Horton, 2000).

C_4 Pathway in C_3 Plants

A majority of plants, including important crops such as rice and wheat, fix atmospheric carbon dioxide by the C_3 photosynthetic pathway. An inherent feature of the C_3 pathway that reduces overall photosynthetic efficiency is the oxygenase reaction of Rubisco. Oxygen that is released by this activity inhibits the carbon fixation reaction and leads to a loss of CO_2 through photorespiration. Environmental stresses such as high temperature and water limitation could further decrease photosynthetic efficiency. Efforts to engineer Rubisco such that its carboxylase activity dominates over oxygenase activity may not be fruitful (Somerville, 1990).

C_4 plants such as maize, on the other hand, have evolved mechanisms by which a reduction in oxygen-mediated photosynthetic efficiency could be overcome. The C_4 pathway enables increased local CO_2 concentration around Rubisco, thus inhibiting its oxygenase activity, which is also reflected in reduced photorespiration. Phosphoenolpyruvate carboxylase (PEPC) mediates the fixing of CO_2 into the C_4 acid oxaloacetate. PEPC is insensitive to oxygen and has a very high affinity for bicarbonate, which is its substrate. These features of C_4 plants confer higher photosynthetic capacity and better water- and nutrient-use efficiency than in C_3 plants, especially under low CO_2 conditions.

Efforts to constitutively overexpress C_4 PEPC in C_3 plants did not yield a significant increase in photosynthetic capacity when the CaMV 35S promoter was used to drive gene expression (Hudspeth *et al.*, 1991; Kogami *et al.*, 1994; Gehlen *et al.*, 1996). The expression of the transgene was low in these studies. The introduction of a native maize PEPC gene into rice resulted in high expression of the gene and the accumulation of high levels of the enzyme. Transgenic rice plants with high levels of PEPC were characterized by a reduced quantum of photosynthesis inhibition by oxygen (Ku *et al.*, 1999). This is a classical example that shows the installation of a new pathway in a plant using a single gene from a different source. The maize PEPC gene was also introduced into an indica rice and the photosynthetic efficiency of the primary transgenic plants is under evaluation (Datta *et al.*, unpublished). Therefore, efforts are being made to install all three key enzymes involved in the C_4 pathway (PEPC: pyruvate orthophosphate dikinase, PPDK: and NADP-malic enzyme, NADP-ME) in rice to realize maximum photosynthetic efficiency. Moreover, further work needs to be done to study whether, in the field, better photosynthesis would be seen in such transgenic rice plants under stress conditions.

It is also important to consider the effect of transgenesis on transgenic plants. In work carried out by Ku *et al.* (1999), even though the protein encoded by the transgene accounted for 12 % of the total soluble protein in the leaves, no loss of fertility or phenotypic abnormalities were encountered. It is known that NADP-ME is induced under low CO_2 conditions and thus might help improve photosynthetic capacity under stress. Takeuchi *et al.* (2000) and Tsuchida *et al.* (2001) overexpressed maize NADP-ME in the chloroplast of rice. Transgenic rice plants showed abnormal chloroplasts lacking the thylakoid-stacking characteristic of the chloroplasts in the bundle sheath cells of C_4 plants. These observations suggest that the enzyme itself affects the development of chloroplasts.

Exploiting Heterosis

Heterosis breeding is a proven way of increasing productivity in many crop species. However, in many grain crops, the exploitation of heterosis hinges on the availability of a good male sterility and fertility restoration system. Over the past several decades, many cytoplasmic male sterility (CMS) sources have been successfully developed and used for hybrid seed production in various crop species, including rice. In many cases, the non-availability of proper CMS sources and/or restoration systems has been a major limitation to the development of commercial F_1 hybrid varieties. Sometimes the lack of diverse cytoplasmic sources has given rise to concerns about cytoplasmic uniformity and the consequent genetic vulnerability to pest and disease outbreaks. This points to a need to develop alternative male sterility sources and fertility restoration systems. Transgenic plants can address this need as has been demonstrated in at least a few instances.

One of the earliest and most successful attempts to induce male sterility by genetic engineering involved the transfer and tissue-specific expression of a "toxin" gene that disrupted normal pollen development. The toxin gene in this case was an RNAse from a fungal source, *Barnase*, which was made to express itself specifically in the tapetal tissue of developing tobacco anthers by using the tapetum-specific promoter TA29 (Mariani *et al.*, 1990). In another study, the pollen-specific promoter from rice, PS1, was fused to the *Barnase* gene from *Bacillus amyloliquefaciens*, which encodes a secreted ribonuclease and was introduced into tobacco using *Agrobacterium*-mediated transformation (Zhan *et al.*, 1996). The transgenic tobacco plants obtained showed normal vegetative and floral development, but displayed a range of reproductive properties from slightly reduced fertility to complete sterility. Several studies have been reported in different crops. Recently, in rice, expression vectors carrying the male sterility and fertility restorer genes (*barnase* and *barstar*) fused to a tapetum-specific promoter have been successfully transferred (Zhang *et al.*, 1998).

Nutrition-rich Crops

Since agriculture is the primary source of nutrients and poor diets are a fundamental cause of malnutrition, it is indispensable to have nutrition-rich crops for solving the problems caused by such deficiencies. Nutritional genomics will have a tremendous impact on the improvement of foods for human health (DellaPenna, 1999). Datta and Bouis (2000) have discussed the potential of biotechnology to improve human nutrition through the recent breakthroughs in genetic engineering, such as the development of golden rice, iron-fortified crops, etc.

Three different genes involved in the metabolic pathway, phytoene synthase and lycopene cyclase (*psy* and *lyc* cloned from *Narcissus pseudonarcissus*), and phytoene desaturase (*crtI* cloned from *Erwinia uredovora*), have been introduced into rice through the *Agrobacterium*-mediated transformation method, which resulted in what is called "golden rice" which produces beta carotene, which is ultimately converted into vitamin A in the human body (Ye *et al.*, 2000). This is a remarkable achievement in the transgenic research that could help eradicate blindness caused by vitamin-A deficiency. Research is in progress at the International Rice Research Institute to engineer these genes for the production of transgenic indica rice with beta-carotene in different elite IRRI-bred cultivars as well as other cultivars adapted to different developing countries (IRRI, 2000).

Similarly, iron deficiency leading to anemia, which affects around 30 % of the world's population, is prevalent in developing countries. Most people get their daily iron requirement from vegetables. So, increasing the iron content in plants by genetic manipulation would have a significant effect on human health. An iron storage protein gene, *ferritin*, driven by the constitutive 35S promoter, was successfully transferred to tobacco, in which the leaves of transgenics had a maximum of 30 % more iron than the non-transformed control plants (Goto *et al.*, 2000b). In lettuce, transgenics showed 1.2-1.7 times more iron content, enhanced early developmental growth, and superior photosynthesis compared with the control plants (Goto *et al.*, 2000a). In rice and wheat, also, the use of a constitutive promoter resulted in higher iron content in the vegetative tissues, but not in the seeds (Drakakaki *et al.*, 2000). However, the *ferritin* gene placed under an endosperm-specific promoter, glutelin, was expressed in the seeds (the target tissue) in rice (Goto *et al.*, 1999). The iron content in the endosperm of transgenic rice was significantly higher than that of the non-transformant. All these studies used the *ferritin* gene cloned from soybean. Recently, Lucca *et al.* (2001) also observed increased iron content in rice seeds transformed with the *Phaseolus ferritin* gene.

Furthermore, since the endogenous phytic acid inhibits iron bioavailability, a heat-stable phytase from *Aspergillus fumigatus* was introduced into rice (Lucca *et al.*, 2001), which increased the level of phytase 130-fold, enough to hydrolyze the phytic acid. In addition, as cysteine peptides are considered to be a major enhancer of iron absorption, an endogenous cysteine-rich metallothionine-like protein was overexpressed in rice, which increased the cysteine residues seven-fold (Lucca *et al.*, 2001). This demonstrates that high-phytase rice, with increased iron content and rich in cysteine peptide, could potentially improve the iron nutrition in the rice-eating population. This technique for iron content enhancement has potential for other crops also.

Another gene, ferric chelate reductase (*FRO2*), which allows plants to uptake more iron in iron-deficient soil (Robinson *et al.*, 1999), may make it possible to co-integrate this gene with *ferritin, phytase*, and *cysteine* peptide genes for high iron uptake and storage in the plant, and bioavailability and absorption in the human body. At IRRI, transgenic rice has also been produced with *FRO2* and further work is in progress (Datta *et al.*, 2000).

Another important, essential but limiting amino acid in rice, lysine, which promotes the uptake of trace elements, is a potential candidate for the nutritional improvement of crops. Two bacterial genes, *dapA* and *lysC* from *Corynebacterium*, have been shown to enhance lysine about fivefolds in canola, maize, and soybean seeds (Falco *et al.*, 1995; Mazur *et al.*, 1999). To introduce these genes into rice, IRRI is collaborating with DuPont to develop lysine-rich rice.

Conclusions

Biotechnology, especially transgenics genetically modified (GM crops), can improve farmer and consumer well-being in many ways. Sustaining crop productivity by developing in-built disease- and insect-resistant crops, and/or increased crop productivity through novel genes, and value-added crops could significantly increase the food supply, thereby helping reduce food prices for poor farmers. A brief survey of some classic developments in transgenic research during the last two decades are presented in Table 1. Recent significant achievements in plant genetic engineering for nutrition-rich crops could reduce human malnutrition. Farmers will always benefit from growing these crops because of a guaranteed higher price. However, the complete realization of such potential of biotechnology could take time because of the large investment required in agricultural research and other public and on-farm infrastructure (Datta and Bouis, 2000).

Nonetheless, like any other technology, the controversy surrounding GM crops is no exception in a heterogeneous society. Several concerns on different issues, such as

Table 1. Some classic developments in transgenic research

Year	Achievement in transgenics	Transgene	Method	Reference
1983	First transformation event demonstrated in tobacco	*gusphaseolin*	*Agrobacterium*	Bevan *et al.*, (1983) Fraley *et al.* (1983),
	Bean phaseolin transferred to sunflower and tobacco		*Agrobacterium*	Herrera-Estrella *et al.* (1983), Murai *et al.* (1983)
1984	First report of a bacterial gene expression in tobacco	*NPTII*	*Agrobacterium*	Horsch *et al.* (1984)
1985	First transient expression in maize	*gus*	Protoplast	Fromm *et al.* (1985)
1987	Biolistic transformation through particle gun bombardment			Sanford et al. (1985)
	established	*NPTII*	Agrobacterium	Firoozabady *et al.* (1987)
	First cotton transgenic plants produced			
1988	Transient expression in *japonica*	*gus*	Protoplast	Zhang and Wu (1988)
	rice	*aph*	Protoplast	Toriyama *et al.* (1988)
	Rice plants with Gus expression			
	in T_0 only; no molecular data	*Glyophosphate*	*Agrobacterium*	Hinchee *et al.* (1988)
	First stable soybean transgenics developed	*tolearance*		
1989	First stable transgenic *japonica* rice	*hph*	Protoplast	Shimamoto *et al.* (1989)
1990	First stable fertile transgenic *indica* rice	*hph*	Protoplast	Datta *et al.* (1990)
	First stable transgenic maize produced	*gus*	Protoplast	Fromm *et al.* (1990)
	First success of *Agrobacterium*-mediated transformation in *japonica* rice	*gus, NPTII*	Agrobacterium	Raineri *et al.* (1990)
	Transgenic cotton with insect resistance developed			
	First report of transient gus expression in wheat	*Bt*	*Agrobacterium*	Perlak *et al.* (1990)
		gus	Biolistic	Londsale *et al.* (1990)
1991	Stable transformation in Sorghum	*NPTII*	Protoplast	Battraw and Hall
	Successful stable transformation in *Brassica*	*gus*	*Agrobacterium*	(1990) Barfield and Pua (1991)
1992	Herbicide tolerant *indica* rice developed	*bar*	Protoplast (PEG) Biolistic	Datta *et al.* (1992)
	First successful stable herbicide resistant wheat developed	*bar*	Protoplast	Vasil *et al.* (1992)
	Herbicide tolerant *japonica* rice developed	*bar*	(electroporation) Biolistic	Toki *et al.* (1992)
	First transgenic sugarcane	*gus*		Bower and Birch (1992)

1993	First transformation of chickpea			Fontana *et al.* (1993)
	First field testing of virus-resistant transgenic crops	*gus*	*Agrobacterium*	Kaniewski and Thomas (1993)
	Field performance of sulfonylurea-resistant transgenic flue-cured tobacco			Brandle and Miki (1993)
1994	Transgenic *japonica* and *indica* rice	*hph*	*Agrobacterium*	Hiei *et al.* (1994)
	First stable independent barley transgenics developed	*bar, gus*	Biolistic	Wan and Lemaux (1994)
	First Basta-resistant transgenic triticale	*bar, gus*	Biolistic	Zimmy *et al.* (1994)
1995	Wheat transgenics for stress tolerance	*CP4, Gox*	Biolistic	Zhou *et al.* (1995)
1996	Mild mosaic virus-resistant barley developed	*CP*	Biolistic	Hagio *et al.* (1996)
1997	Stem borer-resistant sugarcane developed	*cry1A(b)*	Biolistic	Arencebia *et al.* (1997)
	First successful report on Agrotransformation of barley	*bar, uidA*	*Agrobacterium*	Tingay (1997)
1998	First detailed report on the comparative efficiency of different promoters driving agronomically important genes	*cry1A(b), cry1A(c), cry1A(b)/ cry1A(c)*	Biolistic, Protoplast	Datta *et al.* (1998)
1999	First iron-rich rice	*Ferritin PEPc*	*Agrobacterium*	Goto *et al.* (1999)
	First report of C_4 gene transferred to rice		*Agrobacterium*	Ku *et al.* (1999)
	Increased iron uptake by transgenic tobacco	*FRO2*	*Agrobacterium*	Robinson *et al.* (1999)
	Increased Provit A and Vitamin E in transgenic canola	*crtB*	*Agrobacterium*	Shewmaker *et al.* (1999)
2000	First field testing of transgenic rice with agronomically important genes	*Bt*	Biolistic	Ye *et al.* (2000)
		Xa21	Biolistic	Tu *et al.* (2000)
	Beta carotene-rich (golden) rice	psy, lyc, crtI	*Agrobacterium*	Ye *et al.* (2000)
2001	Development of homozygous transgenic in less than a year	*chi11*	Biolistic	Baisakh *et al.* (2001)

horizontal gene flow by outcrossing, food safety, beneficiaries of the technology, and the conflict between plant varietal protection and intellectual property rights for genes and technology discovery, need to be addressed. Finally, successful product development needs extensive field trials and public understanding. Based on the experience of several successful field evaluations of GM crops, it is predicted that gene technology combined with precise plant breeding and efficient crop management could provide the benefits people need. And, once the fruits of genetic engineering in agriculture reach small farms and industrial operations, everyone can benefit from such development (Datta, 2000).

References

Alam M.F., Datta K., Abrigo E., Vasquez A., Senadhira D. and Datta S.K. 1998. Production of transgenic deep water indica rice plants expressing a synthetic *Bacillus thuringiensis cryIA(b)* gene with enhanced resistance to yellow stem borer. *Plant Sci.*, **135**: 25-30.

Alam M.F., Datta K., Abrigo E., Oliva N.P., Tu J., Virmani S.S. and Datta S.K. 1999. Transgenic insect-resistant maintainer line (IR68899B) for improvement of hybrid rice. *Plant Cell Rep.*, **18**: 572-575.

Alia, Kondo Y., Sakamoto A., Nonaka H., Hayashi H.A., Saradhi P.P., Chen T.H.H. and Murata N. 1999. Enhanced tolerance to light stress of transgenic *Arabidopsis* plants that express the codA gene for a bacterial choline oxidase. *Plant Mol. Biol.*, **40**: 279-288

Alia, Hayashi H., Chen T.H.H. and Murata N. 1998. Transformation with a gene for choline oxidase enhances the cold tolerance of *Arabidopsis* during germination and early growth. *Plant Cell Environ.*, **21**: 232-239.

Andersen S.E., Bastola D.R. and Minorcha S.C. 1998. Metabolism of polyamines in transgenic cells of carrot expressing a mouse ornithine decarboxylase cDNA. *Plant Physiol.*, **116**: 299-307.

Angenon G., Dillen W. and Van Montagu M. 1994. Antibiotic resistance markers for plant transformation. *Plant Mol. Biol. Manu.*, **C1**: 1-13.

Anuratha C.S., Zen K.C., Cole K.C., Mew T. and Muthukrishnan S. 1996. Induction of chitinase and beta-1, 3-glucanases in *Rhizoctonia solani* infected rice plants: isolation of an infection-related chitinase cDNA clone. *Physiol. Plant*, **97**: 39-46.

Atkins D., Young M., Uzzell S., Kelly L., Fillatti J. and Gerlach W.L. 1995. The expression of antisense and ribozyme genes targeting citrus exocortis viroid in transgenic plants. *J. Gen. Virol.*, **76**: 1781-1790.

Arencibia A., Vazquez R., Preto D., Tillez P., Carmona E., Coegeo A., Hernandez L., De La Riva G. and Selman-Housein G. 1997. Transgenic sugarcane plants resistant to stem borer attack. *Mol. Breed.*, **3**: 247-255.

Baisakh N. 2000. Improvement of rainfed lowland *indica* rice through *in vitro* culture and genetic engineering. Ph. D. Thesis (Utkal Univ.), India.

Baisakh N., Datta K., Oliva N.P. and Datta S.K. 1999. Comparative molecular and phenotypic characterization of transgenic rice with *chitinase* gene developed through biolistic and *Agrobacterium*-mediated transformation. *Rice Genet. Newsl.*, **16**: 149-152.

Baisakh N., Datta K., Oliva N., Ona I., Rao G.J.N., Mew T.W. and Datta S.K. 2001. Rapid development of homozygous transgenic rice using anther culture harboring rice *chitinase* gene for enhanced sheath blight resistance. *Plant Biotechnol.*, **18**: 101-108.

Baisakh N., Datta K., Rashid H., Oliva N. and Datta S.K. 2000. *Agrobacterium tumefaciens*-mediated transformation of an elite *indica* rice maintainer line IR68899B with a reconstructed T-DNA carrying multiple genes. *Rice Genet. Newsl.*, **17**: 122-125.

Barfield D.G. and Pua E.C. 1991. Gene transfer in plants of *Brassica juncea* using *Agrobacterium tumefaciens*- mediated transformation. *Plant Cell Rep.*, **10**: 308-314.

Barton K.A. and Miller M.J. 1991. Insecticidal toxins in plants. EPA 0431829 A1 910612.

Barton K.A., Whiteley H. and Yang N.S. 1987. *Bacillus thuringiensis* δ-endotoxin in transgenic *Nicotiana tabacum* provides resistance to lepidopteran insects. *Plant Physiol.*, **85**: 1103-1109.

Battraw M.J. and Hall T.C. 1990. Histochemical analysis of CaMV 35S promoter β-glucouronidase gene expression in transgenic rice plants. *Plant Mol. Biol.*, **22**: 1067-1085.

Beachy R.N., Loesch-Fries S. and Tumer N.E. 1990. Coat protein- mediated resistance against virus infection. *Annu. Rev. Phytopathol.*, **28**:451-474.

Beffa R. and Meins F. Jr. 1996. Pathogenesis-related functions of plant β1,3-glucanases investigated by antisense transformation: *A review. Gene*, **179**: 97-103.

Benfey P.N. and Chua N.H. 1990. The cauliflower mosaic virus 35S promoter: combinatorial regulation of transcription in plants. *Science*, **250**: 959-966.

Bevan M.W., Flavell R.B. and Chilton M.D. 1983. A chimeric antibiotic resistance marker gene as selectable marker for plant transformation. *Nature*, **304**: 184-187.

Bidney D., Scelonge J., Martich M., Burrus L.S. and Huffman G. 1992. Microprojectile bombardment of plant tissues increases transformation frequency by *Agrobacterium tumefaciens*. *Plant Mol. Biol.*, **18**: 301-303.

Blochinger K. and Diddelmann H. 1984. Hygromycin B phosphotransferase as a selectable marker for DNA transfer experiments with higher eukaryotic cells. *Mol. Cell Biol.*, **4**: 2929.

Bower R. and Birch R. 1992. Transgenic sugarcane plants via microprojectile bombardment. *Plant J.*, **2**: 409-416.

Bowler C., Slooten L., Vandenbranden S., De Rycke R., Botterman J., Sybesma C., Van Montagu M. and Inze D. 1991. Manganese superoxide dismutase can reduce cellular damage mediated by oxygen radicals in transgenic plants. *EMBO J.*, **10**: 1723-1732.

Brandle J.E. and Miki B.L. 1993. Agronomic performance of sulfonylurea-resistant transgenic flue-cured tobacco grown under field conditions. *Crop Sci.*, **33**: 847-52.

Broglie K., Chet I., Holliday M., Cressman R., Biddle P., Knowlton S., Mauvis C.J. and Broglie R. 1991. Transgenic plants with enhanced resistance to the fungal pathogen *Rhizoctonia solani*. *Science*, **254**: 1194-1197.

Cao J., Duan X.L., McElroy D. and Wu R. 1992. Regeneration of herbicide resistant transgenic rice plants following microprojectile-mediated transformation of suspension culture cells. *Plant Cell Rep.*, **11**: 586-591.

Chalfie M., Tu Y., Euskirchen G., Ward W.W. and Prasher D.C. 1994. Green fluorescent protein as a marker for gene expression. *Science*, **263**: 802-805.

Cho M.J., Choi H.W., Buchanan B.B. and Lemaux P.G. 1999. Inheritance of tissue-specific expression of barley hordein promoter-*uidA* fusions in transgenic barley plants. *Theor. Appl. Genet.*, **98**: 1253-1262.

Christensen A.H., Sharrock R.A. and Quail P.H. 1992. Maize polyubiquitin genes: structure, thermal perturbation of expression and transcript splicing, and promoter activity following transfer to protoplasts by electroporation. *Plant Mol. Biol.*, **18**: 675-689.

Christou P., Ford T.M. and Kofron M. 1991. Production of transgenic rice (*Oryza sativa* L.) plants from agronomically important indica and japonica varieties via electric discharge particle acceleration of exogenous DNA into immature zygotic embryos. *Bio/Technology*, **9**: 957-962.

Chung M.H., Chen M.K. and Pan S.M. 2000. Floral spray transformation can efficiently generate *Arabidopsis* transgenic plants. *Transgenic Res.*, **9**: 471-476.

Cooper B., Lapidot M., Heick J.A., Dodds J.A. and Beachy R.N. 1995. A defective movement protein of TMV in transgenic plants confers resistance to multiple viruses whereas the functional analog increases susceptibility. *Virology*, **206**: 307-313.

Corbin D.R., Greenplate J.T., Wong E.Y. and Purcell J.P. 1994. Cloning of an insecticidal cholesterol oxidase gene and its expression in bacteria and plant protoplasts. Appl. Environ. Microbiol. *Biotechnol.*, **44**: 133-138.

Datta K. and Datta S.K. 2002. Plant transformation. *In*: *Molecular Plant Biology*, Vol. 1. A Practical Approach, (eds.) Gilmartin P.M. and Bowler C., Oxford University Press, UK, pp. 13-32.

Datta K., Tu J., Oliva N., Ona I., Velazhahan R., Mew T.W., Muthukrishnan S. and Datta S.K. 2001. Enhanced resistance to sheath blight by constitutive expression of infection-related rice chitinase in transgenic elite *indica* rice cultivars. *Plant Sci.*, **160**: 405-414.

Datta K., Koukolikova-Nicola Z., Baisakh N., Oliva N. and Datta S.K. 2000. *Agrobacterium*-mediated engineering for sheath blight resistance of indica rice cultivars from different ecosystems. *Theor. Appl. Genet.*, **100**: 832-839.

Datta K., Velazhahan R., Oliva N., Mew T., Khush G., Muthukrishnan S. and Datta S.K. 1999a. Over-expression of cloned rice thaumatin-like protein (PR-5) gene in transgenic rice plants enhances environment-friendly resistance to *Rhizoctonia solani* causing sheath blight disease. *Theor. Appl. Genet.*, **98**: 1138-1145.

Datta K., Muthukrishnan S. and Datta S.K. 1999b. Expression and function of PR-protein genes in transgenic plants. *In*: *"Pathogenesis-related proteins in plants"*, (eds.) Datta S.K. and Muthukrishnan S., CRC Press LLC, Boca Raton, Florida, USA. pp. 261-277.

Datta K., Vasquez A., Tu J., Torrizo L., Alam M.F., Oliva N., Abrigo E.M., Khush G.S. and Datta S.K. 1998. Constitutive and tissue-specific differential expression of *cryIA(b)* gene in transgenic rice plants conferring enhanced resistance to insect pests. *Theor. Appl. Genet.*, **97**: 20-30.

Datta S.K. 2000. Potential benefit of genetic engineering in plant breeding: Rice, a case study. *Agric. Chem. Biotechnol.*, **43**: 197-206.

Datta S.K. 1999, Transgenic cereals: *Oryza sativa* (rice). *In*: *Molecular Improvement of Cereal Crops*, (ed.) Vasil I.K., Kluwer Acad. Publ., UK. pp. 149-187.

Datta S.K., Abrigo E.M., Arboleda M.D., Baisakh N., Dey M., Disonglo A.V., Thet K.M., Torrizo L.B., Tu J., Oliva N. and Datta K. 2000. Transgene pyramiding: Expression and function. *In*: *4th International Rice Genetics Symposium*, October 22-27, IRRI, Los Baqos, Philippines.

Datta S.K. and Bouis H.E. 2000. Application of biotechnology to improving the nutritional quality of rice. *Food Nutr. Bull.*, **21**: 451-456.

Datta S.K. and Muthukrishnan S. 1999. *Pathogenesis-related proteins in plants.* CRC Press, Boca Raton, Florida, USA.

Datta S.K., Torrizo L.B., Tu J., Oliva N. and Datta K. 1997. Production and molecular evaluation of transgenic rice plants. *IRRI Discussion Paper Series No. 21.* International Rice Research Institute, Los Banos, Philippines.

Datta S.K., Datta K., Soltanifar N., Donn G. and Potrykus I. 1992. Herbicide-resistant *indica* rice plants from IRRI breeding line IR72 after PEG-mediated transformation of protoplast. *Plant Mol. Biol.*, **20**: 619-629.

Datta S.K., Peterhans A., Datta K. and Potrykus I. 1990. Genetically engineered fertile *indica*-rice plants recovered from protoplasts. *Bio/Technology*, **8**: 736-740.

De Cosa B., Moar W., Lee S.B., Miller M.J. and Daniell H. 2001. Overexpression of the *Bt cry2Aa2* operon in chloroplast leads to formation of insecticidal crystals. *Nature Biotechnol.*, **19**: 1-4.

De Feyter R., Young M., Schroeder K., Dennis E.S. and Gerlach W.L. 1996. A ribozyme gene and an antisense gene are equally effective in conferring resistance to tobacco mosaic virus on transgenic tobacco. *Mol. Gen. Genet.*, **250**: 329-338.

Della Penna D. 1999. Nutritional genomics: Manipulating plant micronutrients to improve human health. *Science*, **285**: 375-379.

Dennis E.S., Gerlach W.I., Pryor A.J., Bernetzen J.L., Inglis A., Llewellyn D., Sachs M.M., Ferl R.J. and Peacock W.J. 1984. Molecular analysis of the alcohol dehydrogenase (*Adh 1*) gene of maize. *Nucl. Acids. Res.*, **12**: 3983-4000.

Dodds P.N., Ferguson C., Clarke A.E. and Newbigin E. 1999. Pollen-expressed s-RNAs are not involved in self-incompatibility in *Lycopersicon peruvianum*. Sex. *Plant Reprod.*, **12**: 76-87.

Drakakaki G., Christou P. and Stoger E. 2000. Constitutive expression of soybean ferritin cDNA in transgenic wheat and rice results in increased iron levels in vegetative tissues but not in seeds. *Transgenic Res.*, **9**: 445-452.

Ebskamp M.J.M., Spronk B.A., Weisbeek P.J., Smeekens S.C.M. and Van-der Meer I.M. 1994. Accumulation of fructose polymers in transgenic tobacco. *Bio/Technology*, **12**: 272-275.

Edington B.V., Dixon R.A. and Nelson R.S. 1993. Ribozymes: descriptions and uses. *In*: *Transgenic plants : fundamentals and application*. (ed.) Dekker M., New York. pp. 301.

Estruch J.J., Warren G.W., Mullins M.A., Nye G.J., Craig J.A. and Koziel M.G. 1996. Vip3A, a novel *Bacillus thuringiensis* vegetative insecticidal protein with a wide spectrum of activities against lepidopteran insects. *Proc. Natl. Acad. Sci.*, USA, **93**: 5389-5394.

Falco S.C., Guida T., Locke M., Mauvais J., Sanders C., Ward R.T. and Webber P. 1995. Transgenic canola and soybean seeds with increased lysine. *Bio/Technology*, **13**: 577-582.

Faustmann O., Kern R., Sanger H.L. and Muhlbach H.P. 1986. Potato spindle tuber viroid (PSTV) RNA oligomers of (+) and (–) polarity are synthesized in potato protoplasts after liposome-mediated infection with PSTV. *Virus Res.*, **4**: 213-227.

Firoozabady E., DeBoer D.L., Merlo D.J., Jalk E.L., Amerson L.N., Rashka K.E. and Murray E.E. 1987. Transformation of cotton (*Gossypium hirsutum* L.) by *Agrobacterium tumefaciens* and regeneration of transgenic plants. *Plant Mol. Biol.*, **10**: 105-116.

Fischoff D.A., Bowdish K.S., Perlak F.J., Marrone P.G., McCormick S.M., Niedermeyer J.G., Dean D.A., Kusano-Kretzmer K., Mayer E.J., Rochester D.E., Rogers S.G. and Fraley R.T. 1987. Insect tolerant transgenic tomato plants. *Bio/Technology*, **5**: 807-813.

Flavell R.B. 1994. Inactivation of gene expression in plants as a consequence of specific duplication. *Proc. Natl. Acad. Sci.*, USA, **91**: 3490-3496.

Fontana G.S., Sentini L., Caretto S., Frugis G. and Mariotti D. 1993. Genetic transformation in the grain legume *Cicer arietinum* L. (chickpea). *Plant Cell Rep.*, **12**: 194-198.

Fraley R.T., Rogers S.G., Horsch R.B., Sanders P., Flick J., Adams S., Bittner M., Brand L., Fink C., Fray J., Gallupi G., Goldberg S. and Woo S. 1983. Expression of bacterial genes in plant cells. *Proc. Natl. Acad. Sci.*, USA, **80**: 4803-4806.

Fromm M.E., Taylor L.P. and Walbot V. 1985. Expression of genes transferred into monocot and dicot plant cells by electroporation. *Proc. Natl. Acad. Sci.*, USA, **82**: 5824-5828.

Fromm M.E., Morish F., Armstrong C., Williams R., Thomas J.C. and Klein T.M. 1990. Inheritance and expression of chimeric genes in the progeny of transgenic maize plants. *Bio/Technol.*, **8**: 389-396.

Fu X., Duc L.T., Fontana G.S., Bong B.B., Tinjuangjun P., Sudhakar D., Twyman R.M., Christou P. and Kohli A. 2000. Linear transgene constructs lacking vector backbone sequences generate low-copy-number transgenic plants with simple integration patterns. *Transgenic Res.*, **9**: 11-19.

Fujimoto H., Itoh K., Yamamoto M., Kyozuka J. and Shimamoto K. 1993. Insect resistant rice generated by introduction of a modified delta-endotoxin gene of *Bacillus thuringiensis*. *Bio/Technology*, **11**: 1151-1155.

Galston A.W., Kaur S.R., Altabella T. and Tiburcio A.F. 1997. Plant polyamines in reproductive activity and response to abiotic stress. *Acta Bot.*, **110**: 197-207.

Gehlen J., Panstruga R., Smets H., Merkelbach S., Kleines M. and Porsch P. 1996. Effects of altered phosphoenolpyruvate carboxylase activities on transgenic C_3 plant *Solanum tuberosum*. *Plant Mol. Biol.*, **32**: 831-848.

Gianessi L.P. and Carpenter J.E. 1999. Agricultural Biotechnology: Insect control benefits. National Center for Food and Agricultural Policy, Washington D.C. USA (www.ncfap.org).

Goldberg R.B. 2001. From Cot curves to genomics. How gene cloning established new concepts in plant biology? *Plant Physiol.*, **125**: 4-8.

Goto F., Yoshihara T. and Saiki H. 2000a. Iron accumulation and enhanced growth in transgenic lettuce plants expressing the iron-binding protein ferritin. *Theor. Appl. Genet.*, **100**: 658-664.

Goto F., Yoshihara T. and Saiki H. 2000b. Iron accumulation in tobacco plants expressing soybean ferritin gene. *Transgenic Res.*, **7**: 173-180.

Goto F., Yoshihara T., Shigemoto N., Toki S. and Takaiwa F. 1999. Iron fortification of rice seed by the soybean ferritin gene. *Nature Biotechnol.*, **17**: 282-286.

Guilley H., Dudley R.K., Jonard G., Balazs E. and Richards K.E. 1982. Transcription of cauliflower mosaic virus DNA: Detection of promoter sequences, and characterization of transcription. *Cell*, **30**: 763-773.

Gupta A.S., Heinen J.L., Holaday A.S., Burke J.J. and Allen R.D. 1993. Increased resistance to oxidative stress in transgenic plants that overstress chloroplastic Cu/Zn superoxide dismutase. *Proc. Natl. Acad. Sci.*, USA, **90**: 1629-1633.

Hagio T., Kasiawazaki S. and Hirabayashi T. 1996. Genetic transformation of barley (*Hordeum vulgare* L.) using coat protein genes of barley mild mosaic virus (BaMMV) and barley yellow mosaic virus (BaYmv). *In vitro Plant, J. SIVB.*, **32**: 309.

Hamilton C.M., Frary A., Leewis L. and Tanksley S.D. 1996. Stable transfer of intact high molecular weight DNA into plant chromosomes. *Proc. Natl. Acad. Sci.*, USA, **93**: 9975-9979.

Hammond J. and Kamo K. 1995. Effective resistance to potyvirus infection conferred by expression of antisense RNA in transgenic plants. *Mol. Plant-Microbe Interact.*, **8**: 674-682.

Hammond-Kosack K.E. and Jones J.D.G. 1996. Resistance gene-dependent plant defense responses. *Plant Cell*, **8**: 1773-1791.

Hansen G.V. and Chilton M.D. 1996. "Agrolistic" transformation of plant cells: Integration of T-strands generated *in planta*. *Proc. Natl. Acad. Sci.*, USA, **93**: 14978-14983.

Hayashi H.A., Mustardy L., Deshnium P., Ida M. and Murata N. 1997. Transformation of *Arabidopsis thaliana* with the *codA* gene for choline oxidase: accumulation of glycine betaine and enhanced tolerance to salt and cold stress. *Plant J.*, **12**: 133-142.

Hayenga M.L. 1999. "Structural Change in the Biotech Seed and Chemical Industrial Complex." AgBioForum 1(2), 43-55. Retrieved 1 January 1999 from the World Wide Web: http://www.agbioforum.missouri.edu.

Herrera-Estrella L., De Block M., Messens E., Hernalsteens J.P., Van Montagu M. and Schell J. 1983. Chimeric genes as dominant selectable markers in plant cells. *Embo J.*, **2**: 987-995.

Hiei Y., Ohta S., Komari T. and Kumashiro T. 1994. Efficient transformation of rice (*Oryza sativa* L.) mediated by *Agrobacterium* and sequence analysis of the boundaries of the T-DNA. *Plant J.*, **6**: 271-282.

Hilder V.A. and Boulter D. 1999. Genetic engineering of crop plants for insect resistance: a critical review. *Crop Prot.*, **18**: 177-191.

Hilder V.A., Gatehouse A.M.R., Gatehouse J.A., Gatehouse L.N., Shi Y., Hamilton W.D.O., Merryweather A., Newell C.A., Timans J.C., Peumans W.J., Van Damne E. and Boulter D. 1995. Expression of snowdrop lectin in transgenic tobacco plants results in added protection against aphids. *Transgenic Res.*, **4**: 18-25.

Hinchee M.A.W., Connor-Ward D.V., Newell C.A., McDonnell R.E., Sato S.J., Gasser C.S., Fischoff D.A., Re D.B., Fraley R.T. and Horsch R.B. 1988. Production of transgenic soybean plants using *Agrobacterium*-mediated transformation gene transfer. *Bio/Technology*, **6**: 915-922.

Hodgson J. 1999. Monarch *Bt*-corn paper questioned. *Nature*, **17**: 627.

Holmberg N. and Bulow L. 1998. Improving stress tolerance in plants by gene transfer. Trends *Plant Sci.*, **3**: 61-66.

Horsch R.B., Fraley R.T., Rodges S.G., Sanders P.R., Lioyd A. and Hoffman N. 1984. Inheritance of functional foreign genes in plants. *Science*, **223**: 469-498.

Horton P. 2000. Prospects for crop improvement through the genetic manipulation of photosynthesis: morphological and biochemical aspects of light capture. *J. Exp. Bot.*, **51**: 475-485.

Hossain M., Bennett J., Datta S.K., Leung H. and Khush G.S. 2000. Biotechnology research in rice for Asia: priorities, focus and directions. *In*: *Agricultural biotechnology in developing countries: towards optimizing the benefits for the poor.* (eds.) M. Qaim, A. Kratigger, and J. von Braun, Kluwer Academic Publ.

Huang J., Hirji R., Adam L., Rozwadowski K.L., Hammerlindl J.K., Keller W.A., Selvaraj G. and Huang J. 2000. Genetic engineering of glycinebetaine production toward enhancing stress tolerance in plants: metabolic limitations. *Plant Physiol.*, **122**: 747-756.

Hudspeth R.L., Grula J.W., Dai Z., Edwards G.E. and Ku M.S.B. 1991. Expression of maize phosphoenolpyruvate carboxylase in transgenic tobacco: Effects on biochemistry and physiology. *Plant Physiol.*, **98**: 458-464.

IRRI 2000. Golden rice: The eyes of the world are watching. *IRRI Annual Report.* International Rice Research Institute, Los banos, Philippines.

Ishimoto M., Sato T., Chrispeels M.J. and Kitamura K. 1996. Bruchid resistance of transgenic azuki bean expressing seed alpha amylase inhibitor of common bean. *Entomol. Exp. Appl.*, **79**: 309-315.

Jaglo-Ottosen K.R., Gilmour S.J., Zarka D.G., Schabenberger O. and Thomashow M.F. 1998. *Arabidopsis* CBF1 overexpression induces COR genes and enhances freezing tolerance. *Science*, **280**: 104-106.

Jain R.K. and Wu R. 1999. Construction of vectors for constitutive and stress-inducible expression of salt and drought tolerant genes in rice. *In*: *Abstracts of the Rockefeller Foundation General Meeting of Int. Pro. on Rice Biotechnol.*, Phuket, Thailand. pp. 342.

James C. 2000. Global review of commercialized transgenic crops: 2000. *International Service for Acquisition of Agri-Biotech Applications Publ.*, New York, USA.

Jefferson R.A., Kacanagh T.A. and Bevan M.W. 1986. β-glucouronidase from *Escherichia coli* as a gene-fusion marker. *Proc. Natl. Acad. Sci.*, USA, **83**: 8447-8451.

Joersbo M., Donaldson I., Kreiberg J., Petersen S.G., Brunstedt J. and Okkels F.T. 1998. Analysis of mannose selection used for transformation of sugar beet. *Mol. Breed.*, **4**: 111-117.

Kaniewski W.K. and Thomas P.R. 1993. Field testing of virus resistant transgenic plants. *Seminars in Virology*, **4**: 389-396.

Kaniewski W.K. and Thomas P.E. 1999. Field testing for virus resistance and agronomic performance in transgenic plants. *Mol. Biotechnol.*, **12**: 101-115.

Karakas B., Ozias-Akins P., Stushnoff C., Suefferheld M. and Rieger M. 1997. Salinity and drought tolerance of mannitol accumulating transgenic tobacco. *Plant Cell Environ.*, **20**: 609-616.

Kasuga M., Liu Q., Miura S., Yamaguchi-Shinozaki K. and Shinozaki K. 1999. Improving plant drought, salt, and freezing tolerance by gene transfer of a single stress-inducible transcription factor. *Nature Biotech.*, **17**: 287-291.

Kaufmann S., Legrand M., Geoffroy P. and Fritig B. 1987. Biological function of 'pathogenesis-related' proteins: four PR proteins of tobacco have β-1, 3-glucanase activity. *EMBO J.*, **6**: 3209-3212.

Kharb P., Dong J., Islam-Faridi M.N., Jiang Y., Teng W., Yong X., Rasdale P., Stelly D.M. and Hall T.C. 1999. Tissue specific expression, gene silencing and genomic location of a rice root-specific transgene. *In*: *Abstracts of the Rockefeller Foundation General meeting of Int. Pro. on Rice Biotechnol.* Phuket, Thailand, pp. 356.

Khush G.S., Mackill D.J. and Sidhu G.S. 1989. Breeding rice for rediatance to bacterial blight. *In*: "*Bacterial Blight of Rice*" International Rice Research Institute, Los Banos, Philippines. pp. 207-217.

Kishor P.B.K., Hong Z., Miao G.H., Hu C.A.A. and Verma D.P.S. 1995. Overexpression of 1-pyrroline-5-carboxylate synthase increases proline production and confers osmotolerance in transgenic plants. *Plant Physiol.*, **108**: 1387-1394.

Klee H., Hayford M., Kretzmer K., Barry G. and Kishore G. 1991. Control of ethylene synthesis by expression of a bacterial enzyme in transgenic tomato plants. *Plant Cell*, **3**: 1187-1193.

Knight D.M., Flomerfelt F.A. and Ghrayed J. 1987. Expression of the art/trs protein of HIV and study of its role in viral envelope synthesis. *Science*, **236**: 837-840.

Kogami H., Shono M., Koike T., Yanagiswa S., Izui K. and Sentoku N. 1994. Molecular and physiological evaluation of transgenic tobacco plants expressing a maize phosphoenolpyruvate carboxylase gene under the control of the cauliflower mosaic virus 35S promoter. *Transgenic Res.*, **3**: 287-296.

Kong S.G., Jang I.C., Kim J.K. and Kim I.S. 1999. Transgenic rice plants overexpressing the *Arabidopsis* phytochrome A gene to suppress shade-avoidance response. *In*: *Abstracts of the Rockefeller Foundation General meeting of Int. Pro. on Rice Biotech.*, Phuket, Thailand. pp. 344.

Koziel M.G., Beland G.L., Bowman C., Carozzi N.B., Crenshaw R., Crossland L., Dawson J., Desai N., Hill M., Kadwell S., Launis K., Lewis K., Maddox D., McPherson K., Meghji M.R., Merlin E., Rhodes R., Warren G.W., Wright M. and Evola S.V. 1993. Field performance of elite transgenic maize plants expressing an insecticidal protein derived from *Bacillus thuringiensis*. *Biotechnology*, **11**: 194-200.

Ku M.S.B., Agarie S., Nomura M., Fukayama H., Tsuchida H., Ono K., Hirose S., Toki S., Miyao M. and Matusuoka M. 1999. High-level expression of maize phosphoenolpyruvate carboxylase in transgenic rice plants. *Nature Biotech.*, **17**: 76-80.

Lee S.B., Jeon J.S., Jung K.H. and An G. 1999. Binary vectors for efficient transformation of rice. *J. Plant Biol.*, **42**: 310-316.

Li Z., Upadhyaya N.M., Meena S., Gibbs A.J. and Waterhouse P.M. 1997. Comparison of promoters and selectable markers genes for use in *indica* rice transformation. *Mol. Breed.*, **3**: 1-14.

Lin W., Anuratha C.S., Datta K., Potrykus I., Muthukrishnan S. and Datta S.K. 1995. Genetic engineering of rice for resistance to sheath blight. *Bio/Technology*, **13**: 686-691.

Lindbo J.A. and Dougherty W.G. 1992. Untranslatable transcripts of the tobacco etch virus coat protein gene sequence can interfere with tobacco etch virus replication in transgenic plants and protoplasts. *Virology*, **189**: 725-733.

Londsale D., Onde S. and Cuming A. 1990. Transient expression of exogenous DNA in intact viable wheat embryos following particle bombardment. *J. Exp. Bot.*, **41**: 1161-1165.

Losey J.E., Rayor L.S. and Carter M.E. 1999. Transgenic pollen harms monarch larvae. *Nature*, **399**: 214.

Lucca P., Hurrel R. and Potrykus I. 2001. Genetic engineering approaches to improve the bioavailability and the level of iron in rice grains. *Theor. Appl. Genet.*, **102**: 392-397.

Malca I., Endo R.M. and Long M.R. 1967. Mechanism of glucose counteraction of inhibition of root elongation by galactose, mannose and glucosamine. *Phytopathology*, **57**: 272-278.

Mariani C., de Beuckeleer M., Truettner J., Leemans J. and Goldberg R.B. 1990. Induction of male sterility in plants by a chimeric ribonuclease gene. *Nature*, **347**: 737-741.

Matousek J., Schroder A.R.W., Trnena L., Reimers M., Baumstark T., Dedic P., Vlasak J., Becker J., Kreuzaler F., Fladung M. and Riesner D. 1994. Inhibition of viroid infection by antisense RNA expression in transgenic plants. *Biol. Chem.*, **375**: 765-777.

Mazur B., Krebbers E. and Tingey S. 1999. Gene discovery and product development for grain quality traits. *Science*, **285**: 372-375.

McElroy D., Zhang W., Cao J. and Wu R. 1990. Isolation of an efficient actin promoter for use in rice transformation. *Plant Cell*, **2**: 163-171.

McKersie B.D., Murnaghan J., Jones K.S. and Bowley S.R. 2000. Iron-superoxide dismutase expression in transgenic alfalfa increases winter survival without a detectable increase in photosynthetic oxidative stress tolerance. *Plant Physiol.*, **122**: 1427-1437.

Meer I.M. van der, Ebskamp M.J.M., Visser R.G.F., Weisbeek P.J. and Smeekens S.C.M. 1994. Fructan as a new carbohydrate sink in transgenic potato plants. *Plant Cell*, **6**: 561-570.

Moore A.E. and Stone B.A. 1972. Effect of senescence and hormone treatment on the activity of a β-1, 3-glucan hydrolase in *Nicotiana glutinosa* leaves. *Planta*, **104**: 93-109.

Moran J.F., Becana M., Iturbe-Ormaetxe I., Frechilla S., Klucas R.V. and Aparicio-Tejo P. 1994. Drought induces oxidative stress in pea plants. *Planta*, **194**: 346-352.

Murai N., Sutton D.W., Murray M.G., Slighton J.L., Merlo D.J., Reichert N.A., Sengupta-Gopalan C., Stock C.A., Barker R.F., Kemp J.D. and Hall T.C. 1983. Phaseolin gene is expressed after transfer to sunflower by tumor-inducing plasmid vectors. *Science*, **22**: 476-482.

Muthukumar B. and Narayanan K.K. 1998. Biotechnology for insect resistant plants. *In*: *Biopesticides in insect pest management*, (eds.) Ignacimuthu S., Sen A., Phoenix Publishing House, New Delhi. pp. 16-24.

Nanjo T., Kobayashi M., Yoshiba Y., Sanada Y., Wada K., Tsukaya H., Kakubari Y., Yamaguchi S.K. and Shinozaki K. 1999. Biological functions of proline in morphogenesis and osmotolerance revealed in antisense transgenic *Arabidopsis thaliana*. *Plant J.*, **18**: 185-193.

Napoli C., Lemieux C. and Jorgensen R. 1990. Introduction of chimeric chalcone synthase gene into *Petunia* results in reversible cosuppression of homologous genes in *trans*. *Plant Cell*, **2**: 279-289.

Ogawa T., Hori T. and Ishida I. 1996. Virus-induced cell death in plants expressing the mammalian 2'5' oligoadenylate system. *Nature Biotech.*, **14**: 1566-1569.

Ow D.W., Wood K.V., DeLucca M., DeWet J.R., Helinski D.R. and Howell S.H. 1986. Transient and stable expression of firefly luciferase gene in plant cells and transgenic plants. *Science*, **234**: 856-859.

Oxtoby E. and Hughes M.A. 1990. Engineering herbicide tolerance into crops. *Trends Biotech.*, **8**: 61-65.

Pandey S. and Velasco L. 1999. Economics of direct seeding in Asia, patterns of adoption and research priorities. *IRRN*, **2**: 6-11.

Perlak F.J., Fuchs R.L., Dean D.A., McPherson S.L. and Fischhoff D.A. 1991. Modification of coding sequences enhances plant expression of insect control protein genes. *Proc. Natl. Acad. Sci.*, USA, **88**: 3324-3332.

Perlak F.J., Deaton W.R., Armstrong T.A., Fuchs R.L., Sims S.R., Greenplate J.T. and Fischoff D.A. 1990. Insect-resistant cotton plants. *Bio/Technology*, **8**: 939-943.

Peterhans A., Datta S.K., Datta K., Goodall G.J., Potrykus I. and Paszkowski J. 1990. Recognition efficiency of *Dicotyledoneae*-specific promoter and RNA processing signals in rice. *Mol. Gen. Genet.*, **222**: 361-368.

Pilon-Smits E.A.H., Terry N., Sears T., Kim H.S., Zayed A., Hwang S.B., van Dun K., Voogd E., Verwoerd T.C., Krutwagen R.W.H.H. and Goddijn O.J.M. 1998. Trehalose-producing transgenic tobacco plants show improved growth performance under drought stress. *J. Plant Physiol.*, **152**: 525-532.

Pilon-Smits E.A.H., Ebskamp M.J.M., Paul M.J., Jeuken M.J.W., Weisbeek P.J. and Smeekens S.C.M. 1995. Improved performance of transgenic fructan-accumulating tobacco under drought stress. *Plant Physiol.*, **107**: 125-130.

Potrykus I. 1990a. Gene transfer to cereals: an assessment. *Bio/Technology*, **8**: 535-542.

Potrykus I. 1990b. Gene transfer to plants: assessment and perspective. *Physiol. Plant.*, **79**: 125-134.

Rai M. and Prasanna B.M. 2000. *Transgenics in agriculture*. ICAR Publ. India. pp.

Raineri D.M., Bottino P., Gordon M.P. and Nester E.W. 1990. *Agrobacterium*-mediated transformation of rice (*Oryza sativa* L.). *Bio/Technology*, **8**: 33-38.

Ramanathan V. and Veluthambi K. 1995. Transfer of non-T-DNA portions of *Agrobacterium tumefaciens* Ti plasmid pTiA6 from the left terminus of T-DNA. *Plant Mol. Biol.*, **28**: 1149-1154.

Rathore K.S., Chowdhury V.K. and Hodges T.K. 1993. Use of *bar* as a selectable marker gene and for the production of herbicide-resistant rice plants from protoplasts. *Plant Mol. Biol.*, **21**: 871-884.

Rigev A., Keller M., Strizhov M., Sneh B., Prudovsky E., Chet I., Ginzberg I., koncz Z., Schell J. and Zilberstein A. 1996. Synergistic activity of a *Bacillus thuringiensis* δ-endotoxin and bacterial endochitinase against *Spodoptera littoralis* larvae. *Appl. Environ. Microbiol.*, **62**: 3581-3586.

Robinson N.J., Procter C.M., Connolly E.L. and Guerinot M.L. 1999. A ferric-chelate reductase for iron uptake from soils. *Nature*, **397**: 694-697.

Romero C., Belles J.M., Vaya J.L., Serrano R. and Culianez-Macia F.A. 1997. Expression of the yeast trehalose-6-phosphate synthase gene in transgenic tobacco plants: pleiotropic phenotypes include drought tolerance. *Planta*, **201**: 293-297.

Roxas V.P., Smith R.K. Jr, Allen E.R. and Allen R.D. 1997. Overexpression of glutathione S-transferase/glutathione peroxidase enhances the growth of transgenic tobacco seedlings during stress. *Nature Biotech.*, **15**: 988-991.

Sakamoto A., Alia A. and Murata N. 1998. Metabolic engineering of rice leading to biosynthesis of glycinebetaine and tolerance to salt and cold. *Plant Mol. Biol.*, **38**: 1011-1019.

Sanford J.C. and Johnston S.A. 1985. The concept of parasite- derived resistance-deriving resistance genes from the parasite's own genome. *J. Theor. Biol.*, **113**: 395-405.

Sanford J.C., Klein T.M., Wolf E.D. and Allen N. 1987. Delivery of substances into cells and tissues using a particular bombardment process. *J. Part. Sci. Tech.*, **6**: 559-563.

Schellenbaum L., Sprenger N., Schuepp H., Wiemken A. and Boller T. 1999. Effects of drought, transgenic expression of a fructan synthesizing enzyme and of mycorrhizal symbiosis on growth and soluble carbohydrate pools in tobacco plants. *New Phytol.*, **142**: 67-77.

Sheehy R.E., Kramer M. and Hiatt W.R. 1988. Reduction of polygalacturonase activity in tomato fruit by antisense RNA. *Proc. Natl. Acad. Sci.*, USA, **85**: 8805-8809.

Shen B., Jensen R.G. and Bohnert H.J. 1997. Increased resistance to oxidative stress in transgenic plants by targeting manitol biosynthesis to chloroplasts. *Plant Physiol.*, **113**: 1177-1183.

Sheveleva E., Chmara W., Bohnert H.J. and Jensen R.G. 1997. Increased salt and drought tolerance by D-ononitol production in transgenic *Nicotiana tabacum* L. *Plant Physiol.*, **115**: 1211-1219.

Sheveleva E.V., Jensen R.G., Bohnert H.J. and Parry M. 2000. Disturbance in the allocation of carbohydrates to regenerative organs in transgenic *Nicotiana tabacum* L. *J. Exp. Bot.*, **51**: 115-122.

Shewmaker C.K., Sheehy J.A., Daley M., Colburn S., Ke D.Y. 1999. Seed-specific expression of phytoene synthase: increase in carotenoids and other metabolic effects. *Plant J.*, **20**: 401-412.

Shewry P.R. 1993. Barley seed proteins. *In*: *Barley: Chemistry and technology*, (eds.) A.W. MacGregor and R.S. Bhatty, Amer. Assoc. Cereal Chemists Inc., St. Paul, Minnesota. pp. 131-197.

Shimamoto K., Terada R., Izawa T. and Fujimoto H. 1989. Fertile transgenic rice plants regenerated from transformed protoplasts. *Nature*, **338**: 274-276.

Shukle R.H. and Murdock L.L. 1993. Lipoxygenase, trypsin inhibitor, and lectin from soybeans: Effect on larval growth of *Manduca sexta* (Lepidoptera: Sphingidae). *Environ. Entomol.*, **12**: 787-791.

Smith C.J.S., Watson C.F., Ray J., Bird C.J., Morris P.C., Schuch W. and Grierson D. 1988. Antisense RNA inhibition of polygalacturonase activity in transgenic tomatoes. *Nature*, **334**: 724-726.

Snow A.A. and Palma P.M. 1997. Commercialization of transgenic plants: Potential ecological risks. *Bioscience*, **47**: 86-96.

Somerville C.R. 1990. The biochemical basis for plant improvement. *In*: *Plant Physiology and Plant Molecular Biology*, (eds.) Dennis D.T. and Turpin D.H., Longman Group, Essex, UK. pp. 490-501.

Song W.Y., Wang G.L., Chen L.L., Kim H.S., Pi L.Y., Holsten T., Gardner J., Wang B., Zhai W.X., Zhu L.H., Fauquet C. and Ronald P.C. 1995. A receptor kinase-like protein encoded by the rice disease resistance gene, *Xa21*. *Science*, **270**: 1804-1806.

Sprenger N., Schellenbaum L., Dun K. van, Boller T., Wiemken A. and Van Dun K. 1997. Fructan synthesis in transgenic tobacco and chicory plants expressing barley sucrose: fructan 6-fructosyltransferase. *FEBS Lett.*, **400**: 355-358.

Sudhakar D., Fu X., Stoger E., Williams S., Spence J., Brown D.P., Bharathi M., Gatehouse J.A. and Christou P. 1998. Expression and immunolocalization of the snowdrop lectin, GNA, in transgenic rice plants. *Transgenic Res.*, **7**: 371-378.

Tabashnik B.E., Patin A.L., Dennehy T.J., Liu Y.B., Carriere Y., Sims M.A. and Antilla L. 2000. Frequency of resistance to *Bacillus thuringiensis* in field populations of pink bollworm. *Proc. Natl. Acad. Sci.*, **97**: 12980-12984.

Tacke E., Salamini F. and Rohde W. 1996. Genetic engineering of potato for broad-spectrum protection against virus infection. *Nature Biotech.*, **14**: 1597-1601.

Tada Y., Sakamoto M., Matsuoka M. and Fujimura T. 1991. Expression of a monocot LHCP promoter in transgenic rice. *EMBO J.*, **10**: 1803-1808.

Takeuchi Y., Akagi H., Kamasawa N., Osumi M. and Honda H. 2000. Aberrant chloroplasts in transgenic rice plants expressing a high level of maize NADP-dependent malic enzyme. *Planta*, **211**: 265-274.

Tarczynski M.C., Jensen R.G. and Bohnert H.J. 1993. Stress protection of transgenic tobacco by production of the osmolyte mannitol. *Science*, **259**: 508-510.

Tepperman J., Katayama C. and Dunsmuir P. 1988. Cloning and nucleotide sequence of a *Petunia* gene encoding a chloroplast localized superoxide dismutase. *Plant Mol. Biol. Int. J. Fund. Res. Genet. Eng.*, **11**: 871-872.

Thomas J.C., Sepahi M., Arendall B. and Bohnert H.J. 1997. Enhancement of seed germination in high salinity by engineering mannitol expression in *Arabidopsis thaliana*. *Plant Cell Environ.*, **18**: 801-806.

Tingay S., McElroy D., Kalla R., Fieg S., Wang M., Thornton S. and Brettell R. 1997. *Agrobacterium tumefaciens*-mediated barley transformation. *Plant J.*, **11**: 1369-1376.

Tinland B. 1996. The integration of T-DNA into plant genomes. *Trends Plant Sci.*, **1**: 178-184.

Toriyama K., Arimoto Y., Uchimiya H. and Hinata K. 1988. Transgenic rice plants after direct gene transfer into protoplasts. *Bio/Technol.*, **6**: 1072-1074.

Toki S., Takamatsu S., Nojiri C., Ooba S., Anzai H., Iwata M., Christensen A.H., Quail P.H. and Uchimiya H. 1992. Expression of a maize ubiquitin gene promoter-bar chimeric gene in transgenic rice plants. *Plant Physiol.*, **100**: 1503-1507.

Tsuchida H., Tamai T., Fukayama H., Agarie S., Nomura M., Onodera H., Ono K., Nishizawa Y., Lee-ByungHyun, Hirose S., Toki S., Ku M.S.B., Matsuoka M., Miyao M. and Lee B.H. 2001. High level expression of C4-specific NADP-malic enzyme in leaves and impairment of photoautotrophic growth in a C3 plant, rice. *Plant Cell Physiol.*, **42**: 138-145.

Tu J., Datta K., Alam M.F., Fan Y.L., Khush G.S. and Datta S.K. 1998a. Expression and function of a hybrid *Bt* toxin gene in transgenic rice conferring resistance to insect pest. *Plant Biotechnol.*, **15**: 195-203.

Tu J., Ona I., Zhang Q.F., Mew T.W., Khush G.S. and Datta S.K. 1998b. Transgenic rice variety IR72 with *Xa21* is resistant to bacterial blight. *Theor. Appl. Genet.*, **97**: 31-36.

Tu J., Zhang G., Datta K., Xu C., He Y., Zhang Q., Khush G.S. and Datta S.K. 2000a. Field performance of transgenic elite commercial hybrid rice expressing *Bacillus thuringiensis* δ-endotoxin. *Nature Biotechnol.*, **18**: 1101-1104.

Tu J., Datta K., Khush G.S., Zhang Q. and Datta S.K. 2000b. Field performance of *Xa21* transgenic *indica* rice (*Oryza sativa* L.), IR72. *Theor. Appl. Genet.*, **101**: 15-20.

Vaeck M., Reynaerts A., Hofte H., Jansens S., Beukeleer M.D., Dean C., Zabeau M., Montagu M.V. and Leemans J. 1987. Transgenic plants protected from insect attack. *Nature*, **327**: 239-247.

Van Breusegem F., Slooten L., Stassart J.M., Moens T., Botterman J., Van Montagu M. and Inze D. 1999. Overproduction of *Arabidopsis thaliana* FeSOD confers oxidative stress tolerance to transgenic maize. *Plant Cell Physiol.*, **40**: 515-523.

Van Camp W., Capiau K., Van Montagu M., Inze D. and Slooten L. 1996. Enhancement of oxidative stress tolerance in transgenic tobacco plants overproducing Fe-superoxide dismutase in chloroplasts. *Plant Physiol.*, **112**: 1703-1714.

Van Rensburg L. and Kruger G.H.J. 1994. Applicability of abscisic acid and (or) proline accumulation as selection criteria for drought tolerance in *Nicotiana tabacum. Can J. Bot.*, **72**: 1535-1540.

Vasil V., Castillo A.M., Fromm M.E. and Vasil I.K. 1992. Herbicide resistant fertile transgenic wheat plants obtained by microprojectile bombardment of regenerable embryogenic callus. *Bio/Technol.*, **10**: 667-674.

Waldron C., Murphy E.B., Roberts J.L., Gustafson G.D., Armour S.L. and Malcolm S.K. 1985. Resistance to hygromycin B: A new marker for plant transformation studies. *Plant Mol. Biol.*, **5**: 103-108.

Wan Y. and Lemaux P.G. 1994. Generation of large numbers of independently transformed fertile barley plants. *Plant Physiol.*, **104**: 37-48.

Wang G.L., Song W.Y., Wu R.L., Sideris S. and Ronald P.C. 1996. The cloned gene *Xa21* confers resistance to multiple *Xanthomonas oryzae* pv. *oryzae* isolates in transgenic plants. *Mol. Plant-Microbe Interact.*, **9**: 850-855.

Ward E.R., Uknes S.J., Williams S.C., Dincher S.S., Wiederhold D.L., Alexander D.C., Ahl-Goy P., Mitraux J.P. and Ryals J.A. 1991. Coordinate gene activity in response to agents that induce systemic acquired resistance. *Plant Cell*, **3**: 1085-1094.

Wenck A., Czako M., Kanevski I. and Marton L. 1997. Frequent colinear transfer of DNA inclusive of the whole binary vector during *Agrobacterium*-mediated transformation. *Plant Mol. Biol.*, **34**: 913-922.

Wenzel G. 1985. Strategies in unconventional breeding for disease resistance. *Ann. Rev. Phytopathol*, **23**: 149-162.

Wu C., Fan Y.L., Zhang C., Oliva N. and Datta S.K. 1997. Transgenic fertile japonica rice plants expressing modified *cry1A(b)* gene resistant to yellow stem borer. *Plant Cell Rep.*, **17**: 129-132.

Xu D.P., Duan X.L., Wang B.Y., Hong B.M., Ho T.H.D. and Wu R. 1996. Expression of a late embryogenesis abundant protein gene, HVA1, from barley confers tolerance to water-deficit and salt stress in transgenic rice. *Plant Physiol.*, **110**: 249-257.

Xu D., Xue Q., McElroy D., Mawal Y., Hilder V.A. and Wu R. 1996. Constitutive expression of a cowpea trypsin inhibitor gene, *CpTi*, in transgenic rice plants confers resistance to two major rice insect pests. *Mol. Breed.*, **2**: 167-173.

Xu D., McElroy D., Thornburg R.W. and Wu R. 1993. Systemic induction of a potato *pin2* promoter by wounding, methyl jasmonate and abscisic acid in transgenic rice plants. *Plant Mol. Biol.*, **22**: 573-588.

Yang X., Yie Y., Zhu F., Liu Y., Kang L., Wang X. and Tien P. 1997. Ribozyme-mediated high resistance against potato spindle tuber viroid in transgenic potatoes. *Proc. Natl. Acad. Sci.*, USA, **94**: 4861-4865.

Ye G.Y., Tu J., Hu C., Datta K. and Datta S.K. 2001. Transgenic IR72 with fused *Bt* gene *cry1Ab/cry1Ac* from *Bacillus thuringiensis* is resistant against four lepidopteran species under field conditions. *Plant Biotechnol.*, **18**: 125-133.

Ye X., Al-Babili S., Kloti A., Zhang J., Lucca P., Beyer P. and Potrykus I. 2000. Engineering the Provitamin A (β-carotene) biosynthetic pathway into (carotenoid free) rice endosperm. *Science*, **287**: 303-305.

Yeh S.D., Bau H.J., Cheng Y.H., Yu T.A., Yang J.S., Yeh S.D., Bau H.J., Cheng Y.H., Yu T.A., Yang J.S. and Drew R.A. 1998. Greenhouse and field evaluations of coat-protein transgenic papaya resistant to papaya ringspot virus. *Acta Hortic.*, **461**: 321-328.

Yeo E.T., Kwon H.B., Han S.E., Lee J.T., Ryu J.C. and Byun M.O. 2000. Genetic engineering of drought resistant potato plants by introduction of the trehalose-6-phosphate synthase (TPS1) gene from *Saccharomyces cerevisiase*. *Mol. Cells.*, **10**: 263-268.

Yoshiba Y., Kiyosue T., Nakashima K., Yamaguchi-Shinozaki K. and Shinozaki K. 1997. Regulation of levels of proline as an osmolyte in plants under water stress. *Plant Cell Physiol.*, **38**: 1095-1102.

Yoshikawa M., Tsuda M. and Takeuchi Y. 1993. Resistance to fungal diseases in transgenic tobacco plants expressing phytoalexin elicitor-releasing factor, β-1, 3-endoglucanase from soybean. *Naturwissenschaften*, **80**: 417-423.

Zhan X.Y., Wu H.M. and Cheung A.Y. 1996. Nuclear male sterility induced by pollen-specific expression of a ribonuclease. Sex. *Plant Reprod.*, **9**: 35-43.

Zhang S., Song W.Y., Chen L.L., Ruan D., Taylor N., Ronald P.C., Beachy R.N. and Fauquet C. 1998. Transgenic elite *indica* rice varieties, resistant to *Xanthomonas oryzae* pv. *oryzae*. *Mol. Breed.*, **4**: 551-558.

Zhang W. and Wu R. 1988. Efficient regeneration of transgenic rice plants from protoplasts and correctly regulated expression of foreign genes in the plants. *Theor. Appl. Genet.*, **76**: 835-840.

Zheng Z., Hayashimoto A., Li Z. and Murai N. 1991. Hygromycin resistance gene cassettes for vector construction and selection of transformed rice plants. *Plant Physiol.*, **97**: 832-835.
Zhou H., Arrowsmith J.W., Fromm M.E., Hironaka C.M., Taylor M.L., Rodiguez D., Pajeau M.E., Brown S.M., Santino C.G. and Fry J.E. 1995. Glyphosphate-tolerant CP4 and GOX genes as a selectable marker in wheat transformation. *Plant Cell Rep.*, **15**: 159-163.
Zhu B.C., Su J., Chang M.C., Verma D.P.S., Fan Y.L. and Wu R. 1998. Over-expression of a DELTA1-pyrroline-5-carboxylate synthase gene and analysis of tolerance to water- and salt-stress in transgenic rice. *Plant Sci.*, **139**: 41-48.
Zimmy J., Becker D., Brettschneider R. and Lvrz H. 1994. First transgenic triticale (x *Triticosecale* Wittmack). (Abstr) *In*: *Seventh Intl. Cong. Plant Tiss. Cell Cult.*, June 12-17, Firenze. S7-48.
Zupan J.R. and Zambryski P. 1995. Transfer of T-DNA from *Agrobacterium* to the plant cell. *Plant Physiol.*, **107**: 1041-1047.
Zupan J., Muth T.R., Draper O. and Zambryski P. 2000. The transfer of DNA from *Agrobacterium tumefaciens* into plants: a feast of fundamental insights. *Plant J.*, **23**: 11-28.

Plant Breeding - Mendelian to Molecular Approaches
H. K. Jain and M. C. Kharkwal (eds.)

15

Selection for Simple and Complex Traits

V.P. Gupta[1], G.S. Nanda[2] and Darbeshwar Roy[3]

Abstract

Selection is the essence of plant and animal breeding. Selection has been practiced for productive plant type ever since the potential of certain plant species was realized as food source. Before the rediscovery of Mendel's laws, selection was practiced without considering the genetics of the trait. After the rediscovery and end of controversy surrounding the nature of inheritance of quantitative variation, genetic architecture of qualitative, quasi-quantitative and quantitative traits was worked out. Selection procedures in case of different types of traits, and under various systems of matings were developed, applied and their efficiencies were compared. Expected response to selection under various selection schemes was worked out and the gap between the observed and the expected response was narrowed down by obtaining precise estimates of genetic parameters. With the advancement in genetical research isozymes, the gene products were suggested for use as markers for selecting other linked trait(s), but the limitation is that the numbers of markers available are small. Now with the advent of molecular marker technology, marker assisted selection (MAS) may be practiced for each of the QTLs (quantitative trait loci) determining the polygenic trait. At present, the limitation with this technology is that various methods detect QTL with poor precision (5-10 cM), but it is hoped that in future the QLTs will be located with very high precision, and this inturn will further improve the efficiency of selection and prediction of response with higher accuracy.

Introduction

Selection is one of the most important evolutionary forces. Evolution (via natural selection) and domestication (via artificial selection) has created and improved the plant species. Evolution is defined as the change in the gene frequency. Considering most changes are quantitative in nature, polygenes are important in evolution. During selection individuals are chosen to become parents for the next generation. In natural selection, individuals with more fitness (viability and fertility) value become parents for the next generation; whereas in human guided selection, individuals with high economic worth are

[1]Rajendra Agricultural University, Pusa 848 125, India
[2]Punjab Agricultural University, Ludhiana 141 004, India
[3]G.B. Pant University of Agriculture & Technology, Pantnagar 263 145, India

selected as parents. Ever since the potential of certain plant species as food crop was recognized, selection has been practiced for more productive plant type. Although accounts of the extent to which selection is effective appeared in the late eighteenth century, the nineteenth century contributed to the understanding of the effect of selection in self- and cross- pollinated crops. Selection was systematically applied only after the rediscovery of Mendel's laws of heredity and end of controversy about the nature of quantitative variation, and Johannsen's (1903) work with seed yield in bean, which led to the development of pure line theory of selection. Geneticists and breeders started working out the genetics of the traits on which selection applied was based.

Simple *versus* Complex Trait

Plant breeders manipulate traits such as morphological (plant height, maturity, etc.) anatomical, physiological, biochemical or chemical, resistance to biotic (pathogens, insects, nematodes) and abiotic (drought, flood, toxicity, alkalinity, salinity, chilling, etc.) stress besides stability and adaptability. Considering the underlying genetics, these traits can be grouped into qualitative and quantitative. The simple or simply inherited traits are Mendelian traits that are under the control of one or a few genes. These traits are less influenced by the environment. The genes controlling such traits may show complete dominance, partial dominance, over dominance or co-dominance. Gene at one locus may either act independently of the gene at another locus. The phenotypic effects of gene(s) can be changed or modified in a quantitative fashion by gene(s) at other loci called modifiers. The two genes may be linked (tightly or loosely), and finally, the gene(s) determining a trait may show multiple allelism as in case of self-incompatibility (S locus) gene and Rp locus in maize conferring resistance to rust. Economically important, simply inherited traits are resistances to pathogens and insects, for example, vertical resistance to leaf blight caused by *Phytophthora infestans* in potato and resistance to yellow rust, *Puccinia striiformis*, in wheat.

The complex traits are quantitative traits, that are under control of polygenes. These traits show continuous variation and take a bell shaped curve in contrast to discontinuous variation shown by a qualitative trait. These traits are much more influenced by the environment. They are also inherited in Mendelian fashion. Polygenes controlling these traits also show epistasis, linkage and pleiotropy like Mendelian factors. Most of the economically important traits such as yield and yield components are polygenic in nature.

Selection for Qualitative Trait

It is easier to score a qualitative trait and practice selection. Once an individual with a desirable qualitative trait is found, it can be easily fixed. The selected individual is selfed, and if no segregation occurs, then that trait is assumed fixed. But if segregation occurs in the selfing generation, then one will have to practice selection again, and distinguish between segregating and non-segregating selections by further selfing. Selection for a desirable qualitative trait is applied in F_2 generation in the pedigree method, and by F_4 one can isolate lines breeding true for that qualitative trait. If the qualitative trait is determined by a few genes, then selection is applied late in the inbreeding generations. In case of bulk population, qualitative trait will be fixed in the later generation population, and the individual plant selected will breed true for a qualitative trait; whereas, in case of single seed descent method, the individual line selected will breed true for a qualitative trait. In case of clonally propagated crop, the individual clone selected for a qualitative trait from F_1 generation will breed true, irrespective of whether it is fixed or not.

Selection for Qualitative Trait in Natural Population

Whether selection is for qualitative or quantitative trait, in a random mating population it increases the frequency of genes determining that trait. The population can be characterized by the gene and genotypic frequencies. The gene and genotypic frequency will remain constant in a H.W. equilibrium population provided the population is a large and undergoing random mating and there is no selection, mutation, migration and drift. Considering one locus with 2 alleles (A, a) with frequencies $f(A) = u$ and $f(a) = v$ and $u + v = 1$, the three genotypes, *AA*, *Aa* and *aa* with frequencies u^2, $2uv$ and v^2, respectively, will appear in the population. It can be shown mathematically that selection is most effective at intermediate gene frequency ($u = v = 0.5$) and it is least effective when either u or v is very large or small. Further, it can be shown that when a favourable gene appears in the population, it spreads very slowly initially, but becomes faster at intermediate gene frequency (Falconer, 1960). The best example of selection comes from human inherited disease achondroplasia, a dominant mutation, resulting in dwarfism and phenylkaptonuria, a recessive mutation causing mental retardation. In case of dominant mutation the homozygote (aa) is usually lethal, heterozygote (*Aa*) is defective and less fit than *aa*, and homozygote (*AA*) is normal; whereas in case of recessive mutation *AA* and *Aa* are normal, whereas *aa* has lower fitness and there exists a balance between mutation and selection. In the second example, sickle cell anemia another human inherited disorder is caused by a single recessive gene (*h*) and recessive homozygotes (*hh*) suffer from sickle cell anemia but show resistance to malaria. The dominant homozygotes (*HH*) suffer from malaria but show resistance to anemia and the heterozygotes (*Hh*) show resistance against both anemia and malaria. Heterozygotes are thus more fit than homozygotes and selection keeps the frequency of deleterious gene at higher frequency by way of removing recurrent mutant and the segregants from the favoured heterozygote.

Selection for Quantitative Traits

The success of selection in case of quantitative traits depends on (*i*) variation in the trait and (*ii*) the underlying genetic causes of variation. The total phenotypic variation can be partitioned into genotypic variation and environmental variation (Fisher, 1918), given by

$$\sigma_P^2 = \sigma_G^2 + \sigma_E^2$$

The genotypic variation (σ_G^2) can be further partitioned into additive, dominance and interaction variation, as

$$\sigma_G^2 = \sigma_A^2 + \sigma_D^2 + \sigma_I^2$$

Thus the total phenotypic variation (σ_p^2) becomes

$$\sigma_p^2 = \sigma_A^2 + \sigma_D^2 + \sigma_I^2 + \sigma_E^2$$

In the presence of genotype × environment interaction

$$\sigma_p^2 = \sigma_G^2 + \sigma_{G \times E}^2 + \sigma_E^2$$

$$= \sigma_A^2 + \sigma_D^2 + \sigma_I^2 + \sigma_{A \times E}^2 + \sigma_{D \times E}^2 + \sigma_{I \times E}^2$$

The estimates of different components of variation can be obtained using means, variances and covariances of basic generations and multiple mating designs (Comstock and Robinson, 1952; Falconer, 1960; Mather and Jinks, 1971; Kearsey and Jinks, 1968).

Types of Variability

There are two types of variability: free or expressed variability and potential or hidden variability. Selection acts on the free variability and the potential variability must be converted into free variability through recombination and segregation. The concept of free and potential variability can be understood by considering the following example. In case of two loci (*A*, *B*) with two alleles system (*A*, *a* and *B*, *b*) and assuming strictly additive model, the two pure breeding homozygous lines will be AABB and aabb with genotypic values as +2*d* and -2*d*, where d is the additive genetic effect and measured as deviation from the mid-parental value. The free variability results from the phenotypic differences between the homozygous parents and is measured by the variance, the additive genetic variance whereas, the potential variability results from additive balance per locus (heterozygosity) heterozygotic potential, for example AaBb (0), AaBB (+*d*), Aabb (–*d*), AABb (+*d*), aaBB (–*d*) and additive balance between loci (balanced homozygote) for example, aaBB (0), AAbb (0), the homozygotic potential. Thus, the F_1 (AaBb) of the two parents AABB and aabb will show heterozygotic potential variability and will not be acted upon by selection. In the F_2 and subsequent inbreeding generations, the proportion of free and potential variability will take the following form (Mather, 1973).

	Variability		
	Free	Potential	
Generation		Heterozygotic	Homozygotic
Parents (AABB,aabb)	$4d^2$	0	0
F_1	0	$4d^2$	0
F_2	d^2	$2d^2$	d^2
F_3	$3/2d^2$	d^2	$3/2d^2$
F_4	$7/4d^2$	$1/2d^2$	$7/4d^2$
F_∞	$2d^2$	0	$2d^2$

It shows that all the variability in the F_1 will be of heterozygotic potential type. In the F_2 generation the usable free variability is 1/4 of the original variability of parental generation and the remaining 3/4 variability is of potential form, out of which 1/2 is heterozygotic potential and 1/4 is homozygotic potential. Considering n loci the ratio of free, heterozygotic and homozygotic potential variability in F_2 will be in the ratio of 1 : n heterogygotic : n–1 homozygotic. Extension of this additive model to a polygenic case shows that out of a large total variation, only a small amount of variation exists as free and sensitive to selection in any one generation and that a large amount of variation exist as potential. The potential variability locked into heterozygotic form (AaBb) is released as free variability upon recombination and segregation and the homozygotic potential variability locked into AAbb and aaBB will then be converted into heterozygotic potential form by way of crossing AAbb with aaBB, which will then be converted into free variability through recombination and segregation.

If the genes determining polygenic trait are linked, then the flow of variation between free and potential forms will be significantly affected. Theory favours the evolution of repulsion phase linkage, in which increasing (+) and decreasing (–) alleles alternate along a given chromosome, which ensures a moderate release of free variation and a large store of potential variation. The formation of homozygotes contributing to free variability will be largely dependent on the rate of recombination, and further multiple crossing over will be required to release all the variability. This shows that the rate of release of free variability will be slow but continuous, which is supported from observation in case of selection for oil and protein content in maize. Significant gains were recorded in oil and protein per cent even after 95 cycles of selection even though the population was closed, i.e., with no introduction of additional genetic variability and with narrow genetic base. Recombination *per se* contributed to the creation of genetic variability upon which selection acts. Two explanations were put forward : One, a vast amount of genetic diversity have been present in the founding population and two, that these reservoirs have not been exhausted or the novel genetic diversity is being generated during the course of selection. Intragenic recombination serves as a mechanism for creation of novel alleles that contribute to the genetic diversity. A large proportion of variation occurs within genetic sequences and these events can create novel alleles. It could probably account for a very large proportion of the total recombination that occurs in maize genome. Hence, meiotic recombination may be one of the mechanisms that generate vast reservoir of genetic diversity.

Linkage Disequilibrium

If the two loci A and B are in coupling phase and the rate of recombination is p then the F_1 (AaBb) of the cross AABB × aabb will produce gametes aB, Ab, AB and ab in the ratio of $1/2p$, $1/2p$, $1/2(1\text{-}p)$ and $1/2(1\text{-}p)$, respectively. In case of repulsion phase linkage the F_1 of the two parents AAbb and aaBB will produce four types of gametes in the proportion $1/2(1\text{-}p)$, $1/2(1\text{-}p)$, $1/2p$ and $1/2p$ with p as the rate of recombination. Thus, the proportion of free and potential (homozygotic) variability in the F_2 generation population will be as follows :

Linkage phase	Free	Homozygotic potential
Coupling (AABB × aabb)	$2(1\text{-}p)\ d^2$	$2pd^2$
Repulsion (AAbb × aaBB)	$2pd^2$	$2(1\text{-}2p)\ d^2$

It shows that the free or potential variability is dependent on the rate of recombination. In case of large random mating population at equilibrium, the frequencies of four types of gametes produced would be independent of the recombination value and depend only on the gene frequencies. Assuming p, q, r and s being the frequencies of alleles A, a, B and b the gametic frequencies of the randomly produced gametes AB, Ab, aB and ab would be pr, ps, qr and qs, respectively and it will remain constant, generation after generation provided there is no selection. Selection is one of the causes of linkage disequilibrium and thus the gametic disequilibrium which is enhanced by linkage, is a potential means of detecting selection in the natural population. In case of epistasis, selection will act on the additive × additive type of epistasis, which is fixable.

Types of Selection in Natural Population

There are three types of selection: (*i*) Stabilizing selection, (*ii*) Directional selection and (*iii*) Disruptive selection.

Stabilizing Selection: When the selection favours optimal phenotypes approximately close to the mean of the expressed variation, it is called stabilizing selection. Here optimal phenotype does not imply a single optimal genotype. In a polygenic system, a number of genotypes can have very similar, if not identical, phenotypes and the selected individual will give rise to progeny with the same parental mean on an average and with minimal variation among progenies. So the stabilizing selection will stabilize the population at this mean with least variation. The main effect of stabilizing selection is a decrease in phenotypic variance. Considering A and B loci with two alleles system, if selection is for intermediate phenotype, i.e. if there is stabilizing selection then the gametic frequencies of aB and Ab will be greater than AB and ab, and it will lead to the development of dispersion/repulsion linkage. Most traits in natural population exist under stabilizing selection, but changes in environmental condition will lead to directional selection.

Directional Selection: If the selection favours some phenotype and hence the associàted genotypes which deviates from the mean of the population, directional selection is said to be operative. Here the mean of offspring produced by the selected phenotype will result in a higher mean as compared to that of the original population. Directional selection is applied in practical plant and animal breeding programmes. There is directional selection for yield in crop plants and milk yield in animals. In case of traits where lower mean values are desirable, individuals with lower mean values than the mean of the population are selected, with the objective of developing a population with lower mean than the original population mean. In comparison to stabilizing selection which favours genetical invariance, directional selection favours flexibility of genetical variation. This is because the response to directional selection implies change in mean, but continuous change in mean requires variation, the presence of free heritable variation.

Fisher (1930) gave a mathematical treatment of the response of a population to directional selection in case of fitness trait and he proposed the fundamental theorem of natural selection, which states that the rate of increase in fitness of any organism at any time is equal to its genetic variance in fitness at that time where genetic variance is of additive genetic type (D).

Considering 2 loci with 2 alleles system, the directional selection goes on in the direction of AB from aa to Ab or aB to AB. Occurrence of ab will be rare, Ab or aB will be intermediate and AB will be more common and there is thus development of a negative linkage disequilibrium. Yield and other economically important traits are under directional selection in case of artificial selection. Fitness traits such as viability and fertility are under directional selection in natural population.

Disruptive Selection: When a population is raised in two or more ecological niches, the selection will be towards two or more optimal phenotypes, each characteristic of one of the niches within one distribution, and thus will lead to development of a bimodal or multimodal distribution. Such a selection is called disruptive selection. It will have a disruptive effect on the continuity of variation in the population. The presence of two or more optima leads to an increase in variance. With 2 loci and 2 alleles system, the two extremes AB and ab are favoured and the intermediate aB and Ab are disfavoured, and thus there is development of a positive linkage disequilibrium. The development of an individual is thus channelled into one optimal phenotype or another by a developmental switch mechanism. The switch mechanism may involve environmental cues (spatial, seasonal, nutritional) or be genetic (nuclear gene, fertilization, cytoplasm). When the optimal alternate in time or space, a common genotype may characterize a population over several generations but produce distinct phenotypes under different environmental conditions. In practical plant breeding programme, such disruptive selection is being applied in a breeding method called shuttle breeding for developing a variety with higher stability and adaptability. Where optima coexist, their relative frequencies may be critical to the function of each. Environmental cues may also operate in a few such systems. Where genetic switch mechanism operates, it has the capacity to adjust more finely to given frequencies.

Response to Selection

Suppose there is a population with the mean as $\overline{Y}_P$. Let all the individuals having phenotypic values $\geq X$ be selected as parents for the next generation. Assuming the mean of the selected parents as $\overline{Y}_s$ and the mean of the progeny as $\overline{Y}_R$

$\overline{Y}_s - \overline{Y}_P = S$ = Selection differential

And the response to selection (R) is

$R = \overline{Y}_s - \overline{Y}_P$

Theoretically expected response to selection (R) is

$R = h^2 \times S$ (Falconer, 1960)

where h^2 is the heritability of that trait and calculated as σ_A^2/σ_P^2. Heritability calculated as

$$h^2 = R/S$$

is called the realized heritability. If $R/S = 1.0$ it indicates that all the phenotypic variation consists wholly of additive genetic variance and there is 100 % transfer of the phenotypic values from selected parents to the progeny. A ratio of < 1.0 shows the mean degree to which additive genetic variation contributes to the total phenotypic variation.

To make R unitless in order to facilitate comparison of response between different traits and different populations, both R and S are expressed in terms of phenotypic standard deviation, σ_P, R/σ_P and S/σ_P. Thus the generalized response will be

$$R/\sigma_P = (S/\sigma_P)\, h^2 = ih^2, \text{ where } i = S/\sigma_P \text{ and } R = \sigma_P i\, h^2$$

where $i(Z/p)$ is called the intensity of selection. Here p is the proportion of selected individuals and Z is the height of the ordinate at the point of truncation X. The value of i depends on the proportion of the total selected as parents.

The speed of selection thus depends on (*i*) selection intensity, (*ii*) number of genes/effective factors, (*iii*) additive genetic and dominance variance, (*iv*) linkage, (*v*) magnitude of environmental component, (*vi*) initial amount of free and potential variability, and (*vii*) sampling error. This shows that the efficiency of selection will depend on how precisely we have worked out the genetic architecture of a trait. Theoretically response predicted is valid only for one generation, though in practice its predictive value can hold for several generations. Selection will result in change of heritability estimate, which decreases as the selection accumulates more and more homozygotes and variation becomes fixed.

The response will be lower if a trait is correlated with another trait. Furthermore, as different phenotypes may achieve a given economic objective, although the paths of achieving the objective are different, selection response will be different for different paths adopted. For example, high yield in maize can be achieved through either increasing the cob size or increasing the number of ears, but the response to selection for these two paths will be different.

Correlated Response and Selection Limits

Response to selection in one trait often results in response in other trait(s). This could be as a result of linkage and or pleiotropy. The most typical correlated response consists of reduction in fertility, viability and other fitness traits. The correlated response of trait Y (CR_Y) as a result selection for trait X is

$$CR_Y = ih_x\, h_Y\, r_A\, \sigma_{PY}$$

where, r_A is genetic correlation and σ_{PY} is the standard deviation of trait Y. Thus the response of a correlated character can be predicted if genetic correlation and heritability of the two traits are known. The genetic correlation between traits will change as selection is continued.

Selection Plateau

Plateau in population can arise as a result of a variety of reasons, which may be broadly divided into physiological and genetic limits. Selection limit is based on the assumption that selection will lead to the fixation of all favourable alleles with the consequent loss of genetic variance. Response to selection will be rapid in the beginning but soon plateau, which will be judged by the observed fall in the genetic variance, suggesting that genetic limit has been reached. When selection has been applied to two or more traits simultaneously, then like heritability the positive correlation coefficient between traits initially will change and eventually the genetic correlation become negative at plateau. Plateau can appear for some traits because of physiological limits. These are the traits which are affected by a number of developmental processes.

Methods of Selection

There are three methods of selection. (*i*) Individual or mass selection, (*ii*) Family selection and (*iii*) Within family selection.

In case of mass or individual selection, individual's own performance is the source of information about its breeding value and thus individuals with high phenotypic values are selected. It performs better for traits with intermediate to high heritability. Family selection can be of three types : (*a*) full-sib family selection, (*b*) half-sib family selection and (*c*) S_1 or S_2 family section. Family selection performs better for traits having low heritability. When a family's performance is high because of common environmental effect or maternal effect and the heritability is low for that, then within family selection is practised. In the combined selection two or more selection methods can be used, either simultaneously or alternatively in the same breeding programme. For example, in pedigree method individual selection is practised in F_2 followed by family selection in F_3. The selection procedures employed in self-and-cross-fertilizing crops for improving complex polygenetic traits are given in Table 1.

These selection procedures are discussed in detail elsewhere (Hallauer and Miranda, 1981 and Roy, 2000). In the population improvement method, the objective is to increase the frequencies of desirable alleles in the population while maintaining a high degree of heterozygosity and recurrent selection is practised to attain such a goal. Recurrent selection has also been suggested as a means of improving quantitative trait in self-fertilizing crops using male sterility (Brim and Stuber, 1973). Gilmore (1964) suggested use of reciprocal recurrent selection in some naturally self-pollinated species. If two populations are to be improved simultaneously, than one can employ reciprocal recurrent selection.

Table 1. Selection procedures in self- and cross-pollinated crops

I. Selection procedure in self-pollinated crops

	Selection procedure	Method of selection
1.	Pedigree	Individual selection, family selection, within family selection
2.	Bulk	Natural selection, individual selection
3.	Single seed descent	Individual selection
4.	Dihaploidy	Individual selection
5.	Back crossing	Individual selection
6.	Diallel selective mating (Jenson, 1970)	Individual selection, family selection
7.	Composite cross breeding (Suneson, 1956)	Individual selection, family selection

II. Selection methods in cross-pollinated crops

Population improvement		Development of inbreds
Intra-	Inter-	
1. Individual or mass selection	1. Half-sib reciprocal recurrent selection (Comstock *et al.*, 1949)	1. Pedigree
2. Family (*a*) Full-sib (*b*) Half-sib	2. Full-sib reciprocal recurrent selection (Hallauer and Eberhart, 1970)	2. Backcross
(*i*) Ear-to-row, Modified ear-to-row (Hopkins, 1899; Lonnquist, 1964)	3. Modified reciprocal recurrent selection (Paterniani, 1973)	3. Single seed descent
(*ii*) Top-cross progeny (*a*) Recurrent selection for general combining ability (Jenkins, 1935) (*b*) Recurrent selection for specific combining ability (Hull,1945)		4. Gamete selection (Stadler, 1944) 5. Dihaploidy
3. Selfed progeny (S_1 or S_2) selection		

Selection Criterion

In the above selection procedures, the selection can be practised for one, two, or more traits at a time and thus selection can be of two types: (*i*) Single trait selection and (*ii*) Multiple trait selection. In the single trait selection one character is selected at a time and selection is practised for that trait. Once a desired level of improvement is made, selection is abandoned for that trait and another trait is selected for improvement, and thus different traits are improved in succession. This tandem method of selection takes longer time to improve the different desired traits and there is problem with the negatively correlated traits. In the independent culling level different limits are set for different traits, and individuals having phenotypic values equal or greater than the set limits are selected to become the parents for the next generation. This type of selection is being practised by most plant and animal breeders. In case of selection index (Fisher, 1936; Smith, 1936), an index value is given to each individual or family and only those individuals or families which have index values ≥ the set level of index value are selected to produce the progenies. The index is constructed as follows:

$$I = b_1 X_1 + b_2 X_2 + \ldots + b_n X_n$$

where, I is the aggregate phenotypic worth of an individual, $X_1, X_2, \ldots X_n$ are the phenotypic scores of n characters on the individuals and $b_1, b_2, \ldots b_n$ are the respective coefficients, called weights. The aggregate genetic worth (T) of an individual can be written as

$$T = V_1 Y_1 + V_2 Y_2 + \ldots\ldots + V_n Y_n$$

where V_i 's are the economic weights of the n traits, respectively, Y_i 's are the additive genetic values for the n traits. The maximization of correlation between I and T leads to the following set of equations, given in the matrix form.

$$Pb = GV$$

where, P is the matrix of phenotypic variances and covariances, G is the matrix of genotypic variances and covariances, and V is the vector of relative weights and b is the vector of unknown b values. This is the best method for improving the aggregate phenotype and will be most useful in population improvement programme. The efficiency of selection index increases with increase in the number of traits and decreasing differences in the relative importance of traits. Kempthorne and Nordskog (1958) modified the standard index and called restricted selection index for a situation in which changes are required in n_1 traits of n ($= n_1 + n_2$) traits, and no change is required in n_2 traits. Tallis (1962) modified the restricted selection index for cases where it was desirable to improve n_1 traits without any limits but n_2 traits only to a pre-determined level. Williams (1962) suggested the construction of an index called base index, in which the traits are weighed directly by their economic values. Pesek and Baker (1969) suggested the use of "desired gains" for obtaining weights, where assigning relative economic weight is a problem.

Increasing Efficiency of Selection

In order to improve the efficiency of selection, it is necessary to accurately estimate the genetic worth of an individual or family. This can be achieved by controlling and measuring the environmental sources of variation. Various field designs such as grids, rhombic grids and honeycombs were developed in this regard (Gardner, 1961; Fasoulas, 1983) to reduce the effect of soil heterogeneity, competition and $g \times e$ interaction. Moving block method (Fasoulas, 1987) was suggested to reduce the $g \times e$ interaction in case of family.

Use of Isozymes/proteins in Selection

Isozymes/proteins are the gene products which can be assayed electrophoretically. Isozyme as a trait can be used for determining the degree of genetic differentiation between taxa, for detecting abnormal segregation and for comparing recombination rates (Tanksley and Rick, 1980). Isozymes/protein profiles can be used for characterization of a genotype. Isozymes have shown linkages to important agronomical traits. Thus, where this biochemical trait is tightly linked with monogenic, morphological, physiological, phytopathological traits, it can be used as a marker for selection of that trait. If the two parents and their F_1 differ in their isozyme profile, isozymes can be used as marker in selection in backcrossing, pedigree, bulk and single seed descent method in self-pollinated crops. Isozyme is particularly useful when the gene(s) to be introduced into a cultivated crop plant is from a wild species, i.e. it will be useful in interspecific crosses or crosses involving distant relatives. Within cultivars isozyme selection offers little or no help, since most cultivars and land races have almost identical genotype for isozymic loci. Whether heterozygotes have higher competitive ability and are more fit than homozygotes can be checked in case of bulk population. In case of pedigree method it will help identifying plants with superior performance because of heterozygosity. In case of cross pollinated crops, isozymes can be studied to estimate allozymes frequencies for calculating the amount of heterozygosity present in the population, as the objective in recurrent selection scheme is to increase the frequency of desirable alleles while maintaining a high degree of heterozygosity. In case of synthetics the rate of out-crossing between individual inbreds and overall crossing rate can be estimated.

In case of clonally propagated crops, the superior clone(s) representing highly heterotic combination or maximum heterozygosity can be isolated in the F_1 generation. The use of this trait in selection will result in saving of time, space and effort in selecting desirable plant type at the seedling stage.

Variation at gene product level has not produced visible morphological variation and there are limitations with isozymes in assessing quantitative genetic variation. However, isozymes markers have been used to identify quantitative trait loci (QTL) in maize, tobacco, wild oats, soybean, etc. (Stuber, 1992). Stuber and Edwards (1986) demonstrated the effectiveness of isozyme marker in marker assisted selection in maize.

Application of Molecular Marker in Selection

In the marker assisted selection, the linkage of a molecular marker with QTLs (quantitative trait loci) determining a quantitative trait is used to identify and locate the QTLs. The QTL mapping thus allows the estimation of various genetic parameters, such as genetic effect, the number of QTLs, heritability, etc., on which the response to selection depends. Once a QTL is precisely located, its effect can be measured very accurately, and this will help reduce the gap between the expected and the observed response to selection. The various molecular marker technologies that are being used are restriction fragment length polymorphism (RFLP), random amplified polymorphic DNAs (RAPD) (Williams *et al.*, 1990), amplified fragment length polymorphism (AFLP) and mini- and micro-satellites (Rafalski and Tingey, 1993). Also, the different methods for estimating genetic effects are single marker (Weller, 1986), interval mapping (Lander and Botstein,

1989), composite interval mapping (Haley and Knot, 1992) and marker regression (Kearsey and Hyne, 1994). At present the problem with these methods is that they locate QTL with poor precision (5-10 cM). The estimate of QTL effects, particularly the dominant genetic effect is high and estimate of number of QTLs per trait is generally low (< 8). There is no large correlation between the numbers of QTL found and the amount of genetic variation explained (Kearsey and Farquhar, 1998). The response due to marker aided selection, which is a function of heritability and the percentage of additive genetic variance accounted for by the marker ($\sigma^2 M/\sigma^2 A$) or QTL ($\sigma^2 Q/\sigma^2 A$) (Smith, 1967; Lande and Thompson, 1990) has been maximized but the gap between the observed and the expected response still remains.

Molecular markers have shown linkages with agronomically important monogenic as well as QTLs determining polygenic traits (Dudley, 1993; Lee, 1995; Paterson *et al.*, 1993), and thus can be used as marker for improvement of these traits (Table 2) in the breeding procedures such as pedigree, backcross, population improvement and hybrid breeding (Beckmann and Soller, 1983; Soller and Beckmann, 1986).

Selection Gains in Plant Breeding

There are numerous examples of gains due to selection but only a few are mentioned here. Selection for monogenic traits has resulted in the development of stem rust (*Puccinia graminis* f. sp. *tritici*) and leaf rust (*Puccinia recondita*) resistant varieties in wheat in India. The genes used were *Sr2* and *Lr34*, respectively. Development of a series of potato varieties with *R* genes against races of leaf blight (*Phytophthora infestans*) is well known. Genes H_{13}, H_{22}, H_{23}, H_{24} and H_{26} showing risistance to Hessian fly have been transferred to hexaploid wheat.

With regard to improvement in polygenic traits, wherever selection has been applied, significant improvements have been made. Whether it is cereals, legumes, oilseeds, fodder crops, sugar crops or any other crops and whether it is pure breeding varieties, hybrids, inbreds, composites or synthetics, yield levels have increased tremendously as a result of man guided selection. For example, in maize, during the period 1890-1930, the rate of gain in yield was about 1/50 bushel/acre/year. The rate of gain reached to slightly more than one bushel/acre/year (50 times higher during the 30-year period of double-cross hybrid and in 25-year period of single-cross hybrid). The annual rate of gain rose to nearly 1.76 bushel/acre/year. The well known examples of successful selection was development of wheat varieties such as Larma Rojo 64, Sonora 64, Kalyan Sona and Sonalika, which were the basis for green revolution that began in 1968-69 in India. These varieties are still popular today. Similarly, in case of rice, TN1, IR-8, Jaya, IR-20, IR-24, IR-36 and IR-72 varieties developed at IRRI were the result of successful selection. The development of quality protein maize (QPM) varieties with high lysine content developed by Vasal and his colleagues (1979) of CIMMYT, is the result of selection. Finally the development of '0' erucic acid varieties in *Brassica* sp. is an example of selection for oil quality (Downey, 1964). Even our present day varieties are having higher adaptability in comparison to local ones, which has been made possible because of selection.

Table 2. The associations between molecular markers and agronomically important traits in different crops such as wheat, rice, maize and tomato

Trait	Gene
Wheat	
Dwarfing genes	*Rht 12, Rht 8, Rht-B1, Rht-D1*
Kernel hardness	*Ha*
Al tolerance	*Alt 2, Alt BH*
Fertility restoration	*Rf 4, Rf 3, 5DRF*
Leaf rust resistance	*Lr1,3, 9, 10, 13, 18, 19, 23, 24, 27, 28, 29, 31, 32, 34, 35*
Powdery mildew resistance	*Pm, Pm 1, 2, Pm 3b, Pm 4a, 4b, 6, 12, 21*
Grain protein content	QTL
Pre-harvest sprouting tolerance	QTL
Haploid formation	QTL
Green plant formation	QTL
Rice	
Semi dwarf character	*Sd-1*
Brown plant hopper resistance gene	*Bph-1*
Bacterial leaf blight resistance	*Xa-21, Xa-1*
Blast resistance	*Pi-2 (t), Pi-4(t)*
Maize	
Viral resistance	
Maize dwarf mosaic virus	Major gene
Maize streak virus	Major gene
Maize mosaic virus	Major gene
Insect resistance	
European corn borer	QTLs
South eastern corn borer	QTLs
Corn ear worm	QTLs
Fungal resistance	
Northern leaf blight	QTLs
Fusarium ear rot	QTLs
Agronomic and morphological factors	
Yield	QTLs
Plant height	QTLs
Anthesis	QTLs
Silking	QTLs
Abiotic stresses	
Drought	QTLs
Tomato	
Resistance to TMV	*Tm-2a*
Resistance against root knot nematode	*Mi*
Resistance to *Pseudomonas syringae*	*Pto*

With the advances in molecular marker technology and other biotechnological tools and improvement in field designs, a fine characterization of trait and the genotype would be possible in future. This in turn will increase the efficiency of selection, and the prediction of response to selection will be with much higher accuracy. The yield plateau is expected to break and further improvement is expected.

References

Allard R.W. 1960. *Principles of Plant Breeding*. John Wiley and Sons, New York.

Beckmann J.S. and Soller M. 1986. Restriction fragment length polymorphisms in plant genetic improvement: Methodologies, mapping and costs. *Theor. Appl. Genet.*, **67**: 35-43.

Brim C.A. and Stuber C.W. 1973. Application of genetic male sterility to recurrent selection schemes in soybean. *Crop Sci.*, **13**: 528-530.

Comstock R.E., Robinson H.F. and Harvey P.H. 1949. A breeding procedure designed to make use of both general and specific combining ability. *Agron. J.*, **41**: 360-367.

Comstock R.E. and Robinson H.F. 1952. Estimation of average dominance of genes. p. 494-516. *In*: *Heterosis*, (ed.) J.W. Gowen, Iowa State College Press, Ames, Iowa.

Downey R.K. 1964. A selection of *Brassica campestral* L. containing no erucic acid in its seed oil. *Can. J. Plant Sci.*, **44**: 295.

Dudley J.W. 1993. Molecular markers in plant improvement : Manipulation of genes affecting quantitative traits. *Crop Sci.*, **33**: 661-668.

Empig L.T., Gardner C.O. and Compton W.A. 1972. Theoretical gains for different population improvement procedures. *Neb. Agr. Exp. Sta. Misc. Publ.*, 26.

Falconer D.S. 1960. *Introduction to Quantitative Genetics*, Ist Edn. Oliver and Boyd, Edinburgh.

Fasoulas A.C. 1983. Rating cultivars and traits in approved plant breeding. *Euphytica*, **32**: 939-943.

Fasoulas A.C. 1987. A moving black evaluation technique for improving the efficiency of pedigree selection. *Euphytica*, **36**: 477-478.

Fisher R.A. 1918. The correlation between relatives on the supposition of Mendelian inheritance. *Trans. R. Soc. Edinb.*, **52**: 399-433.

Fisher R.A. 1930. *The Genetical Theory of Natural Selection*. Calrendon Press, Oxford.

Fisher R.A. 1936. The use of multiple measurements in taxonomic problems. *Annals of Eugenics*, **7**: 179-188.

Gardner C.O. 1961. An valuation of effects of mass selection and seed irradiation with thermal neutrons on yield of corn. *Crop Sci.*, **1**: 241-245.

Glimore E.C. Jr. 1964. Suggested methods of using reciprocal recurrent selection in some naturally self-pollinated species. *Crop Sci.*, **4**: 323-325.

Goulden C.H. 1939. Problems in plant selection. *In*: *Proc. 7th Int. Genet. Congr.*, (ed.) R.C. Burnett, Edinburgh, Cambridge Univ. Press, pp. 132-133.

Haley C.S. and Knot S.A. 1992. A simple regression method for mapping quantitative trait loci in line crosses using flanking markers. *Heredity*, **69**: 315-324.

Hallauer A.R. and Eberhart S.A. 1970. Reciprocal full-sib selection. *Crop Sci.*, **10**: 315-316.

Hallauer A.R. and Miranda F.O., J.B. 1981. *Quantitative Genetics in Maize Breeding*, Iowa State Univ. Press/Ames.

Hardy G.H. 1908. Mendelian proportions in mixed populations. *Science*, **28**: 49-50.

Hazel L.N. 1943. The genetic basis for constructing selection index. *Genetics*, **28**: 476-490.

Hopkins C.G. 1899. Improvement in the chemical composition of the corn kernel. *Illinois Agric. Exp. Stn. Bull.*, **55**: 205-240.

Hull F.H. 1945. Recurrent selection for specific combining ability. *Agron. J.*, **37**: 134-145.

Jenkins M.T. 1935. The effect of inbreeding and selection within lines of maize upon hybrids made after successive generations of selfing. *Iowa State Coll. Jour. Sci.*, **9**: 429-450.

Jenson N.F. 1970. A diallel selective mating system for cereal breeding. *Crop Sci.*, **10**: 629-635.

Johannsen W.L. 1903. Uber Erblickeit in populationen und in reinen cf. chapt., Leninen. Gustav Fischer, Jena. (Heredity in populations and pure lines. In: *Classic Papers in Genetics*. (ed.) Peters J.A., Prentice Hall, Englewood Cliffs, New Jersy, 1959, pp. 20-26).

Kearsey M.J. and Farquhar A.G.L. 1998. QTL analysis, where are we now?. *Heredity*, **80**: 137-142.

Kearsey M. J. and Jinks J. L. 1968. A general method of detecting additive, dominance and epistatic variation for medical traits. I. *Theory Heredity*, **23**: 403-408.

Kearsey M.J. and Pooni H.S. 1996. *The Genetical Analysis of Quantitative Traits*. Chapman and Hall, London.

Kearsey M.J. and Hyne V. 1994. QTL analysis : a simple marker regression approach. *Theor. Appl. Genet.*, **89**: 698-702.

Kempthorne O. and Nordskog A.W. 1958. Restricted selection index. *Biometrics*, **15**: 10-19.

Lande R. and Thompson R. 1990. Efficiency of marker assisted selection in the improvement of quantitative traits. *Genetics*, **124**: 743-756.

Lander E.I. and Botstein D. 1989. Mapping Mendelian factors underlying quantitative traits using RFLP linkage maps. *Genetics*, **121**: 185-199.

Lee M. 1995. DNA markers and plant breeding programs. *Advances in Agron.*, **55**: 265-341.

Lonnquist J.H. 1964. A modification of ear-to-row procedure for the improvement of maize populations. *Crop Sci.*, **4**: 227-228.

Mather K. 1973. *Genetical structures of population*. Chapman and Hall, London.

Mather K. 1987. Consequences of stabilizing selection for polygenic variation. *Heredity*, **58**: 267-277.

Mather K. and Jinks J.L. 1971. *Biometrical Genetics*. Chapman and Hall, London.

Paran I. and Michelmore R.W. 1993. Development of reliable PCR-based markers linked to downy mildew resistance gene in lettuce. *Theor. Appl. Genet.*, **85**: 985-993.

Paterniani E.A. and Vencovsky R. 1977. Reciprocal recurrent selection based on half-sib progenies and prolific plants in maize. *Maydica*, **22**: 141-152.

Paterniani E.A. 1967. Inter-population improvement : Reciprocal recurrent selection variations. *Maize*. 8. CIMMYT.

Paterniani E.A. 1973. Recent studies on heterosis. *In*: *Agricultural Genetics*, (ed.) R. Moav, pp. 1-22. Natl. Counc. Res. Dev., Jerusalem.

Paterson A.H., Tanksley J.D. and Sorrells M.E. 1993. DNA markers in plant improvement. *Adv. Agron.*, **46**: 36-90.

Pesek J. and Baker R.J. 1969. Desired improvement in relation to selection indices. *Canadian J. Plant Sci.*, **49**: 803-804.

Rafalski J.A. and Tingey S.V. 1993. Genetic diagnostics in plant breeding. RAPDs, Microsatellites and machines. *Trends Genet.*, **9**: 275-279.

Roy Darbeshwar. 2000. *Plant Breeding - Analysis and Exploitation of Variation.* Narosa, New Delhi.

Smith C. 1967. Improvement of metric traits through specific genetic loci. *Anim. Prod.*, **9**: 349-358.

Smith H.F. 1936. A discriminate function for plant selection. *Ann. Eugen*, London. **7**: 240-250.

Soller M. and Beckmann J.S. 1983. Genetic polymorphisms in varietal identification and genetic improvement. *Theor. Appl. Genet.*, **67**: 25-33.

Sprague G.F. and Eberhart S.A. 1977. Corn breeding. *In*: *Corn and Corn Improvement*, G.F. Sprague, ed. pp. 305-362. Am. Soc. Agron. Madison, Wis.

Stadler L.J. 1944. Gamete selection in plant breeding. *J. Am. Soc. Agron.*, **36**: 988-989.

Stuber C.W. 1992. Biochemical and molecular markers in plant breeding. *Plant Breeding Rev.*, **10**: 37-61.

Sunesen C.A. 1956. An evolutionary plant breeding method. *Agron. J.*, **48**: 188-191.

Tanksley S.D. and Rick M. 1980. Isozyme linkage map of tomato: applications in genetics and plant breeding. *Theor. Appl. Genet.*, **57**: 161-170.

Tanksley S.D. and Orton T.J. 1983. *Isozymes in Plant Genetics and Breeding.* Elsevier. Amsterdam.

Vasal S.K., Villegas E. and Bauer R. 1979. Present status of breeding quality protein maize, pp. 127-48. *In*: *Proc. Symp. On Seed Protein Improvement in Cereals and Grain Legumes.* Neuherberg, Fed. Repub. Ger., 4-8 Sept., 1978.

Weinberg W. 1908. Uben Den Nachweis der Verebung bein Menschen.

Weller J.L. 1986. Maximum with the aid of genetic markers. *Biometrics*, **42**: 627-640.

Williams J.G.K., Kubelik A.R., Livak K.J., Rafalski J.A. and Tingey S.V. 1990. DNA polymorphisms amplified by arbitrary primers are useful as genetic markers. *Nucleic Acid Res.*, **18**: 6531-6535.

Williams J.S. 1962. The evaluation of a selection index. *Biometrics*, **18**: 120-122.

Plant Breeding - Mendelian to Molecular Approaches
H. K. Jain and M. C. Kharkwal (eds.)

16

Population Improvement Strategies for Crop Improvement

S.K. Vasal[1], N.N. Singh[2], B.S. Dhillon[3], S.J. Patil[4]

Abstract

Recurrent selection is an important breeding method employed to improve the populations of crop plants particularly those of cross-pollinated species. It may be undertaken in one (intra-population improvement) or two populations (inter-population improvement) to upgrade the performance *per se* or combining ability. Intra-population improvement may involve evaluation of individuals (mass selection) or of progenies (full-sibs, half-sibs or selfed progeny selection) that have been developed within a population or test-cross progenies developed by using a tester (related or unrelated, narrow or broad-genetic base). Various methods of intra-population improvement aim at enhancement of the performance of the population *per se*, random mated or selfed generation, except that in case of test-cross progenies evaluation, the emphasis is on improving combining ability. Inter-population improvement involves simultaneous improvement of two heterotic populations with a focus on their combining ability. In inter-population improvement methods, also known as reciprocal recurrent selection, there are two important alternatives, based on the evaluation of half-sib or full-sib progenies developed by using the other heterotic population or inbred line from that as a tester. There are several possible modifications and additional features of various intra- and inter-population methods that can be incorporated to meet specific objectives of a programme. The integration of population improvement and hybrid research activities, and the improvement of traits related to stress have also been discussed.

Introduction

Breeding methods and selection strategies for improving populations in crop plants have developed and evolved continuously over the past sixty years. A few cross-pollinated plants such as maize have given a lead in developing new methodologies, which subsequently have been deployed in other crop plants. Practically, all the known breeding methods are flexible and can be modified appropriately to suit crop needs and tailored for

[1]International Maize and Wheat Improvement Center (CIMMYT), Mexico
[2]Indian Agricultural Research Institute, New Delhi 110 012, India
[3]National Bureau of Plant Genetic Resources, New Delhi 110 012, India
[4]University of Agricultural Sciences, Dharwad 580 005, India

special emphasis on specific traits. In literature, one will find much greater emphasis on cross-pollinated crops compared to the self-pollinated ones. Difficulties in pollination and recombination in self-pollinated plants make population improvement more cumbersome, less cost-effective and difficult to pursue for long-term selection objectives and gains. Nevertheless, conscious efforts have been made to use relevant procedures and strategies to a limited extent. The paper is intended to present important procedures that are reported and used in different crops. A critique of some of the most commonly used procedures has been presented. Also some possible modifications have been suggested to make the methods more relevant to crop and trait-oriented goals.

Recurrent Selection in Self-pollinated Crops

The use of recurrent selection programs to improve populations in self-pollinated crops has been limited. Breeders have proposed the use of recurrent selection programmes in several crops (Hanson *et al.*, 1967; Miller and Rawlings, 1967; Verma *et al.*, 1977; Gill 1980; Yunus and Paroda 1982 and many others). Most authors point out some major hurdles in achieving recombination and difficulty in making pollinations by hand. Data from systematic studies are limited and convincing evidence still lacking. Several considerations including time factor, laborious and expensive pollinations, unavailability of off-seasons/contrasting sites, either unknown or unsatisfactory male sterility systems, lack of good chemical emasculents and emphasis on relatively short-term goals prevent use of such breeding approaches.

Greater use of recurrent selection programmes becomes feasible if satisfactory and appropriate cytoplasmic male sterility (CMS) systems or genic male sterility, are available. Availability of suitable gametocides or chemical emasculants is another alternative to facilitate pollination and recombination process. To exploit fully the benefits of recurrent selection, continuous intermating generations are desirable and need to be obtained between selected plants or families using CMS system(s), chemical emasculents and perhaps thermal and photosensitive systems. The genic male sterility system is preferred over cytoplasmic system. Genic male sterility has been used by breeders of several crops (Suneson, 1956; Brim and Stuber, 1973; Sorrels and Fritz, 1982). Through the use of genic male sterility system, half-sib progenies can be generated. As expected this brings about cost-effectiveness and saves difficulties in large scale, cumbersome hand pollinations.

Selection methodologies commonly used in self-pollinated crops include pedigree method, bulk method, and a combination or variants of the two methods. Backcross breeding methods with some modifications have also been used to a limited extent. Diallel selective mating has also been suggested (Jenson, 1970, 1978). It is considered a dynamic approach, which involves selection in single and multiple crosses and permits introgression of new germplasm, besides natural spin off of F_3 families. The procedure involves diallel mating among selected plants to develop F_1's followed by crossing again in either diallel or partial diallel manner, depending on the number of F_1s.

These methods have some obvious limitations. The recombination among genes is restricted, as selfing leads to rapid fixation of genes. Thus, breaking undesirable linkages is difficult. Conscious efforts are required to introduce some sort of intercrossing/intermating rather than selfing, to slow down the rapid fixation of alleles and help facilitate obtaining desirable recombinants. Another problem of common concern to breeders of self-pollinated crops is the narrowing down of genetic variability. Selection strategies are critical but may not necessarily be a limiting factor for genetic advance (Rajaram, 1987). The initial genetic variability is, however, very important to make genetic advance and can be eroded very fast in the absence of appropriate methodology. Thus, the efficacy and success of pedigree or other methods can be enhanced by developing several types of breeding populations, with enhanced genetic variability involving three or four progenitors as is the case with 3-way and 4-way crosses.

Intra-population methods have been advocated and used to a limited extent in self-pollinated crops to improve various traits (Gill, 1980; Verma, 1986). The use of inter-population methods is practically absent. The use of reciprocal recurrent selection for inter-population improvement in self-pollinated crops using genic sterility has been described by Gilmore (1964).

Recurrent Selection in Cross-Pollinated Crops

Recurrent selection procedures have been used extensively and have witnessed continuous evolution from individual plant selection to more family-based selection schemes relying on test-cross performance. Practically all reported methods in literature have either been utilized in their original form or modified somewhat in more than one way to meet the general and specific breeding goals. Institutions and programmes that have adopted two-tier or multi-tier germplasm management system to facilitate gene flow from one tier to another, have practised different methodologies, generally a simple one, in pre-breeding stages and more sophisticated ones in front-line advanced elite materials. Recurrent selection procedures to improve populations can be classified in several ways (Vasal *et al.*, 1996) and are presented in Fig. 1. The methods are listed but detailed discussions are limited to highlighting only some of their important features. Modifications known/reported have been discussed along with other possible ones to achieve specific objectives. The paper also emphasizes more recent trends in population improvement, especially on stress tolerance related traits and some specialty traits. Considering current trends and importance of hybrid cultivars in recent years, a critique of various methodologies to enhance or improve hybrid-oriented features through recurrent selection is discussed. Examples cited and methodologies discussed are mostly from maize, but have relevance to other crops also.

Several researchers have discussed recurrent selection procedures to improve maize populations (Moll and Stuber, 1974; Sprague and Eberhart, 1977; Hallauer and Miranda, 1988; Hallauer 1992; Dhillon and Malhi, 1996; Vasal *et al.*, 1996 and Vasal *et al.*, 1997). The methods, in general, can be classified under two categories, namely intra-population and inter-population methods. Within each category various options and their modifications are discussed.

Intra *versus* inter-population improvement

- Intra-population
  - Individual *versus* family structure
  - Non-inbred *versus* selfed progenies
  - *Per se versus* test-cross performance
  - Broad-based *versus* narrow-based testers
  - Parental *versus* non-parental testers
- Inter-population
  - Test-cross individual *versus* families
  - Half-sib *versus* full-sib test-cross progenies
  - Non-parental tester type
    - Population
    - Derived synthetic
    - Single cross
    - Inbred

Choosing among intra-population alternatives

- Individual plant selection
  - Simple mass selection
  - Stratified mass selection
- Family-based selection
  - Half-sib selection and its variations
  - Full-sib selection and its variations
  - Selfed-progeny selection
    - S_1 selection and its variations
    - S_2 selection and its variations
  - Combination of selfed and non-inbred families
    - S_1 and half-sib families
    - S_1 and full-sib families

Choosing among inter-population alternatives

- Reciprocal recurrent selection - half-sib (RRS-HS)
  - Test-cross individuals - RRS-HS
  - Test-cross families - Modified RRS-HS
- Reciprocal recurrent selection - full-sibs (RRS-FS)
  - $S_0 \times S_0$ - RRS-FS
  - $S_1 \times S_1$ - Modified RRS-FS

Fig. 1. Breeding alternatives for population improvement

Intra-Population Recurrent Selection

Intra-population improvement methods are designed to improve *per se* performance of populations for quantitative traits of economic importance. The cyclic improvement process accumulates favourable alleles through a slow and gradual process, in a manner that improves the mean performance of the population. The recurrent selection procedures may involve individuals, families and in some instances, even test-crosses. As will be seen later, some of the recurrent selection procedures are based on the evaluation of individuals and even families. In family based methods, the breeders have several options of evaluating different types of families, such as half-sibs (HS), full-sibs (FS), selfed progenies (S_1–S_n), and test-cross progenies (TC). The evaluations can be done in replicated or unreplicated experiments in single or multiple environments. Family-based selection schemes involve three important steps, namely development of families, evaluation of families, and recombination of selected families. Modifications of some of the standard schemes have been suggested, as also the combinations of different families as a part of a given scheme. The following sections discuss some of the important schemes that are commonly used by the maize breeders.

Mass Selection: It is the simplest, oldest and least expensive method. The procedural details are simple. The method involves planting balanced bulk seed of the population to be improved, in isolation. During harvest, the ears are saved from selected plants. A balanced bulk of the selected ears is made for planting the next cycle. The method exploits additive gene effects plus additive × additive interactions. The efficiency of this method depends greatly on the heritability of the trait under selection. Selection results using mass selection have been quite effective for less complex traits (Gardner, 1969; Jugenheimer, 1985; Hallauer and Miranda, 1988; Hallauer *et al.*, 1988; Paterniani, 1990; Hallauer, 1992; Pandey and Gardner, 1992). Mass selection could be very effective for those traits that can be identified or selected prior to, or at the time of flowering. A few examples of such traits in maize could be flowering, leaf angle, prolificacy, resistance to thrips, fall armyworm, reduced plant and ear height, reduced anthesis-silking interval & barrenness, and tolerance to pre-flowering diseases. Greater response to mass selection has been reported for prolificacy (Lonnquist, 1967; Torregroza and Harpstead, 1967 and Arboleda-Rivera and Compton, 1974). Mass selection is also reported to be successful for early flowering (Hallauer and Sears, 1972).

Gains for adaptation have been impressive in some of the studies that have been reported (Hallauer and Sears, 1972; Troyer and Brown, 1976; Compton *et al.*, 1979; Hallauer, 1994). Adaptation of exotic tropical germplasm to temperate conditions was successfully accomplished in five to six cycles of mass selection for early flowering in temperate environment (Hallauer and Sears, 1972; Troyer and Brown, 1976 and Compton *et al.*, 1979). Similarly, remarkable results have been achieved in improving adaptation of tropical and subtropical maize germplasm under temperate long day conditions in China (Zhang and Shi, 2000). The Chinese Academy of Agricultural Sciences introduced two subtropical quality protein maize (QPM) gene pools from International Maize and Wheat Improvement Center (CIMMYT), that were relatively late in flowering and maturity, in addition to having the problems of unsynchronization, barrenness, and very low productivity. Four cycles of bi-parental mass selection reduced the photoperiod sensitivity and improved the yield performance under long day conditions.

Many earlier reports in literature point out lack of effectiveness of mass selection, because of several reasons including lack of control on soil variability, no control on post-flowering traits, and low heritability of traits such as grain yield (less than 10 % on individual plant basis). Mass selection may be effective for individual or a combination of a few traits at a time rather than a whole series of characters. Also, experience in maize has shown that if field execution is not appropriate, it may result in taller and later maturing plants. Despite some limitations, mass selection can be a very relevant procedure in many situations characterised by resource limitations, non-availability of additional test sites, lack of seed storage facilities even for short term, lack of trained scientific manpower, facilities for proper record keeping, and continuity of technical personnel.

To overcome soil heterogeneity problems, a modification of simple mass selection has been suggested, which is commonly known as stratified mass selection (Gardner, 1961). The basic details are the same as in simple mass selection, except that the field research area is divided into a number of grids and that superior plants are selected from each grid to reduce environmental variation. Simple or modified mass selection can also be used in small grain cereals to increase the frequency of favourable alleles in the segregating generations.

From the above, it is apparent that despite some limitations, mass selection can play an important role in crop improvement programmes, particularly for those traits that are governed by simple genetic system and are highly heritable. Also, for complex traits mass selection can be a useful tool, provided some secondary traits can be identified that have high heritability and are known to contribute to the improvement of the trait, as the case with drought tolerance in maize. Mass selection can be deployed to select successfully for photo-period insensitive genes, to permit use of exotic germplasm and to broaden the germplasm base in a breeding programme. Under resource constraint situation, mass selection will continue to be a preferred method. In cross-pollinated crops with little to drastic inbreeding effects, this procedure will not help to improve trait(s) of interest for hybrid breeding work. Plant density and stress manipulations can be deployed to improve behaviour under such stresses. The products resulting from mass selection will be improved populations and the possibilities of integrating this method with hybrid related activities are minimal. Dhillon (1991a) proposed mass selection in a selfed population.

Family Based Intra-Population Improvement

In family based schemes the selection is based on the performance of progenies which could either be inbred or non-inbred. Several possibilities exist of structuring populations in the form of families as full-sibs, half-sibs, selfed progenies and test-crosses. Several factors will guide the choice of method and progeny, including the availability of off-seasons, test sites, ability to store remnant seed, need for germplasm products (variety, hybrid or both), selection traits, trait heritability, seed quantities needed, degree of control over sexes (both parents or only one parent), and time required to complete a cycle. As opposed to individual plant selection alternatives, the family based selection methods result in greater gains, especially for complex traits with low heritability but are more demanding in resources, record keeping, data collection and analysis, and overall management of the trials. It is needless to emphasize that good field execution is important, irrespective of the choice of the breeding method. Also it is important in schemes requiring more steps in progeny regeneration, where efforts should be directed to utilize the intervening time between cycles more efficiently by emphasizing selection for those traits, which are deficient in the population or for building up stress tolerance. For selection purposes it is also important to know the composition of genetic variance in different types of families. For half-sibs, the $\sigma^2 G = 0.25\ \sigma^2 A$; for FS, $\sigma^2_G = 0.5\ \sigma^2 A + 0.25 = \sigma^2 D$; for S1, $\sigma^2_G = \sigma^2\ A + 0.25 = \sigma^2\ D$; and for S2, $\sigma^2 A + 0.19\ \sigma^2 D$; where σ^2_G, σ^2_A and σ^2_D are genotypic variance, additive-genetic variance and dominance variance, respectively. Variance among and within different families is important to know as described below:

variance among full-sib families : $0.5\ \sigma^2 A + 0.25\ \sigma^2 D$

variance within full-sib families : $0.50\ \sigma^2 A + 0.75\ \sigma^2 D$

variance among half-sib families : $0.25\ \sigma^2 A$

variance within half-sib families : $0.75\ \sigma^2 A + \sigma^2 D$

It is also important to point out that the additive genetic variance among S_1 progenies is two and four times of that among full-sib and half-sib families, respectively. All schemes involve three essential steps of developing, evaluating and recombining of superior progenies. Salient features and some critical points about different schemes are described below.

Full-Sib Recurrent Selection

The method involves developing full-sibs by making plant-to-plant crosses. Direct and reciprocal crosses can be attempted depending on the need for the seed quantity. Multi-location evaluation of progenies is done followed by the recombination of the superior families based on across-location performance data to complete a cycle of selection. A cycle of selection can be completed in one year provided two seasons are available and the recombination and generation of new families is done simultaneously in the same season. This method has been used extensively at CIMMYT to improve populations and to develop experimental varieties. The selection intensity for the development of experimental varieties is stringent, not exceeding 4 per cent. To take full advantage of data from all the locations, a two-year cycle was actually practised and a unique feature of intra-family improvement was introduced to include inbreeding and to take full advantage of intervening period available between the two cycles. The average gain from different studies has been 3.81 per cent per cycle (Hallauer and Miranda, 1988). The results from CIMMYT have been variable with different populations (Vasal *et al.*, 1982; Pandey *et al.*, 1991). The full-sib method completing one cycle of selection in one year has also been used extensively in the Indian maize program, to improve populations and to develop cultivars (Dhillon *et al.*, 1987). Singh *et al.* (1986) carried out successful selection for prolificacy in a composite population and completed two cycles in a year by staggered sowing of evaluation experiments and breeding nurseries.

Maize breeders and physiologists at CIMMYT have also used this method in improving drought tolerance of Tuxpeno Sequia population (Bolanos and Edmeades, 1993). The method has been used successfully in various institutions in several countries to improve populations for release of open-pollinated varieties and to serve as a better source of germplasm for hybrid related activities. The method in its standard form is not conducive for hybrid research and to improve germplasm for hybrid-oriented traits. Conscious efforts need to be made in this direction. As pointed out earlier, inbreeding can be introduced as was done at CIMMYT to affect intra-family improvement. Additional modifications are possible to aid hybrid research, provided full-sibs are developed using $S_1 \times S_1$ or $S_2 \times S_2$ in contrast to $S_0 \times S_0$. Such modifications can help improve populations for inbreeding tolerance, combining ability and to identify potential early generation lines and intra-population inter-line hybrids. Such a scheme has been used in the case of a CIMMYT's population 31. Additional modification that will aid hybrid research, will be to develop full-sibs using special mating designs and partial diallel mating system to identify better combining families (Dhillon *et al.*, 1987).

Half-Sib Recurrent Selection

Half-sib breeding procedure and its various modifications have gained prominence beginning mid-sixties, with the suggestion of modified ear-to-row (MER) method by Lonnquist (1964). Ear-to-row method was introduced by Hopkins in 1896 to improve chemical composition of maize kernel for oil and protein content. The method was subsequently used by some researchers to improve yield and agronomic traits by selecting superior ears in good performing families. The results, however, were not encouraging as these experiments suffered from various limitations including inbreeding and no pollen control. Several improvements and modifications in the use of this procedure have been made at CIMMYT and elsewhere, to form new genetically broad-based gene pools, composites and to improve existing or newly formed populations (Vasal *et al.*, 1982: Pandey *et al.*, 1984; Pandey and Gardner, 1992; Vasal *et al.*, 1997). The procedure has also been used to improve specific traits, such as early flowering and maturity, prolificacy, kernel modification and agronomic performance in quality protein maize gene pools at CIMMYT. Various modifications of half-sib procedure are discussed below.

(i) ***Half-Sib Selection***: The procedure involves development of half-sib progenies either through controlled bulk sib pollination, or by selecting ears from randomly pollinated bulk planting of maize population in isolation. Half-sib progenies are evaluated at different locations. The superior performing progenies based on across-location performance data are selected for recombination to complete a cycle of selection and to generate a new set of progenies for the next cycle of selection. Using this procedure a cycle of selection can be completed in one year, provided two seasons are available. The method is simple, resource effective and involves no hand pollination, if recombination and new progenies are developed in a half-sib recombination crossing block. Since selected progenies are recombined, the method provides control on both sexes.

(ii) ***Modified ear-to-row Selection***: This procedure was suggested by Lonnquist (1964). In this half-sib progeny tests are conducted at 3-4 appropriate sites using single replicate. In the same season, a half-sib crossing block is planted at a different location in isolation. In the crossing block half-sib ears selected in the previous cycle are planted as separate families as female rows and the male rows are planted with a balanced male composite comprised of all selected ears. The ratio of female to male is kept at 2:1. Female rows are detaselled at flowering to permit pollination by bulk pollen from the male rows. Between and within row selection is practiced to select half-sib ears for the ensuing cycle of selection. The data from progeny test sites are used to select the best performing progenies in the crossing block. Within row selection is performed in the selected rows, to reach the number required for the next cycle of selection. The MER is a unique method based on family performance, permitting one cycle of selection in one season. Though families are selected on the basis of progeny tests, the male bulk in the crossing block provides no control, as it is constituted from both selected and non-selected progenies. Selection gains using this method for grain yield have been quite good, averaging 6.5 per cent per cycle (Hallauer and Miranda, 1988). Modification of MER has been suggested wherein only selected families are recombined, so that male bulk is made up from the selected families as well (Compton and Comstock, 1976). This, of course, will take an additional season to complete selection cycle.

(iii) ***Modified ear-to-row Selection (unreplicated evaluation)***: Lonnquist's MER having only half-sib crossing block without replication has been extensively used at CIMMYT in forming and subsequently improving gene pools. Mild selection has generally been practiced in male rows for tall and diseased plants. Artificial manipulations for diseases and insect-pests can be deployed as necessary. Using this procedure one cycle of selection can be completed in one year. Progress made and reported in several gene pools has been encouraging (Pandey *et al.*, 1991). Similarly good progress has been achieved in improving kernel modification and agronomic performance in tropical and subtropical gene pools (Vasal, 1994).

A detailed discussion of recurrent selection methods based on evaluation-cum-recombination block has been published by Vasal *et al.* (1997). The paper describes many modifications to meet programme objectives and to enhance genetic gains. Several additional suggestions include making balanced male composite from elite fraction of selected half-sib ears, practicing selection in the male rows for those traits that express before flowering, high plant density planting in the male rows to enable more intense selection and practising between family selection in comparison with the adjacent male rows to minimize environmental effects. Gain per cycle for grain yield and other traits is presented from several tropical and subtropical gene pools. It is important to note that yield improvement has been achieved concurrent with reduction in maturity and plant height.

The foregoing half-sib procedures are simple and resource effective. Continuous long-term improvements are possible while still maintaining broad genetic variability. Continuous selection and recombination provide opportunities for generating new and novel genetic combinations and to break linkages among traits that have negative associations. Evaluation and improvement for combining ability are inherent in the system. Germplasm undergoing selection can be kept open-ended and introgression of new and superior accessions can be done in an appropriate and systematic manner. The system provides tremendous flexibility in deploying varied manipulations and emphasizing good field execution at various stages of crop growth. In programmes using management system for continuous flow of germplasm from backup to advanced stages, this procedure is considered highly appropriate and has been used quite successfully at CIMMYT for the past thirty years or so. The procedure provides improved source germplasm for hybrid work, but is difficult to integrate with hybrid research until and unless modifications are made to include inbreeding, combining ability, and improved cross-bred phases in the on-going half-sib recurrent selection programs.

Selfed Progeny Selection

Such selection schemes are based on selfed progenies and take a longer time to complete a cycle of selection depending on whether S_1, S_2 or more advanced selfed progenies are involved in the evaluation process. Following evaluation of selfed progenies, the selected ones are recombined using the remnant seed. Thus, a cycle of S_1 selection can be completed in three seasons, and of S_2 in four seasons. A minimum of two years would thus be required, provided two seasons are available in a year. Some researchers have practiced S_3 recurrent selection (Vasal *et al.*, 1995), but it is not recommended as a general procedure because of time factor and achievable gains on per-year basis. Selfed progeny selection schemes contribute to inbreeeding tolerance, which is considered desirable in developing vigorous and productive inbreds. Such schemes integrate better with inbred development efforts, as inbreeding can be continued in superior performing selfed progenies. Selfed progeny selection schemes are particularly suited for disease and insect-pests resistance research work. These schemes in general give higher gains, but the gains over time are adversely affected when contribution of non-additive genetic variation is increased. In the initial phases of the selection process, it is recommended to pre-screen large number of progenies in unreplicated observational nurseries followed by evaluation of the selected fraction in replicated multi-locational trials. This does take an additional season, but saves resources by not testing obvious unwanted and diseased families. Seed quantities can be a limiting factor in selfed progeny selection procedures, but can be easily resolved by using S_2 bulk seed developed from each S_1 progeny. To maintain selection gains over longer periods, it is suggested that no fewer than 20-50 inbred progenies be recombined (Hallauer, 1992). Selfed progeny selection procedures tend to increase plant height and, thus, conscious efforts are needed to guard against this trend.

A modified version of S_1 selection has been suggested by Dhillon and Khehra (1989). In this procedure the evaluation and recombination phases are combined and a cycle of selection is completed in two seasons or one year. The evaluation test in one of the environments is planted as a half-sib crossing block in isolation. Superior progenies in the crossing block are identified based on across-location performance data and in these selected progenies, within- progeny selection is practiced.

Combined Family Based Selection Schemes

Selection schemes can be combined using different types of families. Also sometimes families evaluated and genotype recombined are different. In the first scheme the half-sibs or top-crosses are evaluated and the remnant S_1 seed of superior families is recombined. This scheme is particularly suited in maize populations having prolific tendency. The bulk seed is planted in isolation. One of the ears is left open for open pollination to develop top-crosses and the second ear is manually self-pollinated in the same season. In the next season the top-crosses are evaluated and in the following season the remnant S_1 seed of superior performing top- crosses is recombined.

The half-sib and selfed progeny procedures can also be combined as a two-stage recurrent selection program (Dhillon, 1991b, 1993). The S_1 progenies developed from half-sib families are evaluated and recombined simultaneously in a half-sib crossing block in isolation. Saved half-sib ears are planted for evaluation and selves are made and saved only from superior performing half-sib families. Thus, the selfed and half-sib progeny selections alternate each other. Such a procedure has been used at CIMMYT in improving tropical late maturing gene pools. This procedure incorporates both inbreeding and combining ability and can thus help hybrid breeding efforts in addition to cyclic improvement of the pools. Dhillon (1998) gave comprehensive recurrent selection based on the performance *per se* of selfed and crossed progenies and of test-crosses.

Test-Cross Selection

Selection schemes of this type are based on combining ability evaluation. Here the test-crosses of S_0 plants (Jenkins, 1940; Hull, 1945) or of S_1 or S_2 progenies are developed and evaluated. The procedures are described in literature as recurrent selection for general (*gca*) and specific combining ability (*sca*). The main difference lies in the type of testers that are used; that broad and narrow genetic base testers are used for improving *gca* and *sca*, respectively. The time required to complete a cycle of selection will, thus, depend on what type of selfed progenies (S_1 or S_2) are test-crossed. Such schemes are useful when emphasis in the programme is shifted towards combining ability, hybrid-oriented germplasm and integration of hybrid and population development efforts. Such procedures can also be recommended where the need is to identify superior early generation good combiner lines for further inbreeding or improving a population *per se*, or to derive synthetics as byproducts of the selection process. Selection of stress related traits can be emphasized during the formation and screening of selfed progenies as well as during test-cross evaluation.

Inter-Population Improvement

Inter-population improvement schemes are dual purpose selection alternatives to aid both population and hybrid development efforts. Also two heterotic populations are improved simultaneously and emphasis is placed on combining ability. It is preferable that only agronomically superior populations are subjected to this type of selection programme. Populations that are deficient in agronomic traits should wait until those traits have been improved. Also it is suggested that populations which have not previously been improved through some sort of selfed progeny selection scheme should be subjected to inbreeding pressure to expose and eliminate deleterious alleles. Emphasis in such schemes should not only be on improving combining ability and cross-bred performance, but also on improving tolerance to inbreeding, as this has been a serious drawback when dealing with tropical and subtropical germplasm for the development of vigorous and productive maize inbreds. It is important that conscious efforts be made to make such schemes efficient in terms of resources and types of products that are generated to benefit both open pollinated varieties and hybrid research efforts. Selection of appropriate testers should receive utmost attention to meet programme objectives and goals. Stress tolerance related traits should be integrated in various steps during the progeny regeneration and in the evaluation of test-crosses. The germplasm products resulting from such selection programmes include improved populations, early generation lines for further inbreeding, intra-population synthetics, inter-population synthetics, potential hybrids depending on testers used, new potential testers from each population and providing opportunities for recycling early generation lines.

Two commonly used inter-population improvement schemes are reciprocal recurrent selection - half-sibs (RRS-HS) (Comstock *et al.*, 1949) and reciprocal recurrent selection/full-sibs (RRS-FS) (Hallauer and Eberhart, 1970; Hallauer, 1973). As originally described, the schemes are not particularly suitable if the populations lack tolerance to inbreeding. The schemes also ignore *per se* performance of the lines during the selection process. The original schemes also recommend evaluating S_0 test crosses and recombining the parental S_1 seeds of good performing plants. To meet practical needs and objectives of the breeding programmes, the schemes have been modified in recent years. The modified schemes attempt test-crosses (HS or FS) on S_1 or S_2 progenies and also permit selfed progeny evaluation for elimination of undesirable progenies (Dhillon *et al.*,

1986). The RRS-FS schemes have an added advantage over RRS-HS, in that, only 50 % of the resources are spent on test-cross progeny evaluation trials. In recent years CIMMYT breeders have switched over to such schemes. Also stress tolerance related traits have been integrated during various steps of progeny regeneration. These types of schemes are not necessary requirements for hybrid development, but from long-term perspectives, the populations will become genetically more divergent with improved cross-bred performance and will certainly aid in the development of better high yielding hybrids.

References

Arboleda-Rivera Fernando and Compton W.A. 1974. Differential response of maize to mass selection in diverse selection environments. *Theor. Appl. Genet.*, **44**: 77-81.

Brimm C.A. and Stubers C.W. 1973. Application of genetic male sterility to recurrent selection schemes in soybean. *Crop Sci.*, **13**: 528-530.

Bolanas J. and Edmeades G.O. 1993. Eight cycles of selection for drought tolerance in lowland tropical maize, II : Responses in reproductive behaviour. *Field Crops Res.*, **31**: 253-268.

CIMMYT. 1984. Development, maintenance and seed production of open-pollinated varieties. Mexico, D.F.

Compton W.A., Mumm R.F. and Mathema B. 1979. Progress from adaptive mass selection in imcompletely adapted maize populations. *Crop Science*, **19**: 531-533.

Compton W.A. and Bahadur K. 1977. Ten cycles of progress from modified ear-to-row selection in corn. *Crop Sci.*, **17**: 378-380.

Compton W.A. and Comstock E.E. 1976. More on modified ear-to-row selection in corn. *Crop Sci.*, **16**: 122.

Comstock R.E., Robinson H.F. and Harvey P.H. 1949. A breeding procedure designed to make maximum use of both general and specific combining ability. *Agron J.*, **41**: 360-367.

De Leon C. and Pandey S. 1989. Improvement of resistance to ear and stalk rots and agronomic traits in tropical maize gene pools. *Crop Sci.*, **29**: 12-17.

Dhillon B.S. 1991a. Alternate recurrent selection of S_1 and half-sib families for intra-population improvement. *Maydica*, **36**: 45-48.

Dhillon B.S. 1991b. Recurent mass selection based on selfed-plant evaluation in allogamous species. *Crop Sci.*, **31**: 1075-1077.

Dhillon B.S. 1993. Alternate recurrent selection. *Crop Sci.*, **33**: 650.

Dhillon B.S. 1998. Recurrent selection for combining ability and performance *per se* of cross-bred and selfed families. *Maydica*, **43**: 155-160.

Dhillon B.S. and Khehra A.S. 1989. Modified S_1 recurrent selection in maize improvement. *Crop Sci.*, **29**: 226-228.

Dhillon B.S. and Malhi N.S. 1996. Recent advances in recurrent selection for intra-population improvement. *Crop Improv.*, **11**: 88-91.

Dhillon B.S., Khehra A.S. and Singh M. 1987. Modified full-sib selection and estimation of genetic parameters. *Theor. Appl. Genet.*, **73**: 672-674.

Dhillon B.S., Malhi N.S., Saxena V.K. and Khehra A.S. 1986. Hico breeding in maize. p. 279-285. *In*: *New Frontiers in Breeding Researches*, (eds.) B. Napompeth and S. Subhadrabandhu, Kasetsart Univ., Bangkok, Thailand.

Eberhart S.A. 1970. Factors affecting efficiencies of breeding methods. *Agri. Soils*, **15**: 669-680.

Eberhart S.A., Harrison M.N. and Ogada F. 1967. A comprehensive breeding method. *Zuchter*, **37**: 169-174.

Empig L.T., Gardner C.O. and Compton W.A. 1972. Theoretical gains for different population improvement procedures. *Nebr. Agri. Exp. Sta. Bull.*, MP26 (rev.), Lincoln, NE.

Gardner C.O. 1961. An evaluation of mass selection and seed irradiation with thermal neutrons on yield of corn. *Crop Sci.*, **1**: 241-245.

Gardner C.O. 1969. The role of mass selection and mutagenic treatment in modern corn breeding, p 15-21. *In*: *Proc. 24th Corn Res. Conference*, American Seed Trade Association.

Gardner C.O. 1977. Quantitative Genetic Studies and population improvement in maize and sorghum. pp. 475-489. *In*: *Proc. Int. Conf. Quant. Genetics*, (eds.) E. Pollak, O. Kempthorne and T.B. Bailey Jr., Iowa State Univ. Press, Amer.

Gardner C.O. 1978. Population improvement in maize. Pp 207-228. *In*: *Maize Breeding and Genetics*, (ed.) D.B. Walden, John Wiley and Sons, New York, USA.

Gill K.S. 1980. Recent concepts in breeding methodology of self-pollinated crops, pp. 39-58, *In*: *Breeding methods for the improvement of pulse crops*, (ed.) K.S. Gill, Indian Council of Agricultural Research, New Delhi.

Gilmore E.C. Jr. 1964. Suggested method of using reciprocal recurrent selection in some naturally self-pollinated species. *Crop Sci.*, **4**: 323-325.

Good R.L. and Hallauer A.R. 1977. Inbreeding depression in maize by selfing and full-sibling. *Crop Sci.*, **17**: 925-940.

Granades G., Pandey S. and Ceballas H. 1993. Response to selection for tolerance to acid soils in a tropical maize population. *Crop Sci.*, **33**: 936-940.

Hallauer A.R. 1970. Zygote selection for the development of single cross hybrids in maize. *Adv. Front. Plant Sci.*, **25**: 75-81.

Hallauer A.R. 1973. Hybrid development and population improvement in reciprocal full-sib selection. *Egyptian J. Genetic. Cytology*, **1**: 84-101.

Hallauer A.R. 1975. Relation of gene action and types of testers in maize breeding procedures. *Corn Sorghum Res. Corp.*, **30**: 150-165.

Hallauer A.R. 1981. Selection and breeding methods. pp. 3-55. *In*: *Plant Breeding II*. (ed.) K.J. Frey, Iowa State University Press, Ames., Iowa.

Hallauer A.R. 1984. Reciprocal full-sib selection in maize. *Crop Sci.*, **24**: 755-759.

Hallauer A.R. 1992. Recurrent selection in maize. *Plant Breed. Rev.*, **9**: 115-179.

Hallauer A.R. 1994. Registration of BS28 and BS29 maize germplasm. *Crop Sci.*, **34**: 544-545.

Hallauer A.R. and Eberhart S.A. 1970. Reciprocal full-sib selection. *Crop Sci.*, **10**: 315-316.

Hallauer A.R. and Miranda J.B. 1988. *Quantitative Genetics in Maize Breeding*. Iowa State University Press, Ames., Iowa.

Hallauer A.R., Russell W.A. and Lamkey K.R. 1988. Corn breeding. *In*: *Corn and Corn improvement*. (eds.) G.F. Sprague and J.W. Dudley, Am. Soc. Agron., Madison, Wisconsin.

Hallauer A.R. and Sears J.H. 1972. Integrating exotic germplasm into corn belt maize breeding programs. *Crop Sci.*, **12**: 203-206.

Hanson W.D., Probst A.H. and Caldwell B.E. 1967. Evaluation of a population of soybean genotypes with implications for improving self pollinated crops. *Crop Sci.*, **7**: 99-103.

Hopkins C.G. 1896. Improvement in the chemical composition of the corn kernel. *Illinois Agr. Expt. Sta. Bull.*, **55**: 205-240.

Hull F.H. 1945. Recurrent selection and specific combining ability in corn. *J. Am. Soc. Agron.*, **37**: 134-145.

Jenkins M.T. 1940. The segregation of genes affecting yield of grain in maize. *J. Am. Soc. Agron.*, **32**: 55-63.

Jenson N.F. 1970. A diallel selective mating system for cereal breeding. *Crop Sci.*, **10**: 629-635.

Jenson N.F. 1978. Composite breeding methods and the DSM system in cereals. *Crop Sci.*, **18**: 622-626.

Johnson E.C., Fischer K.S., Edmeades G.O. and Palmer A.F.E. 1986. Recurrent selection for reduced plant height in low land tropical maize. *Crop Sci.*, **26**: 253-260.

Jugenheimer R.W. 1985. *Corn improvement, seed production and uses.* Robert E. Krieger Publishing, Malabar, FL, USA.

Lonnquist J.H. 1964. Modification of the ear-to-row procedure for the improvement of maize populations. *Crop Sci.*, **4**: 227-228.

Lonnquist J.H. 1967. Mass selection for prolification in maize. *Zuchter*, **37**: 185-188.

Miller P.A. and Rawlings J.O. 1967. Break-up of initial linkage blocks through inter-mating in a cotton breeding population. *Crop Sci.*, **7**: 199-204.

Moll R.H. and Stuber C.W. 1974. Quantitative genetics: Emperical results relevant to plant breeding. *Adv. Agron.*, **26**: 277-313.

Pandey S. and Gardner C.O. 1992. Recurrent selection for population, variety and hybrid development in tropical maize. *Adv. Agron.*, **48**: 1-87.

Pandey S., Vasal S.K. and Deutsch J.A. 1991. Performance of open-pollinated maize cultivars selected from 10 tropical maize populations. *Crop Sci.*, **31**: 285-289.

Pandey S., Vasal S.K., De Leon C., Ortega A.C., Granades G. and Villegas E. 1984. Development and improvement of maize populations. *Genetika*, **16**: 23-42.

Paterniani E. 1990. Maize breeding in the tropics. *Crit. Rev. Plant Sci.*, **9**: 125-154.

Paterniani E. 1967. Selection among and within half-sib families in a Brazilian population of maize (*Zea mays* L.). *Crop Sci.*, **7**: 212-216.

Rajaram S. 1987. Objective and methodology of CIMMYT wheat breeding programme in global context. *Proc. First Symp. Crop Improv.*, **1**: 131-146.

Russell W.A., Blackburn D.J. and Lamkey K.R. 1992. Evaluation of a modified reciprocal recurrent selection procedure for maize improvement. *Maydica*, **37**: 61-67.

Russell W.A., Eberhart S.A. and Vega U.A. 1973. Recurrent selection for specific combining for yield in two maize populations. *Crop Sci.*, **13**: 257-261.

San Vicente F.M. and Hallauer A.R. 1993. Mass selection for adaptation in Antigua maize (*Zea mays* L.) composite. *J. Iowa Acad. Sci.*, **100**: 9-12.

Singh M., Khehra A.S. and Dhillon B.S. 1986. Direct and correlated response to recurrent full-sib selection for prolificity in maize. *Crop Sci.*, **26**: 275-278.

Sorrels M.E. and Fritz S.E. 1982. Application of a dominant male-sterile allele to the improvement of self-pollinated crops. *Crop Sci.*, **22**: 1033-1035.

Sprague G.F. and Eberhart S.A. 1977. Corn breeding. pp. 305-362. *In*: *Corn and Corn Improvement.* (ed.) G.F. Sprague, Am. Soc. Agron. Madison, Wisconsin, USA.

Torregroza M. and Harpstead D.D. 1967. Effects of mass selection for ears per plant in maize. *Agron. Abstr.*, P. 20.

Troyer A.F. and Brown W.L. 1972. Selection for early flowering in corn. *Crop Sci.*, **12**: 301-304.

Troyer A.F. and Brown W.L. 1976. Selection for early flowering in corn. Seven late synthetics. *Crop Sci.*, **16**: 695-697.

Vasal S.K. 1994. High quality protein corn. P. 80-121. *In*: *Specialty Corns*, (ed.) A.R. Hallauer, CRC Press, Boca Raton.

Vasal S.K., Ortega A.C. and Pandey S. 1982. CIMMYT's maize germplasm management, improvement and utilization program. CIMMYT, Mexico, DF.

Vasal S.K., Dhillon B.S. and Srinivasan G. 1995. Changing scenario of hybrid maize breeding and research strategies to develop two-parent hybrids. pp. 19-36. *In*: *Hybrid Research and Development.* (eds.) M. Rai and S. Mauria, Ind. Soc. Seed Tech., New Delhi.

Vasal S.K., Dhillon B.S. and Pandey S. 1997. Recurrent selection methods based on evaluation-cum-recombination block. *Plant Breed. Rev.*, **14**: 39-63.

Vasal S.K., Cordora H.S., Beck D.L. and Edmeades G.O. 1997. Choices among breeding procedures and strategies for developing stress tolerant maize germplasm. *In*: *Proc. Developing Drought and Low N-tolerant Maize*, (eds.) G.O. Edmeades *et al.*, 25-29 March, 1996. CIMMYT, El Batan, Mexico.

Vasal S.K., Cordora H.S., Pandey S. and Srinivasan G. 1999. Tropical maize and heterosis. pp. 363-375. *In*: *Genetics and Exploitation of Heterosis in Crops*, (eds.) J.G. Coors and S. Pandey, Amer. Soc. Agron., Madison, Wisconsin, USA.

Verma M.M. 1986. Recurrent selection for improvement in self-pollinated crops, pp. 39-54, *In*: *Genetics and Crop Improvement*, (eds.) P.K. Gupta and J.R. Bahl, Rastogi and Co., Meerut, India.

Verma M.M., Kochhar S. and Kapoor W.R. 1977. The assessment of biparental approach in a wheat cross. *Z. Pflanzenzuchtg*, **82**: 174-181.

Webel O.D. and Lonnquist J.H. 1967. An evaluation of modified ear-to-row selection in a population of corn (*Zea mays* L.). *Crop Sci.*, **7**: 651-655.

Yunus M. and Paroda R.S. 1982. Impact of biparental mating on correlation coefficients in bread wheat. *Theor. Appl. Genet.*, **62**: 337-343.

Zhang Shihuang and Shi Dequan. 2000. QPM breeding program in China. *In*: *Proc. Seventh Asian Regional Maize Workshop.* PCARRD, Las Banos, Philippines. (eds.) Vasal S.K., Gonzalez Cericeris F. and Fan Xing Ming.

Plant Breeding - Mendelian to Molecular Approaches
H. K. Jain and M. C. Kharkwal (eds.)

Classical and Molecular Concepts of Heterosis

S.S. Virmani[1], M.P. Pandey[2], I.S. Singh[2], Wei Jun Xu[1]

Abstract

Heterosis has contributed significantly towards increased crop production and it is the basis of multi-billion dollar agri-business in the world. Although extensive studies have been conducted yet the physiological, biochemical and molecular explanation of this phenomenon remains largely unexplained. During 1940s through 1970s, advances made in quantitative genetics led to development of dominance, over-dominance, and epistasis (classical concepts) theories of heterosis. Consequently, breeding strategies led to establishment of heterotic group for hybrid breeding. Despite these successes, plant breeders and geneticists are not able to predict the extent of heterosis. Over the years, several physiological as well as biochemical concepts have been put forward but these are unable to predict heterotic combinations. Molecular marker technologies are now being used to identify and locate heterotic gene blocks which can be incorporated in the selected parental lines to develop heterotic hybrids which result from the mechanisms that are not likely to be captured in an inbred *viz.*, loci with dosage divergence between inbreds, and differential DNA methylation in hybrids *vs.* inbreds. With the help of transformation technology in future it may also be possible to transfer heterotic chromosome blocks or QTL across species in parental lines. So far, utilization of heterosis has gone ahead of the understanding of this phenomenon, however, in future, use of molecular and transformation tools should increase our understanding of heterosis such that it may lead the path to better utilization of heterosis.

Introduction

Heterosis, or hybrid vigor, usually refers to superior performance of F_1 over its parents. Although Koelreuter, Darwin, Knight, and Burbank observed this phenomenon in late 19th century, the concept for utilizing heterosis for crop improvement was not developed, until rediscovery of Mendel's laws in 1900, and subsequent work of Shull (1908), East (1908) and Jones (1917).

Currently, heterosis is a major factor for increased crop production in maize, rice, sorghum, pearl millet, cotton, sunflower, tomato, eggplants, chillies, onion and sugar beet (Virmani, 1996, 1997 and Hallauer, 1999). It has become the basis of multi-billion dollar agri-business in the world (Philips, 1999).

[1]International Rice Research Institute, Metro Manila, Philippines
[2]G.B. Pant University of Agriculture & Technology, Pantnagar 263 145, India

Commercial hybrid rice cultivars have been successfully released in China, India, Vietnam and the Philippines, giving estimated yield increases of 15-20% (Rangaswamy and Jayamani, 1996; Tan, 1994). Duvick (1999) reported that hybrid varieties in maize, sorghum, sunflowers and rice added 90 million tonnes to global food production, which spread cultivation of about 34 million hectare land from these crops. Thus heterosis has not only helped to achieve food security, but it also contributes to environmental protection.

Although extensive studies have been conducted, yet the physiological, biochemical and molecular explanation of this phenomenon still remains largely unexplained (Stuber, 1994; Philips, 1999). So far utilization of heterosis has stayed ahead of its understanding. With rapid development of molecular biology, molecular approaches are being extensively used to understand and utilize heterosis.

Historical Resume of Heterosis

Although workable methods of utilizing hybrid vigour in crop production were known (Koelreuter, 1766; Johnson, 1891; Shamel, 1905), yet its genetic concepts were not recognized until the rediscovery of Mendel's laws in 1900. With the discovery of Mendel's long forgotten paper "Versuche Ueber Pflanzen Hybriden" and the monumental work of East (1911-1915), East and Hayes (1912) and Shull (1904-1914), the scientific concept of heterosis phenomenon emerged. In 1908, Shull published the famous paper on "The composition of a field of maize" and made the following landmark conclusions:

- Self fertilization tended to purify strains and released new variation
- Selfed plants were weaker than corresponding cross fertilized ones
- Each successive generation of close inbreeding further reduced the strain to other simple constituent biotypes which were weaker than the hybrid combination and it was possible to completely cancel in a single year, the accumulated deterioration.

East and Hayes (1912) explained the concept as 'the decrease in vigour due to inbreeding naturally cross-fertilizing species and increase in vigour due to crossing naturally self-fertilizing species are manifestations of one and the same phenomenon termed 'heterozygosis'. Crossing produced heterozygosis in all the characters for which the parents differ while inbreeding tended to provide homozygosis automatically. It was in 1914 that Shull proposed the term 'heterosis' in a lecture delivered in Goettingen, Germany to represent both the phenomenon (*viz.* inbreeding and hybrid vigor) and suggested it as a "potent source of practical gains" for utilization in animals and plants.

Rapid development in the heterosis concept occurred in the post-Mendelian period with the advent of the knowledge in genetics. Shull (1952) stated that increased vigour, size, fruitfulness, speed of development, resistance to disease and insect-pests or to climatic rigors of any kind, manifested by cross-bred organisms as compared with corresponding inbred could be the result of unlikeness in the constitution of the uniting parental gametes. He recognized heterosis as the result of interaction of unlike gametes and stated that it was closely related to the well-known cases of the interaction of complementary genes. In the absence of knowledge of quantitative genetics at that time, the heterosis concept was conceived as a simply "diffused phenomenon", incapable of analysis into the interactions of specific individual genes and individual gene interactions.

Biometrical Concepts of Heterosis

The simplest genetical model for estimating heterosis is based on the difference between the hybrid and parental means, which is called midparent heterosis and is often expressed as a percentage of midparent. Heterosis may also be defined as the amount by which the mean of F_1 family exceeds its better parent (Mather and Jinks, 1982).

Quantitative genetic theory of heterosis is often developed on the basis of single locus theory. Falconer (1960) demonstrated that midparent heterosis, also called panmictic mid parent heterosis (Lamkey and Edwards 1999), could be expressed as $H_{mp} = y^2 d$, where d is the level of dominance, and y is the difference in allele frequency of the parents. Therefore, the level of heterosis expressed in a hybrid depended largely on allelic frequency differences in two parents and presence of certain level of dominance. If two inbred lines instead of populations, as in the case above, are crossed, y can only be 0 or 1. Therefore, heterosis in a cross of two inbreds is a function of level of dominance at loci with different alleles only. Heterosis shown by F_2 population derived from inbreeding F_1 (also called inbred mid parent heterosis) is expressed as $H_{imp} = \frac{1}{2} y^2 d$.

The difference between panmictic and inbred midparent heterosis reflects the vigor lost due to inbreeding. Genetic divergence and dominance are involved in both inbred and panmictic midparent heterosis.

Single locus theory has directly led to two prominent theories of heterosis called dominance and overdominance hypothesis. Heterosis under dominance hypothesis is due to sheltering of deleterious recessives by their dominant counterpart alleles (Davenport, 1908; Bruce, 1910; Keeble and Pellew, 1910). Heterosis under overdominance assumes that heterozygosity *per se* is important for hybrid vigor (Shull, 1908; East, 1908, 1936; Hull, 1945 and Crow, 1948). Both the theoretical expectations and the experimental evidence suggest absence of genuine overdominance.

Non-allelic interaction and linkage disequilibrium appears to be misinterpreted as overdominance. Beyond single locus theory, one can easily recognize that a complex trait is actually due to several underlying multiplicatively acting traits, which are controlled by a number of genetic loci. Therefore, epistasis may actually contribute a great deal to the heterosis expressed in a cross. Several studies (Bauman, 1959; Gorsline, 1961; Gamble 1962 and Sprague *et al.*, 1962) have conclusively proved that epistasis contributes a part of total hereditary variance in case of quantitative characters such as grain yield in maize.

Cheverud and Routman (1995) have identified two concepts of epistasis, namely, physiological and statistical epistasis. Further, they demonstrated that physiological epistasis could either suppress or enhance statistical epistasis. For understanding the biometrical genetics of heterosis and especially the role of non-allelic interaction in heterosis, the reader is referred to Jinks (1983).

For self pollinated crops, if there were midparent heterosis but no inbreeding depression, then there would be a good evidence for the existence of additive by additive epistasis. Recent studies in rice, a self pollinated crop, indicate a widespread epistatic loci (Yu *et al.*, 1997 and Li *et al.*, 2000). Basic generations and multiple mating designs can be used to obtain information regarding gene action and interaction.

Despite tremendous efforts made during several decades on studying gene action involved in heterosis, we do not know the exact number of loci involved in heterotic expression or their distributions for the effect or interactions. It appears to be extremely difficult to solve these problems with only classical methods of quantitative genetics. Developments in biochemical and molecular techniques in recent decades have opened new avenues to study this phenomenon more efficiently.

Physiological and Bio-chemical Concepts of Heterosis

Grain yield is a complex trait resulting from the action and interaction of essential metabolic processes over a period of time. The increase in yield for many crop species is generally attributed to changes in allocation of assimilates rather than changes in total production (Ziegler, 1990; and Evans, 1993). It is obvious that harvest index (HI), which is the function of changes in pattern of partitioning the accumulated dry matter has been significantly improved in several important crop species, including rice and wheat. However, these partitioning allocation differences must eventually be traceable to metabolic reactions (Eastin *et al.*, 1999). Metabolic studies in sorghum have revealed that metabolic differences existing in crop varieties almost certainly contributed to yield differences (Rice and Eastin, 1986; Eastin *et al.*, 1983). The metabolic efficiency may substantially relate to heterozygosity and heterosis. Significant heterosis for HI has been reported in rice (Ponnuthurai *et al.*, 1984). In view of these results, it seems likely that advantage in metabolic establishment of heterotic hybrid should be a major reason for high heterosis and perhaps has been manipulated subconsciously by plant breeders.

Hageman *et al.* (1967) proposed the metabolic balance concept for determining the hybrid response. The elements of this concept (Stuber, 1999) are: (1) inbreds tend to have unbalanced metabolic systems with enzymes controlled at variable levels of activities from low to high, (2) highly homozygous lines may have some important enzyme reactions that are severely limiting metabolism, and (3) the specific limitations probably differ among individual lines, which may be overcome in a heterozygote by a proper choice of complementary inbred parents, resulting in better balanced metabolic system in the resulting hybrids.

The metabolic balance concept is largely analogous to the concept of limiting factors or physiological bottlenecks, in which the limiting factors may reside at different loci of different parental inbreds. The mutation of any gene that controls any of the key metabolic steps or functions will certainly be reflected in the resulting systems. Srivastava (1981) stated that it was likely that heterotic hybrids were endowed with superior systems of chloroplasts and mitochondria, and such superiority was provided by genomic and intergenomic interactions.

Extensive studies on comparisons between hybrid and inbred cultivars have indicated that the yield advantage of heterotic hybrids may logically result from their higher photosynthetic efficiency (McDonald *et al.*, 1974 and Murayama *et al.*,1984 and 1987) and lower respiration rate (Murayama *et al.*, 1984) as well as higher heterosis in other related physiological traits, namely, leaf area index (Ponnuthurai *et al.*, 1984) and cumulative growth rate (Blanco *et al.*, 1990). Capability of dynamic response to environmental changes of heterotic hybrids should contribute substantially to the average yield increases. It has also been reported that hybrids tend to be superior in drought and salt tolerance (Akaba and Yabuno, 1975 and Senadhira and Virmani, 1987), in ratooning ability (Chauhan *et al.*, 1983) as well as submergence tolerance.

Metabolic processes in living organisms are regulated by both regulatory proteins (and their coding genes), and although there are various mechanisms of regulation of gene activity, several studies, however, suggest that they are significantly related to manifestation of heterosis (Tsaftaris, 1990 and Tsaftaris *et al.*, 1999). The information concerning the quantitative expression for regulatory proteins and their coding genes is largely obtained from the analysis of RNA amount polymorphism (RAP) and Protein amount polymorphism (PAP), respectively. Overall results suggested that quantitative expression of certain genes is responsible for heterosis manifestation in hybrids (Cherry *et al.*, 1961; Nebiolo *et al.*, 1983; Romagnoli *et al.*, 1990; Tsaftaris and Polidoros, 1993). One mechanism involved in regulating expression output of genes is methylation of cytosine residues in their DNA (Cedar, 1984). Although it has been used in studies related to functions of molecular significance, such as modulation of gene expression, many questions as to its function still remain unsolved (Hepburn *et al.*, 1987; Hollick et al., 1997; and Richards, 1997). By extrapolating these observations, Stuber (1999) hypothesized that degree of DNA methylation may be regulating heterosis such that there was a gradual accumulation of methylation during selfing, which is then released/repatterned in hybrids. Tsaftaris *et al.*, (1997) reported that maize DNA is highly methylated and the methylation was found to be genotype, tissue and developmental stage specific as well as affected by the environmental conditions. Hybrids were less methylated than inbreds. Inbreds varied in their methylation status and improved lines were less methylated than older and low yielding ones.

Although a number of physiological and biochemical concepts of heterosis have been proposed to explain the phenomenon of heterosis, yet, to date none of these concepts, either individually or jointly, have shown the potential to predict heterosis, such that plant breeders can choose the parents with confidence to develop hybrids.

Molecular Concepts of Heterosis

Molecular markers can be used to determine molecular basis of heterosis by identifying individual QTLs and understanding their actions and interactions. Molecular marker polymorphism can help to measure genetic distance or genomic similarities between pair of parental lines of a hybrid. Smith and Smith (1991) reported that RFLP diversity among F_1 hybrids were shown to be highly correlated with distance based on multigenic traits such as pedigree, F_1 yield and yield heterosis, however, Dudley's (1991) Modified Roger's distance (MRD) values (a measure of genetic distance between inbreds based on marker data) were not significantly correlated with hybrid yields. Zhang *et al.* (1995) reported very high correlation between midparent heterosis and specific heterozygosity based on the markers that detected significant effects for heading date, plant height, straw weight, grain yield and biomass in rice. Saghai Maroof *et al.* (1997), however, concluded that the level of correlations between RFLP marker distance and hybrid performance was dependent on the germplasm used. Bernardo (1992) suggested that molecular markers would be useful for predicting hybrid performance only when a significant portion (> 50 %) of the selected markers were linked with quantitative trait loci (QTLs).

Studies on QTLs contributing to grain yield, yield components, and heterosis in maize (Stuber *et al.*, 1992) and rice (Xiao *et al.*, 1995; Yu *et al.*, 1997 and Li *et al.*, 2000) have indicated that the degree of correlation between overall genomic heterozygosity and phenotypic performance largely depends on genetic composition of the traits. Theoretically, a trait controlled by few loci should show little correlation between phenotype and overall heterozygosity across genome, whereas a trait controlled by many loci should show a high correlation. The theory has provided reasonable explanation that a high correlation was detected in maize population derived from an elite cross between B73 and Mo17, in which QTLs affecting grain yield were found through out the genome, except chromosome 6 (Stuber *et al.*, 1992). In contrast, no significant correlation between overall heterozygosity and heterosis was detected in rice (Xu *et al.*, 1999 and Yu *et al.*, 1997), since few QTLs were found affecting grain yield in limited rice genome and epistasis between QTLs was common and significant (Yu *et al.*, 1997; Li *et al.*,2000 and Xiao *et al.*, 1996).

QTLs mapping studies have further indicated the importance of overdominance and pseudo-overdominance as well as epistasis in expression of the heterosis. In maize, Stuber *et al.* (1992) reported that whenever a QTL for grain yield was detected, the heterozygote always had higher mean performance. QTL alleles causing high grain yield showed a strong tendency towards overdominance. Similarly, overdominance was observed for most of the QTLs for yield in rice (Yu *et al.*, 1997). Recent evidence for overdominance in rice was reported by Li *et al.* (2000). Fourteeen of the 24 main-effect QTLs detected in the backcross and testcross populations appeared to be overdominant, and the dominant effects at all main-effect QTLs detected in the two BCF_1 and $413F_1$ populations were positive (> 0), resulting in increased grain yield. Heterosis may result from either true overdominance at single loci or pseudo-overdominance due to repulsion-phase linkage between advantageous alleles (Crow, 1952). It may be impossible to distinguish overdominance from pseudo- overdominance without extensive recombinational separation of closely linked loci or cloning of the QTLs. In a recent fine-mapping study in maize, Graham *et al.* (1997) successfully partitioned a QTL at chromosome 5, which appeared to show overdominance into two loci linked in repulsion phase with dominant gene action at each. This shows that dominance rather than overdominance is the cause of heterosis.

In an experiment designed to identify QTLs for heterosis in maize by using recombinant inbred lines (RILs) tested with three elite inbreds, genomic locations of QTLs with significant effect on yields differed greatly among testers (Guffy *et al.*, 1988). It was concluded that the interaction of QTL effect with genetic background was highly significant. However, Stuber *et al.* (1992) did not find convincible evidence for epistasis in maize. Most of evidences for epistasis affecting heterosis come from rice. In F_3 families derived from a elite rice hybrid - Zhenshan 97/Minghui 63, Yu *et al.* (1997) detected 32 QTLs for the four traits related to grain yield and showed strong evidence for widespread epistatic interactions in the population, which included all types of digenic interactions, such as additive by additive, additive by dominance and dominance by dominance. Recently, Li *et al.* (2000) indicated that: (1) in each of their five mapping populations, the majority of the phenotypic variation in the F_1 mean and heterosis values was due to epistatic QTLs, while only a small portion of the vàriation was due to main effect QTLs, and (2) 25 of the 29 QTLs with significant main-effects were involved in epistasis detected in one or more populations. Only four out of at least 54 QTLs identified in their study were not involved in epistasis in any of the mapping populations.

Although findings from mapping studies showed that QTLs were generally distributed throughout genome, certain chromosome region appears to contribute more effect than others to trait expressions (Stuber, 1994a). In population derived from 9024 (*indica*)/LH422 (*japonica*) cross with high heterotic performance, Xiao *et al.* (1995) reported that not all of the QTLs detected for traits had similar phenotype effects. Stuber *et al.* (1992) found that two chromosome regions in maize, the long arm of chromosome 4 and the short arm of chromosome 5, including the centromeric region, contributed significantly to the expression of grain yield. Using marker assisted selection, Stuber (1994a) transferred chromosome segments of TX 303 and Oh43 into B73 and Mo17 respectively. Several enhanced hybrids were obtained.

Genetic dissection of inbreeding depression in rice has indicated a different, however, partial overlap of genes with heterosis (Li *et al.*, 2000). Most of QTLs associated with inbreeding depression in rice appeared to be involved in epistasis, providing further evidence of importance of epistasis as a genetic basis of rice.

Future Outlook on Heterosis Breeding

Initially, the exploitation of heterosis was confined to corn and a few other open pollinated species but presently heterosis is being exploited in practically all species cutting across mode of reproduction. More and more evidences were showing that heterotic expression in a cross is a function of genetic backgrounds and genes. Heterotic grouping would remain as a key strategic method in management of diverse genetic backgrounds and genes. A perfect grouping method should be built on both elements. Presently, heterotic grouping based on molecular techniques like RFLPs, AFLPs, SSRs essentially belong to type of gene classification. The method might be more effective in certain cases of maize, in which epistasis was found less important for heterotic expression. However, it might be less effective for crop species, such as rice, in which epistasis has played a key role in heterotic expression. Therefore, it is imperative for plant breeder to develop new methodologies to make heterotic grouping more effective and efficient.

New molecular tools such as expressed sequence tags (ESTs) and microarray technologies and support from genomics or functional genomics may help in identifying genes, associated with heterosis in future (Lee, 1999). This may help to identify universal or unique features of heterosis in crop species. Thus, in future, both classical and molecular tools would very much help us in improving our understanding and manipulation of heterosis.

References

Akaba M. and Yabuno T.T. 1975. Breeding for saline-resistant varieties of rice. III. Response of F_1 hybrids to salinity in reciprocal crosses between Jhona 349 and Magnolia. *Jpn. J. Breed.*, **25**: 215-220.

Bauman L.F. 1959. Evaluation of non-allelic gene interaction in determining yield, ear height, and kernel row number in corn. *Agron. J.*, **51**: 531-534.

Bernardo R. 1992. Relationship between single cross performance and molecular marker heterozygosity. *Theor. Appl. Genet.*, **83**: 628-634.

Blanco L., Casal C., Akita S. and Virmani S.S. 1990. Biomass, grain yield and harvest index of F_1 rice hybrids and inbreds. *IRRN*, **15**: 9-10.

Bruce A.B. 1910. The Mendelian theory of heredity and augmentation of vigour. *Science*, **32**: 627-628.

Cedar H. 1984. DNA methylation and gene expression. *In*: *DNA methylation, biochemistry and biological significance.* (eds) A. Razin *et al.* Springer Verlag, New York, pp 147-164.

Chauhan J.S., Virmani S.S., Aquino R., Vergara B.S. 1983. Evaluation of hybrid rice for ratooning ability. *IRRN*, **8**: 6.

Cheverud J.M. and Routman E.J. 1995. Epistasis and its contribution to genetic variance components. *Genetics*, **139**: 1455-1461.

Cherry J.H., Hagerman R.H., Rutger J.N. and Jones B.J. 1961. Acid-soluble nucleotides and ribonucleic acid of different corn inbreds and single cross hybrids. *Crop Sci.*, **1**: 133-137.

Crow J.F. 1948. Alternative hypothesis of hybrid vigour. *Genetics*, **33**: 477-487.

Crow J.F. 1952. Dominance and overdominance. pp 282-297. *In*: *Heterosis*, (ed.) J. W. Gowen Iowa State college Press, Ames

Crow J.F. 2000. The rise and fall of overdominance. *In*: *Plant Breeding Reviews*, **17**: 225-257.

Davenport C.B. 1908. Degeneration, albinism and inbreeding. *Science*, **28**: 454-455.

Dudley J.W., Saghai Maroof M.A. and Rufener G.K. 1991. Molecular markers and grouping of parents in maize breeding program. *Crop Sci.*, **31**: 718-723.

Duvick N.D. 1999. Heterosis: Feeding people and protecting natural resource. *In*: *Genetics and Exploitation of Heterosis in Crop.* (eds.) J. G. Coors and S. Pandey, 19-29.
East E.M. 1908. Inbreeding in corn. *Rept. Connecticut Agric. Expt. Sta.* For 1907. pp. 419-428.
East E.M. and Jones D.F. 1919. *Inbreeding and outbreeding.* J. B. Lippincott Co., Philadelphia, PA
East E.M. 1936. Heterosis, *Genetics*, **21**: 375-397.
East E.M. and Hayes H.K. 1912. Heterozygosis in evolution and in plant breeding. U S Dept. Agric. *Bur. Plant Indust. Bull.*, 243, pp. 58.
Eastin J.D., Petersen C.L., Zavala-Garcia F. and Dhopte A. 1999. Potential heterosis associated with developmental and metabolic processes in sorghum and maize. *In*: *Genetics and Exploitation of Heterosis in Crop.* (eds.) J. G. Coors and S. Pandey. 205-220.
Eastin J.D., Castleberry R.M., Gerik T.J., Hultwuist J.H., Mahalakshmi V., Ogunlela V.B. and Rice J.R. 1983. Physiological aspects of high temperature and water stress. *In*: *Crop reactions to water and temperature stresses in humid, temperate climates.* (eds.) C. D. Raper Jr. and P. J. Kramer. Westview Press, Boulder, Co. pp 91-112.
Evans L.T. 1993. *Crop evolution, adaptation and yield.* Cambridge University Press, New York.
Falconer D.S. and Mackey T.F.C. 1996. *Introduction to quantitative genetics.* 4th ed. Longman, Essex, England.
Gamble E.E. 1962. Gene effects in corn (*Zea mays* L.) I. Separation and relative importance of gene effects for yield. *Canad. J. Pl. Sci.*, **42**: 339-348.
Gartner C.F. 1849. Versuche und Beobachtungen uber die Bastarderzeugung im *Pflanzenreich.* 791 pp Stuttgart
Gorsline G.W. 1961. Phenotypic epistasis for ten quantitative characters in maize. *Crop Sci.*, **1**: 55-58.
Gowen G.W. 1952. Hybrid vigour in *Drosophila*. pp. 474-493. *In*: *Heterosis.* (ed.) J. W. Gowen. Iowa State College Press, Ames
Graham G.I., Wolff D.W. and Stuber C.W. 1997. Characterization of a yield quantitative trait locus on chromosome five of maize by fine mapping. *Crop Sci.*, **37**: 1601-1610.
Hageman R.H., Leng E.R. and Dudley J.W. 1967. A biochemical approach to corn breeding. *Advan. Agron.*, **19**: 45-86.
Hallauer A.R. 1999. Heterosis: what have we learnt What have we done? Where are we headed? *In*: *Genetics and Exploitation of Heterosis in Crop.* (eds.) J. G. Coors and S. Pandey. pp.483-492.
Hallauer A.R. and Miranda J.B. 1988. *Quantitative genetics in maize breeding.* 2nd ed., Iowa State Univ. Press, Ames.
Hepburn A.G., Belanger F.C. and Mattheis J.R. 1987. DNA methylation in plants. *Developmental Genetics*, **8**: 475-493.
Hollick J.B., Dorweiler J.E. and Chandler V.L. 1997. Paramutation and related allelic interactions. *Trends Genet.*, **13**: 302-308.
Hull F.H. 1945. Recurrent selection for specific combining ability in corn. *J. Am. Soc. Agron.*, **37**: 134-145.
Jinks J.L. 1983. Biometrical genetics of heterosis. *In*: *Heterosis-Reappraisal of the theory and practice.* (ed.) Frankal R. Springer-Verlag pp 1-46.
Johnson S.W. 1891. *How crops grow*? Orange Judd & Co., New York.
Jones D.F. 1917. Dominance and linked factors as a means of accounting for heterosis. *Genetics*, **2**: 466-479.

Keeble F. and Pellew C. 1910. The mode of inheritance of stature and time of flowering in peas (*Pisuem sativum*), *J. Genet.*, **1**: 47-56.

Koelreuter J.G. 1766. Vorlaufigennachircht von einigen das Geschlecht der Pflanzen betreffenden versuchen und beobachtungen. 266 pp. Leipzig.

Lamkey K.R. and Edwards J.W. 1999. Quantitative genetics of heterosis. *In*: *Genetics and Exploitation of Heterosis in Crop.* (eds.) J. G. Coors and S. Pandey. pp 31-48.

Lee M. 1999. Towards Understanding and manipulating heterosis in crop plants-can molecular genetics and genome projects help? *In*: *Genetics and Exploitation of Heterosis in Crop.* (eds.) J. G. Coors and S. Pandey. pp. 185-194.

Li Z.K., Luo L.J., Mei H.W., Wang D.L., Shu Q.Y., Tabien R., Zhong D.B., Ying C.S., Stansel J.W., Khush G.S. and Paterson A.H. 2000. Genetic basis of inbreeding depression and heterosis in rice (*Oryza sativa* L): I. Biomass and grain yield. *Genetics.*

Mackill D.J. 1995. Classification of *japonica* rice cultivars with RAPD markers. *Crop Sci.*, **35**: 889-894.

Mather K. and Jinks J.L. 1982. *Biometrical Genetics*. 3rd ed. Chapman and Hall, London.

McDonald D.J., Stansel J.W. and Gilmore E.C. 1974. Breeding for high photosynthetic rate in rice. *Indian J. Genet.*, **34**: 1068-1073.

Moser H. and Li M. 1994. RFLP variation and genealogical distance, multivariate distance, heterosis and genetic variation in oats. *Theor. Appl. Genet.*, **87**: 947-956.

Mumm R.H. and Dudley J.W. 1994. A classification of 148 U.S. maize inbreds. I. Cluster analysis based on RFLPs. *Crop Sci.*, **34**: 842-851.

Murayama S., Miyasato Y. and Nose A. 1982. Studies on dry matter production of F_1 rice hybrids. I. Heterosis in photosynthetic activity. *Jpn. J. Crop Sci.*, (Spe. Issue) **2**: 85-86 (in Japanese)

Murayama S, Ogasawara N., Miyasato S. and Nose H. 1984. Dry matter production of F_1 rice hybrids. III. Heterosis in photosynthetic activity. *Jpn. J. Crop Sci.*, (Spe. Issue) **53**: 100-101 (in Japanese).

Murayama S, Miyasato K. and Nose A. 1987. Studies on dry matter production of F_1 rice hybrids. 1. Heterosis in the single leaf photosynthetic rate. *Jpn. J. Crop Sci.*, **56**: 198-203.

Nebiolo C.M., Kaczazczyk W.J. and Ulrivh V. 1983. Manifestation of hybrid vigor in RNA synthesis parameters by corn seedling protoplasts in the presence and absence of gibberellic acid. *Plant Sci. Lett.*, **28**: 95-206.

Ponnuthurai S., Virmani S.S. and Vergara B.S. 1984. Comparative studies on the growth and grain yield of some F_1 rice (*Oryza sativa* L.) hybrids. Philipp. *J. Crop. Sci.*, **9**: 183-193.

Philips R.L. 1999. Research needs in heterosis. *In*: *Genetics and Exploitation of Heterosis in Crop.* (eds.) J. G. Coors and S. Pandey. pp. 501-507.

Pray L.A. and Goodnight C.J. 1995. Genetic variability in inbreeding depression in the flour beetle, *T. castaneum. Evolution*, **49**: 179-188.

Rangaswamy M. and Jayamani P. 1996. Hybrid rice research in Tamil Nadu. *In*: *Hybrid Rice Technology*, (ed.) M. Rangaswamy, School of Genetics, Tamil Nadu Agric. Univ. Coimbatore, India.

Rice J.R. and Eastin J.D. 1986. Grain sorghum root responses to water and temperature during reproductive development. *Crop Sci.*, **26**: 547-551.

Richards E.J. 1997. DNA methylation and plant development. *Trends Genet.*, **13**: 319-323.

Romagnoli S., Maddaloni M., Livini C. and Motto M. 1990. Relationship between gene expression and hybrid vigor in primary root tips of young maize (*Zea mays* L.) plantlets. *Theor. Appl. Genet.*, **80**: 767-775.

Senadhira D. and Virmani S.S. 1987. Survival of some F_1 rice hybrids and their parents in saline soil. *IRRN*, **12**: 14-15.

Shamel A.D. 1905. The effect of inbreeding in plants. U S Dept. Agric. Year book for 1905. pp 377-392.

Shull G.H. 1908. The composition of a field of maize. Rept. Amer. *Breeders Assoc.*, **4**: 296-301.

Shull G.H. 1914. Duplicate genes for capsule form in Bursa bursa-pastoris. *Zeitschr. Induk. Abstamm. U. verebungsl.*, **1**: 947-149.

Shull G.H. 1952. Beginning of the heterosis concept. pp 14-48. *In*: *Heterosis*. (ed.) J. W. Gowen. Iowa State Cllege Press.

Smith J.S.C. and Smith O.S. 1992. Fingerprinting crop varieties. *Adv. Agronomy*, **47**: 85-139.

Smith J.S.C. and Smith O.S. 1991. Restriction fragment length polymorphisms can differentiate among U.S. maize hybrids. *Crop Sci.*, **31**: 893-899.

Sprague G.F., Russel W.A., Penny L.H., Horner R.W. and Hanson W.D. 1962. Effect of epistasis on grain yeld in maize. *Crop Sci.*, **2**: 205-208.

Srivastava H.K. 1981. Intergenomic interaction, heterosis, and improvement of crop yield. *Adv. Agron.*, **34**: 117-195.

Stigfield G.H. 1950. Heterozygosis and hybrid vigour in maize. *Agron. J.*, **42**: 145-152.

Stuber C.W., Lincoln S.E., Wolff D.W., Helentjaris T. and Lander E.S. 1992. Identification of genetic factors contributioing to heterosis in a hybrid from two elite maize inbred lines using molecular markers. *Genetics*, **132**: 823-839.

Stuber C.W. 1994a. Enhancement of grain yield in maize hybrids using marker facilitated introgression of QTLs. *In*: *Analysis of Molecular Marker Data*. Corvallis. Oregon.

Stuber C.W. 1994b. Heterosis in plant breeding. *Plant Breeding Rev.*, **12**: 227-251.

Stuber C.W. 1999. Biochemistry, molecular biology, and physiology of heterosis. *In*: *Genetics and Exploitation of Heterosis in Crop*. (eds.) J. G. Coors and S. Pandey. pp. 173-183.

Stuber C.W., Williams W.P. and Moll R.H. 1973. Epistasis in maize (*Zea mays*). III. Significance in predictions of hybrid performances. *Crop Sci.*, **13**: 195-200.

Tsaftaris A.S. 1990. Biochemical analysis of inbreds and their heterotic hybrids in maize. *In*: *Isozymes: Structure, function and use in biology and medicine*. (eds.) A. Ogita and C. Markert. Wiley-Liss, New York.

Tsaftaris A.S., Kafka M. and Polidoros A. 1997. Epigenetic modifications of total genomic maize DNA. the role of growth conditions. *In*: *Genetics, biochemistry and breeding of maize and sorghum*. (ed.) A. S. Tsaftaris. The Royal Soc. of Chem. Press. Cambridge, England.

Tsaftaris A.S., Kafka M., Polidoros A. and Tani E. 1999. Epigenetic change in maize DNA and heterosis. *In*: *Genetics and Exploitation of Heterosis in Crop*, (eds.) J. G. Coors and S. Pandey. pp. 195-203.

Tsaftaris A.S. and Polidoros A. 1993. Studying the expression of genes in maize parental inbreds and heterosis and non-heterotic hybrids. *In*: *Proc. XII Eucarpia Maize and Sorghum Conf.*, Bergamo, Italy., (eds) A. Bianci *et al.*

Tan N.C. 1994. Progress in hybrid rice program in Vietnam. *International Rice Commission*, **43**: 23-28.

Virmani S.S. 1996. Hybrid rice. *Adv. Agron.*, **57**: 377-461.

Virmani S.S. 1997. Hybrid rice research and development in the tropics. *In*: *Proc. 3rd Int. Symp. on Hybrid Rice*. Hyderabad, India 14-16 Nov. 1996.

Whaley W.G. 1944. Heterosis. *Bot. Rev.*, **10**: 461-498.

Yu S.B., Li J.X., Xu C.G., Tan Y.F. and Gao Y.J. 1997. Importance of epistasis as the genetic basis of heterosis in an elite rice hybrid. *Proc. Natl. Acad. Sci.*, USA, **94**: 9226-9231.

Ziegler H. 1990. Role of plant physiology in assessing productivity potential under stress environment. *In*: *Proc. of the Int. Cong. of Plant Physiology*. Soc. Plant Phys. and Biochem. Water technology Center and Indian Agricultural Res. Inst. New Delhi. (eds.) S. K. Sinha *et al.*

Xiao J., Li J.X., Yuan L. and Tanksley S.D. 1995. Dominance is the major genetic basis of heterosis in rice as revealed by QTL analysis using molecular markers. *Genetics*, **140**: 745-754.

Xiao J., Li J.X., Yuan L., McCouch S.R. and Tanksley S.D. 1996. Genetic diversity and its relationship to hybrid performance and heterosis in rice as revealed by PCR-based markers. *Theor. Appl. Genet.*, **92**: 637-643.

Xu W., Virmani S.S., Hernandez J.E., Li Z.K. and Redona E.D. 1999. Association between simple sequence repeat (SSR) marker diversity, pedigree record, quantitative variation, and hybrid performance in rice. *IRRN*, **24**: 10-11.

Zhang Q., Gao Y.J., Yang S.H., Ragab R., Saghai Maroof M.A. and Li Z.B. 1994. A diallel analysis in elite hybrid rice based on RFLPs and microsatellites. *Theor. Appl. Genet.*, **89**: 185-192.

Plant Breeding - Mendelian to Molecular Approaches
H. K. Jain and M. C. Kharkwal (eds.)

18

Advances in Hybrid Breeding Methodology

B.S. Dhillon[1], A.K. Singh[2], B.P.S. Lather[3] and G. Srinivasan[4]

Abstract

Hybrid cultivars are the first generation (F_1) progeny of a cross between two or more selected and genetically diverse parents, and these exploit heterosis. Heterosis is defined as increased vigour, size, fruitfulness, speed of development, resistance to disease and insect-pests or to climatic rigours of any kind in F_1 generation. Dominance of linked loci, over dominance, and epistatic gene action were proposed to explain the genetic basis of heterosis. Logically all types of gene action controlling the inheritance of a trait should be expected to contribute to heterosis. Hybrid breeding started in maize, and experiences in this crop have greatly contributed to the development of hybrid breeding methodology which is continuously evolving and expanding. In sorghum, pearl millet, sunflower etc., hybrid breeding started with the development of male sterility and fertility restoration system for pollination control. The technology was largely confined to cross-pollinated crops with a few exceptions like tomato and cotton. In the late 1970s, the success of hybrid rice, a strictly self-pollinated crop in China provided an impetus to hybrid breeding. The possibility to use environment-sensitive genetic male sterility and chemical hybridizing agents has opened the avenue of widening the base of parental germplasm and enhancing the magnitude of heterosis; and it is likely to simplify hybrid breeding and seed production. Further, the use of recurrent selection to develop hybrid breeding oriented source germplasm needs greater emphasis. The application of biotechnological techniques such as doubled haploidy, somatic hybridization, analysis of molecular diversity and marker assisted selection are expected to provide cutting edge to hybrid breeding; for which biotechnology needs to be integrated with hybrid breeding. Another exciting area is the development of apomictic hybrid that should greatly expand the coverage by hybrid cultivars.

Introduction

Hybrid cultivar is the first generation (F_1) progeny of a cross between two selected and genetically diverse parents; and the superiority of the performance of a hybrid over its

[1]National Bureau of Plant Genetic Resources, New Delhi 110 012, India
[2]Indian Agricultural Research Institute, New Delhi 110 012, India
[3]Ch. Charan Singh Haryana Agricultural University, Hisar 125 004, India
[4]International Maize and Wheat Improvement Center (CIMMYT), El Batan, Mexico

parent(s) is known as heterosis. The development of hybrids and their commercial cultivation is one of the major achievements in plant breeding. It was first accomplished in maize (*Zea mays* L.). The research conducted by Shull (1908, 1909) on inbreeding and outbreeding, development of single cross hybrids and suggestion made by Jones (1918) to commercially exploit double cross hybrids, laid the foundation of the success story of hybrid maize. In addition, top-cross test for preliminary screening of inbred lines (Davis, 1927), early testing of inbred lines (Jenkins, 1935), prediction of double-cross performance (Jenkins, 1934) and studies on variability among different types of hybrids (Cockerham, 1961) greatly facilitated hybrid maize breeding. Sprague and Tatum (1942) expounded the concept of combining ability, which was used widely and proved very useful in hybrid breeding.

The first commercial planting of a double-cross hybrid was made in 1921 in the USA. The first series of hybrids, in the 1920s, yielded about 15 per cent higher than the better open-pollinated (OP) varieties (Duvick, 1999). The adoption of maize hybrids was initially quite slow, but began to increase rapidly thereafter, and after 15 years 95 per cent of the maize acreage in the US was under hybrids. By about 1965, maize plantings were essentially 100 per cent under hybrids.

By the late 1950s, a few single crosses were put under cultivation due to concern about the plateauing yield trend for maize in the USA; and there was a rapid shift from the cultivation of double-cross to single-cross hybrids. By 1988, single crosses, modified single crosses and three-way crosses accounted for approximately 90, 10 and < 1 per cent, respectively, of hybrid seed sold in the USA and Canada (Wych, 1988). The development of inbred lines having better yield potential and vigour, improved seed-crop production-protection technologies and a well-developed seed industry made single-cross hybrid seed production commercially feasible. Earlier single crosses were not accepted, due to non-availability of vigorous and productive inbred lines capable of producing highly heterotic single crosses, and good quality seed at reasonable price.

In China, single-cross hybrids occupied about 90 per cent of the maize acreage in 1993. (S.H. Zhang, 1994, pers. commun.). In India, double-crosses and double top-crosses were first released in early 1960s, whereas the first single-cross hybrid was released in 1995 (Dhillon *et al.*, 1995). All these types of hybrids are still under cultivation.

A landmark in hybrid breeding is the exploitation of cytoplasmic male sterility in onion (*Allium cepa* L.), also a cross-pollinated crop, in 1940s in the USA (Jones and Clarke, 1943). The discovery of cytoplasmic male sterility and genes for fertility restoration in sorghum [*Sorghum bicolor* (L.) Moench], a predominantly self-pollinated species, in 1954 (Stephens and Holland, 1954) enabled hybrid breeding and commercial hybrid seed production. The male sterility is being currently used in an array of crops. Another important achievement in hybrid breeding was the development of first commercial tomato (*Lycopersicon esculentum* Mill.) hybrid in Japan in 1940. Tomato is a self-pollinated plant with hermaphrodite flower wherein hybrid seed is earily produced on commercial scale by hand emasculation and pollination owing to large number of seeds set per pollination.

Hybrid breeding is now being practised in a large number of cross-pollinated crops such as maize (*Zea mays* L.), pearlmillet [*Pennisetum glaucum* (L.) R. Br.], sunflower (*Helianthus annuus* L.), castor (*Ricinus communis* L.), muskmelon (*Cucumis melo* L.), watermelon [*Citrullus lanatus* (Thunb.) Mats. & Nakai), onion (*Allium cepa* L.); predominantly self-pollinated crops as in sorghum (*Sorghum bicolor* L.), cotton (*Gossypium* spp.), pigeonpea [*Cajanus cajan* (L.) Millsp.), Safflower (*Carthamus tinctorius* L.), oilseed rape (*Brassica napus* L.), chilli (*Capsicum annuum* L.), brinjal (*Solanum melongena* L.); and self-pollinated crops, for example rice (*Oryza sativa* L.), tomato (*Lycopersicon esculatum* Mill.) etc. India has the distinction of being the world leader in producing commercial hybrids in cotton, grain pearlmillet, pigeonpea, castor and safflower.

Hybrid Breeding Methodology

Shull (1909), on the basis of his experiments initiated in 1905, gave a 'pure-line method of corn breeding', which laid the foundation of present-day hybrid breeding. The method involved three steps: (*i*) large-scale inbreeding to obtain many homozygous or nearly homozygous inbred lines, (*ii*) testing the selected inbred lines in all possible crosses, and (*iii*) practical utilization of selected inbred lines in seed production of single-cross hybrids. Shull recognized that the hybrid produced from inbred lines, which are homozygous and homogenous, would be uniform and true to type. The procedure is, however, modified for deployment of male sterility as discussed later.

Types of Hybrids

A single cross is a hybrid between two inbred lines ($I_1 \times I_2$, I denoting an inbred line). A double cross is a hybrid between two single crosses [$(I_1 \times I_2) \times (I_3 \times I_4)$]; and a three-way cross is a hybrid between a single-cross parent and an inbred parent [$(I_1 \times I_2) \times I_3$]. Further, there are modified single cross [$(I_1 \times I'_1) \times I_2$], double modified single cross [$(I_1 \times I'_1) \times (I_2 \times I'_2)$], modified three-way cross [$(I_1 \times I_2) \times (I_3 \times I'_3)$], the prime being used to indicate sister lines. With respect to non-inbred parent(s), a cross between two varieties is a varietal hybrid ($V_1 \times V_2$, V denoting a variety), between an inbred and a variety ($I_1 \times V_1$) a top-cross and between a single cross and a variety is a double top-cross hybrid [$(I_1 \times I_2) \times V_1$].

The single, double and three-way crosses have mostly been used commercially. The most striking advantage of single crosses over double and three-way crosses is that single-cross breeding is simpler and faster.

All hybrids other than single crosses are heterogeneous. Being heterozygous and heterogeneous, these possess individual as well as population buffering (Allard and Bradshaw, 1964). On the other hand, single-cross hybrids are homogeneous and possess only individual buffering. Thus, there was some concern about their stability of performance. But experience showed that stable single-cross hybrids could be identified (Troyer, 1996). The uniformity of single crosses has been an important factor in their spread. Seed production in single crosses is also less cumbersome and quicker than multi-parent hybrids. The number of seasons and isolations required for seed production of single crosses is less. But single crosses must have vigorous parents.

Development and Improvement of Parental Lines

Inbreeding System: In cross-pollinated crops inbred lines are generally developed through self-pollination, which is the extreme form of inbreeding and leads to the most rapid fixation of genes. Usually, at least five selfings are needed to obtain nearly homozygous inbred lines. In case of extreme inbreeding depression, early generation lines, S_2 or so, may be recycled. In case of self-incompatible crops, selfing is carried out using bud pollination and other techniques (Shivanna and Sawhney, 1997). In self-pollinated crops, natural self-pollination leads to the development of pure lines.

Methods of Line Development: The pedigree method is most widely used to develop inbred lines. Initially pedigree selection was practised in adapted OP landrace cultivars (pedigree method when used in an OP population is referred to as the standard method; Sprague and Eberhart, 1977). Later it was observed that repeated sampling of the same landrace cultivar was not productive. This led to the approach of developing planned single crosses of elite lines that complemented each other for different traits; but depending on the objectives and availability of inbred lines, complex crosses and backcrosses were also developed (Darrah and Zuber, 1986).

In pedigree method, seeds of selected plants are planted ear to row. Phenotypic selection is carried out both among and within progenies. As selfing progresses, the differences among progenies increase and those within progenies decrease, and the selection pressure is adjusted accordingly. It seems that pedigree selection will continue to be an important method and may get further emphasized in populations with restricted genetic base, which is a matter of concern.

The backcross method proposed by Harlan and Pope (1922), is the second most important method. It is a very easy and effective method to handle one or two genes and has been widely used to incorporate resistance to disease and insect-pests. For quantitative traits having complex inheritance, the results have been variable (Rinke and Sentz, 1961; Duvick, 1974; Geadelmann and Peterson, 1978). The important factor affecting success of backcross method is the heritability of the trait. Normally, three to five backcrosses are made, but the number may vary depending on the recovery of the characters of the recurrent parent, inheritance pattern of the trait under transfer, selection during backcrossing, genetic similarities between recurrent and non-recurrent parent and linkages. When a parent is a population, a sufficient number of plants needs to be sampled to represent the population. Richey (1927) proposed convergent improvement that aims at parallel improvement of two inbred lines by the reciprocal addition of dominant favourable genes present in one line but lacking in the other, though backcrossing.

Brim (1966) suggested single-seed descent method in which single seed harvested from each plant in F_2 and subsequent generations is bulked to raise the next generation. This method enables use of off-season breeding nursery/green house but no selection can be done during inbreeding. It has been used on a limited scale. Furthermore, several innovative techniques have been proposed for rapid development of inbred lines and these are discussed later under biotechnological applications.

Pedigree versus Backcross Method: Bailey and Comstock (1976), Bailey (1977), Ho and Comstock (1980) and Johnson (1980) conducted simulation studies on intra-line selection within an F_2 population, developed from a pair of inbred lines, followed by selection amongst pure lines. They showed that in the cross having two parental lines that had equal number of loci with favorable alleles, the probability of concentrating desirable alleles in one derivative line was relatively higher. Johnson (1980) also concluded that little or no progress could be expected from pedigree selection unless the parents were nearly equal in value. The simulation studies support the general observation that selection is more productive within a population developed from a cross of good line × good line than from a cross of good line × poor line. The latter situation is reserved primarily for transferring specific traits using backcross method

Dudley (1982) developed a theory regarding the type of segregating populations in which selection can be initiated. He suggested selection in the F_2 or backcross population prior to selfing. Melchinger *et al.* (1988) showed that, generally, the F_2 population was superior when heritability and selection intensity were high. Lamkey *et al.* (1995) showed that, in the presence of epistasis, the choice between F_2 and backcross populations was primarily a function of selection intensity, the F_2 generation being better under high selection intensity.

Sampling to Initiate Inbreeding: Most plant breeders prefer to have a smaller sample size and handle a larger number of populations rather than sampling a large number of individuals in a smaller number of populations. Same practice is followed with respect to the number of families and their sample size. Bauman (1981) reported that 500 was the most common sample size among US maize breeders.

Source Populations: Initially, adapted OP landrace varieties were used as source populations, but now populations developed from planned crosses have become more important (Zuber and Darrah, 1980; Bauman, 1981; Darrah and Zuber, 1986). Darrah and Zuber (1986) reported that for the development of new lines, US maize breeders mostly use single crosses (42%), followed by backcrosses (20%), synthetics and composites (11%), populations improved by recurrent selection (8%), adapted exotic varieties (5%), exotic × adapted varieties (4%), three-way crosses (4%), open-pollinated varieties (2%) and double crosses (1%). Recycling of inbreds has been highly effective (Duvick, 1977; Russell, 1986). The improved synthetics and populations should gain more emphasis in future (Russell, 1985; Hallauer, 1990, 1992) in order to overcome the problem of narrow genetic base. Dudley (1984, 1987) and Gerloff and Smith (1988) developed a method to identify germplasm carrying new and useful alleles. This method has been evaluated in a number of recent studies and should help breeders to select useful donor lines.

With single-cross hybrids gaining importance, however, yield testing of the lines *per se* has become important. Vigorous parental lines are required and greater attention is given to seed yield, pollen production and other traits. The lines do not need to be tested as extensively as test-crosses. Observation rows grown at least at two location for 2 years may be adequate to determine if the line can be used in seed production. While evaluating inbred lines it is necessary to keep the hybrid seed production environment (agro-climatic region, season) in view. Furthermore, it is important to subject the inbred lines to agronomic manipulations so as to develop crop production technology package that enhances seed production. Evaluation of the lines *per se* generates information that can be used to predict hybrid performance for various traits except for grain yield.

Combining Ability and its Evaluation

Combining Ability: The value of an inbred line in hybrid breeding ultimately depends on its ability to combine well with other lines to produce superior hybrids. Very few productive lines prove to be good combiners. Hallauer and Miranda (1988) estimated that in the USA, one in 10,000 lines (0.01%) of those tested in the S_2 or S_3 generation and found productive has been used to some extent ($\geq 0.1\%$ of the seed requirement) in commercial hybrid seed production.

The concept of combining ability has been very useful in hybrid breeding. Sprague and Tatum (1942) coined the terms general combining ability (GCA) and specific combining ability (SCA). GCA is the average performance of an inbred line in a series of hybrid combinations, whereas SCA refers to the deviation of the hybrid from the performance expected on the basis of GCA. Griffing (1956) gave procedures for estimation of GCA and SCA effects and genetical components of variation.

Top-cross Test: Initially the lines were selected on the basis of their performance in single crosses produced generally by diallel matings and sometimes by factorial matings. With the development of breeding programmes, the number of test lines increased and it became very difficult to develop and evaluate all possible single crosses. Further, the performance of inbred lines *per se* also did not prove to be a useful guide to select lines with good combining ability for grain yield. Davis (1927) suggested the use of the top-cross method for preliminary evaluation of a large number of inbred lines for GCA. Jenkins and Brunson (1932) also showed its effectiveness. The lines identified as promising through top-cross test were evaluated in single crosses to identify the most productive specific combinations.

Early Generation Testing: In the beginning, top-cross evaluation for combining ability was carried out after five or six generations of selfing. However, as breeding programmes expanded, the need for early evaluation for combining ability of inbred lines became important. In 1935, Jenkins suggested early testing. It is based on the assumption that S_0 plants or S_1 lines differ in their combining ability and these differences can be detected by a top-cross test and tend to remain fairly stable in subsequent generations of inbreeding. Bauman (1981) reported that majority of the US maize breeders (78%) conducted test-cross evaluation in S_2 to S_4 generation, S_3 being the most common (33%).

The source population may also be a decisive factor. If the source population is elite material, hybrid testing can be started in early generations. Conversely, if the source population is relatively unselected/unadapted material, visual selection for two or three selfing generations is useful to select for adaptation and certain agronomic traits, and hybrid evaluation may be delayed to about S_4.

Nature and Number of Testers: The correct choice of a tester is very important. The use of narrow genetic-base (inbred line or single cross) *versus* broad genetic-base (synthetic, composite or other OP variety), related (parental population) *versus* unrelated and low *versus* high performance (low or high gene frequency) testers have been evaluated in a number of studies.

Rawlings and Thompson (1962) stated two requisites for a good tester to evaluate inbred lines. The tester must discriminate effectively among the materials under evaluation, and also must classify them correctly. Hallauer (1975) suggested that a tester should be simple to use, should correctly classify merit of the lines and maximize genetic gain.

For maximum efficiency, choice of a tester depends on the objective of the breeding programme. For example, if the objective is to identify good combiners for a given line or single cross, the appropriate tester is that line or a single cross itself, and SCA is of prime importance. The other factors, which have bearing on the choice of tester are the stages of development of a breeding programme, availability of the testers, type of germplasm under evaluation and type of hybrids to be developed. In hybrid breeding, on the basis of empirical information, a tester from the opposite heterotic population is used.

Initially, top-cross testing usually involved a broad genetic-base tester. The objective was to have a preliminary evaluation of GCA. Green (1948) suggested the use of a synthetic tester based on inbred lines in current use. Hull (1945) suggested that a tester with narrow genetic base should be employed. His suggestion was based on the presumption that heterosis was due to over-dominance. He considered the variance among test crosses and concluded that a tester which is homozygous recessive at all loci would be the most effective. Rawlings and Thomspon (1962) extended Hull's idea, and showed that the low performance tester, presumably the one having a low frequency of favourable alleles, would be the most effective tester. Allison and Curnow (1966) considered the expected gain from selection in a population and arrived at a conclusion similar to that of Rawlings and Thompson (1962). They emphasized that a low performance tester must have a high frequency of alleles at the loci under selection, advocated the use of parental population as a tester since it ensures low frequency of genes at the loci that need improvement, and suggested repeated selection for low yield within the population for developing a good tester.

Although theoretical considerations show that a low performance tester is better than a high performance tester, there is a problem in identifying a tester having a low frequency of favourable alleles that are under improvement. Another limitation is the low seed yield of poor testers. Even if the problems associated with low yield are overcome, the use of such a tester would not be acceptable in applied breeding programmes. Hallauer and Lopez-Perez (1979) reported that an unrelated elite tester may be as effective as related low performance tester. An elite inbred tester has the advantage of identifying the lines with good combining ability, which can be directly used in a breeding programme. Theoretical considerations, however, restrict general recommendation of the use of an unrelated elite line tester.

It was believed that the use of narrow genetic-base tester would improve SCA rather than GCA. However, empirical results had shown that a narrow genetic-base tester was effective in improving the populations *per se* and their combining ability with testers used in the experiment as well as other testers (Horner *et al.*, 1973, 1976; Russell *et al.*, 1973; Russell and Eberhart, 1975; Walejko and Russell, 1977; Zambezi *et al.*, 1986). In some studies, testers with narrow and broad genetic base were compared, and the former was observed to be more effective in improving the population. These results indicated a predominance of additive-genetic variability and partial to complete dominance. On the basis of studies on narrow *versus* broad genetic-base testers, Russell and Eberhart (1975) proposed that in reciprocal recurrent selection, rather than the opposite population, an inbred line from the opposite population should be used as a tester. Russell *et al.* (1992) and Rademacher *et al.* (1999) evaluated this modified reciprocal recurrent selection (inbred tester) and observed that it was either not effective or less effective than HS reciprocal recurrent selection of Comstock *et al.* (1949). It seems that the efficiency of selection based on inbred tester is highly population- and tester-specific. Comstock (1979) considered expected change in gene frequency due to selection based on the performance of test-crosses with inbred and population testers in reciprocal recurrent selection, and concluded that a population is a better tester.

Bauman (1981) reported that 89 per cent of US maize breeders used an inbred tester and 11 per cent, a single-cross tester. The use of inbred lines enables breeders to exploit SCA. Though considerable information has been generated on the basis of theoretical considerations and experimental data, the choice of tester is still enigmatic. The tester varies from one programme to another.

The use of partial diallel crosses has been suggested to evaluate a large number of lines for GCA (Kempthorne and Curnow, 1961). The studies on partial diallel crosses suggested that seven crosses per parent (equivalent to 3.5 testers in terms of the number of crosses to be tested) provided reliable estimates of GCA (Dhillon, 1975). Some workers have also used partial factorial crosses. Partial diallel or partial factorial crosses, however, have not become common, possibly due to the comparatively difficult schemes of mating and analyses of data, and limited information on SCA effects. Bauman (1981), based on a survey, found that 71 per cent of US maize breeders used two testers and 29 per cent used one tester. He suggested the use of two testers. Some breeders, however, use as many as six testers for the first evaluation of fixed lines. Moreover, the number of testers depends also on resources.

Maintenance of Parental Lines

Inbred lines are generally maintained by a system of self-pollination and growing progenies in ear-to-rows, so as to observe changes for various traits. One purpose of developing inbred lines is to obtain genotypes whose genetic constitution will be maintained without change. Various workers have reported mutations (Sprague *et al.*, 1960; Russell *et al.*, 1963). Relic heterozygosity from either an S_0 plant or a mutation in some later generation may cause changes over generations of reproduction. Variation has been shown to occur in inbred lines after they have been interchanged among experimental stations and reproduced in different environments. It is, therefore, desirable to keep a stock culture in cold storage.

Significant changes have been reported over successive generations of long-term inbred line maintenance. Bogenschutz and Russell (1986) observed a definite trend towards reduction in line performance. Further, experience shows that some lines are inherently stable, whereas others are inherently unstable.

Prediction of Hybrid Performance

The number of crosses, particularly multi-parent crosses, increases very rapidly with an increase in the number of parental lines. For n inbred lines, there are $n(n-1)/2$ single crosses, $3n(n-1)(n-2)/6$ three-way crosses and $3n(n-1)(n-2)(n-3)/24$ double crosses. Evidently, it is not possible to develop and evaluate all possible multi-parent crosses among even a small number of lines. Therefore, methods were developed to predict the performance of double crosses. Jenkins (1934) evaluated different prediction methods and reported that the most accurate estimate for a double-cross performance was provided by the mean yield of four non-parental single crosses of a double cross. Jenkin's method was routinely used to predict the performance of double as well as three-way crosses by maize breeders, but its importance has decreased with a shift from multi-parent to single crosses.

Efforts to predict yield potential of single crosses based on the performance of inbred parents *per se* and molecular marker divergence among the parents have been either unsuccessful in general or of limited utility in some cases (Gama and Hallauer, 1977; Melchinger *et al.*, 1992). Animal breeders have been utilizing the Best Linear Unbiased Prediction (BLUP) method for decades for evaluating the genetic merit of animals (Henderson, 1975). Recently, Bernardo (1994, 1996, 1999) has paid attention to the prediction of single-cross performance based on BLUP using genetic models involving both GCA and SCA. Further studies must be conducted to confirm the utility of BLUP.

Genetic Diversity

The expression of heterosis (H) over mid-parent depends on the difference in allele frequency (y) of the parents and dominance effect (d) at various loci, that is $H = dy^2$ (Falconer 1988). Therefore, some level of dominance and genetic diversity are necessary for expression of heterosis. The importance of genetic diversity is kept in mind while recycling inbred lines and developing experimental hybrids (Hallauer and Miranda 1988;

Smith 1988). The general guideline is to cross related lines to derive new lines and to use the inbred lines derived from different heterotic populations as immediate parents of commercial hybrids.

Several studies have shown that inbred lines from diverse populations tend to be more productive than crosses between inbred lines from the same variety. Genetic diversity was inferred from geographical origin and ancestral relationships. Heterosis increased as presumed genetic diversity increased (Moll *et al.*, 1962), but it decreased in the crosses of varieties that were assumed to be extremely diverse (Moll *et al.*, 1965). There seems to be an optimal level of diversity, beyond which heterosis does not increase, or may even decrease. This may be due to unfavourable interaction of coadapted gene complexes, epistasis, or physiological incompatibility.

The order of parents in the complex hybrids is very important and is related to genetic diversity among parents. In double-cross hybrid breeding, if lines I_1 and I_2 have been derived from one source and lines I_{11} and I_{12} from another, the highest-yielding double cross of these lines is likely to be $(I_1 \times I_2) \times (I_{11} \times I_{12})$ rather than the other two double crosses. The underlying idea is to combine related inbreds to develop parental single crosses, so that the differences between parental single cross and the expression of heterosis in double cross are maximized.

Estimates of genetic diversity can be based on ancestral relationships, geographical distribution, morphological differences, or molecular traits. Contrary to expectations, the results of the studies conducted so far on the use of biochemical and molecular marker-based genetic distance among inbred lines for the prediction of heterosis have not been consistent as discussed later.

Heterotic Patterns and Development of Heterotic Pools

The choice of germplasm is the easiest and perhaps most critical component in plant breeding. In hybrid breeding, germplasm should possess good performance *per se*, inbreeding tolerance and desirable agronomic traits; and the material should show a high level of heterosis in cross combinations. The probability of identifying superior crosses is greater between populations of known heterotic patterns than within a population. A number of scientists have identified, developed or described heterotic populations in maize (see Hallauer *et al.*, 1988; Dhillon and Prasanna, 2001).

The heterotic patterns have a strong impact on a breeding programme and, once established, are not easy to change. Yet heterotic patterns in maize seem to have been established empirically by relating the heterosis observed in crosses with the origin of the parents in the crosses. An ideal approach would be to cross all potential germplasm sources in diallel crosses and select the most promising heterotic patterns. This, however, is not possible due to the large number of materials involved.

The most promising germplasm should be evaluated in a series of diallel crosses, on the basis of which two (or more) heterotic stocks should be identified for use as testers. These should be crossed with diverse germplasm and those showing heterosis with a tester be assigned to the opposite heterotic pool. Unless dealing with very large and diverse germplasm, it would be desirable to work with inbred lines. Further, no germplasm should be put in both pools. The pools should be broad based, genetically diverse and highly heterotic with acceptable performance *per se* and inbreeding tolerance.

Heterotic populations have generally been treated as closed, in the experiments reported in the literature. In applied plant breeding, having open-ended populations is a preferable approach. It has the advantage of introgressing new germplasm to maintain or increase within-pool genetic variation, enhance inter-pool heterosis and rectify specific defects.

Heterotic pools need to be made as divergent as possible, because heterosis depends on the differences in allele frequency between two populations. Genes should be concentrated for contrasting traits in two heterotic pools. For example, in maize, the seed parent should possess high yield potential, large bold kernels, good silk emergence, small tassel and other desirable agronomic features, whereas the pollen parent should provide enough pollen at the level of ear placement or above, and have good pollen spread.

Development of Hybrid Breeding-oriented Source Germplasm

The probability of deriving superior inbred lines is dependent on the proportion of superior genotypes in the base population. During 1940s, a breeding approach called recurrent selection was developed, which involves selection, generation after generation with intermating of selects (Jenkins, 1940; Hull, 1945; Comstock *et al.*, 1949). The objective is to gradually increase the frequency of favourable alleles while maintaining genetic variability within the population through different methods of recurrent selection.

The improvement of the population cross between breeding populations through recurrent selection is the key to the more effective development of improved hybrids. Empirical data and statistical genetic theory indicate that the improvement of the breeding populations results in a corresponding improvement in hybrids developed by crossing inbred lines derived from these populations (Sprague and Eberhart, 1977). Experimental results also indicate better chances of developing superior source germplasm by using reciprocal recurrent selection (Moll and Hanson, 1984; Odhiambo and Compton, 1989; Chang-Jian *et al.*, 1990; Eyherabide and Hallauer, 1991; Moll *et al.*, 1994; Landi and Frascaroli, 1995).

There are basically three methods of reciprocal recurrent selection, namely HS reciprocal recurrent selection (Comstock *et al.*, 1949), FS reciprocal recurrent selection (Hallauer and Eberhart, 1970; Hallauer, 1973) and reciprocal recurrent selection using inbred tester (Russell and Eberhart, 1975). In these methods, two heterotic populations are simultaneously improved, based on the performance of inter-population HSs, inter-population FSs or inter-population test-crosses using inbred tester. The objective is to improve the performance of inter-population cross and heterosis therein, and consequently, the performance of hybrids formed by crossing inbred lines derived from improved populations.

Reciprocal recurrent selection is a dynamic system that permits continuous improvement of source population and introgression of new germplasm into populations under improvement. However, the introgression needs to be done with utmost care. The new materials must have acceptable level of performance to qualify for incorporation into heterotic pools. If required, unadapted germplasm, particularly exotics, may be subjected to selection for adaptation using simple recurrent selection methods such as modified mass selection (Gardner, 1961) and selfed-plant mass selection (Dhillon, 1991a). With performance of inbred lines *per se* gaining importance, tolerance to inbreeding stress has become more relevant. Recurrent selection based on selfed plants or families or related methods (Eberhart, 1970; Dhillon and Khehra, 1989; Dhillon, 1991a, b) may be undertaken to enhance inbreeding stress tolerance. The introgression of new germplasm needs to be closely monitored to avoid dilution of previous gains. In fact only a small proportion (≤ 10 per cent) of germplasm should be introgressed.

Cramer and Kannenberg (1992) presented a comprehensive germplasm management system for maize named Hierarchical Open-ended Population Enrichment (HOPE). It has dual objectives of developing useful inbred lines and diversifying base germplasm. It involves two heterotic populations at four levels of performance; and the selection methods change from simple to complex (mass selection to reciprocal recurrent selection) as the level of performance improves. Popi *et al.* (2000) evaluated the improved populations after 15 years of operation of HOPE system and reported that the system has been successful in offering new variability.

Population improvement programmes are generally long-term projects on germplasm enhancement. But reciprocal recurrent selection offers excellent opportunities of extracting hybrids for immediate use. Reciprocal recurrent selection, however, has not been widely used in hybrid breeding, probably because it is complex, resource consuming and not as efficient at recovering inbred lines as other methods of inbred line development. Most of the inbred lines, are being developed by recycling inbred lines but it may have undesirable consequences in narrowed genetic base and accompanying genetic vulnerability.

Seed Production Technology

The hybrid seed chain includes the maintenance and increase of inbred lines and seed production of parental crosses and commercial hybrids. Seed production at various stages must be undertaken under rigid control and stringent isolation to ensure a high degree of genetic purity in the seed produced and satisfactory performance of the final product. Seed production plots have to be regularly monitored.

Breeder seed of inbred lines is normally derived from bulked self-pollinated seed and this should adequately represent the genetic constitution of the inbred lines. Inbred seed is increased by natural random sib-mating. In maize hybrid seed production plots, common planting patterns include 4 : 1 (four female rows : one male row), 4 : 2, 4 : 1 : 4 :2, 4 : 1 : 2 : 1 and 6 : 2. Occasionally, there is solid planting of female with interplanted male, either in every other or every fourth between-row space. In many cases, the male parent is destroyed after pollination is complete.

Studies have shown that the depth of the field in the direction of the contaminating source is important, since there is greatest contamination in the 50-75 m distance from the source. It is often believed that the best isolation is the perfect nick, that is, when a male parent starts shedding pollen just before the silks start emerging from the husks of the ear shoot of the female parent. Split-date planting of female and male parents is restored to achieve timely nick of parents and is the most popular method of making gross alterations in flowering date (Shoultz 1985). Other methods to obtain small adjustments in pollen shed are variable fertilizer rates, variable planting depths, treatments to delay germination (Bauman, 1967; Bansal *et al.*, 1993), clipping (Cloninger *et al.*, 1974) or flaming (Fowler, 1967).

Minimum standards for isolation of seed production blocks are well established in maize (Anonymous, 1971; Cowan, 1972). These take into consideration seed colour, endosperm type and class of seed. Minimum isolation distances are modified by additional border rows, size of production block, differential flowering dates and adequate natural barriers. Border rows have been reported to be more effective than natural barriers (Jones and Brooks, 1952). The male parent is often double-planted at a spaced interval to extend the pollen-shedding period. An abundant amount of male pollen being beneficial, the male parent is sometimes planted under high plant population density to increase pollen load. Appropriate rotations are followed to avoid volunteer maize plants.

Pollination Control Mechanisms

Manual Pollination Control

Hand emasculation and pollination method of hybrid seed production is economically viable in those crops in which single pollination results in large number of seeds, seed price is high and seed rate in commercial planting is low. Maize has unique floral morphology that allows easy emasculation (detasseling), control pollination including cross-pollination. In other crops, except for a few, it was the deployment of male sterility that enabled commercial exploitation of hybrid cultivars.

In maize, detasseling, that is, physical removal of the tassel from the female plant, is done either manually or in combination with mechanical detasselers; and there is natural pollen transfer. In cotton, tomato and brinjal, hand emasculation and pollination is done. In cotton, the *Dapog* method, which involves one step removal of calyx and corolla using nail is commonly used. In maize and cotton, major seed production is though manual or mechanical detasseling/emasculation, but cytoplasmic-nuclear male sterility (CNMS) and nuclear male sterility (NMS) are also being used respectively.

Genetically Controlled Male Sterility

Male sterility is defined as the inability of a plant to produce or release functional pollen. The occurrence of male sterility was widely documented as early as 1930s (Kaul, 1998). It was first deployed in hybrid onion breeding, and now is being used in a number of crops. It is of different types: cytoplasmic male sterility (CMS), which is due to cytoplasmic factor; nuclear or genetic male sterility (NMS or GMS), which is controlled by nuclear gene(s); cytoplasmic- nuclear male sterility (CNMS or CGMS) system, which results from nuclear-cytoplasmic interaction; and environment-sensitive male sterility (EGMS), which is induced by the interaction of genetic and environmental factors.

In addition to the methodology outlined by Shull (1909) for hybrid breeding, the deployment of male sterility involves identification of male sterility (may be cytoplasmic or nuclear) and nuclear gene(s) for fertility restoration in case of CNMS system, transfer of male sterility and fertility restoration to appropriate parents through backcrossing, and adoption of special procedure to maintain male-sterile parent. Different systems of genetically controlled male sterility are briefly discussed hereunder.

Cytoplasmic Male Sterility: CMS is used in those crops where vegetative parts rather than seed is of commercial value, such as many vegetables and ornamentals. Male-sterile female parent is crossed with male parent, which carries fertile cytoplasm. The F_1, thus produced, has high hetrosis for vegetative growth and vigour but is male-sterile.

Nuclear Male Sterility: In case of NMS, female and male parents of a hybrid, carry gene(s) for male sterility and fertility, respectively. Male-sterile female parent may be maintained and multiplied as a population having an equal proportion of male-fertile heterozygote and male-sterile homozygote, which may be designated as *Msms* and *msms*, respectively, assuming that there is one gene control, and that the allele responsible for male fertility (*Ms*) is dominant over that for male sterility (*ms*). For multiplication, this population is grown in isolation, *msms* plants identified, and their seed produce harvested, which again shows segregation of male-fertile heterozygote and male-sterile homozygote in an equal proportion. For hybrid seed production, this seed of female parent is interplanted with male parent, male-fertile (*Msms*) plants in female rows are removed before pollination, and the produce of female parent harvested as hybrid seed. NMS is being used in pigeonpea, safflower, cotton, muskmelon and chilli.

Hybrid seed production using NMS is cumbersome and costly. However, the seed production is greatly aided if appropriate male-sterility/fertility allele is closely linked to a morphological marker locus that expresses at early seedling stage, or at least before anthesis and helps in the identification of male-sterile/fertile plants.

Cytoplasmic-Nuclear Male Sterility: In case of CNMS system, the female parent carries cytoplasm as well as genes for male sterility, and is commonly known as CMS or *A*-line. *A*-line has a counterpart maintainer, commonly known as maintainer or *B*-line, which has same genotype as *A*-line except that it carries normal fertile cytoplasm. *A*-line, is maintained by crossing it as female, with *B*-line; whereas *B*-line is maintained in isolation. Hybrid seed is produced by crossing *A*-line with its partner, commonly known as restorer or *R*-line, which carries genes for fertility restoration. The *R*-line is also maintained in isolation. The F_1 hybrid between *A* and *R* lines is male fertile and shows hybrid vigour. The CNMS is being used in many crops including pearlmillet, sunflower, sorghum, rice and maize.

Many workers, starting with Rhodes (1931), have reported CNMS in maize. Duvick (1965) and Beckett (1971) defined three types of cytoplasm: T (Texas), S (USDA) and C(Charrua). The T-cytoplasm has proven to be the most satisfactory, because it imparts full sterility to most inbred lines and its fertility restoration is relatively easy. Before 1970, the use of CNMS was thought to be an economical, very satisfactory and practical method of large-scale hybrid seed production. The T-cytoplasm was used extensively during the 1950s. In 1970, the southern corn leaf blight epidemic severely damaged the crop in the US Corn Belt (Tatum, 1971). The epidemic was due to a new race (T-race) of *Drechslera maydis*, which was virulent on T-cytoplasm. Since 1974, the use of cytoplasms other than T started again on a limited scale and C-cytoplasm is being favoured over S-cytoplasm (WT Schapaugh, 1987, cf. Wych, 1988). The southern corn leaf blight epidemic was instrumental in enlightening plant breeders about genetic homogeneity and associated genetic vulnerability to disease and insect-pest epidemics.

There are four cytoplasmic sources of male sterility in sorghum (A_1 or milo, A_2, A_3, A_4) as well as in pearlmillet (A_1, A_2, A_3, A_4). But only one source, A_1, in both crops is being used in commercial hybrid seed production. Similarly, only one cytoplasmic source of male sterility is being used in rice (*WA*), sunflower (*PET*1) and most other crops, though many sources are available. There is an urgent need to develop additional usable sources of cytoplasmic male sterility.

Another limitation of CNMS system is that, generally, not many male-sterile lines having good combining ability are available. This limits the chances of selection of female parent. The usual approach is to cross these CMS lines with a large number of pollinators to produce test-cross F_1s for evaluation. In such a situation, three-way cross hybrid breeding may be more rewarding than single-cross hybrid breeding (Dhillon, 1998b). This approach provides an opportunity to manipulate at least 75 per cent of the parentage, in contrast to 50 per cent in single crosses. But breeding and seed productions of multi-parent hybrids are more complex than single-cross hybrids as discussed earlier.

The exploitation of CNMS depends on the identification of potential maintainers and restorers. Potential maintainers must also be good general combiners, to be a candidate for their conversion into CMS lines for use in hybrid breeding. This limits the choice of parents, particularly female parent. A large germplasm are observed to be partial restorers which are often ignored. The partial restorers, however, can be bred to complete restorers through selection and intermating. Thus, large germplasm comprising partial restorer can be channelled back into hybrid breeding programme.

Environment-Sensitive Genetic Male Sterility (EGMS): In EGMS, male sterility is induced when the plant, at a sensitive stage of its development, is exposed to a particular environmental factor(s); whereas fertility is restored at some other level of environmental factor(s). During the sterile phase, EGMS line can be used as a female parent to produce hybrid seed. During the fertile phase, the line is multiplied through self-fertilization. Thus, only two lines are required for the production of hybrid seed, including maintenance/multiplication of male-sterile line. The EGMS-based system is known as two-line system of hybrid breeding as opposed to three-line system involving CNMS (see Virmani and Ilyas Ahmed, 2001).

If the EGMS is induced in response to variation in photoperiod, the resulting male sterility is known as photoperiod-sensitive genetic male sterility (PGMS). Shi (1981) observed such a male-sterile plant in a *japonica* rice cultivar in 1973. The plant appeared male sterile, when headed under long day condition and turned male fertile, under short day condition. Since then the PGMS trait has been transferred to several *indica* and *japonica* rice cultivars. On the other hand, if EGMS is induced in response to variation in temperature, it is known as temperature-sensitive genetic male sterility (TGMS). The first TGMS line of rice was isolated as a spontaneous mutant (Tan *et al.*, 1990). Following this, Maruyama *et al.* (1991) induced TGMS mutant in a variety Remei, which exhibited no seed set under 31/24 °C, partial fertility under 28/21 °C and complete fertility under 25/15 °C. A reverse TGMS system where sterility occurs under low temperature (24 °C) and transformation to fertility takes place at high temperature (27 °C) has also been reported (Zhang *et al.*, 1991). The report of EGMS in rice and their subsequent utilization in hybrid rice breeding in China has paved the way for its intensive search and utilization in other crops too.

The system based on EGMS offers several advantages. The choice of parental lines is greatly widened and the chances of development of heterotic hybrid are enhanced. Hybrid breeding and seed production are greatly simplified. In addition, the field area ratio between male-sterile multiplication, hybrid seed production and commercial cultivation of hybrid becomes more economical. Keeping in view the availability of temperature ranges in India, the TGMS system can be effectively used. Unlike the PGMS, wherein fertility-sterility transformation process is greatly influenced by temperature, TGMS is less affected by day length in rice. Hence, intensive efforts should be made for identifying TGMS sources in various crops to utilize them in hybrid breeding.

Gametocide Induced Male Sterility

Male gametocide, also known as chemical hybridizing agent (CHA), also offers the advantages like that of EGMS in terms of widening the choice of parental lines, and simplifying hybrid breeding and seed production. Any two parents which are heterotic in a cross can be directly used for hybrid breeding and seed production. An ideal CHA should result into complete male sterility without affecting female fertility; should have wide spectrum to induce male sterility in successively emerging tillers; least phytotoxicity; no carcinogenicity and no residual toxicity to human being and animals. At the same time CHA-application should be easy and economical. CHAs such as etherel (Perez *et al.*, 1973; Parmar *et al.*, 1979), oxinilates and aresenates (Ali *et al.*, 1990) have been found effective in rice. A proprietary chemical formulation is being used for hybrid seed production in wheat. However, intensive research efforts are required on this line to make the CHA technology effective and commercially applicable.

Hybrid Breeding in Self-pollinated Crops : Constraints and Opportunities

Although appreciable heterosis was reported, it remained commercially unexploited in these very important crops owing to their strict self-pollinating behaviour, which posed serious problems particularly in hybrid seed production. The commercial exploitation of hybrid rice in China using CNMS system (Yuan, 1977) motivated the breeders of self-pollinated crops, particularly those working in rice, to intensively research on hybrid breeding. These efforts resulted in development/identification of effective male sterility systems, parental lines with desired plant type, heterotic cross-combinations, and standardization of hybrid seed production technology. However, there are some constraints such as low magnitude of heterosis and limited cross-pollination, which make it difficult to develop hybrids and undertake commercial seed production.

Hybrid breeding in these crops is facilitated by the deployment of male sterility, mostly CNMS. In the approach followed so far, the main criteria for selection of parental lines is the maintenance/restoration behaviour and performance *per se*. Many CMS lines have been developed without evaluating combining ability. On the whole, systematic breeding approach for the development of heterotic parents and pools has been lacking. The major constraints and approaches to overcome them are briefly discussed here.

Limited Choice of Parental Lines

The choice of parental lines in CNMS system is highly limited; in case of female parents by their maintenance behaviour and combining ability and in case of pollen parents by restoration ability. Joint study by International Rice Research Institute (IRRI), Philippines and China indicated that only about 5 and 10 per cent of the lines were effective maintainers and 24 and 15 per cent were effective restorers at IRRI and in China, respectively (Virmani, 1994), and thus the bulk of materials are partial restorers.

It is, however, possible to utilize the partial restorers to develop perfect restorers. In Basmati rice, effective restorers of WA cytoplasm have been developed through intermating of plants having enhanced fertility restoration in F_2 and subsequent generations of partially restored F_1 hybrids (Virmani and Zaman, 1998). In these studies, there is a need for utmost caution with respect to choice of *A* line for developing $A \times R$ crosses to improve restorer lines. *A* lines which have potential for commercial exploitation or even their related lines should not be used. It is necessary as *A* line used in restoration ability improvement programme are bound to contribute some genes to improved restorers, which will have an adverse effect on the ability of improved restorer to produce heterotic hybrids with CMS lines.

The choice of parental lines can be greatly broadened by using EGMS or CHA as discussed earlier. The occurrence of EGMS in different crops and its utilization has been reviewed by Virmani and Ilyas-Ahmed (2001).

Low Magnitude of Heterosis

The magnitude of heterosis generally observed in self-pollinators is much lower than that in cross-pollinators. This is a limiting factor in hybrid breeding in some self-pollinated crops, such as rice, wheat and *Brassicas*. To make hybrid technology commercially viable, relatively higher magnitude of heterosis would be required. An alternative is to channel more diverse germplasm; for example, development of inter-subspecific (*indica* × *japonica*) hybrids in rice using wide compatibility gene(s) (Ikehashi and Araki, 1984), and inter-specific hybrids in cotton. Similarly, synthetic wheat lines developed at International Maize and Wheat Improvement Center (CIMMYT) may be used in hybrid wheat breeding. Introgression of diverse materials should be taken up as a long-term objective, but while doing so care should be taken to maintain heterotic patterns.

Need to Develop Heterotic Pools

There is an urgent need to systematize breeding programmes to develop and improve heterotic pools for use as source populations for hybrid breeding. It is as important in self-pollinated crops as in maize and other cross-pollinated crops, though it may not be as easy and effective in these crops owing to their self-pollinating behaviour.

The development and improvement of maintainer and restorer heterotic pools through reciprocal recurrent selection requires extensive intermating, which is a limitation in self-pollinated crops. The NMS can be used to overcome it and facilitate intermating (Jain and Suneson, 1963; Brim and Stuber, 1973). For the development of a pool, say maintainer pool, a recessive gene for male sterility is transferred to a perfect maintainer line, which is then crossed with other perfect maintainers. An equal quantity of seed of all these F_1 crosses is bulked to constitute a maintainer pool. All plants in F_1 generation are heterozygous and fertile. These are planted and harvested in bulk. The harvest constitutes F_2 generation having three genotypes (1 *MsMs* : 2 *Msms* : 1 *msms*). In the next generation, the produce of *msms* plants, which have been cross-pollinated with *Ms* and *ms* pollen, is harvested, bulked and planted. This bulk constitutes 2 *Msms* : 1 *msms* plants. The produce of *msms* plants is harvested which is expected to show the classical segregation of a test-cross (1 *Msms* : 1 *msms*). Each time the pool is to be planted in isolation and supplementary pollination undertaken for enhancing outcrossing. Selection for desirable traits is conducted and the process continued till desired segregants are obtained and used to develop maintainer lines (*MsMs*), which are then converted into CMS line. In an applied breeding programme, it would be desirable to incorporate new maintainer lines into the pool, as and when identified. Same approach is to be followed for development and improvement of the other pool, that is restorer pool. This approach has been used at IRRI to develop maintainer and restorer pools in rice, which are being further improved (Virmani, 1994).

Care should be taken that materials are assigned to the pools on the basis of heterotic affinities and not just on the basis of maintenance or restoration reaction and performance *per se*. There is an array of desirable traits (agronomic, quality, stress tolerance/resistance, adaptation, etc.), which should be considered with respect to both pools, but most important ones are seed yield, size and viability in maintainer pool, pollen production ability in restorer pool, and respective characters that enhance cross-pollination in both pools.

Recycling Maintainer and Restorer Lines

Development and improvement of pools is a long-term project. To meet immediate needs B × B and R × R crosses should be attempted by using carefully selected lines that complement each other. The desired types may be selected in segregating generations. For recycling, pedigree as well as backcross methods may be used depending upon the objectives and materials. However, there is a need for utmost caution not to cross maintainers and restorers except when specific gene(s) from a specific source have to be incorporated, presumably through backcrossing.

Low Seed Yield

Efficient and economic production of hybrid seed on large scale is very vital for popularization of hybrids. In self-pollinating species, restricted outcrossing is a major barrier, adversely affecting the hybrid ($A \times R$) and *A*-line ($A \times B$) yield in seed production plots and, thus, leading to increased seed cost. Therefore, breeding parental lines for plant and floral traits that promote outcrossing is extremely important. Virmani and Edwards (1983) reported variation for various floral traits that influence outcrossing in cultivated rice *vis-a-vis* wild species. Taillebois and Guimaraes (1988) transferred long stigma of wild species *Oryza longistaminata*, a dominant trait, to *Oryza sativa*. This transfer was associated with partial to complete male sterility and seed shattering, which was considerably reduced though backcross breeding. Like rice, many of the wild relatives of wheat have 100 per cent cross-pollination and can serve as useful sources for floral traits promoting outcrossing in wheat.

The yields in *A*-line seed multiplication plot and hybrid seed production plot can be enhanced by optimization of seed-crop production technology. In rice, for example, where CNMS system is being widely used for hybrid seed production, various aspects, such as, sparse nursery seeding, differential row-to-row spacing (male : male, male : female, female : female), differential number of plants per hill, differential row ratio [8 (*A*-line) : 2 (*B*-line) for *A*-line multiplication and 10 (*A*-line) : 2 (*R*-line) for hybrid seed production], GA3 application, supplementary pollination by rope pulling or shaking by stick, and synchronization of flowering between female and male lines by differential use of nutrients has helped in enhancing hybrid seed yield.

Hybrid Breeding: Biotechnological Applications

The new techniques of biotechnology have several potential applications in hybrid breeding as briefly discussed here.

Doubled Haploids

Many innovative techniques have been proposed for rapid development of homozygous lines. These include induction of haploids through maternal or paternal gametes in maize (Chase, 1952; Goodsell, 1961; Kermicle, 1969; Wu, 1986), haploid initiator *hap* gene in barley (*Hordeum vulgare* L.), and selective chromosome elimination in barley (Kao and Kasha, 1969) and wheat (Barclay, 1975; Laurie and Bennett, 1988); and the most common method is through culture of gametophyte tissues (anther/microspore, ovary/ovule) and regeneration of haploid plants followed by chromosome doubling of haploid plants through colchicine treatment of apical meristem or axillary buds. These methods have the advantages of rapid development of pure lines, and in the conversion of inbred lines from normal to sterile cytoplasms. In maize, the genetic marker technique to identify haploids has been used. The disadvantage of these methods are low frequency of production of haploids. These techniques have been tested in some breeding programmes, but did not prove to be important in inbred line development in maize (Sprague and Eberhart, 1977; Hallauer *et al.*, 1988). Doubled haploidy approach may also be used to maintain genetic purity of parental lines.

Somatic Hybridization

Somatic hybridization can be used for the diversification of male sterility inducing cytoplasm. Development of alloplasmic CMS lines carrying cytoplasm of sexually incompatible wild species and nucleus of cultivated species can be effectively achieved by somatic hybridization. In *Brassicas*, this approach has been used to develop several diverse sources of cytoplasmic male sterility (Kirti *et al.*, 1995a, b). It also offers a unique opportunity for recombination in organelle genomes particularly mitrochondrial genome. The approach has been used to correct CMS-associated deformities in alloplasmic CMS lines in *Brassica* (Kirti *et al.*, 1995b). Since the wild species contributing cytoplasm for development of alloplasmic CMS lines are the most probable source of restorer genes, the somatic hybridization can also be simultaneously used for transfer of restorer gene(s).

Molecular Markers

Molecular markers can be used for a variety of purposes in hybrid breeding programme, which include assessment of genetic diversity among parental lines, establishment of heterotic pools, prediction of hybrid performance, assessment and maintenance of genetic purity of parental lines and hybrid seed and marker assisted transfer of specific genes/QTLs controlling the heterosis for desirable traits, such as high yield, other agronomic traits, resistance/tolerance to biotic and abiotic stresses, EGMS, fertility restoration, floral traits promoting outcrossing, wide compatibility, etc. particularly from wild species. Several workers have estimated correlation between isozyme allelic diversity among inbred parents and grain yield of single crosses in maize (Hunter and Kannenberg, 1971; Heidrich-Sorbrinho and Corderio, 1975; Hadjinov *et al.*, 1982; Kahler *et al.*, 1986; Smith and Smith, 1989). These studies indicated limited value of isozyme diversity in the prediction of hybrid performance.

Molecular diversity study among parental lines based on restriction fragment length polymorphism (RFLP) analysis in relation to grain yield of maize hybrids has been evaluated by many workers (Lee *et al.*, 1989; Godshalk *et al.*, 1990, Melchinger *et al.*, 1990, 1992; Dudley *et al.*, 1991; Boppenmair *et al.*, 1992; Dhillon *et al.*, 1993; Bernardo, 1994; Dubreuil *et al.*, 1996). They observed a positive correlation, but the magnitude of correlation was too low to have predictive value. Molecular divergence based on random amplified polymorphic DNA (RAPD), amplified fragment length polymorphism (AFLP) and simple sequence repeats (SSR) in relation to hybrid performance in rice (Xiao *et al.*, 1996 and Zhang *et al.*, 1996; Saghai Maroof *et al.*, 1997), wheat (Martin *et al.*, 1995), Indian mustard (Jain *et al.*, 1994) and chickpea (Sant *et al.*, 1999) has shown similar results. However, it is hoped that an evaluation of molecular divergence based on a large number of markers that probe into the coding regions, such as isozymes, RFLP and EST, may prove helpful.

Marker assisted transfer of selected QTLs for enhancing performance of maize hybrids was achieved by Stuber *et al.* (1992). They identified heterotic QTLs in maize inbreds (Tx 303, Oh 43) and mobilized them using marker assisted selection to improve the inbred parents (B 73, Mo 17) of a commercial hybrid. The F_1 hybrid from improved inbred lines showed significant gain over the commercial hybrid. To achieve the marker assisted gene introgression in breeding programs in a precise manner, Humberto Reyes-Valdes (2000) proposed a model to predict the probability of donor parent genetic material being present in the specific region of the genome and its proportion at the chromosome specific or whole genome level.

Molecular markers linked to genes controlling the expression of traits which have significance in hybrid breeding such as PGMS (Zhang *et al.*, 1994), TGMS (Wang *et al.*, 1995; Subudhi *et al.*, 1997), fertility restoration (Ichikawa *et al.*, 1997; Zhang *et al.*, 1997; Yao *et al.*, 1997; Liu *et al.*, 1997; Mishra *et al.*, 2001) in rice have already been identified and can be effectively transferred to desired parental lines.

Male Sterility/Fertility-Restoration System

Recombinant DNA techniques have made it possible to engineer entirely new systems of male sterility and fertility restoration by manipulating one or more developmental steps of functional pollen production. Induction of male sterility by using barnase-bastar system (Mariani *et al.*, 1990; 1992), modified glucanase gene (Worrall *et al.*, 1992), flavonoid biosynthesis inhibition (Van der Meer *et al.*, 1992) and hormone engineering (Schmulling *et al.*, 1988; 1993) are some of the exciting developments in this area. Although, genetically engineered male sterility systems are not currently in commercial use except barnase-barstar system, these are likely to have a significant impact in future hybrid breeding programme (Williams *et al.*, 1997).

Evidently, the applications of novel techniques of biotechnology are expected to help in better understanding and efficient execution of plant breeding programme including those of hybrid breeding.

Apomixis

Apomixis is a genetically controlled reproductive mechanism that can be used to vegetatively propagate a genotype through seed. Through various gametophytic apomictic mechanisms (Asker and Jerling, 1992), an egg cell with an unreduced chromosome number develops into an embryo without fertilization by sperm cell; however, fertilization of one or more nuclei of central cell by a sperm cell is usually necessary for endosperm development. Endosperm development is especially important in grain crops. The discovery of sexual plants in apomictic species to produce apomictic hybrids (Bashaw and Hussey, 1992), progress in transferring apomixis from wild to cultivated species, and information on genetics of apomixis (Asker and Jerling, 1992), progress made in molecular mapping of gene(s) controlling apomixis (Hanna *et al.*, 1996) and rapid advances being made in molecular technologies have all contributed to renewed interest and efforts to use apomixis to produce true breeding hybrids (Hanna *et al.*, 1999). CIMMYT is working on the development of apomictic maize. Genes for apomixis have been transferred from *Tripsacum* into maize. Some laboratories are already working on the identification, cloning and transfer of gene(s) responsible for apomixis. This reproductive mechanism introduced in a hybrid could make it breed true and help in production of high quality pure seed. Development of apomictic hybrids would enable poor farmers in developing countries to adopt hybrid cultivars more easily.

Partial Utilization of Heterosis

Hybrids, though, superior to OP cultivars have not become popular in many countries or regions due to lack of technical and financial resources, lack of well developed seed industry and farmers' awareness. One can exploit hybrid vigour through the use of composites and synthetics by advising farmers to go for such variaties. Development of synthetics is particularly useful in forage crops in which the flowers are small and it is difficult to make crosses. This approach can be followed in other crops also. In the crops having high inbreeding depression, the parents should be popular, but in inbreeding tolerant species, parents with narrow genetic-base can also be used.

As discussed earlier, hybrid cotton seed is mainly produced though manual emasculation and pollination, and it requires trained manpower, so that hybrid seed production is economically viable. In India, hybrid cotton seed is being produced and even exported. However, in other countries, many studies have been conducted on heterosis in F_1 and F_2 generation; and both F_1 and F_2 generations are being commercially cultivated (Bauer and Green, 1996; Davis, 1998). Parents of such hybrids should not differ widely in their fibre parameters and F_2 should not show large inbreeding depression.

Conclusions

There are several opportunities and challenges in respect to development of hybrid breeding methodology, which need to be harnessed. These include, identification of heterotic patterns, development and improvement of heterotic pools (including maintainer and restorer pools), integrating hybrid breeding with recurrent selection, resolving nature of tester, selecting inbreds for combining ability and agronomic performance simultaneously, precise prediction of single-cross performance and broadening genetic-base of germplasm. In addition, diversifying cytoplasmic sources of male sterility, developing new CNMS systems and using EGMS and CHAs to simplify the seed production programme need special attention. Development and use of doubled haploids, marker based classification of germplasm into heterotic pools, transfer of heterotic QTLs and other traits related to hybrid breeding from within and related species through integration of biotechnology with hybrid breeding should be taken as priority area of research. Fixation of heterosis through development of true breeding hybrids using apomixis is the most exciting area, so far as the development of hybrid breeding methodology is concerned. Though, it may not be easy and quick to overcome the challenges and exploit the opportunities, yet, conventional hybrid breeding, refined by systemization of methodology, and augmented by innovative biotechnologies seems to hold great promise in making hybrid breeding a universal approach.

References

Ali J., Devakumar C., Zaman F.U. and Sadananda A.R. 1990. Relative gametocidal potency of oxanilates and arsenates in rice. p. 18-19. *In*: *Extended Summary, Proc. Intern. Symp. on Rice Research: New Frontiers*, Dte. of Rice Research, Hyderabad, India.

Allard R.W. and Bradshaw A.D. 1964. Implications of genotype-environment interactions in applied plant breeding. *Crop Sci.*, **4**: 503-508.

Allison J.C.S. and Curnow R.N. 1966. On the choice of tester parent for the breeding of synthetic varieties of maize (*Zea mays* L.). *Crop Sci.*, **6**: 541-544.

Aman M.A. and Sarkar K.R. 1978. Selection for haploidy inducing potential in maize. *Indian J. Genet.*, **38**: 452-457.

Anonynmous. 1971. *Certification Handbook.* Publ., 23. Rv. 1984. Assoc. Off. Seed. Cert. Agencies. Raleigh, USA.

Asker S.E. and Jerling L. 1992. *Apomixis in Plants*. 298. CRC Press, Boca Raton, Florida.

Bailey T.B. Jr. 1977. Selection limits in self-fertilizing population following the cross of homozygous lines. p. 399-412. *In*: *Proc. Int. Conf. Nat. Genet.* (eds.) E. Pollak, O. Kempthorne, and T. B. Bailey Jr., Iowa State Univeristy Press, Ames, USA.

Bailey T.B. Jr. and Comstock R.E. 1976. Linkage and synthesis of better genotypes in self-fertilizing species. *Crop Sci.*, **16**: 363-370.

Balko L.G. and Russell W.A. 1980. Effects of rate of nitrogen fertilizer on maize inbred lines and hybrid progeny. II. Correlations among agronomic traits. *Maydica*, **25**: 81-94.

Bauer P.J. and Green C.C. 1996. Evaluation of F_2 genotypes of cotton for conservation tillage. *Crop Sci.*, **36**: 655-658.

Bansal U.K., Dhillon B.S. and Saxena V.K. 1993. Manipulation of flowering date in maize. p. 88-89. *In*: *Heterosis Breeding in Crop Plants-Theory and Application.* (eds.) M. M. Verma, D. S. Virk and G. S. Chahal. Crop Improv. Soc. of India, Ludhiana.

Barclay I.R. 1975. High frequencies of haploid production in wheat (*Triticum aestivum*) by chromosome elimination. *Nature*, **256**: 410-411.

Bashaw E.C. and Hussey M.A. 1992. Apomixis in *Cenchrus*. p 1-4. *Proc. Apomixis Workshop*, Atlanta, G.A. 11-12 Feb. 1992. Nat. Technical Inform. Serv. Springfield, V.A.

Bauman L.F. 1967. Seed coating to delay emergence. *Hyb. Corn Ind. Res. Conf.*, **22**: 49-52.

Bauman L.F. 1981. Review of methods used by breeders to develop superior corn inbreds. *Corn Sorghum Res. Conf.*, **36**: 199-208.

Beal J.F. 1878. Report of the professor of botany and horticulture. p. 445-457. *Report of Michigan State Board of Agric.*, Lousing, USA.

Beckett J.B. 1971. Classification of male sterile cytoplasm in maize (*Zea mays* L.). *Crop Sci.*, **11**: 724-727.

Bernardo R. 1992. Retention of genetically superior inbred lines during early generation test crossing in maize. *Crop Sci.*, **32**: 923-927.

Bernardo R. 1994. Prediction of maize single-cross performance using RFLPs and information from related hybrids. *Crop Sci.*, **34**: 20-25.

Bernardo R. 1996. Best linear unbiased prediction of maize single-cross performance. *Crop Sci.*, **36**: 50-56.

Bernardo R. 1999. Best linear unbiased predictor analysis. p. 269-276. *In*: *Genetics and Exploitation of Heterosis in Crops.* (eds.) J. G. Coors and S. Pandey. Am. Soc. Agron., Madison, Wisconsin, USA.

Bogenschutz T.G. and Russell W.A. 1986. An evaluation for genetic variation within inbred lines maintained by sib-mating and self-pollination. *Euphytica*, **35**: 403-412.

Boppenmair J., Melchinger A.E., Brunklaus-Jung E., Geiger H.H. and Herrmann R.G. 1992. Genetic diversity for RFLPs in European maize inbreds: I. Relation to performance of Flint × Dent crosses for forage traits. *Crop Sci.*, **32**: 895-902.

Briggle L.W. 1963. Heterosis in wheat - a review. *Crop Sci.*, **3**: 407-412.

Brim C. A. 1966. A modified pedigree method in soybean. *Crop Sci.*, **6**: 220.

Brim C.A. and Stuber C.W. 1973. Application of genetic male sterility to recurrent selection schemes in Soybean. *Crop Sci.*, **13**: 528-530.

Chang-Jian J., Cockerham C.C. and Moll R.H. 1990. Inter and intra-cultivar effects of selection on heterosis. *Crop Sci.*, **30**: 44-49.

Chase S.S. 1952. Production of homozygous diploids of maize from monoploids. *Agron. J.*, **44**: 263-267.

Cloninger F.D., Zuber M.S. and Horrocks R.D. 1974. Synchronization of flowering of corn (*Zea mays* L.) by clipping young plants. *Agron. J.*, **66**: 270-272.

Cockerham C.C. 1961. Implications of genetic variance in hybrid breeding programme. *Crop Sci.*, **1**: 47-52.

Cockerham C.C. 1967. Prediction of double crosses from single crosses. *Zuechter*, **37**: 160-169.

Comstock R.E. 1979. Inbred line *vs* the population as tester in reciprocal recurrent selection. *Crop Sci.*, **19**: 881-886.

Comstock R.E., Robinson H.F. and Harvey P.H. 1949. A breeding procedure designed to make maximum use of both general and specific combining ability. *Agron. J.*, **41**: 360-367.

Cowan R.J. 1972. Seed certification. p. 371-397. *In*: *Seed Biology*. (ed.) T.T. Kozlowski. Academic Press, New York, USA.

Cramer M.M. and Kannenberg L.W. 1992. Five years of HOPE: the hierarchical open-ended corn breeding system. *Crop Sci.*, **32**: 1163-1171.

Darrah L.L. and Zuber M.S. 1986. 1985 United States farm maize germplasm base and commercial breeding strategies. *Crop Sci.*, **26**: 1109-1113.

Davis D.D. 1998. Cotton. p. 357-380. *In*: *Hybrid Cultivar Development.* (eds.) S.S. Banga and S.K. Banga, Narosa Publishing House, New Delhi.

Davis R.L. 1927. Report of the plant breeder. *Rep. Puerto Rice Agric. Exp. Stn.*, Puerto Rico, pp. 14-15.

Dhillon B.S. 1975. The application of partial-diallel crosses in plant breeding - a review. *Crop Improv.*, **2**: 1-8.

Dhillon B.S. 1998a. Recurrent selection for combining ability and performance *per se* of cross-bred and selfed families. *Maydica*, **43**: 155-160.

Dhillon B.S. 1998b. Quantitative-genetic implications of the use of cytoplasmic-genic male sterility in relation to the types of hybrids in crop plants. *Crop Improv.*, **25**: 1-5.

Dhillon B.S. 1991a. Recurrent mass selection based on selfed-plant evaluation in allogamous species. *Crop Sci.*, **31**: 1075-1077.

Dhillon B.S. 1991b. Alternate recurrent selection of S_1 and half-sib families for intrapopulation improvement. *Maydica*, **36**: 45-48.

Dhillon B.S. and Khehra A.S. 1989. Modified S1 recurrent selection in maize improvement. *Crop Sci.*, **29**: 226-228.

Dhillon B.S. and Prasanna B.M. 2001. Maize. pp. 147-185. *In*: *Breeding Field Crops*. (ed.) V.L. Chopra, Oxford & IBH Publishing Co. Pvt. Ltd., New Delhi.

Dhillon B.S., Boppermaier J., Pollmer W.G., Herrmann R.G. and Melchinger A.E. 1993. Relationship of restriction fragment length polymorphism among European maize inbreds with ear dry matter yield of their hybrids. *Maydica*, **38**: 245-248.

Dhillon B.S., Malhi N.S. and Saxena V.K. 1997. Development and improvement of heterotic pools in maize. p. 74-75. *In*: *Proc. Intl. Symp. on Genetics and Exploitation of Heterosis in Crops*, CIMMYT, Mexico.

Dhillon B.S., Malhi N.S., Saxena V.K., Kapoor W.R., Pal S.S., Singh M. and Khehra A.S. 1995. Paras - first single cross hybrid of maize in India. *J. Res. Punjab Agric. Univ.*, **32**: 238-239.

Dubreuil P., Dufour P., Krejci E., Causse M., Vienne D., Gallais A. and Charcosset A. 1996. Organization of RFLP diversity among inbred lines of maize representing the most significant heterotic groups. *Crop Sci.*, **36**: 790-799.

Dudley J.W. 1982. Theory of transfer of alleles. *Crop Sci.*, **22**: 631-637.

Dudley J.W. 1984. Identifying parents for use in a pedigree breeding programme. *Corn Sorghum Res. Conf.*, **39**: 176-188.

Dudley J.W. 1987. Modification of methods for identifying inbred lines useful for improving parents of elite line crosses. *Crop Sci.*, **27**: 944-947.

Dudley J.W., Saghai Maroof M.A. and Ruefner G.K. 1991. Molecular markers and grouping of parents in maize breeding programs. *Crop Sci.*, **31**: 718-728.

Duvick D. N. 1999. Heterosis: feeding people and protecting natural resources. p. 19-29. *In*: *Genetics and Exploitation of Hectorsis in Crops*. (eds.) J.G. Coors and S. Pandey, *Am. Soc. Agron.*, Madison, Wisconsin, USA.

Duvick D.N. 1965. Cytoplasmic pollen sterility in corn. *Adv. Genet.*, **13**: 1-56.

Duvick D.N. 1966. Influence of morphology and sterility on breeding methodology. p. 85-138. *In*: *Plant Breeding*, (ed.) K.J. Frey, Iowa state university press, Ames, IA. Vol. I.

Duvick D.N. 1974. Continuous backcrossing to transfer prolificacy to a single inbred line of maize. *Crop Sci.*, **14**: 69-71.

Duvick D.N. 1977. Genetic rates of gain in hybrid maize yields during the past 40 years. *Maydica*, **22**: 187-196.

Eberhart S.A. 1964. Theoretical relations among single, three-way and double-cross hybrids. *Biometrics*, **20**: 522-539.

Eberhart S.A. 1970. Factors affecting efficiencies of breeding methods. *Afr. Soils*, **15**: 669-680.

Eberhart S.A., Russell W.A. and Penny L.H. 1964. Double-cross hybrid prediction when epistasis is present. *Crop Sci.*, **4**: 363-366.

Eyherabide G.H. and Hallauer A.R. 1991. Reciprocal full-sib recurrent selection in maize. I. Direct and indirect responses. *Crop Sci.*, **31**: 952-959.

Falconer D.S. 1988. *Introduction of Quantitative Genetics*. Longman, London, UK.

Fowler W. 1967. Cultural practices for today's seed fields. *Hybrid Corn Ind. Res. Conf.*, **22**: 53-58.

Gama E.E.C. and Hallauer A.R. 1977. Relation between inbred and hybrid traits in maize. *Crop Sci.*, **17**: 703-706.

Gardner C.O. 1961. An evaluation of effects of mass selection and seed iradiation with thermal neutrons on yield of corn. *Crop Sci.*, **18**: 387-390.

Geadelman J.L. and Peterson R.H. 1978. Effects of two yield component selection procedures on maize. *Crop Sci.*, **18**: 387-390.

Gerloff J.W. and Smith O.S. 1988. Choice of method for identifying germplasm with superior alleles. I. Theoretical results. *Theor. Appl. Genet.*, **76**: 209-216.

Godshalk E.B., Lee M. and Lamkey K.R. 1990. Relationship of restriction fragment length polymorphism to single cross hybrid performance in maize. *Theor. Appl. Genet.*, **80**: 273-280.

Goodsell S. 1961. Male sterility in corn by androgenesis. *Crop Sci.*, **1**: 227-228.

Green J.M. 1948. Relative value of two testers for estimating topcross performance in segregating maize populations. *J. Am. Soc. Agron.*, **40**: 45-57.

Griffing B. 1956. Detasseling responses in corn. *Agron. J.*, **48**: 247-249.

Hadjinov M.I., Sherbak V.S., Benko N.I., Gusev V.P., Sukhorzheus-Kaya T.B. and Voronova L.P. 1982. Interrelationships between isozyome diversity and combining ability in maize lines. *Maydica*, **27**: 135-139.

Hallauer A.R. 1973. Hybrid development and population improvement in reciprocal full-sib selection. *Egyptian J. Genet. Cytol.*, **1**: 84-101.

Hallauer A.R. 1975. Relation of gene action and types of testers in maize breeding procedures. *Corn Sorghum Res. Conf.*, **30**: 150-165.

Hallauer A.R. 1990. Methods used in developing maize inbreds. *Maydica*, **35**: 1-16.

Hallauer A.R. 1992. Recurrent selection in maize. *Plant Breed. Rev.*, **9**: 115-179.

Hallauer A.R. and Lopez-Perez E. 1979. Comparisons among testers for evaluating lines of corn. *Corn Sorghum Res. Conf.*, **34**: 57-75.

Hallauer A.R. and Miranda J.B. 1988. *Quantitative Genetics in Maize Breeding*. Iowa State University Press, Ames, Iowa.

Hallauer A.R. and Eberhart S.A. 1970. Reciprocal full-sib selection. *Crop Sci.*, **10**: 315-316.

Hallauer A.R., Russell W.A. and Lamkey K.R. 1988. Corn breeding. p. 463-564. *In*: *Corn and Corn Improvement*. (eds.) G.F. Sprague and J.W. Dudley, Am. Soc. Agron., Madison, Wisconsin, USA.

Hanna W.P., Ozias-Akins and Roche D. 1999. Apomixis and heterosis. p. 335-342. *In*: *Genetics and Exploitation of Heterosis in Crops*. (eds.) J.G. Coors and S. Pandey. Am. Soc. Agron., Madison, Wisconsin, USA.

Hanna W.P., Roche D. and Ozias-Akins P. 1996. Use of apomixis in crop improvement-traditional and molecular approaches. *In*: *Proc. Third International Symp. on Hybrid Rice*, Hyderabad, India, 14-16 Nov. 1996.

Harlan H.V. and Pope M.N. 1922. The use and value of backcrosses in small grain breeding. *J. Hered.*, **13**: 319-322.

Hayes H.K. 1963. *A Professor's Story of Hybrid Corn*, Burgess, Minneapolis, USA.

Heidrich-Sobrinho E. and Corderio A.R. 1975. Codominant isoenzymic alleles as markers of genetic diversity correlated with heterosis in maize (*Zea mays* L.). *Theor. Appl. Genet.*, **46**: 197-199.

Henderson C.R. 1975. Best linear unbiased estimation and prediction under a selection model. *Biometrics*, **31**: 423-447.

Ho Y.T. and Comstock R.E. 1980. Combining superior alleles from two homozygous populations in a cross-fertilizing species. *Genet. Res.*, **36**: 277-287.

Horner E.S., Lundy H.W., Lutrick M.C. and Chapman W.H. 1973. Comparison of three methods of recurrent selection in maize. *Crop Sci.* ,**13**: 485-489.

Horner E.S., Lutrick M.C., Chapman W.H. and Martin F.G. 1976. Effect of recurrent selection for combining ability with a single cross tester in maize. *Crop Sci.*, **16**: 5-8.

Hull F.H. 1945. Recurrent selection and specific combining ability in corn. *J. Am. Soc. Agron.*, **37**: 134-45.

Humberto Reyes-Valdes M. 2000. A model for marker-based selection in gene introgression breeding programs. *Crop Sci.*, **40**: 91-98.

Hunter R.B. and Kannenberg L.W. 1971. Isozyme characterization of corn (*Zea mays* L.) inbreds and its relationship to single cross hybrid performance. *Can. J. Genet. Cytol.*, **13**: 649-655.

Ichikawa N., Kishimoto N., Inagaki A., Nakamura A., Koshino Y., Yokozeki Y., Oka M., Samoto S., Akagi H., Higo K., Shiniyo C., Fujimura J. and Shimada H. 1997. A rapid PCR-aided selection of rice line containing the *Rf-1* gene which is involved in restoration of cytoplasmic male sterility. *Mol. Breeding*, **3**: 195-202.

Ikehashi H. and Araki H. 1984. Varietal screening for compatibility type revealed in F_1 fertility of crosses in rice. *Jpn. J. Breed*, **34** : 304-312.

Jain A., Bhatia S., Banga S.S., Prakash S. and Lakshmikumaran M. 1994. Potential use of random amplified polymorphic DNA (RAPD) technique to study the genetic diversity in Indian Mustard (*Brassica juncea*) and its relationship to heterosis. *Theor. Appl. Genet.*, **88**: 116-122.

Jain S.K. and Suneson C.A. 1963. Male sterility for increased outbreeding in populations of barley. *Nature*, **199**: 407-408.

Jenkins M.T. 1934. Methods of estimating the performance of double crosses in corn. *J. Am. Soc. Agron.*, **26**: 199-204.

Jenkins M.T. 1935. The effect of inbreeding and of selection within inbred lines of maize upon hybrids made after successive generations of selfing. *Iowa State J. Sci.*, **9**: 429-450.

Jenkins M.T. 1940. The segregation of genes affecting yield of grain in maize. *J. Soc. Agron.*, **32**: 55-63.

Jenkins M.T. and Brunson A.M. 1932. Methods of testing inbred lines of maize in crossbred combination. *J. Am. Soc. Agron.*, **24**: 523-530.

Johnson G.R. 1980. Simple quantitative theory for selections during inbred line development. III. *Corn Breeder's School*, **16**: 1-18.

Jones D.F. 1918. The effects of inbreeding and crossbreeding upon development. *Conn Agric. Exp. Stn. Bull.*, **207**: 5-100.

Jones H.A. and Clarke A.E. 1943. Inheritence of male sterility in the onion and the production of hybrid seed. *Proc. Am. Soc. Hort. Sci.*, **43**: 189-194.

Jones J.W. 1926. Hybrid vigour in rice. *J. Am. Soc. Agron.*, **18**: 424-428.

Jones M.D. and Brooks J.S. 1952. Effect of tree barriers on outcrossing in corn. *Oklahoma Agric. Exp. Stn. Tech. Bull.*, **T-45**: 3-11.

Kahler A.L., Hallauer A.R. and Gardner C.O. 1986. Allozyme polymorphisms within and among open-pollinated and adapted exotic populations of maize. *Theor. Appl. Genet.*, **72**: 592-601.

Kao K.N. and Kasha K.J. 1969. Haploidy from interspecific crosses with tetraploid barley. p. 82-88. *In*: *Barley Genetics*. (ed.) R.A. Nilan, Washington State Univ. Press, Washington.

Kaul M.L.H. 1998. Male sterility: classification and concept. p. 17-45. *In*: *Hybrid Cultivar Development*. (eds.) S.S. Banga and S.K. Banga, Narosa Publishing House, New Delhi.

Kempthorne O. and Curnow R.N. 1961. The partial diallel cross. *Biometrics*, **17**: 229-250.

Kermicle J.L. 1969. Androgenesis conditioned by a mutation in maize. *Science*, **166**: 1422-1424.

Kim C.H. and Rutger J.N. 1988. *Heterosis in rice*. p. 39-54. In Hybrid Rice. IRRI, Manila, Philippines.

Kirti P.B., Banga S.S., Prakash S. and Chopra V.L. 1995a. Transfer of *ogu* cytoplasmic male sterility to *Brassica juncea* and improvement of male sterility line through somatic cell fusion. *Theor. Appl. Genet.*, **91**: 517-521.

Kirti P.B., Mohapatra T., Khanna H. and Prakash S. 1995b. *Diplotaxis catholica* + *Brassica juncea* somatic hybrids: molecular and cytogenetic characterization. *Plant Cell Rep.*, **14**: 593-597.

Lamkey K.R., Schnicker B.J. and Melchinger A.E. 1995. Epistasis in an elite maize hybrid and choice of generation for inbred line development. *Crop Sci.*, **35**: 1272-1281.

Landi P and Frascaroli E. 1995. Response to a modified reciprocal recurrent selection in two maize synthetics. *Crop Sci.*, **35**: 791-797.

Laurie D.A. and Bennett M.D. 1988. The production of haploid wheat plants from wheat × maize crosses. *Theor. Appl. Genet.*, **76**: 393-397.

Lee M., Godshalk E.B., Lamkey K.R. and Woodman W.W. 1989. Association of restriction fragment length polymorphisms among maize inbreds with agronomic performance of their crosses. *Crop Sci.*, **29**: 1067-1071.

Liu K., Wang J., Li H., Xu C., Liu A., Li X. and Zhang Q. 1997. A genome-wide analysis of wide compatibility in rice and the precise location of S_5 locus in the molecular map. *Theor. Appl. Genet.*, **95**: 809-814.

Mariani C., Beukeber M.D., Truettever J., Leemans J. and Goldberg R.B. 1990. Induction of male sterility in plants by a chimeric ribonuclease gene. *Nature*, **347**: 737-741.

Mariani C., Gossele V., De Beuckeber M., De Block M., Goldberg R.B., De Greef W. and Leemans J. 1992. A chimeric ribonuclease-inhibitor gene restores fertility to male sterile plants. *Nature* (London), **356**: 385-387.

Martin J.M., Talbert L.E., Lanning S.P. and Blake N.K. 1995. Hybrid performance in wheat as related to parental diversity. *Crop Sci.*, **35**: 104-108.

Maruyama K., Araki H. and Kato H. 1991. Thermosensitive genetic male sterility induced by irradiation. *In*: *Rice Genetics II*. IRRI, Manila, Philippines. p. 227-235.

Melchinger A.E. 1987. Expectation of means and variances of testcrosses produced from F_2 and backcross individuals and their selfed progenies. *Heredity*, **59**: 105-115.

Melchinger A.E., Boppenmaier J., Dhillon B.S., Pollmer W.G. and Herrmann R.G. 1992. Genetic diversity for RFLPs in European maize inbreds. II. Relation to the performance of hybrids within *versus* between heterotic groups for forage trait. *Theor. Appl. Genet.*, **84**: 672-681.

Melchinger A.E., Lee M., Lamkey K.R., Hallauer A.R. and Woodman W.L. 1990. Genetic diversity for restriction fragment length polymorphism and heterosis for two diallel sets of maize inbreds. *Theor. Appl. Genetics*, **80**: 488-496.

Melchinger A.E., Schmidt W. and Geiger H.H. 1988. Comparison of testcrosses produced from F_2 and first-backcross populations in maize. *Crop Sci.*, **28**: 743-749.

Mishra G.P., Singh R.K., Mohapatra T., Singh A.K., Prabhu K.V. and Zaman F.U. 2001. Molecular mapping of a fertility restorer gene in Basmati rice using microsatellite markers. *Indian J. Genet.*, **61**: 348-349.

Moll R.H. and Hasnon W.D. 1984. Comparisons of effects of intrapopulation vs. interpopulation selection in maize. *Crop Sci.*, **24**: 1047-1052.

Moll R.H., Jackson W.A. and Mikkelson R.L. 1994. Recurrent selection for maize grain yield: Dry matter and nitrogen accumulation and partitioning changes. *Crop Sci.*, **34**: 874-881.

Moll R.H., Lonnquist J.H., Fortuna J.V. and Johnson E.C. 1965. The relation of heterosis and genetic divergence in maize. *Genetics*, **52**: 139-144.

Moll R.H., Salhuana W.S. and Robinson H.F. 1962. Heterosis and genetic diversity in variety crosses in maize. *Crop Sci.*, **2**: 197-198.

Otsuka Y., Eberhart S.A. and Russell W.A. 1972. Comparison of prediction formulae for maize hybrids. *Crop Sci.*, **12**: 325-331.

Parmar K.S., Siddiq E.A. and Swaminathan M.S. 1979. Chemical induction of male sterility in rice. *Indian J. Genet.*, **39**: 529-541.

Perez A.T., Chang T.T., Beachell H.M., Vergara B.S. and Marciano A.P. 1973. Induction of male sterility in rice with ethrel and RH-531. *SABRAO News Lett.*, **5**: 133-139.

Popi J., Rajnpreht J., Kannenberg L.W. and Pauls K.P. 2000. Random amplified polymorphic DNA-based evaluation of diversity in the hierarchical, open-ended population enrichment maize breeding system. *Crop Sci.*, **40**: 619-625.

Rademacher M.A.M., Hallauer A.R. and Russell W.A. 1999. Comparative response of two reciprocal recurrent selection methods in BS21 and BS22 maize population. *Crop Sci.*, **39**: 87-97.

Rawlings J.C. and Thompson D.L. 1962. Performance level as criterion for the choice of maize testers. *Crop Sci.*, **2**: 217-220.

Rhodes M.M. 1931. Cytoplasmic inheritance of male sterility in *Zea mays*. *Science*, **73**: 340-341.

Richey F.D. 1922. The experimental basis for the present status of corn breeding. *J. Am. Soc. Agron.*, **14**: 1-17.

Richey F.D. 1927. The covergent improvement of selfed lines of corn. *Am. Nat.*, **61**: 430-449.

Rinke E.H. and Sentz J.C. 1961. Moving Corn-Belt germplasm northward. *Hybrid Corn Ind. Conf.*, **16**: 53-56.

Russell W.A. 1985. Comparison of hybrid performance of maize inbred lines developed from the original and improved cycles of BSSS. *Maydica*, **30**: 407-419.

Russell W.A. 1986. Contributions of breeding to maize improvement in the United States, 1920s-1980s. *Iowa States J. Res.*, **61**: 5-34.

Russell W.A. and Eberhart S.A. 1975. Hybrid performance of selected maize lines from reciprocal recurrent selection and testcross selection programmes. *Crop Sci.*, **15**: 1-4.

Russell W.A., Blackburn D.J. and Lamkey K.R. 1992. Evaluation of a modified reciprocal recurrent selection procedure for maize improvement. *Maydica*, **37**: 61-67.

Russell W.A., Sprague G.F. and Penny L.H. 1963. Mutations affecting quantitative characters in long-term inbred lines of maize. *Crop Sci.*, **3**: 175-178.

Russell W.A., Eberhart S.A. and Vega U.A. 1973. Recurrent selection for specific combining ability for yield in two maize populations. *Crop Sci.*, **13**: 257-261.

Saghai Maroof M.A., Yang G.P., Zhang Q. and Gravois K.A. 1997. Correlation between molecular marker distance and hybrid performance in US long grain rice. *Crop Sci.*, **37**: 145-150.

Sant V.J., Patankar A.G., Sarode N.D., Mhase L.B., Sainani P.K., Deshmukh R.B., Ranjekar P.K., and Gupta V.S. 1999. Potential of DNA markers in detecting divergence and in analyzing heterosis in Indian elite chickpea cultivars. *Theor. Appl. Genet.*, **98**: 1217-1225.

Schmulling T., Rohrig H., Pilz S., Walden R. and Schell J. 1993. Restoration of fertility by antisense RNA in genetically energineered male sterile tobacco plants. *Mol. Gen. Genet.*, **237**: 385-394.

Schmulling T., Schell J. and Spena A. 1988. Single gene from *Agrobacterium rhizogenes* influence plant development. *EMBO J.*, **7**: 2621-2629.

Shi M.S. 1981. Preliminary report of breeding and utilization of late *japonica* natural double purpose line. *J. Hubei Agric. Sci.*, **7**: 1-3.

Shivanna K.R. and Sawhney V.K. 1997. Pollen Biotechnology for Crop Production and Improvement. Cambridge University Press, Cambridge, UK.

Shoultz D. 1985. An evaluation of parent delay techniques. *Corn Sorghum Res. Conf.*, **440**: 151-160.

Shull G.H. 1908. The composition of a field of maize. *Am. Breeder's Assoc. Rep.*, **4**: 296-301.

Shull G.H. 1909. A pure-line method of corn breeding. *Am. Breeder's Assoc. Rep.*, **5**: 51-59.

Shull G.H. 1952. Beginnings of the heterosis concept. p. 14-48. *In*: *Heterosis*. (ed.) J.W. Gowen, Iowa State College Press, Ames, USA.

Smith J.S.C. 1988. Diversity of United States hybrid maize germplasm. Isozymic and chromatographic evidence. *Crop Sci.*, **28**: 63-69.

Smith J.S.C. and Smith O.S. 1989. The description and assessment of distances between inbred lines of maize. II. The utility of morphological, biochemical, and genetic descriptors and a scheme for the testing of distinctiveness between inbred lines. *Maydica*, **34**: 151-161.

Sprague G. F. 1946. Early testing of inbred lines. *J. Amer. Soc. Agron.*, **38**: 108-117.

Sprague G.F. and Tatum L.A. 1942. General *vs* specific combining ability in single crosses of corn. *J. Am. Soc. Agron.*, **34**: 923-932.

Sprague G.F. and Eberhart S.A. 1977. Corn breeding. p. 305-362. *In*: *Corn and Corn Improvement*. (ed.) G.F. Sprague, Am. Soc. Agron., Madison, Wisconsin, USA.

Sprague G.F., Russell W.A. and Penny L.H. 1960. Mutation affecting quantitative traits in the selfed progeny of doubled monoploid stocks. *Genetics*, **45**: 855-866.

Stephens J.C. and Holland R.F. 1954. Cytoplasmic male sterility for hybrid sorghum seed production. *Argon. J.*, **46**: 320-323.

Stuber C.W., Lincoln S.E., Wolff D.W., Helentjaris T. and Lander E.S. 1992. Identification of genetic factors contributing to heterosis in a hybrid from two elite maize inbred lines using molecular markers. *Genetics*, **132**: 823-839.

Stuber C.W., Williams W.P. and Moll R.H. 1973. Epistasis in maize (*Zea mays* L.), III. Significance in predictions of hybrid performance. *Crop Sci.*, **13**: 195-200.

Subudhi P.K., Borkakati R.P., Virmani S.S. and Huang N. 1997. Molecular mapping of a thermosensitive genetic male sterlity gene in rice using bulked segregant analysis. *Genome*, **40**: 188-194.

Taillebois J. and Guimaraes E.P. 1988. Improving out crossing rate in rice [*Oryza sativa* L.]. p. 175-180. *In*: *Hybrid Rice*. IRRI, Manila, Philippines.

Tan Z.C., Li Y.Y., Chen L.B. and Zhou G.S. 1990. Studies on ecological and adaptability of dual purpose line An-Nong S-1. *Hybrid Rice*, **3**: 35-38.

Tatum L.A. 1971. The southern corn leaf blight epidemic. *Science*, **171**: 1113-1116.

Troyer A.F. 1996. Breeding widely adapted, popular maize hybrids. *Euphytica*, **92**: 163-174.

Van der Meer I.M., Stam M.E., Van Tunen A.J., Moll J.M.N. and Stuitje A.R. 1992. Antisense inhibition of flavonoid biosynthesis in petunia anthers results in male sterility. *Plant Cell*, **4**: 253-262.

Virmani S.S. 1994. *Heterosis and hybrid rice breeding.* Monographs on Theoretical and Applied Genetics. Springer-Verlag.

Virmani S.S. and Edwards I. 1983. Current status and future prospects of breeding hybrid rice and wheat. *Adv. Agron.*, **36**: 145-214.

Virmani S.S. and Zaman F.U. 1998. Improving grain quality of hybrid rice: challenges, strategies and achievements. p. 177-186. *In*: *Advances in Hybrid Rice Technology.* (eds.) S.S. Virmani, E.A. Siddiq and K. Murlidharan, Proc. 3rd Int. Symp. Hybrid Rice, Nov. 14-16, 1996, Hyderabad, India, IRRI, Los Banos, Philippines.

Virmani S.S. and Ilyas-Ahmed M. 2001. Environment-sensitive genetic male sterility (EGMS) in crops. *Adv. Agron.*, **72**: 139-195.

Walejko R.N. and Russeell W.A. 1977. Evaluation of recurrent selection for specific combining ability in two open-pollinated maize cultivars. *Crop Sci.*, **17**: 647-651.

Wang Z., Zheng F., Shen G., Gao J., Snusted D.P., Li M., Jihaung J. and Hong M. 1995. The amylose content in rice endosperm of the waxy gene. *Plant J.*, **7**: 613-622.

Williams M.E., Leemans J. and Michiels F. 1997. Male sterlity through recombinant DNA technology. p. 237-258. *In*: *Pollen Biotechnology for Crops Production and Improvement.* (eds.) K.R. Shivanna and V.K. Sawhney, Cambridge Univ. Press, Cambridge, UK.

Worrall D., Hird D.L., Hodge R., Paul W., Draper J. and Scott R. 1992. Premature dissolution of the microsporocyte callose wall causes male sterility in transgenic tobacco. *Plant Cell*, **4**: 759-771.

Wu J.L. 1986. Breeding haploid corn by anther culture. p. 149-161. *In*: *Haploids of Higher Plants In vitro.* (eds.) H. Hu and H.Y. Yang, Springer, Berlin, Germany.

Wych R.D. 1988. Production of hybrid seed corn. p. 565-607. *In*: *Corn and Corn Improvement.* (eds.) G.F. Sprague and J.W. Dudley, Am. Soc. Agron., Maidson, USA.

Xiao J., Li J., Yuan L., Mc Couch S.R. and Tanksley S.D. 1996. Genetic diversity and its relationship to hybrid performance and heterosis in rice as revealed by PCR-based marker. *Theor. Appl. Genet.*, **92**: 637-643.

Xiao J., Grandillo S., Ahri S.N., Mc Couch S.R., Tanksley S.D., Li J. and Yuan L.P. 1996. Genes from wild rice improve yield. *Nature*, **384**: 223-224.

Yao F. Xu C.G., Yu S.B., Li X.J., Gao Y.J., Li X.H. and Jing O. *et al.* 1997. Mapping fertility-restoring genes of rice WA cytoplasmic male sterility using SSLP markers 171 Zhang. Mapping and genetic analysis of two fertility restorer loci in the wild-abortive cytoplasmic male sterility system of rice (*Oryza sativa* L.), *Euphytica*, **98**: 183-187.

Yuan L.P. 1977. The execution and theory of developing hybrid rice. (In Chinese.) Zhonggue. Nongye Kexue (*Chinese Agric. Sci.*) **1**: 27-31.

Zambezi B.T., Horner E.S. and Martin F.G. 1986. Inbred lines as testers for general combining ability in maize. *Crop Sci.*, **26**: 908-910.

Zhang G., Bharaj T.S., Lu Y., Virmani S.S. and Huang N. 1997. Mapping of the Rf-3 nuclear fertility-restoring gene for WA cytoplasmic male sterility in rice using RAPD and RFLP markers. *Theor. Appl. Genet.*, **94**: 27-33.

Zhang Q.F., Shen B.S., Dai X.K., Mei M.H., Saghai Maroof M.A. and Li Z.B. 1994. Using bulked extremes and recessive class to map genes for photoperiod-sensitive genetic male sterility in rice. *Proc. Natl. Acad. Sci.*, USA. **91**: 8675-8679.

Zhang Q.F., Zhou Z.Q., Yang G.P., Xu C.G., Liu K.D. and Saghai Maroof M.A. 1996. Molecular marker heterozygosity and hybrid performance in *indica* and *japonica* rice. *Theor. Appl. Genet.*, **93**: 1218-1224.

Zhang Z.G., Yuan S.C., Zen H.L., Li Y.Z., Li Z.C. and Wei C.L. 1991. Preliminary observation of fertility changes in the new type temperature sensitive male-sterile rice. IV A. *Hybrid Rice*, **1**: 31-34.

Zuber M.S. and Darrah L.L. 1980. 1979 US corn germplasm base. *Corn Sorghum Res. Conf.*, **35**: 234-249.

Plant Breeding - Mendelian to Molecular Approaches
H. K. Jain and M. C. Kharkwal (eds.)

19

Heterosis in Crop Improvement

E.A. Siddiq[1], O.P. Govila[2] and S.S. Banga[3]

Abstract

An effort has been made in the present paper to critically assess the level of success and future strategies in the development and adoption of hybrid technology in selected field crops under three categories: (*i*) extention of hybrid technology to new niches and situations (maize, sorghum and pearl millet); (*ii*) sustenance of hybrid technology in new crops of proven potential (rice, cotton and sunflower), and (*iii*) extending hybrid technology to new prospective crops (*Brassica*, wheat, pigeonpea and castor). Maize closely followed by sorghum and pearl millet have been the pioneer crops in India since mid fifties. However, lack of single cross short duration hybrids in maize; lack of higher yielding dual purpose hybrids suitable for *rabi* season in sorghum and narrow cytoplasmic base of CMS system in pearl millet are certain limitations which we need to overcome. Failure to produce and supply targeted quantity of hybrid seed, specially of public bred hybrids warrants urgent attention. High yielding cotton hybrids combining early maturity and resistance to leaf curl virus and boll worm complex for irrigated north and north-west regions having limited growing season are yet to be developed. Dissemination of transgenic *Bt.* cotton hybrids/varieties all over to effectively manage boll worm complex and thereby stabilise yield and reduce cost on plant protection is required. One of the major concerns in sunflower hybrids is slow pace of development of heterotic gene pools, exclusive dependence on a single CMS source (*PET-1*) and narrow genetic diversity for seed yield and oil content. Broadening of genetic base through enrichment of source nursery with germplasm for yield vigour is important keeping in view the fact that available genetic variability in Indian mustard is too inadequate to realize the desired level of yield heterosis; at the same time, it is worth searching for cytoplasmic-genetic male sterility source. First wheat hybrid "Pratham" is being commercialised. However, viable seed production technology either through CMS/chemical based system is yet to be perfected.

[1]Directorate of Rice Research, Hyderabad 500 030, India
[2]Indian Agricultural Research Institute, New Delhi 110 012, India
[3]Punjab Agricultural University, Ludhiana 141 004, India

Introduction

Since prehistoric days man has been aware of the importance of hybrid vigour in plants and animals, without knowing the underlying genetic principles. Native Indians in the Americas were practicing for long 'elite plant selection' for seed purpose to sustain their corn yields. In a way it was a forerunner to the scientific exploitation of heterosis through hybrids, composites and synthetics in a variety of crop plants especially open pollinated corn and pearlmillet and often cross pollinated sorghum and cotton. Knowledge of existence of exploitable heterosis and discovery of genetic and cytoplasmic-genetic male sterility sources enabled crop breeders develop and exploit hybrid technology extensively. Over 70 year long history of hybrid breeding in corn and 50 year experience with sorghum and pearlmillet have established that hybrids have (*a*) distinct yield advantage over pure line and open pollinated varieties (*b*) resilience to perform better than varieties under stress situations and (*c*) potential for progressive enhancement of genetic yield level. Equally rewarding experience gained during the last two decades with a variety of crops has prompted the country to identify 'hybrid technology' as one of the thrust areas for sustained research since the 9th Plan for achieving future production targets. The mega network project launched accordingly on hybrid technology in field crops has enabled the country to adopt the technology over 15.63 m ha.

Prospects of and Problems Constraining Hybrid Technology

India has made spectacular advance in the agricultural sector culminating in self-sufficiency in food and many of the essential commodities during the last four decades. Yet, the country can not remain complacent, given the size of expected demands of food, feed, fibre, etc., in the face of shrinking/ absence of many of the favourable growth factors of the seventies and eighties and limitations of the currently available technology to find solutions to many of the yield depressing factors (Table 1).

Table 1. Present status and demand projection of agricultural Commodities by 2007-08

Crop/Commodity	1998-99		Production demand by 2007-08 (m tons)	Target annual growth of yield (%)
	Production (m tons)	Yield (kg/ha)		
Rice	88.6	1985	97.0	2.12
Wheat	70.8	2583	72.1	0.69
Millets	31.5	1065	34.7	0.91
Pulses	13.6	608	18.7	3.93
Oilseeds	21.5	855	---	1.91
Cotton	12.2*	223	---	---
Vegetables	74.5	---	102.6	3.54

*Mill.bales; **Pertaining to low income growth (3.5% growth in per capita GDP)

The contemplated technological interventions have to be such that they not only help enhance the level of productivity on a sustainable basis but also generate more income and employment opportunities in the rural areas. Hybrid technology is the readily adoptable option to break the decades long yield stagnation at low levels in pulses (pigeonpea) and oilseeds (*Brassica* and sesame) and find a new yield threshold in already high yielding dwarf varieties of rice and wheat. Whereas distinct yield advantage possible through hybrids can help release sizeable area under a given crop for other high value crops, as China, following extensive adoption of hybrid rice, could release prime 3-4 m ha of rice area for more remunerative non-rice crops. Thus, hybrid technology can help to diversify cropping patterns in intensively cropped areas. Impact analysis of the technology across crops has shown enhanced rural income and employment opportunities respectively, by high net return realizable through the yield advantage and engagement of rural folk, especially women in hybrid seed production and processing. Production in excess of domestic need as a result of extensive adoption of hybrid technology would help the country build sustainable export market for rice, maize, cotton, etc. Taking advantage of the large and well organized seed industry in both public and private sectors, scope for export of hybrid seed of various crops could as well be explored and exploited.

Despite such advantages, pace of adoption of the technology is much less than desired. Problems constraining the spread varies with the crop. Often, lack of genetic purity of seed and high seed cost are the major constraints common to all crops, besides crop specific drawbacks. Whereas it is lack of short duration hybrids suited to over 45 % of the rainfed area in the states of Madhya Pradesh, Rajasthan and Uttar Pradesh in the case of maize, lack of higher yielding dual purpose hybrids suitable for *rabi* season in sorghum, narrow genetic/cytoplasmic base of male sterility system in pearlmillet and still unacceptable cooking quality and duration in rice are proving to be serious constraints. In Indian mustard (*Brassica juncea*) it is still absence of strong sources of fertility restoration to already available *cms* sources; while lack of stable pollination control system continues to be the major hurdle in safflower, pigeonpea, cotton, sesame and vegetable crops. Nevertheless, if hybrid technology is regarded a success story in cotton and some of the vegetable crops like cabbage, brinjal, chillies, tomato, etc., it is because of their amenability to manual emasculation/pollination, taking advantage respectively of large flower and large number of seeds obtainable with single pollination for seed production on commercial scale.

An effort has been made in the present exercise to critically assess the level of success in the development and adoption of hybrid technology in selected field crops under the following three strategies:

(*i*) Extension of hybrid technology to new niches and situations in traditional hybrid crops

(*ii*) Sustenance of hybrid technology in hybrid crops of proven potential like rice and cotton

(*iii*) Extension of hybrid technology to other prospective crops like wheat

Extension of Hybrid Technology in Millets to New Niches and Situations

Maize closely followed by sorghum and pearlmillet have been the pioneer hybrid crops in

India since mid-fifties. But for the development and extensive adoption of hybrid technology in these crops, it would not have been possible to sustain millet production to around 32 million ha, in the wake of fast declining area under millets.

Maize: Mainly a food grain in many parts of the country, maize (*Zea mays* L.) has become equally an important feed grain following rapid growth of livestock-poultry industry during the last two decades. Given the growing feed grain demand from east Asian countries, maize would, in the coming years, be one of potential export commodities. Breeding emphasis on development of hybrids and composites has helped raise the yield level from < 1.0 to 1.7 t/ha during this period. Sadly, however, the yield level is far below that of the world average (4.2 t/ha), warranting diagnosis and correction of factors that restrain the productivity. Very large area under rainfed condition and motivated shift of maize from its traditional high productivity areas to marginal lands are the two major factors among many for low national yield average.

In the last 40 years, over 60 hybrids have been released for general cultivation, and of these, a few such as Deccan 109 and KH 510 in the medium maturity (85-90 d) group and Pusa Early Hybrid Makka 1 (PEHM1), PEHM2, and Him 129 in the early/extra early maturity (75-85 d) are only popular but account for not more than 10-15 % of the total area planted to hybrids. Equal number of hybrids bred by the private sector, however, account for more than 80 % of the area under hybrids (Singh *et al., 1996).*

Maize is traditionally a *kharif* crop in India. Demonstration in the mid-sixties of the possibilities of growing it in winter marked the beginning of *rabi* planting of the crop in the northern and eastern plains (Bihar and UP) and peninsular India (AP, Karnataka and Tamil Nadu). *Rabi* maize largely raised as irrigated full season crop, yields in the range of 6-8 t/ha, as against less than 2 t/ha by *kharif*/summer maize. Mild temperature, high response to applied fertilizer nutrients and low pest/disease pressure, make *rabi* maize more productive. There is immense scope to further raise the *rabi* yields, if learning from the decades long experience of the USA, the ongoing process of shifting to the more productive single cross hybrids from the traditionally adopted double and double top cross hybrids is accelerated.

The region comprising the states of Rajasthan, Gujarat, Madhya Pradesh and Uttar Pradesh accounts for over 45 % of the total maize area in the country. It is totally rainfed with lowest yields. The high yield hybrid technology has bypassed such a large environmentally handicapped corn belt. There are hardly any high yielding hybrid/composite of very early maturity for this region. On the pretext of low seed yield and hence high seed cost and wrongly perceived low seed demand, the region/ecology was not among the target areas of maize hybrid technology in the private sector seed agencies. Possibly the same reason has made the public sector institutions as well to confine to open pollinated varieties. Augmented research following identification of this ecology as one of the potential niches for extending hybrid technology, has led to the identification of commercially viable extra early maturing hybrids such as PEHM1, PEHM2 and a few by the private sector institutions.

Largely, populations that excessively depend on rainfed food crops remain insecure about food as well as nutrition. While addressing the chronic problem of calorie deficiency through productivity enhancement, scope for improvement of the nutritional value should not be overlooked. Maize, though has the distinction of being the only cereal with rich β-carotene (source of VitA), suffers from poor protein quality. Successful recombination of high quality protein with high yield recently by CIMMYT, has resulted in development of quality protein maize (QPM) hybrids. The composites are also found to be bright. In China, over two lakh hectares have been planted with QPM during the last two years. India has already directed its breeding research towards development of nutritionally rich hybrids/composites, especially for rainfed areas.

Accordingly, the following have been identified as the thrust areas of research on maize.

- Development of extra early hybrids/composites, tolerant to moisture stress for predominantly rainfed regions comprising the states of Rajasthan Gujarat, Madhya Pradesh and Uttar Pradesh.
- Expansion of area under irrigated *rabi* maize for very high productivity in the prospective regions comprising north Bihar and Uttar Pradesh, districts adjoining Bihar, Andhra Pradesh, Tamil Nadu and Karnataka with emphasis on single cross hybrids.
- Popularization of single cross hybrids adapted to temperate conditions in the hill states of Himachal Pradesh and Kashmir.
- While accelerating breeding research giving priority attention to what have been identified above, efforts have to be made, (*a*) to assess the scope for various maize-based products and demands from food/feed industry, so as to diversify the varietal composition, (*b*) to enhance nutritional enrichment of maize in general and rainfed maize in particular and (*c*) to explore the prospects of developing quality protein maize (QPM) and other specialty maize.

Sorghum: Amongst the food grain crops Sorghum ranks third [*Sorghum bicolor* (L.) Moench] in terms of area and production. Between early 1960s and mid 1990s, when area under millets registered gradual decrease due to more remunerative crops like rice, pulses and oilseeds, sorghum suffered the most, losing 30 % of the area (from 18.0 to 12.3 m ha). Of the 5.61 m ha lost, *kharif* sorghum alone accounts for 4.5 m ha. Nevertheless, production growth has been on the increase due to steady increase in the productivity level, especially of *kharif* sorghum (110 %) from 497 kg in 1964-67 to 1013 kg/ha in 1993-99. Though *rabi* sorghum also registered an increase, it was slow and marginal (32 %). Increased productivity is attributed to adoption of progressively higher yielding hybrids and their extensive adoption. The area under the public sector bred hybrids like CSH1 and CSH2 started at a moderate 0.18 m ha in 1966-67, rose to 7.1 m ha (49.2 %) by early 1990s, when more productive hybrids like CSH5, CSH6 and CSH9 were available. Today there are as many as 18 nationally produced hybrids, which include early maturing

(< 100 d) CSH1, CSH6, CSH14 and CSH17 and medium maturing (100-115 d) CSH5, CSH9, CSH13, CSH16 and CSH18. Adoption of the hybrids has been relatively fast in the states of Maharashtra and Karnataka, where moisture stress is a major constraint. The percentage coverage of hybrids as of today, is the highest in Tamil Nadu (90 %) closely followed by Maharashtra (80 %) and Madhya Pradesh (65.2 %) (Rana *et al.*, 1999).

Survival of sorghum in general, *kharif* sorghum in particular, is under threat, because of its poor return to the grower as compared to competitive crops like soybean, sunflower, cotton and groundnut. One sure way of improving the net return is by realization of the yield potential of hybrids. Studies indicate yield gap between potential and actual farmers' yields to be 50-70 % and the percentage gap to vary widely within the same ecology and state. Among the remedies, development of more stable hybrids combining resistance to grain mould and charcoal rot, tolerance to moisture stress and other yield limiting factors, adding value and finding more remunerative alternate uses including livestock/poultry feed, industrial alcohol, etc., are important.

Unlike *kharif* sorghum, *rabi* sorghum, accounting for nearly an equal area and confined largely to Maharashtra and Karnataka, goes largely for human consumption. Being dual purpose like the locally popular variety Maldandi 35, the hybrids provide fodder too. The yield level of *rabi* hybrid is just one half of *kharif* hybrid due to various biological and environmental limitations. Yet, *rabi* sorghum is considered economically more competitive than *kharif* sorghum.

The need to survive on residual moisture under shallow to medium soil depths, difficulties still experienced in satisfactorily recombining high grain and fodder yields with consumer preferred grain quality and desired level of resistance to shoot fly, the killer pest and lack of parental lines ideally adapted to rabi conditions continue to be the major constraints to high productivity. Largely on account of these reasons, very few hybrids (CSH12R, CSH13R, CSH15R) could be released so far. Among them, CSH15 bred on bold grain *rabi* based parental lines offers better opportunity to replace the widely grown Maldandi type of land races, which are known for superior grain quality and higher survival value.

The following are some of the areas/issues that need research attention for sustenance of the technology:

- Parental line improvement by diversification of *cms* and restorer source involving more and more exotic sources for higher yield vigour in *kharif* sorghum.
- Improvement of parental lines for evolving more heterotic hybrids combining desired consumer quality and adaptation to the harsh environment in *rabi* sorghum.
- Development of hybrids satisfying market demands for varied alternate uses.
- Development of future hybrids with high level of resistance to biotic (shootfly, stem borer, grain mold and charcoal rot) and abiotic (drought) stresses by genetic engineering techniques using novel genes and

- Exploring the prospects of breeding for high β carotene content, taking advantage of constructs already available for rice.

Pearl millet: Fourth most important food grain crop pearl millet [*Pennisetum glaucum* (L.) R.Br.] is largely a crop of arid and semi-arid regions, particularly the north western part of India comprising the states of Rajasthan, Maharashtra, Gujarat, Uttar Pradesh and Haryana, which account for 95 % of the area under the crop. Like sorghum, pearl millet area giving place to more productive and remunerative crops, declined by about 2 m ha during the last 20 years (12 m ha in 1978-80 to 9.7 m ha in 1996-98). But production was on the increase from 4.75 to 7.68 million tonnes due to substantial increase in productivity (400 to 789 kg/ha) during the corresponding period. The growth is largely on account of extensive adoption of hybrids/composites since their introduction. Percentage adoption of hybrids is estimated to be around 42 % during 1986-87. For a comprehensive account aspect of pearl millet reader may refer to Khairwal *et al.* (1999).

Being a crop of most risky arid/semi-arid ecolôgy, breeding emphasis has been for extra earliness coupled with tolerance to moisture stress, resistance to major diseases like downy mildew and reasonable fodder and grain yields. Discovery of cytoplasmic male sterility (CMS) marked the beginning of hybrid technology in pearl millet. Since the release of the first hybrid HB 1 in 1965 based on the cms line Tift23A, as many as 50 hybrids have been released so far for general cultivation. Private sector role is no less in this regard. Inspite of its overall success, pearl millet hybrid programme has not been free from occasional setbacks, largely due to breakdown of resistance to the disease (Govila, 2001).

Hybrids HB3, HB4, BJ104, MB104 and MBH110 that were popular for 5-10 years in several parts of the country had to be successively withdrawn from cultivation, when they proved highly susceptible to downy mildew. Narrow genetic base of male sterility source *viz.*, Tift 23A is attributed to the vulnerability to the disease, though there is no convincing evidence to suggest association between male sterility and susceptibility to the disease. Equally, excessive dependence on very few restorers such as J104, K560, K230, etc. could have added to the vulnerability. Linkage or no linkage, experience shows that all the hybrids based on Tift 23A cytoplasm, become susceptible to the disease after a few years (Govila, 2001; Dave 1987). If this is true, then the presently popular hybrids such as ICMH451, Pusa 23, HHB67 and MLBH 104, again based on the same CMS source would also succumb to the disease some time later. Realizing that narrow genetic base of CMS-restorer sources is the cause of the recurrent problem of downy mildew, there is a serious effort to diversify them. Alternate sources like A_2 and A_3 characterized by Athwal at PAU, Ludhiana (Burton and Athwal, 1967) and A_4 by Hanna, USA (Hanna, 1989) are being evaluated for their stability of sterility expression and association of the disease susceptibility, though no hybrid based on them has as yet been released for commercial cultivation.

Another aspect that deserves serious breeding attention is adaptability of hybrids to availability of moisture for a very short period. Area characterized by such a situation is quite sizeable, especially western parts of Rajasthan, which receives less than 400 mm rainfall. To take advantage of resilience of hybrids under stress situations, there is need for concerted effort to develop extra early hybrids/composites (< 70 d) combining tolerance to moisture stress and resistance to downy mildew.

The following are some of the areas and issues that need focussed research

- Diversification of CMS - restorer lines insulated with resistance to downy mildew
- Development of appropriate hybrids for summer crop in Gujarat, which yield double that of main *kharif* crop.
- Development of extra early hybrids/composites combining high level of tolerance to moisture stress and resistance to downy mildew for large but very low rainfall (< 400 mm) semi-arid areas of Rajasthan.
- Enrichment of gene pools now in use by involving increasingly exotic germplasm and sources of value from secondary and tertiary gene pools. Besides natural variability and novel genes of interest be exploited through genetic engineering approaches.
- Improvement of nutritional value with special reference to β-carotene, exploiting both natural variability and novel gene(s).
- Development of true breeding hybrids exploiting the phenomenon of apospory form of apomixis known in pearl millet and its wild allies.
- Development of more productive fodder hybrids utilizing A_4 source of CMS.

Sustenance of Hybrid Technology in New Crops of Proven Potential

Hybrid technology is a commercial success in rice, cotton, sunflower and castor. Its sustenance vital for achieving future production targets warrants further improvement, keeping in view the factors that impede wide adoption, as discussed cropwise below:

Rice: Contributing 55 % to cereal production, rice is the staple food for more than 65 % of our population and source of livelihood for 100 million rural households. A chronically food and rice deficit country, India became self-sufficient following development and extensive adoption of high yielding dwarf varieties since mid-sixties. The rice scenario in the nineties, when productivity growth declined to far below the required level and actual yield nearly approached potential yield in high productivity states, need has arisen for technologies capable of raising further the genetic yield level. Hybrid technology is one option. But research efforts were academic for long until the discovery of the stable CMS source in the wild species *O. sativa, spontanea* designated as 'wild abortive' or 'WA type' that made hybrid technology a commercial feasibility (Yuan, 1977). It is a landmark in the history of rice breeding. Using WA-based CMS lines such as V20A Chinese breeders started exploiting hybrid vigour successfully on commercial scale since late seventies. The yield advantage of a ton more than the best pure line variety in a given region has led China to plant hybrids over 55 % (18 m ha) of its rice area. The success story prompted IRRI and countries like India to develop appropriate WA-based CMS lines well adapted to typical tropical conditions. Two CMS lines developed by IRRI, *viz.* IR58025A and IR62829A provided the base for India to launch hybrid breeding programme on a network mode in 1989-90. By 1995, the network came up with half a dozen of first generation hybrids (APRH1, APRH2, TNRH1, KRH1, CNRH1, DRRH1) with reasonably productive package of hybrid seed production (Siddiq *et al.*, 1995). Convinced of their yield advantage of a ton over the best variety, the country envisaged to take the technology to a million hectares by the year 2000. Another six heterotic hybrids including a couple from the private sector (KRH2, Pant Sankar Dhan1, CORH2, Sahyadri, Narendra Sankar Dhan2, PHB71) added subsequently, however, could not help achieve the targeted level of adoption. Going by seed production estimated for 2001-2002 by the private sector, which account for more than 90 % of the hybrid seed being produced and marketed, area coverage under hybrids is expected to be around 0.2 m ha in *kharif* 2002.

Inspite of sizeable yield advantage the pace of adoption of hybrids is quite slow, because of the following among many reasons:

- Focus on 'target environment' instead of 'target area' approach for intensive research and development efforts (area specific material generation, intensive testing, development of package of cultivation and effective transfer of the technology).
- Lack of hybrids of appropriate maturity to major rice belts (coastal Andhra Pradesh, Tamil Nadu, north and west India and *Boro* season in the eastern region), nearly all the first and second generation hybrids being early or medium early.
- Failure to produce and supply targeted quantity of hybrid seed.
- Non-maintenance of genetic purity of seed.
- Consumer resistance to quality in traditional areas owing to stickiness and aroma (the widely used male sterile line *viz*., IR 58025 A being of a low amylose aromatic rice).
- High quality rices scoring over hybrids on account of higher premium and hence higher return to the grower (The variety BPT 5204 gets premium price in Andhra Pradesh nullifying the yield advantage of hybrids) and
- Susceptibility to major diseases and insect pests.

Keeping these deficiencies in view, corrective research/development measures have been undertaken during the last three years. In respect of focussed area approach, two most representative districts in each of the target states have been chosen for intensive research, testing, training and dissemination of technology. Research efforts to evolve appropriate parental lines to fulfill the requirement of medium and medium late hybrids combining preferred aroma-free non-sticky cooking quality have been quite rewarding. Medium duration CMS lines having high amylose content such as PMS3A, PMS10A, PMS12A, APMS5A, CRMS31A, CRMS32A, IR68888A and IR68897A are now being widely used at southern and eastern network centres (Table 2). Several hybrids in the pipeline are expected to help correct to a great extent duration-quality related deficiencies soon. One of the achievements in this regard is the development of PRH-10, the world's first hybrid of Basmati quality. Maturing in 110 days, the hybrid outyields Pusa Basmati-1 by over 30 per cent.

Table 2. Promising CMS lines in rice

Early duration - non aromatic (DFF-90-95d)		Medium & Medium late - non aromatic (DFF-98-105d)	
IR 68888A	Irrigated	H*CMS* 2-1A	Irrigated
IR 68897A	Irrigated	H*CMS* 1-2A	Irrigated
IR 68886A	Irrigated	CRMs 31A	Shallow lowland
DRR 2A	Irrigated	CRMs 32A	Shallow lowland
Pusa 5A	Irrigated	PMS 11A	Irrigated
COMs 9A	Irrigated	PMS 14A	Irrigated
	Irrigated	APMS 6A	Mainly rainfed

Late - non aromatic (DFF > 110 d)

As far as the maintenance of genetic purity of seed is concerned, an integrated approach involving traditional nucleus seed production manually or by paired crossing, maintenance of proper isolation and grow out test for production of breeder/foundation seed and adoption of molecular marker technology for ensuring purity of cms, maintainer and restorer lines is required. Inadequate seed supply is a problem with only public bred hybrids. A consortium of seed producers being contemplated for production and supply of foundation seed of public hybrids to small private sector (small seed companies, NGOs, farmer cooperatives) and to large public sector (NSC, SFCI), seed production agencies become effective, seed supply would cease to be a constraint to wide adoption of the technology.

Besides the correction of above listed technological deficiencies, sustenance of the technology would also depend on the following:

- Enhancement of the magnitude of yield heterosis
- Development of hybrids for favorable shallow lowland ecology (3 m ha) and irrigated *boro* season (2 m ha) in eastern India
- Improvement of grain quality keeping in view regional preference and export demand.
- Insulation of all future hybrids with desired level of resistance to major insect pests and diseases
- Ensure reasonable seed cost through enhanced seed yield and reduced cost of seed production.

Enhancement of yield heterosis is the foremost condition for sustenance of the technology. Among the breeding efforts underway to evolve progressively more heterotic hybrids, structured parental line improvement and intervarietal (*indica/japonica*) combinations exploiting wide compatibility gene through conventional 3-line as well as temperature sensitive male sterility-based 2-line approaches are important (Siddiq and Ali, 2000; Hoan *et al.*, 1999). Intervarietal 2-line hybrids now under advanced stages of testing like MLTG4/IR50, MLTG4/K Hamsa have an yield advantage of 15-20 % over Intervarietal (*indica/indica*) hybrids evaluated so far.

Seed cost is yet another determinant of the pace of technology. Research to reduce cost of seed production and package to increase seed yield are important for reducing the seed cost. Identification and use of locations/seasons ideal for realizing higher seed yield, emphasis on development of well nicking hybrids and package to minimize the use of GA3 are receiving priority to reduce the seed cost. Hybrid seed production by environment sensitive genic male sterility based two-line approach now under development and testing might help reduce the seed production cost substantially (Siddiq and Ali, 2000).

Cotton: Providing directly and indirectly livelihood to over 60 million people and accounting for about 30 % of our export earnings, cotton is the number one commercial crop of the country. India has the distinction of growing all the four spinnable lint-bearing species of *Gossypium*, *viz.*, tetraploid *G. hirsutum* and *G. barbadense* and diploid *G. arboreum* and *G. herbaceum*. Cotton is grown in three agroclimatic zones, *viz.*, as irrigated crop in the northern region, as rainfed crop in the central region and as mainly rainfed crop in the southern region. Rainfed cotton accounts for 65 % of the total area under the crop. Grown over 1.25 m ha producing 16.9 m bales, cotton productivity, though increased significantly since mid-sixties, is still far lower than that in many major cotton growing countries. Whatever increase was achieved, had been after the advent of hybrid cotton of long and extra long staple.

Existence of exploitable hybrid vigour in inter and intraspecific crosses has been known since long. Yet, commercial exploitation of it was not possible for long due to lack of male sterility- based hybrid seed production system. This situation continued till Indian breeders demonstrated the commercial feasibility of producing hybrid seed by manual emasculation- pollination. Surat centre under the leadership of C.T. Patel came out with 'Hybrid 4', the first ever released hybrid cotton in the world (Patel, 1971). The hybrid between the popular Indo-American extra long staple variety Gujarat 67 and the exotic *G. hirsutam* variety 'American Nectariless' yielded 90-100 q/ha under irrigated condition and 15-20 q/ha under rainfed condition. Following this several inter and intraspecific hybrids have been released by both public and private sector institutions. Among them, intra *hirsutum* hybrids JKHY1, NHH1, H6, PK1, Hy2, G.cot, Hs, MECH1 MECH11, Ankur 51, CICRNH1, NHH44, HH712, Somnath and interspecific hybrids Varalaxmi, OBS156, RHR 253, NHB1, KCH1, JK Hy1, DCH32, HB224, MECH12, MECH, NHB12 etc., are important. To take advantage of wider adaptability of *desi* cotton to biotic/abiotic stresses and suitability for open end spinning same strategy was employed to develop interspecific hybrids in diploid cotton as well. Hybrids G.cot DH7, G cot DH9 and three more have been released for central India and other parts of the country (Singh *et al.*, 1980; Bhale, 1987, Narayanan *et al.*, 1989; Joshi, 1997 and ICAC, 1997).

Today hybrids account for 45 % of the total area planted to the crop. Apart from high productivity, hybrids combining better fibre quality have helped greatly meet the country's need for long and extra long staple cotton to start with and medium and superior medium staple later, thus correcting the imbalances in the availability of fibre categories ranging from 20s to 80s counts. While meeting adequately the domestic need, higher production has enabled the country export sizeably cotton in various forms, especially as yarn, fabric and fibre.

Sustenance of hybrid technology in cotton requires addressing of factors that constrain production as well as marketing. High seed cost on account of excessive dependence on manual crossing for hybrid seed production is one serious problem. Use of male sterility system, either genetic or cytoplasmic-genetic, is one option to reduce the cost of seed production. Although both the sources of genetic and cytoplamic male sterility were known to exist, the former was exploited first. Double recessive male sterile source (*ms5, ms5, ms6 ms6*) developed by Weaver has been used first to reconstitute successfully the popular hybrids H4 and Varalaxmi followed by CPH2, MECH1, Ankur 15 and a few more. Subsequently, cytoplasmic-genetic male sterility system based on *G. harknessii* cytoplasmic sterility source was used taking advantage of restorer sources available in exotic genotypes of *hirsutum - barbadense* background. MECH-4 released for the central zone, CAHH 468 and a few more from the private sector are all CMS-based hybrids.

The second major problem is the slow pace of development and adoption of hybrid technology for the irrigated north and north-west India. Even a few early maturing hybrids such as PCHH1, HHH81 and Raj 16 developed under the special thrust programme of the ICAR, are not popular among farmers, as the yield advantage is not convincing enough in comparison to the popular high yielding varieties. Coupled with narrow growing season, the effective flowering period is limited to 45-50 days between 15th August and 25th September.

The third and persistent problem with no solution as yet in sight is heavy crop losses due to bollworm complex. Indiscriminate and excessive use of chemical pesticides are endangering the fauna and flora around, besides failing to protect the crop from the pest. The only option left today to manage the pest and minimize the crop losses is resorting to transgenic strategy. Crystal protein *Cry 1A(c)* based *Bt* transgenic cotton in the background of already popular hybrids like MECH12, MECH162 and MECH184 are under extensive field testing for efficacy, biosafety and agronomic performance in comparison to standard check hybrids. Their approval for general cultivation granted by the GEAC/ICAR might help substantially minimize the crop losses, while reducing the cost of production, over 45 % of which is on account of chemical pest control. The following are some of the thrust areas of research for sustained exploitation of hybrid technology.

- Strengthening of research for development of high yielding hybrids combining early maturity and resistance to leaf curl virus and bollworm complex for irrigated north and north-west India (limited growing season).
- Development of extra early hybrids well adapted to shallow and medium deep soils and moisture stress conditions for rainfed central and western India.
- Development and dissemination of transgenic *Bt* cotton hybrids all over to manage effectively the bollworm complex and thereby stabilize yield and reduce cost on plant protection.
- Augmentation of research to develop and use cytoplasmic-genetic male sterility-fertility restoration based 3-line hybrid technology and
- Domestic and export market driven research for development of future hybrids.

Sunflower: Cytoplasmic male sterility-based hybrid technology became a reality in the 1970s (Enns *et al.*, 1970; Kinman, 1970; Vranceanu and Stoenescu, 1970 and Leclercq, 1971). Extensive adoption of hybrids in preference to open pollinated varieties placed sunflower (*Helianthus annus* L.) among the major oilseed crops in the world. Though it is a recent introduction to India, its impressive yield advantage and suitability to different cropping systems, facilitated its rapid adoption, resulting in steady and significant increase in oilseed production and productivity during the last two decades. The hybrid technology that led to the release of several hybrids, notably KBSH1, PSFH67, PKVSH27, etc., by the public sector and DSH1, LDMRSH1 and LDMRSH3 by the private institutions was based on the widely used CMS source *PET1* (Virupakshappa *et al.*, 1998). Unlike other crop plants, sunflower has many diverse CMS sources of both spontaneous occurrences as in the cultivar *H. annuus* and alien origin (Miller and Fick, 1997). Besides alien species, wild ecotypes of *H. annuus* have also been found to be valuable sources of CMS (Tables 3 and 4).

Table 3. CMS lines of euplasmic origin in sunflower

Origin	FAO code
Spontaneous occurrence	
H. annuus - 397	ANN 1
H. annuus - 517	ANN 2
H. annuus - 519	ANN 3
NS- ANN 81	ANN 5
Interspecific crosses	
H. annuus/H. lenticularis	ANL 1
H. annuus/H. texanus	ANT 1

Table 4. CMS lines of alloplasmic origin in sunflower

Cytoplasm	FAO code
H. argophyllus	ARGI, ARG2, ARG4
H. anomalus	ARO1
H. bolanderi	BOL1, BOL2
H. exilis	EXI1
H. giganteus	GIG1
H. grosseserratus	GRO1
H. maximiliani	MAX1
H. niveus canescans	NIC1
H. petiolaris petiolaris	PEP1
H. petiolaris fallax	PEF1
H. petiolaris	PET1
H. praecox praecox	PRP1
H. rigidus	RIG1

The strength of the hybrid technology in this crop lies in the availability of four distinct heterotic groups, *viz*., (*a*) open pollinated varieties of Russia, (*b*) USA restorer group, (*c*) Romanian female and South African derivatives and (*d*) Argentina INTA-OP varieties. The Russian and Argentina INTA open pollinated varieties constitute a good source of maintainers, while the USA restorer stock is a potential source of restorers and disease - insect pest resistance. Study of crosses between subgroups reveal them to represent distinct heterotic groups. Efforts in India in this direction have been limited to grouping of germplasm based on oil yield, resistance to stresses, etc. and such groups do not constitute heterotic pools.

Among the crop introductions, sunflower, largely as hybrids, has been contributing sizeably to India's edible oil production. Promotion of the technology is therefore, vital for sustained self-sufficiency in edible oil. The following are some of the areas/issues that need due research attention:

- Stagnation and inconsistency of yield performance over locations.
- Continuous decline of average yield in most of the coordinated yield trials.
- Vulnerability to *Alternaria* and sunflower necrosis (Tobacco leaf streak virus).
- Exclusive dependence still on a single (CMS) source (*PET1*) and relatively narrow genetic diversity for seed yield and oil content.
- Slow pace of development of heterotic gene pools.
- Poor maintenance of genetic purity of parental lines of all popular hybrids.

Extending Hybrid Technology to New Prospective Crops

Although exploitable heterosis for grain yield is available in many more field crops, hybrid technology is yet to become a commercial reality in them either due to lack of suitable pollination control system or poor outcrossing affecting seed production. Prospective among them are Indian mustard (*Brassica juncea*), pigeonpea (*Cajanus cajan*) bread wheat (*Triticum aestivum*) and castor (*Ricinus communis*). While efforts continue to develop proven and widely adapted cytoplasmic-genetic male sterility system, feasibility of various other approaches including genetic male sterility, chemical hybridizing agents and genetically engineered male sterility are being explored as detailed below cropwise.

Indian Mustard: Oilseed *Brassica* is the second major oilseed crop of India next to groundnut and Indian mustard (*B. juncea*) accounts for the largest area under cultivation. Unlike other oilseed crops, although there has been a steady increase in its productivity level, the growth has been largely due to factors other than genetic enhancement. For over several decades, in spite of many released in the name of progressively higher yielding varieties, yieldwise they are not significantly superior to one another. It was this genetic yield stagnation that prompted study of the prospects of heterosis exploitation as demonstrated and commercialized in *Brassica napus* using CMS-based male sterility system. Yield heterosis is known to be as high as 100 % in Indian mustard and reportedly it is positively associated with the geographic diversity of germplasm. Constraint to heterosis exploitation, however, has been the lack of effective pollination control system. Theoretically, self-incompatibility (SI) found in some of the widely cultivated *brassicas*, is the best pollination control system and more economical than the CMS-based system of hybrid seed production. SI, although, a dominant trait over self-compatibility, can not be used in crops, where grain is the economic product, as F_1's will be self-incompatible. Among the various strategies being contemplated for exploitation of heterosis, CMS-based system and genetically engineered male sterility-fertility restoration system are considered promising.

Diverse CMS sources have been identified and usable CMS lines developed in *Brassica* oilseeds by substituting *Brassica* nuclear genome in alien cytoplasm (Table 5).

Table 5. Cytoplasmic male sterility systems in *Brassica* oilseeds

Cytoplasm	Crop
Ogura	*B. napus*; *B. juncea*
Polima	B. napus
Tournefortii	B. napus; *B. juncea*
Oxyrrihna	B. napus; *B. juncea*
Sifolia	B. napus; *B. juncea*
Moricandia	B. juncea
Trachistoma	B. juncea
Lyratus	B. juncea
Canariense	B. juncea; *B. napus*

Gene pools are defined mainly on the basis of geographic diversity. There are in all four geographic groups, *viz*., Indian, European, Asian and Canadian. Though geographic diversity and level of heterosis appear to be associated, there has been no attempt to develop hybrid breeding oriented heterotic gene pools. The first ever made attempt in this direction has been the programme underway to develop maintainer and restorer gene pools.

Alien cytoplasm-based *cms* lines were developed by both sexual and somatic hybridization (Kirti *et al*., 1991; 1995). Often CMS lines based on alien cytoplasm are handicapped with associated floral deformities, absence of nectaries, chlorosis, etc. because of incompatibility between the organelle genomes (*mt* and *ct*) of the allied species and nuclear genome of the cultivars. It is only a small fraction of the *mt* genome of the wild species that induces male sterility in interaction with the nuclear genome. However, while developing alloplasmic CMS lines inclusion of entire organellic genome of the wild species results in the defective and poor performing CMS lines. It would be worthwhile, if only the small sterility inducing portion of the genome of the *mt* of wild species is retained, substituting the rest of the *mt* with that of the cultivar, keeping the entire *ct* genome of the cultivar intact. This has been successfully achieved by somatic hybridization, which facilitates recombination between *mt* genomes and selection of recombinant mitochondria, having the region of interest from the wild species [Kirti *et al*.,

1992(a); 1992(b)]. Recombination between the *ct* genomes is a rare event and as a result somatic hybrids retain *ct* genome of the cultivar. In developing alloplasmic CMS lines devoid of associated floral deformities and chlorosis, somatic hybridization has been found handy.

Exploitation of hybrid vigour in *Brassicas* is so far restricted to only *B. napus*. Napus hybrids based on *polima* cytoplasm have been released for general cultivation in China, Europe, Canada and Australia, while *tournefortii* CMS-based hybrid PGSH51 is the first to be released in India. In the hybrids, in spite of yield advantage of 15-25 % over the best inbred varieties, their adaptability to a restricted area is proving to be a serious constraint to wide adoption. Hybrids in Indian mustard are not as yet a reality, largely due to lack of strong sources of fertility restoration to many of the CMS sources. Recent reports on the identification of restorer sources for CMS lines based on *moricandia*, *lyratus* and *canariense* cytoplasm are quite encouraging (Prakash *et al.*, 2001 and Banga and Banga, 1997). These lines were developed by selective introgression of fertility restorer gene(s) from respective cytoplasm donor species through homoeologous pairing. The chloloroplast deficiency associated with *mori* CMS could be rectified through protoplast fusion and genetic introgression from *moricandia* genome. Initial results on the performance of *mori* CMS-based hybrids have indicated stable expression of fertility restoration over a wide range of environments. The breakthrough has paved the way for accelerated hybrid breeding programme in Indian mustard.

Recombinant DNA technology has introduced a novel means of inducing male sterility across plant species by selectively disturbing one or the other stages of pollen development. Mariani *et al.* (1990) were the first to demonstrate in tobacco that plants spliced with a chimeric dominant gene *barnase* (bacterial RNAse from *Bacillus amyloliquefaciens*) driven by the tapetum specific promoter TA 29 from tobacco induce male sterility. They also demonstrated the possibility of restoration of fertility by a dominant gene *barstar* extracted from the same bacterium. The protein encoded by *barstar* suppresses the cytotoxic effect of *barnase* and restores thereby pollen fertility. Maintenance of male sterility requires, as in GMS system, a strategy to selectively remove fertiles, when it segregates in the ratio of 1F:1S. Linking of a dominant herbicide resistance gene '*bar*' with *barnase* (*Ms*) under a constitutive promoter (CaMV 35S) helps to eliminate fertiles when sprayed with the herbicide.

While *B. napus* hybrids based on the *barnase-barstar* system have been registered in Europe for commercial cultivation, mustard (*B. juncea*) hybrids similarly developed by Pro Agro are undergoing field evaluation at present in India. Modified *barnase-barstar* system in Indian varietal background such as *Varuna* is in the advanced stages of development at the Delhi University.

Commercial realization of hybrid mustard requires more intensive research and testing. The following are some of the aspects that need attention.

- Identification of strong restorer sources to stable cms lines now available.
- Broadening of genetic base through enrichment of source nursery with exotic

germplasm for yield vigour, keeping in view the fact that the available genetic variability in Indian mustard is too inadequate to realize the desired level of yield heterosis and

- Development of '00' hybrids either by conventional CMS-based male sterility or genetically engineered male sterility in Indian mustard with emphasis on seed yield and percentage oil content

Wheat: Being the second most important food crop next to rice, steady production growth of wheat is vital for the country to remain food secure. With the present pace of annual yield growth at around 1 %, achievement of the targeted 110 m tonnes by 2020 would not be an easy task. Since breaking of yield barrier twice earlier — first through Mexican dwarf varieties like 'Sonara 64' and later by '*Veery*' varieties, further rise in genetic yield level requires a different strategy — either a new plant type or hybrid technology (Mahajan and Nagarajan, 1998 and Mahajan *et al.*, 1999). Wheat is one of the earliest crops, wherein prospects of hybrid vigour exploitation was explored way back in the fifties. Identification of cytoplasmic male sterility sources like *T. timopheevi*, no doubt, had enabled development of 3-line hybrid breeding long back. But on account of poor seed yield attributable to low percentage outcrossing and high seed rate required, the technology could not be commercialized for long. In the seventies the strategy of use of gametocides was explored again with limited success. In the last 10 years, however, significant progress has been made under the Network Research on Hybrid Wheat, towards development of gametocide-based hybrid technology. Among 50 chemical hybridizing agents (CHA) developed and evaluated by the Directorate of Wheat Research, Karnal in collaboration with the National Chemical Laboratory, Pune, CH9701, CH9702, CH9078, CH9831 and CH9832 have been found promising (Mahajan *et al.*, 1997; 2000). Effective package for hybrid seed production comprising optimum dose and crop stage for CHA application has been developed. Evaluation of about 50 test hybrids produced using the CHA, revealed three (HM 99168, HM 99160 and HM 9997) to yield 15 % higher than the standard variety (Ganga Rao *et al.*, 2000).

Indian Agricultural Research Institute, Delhi, Punjab Agricultural University, Ludhiana and Haryana Agricultural University, Hisar as well as Mahyco Seed Company in the private sector are engaged in hybrid breeding through development and use of cytoplasmic male sterility-fertility restoration system. As many as 10 CMS lines (Lok 1 KMS 9A, 2009 KMS 9A, 2038A, 2046 KMS 9A, 2041 KMS 9A, 2042 A, 2046 KMS 20A, 2019 KMS 19A and 2160 A) based on diverse cytoplasmic male sterility sources viz., *T. timopheevi*, *T. araraticum*, *T. zhukovskyi*, *Aegilops speltoides*, *Ae. squarrasa* and *Ae. caudata* have been developed at IARI, New Delhi and using most stable CMS lines among them a large number of test crosses have been effected for yield assessment and seed production (S.M.S. Tomar, personal communication). Exploitable yield heterosis has been found to be as high as 41 % (Zehr *et al.*, 1997). Significantly, Mahyco has already come out with India's first commercialized hybrids 'Pratham 7050' and 'Pratham 7070' for general cultivation. They are reported to yield one ton more per hectare than the best variety in the central and eastern wheat belts. Added to higher yield combined with tolerance to brown rust disease they enjoy 3 % higher premium in the grain market because of their lustrous amber colour grains and highly preferred *chapati* making quality. Wide adoption in the west and east has now convinced the traditional north and north-western regions also to gradually adopt them.

The momentum gained in the last five years towards development of hybrid technology by both CHA and CMS approaches may be strengthened, given the encouraging findings. The notion that exploitable hybrid vigour may not be high enough, as bread wheat is itself a natural hybrid (amphidiploid) does not seem to be valid, if reports of as much hybrid vigour as 40 % are any indication. Resynthesized strains of hexaploid wheat and transfer of still not exploited yield genes from winter wheat can form a potential source nursery for developing hybrids with higher yield advantage.

Pigeonpea: One of the major pulse crops, Pigeonpea [*Cajanus cajan* (L.) Millsp.] is cultivated over a wide range of cropping systems in the tropics and subtropics. India alone accounts for 90 % of the world acreage and production. Inspite of decades of breeding research, yield remains stagnant for long at very low levels. Most of the popular cultivars are pureline selections from the local varieties.

Exploitation of heterosis as a means to raise the genetic yield level started with the isolation of genetic male sterile lines *viz*., *Ms3A* and *Ms4A* from the germplasm accessions ICP 1555 and ICP 1696, respectively (Reddy *et al.*, 1978). The single recessive gene controlled male sterility (*ms1*) associated with the visible marker 'white translucent anther' with no pollen enables easy identification and removal of fertile segregants at flowering. *B 15 B* is yet another genetic male sterile source identified in Australia (Wallis *et al.*, 1981). Though it is also single recessive gene controlled, it is nonallelic (*ms2*) to *ms1*, capable of masking the expression of *ms1*. The third source identified as a spontaneous mutant in the popular variety UPAS 120 proved valuable in developing the first ever genetic male sterility-based pigeonpea hybrid ICPH8 (Ms Prabhat DT/ICPH 161) by ICRISAT (Verulker and Singh, 1997). Extensive yield trials conducted in the three zones of India during 1983-1987 revealed ICPH 8 to show an yield advantage of 22-53 % over the best check variety. The heterotic response is attributed to higher biomass, resulting in consistently higher growth rate. The extent of natural outcrossing being adequate, hybrid seed production is considered viable from field, operational and economic angles. Since its release for general cultivation, several genetic male sterility based hybrids have been commercialized. But their genetic base being quite narrow no further progress in yield could be achieved.

Research efforts to identify cytoplasmic male sterility sources are on since the beginning. The strategy has been to substitute the nuclear genome of euplasmic pigeonpea into the cytoplasm of wild relatives through backcrossing. Wild species like *C. sericeus, C. acutifolius, C. cajanifolius* and *C. albicans* have been used for developing alloplasmic lines of *C. cajan*. CMS lines have been developed by two methods, *viz.(a)* wide hybridization involving *C. sericeus* and *C. cajan* and substitution backcrossing (Van der

Maesen, 1990) and (*b*) multiple cross genome transfer. The male sterility is maintainable by pigeonpea genotypes ICPL 85030 and ICPL 90035 of ICRISAT and GT 288A, the first CGMS stable line developed by using *Cajanus scarabaeoides* at GAU, Gujarat (Tikka *et al.*, 1997). To make CMS-based hybrid technology economically sustainable, accelerated research for parental line improvement for high combining ability for yield heterosis and plant type to facilitate easy hybrid and CMS seed production is inevitable. Development of heterotic gene pools using genetic male sterility would help greatly broaden the genetic base of source nursery for yield and tolerance to biotic/abiotic stresses.

Castor: A minor oilseed crop, castor (*Ricinus communis* L.) is the most preferred one in drought prone regions of the country. It was one among the very low yielding crops until the advent of hybrid technology. Release of the Pioneer hybrids GCH 4 and CGH 5 has been a breakthrough in castor production and productivity in the irrigated areas of Gujarat and Rajasthan. Unlike in other field crops, wherein either cytoplasmic-genetic or genetic or manual pollination control system is the practice, hybrid technology in castor is based on pistillate sources available in the otherwise monoecious crop. Multiplication of pistillate line and hybrid seed production are done by adjusting flowering phase with periods of conducive temperature.

There are three different pistillate sources, *viz.* N, S and NES types. The N type pistillate line governed by a recessive sex-switching gene (*ff*) is maintained by sib-mating in a progeny of pistillate plants (seed set by interspersed staminate flowers), which segregate in the ratio of 1 pistillate to 1 monoecious plant. (Claussen and Hoffman, 1950 and Katayama, 1957). The S-type pistillate line is characterized by the phenomenon of sex reversal, i.e. female to monoecious at any time after the appearance of first receme (Shifriss, 1960 and Brigham and Spears, 1961). Repeated inbreeding of the sex reversal types results in a spectrum of types, each reverting at different times of development and monoecists. The third type *viz.*, NES is homozygous for N-pistillate gene along with environment sensitive gene(s) for interspersed staminate flowers (Zimmerman and Smith, 1966 and Ankineedu and Rao, 1973). Yet another system is an induced mutant having male flowers with fused parianth (Subramanian and Sivasubramanian, 1975), which prevents pollen shedding and hence usable as a stable female line in commercial seed production. In India, S-type pistillate lines are the widely used female parents in many of the commercial hybrids. Wilt resistant pistillate lines based on S-type, NES type and the mutant VP-1 that are more stable in their sex expression are increasingly used in commercial hybrid seed production. Many hybrids based on them are under extensive evaluation. Besides adjustment of planting of parental lines to periods of ideal temperature, application of GA3 at 2-4 leaf stage is reported to increase female tendency in primary, secondary and tertiary recemes (Shifriss 1961), while use of ethyl hydrogen-1-1propyl phosphate increases male tendency (Philipos and Narayanaswamy, 1976).

Hybrids significantly out yield OP varieties. The yield advantage is not exclusively due to heterosis. Improvement of inbred lines has as well contributed substantially to the yield increase. Popular hybrids like GAUCH 1, DCH 32 and DCH 177 besides the first generation hybrids GAUCH 4 and GAUCH 5 have contributed immensely to production and productivity of castor. Nevertheless, the fact remains that hybrid technology is not as yet perfect enough. The following are some of the issues that need research attention to sustain the hybrid technology.

- The yield advantage in hybrids is still largely on account of potential of the inbred lines, warranting development of more heterotic pools for realizing higher yield heterosis.
- The gains of hybrid technology are impressive under irrigated conditions only. The technology is yet to make its impact in the drought-prone rainfed areas.
- Pistillate line multiplication and hybrid seed production have serious problems of proper seed set, dependence on interspersed staminate flowers for multiplication of pistillate lines and pollen source for hybrid cultivation. Often hybrids are not 100 % hybrids. Serious research is warranted to prefect the pollination control systems in use.
- It is worth searching for cytoplasmic-genetic male sterility source.

Conclusion

Hybrid technology is one proven approach India has experienced over the last five decades. Inspite of all research and development efforts, beyond millets, the technology has not made its impact to the desired extent, given the low and fluctuating percentage adoption. Continued dependence on the use of genetic diversity started decades back in heterosis breeding, excessive reliance invariability on a single source of cytoplasmic male sterility in all crops, weak mechanism for maintenance of genetic purity of seed and lack of well demarcated target area/ecology for focused research and development are broadly the reasons that stand in the way of reaping the full potential of hybrid technology. Integrated research involving conventional and innovative approaches and improvement of seed production supply system are the means for sustained exploitation of the technology.

References

Ankineedu G. and N. Ganga Prasad Rao. 1973. Development of pistillate castor, *Indian J. Genet.*, **33**: 416-422.

Banga S.S. and Banga S.K. 1997. *Enarthrocarpus lyratus* cytoplasm causes male sterility in oilseed rape. *In*: *Abstracts Int. Symp. Heterosis in Crops*, Mexico, pp. 120-121.

Bhale N.L. 1987. Hybrid cottons in India. *FAO Consultancy Report*, CICR Nagpur, India.

Brigham R.D. and Spears B.R. 1961. Castorbeans in Texas. *Texas Agr. Expt. Sta. Bul.*, 954.

Burton G.W. and Athwal D.S. 1967. Two additional sources of cytoplasmic male sterility in pearlmillet and their relationship to Tift 23A. *Crop Sci.*, **7**: 209-211.

Classen E. and Hoffman A. 1950. The inheritance of pistillate character in castor and its possible utilization in the production of hybrid seed. *Agron. J.*, **42**: 79-82.

Dave H.R. 1987. Pearlmillet hybrids. *In*: *Proc. of the International Pearl millet Workshop*, 7-11 April, 1986, pp. 139-142, ICRISAT, Patancheru, A.P. 502324, India.

Enns H., Dorrell D.C., Hoes J.A. and Chubb W.O. 1970. Sunflower research. A progress report. *In*: *Proc. 4th Int. Sunflower Conference*, Memphis, Tennessee, pp. 162-167.

Ganga Rao N.V.P.R., Mahajan V. and Nagarajan S. 2000. Effect of chemical hybridizing agent on floret opening and seed set in wheat (*Triticum aestivum*). *Indian J. Agric. Sci.*, **70**: 689-670.

Govila O.P. 2001. Pearl millet. *In*: *Breeding Field Crops*. (ed.) V.L. Chopra. Oxford and IBH Publishing Co. Pvt. Ltd., pp. 211-234.

Govila O.P., Rai K.N., Chopra K.R., Andrews D.J. and Stegmeier W.D. 1997. Breeding pearlmillet for developing countries : Indian Experience. *In*: *Proc. Int. Conference on Genetic Improvement of Sorghum and Pearl millet*, Sept. 23-27, Lubbock, Texas, pp. 97-118.

Hanna W.W. 1989. Characteristics and stability of a new nuclear cytoplasmic male sterile source in pearl millet. *Crop Sci.*, **29**: 1457-1459.

Hoan N.T., Sarma N.P. and Siddiq E.A. 1999. Identification and characterization of new sources of cytoplasmic male sterility in rice. *Plant Breeding*, **116**: 547-551.

ICAC 1997. Commercial cotton hybrids. *ICAC Recorder*, June 3-14.

Joshi M. 1997. *Hybrid cotton in India*. Kalyani Publisher, Ludhiana, India, pp. 1-18.

Katayama Y. 1957. The progeny of female individuals in castor beans. *Jpn. J. Genet.*, **23**: 19.

Khairwal I.S., Rai K.N., Andrews D.J. and Harinarayana G. 1999. *Pearl millet Breeding*. Oxford and IBH Publishing Co. Pvt. Ltd., New Delhi.

Kinman M.L. 1970. New developments in the USDA and state experiment station sunflower breeding programmes. *In*: *Proc. 4th Int. Sunflower Conference*, Memphis, Tennessee, pp. 181-183.

Kirti P.B., Banga S.S., Prakash S. and Chopra V.L. 1995. Transfer of *ogu* cytoplasmic male sterility to *Brassica juncea* and improvement of male sterile line through somatic cell fusion. *Theor. Appl. Genet.*, **91**: 517-521.

Kirti P.B., Narsimhulu S.B., Prakash S. and Chopra V.L. 1992a. Production and characterization of somatic hybrids of *Trachystoma ballii* and *Brassica juncea*. *Plant Cell Rep.*, **11**: 90-92.

Kirti P.B., Narsimhulu S.B., Prakash S. and Chopra V.L. 1992b. Somatic hybridization between *Brassica juncea* and *Moricandia arvensis* by protoplast fusion. *Plant Cell Rep.*, **11**: 318-321.

Kirti P.B., Prakash S. and Chopra V.L. 1991. Interspecific hybridization between *Brassica juncea* and *B. spinencens* through protoplast fusion. *Plant Cell Rep.*, **9**: 639-642.

Leclercq P. 1971. La sterilite male cytoplasmique du la fertilite. *Ann. Amel. Plant*, **21**: 45-54.

Mahajan V. and Nagarajan S. 1998. Opportunities in hybrid wheat - a review. *PINSA*, **B64**: 51-58.

Mahajan V., Nagarajan S., Deshpande V.H. and Kelkar R.G. 2000. Screening chemical hybridizing agent for development of hybrid wheat. *Current Sci.*, **78**: 235-236.

Mahajan V., Nagarajan S., Srivastava M., Kumar V. and Ganga Rao N.V.P.R. 1999. Commercial heterosis in wheat - an overview. *Rachis*, **18**: 13-16.

Mahajan V., Singh K. and Kelkar R.G. 1997. Possibility of using CHA route in developing wheat hybrids, pp. 101-106. *In*: *Wheat Research Needs Beyond 2000 AD*, Narosa Publishing House, New Delhi.

Mariani C., Beuckeleer M.D., Truettner J., Leemans J. and Goldberg R.B. 1990. Induction of male sterility in plants by chimeric ribonuclease gene. *Nature*, **347**: 737-741.

Miller J.F. and Fick G.N. 1997. The genetics of sunflower. *In*: *Sunflower Technology and Production*. (ed.) A.A. Schneiter, ASA, CSSA, SSSA; Madison, WI, USA, pp. 441-495.

Narayanan S.S., Singh P. and Bhat M.G. 1989. Heterosis breeding in cotton. *In*: *Heterosis Breeding*, (ed.) R.C. Rakshit, pp. 37-38. BCKV, Kalyani, West Bengal, India.

Patel C.J. 1971. Hybrid 4 - A new hope towards self sufficiency in cotton in India. *Cot. Dev.*, **1**: 1-6.

Philipos H.V. and Narayanswamy S. 1976. Induced potential for male sex expression in *Ricinus Communis* L. By ethyl hydrogen-1-propythophonate. *Protoplasma*, **87**: 71-77.

Prakash S., Ahuja I., Bhatt S.R., Kirti P.B. and Chopra V.L. 2001. Expression of male sterility in alloplasmic *Brassica juncea* with *Erucastrum canariense* cytoplasm and development of a system for fertility restoration. *Plant Breeding*, **120**: 1-5.

Rana B.S., Rao M.H., Indira S., Singh B.U., Chari Appaji and Vilas Tonapi. 1999. Technology for increasing sorghum production and value addition. *National Research for Sorghum (ICAR)*, Rajindranagar, Hyderabad 500030, pp. 1-67.

Reddy B.V.S., Green J.M. and Bisen S.S. 1978. Genetic male sterility in pigeonpea. *Crop Sci.*, **18**: 362-364.

Shifriss O. 1960. Conventional and unconventional system controlling sex variations in *Ricinus. J. Genetics*, **57**: 361-368.

Siddiq E.A. and Ali Jauhar. 2000. Innovative male sterility system and exploitation of hybrid vigour in crop plants I. Environment sensitive genic male sterility system. *PINSA*, **65**: 320-339.

Siddiq E.A., Ahmed M.I., Viraktamath B.C., Ali J. and Hoan N.T. 1995. Status of hybrid rice research in India. *In*: *Hybrid Rice Research and Development*. Indian Society of Seed Technology, New Delhi, pp. 139-158.

Singh N.N., Gupta N.P. and Mauriya S. 1996. Maize. *In*: *50 Years of Crop Science Research in India*. (eds.) R.S. Paroda, K.L. Chadha, ICAR, New Delhi, pp. 252-264.

Singh T.H., Chahal G.S., Bharadwaj H.C. and Tikku P.L. 1980. Exploitation of heterosis in cotton - A review. *Indian J. Soc. Cotton Improvement*, **19**: 113-123.

Subramanian M. and Sivasubramanian P. 1975. X-ray-induced fused perianth in *Ricinus communis* L. *Radiate. Bot.*, **15**: 397-408.

Tikka S.B.S., Parmar L.D. and Chauhan R.M. 1997. First record of cytoplasmic-genic male sterility system in pigeonpea [*Cajanus cajan* (L.) Millsp.] through wide hybridization. *GAU Res. J.*, **22**: 160-162.

Van der Maesen. 1986. Pigeonpea: Origin, History, Isolation and Taxonomy. *In*: *The Pigeonpea*. (eds.) Y.L. Nene, S.D. Hall and V.K. Sheila. CAB Internl., ICRISAT, pp. 15-46.

Verulker S.B. and Singh D.P. 1997. Inheritance of spontaneous male sterility in pigeonpea. *Theor. Appl. Genet.*, **94**: 1102-1103.

Virupakshappa K., Ranganatha A.R.G. and Reddy B.N. 1998. Hybrid sunflower seed production. Directorate of Oilseed Production Technology. Directorate of Oilseed Research, Hyderabad, pp. 1-99.

Vranceanu A.V. and Stoenescu F.M. 1970. Testing sunflower cytoplasm in relation to genes for male sterility. *Plant Breed. Abstr.*, **43**: 3837.

Wallis E.S., Saxena K.B. and Byth D.E. 1981. A new source of genetic male sterility in pigeonpea. *In*: *Proceedings of the International workshop on pigeonpea*, Vol. 2, 15-19 December 1980, ICRISAT Centre, Patancheru, India, ICRISAT, pp. 105-108.

Yuan L.P. 1977. The execution of theory of developing hybrid rice. *Chinese Agric. Sci.*, **1B**: 27-31 (In Chinese).

Zehr D.E., Ratnalikar V.P., Reddy L.M.M. and Pandey L.V. 1997. Strategies for utilizing heterosis in wheat, rice and oil seed *Brassica* in India. pp. 232-233. Book of abstract: The genetics and exploitation of heterosis in crops. *In*: *An International Symposium on Heterosis in Crops*, 17-22 August, 1997, Mexico City, Mexico.

Zimmerman L.H. and Smith J.D. 1996. Production of F_1 hybrid seed in castor beans by use of sex genes sensitive to environment. *Crop Sci.*, **6**: 406-409.

Plant Breeding - Mendelian to Molecular Approaches
H. K. Jain and M. C. Kharkwal (eds.)

20

Cytoplasmic-Nuclear Male Sterility: Origin, Evaluation and Utilization

B.V.S. Reddy[1], K.N. Rai[1], N.P. Sarma[2], Ish Kumar[3], and K.B. Saxena[1]

Abstract

Cytoplasmic-nuclear male sterility (CMS), a potential system for economical hybrid seeds production, results from disharmonious interaction between the cytoplasmic and nuclear genetic factors. Evaluation of CMS system for its utility is determined by many factors. Unstable male sterility increases cost and affects quality of hybrid seed production. Cooler weather during panicle development promotes instability in maize and pigeonpea, high humidity in pearl millet and hot temperature of 42 ^{o}C and above in sorghum. Effect of genetic background of B-lines on male sterility has been found in case of A_1 CMS system in pearl millet and sorghum, and fertility restorer gene expressivity in rice. Maintainer gene frequency in germplasm is another factor to be considered which has bearing on genetic diversification of A-lines. In case of A_1 CMS system in sorghum and pearl millet landraces have greater proportion as restorers. Character association influences the commercial viability of a CMS system. Approaches to estimate the diversity of CMS systems to enhance the exploitation of outstanding inbred lines and also to overcome the risk of disease and pest epidemics, include both traditional field evaluation of hybrids for differential patterns of male fertility restoration and RFLP analysis of mitochondrial DNA. Steps involved in CMS utilization in hybrid cultivar breeding including (*i*) identification of potential CMS sources, (*ii*) development of seed parents (A lines) exhibiting complete sterility, (*iii*) maintenance of seed parents by taking appropriate steps to prevent contamination with foreign pollen, (*iv*) development of restorer parents through identification of promising restorers, genetic improvement of existing restorers and transfer of restorer genes into high yielding inbred lines and (*v*) hybrid seed production keeping in view the four key elements - isolation distance, off type removal, flowering synchrony between A- and R-lines, and A:R lines ratio are discussed.

---

[1]International Crops Research Inst. for the Semi-Arid Tropics, Patancheru 502 324, India
[2]Directorate of Rice Research, Hyderabad 500 030, India
[3]Hybrid Rice International Ltd., Hyderabad 500 008, India

Introduction

Heterosis for grain and biomass yield has been observed in all crops regardless of their breeding systems. In a highly cross-pollinated crop such as pearl millet [*Pennisetum glaucum* (L.) R. Br.] (85 % outcrossing), better-parent heterosis of up to 424 % has been reported for grain yield of single-cross hybrids (Virk, 1988). Such a high degree of heterosis for grain yield in pearl millet is a consequence of high degree of inbreeding depression (Rai *et al.*, 1985). Hybrids as a cultivar option are determined more by their yield advantage over improved varieties than by yield advantage over their inbred parents. It has been observed that highest yielding single-cross hybrids of pearl millet have 20-30 % grain yield advantage over the highest yielding open-pollinated varieties of comparable maturity. In predominantly self-pollinated crops such as pigeonpea, *Cajanus cajan* (L.) Millsp. (15-30 % outcrossing) and sorghum, *Sorghum bicolor* (L.) Moench, the natural cultivars are homozygous. In cultivated sorghum, depending on cultivars and climatic conditions (particularly wind), the outcrossing ranges from 2 to 15 % (House, 1985). In wild sorghum, it is higher than in the cultivated sorghum (Reddy, 1997). In pigeonpea, the heterosis in hybrids has been reported to be 20-100 % over the best parent of similar maturity (Saxena, 2001a) while in sorghum, it is 15-48 % (Maunder, 1972). Asian cultivated rice (*Oryza sativa* L.) is strictly self pollinated (highly autogamous). Natural outcrossing is extremely low (1 %). Depending on environmental conditions and cultivars, the extent of outcrossing may vary from 0 to 6.8 % (Sahadevan and Namboodri, 1963). Through natural cross-pollination, percentage of florets setting seed vary between 0 to 44 % (Athwal and Virmani, 1972; Carnhan *et al.*, 1972). Commercial rice hybrids are reported to show 20-30 % standard heterosis for grain yield (Lin and Yuan, 1980; Yuan *et al.*, 1989). The yield advantage in hybrids as percent over high yielding variety has been reported to be 32 % (Virmani, 1996). This order of yield advantage in hybrids is of considerable economic significance, providing the basis for enhanced hybrid research and development efforts.

Hybrid cultivar development requires additional research investment over those needed for non-hybrid cultivars for developing suitable parental lines. The advantage with hybrid technology, however, is that relatively greater number of players find it attractive to invest in it. This is particularly true of private sector, which has been emerging as an important partner in hybrid research and development. Besides the yield advantage, there are several other factors that determine the success of hybrid programs in any crop. The foremost of these is the identification and development of a biological system that helps in economic production of hybrid seed on a commercial scale. Systems such as genetic male sterility in sorghum and pigeonpea, and temperature-sensitive genetic male sterility (TGMS) in rice, have been developed, but at present these do not provide either the economy of scales or the flexibility of choice of locations and seasons during the hybrid seed production in tropical countries. Cytoplasmic-nuclear male sterility (CMS) provides a system that potentially overcomes the above problems, leading to a more reliable and economical hybrid seed production technology. CMS systems and hybrid seed production technologies in crops like maize (*Zea mays* L.), pearl millet and sorghum were developed in early 1960s and became a matter of routine commercial application. Development of such technologies in other crops such as rice (Lin and Yuan, 1980) and pigeonpea (Saxena, 2001a), is a relatively recent effort. The objective of this paper is to review the CMS research relevant to the development of high yielding hybrids in pearl millet, sorghum, rice and pigeonpea.

Origin

CMS is a physiological abnormality, resulting from a disharmonious interaction between the cytoplasmic factors (now widely identified as mitochondrial genetic factors) and nuclear genetic factors, leading to the production of degenerated or non-viable pollen grains or non-dehiscent anthers with or without functional pollen grains. Understandably, this disharmonious interaction is likely to be more pronounced in populations incorporating divergent sources of cytoplasm and nuclear genes. Thus, in natural landrace populations, CMS is likely to occur more in cross-pollinated crops than in self-pollinated crops. Also, biologically speaking, male-sterility is a negative trait, and as such natural selection pressure works against this trait. As a result, it is more likely to be lost in self-pollinated crops than in cross-pollinated ones.

Cultivated Germplasm

Segregating populations derived from hybridization programs or composites of hybrid origins serve as useful materials to search for CMS sources, regardless of the breeding system of the crop. Several CMS sources in maize (Duvick, 1965) and the first commercially viable A_1 CMS sources in both sorghum (Stephens and Holland, 1954) and pearl millet (Burton, 1958) were identified in F_2 populations derived from crosses between inbred lines of diverse origin. For instance, the A_1 CMS source in pearl millet was identified in a F_2 population derived from a cross Tift 556 × Tift 23. Similarly, the A_1 CMS source in sorghum was identified in the F_2 population of cross Double Dwarf Yellow Sooner Milo × Texas Blackhull Kafir by Stephens and Holland (1954), where the milo inbred belongs to durra race from Sudan and Ethiopia border (Duncan *et al.*, 1991) and the *Kafir* inbred from Eastern Africa (House, 1985).

A partial listing of 54 CMS sources in maize (Duvick, 1965) showed them having occurred in a wide range of cultivated maize germplasm, mostly open-pollinated varieties. The first two discoveries of CMS sources (Peruvian and Argentinian) in 1930s and the once most commercially used source (Texas A&M) in maize were all derived from open-pollinated varieties. Five sources out of 32 tested (4 USA + 1 Mexico) by Duvick (1965) resembled the Texas CMS source, and the remainder the USDA - *iojap* source. They could not distinguish the two other sources tested. A similar search in pearl millet led to the identification of various CMS sources, which is not unexpected as pearl millet is also an outbreeder (similar to maize) and it is endowed with as much variability, both within and between populations, especially in the western African region. Madhava Menon (1959) identified a CMS source in an open-pollinated landrace variety (PT 819) from Bellary district of the then Mysore state. Athwal (1961) identified a CMS source (later designated as A_2) in IP 189, a landrace population from western Africa. Later, several CMS sources were identified in germplasm collections from Ghana and Botswana (Appa Rao *et al.*, 1989) and in broad-based genepools (Rai and Hash, 1993; Rai, 1995). This shows that open-pollinated varieties and broad-based genepools in cross-pollinated crops provide promising materials, where search for CMS sources could be made. In sorghum, a predominantly self-fertilizing crop, Schertz and Pring (1982) listed more than 40 CMS sources - all from cultivated sources. Nagur and Menon (1974) studied several of these sources and recognized four different classes based on fertility/sterility response in hybrids. These were further studied by Reddy (1992) in an effort to classify them and to find minimum differential testers. These were designated as A_1 (CK 60; origin East Africa), A_2 (IS 12662C; origin Ethiopia), A_3 (IS 1112C; origin India) and A_4 (M35-1, VZM 2 and G1; origin India). Those in the A_4 CMS system were further classified, based on RFLP analysis, into three separate cytoplasms (Sivaramakrishnan *et al.*, 1997). Further, a male-sterile plant was discovered by Webster and Singh (1964) in 9E, a selection made in Ghana. Schertz and Pring (1982) described several analogues of these cytoplasms in sorghum. In pigeonpea, a predominantly self-fertilizing crop as sorghum, no CMS source has yet been reported in the cultivated or landrace varieties. In rice, the role of cytoplasm causing male sterility was first reported by Sampath and Mohanty (1954). The first cytoplasmic male-sterile line used to develop commercial F_1 rice hybrids was developed in China in 1973 from a male-sterile plant in a population of wild rice - *Oryza sativa f. spontanea* (Yuan, 1977). The sterile plant designated as wild rice with abortive pollen (WA) served as source for a number of CMS lines (Lin and Yuan, 1980; Virmani *et al.*, 1986). Later on, *indica* × *japonica* crosses of the cultivated rice yielded several new CMS sources in China. Lin and Ming (1991) (C.f. Virmani, 1996) described four such new sources: Seng Qi, Nan Guang Zhan, Gui Hua Huang and Zhao Tong Beizigu. Other CMS sources identified are: V20B, Kalinga-1, Ptb 10, Co 41 (Pradhan *et al.*, 1990) and Lalruma (Pradhan and Jachuck, 1998). Among them V20B maintains CMS-WA (wild source), but it itself is a *japonica* cultivar. In Japan, CMS system in rice was reported by Shinjyo and Omura (1966) and Shinjyo (1969) in *indica* var. Chinsurah Boro II, which had S cytoplasm and R gene and variety T-65 having a maintainer gene. This cytoplasm was named as BT or BO. Similarly, Athwal and Virmani (1972) identified the variety TN 1, which had both CMS cytoplasm and R gene, whereas variety Pankhari 203 was found to be a maintainer of sterility. No stable CMS lines from this could, however, be established. Li Zhengyon (1980) identified CMS plants in variety Teibei 8 and later developed a *japonica* male-sterile line of Dien 1 type. Several other CMS sources identified in rice are given in Table 1.

Table 1. CMS sources in cultivated and wild germplasm of sorghum, pearl millet, rice, and pigeonpea.

Germplasm type	Crop	CMS sources
Cultivated	Sorghum	A_1, A_2, A_3, A_4 (G_1, Maldandi and VZM) and 9E and several analogues
	Pearl millet	A_1, A_2, A_3, A_5, Aegp, PT 732, DSA 59-1A, DSA 105A, DSA 118A, DSA 134A, DSA 1444-1A, PMC 23A, PMC, 30A
	Rice	Chinsurah Boro II, Lead rice, Tadukan, Tian Dong, Dong Pu, TN 1, Gambiaca, Birco, Tian Ji Du, IR 24, Jing Chuan Nao, Sheng Qi, Li Up, Zhao Jin Feng, Zhao Tong Bei, Dissi Hatif, V20B, Kalinga-I, Lalruma, Khiaboro
	Pigeonpea	Nil
Wild	Sorghum	Nil
	Pearl millet	Av, A_4
	Rice	*Oryza sativa f. spontanea* (WA), Red-awned wild, *O. sativa f. spontanea* (Dwarf WA), ARC 13829-16, *O. perennis*, *O. glumaepatula*, *O. rufipogon*, *O. nivara*
	Pigeonpea	*Cajanus sericeus* and *C. scarabaeoides*

Wild Germplasm

Based on the premise that cytoplasm × nuclear interactions are likely to be more disharmonious in crosses involving divergent materials, the probability of identifying CMS will be more in populations derived from wild × cultivated crosses than in those derived from cultivated × cultivated crosses, as the former offers far greater diversity. In such crosses, it is important that the wild germplasm is used as female parent, as it is likely to have a higher frequency of the wild alleles (i.e. dominant restorers) than mutant alleles (i.e., recessive maintainers). Two CMS sources (designated as Av and A_4) in pearl millet were identified in the F_2 populations derived from crosses between germplasm accessions of *P. glaucum* ssp. *monodii* (= *violaceum*) (used as female parents) and cultivated pearl millet germplasm used as male parents (Marchais and Pernes 1985; Hanna, 1989). Several other CMS sources recently identified from *monodii* × cultivated crosses, are being evaluated for their cytoplasmic diversity (W.W. Hanna, personal communication). Stable CMS sources in sorghum have not been identified for use in hybrid program so far from the wild germplasm. In pigeonpea, CMS lines have been derived from inter-specific crosses involving cultivated types as male and its wild relatives as female parents. At ICRISAT, *C. sericeus* and at Gujarat Agricultural University, *C. scarabaeoides* were used as cytoplasm donors (Saxena, 2001a). In rice, the real breakthrough in developing commercial F_1 hybrid rice came when a natural male-sterile plant was found in wild rice *O. sativa f. spontanea* in China. This CMS source, as indicated earlier, is designated as CMS-WA. More than 20 cytoplasms of wild rice have been identified and several CMS lines were derived by using distant nucleus substitution backcrosses. The CMS lines have also been developed involving inter-specific crosses using wild rice (*O. perennis, O. glaberrima* or *O. rufipogon*) as female parent, and cultivated rice like *O. sativa* as male parents. In 1973, Sichuan Agriculture College reported development of CMS line using Gambiaca (an *indica* variety of West Africa) as source of cytoplasm. In 1977, Wuhan University reported development of *indica* CMS line of Honglien type in which sterile cytoplasm was from red-awned wild rice. Pradhan *et al.* (1990) identified two new CMS sources, V20B and sattari, through *indica* × *japonica* hybridization. Recently, two CMS lines viz., IR 66707A with cytoplasm from *O. perennis* and nuclear genome of IR 64; and IR 69700A with cytoplasm of *O. glumaepatula* and nuclear genome of IR 64 have been developed. Genetic tests show that these lines possess cytoplasm that are different from WA cytoplasm (Dalmacio *et al.*, 1995; 1996). Recently, new CMS lines viz., RPMS 1 (*O. rufipogon*), RPMS 2 (*O. nivara*) and RPMS 4 (*O. rufipogon*) have been developed at the Directorate of Rice Research (Hoan *et al.*, 1997). There are other systems such as thermosensitive genetic male sterility (Maruyama *et al.*, 1991), but these are outside the scope of this review.

Further details of various CMS systems that have potential use in hybrid development are given in Table 2.

Table 2. Potential CMS systems for commercial utilization in sorghum, pearl millet, rice, and pigeonpea

Crop	CMS system	Origin	Reference
Sorghum	A_1	Milo (CK 60A) A_1	Stephens and Holland (1954)
	A_2	Caudatum-Nigricans (IS 12662C)	Schertz and Ritchey (1977)
Pearl millet	A_1	Tift 556	Burton (1958)
	A_4	*P. glaucum* subsp. *monodii*	Hanna (1989)
	A_5	Large-seeded genepool	Rai (1995)
Rice (*Indica*)	WA-CMS	*O. sativa f. spontanea*	Lin and Yuan (1980)
	Gam	Gambiaca	Lin and Yuan (1980)
	Dwarf WA	Dwarf MS wild rice	Virmani and Wan (1988)
Pigeonpea		*C. sericeus* and *C. scarabaeoides*	Saxena (2001a)

Evaluation

Although numerous CMS sources have been found in the crops under discussion, all have not been found commercially useful. There are various factors that determine the utility of a CMS system. These include stability of male sterility, effect of genetic background on male sterility, maintainer gene frequency in germplasm, character association, and male fertility restoration behaviour.

Stability of Male Sterility

Instability of male sterility in A-lines increases the problem of roguing of pollen shedders from seed production plots (i.e., higher seed production cost). Further, hybrid seed yields are also reduced. Such an unstable CMS system also reduces breeding efficiency as the backcross progenies found fully sterile initially may not be necessarily so during the subsequent generations, leading to their rejection. Stability of male sterility also has a direct bearing on the cost and quality of hybrid seed production. Ideally, a commercial male sterile line should neither shed pollen nor should it set seed when selfed, regardless of the location and the season. This, however, is seldom possible. For instance, several A-lines based on the A_1 CMS systems in both sorghum and pearl millet, have been extensively used to breed hybrids, which are planted on millions of hectares in India alone. Yet, most of these A-lines produce, albeit low frequency (< 1 %) of pollen shedders, depending on the environment. In maize, cooler weather promotes instability in male sterility in both USDA and Texas CMS sources, and hot dry weather promotes stability of male-sterility (Duvick, 1965).

In pearl millet, it has been observed that A_1 CMS lines produce a relatively higher frequency of pollen shedders during the rainy season than in the dry season. Three new CMS systems (A_4, A_5 and Aegp) have been identified in recent years in pearl millet which produce A-lines with no or much less pollen shedders. In case of sorghum, several workers reported the role of temperature on the expression of male sterility and fertility restoration (Downes and Marshall, 1971; Li *et al.*, 1981). It affects some cytoplasms more than the others (Schertz *et al.*, 1997). The work at ICRISAT showed that fertility restoration is poor when night temperature falls below 10 °C just before flowering, during post rainy season, and that the male sterility in CMS lines breaks down when the day temperature rises above 42 °C, before flowering (Reddy and Stenhouse, 1994). This evidently increases the need to screen the CMS lines in areas where the temperature rises above 42 °C before flowering for the absence of seed setting under bag to ensure stability of male sterility. The hybrids need to be screened in areas where night temperatures are low for seed setting under bags to identify stable fertility restorers. We compared A_1, A_2, A_3 and A_4 male-sterile lines for seed setting under selfing during summer (when temperatures exceed 42 °C) at Bhavanisagar, India. It was found that the A_1 is more stable in maintaining male sterility than A_2, A_3, and A_4 (Maldandi); A_3 is more stable than A_2 and A_4; and A_2 is more stable than A_4 (Reddy and Stenhouse 1994). Devi and Murthy (1993) showed that the tapetum was intact and pollen was sterile in A_2 male-sterile lines in winter (low temperature), while partial or complete degeneration of tapetum occurred and viable pollen grains were found in summer (high temperature), indicating the unstable nature of male sterility in A_2 CMS system. At ICRISAT, the low temperature-induced female sterility in A_1 CMS female lines (e.g. 296A) indicated that female sterility may be reduced significantly by using their non-parental single cross F_1 male-sterile lines (Reddy, 1992). At ICRISAT, in maintainer lines development programme, the frequency of maintainer lines observed in A_1 (Table 3) and A_2 CMS systems (Table 4) was higher in post-rainy season (cooler temperature) than in the rainy season (warmer temperature). Others in Indian National programs reported that fertility restoration frequency in A_2 CMS system is higher in post-rainy season than in rainy season (U. R. Murthy, personal communication). However, the data presented in Table 4 do not support the above observation. Further data are required involving same females in both seasons.

Table 3. Maintainer and restorer frequencies in sorghum A_1 cytoplasm in rainy and postrainy seasons at ICRISAT, Patancheru

Season	A line	Total testcrosses	Frequency	
			Maintainers	Restorers
Rainy season 2000	ICSA 56	75	0.80	0.20
	ICSA 84	66	0.82	0.18
	ICSA 101	95	0.84	0.16
	CK 60 A	49	0.86	0.14
	Total	**285**	**0.83**	**0.17**
Post-rainy season 2000	ICSA 1	39	0.62	0.38
	ICSA 9	39	0.87	0.13
	ICSA 101	200	0.95	0.06
	ICSA 88005	21	0.90	0.10
	Total	**299**	**0.89**	**0.11**

Table 4. Maintainer and restorer frequencies in sorghum A_2 cytoplasm in rainy and postrainy seasons at ICRISAT, Patancheru.

Season	A_2 line	Total test-crosses	Frequency	
			Maintainers	Restorers
Rainy season 1999	MR 750	215	0.84	0.16
	ICSA 94003	55	0.95	0.05
	Total	**270**	**0.86**	**0.14**
Post-rainy season 1999	MR 750	19	0.84	0.16
	ICSA 88004	110	0.86	0.14
	ICSA 94001	20	0.85	0.15
Post-rainy season 2000	ICSA 38	133	0.97	0.03
	ICSA 743	72	1.00	0.00
	ICSA 88001	34	0.85	0.15
	Total	**388**	**0.92**	**0.08**

In pigeonpea, breakdown of male sterility is common at low temperatures and short photoperiod, as observed in some CMS lines at ICRISAT, (Saxena, 2001b). In pearl millet, breakdown in male sterility occurs in rainy season due to high humidity. Unlike in cross-pollinated crops, absolutely complete stability of male sterility is of high significance in self-pollinated crops like rice. Initially, in hybrid rice program, a male sterile line, *viz*. IR 58025A (WA cytoplasm) was used in developing hybrids. This line, on account of its instability for male sterility, always resulted into some female type sterile plants in F_1 hybrids. A similar difficulty was encountered with another CMS line IR 62829A (WA cytoplasm). However, now both these lines are in use. Most commonly used female line now is IR 58025A (WA cytoplasm). The previously suspected instability in male-sterility in female line IR 58025A does not hold true any more, as many completely genetically homogeneous pure hybrids have been commercialized with this line. Thus, any pollen shedders observed by some breeders in this line could be attributed to contamination resulting from incomplete isolation during the maintenance of this line, and not due to instability of the line. Various newly developed WA-CMS lines have shown complete male sterility under different climatic conditions, and are perfectly usable for development of new hybrids.

Effect of Genetic Background on Male Sterility

Nuclear genetic diversity of A-lines provides opportunity for diversifying the genetic base of hybrid cultivars with higher yield potential. The effectiveness of such genetic diversification efforts is determined by whether or not the genetic background of maintainer lines (B-lines) influences the stability of male sterility of A-lines produced from them. In case of the A_1 CMS system in pearl millet, it has been observed that a good maintainer of an A-line is not necessarily a good maintainer of another unrelated A-line carrying the same cytoplasm (Table 5), indicating the effect of nuclear genetic background of B-lines on male sterility. In sorghum, the *kafir*-based crosses with CK 60B produce higher frequency of B-lines than the *caudatum*-based B-lines with A_1 CMS source, which essentially derives its male-sterility maintainer genes from *kafir*.

Table 5. Average pollen shedding scores of male sterile × maintainer pearl millet hybrids for two seasons in 2 yr at ICRISAT Center.

	B-lines																			
	5141B				Pb 111B				81B				842B				843B			
	Average pollen shedding scores																			
A-line	A†	B	C	D	A	B	C	D	A	B	C	D	A	B	C	D	A	B	C	D
5141A	1‡	1	1	1	4	4	4	4	1	1	1	1	4	4	4	4	3	3	4	3
Pb 111A	3	3	4	4	1	1	1	1	1	1	1	1	4	4	4	4	4	4	4	4
81A	1	1	1	1	1	1	1	1	1	1	1	1	1	1	1	1	1	2	1	1
842A	3	4	4	4	4	3	4	1	1	1	1	1	1	1	1	1	4	4	4	4
843A	$	-	-	-	4	4	4	4	1	1	1	1	2	2	1	1	1	1	1	1
834A	4	4	4	4	3	4	4	3	3	4	1	1	4	4	3	1	4	4	4	4
PT 32A	4	4	2	3	4	4	4	4	3	-	1	-	4	4	4	4	4	4	4	4

†A-Rainy season 1985; B-Rainy season 1986; C-Dry season 1986; D-Dry season 1987; ‡1-All plants having shrunken anthers and shedding no pollen; 2-Most of the plants having shrunken anthers and shedding no pollen; 3-Most of the plants having plump anthers and shedding pollen; 4-All plants having plump anthers and shedding pollen; $-Not tested; Adopted from Rai and Hash (1993).

At ICRISAT, the maintainer gene frequency ranged from 0.5 to 1.0 in the derivatives of various sets of crosses with A_1 CMS system (Table 6), although both the parents involved in such crosses were maintainers.

Table 6. Maintainer and restorer frequencies in various sorghum A_1 CMS maintainer lines derived from B × B crosses at ICRISAT, Patancheru

S. No.	B-lines used as female	No. of B-lines used as males	No. of test-crosses	Frequency	
				Maintainers	Restorers
1	ICSB 11	4	6	0.3	0.7
2	ICSB 73	5	31	0.9	0.1
3	ICSB 89	5	18	0.9	0.1
4	ICSB 101	5	21	1.0	0.0
5	ICSB 443	3	18	0.6	0.4
6	ICSB 463	4	4	0.5	0.5
7	ICSB 568	1	4	0.5	0.5
8	ICSB 203	2	9	0.3	0.7
Total			111	0.8	0.2

In rice, the expressivity of fertility restorer (R) gene varies depending on the nuclear background of the female parent, even though they possess the same cytoplasm (Kumari *et al.*, 1998). To cite an example, hybrids involving varieties C 29 and PMK 2 with WA-cytoplasmic male sterile (CMS) lines IR 58025A, IR 62829A and PMS 3A - showed marked differences in pollen and spikelet fertility. C 29 behaved as a partial restorer with IR 62829A and PMS 3A (21-90 %), but with IR 58025A it behaved as maintainer, though both females belong to the same CMS system. Similar is the effect of PMK 2. This type of variation may be due to presence of modifier genes present in the R parents (Ganesan *et al.*, 1998). Govindraj and Virmani (1988) presented evidence for inhibitory genes present in a CMS line of rice causing incomplete fertility restoration by some established restorer lines. Similarly, CMS lines V20A, IR 58025A, and IR 62829A which have the same cytoplasm but different nuclear genotypes, showed differences in maintainer frequency (Virmani, 1994). In such cases, some of the inbred lines derived from B × B crosses are likely to be poor maintainers, leading to reduced A-line breeding efficiency. Stability of male fertility is another important factor in rice. It has been observed that some hybrids which show complete spikelet fertility in northern India, show only partial fertility in southern Indian region. Variations in fertility restoration across locations have also been observed within the region. In a typical case (Ish Kumar, unpublished) one hybrid showed complete fertility in Mandya (Karnataka), whereas, it was observed to have lowest fertility (partially sterile) in Hyderabad (Andhra Pradesh) in India. Use of such male parents must be discouraged, as they do not provide widely adapted fertile hybrids.

Maintainer or Restorer Gene Frequency

Frequency of maintainer gene in a diverse range of improved populations and breeding lines has a direct bearing on the success of genetic diversification of A-lines. Conversely, the frequency of restorers influences directly the use of such diversified male-steriles in grain hybrid development. It has been observed that only 10-15 % of the inbred lines behave as complete and stable maintainers of the A_1 CMS system in pearl millet, and relatively large proportions behave as partial and unstable maintainers. In contrast, more than 60 % of the inbred lines behave as complete and stable maintainers of the A_4 CMS system and more than 99 % of the inbred lines behave as complete and stable maintainers of the A_5 CMS system. Thus, the A_4 and A_5 CMS systems provide much greater opportunities for genetic diversification of A-lines in pearl millet.

The very low frequency of restorers of the A_5 CMS system in pearl millet poses problems in its utilization for grain hybrid program, Therefore, based on the premise that the same germplasm source where the A_5 CMS was identified should also be a source of its restorers, an extensive search was conducted in the Large-seeded genepool 1 (the source of A_5 cytoplasm), four other diverse and broad-based genepools and two open-pollinated varieties. Low frequency of restorers were identified in each population (Table 7), although large variations occurred in the level of fertility restoration ability across the populations. A highly male fertile restorer stock in the A_5 cytoplasmic background was developed from a fertile plant in the hybrid made with the Large-seeded genepool 1, which produces profuse pollen and gives more than 90 % seed set under selfing, both *per se* as well as in the hybrid made on 81A_5. In general, finding commercially viable restorers is much less of a problem than finding maintainers in cross-pollinated crops. This is because, even a lower level of fertility restoration in hybrids of cross-pollinated crops may have no adverse effect on grain yield. In fact, reduced male fertility level in hybrids appears to be associated with higher grain yield in maize (Duvick, 1965) and pearl millet (KN Rai, unpublished). Thus, stability of male fertility is more important than the lower level of stable male fertility. For instance, the otherwise male fertile pearl millet A_1-system hybrids in the midwest of the United States have been found to turn sterile in the event of temperatures falling below 10 °C during flowering. Temperature sensitivity of fertility restoration in other CMS systems of pearl millet is yet to be investigated.

Table 7. Pollen-fertile plants in topcross hybrids of four pearl millet genepools and two open-pollinated varieties of OPVs crossed onto 81A_5, Patancheru.

Genepool/OPV	Season	No. of hybrid plants	Pollen-fertile hybrid plants	
			Number	Per cent
Genepool	1995 rainy			
Early		694	9	1.3
Large-seeded (LSGP 1)		645	6	0.9
Large-seeded (LSGP 2)		723	1	<0.1
Large spike		661	9	1.4
High tillering		721	11	1.5
OPV	1997 rainy			
ICMR 312		541	76	14.0[1]
IAC-ISC TCP 2		544	42	0.7[2]

1. Includes 61 plants (80.3 % of the pollen-fertile plants) that had poor pollen shed; 2. All plants had poor pollen shed; Source: Rai *et al.* (1999)

In sorghum, Scheuring and Miller (1978) found a restorer frequency of 0.62 and maintainer frequency of 0.23 on milo (A_1) cytoplasm in the world collection of 3,507 sorghum accessions. The work carried out at ICRISAT showed a restoration frequency of 0.9 on A_1, 0.5 on A_2, 0.1 on A_3, and 0.3 on A_4, when 48 germplasm lines were test-crossed onto A_1, A_2, A_3, and A_4 CMS systems (Belum Reddy, unpublished data). Senthil *et al.* (1998) found that the frequency of restorers was 0.15 on A_1, 0.04 on A_2, 0.01 on A_3, and 0.03 on A_4 CMS systems. This suggests that the restorer frequency is high on A_1 and low on A_3 system. Considering the restoration frequency, A_1 CMS system provides the widest possible choice in selecting restorers. In B-line development programme, the restorer gene frequency is also higher in A_1 CMS system than in A_2, A_3, and A_4 (Table 8).

Table 8. Maintainer and restorer frequency in various cytoplasms of sorghum in rainy and post-rainy seasons, 1999 and 2000 at ICRISAT, Patancheru

Season	Cytoplasm	A-lines used in test-crossing	Total no. of pollinators	Frequency	
				Maintainers	Restorers
Rainy	A_1	4	285	0.83	0.17
	A_2	2	270	0.86	0.14
Postrainy	A_1	4	299	0.89	0.11
	A_2	6	388	0.92	0.08
Variable	A_3	4	94	0.86	0.14
Variable	A_4	4	82	0.84	0.16

In case of completely self-pollinated crops like rice, high level of fertility restoration is very important. Low level of fertility restoration in rice has been found to give hybrids with partial sterility, which is due to lack of out-pollination. The commercial acceptability of such hybrids in the market is very low, as the lower fertility reduces the heterotic effect for grain yield. Studies carried out at the International Rice Research Institute (IRRI), Philippines showed that the frequency of maintainer lines was higher in elite lines from Korea (*japonica* rice) than those from India, Pakistan, Philippines, Indonesia and Vietnam. It is generally observed that effective restorers were concentrated in races cultivated in south and south-east Asia, and southern China while non-restorers were concentrated in northern China and far-eastern Asia. Frequency of restorer lines in rice has been reported to be generally higher in varieties of geographic origin from lower latitudes compared to those from higher latitudes. Furthermore, restorer frequency is higher in *indica* races (18 %) compared to *japonica*. In China, long duration *indica* races showed higher frequency of restorers than the early maturing *indica* races (Yuan, 1985). Such a correlation may not hold good outside China, because of extensive hybridization between late and early rice cultivars (Yuan and Virmani, 1988). However, in Indian subcontinent, aman, Assam and boro races of *indica* showed high frequency of restorers (Lin and Zhu, 1988). The frequency of restorer gene in breeding material varied from 9 to 33 % at different centers in India (Anonymous, 1998).

In pigeonpea, male fertility restorer frequency is very low. Only two varieties, HPL 24 and ICPL 129-3 have been found to restore the fertility of CMS lines based on *C. sericeus* cytoplasm (Saxena, 2001b).

Effects of Male Sterility Inducing Cytoplasm

Commercial viability of a CMS system depends on the extent of its association with characters of economic importance. The once most widely used Texas CMS source in maize is a historical example, which had to be withdrawn from field deployment because of its association with southern leaf blight caused by *Bipolaris maydis* (Nisikado) Shoemaker (Scheifele *et al.*, 1970). A downy mildew (*Sclerospora graminicola* (Sacc.) Schroet) epidemic on a pearl millet hybrid HB-3 (developed on male-sterile line, Tift 23 A_1) that occurred in India during 1971, gave an initial suspicion that the A_1 cytoplasm was associated with this disease. Later studies showed that the A_1 cytoplasm in pearl millet had nothing to do with the susceptibility to downy mildew (Anand Kumar *et al.*, 1983; Yadav *et al.*, 1993). Although pearl millet hybrids based on the A_1 cytoplasm are more susceptible than open-pollinated varieties to both ergot (*Claviceps fusiformis* Loveless) and smut (*Tolyposporium penicillariae* Bref), it has been found to be due to the CMS-mediated reduced male fertility rather than due to the cytoplasm *per se* (Rai and Thakur, 1995; 1996). Similarly, A-lines and their hybrids have been found to give significantly higher grain yield than their counterpart B-lines and hybrids, but this also appears to be associated with CMS-induced male sterility rather than due to the effect of cytoplasm *per se* (KN Rai, unpublished). The A_1 cytoplasm has been observed to cause a day or two of earliness, but this difference is of no practical significance.

In sorghum, Sharma *et al.* (1994) found that spikelet damage and adult emergence of midge was significantly lower on midge resistant B-lines (PM 7061 and PM 7068) of A_1 CMS source than their corresponding A-lines and *vice versa* in the midge susceptible-parental lines (296A/B and ICSA/B 42). At ICRISAT, considerable variation was observed between male-sterile lines and maintainer lines in A_1 CMS system for flowering. In early group, a few A-lines tend to be late by a day or two, which is not of much significance. But in the medium and late maturity groups, A-lines tend to be late in flowering in significant number and there is a tendency of increased delay in A-lines flowering with increased maturity period (Table 9). B-lines have more open panicles than their A-lines.

Table 9. Frequency of sorghum male-sterile and maintainer lines differing in days to 50 % flowering at ICRISAT, Patancheru

Differences in days (A-B)	Frequency of A/B lines in maturity groups		
	Early (<67 days)	Medium (67-74 days)	Late (>74 days)
-2	0.00	0.02	0.04
-1	0.00	0.08	0.08
0	0.74	0.41	0.31
1	0.24	0.40	0.35
2	0.03	0.09	0.19
3	0.00	0.00	0.04
Total number tested	34	108	26

Further, A_1 cytoplasm is more susceptible to shoot fly than the maintainer line cytoplasm, while the reverse is true for stem borer resistance (Belum Reddy, unpublished data). This finding has significance in developing shoot fly resistant hybrids.

In rice, CMS has been observed to affect panicle exertion. The panicle exertion of CMS lines is not complete, when compared to the B-line with normal cytoplasm. The hybrid, however, does not face incomplete exertion, as the negative effect of male sterility inducing cytoplasms is rectified due to restoration of fertility and heterosis for peduncle length. The hybrids, thus, have complete panicle exertion. Yang (1980), Pan (1982) and Shen (1982) found cytoplasmic effects on grain weight, but through right combinations, the cytoplasmic effects could be overcome in the hybrids of parents with high combining ability and high fertility restoration ability. Another effect of WA cytoplasm is that it delays flowering by 4-6 days compared to its counterpart B-line cytoplasm. This difficulty can, however, be overcome by staggered sowing of A- and B-lines. So far, CMS cytoplasms have not been found to be associated with susceptibility to any major diseases or insects in rice.

CMS Diversity

The need for cytoplasmic diversification of A-lines (and hybrids) to reduce the potential risk of disease and insect pest epidemics associated with cytoplasmic uniformity is a common knowledge. What has not generally been appreciated is that cytoplasmic diversification also enhances the ability of a plant breeder to diversify the genetic base of A-lines, as some of the outstanding inbred lines not found to be maintainers of one CMS system may be maintainers of the other CMS systems.

There are two approaches of CMS diversity analysis which serve two different but overlapping purposes. The first one deals with the traditional field evaluation of hybrids for differential patterns of male fertility restoration, while the second one deals with the mitochondrial DNA (mtDNA) analysis. The differential fertility restoration patterns of hybrids provides an evidence that the CMS sources involved in the study are different from each other, both for nuclear as well as mitochondrial genes. Schertz and Pring (1982) reviewed sorghum CMS and provided a summary of various cytoplasm sources studied through fertility restoration patterns i.e. 42 in India, 24 in USA, and one in Africa. Some of the cytoplasms were reported to be similar in reaction, considering their restoration patterns. For example, Schertz and Pring (1982) (quoting U.R. Murthy, 1996, personal communication) indicated that cytoplasms of G1 (G1-S, ms G1, G1-G, G1-A) are analogous to IS 1112C of USA. Over the years, many of these cytoplasm sources were either lost and/or not widely available. The most commonly available ones include: A_1 (milo source), A_2 (IS 12662C or TAM 428), A_3 (IS 1112C) and 9E (a selection made in 9E) of USA source, and A_4 (Guntur, VZM, and Maldandi) of Indian source. These cytoplasms have been categorized based on the pattern of fertility restoration. Reddy and Stenhouse (1994) reported the identification of minimum differential testers for A_1 to A_4 cytoplasms as follows:

- TAM 428B (A_2) gives fertile F_1s only on A_1 cytoplasm,
- IS 84B (A_4-Maldandi) gives fertile F_1s on A_1 and A_2 cytoplasms,
- IS 5767R (A_4-Maldandi) gives fertile F_1s on all cytoplasms, except A_3, and
- CK 60B (A_1) gives male-sterile F_1s on all cytoplasms

Based on anther morphology and pollen development, these A_1 to A_4 (Guntur, VZM, Maldandi) and 9E cytoplasms were further subdivided into two distinct groups: (*i*) those with small anthers but without fertile pollen, which degenerates during microsporogenesis (A_1 and A_2) and (*ii*) those with large non-dehiscent anthers that may contain some viable pollen (A_3, A_4 and 9E) (Schertz *et al.*, 1989). ICRISAT is maintaining A_1 to A_4 cytoplasms.

The lack of differential restoration patterns, however, does not provide conclusive evidence that the CMS sources involved are necessarily similar, as it is possible that the pollinator parents used in developing the testcrosses were not adequate in number and different enough in diversity to pick up the CMS differences. It is also important in such field studies that testcrosses to be evaluated are made on isonuclear A-lines to ensure that genotypic differences of the female parents are not confounded with their cytoplasmic differences in determining fertility restoration of testcrosses.

The RFLP analysis of mtDNA based on CMS-specific enzyme-probe combinations provides information on the similarity or dissimilarity of cytoplasms. Differences in the RFLP patterns indicate cytoplasmic differences, although such differences may arise due to mitochondrial genomic differences that may or may not have anything to do with male fertility restoration. For instance, Chhabra *et al.* (1998) evaluated nine pearl millet CMS sources, identified from the Large-seeded genepool of different male sterile plants. Based on RFLP patterns of mtDNA, seven cytoplasmic groups were identified. A field study of male fertility restoration of testcrosses, developed on five of these (in isonuclear genetic background), however, grouped three sources along with A_1 system and two with the A_4 CMS system (K.N. Rai, unpublished). In sorghum, hybrids of three A_4 CMS sources (Guntur, VZM and Maldandi) had a similar male fertility restoration pattern and were designated as an A_4 CMS system. RFLP patterns of mt DNA of these three sources, however, revealed them to be different from each other (Sivaramakrishnan *et al.*, 1997). Thus, both traditional restoration pattern study and molecular approach are required to provide a conclusive evidence of whether the new CMS sources differ among themselves as well as from the existing ones.

Diversification of CMS sources is an important consideration in rice too, to avoid any potential risk of breakdown to any serious disease or pest. Virmani and Shinjyo (1988) mentioned several CMS sources, but it is not clear whether all these are distinct from each other. So far, about 95 % of the hybrids commercialized in China and elsewhere possess only one type of cytoplasm, i.e. WA (Brar *et al.*, 1998). CMS-WA cytoplasm gives stable CMS lines for which high frequency of restorers are available. Efforts have been made to identify new sources of cytoplasm from wild species belonging to AA genome species, eg. *O. nivara, O. rufipogon,* and *O. perennis*. Two CMS lines have been developed at IRRI from interspecific hybridization and substitution backcrossing (Dalmacio *et al.*, 1995; 1996). One line IR 66707A has the cytoplasm of *O. perennis* and the nuclear genome of IR 64. Another line, IR 69700A has the cytoplasm of *O. glumaepatula* with nuclear genome of IR64. At the Directorate of Rice Research, Hyderabad, sources of male sterile cytoplasms were identified from accessions of *O. nivara* and *O. rufipogon,* and using these, six new CMS lines were developed that could be grouped into four classes (Hoan *et al.*, 1997; Brar *et al.*, 1998). Two CMS lines Pusha A and Mangala A with MS 577A cytoplasm from *O. rufipogon* have also been developed. However, effective restorers for these are yet to be identified.

At the Central Rice Research Institute, Cuttack, new CMS sources have been identified from improved lines like Kalinga 1, V20B and Lalruma. Although many restorers have been identified for Kalinga 1 cytoplasm, only one restorer each has been identified for V20B and Lalruma cytoplasms.

Utilization

CMS utilization in breeding hybrid cultivars involves development and maintenance of seed parents (female parents and their maintainers) and male parents. The latter should have an additional trait of fertility restoration ability, if hybrids are developed for grain production. Depending on the type of hybrids envisaged and possible, both seed parents as well as restorer parents could be either inbred lines (A-line, B-line and R-line) or heterozygous and heterogeneous populations (male-sterile, maintainer and restorer population) in cross-pollinated crops. We shall, however, confine ourselves to single-cross grain hybrids produced by crossing an inbred male sterile line (A-line) with a restorer line (R-line).

Potential CMS Sources

Although several CMS systems have been identified in several crops, essentially only one CMS system is being mostly used on large-scale development of commercial hybrids [e.g., Texas CMS system in maize, (but abandoned later), A_1 CMS system in sorghum and pearl millet, and WA CMS system in rice]. This may have largely to do with the practical concern that an effective CMS utilization strategy entails parallel research and development efforts for both seed parents as well as restorer parents. Thus, several established restorers of the existing commercial CMS system may become obsolete due to their inability to restore male fertility in hybrids of new CMS systems. In addition to fertility restoration, other problems associated with the use of alternate CMS systems include stability of male sterility, extent of heterosis for economic traits and ease with which these male-steriles can be identified from male fertiles in the field. All these should be analyzed to make a rational judgment about CMS diversification. Often such analyses are not carried out, leaving new useful CMS systems unused or under-used.

In sorghum, as indicated earlier, A_1 cytoplasm is more stable than other alternative cytoplasms, and restorer frequency with A_1 CMS is higher than with others. The heterosis estimates reported for grain yield in A_1 CMS system have varied. For example, the Indian National Program Testing results showed that the standard heterosis with A_1 CMS system for grain yield ranged from 18 to 31 % in the rainy season, and from 19 to 29 % in the post-rainy season during the years 1999 and 2000. The heterobeltiosis estimates in the same studies ranged from 15 to 26 % in the rainy season and from 1.5 to 11 % in the post-rainy season. Siddiq *et al.* (1993) reported that heterobeltiosis was 38 % for grain yield in rainy season. Similar studies with alternative CMS systems are limited.

Comparison of heterosis with A_1 CMS, which is commonly used, with those obtained on other systems will be useful. Senthil *et al.* (1998) reported that the A_1 CMS system in sorghum produced higher number of heterotic combinations than A_2, A_3 or A_4 systems. Kishan and Borikar (1989a) observed that A_2-based hybrids had larger grains and higher yields than A_1- and A_4-based hybrids. Based on testing 15 hybrids derived from three isonuclear male-sterile lines and five common restorers, the A_4-based hybrids were found inferior to others for grain yield in the rainy season. However, in another study, Kishan and Borikar (1989b) noticed that A_4-based hybrids had higher grain yield and larger grain size than A_1 hybrids during the post-rainy season. Murthy (1986) observed that the frequency of segregation of plants for tall and dwarfs conformed to the known theory that short height is controlled by four recessive non-linked genes in crosses of two dwarf isocytoplasmic lines carrying A_1 cytoplasm and two tall tropical landraces (IS 2317 and IS 35613). In the crosses of dwarf isocytoplasmic lines with the A_2 cytoplasm and two landraces, the segregation pattern of dwarf and tall plants deviated significantly from the four gene theory, indicating the effect of A_2 cytoplasm on plant height. Considering the restoration frequency and hybrid performance, it is advantageous to use A_2 CMS system among the alternative cytoplasms available in sorghum. Accurate comparisons of CMS systems are only possible if the hybrids are developed from crosses between isonuclear lines and common restorers. The above studies, however, did not involve such materials. Nevertheless, A_2 CMS system is not popular, as the anthers in A_2 male-steriles, unlike the A_1 male-steriles, mimic the fertile or maintainer lines and lead to difficulties in monitoring the purity of hybrid seed production. Only one hybrid based on A_2 CMS system has been released so far (in China) for commercial cultivation in the world.

In rice, CMS line IR 66707A having cytoplasm of *O. perennis* (Acc.104823) and nuclear genome of IR 64 has been subjected to molecular analysis. The results confirmed that it is different from WA source. Molecular characterization of mitochondrial genome of rice lines of WA origin has also been carried out by Seth *et al.* (1996). Research is underway to use polymorphism in male-sterile genes and in the expression of male sterility to differentiate maintainers and restorer lines, as also to distinguish WA cytoplasm from Chinsurah boro II cytoplasm.

Seed Parents Development

Development of maintainers is an important aspect of seed parents development and it encompasses various issues, such as what characters to consider in B-lines, when and how to test for combining ability and when to initiate their conversion into A-lines. But these are outside the scope of our paper, and hence will not be presented here. Rather, we shall confine ourselves to issues related to utilization of a specific CMS system and cytoplasmic diversification of A-lines.

During the process of conversion of a B-line into an A-line with a given cytoplasm, it is important to ensure that all the plants in the F_1 and backcross progenies are fully sterile. Occurrence of partial sterility (i.e., very low degree of seedset under selfing) in plants in the initial generations is an indication that either the CMS system itself has a problem of poor sterility induction (as in the case of the A_2 and A_3 CMS systems in pearl millet), or that the specific B-line itself is a poor maintainer.

If the latter is true, then a wider search among inbred lines should lead to the identification of good maintainer lines. In completely self-pollinated crops like rice, any maintainer line which does not offer 100 % sterility from early generation itself in wet and dry seasons, should not be converted into CMS line and used in hybrid development. Another situation has been found to be very common in crops like rice, where no selection for maintainer or restorer is being done by the conventional breeders. One may find fertile, sterile and partially sterile plants to occur in the F_1 and backcross progenies, but all sterile plants neither shed any pollen nor do they set any seed under selfing. In general, this is an indication of residual heterozygosity at loci responsible for male sterility, and it should be first noticeable in the F_1 itself, discounting any role of contamination in the succeeding (backcross) generations. Should this situation arise, resort should be made to making paired crosses between sterile plants of F_1 (or backcross generations) and the recurrent parent (B-line), and using the selfed seed of the B-line for backcrossing in the subsequent generation. Two generations of this backcross scheme should normally lead to fully sterile and stable backcross progenies. Plant-to-plant backcrossing coupled with testing the male-sterile line for stability has been used successfully at ICRISAT to develop stable male-sterile lines with A_1 CMS system, whenever fertility segregation is found. In some cases, segregation for fertility (rather occurrence of pollen shedder) may continue longer and for ever (as in case of male sterile line 834A in pearl millet), requiring exclusion of such lines from the conversion programme as early as possible.

Utilization of alternative CMS sources would require complementary efforts in developing corresponding restorers. Because such activities do not take much resources, new and alternative CMS sources should be used, only if they provide some overall advantage in terms of breeding efficiency and A-line maintenance. Until that has been demonstrated, the only larger interest they serve is their availability as standby in case the CMS system under commercial use succumbs to some unforeseen disease or insect pest.

Commercial rice hybrids released in India, Vietnam, and Philippines are based on CMS lines derived from WA cytoplasm. Attempts are being made to breed A-lines with more than one cytoplasm, to reduce the risk of genetic vulnerability. Many programmes in various countries are utilizing particularly A_1 CMS systems in sorghum and pearl millet in developing commercial hybrids. In pigeonpea, the systems are yet to be operationalized and it is likely that both the sources, *C. sericeus* or *C. scarabaeoides* may work. In maize, (1) Texas and the other CMS systems are associated with yield reduction and susceptibility to diseases in hybrids, and (2) the detasseling system lends itself to produce hybrids cheaply on a large scale, and (3) just lines can be developed faster than CMS lines. Therefore, the detasseling method involving high yielding inbred lines is used widely in hybrid production.

Seed Parents Maintenance

Occurrence of low frequency of pollen shedders in A-lines is not uncommon, although it varies a great deal with crops, CMS systems, and seed production season. It is significant to note that male sterility makes the lines more prone to contamination with foreign pollen, although the extent of it is determined by several factors such as isolation distance, wind and insect borne pollen load, pollen viability and genotype. For instance, pearl millet produces profuse pollen, which are robust by nature and can remain viable for more than a day at normal room and air temperature below 38 °C. This creates conditions for greater contamination of A-lines in pearl millet, requiring that even breeders and technicians do not move freely from one seed production field to the other on the same day. However, in other crops such as sorghum, pollen is viable for 1-2 h at room temperature and hence there are less chances for contamination. On the other extreme, pollen viability in rice lasts only 5-10 minutes. The purity of seed parents in this crop can well be maintained by keeping recommended isolation distance of about 250 meters.

Regardless of the care taken to prevent contamination with foreign pollen, an A-line does produce pollen shedders, and these plants are morphologically identical to those of the counterpart B-line. Such pollen shedders arise from mutation, discounting mechanical mixture from the B-line. Their frequency may vary depending on the cytoplasm, the nuclear genetic background, and the environmental factors. Also, mutation giving rise to these pollen shedders, which are called fertile revertants, may occur either in the cytoplasm or in the nucleus. In pearl millet and sorghum, a majority of the mutations producing fertile revertants are in the cytoplasm, in which case they genetically resemble the counterpart B-lines. In such a situation, maintenance of the purity of A-lines is an easy task provided these fertile revertants are rogued out of the A-line rows any time before harvesting. The roguing is easier at the time of flowering, when the fertile plants can be more easily distinguished from the sterile ones. In rice, B-lines flower 4-5 days earlier than A-lines and they can be rogued out easily before flowering. If a nuclear gene mutation (from rf to Rf allele) is the cause of reversion to male fertility, then the situation becomes more complicated, in that any contamination of the A-line with B-line from this plant will lead to a spread of the Rf allele in the lines. Carriers of this allele in the A-line cannot be detected until flowering when they shed pollen (by which time they would already have contaminated the A-line), while such carriers in the B-line cannot be detected unless evaluated for their test-cross performance. In this case, both A-line and

B-line will have to be purified through paired crossing and test-cross evaluation for male sterility for two consecutive generations. Bulking the selfed seed of those B-line rows that produced fully sterile test-cross progenies will reconstitute the B-line. A similar bulking of the seed of plant-by-plant crosses produced between those B-line and test-cross rows will reconstitute the A-line.

A diversified (nuclear genes) set of seed parents have been developed in sorghum and pearl millet at ICRISAT and in national programs in the USA, China, India and Australia using the A_1 CMS systems. In rice, a large number of male-sterile lines have been developed in programs at IRRI, China and India.

Restorer Parents Development

Identification of promising restorer lines with respect to agronomic and adaptation traits and high general combining ability for grain yield is an important breeding activity. Equally important is genetic improvement of existing restorer lines for these traits. But these are outside the scope of this paper and hence will not be presented here. A situation may arise where an inbred line has been found to have produced a high-yielding, but a completely or partially male-sterile hybrid. This is likely to occur more within those CMS systems for which there is low frequency of restorers in the general breeding materials, *viz.* A_5 CMS system in pearl millet and A_3 CMS system in sorghum. In such cases, the most effective approach is to identify best restorer source (s) and undertake backcross transfer of restorer gene(s) in these lines. Considering the paucity of restorers of the A_4 and A_5 CMS systems in pearl millet and A_2, A_3 and A_4 CMS system in sorghum, a backcross breeding scheme (Fig. 1) for development of restorers of these two CMS systems was recently initiated at ICRISAT. In this breeding scheme, an A-line is crossed with a restorer gene donor, which produces a male-fertile F_1 hybrid. This F_1 is crossed (as a female parent) with that inbred line (as a recurrent parent) which is to be converted into a restorer version. The three-way hybrid so produced will segregate for fertile and sterile plants (in the sterile cytoplasmic background). In case there is segregation for varying fertility levels, the most fertile plants are selected and crossed (again as a female) with the recurrent parent to produce BC_1. This process of selection of most fertile plants and crossing them with the recurrent parent is continued until BC_7 or till the BC progeny becomes indistinguishable from the recurrent parent. After the final backcross, 4-5 most fertile plants should be selected and selfed to produce BC-derived F_2 progenies, which should be grown head-to-row. All will segregate for both fertile and sterile plants. Some of the fertile plants will be homozygous for restorer gene (s), and hence will produce uniformly fertile F_3 progenies. Bulked seed from the selfed plants of these progenies will produce restorer version of the recurrent parent (essentially a non-restorer line).

A diversified set of restorers for A_1 and A_2 and the common restorer for A_1 and A_2 are developed at ICRISAT and in other national programs in sorghum. Similarly, a large number of restorers for A_4 and A_5 CMS systems in pearl millet are being developed at ICRISAT. However, work is at initial stage in developing restorers for the two CMS

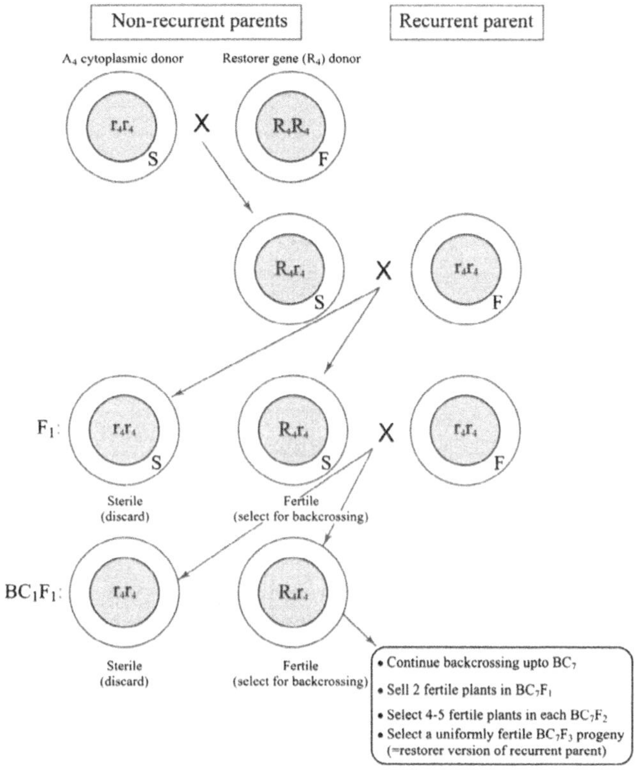

Fig. 1. Schematic back-cross procedure for breeding restorer lines of A_4 cytoplasmic-nuclear male sterility

systems in pigeonpea. Several restorers in rice have been developed for CMS lines derived from WA CMS system in IRRI, and national programs in China and India.

The maintenance of genetic purity of parents is an important problem in large scale seed production of parental lines and hybrids. Conventionally, seed purity of hybrids and parental stocks is done by morphological assay called "Grow Out Test" (GOT). Isozymes, and more recently micro satellite DNA markers, are being used in rice as a substitute to GOT (Sarma *et al.* 2001).

Hybrid Seed Production

In most crops, the four key elements in hybrid seed production are: (*i*) isolation distance, (*ii*) removing off types, (*iii*) flowering synchrony between A- and R-lines and (*iv*) A:R lines ratio. In crops like rice, exsertion of panicles from leaf sheath and angle and duration of flower opening are the important factors. Isolation distance is determined by the breeding system of the crop and the class of seed. Thus, for the highly cross-pollinated pearl millet, recommended isolation distance is 1000 m for breeder and foundation seed of parental lines and 200 m for the certified seed production of hybrids. The isolation distance is 400 m for hybrid parents and 200 m for hybrids for producing breeder and foundation seed in sorghum. In all crops, it is highly essential to remove the off-types before flowering in parents and the pollen shedders at flowering in the female parents. For rice, the isolation distance should be 150 m for breeder and foundation seed and 100 m for certified seed production.

The large flowering synchrony problem arises, but only in the case of certified hybrid seed production, for two reasons: (*i*) A- and R-lines of the highest yielding hybrids may have different flowering time (and this is not uncommon), and (*ii*) both parents that appeared to be synchronous for flowering in the environments where they were bred, may display different flowering time in some of the seed production environment due to differences in their sensitivity to climatic factors such as temperature and photoperiod. Flowering differences of up to 20 days have been effectively managed in pearl millet by staggered sowing of A- and R-lines. In self-fertilized crops like sorghum and rice, hybrids with A- and R-lines differing in flowering by more than 10 days are discouraged. In sorghum, differences in flowering of A- and R-lines up to 10 days have been managed with staggered planting and/or with transplanting by several seed companies.

In case of rice, a CMS line takes 4-6 more days to flower than that of the B-line, on account of cytoplasmic effects. Since transplanting is commonly followed in rice, flowering synchrony of A- and B-lines can be managed by sowing A-line 4-6 days earlier than B-line in nursery and transplanting of both A- and B-lines can be done on the same day in the main field.

In hybrid seed production of rice, the difference between A- and R-lines of currently cultivated hybrids is 4-10 days in India. A difference of 7-10 days between A- and R-line has ideally been managed by many companies. Even 15-20 days difference can be managed through the current technology in India. In countries like China, a difference of up to 40 days in A- and R-line has been very well managed by sowing and transplanting of A- and R-line on different dates on one hand and managing synchrony of flowering through various management practices on the other.

The parental lines ratio will also depend on the breeding system of the crop, mode of pollination and pollen-producing ability. In case of pearl millet that generally produces profuse pollen, which become wind borne, the recommended ratio is 4:2 or 6:2 between A- and B-lines and 4:1 or 6:1 between A- and R-lines. If R-line is very vigorous and has height much taller than A-line (as in case of the parental lines of hybrid ICMH 451), farmers have found even 10-12A:1R ratio most economical. In sorghum, pollination is by wind and generally the R-lines produce profuse pollen. A- and B-lines flower almost simultaneously. The recommended ratio, as in pearl millet is 4:2 or 6:2 between A- and B-lines, and 4:1 or 6:1 between A- and R-lines, and these are followed generally in India and China. However, in countries like the USA, Australia, Colombia, Brazil, 8:1 or 12:1 ratio between A- and R-lines, is followed for hybrid seed production. In rice, the parental line ratio of 8-10:2 has been found to be ideal in India but in countries like China, where specific parental lines have been developed for hybrids, i.e. A-lines have been bred with very high outpollination potential and R-line with high pollen production potential, a ratio of from 12 : 2 to 16 : 2 is commonly followed. In order to enhance panicle exsertion from leaf sheath and stigma from lemma and palea, seed producers in China use GA_3 up to 200 g ha^{-1}. In India, GA3 dose of up to 50-60 g ha^{-1} is recommended, on account of high cost of GA_3. Technological innovations to lower the cost of hybrid seed production have been given by Mao *et al.* (1998).

In pearl millet and sorghum, where A-line and hybrid seed production is relatively easy, seed yields of 1.5-2.0 t ha^{-1} have been reported on CMS lines. Based on several experimental studies carried out on seed technological aspects in rice, package of practices have been developed for seed production of hybrids and CMS lines multiplication in China and in India (Virmani *et al.*, 1997). Seed yield of 1.5-2.0 t ha^{-1} of A-lines and hybrids can be obtained in India. Seed yields can be further enhanced through increasing the outcrossing potential of parental stock and refinement of seed production technology.

References

Anand Kumar K., Jain R.P., and Singh S.D. 1983. Downy mildew reactions of pearl millet lines with and without cytoplasmic male sterility. *Plant Dis.*, **67**: 663-665.

Anonymous. 1998. Hybrid Rice Progress Report 1989-98. ICAR project on promotion of Research and development efforts on hybrids in selected crops - Rice. Directorate of Rice Research, Rajendranagar, Hyderabad, India. 131 pp.

Appa Rao S., Mangesha M.H. and Rajagopal Reddy C. 1989. Development of cytoplasmic male-sterile lines of pearl millet from Ghana and Botswana germplasm. Pp 817-823. *In*: *Perspectives in Cytology and Genetics*, (eds.) G.K. Manna and U. Sinha, All India Congress of Cytology and Genetics, Kalyani Publishers, India.

Athwal D.S. 1961. Recent developments in the breeding and improvement of bajra (pearl millet) in the Punjab. *Madras Agric. J.*, **48**: 18-19 (Abstract).

Athwal D.S. and Virmani S.S. 1972. Cytoplasmic male sterility and hybrid breeding in rice. *In*: *Rice Breeding*. pp. 615-620. Int. Rice Res. Inst. Manila, Philippines.

Burton G.W. 1958. Cytoplasmic male-sterility in pearl millet [*Pennisetum glaucum* (L.) R.Br.]. *Agron. J.*, **50**: 230.

Brar D.S., Zhu Y.G., Ahmed M.I., Jachuck P.J. and Virmani S.S. 1998. Diversifying the CMS system to improve the sustainability of hybrid rice technology. *In*: *Hybrid Rice Technology*, p. 443. Proc. 3rd Intl. Symp. on Hybrid Rice, 14-16 Nov, 1996, Hyderabad, India.

Carnhan H.L., Erickson J.R., Tseng S.T. and Rutger J.N. 1972. Outlook for hybrid rice in USA. *In*: *Rice Breeding*, pp. 603-607, Int. Rice Res. Inst., Manila, Philippines.

Chhabra A.K., Rai K.N., Khairwal I.S., Sivaramakrishnan S. and Hash C.T. 1998. Mitochondrial DNA-RFLP analysis distinguishes new CMS sources in pearl millet (*Pennisetum glaucum* (L.) R.Br.). *J. Plant Biochem. and Biotech.*, **7**: 85-92.

Dalmacio R.D., Brar D.S., Ishii T., Sitch L.A., Virmani S.S. and Khush G.S. 1995. Identification and transfer of a new cytoplasmic male sterility sources from *Oryza perennis* into *indica* rice (*Oryza sativa* L.) *Euphytica*, **82**: 221-225.

Dalmacio R.D., Brar D.S., Virmani S.S. and Khush G.S. 1996. Male sterile line in rice (*Oryza sativa*) developed with *O. glumaepatula* cytoplasm. *IRRN.*, **21**: 22-23

Devi D.R. and Murthy U.R. 1993. Genotype environment interaction in tapetal development A_2 based sorghum genotypes. *Sorghum Newsl.*, **34**: 41.

Downes R.W. and Marshall D.R. 1971. Low temperature induced male sterility in Sorghum bicolor. *Aust. J. Exp. Agric. Anim. Husb.*, **11**: 352-356.

Duncan R.R., Bramel-Cox P.J. and Miller F.R. 1991. Contributions of introduced sorghum germplasm to hybrid development in the USA. *In*: *Use of plant introductions in cultivar development.* Part 1. (eds.) Shands H.L. and L.E. Wiesner, Proceedings of a symposium sponsored by Division C-1 of the Crop Sci. Society of America in Las Vegas, Nevada, USA, pp. 69-102, Crop Sci. Society of America, Madison, Wisconsin, USA.

Duvick D.N. 1965. Cytoplasmic pollen sterility in corn. *Adv. Genet.*, **13**: 1-52.

Ganesan K.N., Thiyagarajan K. and Rangaswamy M. 1998. Effect of minor genes in restoration fertility in CMS lines of rice. *International Rice Res. Notes.*, **23**: 9.

Govindraj K. and Virmani S.S. 1988. Genetics of fertility restoration of WA type cytoplasmic male-sterility in rice. *Crop Sci.*, **28**: 787-792.

Hanna W.W. 1989. Characteristics and stability of a new cytoplasmic-nuclear male-sterile source in pearl millet. *Crop Sci.*, **27**: 1457-1459.

Hoan N.T., Sarma N.P. and Siddiq E.A. 1997. Identification and characterization of new sources of cytoplasmic male sterility in rice. *Pl. Breed.*, **116**: 547-551.

House L.R. 1985. *A Guide to Sorghum Breeding.* (2nd Edn.). International Crops Research Institute for the Semi-Arid Tropics (ICRISAT), Patancheru 502 324, A.P., India. 206 pp.

Kishan A.G. and Borikar S.T. 1989a. Genetic relationship between some cytoplasmic male sterility systems in sorghum. *Euphytica*, **42**: 259-269.

Kishan A.G. and Borikar S.T. 1989b. Comparative performance of Maldandi *vs* Milo cytoplasm in sorghum. *J. Maharashtra Agric. Univ.*, **14**: 192-195.

Kumari S.L., Mahadevappa M. and Kulkarni R.S. 1998. Fertility restoration study in four WA CMS lines of rice. *Int. Rice Res. Notes*, **23**: 9.

Li Zhengyon. 1980. *Dian type hybrid rice.* Yuan People's Press, China.

Li Z.J., Zhang K.T., Oing Y.X. and Xo J.X. 1981. Effect of climate on male sterility in sorghum. *Agron. Sinica*, **7**: 129-133.

Lin S.C. and Yuan L.P. 1980. Hybrid rice breeding in China. *In*: *Innovative approaches in Rice Breeding.* pp. 35-51, Int. Rice Res. Inst. Manila, Philippines.

Lin S.C. and Ming S.K. 1991. *Rice varieties and their geneology in China.* Shanghai Scientific and Technical Publishers, Shanghai, China.

Lin Z. and Zhu Y. 1988. Rice male-sterile cytoplasm and fertility restoration. *In*: *Hybrid Rice.* pp. 85-102, Int. Rice Res. Inst. Manila, Philippines.

Madhava Menon P.M. 1959. Occurrence of cytoplasmic male- sterility in pearl millet (*Pennisetum typhoides* Stapf & Hubb.), *Curr. Sci.*, **28**: 165-167.

Mao C.X., Virmani S.S. and Ish Kumar. 1998. Technological innovations to lower the costs of hybrid seed production. *In*: *Advances in hybrid rice technology.* Proceedings of 3rd International Symposium on Hybrid Rice. November 14 -16, 1996 pp. 111-128.

Marchais L. and Pernes J. 1985. Genetic divergence between wild and cultivated pearl millets (*Pennisetum typhoides*). I. Male sterility. *Z. Pflanzenzuchtg*, **95**: 103-112.

Maruyama K., Araki K. and Keto H. 1991 Thermosensitive genetic male sterility induced by irradiation. *In*: *Rice Genetics II*, pp.227-232, Int. Rice Res. Inst. Manila.

Maunder A.B. 1972. Objectives and approaches to grain and forage sorghum improvement in the Americas. *In*: *Sorghum in Seventies*, (eds.) N. Ganga Prasada Rao and L.R. House. Oxford and IBH publishing Co., New Delhi, India. pp. 60-99.

Murthy U.R. 1986. Effect of A_2 cytoplasm on the inheritance of plant height in temperate × tropical sorghum crosses. *Sorghum Newsl.*, **29**: 77.

Nagur T. and Menon P.M. 1974. Characterization of male sterility-inducing cytoplasms in sorghum. *Sorghum Newsl.*, **17**: 18.

Pan Xigan. 1982. *Analysis of the effect of cytoplasm on heterosis*. Agricultural Science & Technology, Jiangxi.

Pradhan S.B. and Jachuck P.J. 1998. Krishna CMS line in the background of four different cytoplasmic sources in rice. *Int. Rice. Res. Notes*, **23**: 6.

Pradhan S.B., Ratho S.N. and Jachuck P.J. 1990. New sources of cytoplasmic-genetic male-sterility in cultivated *indica* rices. *Euphytica*, **48**: 215-218.

Rai K.N., Andrews D.J. and Suresh Babu V. 1985. Inbreeding depression in pearl millet composites. *Z. Pflanzenzuchtg*, **94**: 201-207.

Rai K.N. and Hash C.T. 1993. A new source of cytoplasmic-nuclear male-sterility in pearl millet. Cereals Program, ICRISAT. *Annual Report* (1992).

Rai K.N. 1995. A new cytoplasmic-nuclear male-sterility system in pearl millet. *Plant Breed.*, **114**: 445-447.

Rai K.N. and Thakur R.P. 1995. Ergot reaction of pearl millet hybrids affected by fertility restoration and genetic resistance of parental lines. *Euphytica*, **83**: 225-231.

Rai K.N. and Thakur R.P. 1996. Smut reaction of pearl millet hybrids affected by fertility restoration and genetic resistance of parental lines. *Euphytica*, **90**: 31-37.

Rai K.N., Andrews D.J., Rao A.S., Rajewski J.F. and Du R.H. 1999. Restorer sources of A_5 cytoplasmic-nuclear male sterility in *Pennisetum* germplasm and its implications in pearl millet hybrid breeding. *Plant Genetic Resources Newsletter*, 1999, **120** : 20-24.

Reddy Belum V.S. 1992. Varietal improvement: Three-way single cross hybrids. *In*: *Annual Report* 1991, pp. 66-67, International Crops Research Institute for the Semi-Arid Tropics (ICRISAT), Patancheru 502 324, Andhra Pradesh, India.

Reddy Belum V.S. 1997. Production of sorghum hybrids. *In*: *Training Manual on Development of Cultivars and Seed Production Techniques in Sorghum and Pearl Millet*, (eds.) Singh, Faujdar, K.N. Rai, Belum V.S. Reddy & B. Diwakar. pp. 31-33, International Crops Research Institute for the Semi-Arid Tropics (ICRISAT), Patancheru 502 324, Andhra Pradesh, India.

Reddy Belum V.S. and Stenhouse J.W. 1994. Sorghum improvement for the semi-arid tropic region: past, current and future research thrusts in Asia. *Punjabrao Krishi Vidyapeeth Res. J.*, **18**: 155-170.

Sahadevan P.C. and Namboodri K.M.N. 1963. National outcrossing in rice. *Proc. Indian Acad. Sci.*, Sect. B., **58**: 176-185.

Sampath S. and Mohanty H.K. 1954. Cytology of semi sterile rice hybrid. *Curr. Sci.*, **23**: 182-183.

Sarma N.P., Sundaram R.M., Thirumurugan T., Viraktamath B.C., Ramesha M.S., Yashitola J. and Sonti R. 2001. DNA marker technology for purity assessment of rice hybrids and parental stocks. *ICAR Science and Technology Newsletter*, **7**: 1-2.

Saxena K.B. 2001a. Heterosis breeding in Pulses - Problems and Prospects, pp. 61-72. *In*: *Pulses for Sustainable Agriculture and Nutritional Security.* (eds.) Ali, Masood, Chaturvedi S.K. and Gurha S.N. Indian Institute of Pulses Research, Kanpur.

Saxena K.B. 2001b. Prospects for hybrid pigeopea. Paper presented at National Symposium on Pulses and Oil Seeds on sustainable agriculture. Tamil Nadu Agricultural University, Coimbatore. 29-31 July 2001.

Scheifele G.L, Whitehead G.W. and Rowe C. 1970. Increased susceptibility to southern leaf spot (*Helminthosporium maydis*) in inbred lines and hybrids of maize with Texas male-sterile cytoplasm. *Plant Dis. Rep.*, **54**: 501-503.

Schertz K.F. and Ritchey J.R. 1977. Cytoplasmic-genetic male-sterility systems in sorghum. *Crop Sci.*, **18**: 890-893.

Schertz K.F. and Pring D.R. 1982. Cytoplasmic sterility systems in sorghum. *In*: *Proceedings of an International Symposium on Sorghum in the Eighties*, (eds.) Rao, N.G.P. & L.R. House. pp. 373-383, International Crops Research Institute for the Semi-Arid Tropics (ICRISAT) center, Patancheru 502 324, Andhra Pradesh, India.

Schertz K.F., Sotomayor-Rios A. and Torraco-Cardona S. 1989. Cytoplasmic-nuclear male-sterility opportunities in breeding and genetics. *Proceedings of the Grain Sorghum Research and Utilization Conference*, **16**: 175-186.

Schertz K.F., Sivaramakrishnan S., Hanna W.W., Mullet J., Yi Sun, Murty,U.R., Pring D.R., Rai K.N. and Reddy Belum V.S. 1997. Alternate cytoplasms and apomixis of sorghum and millet. *In*: *Proceedings of the International Conference on Genetic Improvement of Sorghum and Pearl Millet*, 22-27 September 1996, Holiday Inn Plaza, Lubbock, Texas, USA. pp.213-223, Texas Agricultural University

Scheuring J.F. and Miller F.R. 1978. Fertility restorers and sterility maintainers to the milo-*kafir* genetic cytoplasmic male sterility system in the sorghum world collection, *Misc. Publ., Texas A&M Agricultural Experiment Station*, College Station, Texas, USA, 1367:47 pp.

Senthil N., Ramasamy P. and Khan A.K.F. 1998. Fertility restoration and heterosis involving different cytoplasms in sorghum [*Sorghum bicolor* (L.) Moench] hybrids. *J. Genet.& Breeding*, **52**: 339-342.

Seth P., Sane A.P., Nath P. and Sane P.V. 1996. Molecular characterization of mitochondrial genomes of rice lines containing wild abortive (WA) male-sterile and fertile cytoplasms. *J. Plant. Biochemistry & Biotechnology*, **5**: 75-82.

Sharma H.C., Vidyasagar P., Abraham C.V. and Nwanze K.F. 1994. Effect of cytoplasmic male-sterility in sorghum on host plant interaction with sorghum midge, *Contarinia sorghicola*. *Euphytica*, **74**: 35-39.

Shen Xiaobang. 1982. Preliminary study on the genetic effect of rice sterile cytoplasm. I. The effect of sterile cytoplasms on the agronomic characters of hybrid rice. *J. Huazhong Agric. Coll.*, 3.

Shinjyo C. and Omura T. 1966. Cytoplasmic male sterility in cultivated rice, *Oryza sativa* L.I. Fertiles of F_1, F_2, and offsprings obtained from their mutual reciprocal backcrosses and segregation of completely male sterile plants. *Jpn. J. Breed.*, **16**: 179-180.

Shinjyo C. 1969. Cytoplasmic-genetic male-sterility in cultivated rice *Oryza Sativa* L. II. The inheritance of male sterility. *Jpn. J. Genet.*, **44**: 149-156.

Siddiq M.A., Madrap I.A. and Ambekar S.S. 1993. Heterosis studies in sorghum. *In*: *The Indian Journal of Genetics & Plant Breeding*. Abstract of papers. pp. 123-124. National symposium on Plant Breeding Strategies for India 2000 A.D. and Beyond, Dec 25-27, 1993.

Sivaramakrishnan S., Seetha K. and Reddy Belum V.S. 1997. Characterization of the A_4 cytoplasmic male-sterile lines of sorghum using RFLP of mt DNA. *Euphytica*, **93**: 301-305.

Stephens J.C. and Holland P.F. 1954. Cytoplasmic male sterility for hybrid sorghum seed production. *Agron. J.*, **46**: 20-23.

Virk D.S. 1988. Biometrical analysis in pearl millet: a review. *Crop Improv.*, **15**: 1-9.

Virmani S.S., Govind Raj K., Casal C., Dalmacio R.D. and Aurin P.A. 1986. Current knowledge and outlook on cytoplasmic-genetic male-sterility and fertility restoration in rice. *In*: *Rice Genet.*, pp. 633-647, Int. Rice Res. Inst., Manila, Philippines.

Virmani S.S. and Shinjyo C. 1988. Current status of analysis and symbols for male sterile cytoplasm and fertility restoring genes. *Rice Genet. Newsl.*, **5**: 9-15.

Virmani S.S. and Wan B.H. 1988. Development of CMS lines in hybrid rice breeding. *In*: *Hybrid rice*. pp. 103-114. Int. Rice Res. Inst. Manila, Philippines.

Virmani S.S. 1994. Prospects of hybrid rice in the tropics and sub-tropics. *In*: *Hybrid Rice Technology - New Developments and Future Prospects*. pp.7-20. Int. Rice Res. Inst, Los Banos, Philippines.

Virmani S.S. 1996. *Hybrid Rice*. Adv. Agron., **57**: 377-463.

Virmani S.S., Viraktamath B.C., Casal C.L., Toledo R.S., Lopez M.T. and Manalo J.O. 1997. *Hybrid Rice Breeding Manual*. Int. Rice Res. Inst., Las Banos, Laguna, Philippines. P.151.

Webster O.J. and Singh S.P. 1964. Breeding behavior and histological structure of a nondehiscent anther character in *Sorghum vulgare*. *Crop Sci.*, **4**: 656-658.

Yadav O.P., Manga V.K. and Gupta G.K. 1993. Influence of A_1 cytoplasmic substitution on the downy mildew incidence in pearl millet. *Theor. Appl. Genet.*, **87**: 558-560.

Yang Rencui. 1980. Effect of sterile cytoplasm of wild abortive rice on F_1 generation *J. Fujian Agric. Coll.*, **12**: 14-21.

Yuan L.P. 1977. The execution theory of developing hybrid rice. Zhonggue Nongye Kexue. *Chinese Agric. Sci.*, **1**: 27-31.

Yuan L.P. 1985. *A concise course in hybrid rice*. Human Technological press, China. p. 168.

Yuan L.P. and Virmani S.S. 1988. Organization of hybrid rice breeding program. *In*: *Hybrid Rice*. pp. 33-37. Int. Rice Res. Inst., Manila, Philippines.

Yuan L.P, Virmani S.S. and Mao C.X. 1989. Hybrid Rice - Achievements and future outlook. *In*: *Progress in Irrigated Rice Research*. pp. 219-235, Int. Rice Res. Inst., Manila, Philippines.

Plant Breeding - Mendelian to Molecular Approaches
H. K. Jain and M. C. Kharkwal (eds.)

Quantitative Trait Improvement: Classical and Molecular Concepts

V. Arunachalam[1], K.V. Prabhu[2] and V. Sujata[2]

Abstract

A desired end product like grain yield is a complex function of a number of component traits. They are, in general, governed by a number of interacting genes following Mendelian inheritance. Such traits are quantitative in nature and are expressed in an environment. Methods of improving such quantitative traits (QTs), are influenced by environmental variation. Mendelian methods can, therefore, be time-ineffective in achieving a desired level of QT improvement. In contrast, recent progress in identifying molecular markers which behave like Mendelian genes in inheritance pattern are observed to open up new pathways. Marker-assisted QT improvement methods are viewed as an option to obtain targeted gains in a short-term. At a time when quantum increases in yield and quality are desired with a genetic control of biotic and abiotic stresses, molecular methods are projected to become dominant alternatives to proven Mendelian methods of breeding. The possibility to locate a number of markers closely linked to QTs is seen as a point in favour of efficient QT dissection and repair of deficient section. This paper underlines, therefore, the basic principles behind Mendelian and molecular methods of QT improvement and attempts to evaluate their utility and efficiency. After reviewing a range of published studies, the possibility and feasibility of QT improvement are brought to light.

Introduction

Quantitative analysis is commonly resorted to by researchers in many fields such as plant breeding, genetics, molecular biology, agronomy, economics, statistics, psychology, medicine and the like. In most cases, the analysis is carried out on some measurable attributes. Most often they are called metric characters. Even a qualitative measurement like, the intensity of a disease expression can be graded into categories from no expression to low, moderate, high or terminal expression. In fact, the qualitative description admits of a quantitative score, thereby making it appropriate for a quantitative analysis. Obviously, the possible control of the trait by more than a few Mendelian genes is underlined.

[1]M.S. Swaminathan Research Foundation, Chennai 600 113, India
[2]Indian Agricultural Research Institute, New Delhi 110 012, India

Plant breeding deals with traits that are associated with economic gain. Such traits require constant genetic enhancement and in turn, appropriate quantitative analysis. Therefore, plant breeding paradigms are often pioneered on principles of Mendelian genetics and deal with quantitative trait (QT) improvement. The area of breeding for high performance is vast and varied. Over the last century, ever since the understanding of Mendelian genetic principles led to scientifically designed plant breeding, gradual but significant improvement of QTs has been achieved in most crops by careful analysis of data on breeding populations. Over the last decade, techniques based on molecular biology incorporating isozyme/protein and DNA polymorphism, have added further strength to the analytical processes behind efficient plant breeding methods. Whatever the method, classical or molecular, it should be able to scan the total variation in the breeding population and partition it into genetic (heritable) and environmental components. In this review, we focus on two of the common plant breeding processes, genetic characterization and QT improvement, in the light of Mendelian and molecular approaches.

Mendelian Genes Governing Quantitative Trait

Quantitative Trait (QT)

By definition, QT is one whose measurements on individuals, over space (locations, environments) and time (years, seasons) are supposed, in principle, to follow a continuous distribution. In plant improvement context, it is assumed that the continuous distribution would be a normal distribution. In cases when the population does not follow normal distribution for the QT, the variable is transformed, wherever feasible, to restore normality. Based on a normal distribution of QT values, some useful basics relevant to plant breeding are as follows:

- QTs are measured on phenotypes, that is the genotypes expressed in an environment. It is hardly feasible to separate the genotypic and environmental values from the phenotypic value.
- Assuming therefore a genotype × environment interaction, the phenotypic variance, P^2 can be given as $P^2 = G^2 + E^2 + G \times E$ where P^2 represents the genetic, E^2, the environmental and G × E, the genotype × environment interaction, variance.
- Considering the genotypes ZZ, Zz, zz given by the gene Z with alleles Z and z to be explained by successive gene substitutions, the genetic variance G can further be partitioned into additive (A) and dominance (D) variance, so that $G^2 = A^2 + D^2$. The additive variance A^2 can be computed by assigning the values 1, 0 and –1 to the genotypes ZZ, Zz and zz.
- QTs, in general, are governed by many genes, each producing its own effect, small or big, positive or negative, but together, producing a pronounced inter-genic interaction (epistasis) effect. Because of the environmental influence as explained above, the QT values vary across environments and across genotypes. The QT variation across individuals (genotypes) would then reflect genetic variation influenced by the environmental effects peculiar to that environment. Hence it is feasible to use QT

variance for a comparative evaluation of the genetic potential of varieties in a specific environment.

- The genes governing QTs follow Mendelian inheritance; this inheritance, however, is not always distinguishable in every individual gene, as often the trait expression showing a range of results by the action of more than one gene and the effect of the environment. With these provisions, we may call the genes governing QTs as Mendelian genes.

Molecular Markers

In contrast, the advent of the application of concepts of molecular biology gave rise to a set of molecular marker loci such as restriction fragment length polymorphism (RFLP), randomly amplified polymorphic DNA (RAPD), amplified fragment length polymorphism (AFLP), mini- and micro-satellites, etc., with analogous properties of Mendelian genes. The salient difference is that, the marker loci lack an expression of their own and are distinguishable only through their nucleotide sequence size. The logic behind marker loci has been distinctly developed and would not be repeated here.

Molecular markers are insensitive to the environment (contrasting Mendelian genes controlling a QT) validating the model P = G as an initiating step. So it is affirmed that molecular markers provide genetic values directly. But the marker loci do not have an expression. Hence a very tight linkage between a marker locus and a Mendelian gene governing a QT is invoked to measure P (= G). Where such tight linkages are unavailable, it is not possible to estimate G (as = P). But it must be recognized that P of a QT has already an environment effect in it. Regardless of the fact that marker loci are insensitive to E, when a measure of P is used to assign a G value (P value) to a molecular genotype, E effect gets automatically injected into it.

The Contrast Between Mendelian Genes and Molecular Markers

At best therefore, when molecular techniques recognise a genotype carrying either 2, 1 or zero dominant genes, they are assigned as genetic values to the respective genotypes. Those values are fixed and remain constant for every molecular genotype regardless of organisms (be it a crop plant, animal, human being or whatever) or QTs (be it flowering time, seedling or plant height, grain yield or whatever) (see, for example, Martin *et al.*, 1989; Jansen and Stam, 1994 and Zhang *et al.*, 1994). Genetic variance calculated on this basis is, with respect, to a single marker. When multiple markers are available, at best such individual genetic variances can be added up. The important epistatic variance associated with polygenes cannot be calculated and included.

The important drawbacks in the formulation are, namely, the impossibility of accounting for epistatic variation of a QT and the silent inclusion of environmental contribution when measuring P as G, which are highlighted as strengths [independence of epistatis, insensitivity to environment] of molecular markers, a fact that needs careful recognition.

The contrasting conceptualisation of Mendelian and marker genes diverges further when inferences based on QTs are invoked to synergise with those based on molecular genetic parameters. The areas of such inferences are widely different and it is not possible to deal with every one of them. We restrict our attention, therefore, only to those which concern breeding for QT improvement. While we cruise through important breeding strategies based on Mendelian QTs, we would review relevant developments of molecular applications through related concepts and case studies from published literature.

Genetic Characterization

We saw that QTs are the expressed effects of underlying genotype in association with its environment. A genotype can then be effectively defined by a set of QTs. How many QTs and what type of QTs are live questions. Over the years, it has been generally accepted that those related to the fitness of an organism (Murty and Arunachalam, 1966), particularly those related to survival and those that describe life history characteristics (Storfer, 1996), deserve preference. In valid genetic distinction and conservation programmes, a host of biological and economic factors also deserve weightage in the choice of traits (Avise, 1989), in addition to cultural traditions altering genetic selection (Spinney, 1997).

Genetic Divergence

Genetic distinction or relatedness is an important concept in plant breeding, for the selection of parents as well as for distinguishing among possible recombinants in the breeding populations. Such distinction is made using a measure of genetic distance or genetic divergence. In essence, a set of genotypes is sought to be classified into groups on their *inter-se* genetic distances. Ideally, a group should be such that the 'within-group' distance is far smaller than the 'between-group' distance. It is also understood that it would be of little use if genotypes are grouped on just one QT, because plant breeding almost never is practiced with just one trait in mind. Equally well understood is the fact that genetic component of variance is trait-specific and is relevant only to the particular trial. Thus a divergence analysis needs to include as many important traits as possible to circumscribe, to a large extent, the total genetic information based on which genetically divergent groups can be distinguished. A number of traits would jointly assess the genetic potential of an individual, giving appropriate weightage to the heritable and environmental contributions of trait expression.

Classical Concepts: Measures of genetic divergence and methods of grouping are varied in number, application and underlying logic. One divergence measure is given by multivariate distance (D^2).

As mentioned earlier, a variety can be visualised to be firmly characterised by a set of selected QTs; each QT constitutes an axis in a multi-dimensional plane and the QT values give the co-ordinates of the variety and delineate it on the plane. QTs being governed by a constellation of genes, the multivariate distance between two varieties would measure the genetic distance or divergence between them. The divergence measure is then provided by the multivariate distance (given by Mahalanobis' D^2 statistic). The possible divergence values between pairs of varieties are computed using the distance statistic (see, for example, Rao, 1952; Arunachalam, 1981). Scanning the pattern of D^2 values, a raw grouping is obtained which is then fine-tuned following a method due to Tocher, explained in Rao (1952). A norm is set to decide the difference between any two D^2 (or alternatively the divergence between any two varieties) within a group that can be defensibly tolerated. That norm cannot be set following a fixed rule (as was done by Singh and Choudhary, 1983) as it is dependent on the range and distribution of observed D^2 values. Once the groups are firmed up, intra- and inter-group average D^2 are calculated; they delineate the varieties into clearly defined groups.

The method first developed and applied in 1965 (Murty *et al.*, 1965) was adopted widely across crops and varied genetic material for the next decade-and-a-half. In many instances, the grouping was confirmed by principal component analysis also. Computer software for calculating D^2-values, which otherwise involve complex computations, was also developed (Murty and Arunachalam, 1967) enabling grouping based on a large set of QTs. Yet dealing with a large number of varieties renders scanning a large number of distances (between all possible pairs of varieties) quite arduous. Therefore a modified method of grouping was developed (Vairavan *et al.*, 1973; Durga Prasad *et al.*, 1985). As the utility of D^2-statistic in genetic classification became more and more pronounced, further modifications were made (Arunachalam and Bandyopadhyay, 1984a; Arunachalam *et al.*, 1998) to enable grouping of varieties in four divergence classes uniformly. This enabled an unbiased comparison across experiments (or even crops where such a process is justifiable). One of the utilities of the genetic grouping is efficient selection of parents for a hybridisation programme. The relevant hypothesis was that F_1 hybrids between parents, chosen from widely divergent groups, would more often be heterotic than F_1s between mildly divergent parents. However, there were no clear guidelines as to how divergent the parental groups need to be and whether the divergence norm is the maximum of inter-group distances, or should be based on a comparative scale irrespective of such magnitudes.

The modified grouping procedure provides four divergence classes (DC), DC1 containing genotypes with large *inter-se* distances and DC4 with small distances. The association between parental genetic divergence and F_1 heterosis was then examined in a range of experiments to arrive at a focused hypothesis, namely, the F_1s between parents chosen from intermediate divergence classes, DC_3 and DC_2 have a higher probability of being heterotic than other F_1 combinations. This hypothesis has been test-verified in a range of crops (Arunachalam and Bandyopadhyay, 1984b; Arunachalam *et al.*, 1984 and Durga Prasad *et al.*, 1985). Thus we see that proven genetic concepts concerning breeding for QT improvement can continuously be improved to become more powerful and efficient by continuing basic and strategic research consistently.

Molecular Concepts: Refined and developing techniques of a vast kind are now available to identify molecular marker loci spanning the entire stretch of DNA across the coding and non-coding regions. As has been pointed out, in contrast to a phenotypic expression of QT genes, the marker loci express only through their presence or absence in linkage with the closely associated QT alleles. Therefore, molecular genetic distance can only be defined on the frequency of genes/alleles at a segregating locus. Such distances were initially developed to study the affinity between human populations where, unlike in plants, it is not easy to locate quantitative trāits that define an individual satisfactorily. Under a set of conditions, the genetic distance concept was first developed (Nei, 1972) to deal with local populations within species.

A coefficient of identity, I was defined based on the probability of a gene from one individual being identical to another from another individual. The genetic distance between the two individuals was defined to be $D = -\log_e I$.

It was recognised, however, that Nei's measure of D is affected by a number of factors such as detection of gene differences, varying rate of nucleotide substitution at different loci, etc. The applicability of the measure is extended to any population, including selfing populations. Further I and D depend solely on gene frequencies than on genotypic frequencies. This work was extended to measure "nucleotide diversity" under a restrictive assumption that all nucleotides are distributed at random over the DNA sequences with a given $G + C$ content (Nei and Li, 1979). The concepts underlying genetic distance measure, D have been further refined (Nei, 1987), though the basic definition given earlier remained the same. It is now used almost exclusively in molecular genetic studies as a software incorporating it, is readily available for molecular data analysis (see, for example, Wang *et al.*, 1992).

An additional dimension to molecular genetic divergence was added by Hamrick and Godt (1990), especially in plant species analyses, when dealing with isozyme data which are data on actually expressible traits. It was suggested that the percentage polymorphic loci, the mean number of alleles per locus, the effective number of alleles per locus and the mean genetic diversity over all the loci be calculated. While the first parameter was sufficient to distinguish among species, the latter three were essential for an analysis of inter- and intra- population level genetic divergence. Further, accounting for the quantitative differences in the allelic expression of isozymes through density measurements on gels run with *per-se* protein extracts, Arunachalam *et al.* (1996) added the parameters variance for relative mobilities and variance for relative densities among isozyme loci. These parameters put together make the divergence analysis with isozymes (as expressions, not merely as inert markers scored for their presence or absence) more comparable to that based on the QTs of interest.

A Comparative Analysis of Genetic Distance Based on QTs and on Marker Allele Frequency

The fundamental aim in measuring genetic distances is to identify individuals (essentially phenotypes, equivalently, genotypes expressing in an environment) who share a similar performance. This definition is purposeful; it springs immediately the following doubt: In view of environmental effects, phenotypic performance may be deceptive in identifying genetic affinity. But, in contrast, identifying genetic affinity through the frequency of marker genes which are independent of environmental effects may not help either. In reality, we face a different type of problem. It is known that molecular techniques can identify only single diallelic marker loci. QTs are usually polygenic with high epistasis. Even when more number of marker loci are identified, they are independent and free from epistatic effects. It is possible, therefore, that two genotypes decided to be genetically close on one or more marker genes show completely differing phenotypes when stable and uniformly high performance is the ultimate target of a breeding process. Recognising that the genetic distance given by I or D depend solely on gene and not on genotypic frequencies (Nei, 1972), it is a moot question whether such a measure can identify genotypic affinity. In experiments on potato, it was found that divergence estimates based on quantitative measurements of proteins and esterases did not correlate with genetic divergence estimated on morphometric traits (Loiselle *et al.*, 1991). In this case, divergence was measured by Mahalanobis' distance (D^2). While divergence (M) on proteins and esterases, separated by PAGE by three methods - a buffer of pH 7.9 and another of pH 8.9 for proteins, and of pH 8.9 again for esterases - was based on the number of bands and band intensities, that on QTs included agronomic data (A) on seven traits, three of which are scores, and general combining ability effects (G). In addition inbreeding coefficients of parents were calculated on 12 generation data (F). The divergence estimates are made on principal component estimates. We see, in this study, the QT measurements were put to complex transformations before calculating D^2s. For example, the per cent contribution to total variation of the six or seven principal components would vary across the three cases, A, M and G. D^2 based on such components may not provide a valid comparison. In contrast, in a detailed study with isozyme variation converted into 5 QTs (Arunachalam *et al.*, 1996), a high agreement of up to 78 % between the grouping made on six morphometric traits and that made on 10 isozyme QTs (5 for peroxidase + 5 for esterase) was recorded (Bharti, 2000). The fundamental and key difference in the above studies is a novel characterisation of isozyme variation into QTs which was absent in the study of Loiselle *et al.* (1991).

The concept of intelligent gene search replacing systematic gene research (Antonarakis, 1994), and markers strategically located in gene-rich (or GC rich) regions instead of the entire genome scan, may be of relevance in genetic identification studies too. Such a clue gains ground if we consider the study of 93 accessions of rice representing 21 species (Wang *et al.*, 1992). It was found that RFLP analysis using up to 25 single copy probes showed segregation of polymorphic accessions for only one or two probes. A majority (78 %) of accessions turned out to be monomorphic. Most of the probes used in the study detected polymorphism between cultivated and wild rices only. In addition, a study on phylogenetic classification has cautioned that it is not logical to equate gene phylogeny and organismal (including genotypic) phylogeny (Doolittle, 1999).

In the light of the above analogy, a comparative sketch of molecular and morphometric distances may help to consolidate our viewpoints.

A basic comparison (Table 1) of the distance properties reveals that morphometric distance is more directly relevant to strategies for QT improvement in terms of parental choice and F_1 heterosis than molecular distance.

Table 1. The salient differences between molecular (Nei's D statistic) and morphometric (Mahalanobis' D^2 statistic) distances

	Molecular distance	Morphometric distance
Based on	allelic concentration in the discovered marker genes expressed	expressed QTs in the concerned environment
Individuals defined on	marker genes	a set of identified QTs distinguishing performance
Genetic constitution	a number of non-interacting independent genes	a number of interacting genes displaying epistatic effects
Environment	independent	highly dependent
Formula	function of probabilities of a paternal and maternal gene being identical	multivariate distance statistic using environmental variation to correct genotypic variation
Genetic classification	usually dendrogram allowing flexibility of inferences	grouping devoid of bias into 4 divergence classes
Choice of parents for realising a heterotic F_1	not targeted	specifically targeted

Marker - QT Linkage

The disparities between molecular and morphometric evaluation would be narrowed if (a) there are efficient ways of determining marker - QT linkage and (b) an estimate of linkage can be obtained satisfying consistency, efficiency and sufficiency properties. Assuming that such a marker, tightly linked to a QT is found, it should imply that any genetic alteration in the marker should reflect in the performance of the associated QT.

One can visualize the association between a molecular marker that is qualitative in its expression and a QT with a possible analogy that can be drawn on the association between a morphological marker that is qualitative and a QT. The first example of an association between marker locus and QT was reported as early as 1923 by Sax between a pigment locus and seed size in *Phaseolus vulgaris*. One of the parental lines was homozygous for dominant pigment allele P and had seeds that weighed on an average 48 cg. The other parent was homozygous for the recessive allele p that codes for lack of pigmentation with an average seed weight of 21 cg. The F_2 of the cross is (as adapted from Falconer and Mackay, 1996) detailed below:

Marker genotype	PP	Pp	pp
Seed weight genotype	SS	Ss	ss
Seed weight (cg)	30.7	28.3	26.4

If we assign genotype PPSS to parent 1 and ppss to parent 2 based on pigmentation and seed weight, then as explained above under heading QT, genetic effects can be computed for the QT, seed weight. We can now estimate the mean of each marker genotype class by multiplying the frequency of each genotype by its genotypic value and summing within the marker class. The F_2 mean can then be calculated using the Mendelian frequencies of 1/4, 1/2, 1/4 for the marker genotypes PP, Pp and pp as usual. If the SS locus is linked to the PP locus, we can see, without going into intricacies of linkage analysis between the two loci, that there is a significant deviation in the (PPSS - ppss) genotypes equal to 1/2 (30.7 – 26.4) = 2.15 cg. This gives us the additive effect in the population. Similarly the differences between the PpSs genotype and average of the two homozygotes equal to 28.3 - 1/2 (30.7 + 26.4) = – 0.25 cg. We note that the difference between the parents for seed weight was 27 cg and that in the F_2 for the same genotypes is 4.3 cg. The latter is about 16 % of the original parental differences, implying that 16 % is the extent of association between the marker locus PP and the associated locus SS. In dealing with the above example, we assumed that the seed weight trait is also controlled by a single gene. Had it been only one major locus, what is now known as QTL or quantitative trait locus, and had there been complete linkage between PP and this QTL locus, the marker class difference given by SS - ss in the F_2 would have been equal to the parental difference. However, it can rarely be so, as QTs are obviously under the influence of several genes as we could infer from this example. But it is clear from the the above simple experiment by Sax (1923) that a marker that segregates in simple Mendelian pattern can be linked to a QT and at the same time, the particular marker may not associate itself with all the loci (QTLs) that jointly express as the QT. One has to therefore explore many markers for establishing linkage with the remaining QTLs, so that together the markers can explain the total variation in the QT. Although in the above example, we have used a morphological marker, the analogy is identical to that used in associating a molecular marker with a QT.

The above analogy may look simple to explain the association between a molecular marker and QTLs. But many problems arise in practice regarding optimal methods of statistical analysis and genetic interpretation. For example, in the above case, we did not consider how much of the expression in the seed weight in the parental, F_1 and F_2 generations was influenced or modified by the environment in the three seasons, when Sax conducted his experiments. After having understood how a molecular marker can be linked to a QT, one has to be equally cautious in extending information known on one set of reference population to another.

Thus, in case a marker is used to identify a QT like insect resistance in a known genetic source, it should imply that an unknown genetic source, carrying the marker should reveal insect resistance. Clear-cut evidence of these null hypotheses has yet to become available. Most of the studies reported in the literature deal with markers with their locations in a linkage map linked to QTs based on already developed genetic material carrying the QTs.

Linkage Mapping with Molecular Markers for QTs

Another area concerns new approaches to increasing the power of QTL detection and mapping. For instance, the approach of Jansen and Stam (1994) uses markers as cofactors (as a working substitute for simultaneous mapping of multiple QTLs) and parental and F_1 data to fix the joint QTL effects and the environmental error. According to them, their procedure decomposed more powerfully the phenotypic variation into genetic and environmental variation and thus improved the accuracy of QTL mapping. But in the case of multiple loci, recombination fraction among the marker loci is taken from an already available linkage map of the mapping population. Thus, the genetic linkage map of the markers is fixed first and a putative QTL is moved along the genetic map, so that for a given map position of each QTL, all recombination frequencies are fixed. The interval mapping procedures and the regression models used in this approach assume additivity of effects over QTLs. Soller *et al.* (1976) while handling isozyme loci as markers linked to a quantitative trait have indicated that depending on additivity or dominance among the QTs, one has to resort to either back cross, F_2 or other combinations. They also suggest that depending on the linkage phase, different mapping populations of appropriate size need to be used for mapping the QTLs. Similar considerations were explained further by Beckmann and Soller (1988) on matings and the size of population required to detect linkage between marker loci and loci affecting QTs. Thus all the assumptions imbedded in the process of QTLs dissection are not realistic to match practical situations.

In the usual methods, a large sample size is required to detect marker - QTL linkage. Since this would increase cost and time, Weller and Wyler (1992) have suggested selective genotyping in place of random sampling. Motro and Soller (1993) have systematically explored the role of sequential sampling in determining marker - QTL linkages. They concluded that a combination of sequential sampling and selective genotyping can be expected to reduce the overall number of marker evaluations by an 8-fold factor as compared to the classical, fixed-size sampling from an unselected population. However, how successful it would be in practice is not evident, particularly when the yardstick to identify the sample size of the 'selective genotyping' is not unique and a definite procedure to combine sequential sampling with selective genotyping is unavailable.

The concept of QTL dissection gets more complicated when recent investigations indicate that moderate variation in Simple Sequence Repeat (SSR) number can affect phenotype (Kashi *et al.*, 1997). The observation that there are substantial statistical problems that need to be overcome before inferences of epistasis based on the molecular marker approaches can be made (Fenster *et al.*, 1997) adds another note of caution. Greater stress has been made by Ellis (1986), that ignoring gene interactions as also the effect of environmental factors on QTs would severely limit the utility of RFLP (or for that matter any molecular marker) approach to QT improvement. In an illustrative paper titled 'character dissection', Michelmore and Shaw (1988) echo similar views and state that biometrical approaches have been exceptionally successful in the absence of precise knowledge about the number or location of the genes involved. Marker-based selection is likely to prove highly effective in specific cases; however, careful theoretical consideration of direct and indirect selection alternatives may save considerable time and resources. In a recent review, Simmonds (2000) has reiterated that biometrical approaches are still robust in helping plant breeders for effective improvement of quantitative traits.

Practical Use of Marker - QT Association

There are also reports that complex QTLs have been detected by molecular markers. Using usual procedures of interval mapping and an RFLP map earlier generated as a base, Bagali *et al.* (1999) observed that they could identify two major and five minor QTLs for GS3, grain filling duration (between flowering and maturity) in rice. They identified that the chromosomes 1, 3, 7, 8, 11 were involved in the QT, GS3. But it is intriguing to note some unproved conclusions of the study: (*a*) some of the QTLs identified in this study remained stable across rice populations and across locations, (*b*) marker - assisted selection "will facilitate genetic metamorphosis ameliorating fundamental processes of yield formation". Gura (2000) has listed a number of methods and techniques becoming available to 'dig up' plant genes. Some of them are BLAST search (computer scan of databases), synteny (comparison of chromosomal gene maps from different species), chimeplasty (inserts DNA/RNA hybrids into cells), TUSC (trait utility system for corn; inserts DNA that jumps into genes), activation tagging (inserts DNA enhancers via a plant cell-infecting bacterium), RNA silencing/gene over- expression (infects tobacco plants with genetically altered TMV), microarray analysis (DNA snippets on chips), proteomics (two dimensional gels of protein expression) and metanomics (gas chromotography and mass spectrometry profiles). With all the sophisticated tools now available it has been shown that the ultimate goal of gene manipulation for optimal benefits continues to be an evasive outcome. An example is cited where the starch content of potatoes was sought to be increased and the sugar levels reduced as a requirement for potato chips industry. Metanomics interventions ended up with transgenic tubers losing both sugar and starch (Gura, 2000). Likewise, in the case of golden rice, a question was raised whether production of carotenoids in rice endosperm will entail any metabolic trade- off. Shunting more of the common precursor GGPP into carotenoid production might result in a decrease in other compounds, whose synthesis is dependent on GGPP. In tomatoes a similar problem was encountered. It is to be seen whether the golden rice is free from those problems which can be ensured only from vast field testing (Guerinot, 2000).

Detailed studies on linkage between RFLP markers and soluble solid content (SS), an important QT in tomato, were made by a group of workers during the late 80s (see, for example, Osborn *et al.*, 1987; Tanksley and Hewitt, 1988; Paterson *et al.*, 1991). The conceptual utility of marker-assisted selection (MAS) for the QT was tested across different crosses between tomato species, different generations up to F_3 and different locations and year environments. A question of grave importance was whether SS genes in the genus *Lycopersicon* would map to the same or similar orthologous genomic regions regardless of the genetic background or the environment. When searched across data published in 4 papers, out of a total of 28 possible positions for 7 SS loci, only 5 were common to at least two (Arunachalam and Chandrashekharan, 1993). The data of Paterson *et al.* (1991) also showed clearly that the positions of SS genes varied even with environment. These data demonstrate that even for crosses between related species, QT polygenes will have to be located everytime a new species or even a new variety is employed as a parent in a breeding programme and in every environment. This requirement is highly expensive in terms of cost and time and logically impractical even in elementary backcross and F_2 breeding schemes. The magnitude of cost would escalate several fold in current breeding schemes employed to evolve high yielding varieties, particularly with complex pedigree involving multiple mating and exhaustive selection methods.

Phylogenetic studies is another important area which conventionally advocates the use of an 'exhaustive' set of QTs 'total evidence' approach, since it is supposed to maximize 'informativeness' and 'explanatory power' of the character data used in the analysis (Huelsenbeck *et al.*, 1996). After examining three methods of analysis i.e. total evidence, separate analysis and conditional combination approaches, it was suggested that total evidence would be a good rule of thumb to use when data heterogeneity is absent or uncommon. Obviously, if this suggestion is to be implemented, molecular marker identification and MAS would become too huge and inoperative compared to improved Mendelian approaches. Further, recent studies using natural populations of *Drosophila* show the need for extreme care in sorting out the large number of molecular polymorphisms (or marker loci) found at most loci to identify the nucleotide changes responsible for phenotypic variation in complex traits. Some pointers to the need for extreme care are provided by the identification of three different regions within Adh locus governing alcohol dehydrogenase activity in *D. melanogaster* forming superalleles with large effects on transcription rates, the possibility of non-coding regions playing an active role in quantitative variation, tracking marker associations directly to the actual nucleotides leading to the trait differences (quantitative trait nucleotides, QTN), and the problems of identifying the QTN within the QTL (Phillips, 1999).

Isozyme Marker Variation as Associated With QT Variation: Without dissecting the QT into QTLs through complex assumptions with the help of DNA markers, it is worthwhile to consider, as an effective measure, isozyme markers whose expression is quantifiable for exploring their possible association with the target QTs. In the backdrop of the debility of marker-QT association, protein-based markers like isozymes have a higher potential to understand the nature and magnitude of QT variation. In analysing the question whether proteins predate DNA, Freeland *et al.* (1999) endorsed the view that 'modern organisms process genotype into phenotype through two distinct stages. First DNA genes are transcribed into RNA messages (mRNA); these messages are then translated into proteins' which subsequently give rise to expression as QTs. In the process of transcription and translation, environment plays a role. Therefore, if the isozyme variation which is the expression of protein products of the gene/genes, can be quantified efficiently, can logically reflect QT variation, being a precocious precursor of the QTs. It is then logical to postulate an association between isozyme and QT variation.

Exhaustive studies in establishing such association and utilising it for QT improvement in mustard (*Brassica* spp.), sought to answer the following major questions:

(*a*) Can isozyme variation be quantified as a set of QTs uniquely?

(*b*) Can the association between isozyme and QT variation be used to plan hybridisation programmes?

(*c*) Can the association between isozyme and QT variation be used to detect heterotic F_1s and transgressive segregants in F_2 generation ?

The answers to the questions turned out to be firm positives. As mentioned earlier, five parameters exhausting the information in a gel pattern were identified as QTs (Arunachalam *et al.*, 1996). The grouping of a number of varieties from a few species of *Brassica* using variation of a few isozyme markers such as esterase, peroxidase and GOT measured by five isozyme QTs for each enzyme was found to be quite close to that based on six morphological QTs which have earlier been identified to sufficiently exhaust the total QT variation (see for example, Arunachalam *et al.*, 1998; Bharti, 2000). Efficiencies of individual enzymes alone and in combination in predicting the variation given by the six diagnostic morphological QTs varied from 32.3 % for the enzyme, esterase to 72.9 % for the combination, esterase + peroxidase (Arunachalam *et al.*, 1998). These results were based on isozyme data collected using a laser densitometer (Pharmacia Biotech) and an associated GELSCAN software. The efficiency of enzymes was evaluated on a range of genetic material - species and varieties of *Brassica* grown in field in a replicated design, a large number of plant samples and repeated across 3 years. Taking into account that only three enzymes could be used in these experiments, it was possible to classify the genotypes within and among species for the QTs like plant height, number of primary and secondary branches, yield of primary and secondary branches, and harvest index with high efficiency on the basis of isozyme QTs in a breeding material that comprised advanced lines from four *Brassica* species and in a material having 106 germplasm accessions not characterized for their useful QTs. The efficiency varied among the enzymes. The polymorphism in esterase isozymes had the maximum efficiency, while the one with peroxidase had the least efficiency. But esterase and glutamine oxaloacetate transaminase (GOT) isozymes marked the above six QTs of breeders' interest with over 75 % and 66 % efficiency respectively, in different materials in repeated trials. This clearly indicated that with more polymorphic enzymes, an improved efficiency could be achieved in *Brassica* (Bharti, 2000).

In another experiment, parents of potential heterotic F_1s and F_2 segregants transgressing of heterotic F_1 performance for yield contributing traits were identified on the isozyme data alone, successfully using as early as 28-35 days old seedlings in six

Brassica juncea × *Brassica juncea* crosses. Depending on the extent of polymorphism between parents for the expression of isozymes, a success rate of above 90 % was recorded for the early detection of transgressive segregants for yield contributing QTs in four of the six crosses with esterase, GOT, peroxidase A and peroxidase B (Aruna Kumari, 1999). Details of the logic behind the high association between isozyme and morpho-QTs and the efficiency of the enzymes would be found in Arunachalam *et al.* (1998). However one vital reason has to be re-emphasised; isozymes like QTs have an expression, and are, like QTs, sensitive to environmental variation. Further a method was innovated to measure isozyme variation through five QTs. In contrast, fixed values are associated with any molecular genotype –1 for ZZ, o for Zz and –1 for zz as was seen earlier; since they do not have an expression on their own, they have to lean on an associated QT when even the basics of an association are not above question.

In an exhaustive review of biochemical and molecular markers in plant breeding, Stuber (1992) reported that selections based solely on the manipulation of allelic frequencies at seven isozyme loci in an open-pollinated population of maize increased grain yield significantly. Based on further studies with F_2 populations of maize, it was observed that marker-based selection on 15 isozyme marker loci (which probably represented no more than 30-40 % of the genome) was as effective as phenotypic selection, which would be expected to encompass the entire genome. At the same time, another study from maize which used 230 RFLP marker loci and 31 isozyme loci in predicting F_1 grain yield of more than 100 hybrids from 37 elite inbreds gave an R^2 of 0.87 when F_1 yields were plotted against RFLP diversity. The corresponding R^2 was 0.37 with isozyme diversity. Since contribution of each marker is not compared, it has been reported that the results are not conclusive. Just like identifying QTs with high potential to characterise a genotype, one could identify potential enzymes in predicting grain yield. In this context, it has also been realized that 'providing adequate recombination so that single QTLs (with major effects) can be distinguished from clusters of QTLs (each with small effects) will be difficult to achieve.' (Stuber, 1992). Further, mere number of markers alone cannot justify the efficiency of one over the other; their individual contributions would provide a better platform to judge the relative efficiency.

The theory that selective genotyping will serve to save large samples and economize the number of markers needed would be appropriate only if a single QT is evaluated, because the trait distribution and thus the individuals selected for genotyping will differ for each trait.

The simulation study used by Simpson (1989) to counter the reported weaknesses of RFLP markers in detecting linkage with inbred lines similarly lacks strength in its arguments. A somewhat different study with 114 single-cross hybrids of maize classified parental diversity with 21 isozyme loci (genotypes) as similar and dissimilar. The hybrids between dissimilar isozyme group showed 10 % significantly higher average grain yield of hybrids than those between similar isozyme group (Frei *et al.*, 1986). This result is in consonance with known relationship between parental diversity and heterosis.

Realising the fact that the interaction of QTL effects with genetic background can be highly significant (Guffy *et al.*, 1988) and marker-based selection (based on 15 isozyme marker loci, which probably represent no more than 30-40 % of the genome) was as effective as phenotypic selection, which would be expected to encompass the entire genome (Stuber, 1992) the projected insufficiency of isozyme markers does not in any way debilitate their potential to mark phenotypic performance. It seems justified to postulate therefore the following:

- More than marker types, the possibililty to locate them freely and in high number across the entire genome and the proposition that the sheer number of markers has to trap one or other QTs as closely as possible, it is the logical reasoning to conceive of a correlation between markers (environment-independent) and QTs (highly environment-dependent) that is the most important. On that count, isozymes would definitely score over other markers until environment-consonant molecular markers are discovered.
- Regardless of this basic inadequacy, even if methods to measure molecular markers exhaustively, as we have done in isozymes, are developed, the basis of relating molecular markers with QTs would derive special strength.
- It is essential to test the repeatability of marker efficiency across genetic material and environments. In case of need, separate sets of markers for each crop species would be in order. Most crucial is the need to establish unequivocally the converse hypothesis that material carrying specific markers would show desired level of QT performance.

Gene product variations in the metabolic pathway leading to the QTs are now more practically and realistically seen as possible indicators with a potential to quantify genetic distances and thereby predict promising hybrids. Rhodes *et al.* (1992) observed that a small difference in the metabolic flux could lead to substantial heterosis despite additivity at the enzyme level. Kacser and Burns (1983) had earlier established that a quantitative relationship between metabolic flux and enzyme quantity is determined by gene dosage and allelic constitution. Keightley (1997) provided an insight into the metabolic basis for dominance and recessivity, and the allelic relationships exploited in handling the F_1 and segregating generations, using protein expression as the metabolite corresponding to the genotypic constitution in the population. Crow (2000) stated in his review on the use of molecular markers for heterosis with QTLs, that it was not possible to associate yield changes with specific QTLs which however, could be achieved by possibly identifying molecular markers associated with each QTL. But a realistic proposition is to establish associations of specific QTLs with specific protein (solute, enzyme) pool and then associate this metabolic quantity with grain yield expression. The isozyme based quantitative trait exploration detailed earlier is one step closer to the above observations.

Latest revelations from the human genome project has confirmed, in the words of Craig Venter, one of the most important scientists in the effort to map the human genome, that 'the notion that one gene equals one disease, or that one gene produces one key protein, is flying out of the window,' (Banerji and Shah, 2001). Further, in more than 98 % of human disease, one gene - one disease model does not hold; many genes interacting with each other appear to play a role. External environment of the organism as a whole, has a definitive role in gene expression. The human genome project results confirmed that DNA sequences within one gene may be used in coding many proteins. Clearly 'the control pathway of gene expression is not closed and linear, but dynamic and circular', leading to 'change patterns of gene expression in a context-dependent manner' (Banerji and Shah, 2001).

Quantitative Trait Improvement — Where are we?

In the current context of advocacy of approaches intersecting molecular and Mendelian theory, it is primarily essential to take stock of relevant improvement in the methodology based on Mendelian theory, which adds strength to earlier proven approaches to breeding for QT improvement. Without elaborating the methodological improvements, it would be appropriate to observe the following:

- The infirmities of the established classification procedures of genetic entries on the basis of genetic divergence computed on morphometric traits have been removed. It is possible now to cast genetic entries unequivocally into four divergence classes on self-set norms. Theory of heterosis taking into account the role of epistatic interaction has been set on firm grounds (Jinks, 1955). It has been established that hybrids between parents chosen from intermediate divergence classes have a higher chance to show heterosis for a number of QTs simultaneously. Parents with intermediate mean performance will have genes in repulsion phase (Mather and Jinks, 1982).
- Having set the norms for parental choice, a theory of early generation selection, particularly of heterotic F_1s, has been developed. This would be cost-effective if large F_2 population is to be raised to locate recombinants and transgressive segregants.
- A method to rank F_2 plants on the basis of a diagnostic set of QTs for every crop (that can be selected through stepwise multiple regression) in the order of their potential performance into four equal segments has been developed. In most cases, the top segment is enough to be selected. Their F_3 would have a high probability of containing recombinant plants. It may be mentioned that F_2 plant performance is computed as the value of a regression index using the diagnostic traits (Aruna Kumari, 1999, Mithra 1999).
- It has been established theoretically Pooni and Jinks (1986) and experimentally that selection based on F_3 family mean would be very effective in producing superior recombinant pure breeding lines. Further Pooni and Jinks (1976) proposed a method for predicting the proportion of superior recombinant pure breeding lines derived from F_2 obtained from the F_1 of the two pure breeding parents.
- From F_3 onwards, established procedures of within-, between- and combined family selection (Falconer and Mackay, 1996) would be able to identify the potential families which can be advanced to homogeneity.

- In all the decision processes mentioned above, the developed methods of assessing performance across a set of interacting QTs (Arunachalam and Bandyopadhyay, 1984a; Arunachalam, 1993) could optimise efficiency.

Next in merit is the independent role of MAS and its complementarity to proven methods of QT improvement. We only demonstrated a method of quantifying isozyme variation as QTs and applying procedures analogous to those established for QT improvement. The method is relevant to the following highly specific objectives:

1. Can marker assays produce comparative results consonant with those produced by QTs?
2. If so, can we identify areas of their specific and defensible applications in the field?

It was possible to establish high commonality between genetic classification of species and varieties by marker assays (isozymes) with that based on QTs. This would mean that in *Brassica* (which was our test crop), we can identify parents, heterotic F_1s and desirable F_2s in advance. *Brassica* is a rabi crop. It would take one season to evaluate and select parents and another season to make crosses. Though an off-season is generally used, its environmental effects can confound performance. Isozyme assays can be done using plants raised in growth chambers. This would be season-independent and disease-free. A one-to-one correspondence was established between the assay made using growth chamber-raised plants and field-grown ones earlier. Parents could be identified before the crop season using isozyme assays and in the subsequent field season, crosses can be made. Such procedures can also select heterotic F_1s and desired F_2 recombinants. Particularly in the latter, this process would help in saving a large number of F_2 plants from being grown to maturity and evaluated for a number of QTs. Pilot checking of these concepts have shown highly encouraging results (Aruna Kumari, 1999; Mithra, 1999).

Softwares have been developed starting right from gel assay until selection of desired plants. This would take the load of analysis of data and inferences from out of breeders.

What is Generally Done in QT Improvement Research?

As was mentioned earlier, there was a decade period in 1970s in India when all the research dissertations were on grouping of varieties on genetic distances measured by Mahalanobis D^2 statistic. Such decades continued with the other major areas of development in Quantitative Genetics such as combining ability (Jink's approach and Griffing's approach) using diallel crosses, selection of parents on fractional diallels, selective mating systems in self-pollinated crops, complex mating systems and selection and so on. Likewise the current approach in MAS is processing data on Nei's distance statistic and dendrograms, and mapping new marker genes on the already developed/developing molecular maps. It has been pointed out that researchers in molecular biology 'are tempted by turnkey instrument packages that include hardware and software rather than trying to define their very specific needs and engineer an original solution'. (Peccoud, 1995). It is equally true with breeding practitioners. Therefore we are led, though belatedly, by investigations done elsewhere and demonstrate far limited capacity to lead by innovation and cutting-edge technologies. This way, we cannot, for an indefinite period, come to grips with present or future problems peculiar to our environment and attempt cost- and human resource-effective solutions.

What is the Need of the Day?

With the sprinting developments in digital revolution, horizons, though unreachable once, are in our choice to reach. At the high end of the scale we have a modern communication avenue of "virtual instrumentation". A programming language called LabVIEW permits manipulating of simulated knobs and switches on the computer screen to observe data on indicators, plots, charts and so on. Having designed the DNA molecule, the molecular biologist could leave the design of recombination protocol to vector NTI and its synthesis to CLARA, both of which are robotics based (Peccoud, 1995). At the other end of the scale, several moot questions are to be recognised:

- Whether the method used in commercial software to compute linkages of markers with QTs is understood
- Even when understood, whether capability exists to understand its dire limitations and to modify them to the needs
- Whether the basis behind Nei's genetic distance conception is valid to the experiments we conduct and
- Whether alternative classification procedures exist or can be developed

Identical would be the situation with breeders attempting QT improvement. The design of mating and methods of selection which hold the key to targeted QT improvement are directed by experiments done and published elsewhere and limited to the availability of software for analysis of data. Priorities are thus set not by the crucial problems in a crop or a site or an environment, but by availability of means to evaluate results and arrive at inferences and applicable protocols.

Thus the big hurdle in the path of innovative and problem-specific research is the highly limited 'techniracy' (technological literacy) of researchers. Most often it is limited to matters of biology and hardly, if ever, there are inclinations to delve into at least concerned knowledge areas of other fields (say methods of linkage estimation, knowledge and operational skill of existing computer softwares, knowledge of inter-faces between computer and laboratory instruments, etc.). There are no motivating incentives or compulsions to extend the horizon of already-gathered knowledge. There are no incentives either to attract bright students to innovative and future-oriented biological research or to retain existing bright scientists in active and innovative research pursuits. In the existing environment, inter-disciplinary interaction remains a far-cry. Hence teachers to mould incoming young minds towards frontier technologies including associated skills in allied fields are hard to come by. Path-breaking, bold policy decisions are the need of the day to integrate agriculture with other fields of relevance. Until such time, the minimum we could do is to compulsorily introduce undergraduate and graduate students in biology and agriculture to such inter-disciplinary education and skill orientation, so that they can handle modern biology and its developments in a modern way.

Environment-genotype symbiosis is the key to sustainable performance. The path-breaking February 2001 research finding by the group of scientists from U.S.'s Celera Company and the U.K. Wellcome Trust earlier involved in the human genome project is that more than genes, there are powerful environmental influences "vastly" more crucial in determining human behaviour. Moulding environment would then be a better short-term option than moulding genotypes. A conciliating approach towards this goal should be the right path to a rapid progress.

References

Antonarakis S.E. 1994. Genome linkage scanning: systematic or intelligent? *Nature Genetics*, **8**: 211-212.

Aruna Kumari K. 1999. Marker-aided selection for quantitative trait improvement in *Brassicas*, Ph.D. Thesis, Indian Agricultural Research Institute, New Delhi.

Arunachalam V. 1981. Genetic distance in plant breeding. *Indian J. Genet.*, **41**: 226-236.

Arunachalam V. 1993. A method to rank multiple trait performance of biological entities. *Biometrical Journal*, **35**: 959-966.

Arunachalam V. and Bandyopadhyay A. 1984a. A method to make decisions jointly on a number of dependent characters. *Indian J. Genet.*, **44**: 419-424.

Arunachalam V. and Bandyopadhyay A. 1984b. Limits to genetic divergence for occurrence of heterosis - Experimental evidence from crop plants. *Indian J. Genet.*, **44**: 548-554.

Arunachalam V. and Chandrashekharan S. 1993. RFLP approach to breeding for quantitative traits in plants - a critique. *J. Genet.*, **72**: 73-83.

Arunachalam V., Prabhu K.V. and Sujata V. 1996. Isozyme parameters in genetic evaluation in *Brassica* - A tribute to Haldane's vision on the synergy between biochemistry and genetics. *Curr. Sci.*, **70**: 1062-1065.

Arunachalam V., Prabhu K.V. and Sujata V. 1998. Multivariate methods of quantitative evaluation for crop improvement: conventional and molecular approaches revisited. pp.793-807. *In*: *Crop Productivity and Sustainability - Shaping the Future*. Proc. 2nd International Crop Science Congress, (eds.) V.L. Chopra, R.B. Singh and Anupam Varma, Oxford & IBH Publishing Co., New Delhi, India, 1111 pp.

Arunachalam V., Bandyopadhyay A., Nigam S.N. and Gibbons R.W. 1984. Heterosis in relation to genetic divergence and specific combining ability in groundnut, *Arachis hypogaea* L. *Euphytica*, **33**: 33-39.

Avise J.C. 1989. A role for molecular genetics in the recognition and conservation of endangered species. *TREE*, **4**: 279-281.

Bagali P.G., Shasidar H.E., Hittalmani S. and Huang N. 1999. Identification of quantitative trait loci associated with grain filling durationin rice (*Oryza sativa* L.). *Curr. Sci.*, **77**: 743-746.

Banerji D. and Shah M. 2001. Fluid genome - a paradigm shift. *The Hindu*, 6 August 2001, p.12.

Beckmann J.S. and Soller M. 1988. Detection of linkage between marker loci and loci affecting quantitative traits in crosses between segregating populations. *Theor. Appl. Genet.*, **76**: 228-236.

Bharti Shefali. 2000. Quantitative trait fingerprinting of genetic accessions in *Brassica*, Unpublished Ph.D. Thesis, Agra University, Agra.

Crow J.F. 2000. The rise and fall of overdominance hypothesis. *Plant Breeding Reviews*, **17**: 225-258.

Doolittle W.F. 1999. Phylogenetic classification and the universal tree. *Science*, **284**: 2124-8.

Durga Prasad M.M.K., Arunachalam V. and Bandyopadhyay A. 1985. Diversity pattern elucidating choice of parents for hybridization in varieties of groundnut, *Arachis hypogaea* L. *Trop. Agric.*, **62**: 237-242.

Ellis T.H.N. 1986. Restriction fragment length polymorphism markers in relation to quantitative characters. *Theor. Appl. Genet.*, **72**: 1-2.

Falconer D.S. and Mackay T.F.C. 1996. *Introduction to Quantitative Genetics*. Fourth Edition. Longman, U.K.

Fenster C.B., Galloway L.F. and Chao L. 1997. Epistasis and its consequences for the evolution of natural populations. *TREE*, **12**: 282-286.

Freeland S.J., Knight R.D. and Landweber L.F. 1999. Do proteins predate DNA? *Science*, **286**: 690-692.

Frei O.M., Stuber C.W. and Goodman M.M. 1986. Use of allozymes as genetic markers for predicting performance in maize single cross hybrids. *Crop Sci.*, **26**: 37-42.

Griffing B. 1956. Concept of general and specific combining ability in relation to diallel crossing systems. *Austr. Journ. Bio. Sci.*, **9**: 463-493.

Guerinot M.L. 2000. The green revolution strikes gold. *Science*, **287**: 241-243.

Guffy R.D., Stuber C.W. and Helenkjaris T. 1988. Molecular markers for evaluating quantitative traits across varying genetic backgrounds and environments in maize. *Agron. Abstracts*, p.82 [Abstr.].

Gura T. 2000. Reaping the plant gene harvest. *Science*, **287**: 412-4.

Hamrick J.L. and Godt M.J.W. 1990. Allozyme diversity in plant species. *In*: "*Plant Population Genetics, Breeding and Genetic Resources*". (eds.) A. H. D. Brown, M.T. Clegg, A. L. Kahler and B.S. Weir, Sinauer Associates, Sunderland, USA. Pp 43-63.

Huelsenbeck J.P., Bull J.J. and Cunningham C.W. 1996. Combining data in phylogenetic analysis. *TREE*, **11**: 152-8.

Jansen R.C. and Stam P. 1994. High resolution of quantitative traits into multiple loci via interval mapping. *Genetics*, **136**: 1447-55.

Jinks J.L. 1955. A survey of the genetic basis of heterosis in a variety of diallel crosses. *Heredity*, **9**: 223-229.

Jinks J.L. and Pooni H.S. 1976. Predicting the properties of recombinant inbred lines derived by single seed descent. *Heredity*, **36**: 253-266.

Kacser H. and Burns J. A. 1983. The molecular basis of dominance. *Genetics*, **97**: 639-646.

Kashi Y., King D. and Soller M. 1997. Simple sequence repeats as a source of quantitative genetic variation. *Trend. Genet.*, **13**: 74-78.

Keightley P.D. 1997. A metabolic basis for dominance and recessivity. *Genetics*, **143**: 621-625.

Loiselle F., Tai G.C.C. and Christie B.R. 1991. Pedigree, agronomic and molecular divergence of parents in relation to progeny performance in potato. *Potato Research*, **34**: 305-316.

Mahalanobis P.C. 1936. On the generalized distance in statistics. *Proc. Natl. Sci. India*, **2**: 49-55.

Mather K. and J. L. Jinks. 2982. *Biometrical Genetics*, 3rd Edition. Chapman and Hall, London.

Martin B., Nienhaus J., King G. and Schaffer A. 1989. Restriction fragment length polymorphisms associated with water use efficiency in tomato. *Science*, **243**: 1725-1728.

Michelmore R.W. and Shaw D.V. 1988. Character dissection. *Nature*, **335**: 672-673.
Mithra A.C.R. 1999. Evaluation of yield performance of early generation selections in *Brassica*. M.Sc Thesis, Indian Agricultural Research Institute, New Delhi.
Motro U. and Soller M. 1993. Sequential sampling in determining linkage between marker loci and quantitative trait loci. *Theor. Appl. Genet.*, **85**: 658-664.
Murty B.R. and Arunachalam V. 1966. The nature of divergence in relation to breeding systems in some crop plants. *Indian J. Genet.*, **26A**: 188-198.
Murty B.R. and Arunachalam V. 1967. Computer programmes for some problems in Biometrical Genetics. I. Use of Mahalanobis' D-square in classificatory problems. *Indian J. Genet.*, **27**: 60-69.
Murty B.R., Mathur J.B.L. and Arunachalam V. 1965. Self-incompatibility and genetic divergence in *Brassica campestris* var. brown sarson. *Sankhya B*, **27**: 271-278.
Nei M. 1972. Genetic distance between populations. *Am. Nat.*, **106**: 283-291.
Nei M. 1987. *Molecular Evolutionary Genetics*. Columbia Univ. Press, New York. pp. 106-107.
Nei M. and Li W.H. 1979. Mathematical model for studying genetic variation in terms of restriction endonucleases. *Proc. Natl. Acad. Sci.*, **76**: 5269-5273.
Osborn T.C., Alexander D.C. and Fobes J.F. 1987. Identification of restriction fragment length polymorphisms linked to genes controlling soluble solids content in tomato fruit. *Theor. Appl. Genet.*, **73**: 350-356.
Paterson A.H., Damon S., Hewitt J.D., Zamir D., Rabinowitch H.D., Lincoln S.E., Lander E.S. and Tanksley S.D. 1991. Mendelian factors underlying quantitative traits in tomato: comparison across species, generation and environments. *Genetics*, **127**: 181-197.
Peccoud J. 1995. Automating molecular biology: A question of communication. *Biotechnology*, **13**: 741-745.
Phillips P.C. 1999. From complex traits to complex alleles. *Trend. Genet.*, **15**: 6-8.
Pooni H.S. and J.L. Jinks. 1986. Estimation of true additive genetic variance SIGMK-k-i equals Id-2-i in the presence of linkage disequilibrium. *Heredity*, **57**: 341-344.
Rao C.R. 1952. *Statistical Methods in Biometrical Research*. John Wiley & Sons, New York.
Rhodes D., Ju G.C., Yang W.J. and Samaras Y. 1992. Plant metabolism and heterosis. *Plant Breeding Reviews*, **10**: 53-91.
Sax K. 1923. The association of size differences with seed coat pattern and pigmentation in *Phaseolus vulgaris*. *Genetics*, **8**: 522-560.
Simpson S.P. 1989. Detection of linkage between quantitative trait loci and restriction fragment length polymorphisms using inbred lines. *Theor. Appl. Genet.*, **77**: 815-819.
Singh R.K. and Choudhary B.D. 1983. *Biometrical Methods in Quantitative Genetic Analysis*. Kalyani Publishers, New Delhi.
Soller M., Brody T. and Genizi A. 1976. On the power of experimental designs for the detection of linkage between marker loci and quantitative loci in crosses between inbred lines. *Theor. Appl. Genet.*, **47**: 35-39.
Spinney L. 1997. The unselfish gene. *New Scientist*, **25**: 28-32.
Storfer A. 1996. Quantitative Genetics: a promising approach for the assessment of genetic variation in endangered species. *TREE*, **11**: 343-348.
Stuber C.W. 1992. Biochemical and molecular markers in plant breeding. *Plant Breeding Rev.*, **10**: 37-61.

Simmonds N.W. 2000. An informal history of statistics. *Plant Breeding Reviews*, **17**: 259-314.

Tanksley S.D. and Hewitt J. 1988. Use of molecular markers in breeding for soluble solids content in tomato - a re-examination. *Theor. Appl. Genet.*, **75**: 811-823.

Vairavan S., Siddiq E.A., Arunachalam V. and Swaminathan M.S. 1973. A study of the nature of divergence in rice from Assam and NE Himalayas. *Theor. Appl. Genet.*, **43**: 213-221.

Wang Z.Y., Second G. and Tanksley S.D. 1992. Polymorphism and phylogenetic relationships among species in the genus *Oryza* as determined by analysis of nuclear RFLPs. *Theor. Appl. Genet.*, **83**: 565-581.

Weller J.I. and Wyler A. 1992. Power of different sampling strategies to detect quantitative trait loci variance effects. *Theor. Appl. Genet.*, **83**: 582-588.

Zhang Q., Gao Y.J., Yang S.H., Ragab R.A., Maroof M.A.S. and Li Z.B. 1994. A diallel analysis of heterosis in elite hybrid rice based on RFLPs and microsatellites. *Theor. Appl. Genet.*, **89**: 185-192.

Plant Breeding - Mendelian to Molecular Approaches
H. K. Jain and M. C. Kharkwal (eds.)

22

Mating Designs and Their Implications for Plant Breeding

R.K. Singh[1], H.S. Pooni[2], M. Singh[3] and A. Bandopadhyaya[4]

Abstract

The traits that Mendel chose to study could be classified into clearly defined, discrete groups. Thus, by following the ratios from a cross between two distinct pure breeding lines in different generations, one could easily understand the genetics of such traits. However, there were a large number of traits where the variation was more of a continuous nature. To study inheritance of these so called quantitative traits, biometrical approaches were developed. Subsequently, mating designs were developed to estimate different genetic components of variation. Based on information generated through mating designs, methods were developed to predict performance of hybrids and populations, and identify breeding methods designed to utilize different types of gene actions. Molecular markers technologies have helped identifying QTLs (quantitative trait loci). Interval mapping, based on maximum likelihood methods and implemented in the software package MAPMAKER, is now widely used in QTL mapping experiments and information regarding the minimum number of genes, their additive/dominance effects, and the extent of variation explained by the involved QTLs can be obtained. This paper reviews the development and role of mating designs in elucidating the nature of quantitative variation for their use in plant breeding, during the past, and how is this being now integrated with the molecular techniques for more precision and efficiency in plant breeding.

Introduction

Mendel's discoveries of the laws of inheritance laid the foundation of the modern genetics and its application to plant breeding. Mendel's work published in 1865 was rediscovered in 1900. Mendel's laws of heredity apply to characters which are easily classified into distinct groups and whose expression depend upon one or a few major genes. Johannsen's work on beans led to the idea of heritable and non-heritable variation (Johannsen, 1913), while Nilsson-Ehle (1909) and East (1910) unified the idea of Mendelism and pure-line theory and concluded that the multiple factors (Mather's polygenes) controlling quantitative traits showing continuous variation segregate in Mendelian fashion. Sax (1923) showed linkage of polygenes with Mendelian genes.

[1]International Rice Research Institute, India-Office, New Delhi 110 012, India
[2]University of Birmingham, Birmingham, U.K.
[3]National Bureau of Plant Genetic Resources, New Delhi 110 012, India
[4]Indian Council of Agricultural Research, New Delhi 110 012, India

Fisher (1918) was the first to partition the variation into heritable and non-heritable components, and further to subdivide the former into additive, dominance and epistatic effects. In 1949, Mather published his first edition of the 'Biometrical Genetics' summarizing what was then known about the inheritance of quantitative traits.

Based on the principles of correlation between relatives and how they could help partition the variation into different genetic components, as well as by using the second-degree statistics, various experimental mating designs were proposed (Mather, 1949; Comstock and Robinson 1952; Hayman 1954; Kearsey and Jinks, 1968; Mather and Jinks 1982). The principle was to generate various levels of relatedness among the relative progenies by effecting a series of crosses among individuals of a random mating population, that generated different statistical components of variation from which genetical components of variations were estimated.

Recently, the development of molecular marker technology has opened up the new vistas in understanding of quantitative traits. The QTL analysis and marker-aided selection now provide more precise information on QTL locations, their contribution to variation and selection using the molecular markers associated with them.

Mating Designs: A Survey

Two schools have contributed the most to the development of the mating designs to study quantitative variation. They are the North Carolina State University, USA, and the University of Birmingham, England (UK). The North Carolina group has concentrated on the analyses of random mating populations and has given NCI, NCII and NCIII designs. These designs are essentially the adaptations/extensions of the one-way, two-way and nested analyses of variance that were proposed by Fisher during early 20th century.

Birmingham group, on the other hand, worked on a series of mating designs that were applicable mainly to autogamous species but also to outcrossing species like maize that can be easily selfed or crossed.

Following two important issues should be considered before choosing any mating design:

- The kind of relatives that will be available for analysis, because certain types of relatives are observed more readily in some species than in others, and some types of phenotypic covariance between relatives are more likely to approximate the desired quantities than others.
- The degree of precision that can be achieved in a quantitative genetic survey is a function of the number of individuals measured and the way in which effort is allocated to number of families versus number of individuals within families - an issue of optimum experimental design.

Some of the designs and approaches described by the two groups of schools and others are briefly discussed here.

Parent-offspring regression: The frequent use of the regression of offspring phenotypes on those of their parents for estimating heritabilities specially in natural/ animal populations has obvious reasons that it provides estimate of additive genetic variance.

When there is significant variation in family size, weighted least square regression should be used. Kempthorne and Tandon (1953) showed that appropriate weights are proportional to the inverse of the residual sampling variances of family means about the parent-offspring regression. Hill (1970) suggested the use of assortative mating to improve accuracy of heritability estimates derived from mid-parent-offspring regression.

Sib Analysis: Analysis of collateral relatives, sibs in particular, provides an attractive alternative to parent-offspring regression in estimating genetic parameters. The common environmental effects are the main drawbacks of any sib analysis. The most reliable way to minimize this effect is to exclusively employ paternal half-sib families. There are three types of sib analysis: full-sib families; half-sib families and; combinations of the half-sib and full-sib families.

Biparental matings or the randomly mated paired crosses called BIPs is the simplest, and perhaps the oldest design that has been used by breeders and geneticists, in which a number (= 2n) of individuals are crossed at random (in pairs) to give n families. The progenies of this design are either full-sibs or unrelated. A comparison of between families and within families variances in this case provides an estimate of V_A (additive genetic variance) assuming no dominance ($V_D = 0$) and no common environment ($V_{EC} = 0$). A test of the non-additive variation is also possible, when the parental data are available, as the parent-offspring regression provides a direct estimate of additive genetic variance and the dominance variation is tested as the remainder mean square. Kearsey (1965) proposed an extension of the BIPs design, and called the new design "Augmented BIPs" as it involved reciprocally produced biparental families and the selfed progenies from each of the parents. The new design provides the experimenter orthogonal tests of additive, dominance and maternal effects.

For out-crossing species, the polycross also provides an estimate of the additive genetic variance, assuming absence of dominance effects (Tysdal *et al.*, 1942). Unlike BIPs, polycross uses covariance of half-sibs (COHS) to estimate V_A.

Comstock and Robinson (1952) proposed three mating designs, called North Carolina (NC) Designs I, II and III. In these, families are produced, some having both parents in common (full-sibs), others with only one parent in common (half-sibs), so generating both full and half-sibs relationships. Traditionally ANOVA using linear models is used to analyze the data.

The North Carolina Design I (NCI) with three statistics provides estimates of both V_A and V_D. It also provides the test for V_A. The design has found widest application in animal and tree breeding. In NC II, the mean squares for males and females supply separate and independent estimates of V_A, while the interaction mean square between males and females provides an independent estimate of the non-additive genetic variance V_D. In some cases, it is also possible to estimate epistatic variation using this design (Foster and Shaw, 1988). Cockerham and Weir (1977b) generalized design II, to incorporate reciprocal crosses. The advantage of this modification is that an explicit partitioning of nuclear and extra nuclear effects is possible.

NC III is by far the most powerful of the NC designs. Subsequently, this design was extended by Kearsey and Jinks (1968) to include a third tester, the F_1 from $L_1 \times L_2$. In this extended form, this design is known as Triple Test Cross (TTC). By the inclusion of the third tester, the power of the design increases, and it provides unambiguous test of non-allelic interactions. It also provides the equally reliable estimates of V_A and V_D components and degree of dominance can be calculated. Both NC III and the TTC have the general utility of investigating any population, irrespective of gene frequency or mating system. In the absence of interaction, both additive and dominance components of variation are estimated with comparable precision in TTC, a property no other design has.

Diallel Analysis: In a diallel, a set of genotypes are crossed in all possible combinations. Griffing's diallel analysis leads to the testing and estimation of the general (GCA) and specific (SCA) combining abilities, and the Hayman/Jinks analysis provides information on the estimates of components of variation (Hayman, 1954 and Jinks 1954). Generalization for Hayman's analysis in F_2 and later generations was given by Singh and Singh (1984a). Griffing (1956a) described four methods of a diallel mating depending upon whether parents and reciprocals are retained or excluded from a particular design. Various forms of half-diallel analysis were found to be interrelated (Singh and Singh 1984b). However, Gardner and Eberhart's model (1966) had some advantages over the others, as it partitioned heterosis into different components (Singh and Singh 1984c). A comparative review of these analyses are given by Christie and Shattuck (1992), while Pooni et al. (1984) developed equations showing similarities between Griffing and Hayman models.

Three levels of the diallel analysis are available, each successive level providing more information but requiring more limiting assumptions (Wright, 1985). First level is the analysis and estimation of GCA and SCA effects. Assuming parents as the random sample, variances of these effects, σ_g^2 and σ_s^2 could also be estimated. The second level analysis provides the estimates of components of genetic variance (V_A; V_D), assuming reference population in Hardy-Weinberg and linkage equilibrium (Baker, 1978; Wright, 1985). The third and final level is the full genetic analysis, devised by Jinks and Hayman (1953). ANOVA of a complete diallel is in fact a combined analysis of reciprocal sums and differences, these being summed over half-sib progeny groups or arrays. In a half-diallel (Method 2 & 4), only the analysis of sums can be performed (Kempthorne and Curnow, 1961). The analysis of sums, which measures heritable variation, can be partitioned into GCA effects (the 'a' item of Hayman, 1954; Hayman and Mather, 1955), and SCA effects (the 'b' item of Jinks and Jones, 1958). When parents are also included, SCA effects can be further partitioned into three contrasts, b_1, b_2 and b_3. The b_1 tests the overall differences between parental and F_1 means. Its significance shows directional dominance. The b_2 item measures consistency of mean dominance deviation over arrays; if significant, it means some parents contain more dominant allele than others, leading to gene asymmetry. The b_3 is the measure of all three non-additive deviations unique to each F_1. The b_3 item, in fact, in method 1 and 2 corresponds to the specific combining ability item of method 3 and 4, respectively. This means, in methods 1 and 2, the significance of SCA effects should be determined from b_3 and not from the b item in the analysis. Like sums, analysis of differences can also be partitioned into two items - c and d (Hayman, 1954). The c item means the average maternal or paternal effect, while 'd' is the estimate of reciprocal effects.

Generation Mean Analysis: Mather (1949) proposed an extensive analysis of family means, universally known as the "basic generation analysis". He utilized six generations (P_1, P_2, F_1, F_2, $F_1 \times P_1$ and $F_1 \times P_2$) to test for and estimate the additive, dominance and epistatic effects of genes at the means level and also provided a rationale to estimate V_A and V_D, the additive and dominance variances attributable to the genes that were segregating in the cross.

The means and variances of these generations provide estimates of genetic and environmental components of variation.

Precision and Power

Power and precision are serious considerations for any experimental design. Standard errors provide rough guides to the accuracy of the variance component estimates.

The optimal experimental design depends upon the h^2 of the trait. However, a relationship between n and E (h^2) indicates the following :

If h^2 is relatively large there is a pronounced increase in SE (h^2) as family size exceeds the optimal value.

When h^2 is small, SE (h^2) is very insensitive to the design, provided that n is greater than 20.

Therefore, in the absence of any prior information on h^2, it would be safer to use family sizes to the order of 5-20.

Power under optimal half-sib design for various values of T and h^2 has been worked out. The optimal design for a nested analysis of variance is defined to be the combination of N males, M females per male, and n progeny per female that minimizes the sampling variance of intraclass correlation of interest. Robertson (1959) has shown that when dominance and common environmental effect are absent, the preferred design for estimating t_{PHS} is to use full-sib families of only single individuals, i.e. to rely on the pure half -sib analysis. If, on the other hand, one desires approximately equal precision in the estimates of t_{PHS} and t_{fs}, it is advisable to allocate at least 3-4 female/male and to maintain full-sib families of approximately $1/(2t_{PHS})$ (but no less than 2) progeny/female.

Although the nested design is often relied on as a means for detecting dominance, it is not particularly powerful in this regard.

For screening a large number of lines, clones or accessions, polycross and top-cross (if appropriate tests are available) would be more appropriate. For estimation of GCA and SCA, the diallel analysis proposed by Griffing (1956b) is used. Similarly, for a complete genetic analysis, Jinks and Hayman's model (1953) provides the best solution. In terms of coverage of population the different designs stand in the following order : BIPs > NCM1 > Polycross > NCM3 > NCM2 > Diallel. However, in terms of information generation, their position is reversed : Diallel > NCM3 > NCM2 > NCM1 > Polycross > BIPS. Kearsey (1965) concluded that, although NCM 1 and 2 were well suited to certain breeding systems, they provide no test for epistasis. The partial diallel offered no advantage over other designs, while half- diallel supplied most information about a small and probably unrepresentative number of parents. For the relative precision of variance component estimates by different mating designs, the evidence suggests that for the same amount of effort, diallel yields more precise estimates for both additive and dominance components of variance than NC II (Pederson, 1972; Namkoong and Roberts, 1974).

Estimates of parameters using any design are based on different assumptions. Most common among them are : diploid behaviour; uncorrelated gene distribution, and absence of epistatis, no multiple allelism, no reciprocal differences and genotype × environment (G × E) interactions. In some designs, (TTC and diallel), the epistatis is detected and its effect is included in the estimates. Similarly, reciprocal differences can be tested in several designs and appropriate measures taken. G × E can also be determined by wide scale testing of the materials. Linkage equilibrium is a serious assumption. Linkage influences only the epistatic components. Except for parent-offspring and mono-zygotic twins relationship, linkage inflates covariance between relatives unless there is no epistasis. The estimate of additive × additive type of covariance is inflated more in half-sib than in full-sib families. Theory for digenic descent was also developed (Schnell, 1961; Weir and Cockerham, 1973; 1977). For more than two loci the solution is too complex and yet to be found.

In conclusion, the use of mating designs and their comparative advantages/ disadvantages depend on various factors like the mating system of the species, coverage of population, types of information, etc.

QTLs Analysis

Biometrical methods are unable to give specific information on the genes as separate entities, as the models are based on the sum of their effects, and the individual gene effects are often too small to be detected. Mather (1949) was the first to determine the number of "effective factors", as he called them (Breeze and Mather, 1957). Precise analysis of the phenotypic effects of individual loci underlying quantitative traits has remained elusive for a purely statistical reason. Limitations introduced by linkage restrict analysis to effective factors, i.e. a linked region of genome that may contain several loci influencing the trait, which can lead to biased estimates of effects of individual loci. Geldermann (1975) referred to a locus underlying a quantitative character as a QTL, for a quantitative locus. Most of the basic principles of "QTL mapping" were demonstrated by the 1960s (Thoday, 1961). Thus, it should now be possible with the help of molecular marker to dissect quantitative traits into their underlying genetic factors, that is estimate the number of loci controlling their expression, localize their position on the chromosome, estimate their genetic effects, detect non-allelic interactions between QTLs, and detect pleiotropy. For QTL mapping experiments the general steps are : detection, location and estimation of effects of QTLs.

For QTL analysis, four approaches are followed:

1. Single marker analysis (Point analysis) (Weller, 1986) - It is a traditional method to detect, locate and estimate the effects of a QTL in the vicinity of a marker. It is a good choice when the goal is simple detection of a QTL linked to a marker other than estimation of its position and effects.
2. Simple interval mapping (SIM) or flanking-marker analysis - Lander and Botstein (1989) proposed an interval mapping approach to locate QTLs. A separate analysis is performed for each pair of adjacent marker loci.
3. Composite-interval mapping (CIM) (Haley and Knott, 1992; Zeng 1994; Jansen 1994, Jansen and Stam 1994) - It considers a marker interval plus a few other well-chosen single markers in each analysis. CIM accounts the variance from other QTL by including partial regression coefficients from markers (co-factors) from other regions of the genome. CIM gives more power and precision than SIM, because effects of other QTLs are not present as residual variance.
4. Marker regression (Kearsey and Hyne 1994; Hyne *et al.*, 1995; Wu and Li, 1994, 1996) - It considers all of the linked markers on a chromosome simultaneously, resulting in a single analysis in each chromosome.

Identification of QTLs requires use of appropriate material, although a segregating F_2 population, F_3, backcrosses, recombinant inbrid line and dihaploid lines can all serve as mapping populations.

Information now available indicates that the detected QTL may represent two or more linked QTLs (Mackey and Tanksley, 1990; Stuber, 1995; Gibson and Hogness, 1996; Huang *et al.*, 1996). Also that polygenes with tiny effects produce quantitative variation, may not seem plausible, since relatively small number of genes account for very large portion of phenotypic variation (Tanksley, 1993; Shrimpton and Robertson, 1988).

Applications of Findings to Plant Breeding

Mating designs developed and applied for investigating inheritance of quantitative traits have helped understand the nature of genetic variation, which in turn, were useful in formulating appropriate breeding methods and improving selection efficiency (Kearsey and Pooni, 1996). For example, if the genetics of a trait indicates true over-dominance then development of hybrid varieties is preferred. Schnell (1982) presented a synoptic study of the methods and categories of plant breeding. Based on various considerations, he identified four categories of plant varieties, each exploiting specific types of genetic variation.

Predicting breeding potential of lines or populations and the performances of varieties - hybrids, synthetics or pure lines - has been another important application of the quantitative genetic analysis. Sets of lines with their proven general combining abilities in polycross, testcrosses and diallel tests, have been used to develop high yielding synthetics in various cross-pollinating species; (Kinman and Sprague, 1945; Becker, 1982; Breese and Lewis, 1960).

Jinks and Pooni (1976) gave a method for predicting the potential of producing superior recombinant inbred lines (RILs). This means making predictions for RILs derived from the F_2, BC_{11}, BC_{12} or any other generation, and comparing their potential for producing superior inbred lines (Pooni and Jinks 1978). The only information required is the parental scores and an estimate of the additive genetic variance (*D*). These prediction methods have been applied to several species (Kearsey and Pooni, 1996), including wheat (Snape, 1982), Barley (Caligari *et al.*, 1985), Spring rape (Engqvist and Becker, 1993), etc. The results have clearly shown that breeding programs need not proceed beyond initial stage without a fair idea of their final outcome, whether in terms of better recombinants or better hybrids. Discrepancies can arise between observed and predicted proportions because of epistasis, linkage or G $\times$ E interactions (Kearsey and Pooni, 1996).

Genetics of hybrid vigour (heterosis) has long been a controversial issue. However, with the gradual improvement in the mating designs and reliability of estimates of genetic parameters, the role of epistasis in heterosis is now accepted (Jinks, 1983; Crow, 2000). Heterosis is normally associated with epistasis and in the absence of epistasis, heterosis appears to be mainly due to dispersed directionally dominant genes. Therefore, it is often possible to produce inbred lines that have higher performance than their heterotic F_1 parent.

The over-dominance in maize or other crops could well be due to associative, over-dominance (Stuber, 1995) rather than true single-locus over-dominance, although over-dominance can not be ruled out at least for some loci (Cockerham and Zeng, 1996). Using a set of different mating populations of maize, Stuber *et al.* (1992) showed that QTLs detected for grain yield exhibited over-dominance (or pseudo-overdominance). Later, fine mapping of a region on chromosome 5 helped dissecting it into two smaller QTLs in repulsion phase, which acted in a dominant manner (Graham *et al.*, 1997). This evidently supports the dominance theory of heterosis. Cockerham and Zeng (1996) analysed the same cross which was studied by Stuber *et al.* (1992) using the modified NCIII design in a later generation and found that the results favoured dominance of multiple-linked increaser genes as the cause of heterosis. In another experiment on rice, similar observations were made (Xiao *et al.*, 1995). Initially QTLs effects suggested overdominance. However, a more detailed study, later on, showed complementation of dominant/partially dominant alleles at different loci in the F_1 hybrid as the major contributor to heterosis. This was strengthened by the fact that some recombinant inbred lines in F_8 population had phenotypic value superior to the F_1 for all the traits evaluated. Further, two-way interaction performed between the marker loci showed absense of any digenic epistasis.

Duvick (1996) made a thoughtful remark that beautifully sums up these developments. "The breeding methods for population improvement that are now available might be as effective as the inbred-hybrid method for making certain kind of genetic improvement. The methods now available to the breeders were not available in the 1930s when the hybrid maize program began". Marker-assisted selection (MAS) can be used to maximize selection efficiency (Lande and Thompson, 1990).

Conclusion

Although quantitative genetic analysis has been in the forefront of the breeding programs in the Western world, it has moved at a much slower pace in the developing countries, including India. Recently, the interest in quantitative genetics has grown due particularly to the advent of modern molecular techniques for dissecting QTLs. With the improved computational facilities, the possibility of integrating quantitative genetic analysis (QTL analysis) with the main breeding program is fast increasing. The effective integration would further enhance breeding efficiency and precision. However, there are problems with QTL mapping. The development of new statistical tools, availability of alternate mapping populations and methodology will lead to more refinement in QTL dissection in the future for understanding the genetics of quantitative traits.

References

Baker R.J. 1978. Issues in diallel analysis. *Crop Sci.*, **18**: 533-536.

Becker H.C. 1982. Zuechtung synthetisher Sorten II. Leistung Vorhersage und Selection der Eltern. Vortr. *Pflanzenzuechg*, **1**: 23-40.

Breese E.L. and Mather K. 1957. The organization of polygenic activity within a chromosome of *Drosophila*. I. Hair characters. *Heredity*, **11**: 373-395.

Breese E.L. and Lewis E.J. 1960. The effect of varying number of parent plants on the behaviour of synthetic varieties. *Annual Report* of the Welsh Plant Breeding Station for 1959. pp. 21-25.

Caligari P.D.S., Powell W. and Jinks J.L. 1985. A comparison of inbred lines derived by double haploidy and seed descent in spring barley (*H. vulgare*). *Annals of Appl. Biol.*, **111**: 667-75.

Comstock R.E. and Robinson H.F. 1952. Estimation of average dominance of genes. *In*: *Heterosis*, (ed.) J.W. Gowen, Iowa State College Press, Avens, Iowa, pp : 494-516.

Cockerham C.C. and Weir B.C. 1977b. Quadratic analyses of reciprocal crosses. *Biometrics*, **33**: 187-203.

Cockerham C.C. and Zeng Z.B. 1996. Design III with marker loci. *Genetics*, **143**: 1437-1456.

Christie B.R. and Shattcuk V.I. 1992. The diallel cross : Design, analysis and use for plant breeder. *In*: *Plant Breeding Reviews*, (ed. J. Janick). John Wiley and Sons, New York, **9**: 9-36.

Crow J.F. 2000. The rise and fall of overdominance. *Plant Breeding Reviews* 17, (ed.) J. Janick. John Wiley & Sons Inc., New York, pp. 225-257.

Duvick D. 1996. Plant breeding, an evolutionary concept. *Crop Sci.*, **36**: 539-548.

East E.M. 1910. A Mendelian interpretation of variation that is apparently continuous. *American Naturalist*, **44**: 65-82.

Engqvist G.M. and Becker H.C. 1993. Prediction of F_6 line performance from F_1 families in spring oil seed rape. Yield and other agronomic traits. *Euphytica*, **69**: 141-147.

Fisher R.A. 1918. The correlation between relatives on the supposition of Mendelian inheritance. *Transactions of the Royal Society of Edinburgh*, **52**: 399-433.

Foster G.S. and Shaw D.V. 1988. Using clonal replicates to explore genetic variation in a perennial plant species. *Theor. Appl. Genet.*, **76**: 788-794.

Gardner C.O. and Eberhart S.A. 1966. Analysis and interpretation of the variety cross diallel and related population. *Biometrics*, **22**: 439-452.

Geldermann H. 1975. Investigation on inheritance of quantitative characters in animal by gene markers. I. Methods. *Theor. Appl. Genet.*, **46**: 300-319.

Gibson G. and Hogness D.S. 1996. Effect of polymorphism in the *Drosophilla* regulatory gene *ultrabithora* on homeotic stability. *Science*, **271**: 200-203.

Graham G.I., Wolff D.W. and Stuber C.W. 1997. Characterization of a yield quantitative trait locus on chromosome five of maize by fine mapping. *Crop Sci.*, **37**: 1601-1610.

Griffing B. 1956a. Concept of general and specific combining ability in relation to diallel crossing systems. *Aust. Jour. Biol. Sci.*, **9**: 463-493.

Griffing B. 1956b. A generalized treatment of the use of diallel crosses in quantitative inheritance. *Heredity*, **10**: 31-50.

Haley C.S. and Knott S.A. 1992. A simple regression method for mapping quantitative trait loci in line crosses using flanking markers. *Heredity*, **69**: 315-324.

Hayman B.I. 1954. The analysis of variance of diallel tables. *Biometrics*, **10**: 235-244.

Hayman B.I. 1960. The theory and analysis of diallel crosses III. *Genetics*, **45**: 155-172.

Hayman B.I. and Mather K. 1955. The description of genic interactions in continuous variation. *Biometrics*, **11**: 69-82.

Hill J., Becker H.C. and Tigerstedt P.M.A. 1998. *Quantitative and ecological aspects of plant breeding*: Chapman and Hall, London.

Hill W.G. 1970. Design of experiments to estimate heritabilityby regression of offspring on selected parents. *Biometrics*, **26**: 566-571.

Huang N., Courtois B., Khush G.S., Liu H., Way G., Wu P. and Zheng K. 1996. Association of quantitative traits loci for plant height with major dwarfing genes in rice. *Heredity*, **77**: 103-137.

Hyne V., Kearsey M.J., Pike D.J. and Snape J.W. 1995. QTL analysis: Unreliability and bias in estimation precesses. *Molecular Breed.*, **1**: 273-282.

Jansen P.C. 1994. Mapping of quantitative trait loci by using genetic markers: An overview of biometrical methods. *In*: *Biometrics in plant breeding*, (eds.) J.W. Van Oijen and J. Jansen, Proc. 9th Meeting of Eucarpia Section. The Netherlands. pp. 116-124.

Jansen R.C. and Stam P. 1994. High resolution of quantitative traits into multiple loci via interval mapping. *Genetics*, **136**: 1447-1455.

Jinks J.L. 1954. The analysis of continuous variation in a diallel cross of Nicotina rustica varieties. *Genetics*, **39**: 767-788.

Jinks J.L. 1983. Biometrical genetics of heterosis. *In*: *Heterosis: Reappraisal of Theory and Practice*, (ed.) R. Frankel, Springer-Verlag, Berlin, pp. 1-46.

Jinks J.L. and Hayman B.I. 1953. The analysis of diallel crosses. *Maize Coop. Newsletter*, **27**: 48-54.

Jinks J.L. and Jones R.M. 1958. Estimation of the components of heterosis. *Genetics*, **43**: 223-234.

Jinks J.L. and Pooni H.S. 1976. Predicting the properties of recombinant inbred lines derived by single seed descent. *Heredity*, **36**: 253-266.

Johannsen W. 1913. *Elemente der exakten Erblichkeitslehere*, 2nd edn. Gustav Fischer, Jena.

Jones R.M. 1965. Analysis of variance of half diallel table. *Heredity*, **20**: 117-121.

Kearsey M.J. 1965. Biometical analysis of a random mating population: A comparison of five experimental designs. *Heredity*, **20**: 205-235.

Kearsey M.J. and Jinks J.L. 1968. A general method of detecting additive, dominance and epistatic variation for metrical traits. II. Application to inbred lines. *Heredity*, **24**: 45-57.

Kearsey M.J. and Hyne V. 1994. QTL analysis: A simple marker regression approach. *Theor. Appl. Genet.*, **89**: 698-702.

Kearsey M.J. and Pooni H.S. 1996. *The Genetical Analysis of Quantitative Traits*. Chapman & Hall, London.

Kempthorne, O. and Curnow, R.N. 1961. The partial diallel cross. *Biometrics*, **17**: 2292-50.

Kempthrone O. and Tandon O.B. 1953. The estimation of heritability by regression of offspring on parent. *Biometrics*, **9**: 90-100.

Kinman M.L. and Sprague G.F. 1945. Relation between number of parental lines and theoretical performance of synthetic varieties of corn. *J. Am. Soc. Agron.*, **37**: 341-351.

Lande R. and Thompson. 1990. Efficienty of marker assisted selection in the improvement of quantitative trait. *Genetics*, **124**: 743-756.

Lander E.S. and Botstein D. 1989. Mapping Mendelian factors underlying quantitative traits using RFLP linkage maps. *Genetics*, **121**: 185-199.

Mackay T.F. and Tanksley C.H. 1990. Molecular and phenotypic variation in the achaete-scente region of *D. melanogaster*. *Nature*, **348**: 64-66.

Mather K. 1949. *Biometrical Genetics*. Ist edn. Methuen, London.

Mather K. and Jinks J.L. 1982. *Biometical Genetics*, 3rd edn. Chapman & Hall, London.

Mendel G. 1865. Versuche ueber Pflanzen-Hybriden. *Naturfi Ver. In Bruenn Verh.*, **4**: 3-47.

Namkoong G. and Roberts J.H. 1974. Choosing mating designs to efficiently estimate genetic variance components for trees. *Silvae Genet.*, **23**: 43-53.

Nilsson-Ehle H. 1909. Kreuzungsuntersuchungen an Hafer und Weizen. *Lunds Univ. Arsskrift*. N. S., Series, **2**: 1-122.

Pederson D.G. 1972. A comparison of four experimental designs for the estimation of heritability. *Theor. Appl. Genet.*, **42**: 371-377.

Pooni H.S. and Jinks J.L. 1978. Predicting the properties of recombinant inbred lines derived by single seed descent for two or more characters simultaneously. *Heredity*, **40**: 349-361.

Pooni H.S., Jinks J.L. and Singh R.K. 1984. Methods of analysis and the estimation of the genetic parameters from a diallel set of crosses. *Heredity*, **52**: 243-253.

Robertson A. 1959. Experimental design in the evaluation of genetic parameters. *Biometrics*, **15**: 219-226.

Sax K. 1923. The association of size different with seed-coat pattern and pigmentation in *Phaseolus vulgaris*. *Genetics*, **8**: 552-560.

Schnell F.W. 1961. Some general formulations of linkage effects in inbreeding. *Genetics*, **46**: 947-957.

Schnell F.W. 1982. A synoptic study of the methods and categories of plant breeding. *Z. Pflanzuechtg*, **89**: 1-18.

Singh M. and Singh R.K. 1984a. Hayman's graphical analysis in segregating generations of a diallel cross. *Biom. J.*, **26**: 69-74.

Singh M. and Singh R.K. 1984b. A comparison of different methods of half diallel analysis. *Theor. Appl. Genet.*, **67**: 323-326.

Singh M. and Singh R.K. 1984c. Some theoretical aspects of analysis in half diallel. *Biom. J.*, **26**: 63-68.

Snape J.W. 1982. Predicting the frequencies of transgressive segregants for yield and yield components in wheat. *Theor. and Appl.Genet.*, **62**: 127-134.

Shrimpton A. and Robertson A. 1988. The isolation of polygenic factors controlling bristle score in *D. melanogaster* II. Distribution of third chromosome bristle effects within chromosome sections. *Genetics*, **118**: 445-459.

Stuber C.W., Lincoln S.E., Wolff D.W., Helentjaris T. and Lander E.S. 1992. Identification of genetic factors contributing to heterosis in a hybrid from two elite maize inbred lines using molecular markers. *Genetics*, **132**: 823-839.

Stuber C.W. 1995. Heterosis in plant breeding. *Plant Breeding Review*, **12**: 227-251.

Tanksley J.M. 1993. Mapping polygenes. *Annual Review Genet.*, **27**: 205-233.

Thoday J.M. 1961. Location of polygenes. *Nature*, **191**: 368-370.

Tysdal H.M., Kieselbach T.A. and Westover H.L. 1942. Alfalfa breeding. *Nebraska Agricultural Experimental Station Research Bulletin*, 124.

Weir B.S. and Cockerham C.C. 1977. Two-locus theory in quantitative genetics. *In*: *Proceedings of the international conference on quantitative genetics*, (eds.) E. Pollak, O. Kempthorne, and T.B. Baily, Jr., pp. 247-269. IOWA State Univ. Press, Ames.

Weller J.I. 1986. Maximum likelihood techniques for the mapping and analysis of quantitative trait loci with the aid of genetic markers. *Biometrics*, **42**: 627-640.

Wright A.J. 1985. Diallel designs, analyses and reference populations. *Heredity*, **54**: 307-311.

Wright S. 1921. Systems of mating. *Genetics*, **6**: 111-173.

Wu W.R. and Li W.M. 1994. A new approach for mapping quantitative trait loci using complete genetic marker linkage maps. *Theor. Appl. Genet.*, **89**: 535-539.

Wu W.R. and Li W.M. 1996. Model fitting and model testing in the method of joint mapping of quantitative trait loci. *Theor. Appl. Genet.*, **92**: 477-482.

Xiao J., Li J., Yuan L. and Tanksley S.D. 1995. Dominance is the major genetic basis of heterosis in rice as revealed by QTL analysis using molecular markers. *Genetics*, **140**: 745-754.

Zeng Z.B. 1994. Precision mapping of quantitative trait loci. *Genetics*, **136**: 1457-1468.

Plant Breeding - Mendelian to Molecular Approaches
H. K. Jain and M. C. Kharkwal (eds.)

23

Genotype-by-Environment Interaction in Crop Improvement

Manjit S. Kang[1], V.T. Prabhakaran[2] and R.B. Mehra[3]

Abstract

In this paper, we discuss the role of genotype by environment interactions (GEI) in crop improvement, especially the following: (1) Implications of GEI in crop improvement, (2) Nature and causes of GEI, (3) Approaches for studying GEI, and (4) Strategies for using GEI in crop improvement. Genetic, biochemical, and metabolic aspects of crop plant-environment interactions also are discussed. The take-home message in this paper is as follow: A lack of GEI could signal a lack of genetic diversity, which causes genetic vulnerability of a crop to disease epidemics, insect infestations, or other stresses. Such stresses provide opportunities for identifying and selecting genotypes.

Introduction

In 2000, the human population of the world reached the six billion mark. Projections are that the world population will almost double (10 billion) by the mid-21st century. The challenge for crop breeders and agronomists is, 'Can we feed four billion more people by 2040?' The key to doubling agricultural production is increased efficiency in utilization of resources (increased productivity per unit of land and of money), and a better understanding and utilization of genotype-by-environment interaction (GEI).

The GEI is a complex phenomenon that undermines the repeatability of experimental results and consequently reduces the selection efficiency (Kearsey and Pooni, 1996). Agricultural researchers have long been cognizant of the various implications of GEI in breeding programs (Yates and Cochran, 1938; Dickerson, 1962 and Allard and Bradshaw, 1964). Understanding the implications of GEI structure/nature is important in crop improvement programs, because a significant GEI can seriously impair selection of superior genotypes in new crop introduction and cultivar development programs. Crop breeders, once equipped with information on the nature and extent of GEI, can determine if cultivars can/should be developed for all environments of interest or if they should develop specific cultivars for specific target environments. Variation among genotypes in phenotypic sensitivity to the environment may necessitate the development of locally adapted varieties (Falconer, 1952). From the standpoint of efficiency, it would be

[1]Louisiana State University Agricultural Center, Baton Rouge, LA 70803-2110, USA
[2]Indian Agricultural Statistics Research Institute, New Delhi 110 012, India
[3]Indian Agricultural Research Institute, New Delhi 110 012, India

desirable, however, if selection progress achieved in one environment could be carried over to another environment.

Genotype-by-environment interaction is a universal issue. The voluminous literature on both GEI and stability analyses reflects the fact that it is an important and intriguing field of study. The GEI and stability of performance are expected to become more relevant issues as greater emphasis is placed on sustainability of agricultural systems. For understanding and efficient utilization of GEI molecular marker technology is being used to study the Quantitative Trait Loci (QTL) by environment interaction (Stuber *et al.*, 1999).

From the standpoint of the role of GEI in crop improvement, we plan to discuss relevant issues in the following broad sections:

- Implications of GEI in crop improvement
- Nature and causes of GEI
- Approaches to study GEI
- Strategies for utilizing GEI in crop improvement

Implications of GEI in Crop Improvement

If the interaction is so large as to cause rank changes among genotypes, then one can speak of rank interaction, which is also termed qualitative or crossover interaction. In this type of interaction the true treatment differences vary not only in magnitude but also in direction. In contrast in quantitative or non-crossover interaction the treatment differences vary only in magnitude. A crop cultivar development program encompasses the breeding phase and performance evaluation phase. Importance and implications of GEI during these phases are discussed below.

The Breeding Phase

Reduced Genetic Gains: As Hill (1975) has rightly pointed out, "plant breeder pins his hopes for crop improvement upon evidence of genetic variation for the character being selected. Accurate estimates of the genetic variance will be obtained only if such estimates are unbiased by variation due to GE interaction". Use of biased estimate, while planning breeding programmes may lead to (*i*) infructuous investment of time and effort not warranted by real genetic worth of the stock, (*ii*) adoption of selection schemes far from optimum, and (*iii*) discrepancy between the realized and expected responses to selection. Thus, for the detection and estimation of GEI it is essential that the genotypes are tested for their performance in a series of environments, multi-location, multi-year or both. Heritability of a trait is a key component in determining genetic advance from selection (Falconer, 1960). The GEI being a component of the total phenotypic variance affects heritability negatively. The larger the GE interaction component, the smaller the heritability estimate; thus, progress from selection is reduced.

For forest tree crops, Matheson and Cotterill (1990) determined loss of potential gain resulting from GEI as follows:

$$C = 1 - [(\sigma_G^2 + \sigma^2)^{1/2}/(\sigma_G^2 + \sigma_{GE}^2 + \sigma^2)^{1/2}]$$

where C = loss of potential gain, σ_G^2 = genetic variance, σ^2 = error variance, and σ_{GE}^2 = GEI variance. They described several ways in which effects of GE interaction on breeding programs could be estimated.

Proliferation of Breeding Stations: The presence of large GEI (cross-over interaction) necessitates the development of varieties suited (or tailored) to different agro-environments based on their adaptability/stability characteristics and/or selecting genotypes that generally perform well over many sites. The first approach would yield greater genetic gains, but costs would likely be higher; the second approach is less expensive but gains also would be less.

Testing in Early Selection Stages: LeClerg (1966) suggested that in the early stages of a testing program, it would be advisable to have as many locations as possible but with only one replication per location, *vis-a-vis* multiple replications at a fewer locations.

Kang (1993a) discussed the disadvantages of discarding genotypes evaluated in only one environment in early stages of a breeding program. Some potentially useful genes could be "lost" from the gene pool due to limited testing. The discarded genotypes could have the potential to do well at another location or in another year.

A large GE interaction reflects the need for testing cultivars in numerous environments (locations and/or years) to obtain reliable results. If the weather patterns and/or management practices differ in target areas, testing must be done at several sites in the target areas. If cultivars are to be used in marginal areas or fringes of a crop (specific adaptation), testing must begin in those areas as early as possible.

The Performance Evaluation Phase

As the magnitude of a significant interaction between two factors increases, the usefulness and reliability of the main effects correspondingly decrease. Since GEI reduces the correlation between phenotypic and genotypic values, the difficulty in identifying truly superior genotypes across environments, especially in the presence of crossover GEI, is magnified.

Multi-environment testing makes it possible to identify cultivars that perform consistently from year to year (small temporal variability) and those that perform consistently from location to location (small spatial variability). Temporal stability is desired by and is beneficial to growers, whereas spatial stability is desired by and is beneficial to seed companies and breeders.

Nature and Causes of Genotype-by-Environment Interaction

An understanding of genotypic responses to individual factors that otherwise detract from genetic gains can aid in interpreting and exploiting GEI (Steiner *et al.*, 1984). When an environmental factor is present at a level other than optimal, it represents a stress. Differences in the rate of increase in response of genotypes at sub-optimal levels reflect differences in efficiency, and differences in the rate of decrease in genotypic response at super-optimal levels reflect differences in tolerance (Baker,1988). For example, water at sub-optimal levels (drought) permits identification of water-use efficient genotypes. At super-optimal water levels (flooding), flood-tolerant plants can be identified.

Plants respond to a variety of environmental cues: nutrients, toxic elements, and salts in the soil solution, gases in the atmosphere, light of different wavelengths, mechanical stimuli, gravity, wounding, pests, pathogens, and symbionts (Crispeels, 1994). The extent of an individual's adaptability to environmental conditions reflects the extent and sophistication of the controls over the synthesis and action of specific proteins (Smith, 1990). Adaptability is a quantitative estimation of the range of environmental conditions to which an individual of a particular genotype can adjust and is determined by the extent and sophistication of its plastic traits (Smith, 1990). Plasticity is simply the expression of variability in the phenotype of individuals of identical genotype (Bradshaw, 1965). Genotypes developed under relatively constant environments are expected to display much less plasticity.

Scandalios (1990) summarized plant responses to environmental stress, pointing out that activated oxygen species (endogenous: byproducts of normal metabolism and exogenous: triggered by environmental factors) were highly reactive molecules that can cause extensive damage to plant cells. The effects of oxidative stress range from simple inhibition of enzyme function to the production of random lesions in proteins and nucleic acids, and the peroxidation of membrane lipids (Scandalios, 1990). Loss of membrane integrity caused by peroxidation, together with direct damage to enzymatic and structural proteins and their respective genes, can cause decreased mitochondrial and chloroplast functions, which, in turn, lower plant's ability to fix carbon and to properly utilize the resulting products. This decrease in metabolic efficiency results in reduced yield.

Plants that have incorporated a variety of environmental signals into their developmental pathways over time have acquired a wide range of adaptive capacities (Scandalios, 1990). Plants use a variety of pathways to communicate information both within and between cells. The diversity of potential pathways indicates that there are a large number of possible ways in which plants can perceive and transduce environmental stresses and changes, be they biological, chemical, or physical in nature (Leigh, 1993).

Biotic Stresses and Mineral Uptake: Biotic stress factors are a major limitation to plant productivity (Higley *et al.*, 1993). Differences in insect and disease resistance among genotypes can be associated with stable or unstable performance across environments. For example, Baker (1990) and Gravois *et al.* (1990) implicated disease resistance as a factor that contributed to GEI in crops.

In the context of GEI, very little attention has been given to the important issue of plants' genetic control of mineral uptake, transport, and metabolism. An understanding of differential ion uptake by plants (Myers, 1960) and physiological genetics of plant nutrition (Epstein, 1972, 1976) is important for crop breeders. Genotype-by-environment interaction could be due to differential survival rates among genotypes. Genetic and environmental factors and their interactions affect the number of seeds each genotype produces and the proportion of seeds of each genotype that reaches maturity (Allard, 1960). To survive in competition (stress), a plant may "reduce" the number of seeds produced but produce a smaller number of viable seeds. Such reproductive adjustment mechanisms may be differential among genotypes. Differential responses among genotypes to herbicides and allelochemicals could result in GEI.

Abiotic Stresses: The major abiotic stresses include atmospheric pollutants, soil stresses, such as salinity, acidity, and mineral toxicity and deficiency, temperature (heat and cold), water (drought and flooding), and tillage operations (Steiner *et al.*, 1984; Blum, 1988; Unsworth and Fuhrer, 1993; Clark and Duncan, 1993; Specht and Laing, 1993). Genetic variation exists for plant responses to many stress factors. Unsworth and Fuhrer (1993) suggested that there was considerable potential for breeding for tolerance to air pollutants. Metabolic processes involve chemical reactions that are mediated by enzymes. Enzymes being products of gene expression (transcription and translation), biochemical and physiological processes are integrated with genetics.

Approaches to The Study of Genotype-Environment Interactions

Four different approaches are available for the statistical analysis of GE interactions (Freeman, 1973; Hill, 1975). These are, the 'Variance component approach', 'regression approach', 'bio-metrical genetics approach' and the 'genetic correlation approach'. The choice among these methods depends on the particular situation in hand and the type of data that are collected by the investigator.

Variance Component Approach

The existence of GE interactions has long been recognized (Fisher and Mackenzie, 1923). Fisher (1926) advocated the use of the factorial design to partition the total variation into components measuring the differences between genotypes, environmental differences and that assessing their joint effects. A galaxy of workers has contributed in the promotion of the analysis of variance to investigate GE interactions, notable among them being Sprague and Federer (1951), Comstock and Robinson (1952), Hanson, Robinson and Comstock (1956) and Comstock and Moll (1963). Unbiased estimates of genetic and GEI components can be readily obtained by equating the expected mean squares with those calculated from the experiment, though obviously the accuracy with which these components can be estimated will depend upon the size of the experiments which can be handled. The precise form taken by the expected values of the mean squares will depend upon the underlying assumptions made in the analysis (fixed, random, or mixed models). The analysis of variance approach is the most fundamental, and is essentially a method for confirming the presence or otherwise of GE interaction as well as estimating the fraction of phenotypic variance attributable to this component. This preliminary analysis must therefore precede any genetic analysis of breeding data, including stability analysis, which the breeder is particularly concerned with.

Notwithstanding its importance, GE interaction is often a distraction in genetic analysis and so a recourse usually taken to overcome this difficulty is by transforming the data to a scale in which there is no interaction. In so doing one fails to recognize that GE interactions are as much a function of the genotypes as they are of the environment and so are partly heritable. A review of previous work emphasizing the importance of GE interaction to plant breeders may be found in Allard and Bradshaw (1964).

The effects of GE interactions upon the variances in non-segregating generations can be examined by considering the model devised by Mather and Jones (1958) and Jones and Mather (1958), a model which expresses Fisher's original ideas in terms of parameters developed for the study of biometrical genetics. Consider two true-breeding lines differing by a single gene *A-a* grown in two environments *X* and *Y*. Four situations are possible as shown in Table 1 and the four phenotypes can be described by considering linear combinations of the parameters *d, e, g* representing the genetic, environmental and interaction deviations, being measured from the overall mean.

Table 1. Four phenotypes of two genotypes grown in two environments expressed as deviations from the overall mean

	Genotype		
Environment	AA	aa	Mean
X	$d+e+g$	$-d+e-g$	e
Y	$d-e-g$	$-d-e+g$	$-e$
Mean	d	$-d$	0

Assuming that each genotype has an equal probability of occurring in all environments, the overall-mean phenotype will be independent of *g*. Not so the variances, however, since the phenotypic variance of the *AA* genotype around its own mean of d is, in effect, $(e+g)^2$, whereas for the *aa* genotype it equals $(e-g)^2$. Variation within non-segregating generations has been used as a measure of developmental stability, in that a lower variance signifies a greater ability to cope with the vagaries of the environment. For out-breeding species, F_1 appears to be more stable than its inbred parents (Mather, 1953), whilst for inbreeding species F_1 is not more stable than its pure breeding parents (Jinks and Mather, 1955).

GE interactions can also affect the variances of segregating generations, though their detection here is less straightforward. Essentially, the method requires the estimation of the genetic components of variation from comparable experiments grown in two or more environments (details of the estimation procedure are given by Mather and Jinks, 1971, Chapter 6). If significant differences exist between the various estimates of the same component, it may be concluded that GE interactions are operative.

Regression Approach

The main drawback of the variance component method described in the previous section is that it does not have the provision for partitioning of GE interaction into components, useful in the analysis of response pattern of genotypes under different environmental conditions. This difficulty is overcome under this procedure. The approach falls conveniently into two parts, a conventional analysis of variance being followed by a joint regression analysis to determine whether the GE interactions are a linear function of the additive environmental component. Unless the initial analysis of the data clearly establishes the significance of the GE interactions, there is no point in proceeding to the joint regression analysis. Although a method of partitioning had been proposed by Yates and Cochran as early as 1938, this was not taken up seriously until Finlay and Wilkinson (1963) rediscovered the same technique and used it in the analysis of adaptation of barley varieties. Since then this method, which uses the regression technique in partitioning the GE variance into linear and nonlinear portions, has been widely adopted in assessing the stability of genotypes over a range of environments (Eberhart and Russell, 1966; Perkins and Jinks, 1968a; Freeman and Perkins 1971; Fripp and Caten, 1971; Tai, 1971). The formal analysis for GEI is undertaken in an identical manner by Yates and Cochran (1938), Finlay and Wilkinson (1963), and Perkins and Jinks (1968). This is based on the basic biometrical genetical model:

$$Y_{ij} = \mu + d_i + (1 + \beta_i)\, e_j + \delta_{ij} + \varepsilon_{ij} \tag{1}$$

for the observed mean yield of the ith genotype ($i = 1, 2, \ldots, t$) in the jth environment ($j = 1, 2, \ldots, s$); where μ is the general mean, d_i is the effect of ith genotype, e_j is the effect of jth environment, β_i is the regression of $(Y_{ij} - \mu - d_i - e_j)$, i.e. g_{ij} on $\mu + e_j$, δ_{ij} is the deviation from regression for the ith genotype in the jth environment and ε_{ij} is the random error. Under usual assumptions model (1) provides for the partitioning of the interaction component with $(t\text{-}1)\ (s\text{-}1)$ degrees of freedom (*d.f.*) into a component for heterogeneity of regression lines with $(t\text{-}1)$ *d.f.* and another one for deviations from regression with $(t\text{-}1)$ $(s\text{-}2)$ *d.f.* In the approach of Eberhart and Russell (1966) the sum of squares for environments and GEI are added together to get the within- genotypes sum of squares, which is partitioned into linear component between environments with 1 *d.f.*, a linear component of GE interaction with $(t\text{-}1)$ *d.f.*, and deviation from linear regression with $t(s\text{-}2)$. Here the regression of Y_{ij} on e_j facilitates this partition.

Both the heterogeneity and residual components are first compared with residual error. If the first component alone is significant, almost all GEI will be of linear type so that performance of each genotype can be predicted from its regression on environment. If the residual alone is significant no useful prediction is possible from this approach. If both components are significant, the heterogeneity item should be re-tested against the residual item to examine whether its contribution is overwhelming. If it is so useful, predictions can still be made.

The joint regression analysis, provides two simple measures of stability — the regression coefficient β_i or b_i (depending on whether g_{ij} or y_{ij} is regressed on to e_j) and the deviation (from regression) which we shall be considering in a subsequent section.

The use of environmental values as independent measures, when in fact they are not independent of the phenotypic values (Y_{ij}), leads to statistically invalid regressions and ambiguity in the assignment of degrees of freedom to certain sums of squares in the joint regression analysis. Although, the non-independence as such is not serious, because marginal means could be regarded as fixed while making inferences about values in the body of the table (Freeman, 1973), the confusion over degrees of freedom remains valid. Freeman and Perkins (1971) avoided this complication by considering an independent environmental index (z) obtained by replication of genotypes or using control or assessment genotypes. Another criticism is regarding the measurement error attached to the environment index. However, as pointed out by Hardwick and Wood (1972) this would not cause any serious bias, provided a large number of genotypes are included in the experiment and the environmental mean square is significantly greater than the error mean square.

Following the criticism by Freeman and Perkins (1971) regarding the non-independence of e_j Fripp and Caten (1971), Fripp (1972) and Jinks and Connolly (1973) conducted extensive studies, involving the comparison of different biological measures among themselves and with physical measures as well as the traditional 'environmental index'. It was found that the overall interpretation of GEI, including comparisons of linear and nonlinear sensitivities, was remarkably similar for reasonable external measures and the environmental mean. The results also demonstrated that the degree of correlation between z_j and e_j values very much depended on how closely the independent assessment set matched with the genotype set under test. Recognizing, however, the need for independent values on statistical ground, there is strong reason for basing the assessment on an appropriately chosen control genotype, wherever possible.

Although, the various models considered earlier lead to the same conclusions in practice, Freeman and Perkins (1971) model is preferable on the ground of statistical soundness and inference value. The results, however, indicates that the Freeman and Perkins (1971) test of adequacy of z_j is too rigid in practice. Accordingly, it is better that the test is done by examining the consistency of difference between z_j and $\hat{e}_j$ values, as proposed by Fripp (1972). The approach of Tai (1971), developed for getting round the logical difficulty of regressing the phenotype on a non-independent environmental measure becomes less relevant when independent biological assessment of environment is feasible. The stability variance approach of Shukla (1972) is perhaps more suited to situations where nonlinear interaction also accounts for a significant portion of the total GE interaction. It is, however, pertinent to note that much of the non-linearity could be removed either by excluding a few aberrant genotypes from the interaction analysis (Hill, 1975) or through grouping of genotypes on the basis of similarities in correlation between observed deviations from regression for each pair of genotypes (Perkins and Jinks, 1968b).

The regression techniques currently employed in the analysis of GE interaction can adequately describe the behaviour of genotypes over different environments only when the genotypic response is fairly linear, a situation characterised by the overwhelming contribution of linear regression component to the total GE interaction variation. In the event of the remainder mean squares (non-linear component), relative to heterogeneity of regression mean square, accounting for a large part of the interaction variation, the characterisation of the genotypes based on the linear regression coefficient will be misleading. Presence of significant non-linear interactions has been reported in different crops such as Carrots and *Nicotiana rustica.* A common approach in non-linear situations is to reduce the complex interaction responses to a series of orderly linear responses. Assuming that the threshold limits differ among genotypes and are subjected to genotypic control, the response of each genotype can be represented by a pair of intersecting straight lines. Looking at the slopes of the regressions representing two distinct environmental categories the genotypes can be classified according to their nature of adaptation in a similar manner as followed in single best fitting line (Verma *et al.*, 1978; Pooni and Jinks, 1980).

Biometrical Genetics Approach

This approach is based on fitting of genetic models to observed generation means, leading to a test of adequacy of genetic and environmental specifications and related inferences on the presence or otherwise of interaction [Bucio Alanis, 1966; Bucio Alanis and Hill, 1966; Bucio Alanis *et al.*, 1969; Hill and Perkins, 1969]. For a pair of inbred lines and the generations derived from an F_1 between them, the generation means can be specified by considering, six parameters, viz. m, $[d]$, $[h]$, e_j, g_{dj} and g_{hj}, the last two being the interactions of e_j with the additive $\{[d]\}$and dominance $\{[h]\}$components, respectively. If additive-dominance model is found satisfactory not only for means averaged over environments, but also for each individual environments, it can be concluded that non-allelic interactions are not present. Accordingly any failure of the m, $[d]$, $[h]$, e on genotype × environment data can be attributed to the presence of GEI.

The linear sensitivities of the parents and F_1 are obtained by regressing $[d] + g_{dj}$ and $[h] + g_{hj}$ respectively on $m + e_j$. This amounts to regressing the observed $(\overline{P}_{1j} - \overline{P}_{2j})/2$ and $\overline{F}_{1j} - (\overline{P}_{1j} - \overline{P}_{2j})/2$ on $m + e_j$ [i.e. $(\overline{P}_{1j} + \overline{P}_{2j})/2$] utilizing the data on P_1, P_2 and F_1 families. After estimating the regression coefficients β_d and β_h, statistical significance of these estimates may be undertaken as test of confirmation of linear GEI. A special feature of the Biometrical method is that the predictions can be extended across several future generations from the values of parents and F_1 hybrid. There is also no problem of 'non-independent environmental index' under this procedure. For an extension of the procedure to an arbitrary number of lines and crosses among them, reference may be made to Perkins and Jinks (1968a).

Genetic Correlation Approach

Genetic correlations between the phenotypic measures of a character from different environments can provide useful information about GEI (Falconer, 1960). The expressions of a single trait in two different environments can be viewed as two separate traits, because the physiological mechanisms in different environments would be, to some extent, different, and consequently, the genes required for high performance of a particular trait could also be, to some extent, different. By regarding the performance in different environments as different traits, the genetic correlation between them can be worked out. The general formula used for this purpose is:

$$r_{genotypic} = \frac{\text{Cov}\,(X, Y)_{genotypic}}{(\text{Var}\,X.\ \text{Var}\,Y)^{1/2}{}_{genotypic}} \tag{2}$$

where Cov $(X, Y)_{genotypic}$ is the genotypic covariance between trait measurement in one environment (X) and that in another environment (Y), and Var X and Var Y are genotypic variances for the trait X and Y, respectively.

A low magnitude of a genetic correlation will mean that the genetic mechanism(s) determining the trait in the two environments are different, which is indicative of the presence of GEI. This idea can be extended (Robertson, 1959) to an indefinite number of environments. For this, an estimate of the average degree of genetic correlation among the environments is obtained under the assumption of equal true genetic correlation between any pair of environments. Since the determination of such a measure from one-way classification involves a lot of computation, a device of expressing the mean squares in a two-way classification (obtained by considering the performances in two environments as separate characters) in terms of variances and covariances in the one-way analysis can be adopted (Robertson, 1959; Dickerson, 1962; Yamada, 1962). If $A_{1,}$ B_1 and C_1 are the mean squares in the two-way ANOVA corresponding to genetic groups, interaction and error sources, Robertson (1959) has shown that the average genetic correlation is

$$\hat{r} = \frac{A_1 - B_1}{A_1 - C_1 + (l-1)\,(B_1 - C_{1)}} \tag{3}$$

where l is the number of environments and assumption has been made that the genotypic standard deviations in the environments are the same.

When the heritability of performance in each environment is available, direct and correlated responses to selection can be worked out and the ratio of the latter response to the former used to measure the gain or loss by carrying out selection in some environment other than the one in which the improved population is destined to live. Often, a reduction in environmental variance that might have been achieved by employing multiple environments in evaluating genotypes might be nullified by GEI (Kang *et al.*, 1984). They also reported genetic, phenotypic, and environmental correlations for several sugarcane traits. The environmental (error) correlations (r_e) determined from the covariance and variance component estimates for the error term is useful in examining whether the mechanisms determining the unaccounted variation (error term) for a trait between two locations are similar. The effect of GEI on genetic correlations was discussed by Aastveit and Aastveit (1993), where they suggested that breeding for stability would be expected to reduce the interactions and thereby genotype-environment correlations. Via and Lande (1987) used the GE correlation approach for a character between two environments to quantify GEI. They referred to r_{ge} as cross-environment genetic correlation. This correlation reflects the extent to which genes pleiotropically affect the trait in two different situations.

When the variance component associated with the GE is not constant across environments, the assumption of homoscedasticity is violated and in this situation genetic correlation between measurements of a character in two environments is estimated from transformed data (Dutilleul and Carriere, 1998).

Exploitation of GEI for Crop Production

Eisemann *et al.* (1990) indicated that future progress in analyzing differences in genotypic adaptation in crop improvement programs would require plant breeders to pay more attention to influences of environmental factors. They listed three ways of dealing with GE interactions in a breeding program: (1) ignore them, i.e., use genotypic means across environments even when GEI exists, (2) avoid them, or (3) exploit them.

Researchers agree that interactions should not be ignored when they are significant and of the crossover type (Tiret *et al.*, 1993). One of the "avoiding GEI" approaches is grouping of similar environments (mega-environments) via a cluster analysis. With environments being more or less homogeneous, genotypes evaluated in them would not show crossover interactions. By clustering environments, potentially useful information may be lost. To identify genotypes with broad adaptation (i.e., stable performance across diverse environments) across many international sites, clustering of test environments or genotypes may not be advisable.

The "exploiting GEI" approach may entail use of stability statistics obtained from performance across diverse environments. This approach allows researchers to identify the causes of GEI and to devise strategies to correct the problem. When the cause for unstable performance of a genotype is known, either the genotype could be improved by genetic means or by providing proper environment (inputs and management) to maximize its productivity.

Broad adaptation or stability of performance (reliability) across environments is desired to conserve resources. To achieve greater success, crop environments need to be characterized as fully as possible. The more the breeders know about the crop environment, the better the job they can do in developing cultivars with wide adaptability or judiciously targeting appropriate cultivars to production environments.

Stability Statistics: Concepts and Usefulness

If possible, the breeder would like to have varieties, which show high performance for yield and other agronomic traits over as wide a range of environmental conditions as possible. However, the wide occurrence of GEI causes difficulty in the identification of superior varieties. In order to overcome this difficulty, an attempt is usually made by the plant breeder to reduce the GE interaction, i.e. dependence of the genotypic ranking on environmental conditions through special breeding techniques like resistance breeding if reduction in yield is due to only diseases and insects. Since only a minor part of the GEI can be attributed to controllable environmental determinants, much reduction in the interaction cannot be achieved. This means that the breeder can no longer hope to find varieties which would excel everywhere and in all the years. The most practical alternative then would be to attempt to produce progressively better-adapted populations to the existing or specific subsets of environments. For the final choice of varieties for general/specific adaptation, apart from the mean performance the stability characteristics of the trial genotypes have to be given due consideration. Various concepts of stability have been advanced.

Biological and Agronomic Concepts of Stability: Stability is an important keyword/concept for plant breeders in analyzing GEI data. Some researchers prefer to use the term 'sensitivity analysis' instead of 'stability analysis'. One seeks performance stability (a lack of GEI) against uncontrollable forces, such as weather, whereas a positive GEI may be sought for controllable factors, such as fertilizer and other inputs.

Depending upon the final goal of the breeder and the character under consideration, two concepts of stability can be introduced, the biological (or static) and agronomic (or dynamic) concepts (Becker, 1981a and Leon, 1985). Under the biological concept a stable genotype is one whose phenotype Y_{ij} shows little deviation from the expected character level $\overline{Y}_{i.}$ when the genotype is performance-tested over a number of environments. This type of yield stability is analogous to the concept of genetic homeostasis, first introduced by Lerner (1954). The biological concept as applied to a character like grain yield would mean a stable genotype performs well under adverse environments, but not so well under favourable environments. But with increased inputs, improved technology, etc., the breeder would prefer a genotype whose performance in a particular environment is at a expected level depending on the level of productivity of the location as measured by the average productivity of all the genotypes grown in that environment. In other words he is interested in a variety, which does not show any genotype-environment interaction. *i.e.* $(Y_{ij} - \overline{Y}_{i.} - \overline{Y}_{.j} + \overline{Y}) \cong 0$, for all i This concept which permits a predictable response $(\overline{Y}_{i.} + \overline{Y}_{.j} - \overline{Y})$ in each environment and no deviation from the amount predicted is known as the agronomic or dynamic concept of stability.

Most of the yield stability statistics in vogue are measures according to the agronomic concept. For the biological concept only two measures are available: the environmental variance $S^2_{Y_i}$ and the environmental coefficient of variation (CV_i). The non-parametric measures, based on rank orders, also belong to the agronomic concept, because the rank orders are essentially a consequence of GEI.

Types of Stability: In Type 1 stability (equivalent to biological stability), a genotype is regarded as stable if its among-environment variance is small; in Type 2 (equivalent to agronomic stability), a genotype is regarded as stable if its response to environments is parallel to the mean response of all genotypes in a test; and in Type 3 stability, a genotype is regarded as stable if the residual mean square, after regressing genotype performance (e.g., yield) on environmental index, is small (Lin *et al.*, 1986). Lin and Binns (1988) proposed a 'Type 4' stability statistic on the basis of predictable and unpredictable non-genetic variation; the predictable component related to locations and the unpredictable component related to years. The mean square for years-within-locations for each genotype as a measure of the unpredictable variation was referred to as 'Type 4 stability statistic.'

Stability Statistics: Lin *et al.* (1986), classified the diferent statistics into four groups depending upon their structure (Table 2).

Table 2. Four types of parametric stability measures and their stability concepts

Group	Type	Stability measure
A	1	Environmental variance
	1	Coefficient variation
B	2	Plaisted and Peterson (1959)
	2	Plaisted (1960)
	2	Wricke (1962)
	2	Shukla (1972)
C	2	b_i of {Finlay and Wilkinson (1963); Eberhart and Russell (1966)}
C	2	β_i of Perkins & Jinks (1968)
D	3	S^2_{di} Perkins and Jinks (1968)
	3	S^2_{di} Eberhart and Russell (1966)
	3	Perkins and Jinks (1968)

As originally defined, the Plaisted and Peterson (1959) measure (θ_i) is computed as the mean of the GE variances for each genotype and indicates the contribution of the *i*th genotype to the total GE interaction and the Plaisted (1960) measure, $\theta_{(i)}$ is estimated as the GE variance obtained after deleting the *i*th genotype from the original data set. These measures as well as the Wricke (1962) and Shukla (1972) measures are closely interrelated through W_i, the ecovalence measure. The statistic W_i is simply the GEI effects, squared and summed across the environments, so it is directly related to GE interactions. Shukla's (1972) 'stability variance'(σi^2), which is the estimate of variance of

$g_{ij} + \bar{e}_{ij}$ in terms of the residuals in a two-way classification, is a useful indicator of the stability of the *i*th genotype. The regression coefficient (b_i) in Finlay and Wilkinson (1963) sense is the regression of $e_j + g_{ij}$ on e_j or equivalently, the regression of $(Y_{ij} - \bar{Y}_{i.})$ on $(\bar{Y}_{.j} - \bar{Y}_{..})$ whereas Perkins and Jinks (1968a) parameter, β_i is computed from the regression of g_{ij} on e_j. The same logic applies to the corresponding deviation mean squares. Other reported statistics include Hanson's (1970) 'genotypic stability' and Pinthus' (1973) coefficient of determination measure (r_i^2). The formula for r_i^2 is, $r_i^2 = b_i^2 \Sigma e_j^2 / [b_i^2 \Sigma e_j^2 + \Sigma \delta_{ij}^2]$ and this approximates to $1 - s_{di}^2 / S_{yi}^2$.

The statistic s_{di}^2 is related to the unpredictable part of the genotypic variability and is therefore considered a stability parameter. The regression coefficient which characterizes the specific response of genotypes to environmental effects is a response parameter rather than a stability parameter (Breese, 1969). Accordingly most authors used it in conjunction with s_{di}^2 for characterising the stability of test genotypes (Becker and Leon, 1988).

Since $CV(y)$ = variance (ln y), approximately the two Group *A* statistics are equivalent except for data scaling. For ranking purposes the 4 statistics in Group *B* and the 2 in Group *C* are equivalent, while Group *D* statistics are identical. When regression of g_{ij} on e_j is strong , the Group *C* statistics are to be preferred over Group *B* because they give both the shape of the response and its variation.

Since statistics of the same type are measuring the same characteristics, rank correlation between them is likely to be high. However, there can be exceptions. As pointed out by Lin (1989) CV_i is S_{yi} weighted by a mean. Therefore, correlation between S_{yi}^2 and CV_i may not be good if the cultivar means are highly variable. The b_i statistics which are type 2, are generally not correlated with S_{yi}^2 and CV_i, which are type1. However, if stable statistics are defined by having $b_i = 0$ (smaller the *b* value greater the stability) type 1 is implied and this results in high correlation.

Tai's (1971) procedure, based on the structural relationship of environmental and GE terms in the linear model for phenotype, gets round the logical difficulties of regressing one variable on to another, which is not independent of the first and extracts two parameters α_i and λ_i representing the linear and non-linear sensitivities of *i*th genotype. He has also shown that $(b_i\text{-}1)$ is always smaller than α_i while the counterpart of λ_i namely deviation mean square/ (error mean square/r) is generally greater than λ_i. For a description of Tai procedure reference may be made to Prabhakaran and Jain (1994).

Hanson (1970) proposed a statistic which combines the information from equivalence and regression into a simple useful measure of yield stability. This measure which includes that part of the variance of environmental effects which could be reduced by breeding and selection was termed as 'genotypic stability' ($D^2_{(i)}$) (Utz, 1972). Suppose, the differences between regression coefficients are completely amenable to breeding procedures, and denote the observed minimum of b by b_{min}. Then a stable genotype is one which does not deviate from the straight line, $Y = \bar{Y}_{i.} + b_{min}\,\hat{e}_j$. The 'genotypic stability' statistic, therefore, becomes:

$$D^2_{(i)} = \sum_i (X_{ij} - \bar{X}_{i.} - b_{min}\,\bar{X}_j + b_{min}\,\bar{X}_{..})^2$$

Environmental variance, ecovalence and deviation mean squares are special cases of 'genotypic stability'; when $b_{min} = 0$, $[D^2_{(i)}/(s\text{-}1)] = S^2_{yi}$, when $b_{min} = 1$, $D^2_{(i)} = W_i$, and for $b_{min} = b_i$, $D^2_{(i)}/(s\text{-}2) = s^2_{di}$.

Hitherto we considered stability parameters when interaction is linear. For the non-linear situation we have the procedures of Verma *et al.* (1978) and Pooni and Jinks (1980). The former method consists of first dividing the environments into two groups, one comprising all the negative environmental indices (below average environment) and the other with all positive environmental indices as well as the negative index with the smallest absolute value (above average environment) and then fitting appropriate curves to the two groups of graded environments. The suitability of various genotypes for general/specific adaptation is decided on the basis of the regression coefficients for the two groups. Pooni and Jinks (1980) refined the two-phase regression technique on the principle that significant part of the non-linearity results from thresholds in response of genotypes to environmental changes. Here the environmental value at which a particular genotype changes its slope is determined as the point of intersection of a pair of best fitting straight lines, which allows for a considerable reduction in the residual mean square.

The Analysis proceeds by fitting linear, quadratic (polynomial) and two-intersecting-lines models for each genotype under test. By comparing the residual mean squares, a best fitting pair of lines (*i*) $y = a_1 + b_1 e_j$, and (ii) $y = a_2 + b_2 e_j$ is selected from all possible pairs corresponding to the observational subsets, (a) $y_1, y_2 \ldots, y_i$ and (b) $y_{i+1}, y_{i+2}, \ldots, y_n$, where i is allowed to take values from 3 to n-3, n being the total number of test environments. A single straight line is rejected in favour of a best fitting pair only if the latter results in significant reduction in the residual mean square and significant difference between b_1 and b_2. Similarly a single-best fitting line will be rejected in favour of a quadratic, only if the quadratic accounts for considerable reduction in residual mean square and the quadratic regression coefficient is significantly different from zero. When the quadratic meets these requirements, the two intersecting lines must account for significant reduction in the mean square, over what is achieved from the quadratic for their acceptance against the quadratic.

The procedure can be presented in a matrix form as:

$$Y_i = a_1\boldsymbol{D}_i + a_2(1-\boldsymbol{D}_i) + [b_1 D_i + b_2(1-D_i)]\,\boldsymbol{X}_i + \boldsymbol{U}_{1i}\boldsymbol{D}_i + \boldsymbol{U}_{2i}(1-\boldsymbol{D}_i) \quad (4)$$

where a_1, a_2, b_1 and b_2 are the regression parameters; $\boldsymbol{Y}_i$ and $\boldsymbol{X}_i$ the dependent and independent variable vectors; $\boldsymbol{U}_{1i}$ and $\boldsymbol{U}_{2i}$ are the remainder vectors for the two regressions and $\boldsymbol{D}_i$ a discrete vector of coefficients with values of 0 and 1, respectively below and above the change-over point between the two sets of environments.

Model (4) can be expressed in an alternate form, by considering the expectation of $\boldsymbol{Y}$ as

$$\boldsymbol{Y} = \boldsymbol{X}\beta \quad (5)$$

where $Y' = (y_1, y_2, ..., y_n)$ is the vector of dependent variable, $\beta' = (\boldsymbol{a}_1, \boldsymbol{b}_1, \boldsymbol{a}_2, \boldsymbol{b}_2)$, the vector of regression parameters and

$$\mathbf{X} = \begin{bmatrix} \mathbf{A} & \mathbf{B} \\ \mathbf{C} & \mathbf{D} \end{bmatrix},$$

a partitioned matrix in which **B** and **C** are null matrices of order $r \times 2$ and $(n-r) \times 2$ respectively $(3 \leq r \leq n-3)$. The first column of **A** and **D** are vectors with r, $(n-r)$ unit elements, respectively. The second column of **A** contains the first r environmental values and that of **D** the remaining $(n-r)$ environmental values. It is easy to see that $\mathbf{X}'\mathbf{X}$ reduces to

$$(\mathbf{X}'\mathbf{X}) = \begin{bmatrix} \mathbf{M} & \mathbf{0} \\ \mathbf{0} & \mathbf{N} \end{bmatrix}$$

where **M** and **N** are non-zero 2×2 matrices and **O** is the null matrix of the same order. Clearly,

$$(\mathbf{X}'\mathbf{X})^{-1} = \begin{bmatrix} \mathbf{M}^{-1} & \mathbf{0} \\ \mathbf{0} & \mathbf{N}^{-1} \end{bmatrix}$$

and the least squares solution of Eq. (5) is

$$\hat{\beta} = (\mathbf{X}'\mathbf{X})^{-1}\,\mathbf{X}'\mathbf{Y}$$

with the corresponding sum of squares due to regression as

$$\hat{\beta}'\mathbf{X}'\mathbf{Y} = \mathbf{Y}'\mathbf{X}(\mathbf{X}'\mathbf{X})^{-1}\,\mathbf{X}'\mathbf{Y}$$

The variance of $\hat{\beta}$ can be obtained as

$$\mathbf{V}\,(\hat{\beta}) = \hat{\sigma}^2\,(\mathbf{X}'\mathbf{X})^{-1}$$

where

$$\hat{\sigma}^2 = \frac{\text{Residual sum of squares}}{(n-4)} = \frac{\mathbf{Y'Y} - \hat{\boldsymbol{\beta}}'\mathbf{X'Y}}{(n-4)}$$

Non-Parametric Measures: There is ample justification for the use of non-parametric measures in the assessment of yield stability of crop varieties. Their chief advantages are : (*i*) No assumptions about the phenotypic observations are needed (*ii*) Sensitivity to measurement errors or to outliers is much less compared to parametric measures. (*iii*) Additions or deletions of one or a few genotypes do not cause distortions to non-parametric measures (*iv*) Most of the time, the breeder, is concerned with crossover interaction, an estimate of stability based on rank-information, therefore, seems more relevant (*v*) These measures are particularly useful in situations where parametric measures fail due the large number of linear GEI. For these reasons non-parametric measures are widely employed in the selection of crop varieties (Huhn, 1979, 1990a,b, 1996; Nassar and Huhn 1987; Nassar *et al.*, 1994; Raiger and Prabhakaran, 2000, 2001].

Consider t genotypes tested in s environments. In non-parametric analysis of GE interaction we deal with ranks of genotypes separately for each of these s environments. The rank of a genotype in a particular environment, r_{ij} cannot be based purely on the phenotypic values Y_{ij}, because the stability has to be measured independently of the genotypic effect. Therefore, r_{ij} the rank of the ith genotype in the jth environment is determined on basis of the corrected phenotypic values, namely $(Y_{ij} - \bar{Y}_{i.})$, $\bar{Y}_{i.}$ being the mean performance of the ith genotype. A few non-parametric measures from these considerations were proposed by Thennarasu (1995).

It is a known fact that the non-parametric methods are less powerful than their parametric counterparts. Raiger and Prabhakaran (2000) showed that when the number of genotypes in the trial is fairly large, the power efficiency of the non-parametric measures will be quite close to those of the parametric measures and in this situation, the risk of selecting inferior genotypes from the use of non-parametric measures is minimal.

Interrelationships Among Stability Measures: The following two relationships are particularly useful in explaining empirical correlations

$$W_i = (b_i - 1)^2 \sum_j \hat{e}_j^2 + \sum_j \hat{\delta}_{ij}^2 \qquad D_{(i)}^2 = (s-1)\, S_{Y_i}^2 + b^*(b^* - 2b_i) \sum_j \hat{e}_j^2 \qquad (6)$$

From (6) it follows that, a high correlation between W_i and $s_{d_i}^2$ is possible only if the covariance between g_{ij} and $\hat{e}_j$ explains only a small part of W_i (Wricke and Weber, 1980). The expression for $D_{(i)}^2$ suggests the possibility of high correlation of b_i and $S_{Y_i}^2$ with $D_{(i)}^2$, if b_i and $S_{Y_i}^2$ are, associated (i.e. smaller b value implies greater stability).

Besides theoretical relationships between the stability statistics empirical correlations are also useful for quantifying the influence of each term of the equation. Thus Rao and

Prabhakaran (2000) examined empirical correlations among common stability statistics for grain yield from multi-location wheat trials. They found to very high correlation of r_i^2 with $S_{d_i}^2$, and $S_{Y_i}^2$ and b_i with $D_{(i)}^2$. The correlation between r_i^2 and W_i (or $\hat{\sigma}_i^2$) was much lower than that observed between r_i^2 and $s_{d_i}^2$. This showed that the influence of the component $b_i^2 \sum_j \hat{e}_j^2$ is very little in r_i^2 whereas it was strong in the case of W_i. In fact, the role of the component $b_i^2 \sum_j \hat{e}_j^2$ or $(b_i - 1)^2 \sum_j \hat{e}_j^2$ becomes important when the number of environments(s) is very limited. This has also lowered the correlation between W_i and $s_{d_i}^2$ to a large extent. The observed high correlation of $S_{Y_i}^2$ and b_i with $D_{(i.)}^2$ was a consequence of the strong association between b_i and $S_{Y_i}^2$. Becker (1981b) reported rank correlations among stability statistics for grain yield in maize, barely and oats. They found strong correlation among W_i, s_{di}^2 and r_i^2 due to relatively large variability of s_{di}^2 compared to b_i. However, with a small (s), and consequently higher variation in b_i correlations of W_i and s_{di}^2 to r_i^2 were only medium and correlation between r_i^2 and b_i were moderate to high (Vasil and Milas, 1984).

Pham and Kang (1988) reported s_{di}^2 to be highly correlated with σ_i^2 and s_i^2, the regression coefficients showed strong association with S_{yi}^2. Kang *et al.* (1987) observed perfect correlation between σ_i^2 and W_i (as expected). These measures (Type 2 stability), providing estimates of contributions of individual genotypes to total GE interaction, are particularly useful to breeders and agronomists. They also can be used to evaluate testing locations by identifying those locations with similar GE interaction pattern (Glaz *et al.*, 1985).

Theoretical relationships of non-parametric measures, among themselves, and to the common parametric measures have not so far been elaborated. These, therefore, have to be judged on the basis of empirical (rank) correlations. From winter wheat trials conducted for 22 years, Leon (1985) observed that two of the Huhn's (1979) statistics, S_i^1 and S_i^4 are highly correlated with S_{di}^2 and $W_{i.}$. The correlations with S_{yi}^2 were, however, quite low signalling that the non-parametric statistics measure stability according to dynamic concept.

Recent Developments: Multivariate statistical procedures that have received current attention are Pattern Analysis (Byth *et al.*, 1976; Cooper and De Lacy, 1994; De Lacy *et al.*, 1996; Williams, 1976), the Additive Main effects and Multiplicative Interaction (AMMI) analysis (Gabriel, 1978; Gauch Jr., 1988 and Gauch and Zobel, 1996) and the

Best Linear Unbiased Prediction (BLUP). When there are a large number of cultivars or locations, identification of the GEI pattern becomes difficult. One solution is to group the data with respect to homogeneous subsets of cultivars or environments using any one of the clustering methods. In pattern analysis we use both cluster analysis and ordination to summarize the data effectively in terms of similarities of mean yield or patterns of response. The AMMI model incorporates both additive and multiplicative components of the two-way structure, that can account more effectively for the underlying interaction (Shafii and Price, 1998). Integrating results obtained from bi-plot graphs with those of the genotypic stability analyses enables clustering of genotypes based on similarity of response and the degree of stability in performance across environments (Shafii and Price, 1998). A brief description of AMMI analysis is given below.

Additive Main effects and Multiplicative Interaction (AMMI) Analysis: Gauch Jr. (1988, 1992) has advocated the use of what he terms AMMI analysis of yield trial data. In many cases this procedure has been shown to increase estimation accuracy, since it fits additive main effects for genotypes and environments by an ordinary ANOVA procedure and then applies PCA to the matrix of residuals that remain after the fitting of main effects. The interaction and the error $[(ge)_{ij} + \varepsilon_{ij}]$ can be decomposed into N space PCA axes.

Zobel *et al.* (1988) compared the performance of AMMI analysis with traditional approaches *viz.*, ANOVA, PCA and linear regression in an analysis of a soyabean yield trial. He found the following to hold true:

(*i*) ANOVA fails to detect a significant interaction component.

(*ii*) PCA fails to identify and separate the significant genotype and environment main effects.

(*iii*) Linear regression many a time accounts for only a small portion of the interaction sum of squares.

The AMMI model for G genotypes and E environments may be written as:

$$Y_{ij} = \mu + d_i + e_j + \sum_{n=1}^{N} \lambda_n \alpha_{in} \gamma_{jn} + \varepsilon_{ij} \qquad (7)$$

$$\varepsilon_{ij} \sim N(0, \sigma^2); \; i = 1, 2, ..., G \qquad j = 1, 2, ..., E.$$

where, Y_{ij}, μ, d_i and e_j are same as defined for model (1), $\sqrt{\lambda_n}\,\alpha_{in}$ and $\sqrt{\lambda_n}\,\gamma_{jn}$ are the principal component scores for *i*th genotype and *j*th environment, respectively. We have $N = \max(G\text{-}1, E\text{-}1)$, i.e. the number of axes retained in the model. Further,

$$\sum_i \alpha_{in}^2 = \sum_j \gamma_{jn}^2 = 1, \; \Sigma_i\, \alpha_{in}\, \alpha_{in'} = \sum_j \gamma_{jn}\, \gamma_{jn'} = 0,$$

and the multiplicative interaction term satisfy the constraints, $\lambda_1 > \lambda_2 > \ldots > \lambda_N > 0$.

Let $Z_{ij} = Y_{ij} - \bar{Y}_{i.} - \bar{Y}_{.j} + \bar{Y}_{..}$ be the estimate of the interaction in the (i, j)th cell and in matrix notation it is possible to write (Z_{ij}) as Z, a $G \times E$ matrix. Now, the estimates of the parameters of the model are:

$\hat{\lambda}_n$ = non-zero eigen values of $Z'Z$ or ZZ' (in descending order)
$\hat{\alpha}_{in}$ = principal component of the row, sum of squares cross product (SSCP) matrix ZZ'
$\hat{\gamma}_{jn}$ = principal component of the column (SSCP) matrix $Z'Z$

Using these we can write

$$z_{ij} = \sum_{n=1}^{N} \hat{\lambda}_n \hat{\alpha}_{in} \hat{\gamma}_{jn}$$

It follows that the PCA scores are

$\sqrt{\lambda_n}\, \hat{\alpha}_{in}$ for the genotype, and $\sqrt{\lambda_n}\, \hat{\gamma}_{jn}$ for the environment.

The least square fit to the AMMI model for two-way balanced data is obtained in two steps.

(*i*) The main effects in the additive parts of the model, i.e. μ, d_i and e_j are analysed by ordinary ANOVA, leaving behind the non-additive residuals.

(*ii*) Interaction term, i.e. $(ge)_{ij}$ is then analysed by PCA. If all the PCA axes are retained in the model, the resulting full model would have as many *d.f.* as the data and would consequently fit the data perfectly. Usually a few PCA axes are good enough to summarise most of the interaction variation, resulting in a reduced AMMI model that leaves a residual.

ANOVA for AMMI analysis takes the following form :

Source	df	SS	MS
Genotypes	G-1		
Environments	E-1		
Interaction	(G-1) (E-1)		
PC (1)			
PC (2)			
PC (3)			
-			
PC(k)	G-E-1-2k		
:			
PC (n)			
Pooled error			

The AMMI model or variations thereof (e.g., factor regression or FANOVA) have been considered for interpreting GEI in various situations (Baril, 1992; Nachit *et al.*, 1992; Gutierrez *et al.*, 1994; Yan *et al.*, 2000). Yan *et al.* (2000) advocate the use of what they call 'GGE' biplot. The GGE includes both the G and GE effects. Hill and Rosenberger (1985) and Stroup and Mulitze (1991) used BLUP to determine the additive main effects for yield trails and Piepho (1994) developed the procedure to include the prediction of GEI. A comparison of BLUP and AMMI procedures is also considered in Piepho (1994).

The Shifted Multiplicative Model (SHMM) (Crossa *et al.*, 1996; Cornelius *et al.*, 1996; Crossa and Cornelius, 2000; Seyedsadr and Cornelius, 1992) is a powerful analytical tool for determining the separability of genotypic effects from environmental effects (implies the absence of genotypic rank changes across environments), separability of environmental effects from genotypic effects and complete separability. Cornelius *et al.* (1992) defined sufficient conditions for the absence of statistically significant: (1) $SHMM_1$ is an adequate model for fitting the data and (2) primary effects of environments have the same sign. Non-significant environmental rank-change interactions is indicated when (1) holds and (3) primary effects of genotypes have the same sign. The absence of significance of both environmental rank-change interactions and genotypic rank-change interactions occurs when (1), (2) and (3) all hold. SHMM with *t* multiplicative terms ($SHMM_t$) is represented by

$$Y_{ij} = \beta + \sum_{k=1}^{t} \lambda_k \alpha_{ik} \gamma_{jk} + e_{ij}$$

where Y_{ij} is the yield of *i*th genotype in the *j*th environment; β the shift parameter; λ_k the singular value for the axis *k*; α_{i1} and γ_{j1} the "primary effects" of the *i*th genotype and the *j*th environment, respectively, α_{i2} and γ_{j2} their "secondary effects", etc; e_{ij} a random error.

Non-parametric methods of Hühn (1996) that are based on cultivar ranks, probability of outperforming a check (Eskridge, 1996), and procedures for selecting cultivars, simultaneously, for yield and stability (Kang, 1988, 1993b, Bajpai and Prabhakaran, 2000) are some of the other methods that have been considered by plant breeders. Gimelfarb (1994) using AMMI demonstrated that GEI may cause a substantial non-linearity in offspring-parent regression and a reversed response to directional selection. He also showed that directional selection might be accompanied by an increase in heritability.

Simultaneous Selection for Yield and Stability: Integration of stability with performance through suitable statistics will go a long way in selecting high yielding, stable cultivars. The development and use of *Yield-Stability statistic* (YS_i) has enabled incorporation of stability in the selection process (Kang, 1993b). A computer program (STABLE) for calculating this statistic is available free of charge (Kang and Magari, 1995). It has been evaluated and found to be useful for recommendation. However, Bajpai and Prabhakaran (2000) observed that Kang's rank-sum method has an inherent weakness in that it is weighing heavily towards yield performance, apart from the arbitrariness in the scoring procedure. Accordingly they proposed three new indices (I_1, I_2, I_3), which were found to be superior to Kang (1993b) indices.

Nassar *et al.* (1994) compared the performance of three parametric and three non-parametric measures based on two criteria, convergence of observed α (type-I error) to the postulated α and the power of the test (1-β). Based on this they recommended the parametric measure, $\frac{1}{S}\sum_{j}\left| x_{ij}-O \right|$ and the non- parametric measure $\frac{1}{S}\sum_{j}\left| r_{ij}-O \right|$ [where x_{ij} is the phenotypic value of the ith genotype in environment j, r_{ij} is the rank based on corrected x_{ij} and O is some measure of optimum performance] for varietal selection. Based on the same criteria, Thennarasu (1995) and Raiger and Prabhakaran (2000) found that their measure *NP*(2) is a useful combined-measure. Annicchiarico (1997b) developed a SAS-based computer program (STABSAS) that performs a complete analysis and computes different measures of genotype stability. Recently, a comprehensive SAS program called 'SASG × ESTAB' was made available by Hussein *et al.* (2000). Their program computes univariate and multivariate stability statistics for balanced data. Specifically, the program calculates: Tai statistics, α_i and λ_i, the measures, $W_i(\sigma_i^2$ and $s_i^2)$, $D_{(i)}^2$, $\bar{\theta}_i$, $\theta_{(i)}$, S_{yi}^2 and CV_i, Kang's rank sum as well as the rank-based non-parametric stability statistics of Hühn, and provides stratified rank analysis of genotypes. The program also calculates Type 4 stability, superiority measure, the desirability index of genotype performance and the pair-wise GEI of genotypes with checks. The program may be downloaded from: http://www.nlh.no/ipf/publikasjoner/hussein/stability/ default.htm.

Contributions of Environmental Variables to Stability: Yield stability is genetically controlled, but the amount of genetic variation is related to the specific statistics used for stability evaluation and to environments. Yield components and other plant characteristics, such as resistance to pests and tolerance to environmental stress factors, affect yield stability. By identifying factors causing GEI, breeders can improve cultivar stability. If instability is caused by susceptibility to a disease, breeding for resistance should reduce losses in disease-prone environments and increase genotype stability.

Methods of assessing contributions of weather variables and other factors (covariates) that contribute to GEI have been developed and employed (Shukla, 1972; Denis, 1988; Van Eeuwijk *et al.*, 1996; Magari *et al.*, 1997). Contributions of different environmental variables to GEI have been estimated (Saeed and Francis, 1984; Kang and Gorman, 1989; Kang *et al.*, 1989; Gorman *et al.*, 1989; Rameau and Denis, 1992; Charmet *et al.*, 1993). Using data from multi-environment maize hybrid yield trials, Kang and Gorman (1989) removed GEI heterogeneity caused by maximum and minimum temperatures, rainfall, and relative humidity.

A combination of two or more environmental variables should remove more heterogeneity from GEI than individual variables. Methods developed by Van Eeuwijk *et al.* (1996) may be useful for this purpose. Recently, Magari *et al.* (1997) determined contributions to total GEI of individual environmental factors and combinations thereof.

They identified precipitation as the most important environmental factor that contributed to GE interaction for ear moisture loss rate in maize. Precipitation + growing degree days from planting to black-layer maturity (GDD-BL) and relative humidity + GDD-BL were the two-factor combinations that explained a larger amount of GE interaction than other combinations did.

Stability Variance for Unbalanced Data: When a set of genotypes are grown in a specific set of environments, a balanced data set results. Often, this ideal situation is not possible, especially, when a wide range of environments, or long-term trials is considered, because hybrids/varieties are continually replaced year after year. Also, the number of replications may not be equal for all genotypes, because experimental plots may be discarded for one reason or another. In such cases, plant breeders must deal with unbalanced data.

Researchers have used different approaches for studying GEI from unbalanced data (Freeman, 1975; Patterson, 1978, 80; Digby, 1979; Patterson and Silvey, 1980; Pedersen *et al.*, 1978; Zhang and Geng, 1986; Gauch and Zobel, 1990; Rameau and Denis, 1992; Piepho, 1994). Usually environmental effects are considered as random and cultivar effects as fixed. Mixed model equations (MME) are useful in such situations (Henderson, 1975).

For calculating stability variances, Shukla (1972) partitioned GE interaction into t, $G_k \times E$ components, one for each genotype. A negative feature of this approach is that values can be negative. Computation of σ_i^2 is impossible from unbalanced data, but $G_k \times E$ variance components ($\sigma^2_{g(k)e}$) can be estimated using the Restricted Maximum Likelihood (REML) approach (Kang and Magari, 1996). Under the mixed model assumptions, the linear model for phenotype can be expressed as:

$$\boldsymbol{Y} = \boldsymbol{l}\mu + \boldsymbol{X}\beta + \boldsymbol{W}\alpha + \boldsymbol{U}\nu + \sum_k Z_k \delta_k + \varepsilon \qquad (8)$$

where $\boldsymbol{Y}$ is a $n \times 1$ vector of observations, $\boldsymbol{l}$ a colum vector of unities; $\boldsymbol{X}$ and β are the design matrix for fixed effects (genotypes) and vector of effects ($1 \leq k \leq g$ where g is the number of genotypes); $\boldsymbol{W}$ and α are, respectively, the design matrix for and vector of environmental effects (α_i), ($1 \leq i \leq s$) where s is the number of environments; U and ν are, respectively, the design matrix for and vector of replications within environment effects (ν_{ij}); Z_k and δ_k are, respectively, the design matrix for and vector of GEI effects of the kth genotype and E is the vector of experimental error effects. Equation (8) can be solved using Henderson's (1975) mixed model equation (MME). The procedure was used in the analysis of maize hybrid yield trial data (unbalanced due to missing observations) on 11 hybrids grown at four locations for four years. For calculating phenotypic stability variances, restricted maximum likelihood (REML) using the EM-type algorithm was employed (Patterson and Thompson, 1971).

REML was used to calculate stability variances, because it can be effectively adapted to unbalanced data (Searle, 1987). The method is translation invariant and gives non-negative estimates of $\sigma^2_{g(k)e}$ (Harville, 1977). The REML methodology is generally preferred to maximum likelihood estimates because it considers the degrees of freedom for fixed effects for calculating error. Calculation of REML stability variances for unbalanced data allows one to obtain reliable estimates of the stability parameter. It also overcomes the difficulties of manipulating unbalanced data (Magari and Kang, 1997).

Balzarini *et al.* (2001) have recently illustrated the use of the BLUP methodology (mixed model approach) to improve genotype performance prediction in multi-environment trials. The use of mixed models to analyze advanced variety trials has the potential to improve predictive precision at virtually no additional cost.

Exploiting Genotype by Environment Interaction

Some of the important strategies for minimizing undesirable aspects of and exploiting beneficial potential of GEI through appropriate breeding, genetic, and statistical methodologies are outlined below.

Breeding for Resistance/Tolerance to Stress Factors

Correct Genetic Cause(s) of GE Interaction: Resistance or tolerance to any type of stress, biotic or abiotic, is essential for stable performance. It is also important to identify the factor(s) that are responsible for GE.

If the causes (e.g., disease resistance/susceptibility) for interaction are traits with monogenic or digenic inheritance, solutions are relatively easy. When breeders know the inheritance of resistance/tolerance to stresses and have resistant germplasm available, resistance/tolerance can be incorporated into commercial products. For traits with more complex inheritance (e.g., quantitative traits), population improvement through recurrent selection to counter one or more stress factors may be necessary before parental lines for hybrids or cultivars are developed. Genetic transformation of otherwise superior cultivars but lacking in a specific trait may be considered.

Whether stability of performance across environments is desirable, depends on whether or not environmental differences are predictable. If GEI is primarily due to variability caused by unpredictable environmental factors, such as year-to-year variation in weather variables, stable varieties that perform reasonably well under a range of conditions should be selected or developed. If GEI is caused by variations in predictable factors, such as soil type and cultural practices, consideration should be given to developing specifically adapted varieties for different environments, or to developing broadly adapted genotypes with reasonably acceptable performance under a range of conditions.

Molecular Markers and QTL-by-Environment Interaction: Economically important characters in crop species are generally quantitative in nature. Development of methodologies using molecular markers such as restriction-fragment length polymorphisms (RFLPs), microsatellites and amplified fragment length polymorphisms (AFLPs) should be beneficial in dissecting and utilizing GEI.

The QTL mapping has led to studies focusing on QTL-by-environment interaction (Paterson *et al.*, 1991; Hayes *et al.*, 1993; Beavis and Keim, 1996; Romagosa *et al.*, 1996; Sari-Gorla *et al.*, 1997; Van Eeuwijk *et al.*, 2000; Stuber and Le Deaux, 2000). Regions of plant genomes that provide stable responses across diverse environments can be identified by determining linkage of QTL to molecular markers, which should enable breeders to manipulate QTL in the same fashion as single genes that control qualitative traits. This could substantially reduce breeding and evaluation time.

The usual method of combined analyses for evaluating QTL × E interaction have been problematic (increased number of parameters) in the mapping model. Approximated methods addressing specific situations have been proposed to ease the problem (Hayes *et al.*, 1993; Tinker and Mather, 1995 and Beavis and Keim, 1996). A method of QTL × E analysis based on classical GEI analysis (Korol *et al.*, 1995; Ronin *et al.*, 1995) allows detection of QTL × E across a large number of environments, without resulting in an increased number of parameters in the mapping model. Romagosa *et al.* (1996) have tried to circumvent the problem of increased number of parameters caused by an increase in the number of environments, by developing an algorithm based on clustering environments using a few detected QTLs with most variable effects across environments. Their method, however, did not allow for the use of environmental covariates to explain GEI. Jiang and Zeng (1995) regarded measurements of the same genotype in different environments as a set of correlated traits to examine QTL × E interaction. This multiple trait analysis limits the number of test environments, because of increased number of parameters associated with them. Jensen *et al.* (1995) proposed a mapping model that included environmental factors, which somewhat reduced the parameter problem.

The use of molecular markers to understand QTL × E interaction also has been successfully made in species other than plants (Gurganus *et al.*, 1998). Significant GE was exhibited by 14 QTLs for sensory bristle number in *Drosophila melanogaster* (Gurganus *et al.*, 1998). Stuber *et al.* (1999) stressed, however, that little was known about the stability of QTL alleles when transferred to different genetic backgrounds and when evaluated in varying environments.

Marker-assisted selection for desirable QTLs should be more reliable than phenotypic selection for quantitative traits. The markers should be distributed throughout the genome. It is highly desirable to identify QTLs for a complex trait (say high yield) that are expressed in a number of environments. From the breeding standpoint, however, a major issue is the possible non- reliability of detecting QTLs that might be used for marker-assisted selection.

Breeding for Stability/Reliability of Performance

Smith *et al.* (1990) pointed out that genetic improvement for low-input conditions would require capitalizing on GE interactions and that slower or limited gains in low-input or stress environments suggested that conventional high-input management of breeding nurseries and evaluation trials might not effectively select genotypes with improved performance under low-input levels. Rosielle and Hamblin (1981) examined the theoretical aspects of selection for yield in stress and non-stress environments, pointing out that selection for tolerance to stress generally reduced mean yield in non-stress environments and that selection for mean productivity generally increased mean yields in both stress and non-stress environments. Annicchiarico and Mariani (1996) explored and advocated the use of artificial environments for selection, as possibly a less expensive alternative to multi-location testing for adaptation, stability, and reliability of yield.

From the standpoint of individual growers, stability across years (temporal) is most important. Breeders should pick up varieties which are in cultivation for over 10-15 years and can be said to have the desirable ability trait for using it in crossing program and the developed line(s) should be taped across years and locations.

Resource allocation: Genotype-by-environment interaction can be employed to judiciously allocate resources in a breeding program (Pandey and Gardner, 1992; Magari *et al.*, 1996). Carter *et al.* (1983) estimated that at a low level of treatment × environment interaction (10 % of error variance), testing in at least two environments was necessary to detect treatment differences of 20 % and it required at least seven environments to detect smaller (10 %) treatment differences for growth analysis experiments in soybean. With a larger magnitude of interaction, a larger number of environments would be needed for a given level of precision in treatment differences.

Magari *et al.* (1996) used multi-environment (different planting dates) data for ear moisture loss rate in maize that exhibited a significant planting date × genotype interaction. The number of planting dates (environments) was found to be a critical factor in determining the precision of an experiment rather than either plot size as number of replications.

Concluding Remarks

Causes of GEI should be identified on a cultivar-by-cultivar basis. Appropriate strategies can then be developed to correct those deficiencies/problems through genetic manipulations and/or proper management/cultural practices. Genotype-by-environment interactions may be minimized by modifying genetic constitution of cultivars by incorporating in them resistance/tolerance to different stresses to which they would likely be exposed. The greater the number of biotic/abiotic stresses that cultivars grown in a region are resistant to, the more stable/reliable their performance would be expected to be. Stresses provide opportunities for identifying and selecting genotypes that are efficient users of sub-optimal levels of inputs or are tolerant of super-optimal levels of inputs.

When crossover GE interactions are encountered, cultivars must be categorized as 'specifically adapted' or 'broadly adapted,' and targeted for suitable environments. Simultaneous selection for yield and stability of performance is an important consideration in breeding programs. Two of the most suitable methods of simultaneously selecting for yield and performance stability are Hühn's S_i^3 and Kang's Rank-sum (1988) statistics.

Genotype-by-environment interactions provide opportunities to better plan breeding programs and allocate resources more efficiently. Integration of DNA-based markers with traditional breeding methodology should be a powerful breeding tool incorporating QTLs determining stability in high yielding genotype(s). A better understanding of signal transduction in plants in response to environmental cues should shed more light on the relationship between crop performance and environmental factors. With the rapid progress being made in the understanding of GEI at the biochemical, physiological, molecular, and DNA levels, prospects of effectively utilizing/exploiting GEI to maximize genetic gains in plant breeding have greatly increased.

References

Aastveit A.H. and Aastveit K. 1993. Effects of genotype- environment interactions on genetic correlations. *Theor. Appl. Genet.*, **86**: 1007-1013.

Aastveit A. H. and Mejza S. 1992. A selected bibliography on statistical methods for the analysis of genotype × environment interaction. *Biuletyn Oceny Odmian*, **25**: 83-97.

Allard R.W. 1960. *Principles of Plant Breeding*. John Wiley & Sons, New York.

Allard R.W. and Bradshaw A.D. 1964. Implications of genotype-environment interactions in applied breeding. *Crop Sci.*, **4**: 503-508.

Annicchiarico P. 1997a. Additive main effects and multiplicative interaction (AMMI) analysis of genotype-location interaction in variety trials repeated over years. *Theor. Appl. Genet.*, **94**: 1072-1077.

Annicchiarico P. 1997b. STABSAS: a SAS computer programme for stability analysis. *Ital. J. Agron.*, **1**: 7-9.

Annicchiarico P. 1999. Variety × location interaction and its implications on breeding lucerne: A case study. P. 35-43. *In*: *Proc. XIII Eucarpia* Medicago *spp. germplasm*. (eds.) F. Veronesi and D. Rosellini, 13-16 Sep. 1999, Perugia, Italy.

Annicchiarico P. and Mariani G. 1996. Prediction of adaptability and yield stability of durum wheat genotypes for yield response in normal and artificially drought-stressed conditions. *Field Crops Res.*, **46**: 71-80.

Annicchiarico P. and Perenzin M. 1994. Adaptation patterns and definition of macro-environments for selection and recommendation of common wheat genotypes in Italy. *Plant Breeding*, **113**: 197-205.

Bajpai P.K. and Prabhakaran V.T. 2000. A new procedure for simultaneous selection for high yielding and stable crop genotypes. *Indian J. Genet.*, **60**: 141-152.

Baker R.J. 1988. Tests for crossover genotype-environmental interactions. *Can J. Plant Sci.*, **68**: 405-410.

Baker R.J. 1990. Crossover genotype-environmental interaction in spring wheat. p. 42-51. *In*: *Genotype-by-environment interaction and plant breeding*. (ed.) M.S. Kang, Louisiana State Univ. Agric. Center, Baton Rouge, LA.

Balzarini M., Milligan S.B. and Kang M.S. 2001. Best linear unbiased prediction: A mixed model approach in multi-environmental trials. p. 353-364 *In*: *Crop Improvement: New Challenges in the 21st Century*. (ed.) M.S. Kang, Food Products Press, Binghamton, NY.

Barah B.C., Binswanger H.P., Rana B.S. and Rao N.G.P. 1981. The use of risk aversion in plant breeding: concept and application. *Euphytica*, **30**: 451-458.

Baril C.P. 1992. Factor regression for interpreting genotype-environment interaction in bread-wheat trials. *Theor. Appl. Genet.*, **83**: 1022-1026.

Beavis W.D. and Keim P. 1996. Identification of quantitative trait loci that are affected by environment. pp. 123-149. *In*: *Genotype-by-Environment Interaction*, (eds.) M.S. Kang and H.G. Gauch, Jr., CRC Press, Boca Raton, FL.

Becker H.C. 1981a. Correlations among some statistical measures of phenotypic stability. *Euphytica*, **30**: 835-840.

Becker H.C. 1981b. Biometrical and empirical relations between different concepts of phenotypic stability. pp. 307-314. *In*: *Quantitative Genetics and Breeding Methods*, (ed.) Gallais, A., Versailles; I.N.R.A.

Becker H.C. and Leon J. 1988. Stability analysis in plant breeding. *Plant Breeding*, **101**: 1-23.

Blum A. 1988. *Plant Breeding for Stress Environments*. CRC ress, Boca Raton, FL.

Boland G.J. 1997. Stability analysis for evaluating the influence of environment on chemical and biological control of white mold (*Sclerotinia sclerotiorum*) of bean. *Biol. Control*, **9**: 7-14.

Bradshaw A.D. 1965. Evolutionary significance of phenotypic plasticity in plants. *Adv. Genetics*, **13**: 115-155.

Bramel-Cox P. J. 1996. Breeding for reliability of performance across unpredictable environments. pp. 309-339. *In*: *Genotype-by-environment interaction*. (eds.) M.S. Kang and H.G. Gauch, Jr. CRC Press, Boca Raton, FL.

Breese E.L. 1969. The measurement and significance of genotype-environment interactions in grasses. *Heredity*, **24**: 27-44.

Bridges W.C. Jr. 1989. Analysis of a plant breeding experiment with heterogeneous variances using mixed model equations. pp 145-151. *In*: *Applications of mixed models in agriculture and related disciplines*. So. Coop. Series Bull. No. 343. Louisiana Agric. Exp. Stn., Baton Rouge, LA.

Bucio Alanis L. 1966. Environmental and genotype-environmental components of variability. I. Inbred lines. *Heredity*, **21**: 387-397.

Bucio Alanis L. and Hill J. 1966. Environmental and genotype-environmental components of variability. II. Heterozygotes. *Heredity*, **21**: 399-405.

Bucio Alanis L., Perkins J.M. and Jinks J.L. 1969. Environmental and genotype-environmental components of variability. V. Segregating generations. *Heredity*, **24**: 115-127.

Byth D.E., Eisemann R.L. and De Lacy I. H. 1976. Two-way pattern analysis of a large data set to evaluate genotypeic adaptation. *Heredity*, **37**: 215-230.

Carter T.E. Jr., Burton J.W., Cappy J.J., Israel D.W. and Boerma H.R. 1983. Coefficients of variation, error variances, and resource allocation in soybean growth analysis experiments. *Agron. J.*, **75**: 691-696.

Ceccarelli S., Erskine W., Hamblin J. and Grando S. 1994. Genotype by environment interaction and international breeding programmes. *Expl. Agric.*, **30**: 177-187.

Charmet G., Balfourier F., Ravel C. and Denis J.B. 1993. Genotype × environment interactions in a core collection of French perennial ryegrass populations. *Theor. Appl. Genet.*, **86**: 731-736.

Clark R.B. and Duncan R.R. 1993. Selection of plants to tolerate soil salinity, acidity, and mineral deficiencies. p. 371-379. *In*: *International Crop Science* I., (eds.) D.R. Bruxton, R. Shibles, R.A. Forsberg, B.L. Blad, K.H. Asay, G.M. Paulsen and R.F. Wilson, Crop Sci. Soc. America, Madison, WI.

Comstock R.E. and Robinson H.F. 1952. Genetic parameters, their estimation and significance. *Proceedings of the Sixth International Grassland Congress*, **1**: 284-91.

Comstock R.E. and Moll R.H. 1963. Genotype-environment interactions. *In*: *Statistical Genetics and Plant Breeding*, (ed.) W.D. Hanson and H.F. Robinson. National Academy of Sciences-National Research Council Publication 982, 164-96.

Cooper M. and De Lacy I.H. 1994. Relationships among analytical methods used to study genotypic variation and genotyp-by-environment interaction in plant breeding multi-environment experiments. *Theor. Appl. Genet.*, **88**: 561-572.

Cooper M. and Hammer G.L. 1996. *Plant adaptation and crop improvement*. CAB International, Wallingford, U.K., ICRISAT, Patancheru, India, and IRRI, Manila, Philippines.

Cornelius P.L., Crossa J. and Seyedsadr M.S. 1996. Statistical tests and estimates of multiplicative models for GE interaction. p. 199-234. *In*: *Genotype-by-environment interaction*, (eds.) M.S. Kang and H.G. Gauch Jr., CRC Press, Boca Raton, FL.

Cornelius P.L., Seyedsadr M.S. and Crossa J. 1992. Using the shifted multiplicative model to search for "separability" in crop vcultivars trials. *Theor. Appl. Genet.*, **84**: 161-172.

Crispeels M.J. 1994. Introduction to 'Signal transduction in plants: A collection of updates.' *Am. Soc. Plant Physiologists*, Rockville, MD.

Crossa J. and Cornelius P.L. 2000. Modelos lineales-bilineales para el analisis de ensayos de genotipos en ambientes multiples. p. 61-88. *In*: *Simposium interaccion genotipo × ambiente*. (eds.) F. Zavala Garcia and N.E. Treviqo Hernandez, SOMEFI-CSSA-UG, Irapuato, Gto, Mexico.

Crossa J., Cornelius P.L. and Seyedsadr M.S. 1996. Using the shifted multiplicative model cluster methods for crossover GE interaction. p. 175-198. *In*: *Genotype-by-environment interaction*, (eds.) M.S. Kang and H.G. Gauch, Jr., CRC Press, Boca Raton, FL.

Dashiell K.E., Ariyo O.J. and Bello L. 1994. Genotype × environment interaction and simultaneous selection for high yield and stability in soybeans (*Glycine max* (L.) Merr.). *Ann. Appl. Biol.*, **124**: 133-139.

De Lacy I.H., Cooper M. and Basford K.E. 1996. Relationships among analytical methods used to study genotype-by-environment interactions and evaluation of their impact on response to selection. p. 51-84. *In*: *Genotype-by-environment interaction*. (eds.) M.S. Kang and H.G. Gauch, Jr., CRC Press, Boca Raton, FL.

Denis J.B. 1988. Two-way analysis using covariates. *Statistics*, **19**: 123-132.

Dickerson G.E. 1962. Implications of genetic-environmental interaction in animal breeding. *Animal Prod.*, **4**: 47-64.

Digby P.G.N. 1979. Modofied joint regression analysis for incomplete variety × environment data. *J. agric. Sci., Camb.*, **93**: 81-86.

Dutilleul P. and Carriere Y. 1998. Among-environment heteroscedasticity and the estimation and testing of genetic correlation. *Heredity*, **80**: 403-413.

Duvick D.N. 1996. Plant breeding, an evolutionary concept. *Crop Sci,*. **36**: 539-548.

Dyke G.V., Lane P.W. and Jenkyn J.F. 1995. Sensitivity (stability) analysis of multiple variety trials, with special reference to data expressed as proportions or percentages. *Expl. Agric.*, **31**: 75-87.

Eberhart S.A. and Russell W.A. 1966. Stability parameters for comparing varieties. *Crop Sci.*, **6**: 36-40.

Eberhart S.A. and Russell W.A. 1969. Yield stability for a 10-line diallel of single-cross and double cross maize hybrids. *Crop Sci.*, **9**: 357-361.

Eisemann R.L., Cooper M. and Woodruff D.R. 1990. Beyond the analytical methodology, better interpretation and exploitation of GE interaction in plant breeding, pp. 108-117. *In*: *Genotype-by-environment interaction and plant breeding*, (ed.) M.S. Kang, Louisiana State Univ. Agric. Center, Baton Rouge, LA.

Epstein E. 1972. *Mineral nutrition of plants: principles and perspectives*. John Wiley, New York.

Epstein E. 1976. Genetic potential for solving problems of soil mineral stress: Adaptation of crops to salinity. p. 73-82. *In*: *Plant adaptation to mineral stress in problem soils*. (ed.) Wright M.L., Cornell University Agricultural Experiment station, Ithaca, NY.

Eskridge K.M. 1996. Analysis of multiple environment trials using the probability of outperforming a check. p. 273-307. *In*: *Genotype-by-environment interaction*. (eds.) Kang M.S. and Gauch H.G. Jr., CRC Press, Boca Raton, FL.

Evans L.T. 1993. *Crop evolution, adaptation and yield*. Cambridge University Press, New York.

Falconer D.S. 1952. Selection for large and small size in mice. *J. Genet.*, **51**: 470-501.

Falconer D.S. 1981. *Introduction to quantitative genetics*. Second edition. The Longman Press, London.

Finlay K.W. and Wilkinson G.N. 1963. The analysis of adaptation in a plant breeding programme. *Aust. J. Agr. Res.*, **14**: 742-754.

Fisher R.A. 1926. The arrangement of field experiments. *Journal of the Ministry of Agriculture*, **33**: 403-13.

Fisher R.A. and Mackknzie W.A. 1923. Studies in crop variation, II. The manurial response of different potato variations. *Journal of Agricultural Science*, Cambridge. **13**: 311-20.

Francis T.R. and Kannenberg L.W. 1978. Yield stability studies in short-season maize. I A descriptive method for grouping genotypes. *Can. J. Plant Sci.*, **58**: 1029-1034.

Freeman G.H. 1973. Statistical methods for the analysis of genotypes-environment interactions. *Heredity*, **31**: 339-354.

Freeman G.H. 1975. Analysis of interactions in incomplete two-way tables. *Applied Statistics*, **24**: 46-55.

Freeman G.H. and Perkins J.M. 1971. Environmental and genotype environmental components of variability. VIII. Relations between genotypes grown in different environments and measurement of these environments. *Heredity*, **27**: 15-23.

Fripp Y.J. 1972. Genotype-environmental interactions in *Schizophyllum commune*. II. Assessing the environment. *Heredity*, **28**: 223-238.

Fripp Y.J. and Caten C.E. 1971. Genotype-environment interactions in *Schirophyllum commune*, I. Analysis and character. *Heredity*, **27**: 393-407.

Fripp Y.J. and Caten C.E. 1973. Genotype-environment interactions in *Schizophyllum commune*. III. The relationship between mean expression and sensitivity to change in environment. *Heredity*, **30**: 341-349.

Gabriel K.R. 1978. Least squares approximation of matrices by additive and multiplicative models. *J.R. Stat. Soc. Ser. B.*, **40**: 186-196.

Gail M. and Simon R. 1985. Testing for qualitative interactions between treatment effects and patient subsets. *Biometrics*, **41**: 361-372.

Gauch H.G. Jr. 1988 Model selection and validation of yield trials with interaction. *Biometrics*, **44**: 705-715

Gauch H.G. Jr. 1992. Statistical analysis of regional yield trials: *AMMI analysis of factorial designs*. Elsevier, Amsterdam.

Gauch H.G. and Zobel R.W. 1990. Imputing missing yield trial data. *Theor. Appl. Genet.*, **70**: 753-761.

Gauch H.G. Jr. and Zobel R.W. 1996. AMMI analysis of yield trials. p. 85-122. *In*: *Genotype-by-environment interaction*. (eds.) M.S. Kang and H.G. Gauch, Jr., CRC Press, Boca Raton, FL.

Gimelfarb A. 1994. Additive-multiplicative approximation of genotype-environment interaction. *Genetics*, **138**: 1339-1349.

Glaz B. and Dean J.L. 1988. Statistical error rates and their implications in sugarcane clone trials. *Agron. J.*, **80**: 560-562.

Glaz B., Miller J.D. and Kang M.S. 1985. Evaluation of cultivar-testing locations in sugarcane. *Theor. Appl. Genet.*, **71**: 22-25.

Gorman D.P., Kang M.S. and Milam M.R. 1989. Contribution of weather variables to genotype × environment interaction in grain sorghum. *Plant Breeding*, **103**: 299-303.

Gravois K.A., Moldenhauer K.A.K. and Rohman P.C. 1990. GEI interaction for rice yield and identification of stable, high-yielding genotypes. p. 181-188. *In*: *Genotype-by-environment interaction and plant breeding*. (ed.) M.S. Kang, Louisiana State Univ. Agric. Center, Baton Rouge, LA.

Gurganus M.J., Fry J.D., Nuzhdin S.V., Pasyukova E.G., Lyman R.F. and Mackay T.F.C. 1998. Genotype-environment interaction at quantitative trait loci affecting sensory bristle number in *Drosophila melanogaster*. *Genetics*, **149**: 1883-1898.

Gutierrez J.C, Lopez M. and El-Zik K.M. 1994. AMMI (additive main effects and multiplicative interactions analysis): A tool to determine adaptability of upland cotton genotypes in Spain. Cotton Improve. Conf. *Beltwide Cotton Conf. Proceedings* Vol. 2 (1994): 688-689.

Hanson C.H., Robinson H.F. and Comstock R.E. 1956. Biometrical studies of yield in segregating populations of Korean Lespedeza. *Agronomy Journal*, **48**: 268-72.

Hanson W.D. 1970. Genotypic stability. *Theor. Appl. Genet.*, **48**: 226-231.

Hardwick R.C. and Wood J.T. 1972. Regression methods for studying gnotype environment interactions. *J. Agric. Sci.*, Camb., **85**: 477-493.

Harville D.A. 1977. Maximum-likelihood approaches to variance component estimation and to related problems. *J. Amer. Stat. Assoc.*, **72**: 320-340.

Hayes P.M., Liu B.H., Knapp S.J., Chen F., Jones B., Blake T., Franckowviak J., Rasmusson D., Sorrells M., Ullrich S.E., Wesenberg D. and Kleinhofs A. 1993. Quantitative trait locus effects and environmental interaction in North American barley germplasm. *Theor. Appl. Genet.*, **87**: 392-401.

Henderson C.R. 1975. Best linear unbiased estimation and prediction under a selection model. *Biometrics*, **31**: 423-447.

Higley L.G., Browde J.A. and Higley P.M. 1993. Moving toward new understandings of biotic stress and stress interactions. p. 749-754. *In*: *International crop science*, (eds.) I.D.R. Bruxton, R. Shibles, R. A. Forsberg, B. L. Blad, K. H. Asay, G. M. Paulsen, and R. F. Wilson, Crop Sci. Soc. America, Madison, WI.

Hill J. 1975. Genotype-environment interaction - a challenge for plant breeding. *J. Agric. Sci.*, Camb., **85**: 477-493.

Hill J. and Perkins J.M. 1969. The environmental induction of heritable changes in *Nicotiana rustica*. Effects of gentoype-environmental interactions. *Genetics*, **61**: 661-675.

Hill R.R. Jr. and Rosenberger J.L. 1985. Methods for combining data from germplasm evaluation trials. *Crop Sci.*, **25**: 467-470.

Hühn M. 1979. Beitrage zur Erfassung der phanotypischen Stabilitat. I. Vorschlag eingier auf Ranginformationen beruhenden Stabilitats parameter. *EDP in Medizin und Biologie.*, **10**: 112-117.

Hühn M. 1990a. Nonparametric measures of phenotypic stability. Part 1: Theory. *Euphytica*, **47**: 189-194.

Hühn M. 1990b. Nonparametric measures of phenotypic stability. Part 2: Application. *Euphytica*, **47**: 195-201.

Hühn M. 1996. Nonparametric analysis of genotype × environment interactions by ranks. p. 235-271. *In*: *Genotype-by-environment interaction.* (eds.) M.S. Kang and H.G. Gauch, Jr., CRC Press, Boca Raton, FL.

Hussein M.A., Bjornstad A. and Aastveit A.H. 2000. SASG×ESTAB: A SAS program for computing genotype × environment stability statistics. *Agron. J.*, **92**: 454-459.

Jenns A.E., Leonard K.J. and Moll R.H. 1982. Stability analyses for estimating relative durability of quantitative resistance. *Theor. Appl. Genet.*, **63**: 183-192.

Jensen R.C., Van Ooijien J.M., Stam P., Lister C. and Dean C. 1995. Genotype-by-environment interaction in genetic mapping of multiple quantitative trait loci. *Theor. Appl. Genet.*, **91**: 33-37.

Jiang C. and Zeng Z.-B. 1995. Multiple trait analysis and genetic mapping for quantitative trait loci. *Genetics*, **140**: 1111-1127.

Jinks J.L. and Connolly V. 1973. Selection for specific and general response to environmental differences. *Heridity*, **30**: 33-40.

Jinks J.L. and Mather K. 1955. Stability in development of heterozygotes and homozygotes. *Proceedings of the Royal Society* Series B, **143**: 561-78.

Jinks J.L. and Perkins J.M. 1970. Environmental and genotype-environmental components of variability. VII. Simultaneous prediction across environments and generations. *Heredity*, **25**: 475-480.

Johnson J.J., Alldredge J.R., Ullrich S.E. and Dangi O. 1992. Replacement of replications with additional locations for grain sorghum cultivar evaluation. *Crop Sci.*, **32**: 43-46.

Jones H.G. 1992. *Plants and microclimate: A quantitative approach to environmental plant physiology*. 2nd edition. Cambridge Univ. Press, Cambridge, U.K.

Jones R.M. and Mather K. 1958. Interaction of genotypes and environment in continuous variation. II. Analysis, *Biometrics*, **14**: 489-98.

Kang M.S. 1988. A rank-sum method for selecting high-yielding, stable corn genotypes. *Cereal Res. Commun.*, **16**: 113-115.

Kang M.S. 1990. *Genotype-by-environment interaction and plant breeding*. Louisiana State Univ. Agric. Center, Baton Rouge, LA.

Kang M.S. 1993a. Issues in GE interaction. p. 67-73. *In*: *Genotype-environment interaction studies in perennial tree crops.* (eds.) Rao V., Hanson I.E. and Rajanaidu N., Palm Oil Research Institute of Malaysia, Kaula Lumpur.

Kang M.S. 1993b. Simultaneous selection for yield and stability in crop performance trials: Consequences for growers. *Agron. J.*, **85**: 754-757.

Kang M.S. 1998. Using genotype-by-environment interaction for crop cultivar development. *Adv. Agronomy*, **62**: 199-252.

Kang M.S. and Gauch H.G. Jr. 1996. *Genotype-by-environment interaction.* CRC Press, Boca Raton, FL.

Kang M.S. and Gorman D.P. 1989. Genotype × environment interaction in maize. *Agron. J.*, **81**: 662-664.

Kang M.S., Harville B.G. and Gorman D.P. 1989. Contribution of weather variables to genotype × environment interaction in soybean. *Field Crops Res.*, **21**: 297-300.

Kang M.S. and Magari R. 1995. STABLE: Basic program for calculating yield-stability statistic. *Agron. J.*, **87**: 276-277.

Kang M. S. and Magari R. 1996. New developments in selecting for phenotypic stability in crop breeding. p. 1-14. *In*: *Genotype-by-environment interaction.* (eds.) Kang M.S. and Gauch H.G. Jr., CRC Press, Boca Raton, FL.

Kang M.S. and Miller J.D. 1984. Genotype x environment interactions for cane and sugar yield and their implications in sugarcane breeding. *Crop Sci.*, **24**: 435-440.

Kang M.S., Miller J.D. and Darrah L.L. 1987. A note on relationship between stability variance and ecovalence. *J. Hered.*, **78**: 107.

Kang M.S., Miller J.D. and Tai P.Y.P. 1984. Clonal and individual repeatability of agronomic traits in sugarcane. *J. Am. Soc. Sugar Cane Techno.*, **3**: 22-27.

Kang M.S. and Pham H.N. 1991. Simultaneous selection for high yielding and stable crop genotypes. *Agron. J.*, **83**: 161-165.

Kearsey M.J. and Pooni H.S. 1996. *The genetical analysis of quantitative traits.* Chapman & Hall, London, UK.

Korol A.B., Ronin Y.I. and Kirzhner V.M. 1995. Interval mapping of quantitative trait loci employing correlated trait complexes. *Genetics*, **140**: 1137-1147.

LeClerg E.L. 1966. Significance of experimental design in plant breeding. p. 243-313. *In*: *Plant breeding.* (ed.) K. J. Frey, Iowa State University Press, Ames, IA.

Leigh R.A. 1993. Perception and transduction of stress by plant cells. p. 223-237. *In*: *Plant adaptation to environmental stress.* (eds.) L. Fowden, T. Mansfield and J. Stoddart, Chapman & Hall, New York.

Leon J. 1985. Beiträge zur Erfassung der phänötypischen stabilität unter besonderer Berückischtigug unterschidilicher Heterogenitäts-und Heterozygotiegrade sowie einer zusammenfassenden Beurteilung von Ertragshöhe and Ertragssicherheit. Dissertation, Christian-Albrechts-Universität Kiel.

Leon J. 1986. Methods of simultaneous estimation of yield and yield stability. *In*: *Biometrics in Plant Breeding.* Proc. 6th Meeting Eucarpia-Section. Birmingham, UK, 299-308.

Lerner I.M. 1954. *Genetic Homeostasis.* Oliver & Boyd, London.

Lin C.S. 1989. (In) Letter to the Editor, *Crop Sci.*, **29**: p. 1334.

Lin C.S. and Binns M.R. 1988. A method of analyzing cultivar × location × year experiments: A new stability parameter. *Theor. Appl. Genet.*, **76**: 425-430.

Lin C.S. and Binns M.R. 1991a. Assessment of a method for cultivar selection based on regional trial data. *Theor. Appl. Genet.*, **82**: 379-388.

Lin C.S. and Binns M.R. 1994. Concepts and methods for analyzing regional trial data for cultivar and location selection. *Plant Breeding Reviews*, 12. John Wiley & Sons, Inc.

Lin C.S., Binns M..R. and Lefkovitch L.P. 1986. Stability analysis: Where do we stand? *Crop Sci.*, **26**: 894-900.

Lin C.S. and Butler G. 1990. Cluster analyses for analyzing tow-way classification data. *Agron. J.*, **82**: 344-348.

Lin C.S. and Morrison M.J. 1992. Selection of test locations for regional trials of barley. *Theor. Appl. Genet.*, **83**: 968-972.

Macchiavelli R.E. and Beaver J.S. 1999. Analysis of genotype-by-environment interaction with AMMI models using SAS PROC MIXED. Appl. Stat Agric., Meeting held at Dep. of Statistics, Kansas State Univ., Manhattan, April 25-27, 1999. (11th): p. 171-183.

Magari R. and Kang M.S. 1997. SAS-STABLE: Stability analyses of balanced and unbalanced data. *Agron. J.*, **89**: 929-932.

Magari R., Kang M.S. and Zhang Y. 1996. Sample size for evaluating field ear moisture loss rate in maize. *Maydica*, **41**: 19-24.

Magari R., Kang M.S. and Zhang Y. 1997. Genotype by environment interaction for ear moisture loss rate in corn. *Crop Sci.*, **37**: 774-779.

Mather K. and Jones R.M. 1958. Interactions of genotypes and environment in continuous variation. I. Description. *Biometrics*, **14**: 343-59.

Mather K. 1953. Genetical control of stability in development. *Heredity*, **7**: 297-336.

Mather K. and Jinks J.L. 1971. *Biometrical Genetics*. 2nd edition. Chapman and Hall, London.

Matheson A.C. and Cotterill P.P. 1990. Utility of genotype × environment interactions. *For. Ecol. Management*, **30**: 159-174.

Matheson A.C. and Raymond C.A. 1984. The impact of genotype × environment interactions on Australian *P. radiata* breeding programs. *Aust. For. Res.*, **14**: 11-25.

Myers W.M. 1960. Genetic control of physiological processes: Consideration of differential ion uptake by plants. p. 201-226. *In*: *Radioisotopes in the biosphere*. (eds.) Caldecott R.S. and Synder L.A., Univ. of Minnesota, Minneapolis, MN.

Nachit M.M., Nachit G., Ketata H., Gauch H.G. Jr. and Zobel R.W. 1992. Use of AMMI and linear regression models to analyze genotype-environment interaction in durum wheat. *Theor. Appl. Genet.*, **83**: 597-601.

Nassar R., Leon J. and Huhn M. 1994. Tests of significance for combined measures of plant stability and performance. *Biom. Journal*, **36**: 109-123.

Nassar R. and Huhn M. 1987. Studies on estimation of phenotypic stability: Test of significance for non-parametric measures of phenotypic stability. *Biometrics*, **43**: 45-53.

Nyquist W.E. 1991. Estimates of heritability and prediction of selection response in plant populations. *Crit. Rev. Plant Sci.*, **10**: 235-322.

Pandey S. and Gardner C.O. 1992. Recurrent selection for population, variety and hybrid improvement in tropical maize. *Adv. Agron.*, **48**: 1-87.

Paterson A.H., Damon S., Hewitt J.D., Zamir D., Rabinowitch H.D., Lincoln S.E., Lander E.S. and Tanksley S.D. 1991. Mendelian factors underlying quantitative traits in tomato: Comparison across species, generations, and environments. *Genetics*, **127**: 181-197.

Patterson H.D. 1978. Routine least squares estimation of variety means in incomplete tables. *J. Natn. Inst. Agric. Bot.*, **14**: 401-413.

Patterson H.D. 1980. Yield sensitivity and straw shortness in varieties of winter wheat. *J. Natn. Inst. Agric. Bot.*, **13**: 142-151.

Patterson H.D. and Silvey V. 1980. Statutory and recommended test trials of crop varieties in the United Kingdom. *J. Roy Stat. Soc.*, **A143**: 219-252.

Patterson H.D. and Thompson R. 1971. Recovery of inter-block information when block sizes are unequal. *Biometrika*, **58**: 545-554.

Pedersen A.R., Everson E.H. and Grafius J.E. 1978. The gene pool concept as basis for cultivar selection and recommendation. *Crop Sci.*, **18**: 883-886.

Perkins J. M. and Jinks J.L. 1968a. Environmental and genotype environmental components of variability Multiple inbred crops. *Heredity*, **23**: 339-356.

Perkins J.M. and Jinks J.L. 1968b. Environmental and genotype environmental components of variability. 4. Non-linear interactions for multiple inbred lines. *Heredity*, **23**: 525-535.

Perkins J.M. 1970. Environmental and genotype-environmental components of variability. VI. Diallel sets of crosses. *Heredity*, **25**: 29-40.

Peschke V.M. and Sachs M.M. 1993. Multiple pyruvate decarboxylase genes in maize are induced by hypoxia. *Mol. Gen. Genet.*, **240**: 206-212.

Pham H.N. and Kang M.S. 1988. Interrelationships among and repeatability of several stability statistics estimated from International Maize trials. *Crop Sci.*, **28**: 925-928.

Piepho H.P. 1994. Missing observations in analysis of stability, *Heredity*, **72**: 141-145. (Correction 73 (1994): (58).

Piepho H.P. 1994. Best linear unbiased prediction (BLUP) for regional yield trials: a comparison to additive main effects and multiplicative interaction (AMMI) analysis. *Theor. Appl. Genet.*, **89**: 647-654.

Piepho H.P. 1998. Methods for comparing the yield stability of cropping systems - A review. *J. Agron. Crop Sci.*, **180**: 193-213.

Pinthus M.J. 1973. Estimate of genotypic value: A proposed method. *Euphytica*, **22**: 121-123.

Plaisted R.L. 1960. A shorter method for evaluating the ability of selections to yield consistently over locations. *Amer. Potato J.*, **37**: 166-172.

Plaisted R.L. and Peterson L.C. 1959. A technique for evaluating the ability of selections to yield consistently in different locations or seasons. *Amer. Potato J.*, **36**: 381-385.

Pooni H.S. and Jinks J.L. 1980. Non-linear genotype × environment interaction. II Statistical models and genetical control. *Heredity*, **45**: 389-400.

Prabhakaran V.T. and Jain J.P. 1994. *Statistical techniques for studying genotype-environment interactions*. South Asian Publishers, New Delhi, India.

Raiger H.L. and Prabhakaran V.T. 2000. A statistical comparison between non-parametric and parametric stability measures. *Indian J. Genet.*, **60**: 417-432.

Raiger H. L and Prabhakaran V.T. 2001. A study on the performance of a few non-parametric stability measures using pearl-millet data . *Indian J. Genet.*, **61**: 7-11.

Rameau C. and Denis J.B. 1992. Characterization of environments in long-term multi-site trials in asparagus, through yield of standard varieties and use of environmental covariates. *Plant Breeding*, **109**: 183-191.

Rao A.R. and Prabhakaran V.T. 2000. On some useful interrelationships among common stability parameters. *Indian J. Genet.*, **60**: 25-361.

Rao V., Henson I.E. and Rajanaidu N. 1993. Genotype × environment interaction in perennial tree crops. *International Society of Oil Palm Breeders* and Palm Oil Research Institute of Malaysia, Kuala Lumpur, Malaysia.

Robbertse P.J. 1989. The role of genotype-environment interaction in adaptability. *So. African For. J.*, **150**: 18-19.

Robertson A. 1959. The sampling variance of the genetic correlation coefficient. *Biometrics*, **15**: 469-485.

Romagosa I., Ullrich S.E., Han F. and Hayes P.M. 1996. Use of additive main effects and multiplicative interaction model in QTL mapping for adaption in barley. *Theor. Appl. Genet.*, **93**: 30-37.

Ronin Y.I., Kirzhner V.M. and Korol A.B. 1995. Linkage between loci of quantitative traits and marker loci. Multi-trait analysis with a single marker. *Theor. Appl. Genet.*, **90**: 776-786.

Rosielle A.A. and Hamblin J. 1981. Theoretical aspects of selection for yield in stress and non-stress environments. *Crop Sci.*, **21**: 943-946.

Saeed M. and Francis C.A. 1984. Association of weather variables in genotype × environment interaction in grain sorghum. *Crop Sci.*, **24**: 13-16.

Sari-Gorla M., Calinski T., Kaczamarek Z. and Krajewski P. 1997. Detection of QTL- environment interaction in maize by a least squares interval mapping method. *Heredity*, **78**: 146-157.

Scandalios J.G. 1990. Response of plant antioxidant defense genes to environmental stress. p. 1-41. *In*: *Advances in genetics*, (eds.) G. Scandalios and T.R.F. Wright. Academic Press, New York.

Searle S.R. 1987. *Linear models for unbalanced data*. John Wiley & Sons, New York.

Searle S.R., Casella G. and McCulloch C.E. 1992. *Variance components*. John Wiley & Sons, New York.

Seyedsadr M.S. and Cornelius P.L. 1992. Shifted multiplicative models for non-additive two-way tables. *Comm. Stat. B. Simul Comp.*, **21**: 807-832.

Seyedsadr M.S. and Cornelius P.L. 1992. Using the shifted multiplicative model to search for "separability" in crop cultivar trials. *Theor. Appl. Genet.*, **84**: 161-172.

Shafii B. and Price W.J. 1998. Analysis of genotype-by-environment interaction using the Additive Main Effects and Multiplicative Interaction model and stability estimates. *J. Agric. Biol. Environ. Stat.*, **3**: 335-345.

Shukla G.K. 1972. Some statistical aspects of partitioning genotype-environmental components of variability. *Heredity*, **29**: 237-245.

Silvey V. 1981. The contribution of new wheat, barley and oat varieties to increasing yield in England and vales 1947-78. *J. National Inst. Agric. Bot.*, **15**: 399-412.

Simmonds N. W. 1981. Genotype (G), environment (E) and GEI components of crop yields. *Expl. Agric.*, **17**: 355-362.

Smith H. 1990. Signal perception, differential expression within multigene families and the molecular basis of phenotypic plasticity. *Plant Cell Environ.*, **13**: 585-594.

Smith M.E., Coffman W.R. and Barker T.C. 1990. Environmental effects on selection under high and low input conditions. p. 261-272. *In*: *Genotype-by-environment interaction and plant breeding*. (ed.) M. S. Kang, Louisiana State Univ. Agric. Center, Baton Rouge, LA.

Specht J.E. and Laing D.R. 1993. Selection for tolerance to abiotic stresses - discussion. p. 381-382. *In*: *International Crop Science* I., (eds.) Bruxton D.R., Shibles R., Forsberg R.A., Blad B.L., Asay K.H., Paulsen G.M. and Wilson R.G., Crop Sci. Soc. America, Madison, WI.

Sprague G.F. and Federer W.T. 1951. A comparison of variance components in corn yield trials. II, Error, year-variety, location-variety and variety components. *Agronomy Jounrnal*, **43**: 535-41.

Steiner K.C, Barbour J.R. and McCormick L.H. 1984. Response of Populus hybrids to aluminium toxicity. *Forest Sci.*, **30**: 404-410.

Stroup W.W. and Mulitze D.K. 1991. Nearest neighbor adjusted best linear unbiased prediction. *Am. Stat.*, **45**: 194-200.

Stuber C.W. and Le Deaux J.R. 2000. QTL × environment interaction in maize when mapping QTLs under several stress conditions. p. 101-112. *In*: *Simposium interaccion genotipo × ambiente*. (eds.) F. Zavala Garcia and N.E. Treviqo Hernandez, SOMEFI-CSSA-UG, Irapuato, Gto, Mexico.

Stuber C.W., Polacco M. and Senior M.L. 1999. Synergy of empirical breeding, marker-assisted selection, and genomics to increase crop yield potential. *Crop Sci.*, **39**: 1571-1583.

Tai G.C.C. 1971. Genotypic stability analysis and its application to potato regional trials. *Crop Sci.*, **11**: 184-190.

Thennarasu K. 1995. On certain non-parametric procedures for studying genotype-environment interactions and yield stability. Unpublished Ph.D. Thesis, P.G. School, IARI, New Delhi.

Tinker N.A. and Mather D.E. 1995. Methods for QTL analysis with progeny replicated in multiple environments. JQTL, **1**: http://probe.nalusda.gov:8000/otherdocs/qtl/Jqtl1995-02/jqtl16r2.html.

Tiret L., Abel L. and Rakotovao R. 1993. Effect of ignoring genotype-environment interaction in segregation analysis of quantitative traits. *Genetic Epidemiology*, **10**: 581-586.

Unsworth M.H. and Fuhrer J. 1993. Crop tolerance to atmospheric pollutants. pp. 363-370. *In*: *International Crop Science* I., (eds.) Bruxton D.R., Shibles R., Forsberg R.A., Blad B.L., Asay K.H., Paulsen G.M. and Wilson R.F., Crop Science Soc., America, Madison, WI.

Utz H.F. 1972. Die Zerlengung der Genotyp × Umwelt Interaktion. *EDV in Medizin und Biologie*, **3**: 52-59.

Van Eeuwijk F.A., Denis J.B. and Kang M.S. 1996. Incorporating additional information on genotypes and environments in models for two-way genotype by environment tables. p. 15-49. *In*: *Genotype-by-environment interaction*. (eds.) M.S. Kang and H.G. Gauch Jr., C.R.C. Press, Boca Raton, Florida, USA.

Van Eeuwijk F.A., Crossa J., Vargas M. and Ribaut J.M. 2000. Modeling QTLs and QTL × E using factorial regression models and partial least squares techniques. p. 43-60. *In*: *Simposium interaccion genotipo × ambiente*. (eds.) F. Zavala Garcia and N.E. Treviqo Hernandez, SOMEFI-CSSA-UG, Irapuato, Gto, Mexico.

Vasil J.D. and Milas S. 1984. Relationship between yield stability parameters estimated with different methods for maize and wheat genotypes. *Vortr. Pflanzenzuchtg*. **7**: 266-279.

Verma M.M., Chahal G.S. and Murty B.R. 1978. Limitations of conventional agression analysis - A proposed modification. *Theor. Appl. Genet.*, **53**: 89-91.

Via S. and Lande R. 1987. Evolution of genetic variation in a spatially heterogeneous environment. *Genet. Res.*, **49**: 147-156.

Virk D.S. and Mangat B.K. 1991. Detection of cross over genotype × environment interactions in pearl millet. *Euphytica*, **52**: 193-199.

Weber W.E., Wricke G. and Westermann T. 1996. Selection of genotypes and prediction of performance by analyzing GE interactions. p. 353-371. *In*: *Genotype-by-environment interaction*. (eds.) M. S. Kang and H. G. Gauch Jr., CRC Press, Boca Raton, FL.

Williams W.T. 1976. *Patter Analysis in Agricultural Science*. Elsevier Scientific Publishing Co., Oxford.

Wricke G. 1962. Uber eine Methode zur Erfassung der Vkologischen Streubreite. *Zeitschrift für Pflanzenzlchtung*, **47**: 92-96.

Wricke G. and Weber W.E. 1980. Erweiterte analyse von wechselwirkungen in versuchsserien. *In*: *Biometrie-heute and morgen*, (eds.) Kopcke, W. and K. Uberla, 87-95. Berlin Heidelberg-New York: Springer-Verlag.

Yamada Y. 1962. Genotype by environment interaction and genetic correlation of the same trait under different environments. *Jap. J. Genet.*, **37**: 498-509.

Yan W., Hunt A.L., Sheng Q. and Szlavnics Z. 2000. Cultivar evaluation and mega-environment investigation based on the GGEI biplot. *Crop Sci.*, **40**: 597-605.

Yates F. and Cochran W.G. 1938. The analysis of groups of experiments. *J. Agric. Sci.*, **28**: 556-580.

Zhang Q. and Geng S. 1986. A method of estimating varietal stability for long-term trials. *Theor. Appl. Genet.*, **71**: 810-814.

Zobel R.W., Wright M.J. and Gauch H.G. Jr. 1988. Statistical analysis of a yield trial. *Agron. J.*, **80**: 388-393.

Plant Breeding - Mendelian to Molecular Approaches
H. K. Jain and M. C. Kharkwal (eds.)

24

Breeding for Wider Adaptability

Darbeshwar Roy[1] and M.C. Kharkwal[2]

Abstract

The success of a variety depends not only on its high yielding potential but also on the stability of its yield over years across environments. If a variety gives high and stable performance over years at a location within an agro-climatic zone, it is called a stable variety. When the same variety is grown in different agro-climatic regions and if it shows high yield with little or no change in performance in comparison to other varieties, it is called a widely adapted variety. The adaptability, thus refers to the reduced variation in performance across locations, whereas stability refers to the reduced variation in performance across years. Before embarking on a breeding programme aiming at developing a widely adaptable variety, it is essential to know the genetics of the trait, the mechanisms leading to the development of this trait and finally the methods of measurement of the trait. This paper describes some of the concepts and methods underpinning the development of high yielding, stable and widely adapted varieties. Mechanisms and measures of stability and adaptability of varieties and their analysis under various agro-climatic conditions have been discussed.

Introduction

The different agro-climatic regions differ with respect to climatic (temperature gradient, photoperiod and rainfall distribution) and edaphic (soil fertility and type) factors and management practices. Phenotype is the product of genotype and environment. In the presence of genotype × environment interaction the phenotype will be the product of genotype, environment and genotype × environment interaction. Same genotype can produce different phenotypes in different environments and different genotypes can produce same phenotype in a particular environment. A stable genotype is one which interacts less with the environment or shows a minimum of genotype × environment interaction. Further, the environment consists of controllable or uncontrollable, predictable or unpredictable factors. So while talking about stability we are considering minimum or less genotype × uncontrollable or unpredictable environmental interaction and further less genotype × unfavorable environment interaction. The predictable environmental conditions are topography, soil type and climatic condition such as day length.

[1]G.B. Pant University of Agriculture and Technology, Pantnagar 263 145, India
[2]Indian Agricultural Research Institute, New Delhi 110 012, India

When the genotype × environment (g × e) interaction is significant, we must further investigate other specific types of interactions such as genotype × location, genotype × specific treatment such as fertility level, irrigation schedule, sowing date etc, genotype × season (year) and genotype × location × year. In case of stability, a variety shows minimum or low variety × season (or year) interaction. Here the environment is unpredictable. In case of higher adaptability, genotype × location × year interaction is low or minimum and here the environments consist of predictable as well as unpredictable factors.

The Need for Wider Adaptability

The need for breeding a variety well adapted over wide areas arises because of the fact that the breeder is already developing a variety for specific adaptation. They are knowingly or unknowingly developing a variety which is highly responsive, as they are selecting material in the best possible environment. Their variety will give stable performance in environment similar to their experimental sites. Variety(ies) with wider adaptability will increase the production quickly and it will be much easier to control the quality of seed. Further, any effect of disease/insect or any other specific factor on the crop can be monitored very precisely and correction measures applied effectively.

Mechanism of Stability

There are two ways in which stabilization of yield over locations and over years can occur (Allard and Bradshaw, 1964).

1. ***Individual Buffering***: In case of pure line variety (homozygous, homogeneous), single cross hybrid (heterozygous, homogeneous) or clones the individuals themselves may be well buffered. Each individual of the population is well adapted to a range of environments.
2. ***Population Buffering***: In case of mixtures in self-pollinated species (made up of homozygous and heterogeneous individuals) and in double or three-way hybrids, improved open-pollinated populations, synthetics and composites in case of cross-pollinated species, where the population is made up of a number of genotypes, each genotype is adapted to a somewhat different range of environments resulting in the stabilization of yield.

Thus what we see is that the variety's yield is stabilized in the face of various environmental influences as a result of buffering, which signifies protection against variability. In other words, this character (stability of yield) is buffered or canalized, so that its development is unaffected by environmental stresses or by underlying genetic variability. Further, it can be said from the above that buffering is not only the property of an individual (homozygous or heterozygous) but also that of a population (homozygous, homogeneous or heterozygous, heterogeneous).

In case of out-breeding species, buffering is a characteristic of heterozygosity. Heterozygous individuals have better buffering ability than homozygotes. Lerner (1954) used the term genetic homeostasis to describe this property. It refers to the tendency of a physiological system to react to external disturbances in such a way that the system is not displaced from its normal values. In other words, it means resistance to change which is just opposite of plasticity. Further it depends on the particular array of gene frequencies built up by a population over a long period of time. Darlington and Mather (1949) called this as co-adapted gene complex or super gene which is of fundamental importance in the adaptation of a population to their environments.

Jinks and Mather (1951) observed that in case of inbreeding species, pure breeding lines differ in their buffering capacity and F_1 does not show increase in stability. That F_1 shows higher stability in comparison to homozygous pure breeding parents has been inferred from the general observation that lower coefficient of variability (CV) value is associated with hybrids but this linear relationship accounts only for a small part of the total variation in environmental sensitivity. They concluded that F_1 shows superiority not because of heterozygosity *per se* but because of gene contents.

The population buffering arises from interaction among different constituent genotypes. Stability in performance of such a population is often associated with genetic diversity (Allard, 1961). Mixtures are more diverse than pure breeding varieties. Three-way and double-cross hybrids are more diverse than single-cross hybrids and have shown higher stability based on estimates of CV, although certain single-cross hybrids have also shown higher stability (Jones 1958; Sprague and Federer, 1951). Thus heterogeneity or diversity provides a higher degree of population buffering. Results from varietal mixtures or multilines, have shown that they yield higher than pure line cultivars in the face of biotic stresses (Barrett, 1978). As the population is made up of a number of genotypes, favourable g × e interaction may be high for some, low for some others and even unfavourable for certain other genotypes. Thus overall the population appears to be giving a low g × e interaction. Reduced g × e interaction can also be due to inter-genotype competition occurring among different coexisting genotypes, but so far little is known about the mechanism underlying population buffering. Whatever may be the mechanism underlying individual or population buffering, it must be measurable in terms of g × e interaction. It is difficult to measure g × e interaction in case of population.

Stability of yield can arise because of plasticity of the individual traits as well. Yield is a function of a number of traits. Almost similar yield can be obtained via slight increase in the value of a component trait of yield and the corresponding decrease in the values of other yield component traits or *vice versa*. Stability in yield thus arises because of component compensation.

Measures of Stability/Adaptability

Various measures of stability proposed are as follows:

1. The contribution of *i*th genotype to the g × e interaction sum of squares is a measure of stability of *i*th genotype (Wricke, 1962 and Plaisted and Peterson, 1959).

2. Finlay and Wilkinson (1963) proposed regression coefficient (b) of *ith* variety yield on the yield of all the varieties for each site and season as a quantitative measure of adaptability of the *i*th genotype. They described the nature of adaptability considering the regression values and the varietal means as follows:

Regression coefficient (b)	Yield level	Nature of adaptability
b = 1.0 (Average stability)	High	Well adapted to all environments
	Low	Poorly adapted to all environments
b < 1.0 (Below average stability)	High	Specifically adapted to favourable environments
b > 1.0 (Above average stability)	Low	Specifically adapted to unfavourable environments
b = 0		Absolute phenotypic stability

Here the varietal means serve only to discriminate between regression coefficients of equal value or to specify performance within a set of environments.
3. Hanson's (1970) measure of stability combines the contribution of the *ith* genotype to g × e interaction sum of squares with its response to environmental change.
4. Eberhart and Russell (1966) and Tai (1971) suggested the use of two parameters, regression coefficient (b) in conjugation with deviation from regression (S^2d) as a measure of stability and a variety is called stable if $b = 1$ and $S^2d = 0$. The two parameters can be estimated following Perkins and Jinks' (1968) analysis as well.
5. Breese (1969) advocated the use of deviation from regression (S^2d) as a measure of stability, which according to him, provides the measurement of unpredictable irregularities in the response to environment. Witcombe and Whittington (1971) observed that more generally S^2d, the non-linear environmental sensitivity will be accounted for by the variance of the genotype's response to all the different environmental variables present, so S^2d is not always analogous to an unpredictable irregularity in response to the environment and it is predictable when the environmental changes, which are not accounted for in the analyses are known. Thus deviations from regression are not due to developmental noise (Waddington, 1957), or related to the concept of developmental homeostasis (Lerner, 1954).

The other measures of stability are Shukla's (1972) stability variance, coefficient of variability (CV) and the relative performance of an entry (Yau and Hamblin, 1994).

The different measures of stability lead to three concepts of stability as follows:

1. A variety is stable if its variance among environmental is small, i.e. the genotype is stable in absolute sense ($b = 0$). This variety is interacting less with the environment or shows minimum g × e interaction.
2. If $b = 1.0$, its response to environment corresponds to the mean response of all the varieties in the experiment. The variety can be said to have average stability.
3. If its deviation mean squares, S^2d is smaller or zero, the variety is called a stable variety.

Thus the definitions of stability are many and varied (Hill, 1975). The relationships between these have been discussed by Witcombe and Whittington (1971), Easton and Clements (1973), Freeman (1973), Marquez-Sanchez (1973) and Lin *et al.*, (1986).

In principle we should look for a variety for which $b = 0$, but in practice we should go for a variety for which $b = 1.0$ and $S^2d = 0$, as we need a variety which is responsive and an ideal variety is one which responds favourably when the environment is favourable (i.e. $b > 1.0$). Such a variety is difficult to find. In other words ideal variety having general adaptability is one with maximum yield potential in most favourable environments and a maximum phenotypic stability (Finlay and Wilkinson, 1963).

Genetics of Stability

As the genotype × environment interactions depend on genotype as well as environment, they are partly heritable. Further the variation in response of genotype to environment is as widespread as variation in yield, so the trait determining the stability of performance is considered a quantitative trait. The genetical architecture of this trait can be worked out using analytical procedures of biometrical genetics (Mather and Jinks, 1982) and further the relative sensitivities of any genotype can be predicted. As the selection programme is carried out in the high fertility environment, selection for high mean performance invariably selects genotypes with high b values and thus there appears positive correlation between mean performance and environmental sensitivity (b_i). That yield and yield stability, are, at least, in part, under independent genetic control was shown by Perkins and Jinks (1968a, 1968b, 1971, 1973) and Jinks and Connolly (1973). So selection of genotype(s) for $b = 1.0$ (average stability) should not be a problem. Further since the low remainder mean sum of squares, S^2d is not correlated with either b_i or mean performance (yield), selection for this character will not be complicated. Thus the aim should be to develop genotype(s) with $b = 1.0$ and showing uniformity in response to environmental effects, i.e. low remainder mean sum of squares ($S^2d = 0$).

Measurement of Stability Parameters

For calculating the stability parameters of different varieties, they are grown in replicated trials in multilocational environments. For detecting g × e interaction and estimating bi's and S^2d_i's either joint regression analysis (Finlay and Wilkinson, 1963 and Perkins and Jinks, 1968), or Eberhart and Russell's (1966) stability analysis is carried out. With s number of varieties and t number of environments, the ANOVAs in Perkins and Jinks and Eberhart and Russell's analyses take the form as given in Table 1.

The joint regression analysis of Finlay and Wilkinson (1963) and Perkins and Jinks (1968) combines the individual regression analysis being carried out for each of the variety for estimating b_i and S^2d_i. The individual regression analysis takes the form as given in Table 2.

Having calculated the bi's and S^2d_i's of the different varieties, testing of their significance is done and those varieties for which $b = 1.0$ and $S^2d = 0$ are classified as stable variety. After that means of the varieties are compared and a variety with high mean

Table 1. ANOVAs in Perkins and Jinks and Eberhart and Russell's analyses

Sources of variation	df
Perkins and Jinks's analysis	
Genotypes	(s-1)
Environments	(t-1)
Genotype × Environment	(s-1) (t-1)
Heterogeneity	(s-1)
Remainder	(s-1) (t-2)
Error	st (r-1)
Eberhart and Russell's analysis	
Genotypes	(s-1)
Environment + g × e	s (t-1)
Environment (linear)	1
G × E (linear)	(s-1)
Pooled deviation	s (t-2)
Genotype 1	(t-2)
Genotype	(t-2)
Pooled error	(s-1) t (r-1)

r is the number of replications

Table 2. Regression analysis for individual variety

Sources of variation	df
Regression	1
Deviation from regression	(t-2)

mean and b = 1.0 and $S^2d = 0$ is finally selected and recommended for commercial cultivation or used as parent in the crossing programme aimed at breeding a high yielding variety with general adaptability.

Multivariate Analysis

As we have seen above, the stability of performance of a variety can be the result of plasticity in different traits and compensation between different component traits. In other words, stability is as a result of balance among responses in different traits and so stability/ adaptability of genotype can also be studied using multivariate techniques (Hardwick and Wood, 1972). The different multivariate analyses that are being used are canonical analysis and factor analysis (Grafius and Kiesling, 1960) and principal component analysis (Freeman and Dowker, 1973). Further, in case of either joint regression analysis by Finlay and Wilkinson (1963), and Perkins and Jinks (1968) or Eberhart and Russell (1966) analysis, the linear model was assumed to be able to explain the genotype-environment interaction. In other words, the phenotype of an individual was assumed to be the result of additive effects of genotype and environment. However, Fisher and Mackenzie (1923) observed that a product formula (multiplicative model) provided a better fit to the yields of varieties in different environments. Considering this Mendel (1971) used principal component analysis to study genotype-environment interaction. In the biplot method (Gabriel, 1971; Gauch Jr., 1988) principal component analysis is applied to genotype-environment interaction to generate multiplicative model. The biplot is a plot which simultaneously displays both the genotypes and the environments, whereas the GGE biplot (Yan and Hunt, 2002) is a plot which displays the genotype main effects (G) plus the genotype-environment interaction (GE). It is constructed by plotting the first two principal components, PC1 and PC2, derived from the principal component analysis of the environment-centered data. This analysis identifies ideal cultivar (s), the ones with a large PC1 score (representing high yielding ability) and a small PC2 score which represents stability. This analysis also identifies the test environment - an environment with one large PC1 score (discriminating genotypes more in terms of genotypic main effect) and a low PC2 score (more representative of the overall environment). The only assumption underlying this graphical analysis is that the genotype and the genotype-environment interaction which are the two sources of variation that are relevant to the cultivar evaluation are sufficiently accounted for by the two principal components, PC1 and PC2.

Non-Parametric Measures of Stability

The various non-parametric measures of phenotypic stability such as ranks and the other two statistics, namely mean absolute rank difference and variance of the ranks for testing and comparing the stabilities of different genotypes have been suggested by Hühn (1979), Nassar and Hühn (1987) and Hühn (1990a&b).

Utility of Stability Analysis

The breeders who generally pick up genotypes with high means and stability on the basis of their performance and ranks in different environments, will have problems if the number of varieties and number of environments in the multilocational trial is large and this is where the stability analysis will be of help. These analyses (univariate or multivariate) will categorise the varieties into the following groups.

1. Varieties with high yield and general stability/adaptability
2. Highly responsive genotypes in favourable environment
3. Genotypes showing better response in poor environment

Besides the analysis will indicate the test environment which can discriminate between genotypes, which is essential for making selection effective.

Proven Adaptable Variety

If a crop variety is in commercial cultivation over wide areas for more than 10-15 years, that variety can be said to have wider adaptability and should be used as one of the parents in the hybridization programme, aiming at developing a high yielding variety having wider adaptability. There are a number of varieties in each crop, which are high yielding and have been in cultivation for a long time and thus can be said to possess wider adaptability, for example, Kalyansona and Sonalika in wheat, IR-24 and Jaya in rice. If we look at the characteristics of the different varieties within a crop species, we find that the wider adaptability is associated with different combinations of traits values. Further, if we compare these high yielding semi-dwarf varieties with the land races which were low yielding but having wider adaptability, we find that the adaptability is associated with both low and high yield, which again confirms that these two traits, yield and adaptability are under different independent genetic control system.

Breeding for Wider Adaptability

Self-Pollinated Crops: In case of self-pollinated crops, a variety of proven adaptability is selected and used as a parent in the crossing programme and the pedigree selection is practiced for extracting recombinant pure breeding line(s) superior in yield and adaptability. The segregating generation population is raised in a number of contrasting environments. This method of breeding uses the principle of disruptive selection. Here the selection is for genotype(s) showing extreme phenotypes at the same time. International Maize and Wheat Improvement Center (CIMMYT), Mexico called this method of breeding a widely adapted variety as "Shuttle breeding". Here the individual/family having plasticity in different traits resulting into stability of yield over different environments is selected. Family/individual is selected on the basis of its mean performance in such environments and advanced. The development of an individual is thus channeled into one optimal phenotype or another by a developmental switch mechanism (genetic, environmental). The general adaptability can thus also be attributed to variation in phenotyptic plasticity. In the later stage of breeding programme stability parameters can also be worked out to see the agreement between the conclusions drawn on the basis of observed mean of a variety and its variance over different environments and the conclusions drawn from the estimates of stability parameters.

As we have seen above the genetic diversity within a variety also leads to population buffering. To exploit this, lines which are similar in height, maturity, grain colour, etc. in $F_6 - F_8$ generation in pedigree or SSD, can be bulked to constitute a variety which should be tested in contrasting environments, in order to examine its adaptability. Similarly, varietal mixtures can be constituted to provide diversity to the variety, and thus different varietal mixtures can be made and tested in different environments and mixtures(s) having high yield and adaptability can be identified.

Cross-Pollinated Crops: In cross-pollinated crops while developing single cross hybrids, inbreds should be thoroughly evaluated and only those inbreds having higher adaptability should be used to produce hybrids. Again as in case of self-pollinated crops while developing inbreds through pedigree method, family/individual should be tested in contrasting environments and selection should be based on the means of inbreds in different environments.

In case of developing improved populations through either inter- or intra-populations, improvement methods, full-sibs or half-sibs or S_1 or S_2 families developed should be tested in markedly different environments. Also, full-sibs or half-sibs or S_1 or S_2 families having higher means over environments should be allowed to intermate to reconstitute the population for starting another cycle of selection. Here it must be noted that the selection on the basis of means in multilocational trial will not work. The environments in which evaluation/testing of lines or families is done must represent contrasting environments. For details the reader is referred to Roy (2000).

Yield at a location or in a season may be affected by a specific factor or factors such as susceptibility of the crop variety to diseases and pest, water logging condition, drought, cooler weather, nutrient deficiency, alkalinity, salinity, acidity, etc. Then these defects should be corrected and thereafter the yield and the adaptability of a variety will improve. Under such unfavourable conditions the *b* value of a widely adapted variety will be low. Further, the S^2d value of a variety susceptible to disease/insect will be very large. So if a gene for resistance to disease/insect is transferred to that variety, its adaptability will improve. If a particular abiotic stress is a characteristic of a location in a particular region, it would be better to breed a variety specifically adapted to such conditions. For example, in case of deep water (flooded) condition, one should develop a deep water rice variety and also one can develop a maize variety specifically suited to flooding condition.

Molecular Approach to Breeding for Wider Adaptability

As there is no mechanism which precisely determines the stability of performance of an individual or population over locations or seasons and with the availability of molecular markers such as Randomnly Amplified Polymorphic DNAs (RAPDs), Amplified Fragment Length Polymorphisms (AFLPs), mini- and micro-satellites (Williams *et al.*, 1990 and Rafalski and Tingey, 1993) it would be worthwhile to detect, locate and estimate the effect of individual Quantitative Trait Loci (QTLs) determining the stability/ adaptability and thus the genetic architecture of this can be precisely worked out. What is required at present is to carry out QTLs analysis in proven or tested, widely adapted varieties in different crops and try to seek the answer to the following questions:

1. How many QTLs are explaining most of the variation in this trait?
2. Whether QTLs determining the adaptability are orthologous or not.
3. Whether or not they show syntenous relationship.

Once the desired QTLs are found they can be transferred to an otherwise desirable genotype. High yielding selections with desired QTLs in the segregating generations can be isolated using molecular markers.

Declining Importance of this Trait

As agricultural research stations are now established in different agro-climatic zones and in regions of specific problems such as flooding, drought/dry areas, high altitude, alkaline, sodic or acidic soil etc., plant breeders are developing varieties for such specific conditions. Further, they are engaged in developing variety with specific adaptation to high fertility, low fertility, irrigated or rain fed condition and there they are more concerned about stability of performance over years/seasons in such conditions rather than about the general adaptability of the variety. They are also engaged in developing variety specifically adapted to multiple cropping systems such as inter-cropping, relay cropping, strip cropping, mixed cropping, ratoon cropping, etc. To fit well in a particular crop rotation system, they are developing early maturing varieties.

With the restriction on free trade vanishing, the breeders are now engaged in developing varieties with specific quality, such as variety with specific level of protein, oil (saturated and unsaturated fatty acid per cent), vitamins, storage/transportation quality, seed size, anti-nutritional factors, infection/infestation of grains with pathogens/insects. There is also a need of variety suitable for mechanization (combine harvesting). Besides these, plant breeders are engaged in developing disease/insect resistant variety, so that there is less or no use of chemicals, which are detrimental to environment as well as human health. Also, here the objective is to control the epidemic through the use of varietal diversity in space and time. Further, they are trying to develop variety which is more suitable to organic farming and is thus less dependent on use of chemical fertilizer, as the people are now more conscious about their health.

So in summary it can be said that the breeders are now engaged in developing quality products which can fetch higher price in the international market. It can also be seen from the above that there is a long list of objectives before plant breeders and they are engaged more in developing varieties which can give stable performance with respect to these quality characteristics, thus, the objective of breeding widely adapted varieties has been lost in this era of globalization. Further, as discussed above, for selection of adaptable/stable genotype(s) the segregating generation population must be grown in the contrasting environments and because of resource crunch the breeder cannot generate enough segregating material and further cannot conduct multilocation trial in the early stage of breeding programme. Therefore, they cannot isolate line(s) with general adaptability and thus they are not making it as one of the objectives. But the international institutes like CIMMYT, International Rice Research Institute (IRRI) have got enough resources and can conduct multilocation/international trials and because of that they are in a position to develop material with wider adaptability.

References

Allard R.W. 1961. Relationship between genetic diversity and consistency of performance in different environments. *Crop Sci.*, **1**: 127-133.

Allard R.W. and Bradshaw A.D. 1964. Implications of genotype- environmental interactions in applied plant breeding. *Crop Sci.*, **4**: 503-507.

Barrett J.A. 1978. A model for epidemic development in varietal mixtures. *In*: "*Plant Disease Epidemiology*", Scott and A., Bainsbridge, Blackwells.

Bradshaw A.D. 1965. Evolutionary significance of phenotypic adaptability in plants. *Adv. Genet.*, **13**: 115-55.

Breese E.L. 1969. The measurement and significance of genotype environment interactions in grasses. *Heredity*, **24**: 27-44.

Darlington C.D. and Mather K. 1949. *The Elements of Genetics*, Allen & Unwin, London.

Easton H.S. and Clements R.J. 1973. The interaction of wheat genotype with a specific factor of the environment. *J. Agr. Sci.*, **80**: 43-52.

Eberhart S.A. and Russell W.A. 1966. Stability parameters for comparing varieties. *Crop Sci.*, **6**: 36-48.

Finlay K.W. and Wilkinson G.N. 1963. The analysis of adaptability in a plant breeding programme. *Australian J. Agr. Res.*, **14**: 742-754.

Fisher R.A. and Mackenzie W.A. 1923. Studies in crop variation 2. The manurial response of different potato varieties. *J. Agric. Sci.*, **13**: 311-320

Freeman G.H. 1973. Statistical analysis for the genotype environment interactions. *Heredity*, **31**: 339-354.

Freeman G.H. and Dowker B.D. 1973. The analysis of variation between and within genotypes and environments. *Heredity*, **30**: 97-109.

Gabriel K.R. 1971. The biplot graphic display of matrices with application to principal component analysis. *Biometrika*, **58**: 453-467.

Gauch H.G. Jr. 1988. Model selection and validation for yield trials with interaction. *Biometrics*, **44**: 705-715.

Grafius J.E. 1956. Components of yield in oats. A geometric interpretation. *Agron. J.*, **48**: 419-423.

Grafius J.E. and Kiesling R.L. 1960. The prediction of relative yields of different oat varieties based on known environmental variables. *Agron. J.*, **52**: 396-399.

Hanson W.D. 1970. Genotypic stability. *Theor. Appl. Genet.*, **40**: 226-231.

Hardwick R.C. and Wood J.T. 1972. Regression methods for studying genotype-environment interactions. *Heredity*, **28**: 209-222.

Hill J. 1975. Genotype-environment interaction a challenge for plant breeding. *J. Agric. Sci.*, **85**: 477-493.

Hühn M. 1979. Beitrage zur Erfassung der phenotypischen Stabilitat. 1. Vorschlag einiger auf Ranginformationen beruhenden stabilitatsparameters. *EDP in Medicine and Biology*, **10**: 112-117.

Hühn M. 1990a. Non-parametric measures of phenotypic stability. Part 1. Theory. *Euphytica*, **47**: 189-194.

Hühn M. 1990b. Non-parametric measures of phenotypic stability. Part 2: Applications. *Euphytica*, **47**: 195-201.

Jinks J.L. and Connolly V. 1973. Selection for general and specific response to environmental differences. *Heredity*, **30**: 33-40.

Jinks J.L. and Mather K. 1955. Stability in the development of hetero-zygotes and homo zygotes, *Proc. Roy. Soc. B.*, **143**: 561-578.

Jones D.F. 1958. Heterosis and homeostasis in evaluation and applied genetics. *Am. Nat.*, **92**: 321-328.

Knight R. 1970. The measurement and interpretation of genotype-environment interactions. *Euphytica*, **19**: 225-235.

Knight R. 1973. The relationship between hybrid vigor and genotype-environment interaction. *Theor. Appl. Genet.*, **43**: 311-318.

Lerner I.M. 1954. *Genetic homeostasis*. Oliver and Boyd, London.

Lin C.S., Binns M.R. and Lefkovitch L.P. 1986. Stability analysis. Where do we stand? *Crop Sci.*, **26**: 894-900.

Mather K. and Jinks J.L. 1982. *Biometrical Genetics*. Champman and Hall, London.

Marquez-Sanchez F. 1973. Relationship between genotype-environmental interaction and stability parameters. *Crop Sci.*, **13**: 577-579.

Mendel J. 1971. Principal component analysis of variance and data structure. *Statist. Neerlandica*, **26**: 119-129.

Nassar R. and Hühn M. 1987. Studies on estimation of phenotypic stability: Test of significance for non-parametric measures of phenotypic stability. *Biometrics*, **43**: 45-53.

Perkins J.M. and Jinks J.L. 1968a. Environmental and genotype-environmental components of variability. III. Multiple lines and crosses. *Heredity*, **23**: 339-356.

Perkins J.M. and Jinks J.L. 1968b. Environmental and genotype-environmental components of variability. IV. Non-linear interactions for multiple inbred lines. *Heredity*, **23**: 525-235.

Perkins J.M. and Jinks J.L. 1971. Analysis of genotype × environment interaction in tripple test cross data. *Heredity*, **26**: 203-209.

Perkins J.M. and Jinks J.L. 1973. The assessment and specificity of environmental and genotype-environmental components of variability. *Heredity*, **30**: 111-126.

Plaisted R.L. and Peterson L.C. 1959. A technique for evaluating the ability of selections to yield consistently in different locations or seasons. *Amer. Potato J.*, **36**: 381-385.

Rafalski J.A. and Tingey S.V. 1993. Genetic diagonestics in plant breeding RAPDs, micro satellites and machines. *Trends Genet.*, **9**: 275-279.

Roy Darbeshwar. 2000. *Plant Breeding - Analysis and Exploitation of Variation*. Narosa, New Delhi.

Shukla G.K. 1972. Some statistical aspects of partitioning genotype-environment components of variability. *Heredity*, **28**: 237-245.

Simmonds N.W. 1962. Variability in crop plants, its use and conservation. *Biol. Rev.*, **37**: 422-465.

Sprague G.F. and Federer W.T. 1951. A comparison of variance components in corn yield trials. II. Error, year × variety, location × variety, and variety components. *Agron. J.*, **43**: 535-541.

Tai G.C.C. 1971. Genotypic stability analysis and its application to potato regional trials. *Crop Sci.*, **11**: 184-190.

Waddington C.H. 1957. "*The Strategy of Genes*". Allen & Unwin, London.

Williams J.G.K., Kubelik A.R., Livak K.J., Rafalski J.A. and Tingey S.V. 1990. DNA polymorphism amplified by arbitrary primers are useful as genetic markers. *Nucleic Acid. Res.*, **18**: 6531-6535.

Witcombe J.R. and Whittington W.J. 1971. A study of the genotype × environment interaction shown by germinating seeds of *Brassica napus*. *Heredity*, **26**: 397-411.

Wricke G. 1962. Uber eine Methode zur Erfassung der Okologischen Streubreite in Feldversuchen. *Zeitschrift fur Pflanzen Zuchtung*, **47**: 92-96.

Yan W. and Hunt L.A. 2002. Biplot analysis of diallel data. *Crop Sci.*, **42**: 21-30.

Yan W., Cornelius P.L., Crossa J. and Hunt L.A. 2001. Two types of GGE biplots for analysing multi-environment trial data. *Crop Sci.*, **41**: 656-663.

Yates F. and Cochran W.G. 1938. The analysis of groups of experiments. *J. Agri. Sci.*, **28**: 556-580.

Yau S.K. and Hamblin J. 1994. Relative yield as a measure of entry performance in variable environments. *Crop Sci.*, **34**: 813-817.

Plant Breeding - Mendelian to Molecular Approaches
H. K. Jain and M. C. Kharkwal (eds.)

25

Plant Ideotype: The Concept and Application

J.P. Tandon[1] and H.K. Jain[2]

Abstract

The plant type concept in crop improvement started to receive major attention with the discovery of dwarfing genes in wheat and rice. The discovery and the analysis that followed showed that yield increases in many crops with the advent of scientific plant breeding are associated with better partitioning efficiency of the total dry matter produced, which on its own may have registered little increase. The two physiological parameters of crop yields - dry matter production and harvest index are now explicitly recognized as targets for future studies as major foodgrains of the world begin to reach a saturation point in their yields. The present paper describes the evolution of the plant type concept and discusses its wider implications. The concept is particularly relevant for modernization of traditional agriculture where genetic diversity for plant types could help to develop improved crop varieties responsive to applications of fertilizers, irrigation and other farm inputs. The plant type genes could help to accelerate the process of crop improvement in many of the developing countries.

Introduction

The process of domestication of crop plants over several thousand years has resulted in profound changes in their morphological traits giving them a very different architecture. Thus, cultivated plants like wheat and rice look very different from their wild progenitors. It is clear that as generations of farmers selected plants from the wild for yield and other desirable characteristics, unconsciously they altered their form and growth habit. This process of transformation of crop plants was accelerated when scientific plant breeding became possible a hundred years ago, following the rediscovery of Mendel's laws of biological inheritance.

This was also the period when vast fields of fossil fuels were being discovered. The associated development of industrialization resulted in the production of inorganic fertilizers, pesticides, weedicides, farm machines and other modern farm inputs. The crop plants with the availability of these inputs were now being grown in a different agronomic environment. Selection of genotypes which would take full advantage of these inputs resulted in significant yield improvements associated with further changes in crop morphology. The discovery of plant type genes and their deployment for a major advance

[1]A-72, Sector 14, Noida 201 301, India
[2]Centre for Science Writing, 40 Surya Niketan, New Delhi 110 092, India

in the yield of wheat and rice in the 1960s gave a new direction to selection for new plant architectures. The dominant concept in the earlier years had been that yield as a quantative trait is determined by a large number of polygenes. While the role of polygenes in determining crop yields is not disputed, it is now widely recognized that oligogenes exist, which offer the possibility of major advances in crop yields under conditions of improved farm management with the application of various inputs. Agriculture world over and more particularly in the developing countries has seen significant advances with the discovery and deployment of these genes in plant breeding programmes. The present paper describes the plant type concept and discusses its implications.

Plant Type and Economic Yield

All plant breeders tend to develop a concept of a plant type which they expect would produce the maximum yield of the desired product. However, Donald (1968) was the first to propose a systematic and scientific concept of plant ideotype based on physiological parameters. The concept was first introduced at the Third International Wheat Genetics Symposium held in Canberra in 1968. He proposed that it is possible to define an ideal plant type which will perform well under field conditions. He noticed that in cereals a plant produces many tillers during its early growth period, all of which do not develop into culms bearing earhead and grains and are thus a waste of resources. He proposed that a plant type which is theoretically efficient based on the knowledge of plant physiology and morphology should be defined first and then the breeders could select for such an ideotype rather than for yield. In case of wheat he proposed a uniculm plant with few small erect leaves, relatively shorter height, strong stem, awned large erect earhead with many florets and high proportion of seminal roots as a basic ideotype designed to produce high grain yield as a plant community.

It is seen that he emphasised primarily on the morphological traits. Later this definition was extended to include physiological, biochemical, anatomical and phenological traits (Rasmusson, 1987). However, before any character is included as a parameter for a plant type, it is important to know its specific contribution to yield. Physiologists have identified a wide range of characters which have a bearing on plant productivity. However, since most of these results emerge from controlled studies, breeders tend to have reservations in using these characters as selection parameters. Moreover, some of these features are cumbersome to measure and the breeders fear that their inclusion will increase the efforts without commensurate gains. Extensive investigations have been carried out by many workers in a wide range of crops to compare the important features of the latest improved high yielding types with the older types and come out with conclusions which show that the concept of 'plant type' selection can help achieve, as also accelerate, yield gains in varieties of the future.

Changes Associated with Improved Plant Varieties

Since the breeders have been extremely successful in enhancing crop productivity, it has been considered appropriate to examine the way the crop varieties have changed over time and the features for which the latter varieties differ from the earlier ones. This is considered the best procedure to plan a strategy for the future. Keeping this in view several researchers have investigated the effect of genetic improvement in crop varieties with superior yield potential against the older types. Such studies have been carried out in several crops such as wheat (Jain and Kulshreshtha, 1976; Austin *et al.*, 1980, 1989; Decherd *et al.*, 1985; Paccaud *et al.*, 1985; Cox *et al.*, 1988; Perry and Antouno, 1989; Siddique *et al.*, 1989; Slafer and Andrade, 1989; Slafer *et al.*, 1990a,b), barley (Rasmusson, 1987), maize (Jain *et al.*, 1976; Crosbie and Moch, 1981), oats (Peltonen, 1990), peanut (Duncan *et al.*, 1978), pulse crops (Jain, 1975). All these workers have concluded that the breeders have been successful in modifying the plants in two respects, first enhancing the production proportion of the part of economic value, mostly grain yield, without any significant alteration in biomass productivity; and second enhance the number of grains produced per unit area. Other specific differences are as follows:

Biomass Production and Harvest Index: Only a fraction of the total biomass produced by a plant during its full growth cycle is of economic value and is harvested as a produce. In grain crops the biomass produced above ground is taken into consideration while that accumulated as roots, although equally important, is left out for practical reasons. Economic (grain) yield (GY) has been defined by Donald and Hamblin (1976) as a product of above ground biomass and harvest index (HI), i.e. GY = Biomass × HI.

Harvest index is defined as the ratio between economic yield and the above ground biomass or the biological yield. It is a measure of the partitioning efficiency of the plant into seed and other vegetative parts (Jain, 1986) and can be expressed as follows:

$$\text{Harvest Index (HI)} = \frac{\text{Economic yield (seed in cereals)}}{\text{Biological yield (total drymatter)}} \times 100$$

It is thus seen that economic yield can be increased by increasing biomass production and/or harvest index. Ideally, high harvest index should be combined with high biomass production.

In most crops genetic increase in grain yield of the improved types has been associated with enhanced harvest index, whereas biomass production has not changed to any significant extent (as referred earlier). The semi-dwarfing genes exploited in wheat and rice have been shown to have a favourable association with harvest index and have helped to achieve a remarkable gain in its expression wherever these have been introduced, thereby making the semi-dwarf types significantly superior yielding over the traditional types. Furthermore, the semi-dwarf plant type is able to resist lodging to a greater extent than the tall types and thus increases its ability to exploit more favourable production environments created by increased use of fertilisers and irrigation. However, some recent studies have shown that the semi-dwarf plant type in these crops may limit canopy photosynthesis and biomass production (Karoda *et al.*, 1989 and Gent, 1995). A taller canopy has superior ventilation, higher carbon dioxide concentration and light penetration. However, even if this is true, their total dry matter production does not appear to suffer in relative terms.

Further increase in harvest index, which is currently around 35-45 per cent in spring wheats and around 50 per cent in winter types, may be feasible, but it appears to be reaching its optimum value. Austin *et al.* (1980) believe that it can be increased up to 60 per cent.

In future, increase in biomass production in combination with appropriate HI would play an increasing role in enhancing productivity of major grain crops. Biomass production has been defined by Biscoe and Gallagher (1977) as the amount of radiation intercepted (RI) and radiation use efficiency (RUE), i.e. the efficiency of conversion of the radiation to dry matter or Biomass = RI × RUE. Biomass production can be increased through agronomic manipulations, such as increased population density and appropriate use of plant nutrient and water as also genetic means such as (*i*) extended photosynthetic duration, i.e. either extended crop duration and/or enhanced duration of the photosynthetic activity, (*ii*) appropriate canopy structure for maximised rate of photosynthesis, (*iii*) improved photosynthetic rate and (*iv*) reduced respiratory losses.

However, increased biomass production does not necessarily mean increased grain yield, unless the plant has the ability to transform it into grain through appropriate HI value.

Evolutionary Basis of Plant Type Concept

Present day crop plants growing in their wild habitats prior to domestication evolved by Darwinian natural selection. Natural selection favours reproductive fitness but the plants must first survive before they can reproduce. In the highly competitive environments of natural habitats, populations of the different plant species are exposed to stress conditions of different kinds - moisture stress during periods of drought; soil fertility stress because of competition with other species forming a part of their ecological community (we now call these plants weeds); and stress due to attacks of pests and pathogens. Natural selection under these unmanaged conditions would favour characteristics like an indeterminate growth habit (the plants would resume growth when the rains come), or early flowering so that they could escape drought; and resistance to diseases and pests.

In the highly managed environment of modern agriculture with the application of various farm inputs, many of these traits like the indeterminate growth habit with excessive vegetative growth, or too little dry matter production before the onset of quick flowering, may not be desirable. The modem plant breeder, therefore, selects those plant types which show an optimum balance between dry matter production and reproductive capacity, in other words between biomass and harvest index. The discovery of plant type genes greatly facilitates selection for these plant types and helps accelerate the process of crop improvement.

The plant type genes are of special value for breeders in the developing countries, where traditional agriculture continues to be widely practised. Human selection has had only a limited role in determining the morphological and physiological characteristics of plants in traditional agriculture. In the absence of applied inputs like inorganic fertilizers natural selection has continued to be an important factor in determining their survival. Many developing countries have come under pressure to modernise their agriculture with their rapidly rising human population, following the advent of modern medicine. The plant type genes are helping the breeders in these countries to genetically reconstruct their land races and local cultivars - a process which the breeders in the industrialized countries achieved in the earlier decades by selecting for polygenes determining yield.

Biomass Production

In the present context only the genetic means of increasing biomass production would be considered.

(*i*) ***Extended Duration of Photosynthetic Activity***: Under appropriate temperature and moisture availability conditions, a longer crop duration means longer duration of photosynthetic activity. It can also be achieved through delaying the process of senescence. The process of senescence is associated with degradation of chlorophyll and is closely associated with breakdown of Rubisco (Mabino *et al.*, 1983) and application of nitrogen late in the crop season helps to protect it from degradation and enhances the duration of photosynthesis. As regard the crop duration, it can be easily increased by genetic means, since considerable variability exists in germplasm of most crops. However, manipulating the crop duration has its limitations, since the available appropriate crop growing season is normally very limited under most situations and quite frequently the crop has to be grown in a specified sequence, particularly under irrigated conditions. Extending the crop duration beyond a certain limit would make it encounter adverse climatic conditions, with negative effect on grain yield. Moreover, extending the duration of photosynthesis may not necessarily result in enhanced yield, unless it is accompanied with appropriate sink size and efficient photosynthate transfer mechanisms.

(*ii*) ***Appropriate Canopy Structure***: The amount of radiation interception is regulated by the leaf size and their appearance and orientation. Immediately after germination as the leaf coverage increases, the rate of biomass accumulation by the crop increases rapidly and reaches an optimum level as the ratio of leaf area to ground surface, i.e. the leaf area index (LAI) reaches an optimum stage where it levels off. An ideal variety should have horizontally displayed canopy during the very early stages of growth to effectively intercept more solar radiation. As the crop grows, a plant population with vertically oriented leaves gives a better light penetration and a higher photosynthetic rate at high LAI. Varieties with erect leaves have higher optimum LAI than varieties with horizontal leaves (Yoshida, 1981). Modern rice and wheat varieties have been specifically selected for erect leaves at flowering stage. V-shape leaf blades are reported to reduce mutual shading and increase canopy photosynthesis (Sasahara *et al.*, 1992). Similarly narrow leaves allow better light penetration and improved canopy photosynthesis. It has already been indicated that a taller plant height allows for better penetration of light in the canopy and results in higher rate of biomass accumulation.

(*iii*) ***Improved Rate of Photosynthesis***: It has been reported that photosynthetic capacity per unit leaf area has not increased in the process of improvement of crops like wheat, sorghum, soybean, cotton, sugarcane and tomato (Evans, 1980; Gefford and Evans, 1981). Comparison of wild diploid and domesticated hexaploid wheats has indicated that the photosynthetic rate per unit leaf area has fallen considerably in the course of domestication and selection (Khan and Tsunoda, 1970; Evans and Dunstone, 1970; Austin *et al.*, 1982). Phosynthetic rate and leaf area have been reported to be negatively correlated (Evans and Dunstone, 1970; Rawson *et al.*, 1983; Morgon and Le Cain, 1991). Crop yield and photosynthesis show a poor correlation (Evans, 1975) and selection for high phosynthetic capacity does not often result in increased RUE and yield (Rawson *et al.*, 1983).

Several researchers have observed that there are hardly any chances of modifying the biochemical mechanism of photosynthesis (Good and Bell, 1980; Waller and Sivah, 1983 and Evans, 1987). However there are some examples where varietal differences in photosynthetic efficiency have been recorded (Ohno, 1976 in rice; Khan and Tsunoda, 1970 and Dantuma, 1973 in wheat; Darinhaff and Shibles, 1970 and Wallace *et al.*, 1972 in soybean).

(*iv*) ***Reduced Respiration***: Respiration accounts for the use of a large part of the photosynthates in the plant. Higher rate of respiration and reduced growth rate in wheat than rye and triticale has been reported (Wingeter *et al.*, 1989). There are no reports of reduced respiration among wheat genotypes. In future there may be possibilities of increasing assimilates available for plant growth by reducing respiratory losses through the use of more efficient pathways.

Other Important Physiological Features

In addition to the biomass production and harvest index there are several physiological features which have been found important for the development of an improved plant type. These are phenology, root system and mobilisation of assimilates stored in the vegetative parts, water use efficiency, etc.

(*i*) ***Phenology***: Phenological development of the plant, i.e. the duration of each of the developmental stages namely, germination, vegetative growth and reproductive stage, determine whether the plant life-cycle is well matched to the resources and constraints of its growing environments. The duration of favourable temperature and moisture supply for plant growth is normally limited in most crop growing regions of the world. In some situations the duration available for growth is determined by the agricultural rotations followed for cultivation of any crop. An unusually long duration of any particular stage may result in unproportionate distribution of plant assimilates and expose a relatively more sensitive stage of development to unfavourable climatic situations leading to poor yield and even failure of the crop. This is particularly crucial in determinate plants like wheat, where there is only one opportunity for the plant for transition from vegetative to the reproductive phase. As such these stages have been very critically studied in wheat and comparisons made between the old and the new plant types.

In wheat five distinct phenological phases are recognised. These are (*a*) seeding to germination, (*b*) germination to ear initation or the double ridge stage, (*c*) ear initiation to appearance of terminal spikelet, (*d*) terminal spikelet appearance to anthesis and (*e*) anthesis to maturity. Duration of these phases has important implications on assimilate partitioning to different organs and other plant traits. Research on these aspects has been reviewed by Kirby and Appleyard (1987), Simmons (1987) and Hay and Kirby (1991). There appears to have been a general trend in spring wheat improvement to reduce the time to reach anthesis through selection for reduced sensitivity to photoperiod and vernalisation. The modern wheat varieties have a faster rate of vegetative development as compared to old types, including faster rate of leaf appearance, shorter vegetative growth, fewer leaves and lesser tillers (Kirby *et al.*, 1989; Siddique *et al.*, 1989 a, b). Modern wheats have a faster rate of spikelet initiation and an extended period of ear development. These also have more tiller survival (51%) as compared to the older types (35%). The duration between terminal spikelet appearance and anthesis is shorter in modern wheats as compared to the old types. Anthesis indicates termination of vegetative growth and start of grain filling. Varieties with short duration from sowing to anthesis tend to have longer grain filling duration (Loss *et al.*, 1989; Austin *et al.*, 1989)

(*ii*) ***Mobilisation of Stored Assimilates***: A high proportion of assimilates accumulated in the wheat grain are produced after anthesis and their proportion varies from 70 to 95 % depending on degree of moisture stress (Rawson and Evans, 1971; Austin *et al.*, 1977; Bidinger *et al.*, 1977; Pheloung and Siddique, 1991; Kobata *et al.*, 1992). These assimilates are derived mainly from the spike, flag leaf and its sheath (Austin and Jones, 1975; Rawson *et al.*, 1983). Awns can also make a substantial contribution to spike photosynthates and yield particularly under dry conditions (Atkins and Norris, 1955; Bremner and Rawson, 1972; Evans *et al.*, 1972 and Olugbeni *et al.*, 1976). Assimilates produced before anthesis in the stem made a significant contribution to grain yield in wheat. These are stored as nonstructural soluble carbohydrates and are rapidly depleted during grain filling. Some other structural substances are also mobilised as senescence progresses. Semi-dwarf types have been found more efficient in remobilising stored assimilates as compared to the tall types (Pheloung and Siddique, 1991).

(*iii*) ***Development of Sink Size***: Sink size in cereals is determined by spikelet/floret number per ear/panicle and number of ears/panicles per m^2. There is a strong negative association between these two yield components. This negative association can be delinked only through increased biomass production during the critical phases of plant development when sink is determined (Slafer *et al.*, 1996). Critical period for sink determination is reported to be 20-30 days before flowering in wheat (Fisher, 1985). Slafer *et al.,* 1996 have proposed the elongation of stem development phase, i.e. from terminal spikelet initiation to flowering to increase biomass accumulation during this phase, since floret number is decided during this period. Richard (1996) has proposed reduction in peduncle length and unproductive tillers to increase supply of photosynthates to the developing ears/panicles.

Grain size is the other component which has large influence on sink size. Here again grain weight and grain number have negative correlation. In rice percentage of filled grains is determined by the activity of the source to the relative size, ability of the spikelets to accept carbohydrates and translocation of assimilates from leaves to spikelets (Yoshida, 1981). A close relationship has been reported between crop growth rate during grain filling and percentage of filled spikelets (Akita, 1989). Prolonging the grain filling duration to increase grain yield has been proposed by Kropff *et al.*, 1994.

(*iv*) ***Root System***: Root development is highly influenced by moisture and nutrient availability, soil type and cultural practices (Hamblin *et al.*, 1990). Under moisture stress conditions roots may comprise as much as 60 % of the total crop biomass at maturity (Gregory *et al.*, 1984; Hamblin *et al.*, 1990; Siddique *et al.*, 1989b). However, under favourable conditions these may contribute as little as 10 % of the total biomass (Lupton *et al.*, 1978). During early growth stages more assimilates are transferred to the root system as compared to the shoots, but after anthesis root growth is reduced. In cereals rooting depth or the rate of elongation has been found better for maximising water uptake than root length, weight or density. Cereal root density in top 30 cm of soil is very high as compared to legumes (Gregory, 1988). However, differences in root morphology and physiology are also important. Water uptake per unit of root length has been found to be greater in grain legumes than cereals (Hamblin and Tenant, 1987).

Modern wheats have been found to have less root dry matter and lower root-shoot ratio than older types (Loss and Siddique, 1994). This was supposedly related to the relatively lesser tillers and the consequent fewer adventitious roots. The new varieties have a root density of 10 gm cm^{-3} in the top ten cm of soil, which is about half the density of the older types. However, under moisture stress conditions where the growth of the two types of wheats is similar, the new types produce deep roots earlier than the old types (Siddique *et al.*, 1989b).

Selection for deeper root system for maximising water update under dry environments has been suggested (Hamblin and Tennant, 1987). However, appropriate techniques for measuring rooting depth on mass scale are not available for use in breeding programmes.

(*v*) ***Water Use Efficiency (WUE)***: This is defined as the amount of dry matter produced per unit of water used. Water use by a crop is usually considered as soil evaporation plus transpiration. Any mechanism that reduces evaporation and increases transpiration would increase WUE, since transpiration is closely associated with photosynthesis and biomass production. In winter cereals spreading growth habit during early stages helps to cover soil surface rapidly and tends to reduce evaporation losses. Early plant vigour allows the plant to take advantage of cooler season for growth and can help to increase water use efficiency. Reduction in xylem diameter of seminal roots is to increase hydraulic resistance in wheat under dry environments and enhanced water use efficiency has been suggested by Richard and Passioura (1989). Epicular wax or glaucausness increases radiation reflectance, reduces leaf temperature and increases transpiration (Johnson *et al.*, 1983). Leaf pubescence has been found to enhance WUE and yield in sorghum under water stress conditions (Baldochi *et al.*, 1983). Accumulation of abscissic acid (ABA) (Turner, 1986) and osmoregulation (Turner and Jones, 1980) for which considerable varietal differences exist, play important roles in regulation of plant water relationship, photosynthesis and accumulation of assimilates under moisture stress conditions.

(*vi*) ***Lodging***: One of the major contribution of the semi-dwarfing genes in wheat and rice was that these reduced the extent of lodging suffered by the tall plants, particularly under favourable yielding conditions and allowed use of appropriate doses of plant nutrients and irrigation to enable the realisation of high yield potential. No further improvements in yield are feasible without enhanced lodging resistance. The types of lodging are bending or breakage of shoot and disturbed roots. It reduces grain yield through reduced canopy photosynthesis, increased respiration, reduced translocation of nutrients and photosynthates for grain filling and greater susceptibility of pests and diseases. The magnitude of damage depends on degree and timing of lodging. Further reduction in plant height of the semi-dwarf varieties to enhance lodging resistance is not a good approach, since it results in reduction of biomass production. Basal internode length, cross sectional area of culm and leaf sheath wrapping have major influence on straw strength in rice (Chang and Vergara, 1972). A similar situation exists in wheat.

(*vii*) ***Efficient Use of Plant Nutrients***: Plants consume large quantities of some of the nutrients like NP and K, while others are required in small amounts. Genetic differences among varieties to utilise these nutrients more efficiently have been observed (Ladha *et al.*, 1995). However more information is needed on these aspects.

(*viii*) ***Abiotic Stress Tolerance***: Atmospheric temperature, soil moisture levels and solar radiation can exert extremely important stresses to plant growth. Similarly abnormal soil reactions (pH levels), excess salt concentration (salinity), nutrient deficiencies (Zn, Fe, Cu, B, Mo) and mineral toxicities (Al, B, Fe, Mn) restrict plant growth. CIMMYT has achieved considerable success in developing varieties of wheat tolerant to acid soils and Al toxicity. Indian wheat breeders have successfully bred alkalinity/salinity tolerant varieties. IRRI has done excellent work on identification of genetic sources of tolerance, inheritance patterns and selection criteria for phosphorus, zinc and iron deficiencies as also iron toxicity in rice. Breeding for high temperature tolerance in wheat during different growth stages is a regular component of varietal improvement in CIMMYT and Indian programme.

Breeding for New Plant Type

The past success in increasing cereal yield potential has mainly been the result of selection for yield *per se* (Loss and Siddique, 1994). Further increase in yield potential would be difficult by the use of this approach, since the levels already are quite high (Slafer *et al.*, 1996). Breeding for a well defined plant type (designated as ideotype by Donald, 1968) which is theoretically efficient based on knowledge of physiology and morphology is developed as a model and selection is made for this specified type rather than for yield. Over the past decade the concept has been extensively worked upon in breeding for a new plant type with markedly enhanced productivity potential in rice at IRRI and wheat at CIMMYT. Some of the features highlighted earlier have formed the basis for the new plant type. It is hoped that the new plant type of wheat and rice would yield 20-30 % higher than the best yielding type under favourable environments.

The model plant type designed by CIMMYT for wheat has thick stems, fewer tillers, large heads, a higher number of grains without commensurate decline in grain weight and high input use efficiency and harvest index. They have already developed a plant type with robust stem, long head (73 cm), multiple spikelets, florets with large glumes, large leaf area and broad leaves and named it 'Buitre'. However, it has been observed that the advantages of the positive traits are counter balanced by unknown physiological imbalances or disorders. This results in spars tillering highly sterile heads, shrivelled grains and high suceptibility to leaf and stripe rusts (Rajaram *et al.*, 1996). Now attempts are being made to improve the 'Buitre' type to develop larger number of spikes with slightly reduced head size and high fertility.

The IRRI model for a superior yielding rice plant type popularly named 'Super Rice' has low tillering capacity (3-4 productive tillers), 200-250 grains per panicle, very sturdy stems, dark thick and erect leaves, vigorous root system and increased harvest index (IRRI, 1989). Strains developed according to this model have been found to have low biomass, poor grain filling and pest susceptibility (Khush and Peng, 1996). Reasons for poor performance of the new plant type have been investigated and attempts are continuing to develop new superior types.

Similar plant type models have been proposed for other crops such as maize, pulses, oilseeds, etc. which are usually based on enhanced harvest index and biomass production and understanding of the yield contributing features as also physiological processes and canopy structure. However, such models have so far remained as ideals and the breeders have restricted themselves to only traditional methods.

Rasmusson (1987) has identified four factors of trait interrelationships that slow progress in breeding for a specified plant type. These are:

(*i*) ***Harmony in Size of Plant Parts***: Symmetry in size among plant parts arising simultaneously from the same meristem would make the development of Donald's plant type with small narrow leaves and large spike very difficult.

(*ii*) ***Compensation Among Plant Parts***: As increase in one yield component in cereals may be accompanied by reduction in the other component, i.e. grain weight and grain number in wheat

(*iii*) ***Pleiotropy***: Certain desirable traits under simple genetic control and easily manipulatable have adverse effects on others through pleitropy. Multiple awned feature in largely, which is governed by a single gene and enhances net photosynthesis in the ear. It is found to reduce kernel number and weight (Rasmusson, 1987).

(*iv*) ***Genetic Background***: Several of the potentially useful traits found suitable for incorporation into the ideal plant types are available in genetic stocks with poor yielding ability and several undesirable features. As such crossing of these inferior stocks with highly improved types (with which breeders normally work) results is segregating populations with huge proportion of undesirable types. Furthermore the genes controlling the desired traits available in inferior stocks may be linked to several undesirable genes which increases difficulties in selecting superior types. There are also possibilities that the desirable trait genes when transferred to the superior yielding background may lose their advantage.

Conclusions

The tremendous success achieved by the use of traditional methods of breeding, primarily based on yield performance and test across environments, may not continue to produce remarkable results in the future, since the yield levels in most extensively cultivated crops have reached high levels and are in fact plateauing. These would require to be supplemented with plant type/ideotype concept based on clearer understanding of the phenological development and physiological and biochemical processes. Traits contributing to enhanced yield potential would require to be clearly identified and their inheritance patterns worked out for their inclusion in the conceptual/model plant type. This would require closer collaboration between breeders, physiologists and modelers.

Breeders are normally reluctant to work with unimproved germplasm since such hybridization leads to high frequency of undesirable types which necessitates the growing of very large segregating populations, increasing the work load tremendously. Incidently most attributes identified by physiologists as desirable tend to be located in unimproved and primitive backgrounds. Furthermore, there is always an apprehension as to how these attributes would behave when brought into the improved backgrounds. Breeders, therefore, concentrate on improved germplasm in their hybridization programmes. It would therefore be appropriate to develop strong 'prebreeding' programmes aimed at transferring the physiologically desirable traits to improved background, so that breeders are encouraged to use them in their programmes. Several of the traits identified by physiologists as useful involve the use of laborious techniques and/or sophisticated costly equipments which makes it difficult to adopt them at many locations. It would, therefore, be desirable to concentrate such work at specified places which are adequately funded to enable them to take up extensive screening of germplasm to identify desirable parental lines which can be used in regular breeding programmes.

Recent developments in molecular biology are expected to assist in selection of traits which are time consuming and expensive to measure and are sensitive to environmental factors. DNA markers which are recognisable in the seedling stage can be used to follow the genes regulating the specified traits which would increase the efficiency of selection.

It is expected that in the future the plant/ideotype concept would be used more extensively in varietal improvement and would involve a close co-operation between breeders, physiologists and molecular biologists.

References

Akita S. 1989. Improving yield potential in tropical rice. *In*: *Progress in irrigated rice research.* International Rice Research Institute, Philippines, pp. 41-73.

Atkins I.M. and Norris M.J. 1955. The influence of awns on yield and certain morphological characters of wheat, *Agron J.*, **47**: 218-220.

Austin R.B. and Jones H.G. 1975. The physiology of wheat. Plant Breeding Inst. *Ann. Rep.*, 1974. pp 20-73 Cambridge U.K.

Austin R.B., Edrich J.A., Ford M.A. and Blackswell R.D. 1977. The fate of dry matter carbohydrates and ^{14}C loss from leaves and stem of wheat during grain filling. *Ann. Bot.*, **41**: 1309-1313.

Austin R.B., Bingham J., Blackwell R.D., Evans L.T., Ford M.A., Morgan C.L. and Taylor M. 1980. Genetic improvement in winter wheat yields since 1900 and associated physiological changes. *J. Agric. Sci.*, **94** 675-689.

Austin R.B. Morgan C.L., Ford M.A. and Bhagwat S.G. 1982. Flag leaf photosynthesis of *Triticum aestivum* and related diploid and tetraploid species. *Ann. Bot.*, **49**: 177-189.

Austin R.B., Ford M.A. and Morgan C.L. 1989. Genetic improvement in the yield of winter wheat. A further evaluation *J. Agric. Sci.*, **112** 295-302.

Baldochi D.D., Verma S.B., Rosenberg N.J., Blad B.L., Garay A. and Specht J.E. 1983. Leaf pubescence effects on the mass and energy exchange between soybean canopies and the atmosphere. *Agron J.*, **75**: 537-539.

Bidinger F., Muscrave R.B. and Fisher R.A. 1977. Contribution of stored parenthesis assimilates to grain yield in wheat and barley. *Nature* (London), **270**: 431-433.

Biscoe P.V. and Gallagher J.N. 1977. Weather, dry matter production and yield. *In*: *Environmental effects of crop physiology*. (eds.) J. J. Lendsberg and C. V. Cutting. Academic Press, London.

Bremner P.M. and Rawson H.M. 1972. Fixation of $^{14}C_2$ by flowering and nonflowering glumes of the wheat ear and pattern of transport of label individual grains. *Aust. J. Biol. Sci.*, **25**: 921-930.

Chang T.T. and Vergara B.S. 1972. Ecological and genetic information on adaptability and yielding ability in tropical varieties. 'Rice Breeding'. *Int. Rice Res. Inst.*, 431-453.

Crosbie T.M. and Moch J.J. 1981. Changes in physiological traits associated with grain yield improvement in three maize breeding programmes. *Crop Sci.*, **21**: 255-259.

Cox T.S., Shroyer J.P., Lui B.H., Sears R.G. and Martin J.J. 1988. Genetic improvement in agronomic traits of hundred winter wheat cultivars from 1919 to 1987. *Crop Sci.*, **28**: 756-760.

Dantuma G. 1973. Rates of photosynthesis in the levels of wheat and barley varieties. *Netherland J. Agri. Sci.*, **21**: 188-189.

Darinhaff G.M. and Shibles R.M. 1970. Varietal differences in net photosynthesis of soybean leaves. *Crop Sci.*, **10**: 42-45.

Decherd E.L., Busch R.H. and Kafoid K. 1985. Physiological aspects of spring wheat improvement, *In*: *Exploitation of Physiological and Genetic Variability to Enhance Crop Productivity*, (eds.) J.E. Harper, L.E. Schrader and R.W. Howel., Ann. Soc. Plt Phy. Rachville Maryland.

Donald C.M. 1968. The breeding of crop ideotype. *Euphytica*, **17**: 385-403.

Donald C.M. 1982. In search of yield. *J. Austral. Inst. Agric. Sci.*, **28**: 171-178.

Donald C.M. and Hamblin J. 1976. The biological yield and harvest index of cereals in agronomic and plant breeding criteria. *Adv. Agron.*, **28**: 361-405.

Duncans W.G., Mellaud R.L.M., Grow and Boote K.J. 1978. Physiological aspects of peanut yield improvement. *Crop Sci.*, **18**: 1015-1020.

Evans L.T. 1975. Wheat. *In*: *'Crop Physiology'*, Cambridge Univ. Press. Camb. U.K.

Evans L.T. 1980. The natural history of crop yield. *Amer. Sci.*, **68**: 388-397.

Evans L.T. 1987. Opportunities for increasing yield potential of wheat. "*Future Development of Maize and Wheat in Third World*", pp 79-96 (CIMMYT, Mexico, D. F.)

Evans L.T. 1993. *Crop Evolution, Adaptation and Yield.* Cambridge Univ. Press, Cambridge.

Evans L.T. and Dunstone R.L. 1970. Some physiological aspects of evolution in wheat. *Aust. J. Biol. Sci.*, **23**: 725-741.

Evans L.T., Bingham J., Jackson P. and Sutherland J. 1972. Effect of awns and drought on the supply of photosynthate and its distribution in winter wheat ears. *Ann. Appl. Biol.*, **70**: 67-78.

Fisher R.A. 1985. Number of kernals in wheat crop and the influence of solar radiation and temperature. *J. Agric. Sci.*, **100**: 447-461.

Gent M.P.N. 1995. Canopy light interception, gas exchange and biomass in reduced height isolines of winter wheat. *Crop. Sci.*, **35**: 1636-1642.

Gefford R.M. and Evans L.T. 1981. Photosynthesis, carbon partitioning yield. *Ann. Rev. Plt. Phy.*, **32**: 485-509.

Good N.E. and Bell D.H. 1980. Photosynthesis, plant productivity and crop yield. *In*: *Biology of Crop Productivity*, (ed.) P.S. Carlson, Acad. Press.

Gregory P.J. 1988. Rust growth of chickpea, faba bean, lentil and pea and effect of water and salt stresses. *In*: *World Crops - cool season food legumes.* (ed.) R. J. Summerfield, Klower Dordrecht. pp 857-867.

Gregory P.J., Sheperd K.P. and Cooper P.J.M. 1984. Effect of fertiliser on root growth and water use of barley in northern Syria. *J. Agric. Res.*, **43**: 555-573.

Hamblin J. and Tennant D. 1987. Root length, density and crop water update. How well are theory correlated. *Aust. J. Agric. Res.*, **38**: 513-527.

Hamblin J., Tennant D. and Perry M.W. 1990. The cost of stress: Dry matter partitioning changes with seasonal supply of water and nitrogen to dryland wheat. *Plant Soil*, **122**: 47-58.

Hay R.K.M. and Kirby E.J.M. 1991. Convergence and synchrony. A review of coordination of development in wheat. *Aust. J. Agric. Res.*, **42**: 661-700.

IRRI. 1989. *Towards 2000 and beyond.* Manila Philippines.

Jain H.K. 1975. Breeding for yield and other attributes in grain legumes. *Ind. J. Genet.*, **35**: 167-187.

Jain H.K. 1986. Eighty years of post Mendelian breeding for crop yield: Nature of selection pressure and future potential. *Ind. J. Genet.*, **46**: (Suppl.) 30-53.

Jain H.K. and Kulshreshtha V.P.. 1976. Dwarfing genes and breeding for yield in bread wheat. *Z. Pflanzenzuchtung*, **76**: 102-112.

Jain H.K., Mukerjee B.K., Singh R.D. and Agrawal K.N. 1976. The present basis and possibilities of breeding for yield in maize. *Z. Pflanzenzuchtg*, **76**: 90-101

Johnson K.D., Richard R.A. and Turner N.C. 1983. Yield water relations and surface reflectance of near isogenic wheat lines differing in glaucousness. *Crop Sci.*, **23**: 318-325.

Khan M.A. and Tsunoda S. 1970. Evolutionary trends in leaf photosynthesis and related leave characters among cultivated wheat species and wild relatives. *Jpn. J. Breeding,* **200**: 133-140.

Khush G.S. and Peng S. 1996. Breeding yield frontiers of rice. *Proceedings of workshop on Revising yield potential in wheat, breeding yield barriers.* CIMMYT, Mexico.

Kirby E.J.M. and Appleyard M. 1987. Development and structure of wheat plant. *In*: "*Wheat Breeding - its scientific Basis*", (ed.) F.G.H. Lipton, Chapman and Hall, London and New York.

Kirby E.J.M., Siddique K.H.M., Perry M.W., Kaesehagen D. and Sterm W.R. 1989. Variation in spikelet initiation and ear development of old and modern wheat varieties. *Field Crops Res.*, **21**: 113-128.

Kobata T., Palta J.A. and Turner N.C. 1992. Rate of development of post-anthesis water deficits and grain filling of spring wheat. *Crop Sci.*, **32** 1238-1242.

Kropff M.J., Crossman K.G., Peng S., Mathows R.B. and Setter T.L. 1994. Quantitative understanding of yield potential. *In*: '*Breeding the Yield Barrier*', (ed.) K.G. Crossman. IRRI Philippines.

Karoda E., Oobawa T. and Ishhara K. 1989 Analysis on difference of dry matter production between rice cultivars with different plant height in relation to gross diffusion inside stands. *Jpn. J. Crop. Sci.*, **58**: 374-382.

Ladha J.K., Kirbs G.J.D., Peng S., Reddy K., Reddy P.M., Singh U. and Bannet J. 1995. Opportunities for increased nitrogen use efficiency from improved rice germplasm in lowland rice ecosystem. IRRI workshop on nutrient use efficiency of cropping systems.

Loss S.P. and Siddique K.H.M. 1994. Morphological and physiological traits associated with wheat yield increases in Mediterranean environments. *Adv. Agron.*, **52**: 269-276.

Loss S.P., Kirby E.J.M., Siddique K.H.M. and Perry M.W. 1989. Grain growth and development of old and modern Australian wheat. *Field Crops Res.*, **21**: 411-416.

Lupton F.G.H., Oliver R.H., Ellis F.B., Barnes B.T., House K.R., Wellonk P.J. and Taylor P.J. 1978. Root and shoot growth of semi-dwarf and taller winter wheats. *Ann. Appl. Biol.*, **77**: 129-144.

Mabino A. and Mal T., Obera K. 1983. Photosynthesis and ribulase-1, 5-biophosphate carboxylase in rice leaves. *Pl. Physics*, **73**: 1002-1007.

Morgan J.A. and Le Cain D.R. 1991. Leaf gas exchange and related leaf traits among 15 winter wheat genotypes. *Crop Sci.*, **31**: 443-448.

Ohno Y. 1976. Varietal differences in photosynthetic efficiency and dry matter production in Indian rice. *Tech. Bull. No. 9.* Tropical Agri. Res. Centre, Min. Agric. and Forestry, Japan.

Olugbeni L.B., Austin R.B. and Bingham J. 1976. The influence of temperature on the contribution of awns to wheat. *Ann. Appl. Biol.*, **84**: 241-250.

Paccaud F.X., Fossats A. and Cao H.S. 1985. Breeding for yield and quality in winter wheat: Consequences for nitrogen uptake and partitioning efficiency. *Z. Pflanzenzucht*, **94**: 89-100.

Perry M.W. and Antouno M. 1989. Yield improvement and associated characteristics of some Australian spring wheats introduced between 1860 and 1982. *Aust. J. Agric. Res.*, **40**: 457-472.

Peltonen Sainio P. 1990. Genetic improvements in the structure of oat stands in northern growing conditions during this century. *Plant Breeding*, **104**: 340-345.

Pheloung P.C. and Siddique K.H.M. 1991. Contribution of stem dry matter to grain yield in wheat cultivars. *Aust. J. Pl. Physiol.*, **18**: 53-64.

Rajaram S., Singh R.P. and Ginhel M. Ven. 1996. Approaches to breeding wheat for wide adaptation. *8th Assembly of wheat breeding society of Australia.* Canbara, Australia.

Rasmusson D.C. 1987. An evaluation of ideotype breeding. *Crop Sci.*, **27**: 1140-1146.

Rasmusson D.C. 1987. Ideotype and yield breeding in barley. *Proc 5th Barley Genetic Symp.*, Obayame, Japan.

Rasmusson D.C. and Gengenbach B.C. 1983. Breeding for physiological traits. *In*: "*Crop Breeding*", (ed.) Wood D.R., Am. Soc. Agron., USA.

Rawson H.M. and Evans L.T. 1971. The contribution of stem reserves to grain development in a range of wheat cultivars of different height. *Aust. J. Agric. Res.*, **4**: 389-397.

Rawson H.M., Hindmarsh J.H., Fischer R.A. and Stochman Y.M. 1983. Changes in leaf photosynthesis with plant ontogeny and relationship with yield per ear on wheat cultivars and 120 progeny. *Aust. J. Plt. Physiol.*, **10**: 503-514.

Richard R.A. 1996. Increasing yield potential source in its strength. *Proc. of workshop on Breeding Yield Barriers in Wheat*, CIMMYT, Mexico.

Richard R.A. and Passioura J.B. 1989. A breeding programme to reduce the diameter of the major xylem vessel in seminal roots of wheat and its effect on grain yield in rainfed environments. *Aust. J. Agric. Res.*, **40**: 943-950.

Siddique K.H.M., Kirby E.J.M. and Perry M.W. 1989a. Ear stem ratio in old and modern wheat varieties: Relationship with improvement in number of grains per ear and yield. *Field Crops Res.*, **21**: 59-78.

Siddique K.H.M., Belford R.K., Perry M.W. and Tennant D. 1989b. Growth, development and light interception of old and modern wheat cultivars in mediterranean environment. *Aust. J. Agric. Res.*, **40**: 473-487.

Simmons S.R. 1987. Growth development and physiology. "*Wheat and Wheat Improvement*", Agron Monogr. No. 13.

Sasahara T., Tabachashi T., Kayaba T. and Tsunoda S. 1992. A new strategy for increasing plant productivity and yield in rice. *Int. Rice Comm. Newsl.*, **41**: 1-4.

Slafer G.A. and Andreade F.H. 1993. Physiological attributes related to the generation of grain yield in bread wheat cultivars released at different ears. *Field Crops Res.*, **31**: 351-367.

Salfer G.A., Andreade F.H. and Feingold S.E. 1990a. Genetic improvement of bread wheat in Argentina: Relationbships between nitrogen and dry matter. *Euphytica*, **50**: 63-71.

Slafer G.A., Andrade F.H. and Satarre E.H. 1990b. Genetic improvement effects on pre-anthesis physiological attributes related to wheat grain yield. *Field Crops Res.*, **23**: 255-264.

Slafer G.A., Calderini D.F. and Mirales D.J. 1996. Generation of yield components and compensation in wheat. Opportunities for further increasing yield potential. *Proceeding of workshop "Breaking Yield Barriers in Wheat"*, CIMMYT. Mexico.

Turner N.C. 1986. Crop water deficits: A decade of progress. *Adv. Agron.*, **39**: 1-51.

Turner N.C. and Jones M. 1980. Turgour maintenance in osmotic adjustment. A review and evaluation, *In*: *"Adoption to Water and High Temperature Stress"*, (eds.) N.C. Turner, P.J. Kramer. pp. 203-215. Wiley, New York.

Wallace D.J., Oyburn J.L. and Munger H.M. 1972. Physiological genetics of crop yield. *Adv. Agron.*, **24**: 92-146.

Waller D. and Sivah M. 1986. Improving photosynthesis by genetic means. *Span*, **2**: 47-49.

Wingeter M., Mc. Cullough D.E. and Hunt L.A. 1989. Leaf gas exchange and plant growth of winter rye, triticale and wheat under contributing temperature extremes. *Crop. Sci.*, **29**: 1256-1260.

Yoshida S. 1981. *Fundamentals of rice crop sciences.* IRRI. Philippines.

Plant Breeding - Mendelian to Molecular Approaches
H. K. Jain and M. C. Kharkwal (eds.)

26

Mutation Breeding for Crop Improvement

M.C. Kharkwal[1], R.N. Pandey[2] and S.E. Pawar[2]

Abstract

Crop improvement programmes through induced mutations were initiated about seven decades ago, immediately after the discovery of mutagenic effects of X-rays on *Drosophila* by Muller in 1927, and barley and maize by Stadler in 1928. During 1950-60, several countries including China, India, the Netherlands, USA and Japan took up the task of crop improvement through mutation breeding approaches. A coordinated programme on mutation breeding was also initiated in rice in south east Asia in 1964 by IAEA. Over 2252 mutant varieties of crop plants including cereals, oilseeds, pulses, vegetables, fruits, fibres and ornamentals have been developed by the end of the 20th century. More than 60% of these mutant varieties were developed and released after 1985. While 1585 varieties were released as direct mutants, the rest were released through cross breeding with mutants. Most of the mutant varieties (around 89 %) have been developed using physical mutagens (X-rays, gamma rays, thermal and fast neutrons), with gamma rays alone accounting for the development of 60 % of the mutant varieties. A wide range of characters which have been improved through mutation breeding include plant architecture, yield, flowering and maturity duration, quality and tolerance to biotic and abiotic stresses. Mutation breeding has made a significant contribution to the national economy of the countries like China, India, Japan, Pakistan and USA. With the release of more than 305 mutant cultivars belonging to 56 plant species, India has also become a major recognised centre for work on induced mutations and the second largest contributor of the mutant varieties in the world. In recent years interest has rekindled in mutation research, since induced mutagenesis is gaining importance in plant molecular biology as a tool to identify and isolate genes and to study their structure and function. These studies will definitely have a major impact on the future crop improvement programmes.

Introduction

Hundred years ago in 1900, Hugo De Vries in the Netherlands, Carl Franz Joseph Correns in Germany and Erich von Tschermak in Austria simultaneously and independently

---

[1]Indian Agricultural Research Institute, New Delhi 110 012, India
[2]Bhabha Atomic Research Centre, Mumbai 400 085, India

rediscovered and verified Mendel's laws of inheritance. De Vries (1901, 1903), who is credited with the discovery of mutations, described them as suddenly arising changes in the organisms, which were inherited and produced relatively large effects on the phenotype. He coined the term 'mutation' and presented an integrated concept concerning the occurrence of sudden, shock-like changes (leaps) of existing traits, which lead to the origin of new species and variation. In his experiments with evening primrose (*Oenothera lamarkiana*), De Vries observed many aberrant types, which he called mutants. The notion of mutation which was used by De Vries to indicate sudden genetic changes as a major cause of evolution quickly became established. Early Mendelians, including De Vries rejected continuous variation as a possible source of evolution. In later years, the occurrance of spontaneous mutations was generally accepted, but it was shown that most mutations did not produce significant phenotypic effects. In contrast, major mutations or sudden discontinuous jumps that were hypothesized by De Vries to be responsible for the origin of new species were shown to be rather infrequent. De Vries in 1904 had referred to the new types of radiations [(X-rays and gamma rays, discovered earlier by scientists like Konrad von Roentgen (1895), Henery Becquerel (1896) and Pierre and Marie Curie (1897)] and suggested that they might be applied to induce mutations artificially. Because of De Vries's concept of mutations as the source of genetic variation and his early ideas about their potential value for plant breeding, his work around the turn of the twentieth century may be marked as the starting point of the discipline of plant breeding (Van Harten, 1998).

Crop improvement programmes through induced mutations were initiated about seven decades ago, immediately after the discovery of mutagenic actions of X-rays on fruit fly (*Drosophila melanggaster*) by Muller in 1927, and in maize, barley and wheat by Stadler in 1928 and 1930. These two discoveries resulted almost immediately in the practical recovery of some economically useful mutants in wheat (Delaunay, 1931 and Sapehin, 1930, 1936) and tobacco (Tollenaar, 1934, 1938). It is significant that an early useful induced tobacco mutant found its utility in the production of F_1 hybrid where the new mutant, *chlorina*, conferred some special quality to the marketed leaf. Later, Freisleben and Lein (1942) reported the induction of mildew resistance in barley by X-irradiation. A major stimulus for much of later work in mutation breeding was the classic paper of Gustafsson (1947) in which the very large number of mutations of barley, especially for chlorophyll characters and stiffness of straw (*erectoides*), was shown to respond in a variety of ways in different genotypes and under different external environments. The X-ray treatment of two-rowed barley by Nilsson-Ehle (1948) in Sweden was observed to produce mutants having characters of dense ears and very stiff straw which could withstand high doses of fertilizers and gave high yield. The Swedish programme covered several crop species and generated a lot of information on mutation breeding.

Auerbach (1941), was the first to report that mustard gas had a mutagenic effect on *Drosophila*, which was similar to that of X-rays on plants. Later, Oehlker (1943),

Auerbach and Robson (1946) and Gustafsson and Mackey (1948) proved that mustard gas was mutagenic in barley as well. Rapoport (1946, 1948) in Russia also demonstrated mutagenic effects of mustard gas and several other chemicals and established that alkylating agents are the most important group of chemical mutagens. Later, a large number of chemical mutagens and radiomimetic substances were discovered. During 1950-70, several countries including China, India, the Netherlands, USA and Japan took up the task of crop improvement through mutation breeding approaches and reported spectacular accomplishments of mutation breeding in evolving several superior crop varieties. In the USA, mutation breeding was employed with some success by Konzak (1954) and Frey (1955) on oats, and by Gregory (1955, 1965) on peanuts. Significant contributions in understanding basic and applied nature of mutation phenomenon on cereal and legume crops were made in Sweden (Gustafsson *et al.*, 1967) and Germany (Gaul *et al.*, 1969; Gottschalk and Wolfe, 1983). International Atomic Energy Agency (IAEA) established Plant Breeding and Genetics Division and a fully functional laboratory at Seibersdorf in Vienna in Austria. A coordinated programme on mutation breeding was initiated in rice in south-east Asia in 1964 by IAEA. In early days of mutation experiments, it was felt that mutation breeding would revolutionize plant breeding and breeders would be able to create new genes at will. However, as the knowledge accumulated (IAEA, 1976), it became clear that mutation breeding was not a magic wand that would create anything at any time. It is because of the fact that none of the mutagenic treatments in a plant system has so far been able to direct a specific change. As a result, following mutagenic treatments, a mixed bag of induced variants is found and many of these may not be of any value. Also, many events take place concurrently in a treated cell to produce variation at several loci within a genome. Because of these reasons it will be unrealistic to expect miracles out of mutation breeding programme. Nevertheless, the accomplishments of mutation breeding in evolving superior crop varieties and its role in basic studies confer it an honorable niche in the ever-going crop improvement programmes is several countries (Kharkwal, 1996).

Development of Crop Varieties

The number of mutant varieties officially released and recorded in the FAO/IAEA Mutant Varieties Database (MVD) before the end of the year 2000 is 2,252 (Table 1) (Muluszynski *et al.*, 2000). Majority of the mutant varieties have been released during the last two decades. The cumulative number of officially released mutant varieties in six continents of the world indicates that Asia tops the regional list closely followed by Europe and North America (Table 2). With more than one hundred mutant varieties each, China, India, Russia, the Netherlands, Germany, USA and Japan are the leading countries among approximately sixty countries engaged in the development and release of mutant varieties. France, Czechoslovakia, Canada, Italy, Pakistan, UK, Bulgaria and Poland are among the other important countries which have released more than thirty cultivars each through mutation breeding (Table 2).

Table 1. Crop varieties developed through induced mutations (till 2000)

Crop	No. of varieties released	Specific crop and number of varieties
Cereals and millets	1072	rice (434), barley (269), wheat (197), corn (68), durum wheat (25), oat (21), sorghum (13), pearl millet (5), minor millets (4), others (36)
Legumes	311	soybean (90), common bean (54), groundnut (48), pea (32), mungbean (19), *Lupinus* sp. (19), fababean (13), chickpea (11), cowpea (9), pigeonpea (5), blackgram (4), lentil (3), alfalfa (1), azuki bean (1), hyacinth bean (1), scarlet runner bean (1)
Oilseeds	59	*Brassica* sp. (27), sesame (16), linseed (7), white mustard (5), castor (4), sunflower (2), others (4)
Vegetables	66	tomato (13), chilli (10), lettuce (6), brinjal (4), chinese cabbage (4), onion (4), potato (4), sweet potato (4), cucumber (2), sugar beet (2), amaranthus (1), bitter gourd (1), kale (1), okra (1), radish (1), ridged gourd (1), snake gourd (1), spinach (1), turnip (1), taro (1), others(4)
Cash industrial crops	81	cotton (24), jute (11), tobacco (11), sugarcane (8), citronella (6), cassava (1), opium poppy (1), tea (1), others (18)
Fruits and forages	111	cherry (12), apple (9), citrus sp. (9), mulberry (7), pear (7), fodder beet (5), banana (2), ber (2), peach (2), pomegranate (2), *Trifolium* sp. (2), almond (1), apricot (1), fig (1), grapes (1), loquat (1), papaya (1), plum (1), raspberry (1), others (44)
Ornamental and Decorative plants	552	chrysanthemum (232), rose (61), dahlia (36), begonia (25), carnation (18), azalea (15), bougainvillea (12), portulaca (11), tulip (9), *Hibiscus* sp. (7), gladiolus (4), lotus (3), wild sage (3), lily (2), polyanthus (2), gerbera (1), others (111)

Most of the released varieties of crop plants belong to seed-propagated species. About 75 % varieties developed and released using induced mutations are of crop plants and 25 % are of the ornamentals. The list of crop and plant species with induced mutant varieties has already reached 175 taxa. This was mainly because of an increase in the application of mutation techniques for the improvement of ornamental plants in the developing countries, where they have become important "cash crops". Of the total 2,252 mutant varieties, 1585 mutant varieties were developed 'directly' after mutagenic treatment and selection in the subsequent generations. The remaining 667 new mutant varieties were developed 'indirectly' through cross breeding of mutants or already released mutant varieties as sources of desired characters in cross breeding programmes. Among the various mutagenic agents used for developing varieties, a great majority (1,411) out of 1,585 directly developed mutant varieties were obtained with the use of radiations, particularly gamma rays (910 mutants) as the mutagen.

Table 2. Mutant cultivars released in six continents and top seven countries of the world

Region wise			Country wise		
Region	No. mutant cultivars	% of total	Country[#]	No. mutant cultivars	% of total
Asia	1142	50.7	China P.R.	605	26.8
Europe	847	37.6	India	259	11.5
N. America	160	7.1	USSR+Russia	210	9.3
L. America	48	2.1	Netherlands	176	7.8
Africa	48	2.1	Germany	138	6.1
Australia	7	0.3	USA	125	5.5
	2252	100.0	Japan	120	5.3

[#]Other countries : France (39), Czech. Rep. (36), Canada (35), Italy (35), Pakistan (32), U.K. (32), Bulgaria (30) and Poland (30)

A perusal of the data on specific crops and number of mutant varieties released in the world (Table 1) indicates that the top ten ranks are occupied by some of the most important food and ornamental plant species in the world agriculture and economics e.g. rice (434), barley (269), chrysanthemum (232), wheat (197), soybean (90), corn (68), rose (61), common bean (54), groundnut (48) and dahlia (36).

Mutation breeding in rice (*Oryza sativa* L.) has been very successful worldwide, probably due to enormous attention of many researchers to this most important food crop, its diploid nature and self-fertilizing character. Out of 1072 mutant varieties of cereals, rice alone accounts for 434 varieties. The first widely known rice mutant cultivar has been cv. Reimei, a short straw (semi-dwarf) mutant from cv. Fujiminori, released in Japan in 1966. Several important cultivars were developed thereafter through cross-breeding with this cultivar (Van Harten, 1998). The mutant cultivar carried an allele that was allelic to the well known sd_1 [semi-dwarf allele in cultivar *Dee-geo-wu-gen* (DGWG), a spontaneous dwarf mutant discovered by Chinese scientists]. This mutant DGWG possessing dwarfness, stiff straw, fertilizer responsiveness, non-lodging and day length insensitivity as its most important traits, was the fore-runner of green revolution. It was used in crossing programmes for developing important cultivars like Taichung Native-1 and IR 8.

In barley (*Hordeum vulgare* L.), 269 mutant varieties for various plant characters have been isolated by various workers. Some of them proved to be superior to best control lines. For instance, the induced barley mutants further used by Swedish scientists in cross breeding gave rise to cultivars which performed very well (Gustafsson *et al.*, 1968). Similarly, a Danish dwarf mutant with remarkably high degree of lodging resistance had very good grain yield performance. Some of the important mutant varieties of barley are 'Pallas' (Sweden), Balber J. (Finland), 'Diamant' (Czechoslovakia), 'Trumpf' (German Democratic Republic), 'Goldspear' (UK), 'Pennrad Luther' (USA) and 'Betina' (France) (Van Harten, 1998). Most of the barley mutant cultivars are characterized with dwarf or semi-dwarf stature, low susceptibility to lodging due to stiff straw, higher tillering and better harvest index.

Plant Characters Improved by Induced Mutations

Plant Architecture

Dwarf and semidwarf mutants with reduced plant height belong to the most frequently arising types in mutation experiments. Micke *et al.* (1990) had reported about 336 cases of induced mutations for changed architecture of the plant among the registered mutant crop cultivars at that time. Reduced plant height having positive effect on yield via improved fertilizer response, increased tillering and lodging resistance was selected in cereals. In grain legumes, Micke (1988) had mentioned 57 examples of improved plant architecture, including 17 dwarf or bushy mutants, 17 with increased lodging resistance, 12 with improved plant architecture, 8 with erect growth habit and three with higher harvest index. The reduced height of most of the mutants was by reduction of internodal length or number of nodes. Semi-dwarfness and earliness are the characters most frequently described in released rice mutant cultivars. In 14 cases, both the characters were mutated simultaneously (Maluszynsky *et al.*, 1995). Dwarf plant type was among the main attributes of several varieties in barley (Kharkwal, 1996). An X-ray induced 'erectoides' mutant was developed into commercial variety 'Pallas' of barley in Sweden which is widely grown in Western Europe. It exhibits a pronounced lodging resistance and high productivity (Gustafsson and Eckman, 1967). In several oilseed crops, plant type was among the most frequently altered character obtained by induced mutations. Dwarf and semi-dwarf plant types were obtained through induced mutations in groundnut, rapeseed mustard, soybean and sesame (Bhatia *et al.*, 1999). Bushy and compact plant type was one of the most desirable attributes in several pulse crops. Dwarf, bushy, compact plant mutants have been induced in chickpea, mungbean, urdbean, cowpea, french bean and other pulse crops (Kharkwal, 2000). Mungbean mutant varieties NM19-19 and NM121-25 have short stature, erect plants with determinate growth habit (Malik, 1988). High yielding dwarf, determinate or semi-determinate mutants of cowpea were obtained through gamma radiation (Anonymous, 1987; Kharkwal *et al.*, 1988)

Maturity

Earliness has been an important objective of breeding in several crops grown under distinct ecological conditions. Changes in maturity period, leading to earliness, have been brought about by induced mutations in several crops. Reduced plant height had often been accompanied with early maturity. However, in several instances, earliness was induced without affecting the plant height. Mutant varieties database of seed propagated crops published in Mutation Breeding Newsletter Number 38, 1991 includes 166 varieties in different crops in which main improved character was earliness. Several additions, however, have been made in each crop during recent years. In most of the crops, earliness in maturity induced through mutagenic treatments has been in the range of 1-3 weeks, though higher range of earliness has also been achieved. Rice variety IIT 60, an EMS-induced mutant of IR 8 matures one month earlier than IR8 with same yield potential (Kharkwal, 1996) and the improved rice variety Yuanfengzao, released in China, matures 45 days earlier than the original variety IR 8 (Wang, 1991). In castor, the mutant variety 'Aruna' developed in India through neutron irradiation matures in 120 days compared to 270 days of the parent cultivar, HC-6 (Ankineedu *et al.*, 1968). In soybean, the mutant variety Raiden matures 25 days earlier than the original variety (Kawai and Amano, 1991). Several mutants showing extra-early maturity have been isolated and used in cross breeding in various crops. An ultra early mutant Ha-6222 of soybean matures in 80 days – 32 days earlier than the original parent, and provides an early parental stock for cross breeding (Wang, 1991).

Grain Yield

One of the most important characters for judging agronomic value of mutants is their yield potential. Therefore, improvement in grain/seed yield has always been the main objective in almost all the crops. The improvement in grain yield through induced mutations has been brought about by alterations in yield contributing traits. Ovule number per ovary, number of carpels per flower, number of spikelets per spike, number of flowers and pods per plant or number of tillers/branches per plant or other yield components have been improved through induced mutations in various crops. Isolation of micromutations or polygenic mutations for higher yield coupled with some other desirable attributes like disease and pest resistance has been reported in chickpea (Kharkwal, 2001). A higher yield/ha of the desired products like oil, protein and starch has also been attempted through induced mutations. A yield breakthrough in the modern high yielding varieties of crops, especially cereals, is attributed to desirable shift in the proportion of grains to straw i.e. higher harvest index. The high harvest index has been achieved in cereals and other crops through induced mutations. The groundnut variety TAG 24, developed at Trombay by using induced mutants in cross breeding, has harvest index as high as 57.5 % (Patil *et al.*, 1995).

Improvement in yield in different crops through induced mutations has been observed to range from marginal to substantially high. Yield increase by about 25 % in comparison to parent or the standard variety may be considered as good gain. However, yield increase of higher magnitude has also been achieved in some crops. Mungbean mutant varieties NIAB Mung 19-19 and 121-25 maturing in 65-70 days and NIAB Mung 20-21 and 13-1 maturing in 55-60 days have 30-40 % higher yield than parental types and 35-65 % higher yield than standard variety 6601 in Pakistan (Awan, 1991).

Grain Quality

The protein calorie deficiency in the diet of people in various areas of the world is one of the most important aspects of the complex socio-economic problems. Therefore, in recent years, there has been greater emphasis on the improvement of seed quality of cereals, pulses, oilseeds and other crops. The objectives have been to improve the protein content and specific amino acids in cereals and pulses, and oil content and fatty acid composition in oilseeds. However, seed quality may encompass all those qualities of seed that may increase its marketability, consumer acceptance and nutritive value. The alteration of seed colour and seed size for better acceptability has been achieved through induced mutations in several crops. The change of wheat seed coat colour from red to amber by gamma radiation resulting in the development of 'Sharbati Sonora' (Swaminathan *et al.*, 1968) is a classical example.

Increase in seed size has been achieved through induced mutations in several crops. Pigeonpea variety TT 6 with large seed size was improved from the small seeded variety T 21 through radiation induced mutation. About 25 % increase in seed size was achieved in this variety (Pawar *et al.*, 1991).

Improvement of seed quality in terms of increase in the quantity of protein and oils and the altered amino acid and fatty acid composition through induced mutations has been achieved in some cereals, pulses and oilseeds. Bhatia *et al.* (1978) reported an increase of 2 % in protein content in the radiation induced wheat mutant TW-1, compared to parent variety Kalyan Sona. However, the grain yield was 80-90 % of the parent cultivar. In oilseed crops, the increased oil yield is achieved by increasing seed or fruit yield and/or increasing the percentage of the oil (Robbelen, 1990). Several cultivars with higher yield and higher oil content have been developed with the help of induced mutations in different oilseed crops. In soybean mutant variety Longfu 81-9825 (Wang, 1991), the negative correlation between protein content and the oil content has been broken. The mutant has about 44 % and 21 % of protein and oil, respectively. Jensen (1991) reported high yielding high lysine mutants in barley; however, no varieties have yet been released. Bhatia *et al.* (1999) recently reviewed mutation breeding work for quality improvement in oilseeds. Mutants with altered fatty acid composition have been isolated in soybean, rapeseed, sunflower, linseed and minor oil crops. New cultivars having altered fatty acid composition have been released in rapeseed, sunflower and linseed. In linseed, fatty acid mutants have been utilized to develop 'linola' cultivars that yield edible oil suitable for human consumption with low (65-76 %) linolenic acid content (Green and Dribnenki, 1994; Dribnenki *et al.*, 1996). In soybean, a number of lines with altered meal and fatty acid contents have been developed by Ethyl-methane sulphonate mutagenesis (Kinney, 1995). Some of these mutant lines containing a low raffinosaccharide meal can be used for animal feed. Mutant lines with an improved fatty acid profile of the oil include high oleic and high stearic soybeans. Elevation of oleic acid levels from 60 to 85 %, with concomitant reduction in linoleic acid from 10 to 3 % has been achieved by haploid technology and mutagenesis in rapeseed (Wong and Swanson, 1991).

An increase in protein content associated with high yield in several chickpea mutant cultivars has been reported by Kharkwal (1998a). Similarly, a high yielding early maturing chickpea mutant with high protein content, named as Hyprosola has been released in Bangladesh (Sheikh *et al.*, 1979). Studies on seed yield and seed protein in induced mutants of mungbean showed that while mutants with higher grain yield were similar to the parent in seed protein content, there were other mutants showing higher seed protein but were lower in seed yield (Bhagwat *et al.*, 1979). Malik *et al.* (1979) selected mungbean mutants exhibiting desirable agronomic attributes and superiority in seed protein. Tickoo and Chandra (1996) reported two mutant lines with high protein content (30.4 % and 30.8 %) accompanied with high yield.

Stress Tolerance

Mutagenic treatments for inducing mutation for a specific trait often result in alteration of several traits. The mutants usually show simultaneous changes in several characters. Such changes may be due to either the pleiotropic effects of a single mutant allele or, more likely, due to simultaneous mutations in other loci (Kawai and Amano, 1991). Some of the crop varieties improved for yield or yield components through induced mutations have also shown improved tolerance to biotic and abiotic stresses. Several mutants selected for different morphological traits were later found to show resistance to certain diseases or insect pests and were therefore used as donor parent for resistance in cross breeding (Table 3). Chickpea mutant varieties, Pusa 408, Pusa 413 and Pusa 417 with resistance to *Ascochyta* blight, *Fusarium* wilt and other diseases and pests have been released for commercial cultivation in India (Kharkwal 1988 and 2001); similarly CM-72, CM-88, NIFA-95 and CM 1918 were released in Pakistan (Haq *et al.*, 1988). Mungbean mutant variety MUM-2 with high yield potential (Bahl and Gupta, 1982) was found to be resistant to MYMV, *Cercospora* leaf spot, leaf-crinkle, bacterial blight and *Macrophomina* blight (Gupta *et al.*, 1996). Rice mutant variety 'Rashmi' released in 1985 was found resistant to gall midge and blast, and was used as a donor for multiple resistance in rice breeding programme (Rao *et al.*, 1998). In the recently released radiation induced mutant cultivar CAZRI Moth-1 of moth bean (*Vigna aconitifolia*), simultaneous improvement in several traits like earliness, large seed size, higher number of pods, higher protein content and high resistance to YMV disease has been observed (Kumar, 2000).

Improvement of crop plants for specific stress tolerance has also been attempted and success to a certain extent achieved through induced mutations. Resistance to pathogens in plants can be complete or partial, durable or non-durable; its genetic control can be dominant or recessive, monogenic or polygenic. In monogenically controlled resistance, new pathotypes often arise spontaneously within short time by mutation and selection among the millions of spores that are produced by the pathogen. Consecutively, the new pathotypes are able to break down the recently introduced defense mechanism in the mutant variety. Protection by a monogenic system often lasts only for a short time. There are however, some very notable exceptions, which showed resistance lasting many years (Micke, 1993). The classical example is the *Mlo* resistance against all pathotypes of powdery mildew (*Erysiphe graminis*) in barley. Powdery mildew in barley has been the first disease against which X-ray induced resistance was obtained in cv. Heine's Haisa in Germany (Freisleben and Lein, 1942). The mutated locus was designated as *Mlo*. In later years, several powdery mildew resistant mutants were obtained. However, so far only one cultivar of agronomic importance with induced source of *Mlo* resistance has been commercialized. This EMS-induced powdery mildew resistant mutant cultivar 'Alexis' was released in 1986 as a malting barley variety in Germany (Van Harten 1998).

Table 3. Disease resistant varieties developed through induced mutations

Crop & Variety	Mutagenic treatment & parent variety	Disease against which improved	Reference
***Avena sativa* L. (Oat)**			
Florad Florida 500	th N, Floriland resistance cross	Victoria blight (*Helminthosporium victorae*)	Konzak *et al.* (1984)
***Cicer arietinum* L. (Chickpea)**			
Pusa 408 (Ajay)	γ-rays, G-130	blight, wilt	Kharkwal (1996)
Pusa 413 (Atul)	γ-rays, G-130	blight, wilt	Kharkwal (1996)
Pusa 417 (Girnar)	γ-rays, BG-203	wilt, blight	Kharkwal (1996)
CM-1918	γ-rays, C-727	blight	Awan, 1991
NIFA-95	γ-rays, line 6151	blight	MBNL No. 44 (1999)
***Oryza sativa* L. (Rice)**			
Biraj	X-rays, OC-1393	*Helminthosporium*	Kharkwal (1996)
Sarla (B-12-4)	T-9	YMV	Kharkwal (1996)
Savitri (Ponmani)	Pankaj × Jagannath	blast & sheath blight	Kharkwal (1996)
Indira	EMS, Tainan-31	blast, bacterial leaf blight	Kharkwal (1996)
UNP-9027	γ-rays, CR-1113	blast (*Pyricularia oryzae*)	Navarro-Alvarez & Solazur-Androvetto (1995)
***Pennisetum glaucum* (L.) R. Br. (Pearl millet)**			
NHB-5	35 kR γ-rays, Male sterile inbred line Tift23A	downy mildew (*Sclerospora graminicola*)	Kharkwal (1996)
Pusa-46	Irradiation of hybrid J104 × K559	downy mildew & *Pyricularia*	Kharkwal (1996)
***Sorghum bicolor* (L.) (Sorghum)**			
SPV-126 (CSV-9)	CS-3541 × CSV-4	charcoal rot, downy mildew, head moulds	Kharkwal (1996)
***Vigna mungo* (L.) Hepper (Blackgram)**			
BINA Mash-1(M-25)	γ-rays,BINA accession No. B-10	*Cercospora* leaf spot	Shaikh and Majid (1995)
***Vigna radiata* (L.) Wilczek. (Mungbean)**			
BINA Moog-2 (MC-246)	γ-rays, MB-55, [Mutant MB-55(4) × V-2273]	MYMV	Ahmed *et al.* (1995)
Pant Moong-2	10 kR γ-rays, ML-26	MYMV	Kharkwal (1996)
BM-4	EMS, T-44	*Macrophomina* blight, powdery mildew, MYMV	Kharkwal (1996)
MUM-2	EMS, K851	MYMV	Kharkwal (1996)
TARM-1,TARM-2 & TARM-18	Crosses with mutants	powdery mildew	Kharkwal (1996)

A number of mutants resistant to specific diseases or insect pests have been isolated in other crops too. The mutants have been directly released as varieties or more often used in cross breeding with elite varieties for transferring the resistance trait. Some of the crop varieties improved for tolerance to diseases are listed in Table 3.

Although mutations for improved resistance against various diseases have been reported in several crops, only a few mutants have shown real practical value. Although Jorgensen (1991) had some doubt on the role of induced mutations in the development of disease resistant varieties, Micke (1991), reported three cases where mutation breeding for resistance has been undisputedly successful. These include *Ascochyta* blight in chickpea, *Sclerospora graminicola* in pearl millet and *Verticillium* wilt in peppermint (a vegetatively propagated crop). The resistance in these crops remained effective for many years (Micke 1993). In recent years there has been more emphasis on breeding for durable resistance.

Induced mutagenesis for resistance to insect pests has met with very limited success. Reports on the mutant cultivars improved for insect resistance are very few. In cowpea, Pathak (1988) obtained two aphid resistant mutants, ICV-11 and ICV-12, following gamma irradiation of the cowpea variety ICV-1. The improved attributes of the mutants included high level of resistance to aphid, an increase in pod length and number of seeds per pod, semi-erect plant type and higher yield (Pathak, 1991).

Induction of mutation for abiotic stresses has been attempted and mutants for aluminium tolerance in banana (Matsumoto and Yamaguchi, 1991), chlorate tolerance in barley and field pea (Kleinhofs *et al.*, 1978), and salt and water stress tolerance in *Nicotiana plumbaginifolia* (Syukur *et al.*, 1991) have been obtained.

Mutants in Recombination Breeding

Although, of the 2252 mutant varieties, 1585 were released as direct mutants and 667 through crosses with various mutants (Nichterlein *et al.* 2000), of late the trend is changing towards more use of induced mutants in recombination breeding (Bhatia, 1991). From our own programme at Trombay, 17 of the 22 released mutant varieties of pulses, oilseeds, cereal and fibre crops were developed by using mutants in cross breeding. Mutants of no value often give promising recombinants when inter-crossed. Nichterlein (1999) enlisted a large number of new and better varieties of common bean (*Phaseolus vulgaris*) that were developed during 1960 - 1988 by using X-ray induced bush type mutants or their derivatives in the pedigree. Mutation breeding programme for durum wheat in Italy involving extensive selection and hybridization work with about 1000 induced mutants led to 11 registered varieties, five of which resulted from direct mutant selection and six from the cross breeding procedures (Scarascia-Mugnozza *et al.*, 1991). Fifteen cultivars of barley, released for cultivation in India were developed using the dwarf gene of RDB-1, a dwarf mutant cultivar derived from neutron irradiation of a tall local cultivar RS-17 (Bhatnagar, 1991). Mungbean variety NIAB Mung 98 was developed through hybridization between an induced mutant NM20-21 and an exotic AVRDC accession VC1482 E (Siddique *et al.*, 1999). Several mutants showing resistance to biotic and abiotic stresses have been isolated and recommended for use in cross breeding with elite cultivars in various crops.

In Vitro - Mutagenesis

With the development of *in vitro* technique for regeneration in several seed as well as vegetatively propagated crops, plant parts were subjected to mutagenic treatments either before or after the establishment of *in vitro* cultures. The subject of *in vitro* mutagenic treatments and isolation of mutants from treated cells or protoplasts in crops like tobacco, potato, tomato, carrot, soybean, rapeseed, sugarcane, maize and rice has been reviewed by Mathews and Bhatia (1983). Novak *et al.* (1986a,b), Novak and Micke (1988) and Cheng *et al.* (1990) have shown that the frequency and spectrum of spontaneous genetic variation (somaclonal variation) that occurs *in vitro* may increase considerably due to mutagenic treatments. Buiatti and Gimelli (1993) obtained mutant cultivars using *in vitro* mutagenesis in different ornamental crops. The work carried out by Novak *et al.* (1990; 1993) using radiation has resulted in the release of banana variety Novarica in Malaysia. The mutant variety Novarica matures 10 weeks earlier than the parent variety, Grande Naine. So far, *in vitro* mutagenesis has not produced results that could make significant economic impact. *In vitro* mutagenesis may help in selecting and developing resistant cultivars for biotic and abiotic stresses with proper screening techniques.

Induced Mutation in Molecular Studies

Induced mutagenesis is gaining importance as a tool in the identification of plant genes using molecular approaches (Shirley *et al.*, 1992). Introduction of recent innovative techniques of induction and detection of mutations have opened a vast area for their useful applications in plant breeding. Recent advances in molecular biology have led to the establishment of several techniques, which have helped researchers to redefine the scope, nature and applications of mutagenesis techniques. The main important established techniques are: (*a*) Site directed mutagenesis (SDM), (*b*) Random mutagenesis, (*c*) Insertional mutagenesis and (*d*) Systematic mutational analysis. Site directed mutagenesis (SDM) technique proves to be most effective for directional mutation. It is a new *in vitro* technique in which a specific change in specific location can be brought about in a gene sequence of interest, so that it has become a basic tool in gene manipulation and protein engineering. Numerous methods and modifications of SDM such as oligonucleotide directed mutagenesis, cassette mutagenesis and PCR based mutagenesis are available, though all have same basic principle of targetting one or few nucleotides. Applications of SDM involves a study of structural and functional changes in enzymes leading to protein engineering. It is also used for improving of storage protein quality, determination of the nature of host-pathogen relationship, development of resistance against pathogens, etc. Besides SDM, other methods like random mutagenesis, insertional mutagenesis and systematic mutational analysis are also available. As a result of improvements in the mutation detection methods in recent years, instead of detecting a mutation at phenotypic level, one can now chase it upto DNA level. Present detection methods use mechanisms like enzymatic cleavage, chemical modification, electrophoretic behaviour study, heteroduplex analysis, denaturing properties, PCR amplification and many more. The ability to detect single base changes is of fundamental importance and have great utilities in plant breeding.

Mutational analysis has been demonstrated as a powerful tool to dissect signaling pathways for plant defense responses (Dangl *et al.*, 1996; Yang *et al.*, 1997). The use of mutational analysis has helped to define the physical size, organization, and the sequence complexity of major clusters of pathogenesis-related genes or the fine gene structure. Such examples include downy mildew resistance genes in lettuce (Anderson *et al.*, 1996) and the *Mlo* locus for powdery mildew resistance in barley (Buschges *et al.*, 1997). Mutational analysis of flavonoid biosynthesis in barley was based on the Carlsberg Collection of 772 induced flavonoid mutants (Jende-Strid, 1993). *Arabidopsis thaliana* has become a model plant for various genetic investigations including defense responses (Glzenbrook *et al.*, 1997; Hutcheson 1998; Shah *et al.*, 1998). Extensive studies using mutants of *Arabidopsis* have indicated that the signal transduction pathways involved in defense mechanism may be similar across species (Yang *et al.*, 1997). Maluszynski *et al.* (1995) referred to molecular evidence for point mutation to 50 bp deletion in case of induced mutations by chemical mutagens, while physical mutagens tend to produce larger alterations from 17 bp to 20 cM in length (Weck *et al.*, 1996). The creation of large deletions may help gene isolation from stocks containing radiation-induced mutations via subtractive hybridization. In addition, insertional mutagenesis by T-DNA integration and transposable elements also represent a powerful alternative to the use of physical or chemical mutagens.

Economic Impact

The economic impact of mutant varieties of rice, in which 434 mutant varieties with improved characters such as semi-dwarfness, earliness, improved grain yield, disease tolerance and improved grain quality, etc. have been released, has been reviewed by Rutger (1992) and Maluszynski (1998). The majority of rice mutant varieties have been developed after selection in segregating mutated populations (direct mutants). The advantage of mutation technique for rapid development of improved varieties from locally well adapted rice germplasm has been recognized by breeders in Vietnam. Within only six years after mutagenic treatment in 1993, two improved varieties, 'TNDB-100' and 'THDB', with earliness and improved grain yield were released for Mekong Delta; these varieties have maintained tolerance to acid sulphate soil or soil salinity. Both varieties are grown on 220,000 ha (Rao *et al.*, 1998).

Van Harten (1998) discussed in detail the economic impact of short straw mutants and some other mutant types in barley. Mutant cultivars, such as 'Pallas', 'Diamant', 'Trumpf', 'Goldspear' 'Pennrad', 'Luther', 'Betina', etc. have been of immense economic value in Sweden, Denmark, Spain, UK, Germany, Czechoslovakia, USA and France.

Wang (1991) had reviewed the mutation-breeding programme in China and concluded that the use of induced mutations for crop improvement was a profitable approach. The estimated output of cereals, fibre and oilseeds had substantially increased and contributed to national economy. Breeders of the mutant varieties of rice, cotton, soybean, wheat, maize, legume forage and mulberry had won National Invention Awards.

Commercial success and economic impact of mutant varieties of grapefruit in Texas and pear in Japan has been highlighted by Nichterlein *et al.* (2000). Mutant varieties 'Star rubi' and 'Rio Red' of grapefruit are grown in almost 75 % of the total grapefruit area in Texas (Sauls, 1999). The mutant variety 'Gold Nijiseeiki' of Japanese pear is more resistant to black spot disease caused by *Alternaria alternata* than its parent. The additional annual income by growing this variety is estimated to be about US$ 50 million (Amano, 1997). In peppermint, the mutant variety, 'Todd's mitcham' forms bulk of the world's production of mint oil.

The Nuclear Institute for Agriculture and Biology (NIAB), Faisalabad, Pakistan, has released four disease resistant varieties of chickpea and nine varieties of mungbean for commercial cultivation. Three of these varieties viz. NM19-19, NM-54 and NM-92 have made significant impact on the total production by doubling the yield of mungbean in Pakistan. The variety NM-92, developed at NIAB in collaboration with the Asian Vegetable Research and Development Centre (AVRDC), Taiwan, having large seed size and durable resistance to yellow mosaic virus (YMV) and *Cercospora* leaf spot (CLS) diseases occupies about 51 % of the cultivated area under mungbean (Ali *et al.*, 1997).

The Indian Contribution to Mutation Breeding

One of the top two countries closely following China and contributing substantially to the number of released mutant varieties in the world, India occupies and an important place among the countries actively engaged in the genetic enhancement of its germplasm through induced mutation techniques. Mutation breeding in India has yielded considerable dividends both in enhancing our knowledge on various mutagenesis processes relevant to crop improvement and for developing improved varieties. In India, sustained efforts for crop improvement through induced mutations were initiated during the later half of 1950s. Some of the major research centres actively engaged in mutation breeding work in several crops and having contributed substantially in the development and release of a large number of mutant varieties are Indian Agricultural Research Institute (IARI), New Delhi; Bhabha Atomic Research Centre (BARC), Mumbai; Tamil Nadu Agricultural University (TNAU), Coimbatore; National Botanical Research Institute (NBRI), Lucknow and several others. Bhatia (1991), while reviewing the economic impact of mutant varieties in India, had put the number of the mutant varieties developed at 205 upto 1990. These varieties belong to cereals, grain legumes, oilseeds, fibre crops, vegetables and ornamentals. The highest number was of ornamentals followed by cereals and grain legumes. In cereals, maximum number of cultivars were released in rice followed by barley and wheat. Kharkwal (1996) enlisted a total of 150 mutant varieties of crops excluding ornamentals. An updated list of the 309 mutant cultivars belonging to 56 plant species approved/released in India by the end of the twentieth century is given in Tables 4 to 8.

Table 4. Number of released mutant varieties in fifty six different species in India

Latin name	Common name	No. var.	Latin name	Common name	No. var.
Ablemoschus esculentus (L.) Moench	Okra	1	*Matricario cammomilla*	German Chamomile	1
Arachis hypogaea L.	Groundnut	14	*Mentha spicata*	Spearmint	1
Bougainvillea spectabilis Wild.	Bougainvillea	10	*Momordica charantia* L.	Bitter gourd	1
Brassica juncea L.	Mustard	6	*Morus alba* L.	Mulberry	1
Cajanus cajan Milsp.	Pigeonpea	5	*Nicotiana tabacum* L.	Totacco	1
Capsicum annum L.	Chilli	1	*Oryza sativa* L.	Rice	41
Carica papaya L.	Papaya	1	*Papaver somaniferum* L.	Opium poppy	1
Chrysanthemum sp.	Chrysanthemum	46	*Pennisietum typhoides* L.	Pearl millet	5
Cicer arietinum L.	Chickpea	7	*Phaseolus vulgaris* L.	French bean	1
Corchorus capsularis L.	White jute	2	*Pisum sativum* L.	Pea	1
Corchorus olitorius L.	Tossa jute	3	*Plantago ovata* L.	Isabgol	1
Curcuma domestica Val.	Turmeric	2	*Polyanthus tuberosa* L.	Polyanthus	2
Cymbopogon winterianus Jowitt.	Citronella	6	*Portulaca grandiflora* L.	Portulaca	11
Cyamopsis tetragonoloba L.	Cluster bean	1	Ricinus communis L.	Castor	4
Dahlia sp.	Dahlia	11	*Rosa* sp.	Rose	15
Dolichos lablab L.	Hyacinth bean	2	*Sachharum officinarum* L.	Sugarcane	9
Eleusine coracana L.	Finger millet	4	*Sesamum indicum* L.	Sesame	3
Gladiolus L.	Gladiolus	2	*Solanum khasianum* Clarke	Khasianum	1
Glycine max L.	Soybean	3	*Solanum melongena* L.	Brinjal	1
Gossypium arborium L.	Desi cotton	1	*Sorghum bicolor* L.	Sorghum	3
Gossypium hirsutum L.	American cotton	8	*Setaria italica* L.	Foxtail millet	1
Hibiscus sinensis L.	Hibiscus	2	*Trichosanthus anguina* L.	Snake gourd	1
Hordeum vulgare L.	Barley	13	*Trifolium alexandrium* L.	Egyptian clover	1
Hyocyamus niger	Indian henbane	2	Triticum aestivum L.	Wheat	4
Lantana depressa L.	Wild sage	3	*Vigna aconitifolia* Jacq. M.	Moth bean	5
Lens culanaris Medik.	Lentil	3	*Vigna mungo* (L.) Hepper	Blackgram	7
Luffa acutangula Roxb.	Ridged gourd	1	*Vigna radiata* (L.) Wilczek.	Mungbean	13
Lycopersicon esculentum M.	Tomato	4	Vigna unguiculata Walp.	Cowpea	7

A perusal of the details of mutant cultivars of agricultural crop plants released or approved for cultivation in India provided in Tables 5 and 6 indicates that majority of these (239) were released between 1970-1990. The largest number of mutant cultivars have been produced in ornamentals (103) followed closely by cereals (69) and legumes (52).

Table 5. Mutant cultivars released between the years and groups of crops and species in India

Year wise			Group wise			
Released between the years	No. of cultivars	% of total	Group	No. of species	No. of cultivars	% of total
1941-1950	4	1.29	Cereals	7	69	22.33
1951-1960	0	0.00	Legumes	11	52	16.83
1961-1970	11	3.56	Oil seeds	5	30	9.71
1971-1980	108	34.95	Vegetables	8	12	3.88
1981-1990	131	42.39	Cash crops	6	24	7.77
1991-2000	47	15.21	Ornamentals	9	103	33.33
NA	8	2.60	Others	10	19	6.15
Total	309	100.0	Total	56	309	100.00

Table 6. Mutant varieties of different crops released for cultivation in India

Crop	Total no. of varieties released	Specific crop and no. of varieties
Cereals	69	rice (39), barley (13), pearl millet (5), finger millet (4), foxtail millet (1), wheat (4), sorghum (3)
Pulses	52	mungbean (13), blackgram (7), chickpea (7), cowpea (7), moth bean (5), pigeonpea (5), lentil (3), lablab bean (2), cluster bean (1), common bean (1), pea (1)
Oilseeds	30	groundnut (14), mustard (6), castor bean (4), sesame (3), soybean (3)
Fibre crops	14	American cotton (8), tossa jute (3), white jute (2), *desi* cotton (1)
Vegetables	12	tomato (4), turmeric (2), bitter gourd (1), brinjal (1), green pepper (1), okra (1), ridge gourd (1), snake gourd (1)
Cash crops	10	sugarcane (9), tobacco (1)
Medicinal crops	16	citronella (8), German chamomile (1), Indian henbane (2), isabgol (1), khasianum (1), opium poppy (2), Spearmint (1)
Fruit trees	2	mulberry (1), papaya (1)
Forage crops	1	Egyptian clover (1)
Ornamentals	103	chrysanthemum (46), rose (16), dahlia (11), portulaca (11), bougainvillea (10), wild sage (3), gladiolus (2), *Hibiscus* sp. (2), tube rose (2)
Total	309	

The success story of mutation breeding in ornamentals and horticultural plants in India, which is at the threshold of a boom in the export market in floriculture and horticulture, is particularly important (Datta, 1997). While chrysanthemum with 46 mutants tops the list of ornamentals, among cereals the leading position with 39 mutant cultivars is occupied by rice, the most important food crop of the world. Keeping in view, the rarity with which mutations occur and the general deleterious nature of induced mutants, the success and achievements of mutation breeding in development of commercially released varieties in such a large and wide range of plant species in India, are thus significant. This is particularly true in Indian situation where all varieties (except ornamental) irrespective of the method through which they are developed have to go through the All India Co-ordinated Trials grid before getting identified/approved for release. As is to be expected, the practical utility of induced mutation technique has been conspicuous only where serious efforts, with a well-defined objectives, have been made. However, inspite of a large number of mutant cultivars belonging to crop species released/approved for cultivation in India, the impression that contributions of mutation breeding have been limited is obviously erroneous particularly when one views it in relation to the budgetary and manpower deployment. Even the most vigorous group engaged in applied mutation breeding at the peak of its activity, has been conspicuously small in comparison with the group engaged in the conventional breeding approach. It is already known that mutation breeding is expected to make a contribution primarily as an important adjunct to the conventional breeding approach.

Mutant varieties like K-84, Jagannath, IIT-48, IIT-60, Sattari and Keshari of rice (Chakrabarti, 1996); Aruna of castor; TG-1, TG-17, TAG-24, Co-2 of groundnut; Pusa-408, Pusa-413, Pusa-417 of chickpea; Co-4, Pant Moong-2, MUM-2, TARM-1 of mungbean; Co-4, TAU-1, TPU-4 of blackgram; and Maru Moth-1 of moth bean (*Vigna aconitifolia*) are among the most important varieties of economic importance released in India (Kharkwal *et al.*, 2001). However, authentic information on the area covered under these cultivars is unfortunately lacking. It is also, however, accepted that some of the mutant cultivars of rice, wheat, castor, groundnut, chickpea, mungbean and urdbean did find wide acceptability and cover a sizable acreage on their release for cultivation in parts of India.

A close examination of the type of mutagens used and the number of mutant cultivars released/developed in India (Table 7), indicates that largest number of mutant varieties (201 or 65 %) have been induced by physical mutagens, gamma rays being the most commonly used and successful (168 cultivars). This does not, however, mean that chemical mutagens are less effective. On the contrary, most of them are very effective, but if high doses are applied the desired mutations may be accompanied by too many undesired ones (Sybenga, 1983). The survey also indicates that major gains have resulted from the use of induced variability when it was fully integrated with crop breeding and more varieties have resulted by using induced mutants in cross breeding programme rather than processing them directly into mutant varieties.

The main improved characteristics of mutant derived cultivars, as reported by the breeders and summarised in Table 7, indicate that improved plant type, higher yield are the most frequent traits in a new variety released for commercial cultivation and marketing. Altered maturity duration, increased to resistance biotic and abiotic stresses and quality characters are some other important traits, which ultimately led to higher productivity and acceptability of these mutant cultivars.

Table 7. Mutagens used/origin and main improved characters reported for mutant cultivars released in India

Mutagens/origin	No. of mutants	Main attribute	No. of occurrence
Gamma rays	169	High yield	86
X-rays	26	Early maturity	65
Neutrons	7	Disease resistance	57
Ethyl methane sulphonate (EMS)	15	Quality characters	39
Dimethyl sulphate (DMS)	4	Grain quality	67
Ethylene imine (EI)	2	Abiotic stress resistance	65
Sodium azide (NaN_3)	2	Improved plant type	181
Other mutagens	25	Other	9
Cross bred	47		
Natural mutants	12		

The mutant cultivars have contributed immensely in augmenting the efforts of Indian plant breeders in achieving the target of self-sufficiency in food production and strong economic growth of the country. Mutation breeding has made significant contribution in increasing the production of rice, groundnut, castor, chickpea, mungbean and urdbean in the Indian subcontinent. In India, ten varieties of mungbean have been released for cultivation for different agro-climatic regions. The varieties Co-4, Pant Mung-2 and TAP-7, though released in early eighties, are still being grown widely in the country. The variety TARM-1, resistant to powdery mildew and YMV diseases, is the first of its kind to be released for *rabi*/rice fallow cultivation. Four of the six mutant varieties of blackgram (urdbean) released in India have been developed at the Bhabha Atomic Research Centre (BARC), Mumbai. One of these mutant varieties, TAU-1, has become the most popular variety in Maharashtra occupying an area of about 5 lakh hectare (over 95 % of the total area under urdbean cultivation in Maharashtra). Since 1990, the Maharashtra State Seed Corporation, Akola, has distributed about 200,000 quintals of certified seeds of TAU-1 to the farmers, which has resulted in an additional production of about 1,29,000 quintals of urdbean annually in Maharashtra. The notional income generated by additional production amounts to Rs. 258 crores annually (Kharkwal *et al.*, 2001).

Future Scope

Bhatia (2000), while presenting his views on the future scope of induced mutation studies, visualised a limited scope in the new millenium for the kind of experiments that were carried out in the past. According to him, future mutation experiments should be aimed at improving the root characters, nodulation in legumes, alteration of fatty acid composition in oilseeds, host pathogen interactions, photo-insensitivity and apomixis in crop plants.

Control of gene silencing in transgenics is another area where induced mutation could be used. Useful induced mutants for these traits can be obtained, if sustained efforts are made using proper screening techniques to identify the mutants. A beginning has already been made in some of these areas. Mutations affecting root length, root hair density, number of seminal roots, root tips and zone of elongation have been reported in barley and maize (Szarejko and Maluszynski, 1980; Feix *et al.*, 1997). Bhatia *et al.* (2000) summarized a large number of spontaneous and induced mutations for nodulation in different grain legumes. Mutants with altered fatty acid composition in soybean, sunflower, *Brassica* species, linseed and other oilseed crops have been reviewed (Bhatia *et al.*, 1999). Day length insensitive mutants in jute (Joshua and Thakre, 1986) and *Sesbania rostrata* (Joshua and Ramani, 1993) have also been reported.

Induced mutations play an important role in plant genome research to understand the function of genes with an aim to improve food security. Recent progress in the sequencing of entire genomes has led to the identification of many genes in model species. A future challenge will be to understand the functions of all these genes. In reverse genetics, radiation, chemical and transposon mutagensis are used to discover the unknown functions of genes by knocking out genes and assigning known DNA sequences to the mutant phenotypes.

In future also, mutation breeding is expected to continue making significant contribution to plant breeding primarily as an important adjunct to the conventional and molecular breeding approaches. The direct use of mutations is a very valuable supplementary approach to plant breeding, particularly when it is desired to improve one or two easily identifiable characters in an otherwise well-adapted variety. The main advantages are firstly, that the basic genotype of the variety is usually altered only slightly while adding the improved character(s); secondly, that the time required to breed the improved variety can be relatively shorter. However, in the long run the use of induced mutations in various cross-combinations in hybridization programmes is probably more important than the direct use of mutants. A survey of the achievements of mutation breeding and related literature during the past 75 years of its history (Kharkwal *et al.*, 2001) suggest that major gains have resulted from the use of induced variability when it was fully integrated with conventional crop breeding programmes. This is evident from the fact that many more mutant varieties have resulted by using induced mutants in crossing programmes in comparison to those, which resulted through the direct release of mutants.

A brief description of the crop cultivars developed, released and/or approved from induced mutants or from the use of such mutants in cross-breeding in 56 plant species for cultivation in India are listed in Table 8.

Table 8. Mutant cultivars of various crops released/approved for cultivation in India

S. No.	Crop / Mutant variety	Year of release	Institution	Mutagen & parent variety	Main characters induced / improved
1.	***Abelmoschus esculentus* L. Moench. (Okra)**				
	MDU-1	1978	TNAU, AC&RI, Madurai	0.04% DES Pusa Sawani	yield
2.	***Arachis hypogaea* L. (Groundnut)**				
	BG-1	1979	BAU, Ranchi	γ-rays	very bold pods and kernels
	BG-2	1983	BAU, Ranchi	450 Gy γ-rays, 41-C	bold kernels
	Co-2	1985	TNAU, Coimbatore	0.2% EMS, Pol-1	yield
	Kaushal	1985		natural mutant of Type-28	resistance for leaf spot and rust diseases
	MH-2	1973	HAU, Hisar	Selection from Gujarat dwarf mutant	yield and resistance for Tikka disease
	TG-1 (Vikram)	1973	BARC, Mumbai	150 Gy X-rays, Spanish Improved	bold kernel and 60 days dormancy
	TG-3	1987	BARC, Mumbai	150 Gy X-rays, Spanish Improved	more number of branches
	TG-17	1985	BARC, Mumbai	dark green leaf mutant × TG-1	medium bold pods and 30 days dormancy
	TGS-1 (Somnath)	1989	BARC, Mumbai	TG18A × M-13	early large kernels and spreading habit
	TKG-19A	1994	BARC, Mumbai	TG-17 × TG-1	bold kernels and 30 days dormancy
	TG-22	1992	BARC, Mumbai	Robut 33-1 × TG-17	medium size kernels and 50 days domrancy
	TAG-24	1991	BARC, Mumbai	(TGS-2 × TGE-1	early- semi-dwarf, drought tolerant and high harvest index (52%)
	TG-26	1995	BARC, Mumbai	BARCG-1 × TG-23	semi-dwarf habit, early, high harvest index (50%) and 20 days dormancy

	TMV-10	1975		A natural mutant from Argentina	large pods and kernels
3.	***Bougainvillea* sp. (Bougainvillea)**				
	Arjuna	1974	NBRI, Lucknow	γ-rays, Partha	variegated leaves
	Jaya	1977		γ-rays, Jayalakshmi	ornamental type
	Jayalaxmi Variegata	1977		γ-rays, Jayalakshmi	ornamental type
	Lady Hudson of C.V.	1979		γ-rays, Lady Hudson of Ceylon	ornamental type
	Los Banos Variegata	1990		γ-rays, Los Banos Beauty	leaf colour
	Mahara Variegata	1990		γ-rays, Mahara	variegated leaves
	Pallavi	1986		γ-rays, Roseville's Delight	variegated leaves
	Poultoni Variegata	1981		γ-rays, Poultoni	variegated leaves
	Silver Top	1978		γ-rays, Versicolour	ornamental type
	Suvarna	1981		γ-rays, Lady Hudson of Ceylon	flower colour
4.	***Brassica juncea* Coss. (Mustard)**				
	RL-1359	1987	PAU, Ludhiana	RLM-514 × Varuna	short duration, high yield, bold seeds, oil content, tolerant to aphids
	RLM-198	1977	PAU, Ludhiana	Irradiation, RL-18	oil content, yield, tolerant to Aphids and *Alternaria* blight
	RLM-514	1984	PAU, Ludhiana	200 kR γ-rays, RL-18	yield, bold grain, early maturity, and high oil content
	RLM-619	1985	PAU, Ludhiana	200 kR γ-rays, RL-18	tolerant to white rust, downey mildew and aphids, early maturing
	TM-2	1987	BARC, Mumbai	X-rays, RL-9	early maturity (90-95 days), black seeded, high oil and yield
	TM-4	1987	BARC, Mumbai	TM-1 × Varuna	early maturity (90-95 days), yellow seeded, high oil & yield

(*Contd.*)

(*Table 8 Contd.*)

S. No.	Crop / Mutant variety	Year of release	Institution	Mutagen & parent variety	Main characters induced / improved
5.	***Cajanus cajan* Millsp. (Pigeonpea)**				
	Co-3	1977	TNAU, Coimbatore	0.6% EMS, Co-1	high yield, resistant to root rot, pod borer and determinate type
	Co-5	1985	TNAU, Coimbatore	16 kR γ-rays, Co-1	early, resistant to pod fly, root rot, sterility mosaic disease and drought tolerant
	TAT-10	1985	BARC, Mumbai	mutant TT-2 × mutant TT-8, T-21	extra early (110-115 days) and medium large seeds
	TT-5	1984	BARC, Mumbai	1.5 kR fast neutrons, T-21	early maturity and bold seed
	TT-6 (T. Vishaka-1)	1983	BARC, Mumbai	1.5 kR fast neutrons, T-21	early maturity (135 days) with bold seeds
6.	***Capsicum annuum frutescens* L. (Chilli)**				
	MDU-1	1976	TNAU, AC&RI, Madurai	30 kR γ-rays, K1	high yielding and compact dwarf type
7.	***Carica papaya* L. (Papaya)**				
	Pusa Nanha	1986	IARI, New Delhi	γ-rays, Ranchi	dwarf plant type
8.	***Chrysanthemum* sp. (Chrysanthemum)**				
	Agnisikha	1987	NBRI, Lucknow	15-25 Gy γ-rays, D-5	flower colour
	Alankar	1982	NBRI, Lucknow	15 Gy γ-rays, D-5	flower colour
	Anamika	1975	NBRI, Lucknow	15 Gy γ-rays, E-13	flower colour
	Aruna	1974	NBRI, Lucknow	15 Gy γ-rays, Ashankit	flower colour
	Asha	1975	NBRI, Lucknow	15 Gy γ-rays, Hope	flower colour
	Ashankit	1974	NBRI, Lucknow	15 Gy γ-rays, Undaunted	flower colour
	Basant	1975	NBRI, Lucknow	10 Gy γ-rays, Paul	flower colour
	Basanti	1979	NBRI, Lucknow	15 Gy γ-rays, E-13	flower colour
	Batik	1994	NBRI, Lucknow	20 Gy γ-rays, Flirt	flower colour
	Colchi Bahar	1985	NBRI, Lucknow	Colchicine, Sharad Bahar	flower colour
	Cosmonaut	1984	NBRI, Lucknow	15-25 Gy γ-rays, Nimrod	flower colour
	Gairik	1974	NBRI, Lucknow	10 Gy γ-rays, Belur Math	flower colour
	Hemanti	1979	NBRI, Lucknow	15 Gy γ-rays, Megami	flower colour

Himani	1974	NBRI, Lucknow	15 Gy γ-rays, E-13	flower colour
Jhalar	1975	NBRI, Lucknow	15 Gy γ-rays, Undaunted	flower colour
Jugnu	1991	NBRI, Lucknow	15 Gy γ-rays, Lalima	flower colour
Kanak	1975	NBRI, Lucknow	15-20 Gy γ-rays, Undaunted	flower colour
Kansya	1974	NBRI, Lucknow	15 Gy γ-rays, Roseday	flower colour
Kapish	1974	NBRI, Lucknow	15 Gy γ-rays, E-13	flower colour
Kumkum	1987	NBRI, Lucknow	20-25 Gy γ-rays, M-71	flower colour
Kunchita	1974	NBRI, Lucknow	15 Gy γ-rays, Undaunted	flower colour
Lohita	1974	NBRI, Lucknow	15 Gy γ-rays, E-13	flower colour
Man Bhawan	1982	NBRI, Lucknow	15 Gy γ-rays, Flirt	flower colour
Navneet	1987	NBRI, Lucknow	15 Gy γ-rays, K. Mauve	flower colour
Navneet Yellow	1993	NBRI, Lucknow	15 Gy γ-rays, Navneet	flower colour
Nirbhaya	1975	NBRI, Lucknow	15 Gy γ-rays, Undaunted	flower colour
Nirbhik	1975	NBRI, Lucknow	15 Gy γ-rays, undaunted	flower colour
Pingal	1974	NBRI, Lucknow	15 Gy γ-rays, Pink Casket	flower colour
Pitaka	1978	NBRI, Lucknow	15 Gy γ-rays, Kansya	flower colour
Pitambar	1978	NBRI, Lucknow	15 Gy γ-rays, Otome Zakura	flower colour
Purnima	1978	NBRI, Lucknow	15 Gy γ-rays, Otome Zakura	flower colour
Raktima	1998	NBRI, Lucknow	15 Gy γ-rays, Shyamal	flower colour
Rohit	1979	NBRI, Lucknow	15 Gy γ-rays, K. Smith	flower colour
Shabnam	1987	NBRI, Lucknow	15 Gy γ-rays, D-5	flower colour
Shafali	1975	NBRI, Lucknow	20 Gy γ-rays, Undaunted	flower colour
Sharad Bahar	1992	NBRI, Lucknow	15 Gy γ-rays, Sharad Mala	flower colour
Sheela	1985	NBRI, Lucknow	20-25 Gy γ-rays, Himani	flower colour
Shukla	1974	NBRI, Lucknow	15 Gy γ-rays, H. Gubby	flower colour
Shveta	1974	NBRI, Lucknow	15 Gy γ-rays, Fish Tail	flower colour
Sonali	1990	NBRI, Lucknow	20 Gy γ-rays, Ratna	flower colour
Subarna	1990	NBRI, Lucknow	20 Gy γ-rays, Flirt	flower colour
Surekha Yellow	1992	NBRI, Lucknow	15 Gy γ-rays, Surekha	flower colour
Swarnim	1975	NBRI, Lucknow	15 Gy γ-rays, Undaunted	flower colour
Tamra	1974	NBRI, Lucknow	20 Gy γ-rays, Goldie	flower colour

(Contd.)

(*Table 8 Contd.*)

S. No.	Crop / Mutant variety	Year of release	Institution	Mutagen & parent variety	Main characters induced / improved
	Taruni	1979	NBRI, Lucknow	25 Gy γ-rays, K. Smith	flower colour
	Tulika	1985	NBRI, Lucknow	15 Gy γ-rays, M-24	flower colour
9.	***Cicer arietinum* L. (Chickpea)**				
	Pusa-408 (Ajay)	1985	IARI, New Delhi	60 kR γ-rays, G-130	high yield, profuse branching, resistant to *Aschochyta* blight, pod borer and nematode
	Pusa-413 (Atul)	1985	IARI, New Delhi	60 kR γ-rays, G-130	high yield, resistant to *Fusarium* wilt, stunt virus, root rot, pod borer and nematode
	Pusa-417 (Girnar)	1985	IARI, New Delhi	60 kR γ-rays, BG-203	high yield, resistant to *Fusarium* wilt, stunt virus, color rot, pod borer and nematode
	RS-11	1970	ARS, Durgapura, Jaipur	RS-10	white flowered and resistant to drought
	RSG-2 (Kiran)	1984	ARS, Durgapura, Jaipur	4.5 × 1012 n/cm^2 fast neutron, RS-10	erect, earlier maturity, salinity tolerant and high yield
	WCG-1 (Sadbhawna)	1997	WC, GBPUA&T, Modipuram	30 kR γ-rays, C-235	yield, early maturity and robust plant
	WCG-2 (Surya)	1999	WC GBPUA&T, Modipuram	30 kr γ-rays, G-130	yield. wilt resistance and white flowered
10.	***Corchorus capsularis* L. (White jute)**				
	Padma (Hyb C)	1983	JRI, Barrackpore	JRC-6165 × JRC-412	tolerance to water logging and less affected by diseases
	Shyamali (JRC-7447)	1974	JRI, Barrakpore	25 kR X-rays, JRC-212	yield and nitrogen fertilizer responsive
11.	***Corchorus olitorius* L. (Tossa jute)**				
	IR-1	1978		γ-rays, JRO 632	plant vigour
	Mahadev (TKG-40)	1983	BARC, Mumbai	th. neutrons, Virescent × involute leathery (JRO-632)	higher fibre yield (10-14%) and early flowering

	Savitri	1985	GBPU, Pantnagar	cross of tobacco leaf × long internode mutant	yield, resistant to aphids and yellow mite
12.	***Curcuma domestica* Val. (Turmeric)**				
	BSR 1	1986	ARS, Bhawanisagar	X-rays, Erode local	rhizome colour
	Co 1	1983	TNAU, Coimbatore	X-rays, Erode local	rhizome colour
13.	***Cymbopogan winterianus* Jowitt. (Citronella)**				
	Bhanumati (OJC-11)	1987		X-rays, Subirrsourav (CKS-CW-S-1)	oil content
	Bibhuti (OJC-5)	1987		X-rays, Subirrsourav (CKS-CW-S-1)	oil content
	Jalpallavi	1998	CIMAP, Lucknow	spontaneous mutant	oil content
	Manjari	1998	CIMAP, Lucknow		oil content
	Niranjan (OJC-6)	1987		X-rays, Subirrsourav (CKS-CW-S-1)	oil content
	Phullara (OJC-22)	1987		X-rays, Subirrsourav (CKS-CW-S-1)	oil content
	Sourav (OJC-3)	1987		X-rays, Subirrsourav (CKS-CW-S-1)	oil content
	Subir (OJC-31)	1987		X-rays, Subirrsourav (CKS-CW-S-1)	oil content
14.	***Cyamopsis tetragonoloba* L. (Cluster bean)**				
	Kanchan Bahar	1996	ARS, Durgapura, Jaipur	0.6% EMS, Durga Bahar	yield
15.	***Dahlia* sp. (Dahlia)**				
	Bichitra	1978	NBRI, Lucknow	γ-rays, Kenya	plant architecture
	Black beauty	1978	NBRI, Lucknow	Black Out	plant architecture
	Happiness	1978	NBRI, Lucknow	Croydon Monarch	plant architecture
	Jayaprakash	1978	NBRI, Lucknow	Croydon Monarch	plant architecture
	Jubilee	1978	NBRI, Lucknow	γ-rays, Kenya	plant architecture
	Jyoti	1978	NBRI, Lucknow	γ-rays, Kenya	plant architecture
	Netaji	1978	NBRI, Lucknow	Eagle Stone	plant architecture
	Pearl	1978	NBRI, Lucknow	Eagle Stone	plant architecture

(Contd.)

(*Table 8 Contd.*)

S. No.	Crop / Mutant variety	Year of release	Institution	Mutagen & parent variety	Main characters induced / improved
	Pride of Sindri	1978	NBRI, Lucknow	γ-rays, Kenya	plant architecture
	Twilight	1978	NBRI, Lucknow	γ-rays, Kenya	plant architecture
	Vivekananda	1978	NBRI, Lucknow	Croydon Master	plant architecture
16.	***Dolichos lablab* L. (Field bean)**				
	Lablab Co-9	1980	TNAU, Coimbatore	natural mutant, Pandal Avarai	erect, bushy, bold seeds and fleshy pods
	Lablab Co-10	1984	TNAU, Coimbatore	24 kR γ-rays	high yielding and photo-insensitive
17.	***Eleusine coracana* L. (Finger millet)**				
	CO-3	1942		(CO-1)	plant height 90-95 cm and maturity 110-115 days
	Dibya Sinha	1976		(AKP-7)	tolerant to blast and stem borer and early maturity
	Hagari-1	1941		(Gidda Aryan)	suitable for *kharif* tracts under irrigated conditions
	K-6	1982		(A natural mutant of a local variety)	very early maturity (75-80 days) and short plant height
18.	***Gladiolus* sp. (Gladiolus)**				
	Shobha	1980	IIHR, Bangalore	γ-rays, Wild Rose	flower colour
	Tambari	1991		γ-rays, Oscar	flower colour
19.	***Glycine max* L. Merr. (Soybean)**				
	Birsa Soybean-1	1983	BAU, Ranchi	mutant of Sepaya Black (fodder variety)	erect, dwarf, early, resistant to bacterial pustules and tolerant to foliage diseases
	NRC-2	1993	NRCG, Indore	radiation induced mutant of Bragg	high yield
	VL Soya-1	1985	VPKAS, Almora	spontaneous mutant of Bragg	tolerant to *Cercospora* leaf spot, determinate, black and bold seeds

20.	***Gossypium arboreum* L. (*Desi* cotton)**				
	DS-1	1985	HAU, Hisar	γ-rays, G-27	resistant to jassids, spotted boll-worm, pink bollworm and bacterial blight
21.	***Gossypium hirsutum* L. (American cotton)**				
	Badnawar 1	1961	IPSI, Indore	cross with Indore-2	yield
	Indore 2	1950	IPSI, Indore		x-rays, MU-4
	Khandwa-2	1971	IPSI, Indore	cross with Indore-2	yield
	MA 9	1948	Mysore	X-rays, Co-2	drought resistant
	MCU-7	1984	AC&RI, Madurai	80 kR X-rays, L 1143 EE	early, yield, increased spinning capacity and resistant to jassids
	MCU-10	1984	AC&RI, Madurai	30 kR γ-rays, MCU-4	resistant to blackarm disease, long staple and drought tolerant
	Pusa Ageti	1978	IARI, New Delhi	γ-rays, Stoneville 213	ginning capacity
	Rasmi	1976	IARI, New Delhi	30 kR γ-rays, MCU-5	day neutral, high yield and superior quality
22.	***Hibiscus* sp. (Hibiscus)**				
	Anjali	1987		γ-rays, Alipore Beauty	flower colour
	Purnima	1979		γ-rays chronic, Alipore Beauty	flower and leaf colour
23.	***Hordeum vulgare* L. (Barley)**				
	BH-75	1983	HAU, Haryana	RD-150 (RDB-1 × EB-795) × Ahor-131/68	early maturity, profuse tillering, resistant to yellow rust and CCN
	DL-253	1981	IARI, New Delhi	20kR γ-rays + 0.1% EMS, Ratna	high tillering, high yield, resistant to covered and loose smut and yellow rust
	K-2578	1980	CSAU, Kanpur	RBD-1 × Vijaya	medium tall, high yield potential and long ears

(*Contd.*)

(*Table 8 Contd.*)

S. No.	Crop / Mutant variety	Year of release	Institution	Mutagen & parent variety	Main characters induced / improved
	Karan-3	1984	IARI, RRS, Karnal	RDB-1 × EB-7576) × Riso Mut. 1508	dwarf, erect leaves, non-lodging and hull-less amber colcoured grains
	Karan-4	1984	IARI, RRS, Karnal	RDB-1 × EB-7576	semi-dwarf, lodging resistant, erect leaves and hull-less amber coloured grains
	Karan-201	1984	IARI, RRS, Karnal	(Azam dl × 1B-65) × RDB-1 × Riso Mut. 1508)	semi dwarf, hullless grain and high protein (16%)
	Karan-265	1989	IARI, RRS, Karnal	RDB-1 × EB-7725) × (Riso Mut 1508)	dwarf, nod-lodging, high tillering, hull-less grain, thick and short leaves
	PL-56	1978	PAU, Ludhiana	0.2% EMS, C-164	high tillering, superior in yield, suited for rainfed areas, semi-tall and bold grains
	RDB-1	1974	ARS, Durgapura, Jaipur	pile neutrons, RS-17	dwarf, early, high yielding, non lodging and less water requirement
	RD-103	1978	ARS, Durgapura, Jaipur	RBD-1 × K-18	stiff straw, dwarf, high yield and high tillering
	RD-137	1981	ARS, Durgapura, Jaipur	RDB-1 × EB-795	medium tall, high yielding and less water requirement
	RD-2035	1988	ARS, Durgapura, Jaipur	RD-137 × PL-101	medium tall, profuse tillering, early maturity and resistant to CCN
	Rajkiran	1982	ARS, Durgapura, Jaipur	RBD-1 × Marocaine-079, CI 8334	dwarf, erect, profusely tillering and immune to Molya disease
24.	***Hyocyamus niger*** **(Indian henbane)**				
	Aekela	1996	CIMAP, Lucknow	20-60 kR γ-rays	an unbranched mutant and high tropane alkaloid

	Aela	1996	CIMAP, Lucknow	20-60 kR γ-rays	yellow flowered mutant with high yield of tropane alkaloid and high biomass
25.	***Lantana depressa* (Wild sage)**				
	L. dep. bicoloured	1986		γ-rays	leaf colour
	L. dep. variagata	1986		γ-rays	flower colour
	Niharika	1986		γ-rays, *L. depressa*	leaf colour
26.	***Lens culinaris* Medic. (Lentil)**				
	B-177 (S-177)			B-77(Asha)	erect and double seeded pods
	PI-77-12 (Arun)	1986		BR-25	spreading growth habit, bigger pods and bold seeds
	S-256 (Ranjan)	1984	PORS, Berhampore	X-rays	spreading type, high yield, suitable for rainfed condition and high protein content
27.	***Luffa acutangula* Roxb. (Ridged gourd)**				
	PKM-1	1984	TNAU, AC&RI, Periyakulum	γ-rays, H160	high yielding, tolerant to pumpkin beetles, fruit fly and leaf spot diseases
28.	***Lycopersicon esculentum* Mill. (Tomato)**				
	Co-3 (Marudham)	1981	TNAU, Coimbatore	EMS-0.1%, Co-1	yield, dwarf, determinate, high Vit C and heat tolerant
	PKM-1	1980	TNAU, AC&RI, Periyakulum	25 kR γ-rays, Annangi	yield
	Pusa Lal Meeruti	1972	IARI, New Delhi	30 kR γ-rays, Meeruti	uniform fruit ripening
	S-12	1969	PAU, Ludhiana	γ-rays	high yield, nematode resistant and reduced plant height
29.	***Matricaria cammomilla* (German chamomile)**				
	Vallery	1994	CIMAP, Lucknow	10-100 kR γ-rays, German Bulk	high chamozalene, high oil and dry herb yield
30.	**Mentha spicata (Spearmint)**				
	Neera	1992	CIMAP, Lucknow		high alkaloid content

(*Contd.*)

(*Table 8 Contd.*)

S. No.	Crop / Mutant variety	Year of release	Institution	Mutagen & parent variety	Main characters induced / improved
31.	***Momordica charantia* L. (Bitter gourd)**				
	MDU-1		TNAU, AC&RI, Madurai	10kR γ-rays, MC 103	tolerant to pumpkin beetle, fruitfly and leafspot disease
32.	***Morus alba* L. (Mulberry)**				
	S-54	1974		EMS, Berhampore	yield
33.	***Nicotiana tabacum* L. (Tobacco)**				
	GSH-3	1979	CTRI, Rajahmundry	(LTH × M4) × CTRI	special quality and yield
34.	***Oryza sativa* L. (Rice)**				
	A.U.-1 (Annamalai Uver Nel)	1976	TNAU, Coimbatore	γ-rays, IR 8	tolerant to alkalinity and salinity, grains short bold with moderate degree of dormancy and maturity 105 days
	Biraj	1984	West Bengal	X-rays, OC-1393	moderate resistance to *Helminthosporium* suitable for rainfed lowlands and better tolerance to submergence
	CNM-20	1980	Rice Res. Station, Chinsurah, W. Bengal	30kR X-rays, IR-8	early maturity by 10-12 days, increased effective tillering, long grain and resistant to BLB, BLS & BPH
	CNM-25	1978	Rice Res. Station, Chinsurah, W. Bengal	30kR X-rays, IR-8	early maturity, high yield, increased effective tillering, long grain and resistant to thrips
	CNM-31	1978	Rice Res. Station, Chinsurah, W. Bengal	30kR X-rays, IR-8	early maturity, increased effective tillering, long grain, high yield and resistant to BLB, BLS, BPH & brown spot
	CRM 49	1999		NaN_3, IR 50	blast resistance
	CRM 51	1999		NaN_3, IR 50	blast resistance
	CRM 53	1999		EMS, IR 50	blast resistance

HM 95	1975	PAU, Ludhiana	50 kR gamma rays, Jhonona 349 × Taichung Native 1	early, dwarf, high yield, photo and thermo insensitive, high protein (12.3%) and high lysine (4.07%)
HPU-8020	1984	HPAU, Palampur	20kR γ-rays, Bala	matures 10-13 days later synchronous tillering and yield potential 7.3 t/ha
HUR 36	1990	BHU, Varanasi	γ-rays, EMS, Mahsuri	earliness
Hari (TR-RNR-21)	1987	BARC, Bombay	IR-8 × TR-5	short culm and higher yield
IIT 48	1972	IIT, Kharagpur	Ethyleneoxide, IR 8	earliness
IIT 60	1973	IIT, Kharagpur	0.5% EMS, IR 8	one month earlier than IR 8
Indira	1980	CRRI, Cuttack	EMS, Tainan-31	tolerant to Blast, BLB and stem borer
Intan Mutant	1988	UAS, Bangalore	EI 0.2%, Intan	early maturing and photoperiod insensitive
Jagannath	1975	OUA&T, Bhubaneswar	30 kR X-rays, IR-8	one month earlier than IR 8
K 84	1968	Jammu & Kashmir	γ-rays, Taichung 65	*Indica* grained mutant and early maturity
Keshari (TET-6215)	1985	Orissa	cross [(T-90 × IR 8) × Jagannath]	early maturity, tolerant to blast, bacterial leaf blight and green leaf hopper
Lakshmi (CNM-6)	1982	Rice Res. Station, Chinsurah, W. Bengal	30kR X-rays, IR-8	early maturity, drought tolerant, increased tillering, long grain and high yield
Mohan (CSR-4)	1980	West Bengal	γ-rays, IR-8	Semi-dwarf (75-80cm), resistant to lodging, salinity and fertilizer responsive
Padmini	1988	CRRI, Cuttack	γ-rays, CR-1014	medium slender grains
PL-56	1975		EMS, C-164	tillering type
PNR-162	1993	IARI, New Delhi	Jaya mut. × Basmati 370	semi-dwarf and early maturity (85-115 days)
PNR-166	1989	IARI, New Delhi	cross	synchronous maturity

(Contd.)

(*Table 8 Contd.*)

S. No.	Crop / Mutant variety	Year of release	Institution	Mutagen & parent variety	Main characters induced / improved
	Rosy Green	1974	NBRI, Lucknow	γ-rays chronic	flower morphology
	PNR-381	1992	IARI, New Delhi	Tainan-2 × Basmati 370	semi-dwarf, early maturity and dual purpose
	PNR-519	1990	IARI, New Delhi	cross	pest resistance
	PNR-550-1-2 (JD-8)	1997	IARI, New Delhi	cross	semi-dwarfness
	PNR-551-4-20 (JD 6)	1997	IARI, New Delhi	cross	semi-dwarfness
	PNR-555-28 (JD 10)	1997	IARI, New Delhi	cross	semi-dwarfness
	PNR-555-5	1990	IARI, New Delhi	cross	earliness
	PNR-555-5 (JD 3)	1998	IARI, New Delhi	cross	yield
	PNR-570-17	1990	IARI, New Delhi	cross	earliness
	PNR-571	1990	IARI, New Delhi	cross	semi-dwarfness
	Prabhavati	1986	MAU, Parbhani	0.2% EMS, Amber mohar local	an iron chlorosis tolerant semi-dwarf mutant with stiff straw, resistant to lodging and responsive to nitrogen
	Rasmi	1985	RARS, Pattambi	22kR γ-rays, Oorpandy	awnless, high yield and tolerant to salinity
	Sattari (CRM-13-3241)	1983	CRRI, Cuttack	γ-rays, (NSJ 200 × Padma)	early maturity, nitrogen responsive and suitable for direct seeding in rainfed upland areas
	Savitri (Ponmani)	1983	Orissa	Pankaj × Jagannath	tolerant to blast and sheath blight
	Vellayani-1	1980	KAU, Kerala	Thermal neutrons, PTB 10	early maturity
35.	***Papaver somaniferum* L. (Opium poppy)**				
	BC-28/9/4 (Vivek)	1992	CIMAP, Lucknow	γ-rays, Shweta	big capsule size and high morphine content

	Sujata	2001	CIMAP, Lucknow	10-100 kr γ-rays, Shweta	opiumless poppy with high seed protein content
36.	***Pennisetum glaucum* (L.) R. Br. (Pearl millet)**				
	ICMH-451	1986	ICRISAT, Patancheru	30 kr γ-rays, Tift 23 DB	mildew resistance
	NHB-3	1975	IARI, New Delhi	cross	*Sclerospora* resistance
	NHB-4	1975	IARI, New Delhi	cross	*Sclerospora* resistance
	NHB-5	1976	IARI, New Delhi	35 kR γ-rays, male sterile inbred line Tift 23A	resistance to mildew disease and bold grains
	Pusa-46	1983	IARI, New Delhi	irradiation of hybrid J104 × K 559	resistance to downy mildew and *Pyricularia*
37.	***Phaseolus vulgaris* L. (French bean)**				
	Pusa Parvati	1970	IARI, New Delhi	X-rays, wax podded	early, resistant to mosaic and powdery mildew
38.	***Pisum sativum* L. (Field pea)**				
	L-116 (Hans)	1979	IARI, New Delhi	EI, P-1163	high yield and semi-erect
39	***Plantago ovata* (Isabgol)**				
	Niharika	1997	CIMAP, Lucknow	10-100 kR γ-rays	long inflorescence and high seed yield
40.	***Polyanthus tuberosa* L. (Tube rose)**				
	Rajat Rekha	1974		γ-rays	single flowered with leaf colour mutant
	Swarna Rekha	1974		γ-rays	double flowered with leaf colour mutant
41.	***Portulaca grandiflora* L. (Portulaca)**				
	Five Petal	1974	NBRI, Lucknow	γ-rays	flower
	Jhumka	1974	NBRI, Lucknow	γ-rays, Karna Pali	flower
	Karna Pali	1974	NBRI, Lucknow	γ-rays	double flower
	Karna Phul	1974	NBRI, Lucknow	γ-rays	flower
	Lalita	1974	NBRI, Lucknow	γ-rays	double flower

(Contd.)

(*Table 8 Contd.*)

S. No.	Crop / Mutant variety	Year of release	Institution	Mutagen & parent variety	Main characters induced / improved
	Mukta	1974	NBRI, Lucknow	γ-rays	double flower
	Pink colour	1974	NBRI, Lucknow	γ-rays chronic	flower colour
	Ratnam	1974	NBRI, Lucknow	γ-rays	flower number
	Semi-double	1974	NBRI, Lucknow	γ-rays	flower
	Vibhuti	1974	NBRI, Lucknow	γ-rays	flower
42.	***Ricinus communis* L. (Castor bean)**				
	Aruna	1970	IARI, Reg. Station Hyderabad	1400 rad, th. neutrons, HC-6	extra early (120 days *vs* 270 days for mother variety) and high yield
	RC-8	1982	UAS, Bangalore	40 kR γ-rays, RC 1188-54	higher TKW and higher yield
	SA-2	1971		A spontaneous mutant of TMV 1	resistant to drought, higher TKW and higher yield
	Sowbhagya (157-B)	1978	IARI Reg. Station, Hyderabad	Aruna dwarf mutant of HC-6 × dwarf mutant of HC-6 Mauthners dwarf	dwarf, non-shattering, suitable for intercropping and long duration maturity
43.	***Rosa* sp. (Rose)**				
	Abhisarika	1971	IARI, New Delhi	γ-rays, Kiss of Fire	flower colour
	Angara	1975	IARI, New Delhi	γ-rays, Montezuma	plant architecture
	Curio	1986	IARI, New Delhi	γ-rays, Imperator	flower colour
	Light Pink Prize	1989	IARI, New Delhi	γ-rays, First Prize	flower colour
	Madhosh	1975	IARI, New Delhi	EMS, Gulzar	flower colour
	Mrinalini Stripe	1991	IARI, New Delhi	γ-rays, Mrinalini	flower colour
	Pink Contempo	1986	IARI, New Delhi	γ-rays, Contempo	flower colour
	Pusa Christiana	1975	IARI, New Delhi	γ-rays, Christiana Dior	flower colour
	Saroda	1984	IARI, New Delhi	γ-rays, Queen Elizabeth	flower colour
	Sharada	1983	IARI, New Delhi	γ-rays, Queen Elizabeth	flower colour
	Striped Christiana Dior	1975	IARI, New Delhi	γ-rays, Christiana Dior	flower colour
	Striped Contempo	1983	IARI, New Delhi	γ-rays, Contempo	flower colour
	Sukumari	1984	IARI, New Delhi	γ-rays, America's Jr. Miss	flower colour
	Tangerine Contempo	1984	IARI, New Delhi	γ-rays, Contempo	flower colour

	Twinkle	1986	IARI, New Delhi	γ-rays, Imperator	flower colour
	Yellow Contempo	1984	IARI, New Delhi	γ-rays, Contempo	flower colour
44.	***Saccharum officinerum* L. (Sugarcane)**				
	Co 312		SBI, Coimbatore	γ-rays	high cane yield
	Co 449		SBI, Coimbatore	γ-rays	resistant to red rot
	Co 527		SBI, Coimbatore	γ-rays	high cane yield
	Co 775		SBI, Coimbatore	γ-rays	high cane yield
	Co 997 mutant	1967	SBI, Coimbatore	γ-rays, Co 997	resistant to red rot
	Co 6608	1966	SBI, Coimbatore	γ-rays, Co 449	resistant to red rot
	Co 8153	1981	SBI, Coimbatore	15 kR γ-rays	Co 6304 × Co 6806, superior juice quality & yield
	Co 85017	1985	SBI, Coimbatore	γ-rays, Co 740	resistant to *U. scitaminea* and high sucrose percentage
	Co 85035	1985	SBI, Coimbatore	γ-rays, Co 740	high cane yield and sucrose percentage
45.	***Sesamum indicum* L. (Sesamum)**				
	Kalika	1985	OUAT, Bhubaneswar	1% EMS, Vinayak	dwarf, high yield, tolerant to *Cercospora* leaf spot, stem, root rot and lodging
	Uma	1990		chemical, Kanak	uniform maturity
	Usha	1990		chemical, Kanak	yield
46.	***Setaria italica* (Foxtail millet)**				
	PS 4	1999	GBPUA&T, Pantnagar	0.2% EMS, SIA 2616	high yield, early maturity and bold seeds with 13-15% protein
47.	***Solanum khasianum* Clarke (Khasianum)**				
	RRL-20-2	1975	RRL, Jammu	γ-rays, Dehradun local	solasodine
48.	***Solanum melongena* L. (Brinjal)**				
	PKM-1	1985	TNAU, AC&RI, Periyakulum	γ-rays, Puzhuthikathiri	high yielding and adapted to rainfed area
49.	***Sorghum bicolor* L. (Sorghum)**				
	Co-21 (SPV-80)	1986	TNAU, Coimbatore	40 kR X-rays, CSV-5	tolerance to *Striga* and high fodder yield (100-107q/ha)

(*Contd.*)

(*Table 8 Contd.*)

S. No.	Crop / Mutant variety	Year of release	Institution	Mutagen & parent variety	Main characters induced / improved
	Mothi (SPV-141)	1978	TNAU, Coimbatore	IS 6928	suitable for late *kharif* and early maturing (105 days)
	SPV-126 (CSV-9)	1983	TNAU, Coimbatore	CS-3541 x CSV-4	resistant to charcoal rot, downey mildew, head moulds and stem borer
50.	***Trichosanthus anguina* L. (Snake gourd)**				
	PKM-1	1982	TNAU, AC&RI, Periyakulum	H 375	high yield
51.	***Trifolium alexandrinum* L. (Egyptian clover)**				
	BL-22	1984		γ-rays, Mescavi	lateness
52.	***Triticum aestivum* L. (Wheat)**				
	NI-5643	1978	Agri. Res. Station, Niphad, Nasik	radiation (New Thatcher × NI-284-S)	earliness
	NP 836	1961	IARI, New Delhi	16 kR X-rays, NP 799	fully awned, higher yield, medium-early maturity with rust resistance
	Pusa Larma	1971	IARI, New Delhi	γ-rays, Lerma Rojo 64-A	grain colour
	Sharbati Sonora	1969	IARI, New Delhi	2600 A UV + 20 kR γ-rays, Sonora 64	amber grain colour, short straw and early maturity
53.	***Vigna aconitifolia* Jacq Marechal (Moth bean)**				
	CAZRI Moth-1	1999	CAZRI, Jodhpur	30 kR γ-rays, Jadia	high yield, resistant to YMV and semi-erect plant having high seed protein
	Maru Moth-1 (IJMM-259)		CAZRI, Jodhpur	30 kR γ-rays, Jadia	drought resistant and early maturity
	RMO-40	1994	CAZRI, Jodhpur	40 kR γ-rays, Jwala	yield and earliness
	RMO-257	1997	CAZRI, Jodhpur	30 kR γ-rays, + 0.6% EMS, Jadia	earliness and high harvest index
	RMO-225	1999	CAZRI, Jodhpur	40 kR γ-rays, + 0.6% EMS, Jadia	earliness and short resistance

54.	***Vigna mungo* (L.) Hepper. (Blackgram)**				
	Co-4	1978	TNAU, Coimbatore	0.02% MMS, Co-1	early maturing, erect, determinate & photoinsensitive
	Manikya	1988	GBPUA&T, Pantnagar	40 kR γ-rays, T-9	high yield & YMV resistance
	Sarla (B-12-4)	1985		T-9	resistant to MYMV and powdery mildew
	TAU-1	1995	BARC, Mumbai & PKV, Akola	T-9 × dark green leaf mutant of var. No. 55	yield, larger seed size and powdery mildew resistant
	TAU-2	1992	BARC, Mumbai & PKV, Akola	T-9 × dark green leaf mutant of var. No. 55	high yield and tolerant to powdery mildew
	TPU-4	1992	BARC, Mumbai & MPKV, Rahuri	γ-rays, UM-201 mutant of var. No. 55 × T-9	high yield and tolerant to MYMV
	TU94-2	1999	BARC, Mumbai	γ-rays, TPU-3 × TAU-5 (early mut. of EC168200)	high yield
55.	***Vigna radiata* (L.) Wilczek (Mungbean)**				
	BM-4	1992	ARS, Badnapur	0.15% EMS, T-44	resistance to *Macrophomina*, tolerant to PM and MYMV
	Co-4	1981	TNAU, Coimbatore	40 kR γ-rays, Co-1	high yield
	Dhauli (TT9E)	1979	OUA&T, Bhubaneshwar	mutant of fixed line of cross T51 × local	high yield
	LGG-407	1993	APAU, Lam	40 kR γ-rays, Pant Mung-2	high yield
	LGG-450	1993	APAU, Lam	40 kR γ-rays, Pant Mung-2	high yield
	ML-26-10-3	1983	PAU, Ludhiana	γ-rays	yield and tolerance to MYMV
	MUM-2	1992	CCSU, Meerut	0.2% EMS, K-851	tolerance to MYMV and early maturity
	PBM-1		PAU, Ludhiana		yield and tolerance to MYMV
	Pant Mung-2	1982	GBPUA&T, Pantnagar	10 kR γ-rays, ML-26	yield and resistance to MYMV
	TAP-7	1983	BARC, Mumbai	30 kR γ-rays, S-8	high yield, earliness, tolerance to PM and leaf spot

(Contd.)

(*Table 8 Contd.*)

S. No.	Crop / Mutant variety			Mutagen & parent variety	Main characters induced / improved
	TARM-1	1995	BARC, Mumbai & PKV, Akola	30 kR γ-rays, S-8	high yield and resistance to MYMV
	TARM-2	1992	BARC, Mumbai & PKV, Akola	30 kR, γ-rays, S-8	high yield and resistance to MYMV
	TARM-18	1996	BARC, Mumbai & PKV, Akola	30 kR γ-rays, S-8 [PDM54 × TARM-2]	high yield and resistance to MYMV
56.	***Vigna unguiculata* Walp. (Cowpea)**				
	Co-5	1986	TNAU, Coimbatore	30 kR γ-rays, Co-1	forage cowpea, high yield and suitable for inter-cropping
	Cowpea-88	1990	PAU, Ludhiana	F_1 seed irradiation (Cowpea-74 × No. H-2)	highly resistant to MYMV, *Anthracnose*, high grain and fodder yield
	Gujarat cowpea-1	1984	GAU	Barsati mutant	high yield, earlyness and resistant to rot knot rematode
	V-16 (Amba)	1981	IARI, New Delhi	DMS, Pusa Phalguni	high yield, resistant to fungal and bacterial diseases
	V-37 (Shreshtha)	1981	IARI, New Delhi	DMS, Pusa Phalguni	high yield and suitable as green fodder
	V-38 (Swarna)	1981	IARI, New Delhi	DMS, Pusa Phalguni	high yield, earlier maturity and resistant to diseases
	V-240	1984	IARI, New Delhi	DMS, Pusa Phalguni	high yield, resistant to fungal, bacterial and viral diseases

References

Ahmed Z.U., Shaikh M.A.Q. and Begum S. 1995. Development of the high yielding and disease resistant mungbean variety, BINA Moong-2 using nuclear and cross breeding techniques. *In*: *Proc. IAEA/FAO Symp.*, on Induced Mutations and Molecular Techniques for Crop Improvement. June 19-23, 1995, Vienna. IAEA, Vienna. pp. 638-640.

Ali M., Malik I.A., Sabir H.M., Ahmad B. 1997. *Technical Bulletin No. 24.* AVRDC, Shanhua, Taiwan, ROC. 66 p.

Amano E. 1997. Mutation breeding in Japan and contribution to the region. *In*: Strategy Paper on *Application of Mutation Techniques for Crop Improvement in East Asia and Pacific Region.* IAEA, Vienna.

Anderson P.A., Okubara P.A., Arroyo-Garcia R., Meyers B.C. and Michelmore R.W. 1996. Molecular analysis of irradiation-induced and spontaneous deletion mutants at a disease resistance locus in *Lactuca sativa. Molecular and General Genetics*, **251**: 316-325.

Ankineedu G., Sharma K.D. and Kulkarni L.G. 1968. Effect of fast neutrons and gamma-rays on castor. *Indian J. Genet.*, **28**: 31-39.

Anonymous. 1977. *Manual on Mutation Breeding* (Second Edition), Technical Reports Series, No. 119. Joint FAO/IAEA Division of Atomic Energy in Food and Agriculture, International Atomic Energy Agency, Vienna, Austria. pp. 288.

Anonymous. 1987. Mutation breeding of cowpea and mungbean in Venezuela. *Mutation Breeding Newsletter*, **29**: 3.

Auerbach C. 1941. The effect of sex on the spontaneous mutation rate in *Drosophila melanogaster*, *J. Genet.*, **11**: 255-265.

Auerbach C. and Robson J.M. 1946. Chemical production of mutations. *Nature*, **157**: 302.

Awan M. A. 1991. Use of induced mutations for crop improvement in Pakistan. *In*: *Proc. IAEA Symp. on Plant Mutation Breeding for Crop Improvement.* June 18-22, 1990. IAEA, Vienna, pp. 67-72.

Bahl J.R. and Gupta P.K. 1982. Promising mutants in mungbean, *Vigna radiata* (L.) Wilczek. *Mutation Breeding Newsletter*, **20**: 13.

Bhagwat S.G., Bhatia C.R., Krishna T.G., Joshua D.C., Mitra R.K., Narahari P., Pawar S.E. and Thakre R.G. 1979. Increasing protein production in cereals and grain legumes. *In*: *Proc. Symp. on Seed Protein Improvement in cereals and grain legumes*, IAEA, Vienna, vol. II.pp. 225-236.

Bhatia C.R. 1991. Economic impact of mutant varieties in India. *In*: *Proc. IAEA Symp. on Plant Mutation Breeding for Crop Improvement.* June 18-22, 1990. Vol. 1. IAEA, Vienna. pp. 33-45.

Bhatia C.R. 2000. Induced mutations for crop improvement - the generation next. *In*: *Proc., DAE-BRNS Symp.*, Dec. 6-8, 2000, Mumbai, India. pp. 1-7.

Bhatia C.R., Desai R.M. and Suseelan K.N. 1978. Attempts to combine high yield and grain protein in wheat. *In*: *Proc. Meeting on Seed Protein Improvement by Nuclear Technique.* Baden, 1997, IAEA, Vienna. pp. 5.

Bhatia C.R., Nichterlein K. and Maluszynsky M. 1999. Oilseed cultivars developed from induced mutations and mutations altering fatty acid composition. *Mutation Breeding Review*, **11**: 1-36.

Bhatnagar S.M. 1991. Dwarf and semi-dwarf barley mutants in cross breeding. *In*: *Proc. FAO/IAEA Symp. on Plant Mutation Breeding for Crop Improvement.* June 18-22, 1990. Vol. 1. IAEA, Vienna. pp. 307-319.

Buiatti M. and Gimelli F. 1993. Somaclonal variation in ornamentals. *In*: *Proc. XVII EUCARPIA Symp. on Creating Genetic Variation in Ornamentals.* (eds.) San Remo. T. Schiva and A. Mercuri San Remo. Instituto Sperimentale per la Floricoltura. pp. 5-24.

Buschges R., Hollrichter K., Panstruga R., Simons G., Wolter M., Frijters A., Van Daelen R., Von der Lee T., Diergaarde P., Groenendijk J., Topsch S., Vos P., Salamini and Schulze-Lefert P. 1997. The barley Mlo gene: a novel control element of plant pathogen resistance. *Cell*, **88**: 695-705.

Chakrabarti S.N. 1996. Role of mutation in improving rice productivity-Retrospect and prospect. *In*: Isotopes & Radiations in Agriculture and Environment Research. (eds.) Sachdev M.S., P. Sachdev and D.L Deb. Indian Society for Nuclear Techniques in Agriculture and Biology. New Delhi. pp. 181-195.

Cheng X.Y., Gao M.W., Liang Z.Q. and Liu K.Z. 1990. Effect of mutagenic treatments on somaclonal variation in wheat (*Triticum aestivum* L.). *Plant Breeding*, **105**: 47-52.

Correns C. 1900. Mendels Regal über das Verhalten der Nach-kommenschaft Rassen Bastarde. *Ber. Deutsch. bot. Ges.*, **18**: 158-168.

Dangl J.L., Dietrich R.A. and Richberg M.H. 1996. Death don't have no mercy: cell death programme in plant-microbe interaction. *The Plant Cell*, **8**: 1793-1807.

Datta S.K. 1997. *Ornamental Plants: Role of Mutation*. Daya Publishing House, Delhi, pp. 220.

Delaunay L.N. 1931. Resultate eines dreijahrigen Rontgen Versuch mit Weitzen. *Der Züchter*, **3**: 129-137.

De Vries H. 1901. *Die Mutationstheorie*. I. Leipzig: Veit & Co.

De Vries H. 1903. *Die Mutationstheorie*. II. Leipzig: Veit & Co

Dribnenki J.C.P., Green A.G. and Atlin G.N. 1996. LinolaTM 989 low linolenic flax. *Can. J. Plant Sci.*, **76**: 329-331.

Feix G., Hochholdinger F. and Wulff D. 1997. Genetic analysis of root formation in maize. *In*: *Developmental Pathways in Plants: Biotechnological Implications*. The Hebrew University of Jerusalem, Rehovot. pp. 10.

Freisleben R. and Lein A. 1942. Über die Auffindung einer mehltauresistenten mutante nach Röntgenbestrahlung einer anfälligen reinen Linie von Sommergerste. *Naturwissenschafen*, **30**: 608.

Frey K.J. 1955. Agronomic mutations in oats induced by X-ray treatment, *Agron. J.*, **47**: 207-210.

Gaul H., Ulonska E., Winkel C.Z. and Braker G. 1969. Micro-mutations influencing yield in barley - studies over nine generations. *In*: *Induced mutations in plants*. Proc. Symp. pullman, 1969, IAEA, Vienna, pp. 375-398.

Glzenbrook I., Rogers E.E. and Ausubel F.M. 1997. Use of *Arabidopsis* for genetic dissection of plant defense responses. *Annual Rev. Genet.*, **31**: 547-569.

Gottashalk W. and Wolff G. 1983. Induced Mutations in Plant Breeding. Monograph on *Theoretical and Applied Genetics*, Springer-Verlag, Berlin, Heidel berg, pp. 323-327.

Green A.G. and Dribnenki J.C.P. 1994. Linola - a new premium polyunsaturated oil. *Lipid Technology*, March/April: 29-33.

Gregory W.C. 1955. X-ray breeding of peanuts (*Arachis hypogaea*) *Agron. J.*, **47**: 396-399.

Gregory W.C. 1965. Mutation frequency, magnitude of change and the probability of improvement in adaptation. *In*: *The use of Induced Mutations in Plant Breeding*, Pergamon Press, Oxford, pp. 429-441.

Gupta P.K., Singh S.P. and Bahl J.R. 1996. A new mungbean variety through mutation breeding. *Mutation Breeding Newsletter*, **42**: 6-7.

Gustafsson A. 1947. Mutations in agricultural plants. *Hereditas*, **33**: 1-100.

Gustafsson A. and Mackey J. 1948. The genetic effects of mustard gas substances and neutrons. *Hereditas*, **34**: 371-386.

Gustafsson A., Lundquist U. and Ekberg I. 1967. Yield reactions and rates of origin of chromosome mutations in barley, *Heriditas*, **56**: 200-206.

Haq M.A., Sadiq M. and Hasan M. 1988. Improvement of chickpea through induced mutations. *In*: *Proc. FAO/IAEA workshop on Improvement of Grain Legume Production using Induced Mutations*. 1-5 July, 1986, Pullman, Washington (USA), IAEA, Vienna, 1988, 75-88.

Hutcheson S.W. 1998. Current concepts of active defense in plants. *Annual Rev. Phytopathol.*, **36**: 59-90.

Jende-Strid B. 1993. Genetic control of flavonoid biosynthesis in barley. *Hereditas*, **119**: 187-204.

Jensen J. 1991. New high yielding, high lysine mutants in barley. *In*: *Proc. IAEA Symp. on Plant Mutation Breeding for Crop Improvement*. June 18-22, 1990. IAEA, Vienna, Vol. II, pp. 31-41.

Jorgensen J.H. 1991. Experience from mutation studies on cereal disease resistance. *In*: *Proc. IAEA Symp. on Plant Mutation Breeding for Crop Improvement*. June 18-22, 1990. IAEA, Vienna, Vol. II. pp. 81-91.

Joshua D.C. and Ramani S. 1993. An induced mutant with extended vegetative phase in stem nodulating *Sesbania rostrata*. *J. Agric. Sci.*, (Cambridge). **120**: 71-73.

Joshua D.C. and Thakre R.G. 1986. A day neutral mutant in jute. *Tropical Agriculture* (Trinidad). **63**: 316-318.

Kawai T. and Amano E. 1991. Mutation breeding in Japan. *In*: *Proc. IAEA Symp. on Plant Mutation Breeding for Crop Improvement*. June 18-22, 1990. IAEA, Vienna, Vol. I. pp. 47-66.

Kharkwal M.C. 1996. Accomplishments of mutation breeding in crop improvement in India. *In*: *Isotopes & Radiations in Agriculture and Environment Research*. (eds.) Sachdev M.S., Sachdev P. and D.L Deb. Indian Society for Nuclear Techniques in Agriculture and Biology. New Delhi. pp. 196-218.

Kharkwal M.C. 1998a. Induced mutations for improvement of protein in chickpea (*Cicer arietinum* L.) *Indian J. Genet.*, **58**: 61-68.

Kharkwal M.C. 1998b. Induced mutations in chickpea (*Cicer arietinum* L.). I. Comparative mutagenic effectiveness and efficiency of physical and chemical mutagens. *Indian J. Genet.*, **58**: 159-168.

Kharkwal M.C. 1998c. Induced mutations in chickpea (*Cicer arietinum* L.). II. Frequency and spectrum of chlorophyll mutations. *Indian J. Genet.*, **58**: 465-474.

Kharkwal M.C. 1999. Induced mutations in chickpea (*Cicer arietinum* L.) III. Frequency and spectrum of viable mutations. *Indian J. Genet.*, **59**: 451-464.

Kharkwal M.C. 2000. Induced mutations in chickpea (*Cicer arietinum* L.) IV. Types of macro-mutations induced. *Indian J. Genet.*, **60**: 305-320.

Kharkwal M.C. 2001. Induced mutations in chickpea (*Cicer arietinum* L.) V. Evaluation of micro-mutations. *Indian J. Genet.*, **61**: 115-124.

Kharkwal M.C., Jain H.K. and Sharma B. 1988. Induced mutations for improvement of chickpea, lentil, pea and cowpea. pp. 89-109. *In*: *Improvement of Grain Legume Production using Induced Mutations, Proc. FAO/IAEA Workshop*. 1-5 July, 1986, Pullman, Washington, USA, IAEA, Vienna, 1988.

Kharkwal M.C., Pawar S.E. and Pandey R.N. 2001. Seventy five years of research on induced mutations with special reference to crop improvement in India. *In*: *Proceedings of NAARRI International Conference on Applications of Radioisotopes and Radiation Technology in the 21st Century*. (eds.) N. Ramamurty, M.Ananthakrishnan and A.N. Nandakumar, December 12-14, 2001, Mumbai. pp. 230-235.

Kinney A.J. 1995. Improving soybean seed quality. *In*: *Proc. IAEA/FAO Symp., on Induced Mutations and Molecular Techniques for Crop Improvement*. June 19-23, 1995, Vienna. IAEA, Vienna. Pp. 101-113.

Kleinhofs A., Warner R.L., Muehlbauer F.J. and Nilan R.A. 1978. Induction and selection of specific gene mutations in *Hordeum* and *Pisum*. *Mutation Research*. **51**: 29-35.

Konzak C.F. 1954. Stem rust resistance in oats induced by nuclear radiation, *Agron. J.*, **46**: 538-540.

Konzak C.F., Kleinhofs A. and Ullrich S.E. 1984. Induced mutations in seed propagated crops. *In*: *Plant Breeding Reviews*, Vol. 2. (ed.) J. Janick. Westport, Connecticut. AVI Publishing Company. pp. 13-72

Kumar D. 2000. A commercial cultivar suited to rainfed arid lands induced in dew bean (*Vigna aconitifolia* Jacq Marechal). *In*: *Proc. DAE-BRNS Symp. on The use of Nuclear and Molecular Techniques in Crop Improvement*. December 6-8, 2000. BARC, Mumbai. pp. 124-130.

Malik I.A. 1988. High yielding and early maturing mutants in mungbean (*Vigna radiata* (L.) Wilczek). *Mutation Breeding Newsletter*, **32**: 7-8.

Malik I.A., Nadeem M.T., Sarwar G. and Ali S. 1979. Evaluation of radiation induced mutant lines of mungbean (*Vigna radiata* (L.) Wilczek) for grain yield and protein content. *In*: *Proc. Symp. on Seed Protein Improvement in cereals and grain legumes*, IAEA, Vienna, Vol. II, pp. 445.

Maluszynski M., Nichterlein K., Van Zanten L. and Ahloowalia B.S. 2000. Officially released mutant varieties – the FAO/IAEA database. *Mutation Breeding Review*, **12**: 1-84.

Maluszynski M., Van Zanten L., Ashri A., Brunner H., Ahloowalia B., Zapata F.J. and Weck E. 1995. Mutation techniques in plant breeding. *In*: *Proc. IAEA/FAO Symp., on Induced Mutations and Molecular Techniques for Crop Improvement*. Vienna, June 19-23, 1995. pp. 489-504.

Mathews H. and Bhatia C.R. 1983. Experimental mutagenesis of *in vitro* cultured plant cells and protoplasts. *Mutation Breeding Newsletter*, **22**: 12-17.

Matsumoto K. and Yamaguchi H. 1991. Induction and selection of aluminium tolerance in banana. *In*: *Proc. FAO/ IAEA Symp. on Plant Mutation Breeding for Crop Improvement*. Vienna, 1990. IAEA, Vienna. Vol. 2. pp. 249-256.

Mendel G. 1865. Versuche über Pflanzen-Hybriden. *Vesh. Naturforschung Ver*. In Brünn Verh., 4: 3-47.

Micke A. 1988. Genetic improvement of grain legumes using induced mutations. *In*: *Proc. FAO/IAEA Workshop on Improvement of Grain Legume Production using Induced Mutations*. 1-5 July, 1986, Pullman, Washington (USA), IAEA, Vienna. 1988. pp. 1-51.

Micke A. 1991. Induced mutations for crop improvement. *In*: *Biotechnology and Mutation Breeding*. Gamma Field symposia 30, 1-21. Ohmiya-Machi, Naka-gun, Ibaraki-ken, Japan. Institute of Radiation Breeding, NIAR MAFF.

Micke A. 1993. Durability of resistance in induced mutants. *In*: *Durability of Disease Resistance*. (eds.) Th. Jacobs and J.E. Parlevliet, Dordrecht: Kluwer Academic Press.

Micke A., Donini B. and Maluszynski M. 1990. Induced mutation for crop improvement. *Mutation Breeding Review*, **7**: 1-41.

Muller H.J. 1927. Artificial transmutation of gene. *Science*, **66**: 84-87.

Murty B.R. 1996. Significance of induced mutations in crop improvement research. *In*: *Isotopes & Radiations in Agriculture and Environment Research*. (eds.) Sachdev M.S., P. Sachdev and D. L. Deb. Indian Society for Nuclear Techniques in Agriculture and Biology. New Delhi. pp. 176-180.

Navarro-Alvarez W. and Salazar-Androvetto R. 1995. Improvement in rice resistance to *Pyricularia oryzae* through mutagenesis. *In*: *Proc. IAEA/FAO Symp., on Induced Mutations and Molecular Techniques for Crop Improvement*. Vienna, June 19-23, 1995. pp. 649-665.

Nichterlein K. 1999. The role of induced mutations in the improvement of common beans (*Phaseolus vulgaris* L.). *Mutation Breeding Newsletter*, **44**: 6-9.

Nichterlein K., Bohlmann H., Nielen S. and Maluszynski M. 2000. Achievements and trends of using induced mutations in crop improvement. *In*: *Proc., DAE-BRNS Symp.*, Dec. 6-8, 2000, Mumbai, India. pp. 27-35.

Nilsson-Ehle H. 1948. *The future possibilities of Swedish barley breeding*. Svalöf 1886-1946, pp. 113-126.

Novak F.J., Afza R., Van Duren M. and Omar M.S. 1990. Mutation induction by gamma irradiation of *in vitro* cultured shoot tips of banana and plantain (*Musa* cvs.). *Tropical Agriculture* (Trinidad), **67**: 21-28.

Novak F.J., Afza R., Daskalov S., Hermelin T. and Lucretti T. 1986a. Assessment of somaclonal and radiation-induced variability in maize. *In*: *Proc. FAO/IAEA Symp. on Nuclear Techniques and* in vitro *Culture for Plant Improvement*. Vienna, 1985. IAEA, Vienna, pp. 29-33.

Novak F.J., Hermelin T., Daskalov S. and Nesticky M. 1986b. *In vitro* mutagenesis in maize. *In*: *Proc. International Symp. EUCARPIA, Genetic Manipulation in Plant Breeding*. Berlin, 1985. (eds.) Horn, W., C.J. Jensen, W. Odenbach, and O. Schieder. Berlin. Walter de Gruyter & Co. pp. 563-576.

Novak F.J. and Micke A. 1988. Induced mutations and *in vitro* techniques for plant breeding. *In*: *Proc. SABRAO Int. Symp. & Workshop on Plant Breeding & Genetic Engineering*. Kuala Lumpur, Malaysia, 1987. (ed.) Zakri A.H., pp. 63-86. City Reprographic Service, Kuala Lumpur.

Novak F.J., Brunner H., Afza R. and Van Duren M. 1993. Mutation breeding of *Musa* sp. (banana, plantain). *Mutation Breeding Newsletter*, **40**: 2-4.

Oehlker F. 1943. Chromosome mutation in meiosis by chemicals. *In*: *Mutation Research — Problems, Results and Prospects*. (ed.) C. Auerbach (1976), Chapman and Hall, U.K.

Pathak R.S. 1988. Induced mutations for resistance to aphid (*Aphis craccivera* Koch) in cowpea. *In*: *Proc. Workshop on Improvement of Grain Legume Production using Induced Mutation*. Pullman, 1-5 July 1986, IAEA, Vienna, 1988, pp. 279-291.

Pathak R.S. 1991. Genetic evaluation of two aphid resistant cowpea mutants in Kenya. *In*: *Proc. IAEA Symp. on Plant Mutation Breeding for Crop Improvement*. June 18-22, 1990. Vol. 2. IAEA, Vienna, pp. 241-247.

Patil S.H., Kale D.M., Deshmukh S.N., Fulzele G.R. and Weginwar D.G. 1995. Semi-dwarf early maturing and high yielding new groundnut variety, TAG-24. J. *Oilseed Research*, **12**: 254-257.

Pawar S.E., Thakre R.G., Reddy K.S. and Bhatia C.R. 1991. Use of induced mutations in the breeding of pulse crops. *In*: *Proc. IAEA Symp. on Plant Mutation Breeding for Crop Improvement*. June 18-22, 1990. Vol. 1. IAEA, Vienna, pp. 413-418.

Rao N.G.P., Singh A., Sivasubramanian V., Murty K.S., Mukhopadhyay A.N. and Abraham C.C. 1998. Rice research and production in India. Present status and future perspective. *Quinquinneal Review Committee Report 1988-96*. Directorate of Rice Research, Hyderabad, India.

Rapoport I.A. 1946. Carbonyl compounds and the chemical mechanism of mutation. *C.R. Doklady Acad. Sci.*, USSR, **54**: 65.

Rapoport I.A. 1948. Alkylation of the gene molecule. *Doklady Acad. Sci.*, USSR., **59**: 1183-86.

Robbelen G. 1990. Mutation breeding for quality improvement: a case study for oilseed crops. *Mutation Breeding Review*, **6**: 1-44.

Rutger J.N. 1992. Searching for apomixis in rice. *In*: *Proc. Apomixis Workshop*. Atlanta, G.A. pp. 36-39.

Sapehin A.A. 1930. Rontgen-mutationen beim Weizen (*Triticum vulgare*). *Der Züchter*, **2**: 257-259.

Sapehin A.A. 1936. X-ray mutants in soft wheat, *Bull. Appl. Bot. Genet. Pl. Breed.* Ser. II., **9**: 3-37.

Sauls J.W. 1999. Texas citrus - Root stock and scion varieties. http://aggie-horticulture.tamu.edu/ citrus/12304.htm. 1-7.

Scarascia-Mugnozza G.T., D'Mato F., Avanzi S., Bagnara D., Belli M.L., Bozzini A., Brunori A., Cervigni T., Devreux M., Donini B., Giorgi B., Martini G., Monti L.M., Moschini E., Mosconi C., Porreca G., Rossi L. 1991. Mutation breeding programme for *durum* wheat (*Triticum turgidum* ssp. *durum* Desf.) improvement in Italy. *In*: *Proc. FAO/IAEA Symp. on Plant Mutation Breeding for Crop Improvement*. June 18-22, 1990. Vol. 1. IAEA, Vienna, pp. 95-109.

Shah J., Kachroo P. and Klessig D. 1998. The *Arabidopsis ssi1* mutation restores pathogenesis related gene expression *npr1* plants and renders defense gene expression salicylic acid dependent. *The Plant Cell*, **11**: 191-206.

Sheikh M.A.Q., Ahmed Z.U., Majid M.A. and Wadud M.A. 1982. A high-yielding and high-protein mutant of chickpea (*Cicer arietinum* L.) derived through mutation breeding, *Environ. Exp. Bot.*, **22**: 483-389.

Shaikh M.A.Q. and Majid M.A. 1995. A disease resistant and high yielding blackgram variety, BINA Mash-1, derived through induced mutation. *In*: *Proc. IAEA/FAO Symp., on Induced Mutations and Molecular Techniques for Crop Improvement*. Vienna, June 19-23, 1995. pp. 635-637.

Shirley B.W., Hanley S. and Goodman H.M. 1992. Effects of ionizing radiation on a plant genome: Analysis of two *Arabidopsis* transparent testa mutations. *The Plant Cell*, **4**: 333-347.

Siddique S.M., Sarwar G., Khattak G.S.S. and Saleem M. 1999. Development of mungbean variety 'NIAB Mung-98' involving induced mutants through conventional breeding. *Mutation Breeding Newsletter*, **44**: 11-13.

Stadler L.J. 1928. Genetic effects of X-rays in maize. *Prod. Natl. Acad. Sci.*, USA, **14**: 69-75.

Stadler L.J. 1930. Some genetic effects of X-rays in plants. *J. Hered.*, **21**: 3-19.

Swaminathan M.S., Siddiq E.A., Savin V.N. and Varughese G. 1968. *Mutations in Plant Breeding II*, IAEA, Vienna, 233.

Sybenga J. 1983. Genetic manipulation in plant breeding: Somatic *versus* generative, *Theor. Appl. Genet.*, **66**: 179-201.

Syukur S., Jacobs M. and Negrutiu I. 1991. Analysis of mutant plants resistant to salt or water stress and to proline analogues obtained from the protoplasts of *Nicotiana plumbaginifolia viviani. In*: *Proc. FAO/IAEA Symp. on Plant Mutation Breeding for Crop Improvement.* 1990, Vienna. IAEA, Vienna, Vol. 2, pp. 265-269.

Szarejko I. and Maluszynski M. 1980. Analysis of usefulness of sodium azide in plant breeding. *Acta. Biol.*, **9**: 60-66.

Tschermak E. Von. Über Kunstliche Kreuzung bei *Pisum sativum*, Ber. der. Bot. Ges., **18**: 232-239.

Tickoo J.L. and Chandra N. 1996. Induced variability and selection for improved quality characters in mungbean, *Vigna radiata* (L.) Wilczek. *DAE-BRNS Symp. on Nuclear Techniques in increasing crop and animal productivity*. BARC, Mumbai. pp. 71.

Tollenaar D. 1934. Untersuchungen über Mutation bei Tabak: I. Entstechungsweise und Wesen Küntslich erzeugter Gene-Mutanten. *Genetica*, **16**: 111-152.

Tollenaar D. 1938. Untersuchungen über Mutation bei Tabak: II. Eigine Künstlich erzeugte Chromosomen-Mutanten. *Genetica*, **20**: 285-294.

Van Harten A.M. 1998. Mutation Breeding: Theory and Practical Applications. Cambridge University Press.

Wang L.Q. 1991. Induced mutation for crop improvement in China. *In*: *Proc. IAEA Symp. on Plant Mutation Breeding for Crop Improvement*. June 18-22, 1990. IAEA, Vienna, pp. 9-32.

Weck E., Maluszynski M., Van Zanten L., Ahloowalia B. and Nichterlein K. 1996. Mutation techniques and related molecular technologies in plant Breeding. *In*: *Proc. Conference on Hundred Years of X-rays and Radioactivity*. BARC, Mumbai. 21-24 February, 1996. (eds.) Sood D.D., H.C. Jain A.V.R. Reddy, K.L. Ramakumar and S.G. Kulkarni. BARC, Mumbai. pp. 489-504.

Wong R.C.S. and Swanson E. 1991. Genetic modification of canola oil: high oleic acid canola. pp. 154-164. *In*: *Fat and Cholesterol Reduced Food.* (eds.) Haberstroh C. and C.E. Morris. Gulf, Houston.

Yang Y., Shah J. and Klessig D.F. 1997. Signal perception and transduction in plant defense responses. *Genes & Development*, **11**: 1621-1639.

Yang X., Yie Y., Zhu F., Kang L., Wang X. and Tien P. 1997. Ribozyme-mediated high resistance against potato spindle tuber viroid in transgenic potatoes. *Proc. Natl. Acad. Sci.*, USA, **94**: 4861-4865.

Plant Breeding - Mendelian to Molecular Approaches
H. K. Jain and M. C. Kharkwal (eds.)

27

Induced Mutations and Selection Techniques for Quantitative Traits

B. Sharma[1] and S.K. Sharma[2]

Abstract

It has been concluded from the theory of quantitative genetics that the polygenic traits are controlled by a multiplicity of minor genes with small but cumulative (additive) effect, and these "minor" genes can mutate with equal probability in both directions, i.e. from the normal (dominant) state to the nonfunctional/dysfunctional (recessive) state and back. The experience gained over the last several decades raises serious questions about the validity of this model to explain continuous variation. Since all genes, major or minor, are made of the same genetic material (DNA), their mutation patterns cannot be different. It is well established that the forward mutations (from dominant to recessive) are induced thousands (or millions) of times more frequently than the back mutations (from recessive to dominant state). Thus, the possibility of shifting population means for a particular character in both directions (plus or minus) through mutations in character-specific genes with equal probability is ruled out. The genes controlling metabolism of a trait (or their mutations) frequently have dosage effect, but they all fall in the category of "major" genes. Therefore, micromutations, which are presumed to be induced in "minor" genes, could be a consequence of minor changes in the functions of many other genes or intergenic interactions which indirectly influence gene expression. Multifactorial inheritance, incomplete dominance, dosage effect, pleiotropy, epistasis, multiple allelism, penetrance and expressivity, conditional (leaky) mutations, various kinds of regulatory mechanisms, and environmental influences on gene expression are very robust phenomena, and each one of them is capable of creating situations that will appear to be a consequence of quantitative variation. Minor changes in character mean and variance — the two parameters to estimate quantitative change — at population level following mutagenic treatment have been reported from a large number of studies. Even varieties carrying the so-called micromutations have been commercialized. Selection can strengthen and stabilize a mutated population. Selection procedures with greater efficiency are discussed.

[1]Indian Agricultural Research Institute, New Delhi 110 012, India
[2]Himachal Pradesh Agricultural University, Palampur 176 062, India

Introduction

There is little need to elaborate on the importance of quantitatively inherited characters in agricultural plants, since most economic traits are treated as polygenic and the major effort of plant breeders is devoted to handling of such characters. The efficiency of a plant breeding programme is determined by the amount of genetic variability available for specific traits in the breeding populations and the efficiency of selection. In conventional plant breeding, variation is generated by hybridization, and selection made in segregating populations. Induced mutations can either generally supplement or partly replace hybridization as the source of variability in specific cases.

Over the past 70 years, numerous attempts have been made to induce beneficial mutations in agricultural plants. Much of this work was concentrated on characters determined by genes with large effects, such as disease resistance, and the more readily observable morphological mutants, like short straw and dense-ear mutants of cereals. Even though a number of potentially useful new strains from mutations in the genes with small and additive effect controlling such characters as yield, maturity time, and plant size were described in literature, it is disappointing to find that little attention has been devoted either to the deliberate selection of such mutants or to the study of mutational behaviour of the characters concerned. Selection based entirely on major phenotypic change practically excludes all the induced variation in quantitatively inherited characters; moreover, such mutations are generally associated with gross pleiotropic effects, leading to reduction in performance compared to their parent varieties. In some studies, induced variation in quantitatively inherited characters was observed more or less by accident; a few workers found it by examining their material critically and then applying selection for such variation, and a small number have deliberately designed experiments to study the nature and magnitude of the induced quantitatively inherited variability.

This review will largely be confined to the last category of studies, for, in these experiments critical data have been accumulated, from which generalizations can be made. It will mostly be restricted to self-fertilizing species in which maximum information has been generated.

The Concept of Micromutations

One of the oldest controversies in the theory of evolution relates to the importance of small variation. In the nineteenth century, Charles Darwin, in particular, discussed this question extensively. Our present concept of mutation is based on the work of Hugo De Vries (1901). Although the term "mutation" has an old history, seldom it was used by Darwin. It was De Vries, who returned to this concept and gave a definite meaning to it, which in most parts is still valid. It is true that De Vries usually thought of mutations as relatively large and conspicuous heritable changes, which gave rise to discontinuous variation. But contrary to what is stated in some text books, De Vries (1901) also clearly included in this term the mutations with small effects. Johannsen (1913), in his bean experiments, described not only spontaneous drastic mutations, but also slight mutations affecting the seed index (the proportion of width *versus* length). This is a character which closely falls into the class of continuous variation, and hence Johannsen may be regarded as the first who really proved the existence of spontaneous small mutations. Baur (1924) emphasized repeatedly the importance of small mutations ("Kleinmutationen") for evolution process resulting from the accumulation of a large number of mutations, majority of which exert only slight effects. Stubbe (1934) described the small mutations for the first time in higher plants such as *Antirrhinum*. Knapp (1950) presented clear suggestions about how to utilize these small mutations in barley. It was Gregory (1955) who, working with peanuts, published the first well-planned experiments on X-ray induced genetic variability for yield. Gaul and Mittelstenscheid (1961) used a systematic approach for the selection of small mutations in barley.

Mutation Breeding: For and Against

Plant breeders with experience in mutation breeding tend to join one of the two groups: (*a*) those who embrace mutation breeding, and (*b*) those who malign it. The former group is of the view that mutation breeding is an established method of crop improvement. It cannot serve as a substitute for the conventional methodologies, but is useful as an addendum to the current procedures. Plant breeders and geneticists of the second category use two rebuttals against the claims of the "mutation breeding enthusiast": (*a*) The amount of natural variability in each crop species is so great that plant breeders are unable to utilize it properly and completely, and (*b*) mutagen-derived populations are loaded with "junky or deleterious" mutations. Nevertheless, to-date, worldwide 2252 mutant varieties have been registered in the FAO/IAEA Mutant Varieties Database of the Plant Breeding and Genetics Section, IAEA, Vienna (Maluszynski *et al.*, 2000). Of these 1585 were released as direct mutants and 667 through cross breeding. Some of them have made a major impact on crop productivity and achieved great economic success. It clearly establishes that mutation breeding still remains an important breeding method of plant improvement. Most of the work on mutation and its application in different crops deals with the induction, identification, isolation, and use of drastic changes in phenotype, brought about by mutational events in the so-called major genes. Lately, the effect of mutagenic factors has been examined in connection with "minor" genes, which are assumed to represent the genetic background for characters with continuous phenotypic variation. The delay in dealing with such quantitative characters, in spite of the fact that majority of production traits in plants and animals show continuous variation, seems to be due to the difficulty in identifying at phenotypic level single micro mutations in a genotype. This is due to the small effect of changes in individual genes and the role of environment which interacts with the genotype to determine quantitative characters.

When Must Selection for Quantitative Traits be Undertaken?

It is generally accepted in breeding of self-pollinated crops that evaluation for quantitative traits affected by multiple loci should be delayed by a few generations after crossing till the population is stabilized. Several studies have suggested that such a breeding method also applies to mutation breeding, that is, selection should be delayed until the M_3 or later generations following mutagenic treatment. In addition, experiments have shown that recurrent mutagenic treatments provide a greater amount and larger range of genetic variability than a single mutagenic treatment. In some cases, depending on the trait in question, selection in later generations after single or recurrent mutagenic treatments may be more rewarding than a single M_3 selection. As predicted by Yonezawa *et al.* (1971), the optimum generation for the initial selection and its efficiency would be influenced by many factors, such as, ontogenetic stage of mutagenic treatment (seed or pollen), character to be improved, and size of the mutagenized population. Therefore, a systematized methodology, by which the stage of delayed selections in breeding schedule can be determined, needs to be developed.

A few attempts have already been made in this direction. Sharma and Sharma (1982a,b) compared the quantum of induced variability for quantitative characters in the M_2 and M_3 generations. Neither varietal nor mutagen-dependent differences were recorded in the genetic variability in both generations. However, characters differ in the manifestation of variability in different generations. It was also reported (Sharma, 1986) that selection for micromutations can be carried out even in the M_2 generation and the efficiency of early generation selection varies from character to character. The M_2 progenies segregating for macromutations are likely to yield greater variability for polygenic traits, which is directly related to mutagenic damage recorded in the M_1 generation. Characters like flowering time, pod bearing branches, peduncles per plant, pods per plant, and seed yield were reported to exhibit more variability than seeds per pod and 1000-grain weight in M_2 generation of lentil following mutagenesis with chemicals and ionizing radiations (Sarker and Sharma, 1988).

Pattern of Induced Quantitative Variation

Several workers have attempted to assess mutagen-induced variation in quantitative characters and have made projections about the progress that can be achieved through subsequent selection. An increase in variance irrespective of character, symmetrical or skewed, has been the general observation while the mean mostly remained unchanged and sometimes even decreased. The hypothesis forwarded by Brock (1965) to explain such behaviour of induced mutations in quantitatively inherited traits is that random mutations are expected to increase the variance and shift the mean away from the direction of previous selection history. However, Gaul and Aasveit (1966) concluded that the mutations for quantitative characters are not related to genotype, and with random mutation a change in the mean value of almost any quantitative character is to be expected, in a direction associated with vitality. Gregory (1968) also stated that mutations with very small phenotypic effect will occur with high frequency and will have an equal probability of being positive or negative in their phenotypic effect.

Methods of Detecting Induced Mutations for Quantitative Traits

As a consequence of the peculiarity of phenotypic manifestation of quantitative characters, the only method available to detect the induction of new variation for quantitative traits following mutagenic treatment is the estimation of mean and variance by statistical methods. The mean values of quantitative traits in populations derived from irradiated gametes (pollen grains) or embryos (dormant seeds) are in most instances lower in the treated than in untreated populations. In a very extensive study, Scossiroli (1966a,b) and Scossiroli *et al.* (1966) reported a similar effect in the same population for a large number of characters. It has been found, however, that the difference between means of the treated and untreated populations can be reduced in subsequent generations through selection. The effect of radiation on the means has been interpreted to be due to detrimental mutations, which are supposed to occur more frequently than the favourable ones, and which can be removed through selection in subsequent generations. The difference sometimes observed in the behaviour of different crop plants has been accounted for by ploidy effect on the selection efficiency against gametes with unfavourable genes (Bhatia and Swaminathan, 1962). The same argument has been used to explain the skewness of distributions for some traits observed in irradiated populations. For other traits, however, no significant deviation from symmetry is found.

The estimates of phenotypic variability for different morphological or production traits are larger in irradiated than in non-irradiated populations. In general, the methods used to identify environmental and genetic components of phenotypic variance are based on the estimates of the expected components of variation obtained in analysis of variance following a hierarchic scheme. Details of such methods may be found in Palenzona (1965) and Scossiroli *et al.* (1966). Also, the methods of analysis of variance devised for diallel crosses (Jinks, 1954; Hayman, 1954, 1958) have been used in some instances (Lawrence, 1965). In general, biometrical analyses of variation have shown that increase in phenotypic variation in generations following irradiation, particularly in self-pollinated plants, is largely due to increase in the genetic component, and it may be accounted for by the effects of mutations in the genes governing quantitative characters. From a plant breeder's point of view, this is an important effect, because larger genetic variation means the possibility of greater response to selection, and higher chances for improvement.

It has been found that genetic variability increases with radiation dose, but the relation between variance and dose is not linear. It has also been found that when a wider range of doses is used, the intermediate dose induces the largest increase in variation and there is a strong decline with the highest dose. The lack of linearity could be due to elimination of multiple deletarious mutations through gametic or zygotic selection and loss of potential genetic variability. Estimates of mutation rate and dose-rate relationship have been obtained in some cases. However, since the number of genes determining a quantitative character is difficult to estimate with reasonable precision, at present identifying only the dose of mutagenic treatment required to enhance genetic variability of a population by a significant margin is of relevance. Only few estimates are available. In rice, Oka *et al.* (1958), assuming a proportional increase of genetic variance with radiation dose, have estimated an increase of 0.0153 units of genetic variance per 1000 R for earliness in heading and 0.0839 per 1000 R for plant height. Estimates of the increase of genetic variance in treated populations were also obtained by Scossiroli (1966a,b and other papers) in durum and bread wheats. But larger dose ranges have to be investigated before conclusions can be drawn regarding the relation between radiation dose and increase of genetic variation.

Relationship between Induction of Macro- and Micromutations

Comparative studies based on macromutation data revealed that the chemical mutagens, particularly alkylating agents, are more effective than ionizing radiations (Brock, 1976; Sharma and Sharma, 1979; Sharma, 1966, 1986; Sarker and Sharma, 1988; Singh and Sharma, 1989; Vinod, 1994; Solanki and Sharma, 1999a,b; Kharkwal, 2001). Nevertheless, mutation induction through the application of radiation was most frequently used (89 % studies) for directly developed mutant varieties, whereas the use of chemical mutagens was relatively infrequent (Nichterlein *et al.*, 2000). Brock (1976) attributed this disparity to later introduction of chemical mutagens. Now it is well established that there is a clear deficiency of information about the influence of chemical mutagens on quantitative characters compared to radiations. It is possible that chemical mutagens may prove to be a better alternative for inducing micromutations, as they induce mutations at a much higher rate and cause less chromosomal disturbances than radiations (Sharma, 1968, 2001).

Effect of Mutagens on Association Between Quantitative Characters

The effect of mutagens on the association between traits in relation to the control has been studied to understand the possible interdependence between characters, their role in the general fitness of the population, the possible occurrence of correlated response to selection, performance of any trait of economic value, and to confirm if any correlation might be changed at random or in a definite direction after induction of new genetic variability (Scossiroli *et al.*, 1966; Sharma and Sharma, 1981, 1984). Comparing the results obtained from the control and irradiated populations, it may be seen that in a few instances the relations between the traits are strengthened in the irradiated population, pointing to a common effect of mutation on the associated traits (pleiotropic effect). In most cases, however, no change is observed and in a very few instances a definite weakening of the phenotypic correlation is observed.

Comparative Induced Variation Through Mutagenesis and Hybridization

A question of obvious practical importance is whether the mutation technique can be effectively used to improve quantitative characters in agricultural plants. Should mutations be considered primarily as an alternative or a supplement to hybridization? To answer this question one has to determine both the magnitude and quality of induced variation. Krull and Frey (1961) made a direct comparison of the magnitude of variation induced by irradiation of parent varieties with the variation generated in the F_2 of a cross between these varieties. The radiation-induced variation generated in the parents was approximately 50 % of that occurring in the F_2. However, it was important that additional variation could be generated in the F_2 by irradiation at F_1 stage. Such a possibility was experimentally demonstrated in pea by Mohan and Sharma (1995).

The most effective way of using the variation for quantitative characters may be to combine the techniques of augmenting variation through hybridization between well-adapted varieties and mutagenesis. Brock (1965) however suggested that variation through mutagenesis will be quickly utilized if induced through cycles of recombination between the selected lines, rather than repeat the induction of variation by successive cycles of radiation alone and selection. Thus, mutation induction is suggested as an alternative to wide hybridization, as a less disruptive method of injecting variation into well adapted genotypes. The effect of these mutations will not be restricted to the character under selection, if selection pressure is maintained for all important characters. The results reported for agricultural species as well as *Drosophila* and yeast support the view that real progress can be achieved. This method in no way precludes the role of mutation as an aid to the domestication of species new to agriculture, or its use in inducing specific mutations (Brock, 1965).

Present View

The published reports available suggest that the small mutations, or micromutations, may affect all morphological and physiological characters, as do the large mutations, or macromutations. These are either minute or drastic mutational changes, respectively. In fact, many mutations result in intermediate types, either as a result of dosage effect (micromutation) or multiple allelism (macromutation). The nature of change in the genetic material must be the same irrespective of whether the phenotypic effect is large or small. There is no reason to assume that polygenes and major genes are basically different types of genetic units. All genes are made of DNA. It is well known that environmental effects may change phenotypic expression of genes. Under one environment a metabolic step affected by a mutation may show a greater change than under another environment.

Mutations with continuous or discontinuous variation of characters must occur as a consequence of similar molecular changes in the genetic material. Moreover, both types of variation depend on the environment for character expression. For instance, under uniform growing conditions (e.g. in growth chambers), segregation for one or a few genes affecting a so-called quantitative character may be realized. In spite of very little individual variations, the phenotypes can still be put into distinct classes. Finally, the classification of characters as continuous or discontinuous may also depend on the nature of a large number of multiple alleles influencing phenotype with minor effects in a population.

It is therefore obvious that the distinction between qualitative (monogenic) and quantitative (polygenic) characters depends on the magnitude of effect. The concept of "minor" genes with slight phenotypic but cumulative effects is highly disputed as the existence of any such genes in large numbers spread all over the genome, thereby giving independent assortment and dosage effect and has no experimental support. On the one hand, there is no evidence for huge amount of genes controlling specific traits and capable of mutating in both directions with equal probability. On the other hand, there are several characteristics for which typical F_2 segregation into discrete classes is never obtained and the entire population shows continuous variation. Almost always, such properties are of a complex nature and can be further split into distinct components with Mendelian inheritance. For example, seed size is made of length, width and thickness. All three properties may behave as monogenic traits. If single-seed weight is to be considered as a genetic property, then seed density and relative proportion of its constituents will also be added. It is beyond imagination that all the constituent characters can be changed simultaneously in the same direction by mutation. If the component characters are changed to variable degree, and randomly in positive as well as negative directions, enough opportunity is created for the "final" composite characteristic (seed weight, for example) to show continuous variation. Multifactorial inheritance, multiple allelism, incomplete dominance, dosage effect, various kinds of regulatory mechanisms, pleiotropic and epistatic effects, penetrance and expressivity, conditional (leaky) mutations and environmental influences on gene expression are very robust phenomena, recognised early in the history of genetics. Each one is capable of modifying character expression and create an impression of continuous variation. All these mechanisms working in tandem further add to the complexity. They can also create a series of phenotypes with a range of character expression in a given population. None of these points towards existence of multiplicity of "minor" genes with additive effect and capable of mutating in forward and reverse directions with equal probability.

In the light of the above discussion, it may be concluded that the so-called micromutations with polygenic control of specific traits (as understood in genetics) are not necessarily changes in the hypothetical genetic elements called "minor" genes. From a plant breeder's point of view, it is not enough to secure a change in mean and variance. More important is to achieve a clearly visible change in an economic trait in the desired direction. Few published reports are available to justify such expectations (e.g. Kharkwal *et al.*, 1988). This could be one possible reason for nearly complete absence of popular varieties developed through the micromutational route except in case of chickpea (Kharkwal *et al.*, 1988), even though a very effective methodology was developed to increase the proportion of plants with minor genetic changes in the mutagenized populations and enhance the chances of success through selection in a short period (Sharma, 1986, 1995).

References

Baur E. 1924. Untersuchungen uber das Wesen, die Entstechung und die Vererbung von Rassenunterschie-den bei *Antirrhinum majus*. *Bibliotheca Genet.*, **4**: 1-70.

Bhatia C.R. and Swaminathan M.S. 1962. Induced polygenic variability in bread wheat and its bearing on selection procedures. *Z. Pflanzenzuchtg*, **48**: 317-326.

Brock R.D. 1965. Induced mutations affecting quantitative characters. *Radiat. Bot.*, **5** (Suppl.): 451-464.

Brock R.D. 1976. Quantitatively inherited variation in *Arabidopsis thaliana* induced by chemical mutagens. *Envir. Exp. Bot.*, **16**: 241-253.

De Vries H. 1901. *Die Mutationstheorie.* Verlag Von Veit und Comp, Leipzig.

Gaul H. and Aastveit K. 1966. Induced variability of culm length in different genotypes of hexaploid wheat following X-irradiation and EMS treatment. *Savremena Polgopr.*, **14**: 253-276.

Gaul H. and Mittelstenscheid L. 1961. Untersuchungen zur selekton von Kleinmutationen bei Gerste. *Z. Pflanzenzuchtg*, **45**: 300-314.

Gregory W.C. 1955. X-ray breeding of peanuts (*Arachis hypogaea* L.). *Agron. J.*, **47**: 396-399.

Gregory W.C. 1968. A radiation breeding experiment with peanuts. *Radiat. Bot.*, **8**: 81-147

Hayman B.I. 1954. The analysis of variance of diallel tables. *Biometrics*, **10**: 235-244.

Hayman B.I. 1958. The theory and analysis of diallel crosses. II. *Genetics*, **43**: 65-85.

Jinks J.L. 1954. The analysis of continuous variation in a diallel cross of *Nicotiana rustica* varieties. *Genetics*, **39**: 767-788.

Johannsen W. 1913. *Elemente der exakten Erblichkeitslehre.* Verlag Gustav Fischer, Jena, 723 p.

Kharkwal M.C. 2001. Induced mutations in chickpea (*Cicer arietinum* L.) V. Evaluation of micro-mutations. *Indian J. Genet.*, **61**: 115-124.

Kharkwal M.C., Jain H.K. and Sharma B. 1988. Induced mutations for improvement of chickpea, lentil, pea and cowpea. pp. 89-109. *In*: *Improvement of Grain Legume Production using Induced Mutations.* Proc. FAO/IAEA Workshop, 1-5 July, 1986, Pullman, Washington, USA, IAEA, Vienna, 1988.

Knapp E. 1950. Grundfragen der experimentellen Mutationsauslosung in ihrer Bedentung fur die praktische Pflanzenzuchtung-Vortrog auf der *Pflanzenzuchtung* 1950 Einbeck: 1-20.

Krull C.F. and Frey K.J. 1961. Genetic variability in oats following hybridization and irradiation. *Crop Sci.*, **1**: 141-146.

Lawrence C.W. 1965. Radiation induced polygenic mutations. *Radiat. Bot.*, **5** (Suppl.): 491-496.

Maluszynski M., Nichterline K., Van Zantan L. and Ahloowalia B.S. 2000. Officially released mutant varieties - the FAO/IAEA database. *Mutation Breeding Review*, **12**: 1-84.

Mohan D. and Sharma B. 1995. Increasing induced polygenic variability in pea. p. 1333-1342. *In*: *Genetic Research and Education: Current Trends and the Next Fifty Years.* (eds.) B. Sharma *et al.*, Proc. Golden Jubilee Symp., 12-15 February, 1991, New Delhi, Indian Soc. Genet. and Plant Breeding., New Delhi.

Nichterlein K., Bohlmann H., Nielen S. and Maluszynski M. 2000. Achievements and trends of using induced mutations in crop improvement. *In*: Proc. DAE-BRNS Symp., Dec. 6-8, 2000, Mumbai, India. pp. 27-35.

Oka H.I., Hayashi J. and Shiojiri I. 1958. Induced mutation of polygenes for quantitative characters in rice. *J. Hered.*, **49**: 11-14.

Palenzona D.L. 1965. Stima della varianza dovuta all 'effetto ambientale nell' analisi di tipo gerarchico. *Genet. Agr.*, **19**: 338-350.

Sarker A. and Sharma B. 1988. Mutagenic effectiveness and efficiency of gamma rays, ethyl methane sulphonate, nitroso-N-ethyl urea and sodium azide in lentil. *Thai J. Agric. Sci.*, **21**: 323-327.

Scossiroli R.E. 1966a. Wheat mutagenesis in quantative traits. *Hereditas*, **2** (Suppl.): 85-101.

Scossiroli R.E. 1966b. Consequenze sui caratteri quantitativi del trattamento dei semi con raggi × in *T. durum*. eterogeneita di progenie derivate de semi trattati. *Atti. Ass. Genet. Ital.*, Paduna, **11**: 147-160.

Scossiroli R.E., Palenzona D.L. and Scossiroli-Pellegrini S. 1966. Studies on the induction of new genetic variability for quantitative traits by seed irradiation and its use for wheat improvement. *In*: *Mutations in Plant Breeding*, (Proc. Panel, Vienna, 1966). IAEA, Vienna: 197-229.

Sharma B. 1966. Influence of some physical and chemical agents on the mutation process in peas (*Pisum sativum* L.) (in Russian). *Genetika* (USSR), **1**: 45-52.

Sharma B. 1968. On the diverse aberration–mutation ratio induced by NMU. *Curr. Sci.*, **37**: 675-676.

Sharma B. 1986. Increasing the efficiency of mutagenesis for micromutations by early generation selection. *Indian J. Genet.*, **46** (Suppl.): 88-100.

Sharma B. 1995. Mutation breeding through induced polygenic variability. *In*: *Genetic Research and Education: Current Trends and the Next Fifty Years.* (eds.) B. Sharma *et al.*, Proc. Golden Jubilee Symp., 12-15 February, 1991, New Delhi, Indian Soc. Genet. and Plant Breeding, New Delhi: 1210-1219.

Sharma B. 2001. Mutagenicity of a new dimethyl nitroso compound in pea (*Pisum sativum* L.). *Indian J. Genet.*, **61**: 235-237.

Sharma S.K. and Sharma B. 1979. Pattern of induced mutability in different genotypes of lentil (*Lens culnaris* Medik.). *Z. Pflanzenzuchtg*, **83**: 315-320.

Sharma S.K. and Sharma B. 1981. Effect of mutagens on character association in lentil. *Indian J. Agric. Sci.*, **51**: 619-622.

Sharma S.K. and Sharma B. 1982a. Differential manifestation of macromutations in different quantitative characters in lentil. *Genetika* (Yugosl.), **14**: 285-293.

Sharma S.K. and Sharma B. 1982b. Induced polygenic variability in lentil (*Lens culinaris* Medic.). *Genet. Agr.*, **36**: 9-18.

Sharma S.K. and Sharma B. 1984. Pattern of induced macro- and micro-mutations with gamma-rays and NMU in lentil. *Env. Expt. Bot.*, **24**: 343-351.

Singh N.P. and Sharma B. 1989. Efficiency of selection for micromutations in M_2 generation in peas. *Pisum Newsletter*, **21**: 61-62.

Solanki I.S. and Sharma B. 1999a. Induction and exploitation of polygenic variability in lentil (*Lens culinaris* Medik.). *J. Genet. & Breed.*, **53**: 79-86.

Solanki I.S. and Sharma B. 1999b. Induction and isolation of morphological mutations in different mutagenic damage groups in lentil. *Indian J. Genet.*, **59**: 479-485.

Stubbe H. 1934. Einige Kleinmutationen von *Antirrhinum majus* L. *Zuchter*, **6**: 299-303.

Vinod. 1994. Study of screening procedures for micromutations in dwarf peas. Ph.D. Thesis. Indian Agricultural Research Institute, New Delhi.

Yonezawa K., Ichii M. and Yamagata H. 1971. Theoretical basis for determining the number of X_2 plants required for improvement of quantitative characters of self-pollinated crops. II. Complement to the theory and numerical computations. *Japan. J. Breed.*, **21**: 160-168.

Plant Breeding - Mendelian to Molecular Approaches
H. K. Jain and M. C. Kharkwal (eds.)

Collection, Conservation and Utilization of Plant Genetic Resources

P.L. Gautam[1], B.B. Singh[2], Sanjeev Saxena[3] and Rajiv K. Sharma[2]

Abstract

Biodiversity conservation is essential for maintaining ecological balance among different living forms of the planet and for self-sustaining growth in crop production processes. Plant genetic resources (PGR) comprise one of the most crucial components of agro biodiversity, which sustains humankind by meeting its demands of food, fodder, fibre and fuel. In addition to ensuring food security, protecting environment and reviving lost resources, the heightened PGR related activities are also fuelling the development of new industrial products as well as new research breakthroughs. Recent developments in molecular biology and biotechnology are adding new dimensions to the utilization pattern of biodiversity as the unit of utilization has been scaled down to DNA level. Even though all the political boundaries of the world are not equally blessed with this most essential life-supporting gift from nature, its just availability for humankind is needed. The immensely rich Indian gene centre with 12 mega bio diversity centres, harbours two Vavilovian centres of origin/centres of diversity of crop plants. The extent of utilization of PGR has not been uniform across different countries. The developed countries are stock piling the rare wild species available anywhere along with hitherto unknown applications of them. These novel PGRs also offer scope for discovering new genes. Current conservation strategies emphasize more on *in situ* conservation including participatory plant breeding. Concerted international efforts are underway to streamline and regulate the accessibility of PGRs at global level. In addition to trade related intellectual property rights system, other recent developments *viz.*, inequitable nature of free access, plant breeder's rights, convention on biological diversity, prior informed consent and global action plans are also discussed.

Introduction

The challenge of creating and maintaining a sustainable environment is the single most pressing issue that confronts the humankind today. The wild and domesticated

[1]G.B. Pant University of Agriculture & Technology, Pantnagar 263 145, India
[2]Indian Agricultural Research Institute, New Delhi 110 012, India
[3]National Bureau of Plant Genetic Resources, New Delhi 110 012, India

components of biological diversity are the primary source from where the humanity derives all of its food, many medicines and industrial raw material. The term biodiversity was introduced by Wilson and Frances (1988) as a contraction of the term biological diversity. Biodiversity refers to the variety of life forms and can be divided into three categories (Banham, 1993):

Genetic diversity	-	the genetic variation within and between populations (all possible gene combinations within one species)
Species diversity	-	the variation among species within one region
Ecosystem diversity	-	the variation of components of a given ecosystem, providing the necessary conditions for populations of species.

Agricultural biodiversity or agro-biodiversity is a subset of biodiversity defined broadly by FAO (1996a) as that part of biodiversity that nurtures people and is nurtured by the people. Qualset *et al.* (1995) defined agro-biodiversity as all crops and livestocks, their wild relatives, and all interacting species of pollinators, symbionts, pests, parasites, predators and competitors. The recent advances in biotechnological research have added a new dimension wherein the genetically modified organisms also form a critical element of the agro-biodiversity. The importance of the functional role of agro-biodiversity has been stressed by Swift and Anderson (1994) in their division of the biotic components of agrosystems into three types:

Productive	-	biota including crop plants and livestocks, producing food, fibre or other products for consumption
Resource	-	biota contributing positively to the productivity of the system *viz.*, pollinators, plants of fallows, and soil biota controlling nutrient cycling
Destructive	-	biota includes weeds, pests, and pathogens

According to this classification, management would require role of farmers and scientists in increasing crop and animal production by encouraging resource biota and discouraging destructive biota.

The term Plant Genetic Resources (PGR) is used to define one of the most crucial components of agro-biodiversity, which directly or indirectly sustains humankind. It is not possible to create these genetic resources artificially except to a very limited extent through mutation breeding or more recently through the application of modern techniques of molecular biology, which facilitate transfer of genetic material between distant relatives. These resources generally referred to as germplasm or genetic material, consist of reproductive or vegetative propagating material of plants. These may include seeds, tissues, cells, pollen, DNA molecule or any other component which contains the functional unit of heredity and can be used in conventional or modern breeding programmes.

Importance of Plant Genetic Resources

The unknown potential of genes, species and ecosystems is of inestimable but certainly high value. The scientists have the task of improving crop production for satisfying the ever increasing demand for food, fodder, fibre and fuel by (i) keeping the modern agricultural crops one step ahead of thousands of pests/diseases and (ii) developing crops which can be grown on marginal lands. They also have the challenge of maintaining sustainable agriculture and protecting the environment. Therefore, the plant breeders in their search for desirable genes have to depend upon the genetic resources.

Food Security

Groombridge (1992) had estimated the existence of 400,000 plant species, while Heywood (1995) estimated the existence of as many as 300,000. Ten per cent (i.e. 30,000) of these plants are edible and about 7,000 have been cultivated or collected by humans for food or other agricultural purposes at one time or another (Wilson, 1992). Only 30 species of all these crops, "feed the world", i.e., they provide 95 % of dietary energy or protein (Virchow, 1999). However, only three crops, *viz.*, rice (26 %), wheat (23 %) and maize (7 %) together account for over 50 % of the global food energy (FAO 1996a).

Potential Development of New Industries

Many organisms have evolved materials whose unusual physical properties make them suitable for making various useful products. They may be obtained from the wild or synthesised by biochemists and in many cases, finding one useful material may lead to the discovery of many more, with subtle differences in physical properties, from related organisms *viz.*, spider silk (with potential to be strong enough to make bullet-proof vests and lightweight materials for the aerospace industry), shellac (used in varnishes, paints, stiffeners and other uses), casein, (used extensively in glue manufacture as well as in plastics and paints), etc.

The potential developments of new industries, based on naturally occurring biological materials, have been increasingly recognised in recent years. Institutions and individuals systematically identify plants and animals with potential commercial value using the scientific practice of "bioprospecting". Moreover, there is an increased realisation that overseas countries have had significant commercial successes in exploiting the Indian bioresources (e.g. turmeric, neem, karela, jamun, Basmati etc.), in many cases with little or no benefit to the Indian community.

Basic Research

Materials in germplasm collections can be used in basic research, for example research on the origins of cultivated plants, the origins of civilisation, the sociology and economics of subsistence societies, and can even test theories of evolution. Advances in molecular biology will greatly expand ways to incorporate germplasm collections into such studies.

For example, molecular markers, cloning and identifying specific genes, someday will allow wide ranging studies of collections from around the globe. Such studies can elucidate crop origin, divergence and relationships. The preceding knowledge paired with sociological studies on how farmer varieties were maintained and used, will help clarify the roles of diversity, utility, and fashion in development of such varieties.

Well-categorised collections with complete passport data can serve experimentalists, who study a range of genotypes or genes. For example, molecular studies of series of pest resistance genes, paired with knowledge of the ecology of the region of their occurrence, can increase our understanding of how resistance, tolerance, and durability interact with pest populations, and with each other.

Reviving Lost Resources

Sometimes genetic enhancement enables production of a crop in an area where it had been abandoned, due to disease or insect problems. In the 1960s, virus and mycoplasm diseases adversely affected maize production in the U.S. mid-south states. Enhancement of adapted inbreds and hybrids with germplasm from tropical and subtropical maize, as well as from non-adapted temperate sources, enabled production of new inbreds and hybrids with disease tolerance. Breeders in this case reacted so quickly that only a dip in maize production occurred in the affected regions. Similarly, the cultivation of local varieties had been banned during the inhuman and coercive regime of Pol Pot in Cambodia. After considerable human misery caused by the war, the country realised the need for increasing agro-biodiversity. At that time, the seeds stored in IRRI's gene bank came to the rescue to rehabilitate the *in situ* diversity (Cunningham, 1987).

Environmental Services

Wild organisms carry out many functions in the environment that are vital to us and would be very difficult to complement ourselves. Their role in pollination (bees, bats), biodegradation (micro- organisms), soil aeration (earthworms), fertilisation (soil bacteria), carbondioxide-oxygen exchange and water storage (plants) has been well recognised.

Bio-remediation or phyto-remediation (if plants do it) refers to the use of living organisms to clean up toxic wastes. Some plant species that live naturally on soils that are rich in heavy metals have evolved biochemical mechanisms for extracting these metals from the soil and accumulating them to very high levels in their tissues. These plants called hyperaccumulators, accumulate copper, nickel, lead, cadmium, chromium, zinc, cobalt, mercury and selenium from toxic waste sites. They can be burned in order to recover the metal in cases where the metal is valuable (copper and nickel). With less valuable metals such as lead, the hyperaccumulating plants are much easier to dispose of than the contaminated soil.

Distribution of Diversity

The diversity in PGR created by thousands of years of agriculture was not equally distributed throughout the globe; some regions are blessed with astonishing diversity, while other areas were relatively impoverished. Vavilov (1926) reasoned that genetic diversity and crop origins were related and by locating the geographic region in which one found the greatest genetic diversity for a crop, its origin can be pinpointed. He theorised that the world's crops originated in eight definable centres located in the third world.

(1) China; (2) India with related centre in Indo-Malaya; (3) Central Asia; (4) The Near East; (5) The Mediterranean; (6) Abyssinia (Ethiopia); (7) Southern Mexico and (8) Central America; and South America (Peru, Ecuador and Bolivia), with two lesser centres (the island of Chiloe off the coast of southern Chile and an eastern secondary centre in Brazil and Paraguay). Basically Vavilov drew lines around areas in which agriculture has been practised for a very long time and in which indigenous civilisations arose. Some concentrations of diversity can be detected, to be sure, but they have little or nothing to do with origins. The concept of secondary centres was invented to account for the fact that centres of diversity are not the same as centres of origin. Also the concept of secondary crops i.e., those derived from weeds of older primary crops was developed.

In recent years, several analyses of world collections or parts of world collections have been made and published, especially in wheat, rice, barley, maize and other major crops. In some as many as 40 isozyme loci have been analysed using electrophoresis. In others flavenoids, seed proteins were studied employing sophisticated computer programs for data analysis. Researchers have subsequently redrawn their boundaries to reflect the growing knowledge of agricultural origins and diversity. The numbers and the size of the centres have been increased and scientists now recognise 12 centres of diversity/origin not all of which lie in the third world.

Patterns of Variation

The Vavilovian concepts of "centre of origin" were too simplistic and when one actually analyses origins crop by crop, it soon becomes apparent that many of them did not originate in Vavilovian centres. Some do not have centres of diversity. Several can be traced to very limited and specific origins and others seem to have originated all over the geographical range of the species. The pattern is much more complex and diffuse than Vavilov visualised. Harlan (1992) has recognised a number of distinct situations which best describe the main patterns of variation. There are five categories as mentioned below:

Endemic: Crops that originated in a limited area and did not spread appreciably. Good examples include the small grain cereals from West Africa, (e.g. *Brachiaria deflexa* and *Digitaria exelis*)

Semiendemic: Crops that originated in a definable centre and with limited dispersal. Examples include two Ethiopian domesticates, *viz*., *Eragrostis tef* and *Guizotia abyssinica*

Monocentric: Crops with a definable centre of origin and wide dispersal without secondary centres of diversity. Crops in this class are mostly plantation and industrial crops such as rubber in the Amazon rainforest and Arabica coffee in Ethiopia.

Oligocentric: Crops with a definable centre of origin, wide dispersal and one or more secondary centres of diversity. The whole Near East complex of barley, emmer-wheat, flax, pea, lentil, oats and chickpea fall in this category; all have secondary centres in Ethiopia and some also have centres in India and/or China.

Noncentric: Crops whose patterns of variation suggest domestication over a wide area. The suggestion may be misleading, of course, but centres are either not apparent or anomalous. Examples include *Sorghum* and common bean (*Phaseolus vulgaris*).

Within the centres of diversity, it is not unusual for a crop to be reasonably uniform over extensive areas and show enormous diversity in very small regions. These relatively small regions (100-500 km across) are defined as microcenters (Harlan, 1992).

Uniqueness of the Indian Gene Centre

Indian subcontinent is one of the 12 mega biodiversity centres (Zeven and Zhokovski, 1975) and represents two of the eight Vavilovian centres of origin and diversity of crop plants (Harlan, 1971). Two of the eighteen globally identified 'hot spots' of biological diversity occur in India, one each in the Western Ghats and the North-eastern Himalayas. There exist almost all the climatic conditions and ecological zones found in different parts of the world. These range from territorial snow cover to equatorial and tropical conditions; from moist humid tropics (with annual rainfall of about 2000 mm in Western Ghats, West Bengal, and Assam) to arid areas (with annual rainfall of 100-300 mm) in the hot desert of Rajasthan and Gujarat, and cold desert of Ladhakh and Lahaul-Spiti and many others in between.

At global level, it is estimated that there are 13-14 million species. As per National Action Plan on Biodiversity (MoEF, 1997), India has 49,219 plant species (12.5 % of global flora). Of these, about 17,000 represent flowering plants (6.8 % of the world flora). India has about 148 endemic genera belonging to over 47 families of higher plants; the largest number (about 2532) of species are located in the Himalayas followed by peninsular region (1,788 species) and Andaman Nicobar islands (185 species). The North Eastern region of India holds more than 7000 species including about 700 species of unique and rare orchids (Arora, 1991, Chandel and Paroda, 2000). It is a treasure house of wild economic plants, particularly wild edible and medicinal plants, which are largely utilised locally or in several Ayruvedic preparations, and can be safely termed as commercially under-utilised.

The origin of Indian agriculture dates back to 2300-1750 BC and till date it retains a great diversity of farming systems and crops, and ranks seventh in the world in number of domesticated plants and animals (Khoshoo, 1995). The preponderance of diseases and pests along with the long association of the host species has evolved a diversity of host-pathogen relationship. About 44 major or minor crop species originated and domesticated here, including some wild relatives of crop species (Gautam and Singh, 1998).

The rich genetic diversity is a result of ethnic and cultural diversity encompassing over 550 tribal communities, belonging to 227 ethnic groups, spread over 5,000-forest villages (Gautam, 1997). These native tribes use about 1,532 wild edible plant species including 145 species of roots and tubers, 521 of leafy vegetable/green, 101 of bulbs and flowers, 647 of fruits and 188 of seeds and nuts. In addition, nearly 9,500 plant species of ethno-botanical uses have been reported from the country, of which around 7,500 are ethno-medicinal and 3,900 multipurpose/edible species. These have often been referred as non-conventional source of food and further bio-prospecting and research on these species can contribute significantly to food security.

India is a primary centre of diversity of crops, such as rice, urd bean, moth bean, pigeon pea, smooth gourd, ridge gourd, pointed gourd, tree cotton, *Capsularis* jute, jackfruit, banana, mango, jamun, large cardamom, black pepper, several minor millets and medicinal plants like *Rauvolfia serpentina* and *Saussurea lappa*. The introduction of germplasm in distant past from the Mediterranean, African and American regions has also significantly enriched the Indian plant genetic resources wealth and the Indian gene centre developed as a secondary centre of diversity for the following crops: African crop: finger millet, sorghum, cowpea, cluster bean, okra, sesame, niger and safflower. Tropical American crops: maize, tomato, pumpkin, chayote, chilli and amaranths (Arora, 1988).

The Spectrum of Plant Genetic Resources Utilisation

The full spectrum of plant genetic resources of any crop consists of several types of collections such as those derived from centres of diversity, those derived from centres of cultivation and those derived from breeding programmes. A functional classification was introduced by the International Biological Programme. They broadly classified the plant genetic resources as land races, modern varieties, wild and weedy relatives of cultivated plants, wild (i.e., non-domesticated) species used or of potential use.

Land Races

The traditional farmers, over the millennia, have given us a priceless heritage of germplasm of major and minor crops. It has been reported that farmers in the areas of crop diversity often plant several crop varieties in one season, especially where traditional agriculture is practised. The land-race populations are often highly variable in appearance, but they are each identifiable, having particular properties or characteristics *viz*., early or late maturing, adaptability to a particular soil type, expected usage and usually have local names (Brush *et al*., 1981, Dennis, 1987).

The genetic variability provides some built-in insurance against disease epidemics as the population contains such an array of genes that no single race of pathogen can build up to epidemic proportions. Some genotypes would be affected each year, but not all of them.

The great variability in land-races makes them good sources for genes for modern plant breeding. Their dependability makes them useful in difficult or marginal situations. Land-races, because they have survived so long among pests and diseases in the centres of diversity, offer a wealth of potential resistance. But some pains must be taken to eliminate the unwanted characteristics as resistance is obtained. Land-races, in addition to furnishing simply inherited useful traits, probably contain numerous favourable linkage blocks.

Indian gene centre has provided valuable genetic diversity for overcoming the yield barriers and increasing yield in a number of crops to meet the challenge of food security not only at national level, but also at international level. In sugarcane, early Coimbatore (Co) varieties like Co 213, 290, 312, 313 and 475 were extensively used, as parents in many countries, like Java (Indonesia), Hawaii (USA), Queensland (Australia) and West Indies. In rice, Intan, Mas and Peta from Indonesia are the products of the cross of Indian variety 'Latisail' with the Chinese variety 'Cina'. Peta is parent of IR 8. Hence, over 80 % of dwarf varieties of rice cultivated in Asia have 'Latisail' germplasm. The multi-resistance lines developed in rice derive the resistance to major pests and diseases, including blast, bacterial blight, tungro virus, gall midge and stem borer from land race material collected from the foothills of Himalayas. In wheat, '*Kharchia* Local'/'*Kharchia* 65' for salt tolerance, 'Hindi 62' for heat tolerance and 'NP 4' for grain quality have been used in several countries of Latin America, Africa, Asia, Arab, USSR and Canada. Similar examples are available in pigeonpea (ICP T6, ICP 6393, ICP 7018, and ICP 7035) and several other crops (Tiwari and Karmakar, 2000).

Modern Varieties

The spread of high yielding varieties (HYV) was more rapid and widespread (dramatic) than anything that ever happened in agriculture before. Within a decade new varieties of wheat and paddy were being grown on nearly 55 million hectares in the third world.

It has been argued that the modern varieties evolved through intensified plant breeding efforts, have a narrow genetic base. All modern North American cultures of soybean could be traced back to a dozen strains from China. Harlan (1987) reported that the majority of hard red winter wheat varieties in the USA originated from just two lines imported from Poland and Russia. However, most of the modern cultivars possessing multiple disease and pest resistance have been evolved from varied sources in the agronomically adapted land-races or varieties and have the plasma of a large number of genetic stocks. Analysing the pedigrees of 258 cultivars of soybean released between 1947 and 1988, Gizlice *et al.* (1994) reported that genetic base was fully defined by 80 ancestors. But now the cause has been identified and the situation is improving in several breeding programmes. Indian wheat programme has so far released over 200 varieties till date during post green revolution phase. About 28 % of all the released varieties are the direct release of foreign introduction or a selection in the introduction and an overwhelming 48 % of all the varieties involve exotic germplasm as a parent (Sharma and Jagshoran, 2000). This shows the utilisation of a judicious mix of exotic and indigenous material in the development of wheat varieties in India. An analysis of the background of CIMMYT wheats indicated that the number of land races used increased from less than 10 in 1950 to over 60 in 1997. Similarly, the mean number of land races in the parentage of selected International Rice Research Institute (IRRI) cultivars increased from four in 1966 to 46 in 1994 (Witcombe, 1999).

In soybean, a majority of the Indian varieties have been developed utilising exotic collections. The contribution of indigenous native collections to the pool of Indian soybean varieties is low; only three varieties (Kalitur, Type 49 and JS 2) have been developed through selection and another three through hybridisation using JS 2 and Kalitur as parent. In the development of other 33 varieties through hybridisation, 39 parents were used. However, only 10 parents (Bragg, Punjab 1, Lee, Hardee, CNS, Improved Pelican, Semmes, UPSM534, EC 39821 and EC7034) have been used widely; the most frequently used parent 'Bragg' has contributed to the development of most of the varieties (Tiwari and Karmakar, 2000). The coefficient of parentage of these most frequently used parents namely, Bragg, Lee, Hardee, CNS and Punjab-1 shows that they themselves are derived from a very narrow genetic base (Lal and Rana, 2000).

Elite Germplasm

This type of germplasm includes elite finished products of the breeding programme that could not become varieties. Additionally, the category also includes the lines coming out of prebreeding programmes. These are the genotypes having traits of economic importance in highly intense form singly or in combination with others but can not become varieties on their own. Breeders preferably and almost exclusively use this elite germplasm to produce new commercial cultivars. But basic breeding methods have not changed for the past century. New approaches are needed, which utilise genetic principles more effectively and develop better ways for identifying and utilising new germplasm (Duvick, 1987).

Registration of Plant Germplasm

In order to recognise the accomplishments and efforts of scientists who identified or developed genetic stocks and improved germplasm for specific traits (resistance/tolerance to biotic and abiotic stresses, quality, male sterility, etc), the Indian Council of Agricultural Research (ICAR) has developed a system of registration of such value-added germplasm with National Bureau of Plant Genetic Research (NBPGR) as the nodal institute. This shall encourage sharing the valuable materials, which otherwise remained under-utilised or was even lost, with other workers in crop improvement programmes. A total of 106 germplasm/genetic stocks belonging to 31 crop species have been registered till date.

Wild and Weedy Relatives of Cultivated Plants

The diversity of the wild relatives has enabled them to survive longer than the oldest cultivated variety and to survive without human assistance. Thus, as sources of resistance they are a treasure and are being used in plant breeding programme in every crop. The estimated strength of wild relatives of crop plants and related taxa occurring in India is about 320 species (Arora and Nayar, 1984). The wild edible plant gene pool includes the naturally occurring wild types, weedy types, protected, semi-protected and domesticated diversity (Hoyt, 1988). Working with them is very difficult, as along with every desirable characteristic that is transferred to cultivated types a number of linked but undesirable characters also come in. In a growing number of crops the resistance genes stored in the wild relatives play an important role, in some instances the only resources available when known sources are used up (Arora and Pandey, 1996).

It is quite common for wild provenances of trees and shrubs to serve as base populations for clonal selection, or to be introduced directly, as landscape ornaments or for forestry production, after adequate evaluation (Spongberg, 1993).

Wild relatives have contributed a great deal to the improvement of cultivated wheat. Table 1 presents a list of wild relatives possessing various important traits which have either already been transferred or are potentially transferable. (Nanda *et al.*, 1998, Gautam *et al.*, 1998 and Friebe *et al.*, 1996).

During the 1970's, grassy stunt virus devastated the rice fields from India to Indonesia, endangering the world's single most important food crop. After a four year search, which screened over 17,000 cultivated and wild rice germplasm, resistance to this disease was found in the species *Oryza nivara*, growing wild in the state of Uttar Pradesh. Today, resistant rice varieties containing the wild Indian gene are grown across the entire Asian rice region (Govindswamy *et al.*, 1966).

Table 1. Desirable traits transferred/available from wild relatives in wheat

S. No.	Species	Trait
1.	*Aegilops comosa*	*Yr8/Sr34*
2.	*A. intermedium*	*Wsm1, Lr38, BYDR*
3.	*A. longgisima*	*Pm13*
4.	*A. speltoides*	*Lr28, Lr35, Lr36, Pm12, Sr32, Lr35/Sr9, Gb5*
5.	*A. squarossa*	*Lr21, Lr22a, Lr32,* a gene on *1DS*
6.	*A. ventricosa*	*Pch1, Sr38/ Lr37/ Yr17*
7.	*A. umbellulata*	*Lr9*
8.	*Agropyron elongatum*	*Lr19/Sr25, Lr24, Lr29, Sr43, Lr19, Sr24/Lr24, Sr26,* WSMR (wheat streak mosaic virus resistant), *Cmc2* (wheat curl mite resistant)
9.	*Haynaldia villosa*	*Pm21*
10.	*Secale cereale*	*Lr25, Lr26, Yr9, Pm7, Pm8, Pm17,* Storage glutenins, *Lr26/Sr31/Yr9/Pm8, Gb2/Sr/Pm17, RWA* (Russian wheat aphid resistance), *Lr25/Pm7, Pm8/Sr31, Lr26/Yr9, Gb2/Pm17, Gb6, Lr45, Sr27, Pm20, H21, H25 11.*
12.	*T. compactum*	*Pm15*
13.	*T. dicoccoides*	*Zur15, Pm16,* Glutenins and many other genes
14.	*T. dicoccum*	*Pm4a, Pm5*
15.	*T. distichum*	*Lr19/Sr25*
16.	*T. macha subletschumicum*	*Pm3b*
17.	*T. spelta*	*Yr5*
18.	*T. spelta duhamelianum*	*Pm10, Pm11*
19.	*T. sphaerococcum*	*Pm3b*
20.	*T. timopheevi*	*Sr36/Pm6, Sr37, Lr18, Sr40*

In another instance, a wild Indian muskmelon provided genes for resistance to downy mildew when the crop in the US was threatened by its outbreak. It ensured the commercial cultivation of the muskmelon crop in the US by reducing the amount of pesticides needed to control downy mildew. Potato was introduced from Peru in the 15th century, and cultivated all over Europe, especially in Ireland, where it became part of their diet. In the 19th century, the disease late blight devastated the crop causing the famous Irish famine when millions of people starved to death. A wild relative of the potato found in Peru, provided the gene for developing disease-resistant variety and revival of potato cultivation. *Abelmoschus tuberculatus* has provided resistance to yellow-vein mosaic virus to the cultivated *A. esculentus* (Arora and Singh, 1973). Likewise *Musa* germplasm from north-eastern region has proved to be a potential source of drought and fire resistance (Arora and Pandey, 1996).

A wild barley plant from Ethiopia provided a gene that protects the $160 million California barley crop from lethal yellow dwarf virus (Briggs, 1978). Wild tomato discovered in the Andes has been used to increase the sugar content of cultivated varieties, increasing their commercial value by $5-8 million per year. In both India and Africa, yields of cassava (tapioca) - one of the most important root crops throughout the tropics, were increased 18-folds because of disease resistance brought in from wild Brazilian cassava. The sugarcane industry in the U.S. was saved from collapse by disease-resistance genes brought in from wild Asian species (Parthasarthy, 1951).

Wild Species Used or of Potential Use

There are many wild species, which are not yet domesticated but are extensively used. Some are widely planted, genetically and culturally, in a near wild state. In India, a number of medicinal plants native to the country occur in the wild. These are of considerable economic importance, for example *Dioscorea deltoidea*, with a distribution in the western and north western Himalayas, is an important source of diosgenin which is a precursor for the synthesis of cortisone and of steroidal sex hormone. Also *Rauvolfia serpentina*, the roots of which contain many alkaloids, notably reserpine, is an antihypertensive drug. It occurs on hilly tracts in various parts of the country. There are many wild species, which are used in pasture and rangelands and as raw materials for chemical industries.

The history of crop production fails to suggest that new major crops are likely to be discovered, but there is a reason for assuming that some species may be taken into cultivation to fulfil requirements arising from the nutritional, medical, chemical or other industrial research. A case in point is the use of berries of certain *Solanum* species by some tribal people in south India, for contraceptive purposes. So far these have not been cultivated and their productivity is not known.

Germplasm Acquisition

Concepts and strategies for plant germplasm utilisation have changed over time. In early years, breeders simply made use of the germplasm they had on hand. Initially sporadic collections of indigenous germplasm of various crops were made, *viz.* rice (Paroda and Malik, 1990), wheat (Howard and Howard, 1910; Pal, 1966), jute (Burkill and Finlow, 1907), tea (Bezbaruah, 1968), sugarcane (Thuljaram Rao and Krishnamurthy, 1968), several legumes (Singh, 1973; Singh *et al.*, 1974), etc.

Later variability in *Portersia coaratata* was collected from the coastal regions. Many joint explorations with other countries have also been carried out. Japanese exploration teams led by H. Kihara in the early 1960s and T. Watabe in the late 1960s and early 1970s made systematic collections in Western Uttar Pradesh, Bihar, Andhra Pradesh, and parts of Maharashtra. H.I. Oka travelled extensively on collection trips in different parts of India. French team from IRAT and ORSTOM collaborated with ICAR for collection particularly of *O. nivara, O. rufipogon,* and *O. officinalis* from Goa, Karnataka, Maharashtra, and Gujarat in 1986. During 1987-89, ICAR and IRRI scientists undertook intensive collection for wild rice in South India and West Bengal (Govindswamy *et al.*, 1966; Saxena and Gautam, 2000).

Genetic markers can aid acquisition by helping to identify 'gaps' and redundancies in collections, and by helping to develop optimal sampling strategies for plant explorations. For example, calyx shape is strongly associated with ploidy in sweet potato, *Ipomoea batatas* (L.) Lem (Bohac *et al.*, 1993). Consequently Austin *et al.* (1993) discriminated 4x sweet potatoes from 6x, efficiently. Similarly, RFLP analysis of chloroplast DNA from *Zea perennis* (Hitch), revealed that some populations of *Teosinte* have a plastid haplotype, that was highly divergent from those in other populations of *Teosinte* (Doebley, 1989). Therefore, collecting more of *Zea perennis* population becomes a priority to fill the gaps in *ex situ* collections.

The Search for New Genes

Although the local germplasm pools were well sampled, and improvement curves began to level, breeders began to look elsewhere for new sources of variability. Perhaps more importantly, when genetic vulnerability became evident as a result of epidemic spread of disease and insect pests, breeders searched with great urgency for novel sources of genetic resistance to these new pests.

The concept of "search for new genes" was developed (Pal, 1937), where in it was stated that:

i) the scientific breeding of cultivated plants has become indispensable in the agriculture of progressive countries to meet the increasing demand for food and raw materials,

ii) the genes present in the contemporary varieties had been mostly utilised and the most desirable combinations had already been explored for improvement through selection and hybridisation, although in many cases the ideal was not fulfilled,

iii) it has became necessary, therefore, to discover or introduce new genes and to devise new ways of combining them with the old genes, and

iv) that there was wide variability in several crop plant species, that wild or little known allied species or varieties existed and in several cases plant species resistant/immune to many devastating diseases, attacked their related species under cultivation. In other cases, they were observed to possess some different but desirable attributes for crop improvement programmes. Many of the most pressing problems of crop production were expected to be solved by incorporating these traits in the commercial cultivated varieties.

For proper management of PGR, work was initiated by ICAR in 1946 with the establishment of a small unit designated as Introduction and Exploration Organisation, in the erstwhile Botany Department of the Indian Agricultural Research Institute (IARI). The unit later functioned as the Plant Introduction Division under IARI from 1961 onwards. In 1976 it evolved into an independent institute - National Bureau of Plant Genetic Resources (NBPGR). The NBPGR acts as a nodal organisation under the ICAR for planning, conducting, promoting, co-ordinating and leading all activities concerning germplasm management including *ex situ* conservation of plant genetic resources. It operates the Indian National Plant Genetic Resources System (INPGRS) with the headquarters at New Delhi. It has a network of 12 regional/base centres in different agro-climatic zones of the country and operates with an active partnership of over 40 ICAR Institutes, National Research Centres, Project Directorates, All India Co- ordinated Projects and State Agricultural Universities designated as the National Active Germplasm Sites (NAGS).

The scientists of the Bureau have, either independently, or in collaboration with other co-operating institutes/centres, have so far conducted 813 explorations in diverse agro-ecological regions and habitats representing different phyto-geographical zones of the country. This has resulted in the assembly of more than 128,611 accessions of crop species, their wild relatives, medicinal, aromatic and other plants of potential economic value. The Bureau is responsible for all germplasm related activities in the country including exploration, collection, export, import, evaluation, conservation, registration, utilisation, etc.

International Efforts

Perhaps, the most significant institution on germplasm in 20th century has been the N. I. Vavilov, All Union Scientific Research Institute on Plant Industry (VIR) in Russia. Vavilov has been the pioneer of scientific, systematic germplasm collection and had assembled over 50, 000 samples of crop plants from over 50 countries in the 1920s and 1930s. After that, American, German, Swedish, British and Latin American collectors made several collecting missions. In the year 1960, developing countries started integrating with global plant genetic resources system, where crop diversity was in abundance. In 1965, a panel of experts on crop germplasm exploration was set up with Sir Otto Frankel as Chair. After several meetings, International Biological Programme (IBP) was evolved. Frankel published a document "FAO/IBP Survey of Crop Genetic Resources in their centres of Origin" in 1973 and 1981 wherein Sir Otto Frankel, Era Bennett, R.O. White, Jack R. Harlan, T.T. Chang, Jack G. Hawkes and others played an important role

(Hawkes, 1983). In 1972, the UN Conference on Human Environment gave responsibility to FAO to establish an international forum and in 1974, International Board of Plant Genetic Resources (IBPGR) was established under the aegis of Consultative Group on International Agricultural Research (CGIAR) to promote and co-ordinate plant genetic resources world wide. Germplasm collection was one of the IBPGR's main activities in collaboration with national PGR programmes. Many International Agricultural Research Centres (IARCs) have also played significant role in several exploration expeditions with IBPGR. Over the years, the IBPGR had sponsored some 650 exploration missions in over 130 countries and a total of over 2,00,000 germplasm samples collected. The IBPGR (now IPGRI) continued to be actively involved in germplasm collection apart from strengthening collection and other conservation activities in national PGR systems in developing countries. Priorities of national programmes depended on needs of national research and development capacity and on the diversity of germplasm within the country.

Evaluation of Germplasm

In the last century, professional plant breeding has become the predominant breeding activity. Nevertheless, in many developing countries selection/breeding by those who grow and use the crops commercially is still important. It is vital to learn more about how the farmer selector/breeders operate and to catalogue their skills and techniques before they disappear. Recently, there has been an increasing interest in studying and applying such indigenous agricultural knowledge, as evidenced by the establishment of the National Innovation Foundation by the Government of India in 2000 (FAO, 1996c).

The diversity of crops has been well-utilised in efforts to solve today's food problems. The crop germplasm collected over several decades became "parents" of the high-yielding, pest-resistant, and well-adapted varieties that resulted in unprecedented increase in yields, ushering in the Green Revolution. The value of collected and conserved crop germplasm has been demonstrated in many screening tests for useful economic traits. Thousands of accessions have been screened for resistance to insect pests and diseases as well as tolerance to different abiotic stresses. Germplasm characterisation and evaluation, complemented by biosystematic studies of the wild species and molecular studies of genetic diversity, are generating the information base for more efficient utilisation of these valuable resources.

At the Directorate of Rice Research (DRR), Hyderabad, a large number of germplasm has been evaluated for agronomic characters for and against major insect pests and diseases and several promising accessions have been identified. In addition, a network project was approved by the ICAR to evaluate available germplasm against major insects and diseases. Up to 1997 more than 15,000 accessions have been screened at various hot spot locations and a catalogue for 12,000 accessions with 21 characters has been published.

Cataloguing and Documentation

The germplasm held by the genetic resource centres remains largely unknown if it is not properly documented. Without basic 'passport,' characterisation and evaluation data, conserved germplasm is not of much value to the stakeholders as also in legal cases under the patenting regime. So far, NBPGR has published 73 crop germplasm catalogues, which describe nearly 82,500 accessions and has compiled computer data files for several more accessions. The NBPGR has compiled Minimal Descriptor list for over a 100 crops, of which descriptors for 43 crops have been published. Similar catalogues have been published by other centres for their mandate crops.

Developing Core Collections

Plant breeders and research workers, working on crop genetic resources have always in their wisdom selected limited sets of accessions from the working collections available. The large size of many collections of plant germplasm has led to increasing concern as to whether the full range of genetic diversity they contain can be effectively utilised. All gene banks have a considerable number of duplicates, the number of truly unique accessions is now so large as to deter their extensive use except for a few, easily identified characters and the management of such collections poses major problems (Holden, 1984).

Recent years have seen the development of rapid chemical methods to measure genetic characteristics, and of biometric methods of handling the resulting data sets. The resulting calculation of the "genetic distance" between accessions can be used to identify a small subset or "core", which contains a large proportion of the total variability of all the accessions. It is then possible to characterise and evaluate the "core" collection, knowing that most of the variability has been sampled. The core collection constituting a proportion of the total germplasm collection, provides genebank managers, plant breeders, and research scientists with a manageable set of accessions to focus on the search for desirable new characters, detailed evaluation, and work on the application of new techniques (Frankel, 1984). Genetic markers impart significant role in determining systematic relationships regarding genetic polymorphism and divergence (Crawford, 1990; Stuessy, 1990). The studies of Schoen and Brown (1993) have indicated that the genetic markers can be used in developing core subset and by employing such markers the allelic richness and genetic polymorphism can be retained. Erskine and Muehlbauer (1991) have used isozyme markers and pigment markers to test the utility of a core subset comprising random samples of lentil.

Conservation Strategies

The need for conservation of plant genetic resources has been mutually accepted; however, there exists a fundamental divergence of views on what objectives should be the focus of the conservation of genetic resources. Ethical aims oriented at preserving all existing biodiversity stand in opposition to anthropocentric objectives, which consider genetic diversity worth maintaining only to the extent that it serves human kind in the present or in the future (Hawkes *et al.*, 1992).

The basic objectives for the conservation of plant genetic resources in the country include:

- Conserving the existing diversity for utilisation in posterity through long term conservation as base collection in the National Genebank
- Ensure proper storage at the National Active Germplasm sites and encourage utilisation of the genetic resources following a network approach with effective linkages between the National Genebank and the active sites.
- Encourage long term conservation of plant genetic resources through conservation of natural habitats and using techniques such as the on-farm conservation and community genebanks.

Two major approaches for achieving the above objectives are: (*i*) *In situ* conservation and (*ii*) *Ex situ* conservation. The choice of conservation strategy depends mainly on the nature of the material to be conserved i.e. the life cycle, mode of reproduction, size and the ecological status (Engels, 1991).

In situ Conservation

In situ conservation refers to protection of plant genetic resources in their natural habitats. The Convention on Biological Diversity has given the highest priority to this approach of conservation, which includes species protected in the wild as well as landraces and other cultivated forms maintained by farmers. This involves the co-ordination of a broad range of economic and social activities within a country. Article 8 of the Convention on Biological Diversity includes the following obligations with respect to *in situ* conservation:

- establish a national system of protected areas and buffer zones
- promote sustainable development in the areas around protected areas
- undertake restoration ecology and species rehabilitation
- protection against the release of living modified organisms
- control alien species
- respect the rights of traditional resource users
- enact endangered species legislation
- manage the underlying "processes" that lead to biodiversity loss
- provide financial support to the less developed countries for conservation activities

In situ conservation requires the conservation of ecosystems and natural habitats and the maintenance and recovery of viable populations of species in their natural surroundings and, in the case of domesticated or cultivated species, in the surroundings, where they have developed their distinctive properties [United Nations Environment Programme (UNEP)/Convention on Biological Diversity (CBD) 1992; UNEP, 1994]. This also includes the preservation of indigenous knowledge (social, cultural and religious status), agro-ecosystems and other wild cultivars. The criteria for site selection for *in situ* conservation within the study areas are: (*i*) wide range of diversity of a single or few crop species within a given site, (*ii*) ecological heterogeneity, (*iii*) possibility to control or monitor the site and (*iv*) easy access for monitoring and management (Tan and Tan, 1998).

The following four basic kinds of multidisciplinary research are needed to successfully run the *in situ* conservation (FAO, 1996b):

(*a*) Ethnobotanical and socio-economic research to understand and analyse farmers knowledge, selection/breeding and, utilisation and management of Plant Genetic Resources for Food and Agriculture (PGFRA) with the approval of the involved farmers with applicable requirements for protection of their knowledge and technologies.

(*b*) Population and conservation biology to understand the dynamics of the local landraces and farmer's varieties (population differences, gene flow, degree of inbreeding and selection pressure, etc.)

(*c*) Crop improvement research in mass selection and simple breeding without significant losses in local biodiversity.

(*d*) Extension studies for lesser known crops including their seed production, marketing and distribution.

There is always a threat for *in situ* collections due to genetic drift, inbreeding, habitat loss or fragmentation, competition from exotic species, pest and over exploitation. However, the rationale of *in situ* conservation arises from its ability to preserve evolutionary processes that generate new germplasm under conditions of natural selection. It helps to maintain important field laboratories for crop biology and biogeography. It also provides a continuous source of germplasm and means for wider participation to conservation exclusively for those countries, which have abundant crop germplasm resource.

The *in situ* conservation of PGR forms an integral part of the biodiversity conservation efforts in the country. Establishment of forest area reserves, national parks, protected areas is being promoted to facilitate their *in situ* conservation. There are at present 85 National Parks, and 448 Wildlife Sanctuaries covering 35,919.03 km^2 and 112,274.45 km^2 area, respectively (approximately 4.2 % of the total land area of the country) and representing the major biogeographic provinces of India. The total extent of protected areas include five designated World Heritage Sites, nine Biosphere Reserves and six Ramsar sites, besides 23 Tiger Reserves (MoEF, 1998). The establishment of sanctuaries in Tura range in Garo Hills of Meghalaya for conservation of rich native diversity of wild *Citrus* and *Musa* species, and for *Rhododendrons* and orchids in Sikkim are examples of *in situ* conservation of economically important species.

Several non-government organisations are engaged in *in situ* conservation of targeted species through national or external assistance or both. For example *in situ* conservation of medicinal plants by Foundation for Revitalisation of Local Health Traditions (FRLHT) in southern states, and *in situ* conservation efforts of Tropical Botanical Garden and Research Institute (TBGRI) in Kerala. Efforts to document and conserve the local landraces/farmers varieties are also being done by M. S. Swaminathan Research Foundation (MSSRF), Chennai and SRISTI, Ahmedabad. On-farm *in situ* conservation of paddy in Bastar area of Madhya Pradesh by NBPGR, Indira Gandhi Agricultural University (IGAU) and IRRI is another interesting activity. Traditionally, various communities have been conserving economic and ecological keystone species (a species that is crucial in determining the nature and structure of the entire ecosystem in which it lives) for medicinal value, in sacred grooves which range from few plants to several acres. The important regions for grooves are Khasi and Jianthia Hills, Western Ghats, Aravalli Hills and Sarguja, Chand and Bastar area in Central India.

On-Farm and Community Efforts

On-farm conservation of PGR is a complementary strategy for *in situ* conservation of valuable plant varieties (developed and protected by farmers for centuries in certain areas). It involves continued cultivation and management by farmers of a diverse set of crop populations in the agrosystems, where the crop has evolved or in secondary centres of diversity (Bellon *et al.*, 1997). The strategy allows for crop populations to continue their evolution in response to natural and human selection (Jackson, 1995; Pham *et al.*, 1996; Engels and Wood, 1999).

Farmers and communities play a pivotal role in on-farm conservation, largely because the food security and self- sufficiency, particularly in the marginal areas, depends largely on the availability of genetic diversity. The communities, usually, have multiple interests or concerns and are confronted with numerous challenges of sustenance. To overcome these, farmers in diversity rich areas often plant several crop varieties in one season and adaptive complex of crop genetic diversity enables them to adopt crops suited to their ecological niches and cultural food production systems and practices. The wider environmental adaptability of agricultural diverse crops and varieties also provides the farmer with a risk aversion measure.

The farmers transform or abandon particular varieties to suit their needs as the conservation of these resources are not the only concern for them. The on-farm conservation of genetic resources can only be achieved if maintaining crop populations and varieties is made advantageous to farmers. Various economic and cultural incentives for farmers to encourage them to continue growing the crop genetic resources are to be developed (Smale and Bellon, 1999; Gupta, 1995).

Participatory Plant Breeding

Green revolution brought about major gains in production potential of wheat and rice. The technology that resulted in green revolution was more market oriented and there perhaps was even no need to consult those resource rich farmers who exploited this new technology. However, decades after the Green Revolution, it has become apparent that these offerings (new varieties) of the non-participatory, non-interactive breeding products did not satisfy the needs of farmers in more marginal agricultural environments (Witcomb, 1998). Additionally, despite very different socio-economic circumstances developing countries adopted from the USA and Europe, a regulatory framework designed to release few, widely adapted cultivars for intensive, mechanised, monoculture cropping systems. In developed countries farmers were regarded only as growers and not as direct consumers, because grain was rarely consumed on farm but was sold to industrial food processors. To set breeding objectives for grain quality, the grain purchasers, rather than the farmers, were consulted. The situation is entirely reversed in a developing country such as India where a great majority of farmers are themselves consumers of what they produce. Farmers may have their own priorities such as straw yield that low resource farmers need for their livestock. This is such a character for which breeders may never worry about. To take care of such needs of farmers, participatory breeding or Participatory Varietial Selection (PVS) is proposed. PVS aims at identifying farmers' needs by discovering what crops and varieties they grow, and what traits they consider important. Farmers visit research stations to select material from the wide range of varieties in breeders' plots. The chosen varieties are given to farmers to grow along side their local varieties with traditional management. Farmers are the unit of replication. Later on farmers decide which variety suits their needs most. All this results in quick adoption of new varieties ultimately leading to higher regional yields.

***Ex situ* Conservation**

The concern for conservation led to considerable worldwide efforts for *ex situ* conservation at national and international level gene banks. Undoubtedly *ex situ* gene banks play an extremely important role in conservation of agro-biodiversity, being the only way to conserve species and genotype diversity for posterity. In some of the extreme cases, the *ex situ* gene banks helped in rehabilitating the *in situ* diversity. *Ex situ* conservation, particularly important for crop gene pools, began on a wide scale about three decades ago and is now practised, to some extent, in almost all countries. Notwithstanding the advantages of *ex situ* conservation, there are limitations of relying only on this approach:

(*a*) Many important species are under-represented because of the recalcitrant nature of the seeds.

(*b*) Genetic shifts or alterations cannot be ruled out due to inappropriateness of storage conditions.

(*c*) Since the crops are grown with external application of fertilisers and pesticides, and use of heavy machinery, the plants slowly get accustomed to more congenial conditions, the roots architecture and assimilatory properties get modified since nutrients are easily available and availability of porous well ploughed soil.

(*d*) *Ex situ* conservation does not maintain evolutionary processes that created the crop germplasm (Hamilton, 1994; Harris, 1989). The genetic resources are not exposed to natural or artificial pressure and therefore no chance exists for further evolution or adaptations.

Over the years India has developed sound scientific management regimes for *ex situ* conservation and access to its genetic resources. Groups of Institutions, scientists, scientific societies, NGO's are addressing the task of protecting the genetic resources while deriving benefit from them through the IN-PGRS. Vigorous efforts have been made in recent years for *ex situ* conservation of plant genetic resources at the NBPGR, New Delhi, which co-ordinates the National Genebank Network in the country.

The National Genebank Network

A major purpose and advantage of the Genebank is convenience of access to the plant genetic resources for the various crop improvement programmes. To achieve the desired aim, the National Genebank operates in the network mode. The germplasm collections conserved are categorised as "base collections" (for long term storage at the NBPGR) and "active collections" (for medium-term storage at the National Active Germplasm Sites (NAGS). These NAGS are responsible for evaluating, cataloguing, distributing and maintaining the active germplasm collections of specific crops (Table 2). The National Genebank thus operates in the network mode. These collections are also linked to the National Genebank through the Jai Vigyan National Science and Technology Mission on Conservation of Agrobiodiversity and also the National Agricultural Technology Project (NATP) on Plant Biodiversity being operated by NBPGR.

The major components of the National Genebank include the seed genebank, field genebank, cryobank and the *in vitro* bank

Seed Genebank: Most of the crops have seeds, which can be dried and stored at low temperatures to increase their longevity and thus the easiest and most economical method of conserving them is in seed genebanks. The *ex situ* seed genebank at NBPGR comprises of 16 long-term modules (total capacity: one million accessions) maintained at –20 °C for housing the base collections. The active collections are distributed in 22 medium term modules maintained at 4 °C for storing the base collections at active sites. At present the genebank holds 2,02,330 accessions in the base collection (Table 3).

The accessions held in the long term stores for 10 or more years are monitored for their viability, seed quantity, etc. as per the genebank standards. According to the monitored results the samples/lots registering viability below the specified standards will be regenerated.

Field Genebank: Other conservation methods become important in cases where seed storage is difficult. A number of species - predominantly tropical and subtropical in origin - such as coconut and cacao, have seeds that are unable to withstand drying and cannot be stored at low temperatures. Even if they are kept under optimum conditions, such "recalcitrant" seeds have life spans that are limited to a few weeks, occasionally months. Some species like coffee, citrus or oil palm can withstand drying but not low temperatures, while others like the potato, cassava or yam are vegetatively propagated. At present the most common way to preserve the genetic resources of these species is in field gene banks, as collections of living plants. However, this leaves the collections exposed to natural disasters and attacks by pests and pathogens. Also, field genebanks are expensive to maintain, because of the land and labour costs involved.

Table 2. Crop germplasm holdings of various National Active Germplasm Sites

Institute/AICRP*/NRC	Crop(s)	No. of Accessions
Agricultural Crops		
Central Institute of Cotton Research	Cotton	8,768
ICAR, Research Complex, NEH Region	Crops of North-East Region	NIL
Indian Grassland & Fodder Research Institute	Fodder Crops	6,267
NRC on Groundnut	Groundnut	6,432
Central Research Institute of Jute & Allied Fibers	Jute & Allied Fibers	3,226
Project Directorate on Maize	Maize	2,500
Directorate of Oil Seeds Research	Safflower, Castor, Sunflower	10,550
AICRP on Pearl Millet	Pearl Millet	2,794
Indian Institute of Pulses Research	Chickpea, Pigeonpea, Lentil, Rajmash, Fieldpea, *Lathyrus*, Urd, Mung	5,021
NRC on Rape Seed & Mustard	Rapeseed & Mustard	8,082
Central Rice Research Institute	Rice	24,000
Indira Gandhi Krishi Vishwa Vidhyalaya, Raipur	Rice & *Lathyrus*	15,000
AICRP on Small Millets	Small millets	8,572
NRC on Sorghum	Sorghum	7,366
NRC on Soybean	Soybean	2,500
Sugarcane Breeding Institute	Sugarcane	5,861
Central Tobacco Research Institute	Tobacco	1,500
NBPGR Headquarter	Under utilised crops	NI
Directorate of Wheat Research	Wheat & Barley	7,000
Horticultural/Agroforestry		
NRC on Agroforestry	Agroforestry spp	40
NRC on Arid Horticulture	Arid fruits & vegetables	1,923
NRC on Banana	Banana & Plantains	907
NRC for Cashew	Cashew	NI
NRC on Citrus	Citrus species	51
NRC for Grapes	Grapes	NI
Central Horticultural Experiment Station, IIHR	Leechi, Bael, Aonla & Jackfruit	
NRC on M & Ac Plants	M & Ac plants	190
Central Institute for Sub Tropical Horticulture	Mango	NI
Silkworm and Mulberry Germplasm Station	Mulberry	NI
NRC on Oil Palm	Oil Palm	119
NRC for Onion & Garlic	Onion & Garlic	NI
NRC for Orchids	Orchids	NI
National Botanical Research Institute	Ornamentals & non-traditional crops	NI
Central Plantation crops Research Inst.	Coconut, Arecanut, Cocoa	522
Central Potato Research Institute	Potato	2,500
Indian Institute of Spices Research	Spices	6,055
Central Institute of Temperate Horticulture, Simla	Temperate horticultural crops	NI
Indian Institute of Horticulture Research, Bangalore	Tropical fruits, Vegetables, Ornamentals and M & Ac plants	11,467
Central Tuber Crops Research Institute	Tuber crops	3,871
Indian Institute of Vegetable Research	Vegetables	16,139

NRC = National Research Centres; M & Ac = Medicinal And Aromatic Plants; AICRP = All India Co-ordinated Research Project; NI = No Information; Source: Singh *et al.*, 2001

Table 3. Status of Base collections in National Genebank (–20 °C)

Crop Group	No. of accessions
Cereals	93,429
Millets and forages	17,081
Pseudo Cereals	2,205
Grain Legumes	32,638
Oilseeds	25,015
Fibre Crops	6,755
Vegetables	10,186
Fruits	139
Medicinal & Aromatic Plants	763
Narcotics	778
Spices & Condiments	1,642
Genetic Stocks	203
Released Varieties	1,404
Duplicate Safety Samples	10,092
Total	**2,02,330**

Source: Singh *et al.*, 2001; modified

Cryobank: Cryopreservation of plant germplasm at ultra-low temperature (–150 °C/–196 °C) using liquid nitrogen is widely being attempted for conservation of mainly germplasm with recalcitrant seeds throughout the world. The cryobank facility presently has 3,076 accessions of varied germplasm of orthodox, intermediate and recalcitrant seed species and also of pollen samples (Chaudhary *et al.*, 2001).

In vitro Bank: A large number of plants native to this region are propagated vegetatively. The technique of *in vitro* culture offers an important additional option for the medium and long-term conservation of these problematic species. The aseptic cultures from the reproductive and somatic tissues of the accessions are raised and maintained under a set of conditions aiming to slow down the growth. About 1138 accessions of various priority crops were maintained under short- to medium-term storage as *in vitro* cultures. These include tuber and bulbous crops (468), fruits (383), spices (251), medicinal and aromatic species (65) (Chaudhary *et al.*, 2001). The cultures are maintained at standardised temperatures and are sub-cultured at 4-24 month's intervals.

Recent Initiative

Realising the importance of sustainable conservation and utilisation of plant genetic resources for present and future, a mission mode programme under NATP on Sustainable Management of Plant biodiversity was undertaken at NBPGR, in 1999. Due to its national importance it has been included under the Prime Ministers programmes as "Jai Vigyan National Science and Technology Mission". The main objective of this programme is the exploration/collection, conservation, evaluation and management of available diversity in India for sustainable utilisation. It is an extensive programme with NBPGR, New Delhi as the lead institute and 10 zonal lead centres identified on the basis of agro-climatic regions of the country. Further, it has 131 collaborators including 38 ICAR institutes, 23 SAUs, 3 laboratories of Defence Research and Development Organization (DRDO), 24 Non Governmental Organizations (NGOs) and other Government Departments like Council for Scientific and Industrial Research (CSIR), Department of Science and Technology (DST) etc. So far, over 600 explorations have been undertaken, which have resulted in acquisition of over 37,000 accessions including the landraces and wild and weedy relatives of crop plants. The voucher samples of all the orthodox seeds are being maintained at the Medium Term Storage Module (4 °C), while the recalcitrant seeds are being maintained at the cryobank. The vegetatively propagated plants are conserved in the in vitro repositories as well as in the field gene banks (Pareek *et al.*, 2001).

Recent Global Developments

The concept of global collections and gene bank-storage of threatened crop genetic resources was initiated in the 1960s with the origin of conservation movement. In 1967 FAO/IBP technical conference on the "Exploration, Utilisation and Conservation of Plant Genetic Resources", was held in Rome. The first UN Conference on the human environment was held in 1972 wherein it was decided that "Both static (seed banks, culture collections, etc.) and dynamic (conservation of populations in evolving natural environments) methods are needed".

There was a time when germplasm was considered as the common heritage of mankind. Consequently, there was free access to plant genetic resources found in tropical centres of diversity, including India. The germplasm was freely collected by scientists/explorers and utilized as the "raw materials" for plant breeding purposes throughout the world. In the process, much of the collected diversity of tropical and sub-tropical origin, thus came to be stored as *ex situ* collections in the industrialised world or in gene banks established by the International Research Centres under the aegis of Consultative Group on International Agricultural Research (CGIAR).

Emergence of the Plant Breeders Rights

Over the past 30 years, plant breeding has become increasingly commercialised, dominated by transnational/multinational seed and agrochemical corporations. Direct outcome of this trend has been the development of "Plant Breeders' Rights", which provided protection (patent) and gave breeders, private monopoly rights over the production, marketing and sale of their varieties. To ensure that the member states acknowledge the achievements of breeders (of new plant varieties), by making available to them exclusive property right, on the basis of a set of uniform and clearly defined principles, an inter-governmental organisation, Union pour le Protection de Obtention Vegetale (UPOV) was established. The UPOV convention was signed in 1961, came into force in 1968 and was revised in 1972, 1978, and 1991.

Inequitable Nature of Free Access

The early 1980s witnessed a tremendous response to the recommendation regarding seed banks. But at the same time, doubts and misgivings also grew. The debate was focused on quality of conservation, on security of collections, and on political issues. The latter involved heated disputes over ownership and control of genetic resources. There was initiation of objections against the inequitable and contradictory nature of free access to plant genetic resources. Concerns were raised about the corresponding compensation for the original donors/innovators of the genetic materials in cases where the patented seeds, based on genes of Third World origin, were fetching profits to transnational seed corporations. Questions were also raised as to who is responsible for conserving plant genetic resources/who controls access to genetic material, and what mechanisms are needed to ensure reciprocal benefits between the "technology rich" countries of the industrialised world and the "gene rich" countries? Late in the period, cultural aspects, such as the role and recognition of indigenous knowledge emerged in the genetic resources conservation debate.

An Ad-hoc Working Group of Experts on Biological Diversity was convened in November 1988 by the United Nations Environmental Programme (UNEP) to explore the need for an international convention on biological diversity. In May 1989, the Ad-hoc Working Group of Technical and Legal Experts was established to prepare an international legal instrument for the conservation and the sustainable use of biological diversity. The experts were to take into account "the need to share the costs and the benefits between the developed and developing countries" as well as "ways and means to support innovation by local people". The Ad-hoc Working Group later known as the Inter-governmental Negotiating Committee, culminated its work on 22 May 1992 with the Nairobi Conference for the Adoption of the Agreed Text of the Convention on Biological Diversity. The convention was opened for signature on June 5, 1992 at the United Nations Conference on Environment and Development (UNCED) (the Rio "Earth Summit") and the Convention on Biological Diversity entered into force on 29 December 1993.

Convention on Biological Diversity (CBD)

The CBD came into force, providing an international legally binding framework for the conservation and sustainable use of biodiversity world-wide. The CBD and UNCED's Agenda 21 recognise that genetic resources for food and agriculture warrant discrete strategies and action within the wider context of plant genetic resources in general. The objectives of the convention as set out in Article 1 are: (*i*) the conservation of biological diversity, (*ii*) the sustainable use of its components, and (*iii*) the fair and equitable sharing of the benefits arising from that use.

Access and Benefit Sharing

The Convention calls on to the parties (refers to national governments) to establish conditions that facilitate access to genetic resources. In other words, source countries are obliged to ensure that their policy, administrative, legal and constitutional activities do not impose restrictions that counter the objectives of the Convention. Along with the requirement of access, the Convention also establishes the principles of fair and equitable sharing of benefits derived from the use of genetic resources. These can be scientific and technological knowledge, skill enhancement, payment on the collection of genetic resources and royalties on product developed, etc.

Access on mutually agreed terms

Article 15(4) of the Convention requires that access to genetic resources should be granted on "mutually agreed terms". It establishes new participatory relationship between provider and user of genetic resources.

Prior Informed Consent (PIC)

Article 15(5) of the Convention states that "access to genetic resources shall be subjected to Prior Informed Consent of the contracting parties providing such resources, unless otherwise determined by that Party". Thereby, the providing Parties have an opportunity to become equal partners and in monitoring the use. In circumstances, where the genetic resources are not in public domain, the PIC is obtained from respective community or individual.

The CBD provisions only apply to those genetic resources that come from the country of origin or that have been acquired in "accordance with this Convention". This specifically excludes *ex situ* holdings acquired prior to CBD from these specific obligations. The issue of *ex situ* collections acquired prior to implementation/the entry into force of the CBD was raised by the gene rich countries, and is an outstanding issue under the CBD. The Nairobi Final Act, adopting the final text of the Convention, in its Resolution 3 recognised the need to seek solutions to outstanding matters concerning plant genetic resources with the FAO, in particular, access to pre-CBD *ex situ* collections. Accordingly, FAO members are currently negotiating the revision of the International Undertaking on Plant Genetic Resources to harmonise the Undertaking with the Convention, including the matter of such *ex situ* collections.

The Designated Germplasm

The 12 centres of the CGIAR have placed their germplasm collections under the control of the FAO as part of International Network of *ex situ* Germplasm Collections. The germplasm called the Designated Germplasm is held in trust for the benefit of the international community. The Centres have also committed not to claim ownership or intellectual property rights over this designated germplasm or related information, but to make it available without restriction to users, provided such users gave similar undertakings on ownership and intellectual property rights. Based on this, the attempt of the Australian Company to patent two chickpea lines obtained from the designated Germplasm supplied by the ICRISAT was foiled. One of these was a landrace from Karnataka, which is listed as "King Kong" in the ICRISAT inventory.

To provide a framework for *ex situ* Plant Genetic Resources for Food and Agriculture (PGRFA), and to resolve questions of legal status of collections that pre-dated the convention, negotiations continue with a number of countries and institutions that have offered to bring their *ex situ* collections of PGRFA into the Network.

Non-mandated *ex situ* Collections

During discussions in CoP-4, it was India who pointed out that not all *ex situ* collections are being covered under the discussions in FAO. *Ex situ* collections refer to collections of seeds, plants, microbes and other forms of life held outside their natural habitat, as seeds in gene banks (most commercially significant, diverse and extensive holdings, of living resources), field plantings (botanical gardens or arboreta), live animals (zoos), pollen (in cold storage), tissue cultures, or microbial holdings (in laboratories). Many countries also have networks of gene banks. The private breeders also hold significant collections.

IUPGR (FAO) is being revised to harmonise with the Convention including the matter of such *ex situ* collections, through the Commission on Genetic Resources for Food and Agriculture (CGRFA). It addresses all components of the mandated biological diversity of relevance to food and agriculture. There is a need to focus on the *ex situ* collections of non-mandated crops of CGRFA. The Conference of the Parties to the CBD continues to debate important issues such as access to genetic resources, intellectual property rights, indigenous knowledge, and biosafety. A parallel process has been underway at FAO to deal with the unique situation facing agricultural biodiversity. This includes revision of the International Undertaking and specifically, action on two critical issues left outside of the Convention: access to *ex situ* collections, and the question of Farmers' Rights.

Global Plan of Action

In June 1996 in the FAO's Fourth International Technical Conference on Plant Genetic Resources for Food and Agriculture (PGRFA), the most important meeting on agricultural biodiversity ever, was held in Leipzig, Germany. The Leipzig Conference adopted the first-ever Global Plan of Action for the Conservation and sustainable Utilisation of PGRFA. The Global Plan represents the input of 158 countries, scientific experts and NGOs, and the synthesis of over 2000 recommendations resulting from regional meetings and country reports. It identifies 20 priority programs for securing and better utilising of PGRFA as a basis for global food security. The Leipzig conference also considered the FAO Report on the State of the World's Plant Genetic Resources, based on report submitted by the countries. The State of the World report provides the first comprehensive assessment of the status of plant genetic resources and existing capacity to conserve and utilise them.

The governments, which met in Leipzig, recognised that the Global Plan of Action (GPA) couldn't be implemented successfully unless Farmers' Rights are realised. At Leipzig, delegates also identified the need for "new and additional" financial support to implement the GPA. The follow-up process now under way requires governments to secure adequate financing to implement the Plan, and realise Farmers' Rights. An International Undertaking, which contains a set of legally binding provisions covering ownership, access to and exchange of plant genetic resources, is now being revised through negotiations between countries. It is this instrument that will establish the rules of the game on access to agricultural biodiversity and Farmers' Rights. Ultimately, the revised International Undertaking may be considered as a protocol to the Convention on Biological Diversity.

Trade Related Intellectual Property Rights System (TRIPS)

The relation between trade and the genetic resources exchange became more intricate under the stringent intellectual property regime imposed by TRIPS agreement under General Agreement Trade and Traffic (GATT), and its successor, the World Trade Organisation. TRIPS and Convention on Biological Diversity (CBD) represent two different approaches to the utilisation of living resources.

Under Article 27.3(b) of the TRIPS, it is obligatory for the member country to "provide protection to plant varieties either by patent or by an effective *sui generis* system, or by any combination thereof". Only a few developing countries have such a system, while most of them are in the process of developing it. In contrast, the developed countries have a well-established mechanism of patent or Plant Variety Protection (PVP). Consequently, the developed countries are capitalising by getting a number of patents and protections, and thereby putting the developing countries into a great disadvantage.

The Indian government has preferred to use *sui generis* options because of three major advantages: (*a*) flexibility, (*b*) better protection of farmers' rights, and (*c*) stronger researchers' exemption. The draft Indian bill is under the consideration of the Parliament. The proposed Indian Plant Varieties and Farmers Rights Bill (PVFRB) has most provisions of UPOV 1978 and some provisions of 1991 also. However, the proposed legislation has many unique features, such as opportunity for registration of extant varieties, registration of farmer's traditional varieties by communities or NGOs on their behalf, constitution of National Gene Fund.

The Geographical Indication

The protection based on Geographical Indication is found in Section 3 of TRIPs, which states: "Geographical Indications are for the purposes of this Agreement, indications which identify the source originating in the territory of a Member, or a region or locality in the territory, where a given quality, reputation or other characteristic of the good is essentially attributable to its Geographical origin". The evident examples of geographic appellation are Champagne for Champagne district in France, Colombian coffee, Havana cigars from Cuba and Scotch whisky exclusively from Scotland. Several for the developing countries including India are yet to enact the legislation on geographic indication. This clause should be exploited as an effective tool to monopolise select sections of the agriculture commodities market for goods like 'Basmati' rice, 'Darjeeling' tea, 'Alfonso' and 'Dasheri' mangoes or the 'Shahi' Leechi. A bill regarding the Geographical Indications is also being discussed in the Indian Parliament.

Future Strategies

Exploration and Collection: Based on past experience and present scenario, the collection and conservation of germplasm deserves urgent attention. Priority collection trips are to be made in the identified areas not sufficiently covered so far and a repeat collection should be made in other areas that showed elaborate diversity in the past. More emphasis should be given in future on trait specific explorations. The exploration and collection activities are to be carried out in a time bound manner, with the participation of the various stakeholders.

There has been a deficiency in sampling the wild germplasm during the earlier missions, when the focus was on the land races and traditional cultivars. In addition, different approaches and strategies are required to be adopted for wild species. There is a need to collect these resources which exhibit rich genetic diversity, as is indicated for example at the molecular level. The improved wide hybridisation techniques, the desire to explore new sources of genetic variability for improvement of the major cultivars and the threat of extinction of the wild populations are the other indicators, which suggest that the demand for wild germplasm will increase markedly. The rapid development in the techniques for utilisation of the germplasm requires the prioritisation of the regions for collections rather than the species.

Collaboration: No country is self-sufficient in PGR, nor can all countries accumulate resources, which would satisfy all their needs. There is a need to collaborate at local, regional and international levels for the acquisition of the germplasm. It is imperative to scrupulously obey the quarantine and biosafety rules for the safe movement of germplasm.

Conservation: A large number of accessions of crop germplasm, available with crop improvement programmes and PGR Centres in the country have yet not been deposited in the National Genebank. There is an urgent need to expand the network to include other centres, which have substantial germplasm holdings, and also to strengthen the existing network. It is to be ensured that a set of all the available PGR with associated database are deposited as base collection with the National Gene Bank in a time bound manner. One set should be maintained as active collections in the NAGs gene bank and should be available with the plant improvement centres to provide the opportunities for improvement, selection and evolution to continue.

Access: The level of utilisation of PGR stored in genebanks is difficult to assess. There is a general lack of feedback documentation on the final use of the distributed material. Genebank managers need to be helped to collect data from the users of their collections.

Evaluation: Characterisation and evaluation are essential to make the resources more immediately useful. A large number of germplasm is awaiting proper evaluation and characterisation. These tasks require substantial inputs. A decentralised well organised systematic evaluation of this genetic wealth in the shortest possible time will help to develop durable and stable multiple resistance varieties. The activity should be conducted in a time bound mission mode manner, with the participation of all the stake holders including ICAR and other public sector research institutes, SAUs and general Universities, NGOs and private research organisations/foundations. There is a need to increase the use of rapid and cheap evaluation tools, using biochemical and molecular techniques for germplasm characterisation. Training in these modern techniques may need to be developed and offered.

Utilisation: The effective utilisation of germplasm is possible only if the potential users of germplasm collections are able to find the information they are looking for. At present, only a few collections have a complete and user-friendly documentation system. With the developments in information technology the necessary tools for collation and dissemination of information are becoming available and need to be effectively pursued.

The information on genetic resources is distributed and almost each genetic resource centre has its own system for documentation of the information. The effective germplasm management can only be done through sharing of information following a network approach, involving all the stakeholders. It is suggested to have a centralised harmonisation of documentation systems involving all the genetic resource centres in the country and also at the Community level. The other option is to adopt a distributed, decentralised model, which offers universal access with decentralised maintenance, and less duplication of effort. The major issues in realising this potential are the development of data standards, and the establishment of quality control procedures. All databases should be able to effectively exchange data and share the information.

Sustainability: The current emphasis of germplasm evaluation is to produce crops less dependent on chemical products. The plant breeders are searching for resistance to abiotic stress (e.g. drought, cold), resistance to biotic stress (e.g. diseases, pests), quality traits, and better use of inputs without detriment to income. However, there is a growing realisation that the farmers demand for a wider choice and variety of horticultural crops and agricultural products, for which diversity in taste, colour, nutritional values and earliness/lateness is highly valued by the market. In addition, the culture and traditions also dictate the requirement of farming communities. There is a need to make the search for the characteristics required by farming communities as quick and efficient as possible for the evaluated germplasm. There is also a need to modify the descriptors for evaluation following bottoms up approach, by documenting and incorporating the requirements of the farming communities.

The adoption of the core collection concept by genebank will likely result in a more structured and efficient approach to identify such limited sets of germplasm and enable workers in different fields to compare results more effectively. This cost-effective approach of documenting collections is to be pursued in a number of crops. The practical pros and cons of the "core" collection approach need to be evaluated. Training in the approach may need to be developed.

Genetic Enhancement: Genetic enhancement in the sense of pre-breeding is a new term, not well known, especially to public funding agencies. Genetic enhancement for crop plants has become necessary in recent years to broaden the relatively narrow genetic base of modern crop cultivars. Such broadening is needed to supply new kinds of pest resistance, to bring in new levels of productivity and stability of performance, and to provide useful new qualities to food and feed products.

Biotechnology will provide essential and innovative support to standard plant breeding in the years to come, bringing in new generic systems, new techniques for selection and identification of genotypes, new ways of making hybrid crops, and, most importantly, deeper understanding of plant gene action, biochemistry and physiology. In addition to broadening the genetic base of established crops, genetic enhancement and plant breeding in the sense of final cultivar development can be used in two other ways: to develop new crops from hitherto uncultivated species, and to change old crops into new crops. Plant breeding utilising genetic enhancement, and sometimes assisted by biotechnology, will be used in the future to develop new intensive-culture crops from wild or weedy species, or from land races. Plant breeding, utilising genetic enhancement, and strongly assisted by biotechnology, also will be used someday to allow old crops to produce new products, such as speciality chemicals.

Genetic Stocks Registration: Registration of genetic stocks and elite germplasm need to be encouraged to ensure their effective utilisation. Genetically enhanced stocks, by public sector breeders for their own use in cultivar production, have been freely available to private breeders as well. In the changing global scenario under the new IPR regimes, modalities for compensation to these public institutions and the breeders will also have to be worked out. Also the public sector institutes are the custodians of the major germplasm holdings, the raw materials for any biotechnology-based enhancement. A partnership of the public and private sector can be developed, envisaging proportionate benefit sharing.

Public Awareness: Public interest in under-utilised crops seems to be increasing. However, it is unclear how large, and just how stable, is the market for any of these new products from minor crops. In order to encourage the sustainable utilisation of genetic resources, and to define market opportunities, costs and benefits, the economic and business aspects of genetic resources need to be studied. Awareness generation in the country or educating the people about the value of plant genetic resources wealth and patenting is essential. There is an urgent need to increase the interaction between breeders, GRU workers, through consultation meetings, workshops, seminars on PGR issues, etc. The emphasis should be on seeking diverse viewpoints of different stakeholders, such as, NGOs, scientists, individuals, etc. Their interface is likely to bring out some more useful management alternatives.

References

Arora R.K. 1988. The Indian Gene centre - Priorities and prospects for collection. *In*: *Plant Genetic Resources - Indian Perspectives*. (eds.) Paroda R.S., Arora R.K. & Chandel K.P.S., NBPGR, New Delhi, pp. 66-75.

Arora R.K. 1991. Plant Diversity in the Indian Gene centre. *In*: *Plant Genetic Resources - Conservation and Management*, (eds) Paroda R.S. & Arora R.K. Intern. Board for Plant Genetic Resources, Regional Office for South Asia and South East Asia, NBPGR Campus, New Delhi, pp. 25-54.

Arora R.K. and Pandey A. 1996. *Wild Edible Plants of India - Diversity, Conservation and Uses*. ICAR, NBPGR.

Arora R.K. and Singh H.B. 1973. *Abelmoschus tuberculatus* and its relation to *A. esculentus* (Malvaceae). *Baileya*, **19**: 90.

Arora R.K. and Nayar E.R. 1984. Wild Relatives of Crop Plants in India. *NBPGR Sci. Mag.* No. **7**: 90p.

Austin D.F., Jarret R.L., Tapia C. and Puente F. de la. 1993. Collecting tetraploid *I. Batatas* (L.) Lam. in Ecuador. FAO/IBPGR *Pl. Genet. Resources Newsl.*, **91/92**: 33-35.

Banham W. 1993. Biodiversity: What is it? Where is it? How and why is it threatened ? *New Series Discussion Paper No. 33.* Development and Project Planning Centre. Univ. of Bradford.

Bellon M.R., Pham L.J. and Jackson M.J. 1997. Genetic conservation - a role for rice farmers, *In*: *Plant Conservation: The* In situ *Approach.* (eds.) Martel N., Ford-Lloyd B.V. & Hawkes J.G. Chapman & Hall, London. pp-263-289.

Bezbaruah H.P. 1968. Genetic improvement of tea in North-East India. Its problems and possibilities. *Indian J. Gen.*, **28A**: 126-134.

Birggs D.E. 1978. *Barley*. Champan & Hall, London

Bohac J.R., Austin D.F. and Jones A. 1993. Discovery of wild tetraploid sweetpotatoes. *Econ. Bot.*, **47**: 193-201.

Brush S.B., Carney H.J. and Huaman Z. 1981. Dynamics of Andean potato cultivars. *Eco. Bot.*, **35**: 75-88.

Burkill I.N. and Finlow R.S. 1907. Resources of Jute. *Agri. Ledger*, **14**: 41-137.

Chandel K.P.S. and Paroda R.S. 2000. Status of Plant Genetic Resources Conservation and Utilization in the Asia-Pacific Region: *Regional Synthesis Report.* Asia-Pacific Association of Agricultural Research Institutions, Bangkok, pp. 18-26.

Chaudhary R., Pandey R., Mandal B.B., Sharma N., Tyagi R.K., Agrawal A., Gupta S., Malik S.K. and Hussain Z. 2001. Germplasm conservation: *In vitro* repository and cryobank. *In*: *National Bureau of Plant Genetic Resources: A Compendium of Achievement.* (eds.) Dhillon B.S., Varaprasad K.S., Srinivasan K., Singh M., Archak S., Srivastava U. & Sharma G.D. NBPGR, New Delhi. pp. 209-223.

Costanza R., d'Arge R., De Groo R., Farber S., Grass O.M., Hannon B., Kimburg K., Naeem S., O'Niell R.V., Paruslo J., Rsaun R.G., Sultan P. and Van de Bett M. 1997. The value of the worlds ecosystem services and national capital. *Science*, **387**: 253-260.

Crawford D. 1990. *Plant molecular systematics: macromolecular approaches.* Wiley, New York.

Cunningham I. 1987. Ethiopian germplasm repatriated by U.S. *Diversity*, **10**: 13.

Dennis S.V. 1987. Farmer management of rice variety diversity in Northern Thailand, Ph.D. Thesis, Cornell Univ. Ithaca, New York.

Doebley J.F. 1989. Molecular evidence for a missing wild relative of maize and the introgression of its chloroplast genome into *Zea perennis. Evolution*, **43**: 1555-1559.

Duvick D.N. 1987. Increasing yield and quality of cereals through breeding. *In*: *Cereals and Legumes in the Food Supply.* (eds.) Dupont J. and Osman E.M. Iowa State University Press, Ames, IA

Engels J.M.M. 1991. Conservation strategies: A holistic approach. A paper presented at the international Okra genetic Resources Workshop Oct 8-12, N. Delhi.

Engels J.M.M. and Wood D. 1999. Conservation of Agrobiodiversity. *In*: *Agrobiodiversity.* (eds) Wood D. & Leeni J.M. Characterization, Utilization and Management, CABI Publishing London. 355-386.

Erskine W. and Muehlbauer F.J. 1991. Allozyme and morphological variability: Outcrossing rate and core collection formation in lentil germplasm. *Theor. Appl. Genet.*, **83**: 119-125.

FAO. 1996a. State of World's Plant Genetic Resources for Food and Agriculture, FAO Rome.

FAO. 1996b. Global Action Plan for the conservation and sustainable utilisation of PGR for food and agriculture: Leipzig declaration by International Tech. Conference on PGR, Leipzig Germany, June 17-23, 1996.

FAO. 1996c. Utilisation of Plant Genetic Resources. International Tech. Conf. on Plant Genetic Resources. FAO Leipzig. 1-4.

Frankel O.H. 1984. Genetic perspectives of germplasm conservation. *In*: *Genetic Manipulation: Impact on Man and Society*. (eds.) Arber W.K., Llimensee K., Peacock W.H. & Straslinger P. Cambridge Univ. Press, Cambridge. pp. 161-170.

Friebe B., Jiang G., Raupp W.J., McIntosh R.A. and Gill B.S. 1996. Characterization of wheat-alien translocations conferring resistance to diseases and pests - Current status. *Euphytica*, **91**: 59-87.

Gautam P.L. and Singh A.K. 1998. Agrobiodiversity and intellectual property rights (IPR) related issues. *Indian J. Pl. Gene. Res.*, **11**: 129.

Gautam P.L. 1997. Agro-biodiversity in the Indian gene centre and conservation strategies- Keynote address. NAAS-NBPGR(ICAR) Workshop on National Concern for management, Conservation and use of Agrobiodiversity. Oct. 15-16, 1997, Shimla. 24p.

Gautam P.L., Mithal S.K. and Kochhar S. 1998. Wheat genetic resources management and its use in National PGR system. *In*: *Wheat: Research Needs Beyond 2000 AD*. (eds.) Nagarajan S., Singh G. & Tyagi B.S. Narosa Publishing House, N. Delhi, pp. 3.

Gizlice Z., Carter T.E. and Burton J.W. 1994. Genetic base for North American soybean cultivars released between 1947 and 1988. *Crop Sci.*, **34**: 1143-1157.

Govindswamy S., Krishnamurthy A. and Sastry N.S. 1966. The role of introgression in the varietal variability in rice in the Jeypore tract of Orissa. *Oryza*, **3**: 74-85.

Groombridge B. 1992. Global Biodiversity: Status of the Earth's Living Resources. A report compiled by the World Conservation Monitoring Centre. Chapman and Hall, London.

Gupta A.K. 1995. Compensating Local Communities for Conserving Biodiversity: How Much, Who Will, How and When, 1995, and published as rewarding Local Communities for conserving Biodiversity: The Case of the Honey Bee in Protection of Global Biodiversity: Converging Strategies, (ed.) Lakshman D. Guruswamy and Jeffrey A McNeely, Durhan and London: Duke University Press, pp. 180-189, 1998, IIMA WP No.99-10-14, October 1999.

Hamilton M.B. 1994. *Ex situ* conservation of wild plant species: time to reassess the genetic assumption and implications of seed banks. *Conserv. Biol.*, **8**: 39-40.

Harlan J.R. 1971. Agricultural origins: Centers and noncenters. *Science*, **174**: 468-474.

Harlan J.R. 1987. Gene Centres and gene utilization in American agriculture. *In*: *Plant Genetic Resources: A Conservation Imperative*. (eds.) Yeotman C.W., Xaflon D. & Wilkes G. AAAS selected symposium 87, Boulde Colorado, 111-129.

Harlan J.R. 1992. Crops and Men. *American Society of Agronomy*, Madison.

Harris D.R. 1989. An evolutionary continum of plant - people interaction. *In*: *Foraging and farming: the evolution of plant exploration*. (eds.) Harris D.R. & Hillman G.C. Unwin -Hyman, London.

Hawkes J.G. 1983. *The Diversity of Crop Plant*. Harvard Univ. Press, Cambridge. 184.

Hawkes J.G., Maxted N. and Ford Lloyd B.V. 1992. *The* Ex Situ *Conservation of Plant Genetic Resources*. Kluwer Acad. Pub, Dordrecht. 32-39.

Heywood V.H. 1995. *Global Biodiversity Assessment.* UNEP. Cambridge Univ. Press. UK.

Holden J.H.W. 1984. The second ten years. *In*: *Crop Genetic Resources: Conservation and Evaluation.* (eds.) Holden J.H.W. & William J.T. George Allen. London UK

Howard A. and Howard G.L.C. 1910. *Wheat in India; its production, varieties and improvement.* Tehacker, Spink and Co. Calcutta. 288 p.

Hoyt E. 1988. *Conserving the Wild Relatives of Crops.* IBPGR- IUCN-WWF, IBPGR, Rome.

Jackson M.J. 1995. Protecting the heritage of rice biodiversity. *Geo Journal,* **35**: 267-274.

Khoshoo T.N. 1994. Biodiversity, bio-productivity and biotechnology. *Agro-biodiversity,* **24**: 251-253.

Khoshoo T.N. 1995. Census of India's Biodiversity: task ahead. *Curr. Sci.,* **69**: 14-17.

Lal S.K. and Rana V.K.S. 2000. Genetic enhancement of soybean. *Indian J. Genet.,* **60**: 483-494.

MoEF 1997. *Draft National Action Plan on Biodiversity,* Government of India, New Delhi.

MoEF 1998. Implementation of article 6 of the Convention on Biological Diversity in India, National Report (interim). National Report to the CBD. Min. of Env. & Forest, New Delhi

Nanda G.S., Rao V.S.P. and Mohan D. 1998. Wheat germplasm- Utilized and underutilized. *In*: *Wheat research needs beyond 2000 AD.* (eds.) Nagarajan S., Singh G. & Tyagi B.S., Proceedings of the International Group Meeting held at DWR, Karnal, India during 12-14 August 1997.

Pal B.P. 1937. The search of new genes. *Agric. Livestocks,* **7**: 372-378.

Pal B.P. 1966. *Wheat.* ICAR, New Delhi. p.300

Pareek S.K., Bhat K.C. and Kumar A. 2001. NATP on sustainable management of plant biodiversity. *In*: *National Bureau of Plant Genetic Resources: A Compendium of Achievement.* (eds.) Dhillon B.S., Varaprasad K.S., Srinivasan K., Singh M., Archak S., Srivastava U. & Sharma G.D. NBPGR, New Delhi. pp. 26-30.

Paroda R.S. and Malik S.S. 1990. Rice Genetic resources - its conservation and use in India. *Oryza,* **27**: 361-369

Parthasarthy N. 1951. Some cytogenetical aspects of the origin of sugarcanes. *Indian J. Genet.,* **11**: 63-66.

Pham J.L., Bellon M.R. and Jackson M.J. 1996. A research program for on-farm conservation of rice genetic resource. *International Rice Res. Notes.,* **21**: 10-11.

Qualset C.O., McGuire P.E. and Warbuton M.L. 1995. Agrobiodiversity - Key to Agricultural Productivity. *California Agriculture,* **49**: 45-49.

Saxena S. and Gautam P.L. 2000. Rice germplasm diversity and conservation. *Rice India,* February: 7-10.

Schoen D.J. and Brown A.H.D. 1993. Conservation of allelic richness in wild crop relatives is aided by assessment of genetic markers. *Proc. Natl. Acad. Sci.,* (USA), **90**: 10623-10627.

Sharma Rajiv K. and Jagshoran. 2000. Contribution of exotic germplasm in wheat breeding in India. *Indian J. of Plant Genetic Resources,* **13**: 75-79.

Singh A.K., Srinivasan K., Jain S.K. and Saxena S. 2001. Germplasm Conservation: Seed Genebank. *In*: *National Bureau of Plant Genetic Resources: A Compendium of Achievement.* (eds.) Dhillon B.S., Varaprasad K.S., Srinivasan K., Singh M., Archak S., Srivastava U. & Sharma G.D., NBPGR, New Delhi. pp. 183-208.

Singh H. 1973. *Survey of crop genetic resources in their centres of diversity.* OH Frankel (ed.) FAO/IBPGR Rome Italy.

Singh H., Joshi B.S., Chandel K.P.S., Pant K.C., Saxena R.K. 1974. Genetic diversity in some asiatic *Phaseolus* species and its conservation. *Indian J. Genet.*, **34A**: 52-57.

Singh H. K. and Lal Banwari. 1998. Ethno-ecological concept of forest management *vis-á-vis* protection and exploitation by tribals in India. *In*: *Biodiversity Conservation Merged Forest and Protected Areas.* (eds.) P.C. Kotwal & Benerjee M.K., Agro Botanica. Bikaner. p. 217-224.

Smale M. and Bellon M.R. 1999. A conceptual framework for valuing on-farm genetic resources. *In*: *Agrobiodiversity: Characterization, Utilization and Management.* (eds.) Wood D. & Lenni J.M., CABI Publishing London, 387-408.

Spongberg S.A. 1993. Exploration and introduction of ornamental and landscape plants from eastern Asia. *In*: *New Crops.* (eds.) Janick J. & Simon J.E., John Wiley & Sons, New York. pp. 140-147.

Stuessy T. 1990. *Plant Taxonomy.* Columbia University Press, NY. 514 p.

Swift M.J. and Anderson J.M. 1994. Biodiversity and ecosystem function in agricultural systems. *In*: *Biodiversity and Ecosystem Function.* (eds.) Schulze E.D. & Mooney H.A., Springer-Verlag, Berlin, 15-41.

Tan A. and Tan A.S. 1998. Data Collection and Analysis in Turkey-GEF Project on *in situ* Conservation of Genetic Diversity. *In*: *Proceedings of Workshop to Develop Tools and Procedures for* in situ *Conservation.* (eds.) Jarvis D.I. & Hodgkins T., On-Farm, 31.

Thulijaram Rao J. and Krishnamurthy T.N. 1968. Accelerating genetic improvement in sugarcane. *Indian J. Genet.*, **28A**: 88-96.

Tiwari S.P. and Karmakar P.G. 2000. Widening the genetic base through germplasm enhancement, pre-breeding and generation of elite breeding materials in soybean. *In* : *The proceedings of 30th annual workshop* held at UAS. Bangalore, March 15-18, 2000-1213.

UNEP. 1994. *Convention on Biological Diversity.* Texts and Annexes. Interim Secretarial for the Convention on Biological Diversity Geneva Executive Centre, Switzerland.

UNEP/CBD. 1992. Global Biodiversity strategy: Guidelines for action to save, study and use Earth's Biotic Wealth Sustainabily and Equitably. World Resource Institute, Washington DC current views on needed policy changes, with a long list of key people in the field.

Vavilov N.I. 1926. Studies on the origin of cultivated plants. *Inst. Appl. Bot. Plant Breed,* Leningrad.

Virchow D. 1999. *Conservation of Genetic Resources. Cost and Implication for a Sustainable Utilizition of Plant Genetic Resources for Food and Agriculture.* Springer - Verlag, Berlin.

Wilson E.O. and Frances M. Peter. 1988. *Biodiversity.* Natl. Acad. Press. Washington DC, USA.

Wilson E.O. 1992. *The Diversity of Life.* Penguin, London.

Witcomb J.R. 1998. Participatory approach to plant breeding and selection. *Biotechnology development and monitor,* **29**: 2-6.

Witcomb J.R., Virk D.S. and Farrington J. 1998. *Seeds of Choice.* Oxford and IBH Publishing company Ltd. New Delhi.

Witcomb J.R. 1999. Does plant breeding lead to a loss of genetic diversity? *In*: *Agrobiodiversity - Characterization,* (eds.) Wood D. & Lenni J.M., Utilization and Management, CABI Publishing London, 245-272.

Zeven S.C. and Zhokovski P.M. 1975. *Dictionary of cultivated plants and their centres of diversity.* Centre for Agric. Pub. & Document. Wageningen, 219.

Plant Breeding - Mendelian to Molecular Approaches
H. K. Jain and M. C. Kharkwal (eds.)

Intellectual Property Rights Related Issues in Plant Breeding

Mangala Rai and S. Mauria

Abstract

International developments during last part of the 20th century brought to the fore the increasing relevance of Intellectual Property Rights (IPRs) in agriculture. Plant breeding has an intimate connection with various IPR-related issues. The paper discusses the various forms of IPRs viz. plant breeders' rights, patent rights, trade secrets, geographical indications and trademarks in relation to agriculture, thereby providing an insight into their various facets as relevant to agricultural research and development. Several aspects like requirements, scope and exemptions of plant breeders' right and patents; the concept of essentially derived varieties, patents in relation to plant breeding, research exemption *vis-á-vis* seed availability of protected varieites, making research exemption really meaningful, concern because of utility patents, salient provisions in Indian Patent Act and Protection of Plant Varieties and Farmers' Rights Act, implementation of plant breeders' rights and farmer's rights, information flow between plant breeders' rights and patent offices, requirements of deposit of protected products, suitability of trade secrets in specific circumstances and as remedy for research-based seed companies, and utilization of geographical indication and trademark concepts are specifically discussed. Subsequently, issues of conceptualization *vis-á-vis* implementation of intellectual property protection of plants, the requirement of supporting policies and programs for public research institutions, material transfer agreements and licensing arrangements for transfer of technology, impact on research agenda in public; and private research institutions, and on research publications and flow of information; and ethical and social issues are discussed for their application to Indian agriculture.

Introduction

Until the arrival of the General Agreement on Tariffs and Trade (GATT) establishing the World Trade Organization (WTO) in 1995, the multilateral and plurilateral treaties administered by the World Intellectual Property Organization (WIPO) constituted the bulk of the international law on intellectual property. The relevant treaties for Intellectual Property Rights (IPRs) related to agriculture were the Paris Convention and related

---

Indian Council of Agricultural Research, New Delhi 110 001, India plurilateral treaties, which dealt with areas such as patents, trademarks, appellations of origin or unfair competition; and the UPOV (The International Union for Protection of New Varieties of Plants, with presently 52 member countries) for protection of new plant varieties. Under the GATT Agreement establishing the WTO, now it is the Agreement on TRIPs (Trade related (aspects of) Intellectual Property Rights) which covers all issues of intellectual property. While TRIPs obliges the adherence to the substantive provisions of the Paris Convention, it goes further in limiting the freedom of countries on several aspects of their intellectual property laws. India is a member of the WTO and is, therefore, obliged to implement the Agreement on TRIPs. Forms of intellectual property described under TRIPs also cover plant materials. The TRIPs Agreement obliges members to provide protection for plant varieties either through patents or through an effective *sui generis* law or through any combination of the two. While TRIPs calls for the institution of an effective *sui generis* system of plant variety protection, there is no reference to UPOV or a call to adhere to any version of it. TRIPs also oblige the patenting of microorganisms and microbiological and non-biological processes for the production of plants and animals. It, however, presently allows the exclusion from patents, of plants and animals and essentially biological processes for their production. The TRIPs Agreement also ensures a universal, minimum level of protection of commercial marks, such as trademarks and geographical indications. For the first time in international law, trade secrets have also been accorded the status of IPRs. Under the TRIPs Agreement, the protection granted for IPRs can be tempered by appropriate provisions in competition law, particularly relating to practices or conditions of licensing of IPRs, which have an adverse effect on trade or transfer and dissemination of technology.

The Convention on Biological Diversity (CBD) is the other important international agreement relevant to a discussion on IPRs and agriculture. It allows sovereign rights on a nation's genetic resources, its stated objectives being: the conservation of biological diversity, the sustainable use of its components, and the fair and equitable sharing of benefits arising out of the utilization of genetic resources. Much is made of the provision on compulsory access to and transfer of technologies relevant to conservation under 'fair and most favorable terms' given in this treaty. However, with the provision that such access and transfer shall be consistent with the adequate and effective protection of IPRs, there is no cause to imagine that this treaty will force transfer of technology on any terms other than those set commercially in the market. Even the provision to cooperate to ensure that IPRs are supportive of and do not run counter to the objectives of the CBD is subject to international law, which now includes the TRIPs Agreement. It is also felt that the fair and equitable sharing of benefits from the commercial use of genetic/biological resources or traditional/indigenous knowledge would remain a good intention, till there are internationally accepted legal instruments to implement these provisions, i.e. there are currently no internationally accepted ways to reward farmers' rights or community IPRs, the other new forms of intellectual property at the stage of initial conceptualization, where the international and national laws are yet to evolve.

Thus, the classical IPRs relevant to plant breeding are: plant breeders' rights, patents (particularly on biotechnological inventions), trade secrets, geographical indications and trade marks; besides the all important farmers' rights and the community IPRs with reference to utilization of genetic resources. These classical IPRs are first introduced in this paper. The discussion then leads to obvious delineation and discussion on several linked issues, including the farmers' rights and community IPRs, which also get appropriately covered under various important sections. The paper has thus attempted to present an overall perspective on the topic. Wherever necessary, provisions available/considered in India and other important countries are mentioned and issues discussed in the Indian context. The conclusion section relates the subject to the overall requirements of agricultural research and development for better growth of Indian agriculture.

Plant Breeders' Rights (PBRs)

(*i*) ***The Requirements, Scope and Exemptions***: To secure PBR on a variety, the variety must be new, distinct (D), uniform (U) and stable (S). DUS characteristics of a new variety require: (*i*) clear distinguishability of the new variety from any other variety of common knowledge in expression of at least one characteristic, (*ii*) uniformity of that particular variety in its relevant characteristic(s) subject to the variation that may be expected from the particular features of its propagation, and (*iii*) stability of that particular variety in its relevant characteristic(s) from generation to generation.

The owner of a PBR Certificate is entitled to prohibit others from selling or offering the variety for sale, importing or exporting the variety, sexually multiplying the variety for marketing, and using it for producing another variety. As against the UPOV's 1978 version, the UPOV 1991 Act also requires authorization of the right-holder for the production of varieties (*i*) whose production requires the repeated use of the protected material, (*ii*) which have been essentially derived from a protected variety, or (*iii*) which are not clearly distinguishable from a protected variety. Since the UPOV 1991 Act also extends the protection to harvested material, there is also the new trend for extending the scope of prohibition under PBRs to harvested products, leading to thinking among breeders to apply for PBRs on crop varieties in countries where that particular crop is not grown. That is thought with a view to take action against imports derived from unlicensed propagation of the variety in countries where plant variety protection (or PBRs) is not available. There are, however, two exceptions: (i) Researchers' exemption - allowing use of any protected variety for conducting experiments or research to develop (as opposed to produce) a variety that is distinct from the protected variety; (ii) Farmers' exemption - to save, sow and re-sow seed on farmers' own holdings. In the USA, the law has even permitted a farmer whose primary occupation is not seed growing, to sell a quantity of seed to other such farmers without any obligation to the PBR certificate owner. India has included in its *sui generis* law to even allow exchanging, sharing or selling unbranded seed by the farmers considering its special situation.

(*ii*) ***EDV Concept in PBRs to Promote Originality in Breeding Researches***: Essentially Derived Varieties (EDVs) are varieties that are virtually entirely constructed upon the basis of the protected varieties from which they are derived. EDVs may be obtained, for example, by the selection of a natural or induced mutant, or of a somaclonal variant, the selection of a variant individual from plants of the initial variety, backcrossing, or transformation by genetic engineering. The arrival of the concept of EDVs within PBRs under UPOV's 1991 version, which many UPOV member countries have already introduced, has made such rights slightly more stringent, requiring authorization of the original breeder for marketing the EDV, and hence, making PBRs closer to the concept of Patent Rights. The EDV concept arrived to discourage plagiarist approaches in plant breeding, its intent is, however, more in the context of biotechnology-based breeding.

A breeder spends much time and effort in the development of a variety using conventional means. A biotechnologist spends time and effort in development and insertion of a gene in a developed variety to further increase its economic worth. If both claim complete rights, it is likely to help the EDV off the market. Thus, incentives for both groups are necessary for uninterrupted development of new improved varieties. However, the arrival of a much-needed reasonable compromise between the two groups is not so simple, particularly if the breeder of the conventional variety and of the EDV belong to two competing groups. The arrival of EDV concept is seen as providing a new balance between the PBR and Patent Rights systems to facilitate the exchange of technology between plant breeders and biotechnologists. In general, however, PBRs are considered to provide a minimum form of protection adapted to the needs of the traditional breeder as well as the farmers, as it can allow scope for exemption for both breeders/researchers and farmers.

(*iii*) ***Research Exemption Provision vis-á-vis Seed Availability of Protected Varieties***: The research exemption, although available in a PBR system with the intention of facilitating uninterrupted crop improvement researches without any hindrances, may again be of little practical consequence because there is no requirement that seed of the protected variety be made available. So, while the research with the variety is allowed, the seed required to conduct the research may not be available. Moreover, the use of trade secrets or contract agreements also permits the restriction of access to, or use of varieties on terms, which are defined solely within the discretion of the owner. It is, therefore, necessary that any PBR system displays an approach, which allays concerns about exchange of germplasm and information. That has to be a critical element for promoting efforts in cultivar development. To encourage crop improvement researches, seed availability of varieties protected through PBRs as well as easy procedures of authorization for developing EDVs are thus necessary.

(*iv*) ***Salient Provisions in Indian Legislation***: As mentioned earlier, under Farmers' exemption, PBRs allow farmers to save protected seed, for his or her own use, besides allowing farmers to sell surplus seed to their neighbours. The provision as it stands in the Indian legislation has rightly taken care of the special situation of farmers in the country and their immense contributions, and allowing the flexibility to Indian farmers to even sell unbranded seed, as their right rather than privilege. The research exemption clause is obviously also kept to facilitate continuity in required crop improvement researches. The clause on 'extant' varieties and provisions for farmers' rights and benefit sharing are unique to Indian legislation. At the same time, the legislation has cared to guard the interests of breeders/breeding institutions by reducing financial harm to them to substantial extent, by providing the EDV Clause. The clauses on compulsory license further safeguard the interest of farmers to ensure uninterrupted availability of quality seed. No doubt, many of these provisions have been included considering the peculiar socio-economic situation in the country, but they have been branded as measures of political necessity. The latter is not considered wrong anywhere in the world. The reasons for exclusion of tuber-propagated plants in the Plant Patent Act of the USA were more political than scientific.

(*v*) ***Implementation of Plant Breeders' and Farmers' Rights System in India***: Indian legislation on Protection of Plant Varieties and Farmers' Rights (PPV & FR) is expected to be implemented with the establishment of an Authority for the purpose. The administrative and legal system with the Authority has to, for example, put in place the mechanism for honouring farmers' rights and benefit sharing, and to draw basic support for granting PBRs from a technical system for DUS testing. To implement the legislation, there are several questions to be addressed to put the overall system in place. Salient among them under discussion are: (*i*) should India become a UPOV member to gain technically as well as administratively; (*ii*) how to go about for collection, conservation and documentation of all 'extant' varieties, which have been given a broader coverage in our legislation; (*iii*) who shall be the agency to do DUS tests, whether it should be a system independent of the All India Coordinated Crop Improvement Programs (AICCIPs) of the Indian Council of Agricultural Research (ICAR), or under the AICCIPs itself; (*iv*) if it has to be under the AICCIPs, will it impact upon other activities of AICCIPs, for which they are meant; (*v*) can any other independent system be established in the present situation in India, for example to work on build, operate and transfer/earn basis; (*vi*) should this statutory requirement be entrusted to breeders themselves, at least in case of minor crops with few varieties, to reduce costs; (*vii*) what kind of infrastructure of international standards is to be developed for such a system, which equally attracts breeders from other countries, and is able to encounter legal implications; (*viii*) should we not have some flexibility in the system to gainfully utilize the services of superannuated crop experts with their vast experience, and particularly when breeders in specific crops are less available; (*ix*) what is the magnitude of costs involved, is the Govt. of India prepared for the same; (*x*) should India be doing DUS test itself on all genera and species or enter into cooperative arrangements as done by many countries to economize on costs;

(*xi*) how to implement provisions on farmers' rights and benefit sharing, for which not only reorientation in many related systems is required but also much more clarity in conceptualization is required; and (*xii*) how to ensure germplasm supply and exchange among components of the National Agricultural Research System (NARS) of India, as has always been the case, in the changed scenario? No doubt, some thinking has been crystallized in recent exercises on the subject, but there are several issues yet to be resolved.

Patent Rights

(*i*) ***The Requirements and Scope***: To be patentable, an invention must be new, industrially applicable, non-obvious, and the subject of an enabling disclosure. The last item requires that the invention be described, so that a person skilled in the art to which the invention relates can reproduce it. In the case of inventions, which consist of living material, it is frequently impossible, by following a written description, to reproduce precisely the same material. To overcome this problem, the general practice is that an applicant may deposit a sample of the relevant living material and the deposit supplements the description in the patent specification, so that description and deposit together are deemed to constitute an enabling disclosure. Unlike in PBRs, there is no question of any exemption to farmers, as the rationale, to consider patents on plant materials (or even PBRs on plant materials!), is to reduce the perceived discrimination between plant developers and industrial inventors. Even the breeders' privilege is theoretically to the extent of exemption for experimental use, permitting only purely academic or non-commercial research. These exemptions are much narrow or less clear than its PBR analogue. Keeping absence in clarity for exemptions to researchers, or making patent rights more stringent than PBRs, was perhaps with increasing realization in the developed world that scientists of biology do create products that were more than mere products of nature, and thus the need to be given parity with industrial inventors in terms of availability of protection regimes.

(*ii*) ***Patents in Relation to Plant Breeding***: Patents on varieties or inbred parental lines and on genes particularly emerge as having substantial significance. The patent protection can obviously have greater effective force for the inventor than a PBR Certificate, as it does not permit farmers' and breeders' privilege in the manner desired. Applications of patents on a gene typically discloses a sequence of a gene, and then typically claims, a gene or protein standing alone corresponding to that sequence, a vector/plasmid incorporating the sequence, and, possibly, a plant (or a particular range of species) that has been transformed by means of such a vector, and even the descendents of the transformed plant. The patent holder thus gains effective control over the use of specified gene in genetic engineering. If, however, the same gene is found naturally in a species, a breeder can use traditional breeding techniques to include the gene in a cultivar without infringing the patent.

(*iii*) ***General Scenario on Patent Protection in Plants***: Under the European Patent Convention, a specific exemption to patentability is created for "plant or animal varieties or essentially biological processes for the production of plants and animals; this provision does not apply to microbiological processes or the products thereof". There has been much discussion about the meaning of this exception; some had speculated that it forbids patent protection to naturally bred plants while permitting it for genetically engineered plants. Other respected commentators suggested that the prohibition is only to varieties *per se*, and is not a bar to patents for generic inventions in plants. The latter view has prevailed, and the European Office now regularly examines patents on plants, at least for claims not directed to a plant variety. In general, the PBRs for plant varieties and patents for microbiological processes or products thereof are the two available systems for protection of plant materials in different developed countries, and essentially biological processes for production of plants or animals are also exempted from patentability. As stated earlier, this provision also exists in TRIPs under WTO.

The situation in the USA is slightly different as having three systems, *viz*. plant variety protection for sexually reproduced plants (with the exception of F_1 hybrids, fungi and bacteria); plant patents for all asexually reproduced plant varieties (in the everyday sense of the word "plant") with the exception of those reproduced by tubers; and utility patents in relation to all plant inventions which fulfill the criteria of the patent system. While the European provisions on patents are more akin to utility patents of the USA, that kind of stringency in provisions is not followed, because activism in favour of consumers of technology is witnessed there in more intense form. In countries like Mexico, Romania and Republic of Korea, plant varieties are protected in hybrid systems, which cover features of both plant variety protection and utility patents. Theoretically, the requirements of novelty, utility and non-obviousness are equally applicable to plant patents and utility patents, although US courts have admitted difficulty in applying the requirement of non-obviousness to plants. Nevertheless, it has been held in a US judgement that a printed publication, which would bar a utility patent application, is not a bar to a plant patent, on the ground that a printed publication is non-enabling for a plant. The US law is also unique as plant patents on asexually produced varieties can be obtained on a "discovered" variety as opposed to an intentionally developed variety, so long as the provisions of "growth in a cultivated area" and the capacity for "stable asexual reproduction" are met, i.e. "sports" are eligible for this type of protection.

(iv) Utility Patents Causing Concern in Gene-rich Countries: Utility patents are granted in relation to products consisting of plant material as well as to inventions concerning processes relating to or uses of plant material, and entitles the patentee to exclude others from making, using or selling the invention. The scope of such protection on a plant material is much broader, and covers a broadly described category of plants or parts of plants, of plant material, and one or more plant varieties. Such broad-based patents severely restrict the receipt and use of potential germplasm. Judicial cases of patents granted on *neem*, turmeric and *basmati* rice in the USA are well known to Indian public concerned with agricultural research and development. These have led our country chasing the patents in US courts, being infringement on our traditional indigenous knowledge with little novelty and non-obviousness in them. No doubt, patents granted in a different country have their limited applicability to that particular country alone; contesting the cases there, even if they involve costs, is perhaps equally necessary, at least for some time in the future, till the Indian public at large and activists on the subject are able to appreciate the Govt. of India's commitment to safeguard our rich heritage, traditions and knowledge. The CBD, considered a big milestone for gene-rich countries, also requires us to assert our sovereignty on our biological resources to convey the right messages to Indian masses.

(*v*) ***Making Research Exemption Really Meaningful***: This issue is perhaps the most complex that one faces in the changed times. A compromise should be possible between the conflicting needs of protecting IPRs and enhancing progress in both economically and socially essential crops - by the use of the research exemption provision for varieties developed and protected by public and quasi-public research activities. The ideal compromise would be to extend the research exemption clause to even include patented varieties. It is not likely that granting such a privilege would create an economic hardship. No doubt, the time allowed for realizing profits would become shorter than at present, but that seems a small price to pay for keeping germplasm openly available for breeding. The legal mechanics of a research exemption for germplasm or a patented variety can be worked out using the public interest clause. In fact, there is need to encourage the use of research exemptions as also to encourage policies, which facilitate exchange of even the products of agricultural molecular biology. In applications for grant of patents, the requirement of disclosure of origin of plant material developed needs be considered positively in countries like the US and others to send the right signals to gene-rich countries. Provisions of equitable sharing and facilitated access to germplasm under CBD thus need implementation in right earnest, by allowing its equity provisions, such as, the realizations in CBD - to ensure facilitated access to germplasm, and that IPRs are supportive of and do not run counter to the objectives of the CBD.

(*vi*) ***Double Protection - Need for Information Flow between PBR and Patent Offices***: The scope for double protection on the same plant species by both patents and PBR in countries like the USA is another cause of concern. Difficulties of this kind may be largely academic if patents protect very few varieties. However, one can envisage a situation, where many hundreds or thousands of varieties are protected in two separate systems with different criteria, as severe difficulties may arise in use of such materials in research. The need for linkages between plant variety and patent offices for easy flow of information will bring another complexity in the exercise of intellectual property protection of plants. This kind of problem does not so far arise in Europe, since patents are not granted for plant varieties, as such.

Also, in the USA, the case law (*Ex parte Hibberd*) provides a possibility of double patenting. The judgement concluded that the existence of the Plant Patent Act did not preclude patenting the same variety under the general (utility) patent statute. However, it did not indicate that more than one patent term should run on the identical product. Thus, in those instances in which a claim is made on the same variety, in a plant utility patent application and in an application under Plant Patent Act, a rejection for double patenting will be made by the examiner, i.e. to the extent that the scope of the claim in the patents and/or applications is identical, two patents would not issue. However, if there is some difference in scope between the two sets of claims, both patents may be issued, provided that common ownership of the two will be maintained throughout the life of both patents, and that the second patent to issue will expire on the same day as the first patent to issue. These requirements must be complied with in a document known as a Terminal Disclaimer.

(*vii*) ***Requirements of Deposit***: Patent rules on the deposit of biological materials, as mentioned earlier, require deposit of materials in either an International Depository Authority (IDA) established under the Budapest Treaty or other suitable depositories, which are specifically recognized by the patent office. While deposit of inherently stable and storable material, such as seeds or lypholized microorganisms is easily possible, there are difficulties in accepting by any deposit authority the living propagatable plant material (ex. a cutting) that is incapable of any cryogenic or other form of suspension. Clearly, if the protection systems do not provide for recognized depositories, the filing for the protection is the only alternative. In such a situation, the description alone without deposit of the material can not be an enabling disclosure, which is the basic purpose in protection of living material, whether by PBRs or patents. Proper depositories are thus a necessary element in such systems.

(*viii*) ***Relevant Provisions in Indian Patent Act***: Section 3 of the Patents Act, 1970 (as amended upto 2002) provides for inventions that are not patentable. Section 3(h) does not allow patenting of a method of agriculture or horticulture. Section 3(j) does not allow patenting of plants and animals in whole, or any part, thereof, other than microorganisms but includes seeds, varieties and species and essentially biological processes for production or propagation of plants and animals. Section 5 allows for patenting of the methods or processes of manufacture of any new substances intended for use, or capable of being used as food or as medicine or drug; or process inventions relating to substances prepared or produced by chemical processes. Non-biological and microbiological processes are thus patentable subject matter, to the extent that they do not fall under the category of non-patentable inventions under Section 3. The Patents (Amendment) Act, 2002 has incorporated a provision in Section 5 for patenting of substances also. Thus, the patent protection can extend to microorganisms and genes, provided the limitations of Section 3 are not attracted. The limitations of Section 3 are, however, not clear in specific clauses. For example, Section 3(h) is vague, as the provision that "a method of agriculture or horticulture" is not a patentable subject matter is too general a description to have any meaning in the present-day times, when there are examples wherein not only the gene patents but patents with much broader scope have been granted. Likewise, Section 3(b) that does not allow patenting of any invention – the primary or intended use or commercial exploitation of which would be contrary to public order or morality, or which causes serious prejudice to human, animal or plant life or health or to the environment – also leaves enough scope for judicial interpretation if judges consider latest scientific developments.

Trade Secrets

(*i*) ***Suitability in Specific Circumstances***: The common law of contract and torts, laws relating to both moveable and immovable property, and confidential and employment relationships, when combined with the physical nature of the plant material in question, have enabled private plant breeding companies to maintain the exclusive use of certain genetic materials and information relating to such materials. This has resulted in certain varieties of plants, particularly F_1 hybrids, only being available from a particular supplier. Component lines of synthetics and composites, and vegetatively propagated varieties based on a limited number of parent clones, provide other examples, where the exclusive possession of elite source materials by a breeder, together with information related to their combination, ensures that only the breeder can offer authentic seed of a particular variety in the market place. If the variety in question is in demand, the breeder in question can expect to secure a premium price to fund his research investment. Trade secret protection, as an alternative, is thus better suited only to plants, which are sold as hybrid seed, and it is not considered an option for inbred plant varieties, where true-to-type seed is to be sold to general public.

(*ii*) ***Remedy for Research-based Seed Companies***: As stated above, protection of this kind led to the development of research-based seed companies, in crops where hybrids are possible or the varieties are based on a limited number of parent clones. Under the law of trade secrets, if information and/or plant material is improperly acquired or misappropriated, the law will, in appropriate circumstances, provide a remedy to the plant breeder. Unlike in most developed countries, where appropriate IP regimes covering plant materials have also been in existence for long to additionally support the growth and expansion of private seed sector, countries like India, with predominance of public R&D in agriculture, have allowed the growth of private seed sector, by either their dependence on public sector for supply of finished plant varieties and their parental lines, or, by their use of the art of trade secrets through minimum strategic research.

(*iii*) ***Trade Secret vis-à-vis Making Seed Certification and PBR a Compulsory Requirement***: In contrast to the system of PBRs, wherein the basic premise for deposit requirement of protected material is that the technology behind the claimed novelty should be reproducible and available to public, breeders opting for the trade secret route consider potential use of their material by any third party, may even be a government agency, as undesirable. Thus, if the importance of public possession of the technology should necessarily outweigh everything else, there is a point for discussion as to whether the deployed seed certification and PBR systems should be voluntary or compulsory requirements for players in this field. One logic for making seed certification a compulsory requirement could be that if seed certification can be compulsory, and not voluntary in European states, why can't it be made so in India with the advent of PBRs, so that the farmer can be sure of the seed quality? This suggestion becomes more relevant as a clause is there in our PPV & FR Act for the performance of the protected variety.

The logic for making PBRs a compulsory requirement is that it might help to reduce the pressures from big private companies for early spread of large number of transgenics, for which even developed countries in Europe are still debating. This is stated, as transgenics, in all likelihood, shall be falling under the category of EDVs, being based on already adapted plant varieties. It is reiterated that the case of transgenics is cited for making the discussion more complete, for taking a well considered view, with absolutely no personal bias against the transgenic technology *per se*. Depending on the advantages of each transgenic, each case should be considered on its own merits, to come as an EDV and not as a patented product. Thus, authorization of and benefit sharing with the breeder of initial variety must, in the least, be a minimum requirement. It is proposed, notwithstanding the essential requirement of testing for biosafety aspects, which, in any case, remains a critical precaution to take on case-by-case basis.

These provisions are suggested for taking a considered view as the still nascent private seed sector in India is jittery of seeing seed certification as a compulsory requirement, and is discussing of keeping itself clear off the PPV & FR Act, in view of the provided provisions on 'extant and essentially derived varieties', and on farmers' rights and benefit sharing.

(*iv*) ***Enforceability of Trade Secrets***: Regarding any action against infringement of trade secrets, there will be need to demonstrate the derivation, i.e. the accused plant material is derived from a material protected with a trade secret. Enforceability of trade secrets may not be difficult in cases of hybrids, but it is not easy when it involves inbred parental lines, since the parent plants must be planted outdoors and over a wide area. This contrasts the traditional trade secret case law, which requires locked doors and security to establish trade secret status. Also, in a sufficiently large quantity of hybrid seed, there will be an occasional seed that was self-pollinated because of incomplete male-sterility. This self-pollinated plant represents the inbred of the maternal parent line. Identification and use of such seed is thus possible. In any case, the enforceability of a trade secret Right in a plant line offers an alternative form of protection, to consider for those plant materials which are not sold as a product in that form. There is, however, a general consensus emerging, that trade secrets are no longer a viable option for protection of plant varieties.

Geographical Indication and Trade Mark

Both these types of intellectual property also impact goods of plant breeding research. Geographical indications (GI) identify the geographic origin of a product. These are indications upon goods, of their origin from a country, region, or locality; where the quality, reputation or other characteristic of the goods are essentially attributable to their geographical origin. The basic objective behind geographical indications is to educate and influence the consumer in his choice of products, to guide him for use of an authentic product. Trade mark is an invented word, a visual symbol, a word, a name, a picture, a device, a label, or a color scheme, applied on goods or services to indicate the consumer that they belong to a particular manufacturer or provider of services, to distinguish them from similar goods manufactured or services provided by others.

A GI does not belong to a particular enterprise. It is used by several enterprises at the same time for the product, which originates in a particular geographical area. As against GI, which is a matter of public law and is non-assignable to any particular owner, trade mark falls in the realm of private law and is framed as per requirements of the owner. With respect to plant breeding, there is need for action on recognition, as GI for products like *basmati* rice and Darjeeling tea, and need for recognition of trade marks for providing identity to products of research of agriculture institutes like the Indian Agricultural Research Institute, New Delhi. It is so felt in view of the Indian experience on contesting the *basmati* patent case of Rice Tech., USA, and the fact that the specific recognition given to wines and spirits under TRIPs calls for concerted efforts for also including special products with origin in developing countries. Indian legislation on GI also needs to be fortified with its subordinate legislation, of making rules, regulations, procedures and systems, besides resolving any attendant difficulties that arise in due course of time.

Concerns of Different Sectors about IPRs

The fears are as diverse as the backgrounds of the concerned groups. Farmers of the developed world have feared that the new trends in plant protection deprive them of the freedom to save and replant seed from their own commercial crop, or to sell some of it to others. Farmers of the third world countries are completely disillusioned with the onslaught of new developments, when amelioration of their present economic conditions is seemingly impossible without major initiative of support from Government sources. Farmer-scale selection and breeding of the small farmers of the developing countries may become meaningless in the midst of expansion of area under productive but protected plant materials. Developing economies, economies in transition and sympathizing organizations with social objectives have feared that intellectual property protection on plants would gradually deprive the third world farmers of better sources of germplasm. Traditional plant breeders have feared that patents on genes and other genetic products of biotechnology would deprive them of sources of germplasm for classical recombination and breeding. People and organizations concerned with crop genetic diversity *per se* have feared that new germplasm would be locked up in private hands, thereby reducing the effective size of the germplasm base available for publicly-supported breeding. Small seed companies fear the increasingly difficult access to new germplasms, and feel that with the arrival of concepts like utility patents, they would certainly go out of business. Commercial breeding companies with strong R&D set-up have feared that public institutions, eager to profit from their research, would restrictively license their protected products to only a few companies, thus depriving other firms of the results of tax-financed research.

In brief, current attempts to increase protection on plant material based research products of any particular group has raised fears among other groups that the exchange and availability of germplasm would be curtailed to their disadvantage. It is in this context that the negotiated text of the Contact Group of the International Undertaking on Plant Genetic Resources for Food and Agriculture was put to the Parties in November 2001 for consideration and adoption. There was unanimity in the Contact Group on the entire text, except the component and part thereof of the genetic resources in the "form" supplied. The country thus needs to do its part to find answers to the questions of concerned groups and individuals, to help identify those fears that are based on fact, and then to help make adjustments that removes or alleviates the problems.

Conceptualizing *vis-à-vis* Implementation of Intellectual Property Protection of Plants

(*i*) ***Implementation Seemingly Not-too-difficult***: Public institutions are invaluable sources of germplasm and knowledge but they can no longer satisfy all the needs of today's increasingly knowledge-based seed industry. Pure necessity, therefore, leads competitors to deal with each other from time to time, to exchange material and knowledge on mutually acceptable terms. In an intellectual property (IP) regime, this shall have its bearing on both within public sector and public-private sector relationships. However, it is also true that to state these concepts of IP protection of plants is relatively easy, but to implement them may be more difficult. To implement, one has to learn from the history of IP protection in developed countries, which perhaps indicates both things. None of the systems developed is perhaps as successful as it was hoped. At the same time, such developed systems have not caused many problems as were predicted initially.

(*ii*) ***Implementation within ICAR-SAU System***: It is imperative that the Indian NARS, which is presently a predominant public sector activity, should deliberate for developing a system while also considering the upcoming private sector in crop science research. Within the ICAR-SAU (Indian Council of Agricultural Research - State Agricultural Universities) system, the approach can be to develop a common memorandum of understanding, as the finances are mainly drawn from public exchequer. For example, Agricultural Research Services of USDA was also suggesting consideration of a research exemption policy for products of plant research conducted or funded by the US Department of Agriculture. Such mechanism may not considerably affect the flow of germplasm within public sector, if monitored and implemented properly.

It should also not be the intent within the ICAR-SAU system to think of implementing the ideal situation, that exchanges should be *quid pro quo*, i.e. equally reciprocal exchanges. This has to be essentially kept in view to ensure the required crop improvement efforts from all those partners of research who depend for funds on public exchequer in large measure. Within the system, there may not be reservations even for informal exchange of germplasm. That further underscores the need for better germplasm documentation systems, and effective monitoring of germplasm flow and use by research leaders.

(*iii*) ***Collaborations with Institutions outside ICAR-SAU System***: In collaborative programs with private or other institutions, which have started operating on comparatively intense form of commercialization, the approach could be through negotiated Material Transfer Agreements (MTAs). Regarding collaborations with the international public CG institutions, these should have consideration for their objective of working in trust for the public at large, giving due consideration to earlier contributions of gene-rich countries and known concerns of the group of developing countries. But it may change, in line with their finalized policy being discussed in the changed situation. If IP protection is to affect even CG institutions and their clients, it would not be surprising that their donors may like to define 'seed' as a purely "private good", and they may deny funding to the variety development efforts in CG institutions, from which developing countries have benefited the most. Equal efforts of this group of developing countries, as a united body, are thus necessary to safeguard their interests in such variety development researches. In fact, collaborative programs on mutually agreed terms could be developed in any public-private or private-private partnership, but such systems would work as long as it is a win-win situation for both parties.

(*iv*) ***Technical Guidance from International Instruments for Simplification of Tasks***: Provisions in international treaties and agreements and technical guidance from established international agencies can help in developing understanding on common protection principles, practices, minimum levels of protection and procedures followed; and thus should also be considered to help in simplification of the tasks. This could help in bringing harmony among different nations and shall thus provide for a reasonable basis for acceptability in international arena. Any country would obviously be considering its best national interests while making programs of implementing IP protection on plants.

Supporting Policies and Programs for Public Research Institutions

(*i*) ***Implementing Policies that Equally Consider IPR Issues***: Hitherto, breeders recognized that the great progress that is made in plant productivity is due in large part to the free exchange of germplasm. It was also felt that protection was not needed for most plants in order to get new varieties used. The change now is the economic stress in public agricultural research system, and consequent intense interest by institutions to examine and exploit income-producing activities. There is thus now more and more interest towards commercialization and commercial potential of research. However, while there is perception among public-supported institutions that there is a commercial potential for facilitating access to germplasm and other technologies, a problem is also seen in commercialization of results arising from research which uses the technology of others. Plant utility patents especially interfere with free and open exchange of biological materials to be used for research, and, therefore, impede future genetic progress. Even though PBRs allows farmers' and researchers' exemption, the issues of minimum distance from a variety protected under PBRs and 'essentially derived varieties' can be of concern to many. Thus, the approach of allaying concerns about exchange of germplasm has to be seen through implementing policies, which appropriately take into account such issues. While the private industry cannot be told about the manner of conducting its business, two important issues primarily concern the public sector, which are as follows.

(*ii*) ***Storage and Access to Germplasm - Necessarily a Public Supported Activity***: There is general agreement, both domestically and internationally, that the collection, evaluation, conservation, utilization and exchange of germplasm is a desirable activity as it provides for the long-term diversity of genes necessary to address continued enhancement of plants to improve yield, quality, resistance to biotic and abiotic stresses etc. Systematic germplasm based research for making collections meaningful, and use significant, is thus a critical necessity, more so for a predominantly agricultural country like India. A strong germplasm research and management and supply system is only possible as a government-supported national activity. In the current trend towards more and more stringent IP regimes, such activities have to care for both economically and socially important crops. Socially important crops mostly include crops of indigenous origin, or crops, which are naturalized in the region to become an integral part of traditional lifestyles in the countryside. Equal efforts on these crops are required not only for social upliftment of rural population using these crops but also for exploiting their possible diversified uses, thus increasing their market potential. In fact, with the advent of biotechnology as a powerful tool, providing opportunities for genes to flow across the plant and animal kingdoms – collection, conservation, evaluation and documentation of all kinds of 'biodiversity' for posterity has become important. Therefore, private sector investment should also be flowing in this direction, if they have to sustain their growing industries on a long-term sustainable basis.

(*iii*) ***A Rational Approach for Continuity in Publicly Funded Crop Technologies***: There is little justification for spending public funds to develop new varieties or improving germplasm if they never reach the intended consumers. While IP protection on plants is justified in some cases in order to ensure use and recovery of public investment, protection can also be justified in other cases where researches have to meet social and supporting objectives of research and development. In order that R&D in public sector is not affected in both kinds of crops and commodities, support of public funds needs to be channelized in a manner that resource generation for public institutions keeps growing steadily, and Government and other possible sources also become more and more interested in allocating budgets for their important activities. This approach alone would help in ensuring the sustainability of such a well thought out, created infrastructure. This would also require rationalization in activities of public institutions.

MTAs and Licensing Arrangements for Transfer of Technology

It has been realized above that the greatest need for all concerned with crop improvement researches is to: (*i*) clearly delineate the advantages of IP protection for sharing of genetic materials and ideas pertaining to them; and (*ii*) identify the present and/or potential problems and disadvantages in use of IP protection for plants. This has to be done with the intention of facilitating correction or amelioration of the problems. In an IP regime, MTAs and licensing arrangements have become important mechanisms for supply of germplasm and technologies. This has also necessitated public-private interface and moving forward together on complementary activities to harness synergies.

(*i*) ***Material Transfer Agreements (MTAs) for Germplasm Exchange and Supply***: The movement toward IP protection of early research results is paralleled by trends in MTAs, which has been referred earlier. This is a concept that has yet to occupy its rightful place in India, but is common in most universities/institutions in developed countries. These agreements require the researchers to require MTAs as part of transferring any genetic materials to other researchers. Typically, in addition to provisions designed to disclaim any possible liabilities, the provisions of these agreements require the recipient to use the material for research purposes only and not to transfer it to third parties. If a commercial invention is derived by the recipient's use, the recipient is committed to negotiate with the supplier about sharing the profits. These agreements have become more and more common, and thus the importance of legal offices attached with institutions.

In such agreements, profit-sharing provisions can help in bringing windfall profits. However, they can be tricky, as actual implementation of sharing, when profits arise, is not an easy task, especially in public institutions. Moreover, MTAs only specify a duty to negotiate, and the negotiation itself may just prove unworkable. Added to such developments is the CBD, wherein the provision of national sovereignty over genetic resources, as against the common heritage concept, virtually undermines the earlier idea of free flow of genetic resources. Here, it must be clearly understood that even in the FAO's 1983 Undertaking on Plant Genetic Resources that espoused the common heritage concept, free access to genetic resources did not mean the access to be free of charge. The contradiction, of the facilitated access of genetic resources provided in the CBD, with the existing and emerging patent regime of WTO, as earlier also referred in the introductory part, has made the situation still complicated. It is hoped that this contradictory situation would be addressed to on WTO platform in the coming months. The obvious implication here is that any future collector or recipient of genetic materials, will have the equivalent of a MTA with provisions for sharing of profits by the donor nations. Such trends raise several issues, like development of codes of conduct for research exemption, codifying what constitutes 'utility' *per se*, avoiding overly extensive MTAs and developing MTAs by attempting a reasonable balance of interests, etc.

(*ii*) ***Licensing vis-à-vis MTAs***: Licensing is one approach, for both the public and private sectors, a primary vehicle, by which IPRs are transferred from the inventor to the farmers and consumer. Licensing agreements are generally private, contractual, negotiated and differ on a case-by-case basis, depending on the technologies and parties involved.

However, there is significant uncertainty in evaluating the usefulness, technical feasibility, and value of individual plant inventions. Uncertainty in assessing usefulness often causes difficulties in negotiating a license in the early research and development stages. This uncertainty has led to the use of MTAs as one type of licensing agreement, which often allows only research activities with the transferred material. Under the terms of these agreements, if a new product is developed subsequently, the owner may reserve the right to grant a license for commercial sale and use, or leave the critical terms of a license arrangement to later negotiation and agreement. At the later stage comes the issue of value of plant intellectual property, and identification of sharing partners in the value. Unfortunately, such issues are left for solution in a competitive environment, allowing product development to continue at its own pace. It seems at present no solution can be attempted.

(*iii*) ***Consistent and Organized Licensing Practices in Public Institutions***: Since it is important to facilitate dissemination of information and to evaluate potential licensing arrangements, public institutions need to be better organized and more business like when handling new inventions and their licensing. The problem is not likely to be there in private institutions for obvious reasons. Procedures defining how licensing decisions are made and by whom, can help in making quicker decision, for which identification of the empowered office with technically competent personnel for such jobs is equally important. This is required, as private companies are generally uninformed about the particular office in a public institution with which they can deal. Such an office should be a kind of facilitation cell, both to publicize about the availability of technology as also offering the basis for licensing and potential commercialization. The latter also requires public institutions to preserve their rights in intellectual property.

(*iv*) ***Questions for Consideration of Licensing in Public Institutions***: The process of licensing IP in public institutions is not a simple task. When considering the potential for licensing plant materials, the following could be some of the questions that need to be answered:

(*i*) Is there a clearly defined ownership position for the plant material, as either intellectual or personal property; is there another property that is critical or contributes to the successful use of the property;

(*ii*) What is the commercial status of the property i.e. does it need further development or is it ready to be marketed and used;

(*iii*) Are there any existing agreements or understandings, written or verbal, involving the property;

(*iv*) Does the potential user industry have any familiarity with the property; for example has the germplasm been grown in state, regional, or national trials;

(*v*) What is the potential competitive position of the property;

(*vi*) What is the potential life expectancy of the property;

(*vii*) Are there any existing standards or accepted norms for the property; to what extent, does the property meet these standards or norms;

(*viii*) Are there existing public/private sector relationships pertinent to this or similar property;

(*ix*) What is the economics of the property *vis-à-vis* the species in question, is it the plant material that is the primary value or is it a downstream product of the plant material;

(*x*) Are there points in the industry or the distribution system, which can be readily accessed or utilized to manage or control the property;

(*xi*) What is the current system(s) for making similar property available i.e. what is the nature of the industry for the species;

(*xii*) What is the amount of material available, how is it generated, how would it be made available for commercialization; and

(*xiii*) What capability is required to bring the property to the market place, does the inventor need to or want to have a role in supplying the property?

Above questions would quickly lead to the conclusion that the licensing of plant material will be a species-by-species and trait-by-trait decision. The type of license to be pursued with plant material is not a generic decision, but must be an individual decision tailored to the unique situation surrounding each property.

(*v*) ***Other Factors to Consider in Public Institutions***: It is perhaps important for public institutions to allow equal access to any protected technology, to ensure avoiding any potential problem of bias, whether perceived or real. While the public be apprised of the entire situation, the licensing procedures should include broad-based notification of full technology when it becomes available. Factors to consider when licensing a new invention include, (*i*) public trust, (*ii*) duty to public at large, (*iii*) duty to taxpayer, and (*iv*) duty to the inventor to protect the invention. This is so, because it must be assumed that the press will examine all institutional decisions. Therefore, the goal for public institutions must be to maximize public benefit; revenue should perhaps not be the highest priority. The latter consideration would, however, require needed encouragement from the government.

Exclusive licensing to any one agency can have effect especially on credibility of public institutions in the eyes of the general public as such arrangements may create tensions with regard to (*i*) access to germplasm and (*ii*) return on investment. It is important to use MTAs whenever possible to clarify the intent of the parties involved. MTAs should be well written, and must be tailored specifically on a case-by-case basis. Guidelines for writing MTAs are also needed, which include several examples. Breeder's Exchange Agreement under the 'Code of Ethics for Germplasm Exchange' may be useful in certain circumstances for defining these guidelines. An initial license agreement through MTA could have a clause stating that any use is for research purposes only, with scope for negotiations in future and having some flexibility for future developments. Provisions for enforcement of a license in case of a breach like absence of 'due diligence' by a licensee who is not marketing the product adequately or when the licensee gives materials to unauthorized third parties could also be there in the MTAs. The concept of 'essentially derived varieties' should also be defined and put into a license agreement.

It is also important to recognize that a new product is most likely the result of a team effort; and it is important to treat plant breeders or plant breeding teams alongside scientists of other disciplines on the basis of 'proportionate creativity'. Consensus on share of profit going to scientists and the institution, including any compensation to other staff, is another aspect for consistency in approach. Also, within a public institution, only certain crop species have the potential and/or opportunity to generate significant revenue; an issue that equally needs a consistent policy for keeping at bay any heart burning among different scientists and ensuring equal enthusiasm in scientists working on non/less-remunerative crops.

Impact on Research Agenda

The use of IPR is only a peripheral issue in the current financial struggles at public institutions, but historically it affects the synergy between public and private researches. The question is: what will be the result of these changing relationships? Equally important is the potential effect of the use of IPR on income, and consequently, on the performance of public institutions. Will the potential for income drive public institutions too strongly toward market concerns? The private sector, however, by its very nature, is focussed on the development of proprietary rights in developed technology, with whatever methods possible in the available scenario.

(*i*) ***Impact on Public Research Agenda***: There is concern that some crop programs and research efforts, with public significance, but little income potential, will be left behind, as the need for income affect research agendas. There is also increased pressure to seek new sources of funding, thereby increasing importance of IPRs as potential sources of support. Government signals to pursue the kinds of research may also be confusing, as it will require public institutions to do research relevant to both social problems and requirements of income generation. Therefore, establishing the right balance between basic and applied research, and the products that would have greater value and lesser value would always be a great challenge, and the external causes that would be influencing this balance would be on the increase. In a developing country like India, the Government would also want emphasis on actual transfer of technologies generated by the public research institutions for addressing the concerns of equity, social justice, and balanced and harmonized growth of crops/commodities/ sectors/sub-sectors/regions.

Research infrastructure in public institutions is, however, fairly attractive to private companies. This situation is the same, both in developed as well as developing countries like India. Sponsored researches from private sector are thus gaining ground, although the agreements underlying such researches may bring with them several restrictions, which complicate research efforts. There could thus be considerable uncertainty about the overall effect of such agreements on research programs. Such trends also lead public institutions to concentrate more on the short-term objectives, and less on long-term objectives of basic and strategic researches and the social objectives for which public institutions are meant. The perception (or reality) of research for profit could have further adverse effect on public funding of research programs. In any case, the overall public good from IPR policies in public institutions will have to justify the costs. Efficient utilization of public R&D infrastructure, in itself, is thus the biggest challenge; and all radical decisions for change should be made on a fully informed basis.

(*ii*) ***Impact on Private Research Agenda***: The importance of IP protection as a part of the overall business strategy has always been part of private sector thinking. Effective lack of proprietary rights' protection created two classes of crops. The first tier, where technology investments were almost always encouraged and research spending as a percentage of sales was relatively higher; and all others, where R&D investments were given careful consideration in terms of the cost-versus-return possibilities and where technology always lagged behind. Examples of the former are hybrid crops, vegetables and ornamentals. Examples of the latter are self-pollinated and food security crops like wheat, rice, peanuts, soybean, pulses, minor millets, etc.

The arrival of IPRs is, however, gradually blurring this distinction between the two classes. For example, where strong breeders' rights and variety registration are possible, companies have invested in non-hybrid crop development. Likewise, the availability of utility patents to protect plant varieties, genes and crop improvement methods makes many companies feel that they need to do more with this particular category of IP to cover a diversity of crops. At the same time, they are confused about the use of such patents for products of their traditional breeding programs. Thus, seed companies, whose research efforts have mostly been traditional breeding, are attempting to meet this threat by increasing research budgets or shifting resources away from traditional programs. In the struggle, the smaller companies may gradually get sidelined from mainstream research and become only subordinate agencies for purely development activities. In the race, the survival depends more on collaborative programs, and a public-private relationship is one necessary alternative for efficiently utilizing the infrastructure of public sector and resources of the private sector.

(*iii*) ***Impact on Research Publications and Flow of Information***: The greatest cost of onslaught of IP regimes is perhaps the decrease in informal flow and exchange of information. Certainly, a significant discovery will be published or patented, but much of the progress made in science comes from free exchange of ideas and germplasm, which occurs well before a discovery ends up in a new process or product. Unlike scientific journal articles, which reveal as much as possible, patent claims tend to reveal only that which is clearly required by the law. Likewise, the quality and quantity of exchange that used to take place at big scientific meetings is not seen in the present times. Especially at plant biotechnology meetings, the presenters have achieved new levels of evasiveness in response to questions.

An important controversy in the area of patenting is the first-to-file *versus* first-to-invent controversy, in which there is good probability of a general trend towards the first-to-file system. It can be exemplified by the situation of two inventors, who make the same invention, independently, but file at different times. In cases where the first-to-invent is not the first-to-file, the patent office or court in countries like USA attempts to reconstruct the actual order in which the firms made the invention. In a typical biotechnology context, this effort may require extremely expensive litigation. There are frequently steps of purification, assay, probe development, sequence identification, actual sequencing, and possibly synthesis, and several institutions may be competing in this process, with different ones leading at different steps. Under the first-to-file system, the first one to the patent office automatically wins. If such a change to first-to-file system becomes a general phenomenon, it will be accompanied by other changes. Thus, in the US system, in a pattern that universities like, one can publish and still file for a patent within a year; elsewhere there is no such grace period. In most other nations, unlike the USA, patents are made public at a fixed time, generally 18 months after filing, even before they have been granted. Therefore, the ideal would perhaps be a compromise, in which the first-to-file system is associated with some form of early publication, so that there is at least some consideration for sharing of new information generated. This can perhaps help in promoting meaningful research in competing institutions. At the same time, it may help in moving towards reduction in inequitable results of IP protection among competing institutions.

Ethical and Social Issues

The ethical and social issues of protecting plant intellectual property rights emanate from the following fundamental questions:

(*i*) What are the principles relating to the virtues and faults of the plant IPR protection process;

(*ii*) What are the appropriate aims of the process, who are its beneficiaries, and who does it neglect;

(*iii*) What social goals should plant IPR protection serve; and

(*iv*) What reasoning supports our judgements and actions on plant IPR protection?

Broadly, these questions imply two things: (*i*) that ethical and social content of the issues is closely tied to important questions about implementation of IPR procedures; and (*ii*) that ethical issues can not always be independent of social issues. In the context of plant breeding and crop improvement, the following issues, with both ethical and social concerns, can lead to identification of recommendations for policy initiatives.

(*i*) ***Access to Germplasm***: Possible questions could be: (*i*) which agencies/institutions/individuals should have access to germplasm; (*ii*) what are the ethical considerations of public versus private access; (*iii*) should compensation be a factor in allowing access; (*iv*) should utility patents restrict access to protected plant materials for only certain type of research; and (*v*) who does this restriction serve, and who does it neglect?

Ethical concerns could be: (*i*) should protection of genetic inventions be allowed when the raw materials of the inventions i.e. undeveloped germplasm, were not the private property of the inventor; (*ii*) should public institutions protect plant inventions with added value; (*iii*) is plant IPR protection by public institutions in the greater public interest; (*iv*) should protection of plant IPR deny, for 15-20 years, access to the new knowledge or technology; (*v*) can the scope of claims for the invention exceed its contribution; and (*vi*) does plant IPR protection allow the public inventor or institution to capture "fruits of labour."

Social concerns are: (*i*) will these methods of protecting plant IPR be adequate to capture social benefits; (*ii*) who shall benefit from access to various methods of plant IPR protection, and who is hindered; (*iii*) who shall decide what to count as benefits; and (*iv*) what external costs are ignored, e.g. environmental problems, social inequities, and geopolitical inequities?

(*ii*) ***Promoting Growth of Biodiversity***: Three basic questions are: (*i*) what are the present and future effects of plant IPR policies on biodiversity; (*ii*) should gene-rich nations be compensated for germplasm obtained from within their borders, if so, how; and (*iii*) how the present day gene-rich but seed-poor countries could be made both gene- and seed-rich by striking an optimum balance? The Convention on Biological Diversity has been a big step in answering these questions. Its importance in the presence of increasing trend towards IP regimes on plants has led to the concept of MTAs, as tools to promote exchange of germplasm among nations, and between private and public parties in different nations.

Ethical concerns are: (*i*) are MTAs appropriate for facilitating the conservation of indigenous genetic resources; (*ii*) whom do they serve, and whom do they neglect; (*iii*) what are the appropriate forms of compensation for access to genetic resources of a developing nation, e.g. cash, forgiving indebtedness, technology transfer, R&D infrastructure; and (iv) what are the external costs of our methods of protecting, or not protecting, plant genetic resources in developing nations?

Social concerns are: (*i*) will the use of MTAs encourage third world nations to develop their natural plant genetic resources; (*ii*) will the use of MTAs to exchange genetic resources increase R&D costs, therefore limiting rate of technology progress; and (*iii*) do MTAs serve the greater good of the global community? Current thinking of developing nations is that all these MTAs have failed to fulfill the aspirations of gene-rich nations. Such a feeling is certainly not a congenial situation for managing the biodiversity, which is so essential for posterity.

(*iii*) ***The Plant as Private Property***: In countries with IP regimes also covering plant materials, plant and plant genetic parts are protected as private property by law or by judicial decisions. However, the fundamental questions are: (*i*) should plant materials be private property; if yes, how should ownership be defined; and (*ii*) what new instruments of ownership should be developed, or are the current systems appropriate?

No doubt, ownership of private property for profit making purposes is a legitimate and primary purpose of the commercial sector. However, ethical concerns with respect to plant as private property are: (*i*) is profit motive the primary objective of the academic sector;

and (*ii*) is it appropriate for the academic sector, especially the public academic sector, to directly compete for profit with the commercial sector in protection and commercialization of plant IPR? At the same time, while the management of plants as private property could encourage investment, increase knowledge, improve living standards, and foster innovation, the social concern is, whether these practices restrict the access of germplasm to some players, narrow the research agenda, and lead to neglect of minor crops and difficult ecosystems.

(*iv*) ***Effects of IP Policies on Social Structures and Agricultural Sustainability***: The use of plant IPRs may be consistent with the paradigm of the market economy, to maximize competitiveness and return on investment, but there are effects of plant IPR policies on the social structure of agriculture, more so in countries like India. Specifically, the question is: how will IPR policies affect the development of sustainable and stable agricultural systems?

Ethical concerns are: (*i*) does protection of plant IPR foster vertical integration in the agricultural sector and displace the farmer as the primary customer of the input industries; (*ii*) who does it benefit, to who is it a disadvantage; and (*iii*) have we adequately considered the external costs of adopting new technologies based upon IP protection on plants?

Social concerns are: (*i*) how will increased adoption of technologies based upon plant IPR protection affect the future of small farm families, of rural communities; (*ii*) will these new technologies be better for the environment; and (*iii*) who has the responsibility for these issues?

(*v*) ***Survival of Public Sector Research and Development***: With erosion of long-term investments in public sector R&D capacity, it is adopting a short-term vision for delivery of results. The academic sector now thinks in terms of increasing its protection of plant IPR to secure new sources of research support. When that is so, the question is about the effects of plant IPR policies on the sustainability, stability and well being of the public sector research systems in the academia and the government.

Ethical concerns are: (*i*) what is the justification for non-uniform implementation of plant IPR policies among academic institutions, and among units within a single institution; (*ii*) what is the justification for non-uniform application of plant IPR policies among the public institutions-government-commercial sector participants in germplasm exchange; and (*iii*) whether the negative public perceptions of contracts of public institutions with the commercial sectors is justified?

Social concerns are: (*i*) does non-standard application of plant IPR policies cause participants or sectors to benefit unequally from advances in science and technology; (*ii*) are public institutions justly perceived as shirking their education mandate; (*iii*) do legislatures react to perceived irresponsibility in teaching by cutting or redirecting scarce research funds; and (*iv*) are public institutions caught up on pedaling the "IPR treadmill" to sustain their plant research programs?

(*vi*) ***Farmers' and Community Rights***: No doubt, there are international efforts to harmonize national policies on farmer-saved seed, but the basic question is: what rights should farmers retain after purchasing and growing protected seed? Also, would it not be fair enough for the farmer to expect the claimed level of harvest from the protected variety in its described set of environment? If it is fair to have a positive answer to the latter question, should he/she then not reserve the right for compensation if the harvested product is not commensurate with the claims made?

Ethical concerns are: (*i*) is the sale of farmer-saved seed "counterfeiting"; and (*ii*) whose interests are being served for prohibiting the sale of farmer-saved seed, and whose are being neglected?

Social concerns are: (*i*) will enforcement of laws prohibiting the farmers to sell seed lead to increased input costs to the farmer; (*ii*) is the farmers' selling seed enough reason for reduction in R&D support to important crops; and (*iii*) should the government not intervene if such critical research areas are abandoned, or given lesser importance?

Regarding implementation of community rights, the basic questions are: (*i*) what should be the possible criteria for identifying community rights; and (*ii*) how to develop a consensus on procedures for according recognition to those who have over millennia conserved and/or improved biodiversity?

Ethical concerns are: (*i*) should political expediency alone determine the interests of traditional communities; and (*ii*) where will the world end up with increasing trend of the rich exploiting the poor?

Social concerns are: (*i*) to what extent critical development inputs alone will satisfy the traditional communities in changing times; and (*ii*) shall any significant movement of traditional communities towards market direction deprive the country of its rich heritage?

(*vii*) ***Education of Students and Public***: Generally, students and public learn about plant IPR protection policy and mechanisms on an ad-hoc basis: students from perhaps their peers in the system, and the general public receive information through the lay media.

Ethical concerns are: (*i*) as mentors with IPR knowledge, do we have a responsibility to the next generation of scientists and educators; and (*ii*) do we have a responsibility to be forthright with the general public?

Social concerns are: (*i*) will an informed public be more likely to make intelligent and rational decisions on plant IPR policy; and (*ii*) will an informed student body and general public foster increased trust of science by the citizen?

In sum and substance, new forms of IPRs for plants raise numerous complex ethical and social issues. While it is necessary that we pause to reflect before rushing headlong into the thicket of issues before us, international commitments require us to adhere to the timeframes given. Many of these ethical and social issues can perhaps be resolved, but there may not be clear-cut answer to specific dilemmas, which many of these issues also create. Therefore, further value judgement will always need to be made in interpretation and application of these principles in the light of particular objectives and priorities. A clear set of ethical principles, however, can serve several useful and practical purposes. Thus, instead of stipulating what ought to be done, ethical and social issues can act as a series of checks before decisions are made; while not supplying substantive answers, they ensure that certain key questions are asked. This process cannot eliminate controversies and disagreements but expose policies and practices to critical scrutiny. An explicit set of principles can encourage rational discussions, informed negotiations and constructive interactions between interested parties in the use of plant genetic resources for socio-economic development. Naturally, a forward movement, duly considering the overall national socio-economic scenario as well as the possibility of fulfilling the minimum requirements of commitments, is needed to serve national interests in the changing global scenario. In general practice that is witnessed, it is easier said than done; and a collective political and administrative "will" of the developing gene-rich countries to succeed in international negotiations, is the only solution.

Conclusion

The inequity arising out of exploitation of biological resources of the gene-rich developing countries and the IPR regimes was voiced by the then India's Minister of Commerce, in the WTO Ministerial Conference at Seattle, USA. He observed, "the TRIPs agreement places the rights of a patent holder on a higher pedestal than obligations. However, it does not confer corresponding rights to countries or indigenous communities whose bio-resources or traditional knowledge are put to use". The failure of the FAO's 1983 International Undertaking on Plant Genetic Resources to forge a truly equitable global access regime, and the increasing trend towards monopolistic tendencies with expansion of IP regimes, has pushed the world towards regulatory mechanisms on a diversity of subjects, which became more possible with adoption of both CBD and WTO. No doubt, an attempt is being made for revision of the FAO International Undertaking on Plant Genetic Resources for Food and Agriculture, with the objective of bringing harmony between these two treaties. But it is also clear that regulatory mechanisms, whether to regulate biodiversity or intellectual property, have come to stay in life of nations on the earth. Therefore, adoption of measures to give effect to provisions of both the treaties has become a necessity. With amendments in the Indian Patent Act, arrival of the Geographical Indications Act and Protection of Plant Varieties and Farmers' Rights' Act, one remaining action is the arrival of our law on Biodiversity. These four legal instruments along with existing legal mechanisms on Trade Secrets and Trade Marks shall complete the requirements of legal instruments related to intellectual property on plants. The subordinate legislations and procedures thus need to be urgently put in place to give effect to these instruments, considered to be in the best national interest.

However, any amount of legal instruments and regulatory mechanisms will remain meaningless unless they positively contribute to national efforts in food and nutrition security, and towards the conservation and improvement of the environment. It is thus imperative that R&D infrastructure already available in the country be best utilized to meet the new challenges, which will be possible only with reorientation in policies and programs in tune with the explained new developments. It would also be inappropriate to accord over-emphasis on IPR issues, and undermining the other more important problems facing agriculture and environment. At this juncture, maintaining continuity in growth of agriculture especially requires research, education and development programs, centered around sustainable management of natural resources, as qualitative and quantitative degradation of land, water and bioresources is assuming unmanageable proportions. Basic and strategic researches considering cropping/farming system perspective, declining factor productivity, nutrient depletion and abiotic/biotic stresses have the potential to meet projected demands in future, and current concerns of environmental deterioration are not to be lost sight of. Integration and use of new science and technology, *viz.* biotechnology, information technology, space and nuclear sciences is equally important, to address the well-known new concerns on productivity, production and sustainability fronts. Many of the new technologies would equally need appropriate regulatory systems to not only streamline the processes involved, but also to keep at bay the attendant controversies. Documentation of traditional wisdom and knowledge is necessary, not only to assert our sovereignty on genetic resources, but also to gainfully use it in the new IP regimes.

Infrastructure reforms in agricultural R&D are essential for: bridging the gaps in research and technology transfer to farmers and industry, executing R&D priorities for sustainable technological change, upgrading skills of rich/poor farmers, reforms in agriculture education system, developing production-to-consumption systems for efficient use of agricultural produce, creating market intelligence systems for exploitation of indigenous/global markets, bringing reforms in trade, increased involvement of private sector, bringing accountability/efficiency in R&D, developing effective systems for acknowledgement and flow of intellectual property, effective role of social sciences and policy research, and participatory mode in research and development. To successfully move towards sustainability in agriculture, there is also a need for: bringing policies and action plans to tackle urbanization and population pressure, identifying clear-cut role of Centre *versus* State in agriculture, restructuring for major state level initiatives, and sustainability of public sector R&D institutions. In this endeavour, while a significant increase in R&D investment is a must, a rational judgement in allocation of resources, keeping an eye on the returns expected, is an equally critical requirement for taking the right decisions.

The agenda for agricultural research and education led development has also to consider both national and global partnerships. A major thrust is needed for development of rainfed, arid, hilly and tribal areas to overcome the earlier left imbalances. Simultaneously, however, there is need for more production from favourable agro-ecologies to move towards all-round progress. Generation of wealth and employment in the agriculture sector would require development, refinement and transfer of crop production technologies in food/fodder/fibre crops, horticulture, roots and tubers, plantation crops, aromatic/medicinal plants, etc. Likewise, development of animal husbandry, poultry, dairy and fisheries is also required for diversification of agriculture, increasing animal protein availability in the food basket and for generating exportable surpluses. Similarly, there would be requirement of soil and environment conservation programs to bring in the necessary element of sustainability in the ultimate objective of developing self-contained production-storage-marketing-consumption systems. Implementation of such self-contained systems should duly consider the various underlying factors, like the issues threatening food security and the earlier and likely impact on resource base to arrive at well-thought specific kinds of technology interventions.

The anxiety and concern about implications of new globalization policies on domestic economy, producer, consumer, food security and long-term growth prospects also need to be duly addressed. Thus, assessment of new risks and opportunities is important to safeguard national agricultural interests. Therefore, the disadvantages of poor infrastructure, poorly developed market and commerce, high energy prices, low capital, low access to credit and farm holdings with low average size are to be addressed without reservations as governmental support is much higher in the developed countries. Certainly, the requirement of a better future demands consideration of all national and international factors and appropriate implementation to successfully thrive in the changed times of competitiveness in all sectors of economic activity. Such a situation demands leadership to lead from the front, and having the six C's viz. credibility, competence, communications, control, caring and commitment, so essential to harness the genic power to steer India as a developed nation in the foreseeable future.

Suggested Reading

Baenziger P. Stephen, Roger A. Klease and Robert F. Barnes. (eds.) 1993. *Intellectual Property Rights: Protection of Plant Materials.* CSSA Spl. Publ. No. 21. Crop Science Society of America, USA. 179p.

Bragdon, Susan H. and David R. Downes. 1998. Recent Policy Trends and Developments Related to the Conservation, Use and Development of Genetic Resources. *Issues in Genetic Resources No. 7*, June 1998. International Plant Genetic Resources Institute, Rome, Italy.

IPGRI. 1996. Access to Plant Genetic Resources and the Equitable Sharing of Benefits: a contribution to the debate on systems for the exchange of germplasm. *Issues in Genetic Resources No. 4*, June 1996. International Plant Genetic Resources Institute, Rome, Italy.

Mauria S. 2000. DUS testing of crop varieties - a synthesis on the subject for new PVP-opting countries. *Plant Varieties and Seeds*, **13**: 69-90.

Rai M. and Kochhar S. 2000. Inter-Government Negotiations for Revision of the International Undertaking of Plant Genetic Resources - an objective assessment of the outstanding issues. *In*: *National Seminar on Inter-Governmental Negotiations*, held at NCDC, New Delhi, July 11, 2000; Ministry of Agriculture, Department of Agriculture & Cooperation, Govt. of India. 11p.

Rai M. 1998. Socio-economic and Ethical Issues in Plant Genetic Resources. *Dr. B.P. Pal Memorial Lecture* 1998. Indian Society of Plant Genetic Resources. National Bureau of Plant Genetic Resources, New Delhi, India. 16p.

Rai M. 1997. Benefit Sharing, Compensation Modes and Financial Mechanisms for Agro-biodiversity Management in India. *In*: *Workshop on National Concern for Management, Conservation and use of Agro-biodiversity.* Organized by National Academy of Agricultural Sciences, Shimla. 8p.

Rai M. 1996. Intellectual property rights in plants - status, issues and strategies. *Crop Improvement*, **23**: 25-30.

Rai M. 1996. Indian Seed Scenario and Protection of New Varieties of Plants. APSA (Asian Pacific Seed Association) *International Workshop*, Sept. 23-24, 1996; Jakarata, Indonesia.

Rai M. 1994. Intellectual Property Rights *vis-á-vis* Hybrid Research and Development in Asia-Pacific Region. *FAO Publ.*, **94/21**: 217-238.

Rai M. 1993. Current situation of plant breeding and seed industry in India and policy on protection of new varieties of plants. *In*: *Proc. Regional Seminar on the Nature and the Rationale for the Protection of New Plant Varieties under the UPOV Convention.* Beijing, China. p.125-131.

Rai M. 1993. Provisions under UPOV and Protection of New Varieties of plants. *In*: Souvenir, National Seminar on *Hybrid Research and Development*, Indian Council of Agricultural Research, New Delhi. p. 1-8.

Santos, Marcio de Miranda and Richard C. Lewontin. 1997. Genetics, plant breeding and patents: conceptual contradictions and practical problems in protecting biological innovations. *Plant Genetic Resources Newsletter*, **112**: 1-8.

Swaminathan M.S. and Jena S. (eds.) 1992. *Biodiversity - Implications for Global Food Security.* Macmillan India Ltd., Madras, India.

Swaminathan M.S. (eds.) 1995. *Farmers' Rights and Plant Genetic Resources - Recognition and Reward.* Macmillan India Ltd., Madras, India.

The Patents Act. 1970 (Act no. 39 of 1970) (As amended by the Patents (Amendment) Act. 2002. Govt. of India Pub.

The Protection of Plant Varieties and Farmers' Rights Act. 2001. Department of Agriculture and Cooperation, Govt. of India.

Plant Breeding - Mendelian to Molecular Approaches
H. K. Jain and M. C. Kharkwal (eds.)

30

Breeding Legumes for Improved Nitrogen Fixation

O.P. Rupela[1] and D.L.N. Rao[2]

Abstract

Legumes can meet about 80 % of their N need under favourable conditions. In addition, they have been estimated to contribute 20 % of the nitrogen needed for global grain and oilseed production. Thus nitrogen (N_2) fixation by legumes plays an important role in maintaining soil fertility in different cropping systems. Methods of measuring N_2-fixation, a pre-requisite to assess its levels, genetic variability in the different traits relevant to N_2-fixation, strategies to select and breed legumes for high N_2-fixation and yield are presented. Where available, examples of successful selection and breeding for high N_2-fixation have been cited. It is argued that any legume breeding program will benefit by growing crops under low soil-N conditions, which will exert a selection pressure in favour of high N_2-fixing plants. Use of simple to screen parameters (having high correlations with yield) for rapid identification of high N_2-fixing parent lines and segregating populations likely to be favoured by breeders are discussed.

Introduction

Legumes are an integral part of different cropping systems. Their role in sustaining crop production through nitrogen (N_2) fixation (a symbiotic process occurring in their nodules) is widely accepted. The fact that legumes fix N_2 was explicitly proved in 1886 by Hellriegel and Wilfarth in Bernburg and published in 1888 (Nutman, 1987). Until then, there was a controversy about the role of nodules. Even today there is a debate on the process of nitrogen fixation, but it is largely centered on methodology of its appropriate quantification (Herridge *et al.*, 1994, Danso *et al.*, 1993) and on methods of optimizing it for maximum harnessing (Herridge and Rose, 2000; Rupela and Beck, 1990). The oil crisis of the early 1970s and the consequent rise in fertilizer N prices resulted in intensive research on biological nitrogen fixation (BNF). Subsequently, the momentum has been sustained due to environmental concerns associated with manufacture of nitrogenous fertilizers and their usage. As people are becoming increasingly environment conscious (Brown *et al.*, 2000), it is predicted that BNF by legumes would gain more importance in the near future.

---

[1]International Crops Research Institute for the Semi-Arid Tropics, Patancheru 502 324, India
[2]Indian Institute of Soil Science, Bhopal 462 038, India

Legumes can potentially fix about 80 % of their own N need and in addition can contribute to the yield of subsequent crops (Saraf *et al.*, 1998; Ahlawat *et al.*, 1998; Lauren *et al.*, 1998). But all these potential benefits can be harnessed under certain conditions. Mere inclusion of a legume in a cropping system does not ensure high BNF. There can be three approaches to harness BNF: Increasing the area of legumes sown by farmers; improved crop, soil and water management to achieve maximum efficiency of BNF including Rhizobium inoculation, and selection of host genotypes to ensure a higher proportion of nitrogen fixation in the plant (P_{fix}). This paper discusses information/knowledge that will help endeavors towards breeding legumes for high BNF. The different sections include nodule formation/functions, methods of measuring N_2-fixation (with emphasis on those potentially useful in breeding), genetic variation for traits relevant to N_2-fixation and strategies for selection and breeding for high N_2-fixation. Where available, examples have been drawn from legumes grown on large areas in India.

Like the other biological processes, N_2-fixation in legumes is sensitive to environmental factors such as temperature, moisture and nutrients. Readers are suggested to refer to reviews on factors affecting N_2-fixation (Howieson *et al.*, 1993; Rawsthorne *et al.*, 1985; Serraj *et al.*, 1999a; Streeter, 1988). A good understanding of the limitations of N_2-fixation should be a help in devising an effective breeding program on N_2-fixation. Such limitations due to factors like mineral nitrogen (N) and water have been discussed.

Partners in the Symbiotic Process

Until 1992, there were four genera of root nodulating bacteria: *Rhizobium, Bradyrhizobium, Sinorhizobium* and *Azorhizobium* (Elkan, 1992). Also, some new species have been identified and the nomenclature of some of the old species have recently been revised (Young, 1996). Four more genera added recently are *Mesorhizobium, Allorhizobium, Methylobacterium* and *Burkholderia.* All the eight genera except *Burkholderia* belong to α-subclass of proteobacteria and the last one belongs to β-subclass of proteobacteria (Young and Haukka, 1996; Van Berkum and Eartly, 1998; De Lajudie *et al.*, 1998; Moulin *et al.*, 2001). It should, however, be noted that these are not new bacteria. These were known or associated with their relevant legumes all along, but due to the advances made in molecular biology techniques, their nomenclature has improved. For scientists interested in developing N_2-fixation technologies for use on farmers' fields, this fast advancing nomenclature of the bacteria presents a complex situation. However, for all practical purposes, the symbiotic relationship between the bacteria forming root nodules in a given host legume remains the same as ever and may not affect field oriented breeding programs on high N_2-fixation. In this paper, we have used "rhizobia" or *Rhizobium* to mean any bacteria/bacterium forming nodules on a given legume.

The process of nodule formation in legumes has been extensively investigated. Comprehensive reviews on different aspects of the subject are available (Caetano-Anolles and Gresshoff, 1006; 1996; Hirsch *et al.*, 2001; Long, 2001; Oldroyd, 2001; Schultze and Kondorosi, 1998; Spaink, 2000). Some reviewers have referred to the infection of legumes roots by the root-nodulating-bacteria (rhizobia) as "a beneficial plant disease" and others have referred to rhizobia as "refined parasite of the legumes". As per present understanding, at least ten steps are involved in the process of formation and functions of a given nodule. Each of these is gene-controlled interaction between the host-plant and the root-nodule bacterium. In response to a variety of substances excreted (e.g. flavonoids) by the host plant roots, rhizobia are attracted to a particular legume species, leading to the formation of specialized root organ, the nodule. A set of bacterial (nod) genes involved in production of lipo-chito-oligosaccharides (nod-factors) that act as signaling molecules are also important for nodulating the specific legumes hosts (Long, 2001). The steps involved in establishment of successful symbiosis are: multiplication and colonization at the root surface (*Roc* root colonization), adhesion of bacteria to the root hair surface (*Roa* root adhesion), curling or branching of host root hairs (*Hab* and *Hac*), root hair branching and curling, formation of infection thread (*Inf*), induction of meristems in the host roots for the nodule initiation and differentiation (*Noi*), bacterioid release from the infection thread (*Bar*), bacterial differentiation (*Bad*), onset of nitrogen fixation (*Nif*), nodule function persistence/maintenance (*Nop*), development of complementary functions associated with N_2-fixation (*Cof*) and its transport (Sprent and Sprent, 1990; Caetano-Anolles and Gresshoff, 1996).

The advances in the understanding of rhizobial genetics has been rapid and been reviewed extensively (Fischer, 1994, Long, 2001; Van Berkum and Eartly, 1998). The symbiotic genes have been located on large plasmids and on the chromosome of rhizobia. Several rhizobial plasmids (such as a 536 kb plasmid NGR234, Freiberg *et al.*, 1997) have been sequenced. An international effort (involving at least 10 institutions) initiated in 1998 resulted in sequencing of the complete genome of *Sinorhizobium meliloti*, comprising 3.65-Mb chromosome, and 1.35-Mb pSymA and 1.68-Mb pSymB megaplasmids (Galibert *et al.*, 2001). Readers are referred to the cited references for details on rhizobial genetics. For this paper, we have chosen to restrict to legumes and legume × *Rhizobium* interactions, as they would apply to developing high nitrogen fixing legume varieties.

Progress in molecular research of BNF has been relatively slow in legume plants. At least 80 genes affecting symbiosis have been identified in different legumes (Tsyganov *et al.*, 1998). Four major types of nodulation variants that have been reported for most legumes are, nod$^-$ (no nodules), nod ± (few or no nodules), fix$^-$ (ineffective nodulation), nod++ (super nodulation or hypernodulation) and nts (nitrate tolerant nodulation). Some salient aspects on these investigations have been listed in Table 1.

Screening Techniques for Improved N_2-Fixation

For a good breeding program on BNF, one would ideally require a high and stable genetic variability for this trait, a high heritability and a high correlation between the given BNF related trait and yield. A rapid and simple to screen method will be an added advantage. There is, however, no single method universally appropriate for measuring N_2-fixation by

Table 1. Nodulation Variants and genes reported for the different legumes

Legume	No. of variants	Genes	Inheritance/Comments
Pea	70	*sym1 to sym40 nod, nod2*	Most of the variants were nod-due to several reasons and monogenic recessive to wild type (mr). Some have nitrate tolerant nodulation but others have semi-dominance (e.g. sym-18). Other interesting traits involve ineffective white nodules, resistance to mycorrhizal infection, premature senescence of nodules, defective root hairs, impaired symbiosome development.
Soybean	20	*nts, rj1-rj8 rj2-Rj5*	Several super nodulating mutants with tolerance to nitrate (nts) (due to absence of auto-regulation). Some were monogenic recessive while others such as RJ2 to RJ5 were monogenic dominant. Most Nod-lines/plants have light green foliage.
Chickpea	12(3)*	*rn1 to rn8*	Several variants developed due to induced mutation. Nodulation variants (from non-nodulation to high nodulation) reported from land races and bred lines and developed by pure line selection. Most mutants studied were monogenic recessive and non-nodulating. *Kabuli* type nod-lines may have light green leaves while the leaves of the *Desi* type lines have symptoms similar to drought or salinity stress.
Common bean	9	*nts, sym1, sym2, Nie, nnd2*	Several nod$^-$ and fix$^-$ (ineffective nodulation) mutants developed. Most were monogenic recessive. Nitrate tolerant supernodulation also reported.
Faba bean	8	*sym1 to sym5 sym2*	Most were Nod$^-$ and monogenic recessive (Sym-3) but some monogenic dominant (Sym-2). Some were fix$^-$ (Sym-1, ineffective nodulation) while others had 3-5 times more nodules than control and tolerant to high nitrate.
Groundnut	7	*GP50 to GP54 GS-6*	All the variants were picked (due to light green foliage) from segregating generations (F_2, F_3) of different crosses involving normal parents. Most were from ICRISAT. Some were reported as having more than one gene, duplicate gene, double recessive or trigenic.
Pigeonpea	(3)	NS (Not studied)	Variants ranging from non-nodulation to high nodulation (higher than parents) were selected from segregating population (F_2) of several crosses. Genetics not studied.

Developed from Bhatia *et al.*, 2001; Rupela, 1992; Rupela and Johansen, 1995; Note: Most of the variants were developed by induced mutations. *Number of variants in parenthesis were developed from segregating population of different crosses of normal parents or were occurring in normal land races or bred varieties.

legumes. N-difference (N yield), ^{15}N, acetylene reduction, xylem solute (ureide) are the most widely used methods. But all these have their own limitations and strengths and cannot provide an accurate measure of N_2-fixation for every legume species grown under all possible variations of soil type and cropping environment. However, some of the methods are more likely than others to provide reliable and quantitative estimates of N_2-fixation. These are summarized in Table 2. However, for preliminary screening of large number of germplasm lines, counting nodule number per plant and taking nodule mass per plant, and even a visual rating scale for nodulation (Rupela, 1990; Rupela and Johansen, 1995) may be sufficient.

Biological yield largely determines N_2-fixation and grain yield (Mytton, 1983; Duc *et al.*, 1988; Kumar Rao and Dart, 1987). Therefore, breeders who operate in low N soils, and select for high grain yield, are highly likely to also select for high N_2-fixation. Thus in breeding programmes a more direct and useful selection trait would be N_2-fixation (or plant N or seed N yield under low soil-N conditions of growth), provided legumes were adequately nodulated and functioning.

Genetic Variability for Nodulation and N_2-Fixation

Presence of a large genotypic variability for traits such as nodule number, nodule mass and acetylene reduction activity (ARA) per plant have been known since early seventies and eighties for chickpea, groundnut and pigeonpea (Nambiar *et al.*, 1988), soybean (Wacek and Brill, 1976), cowpea (Zari *et al.*, 1978), common bean (Graham and Rosas, 1977). Using ^{15}N isotope-based methods, differences among cultivars have been detected in soybean (Hardarson *et al.*, 1989; Rennie *et al.*, 1982), common bean (Rennie and Kemp, 1982; Westermann *et al.*,1981), groundnut (Giller *et al.*, 1987), greengram and blackgram (Sampet and Peoples, unpubl. data cited by Peoples and Crasswell, 1992), pigeonpea (J V D K Kumar Rao, ICRISAT pers. commun.) and chickpea (Rupela *et al.*, unpubl. data). However, an effort to use this variability in breeding for improved N_2-fixation has been limited or non-existent in many of these legumes. Arunachalam *et al.* (1984) found that ARA and nodule mass have good predictive value for plant growth and yield related traits in groundnut. After analysis of a six-parent diallel cross, Nigam *et al.* (1985) observed that non-additive genetic variance for ARA was predominant in groundnut. The groundnut line NC Ac 2821 had the highest general combining ability for ARA, total nitrogen, leaf area and was proposed as a good parent for breeding programs. The crosses made between the high- and low-nodulating chickpea lines to investigate the inheritance of nodulation indicated segregation for nodulation in F_2 populations from non-nodulating to nodulating. But the differences in the extent of nodulation were generally not reflected in plant growth (Fig. 1, O. P. Rupela, unpublished). This may be due to ability of the less nodulated plants to effectively use soil N, which needs to be ascertained by further studies. Fig. 1 shows two plants with large visible difference for nodulation capacity but not for shoot mass. This, however, should not be interpreted to mean that legumes can yield well without nodules.

There is sufficient evidence to show that non-nodulating lines in some legumes (such as chickpea) do not yield at par with nodulated lines, unless provided with high doses of N-fertilizers (Rupela, 1992). In some other legumes (such as groundnut) the non-nodulating lines do not yield at par with nodulating parents even when high doses of N is provided (Nambiar *et al.*, 1986).

Table 2. List of different methods of quantification of nitrogen fixation, their salient features, weakness and strengths

Method	Salient feature	Weaknesses	Strengths	Reference
N-difference	Non-nodulating legume or a suitable cereal is used as reference. Nitrogen in these is deducted from the nitrogen in the test legume	Substantialdifferences between N_2-fixing and non-fixing plants for their physiology and agronomy	-Simple, does not need expensive equipments -When grown on low-N soils, high legume yield or N-yield is well correlated with N_2-fixation. It is therefore a simple trait in breeding for high N_2-fixation.	Boddey *et al.*, 1984 Kumar Rao and Dart, 1987
^{15}N method	-Atmospheric N_2 has a constant 0.3663 atom % ^{15}N (a stable isotope). If ^{15}N enrichment (abundance) in plant-available soil N is different from that in atmospheric N_2, then the proportion of legume N derived from each source can be measured by isotope abundance in the legume and in a non-fixing control wholly dependent on soil N. -Must measure total dry matter and N yield. -Assumption: legume and non-fixing reference plants use soil N with the same isotopic composition	-Requires ^{15}N fertilizer, an expensive item -Requires a very precise massspectrometer	-Widely accepted as accurate -Provides time-averaged estimate of proportion of fixed nitrogen (*Pfix*)	Chalk, 1985; Danso, 1988; Witty *et al.*, 1988
Natural ^{15}N abundance	Soils often have increase in the ^{15}N abundance compared with atmospheric N_2. In regularly cultivated soils ^{15}N values tend to range between 6.0 and 16.0 and are relatively constant with time and depth. The other features are similar to the ^{15}N method.	Require a high accuracy mass spectrometer with ability to measure differences of 0.1 part per thousand (about 0.00004 atom % ^{15}N) Great care in sample preparation to avoid isotopic preparation or contamination	Does not require ^{15}N fertilizers	Shearer and Kohl, 1986; Peoples *et al.*, 1989

	-Accuracy depends on the levels and spatial and temporal uniformity of the ^{15}N in the soil			
Acetylene reduction assay (ARA) method	-The N_2-fixing enzyme, nitrogenase also catalyzes the reduction of acetylene (C_2H_2) to ethylene (C_2H_4). -Generally detached/ disturbed root system is incubated in air tight containers for a known period in an atmosphere containing C_2H_2. The gas sample is analyzed for C_2H_4 using a gaschromatograph. -It is a point in time measurement	-Difficulties in recovering root nodules quantitative -Requiresregular measure-mentthroughout growth of legume. Needs interpolation between the different measurements to obtaintime-integrated measurements. -Diurnal changes in the activity are large -Non-linearity in the rate of C_2H_2 reduction over the period of the assay	-It is rapid, sensitive, simple, and relatively inexpensive (to ^{15}N methods). -Good for screening large number of germplasm lines	Wani *et al.*, 1993 Witty and Minchin, 1988
Xylem N solutes	Principle: N profile in solutes in xylem sap changes from the one dominated by ureide compounds (altantoin and altantoic acid) in N_2-dependentplants that are dominated by nitrate and amino-N in plants using soil N. Relative abundance of ureide N in xylem sap and the proportion of plant N derived from N_2-fixation (*Pfix*) needs to be correlated and is reported extremely strong with regression coefficients close to unity.	- Not all legumes export fixed N_2 as ureides, e.g. crops like chickpea export amides(asparagine, glutamine)- Response of ureide levels to water deficit was very different from that observed for well watered plants. The levels of ureides in the plants and particularly, the petioles, increased as soil dried and N_2-fixation rates decreased (De Silva *et al.*, 1996; Serraj *et al.*, 1999b.	-It is sensitive, simple, and relatively inexpensive (to ^{15}N methods -Good for screening large number of germplasm lines	Peoples *et al.*, 1989; Herridge and Peoples, 1990

Fig. 1. In a cross between two chickpea lines, sixty two days old plants grown on a Vertisol field at ICRISAT Patancheru were uprooted for nodulation studies. Segregation for nodulation at F_2 ranged from few nodules in the plant at right [lower than the low nodulating parent F_3 Partner 4-14-1 (ICRISAT germplasm accession IC-6679)] to normal nodulation in the plant at left [close to that of the normal nodulating parent NEC 802 (ICRISAT germplasm accession IC-5770)]. Few plants nodulated better than the normal nodulating parent (not shown in the figure). However, the plant growth did not seem to correlate with nodulation, suggesting availability of sufficient nitrogen in the field and emphasizing the need of breeding legumes at low soil-N.

Following Mahalanobis D^2 statistic, Dangaria *et al.* (1994) reported large genetic divergence (among 32 chickpea genotypes) for nodule number per plant, nodule mass per plant and nodule size. Cultivar K 850 formed highest nodule mass and clustered separately. These studies thus indicate the complexity of the N_2-fixation related traits. Screening for high N_2-fixation traits for chickpea and groundnut can be made in field. Legumes like pigeonpea offer difficulty for N_2-fixation studies in field, because their nodules are loosely attached to roots and generally fall off during excavation of field grown plants. It is perhaps due to this reason that there were no reports in pigeonpea so far on studies of the type reported above for groundnut.

Intracultivar Variability for N_2-Fixation: Plant-to-plant variability for nodulation within chickpea cultivars, including occurrence of non-nodulating plants in land races and bred cultivars has been reported by Rupela (1992). Consistent variability for nodulation extent was also subsequently detected within pigeonpea cultivars (Rupela, 1994). Unlike in chickpea, however, non-nodulating plants in pigeonpea were found in segregating populations at F_2 (Rupela and Johansen, 1995). In addition to the breeding method used for developing a material, absence of any natural selection pressure for nodulation or N_2-fixation during its development may be responsible for the occurrence of the different nodulation types within a material up to release stage. This view gained strength from the fact that during a screening for high nodulating plants at high mineral N in soil, both high and low nodulating plants were observed in 85 out of 90 advanced breeding lines of chickpea (Rupela, 1994).

Using appropriate screening procedures several different nodulation types [high nodulating (HN), low nodulating (LN), non-nodulating (NN)] have been identified within several chickpea and pigeonpea cultivars (Rupela, 1994). Preliminary studies of Venkateswarlu and Katyal (1994) also indicated plant-to-plant variability within cultivars of groundnut. It is, therefore, likely that intra-cultivars variability is present in other legumes also. Intracultivars variability for a given trait in a crop species is not new. Singh *et al.* (1988) identified downy mildew resistant lines from a highly susceptible parent of pearl millet (open pollinated crop) and M. P. Haware (ex-ICRISAT scientist, personal communication) developed a *fusarium* wilt resistant line from a susceptible cultivar (Annigeri) of chickpea (self-pollinated crop). Obviously, the *Nod*$^-$ (NN) and the low-nodulating (LN) selections are of academic interest and serve as an important reference base in N_2-fixation quantification studies. High-nodulating (HN) selection generally grew better than the NN and LN selections of a given cultivar. Yield trials were conducted only with the LN and HN chickpea selections of G130 and K 850 at five locations in India and a location each in Bangladesh, Nepal and Pakistan (Dudeja *et al.*, 1997; Khanam *et al.*, 1997). At ICRISAT, the HN-selection of cultivar G 130 produced 31 % more grains than its LN-selection at low soil N (N_1) level. The HN-selection of G 130 yielded better even at high soil N (N_2) level. [Note: The two contrasting soil N levels were prepared by applying 0 (N_1) and 100 kg N ha^{-1} (N_2) as urea to the preceding sorghum on Vertisol field after the whole field was depleted for N by growing cover crops for two years]. N is a highly dynamic element and varies through the soil profile due to several factors. At sowing, the available N in the top 15 cm soil was 1.7 times more in the N_2 plots than in N_1 plots (8.7 mg N kg^{-1} soil)]. But the LN and HN selections of another cultivar K 850 yielded the same under N_1 and N_2 levels. This may be due to high root length density of the LN-K 850. In a previous pot trial the root length density of LN-K 850 was 32 m plant^{-1} that was two-times greater than that of the LN-G 130. Perhaps the cultivar K 850 could scavenge the soil N more efficiently than that of G 130 due to its high root length density, and as a result both the HN and LN lines of K 850 yielded similarly. The HN selections, however, yielded higher than the relevant LN selections (Table 3). These studies thus suggest a great scope of enhancing N_2-fixation in legumes through host plant selection.

Table 3. Nodule dry mass (mg plant^{-1}), acetylene reduction activity (µ5M C_2H_4/plant^{-1}-h^{-1}), N_2 fixed (kg ha^{-1}) and grain yield (t ha^{-1}) of chickpea lines selected for different nodulation and N_2-fixation capacity

Chickpea line	Nodule dry mass (mg plant^{-1})*	µM C_2H_4 (plant^{-1} h^{-1})	N_2 fixed (kg ha^{-1})	Grain yield (t ha^{-1})
G 130 HN	163	4.6	23.8	1.34
G 130 LN	93	3.9	13.5	1.08
G 130 (Parent)	117	3.4	17.2	1.22
K 850 HN	219	8.1	30.9	1.49
K 850 LN	88	2.1	23.5	1.33
K 850 (Parent)	197	5.0	24.5	1.36
Rabat-NN (Nod$^-$)	0	0.0	NR	0.88
Annigeri-NN (Nod$^-$)	0	0.0	NR	0.81
Mean	110	3.4	22.2	1.19
SE±	4.5	2.16	1.33	0.04
CV%	16.0	63.00	34.00	20.00

Data are means of four locations (Akola, Badnapur, New Delhi, Sehore) in India. Data from Hisar not considered due to crop heterogeneity. NR=Not relevant; *=Data is from New Delhi and Sehore for the year 95/96; **=Data from New Delhi only; HN=High nodulating selection; LN=Low nodulating selection; Summarized from: Dudeja *et al.*, 1997

Variability for N_2-Fixation Under Stress Conditions: N_2-fixation is sensitive to several environmental factors including soil-N. High nitrogen levels have been reported to suppress N_2-fixation in different legumes (Rawsthorne *et al.*, 1985; Peoples *et al.*, 1987; Buttery and Dirks, 1987). Based on several studies (Wu and Harper, 1991; Buttery *et al.*, 1988; Buttery and Gibson, 1990), nodulation and/or N_2-fixation was reduced by approximately 50 % in different legumes, when N concentration in root environment was between 1.43 and 6.0 mM (approximately equivalent to 20 –90 mg N kg^{-1} soil) in the growth medium. The suppression in N_2-fixation was particularly due to the nitrate fraction in the root growth environment (Streeter, 1988). Groundnut grown on an Alfisol without fertilizer N application in India derived 61 % of its N requirement through N_2-fixation (proportion of N_2-fixed or *Pfix*), and application of 100 kg N ha^{-1} reduced *Pfix* to 47 % (Yoneyama *et al.*, 1990). The suppressive concentration of mineral-N may be reached in the root zone with band application of N-fertilizer, at least for some days after application. Mineral N concentrations of 10-32 mg kg^{-1} soil have been recorded at sowing chickpea in farmers' fields around Hisar in Haryana and Gwalior in Madhya Pradesh (Wani *et al.*, 1997). All these studies indicate the need for developing symbiotic lines of legumes that are tolerant to different stress factors.

Genetic variability for N_2-fixation traits has also been reported under drought conditions. Smith *et al.* (1988) reported that common bean plants reacted to soil dehydration by closing their stomata earlier than soybean plants, which resulted in maintaining high water content. Nodule growth and activity were more sensitive to drought in common bean, and decreased earlier during soil dehydration, compared to soybean. However, Peña-Cabriales and Castellanos (1993) found, in a comparison of common bean cultivars, that the percentage of N derived from N_2-fixation was not affected by drought stress.

Comparisons across species in N_2-fixation response to drought have shown variations that indicate a genetic control. De Vries *et al.* (1989) found that nitrogenase activity per plant and N accumulation in peanut (amide exporter) were less sensitive to drought compared with pigeonpea (*Cajanus cajan* L.) and soybean (ureide exporters). Sinclair *et al.* (1995) has also reported that peanut N_2-fixation was relatively insensitive to soil drying, and they identified cultivar variation in drought. Comparison of the percentage of N derived from N_2-fixation under drought stress by chickpea and pea by Beck *et al.*, 1991 showed differences between the two species. Wakrim and Wery (1995) showed that N_2-fixation in chickpea was depressed under drought conditions, although it had no significant effect on yield.

Salinity in the arid and semi-arid regions of the world is a serious threat to agriculture. Production of grain legumes is particularly vulnerable because of their low tolerance to salinity, combined with the high sensitivity of their BNF system (infection of root hairs by rhizobia and subsequent nodule development). Salinity does not affect colonisation of roots by rhizobia as much as initiation and growth of new nodules (Singleton and Bohlool, 1984). A soil pH of 8.9 was the critical upper limit for high nodulation in most genotypes of chickpea. At pH 9.0-9.2, a genotype selected for high-nodulation outperformed the other four used in the study (Rao *et al.*, 2002). Nodulation was reduced in all the five chickpea genotypes as the electrical conductivity increased from 1.1 to 8.1, but the high nodulating selection CSG 9372, seemed to have more tolerance and formed about three-times more nodules than the salt tolerant line (CSG 8927), even at 6.2 dS m^{-1} (Table 4).

Heritability of Nitrogen Fixation

In a breeding program, establishing the heritability of relevant traits is a pre-requisite. Ronis *et al.* (1985) investigated broad-sense heritability of total and per cent fixed N (using ^{15}N technique), in harvested seed of three F_2 soybean populations each of 110 plants. Broad-sense heritability for fixed N content of seed ranged from 0.53 to 0.60. Heritability estimates for per cent fixed N in seed were moderate at 0.37 and 0.43, suggesting that breeding for improved N_2-fixation of soybean using appropriate parents should be possible. Herridge and Rose (1994) found broad-sense heritability (N_2-fixation assessed by ureide analysis) to range from 0 to 0.31 in F_2 populations of soybean. For F_2 derived F_3 lines it was 0.36 and for F_3 derived F_6 and F_7 lines it ranged from 0.32 to 0.52. Greder *et al.* (1986) estimated broad-sense heritability for nodulation (nodule mass) in soybean, which exceeded 0.55 for each population averaged across sites. Their data, together with correlation analyses of nodulation, agronomic and yield traits indicated that selection for increased nodule mass was warranted.

Table 4. Shoot and root dry mass of five selected genotypes and nodule number and dry mass of nodules of four (the non-nodulating ICC 4918 does not form part of the analysis) selected genotypes of *C. arietinum* grown (60 days) in a saline soil (Rao *et al.*, 2002)

Genotypes	Shoot dry mass (g per pot)				LSD P=0.05	Root dry mass (g per pot)				LSD P=0.05
ECe (dSm^{-1})	**1.00**	**3.20**	**6.20**	**8.10**		**1.00**	**3.20**	**6.20**	**8.10**	
ICC 4918 (non-nodulating)	0.57	0.61	0.50	0.14	0.10 (among ECe)	0.26	0.21	0.15	0.03	0.05 (among ECe)
CSG 8890 (salt-sensitive)	0.76	0.70	0.50	0.32		0.50	0.37	0.19	0.11	
BG 256 (check)	1.11	0.90	0.65	0.47		0.46	0.41	0.24	0.13	
CSG 9372 (high-nodulation)	1.03	1.22	1.14	0.60		0.38	0.46	0.38	0.22	
CSG 8927 (salt-tolerant)	0.93	0.91	0.60	0.54		0.43	0.47	0.31	0.20	
LSD P=0.05	0.12 (among genotypes)					0.06 (among genotypes)				
	Number of nodules per pot					Dry mass of nodules (mg per pot)				
CSG 8890 (salt-sensitive)	23.0	10.0	5.6	7.6	6.5 (among ECe)	8.9	1.7	0.7	0.5	2.7 (among ECe)
BG 256 (check)	22.2	11.8	13.8	7.6		12.8	5.2	3.3	1.2	
CSG 9372 (high-nodulation)	63.2	20.8	21.8	13.0		15.8	5.1	2.8	1.1	
CSG 8927 (salt-tolerant)	25.2	15.6	7.2	2.6		5.9	4.4	1.4	0.3	
LSD P=0.05	6.5 (among genotypes)					2.7 (among genotypes)				

Similarly, Arrendell *et al.* (1985) found in F_5 and F_6 generation progenies from a cross of Virginia and Spanish cultivars of groundnut, that broad sense heritability of nodule number ranged from 0.25 to 0.57; estimates for ARA (acetylene reduction), shoot weight, fruit weight ranged from 0.53 to 0.85. The moderate to high estimates for these traits indicated, that superior nodulating and N_2 fixing genotypes within the populations studied could be readily identified, and that selection for enhanced nitrogen fixation should result in indirect selection for yield.

Singh *et al.*, 1985 reported considerable variation for total nitrogen, total dry matter and nodule fresh weight per plant in 49 purelines of mungbean inoculated with cowpea *Rhizobium* Mo5. Combining ability analysis in an 8 × 8 diallel of diverse purelines showed the importance of both additive and non-additive gene effects in control of the nitrogen fixation traits; non-additive gene effects were more preponderant than additive gene effects. There was considerable transgressive segregation for the three traits in the two F_2 populations used in this study. The three traits showed very high heritability estimates, and total nitrogen showed very high positive correlation with total nitrogen and nodule fresh weight. There was no evidence for an antagonistic relationship between grain yield and nitrogen fixation. It was concluded that selection for total dry weight or nodule fresh weight may be done to improve nitrogen fixation in mung bean without adversely affecting yield and yield traits.

Legume × *Rhizobium* Interactions

There is a range of quantitative variation in nitrogen fixation, which is broadly genetic; but methods of exploiting it have not been fully developed. Difficulties are encountered due to variable amounts of nitrogen fixation when cultivars are nodulated by different *Rhizobium* genotypes, and as different environments are encountered. Rapid genetic advance under selection would result from coincidental selection of plant and *Rhizobium* genotypes (Mytton, 1983). The extent to which these can be induced to reform their specific association is unclear but the mechanism of host strain selectivity as exemplified by the cross-inoculation groupings should be of value in achieving this. These objectives can be pursued by empirical methods, but improvement of legume-*Rhizobium* associations and their systems of management would be assisted by the co-ordinated work of agronomists, microbiologists, physiologists and plant breeders (Mytton, 1983).

How should a field testing program of plant lines be approached? Lie (1981) extended the gene-centre concept of Vavilov (1951) to the area of N_2-fixation, pointing out with examples from the *Pisum* and its *Rhizobium* symbiosis, the variation in response likely to occur among collections from such gene-centres. With soybean, some evidence of that variation is already apparent in the response of Asiatic *versus* American cultivars to cowpea-type rhizobia (Nangju, 1980; Pulver *et al.*, 1982) and in the ineffective response of American cultivars to the fast-growing soybean rhizobia obtained from China. For beans, response differences have been found amongst ancestral lines of *Phaseolus vulgaris* obtained from Mexico (Graham and Temple, 1984) and with other species of *Phaseolus* (Ferrera-Cerrato, 1980; Hohenberg *et al.*, 1982).

Debate on whether to select for increased N_2-fixation in the presence of a single rhizobial strain or with a mixture of strains is likely to continue. Those who advocate a mixed-strain inoculum suggest that selection in the presence of multiple rhizobial genotypes will produce the best host-strain combinations (Barnes *et al.*, 1984). Those advocating the use of a single-strain inoculum believe that because the most competitive rhizobia are not necessarily the most effective at reducing N_2, the presence of multiple rhizobial genotypes may confound the selection of plant genes favouring N_2-fixation (Phillips and Teuber, 1985, Smith *et al.*, 1982). Certainly the degree to which less effective strains from a mixed population nodulate genetically desirable hosts will slow the rate of genetic gain when selecting for host genes that favour increased N_2-fixation. In the case of alfalfa, germplasm selected for increased N_2-fixation in the presence of one *R. meliloti* strain stimulated greater N_2-fixation by other strains (Phillips *et al.*, 1985). Therefore, the imponderable remains - how will the plants screened in the presence of mixed rhizobial conditions interact with various populations of indigenous rhizobia that will be encountered in soils at different sites (Gibson, 1962)?

The reported large legume × *Rhizobium* interaction in *Vicia faba* L. was used to argue that most of the genetic variation was under non-additive genetic control (Mytton, *et al.*, 1977) supporting simultaneous selection of *Rhizobium* and host plant. Other studies showed that significant additive and non-additive genetic variance is present for several traits related to N_2-fixation (Hobbs and Mahon, 1983; Tan, 1981). Thus, some investigators conclude that both legumes and rhizobia should be selected independently (Hobbs and Mahon, 1983). Because rhizobia that nodulate one plant effectively are more likely to perform similarly on related genotypes than on unrelated plants (Hardarson and Jones, 1979; Mytton, 1975), legume selection with one rhizobial strain seems a reasonable approach to increasing N_2-fixation. Scientists following this approach believe that relative rankings of strains in closely related plant materials will not change markedly. Thus, alfalfa plants selected for N_2-fixation with a single strain of *Rhizobium* also improved symbiotic performance of several other *Rhizobium* strains (Phillips *et al.*, 1985).

Another approach that has been followed, ignores host × strain specificity and suggest screening plant genotypes in a field soil containing high numbers of effective rhizobia, because in a field situation and with current inoculant technology, it is not possible to control the mix of rhizobial strain(s) that nodulate a legume crop, making it difficult to establish the highly effective, specific-host-strain combination. The studies of Arunachalam *et al.* (1984), Rao *et al.* (2002), Rupela (1992 and 1994), Dudeja *et al.* (1997) fall under this school. There is also evidence that a superior host genotype selected on the basis of N_2-fixation with one or more highly effective strains will express that superiority with others (Buttery and Dirks, 1987 Phillips and Teuber, 1985; Wiersma and Orf, 1992).

In soybean, indigenous *B. japonicum* strains frequently are only moderately effective at fixing N_2, but they prevent superior strains from forming root nodules. Thus, the advantages of using plant genes to block indigenous strains and permit nodulation by desirable rhizobial strains is obvious. The rj_1 allele is one plant gene that prevents root nodule formation by many indigenous strains (Devine and Weber, 1977). A more complex host plant effect is evident in several primitive soybean introductions (*Glycine max* (L.) Merr.) that restrict, but do not prevent, effective root nodule formation by *B. japonicum* strains in the 123 serocluster (Cregan and Keyser, 1986). A possibly related, but more extreme, host plant effect has been observed with the primitive *G. max* cultivar Peking (Devine, 1984) and the wild soybean (*Glycine soja* Sieb and Zucc) (Cregan and Keyser, 1986), which produce Fix^- nodules with the bacterium *B. japonicum* USDA123 and Fix^+ nodules with the bacterium *R. fredii*. Other "promiscuous" *G. max* lines that apparently form Fix^+ nodules with both *B. japonicum* and *Bradyrhizobium* sp. organisms indigenous to African soils are also known (Kueneman *et al.*, 1984). Genetic relationships among these phenotypes have not yet been reported. It may, however, be possible to use the alleles involved in restricting nodulation by indigenous rhizobia, so that more effective strains could be supplied as inoculants, to ensure greater nodule occupancy and in turn greater legume BNF and yield.

Any environmental factor affecting BNF will most likely influence Legume × rhizobium interactions. A large number of biotic and abiotic factors affect BNF. These include water, temperature, salinity/alkalinity/acidity of soil, soil antagonists and toxic chemicals and have been sufficiently reviewed (Sprent *et al.*, 1983, Streeter, 1988, Vincent, 1988). Allelopathy, a direct or indirect harmful or beneficial effect by one plant on another through production of chemical compounds in the root rhizosphere (Rice, 1974) may be another factor that may influence nodulation and nitrogen fixation. All these factors influence performance of a given plant within a cultivar or a given cultivar. Kharkwal *et al.* (2000) reported that raising two seeds of a given genotype of a legume or of two different legumes sown in a single hill affect nodulation of each other negatively or positively. Of the three legumes (chickpea, lentil and pea) studied, variety HFP4 of pea (the only one studied) showed enhanced nodulation by 20 to 74% compared to the accompanying other legumes and the other varieties/legumes generally reduced nodulation. Any breeding study therefore has to consider influence of such factors and have some in-built safeguards to develop high BNF and high yield materials.

Strategies for Improving N_2-Fixation

Researchers working in this area generally agree that enhanced N_2-fixation by the grain legumes will result from selection and breeding for high N yield, high nitrate tolerance, and specific rhizobial strain × cultivar requirement (Mytton, 1983; Beringer *et al.*, 1988; Bliss and Miller, 1988). This section examines programs that aim to develop cultivars of legumes that incorporate one or more of these characteristics.

Legume N-yield: Agronomic and environmental considerations often limit the biomass yield of a legume crop and therefore the capacity of that crop to fix N_2. Yield will also be determined genetically. Duc *et al.* (1988) evaluated 21 genotypes of faba bean at two sites near Dijon, France, over two years. Their data show the strong correlation between biomass N yield and N_2-fixation. Where comparisons between genotypes could be made over sites and years, they suggested a strong genetic basis for N yield and N_2-fixation. This, however, can be true under low soil-N and not under high soil-N conditions.

With some species, low N yield may be a typical characteristic. In studies over a range of environments and agronomic practices, N yield and N_2-fixation by chickpea were consistently less than for the other cool season food legumes (Rennie and Dubetz, 1986; Evans and Herridge, 1987; Smith *et al.*, 1987; Beck *et al.*, 1991). Average values for N yield were 100 for chickpea, 185 for field pea, 196 for lentil, and 200 kg N ha^{-1} for faba bean. These studies did not indicate that the inherent capacity of chickpea for either nodulation or N_2-fixation was less than for the other species. It seems that increasing N yield of chickpea may result in increased N_2-fixation.

In the common bean, low N yield is the result of low N_2-fixation capacity, rather than *vice versa* (Attewell and Bliss, 1985). A breeding program by Bliss and his co-workers at the University of Wisconsin has produced new genotypes of common bean with higher levels of N_2-fixation, resulting in increased plant vigour and improved N yields (Table 5).

Table 5. Summary of data from two experiments from a breeding program to increase N_2 fixation by the common bean (Attewell and Bliss, 1985)

Parent, cultivar or line	Experiment 1			Experiment 2		
	N_2 fixation (maturity)		Seed yield ($g\ plant^{-1}$)	Maturity (days)	Determinate	N yield at R7 ($mg\ plant^{-1}$)
	($mg\ plant^{-1}$)	% N from air				
Sanilac	76	12	18	85	Yes	591
24-17	583	48	31	110	No	1068
24-21	216	25	19	91	Yes	1045
24-55	192	22	23	94	Yes	668
Puebla	852	57	38	120	No	1429

Until 1980s most scientists assessed N_2-fixation based on the ARA. More recently, ^{15}N methods (enriched and depleted) were used (Pereira *et al.*, 1989; Wolyn *et al.*, 1989). Breeding material was, out of necessity, evaluated under low soil mineral N, because the plant's capacity for N_2-fixation and not the capacity for growth (N yield) was of principal importance. In a soil with moderate to high nitrate N, the capacity of the plant to fix N_2 will not be expressed to the same extent, because of the suppression of N_2-fixation by the soil N. In a species such as chickpea, where N yield rather than N_2-fixation *per se* may be the problem, evaluation would still be more useful in low N soils, because of the added capacity to screen for N_2-fixation as well as for N yield. Because N yield and dry matter yield are generally strongly correlated (e.g., Mytton, 1983), a program to enhance N_2-fixation in chickpea might involve the following:

- Screening a large and diverse germplasm (500-1000 genotypes) of chickpea, inoculated with highly effective rhizobia, for production of dry matter under low N conditions (preferably in the field, but could also be done in a glasshouse);
- Selecting superior genotypes (e.g., top 10 %) for further evaluation;
- A second round-of screening is ideally done in the field on a low N-fertility soil, again with a mixture of highly-effective rhizobia. Assessments should include seed yield and total N yield and
- Comparison of elite genotypes over a range of edaphic (particularly soil N fertility) and environmental (including diverse rhizobial population) conditions for seed yield, N yield and N_2-fixation, and the latter using ^{15}N methods.

Figure 2 outlines a program used in chickpea and pigeonpea (Rupela, 1994; Rupela and Johansen, 1995) for developing high BNF lines. A similar approach was followed by Venkateswarlu and Katyal (1994) in groundnut. Genotypes identified through such a screening protocol are likely to be superior for BNF and seed yield and adapted to the soils and environments for which they were likely to be used. This protocol would have on-farm application (Herridge *et al.*, 1994).

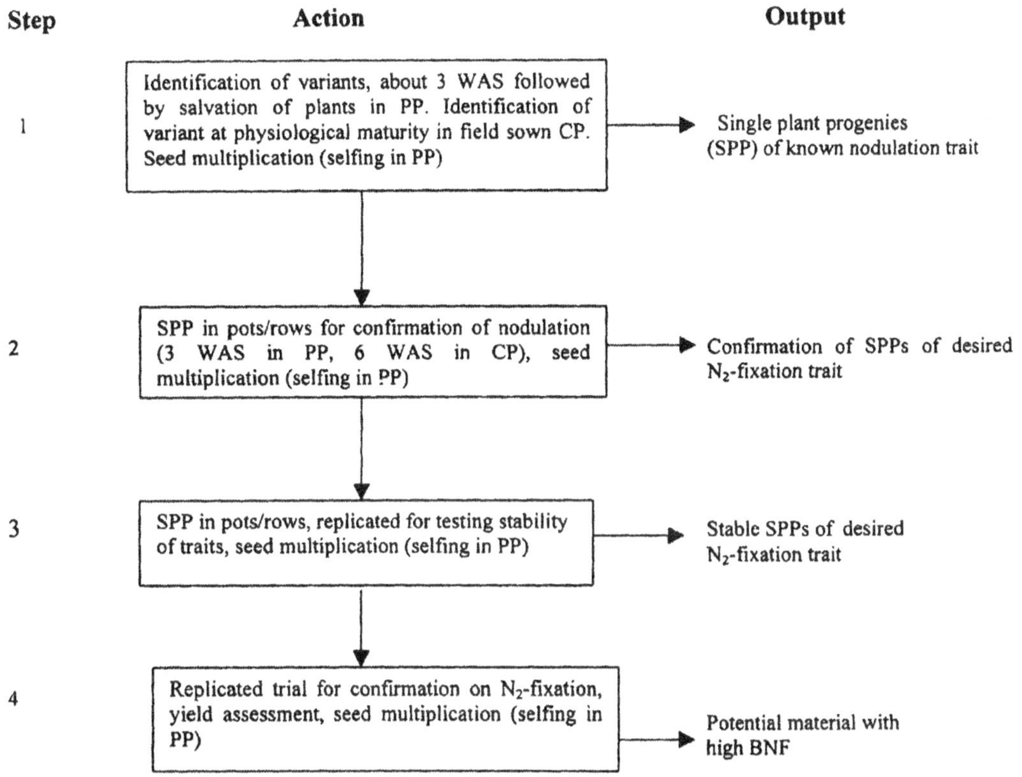

Fig. 2. Protocol for selecting nodulation variants of chickpea (CP) and pigeonpea (PP). The screening nursery for chickpea can be developed in a field while for pigeonpea, evaluation has to be done in a green house. Seed multiplication can be done both in the greenhouse and field. WAS = weeks after sowing

High N_2-fixing genotypes that produced low seed yields or seed of low quality, could be used as donor parents in a breeding program (e.g., as with "Puebla" in the common bean program, Table 5). There may be little scope to adopt this protocol for species such as field pea and faba bean that are already capable of producing high N yields under field conditions, i.e., 300 kg N ha^{-1} and fixing substantial amounts of N_2 (see Brunner and Zapata, 1984; Jensen, 1986). Harvesting this nitrogen into grains may be more important in these legumes.

Brunner and Zapata (1984) compared 19 mutant lines of faba bean with the parent cultivar for seed yield, dry matter and N yield, and N_2-fixation. One line (II-18) was clearly superior in all characteristics, but it was not clear whether the superiority of the line was because of higher growth and N-assimilation rates or because of a longer crop duration. Nevertheless, its superior performance over a range of environments suggested that growth and assimilation rate had been improved by mutagenesis. The authors also concluded that N_2-fixation was determined by N yield, rather than *Pfix*. In a similar study involving pigeonpea, Kumar Rao and Dart (1987) also showed the strong relationship between N yield and N_2-fixation. In that instance, however, N yield appeared to be linked to crop duration.

Traits Associated with N_2-fixation at Process Level: Although sometimes used as criteria for increased N_2-fixation, selection of genotypes on the basis of specific traits that are either directly or indirectly associated with nodulation or nodule (N_2-fixing) activity appears to be of limited value. Heichel *et al.* (1989) concluded that the selection of lines of alfalfa (*Medicago sativa* L.) for activity of various nodular enzymes, including nitrogenase, resulted in experimental populations with enhanced or reduced enzyme activity, but did not result in populations that were different in N_2-fixation. The authors proposed several explanations for their inconclusive findings, including the following:

- Short-term measurements of nitrogenase activity using the ARA may not predict activity over a longer growth period;
- The limitations of the ARA may be too great to even compare treatments on a relative, let alone a quantitative, basis;
- Performance of seedlings in a glasshouse may not predict performance in the field over one or several seasons, and
- Enhanced activities of single or several nodular enzymes may be countered by inadequate or even normal activity of other equally-important enzymes, resulting in no net change in the rate of N_2-fixation.

Thus, any program based on screening at biochemical (enzymetic) levels alone may not be very successful.

Effective Symbiosis with Native Rhizobia: Several researchers have argued that it is a desirable practice to inoculate legumes. It is indeed required for soils where native appropriate rhizobia are lacking. But determination of such needs of inoculation is time consuming. Also, in most soils where a given legume is grown regularly, any further inoculation with laboratory inocula may not enhance nodulation and nitrogen fixation and yield significantly high over non-inoculated control. Exception will be where rhizobia may die due to environmental conditions, such as high summer temperatures and also during rice phase in the rice-legume cropping systems. Most developing countries should select legumes that nodulate effectively with the native rhizobia, because quality of their inoculants may be generally poor (Singleton *et al.*, 1997). Soybean from south-east Asia nodulate successfully with the indigenous rhizobia in Africa, but the USA bred cultivars nodulate poorly (Nangju, 1980). Also, the local soybean (Kali Toor) in Madhya Pradesh has been noted to nodulate more effectively by native rhizobia than some breeding lines involving blood from USA bred cultivars (J.A. Thompson, Tamworth, Australia, ex-ICRISAT scientist, personal communication). Therefore, developing soybean lines that nodulate and fix N_2 quantitatively with native rhizobia may obviate the need for inoculation of such lines, particularly when grown in soils with high population of native rhizobia. Kueneman *et al.*, 1984 reported a selection program for high nodulation/nitrogen fixation in soybean in Nigeria using native rhizobia (i.e. no inoculation) and developed lines with high yield and N_2-fixation.

Large difference for N_2-fixation capacity between cultivars has been reported for several legume species (see section on 'Genetic variability for nodulation and N_2-fixation'). Rupela (1992) reported large variability for nodulation within bred cultivars and landraces of chickpea.

The protocol of developing high N_2-fixation lines is shown in Figure 2. The high nodulating selections developed from two widely used cultivars (G130, K850) using this protocol were evaluated as part of a multilocation (Akola, Badnapur, Hisar, New Delhi, and Sehore) experiment during 1994/95 and 1995/96. Low nodulating selections from both the varieties and the parent lines were used as reference. Non-nodulating selections from two cultivars (ICC4993 = Rabat, a long duration line, and ICC4918 = Annigeri, a short duration line) were used as non-N_2-fixing lines, to allow calculation of amount of N_2-fixed by difference method. Indeed the high nodulating pure line selections from the two cultivars nodulated significantly higher (Table 3), fixed high nitrogen (measured by ARA method at one of the five locations and by difference method at 3 of the 5 locations) besides producing high grain yield. The differences were more pronounced at almost all locations at low soil-N (N_1) than at high soil-N (N_2) conditions. High N_2-fixing selections were also made from advanced breeding lines. The final products developed from six cultivars were evaluated at two soil-N levels, along with their parents. The results suggested 0–293 % improvement in nodule mass, with no or insignificant improvement in total biomass yield (Table 6). It was apparent that this strategy of selecting high nodulating plants and developing their progenies (single plant progenies) into lines (pure line selection) can also be applied for selecting stress tolerant symbiotic lines.

Exclusion of Nodulation by Native Rhizobia: It is generally difficult to displace native rhizobia by inoculation with laboratory multiplied strains. About 10 % or less nodules are formed by inoculant rhizobia when the population of native rhizobia is high (Rupela and Sudarshana, 1990; Devine, 1984; Halliday, 1985). Use of undisturbed cores has been suggested to select rhizobial strains for competitiveness (Sylvester-Bradley *et al.*, 1983). There are indications that selected, highly effective, and competitive strains can be used to increase N_2-fixation and yield, particularly where native rhizobia populations are low or ineffective (Arsac and Cleyet-Marel, 1986; Beck, 1992). Successful nodulation by inoculant strains can often be site-specific depending on soil factors and the composition of the indigenous rhizobial population.

Some groups (Devine, 1984; Cregan and Keyser, 1986) aimed to produce cultivars of the host that bypass the resident rhizobia in the soil to become nodulated by more effective inoculant strains. This strategy has also been applied to pea. Lie (1978; 1981) found that the primitive pea cultivar "Afghanistan" was not nodulated by strains of rhizobia isolated from temperate soils in Europe, but did form nodules with bacterial strain Tom from Turkey. This resistance to nodulation has been ascribed to the recessive gene, *sym*-2 (Holl, 1975). In competition experiments, the European strains caused differential blocking of nodulation by Tom (Lie *et al.*, 1988). Subsequent identification of rhizobial isolates that could nodulate Afghanistan as well as cultivars bred for European conditions, indicated the presence in the rhizobia of a specific genetic region, termed *nodX*. Fobert *et al.* (1991) suggest that the combination of the plant gene, *sym*-2, and the rhizobial gene, *nodX*, may allow ultimate control of nodulation and provide a mechanism for enhancing N_2-fixation, even in the presence of large populations of indigenous rhizobia. This approach assumes that N_2-fixation was limited by the effectiveness of the native rhizobia, which may not always be the case.

Table 6. Mean values on nodule mass and total dry matter of single plant progenies selected for large differences in nodulation capacities within advanced breeding lines of chickpea, ICRISAT, Patancheru, post-rainy season 1995/96.

Parent	No. of selection	Type of means[2]	Nodule mass (mg/plant)			Total dry matter (t ha^{-1})			Nodulation improvement[3] (%)
			N_1	N_2	Mean	N_1	N_2	Mean	
ICCV 89230	2	a	117±15.9	74±15.9	096±10.1	2.2±0.15	2.4±0.15	2.3±0.10	-[4]
		b	114±12.3	55±12.3	085±7.1	2.6±0.12	2.4±0.12	2.5±0.07	-
ICCV 89302	6	a	129±15.9	78±15.9	103±10.1	2.0±0.15	2.4±0.15	2.2±0.10	-
		b	128±9.2	62±9.2	095±4.1	2.1±0.08	2.4±0.08	2.3±0.04	-
ICCV 91016	5	a	076±15.9	21±15.9	049±10.1	2.1±0.15	2.4±0.15	2.3±0.10	64,N_1
		b	079±9.5	39±9.5	059±4.5	2.0±0.09	2.4±0.09	2.2±0.05	129,N_2
ICCV 91019	3	a	036±15.9	21±15.9	029±10.1	1.5±0.15	1.9±0.15	1.7±0.10	293,N_1
		b	092±10.8	52±10.8	072±5.8	2.0±0.10	2.4±0.10	2.2±0.06	273,N_2
ICCV 91026	5	a	072±15.9	47±15.9	060±10.1	1.8±0.15	2.4±0.15	2.1±0.10	88,N_1
		b	085±9.5	26±9.5	056±4.5	1.9±0.09	2.4±0.09	2.2±0.05	-
ICC 4958	2	a	065±15.9	21±15.9	043±10.1	2.3±0.15	2.9±0.15	2.6±0.10	41,N_1
		b	060±12.3	33±12.3	047±7.1	2.0±0.12	2.7±0.12	2.4±0.07	99,N_2
ICC 5003 (Control)			091±15.9	46±15.9	069±10.1	1.3±0.15	2.2±0.15	1.8±0.10	

1. Data for selected parents where selections were significantly different ($P < 0.05$) from parents or among themselves for nodule mass; 2. a = mean values for the parent, b = mean values for selections from the parent; 3. Percentage of increase in nodule mass of a selection at N_1 or N_2 over its parent; 4. – = No improvement; 5. Low (N_1) and high (N_2) soil N level, representing those that can be found in farmers' fields, were created by applying 0 (N_1) and 100 kg N ha^{-1} (N_2) as urea to the preceding sorghum on Vertisol field after the whole field was depleted for N by growing cover crops for two years. N is a highly dynamic element and varies through the soil profile due to several factors. At sowing, the available N in the top 15 cm soil was 1.7 times more in the N_2 plots than in N_1 plots (8.7 mg N kg-1 soil); Source: Rupela 1997.

Induced Mutations: Breeding through induced mutations is a well established strategy to improve legumes (Micke, 1984). High soil-nitrate tolerant (called supernodulating by some authors) lines of pea (Jacobsen and Feenstra, 1984) and soybean (Carroll *et al.*, 1985a) involving mutations have been reported. Supernodulating lines have been reported for soybean cultivars Bragg, Williams, Elgin 87 and Enrei (Herridge and Rose, 2000).

These mutants formed up to 10-20 times more nodulation and AR activity than that of the wild-types in the presence of 5 or 5.5 mM nitrate. However, their yield was 30-40 % lower than the wild types and they had restricted root growth. The super-nodulating soybean mutants seemed to be controlled by a single Mendelian recessive gene operating through the shoot (Delves *et al.*, 1986; Lee *et al.*, 1991). There was no evidence of host × rhizobial strain specificity affecting expression of the supernodulation trait (Carroll *et al.*, 1985b; Gremaud and Harper, 1989). A selection and breeding program for high N_2-fixing and high yield involving supernodulating mutants in Australia suggested that lines with intermediate nodulation (between normal lines and supernodulating mutants) yielded equal to agronomic cultivar Manark and fixed substantially high N_2. At this stage, however, no supernodulating cultivars have been released (Herridge and Rose, 2000).

The Future

There has been limited success in selection and breeding for high N_2-fixation. Development of soybean lines with effective nodulation due to native rhizobia in Africa (Mpepereki *et al.*, 2000), high N_2-fixing cultivars of common bean released in South America (Bliss, 1993), pure lines selections from advanced breeding lines of chickpea with high nitrogen fixation ability (Rupela, 1997) are some examples where materials are available for researchers and/or farmers to evaluate and verify their potential. One likely reason of slow progress may be due to the fact that it would take a multidisciplinary team to achieve success in this complex trait, in addition to the complex trait 'yield' (it may be noted that breeding legumes for high yield, *per se*, has not been as successful as in cereals). Also, any breeding program has to combine several traits (such as pest and drought resistance) along with yield to ensure that the resultant materials are potentially acceptable to farmers. Breeding for high N_2-fixation is feasible and should also be on the research agenda of breeders.

Much of the breeding work is conducted at research stations. Soils at research stations are likely to have higher soil nitrogen than at farmers' fields. High soil-N is known to suppress nitrogen fixation by legumes (Streeter, 1988; Wani *et al.*, 1997). For promoting N_2-fixing traits, breeders should grow their legumes at low soil-N (preferably 10 5g mineral N g^{-1} soil) fields, prepared specially for the purpose. Breeders generally handle large numbers of genotypes and materials. In some materials, genes for N_2-fixation may be co-segregating with genes for the other traits. It is likely, therefore, that trait combinations associated with enhanced N_2-fixation will be identified, if appropriate assessment methods are applied to the segregating populations. If genetic variation for N_2-fixation existed in breeding populations, the high N_2-fixing lines would be produced as a normal outcome. The role of the N_2-fixation specialist in the mainstream breeding program might be to identify sources of genes for particular symbiotic traits and to provide technical guidance and support (Bliss, 1993) during development of desired products. Recent developments in the field of genomics (particularly on *Medicago* and *Lotus*) would provide a better understanding of the expression and regulation of symbiotic genes. It should also open up opportunities for biotech assisted germplasm enhancement and bio-informatics assisted gene mining and utilization. These developments may lead to a better targeted breeding of legumes for high BNF than hitherto possible.

Acknowledgements

We thank Drs. Y.S. Chauhan, H.D. Upadhyaya and L.J. Reddy for comments on the manuscript.

References

Ahlawat I.P.S., Ali M., Yadav R.L., Kumar Rao J.V.D.K., Rego T.J. and Singh R.P. 1998. Biological Nitrogen fixation and residual effects of summer and rainy season grain legumes in rice and wheat cropping systems of the Indo-Gangetic Plain. Pages 31-54 *In*: *Residual Effects of Legumes in Rice and wheat cropping Systems of the Indo-Gangetic Plain*, (eds.) Kumar Rao J.V.D.K., Johansen C. and Rego T.J., Oxford and IBH, New Delhi.

Arrendell S., Wynne J.C., Elkan G.H. and Isleib T.G. 1985. Variation for nitrogen fixation among progenies of a Virginia × Spanish peanut cross. *Crop Sci.*, **25**: 865-869.

Arsac J.F. and Cleyet-Marel J.C. 1986. Serological and ecological studies of *Rhizobium* spp. (*Cicer arietinum* L.) by immunofluorescence and ELISA technique: competitive ability for nodule formation between *Rhizobium* strains. *Plant and Soil*, **94**: 411-423.

Arunachalam V., Pungle G.D., Dutta M., Nambiar P.T.C. and Dart P.J. 1984. Efficiency of nitrogenase activity and nodule mass in predicting the relative performance of genotypes assessed by a number of characters in groundnut (*Arachis hypogaea*). *Exp. Agr.*, **20**: 303-309.

Attewell J. and Bliss F.A. 1985. Host plant characteristics of common bean line selected using indirect measures of N_2 fixation. *In*: *Nitrogen Fixation Research*, (eds.) H. J. Evans, P. J. Bottomley and W. E. Newton. Progress, pp. 3-9. Dordrecht: Martinus Nijhoff.

Barnes D.K., Heichel G.H., Vance C.P. and Ellis W.R. 1984. A multiple-trait breeding program for improving the symbiosis of N_2 fixation between Medicago sativa L. and *Rhizobium meliloti*. *Plant Soil*, **82**: 303-314.

Bhatia C.R., Nichterlein K. and Maluszynki M. 2001. Mutations affecting nodulation in grain legumes and their potential in sustainable cropping systems. *Euphytica*, **120**: 415-432.

Beck D.P. 1992. Yield and N_2-fixation response of chickpea cultivars to inoculation with selected *Rhizobium* strains. *Agronomy Journal*, **84**: 510-516.

Beck D.P., Wery J., Saxena M.C. and Ayadi A. 1991. Dinitrogen fixation and nitrogen balance in cold season food legumes. *Agronomy Journal*, **83**: 334-341.

Beringer J.E., Bisseling T.A. and LaRue T.A. 1988. Improving symbiotic nitrogen fixation through the genetic manipulation of *Rhizobium* and legume plants. pp. 691-702. *In*: *World Crops Cool Season Food Legumes*. (ed.) R. J. Summerfield, Kluwer Academic Publishers, London.

Bliss F.A. 1993. Breeding common bean for improved biological nitrogen fixation. *Plant Soil*, **152**: 71-79.

Bliss F.A. and Miller Jr J.C. 1988. Selecting and breeding grain legumes for enhanced nitrogen fixation. Pages 1001-1012. *In*: *World Crops: Cool Season Food Legumes*, (ed.) R.J. Summerfield, Kluwer, London.

Boddey R.M., Chalk P.M., Victoria R. and Matsui E. 1984. Nitrogen fixation by nodulated soybean under tropical field conditions estimated by the ^{15}N isotope dilution technnique. *Soil biology and biochemistry*, **16**: 583-588.

Brown L., Renner M. and Halweil B. 2000. Vital Signs 2000, The environment trends that are shaping our future. *Worldwatch Institute*. p. 192.

Brunner H. and Zapata F. 1984. Quantitative assessment of symbiotic nitrogen fixation in diverse mutant lines of field bean (*Vicia faba minor*). *Plant and Soil*, **82**: 407-413.

Buttery B.R. and Dirks V.A. 1987. The effects of soybean cultivar, *Rhizobium* strain and nitrate on plant growth, nodule mass and acetylene reduction rate. *Plant Soil*, **98**: 285-293.

Buttery B.R., Park S.J., Finday W.I. and Dhanvantari B.N. 1988. Effects of fumigation and fertilizer on growth, yield, chemical composition and mycorrhizae in whitebean and soybean. *Canadian Journal of Plant Science*, **68**: 677-686.

Buttery B.R. and Gibson A.H. 1990. The effect of nitrate on the time course of nitrogen fixation and growth in *Pisum sativum* and *Vicia faba*. *Plant and Soil*, **127**: 143-146.

Caetano-Anolles G. and Gresshoff P.M. 1996. Advances in positional cloning of nodulation genes in soybean. *Plant and Soil*, **186**: 1-7.

Carroll B.J., McNeil D.L. and Gresshoff P.M. 1985a. Isolation and properties of soybean (*Glycine max*) mutants that nodulate in the presence of high nitrate concentrations. *Proc. Natl. Acad. Sci.*, (USA), **82**: 4164-4166.

Carroll B.J., McNeil D.L. and Gresshoff P.M. 1985b. A supernodulation and nitrate tolerant symbiotic (nts) soybean mutant. *Plant Physiology*, **78**: 34-40.

Chalk P.M. 1985. Estimation of N_2 fixation by isotope dilution: an appraisal of techniques involving ^{15}N enrichment and their appraisal of techniques involving ^{15}N enrichment and their application. *Soil Biol. Biochem.*, **17**: 389-410.

Cregan P.B. and Keyser H.H. 1986. Host restriction of nodulation by *Bradyrhizobium japonicum* strain USDA 123 in soybean. *Crop Sci.*, **26**: 911-916.

Dangaria C.J., Parameshwarappa R., Salimath P.M. and Annigeri B.S. 1994. Genetic divergence for nodulating characters in chickpea. *Legume Research*, **17**: 32-36.

Danso S.K.A. 1988. The use of ^{15}N enriched fertilizers for estimating nitrogen fixation in grain and pasture legumes. Pages 345-358. *In*: *Nitrogen fixation by legumes in Mediterranean agriculture*, (eds.) D.P. Beck and L.A. Materon. Dordrecht, Martinus Nijhoff.

Danso S.K.A., Hardarson G. and Zapata F. 1993. Misconceptions and practical problems in the use of ^{15}N soil enrichment techniques for estimating N_2-fixation. *Plant and Soil*, **152**: 25-52.

De Lajudie P., Laurent Fulele E., Willems A., Torck U., Coopman R., Collins M.D., Kersters K., Dreyfus B. and Gillis M. 1998. *Allorhizobium undicola* gen. nov., sp.nov., nitrogen-fixing bacteria that efficiently nodulate *Neptunia natans* in Senegal. *International Journal of Systematic Bacteriology*, **48**: 1277-1290.

Delves A.C., Mathews A., Day D.A., Carter A.S., Carroll B.J. and Gresshoff P.M. 1986. Regulation of the soybean-*Rhizobium* nodule symbiosis by root and shoot factors. *Plant Physiol.*, **82**: 588-590.

De Silva M., Purcell L.C. and King C.A. 1996. Soybean petiole ureide response to water deficits and decreased transpiration. *Crop Science*, **36**: 611-616.

Devine T.E. 1984. Genetics and breeding of nitrogen fixation. *In*: *Biological Nitrogen Fixation*, (ed.) M. Alexander, pp. 127-154, New York, Plenum.

Devine T.E. and Weber D.F. 1977. Genetic specificity of nodulation. *Euphytica*, **26**: 527-535.

De Vries J.D., Bennett J.M., Albrecht S.L. and Boote K.J. 1989. Water relations, nitrogenase activity and root development of three grain legumes in response to soil water deficits. *Field Crops Research*, **21**: 215-226.

Duc G., Mariotti A. and Amarger N. 1988. Measurments of genetic variability for symbiotic dintrogen fixation in field-grown fababean (*Vicia faba* L.). *Plant and Soil*, **106**: 269-276.

Dudeja S.S., Potdukhe S.R., Namdeo S.L., Datar V.V., Kumar V., Tilak K.V.B.R, Khurana A.L. and Rupela O.P. 1997. Multilocation evaluation of some selected chickpea nodulation variants in India. Pages 261-276. *In*: *Extending nitrogen fixation research to farmers' fields*: Proceedings of an International Workshop on Managing Legume Nitrogen Fixation in the Cropping Systems of Asia, 20-24 Aug 1996, ICRISAT Asia Center, India, (eds.) Rupela O.P., Johansen C. and Herridge D.F., Patancheru 502 324, A.P., India.

Elkan G.H. 1992. Taxonomy of the rhizobia. *Can. J. Microbiol.*, **38**: 446-450.

Evans J. and Herridge D.F. 1987. *In*: *Nitrogen cycling temperate agricultural systems.* (eds.) P. E. Bacon, J. Evans, R. R. Storrier and A. C. Taylor. pp.14-43, Wagga Wagga: Australian Society of Soil Science.

Ferrera-Cerrato R. 1980. Inoculacion de *Rhizobium phaseoli* a differentes especies del genero *Phaseolus* originarias de Mexico. *Rev. Lat-Amer, Microbiol.*, **22**: 175-180.

Fischer H.M. 1994. Genetic regulation of nitrogen fixation in rhizobia. *Microbiol. Rev.*, **58**: 352-386.

Fobert P.R., Roy N., Nash J.H.E. and Iyer V.N. 1991. Procedure for obtaining efficient root nodulation of a pea cultivar by a designed *Rhizobium* strain and preempting nodulation by other strains. *Applied and Environmental Microbiology*, **57**: 1590-1594.

Freiberg C., Fellay R., Bairoch A., Broughton W.J., Rosenthal A. and Perret X. 1997. Molecular basis of symbiosis between *Rhizobium* and legumes. *Nature*, **387**: 394-401.

Galibert F. *et al.* (55 authors). 2001. The composite genome of the legume symbiont *Sinorhizobium meliloti*. *Science*, **293**: 668-672.

Gibson A.H. 1962. Genetic variation in the effectiveness of nodulation of lucerne varieties. *Aust. J. Agric. Res.*, **13**: 388-399.

Giller K.E., Nambiar P.T.C., Srinivasa Rao B., Dart P.J. and Day J.M. 1987. A comparison of nitrogen fixation in genotypes of groundnut (*Arachis hypogaea* L.) using ^{15}N-isotope dilution. *Biol. Fertil. Soils*, **5**: 23-25

Graham P.H. and Rosas J.C. 1977. Growth and development of intermediate bush and climbing cultivars of *Phaseolus vulgaris* L. inoculated with *Rhizobium J. Agric. Sci., Camb.*, **88**: 503-508.

Graham P.H. and Temple S.R. 1984. Selection for improved nitrogen fixation in *Glycine max* (L.) Merr. and *Phaseolus vulgaris* L. *Plant Soil*, **82**: 315-327.

Greder R.R., Orf J.H. and Lambert J.W. 1986. Heritabilities and associations of nodule mass and recovery of *Bradyrhizobium japonnicum* serogroup USDA 110 in soybean. *Crop Sci.*, **26**: 33-37.

Gremaud M.F. and Harper J.E. 1989. Selection and initial characterization of partially nitrate tolerant nodulation mutants of soybean. *Plant Physiol.*, **89**: 169-173.

Halliday J. 1985. Biological nitrogen fixation in tropical agriculture. *In*: *Nitrogen fixation research progress*, (eds.) H.J. Evans, P.J. Bottomley and W.E. Newton, pp. 675-681, Martinus Nijhoff, Boston.

Hardarson G. and Jones D.G. 1979. The inheritance of preference for strains of *Rhizobium trifolii* by white clover (*Trifolium repens*). *Ann. Appl. Biol.*, **92**: 329-333.

Hardarson G., Golbs M. and Danso S.K.A. 1989. Nitrogen fixation in soybean (*Glycine max* L. Merr.) as affected by nodulation patterns. *Soil Biol. Biochem.*, **21**: 783-787.

Heichel G.H., Barnes D.K., Vance C.P. and Sheaffer C.C. 1989. *Journal of Production Agriculture*, **2**: 24-32.

Herridge D.F. and Peoples M.B. 1990. The ureide assay for measuring nitrogen fixation by nodulated soybean calibrated by ^{15}N methods. *Plant Physiol.*, **93**: 495-503.

Herridge D.F., Rupela O.P., Serraj R. and Beck D.P. 1994. Screening techniques and improved biological nitrogen fixation in cool season food legumes. *Euphytica*, **73**: 95-108.

Herridge D.F. and Rose I. A. 1994. Heritability and repeatability of enhanced N_2 fixation in early and late inbreeding generations of soybean. *Crop Sci.*, **34**: 360-367.

Herridge D.F. and Rose I.A. 2000. Breeding for enhanced nitrogen fixation in crop legumes. *Field Crops Research*, **65**: 229-248.

Hirsch A.M., Lum M.R. and Downie J.A. 2001. What makes the Rhizobia-legume symbiosis so special. *Plant Physiology*, **127**: 1484-1492.

Hohenberg J.S., Munns D.N. and Tucker C.L. 1982. *Rhizobium*-host specificities in *Phaseolus coccineus* L. and *Phaseolus vulgaris* L. *Crop Sci.*, **22**: 455-459.

Hobbs S.L.A. and Mahon J.D. 1983. Variability and interaction in the *Pisum sativum* L.-*Rhizobium leguminosarum* symbiosis. *Can. J. Plant Sci.*, **63**: 591-599.

Holl J. 1975. Host plant control of the inheritance of dinitrogen fixation in *Pisum-Rhizobium* symbiosis. *Euphytica*, **24**: 767-770.

Howieson, J.G., Robson A.D. and Ewing M.A. 1993. External phosphate and calcium concentrations and pH, but not the products of rhizobial nodulation genes, affect the attachment of *Rhizobium meliloti* to roots of annual medics. *Soil Biology and Biochemistry*, **25**: 567-573.

Jacobsen E. and Feenstra W.J. 1984. A new pea mutant with efficient nodulation in the presence of nitrate. *Plant Sci. Lett.*, **33** : 337-344.

Jensen E.S. 1986. Symbiotic N_2 fixation in pea and field bean estimated by ^{15}N fertilizer dilution in field experiments with barley as a reference crop. *Plant and Soil*, **92**: 3-13.

Khanam D., Bhuiyan M.A.H., Hossain A.K.M., Ali S., Mustaq N., Bhattarai S. and Maskey S. 1997. Evaluation of selected chickpea nodulation variants in Bangladesh, Nepal, and Pakistan. Pages 277-288. *In*: *Extending nitrogen fixation research to farmers' fields*: Proceedings of an International Workshop on Managing Legume Nitrogen Fixation in the Cropping Systems of Asia, 20-24 Aug 1996, ICRISAT Asia Center, India, (eds.) Rupela O.P., Johansen C. and Herridge D.F., Patancheru 502 324, Andhra Pradesh, India, ICRISAT.

Kharkwal M.C., Pathak S.K. and Nagar J.P. 2000. Effect of Allelopathy on Nodulation in Grain Legumes. *Indian J. Genet.*, **60**: 177-184.

Kueneman E.A., Root W.R., Dashiell K.E. and Hohenberg J. 1984. Breeding soybeans for the tropics capable of nodulating effectively with indigenous *Rhizobium* spp. *Plant and Soil*, **82**: 387-396.

Kumar Rao J.V.D.K. and Dart P.J. 1987. Nodulation, nitrogen fixation and nitrogen uptake in pigeonpea [*Cajanus cajan* (L.) Millsp.] of different maturity groups. *Plant and Soil*, **99**: 255-266.

Lee S.H., Ashley D.A. and Boerma H.R. 1991. Regulation of nodule development in supernodulating mutants and wild-type soybean. *Crop Sci.*, **31**: 688-693.

Lie T.A. 1978. Symbiotic specialisation in pea plants: the requirement of specific *Rhizobium* strains for peas from Afghanistan. *Annals of Applied Biology*, **88**: 462-465.

Lie T.A. 1981. Gene centres, a source for genetic variants in symbiotic nitrogen fixation: host induced ineffectivity in *Pisum sativum* L. ecotype *fulvum*. *Plant Soil*, **61**: 125-134.

Lie T.A., Pijnenborg J. and Timmermans P.C.J.M. 1988. Analysis of the host genes controlling the legume-*Rhizobium* symbiosis: some technical problems and pitfalls. Pages 93-100. *In*: *Nitrogen fixation by legumes in Mediterranean agriculture*, (eds.) D.P. Beck and L.A. Materon. Dordrecht: Martinus Nijhoff.

Lauren J.G., Duxbury J.M., Beri V., Razzaque II M.A., Sattar M.A., Pandey S.P., Bhattarai S., Mann R.A. and Ladha J.K. 1998. Direct and residual effects from forage and green manure legumes in rice-based cropping systems. Pages 55-81. *In*: *Residual Effects of Legumes in Rice and wheat cropping Systems of the Indo-Gangetic Plain*, (eds.) Kumar Rao J.V.D.K., Johansen C. and Rego T.J., India, Patancheru, ICRISAT and New Delhi, Oxford and IBH.

Long S.R. 2001. Genes and signals in the *Rhizobium*-legume symbiosis. *Plant Physiology*, **125**: 69-72.

Micke A. 1984. Mutation breeding of grain legumes. *Plant and Soil*, **82**: 337-357.

Moulin L., Munive A., Dreyfus B. and Boivin-Masson C. 2001. Nodulation of legumes by members of the β-subclass of proteobacteria. *Nature*, **411**: 948-950.

Mpepereki S., Jahaveri F., Davus P. and Giller K.E. 2000. Soyabeans and sustainable agriculture. Promiscuous soyabeans in Southern Africa. *Field Crops Res.*, **65**: 137-150.

Mytton L.R. 1975. Plant genotype × *Rhizobium* strain interactions in white clover. *Ann. Appl. Biol.*, **80**: 103-107.

Mytton L. R., El-Sherbeeny M.H. and Lawes D.A. 1977. Symbiotic variability in *Vicia faba*. 3. Genetic effects of host plant, *Rhizobium* strain and of host × strain interaction. *Euphytica*, **26**: 785-791.

Mytton L.R. 1983. Host plant selection and breeding for improved symbiotic efficiency, *In*: *The Physiology genetics and nodulation of temperate legumes*, (eds.) D.G. Jones and D.R. Daves, pp. 373-393, London: Pitman.

Nambiar P.T.C., Rego T.J. and Srinivasa Rao B. 1986. Comparison of the requirements and utilization of nitrogen by genotypes of sorghum (*Sorghum bicolor* L. Moench), and nodulating and non-nodulating groundnut (*Arachis hypogaea* L.). *Field Crops Research*, **15**: 165-179.

Nambiar P.T.C., Rupela O.P. and Kumar Rao J.V.D.K. 1988. Nodulation and nitrogen fixation in groundnut (*Arachis hypogaea* L.), chickpea (*Cicer arietinum*), and pigeonpea [*Cajanus cajan* (L.) Millsp.]. *In*: *Biological nitrogen fixation: Recent developments*. (ed.) N.S. Subba Rao. pp 53-70. Oxford and IBH Publishers, New Delhi.

Nangju D. 1980. Soybean response to indigenous rhizobia as influenced by cultivar origin. *Agronomy Journal*, **72**: 403-406.

Nigam S.N., Dwivedi S.L., Nambiar P.T.C., Gibbons R.W. and Dart P.J. 1985. Combining ability analysis of N_2-fixation and related traits in peanut. *Peanut Sci.*, **12**: 55-57.

Nutman P.S. 1987. Centenary lecture. Phil. Trans. R. Soc. Lond. B 317:69-106. Reproduced. *In*: *A Century of Nitrogen Fixation Research*, (eds.) F.J. Bergersen and J.R. Postgate, pp 3-40. Royal Society, London.

Oldroyd G.E.D. 2001. Dissecting symbiosis: developments in Nod factor signal transduction. *Annals of Botany*, **87**: 709-718.

Pena-Cabriales J.J. and Castellanos J.Z. 1993. Effects of water stress on N_2-fixation and grain yield of *Phaseolus vulgaris* L. *Plant and Soil*, **152**: 151-155.

Pereira P.A.A., Burris R.H. and Bliss F.A. 1989. N_{15} determined dinitrogen fixation potential of genetically diverse bean lines (*Phaseolus vulgaris* L.). *Plant and Soil*, **120**: 171-179.

Peoples M.B., Faizah A.W., Rerkasem B. and Herridge D.F. 1989. Methods for evaluating nitrogen fixation by nodulated legumes in the Field. *ACIAR Monograph No.11*. ACIAR, Canberra, 76 pp.

Peoples M.B., Sudin M.N. and Herridge D.F. 1987. Translocation of nitrogenous compounds in symbiotic and nitrate-fed amide-exporting legumes. *Journal of Experimental Botany*, **38**: 567-579.

Peoples M.B. and Crasswell E.T. 1992. Biological nitrogen fixation: Investments, expectations and actual contributions to agriculture. *Plant and Soil*, **141**: 13-39.

Phillips D.A., Bedmar E.J., Qualset C.O. and Teuber L.R. 1985. Host legume control of Rhizobium functions. Pages 203-213. *In*: *Nitrogen Fixation and CO_2 metabolism*, (eds.) Ludden P.W., Burris J.E., Elsevier Science Publishing Company Inc.

Phillips D.A. and Teuber L.R. 1985. Genetic improvement of symbiotic nitrogen fixation in legumes. *In*: *Nitrogen Fixation Research Progress*. (eds.) H.J. Evans, P. J. Bottomley and W.E. Newton, pp. 11-17, Martinus Nijhoff Publ., Dordrecht.

Phillips D.A., Cunningham S.D., Bedmar E.J., Sweeney T.C. and Teuber L.R. 1985. Nitrogen assimilation in an improved alfalfa population. *Crop Sci.*, **25**: 1011-1015 .

Pulver E.L., Brockman F. and Wien H.C. 1982. Nodulation of soybean cultivars with *Rhizobium* spp and their response to inoculation with *R. japonicum*. *Crop Sci.*, **22**: 1065-1070.

Rao D.L.N., Giller K.E., Yeo A.R. and Flowers T.J. 2002. The effects of salinity and sodicity upon nodulation and nitrogen fixation in chickpea (*Cicer arietinum*). *Annals of Botany*, **89**: 563-570.

Rawsthorne S., Hadley P., Summerfield R.J. and Roberts E.H. 1985. Effect of supplemental nitrate and thermal regime on the nitrogen nutrition of chickpea (*Cicer arietinum* L.) II. Symbiotic development and nitrogen assimilation. *Plant and Soil*, **83**: 279-283.

Rennie R.J., Dubetz S., Bole J.B. and Muendel H.H. 1982. Dinitrogen fixation measured by ^{15}N isotope dilution in two Canadian soybean cultivars. *Agron. J.*, **74**: 725-730.

Rennie R.J. and Kemp G.A. 1982. N_2-fixation in field beans quantified by ^{15}N isotope dilution. II. Effect of cultivars and beans. *Agron. J.*, **75**: 645-649.

Rennie R.J. and Dubetz S. 1986. Nitrogen-15-Determined Nitrogen Fixation in Field-Grown Chickpea, Lentil, Fababean, and Field pea. *Agron. Journal*, **78**: 654-660.

Rice E.L. 1974. *Allelopathy*, First Edition, Academic Press, New York. pp. 353.

Ronis D.H., Sammons D.J., Kenworthy W.J. and Meisinger J.J. 1985. Heritability of total and fixed N content of the seed in two soybean populations. *Crop Sci.*, **25**: 1-4.

Rupela O.P. 1990. A visual rating system for nodulation of chickpea. *International Chickpea Newsletter*, **22**: 22-25.

Rupela O.P. and Beck D.P. 1990. Prospects for optimizing biological nitrogen fixation in chickpea. Pages 101-103. *In*: *Chickpea in the Nineties*: Proceedings of the Second International Workshop on Chickpea Improvement, 4-8 Dec 1989, ICRISAT Center, India, Patancheru 502 324, India: ICRISAT. (CP 559).

Rupela O.P. and Sudarshana M.R. 1990. Displacement of native rhizobia nodulating chickpea (*Cicer arietinum* L.) by an inoculant strain through soil solarization. *Biology and Fertility of Soils*, **10**: 207-212.

Rupela O.P. 1992. Natural occurrence and salient characters of non-nodulating chickpea plants. *Crop Sci.*, **32**: 349-352.

Rupela O.P. 1994. Screening for intra-cultivaral variability for nodulation of chickpea and pigeonpea. *In*: *Linking Biological Nitrogen Fixation Research in Asia*: Report of a meeting of the Asia Working Group on Biological Nitrogen Fixation in Legumes, (eds.) O.P. Rupela, J.V.D.K. Kumar Rao, S.P. Wani and C. Johansen. pp. 75-83. ICRISAT, Patancheru, A.P., India.

Rupela O.P. and Johansen C. 1995. Identification of non-nodulating, and low and high nodulating plants in pigeonpea. *Soil Biol. Biochem.*, **27**: 539-544.

Rupela O.P. 1997. Field evaluation of some recently developed selections for high nodulation and value of nodulation variants of chickpea. Pages 305-317. *In*: *Extending nitrogen fixation research to farmers' fields*: Proceeding of an International Workshop on Managing Legume Nitrogen Fixation in Cropping Systems of Asia, 20-24 Aug 1996, ICRISAT Asia Center, India (eds.) Rupela O.P., Johansen C. and Herridge D.F., Patancheru.

Saraf C.S., Rupela O.P., Hegde D.M., Yadav R.L., Shivkumar B.G., Bhattarai S., Razzaque II M.A. and Sattar M.A. 1998. Biological nitrogen fixation and residual effects of winter grain legumes in rice and wheat cropping systems of the Indo-Gangetic Plain. Pages 14-30. *In*: *Residual Effects of Legumes in Rice and wheat cropping Systems of the Indo-Gangetic Plain*, (eds.) Kumar Rao J.V.D.K., Johansen C. and Rego T.J., India, ICRISAT, Patancheru, Oxford and IBH, New Delhi.

Schultze M. and Kondorosi A. 1998. Regulation of symbiotic root nodule development. *Annual Review of Genetics*, **32**: 33-57.

Serraj R., Sincliar T.R. and Purcell L.C. 1999a. Symbiotic N_2-fixation response to drought. *Journal of Experimental Botany*, **50**: 143-155

Serraj R., Vadez F., Denison F. and Sinclair T. R. 1999b. Involvement of ureides in nitrogen fixation inhibition in soybean. *Plant Physiology*, **119**: 289-296.

Shearer G. and Kohl D.H. 1986. N_2-fixation in field settings: Estimations based on natural ^{15}N abundance. *Australian Journal of Plant Physiology*, **13**: 699-756.

Sinclair T.R., Leilah A.A. and Schreffler A.K. 1995. Peanut nitrogen fixation (C_2H_2 reduction) response to soil dehydration. *Peanut Science*, **22**: 162-166.

Singh B.D., Singh R.M., Murty B.K. and Singh R.B. 1985. Combining ability analysis of nodulation characters in mungbean (*Vigna radiata* L. Wilczek). *Indian J. Gen.*, **45**: 121-127

Singh S.D., Williams R.J. and Reddy P.M. 1988. Isolation of downy mildew resistant lines from a highly susceptible cutlivar of pearl millet. *Indian Phytopathology*, **41**: 450-456.

Singleton P.W., Boonkerd N., Carr T.J. and Thompson J.A. 1997. Technical and Market Constraints Limiting Legume Inoculant Use in Asia. *In*: *Extending Nitrogen Fixation Research to Farmers' Fields*, (eds.) O.P. Rupela, C. Johansen, D.F. Herridge, ICRISAT, Patancheru. pp.17-38.

Singleton P.W. and Bohlool B.B. 1984. Effect of salinity on nodule formation by soybean. *Plant Physiology*, **74**: 72-76.

Smith D.L., Dijak M. and Hume D.J. 1988. The effect of water deficit on N2 (C2H2) fixation by white bean and soybean. *Canadian Journal of Plant Science*, **68**: 957-967.

Smith G.R., Knight W.E., Peterson H.L. and Hagedorn C. 1982. The effect of *Rhizobium trifolii* strains and crimson clover genotypes on N_2 fixation. *Crop Sci.*, **22**: 970-972.

Smith S.C., Bezdicek D.F., Turco R.F. and Cheng H.H. 1987. Seasonal N_2 fixation by cool-season pulses based on several ^{15}N methods. *Plant and Soil*, **97**: 3-13.

Spaink H.P. 2000. Root nodulation and infection factors produced by rhizobial bacteria. *Annual Review of Microbiology*, **54**: 257-288.

Sprent J.I. and Sprent P. 1990. *Nitrogen fixing organisms: Pure and Applied aspects.* Chapman and Hall, London.

Sprent J.I., Minchin F.R. and Thomas R.J. 1983. Environmental effects on the physiology of nodulation and nitrogen fixation. *In*: *Temperate legumes: Physiology, genetics and nodulation*, (eds.) Jones D.G. and Davies D.R., pp. 269-318. Pitman Books, London, UK.

Streeter J. 1988. Inhibition of legume nodule formation and N_2-fixation by nitrate. *Critical Reviews in Plant Science*, **7**: 1-23.

Sylvester-Bradley R., Ayarza M.A., Mendez J.E. and Moriones R. 1983. Use of undisturbed soil cores for evaluation of *Rhizobium* strains for inoculation of tropical forage legumes in a Colombian Oxisol. *Plant and Soil*, **74**: 237-247.

Tan G.Y. 1981. Genetic variation for acetylene reduction rate and other characters in alfalfa. *Crop Sci.*, **21**: 485-488.

Tsyganov V.E., Morzhina E.V., Stefanov S.Y., Borisov A.Y., Lebsy V.K. and Tikhonovich I.A. 1998. The pea (*Pisum sativum* L.), genes sym33 and sym40 control infection thread formation of root nodule function. *Mol Gen Genet.*, **259**: 491-503.

Van Berkum P. and Eartly B.D. 1998. *The Rhizobiaceae.* Pages 1-24 (eds.) Spaink H.P., A. Kondorosi and P.J.J. Hooykaas. Kluwer Academic Publishers, Dordrecht.

Vavilov N.I. 1951. *The origin, variation, immunity and breeding of cultivated plants*, (English translation), The Ronald Press Co., New York.

Venkateswarlu B. and Katyal J.C. 1994. Improvement of nitrogen fixation in groundout by host-plant selection. *In*: *Linking Biological Nitrogen Fixation Research in Asia*: Report of a meeting of the Asia Working Group on Biological Nitrogen Fixation in Legumes, (eds.) O.P. Rupela, J.V.D.K. Kumar Rao, S.P. Wani and C. Johansen, 6-8 December 1993. International Crops Research Institute for the Semi-Arid Tropics, Patancheru, Andhra Pradesh, India.

Vincent J.M. 1988. The role of legume, *Rhizobium* and environment in nitrogen fixation: Constraints on symbiotic potential and their removal. *In*: *Nitrogen Fixation by Legumes in Mediterranean Agriculture*, (eds.) Beck D.P. and Materon L.A., pp. 275-285.

Wacek T.J. and Brill W.J. 1976. Simple, rapid assay for screening nitrogen-fixing ability in soybean. *Crop Sci.*, **16**: 519-523.

Wakrim M.R. and Wery J. 1995. Analyse des relations entre itat hydrique du sol, rendement en graines et fixation d'azote chez le pois chiche (*Cicer arietinum* L.). *In*: *Facteurs limitant la fixation symbiotique de l'azote dans le bassin mediterraneen.* (ed.) J.J. Drevon, Paris: INRA (Les Colloques No. 77), 257-267.

Wani S.P., Dart P.J. and Upadhyaya M.N. 1983. Factors affecting nitrogenase activity (C_2H_2 reduction) associated with sorghum and millet estimated using the soil core assay. *Canadian Journal of Microbiology*, **29**: 1063-1069.

Wani S.P., Rupela O.P. and Lee K.K. 1997. Soil mineral nitrogen concentration and its influence on biological nitrogen fixation of grain legumes. Pages 183-198. *In*: *Extending nitrogen fixation research to farmers' fields*: Proceedings of an International Workshop on Managing Legume Nitrogen Fixation in the Cropping Systems of Asia, 20-24 Aug 1996, ICRISAT Asia Center, India, (eds.) Rupela O.P., Johansen C. and Herridge D.F., Patancheru 502 324.

Westermann D.T., Kleinkopf G.E., Porter L.K. and Legett G.E. 1981. Nitrogen sources for bean seed production. *Agron. J.*, **73**: 660-664.

Wiersma J.V. and Orf J.H. 1992. Early maturing soybean nodulation and performance with selected *Bradyrhizobium japonicum* strains. *Agron. J.*, **84**: 449-458.

Witty J.F. and Minchin F.R. 1988. Measurement of nitrogen fixation by the acetylene reduction assay; myths and mysteries. *In*: *Nitrogen fixation by legumes in Mediterranean Agriculture.* (eds.) Beck D.P. and Materon L.A., Martinus Nijhoff, Dordrecht. Pp 331-344.

Witty J.F., Rennie R.J. and Atkins C.A. 1988. ^{15}N addition methods for assessing N_2 fixation under field conditions. *In*: *World Crops: Cool season food legumes.* (ed.) Symmerfield R.J., Kluwer Academic Publishers, London. Pp. 715-730.

Wolyn D.J., Attewell J., Ludden P.W. and Bliss F.A. 1989. Indirect measure of N_2-fixation in common bean (*Phaseolus vulgaris* L.) under field conditions. Role of lateral root nodules. *Plant and Soil*, **113**: 181-187.

Wu S. and Harper J.E. 1991 Dinitrogen fixation potential and yield of hyper nodulating soybean mutants: A field evaluation. *Crop Sci.*, **31**: 1233-1240.

Yoneyama T., Nambiar P.T.C., Lee K.K., Srinivasa Rao B. and Williams J.H. 1990. Nitrogen accumulation in three legumes and two cereals with emphasis on estimation of N_2-fixation in the legumes by the natural ^{15}N abundance technique. *Biology and Fertility of Soils*, **9**: 25-30.

Young J.P.W. 1996. Phylogeny and taxonomy of rhizobia. *Plant and Soil*, **186**: 45-52.

Young J.P.W. and Haukka K.E. 1996. Diversity and phylogeny of rhizobia. *New Phytol.*, **133**: 87-94.

Zari K.W., Miller J.C. Jr., Weaver R.W. and Barnes L.W. 1978. Intraspecific variability for nitrogen fixation in southern pea (*Vigna ungiculata*). *J. Am. Soc. Hortic. Sci.*, **103**: 806-808.

Plant Breeding - Mendelian to Molecular Approaches
H. K. Jain and M. C. Kharkwal (eds.)

31

Rising Atmospheric Carbon Dioxide and Crop Responses

D.C. Uprety[1], A.P. Mitra[2], S.C. Garg[2], B. Kimball[3] and D. Lawlor[4]

Abstract

Exponential rise in the atmospheric carbon dioxide (CO_2) concentration anthropogenically has aroused interest to characterize the possible response of crop plants to the elevated carbon dioxide in future. Open Top Chamber (OTC) and Free Air Carbon Dioxide (CO_2) Enrichment (FACE) technologies were developed to study the crop responses. Study revealed that elevated CO_2 significantly ameliorated the adverse stress effect in *Brassica* species. It also demonstrated the possibility of transferring the CO_2 response character in *Brassica oxycamp* hybrid. The stress induced adverse effect on grain composition of *Brassica juncea*, particularly on carbohydrate and oil content was significantly ameliorated due to sequestering of carbon and improvement in water status. Studies on rice cultivars showed that additional carbohydrate contributed to the production of mass tillers. The CO_2 induced increase in the rice grain yield was attributed to the number of tillers and grain per plant. Additional carbohydrate has helped in balancing the profile of photosynthetic proteins to sustain greater photosynthetic activity in rice plant. These studies lead to the establishment of South Asian CO_2 Crop Research Network of Bangladesh, India, Nepal, Pakistan and Sri Lanka and the first Asian FACE facility at the Indian Agricultural Research Institute, New Delhi, India.

Introduction

Anthropogenic activities have lead to the build up of greenhouse gases including water vapour (H_2O), carbon dioxide (CO_2), methane (CH_4), nitrous oxide (N_2O) and chloro fluro carbons (CFC) in the atmosphere, resulting in global climate changes. Global climate changes have posed an unique research challenge to the present day agriculture. The exponential rise in CO_2 concentration is one such important change, which effectively influences the productivity of the crop plants and food security of this region. The rising concentration of atmospheric CO_2 is believed to be due to the burning of fossil fuels such as coal, gas and oil, urbanization, deforestation, and expansion of agriculture.

[1]Indian Agricultural Research Institute, New Delhi 110 012, India
[2]National Physical Laboratory, New Delhi 110 012, India
[3]United States Water Conservation Laboratory, Phoenix, Arizona, U.S.A.
[4]Rothamsted Experimental Station, Harpenden, Herts, U.K.

The Rising Atmospheric Carbon Dioxide

Burning of fossil fuel introduces 5.70 Gt whereas deforestation adds 2.30 Gt of CO_2/a (per annum) in the atmosphere. Of this about 4.7 Gt CO_2/a is being absorbed by oceanic and terrestrial biota. The leftover of about 3.3 Gt CO_2/a in the atmosphere accounts for the rise in its concentration. Measurements of CO_2 in the atmosphere which began in Antarctica in 1957 and Maunaloa (Hawaii) in 1958, indicated clearly that the concentration of CO_2 in the atmosphere is increasing rapidly (Keeling *et al.*, 1982). The ice core studies showed that the CO_2 concentration was about 205 μmol mol^{-1} some 20,000 years ago. During the 10,000 years prior to the Industrial Revolution, its value was 280 μmol mol^{-1}. Prior to 1900 the CO_2 concentration in the atmosphere was 290 μmol mol^{-1}. Maunaloa studies also observed that from 1958 to 1982 there was an increase of 1.0 μmol mol^{-1} CO_2 per year. The 1958 value of atmospheric CO_2 was 316 μmol mol^{-1} and in 1995 (Collette, 1995) and the present, concentrations (Kimball 1997) were recorded to be as high as 360 and 370 μmol mol^{-1}, respectively. It is currently increasing at a rate of about 1.8 μmol mol^{-1} per annum. Measurements in Delhi region (Indian Agricultural Research Institute) also showed a similar trend (Uprety *et al.*, 2000) (Fig. 1). At this rate of increase the concentration of CO_2 in the atmosphere is likely to be doubled (600 μmol mol^{-1}) by the middle of 21st century (Houghton *et al.*, 1990).

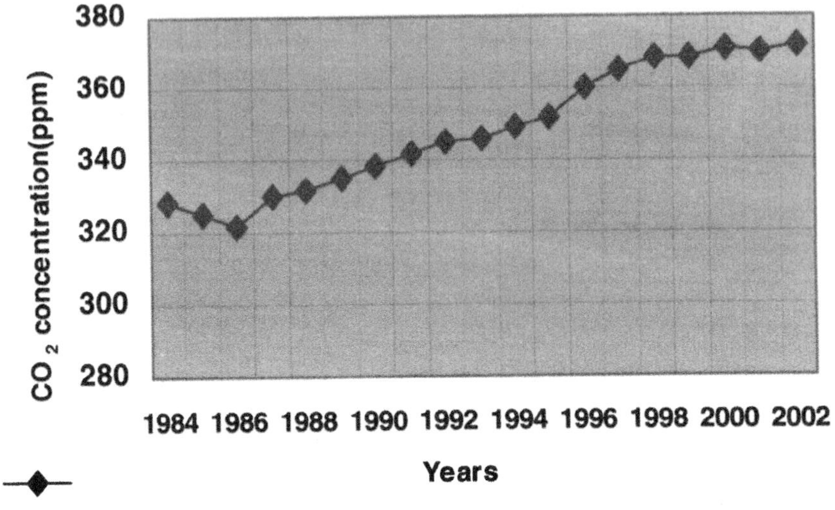

Fig. 1. The rise in atmospheric CO_2 at I.A.R.I., New Delhi, India, 1984-2000 winter crop season

Increase in CO_2 concentration is expected to cause global warming by absorbing long wave heat radiations from earth's surface and may also affect cloudiness and precipitation. This is referred as Climatic Effect. However, according to General Circulation Model (GCM) temperature changes will not be uniform. At polar regions temperature may increase by 4-5 °C and equatorial region by less than 1 °C. CO_2 has

'Biological Effect' as well, because it serves as a substrate for photosynthetic carbon assimilation. There is concomitant decline in photo-respiratory activity and alteration in stomatal aperture. It was reported that C_3 plants (wheat, rice, oilseeds, pulses, etc.) respond to elevated CO_2, since elevated CO_2 reduces oxygenase activity of RuBP carboxylase oxygenase enzyme in plants. C_4 plants (sorghum, maize, sugarcane, etc.) on the other hand show little or no photosynthetic response to elevated CO_2, because C_4 pathway is not competitively inhibited by O_2 and is completely CO_2 saturated. However, there is no consensus on the quantitative effects of increased CO_2 in plant processes and growth due to differences in response at different stages of growth, species of crops and because of growth limiting environmental factors. The ability of farmers to survive/exploit such influences will depend on scientists accurately predicting growth and development responses of crops to such increase in atmospheric CO_2. The selection of genotypes which are likely to be more productive under new environmental conditions is also a major operative. Innovative approaches for conducting long term experiments have been developed to investigate the impact of rising atmospheric CO_2 on crops.

Need for the Plant Response Studies to Atmospheric Carbon dioxide

1. Most of our crops were selected in the past for maximum productivity at 320-330 μmol mol^{-1} CO_2 in the atmosphere and their responses to the higher levels of CO_2 need to be characterized for developing plant types for future.
2. It will be required for tailoring plant species and identifying management practices for their cultivation in a future high CO_2 environment.
3. Most of the CO_2 crop response studies were done on temperate crops (Acock and Allen, 1985) and such studies on tropical and sub- tropical crops were required.
4. Response studies were conducted in closed chambers or short term exposures in the field. They may not reflect the response of crops under natural conditions where other factors such as temperature, moisture, etc. also interact.
5. To generate database on the crop responses for modelling to predict their growth and productivity for future, so that farmers could be able to utilize and exploit this information.

The Government of India and ICAR started impact assessment analysis of crop responses to the rising atmospheric CO_2 through an AP CESS Fund project entitled "Study the effect of rising atmospheric CO_2 on photosynthesis and productivity of crop plants under moisture stress conditions", in 1992 at Indian Agricultural Research Institute, New Delhi, India (Uprety, 1993), with the following objectives :

(*a*) To develop facilities of CO_2 enrichment technology.
(*b*) To study the response of crop plants to the elevated CO_2.
(*c*) Application of crop response data in preparing model and developing ideotype for likely future environment.
(*d*) To characterize the responses to higher CO_2 to tailor plants for future CO_2 environment and identifying the cultivation and nutrient management technologies for them.

Development of Carbon Dioxide Enrichment Technology

Several systems are used to study the response of plant communities to elevated CO_2. There are expensive precisely controlled closed systems that continuously recondition and re-circulate the air in controlled environment of growth chambers. During the current decade there have been new approaches to CO_2 enrichment technology under field conditions. Computer models which predict for field conditions should be based on data collected at the field and give confidence in the conclusions arrived at from these experiments. These experiments are potentially cost effective (Lawlor and Mitchell, 1991). Currently most of the CO_2 experiments utilizing such facilities follow a holistic approach, to design experiments on crops as natural as possible and then to observe their responses.

(*i*) ***Open Top Chamber Technology***: Experimental studies on the crop responses to the elevated CO_2 started with the designing of modified open top chambers at Indian Agricultural Research Institute for South Asian climatic conditions (Uprety *et al.*, 1998). These chambers are cylindrical aluminium frames with clear flexible PVC covering and frustrum to reduce the incursion of external air. CO_2 enriched air is introduced into the chamber with the help of blower to distribute it uniformly using manifold, valves and the flow meters. These chambers have controlled CO_2 enrichment of 580-620 μmol mol^{-1} and facilities to measure dynamic changes in temperature, light and relative humidity without costly and complex environmental controls. These open top chambers are inherently simple and a cost effective means of meeting the requirements of crop response studies under field conditions (Fig. 2).

Fig. 2. Open top chamber CO_2 enrichment technology facility at I.A.R.I., New Delhi, India

(*ii*) ***Free Air CO_2 Enrichment (FACE) Technology***: A simple cost effective Mid-FACE facility was jointly developed by IARI, NPL and Italian group of scientists at Indian Agricultural Research Institute, New Delhi. It consists of 8 m diameter octagon shaped ring wherein CO_2 concentration is regulated on the basis of wind speed and wind direction by PC based controller and PiD valves. It has advantage of enriching a larger area of vegetation with CO_2, without the confounding effects of growth chambers for an extended period of time at lower cost per unit area than any other techniques (Uprety *et al.*, 2000a). (Fig. 3).

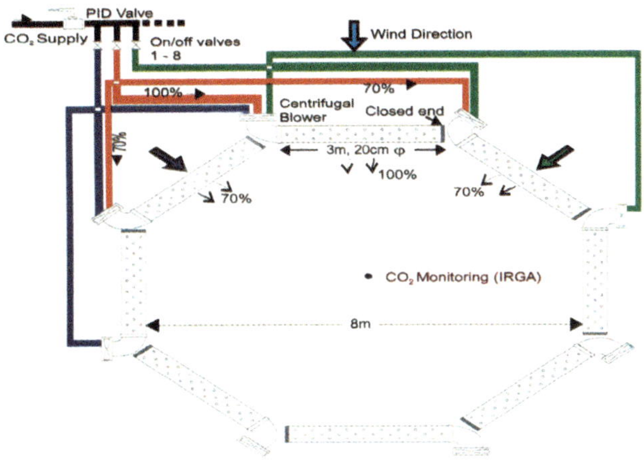

Fig. 3. Mid-FACE ring for South Asia (Structural diagram)

Results

(*i*) ***Amelioration of Moisture Stress Effect***: It was revealed that plants responded to elevated CO_2 significantly under water stress conditions, ameliorating the adverse stress effect on photosynthesis, growth and water relations in *Brassica* species. The elevated CO_2 increased the rate of photosynthesis, promoted greater foliage, more number of siliqua and increased root growth. The carbon need to satisfy the demand for these newly generated sinks was met from new photosynthates produced during the high CO_2 exposure. Plants accumulated increased metabolic reserves in terms of sugars and starch to significantly enhance metabolic flexibility, responding to adverse moisture stress. This flexibility would give plants economic and ecological advantage. The *Brassica* plants would be able to survive under stress condition better in future than their current resistance to drought (Uprety *et al.*, 1995). (Fig. 4).

(*ii*) ***Genetic Transfer of CO_2 Responsive Characters***: Study on the characterization of CO_2 responsiveness in *Brassica oxycamp* hybrid and its parents *B. oxyrrhina* and *B. campestris* showed that the response of *B. oxycamp* hybrid to elevated CO_2 was significantly positive in respect to photosynthesis and growth, similar

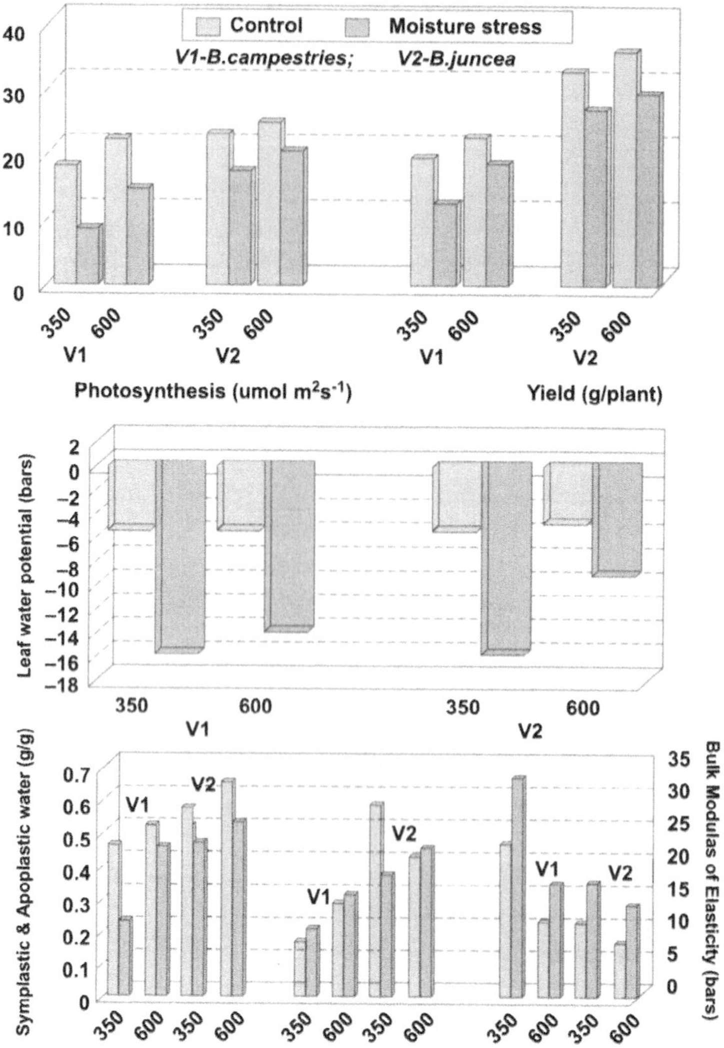

Fig. 4. Effect of elevated CO_2 on various components of *Brassica* species

to that of its parent *B. campestris*. Hybrid *B. oxycamp* and *B. campestris* with greater sink potential responded significantly, whereas, *B. oxyrrhina* with poor sink size did not respond to CO_2 enrichment (Uprety *et al.*, 1998) (Fig. 5).

(*iii*) ***Improvement in Nutrient Constituents in Seeds***: Stress induced adverse effect on grain weight and its composition such as carbohydrate, and oil content in *Brassica* seeds was significantly ameliorated by elevated CO_2 due to sequestration of carbon and improved water status. The reduction in protein level was observed and attributed to the dilution effect caused by excess carbohydrates and fats (Uprety *et al.*, 1997) (Fig. 6).

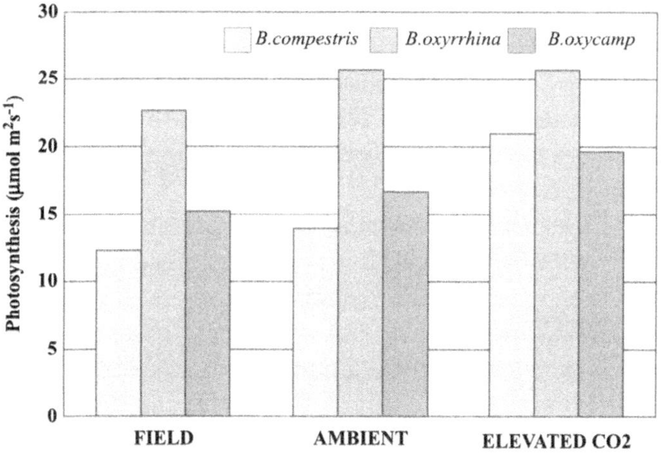

Fig. 5. Effect of elevated CO_2 on the photosynthesis of *B. oxycamp* hybrid and its parents

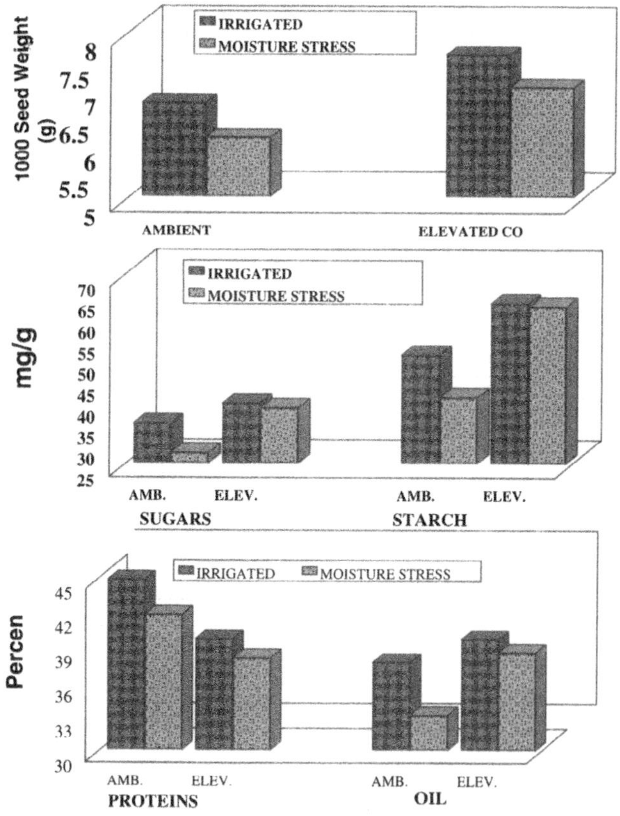

Fig. 6. Interactive effect of elevated CO_2 and moisture stress on the 1000 seed weight, sugars, starch, proteins and oil content of *Brassica juncea* seeds

(*iv*) ***Anatomical Alterations of Leaves***: The Transmission Electron Microscopic (TEM) study of leaf tissues showed a significant increase in the thickness of epidermis, size of mesophyll cells, accumulation of starch and size and number of starch granules per chloroplast. These changes possibly increase storage sites for excessive starch avoiding the disruption of chloroplast ultrastructures by starch overloading (Uprety *et al.*, 2001).

(*v*) ***Response of Rice Cultivars***: Studies on the response of rice cultivars Basmati-1 and Pusa-677 to the elevated CO_2 showed increase in growth, photosynthesis and seed yield. The increase in grain yield of Basmati-1 attributed largely to increased grain number (Uprety *et al.*, 2000b). The increased photosynthesis and greater accumulation of sugar contributed significantly to the accelerated development of leaves and tillers in both the cultivars. The reduction in the low molecular mass proteins including Rubisco and increase in high molecular mass PS2 proteins was observed in both the cultivars. Additional sugars may possibly help in balancing the profile of photosynthetic proteins and sustain greater growth and productivity in rice cultivars, (Uprety *et al.*, 2003).

The response of rice cultivars to the elevated CO_2 with reference to their stomatal characters showed a marked increase in the stomatal resistance, stomatal index, size of guard cells, stroma and epidermal cells. There was marked reduction in stomatal and epidermal cell density. These changes on stomatal characters made rice cultivars adjustable to CO_2 enriched environment. Such adjustments helped in the regulation of gas exchange in rice cultivars in CO_2 enriched environment (Table 1) [Uprety *et al.*, 2002].

Table 1. Effect of elevated CO_2 on the stomatal characters of rice cultivars

		CO_2 treatment	Pusa Basmati-1	P-677	P-834	P-2503-6-693	C.D. at 5% culti.	C.D. at 5% treat.	P
Stomatal resistance (S cm^{-1})		Ambient	1.54	1.72	1.40	1.54	0.85	0.85	n.s.
		Elevated	2.16	2.58	2.25	2.48			
	Upper surface	Ambient	93.56	128.65	105.26	93.56	3.75	3.75	n.s.
		Elevated	81.87	105.26	119.88	87.74			
Stomatal density (n mm^{-2})	Lower surface	Ambient	116.95	149.12	116.95	113.50	4.20	4.20	n.s.
		Elevated	111.11	132.42	131.57	149.12			
Stomatal index (%)	Upper surface	Ambient	28.60	31.90	23.90	24.40	1.20	1.20	n.s.
		Elevated	35.00	35.80	29.50	31.60			
	Lower surface	Ambient	29.80	29.80	28.10	26.60	1.60	1.60	n.s.
		Elevated	35.80	31.80	34.60	40.10			
Size of Stomata (upper surface) (μ m)	Length	Ambient	20.16	17.52	20.19	19.07	2.62	2.62	n.s.
		Elevated	21.07	24.00	22.58	19.29			
	Width	Ambient	10.38	8.87	9.18	8.39	n.s.	1.35	n.s.
		Elevated	10.95	11.69	11.12	11.50			
Stroma (μ m)	Length	Ambient	1.55	0.97	1.98	1.03	0.35	0.35	n.s.
		Elevated	2.09	2.24	2.18	1.66			

These Indian studies relating to the development of CO_2 enrichment technologies and generation of a database on various components of crop responses to the elevated CO_2, have helped in the development of an International CO_2 Crop Research Network of Bangladesh, India, Nepal, Pakistan and Sri Lanka under APN funded CO_2 crop research project. This network has become functional to generate data on crop responses at various locations to be used for developing models of plants for a CO_2 enriched future. This international programme at IARI on global change research provided an inter-disciplinary framework for global change science. Funding from APN for such an effort has also added value to the nationally funded activities, such as training scientists from various Indian and South Asian countries at IARI and conducting multi-country crop CO_2 experiment at various locations in different South Asian countries. These studies will help farmers to choose selected responsive crop cultivars for changed climatic conditions following modified crop management technologies.

Conclusion

From this study, the following conclusions can be arrived at:

1. Basic morpho-physiological data on the crop responses to the elevated CO_2 have been generated and are being processed for modelling.
2. The characterization of these data will help in tailoring plants and identifying management practices to crop cultivation for future CO_2 environment.
3. A south Asian CO_2 crop research network of the participating countries was established.
4. Open top chambers were made and transferred to Bangladesh, Nepal, Pakistan and Sri Lanka for common multi-country, multi-disciplinary experiments to study the responses of crop plants to elevated CO_2.
5. Scientists, students and Government officials from Bangladesh, Nepal, Pakistan and Sri Lanka were trained for CO_2 enrichment research and technology at Indian Agricultural Research Institute and National Physical Laboratory, New Delhi, India under APN funded project.

References

Acock B. ahd Allen L.H. Jr. 1985. Crop responses to elevated carbon dioxide concentration. Pp-53-98. *In*: *Direct effects of increasing carbon dioxide on vegetations.* (eds.) B.R. Strain and J.D. Cure, United States Department of Energy, Carbon dioxide Research Division, DOE/Er-0238, Office of Energy Research, Washington, DC, USA.

Collette E. 1995. Will plants profit from high CO_2?. *Science*, **268**: 654-655.

Houghton J.T., Jenkins G.J. and Ephraumas J.J. 1990. *Climate Change* - The PCC Scientific Assessment. Cambridge Univ. Press. Cambridge, U.K.

Keeling C.D., Bacastow R.B. and Whorf T.P. 1982. Measurements of the concentration of carbon dioxide at Maunaloa Observatory, Hawaii, p 377-385. *In*: *Carbon dioxide review*, (ed.) W.C. Clark, 1982. Oxford University Press, New York, USA

Kimball B.A. 1997. Influence of increasing CO_2 concentration on photosynthetic stimulation of selected weeds. *Photosynth. Res.*, **54**: 199-208.

Lawlor D.W. and Mitchell R.A.C. 1991. The effect of increasing CO_2 on crop photosynthesis and productivity : a review of field studies. *Plant Cell & Environ.*, **14**: 807-818.

Uprety D.C. 1993. Study the effect of rising atmospheric CO_2 on photosynthesis and productivity of crop plants under moisture stress conditions. ICAR AP CESS Fund Project report. pp. 1-23.

Uprety D.C., Dwivedi N., Mohan R. and Paswan G. 2001. Effect of elevated CO_2 concentration on leaf structures of *Brassica juncea* under water stress. *Biologia Plantarum*, **44**: 149-152.

Uprety D.C., Dwivedi N. and Mohan R. 1997. Interactive effect of rising atmospheric carbon dioxide and drought on nutrient constituents of *Brassica juncea* seeds. *Sci. & Culture*, **63**: 291-292.

Uprety D.C., Dwivedi Neeta and Mohan Rajat. 1998. Characterization of CO_2 responsiveness in a *Brassica oxycamp* interspecific hybrid. *J. Agron. & Crop Sci.*, **180**: 7-13.

Uprety D.C., Garg S.C., Tiwari M.K. and Mitra A.P. 2000a. Crop responses to elevated CO_2: Technology and Research (Indian studies). *Global Environ. Res.*, **3**: 155-167.

Uprety D.C., Kumari Sunita, Dwivedi Neeta and Mohan Rajat. 2000b. Effect of elevated CO_2 on the growth and yield of rice variety 'Pusa 834'. *Ind. J. Pl. Physiol.*, **51**(N.S.): 105-107.

Uprety D.C., Mishra R.S. and Abrol Y.P. 1995. Effect of elevated CO_2 and moisture stress on the photosynthesis and water relation in *Brassica* species. *J. Agron. & Crop Sci.*, **175**: 231-237.

Uprety D.C., Dwivedi N., Jain V. and Mohan R. 2002. Effect of elevated carbon dioxide on the stomatal parameters of rice cultivars. *Photosynthetica*, **40**: 315-319.

Uprety D.C., Dwivedi N., Jain V., Mohan R., Saxena D.C., Jolly M. and Paswan G. 2003. Response of rice varieties to elevated CO_2. *Biologia Plantarum*, **46**: 35-39.

The Contributors

Ahuja P.S., Institute of Himalayan Bioresource Technology, Palampur 176 061, India; Email: ahuja@csihbt.ren.nic.in

Arunachalam V., MS Swaminathan Research Foundation, Taramani, Chennai 600 113 , India; Email: v_arunachalam@hotmail.com

Baisakh Niranjan, Plant Breeding, Genetics and Biotechnology Division, International Rice Research Institute, DAPO Box 7777, Metro Manila, Phillipines; Email: n.baisakh@ cgiar.org

Bandyopadhyaya A., National Agricultural Technology Project, (ICAR), KAB-II, Pusa Campus, New Delhi 110 012, India; Email: bandy@icar1.nic.in

Banga S.S., Department of Plant Breeding, Punjab Agricultural University, Ludhiana 141 004 , India; Email: surin11@rediffmail.com

Bharti A.K., Department of Plant Molecular Biology, University of Delhi South Campus, New Delhi 110 021, India; Email: bharti@ waksman.rutgers.edu

Brar D.S., Plant Breeding, Genetics and Biotechnology Division, International Rice Research Institute, DAPO Box 7777, Metro Manila, Phillipines; Email: d.brar@cgiar.org

Datta Swapan K., Plant Breeding, Genetics and Biotechnology Division, International Rice Research Institute, DAPO Box 7777, Metro Manila, Phillipines; Email: s.data@cgiar.org

Dhaliwal H.S., Department of Plant Breeding, Punjab Agricultural University, Ludhiana 141 004, India; Email: hsdhaliwal76@hotmail.com

Dhar M.K., Department of Biotechnology, University of Jammu, Jammu 180 006, India; Email: manojdharm@yahoo.com

Dhillon B.S., National Bureau of Plant Genetic Resources, Pusa Campus, New Delhi 110 012, India; Email: bdhillon@nbpgr.delhi.nic.in

Garg S.C., National Physical Laboratory, New Delhi 110 012, India

Gautam P.L., G.B.Pant University of Agrilculture & Technology, Pantnagar 263 145, India; Email: pl_gautam@yahoo.com

Govila O.P., Division of Genetics, Indian Agricultural Research Institute, New Delhi 110 012, India; Email: govilaop@sify.com

Grover Anil, Department of Plant Molecular Biology, University of Delhi South Campus, New Delhi 110 021, India; Email: grover_anil @hotmail.com

Gupta P.K., Department of Agricultural Botany, Ch. Charan Singh University, Meerut 250 004, India; pkgupta@ndf.vsnl.net.in

Gupta V.P., GH-10/69A, Sunder Apartment, Paschim Vihar, New Delhi 110 067, India;

Gupta V.S., Division of Biochemical Sciences, National Chemical Laboratory, Pashan, Pune 411 008, India; Email: vidya@ems.ncl.res.in

Hittalmani S., Department of Genetics & Plant Breeding, University of Agricultural Sciences, GKVK, Bangalore 560 065, India; Email: shailajah_maslab@rediffmail.com

Hoisington David, Applied Biotechnolgy Centre and Bioinformatics, The International Maize and Wheat Improvement Center (CIMMYT), Mexico DF, Mexico; Email: d.hoisington@cgiar.org

Huerta-Espino J., The International Maize and Wheat Improvement Center (CIMMYT), Campo Experimentale Valle de Mexico-INIFAP, Apdo, Postal 10, 56230, Chapingo, Edo, de Mexico, Mexico; Email: JHuerta@cgiar.org

Jain H.K., Centre for Science Writing, 40, Surya Niketan, New Delhi 110 092, India; E-mail: hkj@vsnl.net

Kang M.S., Department of Agronomy, Louisiana Agricultural Experiment station, Louisiana State University Agricultural Centre,, Baton Rouge, LA 70803-2110, USA; Email: mkang@agctr.lsu.edu

Kapoor Avnish, Department of Plant Molecular Biology, University of Delhi South Campus, New Delhi 110 021, India; Email: avnih@hotmail.com

Kharkwal M.C., Division of Genetics, Indian Agricultural Research Institute, New Delhi 110 012, India; Email: mckharkwal@yahoo.com

Khurana J.P., Department of Plant Molecular Biology, University of Delhi South Campus, New Delhi 110 021, India; Email: khuranaj@ genomeindia.org

Khurana P., Department of Plant Molecular Biology, University of Delhi South Campus, New Delhi 110 021, India; Email: param@ genomeindia.org

Kimball B.A., United States Water Conservation Laboratory, Agricultural Research Service, U.S. Department of Agriculture, Phoenix, Arizona 85040, USA; Email: bkimball@uswcl.ars.ag.gov

Kirti P.B., School of Life Sciences, University of Hyderabad, Hyderabad 500 046, India; Email: pbksl@uohyd.ernet.in

Kole C., Department of Genetics and Plant Breeding, Allahabad Agricultural Institute, Allahabad 211 007, India; Email: kolecr@yahoo.co.in

Kumar D., AICRP on Arid Legumes, Central Arid Zone Research Institute, Jodhpur 342 003, India; Email: dkumar@cazri.raj.nic.in

Kumar Ish, Plant Breeding, Genetics and Biochemistry Division, International Rice Research Institute, DAPO Box 7777, Metro Manila, Philippines; Email: i.kumar@ cgiar.org

Lather B.P.S., Department of Plant Breeding, Ch. Charan Singh Haryana Agricultural University, Hisar 125 004, India

Lawlor D.W., Crop Performance and Improvement Division, Rothamsted Research, Rothamsted Experimental Station, Harpenden, Herts, AL5 2JQ, UK; Email: david.lawlor@ bbsrc.ac.uk

Lakshmikumaran M., Tata Energy Research Institute, India Habitat Centre, Lodhi Road, New Delhi 110 003, India; Email: malks@ teri.res.in

Mauria S., Crops Division, Indian Council of Agricultural Research, Krishi Bhavan, New Delhi 110 001, India; Email: shashankmauria@ yahoo.com

Mehra R.B., Division of Genetics, Indian Agricultural Research Institute, New Delhi 110 012, India; Email: rajbeharimehra@yahoo.com

Mitra A.P., National Physical Laboratory, New Delhi 110 012, India

Mohanty A., Department of Plant Molecular Biology, University of Delhi South Campus, New Delhi 110 021, India; Email: mohanty@cshl.edu

Mohapatra T., National Research Centre on Plant Biotechnology, Indian Agricultural Research Institute, New Delhi 110 012, India; Email: tmnrcpb@mailcity.com

Nanda G.S., Department of Plant Breeding, Punjab Agricultural University, Ludhiana 141 004, India; Email: gsnanda@chd.pau.nic.in

Narayanan K.K., Metahelix Life Sciences Pvt. Ltd., Plot # 3, KIADB, 4th Phase, Bommasandra, Bangalore 562 158, India; Email: k.k.narayanan@ meta-helix.com

Pandey M.P., Department of Genetics and Plant Breeding, G.B.Pant University of Agrilculture & Technology, Pantnagar 263 145, India; Email: matappandey@rediffmail.com

Pandey R.N., Nuclear Agriculture and Biotechnology Division, Bhabha Atomic Research Centre, Mumbai 400 085, India; Email: pandeyrn@yahoo.com

Patil S.J., Univerversity of Agricultural Sciences, Dharwad 580 005, India

Pawar S.E., Nuclear Agriculture and Biotechnology Division, Bhabha Atomic Research Centre, Mumbai 400 085, India; Email: sepawar@magnum.barc.nic.in

Pooni H.S., School of Biosciences, The University of Birmingham, P.O. Box 363, Edgbaston, Birmingham, B 15 2TT, UK; Email: h.s.pooni@bham.ac.uk

Prabhakaran V.T., Department of Biometrics, Indian Agricultural Statistics Research Institute, New Delhi 110 012, India; Email: prabhakaran vt@yahoo.com; vtp@iasri. delhi.nic.in

Prabhu K.V., Division of Genetics, Indian Agricultural Research Institute, New Delhi 110 012, India; Email: kvinodprabhu@ rediffmail. com

Prasanna B.M., Division of Genetics, Indian Agricultural Research Institute, New Delhi 110 012, India; Email: prasanna@ndf.vsnl.net.in

Rai K.N., International Crop Research Institute for Semi Arid Tropics (ICRISAT), Patancheru 500 324, A.P. , India; Email: k.rai@cgiar.org

Rai Mangala, Indian Council of Agricultural Research, Krishi Bhavan, New Delhi 110 001, India; Email: mrai@icar.delhi.nic.in

Rajaram S., The International Maize and Wheat Improvement Center (CIMMYT), Apdo, Postal 6-641, 06600, Mexico DF, Mexico; Email: srajaram@cimmyt.mx

Ramanathan Vai, Metahelix Life Sciences, Plot # 3, KIADB, 4th Phase, Bommasandra, Bangalore 562 153, India; Email: v.ramanathan @meta-helix.com

Ranjekar P.K., Division of Biochemical Sciences, National Chemical Laboratory, Pashan, Pune 411 008, India; Email: pkr@ems. ncl.res.in

Rao D.L.N., AICRP on BNF, Indian Institute of Soil Sciences, Nabi bagh, Bhopal 462 038 , India; Email: dlnrao@iiss.mp.nic.in

Rao N.G.P., HIG (OG) Block 10, Flat 4, Baghlingampally, Hyderabad 500 044, India Email:

Reddy B.V.S., International Crop Research Institute for Semi Arid Tropics (ICRISAT), Patancheru 502 324, A.P., India; Email: b.reddy @cgiar.org

Roy Darbeshwar, Department of Genetics and Plant Breeding, G.B.Pant University of Agriculture & Technology, Pantnagar 263 145, India; Email: beshroy@yahoo.com

Rupela O.P., International Crop Research Institute for Semi Arid Tropics (ICRISAT), Patancheru 500 324, A.P., India; Email: o.rupela@cgiar.com

Saini R.G., Department of Genetics & Biotechnology, Punjab Agricultural University, Ludhiana 141 004, India; Email: sainirg@rediffmail.com

Sarma N.P., Directorate of Rice Research, Rajendranagar, Hyderabad 500 030, India; Email: npsarma@mailcity.com

Saxena K.B., International Crop Research Institute for Semi Arid Tropics (ICRISAT), Patancheru 500 324, A.P., India; Email: k.saxena@cgiar.org

Saxena Sanjeev, Conservation Division, National Bureau of Plant Genetic Resources, Pusa Campus, New Delhi 110 012, India; Email: ssaxena@nifindia.org

Sharma B., Division of Genetics, Indian Agricultural Research Institute, New Delhi 110 012, India; Email: drbrsh@yahoo.co.in

Sharma Rajiv K., Division of Genetics, Indian Agricultural Research Institute, New Delhi 110 012, India; rajivksharma@rediffmail.com

Sharma S.K., Department of Plant Breeding, Himachal Pradesh Agricultural University, Palampur 176 062, India; Email: pbg@hpkv. hp.nic.in

Shashidhar H.E., Department of Genetics & Plant Breeding, University of Agricultural Sciences, GKVK, Bangalore 560 065, India; Email: heshashidhar@yahoo.com

Siddiq E.A., Directorate of Rice Research, Rajendranagar, Hyderabad 500 030, India; Email: eas@drr.nic.in

Singh A.K., Division of Genetics, Indian Agricultural Research Institute, New Delhi 110 012, India; Email: aks_gene@lycos.com

Singh B.B., Division of Genetics, Indian Agricultural Research Institute, New Delhi 110 012, India; Email: head_g@rediffmail.com

Singh I.S., Department of Genetics and Plant Breeding, G.B.Pant University of Agriculture & Technology, Pantnagar 263 145, India; Email: issingh@rediffmail.com

Singh Mahender, Evaluation Division, National Bureau of Plant Genetic Resources, Pusa Campus, New Delhi 110 012, India; Email: eval@nbpgr.delhi.nic.in

Singh N.N., Directorate of Maize Research, Indian Agricultural Research Institute, New Delhi 110 012, India; Email: nns_dmr@ bol.net.in

Singh R.K., International Rice Research Institute, India Office, NASC Complex, DPSM, Pusa, New Delhi 110 012, India; Email: r.k.singh@cgiar.org

Singh Ravi P., The International Maize and Wheat Improvement Center (CIMMYT), Apdo, Postal 6-641, 06600, Mexico DF, Mexico; Email: r.singh@cgiar.org

Srinivasan G., The International Maize and Wheat Improvement Center (CIMMYT), Apdo, Postal 6-641, 06600, Mexico DF, Mexico; Email: g.srinivasan@cgiar.org

Sujata V., Division of Genetics, Indian Agricultural Research Institute, New Delhi 110 012, India; Email: sujatavasudev@ rediffmail.com

Tandon J.P., A-72, Sector 14, Noida 201 301, U.P., India; Email: tandonjp@rediffmail.com

Tyagi Akhilesh K., Department of Plant Molecular Biology, University of Delhi South Campus, New Delhi 110 021, India; Email: akhilesh@genomeindia.org; aktpmb@ hotmail.com

Upadhyaya K.C., School of Life Sciences, Jawaharlal Nehru University, New Delhi 110 067, India; Email: kcupadhyaya@hotmail.com

Uprety D.C., Division of Plant Physiology, Indian Agricultural Research Institute, New Delhi 110 012, India; Email: dc_uprety@ yahoo.com

Vasal S.K., The International Maize and Wheat Improvement Center (CIMMYT), Apdo, Postal 6-641, 06600, Mexico DF, Mexico; Email: s.vasal@cgiar.org

Virmani S.S., Plant Breeding, Genetics and Biotechnology Division, International Rice Research Institute, DAPO Box 7777, Metro Manila, Phillipines; Email: s.s.virmani@ cgiar.org

William M., International Maize and Wheat Improvement Center (CIMMYT), Apdo, Postal 6-641, 06600, Mexico DF, Mexico; Email: mwilliam@cgiar.org

Xu Wei Jun , Plant Breeding, Genetics and Biotechnology Division, International Rice Research Institute, Manila, Phillipines; Email: w.xu@cgiar.org

AUTHOR INDEX

C

D

E

F

G

I

J

M

W

X

Y

Z

SUBJECT INDEX

Q

R

GPSR Compliance

The European Union's (EU) General Product Safety Regulation (GPSR) is a set of rules that requires consumer products to be safe and our obligations to ensure this.

If you have any concerns about our products, you can contact us on ProductSafety@springernature.com

In case Publisher is established outside the EU, the EU authorized representative is:

Springer Nature Customer Service Center GmbH
Europaplatz 3
69115 Heidelberg, Germany

Batch number: 10761760

Printed by Printforce, the Netherlands